Alkoxo and Aryloxo Derivatives of Metals

Alkoxo and Aryloxo Derivatives of Metals

D. C. Bradley, R. C. Mehrotra,
I. P. Rothwell and A. Singh

ACADEMIC PRESS

A Harcourt Science and Technology Company

San Diego   San Francisco   New York
Boston   London   Sydney   Tokyo

This book is printed on acid-free paper.

Copyright © 2001 by ACADEMIC PRESS

Academic Press
A Harcourt Science and Technology Company
Harcourt Place, 32 Jamestown Road, London NW1 7BY, UK
http://www.academicpress.com

Academic Press
A Harcourt Science and Technology Company
525 B Street, Suite 1900, San Diego, California 92101-4495, USA
http://www.academicpress.com

ISBN 0-12-124140-8

Library of Congress Catalog Number: 00-105909

A catalogue record for this book is available from the British Library

Typeset by Laser Words, Madras, India
Printed and bound in Great Britain by MPG Books Ltd, Cornwall.

01 02 03 04 05 MP 9 8 7 6 5 4 3 2 1

Contents

Foreword

The value of a book may well be judged by the number of times a person has to buy it, for, while many books once read gather dust upon a shelf, those more often sought can sometimes be seldom found. Over 20 years ago, I was fortunate to receive a complimentary copy of "*Metal Alkoxides*" by Bradley, Mehrotra and Gaur. As one interested in alkoxide metal chemistry, this proved a valuable reference for me and my research group. In fact, I had to purchase two subsequent copies and probably would have purchased more were it not for the fact that the book became out of print and unavailable except through the library. Now I have received the galley proofs of the second edition entitled "*Alkoxo and Aryloxo Derivatives of Metals*" by Bradley, Mehrotra, Rothwell and Singh. After 20 years, virtually every field of chemistry must have changed to the extent that a new edition would be appropriate. However, it is unlikely that any field of chemistry, save computational chemistry, will have changed as much as that of the chemistry of metal alkoxides and aryloxides during the period 1978–2000. The explosion of interest in metal alkoxides has arisen primarily for two reasons. First and foremost, we have witnessed the tremendous growth of materials chemistry spurred on by the discovery of high temperature superconducting oxides and by the increasingly important role of other metal oxides to technology. Metal alkoxides, mixed metal alkoxides and their related complexes have played an essential role in the development of new routes to these materials either by sol-gel or chemical vapor deposition techniques. In a second area of almost equal magnitude, we have seen the growth of a new area of organometallic chemistry and catalysis supported by alkoxide or aryloxide ancillary ligands. As a consequence of these major changes in chemistry, virtually any issue of a current chemistry journal will feature articles dealing with metal alkoxides and aryloxides. Thus, although the present book owes its origins, and to some extent its format, to the first edition, its content is largely new. For example, while the first edition reported on but a handful of structurally characterized metal alkoxides, this second edition carries a whole chapter dealing with this topic, a chapter with over 500 references to publications. The second edition is therefore most timely, if not somewhat overdue, and will be a most valuable reference work for this rapidly expanding field of chemistry. I only hope that I can hold on to my copy more successfully than I did in the first instance.

Malcolm H. Chisholm FRS
Distinguished Professor of Mathematical and Physical Sciences
The Ohio State University
Department of Chemistry
Columbus, OH 43210-1185 USA
January 2001

1

Introduction

In 1978 the book entitled "Metal Alkoxides" was published.[1] It contained over one thousand references and attempted to summarize most of what was known about metal alkoxides up to that time. A striking feature was the dearth of X-ray crystal structures and so structural aspects necessarily involved speculation based on the results of molecular weight determinations, combined where possible with spectroscopic data.

The intervening years have witnessed a spectacular advance in our knowledge of the chemistry of the metal alkoxides, a development which has been driven primarily by research activity resulting from the realization that these compounds have great potential as precursors for the deposition of metal oxide films for microelectronic device applications and in bulk for producing new ceramic materials. Simultaneously a tremendous advance occurred in X-ray crystallography with the advent of computer-controlled automated diffractometers and with improvements in the techniques for growing and mounting single crystals of the air sensitive metal alkoxides. Consequently the number of structures solved has become so large that in this book a separate chapter with over 500 references has been devoted to crystal structures with much of the data summarized in tabular form. In addition, considerable advances have been made in the synthesis and characterization of a range of new alkoxides of the alkali metals, alkaline earths, yttrium and the lanthanides which together with other new developments has led to a chapter on Homometallic Alkoxides containing well over 1000 references. Similarly the chapter on Heterometallic Alkoxides (previously described as Double Metal Alkoxides) has been expanded to include many novel compounds, with particular emphasis on the recently authenticated species containing two, three and even four different metals in one molecule.

Another area that has expanded in recent years concerns the Industrial Applications of metal alkoxides. Besides the previously mentioned deposition of metal oxides in the microelectronic and ceramics industries there have also been major developments in the catalytic activity of early transition metal alkoxo compounds in several important homogeneous reactions. This has stimulated a growing interest in the mechanisms of reactions catalysed by metal alkoxides.

Metal Oxo Alkoxides are implicated as intermediates in the hydrolysis of metal alkoxides to metal oxides and their importance in the sol–gel process has led to much research activity in this area. Accordingly we have allocated a whole chapter to the Metal Oxo Alkoxides.

In the 1978 book very little space was devoted to metal aryloxides because this area had received scant attention, but the intervening years have seen a resurgence of activity involving the synthesis and characterization of many novel compounds and studies on their catalytic activity. Therefore we have added a separate chapter dealing with this important topic.

In this book we are giving the relevant references at the end of the seven chapters rather than placing them all at the end of the text in the hope that this will be more convenient for the reader. Finally, the authors acknowledge their indebtedness to all of their former research students, postdoctoral assistants, and colleagues for their invaluable contributions to the research which has provided much of the information collected in this publication.

REFERENCE

1. D.C. Bradley, R.C. Mehrotra, and D.P. Gaur, *Metal Alkoxides*, Academic Press, London (1978).

2

Homometallic Alkoxides

1 INTRODUCTION

Metal alkoxides $[M(OR)_x]_n$ (where M = metal or metalloid of valency x; R = simple alkyl, substituted alkyl, or alkenyl group; and n = degree of molecular association), may be deemed to be formed by the replacement of the hydroxylic hydrogen of an alcohol (ROH) by a metal(loid) atom.

Historically, the first homoleptic alkoxo derivatives of elements such as boron and silicon had been described[1,2] as early as 1846, but later progress in the alkoxide chemistry of only half a dozen metals was rather slow and sporadic till the 1950s; since then the chemistry of alkoxides of almost all the metals in the periodic table has been systematically investigated. With a few exceptions, systematic investigations on the structural aspects of metal alkoxides till the mid-1980s[3-10] were limited to studies on molecular association, volatility, chemical reactivity, and spectroscopic (IR, NMR and electronic) as well as magnetic properties. It is only since the early 1980s that definitive X-ray structural elucidation has become feasible and increasingly revealing.

The rapidly advancing applications[11-16] of metal alkoxides for synthesis of ceramic materials by sol–gel/MOCVD (metallo-organic chemical vapour deposition) processes (Chapter 7) have more recently given a new impetus to intensive investigations on synthetic, reactivity (including hydrolytic), structural, and mass-spectroscopic aspects of oxo-alkoxide species.[17-21]

Some of the exciting developments since 1990 in metal alkoxide chemistry have been focussing on the synthesis and structural characterization of novel derivatives involving special types of alkoxo groups such as (i) sterically demanding monodentate (OBu^t, OCHPr$^i_2$, OCHBu$^t_2$, OCMeEtPr^i, OCBu$^t_3$) as well as multidentate (OCR$'$(CH$_2$OPr^i)$_2$) (R$'$ = Bu^t or CF$_3$), OCR$''_2$CH$_2$X (R$''$ = Me or Et, X = OMe, OEt, NMe$_2$) ligands,[21-24] (ii) fluorinated tertiary alkoxo (OCMe(CF$_3$)$_2$, OCMe$_2$(CF$_3$), OC(CF$_3$)$_3$, etc.) moieties,[21-23] and (iii) ligands containing intramolecularly coordinating substituents (OCBu$^t_2$CH$_2$PMe$_2$, OCH$_2$CH$_2$X (X = OMe, OEt, OBu^n, NR$_2$, PR$_2$)).[21,22] Compared to simple alkoxo groups, most of these chelating/sterically demanding ligands possess the inherent advantages of enhancing the solubility and volatility of the products by lowering their nuclearities owing to steric factors and intramolecular coordination.

Solubility and volatility are the two key properties of metal alkoxides which provide convenient methods for their purification as well as making them suitable precursors for high-purity metal oxide-based ceramic materials.

It is noteworthy that the homoleptic platinum group metal (Ru, Rh, Pd, Os, Ir, Pt) alkoxides are kinetically more labile possibly owing to β-hydrogen elimination[9,10,21] type reaction(s) (Eq. 2.1):

$$M—OCHR'R'' \longrightarrow M—H + R'R''C=O \qquad (2.1)$$
$$\downarrow$$
$$M + \tfrac{1}{2}H_2$$

These, therefore, are not generally isolable under ambient conditions unless special types of chelating alkoxo ligands[21] are used.

Although single crystal X-ray studies presented considerable difficulties in the earlier stages,[25] the development of more sophisticated X-ray diffraction techniques has led to the structural elucidation of a number of homo- and heteroleptic alkoxides[17–23] and actual identification of many interesting metal oxo-alkoxide systems (Chapter 4).

In this chapter we shall discuss the synthesis,[3,4,26] chemistry and properties of homometallic alkoxides with more emphasis on homoleptic alkoxides $[M(OR)_x]_n$ and $M(OR)_x.L_n$ with occasional references to metal oxo-alkoxides $MO_y(OR)_{(x-2y)}$ and metal halide alkoxides $M(OR)_{x-y}X_y.L_z$ (where x = valency of metal, L = neutral donor ligand, X = halide, and n, y and z are integers). The discussion will generally exclude organometallic alkoxides and a considerable range of metal-organic compounds containing alkoxo groups, as in these systems the alkoxo groups play only a subsidiary role in determining the nature of the molecule.

2   METHODS OF SYNTHESIS

Metal alkoxides in general are highly moisture-sensitive. Stringent precautions are, therefore, essential during their synthesis and handling; these involve drying of all reagents, solvents, apparatus, and the environment above the reactants and products. Provided that these precautions are taken, the preparation of metal alkoxides, although sometimes tedious and time consuming, is relatively straightforward.

The method employed for the synthesis[3,4,8,17,21] of any metal/metalloid alkoxide depends generally on the electronegativity of the element concerned. Highly electropositive metals with valencies up to three (alkali metals, alkaline earth metals, and lanthanides) react directly with alcohols liberating hydrogen and forming the corresponding metal alkoxides. The reactions of alcohols with less electropositive metals such as magnesium and aluminium, require a catalyst (I_2 or $HgCl_2$) for successful synthesis of their alkoxides. The electrochemical synthesis of metal alkoxides by anodic dissolution of metals (Sc, Y, Ti, Zr, Nb, Ta, Fe, Co, Ni, Cu, Pb) and even metalloids (Si, Ge) in dry alcohols in the presence of a conducting electrolyte (e.g. tetrabutylammonium bromide) appears to offer a promising procedure (Section 2.2) of considerable utility. It may be worthwhile to mention at this stage that the metal atom vapour technique, which has shown exciting results in organometallics, may emerge as one of the potential synthetic routes for metal alkoxides also in future.

For the synthesis of metalloid (B, Si) alkoxides, the method generally employed consists of the reaction of their covalent halides (usually chlorides) with an appropriate alcohol. However, the replacement of chloride by the alkoxo group(s) does not appear to proceed to completion, when the central element is comparatively more electropositive. In such cases (e.g. titanium, niobium, iron, lanthanides, thorium) excluding the strongly electropositive s-block metals, the replacement of halide could in general be pushed to completion by the presence of bases such as ammonia, pyridine, or alkali metal alkoxides.

Another generally applicable method, particularly in the case of electronegative elements, is the esterification of their oxyacids or oxides (acid anhydrides) with alcohols (Section 2.6), and removing the water produced in the reaction continuously.

In addition to the above, alcoholysis or transesterification reactions of metal alkoxides themselves have been widely used for obtaining the targeted homo- and heteroleptic alkoxide derivatives of the same metal. Since the 1960s, the replacement reactions of metal dialkylamides with alcohols has provided a highly convenient and versatile route (Section 2.9) for the synthesis of homoleptic alkoxides of a number of metals, particularly in their lower valency states.

The metal–hydrogen and metal–carbon bond cleavage reactions have also been exploited in some instances (Section 2.10.2).

The following pages present a brief summary of the general methods used for the synthesis of metal and metalloid alkoxides applicable to specific systems. Tables 2.1 and 2.2 in Section 2.1 (pp. 6–14) list some illustrative compounds along with their preparative routes and characterization techniques.

2.1   Reactions of Metals with Alcohols (Method A)

The facility of the direct reaction of a metal with an alcohol depends on both the electropositive nature of the metal and the ramification of the alcohol concerned.

In view of the very feeble acidic character of nonfluorinated alcohols [even weaker than that of water: pK_a values (in parentheses) of some alcohols are $CH_3OH(15.8)$, $CH_3CH_2OH(15.9)$, $(CH_3)_2CHOH(17.1)$, $(CH_3)_3COH(19.2)$, $CF_3CH_2OH(12.8)$, $CH_3(CF_3)_2COH(9.6)$, $(CF_3)_2CHOH(9.3)$, $(CF_3)_3COH(5.4)$], this route is more facile with lower aliphatic and fluorinated alcohols.

2.1.1   s-Block Metals

2.1.1.1   Group 1 metals (Li, Na, K, Rb, Cs)
The more electropositive alkali metals react vigorously with alcohols by replacement of the hydroxylic hydrogen (Eq. 2.2):

$$M + (1+y)ROH \longrightarrow \frac{1}{n}[MOR.yROH]_n + \frac{1}{2}H_2 \uparrow \tag{2.2}$$

$M = $ Li, Na, K, Rb, Cs; R = Me, Et, Pr^i, Bu^t;[3,6,26,27] $y = 0$.

$M = $ Li; R = Bu^t, CMe$_2$Ph;[28] $y = 0$.

$M = $ K, Rb, Cs; R = Bu^t;[29] $y = 1$.

$M = $ K, Rb; R = Bu^t;[29] $y = 0$.

The alkali metals react spontaneously with sterically compact aliphatic alcohols (MeOH, EtOH, etc.) and the speed of the reaction increases with atomic number of the metal, Li < Na < K < Rb < Cs, corresponding to a decrease in ionization potential of the alkali metals. The ramification of the alkyl group is also important, as shown by the

Table 2.1   Examples of some homoleptic alkoxides

Compound[1]	Method of preparation[2]	Characterization techniques[3]	Reference
Group 1			
$[LiOMe]_\infty$	A	X-ray	28a
$[LiOBu^t]_6$	A	IR; ^1H, ^{13}C, ^7Li NMR; MW	28
$[LiOCMe_2Ph]_6$	A	IR; ^1H, ^{13}C, ^7Li NMR; MW; X-ray	28
$[LiOCBu^t_3]_2$	J-2	^1H, ^{13}C, ^7Li NMR; X-ray	396
$[LiOCBu^t_3(thf)]_2$	J-2	X-ray	230
$[LiOCBu^t_2CH_2PMe_2]_2$	J-3	^1H, ^{13}C, ^7Li, ^{31}P NMR; X-ray	22
$[LiOCBu^t_2CH_2PPh_2]_2$	J-3	^1H, ^{13}C, ^7Li, ^{31}P NMR; X-ray	422
$[LiOCBu^t_2CH_2PPh_2]_2(Bu^t_2CO)$	J-3	^1H, ^{13}C, ^7Li, ^{31}P NMR; X-ray	422
$[MOMe]_\infty$ (M = Na, K, Rb, Cs)	A	X-ray	a, b, c, d
$[NaOBu^t]_6$	A	X-ray	e, f
$[NaOBu^t]_9$	A	X-ray	e
$[MOBu^t.HOBu^t]_\infty$ (M = K, Rb)	A	IR; ^1H, ^{13}C NMR; MW; X-ray	29
$[MOBu^t]_4$ M = K, Rb, Cs	A	^1H, ^{13}C NMR; X-ray	29, g
$[Na\{OCH(CF_3)_2\}]_4$	J-2	IR; ^1H, ^{19}F NMR; X-ray	397
Group 2			
$[Be(OMe)_2]_n$	E-3, J-2	IR	214, 385
$[Be(OBu^t)_2]_3$	J-2	IR; ^1H NMR; MW	385
$[Be(OCEt_3)_2]_2$	J-2	IR; ^1H NMR; MW	385
$[Be(OCMe_2CH_2OMe)_2]_2$	I	IR; ^1H NMR; MS	340
$[Be(OCEt_2CH_2OMe)_2]_2$	I	IR; ^1H NMR; MS	340
$[Be\{OC(CF_3)\}_2]_3.OEt_2$	E-2	^1H, ^{19}F NMR; MW	396
$Mg(OMe)_2.3.5MeOH$	A	X-ray	38
$[Ca(\mu\text{-}OR)(OR)(thf)]_2.(toluene)_2$	E-2	IR; ^1H, ^{13}C NMR; X-ray	147
$[Ca(OR)_2(thf)_3].THF$	E-2	IR; ^1H, ^{13}C NMR; X-ray	147
$Ca\{OC(CF_3)_3\}_2$	A	^{19}F NMR	47, 53, 340
$Ca_3(OCHBu^t_2)_6$	I		53, 340
$Ca_2[OCBu^t(CH_2OPr^i)_2]_4$	I	IR; MS; X-ray	53, 340
$Ca[OCBu^t(CH_2OPr^i)(CH_2CH_2NEt_2)]_2$	I	IR; MS	53, 340
$Ca_9(OC_2H_4OMe)_{18}(HOC_2H_4OMe)_2$	A	IR; ^1H, ^{13}C NMR; X-ray	50
$Sr[OC(CF_3)_3]_2$	A	^{19}F NMR	47
$Sr_2[OCBu^t(CH_2OPr^i)_2]_4$	I	IR; MS	53, 340
$Ba(OBu^t)_2$	A	^1H NMR	47
$Ba(OCEt_3)_2$	A	^1H NMR	47
$Ba(OCMeEtPr^i)_2$	A	^1H NMR	47

Table 2.1 *(Continued)*

Compound[1]	Method of preparation[2]	Characterization techniques[3]	Reference
$Ba(OCHBu_2^t)_2$	A	^1H NMR	47
$Ba[OCH(CF_3)_2]_2$	A	^{19}F NMR	47
$Ba[OC(CF_3)_3]_2$	A	^{19}F NMR	47
$[Ba(OBu^t)_2(HOBu^t)_2]_4$	I	^1H, ^{13}C NMR; X-ray	549
$Ba_2[OCBu^t(CH_2OEt)_2]_4$	E-2		53, 340
$Ba_2[OCBu^t(CH_2OPr^i)_2]_4$	A, I	IR; MS	53, 340
$Ba_2(OCPh_3)_4(thf)_3$	A	^1H, ^{13}C NMR; X-ray	48
$Ba[O(CH_2CH_2)_xCH_3]_2$ $(x = 2, 3)$	A	IR; ^1H, ^{13}C NMR; MS	52
Scandium, Yttrium, and Lanthanides			
$[Sc\{OCH(CF_3)_2\}_3(NH_3)_2]_2$	I	IR; ^1H, ^{19}F NMR; MS; X-ray	349
$Ln(OPr^i)_3$ $Ln = Y, Pr, Nd, Sm, Eu,$ $Gd, Tb, Dy, Ho, Et, Tm, Yb, Lu$	A	IR; ^1H NMR (Y, La, Lu); UV-Vis (Pr, Nd, Sm, Ho, Er)	55
$Ln(OPr^i)_3$ $Ln = Y, Dy, Yb$	A	IR; ^1H NMR (Ln = Y)	54
$Ln(OPr^i)_3$ $Ln = Pr, Nd$	E-2	MW	153
$Ln(OR)_3$ $Ln = Pr, Nd;$ $R = Bu^n, Bu^i, Bu^s, Bu^t, Am^n,$ $Am^t, Pr^nCH(Me), Pr^nCMe_2$	G	MW	153
$Gd(OPr^i)_3$	E-2	IR; MW	157
$Er(OPr^i)_3$	E-2	IR; MW	157
$Ln(OMe)_3$ $Ln = Gd, Er$	E-3	IR	157
$Ho(OPr^i)_3$	E-2	MW	158
$[Y\{OCH(CF_3)_2\}_3(thf)_3]$	I	IR; MS; X-ray	349
$[Y\{OCMe_2(CF_3)\}_3]_n$	I	^1H, ^{19}F NMR	349a
$[Y\{OCMe_2(CF_3)\}_3(thf)_{2.5}]$	I	^1H, ^{19}F, ^{89}Y NMR	349a
$[Y\{OCMe(CF_3)_2\}_3]_n$	I	^1H, ^{19}F, ^{89}Y NMR	349a
$[Y\{OCMe(CF_3)_2\}_3(NH_3)_{0.5}]$	I	^1H, ^{19}F, ^{89}Y NMR	349a
$[Y\{OCMe(CF_3)_2\}_3(NH_3)_3]$	I	^1H, ^{19}F, ^{89}Y NMR	349a
$[Y\{OCMe(CF_3)_2\}(thf)_3]$	I	^1H, ^{19}F, ^{89}Y NMR	349a
$[Y\{OCMe(CF_3)_2\}_3(OEt_2)_{0.33}]$	I	^1H, ^{19}F, ^{89}Y NMR	349a
$\{Y\{OCMe(CF_3)_2\}_3(diglyme)\}$	I	^1H, ^{19}F, ^{89}Y NMR	349a
$[Y\{OCMe(CF_3)_2\}_3(HOBu^t)_3]$	I	^1H, ^{19}F, ^{89}Y NMR	349a
$\{Y\{OCH(CF_3)_2\}_3(NH_3)_{0.5}]$	I	^1H, ^{19}F, ^{89}Y NMR	349a
$[Y\{OCH(CF_3)_2\}_3(thf)_3]$	I	^1H, ^{19}F, ^{89}Y NMR	349a
$[Y_3(OBu^t)_9(HOBu^t)_2]$	I	IR; ^1H, ^{13}C, ^{89}Y NMR; MS	345
$[Y_3(OAm^t)_9(HOAm^t)_2]$	I	IR; ^1H, ^{13}C, ^{89}Y NMR; MS	345

(continued overleaf)

Table 2.1   *(Continued)*

Compound[1]	Method of preparation[2]	Characterization techniques[3]	Reference
$[Y(OR)_3]_2$ R = CMe_2Pr^i, $CMeEtPr^i$, CEt_3	I	IR; ^1H, ^{13}C, ^{89}Y NMR; MS	345
$[Y(OC_2H_4OMe)_3]_{10}$	A	IR; ^1H, ^{13}C NMR; X-ray	57
$[La_3(OBu^t)_9(HOBu^t)_2]$	I	^1H, ^{13}C NMR; MS; X-ray	345
$[La(OR)_3]_2$ R = CMe_2Pr^i, $CMeEtPr^i$	I	^1H, ^{13}C NMR; MS	345
$[La_3(OBu^t)_9(thf)_2]$	E-2	^1H, ^{13}C NMR; X-ray	160
$[La(OCPh_3)_3]_2$	I	IR; ^1H, ^{13}C NMR; X-ray	346
$[La\{OCMe(CF_3)_2\}_3(thf)_3]$	I	IR; ^1H, ^{13}C NMR; MS; X-ray	349c
$La_4(OCH_2Bu^t)_{12}$	I	IR; ^1H, ^{13}C NMR; X-ray	348a
$[Ce(OPr^i)_4 \cdot Pr^iOH]_2$	E-1 E-2	MW ^1H, ^{13}C NMR; X-ray	143 460
$Ce(OCBu^t_3)_3$	I	MW	344
$[Ce(OCHBu^t_2)_3]_2$	150°C, vacuum	X-ray	344
$Ce(OR)_4$ R = Me, Et, Pr^n, Bu^n, Bu^i, CH_2Bu^t	G	MW	143
$Ce(OBu^t)_4(thf)_2$	E-2	IR; ^1H, ^{13}C NMR	164
$[Pr\{OCMe(CF_3)_2\}_3(NH_3)_2]_2$	I	IR; MS; X-ray	349
$[Pr\{OCMe(CF_3)_2\}_3(NH_3)_4]$	I	IR; ^1H NMR; X-ray	349b
$[Pr\{OCMe_2(CF_3)\}_3]_3$	I	IR; MS; X-ray	349
$[Nd(OCBu^t_3)_3(thf)]$	I	IR; ^1H NMR; X-ray	959
$Nd_4(OCH_2Bu^t)_{12}$	I	IR; ^1H NMR; X-ray	348a
$Nd_2(OCHPr^i_2)_6(thf)_2$	I	IR; ^1H NMR; X-ray	348
$[Nd(OPr^i)_3 \cdot Pr^iOH]_4$	A	IR	56
$[Eu\{OCMe(CF_3)_2\}_3]_n$	I	IR	349b
$[Eu_2\{OCMe(CF_3)_2\}_6(NH_3)_2]$	I	IR	349b
$[Eu\{OCMe(CF_3)_2\}_3(thf)_3]$	I	^1H, ^{19}F NMR; MS	349c
$[Eu\{OCMe(CF_3)_2\}_3(diglyme)]$	I	^1H, ^{19}F NMR; MS	349c
$[Lu(OCMe_2CH_2OMe)_3]_2$	I	IR; ^1H, ^{13}C NMR; MS; X-ray	355
Actinides			
$[Th(OPr^i)_4]_n$	E-2	MW	141
$[Th(OEt)_4]_n$	G	MW	141
$[Th(OR)_4]_n$	G	MW	143
R = Bu^n, $Pent^n$, CH_2Bu^t			143
R = CMe_3, CMe_2Et, $CMeEt_2$, CMe_2Pr^n, CMe_2Pr^i, CEt_3, $CMeEtPr^n$, $CMeEt,Pr^i$			141a
$Th_4(OPr^i)_{16}(Py)_2$	E-2	IR; ^1H, ^{13}C NMR; X-ray	165
$Th_2(OCHEt_2)_8(Py)$	J-3	IR; ^1H, ^{13}C NMR; X-ray	165
$[Th(OBu^t)_4(Py)_2]$	E-2	IR; ^1H NMR; X-ray	398

Table 2.1 (*Continued*)

Compound[1]	Method of preparation[2]	Characterization techniques[3]	Reference
$Th_2(OBu^t)_8(HOBu^t)$	J-3	IR; ^1H, ^{13}C NMR; X-ray	398
$Th_2(OCHPr^i_2)_8$	J-3	^1H NMR; thermochemical data; X-ray	399
$U(OMe)_4$	E-3, I		213
$U(OR)_4$	I		213
R = Et, Bu^t			
$U_2(OBu^t)_8(HOBu^t)$	J-3	IR; ^1H NMR; UV-Vis	330
$U(OCHBu^t_2)_4$	E-3	^1H NMR; μ_{eff}; MS	229
$U\{OCH(CF_3)_2\}_4(thf)_2$	E-2	^{19}F NMR; μ_{eff}	h
$U\{OC(CF_3)_3\}_4(thf)_2$	E-2	^{19}F NMR; μ_{eff}	h
$U(OEt)_5$	J-1		379
$U(OR)_5$	J-1		289
R = Me, Pr^n, Pr^i, Bu^s, Bu^n, Bu^i			
$U(OBu^t)_5$	G		289
$U(OCH_2CF_3)_5$	E-1		289
$Pu(OPr^i)_4.Pr^iOH$	E-1		144
Group 4			
$[Ti(OEt)_4]_4$	E-1		134
	E-1	^1H NMR	548
		X-ray	434
$M(OR)_4$	G	MW	273
(M = Ti, Zr)			
R = $MeCH_2(CH_2)_2CH_2$,			
$Me_2CHCH_2CH_2$, $MeCH(Et)CH_2$,			
Me_3CCH_2, $CHEt_2$, $CHMePr^n$,			
$CHMePr^i$, CMe_2Et			
$Ti(OR)_4$	G	MW; Lv*; ΔS	274
R = CMe_2Et, $CMeEt_2$			
$[Zr(OR)_4]_n$			
R = Et, Pr^i, Bu^n, Bu^s	E-1	MW	145
R = Pr^i, Pr^n, Bu^n, Am^n	E-1	MW	145a
$[Zr(OPr^i)_4.Pr^iOH]_2$	E-3, I	IR; ^1H, ^{13}C NMR; X-ray	460
$[Hf(OR)_4]_n$	E-1	MW	277
R = Et, Pr^i			
R = Me, Et, Pr^i, Bu^t, Am^t	G	MW	277
$[Hf(OPr^i)_4.Pr^iOH]_2$	E-1	IR; ^1H, ^{13}C NMR; X-ray	460
Group 5			
$V(OR)_4$	E-3, I		331, 333
R = Me, Et, Pr^i, Bu^t		UV-Vis; μ_{eff}; ESR	588, 589, 590
$[Nb(OR)_5]_2$	E-1, G	MW	279, 468
R = Me, Et, Pr^n, Bu^n, n-pentyl			
$[Ta(OR)_5]_n$	E-1, G	MW	280, 312, 469
R = Me, Et, Pr^n, Bu^n, $MeCH_2CH_2CH_2$ (and its isomers), $MeCH_2CH_2CH_2CH_2$ (and its isomers)			

(*continued overleaf*)

Table 2.1   *(Continued)*

Compound[1]	Method of preparation[2]	Characterization techniques[3]	Reference
$[Nb(OPr^i)_5]_2$	B, E-1	IR; MS; X-ray	608
$[M(OEt)_5]_2$	B	MS	79
M = Nb, Ta	E-1	X-ray (M = Nb)	566
$[M(OR)_5]_2$	G	^{1}H NMR	470
(M = Nb, Ta)			
R = Me, Et, Bu^i, Pr^i			
$[Ta(OR)_5]_2$	A	IR; MS	81
R = Me, Et, Bu^n, Pr^i			
$[Ta(OC_2H_4OMe)_5]$	G	MS	81
Group 6			
$[Cr(OCHBu^t_2)_2]_2$	E-3	X-ray	*i*
		UV-Vis; ESR	226
$Cr[OCBu^t(CH_2OPr^i)_2]_2$	E-2	IR; MS	340
$[Cr(OCMe_2CH_2OMe)_3]$	I	IR; MS; X-ray	340
$Cr(OCHBu^t_2)_3(thf)$	E-3	IR; UV-Vis; MS	226
$Cr(OBu^t)_4$	E-2	UV-Vis; μ_{eff}; MW	168
	I		331
	I	Thermochemical data; MS; UV-Vis; μ_{eff}	467, 471
$Cr(OCHBu^t_2)_4$	E-3	IR; UV-Vis; μ_{eff}; MS; X-ray	226
$Mo_2(OR)_6(Mo\equiv Mo)$	I	^{1}H NMR; X-ray (R = CH_2Bu^t)	360, 361, 362
R = Bu^t, CMe_2Ph, Pr^i, CH_2Bu^t			
$Mo_2[OCMe(CF_3)_2]_6$	E-2	^{1}H, ^{13}C NMR; X-ray	170
$Mo_2[OCMe_2(CF_3)]_6$	E-2	^{1}H, ^{13}C NMR	170
$[Mo_2(OCMe_2Et)_6]$	E-3	^{1}H, ^{13}C NMR	170
$Mo_2(OPr^i)_8(Mo=Mo)$	J-1		
	I	^{1}H NMR; MW; X-ray	371, 372
$Mo(OBu^t)_4$	I	^{1}H NMR; MW	371
$Mo_2(OPr^i)_4(HOPr^i)_4(Mo\equiv Mo)$	J-3	IR; ^{1}H NMR; UV-Vis; X-ray	375
$Mo_2(OR)_4(HOR)_4$	J-3	IR; ^{1}H NMR; UV-Vis; X-ray (c-pentyl)	375
R = c-pentyl, c-hexyl			
$Mo_2(OCH_2Bu^t)_4(NHMe_2)_4$	J-3	IR; ^{1}H NMR; UV-Vis; X-ray	375
$Mo_2(OPr^i)_4(Py)_4$	J-3	IR; ^{1}H NMR; UV-Vis; X-ray	375
$W_2(OPr^i)_6$	I	X-ray	*j*
$W_2(OPr^i)_6(Py)_2$	I (+pyridine)	IR; ^{1}H NMR; MS; X-ray	365
$W_2(OBu^t)_6(W\equiv W)$	I	IR; ^{1}H NMR; MS	365
$W_4(OR)_{12}$	I	MW	365
R = Pr^i, CH_2Bu^t			

Table 2.1   (*Continued*)

Compound[1]	Method of preparation[2]	Characterization techniques[3]	Reference
$W_4(OPr^i)_{12}/W_2(OPr^i)_6$	Crystallization of $W_2(OPr^i)_6$ from dimethoxyethane	^1H NMR; MS; X-ray	

$M_4(OCH_2R)_{12}$
  M = Mo, W
  R = c-C_4H_7, c-C_5H_9,
  c-C_6H_{11}, Pr^i
	alcoholysis of $M_2(OBu^t)_6$	IR; ^1H, ^{13}C, ^{95}Mo NMR;
		X-ray (M = Mo; R = c-C_4H_7)
			k

$W_4(OEt)_{16}$	I	^1H NMR; X-ray	366
Group 7			
$[Mn(OR)_2]_n$ R = primary, secondary, and tertiary alcohols	I	Reflectance spectra; μ_{eff}	l
$[Mn(OCHBu_2^t)_2]_2$	I	IR; UV-Vis; ESR; MW	226
	I	X-ray	351
$Mn[OCBu^t(CH_2OPr^i)_2]_2$	I	IR; MS	340
$Re_3(OPr^i)_9$	E-2	^1H NMR; X-ray	173, 174
Group 8			
$[Fe\{OCBu^t(CH_2OEt)_2\}_2]_2$	I	IR; MS	340
Group 9			
$[Co(OCHBu_2^t)_2]_2$	I	IR; UV-Vis; MW	226
$[Co(OCPh_3)_2]_2.n$-C_6H_{14}	I	IR; ^1H NMR; UV-Vis; X-ray	351a
$Co(OCPh_3)_2(thf)_2$	I (THF used as solvent)	IR; ^1H NMR; UV-Vis; X-ray	351a
$Co[OCBu^t(CH_2OPr^i)_2]_2$	I	IR; MS	340
$[Co[OC(CF_3)(CH_2OPr^i)_2]_2]_2$	I	IR; MS	340
Group 10			
$[Ni(OR)_2]_n$ R = Me, Et, Pr^n, Pr^i, Bu^t, Am^t, t-C_6H_{13}	E-3		6, 219
$Ni[OCBu^t(CH_2OPr^i)_2]_2$	E-2	IR; MS	340
$(dppe)Pt(OMe)_2$	E-2	^1H, ^{31}P NMR; X-ray	177, 177a
$Pt(OCH_2CH_2PPh_2)_2$	E-2	IR; ^1H, ^{31}P NMR; MS; X-ray	178
$Pt(OCMe_2CH_2PPh_2)_2$	E-2	IR; ^1H, ^{31}P NMR; X-ray	178a
$Pt(OCMe_2CH_2PPh_2).3.5H_2O$	I	^1H, ^{31}P NMR; X-ray	178b
Group 11			
$[Cu(OCHBu_2^t)]_4$	E-3	MW	226
$[CuOBu^t]_4$	E-3		m
	E-3	IR; ^1H NMR; ESR	n, o
$Cu\{OCH(CF_3)_2\}(PPh_3)_3$	J-2	IR; ^1H NMR; X-ray	p
$Cu(OCHPh_2)(PPh_3)_3$	J-2	IR; ^1H NMR	p
$[CuOCEt_3]_4$	E-2	^1H, ^{13}C NMR; MW	47

(*continued overleaf*)

Table 2.1 *(Continued)*

Compound[1]	Method of preparation[2]	Characterization techniques[3]	Reference
$Cu(OCEt_3)_2$	E-2	^1H, ^{13}C NMR; UV-Vis	47
$Cu[OCH(CF_3)_2]_2(L)$ L = tmeda, teed, bipy, $(py)_2$	E-2, G	UV-Vis; ESR; X-ray (L = tmeda)	175
$Cu[OCMe(CF_3)_2]_2(L)$ L = tmeda, bipy, $(py)_2$	E-2, G	UV-Vis; ESR; X-ray (L = tmeda)	175
Group 12			
$M[OCBu^t(CH_2OPr^i)_2]_2$ M = Zn, Cd	I	IR; MS	340
$[Zn(OR)_2]_n$ R = CEt_3, CEtMe, C_2H_4OMe, $C_2H_4OC_2H_4OMe$, $C_2H_4NMe_2$, $CHMeCH_2NMe_2$, $C_2H_4NMeC_2H_4NMe_2$	I	IR; ^1H NMR	339b
$Cd_9(OC_2H_4OMe)_{18}(HOC_2H_4OMe)_2$	I	IR; ^1H, ^{13}C, ^{113}Cd NMR; X-ray	339c
$[Cd(OBu^t)_2]_n$	I	IR	339c
Group 13			
$[Al(OPr^i)_3]_4$	A	IR; ^1H, ^{13}C, ^{27}Al NMR; MW; MS	6
		X-ray	579, 580
$[Al(OBu^t)_3]_2$	A		71
	I	X-ray	338
$[Ga(OR)_3]_n$ R = Me, Et, Pr^n, Pr^i	E-2	IR; ^1H NMR	187
$[In\{OCMe_2(CF_3)\}_3]_2$	I	IR; ^1H, ^{13}C NMR; X-ray	358
$[In(OPr^i)_3]_n$	E-2	IR; ^1H NMR	187
Group 14			
$Ge(OCBu^t_3)_2$	I	^1H, ^{13}C NMR; X-ray	352
$[Sn(OBu^t)_2]_2$	I	^1H, ^{119}Sn NMR	352
$[Sn(OPr^i)_4.Pr^iOH]_2$	E-2	^1H, ^{13}C, ^{119}Sn NMR X-ray	190 190, 191
$Sn(OBu^t)_4$	I	X-ray	190
$[Pb(OBu^t)_2]_n$	I	IR; ^1H, ^{207}Pb NMR; MS	194
$[Pb(OR)_2]_n$ R = Pr^i, Bu^t, CMe_2Et, CEt_3	I	IR; ^1H, ^{13}C NMR; MW; X-ray (R = Pr^i, C_2H_4OMe; $n = \infty$. R = Bu^t, $n = 3$)	612
CH_2CH_2OMe, $CHMeCH_2NMe_2$			
$[Pb(OPr^i)_2]_x$	E-2		192

Table 2.1   (*Continued*)

Compound[1]	Method of preparation[2]	Characterization techniques[3]	Reference
Group 15			
$Bi(OR)_3$	E-2		202
R = Me, Et, Pr^n, Pr^i			
$Bi(OBu^t)_3$	E-2	IR; ^1H, ^{13}C NMR	203
$[Bi(OC_2H_4OMe)_3]$	E-2; I	IR; ^1H, ^{13}C NMR; MS; X-ray, MW	205, 339
$[Bi(OCH(CF_3)_2]_3(thf)]_2$	E-2	IR; ^1H, ^{19}F NMR; X-ray	206, 206a
$[Bi\{OC(CF_3)_3\}_3]$	Bi + $3(CF_3)_3CCOCl$	IR; ^{19}F NMR	q
Group 16			
$Se(OR)_4$			
R = Me, Et	E-1, E-2	^1H, ^{13}C, ^{77}Se NMR	r, s
R = CH_2CF_3	E-1	^1H, ^{13}C, ^{77}Se, ^{19}F NMR	s
$Te(OR)_4$			
R = Me, Et, Pr^i	E-2		r
R = CH_2CF_3	E-1	^1H, ^{13}C, ^{125}Te, ^{19}F NMR	
R = $C(CF_3)_3$	Te + $4(CF_3)_3CCOCl$	IR; ^{19}F NMR	q

*Lv = Latent heat of vaporization.

[1] bpy = 2,2′-bipyridine; diglyme = bis (2-methoxyethyl) ether (ligand); dppe = 1,2-bis (diphenylphosphino)ethane, Py = pyridine (ligand); teed = N,N,N′,N′-tetraethylethylenediamine (ligand); tmeda = N,N,N′,N′-tetramethylethylenediamine (ligand); [2] Methods A–J (J-1–J-7) as described in text; [3] ESR = electron spin resonance; μ_{eff} = magnetic moment; MS = mass spectrum; MW = molecular weight; UV-Vis = ultraviolet and visible.

[a] E.Weiss, *Helv. Chim. Acta*, **46**, 2051 (1963); [b] E. Weiss and W. Biücher, *Angew. Chem.*, **75**, 1116 (1963); [c] E. Weiss, *Z. Anorg. Allg. Chem.*, **332**, 197 (1964); [d] E. Weiss and H. Alsdorf, *Z. Anorg. Allg. Chem.*, **372**, 2061 (1970); [e] J.E. Davies, J. Kopf, and E. Weiss, *Acta Crystallogr.*, **38**, 2251 (1982); [f] E.Weiss, *Angew. Chem., Int. Ed. Engl.*, **32**, 1501 (1993); [g] E. Weiss, H. Alsdorf, and H. Kühr, *Angew. Chem. Int. Ed. Engl.*, **6**, 801 (1967); [h] R.A. Andersen, *Inorg. Nucl. Chem., Lett.*, **15**, 57 (1979); [i] B.D. Murray, H. Hope, and P.P. Power, *J. Am. Chem. Soc.*, **107**, 169 (1985); [j] M.H. Chisholm, D.L. Clark, J.C. Huffman, and M. Hampden-Smith, *J. Am. Chem. Soc.*, **109**, 7750 (1987); [k] M.H. Chisholm, K.Folting, C.E. Hammond, M.J. Hampden-Smith, and K.G. Moodley, *J. Am. Chem. Soc.*, **111**, 5300 (1989); [l] B. Horvath, R. Moseler, and E.G. Horvath, *Z. Anorg. Allg. Chem.*, **449**, 41 (1979); [m] T. Greiser and E. Weiss, *Chem. Ber.*, **104**, 3142 (1976); [n] T. Tsuda, T. Hashimoto, and T. Saegusa, *J. Am. Chem. Soc.*, **94**, 658 (1972); [o] T.H. Lemmen, G.V. Goeden, J.C. Huffman, R.L. Geerts, and K.G. Caulton, *Inorg. Chem.*, **29**, 3680 (1990); [p] K. Osakada, T. Takizawa, M. Tanaka, and T. Yamamoto, *J. Organomet Chem.*, **473**, 359 (1994); [q] J.M. Canich, G.L. Gard, and J.M. Shreeve, *Inorg. Chem.*, **23**, 441 (1984); [r] N.Temple and W. Schwarz, *Z. Anorg. Allg. Chem.*, **474**, 157 (1981); [s] D.B. Denny, D.Z. Denny, P.T. Hammond, and Y.F. Hsu, *J. Am. Chem. Soc.*, **103**, 2340 (1981).

Table 2.2 Examples of a few selected heteroleptic alkoxides

Compound[1]	Method of preparation[2]	Characterization techniques[3]	References
$[Y_5O(OPr^i)_{13}]$	A	^1H, ^{13}C, ^{89}Y NMR; MW; MS; X-ray	59
$[Ln_5O(OPr^i)_{13}]$ Ln = Sc, Y, Yb	A; E-2	IR; ^1H NMR; MS; X-ray (Ln = Yb)	58
$[Nd_5O(OPr^i)_3(HOPr^i)_2]$	A	IR; X-ray	60
$[Y_3(OBu^t)_7Cl_2(thf)_2]$	E-2	^1H, ^{13}C NMR; X-ray	161
$[Y_3(OBu^t)_8Cl(thf)_2]$	E-2	IR; ^1H, ^{13}C NMR; X-ray	160
$[CeOCBu_3^t)_2(OBu^t)_2]$	$Ce(OCBu_3^t)_2$ + Bu^tOOBu^t	^1H, ^{13}C NMR	383
$[Nd(OCBu_3^t)_2Cl(thf)]_2$	E-3	IR; ^1H NMR; X-ray	959
$[Nd_6(OPr^i)_{17}Cl]$	E-2	X-ray	159
$(Bu_3^tCO)_2UCl_2(thf)_2$	E-3	^1H NMR	227
$UO_2(OBu^t)_2(Ph_3PO)_2$	E-2	IR; ^1H NMR; X-ray	a
$(Bu_3^tCO)_2MCl_2$ M = Ti, Zr	E-3	^1H, ^{13}C NMR	228
$Ti(OPr^i)_2[OCH(CF_3)_2]_2$	G	IR	272
$Ti[OCH(CF_3)_2]_2(OEt)_2(HOEt)$	G	IR, ^1H, ^{19}F NMR; X-ray	272
$[TiCl_3(OPr^i)(HOPr^i)]_2$	D	IR; ^1H, ^{13}C NMR; X-ray	108
$[TiCl_2(OPr^i)_2(HOPr^i)]_2$	D	IR; ^1H, ^{13}C NMR; X-ray	108
$[TiCl_2(OCH_2CH_2Cl)_2(HOCH_2CH_2Cl)]_2$	D	IR; ^1H, ^{13}C NMR; X-ray	108a
$VCl(OMe)_2$	E-3	IR; reflectance spectra; μ_{eff}	592
$VO(OPr^i)_3$	E-3	IR; ^1H, NMR; MW	235
$CrO_2(OR)_2$ R = CH_2CCl_3, CH_2CF_3, CH_2CH_2Cl	E-3	IR; ^1H, ^{19}F NMR; MW	233
$Mo_4(OPr^i)_{10}(OMe)_2$	G	^1H NMR; X-ray	b
$[W(OCH_2CF_3)_2Cl_2(PMe_2Ph)_2]$	$WCl_4(PMe_2Ph)_2$ + $TlOCH_2CF_3$	IR; ^1H, ^{13}C, ^{31}P NMR; X-ray	171
$Re_3(OCHEt_2)_8(H)$	E-2	IR; ^1H, ^{13}C NMR	174
$Re_3(OPr^i)_8(H)$	β-Hydrogen elimination from $Re_3(OPr^i)_9$	IR; ^1H, ^{13}C NMR; X-ray	174
$CoBr(OMe).2MeOH$	E-3	IR; reflectance spectra; μ_{eff}	595
$Ni(OMe)Cl$	E-3	IR; reflectance spectra; μ_{eff}	541
$[Cu(OMe)\{OCH(CF_3)_2\}]_n$	G	IR	175
$Cu_3(OBu^t)_4[OC(CF_3)_3]_2$	G	IR; ^1H, ^{19}F NMR; UV-Vis; μ_{eff}; X-ray	651
$Cu_4(OBu^t)_6[OC(CF_3)_3]_2$	G	IR; ^1H, ^{19}F NMR; UV-Vis; μ_{eff}; X-ray	651
$[Zn(OCEt_3)\{N(SiMe_3)_2)\}]_2$	I	IR; ^1H NMR; MW; X-ray	688
$[Zn_2(OOCMe)_3(OMe)]$	$Zn(OAc)_2$ + $Bu_2^nSn(OMe)_2$	IR; ^1H NMR; X-ray	431a
$[Al(OPr^i)(OBu^t)_2]_2$	G	^1H NMR; MW; MS	492
$[Al(OPr^i)_2(acac)]_2$	$Al(OPr^i)_3$ + acacH	^1H, ^{13}C, ^{27}Al NMR; X-ray	726
$In_5O(OPr^i)_{13}$	E-2	IR; ^1H, ^{13}C NMR; MS; X-ray	58
$[Sn(OBu^t)_3(OAc)(py)]$	$Sn(OBu^t)_4$ + Me_3SiOAc + Py	^1H, ^{13}C, ^{17}O, ^{119}Sn NMR; X-ray	431

[1] acac = acetylacetonate; py = pyridine; thf = tetrahydrofuran(ligand); [2] For methods see text (Section 2); [3] For abbreviations see footnote of Table 2.1.
[a] C.J. Burns, D.C. Smith, A.P. Sattelberger, and H.B. Gray, *Inorg. Chem.*, **31**, 3724 (1992); [b] M.H. Chisholm, C.E. Hammond, M. Hampden-Smith, J.C. Huffman, and W.G. Van der Sluys, *Angew. Chem., Int. Ed. Engl.*, **26**, 904 (1987).

reactivity sequence of alcohols, MeOH > EtOH > Pr^{i}OH > Bu^{t}OH, towards an alkali metal. This order of reactivity is understandable from an electronic viewpoint which predicts a decrease in the acidity of the hydroxyl hydrogen in the same order.

2.1.1.2   Group 2 metals (Be, Mg, Ca, Sr, Ba)

Group 2 metals, being less electropositive than group 1 metals, react sluggishly even with sterically compact alcohols and require a catalyst (iodine or mercury(II) chloride) particularly in cases of lighter group 2 metals (Be and Mg)[30–34] to yield insoluble, polymeric, and nonvolatile metal dialkoxides.

The reaction of magnesium with methanol had been reported[26] to form solvates of different compositions: $Mg(OCH_3)_2.3CH_3OH$ and $Mg(OCH_3)_2.4CH_3OH$,[35,36] which have been shown by X-ray diffraction studies to have the compositions $Mg(OCH_3)_2.2CH_3OH$[37] and $Mg(OCH_3)_2.3.5CH_3OH$,[38] respectively.

With sterically less demanding alcohols, alkoxides of the heavier alkaline earth metals (Ca, Sr, Ba) $[M(OR)_2]_n$ (R = Me, Et, Pr^i) had been prepared by a number of workers[39–44] by reactions of metals with alcohols. These are also oligomeric or polymeric, and nonvolatile.

Interest in the synthesis and chemistry of soluble and volatile alkaline earth metal alkoxides experienced a sudden upsurge in the 1990s,[21–23] owing to the discovery of superconducting ceramics[45,46] containing Ba and Ca.

Reactions of sterically demanding monodentate alcohols[47] with heavier alkaline earth metals (M′) have been reported to yield soluble derivatives:

$$M' + 2R'OH \longrightarrow M'(OR')_2 + H_2 \uparrow \qquad (2.3)$$

M′ = Ba; R′ = CMe$_3$, CEt$_3$, CHMe$_2$, CH(CF$_3$)$_2$. M′ = Ca, Sr; R′ = C(CF$_3$)$_3$.

By contrast, reaction of barium granules with Ph$_3$COH does not appear to take place, even in the presence of I$_2$ or HgCl$_2$ as a catalyst, in refluxing tetrahydrofuran (THF) over three days. However, the same reaction in the presence of ammonia as a catalyst yields X-ray crystallographically characterized dimeric derivative $[H_3Ba_6(O)(OBu^t)_{11}$ $(OCEt_2CH_2O)(thf)_3]$.[48] It may be inferred that ammonia reacts initially with barium to form Ba(NH$_2$)$_2$, which undergoes proton transfer and anion metathesis to yield the desired alkoxide derivative.

Although the reactions of heavier alkaline earth metals with alcohols are generally straightforward, yielding the expected homoleptic derivatives, in some instances it has been reported that the reaction follows a different course to yield an intriguing product as in the case of the formation of X-ray crystallographically characterized[49] oxo-alkoxide cluster of the composition $H_3Ba_6O(OBu^t)_{11}(OCEt_2CH_2O)(thf)$, in the reaction of Ba with Bu^{t}OH in THF. The reasons for the formation of such an unusual product in a simple reaction of the above type (Eq. 2.3) are not yet well understood, but it tends to indicate that either adventitious hydrolysis or alkene/ether elimination may be the main factor. Furthermore, the formation of (OCEt$_2$CH$_2$O) ligated product in this reaction indicates that the diolate ligand is probably formed in a side-reaction involving the solvent tetrahydrofuran molecules.

2-Methoxyethanol (a chelating alcohol) has been shown[50] to react with calcium filings in refluxing n-hexane to yield an X-ray crystallographically authenticated product according to the following reaction:

$$Ca + 4HOC_2H_4OMe \xrightarrow{\text{n-hexane}} \tfrac{1}{9}[Ca_9(OC_2H_4OMe)_{18}].2(HOC_2H_4OMe) + H_2 \uparrow \quad (2.4)$$

By contrast, a similar reaction with barium granules followed a different course[51] to yield $[H_4Ba_6O(OCH_2CH_2OMe)_{14}]$ which has been characterized by single-crystal X-ray diffraction studies.

Recently, it has been reported that monomeric $Ba[O(CH_2CH_2O)_nCH_3]_2$ ($n = 2$ or 3) products are obtained in the reactions of barium granules with an oligoether alcohol[52] in tetrahydrofuran (Eq. 2.5):

$$Ba + 2HO(CH_2CH_2O)_nCH_3 \xrightarrow{\text{THF}} Ba[O(CH_2CH_2O)_nCH_3]_2 + H_2 \uparrow \quad (2.5)$$

where $n = 2$ or 3.

The factor(s) determining the variation in the nature of products in the reaction of Ca/Ba with chelating alcohols obviously require further investigations.

Interestingly, the reaction of barium with a sterically demanding alcohol having donor functionality yields a volatile derivative[53] with excellent solubility (even in n-pentane) (Eq. 2.6):

$$Ba + 2HOCBu^t(CH_2OPr^i)_2 \xrightarrow[(-H_2)]{\text{THF/NH}_3} Ba[OCBu^t(CH_2OPr^i)_2]_2 \quad (2.6)$$

2.1.2   Group 3 and the f-block Metals

The method involving direct reaction of a metal with alcohol was extended by Mazdi-yasni et al.[54] for the formation of scandium, yttrium, and lanthanide alkoxides using mercuric chloride ($10^{-3} - 10^{-4}$ mol per mol of metal) as a catalyst:

$$Ln + 3Pr^iOH \xrightarrow[\text{heat}]{\text{HgCl}_2\,(\text{cat.})} \tfrac{1}{n}[Ln(OPr^i)_3]_n + \tfrac{3}{2}H_2 \uparrow \quad (2.7)$$
$$\text{(excess)}$$

$$Ln = Sc, Y, Dy, \text{ and } Yb.$$

Mercuric chloride appears to form an amalgam with the metal which reacts with isopropyl alcohol to yield the triisopropoxide. Mazdiyasni et al.[54] also noticed that the use of $HgCl_2$ in stoichiometric ratio resulted in the formation of alkenoxide contaminated with chloride. For example, the reaction of yttrium metal, isopropyl alcohol, and mercuric chloride in 1:3:4 molar ratio yielded yttrium isopropeneoxide[55] and hydrogen chloride:

$$Y + 3HOCH(CH_3)_2 + 4HgCl_2 \longrightarrow Y[OC(CH_3)=CH_2]_3 + 4Hg + 8HCl + \tfrac{1}{2}H_2 \quad (2.8)$$

The above route has also been utilized for the synthesis of neodymium[56] and yttrium[57] alkoxides as shown by Eqs (2.9) and (2.10):

$$4Nd + 16Pr^iOH \longrightarrow [Nd(OPr^i)_3.Pr^iOH]_4 + 6H_2 \uparrow \quad (2.9)$$

$$10Y + 30HOC_2H_4OMe \longrightarrow [Y(OC_2H_4OMe)_3]_{10} + 15H_2 \uparrow \quad (2.10)$$

By contrast, interesting oxo-isopropoxides of the type $Ln_5O(OPr^i)_{13}$, where $Ln =$ Sc,[58] Y,[58,59] Nd,[60] and Yb[58] have been isolated from the reaction mixtures resulting from the interaction of metal chips and isopropyl alcohol, out of which, the last three have been characterized by X-ray crystallography.

In restrospect, the isolation of oxo-alkoxide products, in the straightforward reactions of metals with alcohols has generated a new interest in metal oxo-alkoxide products (Chapter 5), a large number of which have been reported in the extensive investigations of Turova et al.[61-63] employing the solubility and vapour pressure studies of $M(OR)_x$–ROH systems.

Bradley et al.[58] have suggested that the formation of $Ln_5O(OPr^i)_{13}$ occurs by the mechanism of metal alkoxide decomposition involving elimination of an ether,[64] according to Eq. (2.11):

$$5Ln(OPr^i)_3 \longrightarrow Ln_5O(OPr^i)_{13} + Pr_2^iO \tag{2.11}$$

Obviously, more quantitative work is essential to explore the extent and course of side-reactions in the interactions of metals with different alcohols under varying experimental conditions.

2.1.3   p-Block Elements

Aluminium alkoxides may be prepared[65-74] by reaction of an alcohol with aluminium activated by I_2, $HgCl_2$, or $SnCl_4$ under refluxing conditions, for example:

$$2Al + 6ROH \xrightarrow[\Delta]{1\%HgCl_2} 2Al(OR)_3 + 3H_2 \uparrow \tag{2.12}$$

(excess)

where R = primary, secondary, or tertiary alkyl groups.

Aluminium triethoxide was first prepared in 1881 by Gladstone and Tribe[65] by the reaction of aluminium metal with ethanol in the presence of iodine as a catalyst. Wislicenus and Kaufman[66] in 1893 reported an alternative method of preparing normal as well as isomeric higher alkoxides of aluminium by reacting amalgamated aluminium with excess of refluxing alcohol. Hillyer[67] prepared aluminium trialkoxides by the reaction of metal with alcohols in the presence of $SnCl_4$ as a catalyst. Tischtschenko[68] in 1899, however, pointed out that the above reactions involving catalysts were useful for the preparation of primary and secondary alkoxides of aluminium, but the reaction of metal with tert-butyl alcohol was very slow even in the presence of catalysts. A successful synthesis of aluminium tri-tert-butoxide described by Adkins and Cox[71] in 1938, involved the reaction of amalgamated aluminium with refluxing tert-butyl alcohol.

By contrast, in a reaction similar to Eq. (2.12), indium forms $In_5O(OPr^i)_{13}$, the structure of which has been established by X-ray crystallography.[58]

Although metallic thallium did not appear to react with alcohols[75] even under refluxing conditions, the reaction of ethyl alcohol with the metal partly exposed to air does occur, resulting in the formation of liquid thallous ethoxide, for which the following course of reactions has been suggested: (Eqs 2.13–2.15):

$$2Tl + \tfrac{1}{2}O_2 \longrightarrow Tl_2O \tag{2.13}$$

$$Tl_2O + EtOH \longrightarrow TlOEt + TlOH \tag{2.14}$$

$$TlOH + EtOH \rightleftharpoons TlOEt + H_2O \tag{2.15}$$

2.2   Electrochemical Technique (Method B)

The possibility of synthesizing metal alkoxides by the anodic dissolution of metals into alcohols containing conducting electrolytes was demonstrated for the first time by Szilard[76] in 1906 for the methoxides of copper and lead. Since then this technique has proved to be most promising. For example, the electrochemical method for the preparation of ethoxides of Ti, Zr, Ta, Si, and Ge[77] was patented by the Monsanto Corporation in 1972, and was later applied by Lehmkuhl et al.[78] for the synthesis of Fe(II), Co, and Ni alkoxides $M(OR)_2$ (R = Me, Et, Bu^n, and Bu^t).

Turova et al.[79] have substantially widened the scope of this technique by the synthesis of a wide variety of homoleptic metal alkoxides and oxo-metal alkoxides: (i) soluble $M(OR)_n$, M = Sc, Y, La, lanthanide,[80] Ti, Zr, Hf, Nb,[79] Ta[81] when R = Me, Et, Pr^i, Bu^n; $MO(OR)_4$, M = Mo, W when R = Me, Et, Pr^i;[82–86] 2-methoxyethoxides of Y, lanthanide, Zr, Hf, Nb, Ta, Fe(III), Co, Ni, Sn(II),[87] and (ii) insoluble metal alkoxides such as $Bi(OMe)_3$;[88] $Cr(OR)_3$, R = Me, Et, MeOC$_2$H$_4$;[89] $V(OR)_3$;[86] $Ni(OR)_2$, R = Me, Pr^n, Pr^i;[90] $Cu(OR)_2$, R = Bu^n, C$_2$H$_4$OMe;[91] $Re_4O_2(OMe)_{16}$.[92]

Besides the above, Banait et al. have also employed the electrochemical reactions of some (including polyhydroxy) alcohols for the synthesis of alkoxides of copper[93] and mercury.[94]

In 1998, the anodic oxidation of molybdenum and tungsten[95] in alcohols in the presence of LiCl (as electroconductive additive) was found to yield a variety of interesting oxo-metal alkoxide complexes, some of which have been authenticated by single-crystal X-ray crystallograpy.

The electrode ionization reactions of alcohols and anode polarized metals in the presence of an electroconductive additive, followed by the interaction of the generated intermediate species and the formation of the final products can by illustrated[96] by the following reactions (Eqs 2.16 and 2.17):

$$M \longrightarrow M^{n+} + ne^- \quad \text{(anode)} \tag{2.16}$$

$$nROH + ne^- \longrightarrow nRO^- + nH\cdot$$

$$nH\cdot \longrightarrow \frac{n}{2}H_2 \quad \text{(cathode)} \tag{2.17}$$

$$M^{n+} + nRO^- \longrightarrow M(OR)_n$$

where M = anode metal and ROH = an appropriate alcohol.

This process has great promise for the direct conversion of the less electropositive metals to their alkoxides owing to its simplicity and high productivity as well as its continuous and non-polluting character (with hydrogen as the major by-product).

The electrochemical technique appears to have been successfully employed in Russia for the commercial production[96] of alkoxides of Y, Ti, Zr, Nb, Ta, Mo, W, Cu, Ge, Sn, and other metals.

2.3 Reactions of Metal Atom Vapours with Alcohols (Method C)

Although the development of metal atom vapour technology over the past three decades has shown tremendous utility for the synthesis of a wide range of organometallic compounds (many of which were inaccessible by conventional techniques),[97] the use of this technique for the synthesis of metal alkoxides and related derivatives does not appear to have been fully exploited.[98] In 1990, Lappert *et al.*[99] demonstrated the utility of this technique for the synthesis of M—O—C bonded compounds by the isolation of alkaline earth metal aryloxides.

2.4 Direct Reactions of Metal Halides with Alcohols (Method D)

By far the most common synthetic technique for metal alkoxides (Eq. 2.18) is the replacement of halides from an appropriate metal halide by alkoxo groups.

$$MCl_n + (x + y)ROH \rightleftharpoons MCl_{n-x}(OR)_x(ROH)_y + xHCl \uparrow \qquad (2.18)$$

Halides of alkaline earth, lanthanide, actinide, and later 3d (Mn, Fe, Co, Ni) metals on interactions with alcohols form crystalline molecular adducts like $MgBr_2.6MeOH$,[100] $CaBr_2.6MeOH$,[100] $LnCl_3.3Pr^iOH$[101–103] where Ln is a lanthanide metal, $ThCl_4.4EtOH$,[104,105] $MCl_2.2ROH$ (M = Mn, Fe, Co, Ni; R = Me, Et, Pr^n, Pr^i).[106] Apart from the alkaline earth metal (Ca, Sr, Ba) halides, all of these undergo alcoholysis in the presence of a suitable base to yield the corresponding homoleptic alkoxide or chloride-alkoxide derivatives (Sections 2.5.1, 2.5.2, and 2.5.3).

Interesting variations in the extent of alcoholysis reactions of tetravalent metal (Ti, Zr, Th, Si) chlorides may be represented[107] by Eqs (2.20–2.23), to which CCl_4 has been added for comparison.

$$CCl_4 + ROH \text{ (excess)} \longrightarrow \text{no reaction} \qquad (2.19)$$

$$SiCl_4 + 4ROH \longrightarrow Si(OR)_4 + 4HCl \uparrow \qquad (2.20)$$

$$TiCl_4 + 3ROH \text{ (excess)} \longrightarrow TiCl_2(OR)_2.ROH + HCl \uparrow \qquad (2.21)$$

$$2ZrCl_4 + 6ROH \text{ (excess)} \longrightarrow ZrCl_2(OR)_2.ROH$$
$$+ ZrCl_3(OR).2ROH + 3HCl \uparrow \qquad (2.22)$$

$$ThCl_4 + 4ROH \text{ (excess)} \longrightarrow ThCl_4.4ROH \qquad (2.23)$$

Depending on the nature of the metal (M), the initial metal chloride (MCl_n) or a product $MCl_{x-y}(OR)_y$ forms an addition complex with alcohol molecules (ROH) without enough perturbation of electronic charges for the reaction to proceed further. The reactions of metal tetrachlorides MCl_4 (M = Ti, Zr, Th) towards ethyl alcohol show a gradation $TiCl_4 > ZrCl_4 > ThCl_4$.[107]

Although no clear explanation is available for the varying reactivity of different metal chlorides with alcohols, it is interesting to note that final products of similar compositions have been isolated in the reactions of tetraalkoxides of these metals with HCl. For example, the reaction of $Ti(OPr^i)_4$ with HCl leads finally to $Ti(OPr^i)_2Cl_2.Pr^iOH$ (Section 4.11.2).

Specific intermediate products according to Eq. (2.18) may be isolated by controlling the conditions (solvent, stoichiometry, or temperature). For example, the equimolar reaction of $TiCl_4$ with Pr^iOH in dichloromethane at room temperature has been shown[108] to yield the dimeric complex $[TiCl_3(HOPr^i)(\mu\text{-}Cl)]_2$. The above reaction in 2–3 molar ratios gives the dimeric complexes $[TiCl_2(OPr^i)(HOPr^i)(\mu\text{-}Cl)]_2$ and $[TiCl_2(OPr^i)(HOPr^i)(\mu\text{-}OPr^i)]_2$ as outlined in Scheme 2.1 on the basis of X-ray structures of the products.

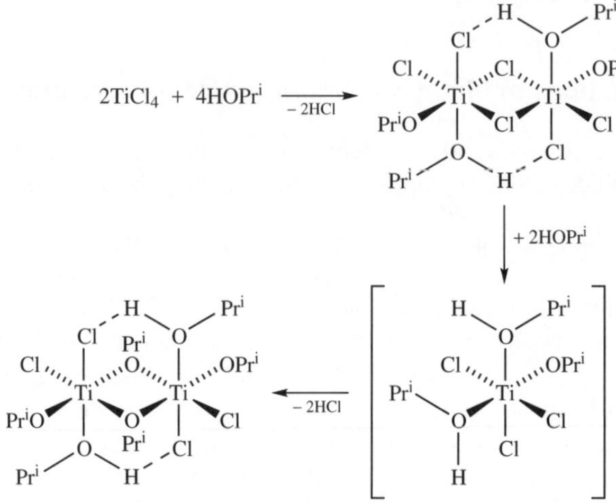

Scheme 2.1

Interestingly, the reaction of metallic uranium in isopropyl alcohol in the presence of stoichiometric amounts of iodine[109] has been shown to afford a mixed iodide–isopropoxide of uranium(IV), $UI_2(OPr^i)_2(Pr^iOH)_2$ (Eq. 2.24):

$$U + 2I_2 + 4Pr^iOH \xrightarrow{Pr^iOH} UI_2(OPr^i)_2(Pr^iOH)_2 + 2HI \qquad (2.24)$$

This type of reaction appears to have considerable promise for the preparation of other polyvalent metal–iodide–isopropoxide complexes.

Out of the p-block elements, anhydrous chlorides of electronegative elements boron,[110–112] silicon,[2,113–117] and phosphorus[118,119] react vigorously with alcohols to yield homoleptic alkoxo derivatives $[M(OR)_x]$ (Eq. 2.25). Although no detailed studies have been made, $AlCl_3$[120,121] and $NbCl_5$[122,123] undergo only partial substitution, while $GeCl_4$[124,125] does not appear to react at all with alcohols.

$$MCl_x + xROH \longrightarrow M(OR)_x + nHCl \uparrow \qquad (2.25)$$

$$(M = B, x = 3; Si, x = 4; P, (As)\ x = 3)$$

The reactions indicated above occur with primary and secondary alcohols only, and have been studied mainly with ethyl and isopropyl alcohols. With a tertiary alcohol (Bu^tOH), silicon tetrachloride yields almost quantitatively $Si(OH)_4$ and Bu^tCl.[126] This has been shown by Ridge and Todd[127] to be due to facile reactivity of HCl initially evolved to yield Bu^tCl and H_2O, which hydrolyses $SiCl_4$.

$$SiCl_4 + 4Bu^tOH \longrightarrow Si(OH)_4 + 4Bu^tCl \uparrow \qquad (2.26)$$

The reaction of $AsCl_3$ with an excess of CF_3CH_2OH[128] gives $As(OCH_2CF_3)_3$, which could be oxidized with chlorine in the presence of CF_3CH_2OH to $As(OCH_2CF_3)_5$ as shown by the following reaction (Eq. 2.27):

$$AsCl_3 + 3CF_3CH_2OH \xrightarrow[-3HCl]{} As(OCH_2CF_3)_3$$

$$\xrightarrow[]{+Cl_2, +2HOCH_2CF_3} As(OCH_2CF_3)_5 + 2HCl \uparrow \qquad (2.27)$$

Following the earlier observations of Fischer,[129] Klejnot[130] observed that the reaction of WCl_6 with ethyl alcohol can be represented by Eqs (2.28) and (2.29):

$$WCl_6 + 2C_2H_5OH \longrightarrow WCl_3(OEt)_2 + \tfrac{1}{2}Cl_2 + 2HCl \uparrow \qquad (2.28)$$

$$Cl_2 + C_2H_5OH \longrightarrow CH_3CHO + 2HCl \uparrow \qquad (2.29)$$

Chloro-alkoxo derivatives of W(V) can be prepared by the direct reactions of WCl_5 with alcohols at $-70°C$.[131,132] The reaction between WCl_4 and the alcohols ROH (R = Me, Et) leads to the $(W=W)^{8+}$-containing derivatives $W_2Cl_4(OR)_4(HOR)_2$,[133] which have been characterized by X-ray crystallography.

2.5 Reactions of Simple and Complex Metal Chlorides or Double Nitrates with Alcohols in the Presence of a Base (Method E)

On the basis of the earlier observations,[3,4,26] it appears that except for a few metal(loid) halides, most of these undergo only partial solvolysis or no solvolysis even under refluxing conditions. Thus in order to achieve the successful preparation of pure homoleptic metal alkoxides, the use of a base such as ammonia, pyridine, trialkyl-amines, and alkali metal alkoxides appears to be essential. While alkali alkoxides provide anions by direct ionization, the role of other bases (:B) could be to increase the concentration of alkoxide anions according to Eqs (2.30)–(2.32):

$$:B + ROH \rightleftharpoons (HB)^+ + OR^- \qquad (2.30)$$

$$OR^- + M—Cl \longrightarrow M—OR + Cl^- \qquad (2.31)$$

$$(HB)^+ + Cl^- \longrightarrow (BH)^+Cl^- \qquad (2.32)$$

Of the commonly employed bases (NH_3, NaOR, KOR) for completion of the reactions and preparation of soluble metal alkoxides, NH_3 appears to have some distinct advantages including: (i) passage of anhydrous ammonia in a reaction mixture of an anhydrous metal chloride and alcohol produces heat by neutralization of the liberated HCl with NH_3; the cooling of the reaction mixture is an index of the completion of the reaction, (ii) precipitated NH_4Cl can be filtered easily, (iii) excess NH_3 can be easily removed by evaporation, whereas (iv) heterobimetallic alkoxides like $NaAl(OR)_4$ and $KZr_2(OR)_9$ tend to be formed with excess of alkali alkoxides.

2.5.1   The Ammonia Method (E-1)

The addition of a base, typically ammonia, to mixtures of metal(loid) halides and alcohols allows the synthesis of homoleptic alkoxides for a wide range of metals and metalloids. Anhydrous ammonia appears to have been employed for the first time by Nelles[134] in 1939 for the preparation of titanium tetra-alkoxides (Eq. 2.33):

$$TiCl_4 + 4ROH + 4NH_3 \xrightarrow{\text{benzene}} Ti(OR)_4 + 4NH_4Cl \downarrow \qquad (2.33)$$
$$\text{(excess)}$$

Zirconium tetra-alkoxides were prepared for the first time in 1950 by the ammonia method,[135] as earlier attempts[136] to use the alkali alkoxide method did not give a pure product, owing to the tendency of zirconium to form stable heterobimetallic alkoxides[137] (Chapter 3) with alkali metals.

The ammonia method has, therefore, been successfully employed[3,4,21,26] for the synthesis of a large number of alkoxides of main-group and transition metals according to the following general reaction (Eq. 2.34):

$$MCl_x + xROH + xNH_3 \xrightarrow{\text{benzene}} M(OR)_x + xNH_4Cl \downarrow \qquad (2.34)$$

Owing to the highly hydrolysable nature of most of the alkoxide derivatives, stringently anhydrous conditions are essential for successful preparation of the alkoxides. Apart from careful drying of all the reagents as well as solvents, gaseous ammonia should be carefully dried by passage through a series of towers packed with anhydrous calcium oxide, followed preferably by bubbling through a solution of aluminium isopropoxide in benzene.

Benzene has been reported to be a good solvent for the preparation of metal alkoxides by the ammonia method, as its presence tends to reduce the solubility of ammonium chloride, which has a fair solubility in ammoniacal alcohols. In addition, the ammonium chloride precipitated tends to be more crystalline under these conditions, making filtration easier and quicker. Although most of the earlier laboratory preparations have been carried out in benzene, the recently emphasized carcinogenic properties of this solvent suggests that the use of an alternative solvent should be explored.

2.5.1.1   Group 3 and f-block metals

To date no unfluorinated alkoxides of scandium, yttrium, and lanthanides[18,21] in the common +3 oxidation state appear to have been prepared by the ammonia method. By contrast, yttrium and lanthanides (Ln) fluoroalkoxide derivatives of the types Ln{OCH(CF$_3$)$_2$}$_3$[138] and Ln{OCH(CF$_3$)$_2$}$_3$.2NH$_3$[139] have been isolated by this route.

It might appear that the ammonia method is applicable to the synthesis of a large number of metal alkoxides, but there are certain limitations. For example, metal chlorides (such as LaCl$_3$)[140] tend to form a stable and insoluble ammoniate M(NH$_3$)$_y$Cl$_n$ instead of the corresponding homoleptic alkoxide derivative. Difficulties may also arise if the metal forms an alkoxide which has a base strength comparable with or greater than that of ammonia. Thorium provides a good example of this type where the ammonia method has not been found to be entirely satisfactory.[141] For example, during the preparation of thorium tetra-alkoxides from ThCl$_4$ and alcohols, Bradley et al.[142] could obtain only thorium trialkoxide monochlorides owing to the partial replacement of chlorides. These workers observed that the alcoholic solutions of $Th(OEt)_4$ or $Th(OPr^i)_4$ were alkaline to thymolphthalein. On the other hand anhydrous ammoniacal alcohols were acidic to this indicator. Thus thorium tetra-alkoxides tend to be more basic than ammonia and the following feasible equilibria (Eqs 2.35 and 2.36) may be responsible for the formation of $Th(OR)_3Cl$ instead of the expected tetra-alkoxides.

$$Th(OR)_4 + NH_4{}^+ \rightleftharpoons Th(OR)_3{}^+ + NH_3 + ROH \qquad (2.35)$$

$$Th(OR)_3{}^+ + Cl^- \rightleftharpoons Th(OR)_3Cl \qquad (2.36)$$

However, it was observed[141] that treatment of alcoholic solutions of thorium tetrachloride with sodium alkoxides gave thorium tetra-alkoxides.

In search of a convenient method for the synthesis of tetra-alkoxides of cerium(IV) and plutonium(IV), which do not form stable chlorides, the complex chlorides $(C_5H_6N)_2MCl_6$ ($M = Ce(IV)$, $Pu(IV)$; $C_5H_6N = $ pyridinium) method proved to be convenient starting materials:

$$(C_5H_6N)_2MCl_6 + 6NH_3 + 4ROH \longrightarrow M(OR)_4 + 6NH_4Cl \downarrow 2C_5H_5N \qquad (2.37)$$

where $M = Ce^{143}$ or Pu^{144} and $R = Pr^i$.

Bradley $et\ al.$[145] had earlier reported that dipyridinium hexachlorozirconate $(C_5H_6N)_2ZrCl_6$, which can be prepared from the commonly available $ZrOCl_2.8H_2O$, also reacted smoothly with alcohol in the presence of ammonia to form the tetra-alkoxides $Zr(OR)_4$.

During an attempt to prepare tetra-$tert$-alkoxides of zirconium and cerium by the reactions of $(C_5H_6N)_2MCl_6$ ($M = Zr$, Ce) with $tert$-butyl alcohol, Bradley and co-workers[143,144] had noticed the formation of $MCl(OBu^t)_3.2C_5H_5N$ as represented by Eq. (2.38):

$$(C_5H_6N)_2MCl_6 + 3Bu^tOH + 5NH_3 \longrightarrow MCl(OBu^t)_3.2C_5H_5N + 5NH_4Cl \downarrow \quad (2.38)$$

As the product reacts with primary alcohols (Eq. 2.39) in the presence of ammonia to give heteroleptic alkoxides, $M(OR)(OBu^t)_3$, steric reasons have been suggested as a possible explanation for the partial replacement reactions with $tert$-butyl alcohol:

$$MCl(OBu^t)_3.2C_5H_5N + EtOH + NH_3 \longrightarrow M(OEt)(OBu^t)_3 + 2C_5H_5N + NH_4Cl \downarrow$$
$$(2.39)$$

It is, however, somewhat intriguing that dipyridinium hexachloro derivatives of zirconium and cerium[146] undergo complete replacement with $Cl_3C.CMe_2OH$, which should apparently be an even more sterically hindered alcohol than Bu^tOH:

$$(C_5H_6N)_2MCl_6 + 4Cl_3C.CMe_2OH + 6NH_3$$
$$\longrightarrow M(OCMe_2CCl_3)_4 + 2C_5H_5N + 6NH_4Cl \downarrow \qquad (2.40)$$

Reactions of MCl_4 ($M = Se$, Te) with a variety of alcohols (MeOH, EtOH, CF_3CH_2OH, Bu^tCH_2OH, Me_2CHOH) in 1:4 molar ratio in THF using Et_3N as a proton acceptor afford corresponding tetra-alkoxides.[146a]

2.5.2   The Sodium (or Potassium) Alkoxide Method (E-2)

This procedure, sometimes referred to as transmetallation or (metathesis or salt-elimination) reaction, is by far the most versatile synthetic method for a wide range of d- and p-block metal alkoxide complexes. The alkali metal (usually sodium or potassium) alkoxide is treated in the presence of excess alcohol with the corresponding metal(loid) halide either in a hydrocarbon (generally benzene) or an ether solvent (Eq. 2.41):

$$MCl_x + xM'OR \longrightarrow M(OR)_x + xM'Cl \downarrow \qquad (2.41)$$

where M = a metal or metalloid and M' = Na or K.

Although this procedure normally results in complete substitution, except for the sterically more demanding alkoxo groups, 100% synthetic predictability is not likely to be achieved. The generality and limitations of Eq. 2.41 for a wide variety of elements may be reflected by the group-wise discussion that follows.

2.5.2.1   s-Block metals

The reaction between CaI_2 and $KOC(Ph)_2CH_2C_6H_4Cl-4$ in (THF) affords a soluble and monomeric alkoxide complex[147] (Eq. 2.42):

$$CaI_2 + 2KOC(Ph)_2CH_2C_6H_4Cl-4 \xrightarrow{THF} [Ca\{OC(Ph)_2CH_2C_6H_4Cl-4\}_2(thf)_n] + 2KI \downarrow$$
$$(2.42)$$

Interaction of BaI_2 and the potassium salt of a donor-functionalized alcohol gives a dimeric product[53] according to Eq. (2.43):

$$BaI_2 + 2KOC(Bu^t)(CH_2OPr^i)_2 \xrightarrow[-2KI]{THF} \tfrac{1}{2}[Ba\{OC(Bu^t)(CH_2OPr^i)_2\}_2]_2 \qquad (2.43)$$

2.5.2.2   Group 3 and f-block metals

The addition of a sodium (or potassium) isopropoxide to an appropriate $LnCl_3.3Pr^iOH$ in a medium of isopropyl alcohol and benzene results in the precipitation of NaCl (or KCl), which is removed by filtration. From the filtrate, quantitative yields of $[Ln(OPr^i)_3]_x$ can be isolated (Eq. 2.44):[103,148-158]

$$LnCl_3.3Pr^iOH + 3MOPr^i \longrightarrow \frac{1}{n}[Ln(OPr^i)_3]_n + 3Pr^iOH + 3MCl \downarrow \qquad (2.44)$$

where Ln = Y,[103,148,149] La,[150,156] Pr,[150-153] Nd,[150-153] Sm,[154,155] Gd,[149,154,156,157] Ho,[158] Er,[149,152,156,157] Yb.[103,148,149]

The above reactions (Eq. 2.44) are frequently not straightforward; a plethora of different (generally unusual) products are generated[18,21] in such reactions even by slight variations in the experimental conditions/manipulations, the order of reactant(s) addition, the stoichiometry of reactants,[159] and the nature of the alkoxide groups.[159-161] Although there appears to be little synthetic control over the nature or structures of the products, an impressive series of structurally novel species have been obtained *via* reactions of alkali metal alkoxides with the appropriate lanthanide chloride or ceric ammonium nitrate, as illustrated below.

Interesting partial substitution reactions between yttrium trichloride and sodium *tert*-butoxide in different (1:2 and 1:3) molar ratios have been reported (Eqs 2.45 and 2.46) by Evans and co-workers:[160,161]

$$3YCl_3 + 7NaOBu^t \xrightarrow{THF} Y_3(OBu^t)_7Cl_2(thf)_2 + 7NaCl \downarrow \qquad (2.45)$$
$$\text{(Complex } \mathbf{A}, 80\%)$$

$$3YCl_3 + 8NaOBu^t \xrightarrow{THF} Y_3(OBu^t)_8Cl(thf)_2 + 8NaCl \downarrow \qquad (2.46)$$
$$\text{(Complex } \mathbf{B}, 80\%)$$

X-ray structural determinations have shown complexes **A** and **B** to have structures which can be represented as $[Y_3(\mu_3\text{-}OBu^t)(\mu_3\text{-}Cl)(\mu_2\text{-}OBu^t)_3(OBu^t)_3Cl(thf)_2]$ and $[Y_3(\mu_3\text{-}OBu^t)(\mu_3\text{-}Cl)(\mu_2\text{-}OBu^t)_3(OBu^t)_4(thf)_2]$, respectively.

These workers further showed[161] that the complex **A** could be converted into **B** by treating with the requisite amount of $NaOBu^t$, but further reaction led to insoluble products:

$$\text{Complex } \mathbf{A} + NaOBu^t \longrightarrow \text{Complex } \mathbf{B} \qquad (2.47)$$

$$\text{Complex } \mathbf{B} + NaOBu^t \longrightarrow \text{Insoluble product} \qquad (2.48)$$

The above findings are rather unusual and intriguing in view of the general trends of metal alkoxide chemistry and are somewhat at variance with the earlier findings (mainly on isopropoxide derivatives) from the research groups of Mehrotra[103,148−150,156−158] and Mazdiyasni.[54,55] Also, although most of the 1:3 reactions of lanthanide trichlorides and alkali alkoxides (mainly methoxides, ethoxides, isopropoxides and even 2-methoxyethoxides[57] have been reported to be quantitative, yet a product (with incomplete chloride substitution) had been reported[159] as $Nd_6(OPr^i)_{17}Cl$ with an interesting structure, in the reaction of $NdCl_3$ with three equivalents of $NaOPr^i$.

By contrast, the reaction of $LaCl_3$ with three equivalents of $NaOBu^t$ in THF was reported[160] to be straightforward, yielding the homoleptic alkoxide complex, $[La(OBu^t)_3]_3 \cdot 2thf$, as represented by Eq. (2.49):

$$3LaCl_3 + 9NaOBu^t \xrightarrow{THF} La_3(\mu_3\text{-}OBu^t)_2(\mu_2\text{-}OBu^t)_3(OBu^t)_4(thf)_2 + 9NaCl \quad (2.49)$$

Starting with the readily available $(NH_4)_2Ce(NO_3)_6$ (CAN), synthesis of Ce(IV) alkoxides, $Ce(OR)_4$, has been reported[162,163] by the reaction represented by Eq. (2.50):

$$(NH_4)_2Ce(NO_3)_6 + 4ROH + 6NaOMe \longrightarrow Ce(OR)_4 + 6NaNO_3 + 2NH_3 + 6MeOH \qquad (2.50)$$

where R = Me, Et, Pr^i, or n-octyl.

The above convenient method was extended by Evans and co-workers[164] in 1989 to the synthesis of a series of ceric *tert*-butoxide complexes with the general formula, $Ce(OBu^t)_n(NO_3)_{4-n}(solvent)$ by the reactions of CAN with $NaOBu^t$ in the appropriate solvent (S = THF or Bu^tOH):

$$(NH_4)_2Ce(NO_3)_6 + (2+n)NaOBu^t$$

$$\xrightarrow{THF \text{ or } Bu^tOH} Ce(OBu^t)_n(NO_3)_{4-n}(S)_2 + 2NH_3 + (n+2)NaNO_3 + 2Bu^tOH \quad (2.51)$$

The reactions in THF have been reported to be cleaner with higher yields in all the cases when $n = 1$–4. The formation of heterobimetallic alkoxides, $Na_2Ce(OR)_6$ and $NaCe_2(OR)_9$ was also reported, when $n = 6$ or 4.5, respectively.

The above novel synthesis of cerium(IV) *tert*-butoxide nitrate complexes appears to be the only report using a nitrate salt as the starting material for the preparation of tertiary butoxo derivatives. It may, therefore, be worthwhile to investigate similar routes for alkoxo (particularly *t*-butoxo) derivatives of other metals (especially in their higher oxidation states) starting with their nitrate salts. The chances of success in these efforts may be higher in cases where the nitrate groups are bonded predominantly in a monodentate manner.

The X-ray crystallographically characterized[165] tetrameric thorium(IV) isopropoxide complex, $Th_4(OPr^i)_{16}(py)_2$, has been prepared in 80% yield by the (1:4) reaction of $ThBr_4(thf)_4$ with $KOPr^i$ in THF followed by addition of excess pyridine (py):

$$ThBr_4(thf)_4 + 4KOPr^i \xrightarrow[\text{(ii) py}]{\text{(i) THF}} \tfrac{1}{4}[Th_4(OPr^i)_{16}](py)_2 + 4KBr \downarrow \qquad (2.52)$$

2.5.2.3 d-Block metals

Titanium tetraethoxide was first obtained pure by Bischoff and Adkins[166] by the reaction of sodium ethoxide with titanium tetrachloride (Eq. 2.53).

$$TiCl_4 + 4NaOEt \longrightarrow Ti(OEt)_4 + 4NaCl \downarrow \qquad (2.53)$$

By contrast, the alkali alkoxide route appears to be inapplicable for the synthesis of zirconium tetra-alkoxides or niobium (tantalum) penta-alkoxides, as these tend to form heterobimetallic alkoxides with alkali metal alkoxides (Chapter 3, Section 3.2.1.1), which volatilize out during final purification, whereas alkali titanium alkoxides, even if formed, dissociate readily to give volatile titanium alkoxides.

The alkali alkoxide method has been extended[167] to the preparation of alkoxides of the hexanuclear niobium and tantalum cluster units, *e.g.*, $[M_6X_{12}](OMe)_2.4MeOH$ (where $M = Nb$ or Ta and $X = Cl$ or Br), and $M_2[Ta_2Cl_{12}](OMe)_6.6MeOH$.

The chromium(IV) alkoxides, $Cr(OR)_4$ (where $R = OBu^t$ or 1-adamantoxide), can be prepared by the reaction of $CrCl_3(thf)_3$ with 4 equivalents of the corresponding K (or Na) alkoxide in the presence of cuprous chloride in THF:

$$CrCl_3(thf)_3 + 4MOR + CuCl \xrightarrow{\text{THF}} Cr(OR)_4 + 4MCl \downarrow + Cu \downarrow \qquad (2.54)$$

$$(M = K,^{168} R = Bu^t; M = Na,^{169} R = 1\text{-adamantyl})$$

The sodium method has been successfully employed for the synthesis of a triply metal–metal bonded fluoroalkoxide of molybdenum (Eq. 2.55).[170]

$$Mo_2Cl_6(dme) + 6NaOCMe_2CF_3 \longrightarrow Mo_2(OCMe_2CF_3)_6 + 6NaCl \downarrow \qquad (2.55)$$

where dme = 1,2-dimethoxyethane.

The reaction of $WCl_4(PMe_2Ph)_2$ with $TlOCH_2CF_3$ in 1:2 molar ratio gives $W(OCH_2CF_3)Cl_2(PMe_2Ph)_2$, which has been studied by spectroscopic and X-ray diffraction methods.[171]

The heteroleptic chloride *tert*-butoxide of rhenium $Re_3(OBu^t)_6Cl_3$ was prepared by Wilkinson *et al.*[172] by the reaction in THF of $Re_3(\mu\text{-}Cl)_3Cl_6(thf)_3$ with $NaOBu^t$ in 1:6 molar ratio:

$$Re_3(\mu\text{-}Cl)_3Cl_6(thf)_3 + 6NaOBu^t \xrightarrow{\text{THF}} Re_3(OBu^t)_6Cl_3 + 6NaCl \downarrow \qquad (2.56)$$

In an attempt to prepare $Re_3(OPr^i)_6Cl_3$, Hoffman *et al.*[173,174] reacted $Re_3(\mu\text{-}Cl)_3Cl_6(thf)_3$ with $NaOPr^i$ in 1:6 molar ratio in THF and isolated the X-ray crystallographically characterized[174] green complex $Re_3(\mu\text{-}OPr^i)_3(OPr^i)_6 \cdot \frac{1}{3}Pr^iOH$ in a very low (18%) yield. Naturally the unreacted $Re_3(\mu\text{-}Cl)_3Cl_6(thf)_3$ had to be removed from the reaction medium.

By contrast, an isopropyl alcohol-free product $Re_3(\mu\text{-}OPr^i)_3(OPr^i)_6$ has been prepared, but again in a low (31%) yield according to Eq. (2.57):

$$Re_3(\mu\text{-}Cl)_3Cl_6(thf)_3 + 9NaOPr^i \xrightarrow{\text{THF}} Re_3(\mu\text{-}OPr^i)_3(OPr^i)_6 + 9NaCl \downarrow \qquad (2.57)$$

The yield of $Re_3(\mu\text{-}OPr^i)_3(OPr^i)_6$ (Eq. 2.57) could be improved (i.e. from 31% to 53%) considerably by the addition of a few drops of acetone to the solvent of crystallization.[174]

The synthesis of simple generally insoluble alkoxides, $M(OR)_2$ (M = Mn, Fe, Co, Ni, Cu, Zn; R = Me, Et, Pr^i), was found not to be feasible owing to the difficulty of separating them from the insoluble alkali metal (Na, K) chlorides (*cf.* preparation through LiOR in Section 2.5.3). However, a soluble copper(II) fluoroalkoxide, $Cu(OR_f)_2(py)_2$ where $R_f = CH(CF_3)_2$ or $C(CF_3)_3$,[175] has been synthesized as shown in Eq. (2.58):

$$CuBr_2 + 2NaOR_f + 2py \longrightarrow Cu(OR_f)_2(py)_2 + 2NaBr \downarrow \qquad (2.58)$$

where $R_f = CH(CF_3)_2$ or $C(CF_3)_3$.

In the absence of ancillary ligands such as C_5H_5, CO, PR_3, and $R_2PCH_2CH_2PR_2$ (R = alkyl or aryl) there are relatively few stable platinum group metal (Ru, Rh, Pd; Os, Ir, Pt) alkoxides[9,10,21] because these metals prefer softer donor ligands (relative to the hard (oxygen) donor alkoxo groups) and the alkoxide ligands when bonded to platinum group metals are labile to thermal decomposition (Section 2.1), typically by a β-hydrogen elimination pathway. However, with the use of some special (fluorinated and/or donor-functionalized) type of alkoxo ligands,[21] the synthesis of hydrocarbon-soluble and monomeric alkoxides of later transition metals including palladium(II) and platinum(II) can be achieved (Eqs 2.59–2.62):

$$cis\text{-}(R_3P)_2MCl_2 + 2NaOCH(CF_3)_2 \longrightarrow cis\text{-}(R_3P)_2M\{OCH(CH_3)_2\}_2 + 2NaCl \downarrow \qquad (2.59)$$

$$(M = Ni, R = Et;^{176} \; M = Pt, R = Ph^{176})$$

$$(dppe)PtCl_2 + 2NaOCH_3 \xrightarrow[\text{slight excess}]{C_6H_6/\text{MeOH}} (dppe)Pt(OCH_3)_2 + 2NaCl \downarrow^{177} \qquad (2.60)$$

$$M^{2+} + 2\,HOC(CF_3)_2CH_2PPh_2 \xrightarrow{2KOH} M\{OC(CF_3)_2CH_2PPh_2\}_2 + 2H^+ \qquad (2.61)$$

where M = Co, Ni, Pd, Pt.[10,24]

$$\text{PtCl}_2(\text{NCBu}^t)_2 + 2\,\text{R}''_2\text{PCH}_2\text{CRR}'\text{OH} \xrightarrow[\text{(ii)}\ x\text{H}_2\text{O}]{\text{(i) NaOH in MeOH}} \text{Pt(OCRR}'\text{CH}_2\text{PR}''_2)_2.(\text{H}_2\text{O})_x$$

$$(2.62)$$

$(\text{R} = \text{R}' = \text{H}, \text{R}'' = \text{Ph}, x = 0 \text{ or } 1;^{178}\ \text{R} = \text{R}' = \text{Me}, \text{R}'' = \text{Ph}, x = 0^{178})$

2.5.2.4 p-Block elements

Meerwein and Bersin[179] investigated the reaction of sodium ethoxide with an alcoholic solution of aluminium trichloride and isolated a product of composition NaAl(OEt)_4.

Bradley et al.[180] attempted to prepare tin tetra-alkoxide by the reactions of tin tetrachloride with sodium alkoxides, but the resultant product was a heterobimetallic alkoxide of tin and sodium:

$$2\text{SnCl}_4 + 9\text{NaOEt} \longrightarrow \text{NaSn}_2(\text{OEt})_9 + 8\text{NaCl} \downarrow \qquad (2.63)$$

However, $\text{NaSn}_2(\text{OEt})_9$ on further treatment with hydrogen chloride or the alcoholate of tin trichloride monoethoxide afforded tin tetraethoxide:

$$3\text{NaSn}_2(\text{OEt})_9 + \text{SnCl}_3(\text{OEt}).\text{EtOH} \longrightarrow 7\text{Sn(OEt)}_4 + 3\text{NaCl} \downarrow + \text{EtOH} \quad (2.64)$$

Interestingly, homometallic alkoxides of gallium,[181–185] indium,[186,187] germanium,[124,188] tin,[189–191] lead,[192–194] arsenic,[195–198] antimony,[195–197,199–201] bismuth,[202–206] and tellurium[207,208] can easily be prepared according to the general reaction (Eq. 2.65):

$$\text{MX}_x + x\text{M}'\text{OR} \longrightarrow \text{M(OR)}_x + x\text{M}'\text{X} \downarrow \qquad (2.65)$$

where M = metal(loid) of groups 13, 14, 15, and 16; X = halide (usually chloride); M′ = Na or K; R = a simple or substituted alkyl group; x = valency of the metal(loid).

The tendency for the formation of heterobimetallic alkoxides with alkali metals,[179,180,195] has generally restricted the applicability of this procedure for the synthesis of homometallic alkoxides of many of these metals, except in cases where the heterobimetallic alkoxides are thermally labile and dissociate to yield the corresponding volatile homometallic alkoxides of p-block metals. However, under suitable conditions, when addition of an excess of alkali metal alkoxide is avoided the chances of the formation of binary alkoxides in an excellent yield is high.

In spite of the above difficulties, this route has been found convenient even for the synthesis of heteroleptic alkoxides such as $\text{M(OBu}^t)\text{Cl}$ (M = Ge, Sn)[209] and AsCl(OEt)_2.[195]

The stoichiometric reactions of alkyltin chlorides with sodium alkoxides generally yield the alkyltin alkoxide derivatives. The method was finally extended to the preparation of $\text{Bu}^n\text{Sn(OR)}_3$,[210,211] in spite of earlier unsuccessful attempts:[212]

$$\text{Bu}^n\text{SnCl}_3 + 3\text{NaOR} \xrightarrow[\text{isopropyl alcohol}]{\text{benzene}} \text{Bu}^n\text{Sn(OR)}_3 + 3\text{NaCl} \downarrow \qquad (2.66)$$

2.5.3   The Lithium Alkoxide Method (E-3)

2.5.3.1 Synthesis of insoluble metal alkoxides (usually methoxides)

Although ammonia and sodium (potassium) alkoxides appear to be conveniently employed as proton acceptors for the preparation of metal alkoxides, the method is not conveniently applicable for the synthesis of insoluble metal methoxides, owing to difficulties in their separation from insoluble ammonium or/and alkali halides. A new method was, therefore, developed by Gilman et al.[213] for the preparation of insoluble uranium tetramethoxide by the reaction of uranium tetrachloride with lithium methoxide in methanol:

$$UCl_4 + 4LiOMe \longrightarrow U(OMe)_4 \downarrow + 4LiCl \qquad (2.67)$$

The advantage of this method over the others described in the preceding sections is that the lithium chloride is soluble in methanol, making the separation of insoluble methoxides easily possible. The method was later extended by other workers for the preparation of methoxides of beryllium,[214] lanthanum,[215] uranium,[216] neptunium,[217] vanadium,[218] iron,[219,220] and copper.[221]

By contrast, pure binary methoxide of zinc[222] and ethoxides[223] of zinc, cadium, and mercury could not be isolated by the lithium alkoxide method possibly because heterobimetallic alkoxide complexes were formed.

Interestingly, Talalaeva et al.[224] have prepared $Zn(OBu^t)_2$ by the reaction of zinc chloride with lithium tert-butoxide in ether solution.

The lithium alkoxide method has been extensively exploited,[7,225] by Mehrotra and co-workers for the synthesis of insoluble alkoxides of later transition metals (Eqs 2.68 and 2.69).

$$CrCl_3(thf)_3 + 3LiOR \xrightarrow{ROH/C_6H_6} Cr(OR)_3 \downarrow + 3LiCl \qquad (2.68)$$

where R = Me, Et, Bu^n; when R = Bu^t, pure product is obtained only when free tert-butyl alcohol is carefully excluded.

$$MCl_2 + 2LiOR \xrightarrow{ROH/C_6H_6} M(OR)_2 \downarrow + 2LiCl \qquad (2.69)$$

(M = Co, R = Me, Et, or Pr^i; M = Ni, R = Me, Et, Pr^n, Pr^i, Bu^s, Bu^t, t-C_5H_{11}, or t-C_6H_{13}; M = Cu, R = Me, Et, Pr^i, or Bu^t)

2.5.3.2   Synthesis of soluble metal alkoxides

The sterically compact metal alkoxides prepared by the lithium alkoxide method are generally insoluble and nonvolatile, owing to the formation of oligomers or polymers involving alkoxo bridging. Minimizing molecular oligomerization, and hence lattice cohesive energies, by saturating the metal coordination sphere with sterically encumbered (mono- and/or multi-dentate) or halogenated (preferably fluorinated) and/or donor-functionalized alkoxo ligands is an attractive strategy for the design and synthesis of hydrocarbon-soluble and volatile derivatives (Table 2.1). In this context the use of lithium derivatives of such ligands,[21] which are conveniently prepared by the interaction of an alcohol with butyllithium, has played an important role. For example, the reactions of LiOR (R = a sterically demanding alkyl, halogenated alkyl, or alkyl with donor functionalities) with metal halides yield derivatives generally with unprecedented structural and reactivity features. The synthesis of such soluble derivatives is generally carried out in Et_2O or THF solvents, in which LiCl tends to be precipitated (Eqs 2.70–2.74):

Homoleptic derivatives

$$MCl_x(thf) + xLiOR \xrightarrow{Et_2O \text{ or } THF(L)} \frac{1}{m}[M(OR)_x(L)_y]_m + xLiCl \downarrow \qquad (2.70)$$

(M = Cr,[226] R = CHBu$_2^t$, $x = 3$, L = THF, $n = 3$, $y = 1$, $m = 1$; M = Cr,[226] R = CHBu$_2^t$, $x = 3$, L = Et$_2$O, $n = 0$, $y = 0$, $m = 1$; M = Y, Nd,[227] R = CBu$_2^t$CH$_2$PMe$_2$, $x = 3$, L = THF, $n = 0$, $y = 0$, $m = 1$; M = U,[228] R = CHBu$_2^t$, $x = 4$, L = THF, $n = 0$, $y = 0$, $m = 1$).

The above reactions when carried out in 1:4 molar ratio with CrCl$_3$(thf)$_3$ and FeCl$_3$[226] result in heterobimetallic alkoxide complexes, respectively of the types [(thf)Li(μ-OCHBu$_2^t$)$_2$Cr(OCHBu$_2^t$)$_2$] and [(HOCBu$_2^t$)Li(μ-OCHBu$_2^t$)$_2$Fe(OCHBu$_2^t$)$_2$].

Heteroleptic derivatives

$$MCl_x + nLiOR \xrightarrow{Et_2O \text{ or } THF/hexane} M(OR)_nCl_{x-n}(L)_y + LiCl \downarrow \qquad (2.71)$$

(M = Ti,[228] R = CBu$_2^t$CH$_2$PMe$_2$,[227a] CBu$_3^t$,[228] $x = 3$, $n = 2$, $y = 0$; M = Ti, Zr,[228] R = CBu$_2^t$, $x = 4$, $n = 1$ or 2, $y = 0$; M = U,[229] $x = 4$, $n = 2$, $y = 2$ (L = Et$_2$O)).

The same reaction (Eq. 2.71) carried out with chromium trichloride or cobalt dichloride with two equivalents of the lithium reagent yields the heterobimetallic species [(thf)$_2$Li(μ-OCBu$_3^t$)Cr(OCBu$_3^t$)]^{230} or [(thf)$_3$Li(μ-Cl)Co(OCBu$_3^t$)$_2$],[231] respectively. The reaction represented by Eq. (2.71) has also been used in the preparation of alkoxide derivatives of vanadium(III), chromium(VI), and molybdenum(III) (Eqs 2.72–2.74):

$$VCl_3 + xLiOCH_2CF_3 \xrightarrow{benzene} VCl_{3-x}(OCH_2CF_3)_x + xLiCl \downarrow \qquad (2.72)$$

where $x = 1$ or 2,[232]

$$CrO_2Cl_2 + xLiOR \longrightarrow CrO_2(OR)_xCl_{2-x} + xLiCl \downarrow \qquad (2.73)$$

where R = CH$_2$CF$_3$, CH$_2$CCl$_3$, CH$_2$CH$_2$Cl when $x = 1$ or 2[233] and R = CH(CF$_3$)$_2$, $x = 2$.[234]

$$Mo_2Cl_6(dme) + 6LiOR \xrightarrow{DME} Mo_2(OR)_6 + 6LiCl \downarrow + DME \qquad (2.74)$$

where R = Bu^t, CMe$_2$Et.[170]

2.6   Preparation of Alkoxide Derivatives from Metal(loid) Hydroxides and Oxides (Method F)

Hydroxides and oxides of non-metals behave as oxyacids or acid anhydrides and therefore react with alcohols to form esters (alkoxides of these non-metallic elements) and water:

$$M(OH)_x + xROH \rightleftharpoons M(OR)_x + xH_2O \uparrow \qquad (2.75)$$

$$MO_x + 2xROH \rightleftharpoons M(OR)_{2x} + xH_2O \uparrow \qquad (2.76)$$

In view of the reversible nature of the reactions, a continuous removal of water during these reactions is necessary to yield the final alkoxide products. This has been accomplished by fractionating out water with organic solvents (benzene, toluene, and xylene) using a Dean–Stark assembly. The preparation of ethoxides by this method has an additional advantage that ethanol forms a minimum boiling ternary azeotrope (water–ethanol–benzene) which helps in the fractionation of water.

The above technique has been quite successful for the synthesis of alkoxides of a number of s- and p-block metals and metalloids such as sodium,[235,236] boron,[237,238] thallium,[239,240] silicon,[241–243] and arsenic.[244–246]

Although synthesis of $VO(OR)_3$ by the reaction of V_2O_5 with alcohols was described as early as 1913,[247] only a few more vanadium oxo-alkoxides[248,249] appear to have been synthesized as shown by Eqs (2.77) and (2.78):

$$V_2O_5 + 6ROH \longrightarrow 2VO(OR)_3 + 3H_2O \uparrow \qquad (2.77)$$

$$2NH_4VO_3 + 6ROH \longrightarrow 2VO(OR)_3 + 2NH_3 \uparrow + 4H_2O \uparrow \qquad (2.78)$$

Interestingly, $Mo_2O_5(OCH_3)_2.2CH_3OH$ has been detected[250] as a co-condensation product of MoO_3 vapour with methyl alcohol, water, and THF at $-196°C$.

The formation of homoleptic alkoxides from oxides and hydroxides appears to be mainly confined to the alkali metals and monovalent thallium as well as boron, silicon, and arsenic. This might be due to the high lattice energies of oxides of higher valency metals.

In view of the importance of alkoxysilanes and alkoxysiloxanes as precursors for glasses and ceramic materials, a process of obtaining these from portland cement and silicate minerals was described in 1990.[251] Under the mild reaction conditions employed, the silicon–oxygen framework in the original mineral tends to be retained in the final alkoxysilane or alkoxysiloxane obtained, e.g.:

$$Ca_3(SiO_4)O + EtOH \longrightarrow (HO)_4Si.xEtOH \xrightarrow{H+} Si(OEt)_4 \qquad (2.79)$$
$$\text{tricalcium silicate}$$
$$\text{(based on } SiO_4 \text{ framework)}$$

$$CaSiO_3 + CaCl_2.2H_2O \xrightarrow[\Delta]{N_2} Ca_8(SiO_3)_4Cl_8 \xrightarrow[+H^+]{+EtOH} [(HO)_2SiO]_4.xEtOH$$
$$\text{wollastonite}$$

$$\xrightarrow[+H^+]{EtOH} [(EtO)_2SiO]_4 \qquad (2.80)$$

Silica in rice-hull ash (92% silica) loaded with 5 wt% potassium hydroxide has been shown to react with dimethyl carbonate in the temperature range 500–600 K to give tetramethoxysilane[252] in 80% yield:

$$SiO_2 + 2(MeO)_2CO \longrightarrow Si(OMe)_4 + 2CO_2 \uparrow \qquad (2.81)$$

Preparation of organometal alkoxides of germanium,[253–255] tin,[256–266] and lead[267,268] have also been described by the reactions of an appropriate organometal oxide or hydroxide with alcohols. The reaction between an organometal oxide and a dialkyl carbonate has also been found to yield the corresponding organometal alkoxides of tin[261] and mercury.[269]

2.7   Alcohol Interchange (or Alcoholysis) Reactions of Metal(loid) Alkoxides (Method G)

One of the characteristic features of metal(loid) alkoxides is their ability to exchange alkoxo groups with alcohols, and this has been widely exploited for the synthesis of new homo- and heteroleptic alkoxide derivatives of various s- , d- , f- , and p-block elements such as beryllium,[219] yttrium,[103,148] titanium,[270–272] zirconium,[135,273–276] hafnium,[277] vanadium,[249,278] niobium,[279–281] tantalum,[280] iron,[282,283] copper,[7,284,285] zinc,[222] cerium,[286,287] praseodymium,[151–153] neodymium,[151–153] samarium,[155] gadolinium,[103,156,157] erbium,[103,156,157] ytterbium,[103,148] thorium,[141,287,288] uranium,[289–291] boron,[111,237,292] aluminium,[293,294] gallium,[184,185] indium,[186] germanium,[113,125,295] tin,[296–300] antimony,[301] and tellurium[208,302] according to the general reaction (Eq. 2.82):

$$M(OR)_x + nR'OH \rightleftharpoons M(OR)_{x-n}(OR')_n + nROH \uparrow \qquad (2.82)$$

If the alcohol R'OH has a higher boiling point than ROH, then the desired product can be easily obtained by shifting the equilibrium of Eq. (2.82) by removing ROH (preferably as an azeotrope with benzene) by fractional distillation.

Many of the final products (particularly those containing sterically congested and chelating alkoxo ligands) prepared by this route assume special significance because of their reduced molecularity, enhanced solubility (in organic solvents), and volatility as well as novelty in structural features (Chapter 4).

There are three important factors that influence the extent of substitution[303,304] in alcoholysis reactions: (i) the steric demands of the alkoxo groups (OR and OR') involved, (ii) the relative O—H bond energies of the reactant and product alcohols, and (iii) the relative bond strengths of the metal–alkoxo bonds of the reactant and product alkoxides, and, in view of their wide applicability, it may be appropriate to discuss briefly the general conditions which govern such equilibria (Eq. 2.82) employed for synthetic purposes.

2.7.1   Interchangeability of Different Alkoxy Groups

In general, the facility for interchange of alkoxy groups increases from tertiary to secondary to primary groups. Verma and Mehrotra[270] tried to determine the extent of such equilibria in the case of titanium alkoxides, $Ti(OR)_4$ and found the following order in the interchangeability of alkoxo groups in alcoholysis reactions: $MeO^- > EtO^- > Pr^iO^- > Bu^tO^-$.

Such an alcoholysis reaction from a more branched alkoxide to a less branched alkoxide is sometimes facilitated even further, if the product is significantly more associated than the reactant alkoxide. For example, Mehrotra[305] has shown that the reactions of monomeric zirconium tetra-*tert*-butoxide with methanol and ethanol in lower stoichiometric ratios are highly exothermic, resulting in almost instantaneous crystallization of the dimeric mixed alkoxide products:

$$2Zr(OBu^t)_4 + 2MeOH \longrightarrow [Zr(\mu\text{-}OMe)(OBu^t)_3]_2 + 2Bu^tOH \qquad (2.83)$$

2.7.2 Steric Factors

The high rate of alcohol interchange has some interesting mechanistic implications especially in view of the strong ($100-110$ kcal mol^{-1}) metal–oxygen bonds[306,307] in titanium alkoxides. The presence of vacant d-orbitals in most of the metals offers a facile initial step in a nucleophilic attack (S_{N2}) of an alcohol molecule on the metal alkoxide (Eq. 2.84), and thus a low activation energy for alcohol interchange involving a four-membered cyclic transition state seems reasonable:

$$\begin{array}{c}
\underset{H}{\overset{R'}{\diagdown}}O \longrightarrow \underset{RO}{\overset{RO}{\diagdown}}M\underset{OR}{\overset{OR}{\diagup}} \longrightarrow \underset{R'}{\overset{H}{\diagdown}}O \longrightarrow \underset{R'O}{\overset{RO}{\diagdown}}M\underset{OR}{\overset{OR}{\diagup}} \longrightarrow \underset{R'O}{\overset{R'O}{\diagdown}}M\underset{OR}{\overset{OR}{\diagup}} \longrightarrow \text{and so on}
\end{array}$$

(+ROH)   (+ROH)

$$(2.84)$$

Obviously, such reactions would be susceptible to steric factors. For example, in an attempt to measure the kinetics of alcohol interchange in titanium and zirconium alkoxides, Bradley[3] observed that an equilibrium was established between the reactants when they were mixed at room temperature. This was later confirmed by ^{1}H NMR spectra which indicated that a mixture of titanium tetraethoxide and ethanol gave only one type of ethoxy signal, thus indicating the rapidity of the exchange of ethoxy groups. A similar observation has been made by Mehrotra and Gaur[308] who found only one type of isopropyl protons in the ^{1}H NMR spectrum of titanium dibromide diisopropoxide in isopropyl alcohol solution indicating a dynamic equilibrium (Section 3.4). On the other hand, a mixture of titanium *tert*-butoxide and *tert*-butyl alcohol showed different types of methyl signals indicating that the rate of exchange is slow, which may be ascribed to steric factors.[26]

2.7.3 Fractionation of More Volatile Product

Even in cases where reactions are rather slow, the equilibrium can be pushed to completion if the alcohol produced in the reaction is continuously fractionated out. For example, alcoholysis of aluminium isopropoxide with primary as well as secondary butyl alcohols can be completed by fractionating out the isopropyl alcohol produced:[293]

$$Al(OPr^i)_3 + 3Bu^nOH \longrightarrow Al(OBu^n)_3 + 3Pr^iOH \uparrow \qquad (2.85)$$

However, in the case of *tert*-butyl alcohol even after careful fractionation, a maximum of only two isopropoxy groups per aluminium atom appears to be replaced:[293]

$$Al(OPr^i)_3 + 2Bu^tOH \longrightarrow \tfrac{1}{2}[Al(OBu^t)_2(OPr^i)]_2 + 2Pr^iOH \uparrow \qquad (2.86)$$

As the product Al(OPr^i)(OBu^t)$_2$ is found to be dimeric it was suggested that aluminium atoms on being surrounded by bulky isopropoxy and *tert*-butoxy groups in a structure of the type (2-I), are shielded so effectively that the lone pair orbital of the oxygen atom of another *tert*-butyl alcohol molecule cannot approach sufficiently close to the 'd' orbitals of aluminium for the interaction to be initiated. A finer difference in susceptibilities to steric factors was further demonstrated by the fact that with aluminium ethoxide, some further (albeit extremely slow) replacement was possible,

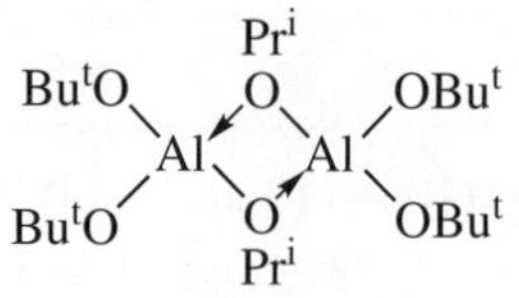

$$
\begin{array}{c}
\text{Pr}^i \\
\text{Bu}^t\text{O} \qquad \text{O} \qquad \text{OBu}^t \\
\diagdown\text{Al}\diagup\diagdown\text{Al}\diagdown \\
\text{Bu}^t\text{O}\diagup \qquad \text{O} \qquad \diagdown\text{OBu}^t \\
\text{Pr}^i
\end{array}
$$

$$(2\text{-}I)$$

resulting in a final product of the composition $Al_2(OEt)(OBu^t)_5$. It appears, therefore, that the corresponding $[Al(OEt)(OBu^t)_2]_2$ may also have (2-II) as one possible structure,[293] in which further replacement at the less shielded of the two aluminium atoms may lead to further reaction leading to a structure of the type (2-III).

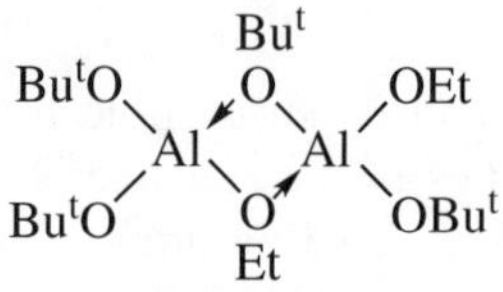

$$
\begin{array}{c}
\text{Bu}^t \\
\text{Bu}^t\text{O} \qquad \text{O} \qquad \text{OEt} \\
\diagdown\text{Al}\diagup\diagdown\text{Al}\diagdown \\
\text{Bu}^t\text{O}\diagup \qquad \text{O} \qquad \diagdown\text{OBu}^t \\
\text{Et}
\end{array}
$$

$$(2\text{-}II)$$

$$
\begin{array}{c}
\text{Bu}^t \\
\text{Bu}^t\text{O} \qquad \text{O} \qquad \text{OBu}^t \\
\diagdown\text{Al}\diagup\diagdown\text{Al}\diagdown \\
\text{Bu}^t\text{O}\diagup \qquad \text{O} \qquad \diagdown\text{OBu}^t \\
\text{Et}
\end{array}
$$

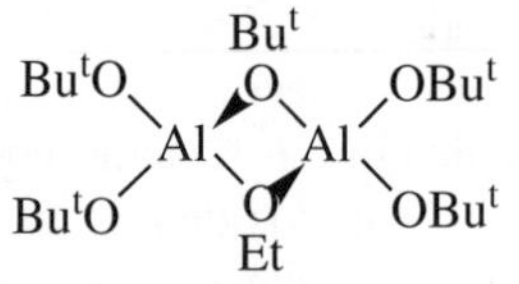

$$(2\text{-}III)$$

2.7.4   Fractionation in the Presence of an Inert Solvent

In order to complete an alcoholysis reaction, an excess of the higher boiling reactant alcohol would be required, so that the lower boiling alcohol could be fractionated out; this can sometimes be achieved by use of a solvent which is higher boiling than even the reactant as well as the product alcohols. When the original alkoxide is ethoxide or isopropoxide, the use of an inert solvent such as benzene offers an added advantage by virtue of the formation of a lower boiling azeotrope with ethyl alcohol or isopropyl alcohol, which facilitates the removal of the liberated alcohol by fractional distillation. For example, zirconium isopropoxide on treatment with alcohols was converted to pure alkoxides by the removal of the isopropyl alcohol–benzene azeotrope continuously on a fractionating column:[135]

$$
Zr(OPr^i)_4 \cdot Pr^iOH + x\,ROH \longrightarrow Zr(OPr^i)_{4-x}(OR)_x + (x+1)Pr^iOH \uparrow \qquad (2.87)
$$

Mehrotra[4,308–310] observed that benzene is a good solvent for the alcoholysis reactions for the following reasons:

(a)   The reactant alcohol can be used economically since only the stoichiometric amount is required for complete replacement.

(b)   When the reaction is carried out with a higher boiling alcohol, the refluxing temperature can be lowered in the presence of benzene and the side reactions are thus minimized.

(c)   The technique has the added advantage that it can be used with different stoichiometric ratios of the reactants. Thus the mixed alkoxides of many elements have been prepared.

(d)   A simple oxidimetric method developed for estimation of ethyl alcohol or isopropyl alcohol present in the azeotrope makes it possible to follow the progress of the reaction quantitatively.

2.7.5   Solubility Factor

The solubility factor has been found to be of some use for the synthesis of insoluble alkoxides. For example, when titanium ethoxide or isopropoxide is treated with methanol, an instantaneous reaction occurs with the separation of insoluble methoxide.[270] The fractionation process is not necessary in cases where such insoluble derivatives are obtained.

The insoluble methoxides are generally not preferred as starting materials for alcoholysis reactions, because they requires a longer refluxing period for completion of the reaction. Bradley *et al.*[274] and Mehrotra,[305] however, made the interesting observation that zirconium methoxide, after prolonged refluxing with excess *tert*-butyl alcohol[274] or *tert*-amyl alcohol[305] in the presence of benzene, finally yielded monomethoxide tri-*tert*-alkoxide only:

$$Zr(OMe)_4 + 3R^tOH \longrightarrow Zr(OMe)(OR^t)_3 + 3MeOH \uparrow \qquad (2.88)$$
$$\text{(excess)}$$

where R^t = *tert*-butyl or *tert*-amyl.

The final product was found to be a stable dimeric species $[(R^tO)_3Zr(\mu\text{-}OMe)_2Zr(OR^t)_3]$, which did not undergo further alcoholysis for steric reasons.

In the light of the above factors, the alcoholysis reactions of a few metal alkoxides may be briefly summarized. In the alcoholysis reactions of boron alkoxides with primary alcohols, Mehrotra and Srivastava[292] observed that reactions proceed to completion conveniently, but with tertiary alcohols, (*tert*-butyl alcohol) mono-alkoxy di-*tert*-butoxide was the final product. The non-replaceability of the last alkoxy group with *tert*-butyl alcohol was ascribed to steric factors. It was observed that the reaction was fast in the beginning but slowed down at the later stages after the formation of the mixed alkoxide. Presumably steric hindrance of the mixed alkoxide $B(OR)(OBu^t)_2$ prevented the close approach of another molecule of *tert*-butyl alcohol.

The alcoholysis reactions of tin(IV) alkoxides are comparatively faster than those of the silicon and germanium analogues and proceed to completion without any catalyst. Bradley[26] thus prepared a number of primary, secondary and tertiary alkoxides by the alcoholysis reactions of tin tetraisopropoxide isopropanolate with various alcohols in the presence of benzene (Eq. 2.89):

$$Sn(OPr^i)_4.Pr^iOH + 4ROH \longrightarrow Sn(OR)_4 + 5Pr^iOH \uparrow \qquad (2.89)$$

However, in the alcoholysis reaction of tin tetraisopropoxide isopropanolate with *tert*-heptyl alcohol in refluxing toluene or benzene, Gupta[311] could isolate only monoisopropoxide tri-*tert*-heptyloxide (Eq. 2.90) which on distillation *in vacuo* disproportionated into tetraisopropoxide and tetra-*tert*-heptyloxide.

$$Sn(OPr^i)_4.Pr^iOH + 3Me_2BuCOH \longrightarrow Sn(OPr^i)(OCBuMe_2)_3 + 4Pr^iOH \uparrow \quad (2.90)$$

In attempts to prepare tantalum pentaisopropoxide and penta-*tert*-butoxide by alcoholysis of tantalum pentamethoxide with isopropyl alcohol or *tert*-butyl alcohol, respectively, Bradley *et al.*[312] could isolate only monomethoxide tetra-isopropoxide or tetra-*tert*-butoxide:

$$Ta(OMe)_5 + 4Pr^iOH(Bu^tOH) \longrightarrow Ta(OMe)(OPr^i/OBu^t)_4 + 4MeOH \uparrow \quad (2.91)$$

Similarly, the reaction of niobium pentaethoxide with isopropyl alcohol yielded monoethoxide tetra-isopropoxide:[312]

$$Nb(OEt)_5 + 4Pr^iOH \longrightarrow Nb(OEt)(OPr^i)_4 + 4EtOH \uparrow \quad (2.92)$$

In contrast to earlier transition metals, some of the later 3d metal (Ni and Co) alkoxides exhibit an interesting variation in their alcoholysis reactions; for example, secondary and tertiary alkoxides of these metals undergo facile alcoholysis with primary alcohols, whereas their primary alkoxides do not appear to undergo alcoholysis with tertiary, secondary, or even other primary alcohols.[7,225]

The alcoholysis reactions of tetra-alkoxysilanes, $Si(OR)_4$, are generally very slow and the reactions of $Si(OEt)_4$ with tertiary alcohols were unsuccessful even in the presence of a variety of catalysts,[313,314] whereas similar reactions with germanium analogues[125,293,315] are quite facile. It may be mentioned that compared to the tetra-alkoxysilanes, the alkylalkoxysilanes appear to offer less steric hindrance to alcoholysis reactions, which have been carried out successfully in a number of cases with the help of catalysts such as sodium, *p*-toluene sulphonic acid, hydrogen chloride, and sulphuric acid.[316] In the alcoholysis reactions of dialkylgermanium dialkoxides with primary, secondary, and tertiary alcohols, Mathur *et al.*[317,318] observed that reactions with primary alcohols proceed easily but those with secondary and tertiary alcohols are completed only in the presence of *p*-toluene sulphonic acid as a catalyst.

The alcoholysis reactions of tellurium isopropoxide with primary, secondary and tertiary alcohols readily yielded the corresponding tetra-alkoxides:[302]

$$Te(OPr^i)_4 + 4ROH \longrightarrow Te(OR)_4 + 4Pr^iOH \quad (2.93)$$

On the other hand when diethyl selenite reacted with primary alcohols it gave dialkyl selenite by alcohol interchange but with secondary and tertiary alcohols only partial replacement was observed.[319]

The differences in solubility of metal isopropoxides and their alcoholates have led to the crystallization of pure isopropoxide isopropanolates of some tetravalent elements. For example, zirconium tetraisopropoxide is a viscous supercooled liquid which dissolves in excess of isopropyl alcohol and crystallizes out in the form of pure white crystals which were characterized as $Zr(OPr^i)_4.Pr^iOH$.[135] Similarly tin tetraisopropoxide and cerium tetraisopropoxide have been obtained as crystalline alcoholates, $Sn(OPr^i)_4.Pr^iOH$[320] and $Ce(OPr^i)_4.Pr^iOH$[287] which were used as the starting materials for the synthesis of a number of new alkoxides.

2.8   Transesterification Reactions of Metal Alkoxides (Method H)

As early as 1938, Baker showed that aluminium alkoxides undergo facile transesterification reactions, which can be represented by Eq. (2.94):

$$Al(OR)_3 + 3CH_3COOR' \rightleftharpoons Al(OR')_3 + 3CH_3COOR \uparrow \qquad (2.94)$$

where R' = a primary or secondary alkyl group.

He also observed that the reactions of $Al(OEt)_3$ with CH_3COOBu^t yielded only the heteroleptic alkoxide, $Al(OEt)(OBu^t)_2$.

Mehrotra[293] in 1954 confirmed Baker's observation and extended the above procedure for the preparation[321] of higher alkoxides of titanium, zirconium, and hafnium; the alcoholysis of the convenient starting alkoxide, $Zr(OPr^i)_4.Pr^iOH$, with Bu^tOH was extremely slow owing to the small difference in the boiling points (in parentheses) of Pr^iOH (82°C) and Bu^tOH (83°C), whereas the much larger difference in the boiling points of CH_3COOPr^i (89°C) and CH_3COOBu^t (98°C) facilitated the reaction considerably, providing the first convenient method for the preparation of $Zr(OBu^t)_4$:

$$Zr(OPr^i)_4.Pr^iOH + 4CH_3COOBu^t \overset{\text{cyclohexane}}{\rightleftharpoons} Zr(OBu^t)_4 + Pr^iOH + 4CH_3COOPr^i$$
$$(2.95)$$

The technique has also been successfully employed for the preparation of heteroleptic alkoxides by carrying out the reaction(s) in the desired stoichiometric ratio in the solvent cyclohexane (b.p. 81°C), which forms a convenient lower boiling azeotrope with CH_3COOPr^i (Eq. 2.96):

$$Zr(OPr^i)_4.Pr^iOH + xCH_3COOBu^t$$

$$\overset{\text{cyclohexane}}{\rightleftharpoons} Zr(OPr^i)_{4-x}(OBu^t)_x + Pr^iOH + xCH_3COOPr^i \qquad (2.96)$$

The method was extended by Mehrotra and co-workers for the preparation of tertiary alkoxides of a number of metals: lanthanides,[142,145,146,150] titanium,[321] hafnium,[321] vanadium,[322] niobium,[323] tantalum,[324] iron,[283] and gallium.[184,185]

Following the transesterification technique, Bradley and Thomas[325,326] extended the technique for a more convenient preparation of trialkylsiloxides of titanium, zirconium, and other metals:

$$M(OPr^i)_4 + 4R_3SiOCOCH_3 \longrightarrow M(OSiR_3)_4 + 4CH_3COOPr^i \uparrow \qquad (2.97)$$

Transesterification reactions have the following advantages over alcoholysis reactions:

(a)   The *tert*-butoxide derivatives of elements can be easily prepared from the corresponding isopropoxides as there is a significant difference in the boiling points of their organic esters (~9°C) compared with the corresponding small difference in the boiling points of the two alcohols (~0.2°C); this makes the fractionation of the more volatile ester much easier.

(b)   In some cases the esters (*e.g.* silyl acetate) are much more stable than the corresponding alcohols (silanol), which sometimes undergo self-condensation (siloxane) or which are oxidized more readily (*e.g.* in the preparation of higher vanadyl alkoxides from the ethoxide).

(c)   The method appears to be less prone to steric factors compared with the alcoholysis technique, and hence tertiary alkoxides may easily be obtained.

The utility of transesterification reactions has been extended considerably by the use of an inert solvent like cyclohexane (b.p. 80.8°C) which forms a convenient azeotrope with ethyl acetate (b.p. 72.8°C) and isopropyl acetate (b.p. 78.9°C). Using this modified technique, it is now possible to prepare mixed alkoxides also by taking the reactants in the desired stoichiometric ratios:

$$M(OEt)_x + nCH_3COOR \longrightarrow M(OEt)_{x-n}(OR)_n + nCH_3COOEt \qquad (2.98)$$

As stated earlier, the method has been extended for the preparation of alkoxides and mixed alkoxides of a large number of metals. However, Mehrotra and Srivastava[327] made the interesting observation that boron alkoxides do not appear to undergo transesterification reactions at all, although they do undergo alcoholysis reactions.

There has been some conjecture about the mechanism of these transesterification reactions. The most obvious mode of reaction could be represented by Eq. (2.99) in which the alkoxy oxygen atom of the ester molecule coordinates to the metal atom:

$$(2.99)$$

Alternatively, the coordination may occur through the carbonyl oxygen atom, giving the metal atom a negative and the carbonyl carbon atom a positive charge, and this type of transition state would probably take the route represented by Eq. (2.100):

$$(2.100)$$

The actual mode of reaction is not yet understood but some support in favour of the latter mechanism may be obtained from the study of the reaction of boron trichloride with organic esters. Lappert[328,329] on the basis of infrared studies has shown that on coordination of an organic ester with Lewis acids like boron trichloride, the electron density on the carbonyl oxygen atom appears to be reduced and, hence, that this should be a more probable donation site for the ester. This mechanism involving coordination of the carbonyl oxygen of the ester rather than its alkoxo oxygen would appear to be less prone to steric hindrance than the alcohol interchange, as actually observed by Mehrotra.[293]

2.9 Reactions of Metal Dialkylamides $M(NR_2)_x$ (R = Me, Et, SiMe$_3$) with Alcohols (Method I)

2.9.1 Derivatives without Metal–Metal Bonds

Metal dialkylamides are reactive toward alcohols, readily eliminating an amine according to Eq. (2.101):

$$M(NR_2)_x + xR'OH \longrightarrow M(OR')_x + xR_2NH \uparrow \qquad (2.101)$$

This method is particularly suitable for those metals which have a greater affinity for oxygen than for nitrogen. The other advantage of this procedure is the generally higher volatility of the liberated dialkylamines, which can readily be volatilized out.

The reactions of the type of Eq. (2.101) have often been employed in the synthesis of metal alkoxides when other routes are either inapplicable or tedious. Historically, this method was first investigated by Jones et al.[213] for the preparation of uranium tetralkoxides U(OR)$_4$ (R = Me, Et) from U(NEt$_4$)$_4$ and alcohols. For the synthesis of U(OBu^t)$_4$ Jones et al.[213] adopted the modified procedure of interacting uranium tetrachloride with ammonia and/or potassium amide and tert-butyl alcohol:

$$UCl_4 + 4KNH_2 \longrightarrow [U(NH_2)_4] \xrightarrow{4Bu^tOH} U(OBu^t)_4 + 4NH_3 \uparrow$$
$$\uparrow_{-2H_2}$$
$$4K + 4NH_3 \text{ (liquid)} \qquad (2.102)$$

Interestingly, the reaction of U(NEt$_2$)$_4$ with excess of tert-butyl alcohol affords the complex $U_2(OBu^t)_8(Bu^tOH)$:[330]

$$2U(NEt_2)_4 + 9Bu^tOH \longrightarrow U_2(OBu^t)_8(Bu^tOH) + 8Et_2NH \uparrow \qquad (2.103)$$

Thomas[331] utilized this method for preparation of a number of difficult-to-synthe-size metal alkoxides M(OR)$_4$ (especially when R = Bu^t), e.g. Zr(OBu^t)$_4$ from Zr(NEt$_2$)$_4$, V(OBu^t)$_4$ from V(NMe$_2$)$_4$, Cr(OBu^t)$_4$ from Cr(NEt$_2$)$_4$, Sn(OBu^t)$_4$ and Sn(OPr^i)$_4$.Pr^{i}OH from Sn(NMe$_2$)$_4$.

Thomas[331] prepared tantalum and niobium penta-alkoxides by the method of Eq. (2.101). For example, the alcoholysis of tris-(dialkylamido) monoalkylimidotan-talum yielded pentaalkoxides very conveniently:

$$RN=Ta(NR_2)_3 + 5ROH \longrightarrow Ta(OR)_5 + 3R_2NH \uparrow + RNH_2 \uparrow \qquad (2.104)$$

However, Nb(NEt$_2$)$_4$ reacted with alcohols to form compounds of the type Nb(OR)$_4$ which were oxidized instantaneously (even after the rigorous exclusion of oxygen) to $Nb(OR)_5$. The probable mode of this reaction can be represented by Eqs (2.105) and (2.106):

$$Nb(NEt_2)_4 + 4ROH \longrightarrow Nb(OR)_4 + 4Et_2NH \uparrow \qquad (2.105)$$

$$Nb(OR)_4 + ROH \longrightarrow Nb(OR)_5 + \tfrac{1}{2}H_2 \uparrow \qquad (2.106)$$

Bradley and co-workers[332–334] have nicely demonstrated the importance of metal dialkylamides as starting materials for preparation of the alkoxides of a wide range of metals.

Reactions of $M(NR_2)_4$ (M = V, Nb, R = Et; Cr, R = Et or Pr^i; and Mo, R = Me) with 1-adamantyl alcohol have been investigated by Wilkinson and co-workers[169] and isolated mononuclear alkoxides of these metals in the tetravalent state.

The reaction represented by Eq. (2.101) is, however, sometimes accompanied by a change in the oxidation state of the metal. For example, $Cr(NEt_2)_4$ reacts with primary and secondary alcohols[335] according to Eq. (2.107). Only tertiary alcohols and the sterically demanding 3,3-dimethyl-2-butanol,[336] which are not prone to this type of oxidation, are known[331] to give chromium(IV) alkoxides.

$$2Cr(NEt_2)_4 + 7R'R''CHOH \longrightarrow Cr(OCHR'R'')_3 + R'R''CO + 8Et_2NH \uparrow \quad (2.107)$$

Similar to the alcoholysis reaction of tetrameric aluminium isopropoxide[293] with *tert*-butyl alcohol, the alcoholysis reaction of dimeric aluminium tris(dimethyl amide)[337,338] with *tert*-butyl alcohol is also slow owing to steric factors. However, the amide route affords finally the *tris* product $[Al(OBu^t)_3]$, (Eq. 2.108) instead of the mixed product of the type $Al_2(OBu^t)_5(OPr^i)$ which was finally obtained in the reaction of $[Al(OPr^i)_3]_4$ with an excess of Bu^tOH.

$$Al_2(NMe_2)_6 + 5Bu^tOH \xrightarrow[-5Me_2NH]{} Al_2(OBu^t)_4(\mu\text{-}OBu^t)(\mu\text{-}NMe_2)$$

$$\xrightarrow[]{Bu^tOH \text{ (excess)}} Al_2(OBu^t)_6 + Me_2NH \qquad (2.108)$$

The reactions of $Bi(NMe_2)_3$ with alcohols afford soluble and volatile alkoxides of bismuth:[339]

$$Bi(NMe_2)_3 + 3ROH \longrightarrow Bi(OR)_3 + 3Me_2NH \uparrow \qquad (2.109)$$

where R = Pr^i (less soluble and non-volatile), CH_2CH_2OMe, $CH_2CH_2NMe_2$, $CHMeCH_2NMe_2$, CMe_2Et.

Although the utility of metal dialkylamides for the synthesis of metal alkoxides (including *tert*-butoxides) had been established for many years, the progress in this direction was rather slow up to the 1980s. Since then the alcoholysis reactions of metal bis(trimethylsilyl) amides involving sterically hindered mono- and multi-dentate (with more recent emphasis on specially designed donor-functionalized) alcohols to reduce the tendency of molecular aggregation and to increase the solubility and volatility of the resulting mono- or di-nuclear alkoxide derivatives of even more electropositive and larger size metals, such as heavier alkaline earths and lanthanide elements, have played a significant role in the development of exciting homometallic alkoxide systems[340,341] as shown by Eqs (2.110–2.112):

2.9.1.1 Alkoxides of Be, Mg, Ca, Sr, and Ba

$$M\{N(SiMe_3)_2\}_2(thf)_y + 2R'OH \xrightarrow{S} \frac{1}{x}[M(OR')_2]_x + 2(Me_3Si)_2NH \uparrow \quad (2.110)$$

(M = Be, R′ = CMe$_2$CH$_2$OEt,[341] $x = 2$, $y = 0$; M = Be, R′ = CEt$_2$CH$_2$OMe,[341] $x = 2$, $y = 0$; M = Mg, R′ = CEt$_2$CH$_2$OMe,[341] $x = 2$, $y = 2$; M = Ca, Sr, Ba, R′ = CBu^t(CH$_2$OPr^i)$_2$,[340] $x = 2$, $y = 2$; M = Ca, R′ = CBu$^t_3$,[53] $x = 2$, $y = 2$; S = n-pentane, n-hexane, or n-heptane).

2.9.1.2 Sc, Y, and the lanthanides

Reactions of Ln$\{N(SiMe_3)_2\}_3$ with a variety of sterically and/or electronically demanding monodentate alcohols (Bu^{t}OH, (CF$_3$)$_2$MeCOH, etc.) as well as sterically hindered donor functionalized multidentate alcohols (HOCBu^t(CH$_2$OR)$_2$ where R = Pr^i or Et, and HOCR$'_2$CH$_2$OR (where R′ = Et, Pr^i, Bu^t; R = Me or Et)[21-23,340-342] afford a plethora of different products by variation of the added alcohol, the metal, the solvent (non-donor or donor), and the stoichiometry of reactants as illustrated by the reactions depicted in Scheme 2.2 and Eqs (2.111) and (2.112):

$$Yb[N(SiMe_3)_2]_2(dme) + 2R'OH \xrightarrow{S} \frac{1}{n}[Yb(OR')_2]_n + 2(Me_3Si)_2NH \uparrow \quad (2.111)$$

where R′ = Bu^t,[354] CBu^t(CH$_2$OPr^i)$_2$,[340] $n = 2$ and S = n-pentane or n-heptane

$$Ln[N(SiMe_3)_2]_3 + 3R'OH \xrightarrow{S} \frac{1}{n}[Ln(OR')_3]_n + 3(Me_3Si)_2NH \uparrow \quad (2.112)$$

(Ln = Sc,[340] R′ = CMe$_2$(OCH$_2$OMe), $n = 2$; Ln = Sc,[340] R′ = CEt$_2$(CH$_2$OMe), $x = 1$; Ln = Y,[355] R′ = CEt$_2$(CH$_2$OMe), $n = 2$; Ln = Y,[356] R′ = CPr$^i_2$(CH$_2$OEt),[356] CMe$_2$CH$_2$NMe$_2$,[355] $n = 1$; Ln = Nd,[355] R′ = CBu^t(CH$_2$OPr^i)$_2$, CPr$^i_2$(CH$_2$OC$_2$H$_4$OMe); $n = 1$; Ln = Nd,[356] R′ = CHBu^t(CH$_2$OEt), CHBu^{t}CH$_2$NEt$_2$, CBu$^t_2$(CH$_2$OEt), CPr$^i_2$(CH$_2$OEt), $n = 1$; Ln = Sm, Yb,[357] R′ = CBu^t(2-CH$_2$NC$_5$H$_3$Me-6)$_2$, $n = 1$; Ln = Lu,[355] R′ = CMe$_2$CH$_2$NMe$_2$, $n = 2$; Ln = Lu,[356] R′ = CMe$_2$CH$_2$OMe, $n = 2$; s = n-pentane, n-hexane, or n-heptane).

The amide–alkoxide exchange reactions has also been successfully applied for the synthesis of many interesting divalent group 12, group 14, and 3d transition metal alkoxides as depicted in Scheme 2.3.

It is now reasonably well established that metal dialkylamides/bis(trimethylsilyl)-amides are valuable starting materials when the more conventional methods for metal alkoxide preparation fail.

In a recent study, Hoffman and co-workers[358] have reported that reactions of indium amides (In[N(Bu^t)SiMe$_3$]$_3$, In(NEt$_2$)$_3$, and In(tmp)$_3$ where tmp = 2,2,6,6-tetramethyl-piperidide) with fluorinated alcohols (Me$_2$(CF$_3$)COH, (CF$_3$)$_2$CHOH, (CF$_3$)$_2$MeCOH) of differing acidic character give different types of X-ray crystallographically characterized interesting fluoroalkoxide derivatives: [In$\{\mu$-OCMe$_2$CF$_3\}\{$OCMe$_2$CF$_3\}_2$]$_2$, [In$\{$OCMe(CF$_3$)$_2\}_3$(H$_2$NBu^t)$_3$], [H$_3$NBu^t][In$\{$OCH(CF$_3$)$_2\}_4$(H$_2$NBu^t)], [In$\{$OCH(CF$_3$)$_2\}_3$(Htmp)], [H$_2$tmp][In$\{$OCR(CF$_3$)$_2\}_4$] (R = H, Me), [H$_2$NEt$_2$][In$\{$OCH(CF$_3$)$_2\}_4$(HNEt$_2$)], and mer-In[OCMe(CF$_3$)$_2$]$_3$(py)$_3$.

$$\xrightarrow{\text{3R'OH}} Ln(OR')_3 + 3(MeSi)_2NH \uparrow$$

$Ln = Y, R' = CEt_3{}^{343}$
$Ln = Ce, R' = CBu_3^t{}^{344}$

$$\xrightarrow{\text{5R'OH (excess)}} \tfrac{1}{3}[Ln_3(OR')_9](R'OH)_2 + 3(Me_3Si)_2NH \uparrow$$

$Ln = Y, La, R' = Bu^t, Am^t{}^{345}$

$$\xrightarrow{\text{3HOCBu}_3^t, \text{ hexane}} Ce\{N(SiMe_3)_2\}_3 + 3(Me_3Si)_2NH \uparrow$$
$$\xrightarrow[{-3Me_2C=CH_2}]{150°C, \text{ vacuum}} \tfrac{1}{2}[Ce_3(OCHBu_2^t)_2(\mu\text{-}OCHBu_2^t)]_2$$

$Ln = Ce^{344}$

$$\xrightarrow{\text{3HOCPh}_3, \text{ toluene}} \tfrac{1}{2}[La(OCPh_3)_2(\mu\text{-}OCPh_3)]_2 + 3(Me_3Si)_2NH \uparrow$$

$Ln = La^{346}$

$$\xrightarrow[\text{hexane CH}_3\text{CN}]{\text{3HOCBu}_3^t} [Nd(OCBu_3^t)_3(CH_3CN)_2] + 3(Me_3Si)_2NH \uparrow$$

$Ln = Nd^{347}$

$$\xrightarrow[\text{hexane/THF}]{\text{3HOCHPr}_2^i} \tfrac{1}{2}[Nd_2(OCHPr_2^i)_6(thf)_2] + 6(Me_3Si)_2NH \uparrow$$
$$\xrightarrow[{-THF}]{C_5H_5N(\text{excess})} \tfrac{1}{2}[Nd_2(OCHPr_2^i)_6(NC_5H_5)_2]$$

$Ln = Nd^{348}$

$$\xrightarrow[\text{benzene}]{\text{3HOCH(CF}_3)_2(\text{excess})} \tfrac{1}{2}[Sc\{OCH(CF_3)_2\}_3(NH_3)_2]_2 + 3(Me_3Si)_2NH \uparrow$$

$Ln = Sc^{349}$

$$\xrightarrow[\text{benzene}]{\text{3HOCMe(CF}_3)_2 \ (\text{excess})} \tfrac{1}{n}[Ln\{OCMe(CF_3)_2\}_3(NH_3)_x]_n + 3(Me_3Si)_2NH \uparrow$$

$Ln = Y,^{349a}\ n = 0, x = 3$
$Ln = La,^{349b}\ n = 2, x = 1 \text{ or } 2$
$Ln = Pr,^{349b}\ n = 2, x = 2$
$Ln = Eu,^{349b}\ n = 2, x = 1$

$$\xrightarrow[\text{THF}]{\text{3HOR}_f} [Ln(OR_f)_3(thf)_3] + 3Me_3Si)_2NH \uparrow$$

$Ln = Y,^{349}\ R_f = CH(CF_3)_2$
$Ln = Y,^{349a}\ La, Eu,^{349b}\ R_f = CMe(CF_3)_2$

$$\xrightarrow[\text{Et}_2\text{O}]{\text{3HOCMe(CF}_3)_2} [Ln\{OCMe(CF_3)_2\}_3(Et_2O)_{0.33}] + 3(Me_3Si)_2NH \uparrow$$

$Ln = Y,^{350}\ La^{349b}$

$$\xrightarrow[\text{benzene}]{\text{3HOR}_f} [Ln(OR_f)_3] + 3(Me_3Si)_2NH \uparrow$$

$Ln = Y,^{349a}\ La, Pr, Eu,^{349a}\ R_f = CMe(CF_3)_2$
$Ln = Y,^{349a}\ R_f = CMe_2(CF_3)$

$$\xrightarrow{\text{3HOCBu}_3^t} [Tm(OCBu_3^t)_3NH_2SiMe_2]_2CH_2$$

$Ln = Tm^{353}$

(left label) $Ln\{N(SiMe_3)_2\}_3$

Scheme 2.2   Reaction of $Ln\{N(SiMe_3)_2\}_3$ with sterically demanding monodentate alcohols.

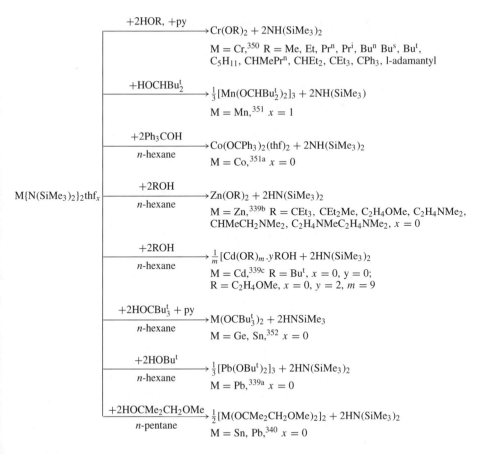

Scheme 2.3 Synthesis of divalent Cr, Mn, Co, Zn, Cd, Ge, Sn, and Pb alkoxides from corresponding metal bis (trimethylsilyl) amides.

The composition of these products is influenced by a number of factors such as (i) the acidic character of the alcohols, (ii) the nature of precursor amide derivatives, (iii) the stoichiometry of the alcohols used, and (iv) the reaction conditions.

2.9.2 Derivatives with Multiple Metal–Metal Bonds

The remarkable utility of the amide route has been established by the synthesis of alkoxides of molybdenum and tungsten (M) in their lower oxidation states ($+3$ or $+4$), e.g. $[M(OR)_3]_n$ and $[M(OR)_4]_n$. Derivatives of the empirical formula $M(OR)_3$ are of special interest as, in these, association occurs primarily through the formation of metal–metal triple (M≡M) bonds[359] rather than through the normal alkoxo bridges (Eq. 2.113)

$$M_2(NMe_2)_6 + >6ROH \longrightarrow (RO)_3M \equiv M(OR)_3 + 6Me_2NH \uparrow \qquad (2.113)$$

$(M = Mo, R = Bu^t,[360] Pr^i, CH_2Bu^t, CMe_2Ph, SiMe_3, SiEt_3,[361,362]$

menthyl;[363] $M = W; R = Bu^t$[364,365]).

With CH_2Bu^t and $SiMe_3$ groups[361,362] the initial complexes were molecular adducts of Me_2NH, e.g. $Mo_2(OR)_6(HNMe_2)_2$, from which dimethylamine-free product could be obtained above 60°C under reduced pressure. However, the addition of excess EtOH to $Mo_2(NMe_2)_6$ affords $Mo_4(OEt)_{12}$.

By contrast, the reactions of $W_2(NMe_2)_6$ with MeOH and EtOH[366,367] give tetra-nuclear products $W_4(OR)_{16}$ containing W—W bonds of order 1 or lower (Eq. 2.114).

$$2W_2(NMe_2)_6 + 16ROH \longrightarrow W_4(OR)_{16} + 12Me_2NH \uparrow + 2H_2 \uparrow \qquad (2.114)$$

Analogous reactions involving cyclohexanol[368] and Pr^iOH[369,370] give $W_2(O\text{-}c\text{-}C_6H_{11})_6$ and $[W_2(\mu\text{-}H)(OPr^i)_7]_2$, respectively.

When $W_2(NMe_2)_6$ and Pr^iOH reacts in pyridine (py), the black crystalline product $W_2(OPr^i)_6(py)_2$ is obtained.[369,370]

A series of molybdenum(IV) alkoxides of empirical formula $Mo(OR)_4$ have been prepared according to Eq. (2.115).[371]

$$Mo(NMe_2)_4 + 4ROH \longrightarrow \frac{1}{n}[Mo(OR)_4]_n + 4Me_2NH \uparrow \qquad (2.115)$$

where R = Me, Et, Pr^i, Bu^t and CH_2Bu^t.

The properties of the above products $[Mo(OR)_4]_n$ depend upon the nature of the group R. For example, of the above only $[Mo(OPr^i)_4]_2$ is diamagnetic. It is a fluxional molecule and the low temperature NMR spectra indicate the structure $(Pr^iO)_3Mo(\mu\text{-}OPr^i)_2Mo(OPr^i)_3$, which appears to be corroborated by X-ray crystallography.[372] The bulky *tert*-butoxide appears to be monomeric ($n = 1$) in benzene, but its structure is not known in the solid state.

Reactions of $Mo(NMe_2)_4$ with two equivalents of alcohol (ROH) (where R = 4-methylcyclohexyl and diisopropyl) afford compounds of formula $Mo(NMe_2)_2(OR)_2$[373] which are dimeric *via* a pair of $\mu\text{-}NMe_2$ ligands both in solution and in the solid state.

Interaction between $1,2\text{-}Mo(Bu^i)_2(NMe_2)_4$ and an excess of isopropyl alcohol in hexane affords[374] $Mo_2(OPr^i)_4(HOPr^i)_4$ (Eq. 2.116):

$$1,2\text{-}Mo_2(Bu^i)_2(NMe_2)_4 + >10Pr^iOH$$

$$\xrightarrow[0°C]{\text{hexane}} Mo_2(OPr^i)_4(HOPr^i)_4 + Me_2C{=}CH_2 + Me_3CH + 4Me_2NH \quad (2.116)$$

In related reactions involving $1,2\text{-}Mo_2(Bu^i)_2(NMe_2)_4$ and both cyclopentanol and cyclohexanol (ROH), thermally unstable compounds of formula $Mo_2(OR)_4(HOR)_4$ have been prepared and characterized by X-ray crystallography when R = Pr^i or c-pentyl.[374]

Interestingly, alcoholysis reactions of $Mo_2(NMe_2)_6$ with fluorinated alcohols such as $Me_2(CF_3)COH$, $Me(CF_3)_2COH$, and $(CF_3)_3COH$ give incomplete or complex reactions.[376,377] This probably arises from the low nucleophilicity of fluoroalcohols and the poor π-bonding characteristics of fluoroalkoxide ligands.[24,378] However, triply bonded *hexakis* (fluoroalkoxo) dimolybdenum derivatives $Mo_2[OCMe_{3-x}(CF_3)_x]_6$ ($x = 1$ or 2) have been conveniently prepared[170] in high yield by metathetical reactions involving $Mo_2Cl_6(dme)_2$ (where dme = 1,2-methoxyethane) and an alkali (Li or Na) metal fluoroalkoxide.

2.10   Miscellaneous Methods of Alkoxide Synthesis (Method J)

2.10.1   By Oxidation Reactions (J-1)

There are other methods of synthesis of metal alkoxides. During the synthesis of uranium tetraethoxide, Gilman and co-workers[379,380] observed that it could be oxidized by an oxidizing agent such as dry air to uranium pentaethoxide. The exact mechanism is not known but it is assumed that uranium(IV) is oxidized to uranium (V); the yield of product was only 65%:

$$5U(OEt)_4 + O_2 \longrightarrow 4U(OEt)_5 + UO_2 \qquad (2.117)$$

Alternatively when uranium tetraethoxide was treated with bromine in 2:1 molar ratio followed by the addition of 2 moles of sodium ethoxide, uranium pentaethoxide was obtained in 90–95% yield:[378,379]

$$2U(OEt)_4 + Br_2 \longrightarrow 2U(OEt)_4Br \qquad (2.118)$$

$$2U(OEt)_4Br + 2NaOEt \longrightarrow 2U(OEt)_5 + 2NaBr \qquad (2.119)$$

It was also observed that uranium pentaethoxide was suceptible to oxidation by air or molecular oxygen to give uranium hexaethoxide, but the yield of the product was very poor. Gilman et al.[380] found that better yields were obtained by oxidizing the double ethoxide of sodium and uranium, $NaU(OEt)_6$, in place of uranium pentaethoxide, $U(OEt)_5$, with bromine, lead tetraacetate or benzoyl peroxide; the yields in these reactions were 5–10, 20, or 40–60%, respectively:

$$U(OEt)_5 + NaOEt \longrightarrow NaU(OEt)_6 \qquad (2.120)$$

$$2NaU(OEt)_6 + Br_2 \longrightarrow 2U(OEt)_6 + 2NaBr \qquad (2.121)$$

$$2NaU(OEt)_6 + Pb(OAc)_4 \longrightarrow 2U(OEt)_6 + 2NaOAc + Pb(OAc)_2 \qquad (2.122)$$

$$2NaU(OEt)_6 + (C_6H_5CO_2)_2O_2 \longrightarrow 2U(OEt)_6 + 2NaO_2CC_6H_5 \qquad (2.123)$$

Chromium trimethoxide was obtained by the photooxidation of aryl tricarbonyl chromium in methanol; the valency of chromium in the product was established by absorption spectra and magnetic studies.[381] Hagihara and Yamazaki[382] synthesized chromium tetra-tert-butoxide by heating a mixture of bis(benzene)chromium and di-tert-butylperoxide in the presence of benzene or petroleum ether in a sealed tube at 90° for about 20 h. The excess of solvent was removed and the green residue was sublimed in vacuo. It was found that this residue was contaminated with a small amount of biphenyl as an impurity which was removed as a complex by the addition of 2,4,7-trinitrofluorenone. The blue product thus obtained showed no characteristic absorption bands due to biphenyl and was characterized as chromium tetra-tert-butoxide $Cr(OBu^t)_4$.

A much more convenient method for synthesis of $Cr(OBu^t)_4$ could be the disproportionation of $[Cr(OBu^t)_3]_n$ (prepared by the lithium alkoxide method with strict exclusion of free Bu^tOH in the reaction mixture) under reduced pressure.[7]

A heteroleptic tert-alkoxide of cerium(IV), $(Bu_3^tCO)_2Ce(OBu^t)_2$, has been reported[383] to be formed according to Eq. (2.124):

$$Ce(OCBu_3^t)_3 + 2Bu^tOOBu^t \longrightarrow (Bu_3^tCO)_2Ce(OBu^t)_2 + \tfrac{1}{2}[Bu_3^tCOOCBu_3^t] \quad (2.124)$$

Pentakis(2,2,2-trifluoroethoxy)arsorane has been prepared by passing chlorine gas into the solution containing $As(OCH_2CF_3)_3$ and excess CF_3CH_2OH:[125]

$$As(OCH_2CF_3)_3 + 2CF_3CH_2OH + Cl_2 \longrightarrow As(OCH_2CF_3)_5 + 2HCl \uparrow \qquad (2.125)$$

2.10.2   From Metal–Carbon Bond Cleavage Reactions (J-2)

The polarity of metal–carbon and metal–hydrogen bonds allows protolysis to occur with alcohols under moderate conditions. Consequently carbanionic metal alkyls react with alcohols to give alkoxides, in excellent yields. The reaction assumes special importance as it forms the basis for the calorimetric measurements of a large number of metal–alkyl bond dissociation energies.[384] This preparative route tends to be more convenient owing to the volatility of the liberated alkane side-products. Organometal alkoxides of beryllium,[385,385a,385b] magnesium,[386,386a,387,388a,389] and zinc,[390-392] were prepared by the equimolar reactions of metal alkyls (MR_2) and an appropriate alcohol ($R'OH$) (*partial alcoholysis*):

$$MR_2 + R'OH \longrightarrow 1/n[RM(OR')]_n + RH \uparrow \qquad (2.126)$$

where M = Be, Mg, Zn; R = Me, Et, Bu^t, Ph; R′ = primary, secondary, and tertiary alcohols; $n =$ is generally 4 with sterically compact R and R′ but may be 2 in cases of sterically demanding R and R′.

Tributylgallium reacts readily with methanol and butanol at low temperatures to yield dimeric dibutylgallium alkoxides in excellent yields.[393]

$$Bu_3Ga + ROH \longrightarrow \tfrac{1}{2}[Bu_2Ga(\mu\text{-}OR)]_2 + BuH \qquad (2.127)$$

By contrast, reactions of $GaMe_3$ or $GaBu_3^t$ with ethanolamine or $HOCPh_3$, respectively, under refluxing conditions afford crystallographically characterized monomeric products (Eq. 2.128):

$$GaR_3 + R'OH \longrightarrow R_2GaOR' + RH \qquad (2.128)$$

$$(R = Me, R' = CH_2CH_2NH_2;^{394} \ R = Bu^t, R' = CPh_3{}^{395})$$

Homoleptic alkoxides of lithium,[230,396] beryllium,[385,385b] magnesium,[397] zirconium,[398,398a] vanadium,[340,399] chromium,[400] rhodium,[401] copper,[402] cobalt,[403] nickel[404] and aluminium[405] have also been prepared by metal–carbon bond cleavage reactions:

$$Bu_3^tCOH \begin{cases} \xrightarrow[n\text{-hexane}]{LiBu^n} \tfrac{1}{2}[LiOCBu_3^t]_2 + Bu^nH \uparrow^{396} & (2.129) \\[2ex] \xrightarrow[THF]{LiBu^n} \tfrac{1}{2}[LiOCBu_3^t(thf)]_2 + Bu^nH^{230} & (2.130) \end{cases}$$

$$MR_2 + 2R'OH \longrightarrow M(OR')_2 + 2ROH \qquad (2.131)$$

$$(M = Be; R/R' = Me, Me,^{385} \ Et/CH(CF_3)_2.^{385b} \ M = Mg;^{397} \ R = Bu^n; R' = CBu_3^t.)$$

$$Zr(CH_2Bu^t)_4 + 4R_fOH \xrightarrow{\text{benzene}} Zr(OR_f)_4 + 4CMe_4 \tag{2.132}$$

where $R_f = CH(CF_3)_2, C(CF_3)Me_2, C(CF_3)_2Me, C(CF_3)_3$.[398,398a]

$$V(C_6H_2Me_3\text{-}2,4,6)_3(thf) \begin{cases} \xrightarrow[\substack{\text{toluene, } (-THF), \\ (-3HOC_6H_2Me_3\text{-}2,4,6)}]{\text{HOCMe}_2CH_2OMe} V(OCMe_2CH_2OMe)_3{}^{340} & (2.133) \\[2em] \xrightarrow[\substack{\text{toluene, } (-THF), \\ (-3HOC_6H_2Me_3\text{-}2,4,6)}]{+3Bu^tOH} \tfrac{1}{2}(Bu^tO)_2V(\mu\text{-}O)(\mu\text{-}OBu^t)V(OBu^t)_2{}^{399} & (2.134) \end{cases}$$

$$Cr(C_5H_5)_2 + 2ROH \longrightarrow \frac{1}{n}[Cr(OR)_2]_n + 2C_5H_6 \tag{2.135}$$

where $R = Me, Et, Pr^i, CH_2Bu^t$.[400]

$$(Me_3P)_3RhMe + ROH \xrightarrow{\text{toluene}} (Me_3P)_3RhOR + CH_4 \tag{2.136}$$

where $R = CH_2CF_3, CH(CF_3)_2$.[401]

$$(Ph_3P)_2CuMe(OEt_2)_{0.5}{}^{402} \begin{cases} \xrightarrow[\substack{+PPh_3}]{+HOCH(CF_3)_2} (Ph_3P)_3Cu[OCH(CF_3)_2] + CH_4 & (2.137) \\[1.5em] \xrightarrow[]{+HOCHPh_2} (Ph_3P)_2Cu(OCHPh_2) + CH_4 & (2.138) \\[1.5em] \xrightarrow[\substack{+PPh_3}]{+HOCHPh_2} (Ph_3P)_3Cu(OCHPh_2) + CH_4 & (2.139) \end{cases}$$

Reactions of alcohols ROH, with $CoMe(PMe_3)_4$ afford polymers of the formula $[Co(OR)_2]_n$ (Eq. 2.140).[403]

$$2CoMe(PMe_3)_4 + 2ROH \longrightarrow \frac{1}{n}[Co(OR)_2]_n + Co(PMe_3)_4 + 4PMe_3 + 2CH_4 \tag{2.140}$$

where $R = Me, Et, Ph, SiMe_3$.

Interactions of $(bpy)NiR_2{}^{404}$ with $HOCH(CF_3)_2$ and $AlEt_3{}^{405}$ with $HOCH_2CCl_3$ give corresponding alkoxide derivatives:

$$(bpy)NiR_2 + 2HOCH(CF_3)_2 \longrightarrow (bpy)Ni\{(OCH(CF_3)_2\}_2 + 2RH \tag{2.141}$$

$$AlEt_3 + 3HOCH_2CCl_3 \longrightarrow Al(OCH_2CCl_3)_3 + 3C_2H_6 \tag{2.142}$$

2.10.3   From Metal–Carbon and Metal–Nitrogen Bond Cleavage Reactions (J-3)

$$[(Me_3Si)_2N]_2M(CH_2SiMe_2NSiMe_3)$$

$$\xrightarrow[-3(Me_3Si)_2NH]{+4Pr^iOH+py} \tfrac{1}{4}M_4(OPr^i)_{16}(py)_2 \qquad (2.143)$$
$$M = Th^{165}$$

$$\xrightarrow[-3HN(SiMe_3)_2]{+Pr^iOH} [M(OPr^i)_4]_n \qquad (2.144)$$
$$M = Th^{165}$$

$$\xrightarrow{+4HOCHPr^i_2} \tfrac{1}{2}M_2(OCHPr^i_2)_8 + 3(Me_3Si)_2NH$$
$$M = Th^{407} \qquad (2.145)$$

$$\xrightarrow{+4HOCHEt_2+py} \tfrac{1}{2}[M_2(OCHEt_2)_8](py)_2 \qquad (2.146)$$
$$M = Th^{165}$$

$$\xrightarrow{>4Bu^tOH/toluene} \tfrac{1}{2}[M_2(OBu^t)_8)](HOBu^t) + 3(Me_3Si)_2NH$$
$$M = Th^{406} \qquad (2.147)$$

$$\xrightarrow[-3(Me_3Si)_2NH]{>4.5Bu^tOH} \tfrac{1}{2}[M_2(OBu^t)_8](HOBu^t) \qquad (2.148)$$
$$M = U^{330}$$

2.10.4   From Metal–Hydrogen Bond Cleavage Reactions (J-4)

The hydridic metal hydrides react with alcohols to yield hydride-alkoxides or binary alkoxides depending on the amounts of alcohols used.

Tetrahydrofuran soluble magnesium dialkoxides $Mg(OR)_2$ (where $R = CPh_3$ or $CMePh_2$) have been prepared[408] by the reactions of magnesium hydride or dimethyl magnesium with the corresponding alcohols:

$$MgH_2 \text{ (or } MgMe_2) + 2ROH \longrightarrow Mg(OR)_2 + H_2 \text{ (or } 2CH_4) \qquad (2.149)$$

The hydrocarbon soluble alkoxides of even the heavier alkaline earth metals (Ba) are accessible[47] according to Eq. (2.150):

$$BaH_2 + 2ROH \longrightarrow Ba(OR)_2 + 2H_2 \uparrow \qquad (2.150)$$

where $R = CH(CF_3)_2$ or $C(CF_3)_3$.

In view of the importance of zinc hydride derivatives as catalysts for methanol production,[409,410] a number of alkoxozinc hydrides have been prepared by the reactions of ZnH_2 with a variety of alcohols:

$$2ZnH_2 + 2ROH \longrightarrow 2HZn(OR) + H_2 \qquad (2.151)$$

where $R = c\text{-}C_6H_{11}$,[411] $CH_2CH_2NMe_2$,[412] and Bu^t.[413]

Aluminium hydride undergoes stepwise substitution[414] by alcohols (Eq. 2.152):

$$AlH_3 + xROH \longrightarrow Al(OR)_xH_{3-x} + x/2\,H_2 \qquad (2.152)$$

2.10.5   By Insertion of Oxygen into Metal–Carbon Bonds (J-5)

The reactivity of molecular oxygen towards metal alkyls is very diverse; for example, some metal alkyls react violently with O_2 and others are inert. There are only a few well-documented examples of reactions which lead to the formation of alkoxides. Since only one oxygen atom is required per metal–alkyl bond, the reactions require a multistep process.

The high reactivity of certain metal alkyls is well known. The autoxidation of dialkylzinc was investigated as early as 1864 by several workers and they showed that dialkylzinc is oxidized rapidly to zinc dialkoxides[415–416]

$$R_2Zn + O_2 \longrightarrow Zn(OR)_2 \qquad (2.153)$$

where $R = Et, Pr^i, C_5H_{11}$.

Abraham[417] studied the oxidation of dialkylzinc and observed that rapid oxidation leads to the formation of bis(alkylperoxy)zinc, $Zn(OOR)_2$. However, when the oxidation was carried out slowly for several days under controlled conditions, zinc dialkoxide was the final product, which on hydrolysis gave alcohol. The reduction of bis(alkylperoxy)zinc with excess of dialkylzinc also afforded zinc dialkoxides:

$$Zn(OOR)_2 + R_2Zn \longrightarrow 2Zn(OR)_2 \qquad (2.154)$$

Similarly the oxidation of trialkylaluminium and trialkylboron with molecular oxygen under controlled conditions also afforded trialkoxides.[418–420]

$$2MR_3 + 3O_2 \longrightarrow 2M(OR)_3 \qquad (2.155)$$

where $M = B, Al$.

When solutions of $(Bu^t_3CO)_2MMe_2$ ($M = Ti, Zr, Hf$) were exposed to either an excess (1 atm) or one equivalent of dry oxygen, the mixed alkoxide derivatives $(Bu^t_3CO)_2M(OMe)_2$ were isolated in >87% yields.[421] The addition of O_2 to $(Bu^t_3CO)TiMe_3$ afforded three different crystalline derivatives[421] as represented by the formula $TiMe_{3-x}(OMe)_x(OCBu^t_3)$ where $x = 1, 2, 3$.

$$(Bu^t_3CO)_2MMe_2 + O_2 \longrightarrow (Bu^t_3CO)_2M(OMe)_2 \qquad (2.156)$$

where $M = Ti, Zr, Hf$.

2.10.6   By Insertion of Aldehydes or Ketones Across a Metal–Carbon or Metal–Hydrogen Bond (J-6)

Coates and Fishwick[385] as early as 1968 observed that dimethylberyllium adds to acetaldehyde or acetone to form $MeBe(OCHMe_2)$ or $MeBe(OBu^t)$, respectively.

In 1986, Lappert and co-workers[422] carried out the reactions of $Li(CH_2PR_2)(tmeda)$ ($R = Me$ or Ph; $tmeda = N,N,N',N'$-tetramethylethylenediamine) with Bu^t_2CO in 1:1 (Eq. 2.157) and 1:2 (Eq. 2.158) molar ratios to yield highly hindered phosphino-alkoxides of lithium:

$$Li(CH_2PMe_2)(tmeda) + Bu^t_2CO \xrightarrow{n\text{-hexane}} \tfrac{1}{2}[Li(OCBu^t_2CH_2PMe_2]_2 + tmeda$$
$$(2.157)$$

$$\text{Li(CH}_2\text{PPh}_2)(\text{tmeda}) + 2\text{Bu}_2^t\text{CO} \xrightarrow{\textit{n}\text{-hexane}} \tfrac{1}{2}\,[\text{Li(OCBu}_2^t\text{CH}_2\text{PPh}_2)_2\text{Li(OCBu}_2^t)]$$

(2.158)

Similarly, acetone adds to the Ga–H bond in dichlorogallane under mild conditions to give gallium dichloride monoisopropoxide[423]

$$\text{HGaCl}_2 + (\text{CH}_3)_2\text{CO} \longrightarrow \text{GaCl}_2\text{OCH(CH}_3)_2$$

(2.159)

Ketones (hexafluoroacetone) insert across a Si–H bond to form the corresponding silicon alkoxide derivative:[424,425]

$$\text{Me}_3\text{SiH} + (\text{CF}_3)_2\text{CO} \longrightarrow \text{Me}_3\text{SiOCH(CF}_3)_2$$

(2.160)

2.10.7   By Exchange Reactions (J-7)

Exchange reactions between two different metal chlorides and alkoxides have been employed for the synthesis of a wide variety of homo- and heteroleptic metal alkoxides. For example, Bradley and Hill[426] observed that a mixture of excess titanium tetrachloride and silicon dichloride diethoxide at 0°C deposited a crystalline product which was characterized as titanium trichloride monoethoxide:

$$\text{TiCl}_4 + \text{SiCl}_2(\text{OEt})_2 \longrightarrow \text{TiCl}_3(\text{OEt}) \downarrow + \text{SiCl}_3(\text{OEt})$$

(2.161)

Similarly titanium trichloride monoethoxide was obtained by reactions of titanium tetrachloride with silicon monochloride triethoxide or silicon tetraethoxide (Eqs 2.162 and 2.163).

$$\text{TiCl}_4 + \text{SiCl(OEt)}_3 \longrightarrow \text{TiCl}_3(\text{OEt}) + \text{SiCl}_2(\text{OEt})_2$$

(2.162)

$$\text{TiCl}_4 + \text{Si(OEt)}_4 \longrightarrow \text{TiCl}_3(\text{OEt}) + \text{SiCl(OEt)}_3$$

(2.163)

Following the use of exchange reactions between TiCl_4 and $\text{Si(OEt)}_4/\text{SiCl(OEt)}_3$, Marks and co-workers[427] have shown that uranium hexamethoxide can be prepared in the direct reaction between UF_6 and an excess of MeSi(OMe)_3 (Eq. 2.164).

$$\text{UF}_6 + 2\text{MeSi(OMe)}_3 \xrightarrow[-76°C]{\text{CH}_2\text{Cl}_2} \text{U(OMe)}_6 + 2\text{MeSiF}_3$$

(2.164)

Jacob[428] has described a variation of this reaction by using $\text{Si(OMe}_3)_4$ for the synthesis of hexamethoxides of uranium, molybdenum and rhenium (Eq. 2.165).

$$\text{MF}_6 + 6\text{Si(OMe)}_4 \longrightarrow \text{M(OMe)}_6 + 6\text{SiF(OMe)}_3$$

(2.165)

where M = U, Mo, Re.

For M = Mo, the reaction proceeds to completion if the product SiF(OMe)_3 in Eq. (2.165) is removed during the reaction by pumping.

Analogous reaction with WF_6 affords WF(OMe)_5, which on reaction with NaOMe yields W(OMe)_6.[428]

First homoleptic dimers of tungsten(v) (Eq. 2.166) and rhenium(v) (Eq. 2.167) have been prepared[429] according to the following reactions:

$$WF_6 + Me_3SiOMe \longrightarrow WF_2(OMe)_4 \xrightarrow[THF]{(i)\ Li} \underset{(x=1-3)}{W_2F_x(OMe)_{10-x}} \xrightarrow[THF]{(ii)\ NaOMe} W_2(OMe)_{10}$$

$$(2.166)$$

$$ReF_6 \xrightarrow[CH_3CN,\ -40^\circ C]{(i)\ Si(OMe)_4} \xrightarrow[THF]{(ii)\ Mg(OMe)_2(HOMe)_2} Re_2(OMe)_{10} \qquad (2.167)$$

An unusual exchange reaction takes place when the organoactinide hydrides $[AnH_2(\eta^5\text{-}C_5Me_5)_2]_2$ are treated with trimethyl phosphite.[430] A methoxo group is transferred to the metal atom while phosphorus is hydrogenated to give a PH bridge:

$$5[AnH_2(\eta^5\text{-}C_5Me_5)_2]_2 + 4P(OMe)_3$$

$$\xrightarrow[25^\circ C]{toluene} \{An(OMe)(\eta^5\text{-}C_5Me_5)_2\}_2PH + 3An(OMe)_2(\eta^5\text{-}C_5Me_5)_2 + 3H_2 \quad (2.168)$$

In contrast to the expected elimination of an ester in the reaction of $Sn(OBu^t)_4$ with $Sn(OAc)_4$ as well as with Me_3SnOAc in a refluxing hydrocarbon solvent, only ligand exchanges of the types shown below have been demonstrated to occur in coordinating solvents such as pyridine:[431]

$$3Sn(OBu^t)_4 + 3Sn(OAc)_4$$

$$\xrightarrow{pyridine} 2Sn(OBu^t)(OAc)_3 + 2Sn(OBu^t)_2(OAc)_2 + 2Sn(OBu^t)_3(OAc) \quad (2.169)$$

$$Sn(OBu^t)_4 + Me_3SnOAc \xrightarrow{pyridine} Sn(OBu^t)_3(OAc)(py) + Me_3Sn(OBu^t) \quad (2.170)$$

The X-ray crystallographically characterized[431a] derivative of zinc, $Zn_2(OAc)_3(OMe)$, has been prepared by the reaction shown in Eq. (2.170a):

$$2Zn(OAc)_2 + Bu_2^nSn(OMe)_2 \longrightarrow Zn_2(OAc)_3(OMe) + Bu_2^nSn(OAc)(OMe)$$

$$(2.170a)$$

3 PHYSICAL PROPERTIES

3.1 Introduction and Structural Characteristics

The physical properties of the metal alkoxides $[M(OR)_x]_n$ are largely influenced by the size and shape of the alkyl (R) group as well as by the valency (x), atomic radius, stereochemistry, and coordination number of the metal. Owing to the high electronegativity of oxygen (3.5 on Pauling scale), the M–OR bonds (in alkoxides of metallic elements) would be expected to possess significant ionic character. Thus metal–oxygen bonds ($M^{\delta+}$—$O^{\delta-}$—C) in metal alkoxides could be expected to possess around 65% ionic character for metals with electronegativity values of the order of 1.5–1.3 (aluminium, titanium, and zirconium) to about 80% ionic character for more electropositive metals with electronegativity values in the range of 1.2–0.9 (alkali, alkaline earth, and lanthanide metals). However, most of these alkoxides show a fair degree of volatility and solubility in common organic solvents, which can be considered as characteristic of covalent compounds. The factors that have been postulated to explain the attenuation in the polarity of the metal–oxygen bond are: (i) the inductive effect (electron release) of the alkyl groups at the oxygen atom, which increases with the branching of the alkyl chain, (ii) the presence of oxygen p to metal d π-bonding for earlier transition metals, and (iii) the formation of oligomeric species through alkoxo bridges of the type:

$$
\begin{array}{c}
\text{R} \\
\text{O} \\
\diagdown\diagup \\
\text{--M} \qquad \text{M--} \\
\diagup\diagdown \\
\text{O} \\
\text{R}
\end{array}
$$

This latter tendency, which leads to coordination polymerization, is a dominant feature of metal alkoxides unless inhibited by steric and electronic factors.[3–8,17–24,342] The extent of aggregation (n) is expected to decrease with the enlargement of the alkyl group (CHMe$_2$, CMe$_3$, CHPr$^i_2$, CHBu$^t_2$, CBu$^t_3$) owing to steric factors and/or the presence of donor functionality in the alkoxo group, e.g. [OC(Bu^t)(CH$_2$OPr^i)$_2$]. Bradley, Mehrotra, and Wardlaw[113,273,274,432] in a series of papers in the 1950s tried to analyse steric and electronic factors for alkoxo derivatives of group 4 metals (Ti, Zr, Hf) and their conclusions have since been found to be broadly applicable to almost all the metals of the periodic table. Results indicate that the extent of aggregation (n) for an appropriate stoichiometry [M(OR)$_x$]$_n$ is dependent on the following considerations: (i) aggregation increases as the metal atom becomes more electron deficient; (ii) the larger the size of the metal atom, the greater the tendency to increase the degree of association (n) by forming alkoxo bridged systems; and (iii) the steric effects of the alkyl substituents, which with increasing steric demand inhibit aggregation and have been found to be of greater importance than the electronic nature of the substituents in determining the ultimate extent of aggregation. The unusually low solubility and volatility of most metal methoxides apparently arise from a combination of these factors as well as a high lattice energy between oligomers arising from the small size of methyl groups on the periphery of the molecules.

The structures of homoleptic alkoxides have been arousing interest for many years and in 1958, long before single-crystal X-ray crystallography became commonplace, Bradley[432] proposed a structural theory for metal alkoxides M(OR)$_x$ based on considerations of molecular weight data, enthalpy, and entropy; the preferred structure is generally one that allows the metal to attain its preferred coordination number and geometry by the minimum degree of oligomerization involving the formation of alkoxo bridges (μ_2 or μ_3) (Fig. 2.1).[8] Within this context, the tetrameric structure of [Ti(OEt)$_4$]$_4$ (Fig. 2.1g) is of historic significance[434] since it provided the first structural test of Bradley's theory for a compound of formula M(OR)$_4$. Each titanium atom achieves an octahedral environment, by the use of four doubly (μ_2) bridging and two triply (μ_3) bridging ethoxo groups. Furthermore, the coordination number of the oxygen atom should not exceed four and, therefore, polymerization can be assumed only on the basis of the stereochemistry of the central metal atom. Bradley[432] tried to predict the

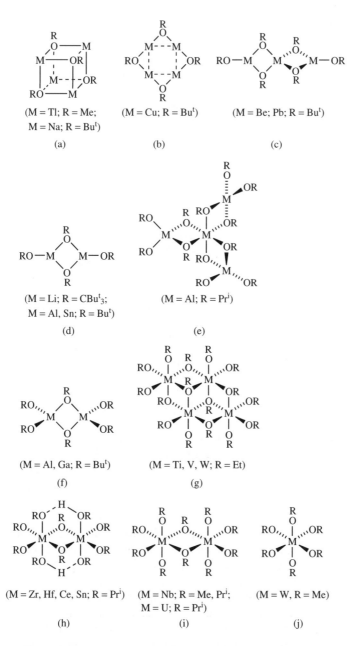

Figure 2.1 Some structural types adopted by homometallic alkoxides.

expected degrees of molecular association for alkoxides of metals and metalloids in different oxidation states, coordination numbers, and geometries (Table 2.3).

The main factors influencing the adoption of one structural type in preference to another[8] are: (i) the empirical formula $M(OR)_x$, (ii) the size and formal oxidation state of the metal atoms, (iii) the steric demand of the alkyl (R) group, and (iv) the d^n

Table 2.3   Proposed correlation amongst empirical formula of a metal alkoxide, coordination number, geometry, and molecular association[432]

Metal alkoxides $M(OR)_x$	Coordination number of M	Stereochemistry of M	Minimum degree of association
MOR	2	OMO $= 180°$	2
	2	OMO $= 120°$	3
	3	Pyramidal	4
$M(OR)_2$	3	Planar One OMO $= 90°$	2
	3	Planar, OMO $= 120°$	3
	4	Tetrahedral	3
	4	Planar	4
	6	Octahedral	Infinite three-dimensional polymer
$M(OR)_3$	4	Tetrahedral	2
	4	Planar	2
	4 and 6	Tetrahedral and octahedral	4
	6	Octahedral	8
$M(OR)_4$	5	Trigonal bipyramidal	2
	6	Octahedral	3
	8	Cubic	4
$M(OR)_5$	6	Octahedral	2
	8	Cubic	4
$M(OR)_6$	8	Cubic	2

configuration for transition metals, leading to pronounced ligand field effects or/and metal–metal bonding.[359]

The data in Table 2.3 show that the values of molecular association predicted by the simple theory of Bradley[432] are in reasonable agreement with those observed experimentally. When the bulk of information collected from spectroscopic and other physicochemical measurements is combined with Bradley's structural theory as well as with the knowledge of structures obtained from the more recent X-ray crystallographic studies, a general classification of the possible structural types (Fig. 2.1) becomes feasible. However, it is not still possible to predict precisely the structure for any given metal alkoxide, and sometimes more than one structural type is possible.

The applications of physicochemical techniques such as IR, NMR, ESR, UV-Vis, and mass spectroscopy as well as magnetic susceptibility measurements have played an important role in the development of metal alkoxide chemistry, by throwing clearer light in the absence of X-ray crystallographic data on the structural features of metal alkoxides up to the 1980s. In spite of some inherent difficulties in the X-ray structural elucidation of metal alkoxide species as pointed out by Bradley (25), the availability of more sophisticated X-ray facilities during the last decade has played a key role in providing solid state structural information on a rapidly increasing number of novel homo- and heterometallic alkoxide systems.

An account of the physical characteristics of homometallic alkoxides is now presented under the following headings:

1. Molecular complexities, volatilities, and thermodynamic data (Section 3.2)
2. Infrared spectra (Section 3.3)

3. Nuclear magnetic resonance spectra (Section 3.4)
4. Electronic absorption, reflectance, and electron spin resonance spectra (Section 3.5)
5. Magnetic susceptibility measurements (Section 3.6)
6. Mass spectra (Section 3.7)
7. EXAFS and XANES studies (Section 3.7).

3.2 Molecular Complexities, Volatilities, and Thermodynamic Data

3.2.1 Introduction

As described in Section 3.1, the observed molecular complexities in nonpolar solvents and the volatility of alkoxides of metals in general appear to be governed by the nature and size of the metal atoms and the inductive effect coupled with the steric demand of the alkoxide groups. This is further confirmed by the extensive data included in Tables 2.4–2.7, 2.9, 2.11, 2.12, and 2.14.

Although studies on the complexity and volatility of alkoxo derivatives of a few elements (B, Si) had been carried out,[1,2] as early as 1846, attention towards these properties was focused[434] in 1950 by the extraordinary volatility of $Zr(OBu^t)_4$ which distilled at about 54°C under 0.1 mm pressure in contrast to the much higher boiling temperature (243°C/0.1 mm) of $Zr(OBu^n)_4$.[135] This marked difference in the volatility of the two isomeric zirconium butoxides was explained on the basis of molecular complexity values of ~3.5 and 1.0, respectively, for the $Zr(OBu^n)_4$ and $Zr(OBu^t)_4$ in refluxing benzene.

This interesting observation led to detailed investigations[113,273,274] during 1950–1952 on the variations in the molecular complexities and volatility of alkoxo derivatives of silicon, titanium, and zirconium. These results are described in detail in Section 3.2.7, but the data on the volatility and molecular complexity of the simplest soluble alkoxides (ethoxides) of the above metals along with those of thorium, germanium, and tin(IV) are presented in Table 2.4, depicting clearly the enhancement in molecular complexity and boiling points with the increasing size and reducing electronegativity of the central metal atom:

Table 2.4   The effect of central atom electronegativity and atomic radius on some physical properties of group 4 and 14 element tetraethoxides, $M(OEt)_4$

	M		$M(OEt)_4$	
Central atom	Electro-negativity	Atomic radius (Å)	B.p. (°C/mm)	Degree of polymerization
C	2.50	0.77	158/760	1.0
Si	1.74	1.11	166/760	1.0
Ge	2.02	1.22	86/12.0	1.0
Sn	1.72	1.41	–	4.0
Ti	1.32	1.32	103/0.1	2.4
Zr	1.22	1.45	190/0.1	3.6
Hf	1.23	1.44	178/0.1	3.6
Th	1.11	1.55	300/0.05	6.0

Table 2.5   Volatility and degree of polymerization (n) of metal amyloxides[273,294]

R in $[M(OR)_x]_n$	$Al(OR)_3$ B.p. (°C/mm)	n	$Ti(OR)_4$ B.p. (°C/mm)	n	$Zr(OR)_4$ B.p. (°C/mm)	n
$CH_3(CH_2)_4$	255/1.0	4.0	175/0.8	1.4	256/0.01	
$(CH_3)_2CH(CH_2)_2$	195/0.1	4.0	148/0.1	1.2	247/0.1	3.3
$(CH_3)(C_2H_5)CHCH_2$	200/0.6	4.1	154/0.5	1.1	238/0.1	3.7
$(CH_3)_3CCH_2$	180/0.8	2.07	105/0.05	1.3	188/0.2	2.4
$(C_2H_5)_2CH$	165/1.0	2.08	112/0.05	1.0	178/0.5	2.0
$(CH_3)(C_3H_7^n)CH$	162/0.5	2.06	135/1.0	1.0	175/0.05	2.0
$(CH_3)(C_3H_7^i)CH$	162/0.6	1.98	131/0.5	1.0	156/0.01	2.0
$(CH_3)_2(C_2H_5)C$	154/0.5	1.97	98/0.1	1.0	95/0.1	1.0

In order to assess the relative effectiveness (Section 3.1) of the two synergetic factors, *i.e.* inductive effect and steric bulk of the alkyl groups, a detailed study was carried out in 1952[273] on the molecular complexities of all the 8 isomeric tertiary amyloxides of titanium and zirconium (Table 2.5). Amongst these, the neopentyloxides are of special significance as the neopentyl alcohol, in spite of being a primary alcohol, had been known to cause steric effects comparable to those of secondary alcohols. The closer resemblance of the neopentoxides of both the metals to their secondary rather than primary amyloxides tended to indicate the greater influence of steric factors in determining the molecular complexity and volatility of these derivatives.

The corresponding data on isomeric amyloxides of aluminium are also included in Table 2.5, indicating the importance of steric factors in their cases also. In fact, detailed investigations were also carried out almost simultaneously[293,294] on aluminium alkoxides as even aluminium tertiary amyloxide exhibits a dimeric behaviour and volatility similar to secondary amyloxides; this has been ascribed to the electron-deficient nature of aluminium, leading to a stronger bridged structure of the type, $(Am^tO)_2Al(\mu\text{-}OAm^t)_2Al(OAm^t)_2$. The strength of this type of bridged structure is also reflected by its presence in the vapour state. An interesting 'ageing' tendency was observed in freshly distilled trimeric $\{Al(OPr^i)_3\}_3$ which slowly changes into a stable tetrameric crystalline form, to which Bradley assigned the interesting structure $[Al\{(\mu\text{-}OPr^i)_2Al(OPr^i)_2\}_3]$, in which the central aluminium atom is hexa-coordinate whereas the three surrounding aluminium atoms are tetra-coordinate.[577]

These early studies by Bradley *et al.*[113,273,274] coupled with those of Mehrotra[293,294] laid the foundation of a clearer understanding for the first time of the molecular complexity and volatility of metal alkoxide derivatives, and led to extensive investigations over several decades on the alkoxides of metals throughout the periodic table, which confirmed the above conclusions; these are described groupwise in Sections 3.2.3–3.2.14, followed by an account of different physico-chemical techniques employed for structural elucidation of metal alkoxides in general.

3.2.2   Special Characteristics of Metal Methoxides

Apart from the monomeric methoxides of the metalloids (B, Si, Ge, As, and Te) which are very volatile and soluble in organic solvents, most of the metal methoxides (except niobium and tantalum pentamethoxides and uranium hexamethoxide) are comparatively nonvolatile and insoluble in common organic solvents. Attempts have been made to explain these special characteristics of methoxides again on the basis of steric and inductive factors. As the molecular weights of the methoxides could not be directly measured, owing to their insolubility and nonvolatility, Verma and Mehrotra,[270] in an attempt to sort out the comparative effect of these two factors, synthesized and studied the properties of heteroleptic ethoxide-methoxides of titanium, $[Ti(OC_2H_5)_3(OCH_3)]$, $[Ti(OC_2H_5)_2(OCH_3)_2]$ and $[Ti(OC_2H_5)(OCH_3)_3]$. From measurements of the molecular weights of these derivatives in benzene, the degree of polymerization of titanium tetramethoxide, $Ti(OMe)_4$ was extrapolated to be four. The single crystal X-ray analysis has confirmed (Chapter 4) the tetrameric[436] nature of $[Ti(OMe)_4]_4$ and this may well explain its low volatility. However, its low solubility suggests that the small size of the peripheral methyl groups leads to a high lattice energy. Incidentally, a sparingly soluble variety of titanium methoxide has been synthesized by Dunn[437] by the reaction between titanium tetrachloride and methanol in the presence of anhydrous ammonia, and this derivative has also been shown to be tetrameric in nature.

3.2.3 Alkoxides of Group 1 Metals

Owing to the strongly electropositive nature of alkali metals, their alkoxides would be expected to be predominantly ionic in character. In fact, sodium ethoxide has been shown to behave as a strong base[136] in ethanol (like sodium hydroxide in water).

Amongst the alkali metals, the lithium derivatives would be expected to be the least ionic as is shown by their solubility in organic solvents and volatility of lithium *tert*-butoxide (110°/0.1 mm);[438] alkoxides of other alkali metals tend to be nonvolatile and decompose on being heated to higher temperatures even under reduced pressure.

Bains[438] reported lithium *tert*-butoxide to be hexameric in cyclohexane. There appears to be some controversy in the literature about the molecular weight of lithium *tert*-butoxide in benzene, the reported degree of polymerization being 4.0,[438] 9.0,[439] and 6.0.[440] The last value appears to be most reliable as these investigators[440] have confirmed their findings by mass spectrometry. X-ray structural findings[28,29] have revealed that tetrameric cubanes and hexameric stacks are the most prevalent structural motifs for molecular alkali metal alkoxides.

In spite of the strongly electropositive character of alkali metals, the importance of the inductive effect of the alkyl group was shown[441] in the comparative molar conductivities of sodium methoxide (92.0 mhos), ethoxide (45.0 mhos), isopropoxide (2.5 mhos) and tertiary butoxides (0.05 mhos) in their parent alcohols. Although the effect could be partially ascribed to the differences in dielectric constants of the alcohols, yet the sharp break between the conductivities of the isopropoxide and *tert*-butoxide could be expected to arise at least partially from the strong +I inductive effect of the tertiary butyl group, making the sodium–oxygen bond much less ionic in character.

In contrast to the insoluble, nonvolatile nature of the simple alkoxo-derivatives of alkali metals, their perfluoro-alkoxo derivatives are comparatively more soluble in polar solvents like acetonitrile, acetone and ether. Thus the perfluoro *tert*-butoxides of lithium and sodium $(MOC_4F_9^t)$ have been found to be high-melting solids which can be distilled under reduced pressures.[442] The corresponding potassium perfluoro *tert*-butoxide could also be sublimed at 140°/0.2 mm pressure.[442] The high volatilities of lithium, sodium, and potassium *tert*-butoxides might well arise from low molecular complexities of these derivatives. In fact, the molecular weight determinations of sodium perfluoro *tert*-butoxides[442] and those of the *tert*-butoxide derivatives[443] of potassium, rubidium, and caesium have indicated these derivatives to be tetrameric in nature. Since the $-I$ effect of perfluoroalkyl groups should enhance the polarity of the M–O bonds, the relatively high volatility and solubility of these perfluoroalkoxides might be due in part to the weak intermolecular forces between the fluorocarbon groups.

During the preparation of trifluoromethoxides of the alkali metals involving reaction of alkali metal fluorides with carbonyl fluoride, Redwood and Willis[444] concluded that because of the ionic character of the fluoromethoxide derivatives $M^+(OCF_3)^-$ the stable trifluoromethoxide derivatives would be formed only with those metals which have a greater atomic size. Thus trifluoromethoxides of potassium, rubidium, and caesium are stable below $-20°C$ but those of lithium and sodium could not be isolated. The trifluoromethoxides of potassium, rubidium, and caesium decompose at $80°C$ under reduced pressure. From the curves of Fig. 2.2, it is clear that the rate of decomposition is slowest for the caesium derivative, which has the largest cation. Thus the order of stability of these alkali metal trifluoromethoxides is $CsOCF_3 > RbOCF_3 > KOCF_3$.

Further, the thermal stability of the product decreases with increasing chain length of the fluoroalkoxide. This is because, although the oxygen atom is negatively charged in the trifluoromethoxide ion, the higher electronegativity of three fluorine atoms would tend to delocalize the charge over the ion and thus bring about comparatively greater stability. However, with pentafluoroethoxide or heptafluoropropoxide, which contain more than one carbon atom, the stability is reduced because the carbon atom is less electronegative than fluorine. Therefore, for the fluoroalkoxide derivative of the same metal the observed stabilities are in the decreasing order:[445] $MOCF_3 > MOCF_2CF_3 > MOCF_2CF_2CF_3 > MOCF(CF_3)_2$ which can be represented graphically by Fig. 2.3.

3.2.4   Alkoxides of Group 2 and 12 Metals

The primary alkoxide derivatives of alkaline earth and other metals of group 2 are generally insoluble nonvolatile compounds whereas their highly branched alkoxides

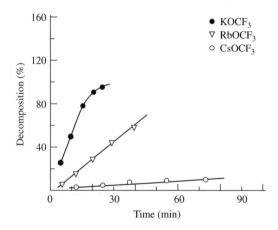

Figure 2.2   Decomposition curve of alkali trifluoromethoxides at $80°C$.

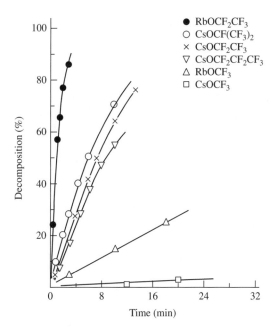

Figure 2.3   Decomposition curves of alkali metal fluoroalkoxides.

tend to be comparatively more volatile and soluble in organic solvents. These characteristics could arise either from their ionic nature or from their high degree of polymerization in the case of the alkaline earth metals. However, the low solubility and volatility in the case of beryllium alkoxides most probably arises from the tendency of beryllium to attain a tetracoordination state through the formation of an almost infinite polymer comparable to the case of beryllium hydride.[446] Thus the dimethoxide, diethoxide, di-n-propoxide and other normal alkoxides of beryllium have been reported to be insoluble products,[214,447] but the trimeric di-*tert*-butoxide, $[Be(OCMe_3)_2]_3$ is soluble in benzene and hexane and can be sublimed at about 100°C under 0.001 mm pressure.[385] Beryllium dialkoxides of higher tertiary homologues are trimeric $[Be(OCMe_2Et)_2]_3$ (subliming at 105°C/0.001 mm), $[Be(OCMeEt_2)_2]_3$ (subliming at 125°C/0.001 mm), and dimeric $[Be(OCEt_3)_2]_2$ (subliming at ~60°C/0.001 mm).[448] The extraordinary volatility of the last product is apparently due to increasing steric bulk as shown by its dimeric nature compared to the trimeric nature of the lower tertiary alkoxides[385] in benzene. In fact, such high volatility appears to indicate dissociation to monomeric species before sublimation and it would be interesting to measure the vapour density of such derivatives at different temperatures.

Compared to the highly polymeric dialkoxides of beryllium, the corresponding alkylberyllium alkoxides have been shown to be volatile and tetrameric in nature. Thus the lowest member of the series, methylberyllium methoxide, is soluble in benzene and ether, and it sublimes at 70°C under 0.001 mm pressure, whereas the sterically more hindered dimeric *tert*-butylberyllium *tert*-butoxide sublimes at a lower temperature, *i.e.* 40–50°C under 0.001 mm pressure.[448,449] The lower degree of polymerization and higher volatility of alkylberyllium alkoxides, $[RBe(OR)]_n$ compared to the corresponding dialkoxides $[Be(OR)_2]_n$ can be partially due to steric factors, but the replacement of one of the electron-withdrawing alkoxo groups by an electron-releasing alkyl group would tend to make the bond much less polar in character.

Similar to beryllium n-dialkoxides, zinc dialkoxides are also insoluble and nonvolatile compounds.[222,447] The alkylzinc alkoxides are, however, less polymeric and exhibit higher volatility. For example, the cryoscopic molecular weight determination in benzene indicates that methylzinc methoxide and *tert*-butoxide as well as ethylzinc *tert*-butoxide are tetrameric[450] with sublimation temperatures of 60, 95, and 105°C, respectively under 0.0001 mm pressure.[390]

Similar behaviour has also been observed in the case of magnesium and other alkaline earth metal alkoxides. The normal dialkoxides of magnesium, calcium, strontium, and barium are insoluble and nonvolatile[39–41] whereas the alkylmagnesium alkoxides are soluble in common organic solvents and some of these (especially methyl and ethyl derivatives) can be volatilized *in vacuo*,[386–391,451] usually accompanied by decomposition.

The properties[22,39–41] of alkaline earth metal alkoxides appear to be dominated by their ionic character and preference for attaining metal coordination numbers ≥ 6, leading to rather large associated species.

3.2.5 Alkoxides of Scandium, Yttrium, and Lanthanides

The alkoxide derivatives of lanthanide elements (which are larger in size and more electropositive) are less soluble than other trivalent (say group 13) alkoxides, probably owing to their ionic nature and higher molecular aggregation *via* alkoxo bridging. Although methoxides and ethoxides of scandium, yttrium, and lanthanides are insoluble and nonvolatile,[6,215] the isopropoxides of yttrium,[54] lanthanum,[150] neodymium,[150,153] samarium,[155] gadolinium,[149,157] dysprosium,[54] holmium,[158] erbium,[157] and ytterbium[54,103,157] could be sublimed under reduced (0.1 mm) pressure in the temperature range 180–280°C. The sublimation temperature appears to be influenced by the size of the lanthanide metal; the smaller the size of the metal atom the greater the volatility. However, the authenticity of these tris-isopropoxides has been questioned (see Ch. 5, p. 385).[158a,b]

Brown and Mazdiyasni[55] carried out thermogravimetric analyses of yttrium, dysprosium, and ytterbium isopropoxides and observed a 30% weight loss up to 200–250°C with the formation of an intermediate hydroxide which was converted to the oxide in the temperature range 750–850°C.

The molecular complexity of lanthanide isopropoxides in solution has been reported to be higher than four,[155,158] although these were reported earlier to be dimeric.[54]

On the basis of mass spectral data, Mazdiyasni *et al*.[54] reported tetrameric nature for neodymium, erbium, and lutetium isopropoxides. The tetrameric structure of $[Ln(OCH_2Bu^t)_3]_4$ (Ln = La, Nd) has recently been confirmed by X-ray crystallography,[452a] whereas the t-butoxide is trinuclear $[La_3(OBu^t)_9(Bu^tOH)_2]$.[452b]

3.2.6 Alkoxides of Actinides

Bradley *et al*.[290] purified uranium tetra-alkoxides by sublimation under reduced pressure and also tried to correlate the volatility of uranium tetra-alkoxides with those of other tetravalent metal alkoxides as shown in Table 2.6 (which includes for comparison relevant data for tetravalent titanium and zirconium derivatives).

Table 2.6   Volatilities of tetravalent metal alkoxides

M(OR)$_4$	Radius (Å)		Sublimation distillation temperature (°C/mm)			
	Atomic	Ionic (M^{4+})	M(OMe)$_4$	M(OEt)$_4$	M(OPr^n)$_4$	M(OPr^i)$_4$
Ti	1.32	0.64	170/0.1	103/0.1	124/0.1	49/0.1
Zr	1.45	0.87	280/10^{-5}	180/0.1	208/0.1	160/0.1
U	1.42	1.05	>300/10^{-4}	220/0.01	240/0.1	160/0.01
Ce	1.65	1.02	>200/*in vac*	>200/*in vac*	>200/*in vac*	160–170/0.05
Th	1.65	1.10	>300/0.05	>300/0.05	—	200–210/0.05–0.1

Uranium pentaethoxide[379] has been found to be thermally stable up to 170°C *in vacuo* and it has been distilled unchanged at 123°C/0.001, 145°C/0.01, and 160°C/0.5 mm pressures, but it decomposes when heated in the range 180–200°C *in vacuo*. Uranium pentamethoxide, penta-*n*-propoxide, and pentaisopropoxide have also been sublimed or distilled respectively[289] at 190–210°C/0.1, 162–164°C/0.001, and 160°C/0.1 mm pressure. The purification of the uranium hexa-alkoxides has also been achieved by distillation *in vacuo*; thus uranium hexamethoxide, hexaethoxide, hexa-*n*-propoxide have been distilled respectively[380] at 87°C/0.1, 93°C/0.18, 105–107°C/0.001 mm pressures. Jones *et al.*[453] have claimed that uranyl diethoxide, UO$_2$(OEt)$_2$ retained three moles of ethanol even on heating at a higher temperature of 200°C/0.004 mm pressure, but Bradley *et al.*[216] observed that uranyl dimethoxide and diethoxide lost their attendant alcohol molecules on being heated to 100°C/0.05 and 80°C/0.05 mm pressures, respectively.

3.2.7   Alkoxides of Group 4 Metals

Titanium and zirconium tetra alkoxides are all volatile derivatives including the methoxides[275] which sublime at 180°C and 280°C, respectively under 0.1 mm pressure. Table 2.7 shows the boiling points and molecular complexities of titanium and zirconium alkoxides, which decrease with increasing branching of the alkoxo groups. Furthermore, the boiling points and complexities (including some data already given in Tables 2.4 and 2.5) are higher for zirconium alkoxides than for their titanium analogues.[113,274,326,432] The molecular complexity of titanium methoxide has been reported to be tetrameric.[270] Caughlan *et al.*[454] measured the molecular weight of the tetraethoxide cryoscopically in benzene and found it to be concentration dependent with equilibrium between monomeric and trimeric species in the concentration range 0–0.3 M. Bradley *et al.*[432,455] on the other hand found the complexities 2.4 and 2.8 ebullioscopically and cryoscopically in benzene, respectively, and these values were found to be concentration independent in the concentration range 2–100 × 10^{-3} M. Higher normal alkoxides of titanium are trimeric in the concentration range 0–0.03 M; the isopropoxide shows an average association of 1.4 whereas tertiary alkoxides are essentially monomeric in refluxing benzene.[113,273,274,432]

Martin and Winter[456] also observed three-fold molecular complexity for titanium tetra-*n*-butoxide and the molecular weight was shown to be concentration dependent. The replacement of butoxy groups by the more electronegative chlorine atoms appears to affect the molecular complexity of tetra-*n*-butoxide considerably. Thus it has been observed that the trimeric titanium monochloride tributoxide does not

Table 2.7   Boiling points and molecular complexities of titanium and zirconium tetra alkoxides

R	B.p. (°C/mm)		Molecular complexity	
	$Ti(OR)_4$	$Zr(OR)_4$	$Ti(OR)_4$	$Zr(OR)_4$
C_2H_5	103/0.1	190/0.1	2.4	3.6
$CH_3CH_2CH_2CH_2$	142/0.1	243/0.1	–	3.4
$CH_3(CH_2)_7$	214/0.1	–	1.4	3.4
$(CH_3)_2CH$	49/0.1	160/0.1	1.4	3.0
$(C_2H_5)_2CH$	112/0.1	181/0.1	1.0	1.0
$(C_3H_7^n)_2CH$	156/0.1	163/0.1	1.0	1.0
$(CH_3)_3C$	93.8/5.0	89/5.0	1.0	1.0
$CH_3(CH_2)_4$	175/0.8	256/0.01	1.4	3.2
$(CH_3)_2CH(CH_2)_2$	148/0.1	247/0.01	1.2	3.3
$(CH_3C_2H_5)CHCH_2$	154/0.5	238/0.01	1.1	3.7
$(CH_3)_3CCH_2$	105/0.05	188/0.2	1.3	2.4
$(C_2H_5)_2CH$	112/0.05	178/0.05	1.0	2.0
$(CH_3C_3H_7^n)CH$	135/1.0	175/0.05	1.0	2.0
$(CH_3C_3H_7^i)CH$	131/0.5	156/0.01	1.0	2.0
$(CH_3)_2C_2H_5C$	98/0.1	95/0.1	1.0	1.0

undergo molecular dissociation at the concentrations at which the tetrabutoxide is almost dissociated.[456] Similarly dimeric dichloride dibutoxide as well as monomeric trichloride monobutoxide also do not undergo any dissociation over a wide concentration range. The above results appear to indicate that the presence of the more electronegative chlorine atom increases the positive charge and acceptor properties of the central titanium atom, and consequently the strength of the alkoxide bridges increases which prevents the molecules from depolymerizing.

Barraclough *et al.*[457] measured the molecular complexities of titanium alkoxides in dioxane solvent and observed that the complexities are concentration dependent except with monomeric *tert*-butoxide which shows concentration-independent molecular weight. For the normal alkoxides, as the concentration is increased, the molecular weight increases and there was no indication of a limiting value of molecular weight up to a concentration of 0.5 M.

Titanium isopropoxide shows an average association of 1.4, whereas its tertiary and higher secondary alkoxides are essentially monomeric in refluxing benzene.[113,273,274,432] Titanium tetrakis(hexafluoroisopropoxide) also shows an average molecular association of 1.5 in boiling benzene.[458]

On the basis of the boiling points measured for various normal alkoxides at different pressures, Cullinane *et al.*[459] observed that the results do not conform to the relation, $\log p = a - b/T$ (p is the pressure in mm at which boiling point T was observed and a and b are constants). However, the latent heats and entropies of vaporization and the Trouton constants indicate that titanium tetraethoxide exhibits anomalous behaviour, *i.e.* it shows a distinctly higher Trouton constant (42.5) than those observed for *n*-propoxide, *n*-butoxide, *n*-amyloxide, and *n*-hexyloxide (29.1, 35.6, 39.4, and 40.5, respectively).

Bradley *et al.*[306,307] measured the heat of formation of liquid trimeric titanium ethoxide and the value was found to be $\Delta H_f^\circ Ti(OC_2H_5)_4(liq) = -349 \pm 1.4$ kcal mol^{-1}, from which the standard heat of formation of monomeric gaseous titanium ethoxide was

deduced to be $\Delta H_f^\circ \mathrm{Ti(OC_2H_5)_4(g)} = -325 \pm 2.4\,\mathrm{kcal\,mol^{-1}}$. The average bond dissociation energy Δ of (Ti—O) bonds in gaseous monomeric titanium ethoxide was calcu-
lated by substituting the known values of $\Delta H_f^\circ \mathrm{(OC_2H_5)(g)} = -8.5 \pm 2\,\mathrm{kcal\,mol^{-1}}$ in
Eq. (2.171) and was found to be $101 \pm 2.1\,\mathrm{kcal\,mol^{-1}}$.

$$4\bar{D} = \Delta H_f^\circ[\mathrm{Ti(g)}] + 4\Delta H_f^\circ(\mathrm{OR})(g) - \Delta H_f^\circ\,\mathrm{Ti(OR)_4(g)} \tag{2.171}$$

Using the well-known alcoholysis reactions of titanium ethoxide in cyclohexane, the
standard heat of formation of various titanium tetra-alkoxides in liquid as well as
gaseous states were determined and the average bond dissociation energy was calculated[306] as given in Table 2.8.

The heats of formation of $\mathrm{Ti(OR)_4}$ appear to depend on the chain length and
branching of the alkoxy group, in the following order: $R = tert\text{-}C_5H_{11} > n\text{-}C_5H_{11} >$
$t\text{-}C_4H_9 > sec\text{-}C_4H_9 > n\text{-}C_4H_9 > iso\text{-}C_3H_7 > n\text{-}C_3H_7 > C_2H_5$.

The boiling points of zirconium alkoxides measured at various pressures conform to
the equation, $\log p = a - b/T$ in the pressure range 2–10 mm.[432] As regards molecular
complexity, zirconium methoxide is an insoluble polymeric derivative. The ethoxide
shows an average association of the order of 3.6 and nearly the same degree of association has also been observed for other normal alkoxides (up to octyloxide). Secondary
alkoxides show variations depending on the chain length: thus the isopropoxide is
trimeric whereas sec-amyloxides are dimeric and these became monomeric as the
chain length increases. Thus the molecular complexities of various secondary alkoxides $[\mathrm{Zr(OCHRR')_4}]_n$ decrease with increasing chain length of R and R' (i.e. when
$R = R' = Me$, $n = 3.0$; $R = Me$, $R' = Et$, $n = 2.5$; $R = R' = Et$, $n = 2.0$; and $R =$
$R' = Pr$, $n = 1.0$). The tertiary alkoxide derivatives are, however, all monomeric irrespective of chain length.

The higher molecular association of zirconium alkoxides compared to those of titanium analogues may be ascribed to the greater atomic radius of zirconium and its
tendency to achieve a higher coordination number. It has also been observed by Bradley
$et\ al.$[113,432] that the molecular complexities of zirconium alkoxides are reduced to a
considerable extent in their parent alcohols, $e.g.$ ethoxide and isopropoxide show association of 2.5 and 2.0 in boiling parent alcohols as against 3.6 and 3.0 observed,
respectively, in benzene. The reduction in molecular complexities in alcohols may be
due to the coordination of free alcohol molecules to zirconium atoms. It has been found

Table 2.8   Thermodynamic properties of titanium alkoxides

R	$\Delta H_f^\circ\ \mathrm{Ti(OR)_4(liq)}$ (kcal mol^{-1})	$\Delta H_f^\circ\ \mathrm{Ti(OR)_4(g)}$ (kcal mol^{-1})	$\bar{D}$ (kcal mol^{-1})	$\bar{E}$ (kcal mol^{-1})
C_2H_5	-349 ± 1.4	-350	101	52
$n\text{-}C_3H_7$	-372 ± 4.2	-354	104	55
$iso\text{-}C_3H_7$	-377 ± 1.6	-360	103	53
$n\text{-}C_4H_9$	-399 ± 1.6	-377	105	56
$iso\text{-}C_4H_9$	-403 ± 1.5	-381	105	54
$sec\text{-}C_4H_9$	-404 ± 1.5	-382	104	53
$tert\text{-}C_4H_9$	-411 ± 1.5	-395	102	51
$n\text{-}C_5H_{11}$	-428 ± 2.2	-403	105	58
$tert\text{-}C_5H_{11}$	-457	-438	109	57

that this type of coordination is quite stable in some special cases, as shown by the crystallization of zirconium isopropoxide in the form of its alcoholate, $Zr(OPr^i)_4 \cdot Pr^iOH$, which has been shown by X-ray crystallography[460] to be dimeric in the solid state also. Thermal studies of zirconium alkoxides have indicated that the highly associated alkoxides show higher values of latent heat and entropies of vaporization than the monomeric tertiary alkoxides, and it therefore appears that the vaporization process appears to be associated with depolymerization from polymeric to monomeric state as shown by Eq. (2.172):

$$[Zr(OR)_4]_n \rightleftharpoons nZr(OR)_4 \qquad (2.172)$$
$$\text{(liquid)} \qquad \text{(vapour)}$$

Table 2.9 indicates that the boiling points of zirconium[113,273,274,432] and hafnium alkoxides[277] are quite similar but differ from those of titanium analogues.[113,273,274,432] An important feature of Table 2.9 is the fact that *tert*-alkoxides of hafnium are more volatile[277] than the analogous zirconium derivatives. The entropies of vaporization of zirconium and hafnium *tert*-amyloxides are also nearly the same (39.5 and 39.8 cal deg^{-1} mol^{-1}, respectively) and these results are in satisfactory accordance with the chemical similarities of zirconium and hafnium.

On the basis of the comparative volatilities of titanium, zirconium, and hafnium tertiary alkoxides measured at various pressures, the following order of volatility may be deduced $P_{Hf} > P_{Zr} > P_{Ti}$. The anomalous order cannot be explained in terms of intermolecular forces between tertiary alkoxides as the heats of vaporization are quite close to each other for these alkoxides. Bradley *et al.*,[461–463] however, concluded that in the case of two species involving the same intermolecular forces, the effect of molecular mass on the entropy of vaporization would be to increase the volatility of the species having the higher mass value. In addition, Table 2.10 shows that the latent heats of vaporization of titanium, zirconium, and hafnium *tert*-alkoxides[462,463] are temperature dependent, which may be ascribed to a slight molecular association in these *tert*-alkoxides in the liquid state.

During the pyrolytic study of zirconium n-, sec-, and *tert*-alkoxides, Bradley and Faktor[464] observed that zirconium ethoxide showed a very slight decomposition when

Table 2.9　Thermodynamic properties and molecular complexities of titanium, zirconium, and hafnium alkoxides

$M(OR)_4$	B.p. (°C/mm)	ΔH_v kcal mol^{-1}	$\Delta S_{15.0}$ cal deg^{-1} mol^{-1}	Molecular complexity
$Ti(OEt)_4$	138.3/5.0	21.6	52.5	2.4
$Ti(OPr^i)_4$	91.3/5.0	14.7	40.5	1.4
$Ti(OBu^t)_4$	93.8/5.0	14.5	39.5	1.0
$Ti(OAm^t)_4$	142.7/5.0	16.7	40.0	1.0
$Zr(OEt)_4$	234.8/5.0	30.2	59.4	3.6
$Zr(OPr^i)_4$	203.8/5.0	31.5	66.1	3.0
$Zr(OBu^t)_4$	89.1/5.0	15.2	42.0	1.0
$Zr(OAm^t)_4$	138.4/5.0	16.3	39.5	1.0
$Hf(OEt)_4$	180–200/0.1	–	–	–
$Hf(OPr^i)_4$	170/0.35	–	–	–
$Hf(OBu^t)_4$	90/6.5	–	–	–
$Hf(OAm^t)_4$	136.7/5.0	16.3	39.8	–

Table 2.10   Vapour pressure of Ti, Zr, and Hf *tert*-butoxides

| | Ti(OBu^t)$_4$ | | Zr(OBu^t)$_4$ | | Hf(OBu^t)$_4$ | | | |
| | | ΔH_v | | ΔH_v | | ΔH_v | | |
$T°$C	P (mm)	(kcal mol^{-1})	P (mm)	(kcal mol^{-1})	P (mm)	(kcal mol^{-1})	$P_{Zr/Ti}$	$P_{Hf/Zr}$
27	0.049	15.8	0.066	15.9	0.069	16.3	1.34	1.05
47	0.25	15.2	0.34	15.3	0.36	15.5	1.34	1.08
67	1.00	14.7	1.35	14.8	1.48	14.7	1.35	1.09
87	3.26	14.1	4.42	14.2	4.80	13.9	1.36	1.09
107	9.04	13.5	12.30	13.7	13.05	13.2	1.36	1.06
127	21.41	13.0	29.77	13.1	30.49	12.4	1.39	1.02

heated at 340°C under 760 mm pressure for about 3 hours; the isopropoxide on the other hand decomposed under these conditions to give volatile products, propylene, isopropyl alcohol, and a solid residue containing 65.8% Zr, which is in the range required for $Zr_2O_3(OPr^i)_2$-ZrO_2. Compared to these derivatives, the *tert*-butoxide and *tert*-amyloxide are thermally unstable. The former is stable for about 70 h at 250°C/760 mm under special conditions whereas the latter is stable indefinitely at 220°C/760 mm, for about 30 h at 250°C/175 mm, and under special conditions for about 6 h at 320°C/760 mm pressure. These alkoxides also, on thermal decomposition, yielded alkene, *tert*-alcohols and solid residues having Zr contents that are also within the range required for $Zr_2O_3(OR^t)_2$-ZrO_2.

The following order of the thermal stability of these alkoxides may thus be assigned in which the ethoxide appears to be thermally more stable than the more volatile secondary and tertiary alkoxides: Zr(OEt)$_4$ $\gg$ Zr(OPr^i)$_4$ > Zr(OBu^t)$_4$ ~ Zr(OAm^t)$_4$. Furthermore, the kinetics of thermal decomposition of zirconium tetra-*tert*-amyloxide[465] studied in a clean glass apparatus in the temperature range 208–247°C indicated that the tertiary amyl alcohol formed during decomposition eliminates water; a molecule of water produced from one molecule of alcohol would produce another two molecules of alcohol by hydrolysis of zirconium *tert*-amyloxide and hence a chain reaction would be set up. Since the dehydration of the tertiary alcohol is a rapid first order reaction, it may be the rate controlling step in the thermal decomposition of tertiary alkoxides.

In comparison with titanium, zirconium, and hafnium alkoxides, the thorium and cerium tetra alkoxides are less volatile; their methoxides and ethoxides are insoluble solids, which could not be sublimed even at higher temperatures *in vacuo*.[143,466] The long chain normal alkoxides are, however, soluble derivatives but these also could not be volatilized *in vacuo* except for cerium tetrapentyloxide which could be sublimed at 260°C/0.05 mm.[143] The soluble isopropoxide and tertiary alkoxides of cerium and thorium could also be volatilized (Table 2.11) under reduced pressure.[141,286,287] The normal alkoxides of cerium, *e.g.* n-propoxide and n-pentyloxide show molecular complexity[143] in the range of 4.2–4.3 in refluxing benzene and this was reduced to a considerable extent in boiling toluene where the observed values fall in the range 3.4–3.44. The analogous thorium alkoxides show higher associations; thus the n-butoxide and n-pentyloxide show associations of the order of 6.44 and 6.20, respectively, in boiling benzene. This appears to indicate that increasing the chain length does not have an appreciable effect on the degree of polymerization of normal alkoxides of cerium and thorium.

The isopropoxides of cerium and thorium also show higher association (3.1 and 3.8, respectively, in boiling benzene) than titanium, zirconium, and hafnium alkoxides.

Table 2.11 Volatilities and molecular complexities of cerium and thorium alkoxides

R	Ce(OR)$_4$		Th(OR)$_4$	
	B.p. (°C/mm)	Molecular complexity	B.p. (°C/mm)	Molecular complexity
CHMe$_2$	160–170/0.5	3.1	200–210/0.1	3.8
CMe$_3$	140–150/0.1	2.5	160/0.1	3.4
CEt$_3$	154/0.05	1.1	148/0.05	1.0
CMe$_2$Et	240/0.1	2.4	208/0.3	2.8
CMeEt$_2$	140/0.06	–	148/0.1	–
CMeEtPr^n	150/0.05	1.0	153/0.1	1.7
CMeEtPr^i	–	–	139/0.05	1.0

In contrast to the monomeric nature of tertiary alkoxides of titanium, zirconium, and hafnium, the corresponding cerium and thorium lower tertiary alkoxides exhibit association, which decreases with increasing chain length of the groups attached to the tertiary carbon atom and finally Th(OCMeEtPr^i)$_4$[286] and Ce(OCMeEtPr^n)$_4$[141] show monomeric behaviour. On the basis of the above observations, the order of volatility of some quadrivalent metal alkoxides may be assigned: Si(OR)$_4$ > Ge(OR)$_4$ > Ti(OR)$_4$ > Hf(OR)$_4$ > Zr(OR)$_4$ > Ce(OR)$_4$ > Th(OR)$_4$. However, for monomeric tertiary alkoxides, the order of volatility is Hf(OR^t)$_4$ > Zr(OR^t)$_4$ > Ti(OR^t)$_4$.

3.2.8 Alkoxides of Group 5 Metals

Vanadium trimethoxide and triethoxide are nonvolatile solids,[218] whereas almost all known vanadium tetra-alkoxides are volatile derivatives with the exception of the tetramethoxide which decomposes under similar conditions.[333] Vanadium tetraethoxide and tetra-*n*-propoxide sublime under 0.5 mm pressure at 100–110°C and 140–150°C bath temperatures, respectively, whereas tetra-*n*-butoxide, tetra isobutoxide, tetra-isopropoxide and tetra-*tert*-butoxide distil at 150–160°C/0.5, 70–80°C/0.1 and 60–70°C/0.1 mm pressures, respectively. The heats of formation of liquid and gaseous vanadium tetra-*tert*-butoxide, determined by reaction calorimetry,[307] were found to be −334 ± 0.8 and −328 kcal mol^{-1}, respectively. Using the relation for the calculation of average bond dissociation energy (used for chromium tetra-*tert*-butoxide)[467] (Section 3.2.9), Bradley *et al.*[307] found the average value of D for V(OBu^t)$_4$ to be 87.5 kcal.

Ebullioscopic molecular weight determinations show that vanadium tetramethoxide is trimeric. The tetraethoxide is dimeric, whereas the tetra-*n*-propoxide, *n*-butoxide and *n*-amyloxide show average degrees of association of 1.38, 1.31, and 1.27, respectively.[333] All secondary and tertiary alkoxides of vanadium are monomeric except the isopropoxide which shows slight association (1.17 complexity) in boiling benzene.

The pentavalent niobium and tantalum alkoxides are reasonably volatile and can be distilled unchanged in the pressure range 0.05–10 mm, *e.g.* pentamethoxides of niobium[468] and tantalum[469] have been distilled at 153°C/0.1 or 200°C/5.5 and 130°C/0.2 or 189°C/10.0 mm pressures, respectively. The data in Table 2.12 indicate that the boiling points are dependent on chain length; furthermore, the methoxides and ethoxides of tantalum are more volatile than the niobium analogues whereas for higher *n*-alkoxides, the reverse is true.[279,280,468,469]

Table 2.12   Boiling points and molecular complexities of niobium and tantalum alkoxides

	$Nb(OR)_5$		$Ta(OR)_5$	
R	B.p. (°C/mm)	Molecular complexity	B.p. (°C/mm)	Molecular complexity
Me	153/0.1	2.11	130/0.2	1.98
Et	156/0.05	2.02	146/0.15	1.98
Pr^n	166/0.05	2.02	184/0.15	1.95
Bu^n	197/0.15	2.01	217/0.15	2.02
n-Pentyl	223/0.15	2.00	239/0.02	2.01
Me_2CH	60–70/0.1	1.00	122/0.1	1.00
Me_3C	110–20/0.3	1.00	96/0.1	1.00
Et_2CH	138/0.1	1.16	153/0.1	1.02
$MePr^nCH$	137.5/0.1	1.03	148/0.1	0.99
$Me_2CHCH_2CH_2$	199/0.1	1.81	210/0.1	1.98

The volatilities of both the niobium and tantalum alkoxides decrease with increasing chain length. The mixed alkoxides of tantalum have been found to be thermally stable: $Ta(OMe)(OPr^i)_4$, $Ta(OEt)(OPr^i)_4$, and $Ta(OMe)(OBu^t)_4$ distil unchanged at 81°C/0.07, 130°C/0.1, and 96°C/0.05 mm, respectively.

With regard to molecular complexity, the primary alkoxides of niobium and tantalum are all dimeric in boiling benzene. Thus, even the methoxides show degrees of association of 2.11 and 1.98, respectively.[468,469] This appears to indicate that chain length has very little effect on the molecular complexity of normal alkoxides of niobium and tantalum. Molecular weight measurements in boiling toluene gave the values 1.90 and 1.83 for niobium and tantalum methoxides which appear to indicate that the higher temperature affects the molecular complexity to a small extent[468] (Table 2.13). Molecular weight measurements in boiling parent alcohols appear to indicate the lowest values for methoxides. This unexpectedly low value for methoxides may be ascribed to the higher dielectric constant of methanol. The isopropoxides of niobium and tantalum are monomeric at room temperature but at lower temperatures show dimeric behaviour as deduced from low temperature NMR data,[312,470] thus indicating an equilibrium of the following type:

$$M_2(OPr^i)_{10} \rightleftharpoons 2M(OPr^i)_5 \qquad (2.173)$$

Table 2.13   Molecular complexity of niobium and tantalum alkoxides, $M(OR)_5$

	Molecular complexity of $M(OR)_5$					
	In benzene		In toluene		In ROH	
R	Nb	Ta	Nb	Ta	Nb	Ta
Me	2.11	1.98	1.90	1.83	1.34	1.20
Et	2.02	1.98	1.89	1.83	1.52	1.78
Pr^n	2.02	1.95	1.79	1.83	1.29	1.70
Bu^n	2.01	2.02	1.74	1.83	1.13	1.40
n-Pentyl	2.00	2.01	–	–	–	–

3.2.9   Alkoxides of Group 6 Metals

The alkoxide derivatives of divalent chromium (as well as manganese, cobalt, and nickel) are all insoluble non-volatile products and their insolubility may be attributed to their polymeric nature.[219] Chromium trialkoxides are also insoluble, except the tri-*tert*-butoxide which is a soluble dimeric compound.[334] Chromium tri-*tert*-butoxide when heated *in vacuo* yielded[7] volatile $Cr(OBu^t)_4$ and nonvolatile $[Cr(OBu^t)_2]_n$:

$$2Cr(OBu^t)_3 \xrightarrow{\text{heat, } vacuo} Cr(OBu^t)_4 + \frac{1}{n}[Cr(OBu^t)_2]_n \qquad (2.174)$$

Chromium tetra-*tert*-butoxide is monomeric.[335,471] Bradley and Hillyer[467] measured the heat of formation of chromium tetra-*tert*-butoxide by means of reaction calorimetry in the presence of 2.0 N aqueous sulphuric acid. The thermochemical equations for chromium tetra-*tert*-butoxide may be assumed as follows:

$$6Cr(OC_4H_9^t)_4(l) + 31H_2O(liq) + 10H^+(aq)$$
$$\longrightarrow 4[Cr(H_2O)_6]^{3+}(aq) + (Cr_2O_7)^{2-}(aq) + 24C_4H_9^tOH(liq) \quad (2.175)$$

$$3Cr(OC_4H_9^t)_4(l) + 16H_2O(liq) + 5H^+(aq)$$
$$\longrightarrow 2[Cr(H_2O)_6]^{3+}(aq) + (HCrO_4)^-(aq) + 12C_4H_9^tOH(liq) \quad (2.176)$$

$$3Cr(OC_4H_9^t)_4(l) + 16H_2O(liq) + 4H^+(aq)$$
$$\longrightarrow 2[Cr(H_2O)_6]^{3+}(aq) + (CrO_4)^{2-}(aq) + 12C_4H_9^tOH(liq) \quad (2.177)$$

The heats of formation, ΔH_f° $Cr(OC_4H_9^t)_4(l)$ from Eqs (2.175)–(2.178) have been obtained as −320.2, −321.5, and −319.4 kcal mol^{-1}, respectively which are in good agreement. The standard heat of formation of gaseous *tert*-butoxide, ΔH_f° $Cr(OC_4H_9^t)_4$ (g) was −305 kcal mol^{-1}. Bradley and Hillyer[467] also calculated the average bond dissociation energy using the relationship:

$$4\bar{D} = \Delta H_f^\circ M(g) + \Delta H_f^\circ (OC_4H_9^t)(g) - \Delta H_f^\circ M(OC_4H_9^t)_4(g) \qquad (2.178)$$

Substituting the Cottrell[472] value for $\Delta H_f^\circ Cr(g) = 94.0$ kcal mol^{-1} and Gray and Williams'[473] value for $\Delta H_f^\circ (OC_4H_9^t)(g) = -25 \pm 2$ kcal mol^{-1} in the above equation, the average bond dissociation energy $\bar{D}$ for chromium tetra-*tert*-butoxide was found to be 73 kcal. This value of $\bar{D}$ is in accordance with the requirement of the ground state gaseous metal atom ($d^5 s^1$) which can be used to obtain the dissociation of chromium tetra-*tert*-butoxide into Cr atoms and $(OC_4H_9^t)$ radicals.

The alkoxides of molybdenum(III) and tungsten(III) of empirical formula '$M(OR)_3$' are associated *via* metal–metal bonding (instead of association through alkoxo bridges), leading to diamagnetic compounds[359-365] of the type $M_2(OR)_6$ (M = Mo; R = Bu^t, Pr^i, CH_2Bu^t, CMe_2Ph, etc. M = W; R = Bu^t) or $Mo_4(OR')_{12}$ (R' = the less bulky Me or Et). Tungsten(IV) methoxide and ethoxide are tetrameric clusters,[366,367] $W_4(OR')_{16}$ (R' = Me, Et). The compounds of the type $M_2(OR)_6$ sublime readily at 70–90°C/ 10^{-3} mm for $Mo_2(OPr^i)_6$ and 100°C/10^{-5} mm for $Mo_2(OBu^t)_6$.

3.2.10 Alkoxides of Later 3d Metals

Iron trialkoxides are volatile compounds, except the methoxide which decomposed on attempted distillation *in vacuo*.[282] Thus ethoxide, *n*-propoxide, *n*-butoxide, isopropoxide, and *tert*-butoxide distilled at 155, 162, 171, 149, and 136°C, respectively under 0.1 mm pressure.[282,283] The ebullioscopic[283] and cryoscopic[474] molecular weight determinations in benzene indicated all these normal alkoxides to be trimeric including the methoxide, whereas the isopropoxide and *tert*-butoxide are dimeric in boiling benzene.[283]

A comparision of the properties of all the alkoxides of trivalent iron with those of aluminium should be interesting as these metals have almost the same covalent radii (1.22 and 1.26 Å, respectively).

It is evident from Table 2.14 that the normal alkoxides of aluminium have generally high boiling points and higher molecular complexities than the iron(III) analogues. However, the boiling points are not all quoted at the same pressure and are therefore not strictly comparable. These differences in molecular complexity are doubtless due to structural differences between aluminium and iron(III) alkoxides and this may be a reflection of stronger intermolecular bonding in the aluminium derivatives.

Alkoxides of divalent Mn, Fe, Co, Ni, and Cu are insoluble and nonvolatile,[7] due to extensive molecular association and hence are not included here.

3.2.11 Alkoxides of Group 13 Elements

The lower alkoxides of boron are highly volatile compounds[475] which can be distilled at atmospheric pressure; these derivatives are essentially monomeric in nature.[238]

In comparing the volatilities of primary, secondary and tertiary alkoxides of boron, it is noteworthy that the primary and secondary butoxides distil unchanged at 128 and 195°C, respectively[292,476] whereas the corresponding *tert*-butoxide undergoes decomposition to produce olefin.[477]

Table 2.14 Boiling points and molecular complexities of iron and aluminium trialkoxides, $M(OR)_3$

R	B.p. (°C/mm)		Molecular complexity	
	Fe	Al	Fe	Al
Me	–	$240/10^{-5}$	2.9	–
Et	155/0.1	162/1.3	2.9	4.1
Pr^n	162/0.1	205/1.0	3.0	4.0
Bu^n	171/0.1	242/0.7	2.9	3.9
Bu^i	173/0.1	–	3.0	–
Bu^s	159/0.1	172/0.5	1.7	2.4
Bu^t	136/0.1	–	1.5	1.95
Bu^nCH_2	178/0.1	255/1.0	3.0	4.0
Bu^iCH_2	200/0.1 (dec)	195/0.1	3.0	4.0
Bu^sCH_2	178/0.1	200/0.6	3.0	4.1
Bu^tCH_2	159/0.1	180/0.8	2.0	2.1
Pr^iMeCH	162/0.1	162/0.6	1.9	2.0

dec = decomposes.

In spite of a considerable amount of work, the mechanism of the thermal decomposition of boron alkoxides is not yet fully understood; it probably occurs *via* a carbonium ion species as the thermal decomposition of boron tris(tetrahydrofurfuroxide) yields 2,3-dihydropyran which is also obtained by vapour-phase dehydration of tetrahydrofurfuryl alcohol.[478] Similarly, decomposition of free *tert*-butyl methylcarbinol has been found to yield the same product as that of its corresponding boron alkoxide derivative;[478] this led Dupuy and King[479] to assume that the thermal decomposition of boron alkoxides might proceed *via* an acid catalysed mechanism rather than *via* a pyrolytic *cis*-elimination mechanism.

The thermochemistry of boron alkoxides was studied by Charnley *et al.*,[480] which led to an estimate of the average B–O bond dissociation energies for boron alkoxides in the range of $110 \pm 5 \, kcal \, mol^{-1}$; these results are consistent with the order: B–F > B–N > B–O > B–Cl, reported earlier by Sidgwick.[481] In another study [482] the reported B–O bond dissociation energy $\bar{D}(B\text{–}OR)$ of $B(OMe)_3$, $B(OEt)_3$, $B(OPr^n)_3$, were 118.0, 117.7, $119.0 \pm 2 \, kcal \, mol^{-1}$, respectively.

Fenwick and Wilson[482] have measured the vapour pressures of boron triphenoxide and substituted triphenoxides manometrically and found that the results were in good agreement with the equation, $\log p$ mm $= a - b/T$ (where p is the pressure in mm at which boiling point T was observed and a and b are constants) over a wide range of temperatures.

Aluminium alkoxides are thermally stable and even the insoluble $[Al(OMe)_3]_n$ may be sublimed at 240°C under high vacuum.[275] The higher alkoxides are all soluble, distillable products and the melting points of the solids increase with increasing complexity of the alkyl chain.[293,294]

In view of the large variations reported in the boiling points and molecular weights of aluminium alkoxides by various authors,[71,484–487] Mehrotra[293] in 1953 measured the boiling points of a number of aluminium alkoxides at different pressures (Table 2.15) (in the range 0.1–10.0 mm pressures) and observed that the boiling points measured in the pressure range 2.0 to 10.0 mm follow the equation, $\log p = a - b/T$. The constants a and b, the latent heats of vaporization, ΔHv, and entropies of vaporization ΔS_5 at 5°C are reported in Table 2.16 for a pressure of 5.0 mm, along with the boiling points and molecular complexities (n) of various alkoxides. Measurements of the same compounds were carried out by Wilhoit[488] isotenoscopically in 1957 and these data are also included in the table.

The data in Table 2.16 show that the latent heats of vaporization, ΔH, and entropies of vaporization, ΔS, decrease regularly with decreasing polymerization of aluminium alkoxides. The entropies of vaporization reported for aluminium alkoxides are slightly higher than those observed for monomeric *tert*-butoxides of titanium and zirconium

Table 2.15   Boiling points of aluminium alkoxides at different pressures

$Al(OR)_3$	B.p. (°C/mm)							
$Al(OEt)_3$	162/1.3	169/1.5	171/2.2	181/4.2	184/5.0	188/6.0	190/6.8	197/10.0
$Al(OPr^i)_3$	106/1.5	111/2.3	118/3.8	122/4.6	124/5.2	131/7.5	132/8.3	135/10.0
$Al(OPr^n)_3$	205/1.0	211/1.9	215/2.0	228/4.0	233/6.9	238/6.9	239/7.5	245/10.0
$Al(OBu^n)_3$	242/0.7	258/2.7	262/3.6	272/6.0	276/7.5	281/8.8	284/10.0	—
$Al(OBu^t)_3$	134/0.25	151/1.3	156/2.0	164/3.5	170/4.6	177/6.6	181/8.0	185/10.0

Table 2.16   Boiling points, latent heats and entropies of vaporization, and molecular complexities of aluminium alkoxides

$[Al(OR)_3]_n$	Physical state	B.p. ($T_{5.0}$)	a	b	ΔH kcal mol^{-1}	ΔS_5 cal deg^{-1} mol^{-1}	n
$Al(OEt)_3$	White solid	184.5 (189.0)	11.8	5100	23.9 (20.2)	52.2 (43.7)	4.1
$Al(OPr)_3$	Colourless liquid	232.5 (222.7)	12.6	6025	27.5 (22.3)	54.5 (45.0)	4.0
$Al(OBu)_3$	Colourless liquid	270.0 (259.6)	12.7	6540	29.9 (24.9)	55.0 (46.7)	3.9
$Al(OPr^i)_3$	White solid	124.0 (139.1)	11.4	4240	19.4 (21.10)	48.9 (51.2)	3.0
$Al(OBu^t)_3$	Colourless solid	172.0 (167.3)	11.3	4270	21.6 (19.5)	48.5 (44.3)	2.4

Values in parentheses are reported by Wilhoit[488] and others by Mehrotra.[293]

(39.5 and 42.0 cal deg^{-1} mol^{-1}, respectively),[274] but these are considerably lower than those of polymerized zirconium primary and secondary alkoxides (the value of entropy of vaporization for zirconium isopropoxide is reported as 66.1–72.1 cal deg^{-1} mol^{-1}).[113,137] In view of the comparatively lower values of entropies of vaporization of aluminium alkoxides, it appears that aluminium alkoxides might be associated in the vapour phase also. In fact Mehrotra[294] has shown that aluminium isopropoxide is dimeric in the vapour phase, whereas it is trimeric in solution when freshly distilled.[293] This special behaviour of aluminium alkoxides and the stability of alkoxo bridges of the type $Al(\mu\text{-}OR)_2Al$, even in the vapour phase, is easily understandable on the basis of the electron-deficient nature of tricovalently bonded aluminium atoms.

A number of investigators[484–487] have reported rather widely varying values of the degree of association of aluminium alkoxides on the basis of molecular weight determinations. These were, therefore, carefully reinvestigated by Mehrotra,[293] who found that freshly distilled aluminium isopropoxide is trimeric in boiling benzene and it changes to a tetrameric form on ageing. With increasing branching of alkoxo groups, the complexity diminished and it was reduced to the dimeric state in the *tert*-butoxide.[293] These results were later confirmed by Shiner *et al.*[489] and Oliver *et al.*[490–492] who made a careful NMR study of these aluminium alkoxides.

The variation in melting points of aluminium alkoxides was ascribed by Wilhoit[493] to various factors such as (i) sporadic hydrolysis, and (ii) the existence of allotropic forms (α and β).

Wilson[494] measured the enthalpy of formation of solid aluminium isopropoxide which was found to be $\Delta H_f^\circ[Al(OPr^i)_3]_4(s) = 1230.8$ kcal mol^{-1}. On the basis of this value, the enthalpy of formation of dimeric gaseous isopropoxide was calculated to be $\Delta H_f^\circ[Al(OPr^i)_3]_2(g) = 584.8$ kcal mol^{-1}. These two factors led to the minimum value of Al–O bridge bond: $\Delta H_{net} = +31.4$ kcal mol^{-1}.

Gallium alkoxides are also volatile compounds. For example, even the lowest member of the series (the insoluble trimethoxide) has been sublimed at 275–280°C under 0.5 mm pressure.[184] The ethoxide and higher normal alkoxides are also volatile and ebullioscopic molecular weight determinations in benzene indicated that these normal alkoxides were tetrameric in character.[184] Mehrotra and co-workers[184] observed gallium isopropoxide and *tert*-butoxide to be dimeric in boiling benzene. The dimeric behaviour of gallium isopropoxide was confirmed in a later study by Oliver and Worrall[495] who found it to be dimeric cryoscopically in dilute benzene solution. However, the cryoscopic molecular weight determination of a freshly prepared concentrated benzene solution of the isopropoxide indicated it to be tetrameric. Evidence in favour of the latter observations was obtained from ^{1}H NMR spectra, which also showed the presence of tetrameric species in a fresh solution which dissociated to the dimeric form on being aged.

3.2.12   Alkoxides of Group 14 Elements

Silicon and germanium alkoxides are highly volatile and most of the derivatives of silicon[316] have been distilled unchanged at atmospheric pressure; for example, silicon tetracthoxide,[316,496] and tetraphenoxide[316,497] distilled at 168°C and 417–420°C, respectively. Similarly, germanium ethoxide[315] and phenoxide[293] could also be volatilized at 85–86°C/12.0 mm and 220–225°C/0.3–0.4 mm, respectively. The low boiling points of these alkoxide derivatives of silicon and germanium are due to their monomeric nature, irrespective of the chain length and branching of alkoxo groups.

Bradley *et al*.[113,273,274,432] measured the boiling points of a number of derivatives under different pressures and observed that the results follow the equation, $\log p = a - b/T$ in the pressure range 2.0–10.0 mm. Although silicon alkoxides are monomeric, the latent heat of vaporization, entropy of vaporization, and Trouton constants increase with the length of the alkyl groups. This unexpected behaviour has been ascribed to the molecular entanglement in the liquid state or to the restricted rotation of the alkoxo group with increasing chain length. It is worth mentioning here that a molecule in the liquid state will require special orientation in order to be disentangled before vaporization, and this may cause an increase in the entropy of vaporization.

The thermal decomposition of silicon tetraethoxide yields silicon oxides at higher temparatures.[475,498] By contrast, tin alkoxides appear to be thermally less stable than the corresponding silicon and germanium analogues.[311,499–501] For example, all normal alkoxides of tin undergo decomposition on being heated under reduced pressure.[311] The sterically hindered isopropoxide, *tert*-butoxide, and *tert*-amyloxide, however, were distilled at 131°C/1.6, 99°C/4.0, and 102°C/0.2 mm, respectively.[311] The ebullioscopic molecular weight measurements in boiling benzene indicate the following order of complexities: methoxide (an insoluble derivative) appears to be highly associated; the normal alkoxides are tetrameric, whereas isopropoxide and *tert*-alkoxides are trimeric and monomeric, respectively.[311]

The alkyl substituents at the tin atom, in view of its larger size have been found to affect the molecular complexities as well as volatility of the derivatives considerably. For example, Gaur *et al*.[210,211] observed that except for highly associated nonvolatile methoxides, all secondary and tertiary trialkoxides of monoalkyltin trialkoxides are highly volatile liquids which could be distilled in good yields under reduced pressure. The complexities of ethoxides, isopropoxides, and *tert*-butoxides of monoalkyltin fall in the order 3.0, 1.5, and 1.0, respectively in boiling benzene. These molecular complexities are in agreement with the ^{119}Sn NMR spectra measured for a number of monoalkyltin trialkoxides.[502] Similarly, dialkyltin dialkoxides and trialkyltin alkoxides also show high volatility but the molecular complexity is comparatively less affected in these derivatives as the branching of alkoxo groups increases.[503] Thus along with dialkyltin disopropoxides and di-*tert*-alkoxides, the normal alkoxides are also monomeric in dilute solution,[262,311,504] but the latter derivatives show some association in concentrated solution.[262,311] Cryoscopic data show that dibutyltin dimethoxide is dimeric in character.[504]

Apart from the smaller sizes of silicon and germanium atoms, the monomeric nature of silicon and germanium alkoxides in contrast to analogous tin derivatives may be ascribed to stronger $p_\pi - d_\pi$ bonding in the former which diminishes in germanium and appears to be insignificant in case of tin alkoxides. It may be assumed that the covalent bonding in the case of silicon would be stabilized more by $p_\pi - d_\pi$ bonding rather than by bridging (σ-bonding) and the shortened bond length in silicon methoxide has been observed experimentally.[505]

3.2.13   Alkoxides of Group 15 Elements

The alkoxides of arsenic and antimony are highly volatile, generally distil easily at low temperatures, and exhibit monomeric behaviour.[244,506,507]

The first homoleptic bismuth trialkoxides, $Bi(OR)_3$ (R = Me, Et, Pr^i), prepared by Mehrotra and Rai[202] in 1966, showed very limited solubility (in benzene or toluene) and low volatility probably became of the formation of polymeric networks involving alkoxo bridges. By contrast, $Bi(OBu^t)_3$ is highly soluble (even in hexane)[203–205] and exhibits higher volatility (sublimed at 80–1000°C/0.01 mmHg)[204,205] and monomeric nature. The derivative $Bi(OCH_2CH_2OMe)_3$ is dimeric in solution although in the solid state a polymeric structure has been established by X-ray crystallography.[203–205] Interestingly, the derivative $Bi(OCMe_2CH_2OMe)_3$ is volatile and appears to be monomeric.[340,508]

3.2.14   Alkoxides of Group 16 Elements (Se, Te)

Selenium and tellurium alkoxides are also highly volatile and tend to be monomeric.[208,509,510] For example, ethoxo-, *n*-butoxo-, and isopropoxo-derivatives of selenium have been volatilized at 76°C/10.0, 110°C/7.0, and 88°C/4.5 mm pressures, respectively.[510] Similarly tellurium tetramethoxide, ethoxide, *n*-butoxide, and isopropoxide[208] distilled at 115°C/9.0, 107°C/5.0, 150°C/0.8, and 76°C/0.5 mm pressure, respectively.

3.3   Infrared Spectra

Infrared spectroscopy has been utilized to verify the identity of metal alkoxides by observing bands (M−O and C−O stretching vibrations) characteristic of the bonded alkoxide group. Owing to the complex structure and often low molecular symmetry of metal alkoxides, the assignment of various types of M−O bands has proved to be rather difficult and the infrared technique has not generally been definitive in structural assignments.

Barraclough *et al.*[511] studied the infrared spectra of a number of metal (aluminium, titanium, zirconium, hafnium, niobium, and tantalum) alkoxides and tentatively assigned the $v(C−O)M$ and $v(M−O)$ bands in these derivatives. It has been observed that ν(C–O)M bands generally appear in the region 900–1150 cm^{-1} and the position of bands is determined by the nature of the alkoxo groups. For example, the ethoxide derivatives show two bands around 1025 and 1070 cm^{-1} which have been assigned to ν_s(C–O)M and ν_{as}(C–O)M stretching vibrations, respectively. However, C–O and C–C coupling complicates the problem. On the other hand isopropoxide derivatives show bands at 1170, 1150, and 950 cm^{-1}. The isopropoxide derivatives also exhibit a strong doublet at about 1375 and 1365 cm^{-1} due to *gem* dimethyl groups. Bell *et al.*[512] had also observed isopropoxide bands at about 1170, 1135 and 1120 cm^{-1} in boron and aluminium isopropoxides.

Guertin *et al.*[513] had earlier assigned the band above 1000 cm^{-1} in aluminium alkoxides to the ν(Al–O) stretching vibration. This observation was later corrected by Barraclough *et al.*[511] who suggested that the band above 1000 cm^{-1} was due to ν(C–O) and a set of five bands observed in the range 539–699 cm^{-1} were due to ν(Al–O) stretching. Wilhoit *et al.*[493] also measured the infrared spectra of pure aluminium ethoxide as well as the hydrolysed product and concluded that on hydrolysis the intensities of infrared bands diminished to a considerable extent except the strong O–H and Al–O–Al bands observed at 3340 and 935 cm^{-1}, respectively. Maijs *et al.*[514] measured the refraction of Al–O bonds of bridging as well as terminal alkoxo groups in aluminium alkoxides and the values derived were 0.5 and 2.5, respectively. The presence of five bands due to Al–O bonds was ascribed to the presence of terminal and bridging groups in these polymeric alkoxides.

Thus the ν(C–O) bands in the metal alkoxides (methoxides, ethoxides, isopropoxides, *n*-butoxides and *tert*-butoxides) appear at 1070, 1025; 1070, 1025; 1375, 1365, 1170, 1150, 980, 950; 1090, 1025; and 940 cm^{-1}, respectively.

The infrared and Raman spectra of group 14 metal alkoxides have been studied in detail by several workers and the νM(C–O) and ν(M–O) bands have been firmly assigned. For example, ν_{as}(Si–O) and ν_s(Si–O) bands in silicon tetra-alkoxides appear in the ranges 720–880 and 640–780 cm^{-1}, respectively.[515–518] The alkyl group appears to influence the ν(Si–O) band position in these silicon alkoxides; thus the bands shift to higher values on increasing the size of the alkoxo groups.[516,518] Newton and Rochow[519] have investigated the infrared spectra of a variety of trialkoxosilanes $(RO)_3SiH$ and the bands observed in the range 840–880 cm^{-1} have been assigned to ν_{as}(Si–O) stretching modes and the very weak band observed due to ν_s(Si–O) in the infrared spectrum gave a highly polarized band in Raman spectra at 700 cm^{-1}. The infrared and Raman spectra of some alkylsilicon alkoxides, phenoxides, and chloride-alkoxides have also been measured and the change in position of ν(Si–O) stretching modes caused by substituents have been interpreted to be due to a change in polarity of σ-bonds.[520–530]

Infrared spectra of germanium tetra-alkoxides showed bands at 1040 and 680 cm^{-1} due to GeO_4 configuration.[531] Studies[532,533] on the spectra of alkyl- and aryl-germanium alkoxides indicate the presence of ν(Ge–O) bands in the range 840–500 cm^{-1}.

The spectra of a variety of alkyltin alkoxides have been measured by several workers[210,265,534–540] and the band around 500 cm^{-1} in these derivatives was tentatively assigned to ν(Sn–O) stretching modes. The bands observed at about 700 and 670 cm^{-1} have been assigned to (Sn–CH$_2$) rocking vibrations arising from *gauche* and *trans* conformations, respectively. The bands due to ν_s and ν_{as}(Sn–C) stretching vibrations, respectively, are at about 500 and 600 cm^{-1}. The strong bands observed in the range $1030-1070\,cm^{-1}$ due to ν(C–O) Sn in primary alkoxides shifted to a lower value $(940-980\,cm^{-1})$ in the isopropoxides and *tert*-butoxides.

Infrared spectral studies on homoleptic methoxides and halide methoxides of the first row divalent transition metals such as chromium, manganese, iron, cobalt, nickel, copper, and zinc have been made by Winter and co-workers.[320,541–543] These workers have also assigned the ν(C–O)M band in the region $1000-1100\,cm^{-1}$ and the ν(M–O) stretching frequencies fall below $600\,cm^{-1}$ in these methoxides. For example the ν(Ni–O) bands in nickel dimethoxide and halide-methoxides appear at 375, $420\,cm^{-1}$, and $360-390$, $420\,cm^{-1}$, respectively[541,543] whereas in copper methoxides, ν(Cu–O) bands appear at $520-545$ and $410-450\,cm^{-1}$.[542]

The earlier study on the infrared spectra of the transition metal alkoxides and the tentative assignments made by Barraclough *et al.*[511] were later confirmed by Lynch *et al.*[544] Table 2.17 indicates the positions of ν(C–O)M and ν(M–O) stretching modes observed in these transition metal alkoxides.

Following the earlier work of Barraclough *et al.*, Bradley and Westlake[545] made a more detailed study of the infrared spectra of the polymeric metal ethoxides. In particular they attempted a definitive assignment of bands to the C–O and M–O vibrations of terminal and bridging ethoxide groups. Also, by measuring relative band intensities of terminal and bridging groups, they attempted to assign structures to the polymeric alkoxides. In assigning bands they used various criteria: thus by comparing the spectra of $[M(OEt)_x]_n$ with those of EtOH (no M–O vibrations and no bridging OEt groups) and $U(OEt)_6$ (an essentially monomeric compound containing no bands due to bridging C–O and M–O groups), it was possible to identify the bands due to M–O and C–O vibrations in terminal and bridging groups as listed in Table 2.18.

Owing to the structural complexity of these molecules and the possibility of coupling of vibrations, the simple criterion of comparison mentioned above was not considered an adequate basis for the assignment, and additional criteria were developed. For example, it was shown that the addition of pyridine as a donor ligand caused the disappearance of some bands but not others. It was reasonable to conclude that only the bands due to bridging ethoxide groups would disappear, due to the replacement of

Table 2.17 Tentative assignments of ν(M–O) stretching frequencies in transition metal alkoxides

Metal alkoxide	ν(C–O)M (cm^{-1})	ν(M–O) (cm^{-1})
Ti(OEt)$_4$	1064, 1042	625, 500
Ti(OPr^i)$_4$	1005, 950	619
Ti(OAm^t)$_4$	1011	615, 576
Zr(OPr^i)$_4$	1011, 958, 945	559, 548
Zr(OBu^t)$_4$	997	540
Zr(OAm^t)$_4$	1010	586, 559, 521
Hf(OPr^i)$_4$	1020, 983	–
Hf(OBu^t)$_4$	990	567, 526
Th(OPr^i)$_4$	996, 973	–
Nb(OEt)$_5$	1063, 1029	571
Ta(OEt)$_5$	1072, 1030	556
Ta(OPr^i)$_5$	1001	557, 540

Table 2.18   **Assignment of M–O and C–O stretching frequencies in metal ethoxides, $[M(OEt)_x]_n$**

M	H	U(VI)	Al(III)[1]	Ti(IV)	Zr(IV)	Hf(IV)[1]	Nb(V)	Ta(V)	U(V)
	–	–	1176 (s)	–	1160 (s)	1168 (vs)	–	–	–
	–	–	–	–	1145 (s)	1145 (s)	1143 (sh)	1148 (sh)	–
Terminal C–O	–	–	–	1138 (s)	1134 (s)	–	–	–	1126 (w)
vibrations									
	1089 (m)	1089 (vs)	1098 (m)	1103 (s)	1095 (m)	1094 (m)	11110 (s)	1120 (s)	1097 (s)
	1049 (s)	1044 (vs)	1079 (m)	1064 (s)	1073 (m)	1075 (m)	1066 (s)	1070 (s)	1056 (s)
	880 (m)	899 (s)	930 (m)	913 (m)	918 (m)	920 (m)	914 (m)	917 (m)	905 (m)
Bridging C–O	–	–	1053 (vs)	1040 (m)	1050 (m)	1039 (s)	1030 (m)	1028 (m)	1024 (m)
vibrations	–	–	895 (s)	882 (m)	894 (m)	878 (m)	880 (w)	878 (w)	874 (m)
Terminal M–O	–	513 (m)	703 (m)	623 (m)	520 (m)[1]	500 (m)[1]	575 (m)	554 (m)	502 (m)
vibrations		392 (m)	512 (m)	520 (m)	412 (m)[1]	425 (m)	–	–	–
Bridging M–O	–	–	646 (m)	577 (m)	463 (m)[1]	470 (m)[1]	485 (m)	499 (m)	390 (m)
vibrations									

[1]Nujol mull; all other spectra in CS_2 solution.

bridging groups by coordinated pyridine:

$$M_2(OEt)_{10} + 2C_5H_5N \longrightarrow 2[C_5H_5N \longrightarrow M(OEt)_5] \qquad (2.179)$$

For example the bands at 1030, 880, and 485 cm^{-1} in $Nb_2(OEt)_{10}$ and at 1028, 878, and 499 cm^{-1} in $Ta_2(OEt)_{10}$ were all eliminated by the addition of pyridine, thus confirming the assignments in Table 2.18. Interestingly the terminal M–O bands at 575 and 554 cm^{-1} in $Nb_2(OEt)_{10}$ and $Ta_2(OEt)_{10}$, respectively, were not diminished in intensity but they were shifted to lower frequencies and to 548 cm^{-1}, respectively, on coordination with pyridine.

In a further independent check on the assignments, a sample of ^{18}O-labelled (72%) $Ta_2(OEt)_{10}$ was synthesized from $Et^{18}OH$ as follows:

$$2Ta(NMe_2)_5 + 10Et^{18}OH \longrightarrow Ta_2(^{18}OEt)_{10} + 10Me_2NH \qquad (2.180)$$

A comparison of the spectra of $Ta_2(^{16}OEt)_{10}$ and $Ta_2(^{18}OEt)_{10}$ should reveal significant shifts in the frequencies of C–O and M–O vibrations but relatively smaller changes for other bands. The data are presented in Table 2.19.

Using a simplified "diatomic molecule" approach for the C–O and M–O vibrations, values were calculated for the frequencies of C–^{18}O and M–^{18}O vibrations and in

Table 2.19   **Oxygen isotope effect on the spectrum of $Ta_2(OEt)_{10}$**

	$Ta_2(^{16}OEt)_{10}$	$Ta(^{18}OEt)_{10}$ (observed)	$Ta_2(^{18}OEt)_{10}$ (calculated)
νC–O (cm^{-1})	1148	~1118	1120
	1120	1098	1092
	1070	1061	1044
	1028	1018	1003
	917	906	895
	878	863	857
νM–O (cm^{-1})	554	542	525 (544)
	499	486	473 (490)

several cases the agreement between observed and calculated was very good. For the M–O vibrations, better agreement (figures in parentheses) was obtained by treating the OEt group as a mass-45 unit attached to tantalum.

Although attempts were also made by Bradley and Westlake[545] to elucidate the ratio of terminal:bridging ethoxide groups by intensity measurements on terminal ($1079-1056\,cm^{-1}$) and bridging ($1053-1024\,cm^{-1}$) C–O bands, yet the findings except in a few cases[546] were not in conformity with the structures of the corresponding metal alkoxides.

Infrared spectra of lanthanide isopropoxides and *tert*-butoxides have been measured by Mehrotra *et al*.[103,157] as well as by Mazdiyasni *et al*.[54,55] In these derivatives also, isopropoxo bands appear in ranges $1175-1160$, $1140-1120$ and at $1380\,cm^{-1}$ due to *gem*-dimethyl structure and a set of bands due to *tert*-butoxo groups arise in the range $1253-945\,cm^{-1}$. It was pointed out by Brown and Mazdiyasni[55] that all the lanthanide isopropoxides exhibit a set of five bands in the region $520-370\,cm^{-1}$. The positions of metal–oxygen bands may vary from 900 to $200\,cm^{-1}$, depending on the mass of the metal and the double bond character in the bond. The νCr–O vibration was assigned by Brown *et al*.[547] at $500-505\,cm^{-1}$ and the δ OCrO vibration at $253-250\,cm^{-1}$ in $Cr(OMe)_3$.

3.4 Nuclear Magnetic Resonance Spectra

3.4.1 Introduction

As NMR spectroscopy has become a well-established technique for elucidation of structural features of organic compounds, it has also been employed quite successfully for throwing light on the structural features of metal alkoxides. All the early NMR work involved proton NMR, which continues to be one of the most useful techniques for the investigation of metal alkoxides. However, metal alkoxide complexes have also been studied by ^{13}C and metal[5,6,18–21] NMR.

Table 2.20 presents some data on the NMR spectra of diamagnetic homo- and heteroleptic metal alkoxides. In comparing chemical shifts it should be kept in mind that the δ values of some metal alkoxides are subject to specific solvent effects.[548] As expected the δ values of the protons on α-carbon atoms are shifted considerably downfield relative to protons on β-carbons. Since the $^1H/^{13}C$ chemical shift is the resultant of several contributing factors, there are no obvious correlations with metal oxidation state, atomic radius, or co-ordination number. However, metal NMR studies have proved to be of considerable importance in indicating the coordination environment of the central metal atom.

Although NMR studies sometimes provide valuable information on the terminal and bridging (μ_2/μ_3) alkoxo groups,[5,6,18–21] yet in a large number of cases it is difficult to derive meaningful conclusions regarding the structures of metal alkoxides owing to various factors such as: (a) intermolecular bridge/terminal or terminal site exchange; (b) proton-catalysed site exchange with the free alcohol molecules present in the system from adventitious hydrolysis; (c) the overlapping of the resonances; (d) the broadening of the signals due to quadrupolar effects, and (e) the oligomeric nature of a number of metal alkoxides. In spite of these limitations, a useful correlation between the (^1H, ^{13}C) NMR chemical shifts and the ligating mode (terminal, μ_2 and μ_3) can be made

Table 2.20   Some data on the ^{1}H NMR spectra of a few selected diamagnetic metal alkoxides

Compound	Solvent	$\delta^1(\beta\text{-CH})$	$\delta^1(\alpha\text{-CH})$	Reference
$[\text{LiOBu}^t]_n$	C_7D_8	1.26 (s)		28
$[\text{LiOCMe}_2\text{Ph}]_6$	C_6D_6	1.34 (s); 7.02 (m, Ph)		28
$[\text{KOBu}^t\text{.Bu}^t\text{OH}]_\infty$	C_7D_8	2.14 (s)		29
$[\text{KOBu}^t]_4$	C_7D_8	1.08 (s)		29
$[\text{RbOBu}^t\text{.Bu}^t\text{OH}]_\infty$	C_7D_8	1.16 (s)		29
$[\text{RbOBu}^t]_4$	C_7D_8	1.00 (s)		29
$[\text{Be(OBu}^t)_2]_3$	Benzene	1.44 (s)		385
	C_6D_{12}	1.38 (μ_2); 1.28 (s)		385
$[\text{Ba(OBu}^t)_2(\text{HOBu}^t)_2]_4$	C_7D_8	1.18 (s)		549
$[\text{Ba(OEt}_3)_2]_n$	C_6D_6	0.81 (t)	1.37 (q)	47
$[\text{Y}_3(\text{OBu}^t)_8\text{Cl(thf)}_2]$	C_7D_8	1.88 (μ_3); 1.50, 1.43 (μ_2); 1.31, 1.24, 122 (s)		160
$[\text{Y}_3(\text{OBu}^t)_7\text{Cl}_2(\text{thf})_2]$	THF-d_8	1.90 (μ_3); 1.51, 1.45 (μ_2); 1.32, 1.26 (s)		161
$[\text{Y(OCEt}_3)_3]$		0.97 (t)	1.54 (q)	345
$[\text{Y}_5\text{O(OPr}^i)_{13}]$	$CDCl_3$	1.50 (μ_3); 1.29 (μ_2); 1.05	4.30 (septet); 4.10 (m)	59
$[\text{La}_3(\text{OBu}^t)_9(\text{HOBu}^t)_2]$	C_7D_8	1.48		345
$[\text{Ce(OCBu}^t_3)_2(\text{OBu}^t)_2]$	C_6D_6	1.38 (s); 1.45 (s, γ-CH)		383
$[\text{Ce(OPr}^i)_4(\text{HOPr}^i)]_2$	C_7D_8	1.41 (d)	4.91 (septet)	460
$[\text{Ce(OBu}^t)_4(\text{thf})_2]$	C_6D_6	1.58 (s)		164
$[\text{Ti(OMe)}_4]_4$	C_6H_6	–	3.90; 3.77; 3.72; 3.46	560
$[\text{Ti(OEt)}_4]_4$	Decalin	1.26 (t)	4.42 (q)	548
$\text{Ti(OPr}^i)_4$	Cyclohexane	1.21 (d)	4.45 (septet)	548
$[\text{Zr(OPr}^i)_4]_n$	Cyclohexane	1.30 (d)	4.50 (septet)	548
$[\text{Hf(OPr}^i)_4]_n$	Cyclohexane	1.31 (d)	4.61 (septet)	548
$[\text{Zr(OBu}^t)_4]$	Cyclohexane	1.24 (s)	–	548
$[\text{Hf(OBu}^t)_4]$	Cyclohexane	1.24 (s)	–	548
$[\text{TiCl}_3(\text{OPr}^i)(\text{HOPr}^i)]_2$	$CDCl_3$	1.59 (d)	5.15 (septet)	108
$[\text{TiCl}_2(\text{OPr}^i)_2(\text{HOPr}^i)]_2$	$CDCl_3$	1.45 (d)	5.00 (br)	108
$[\text{TiCl}_2(\text{OC}_2\text{H}_4\text{Cl})_2] (\text{HOC}_2\text{H}_4\text{Cl})_2$	CD_2Cl_2	3.82 (m) (CH$_2$Cl)	5.09 (m); 4.84 (m); 4.39 (m)	108a
$\text{VO(OPr}^i)_3$	CCl_4	1.43 (d)	5.30 (septet)	235
$[\text{Nb(OMe)}_5]_2$	n-Octane	–	3.04; 4.15; 4.23 (s)	562
$[\text{Nb(OEt)}_5]_2$	n-Octane	1.22 (t)	4.34; 4.45 (q)	562
$[\text{Ta(OMe)}_5]_2$	n-Octane	–	4.03; 4.21; 4.32 (s)	562
$[\text{Ta(OEt)}_5]_2$	n-Octane	1.21 (t)	4.36; 4.46; 4.53 (q)	562
$[\text{Nb(OBu}^t)_5]$	n-Octane	1.37 (s)	–	562
$[\text{Ta(OBu}^t)_5]$	n-Octane	1.36 (s)	–	562
$[\text{CrO}_2(\text{OCH}_2\text{CF}_3)_2]$	Acetonitrile	–	3.95 (s)	233
$[\text{Mo}_2(\text{OPr}^i)_6]$	C_7D_8	1.37 (d)	5.48 (septet)	361
$[\text{Mo}_2(\text{OBu}^t)_6]$	Toluene	1.53	–	5
	C_7D_8	1.48^2	–	360
$\text{Mo}_2(\text{OPr}^i)_4(\text{HOPr}^i)_4$	C_7D_8	1.38 (d)	4.51 (septet)	375
$\text{Mo}_2(\text{OCMe}_2\text{Et})_6$	C_6D_6	1.09 (t)	1.81 (q); 1.48 (s)	170
$\text{Mo}_2(\text{OCMe}_2\text{CF}_3)_6$	C_6D_6	1.37 (s)	–	170
$\text{Mo}_2\text{Cl}_4(\text{OPr}^i)_6$	C_7D_8	0.94 (d); 1.73 (d) (μ_2)	5.58 (septet); 7.21 (septet) (μ_2)	1025

Table 2.20   (*Continued*)

Compound	Solvent	$\delta^1(\beta\text{-CH})$	$\delta^1(\alpha\text{-CH})$	Reference
Mo$_4$Cl$_3$(OPr^i)$_9$	C$_7$D$_8$	1.97, 1.70, 1.58, 1.52, 1.44, 1.38, 1.15, 0.90, 0.89 (doublets)	6.09, 5.91, 5.79, 4.52, 4.45 (septets)	1053
W$_4$(OPr^i)$_{12}$	C$_6$D$_6$	1.2–1.6 (br, overlapping resonance)	6.1 (br), 5.0 (sharp), 4.7 (br), 4.2 (br)	*a*
[W(OMe)$_6$]	Toluene	–	4.51 (s)	5
[Al(OPr^i)$_3$]$_4$	CCl$_4$	1.28 (μ_2); 1.47 (μ_2) doublets; 1.12 (d)		489
[Al(OBu^t)$_3$]$_2$	CCl$_4$	1.23 (s), 1.5 (μ_2)		489
[Sn(OBu^t)$_3$(OAc)(Py)]	C$_5$D$_5$N	1.55 (s)	–	644
Se(OCHMe$_2$)$_4$	C$_7$D$_8$	1.42 (d)	5.41 (septet)	146a
Te(OMe)$_4$	C$_7$D$_8$	–	3.83 (s)	146a
Te(OCH$_2$CH$_3$)$_4$	C$_7$D$_8$	1.26 (t)	4.09 (q)	146a
Te(OCH$_2$Bu^t)$_4$	C$_7$D$_8$	–	3.72 (s)	146a
Te(OCHMe$_2$)$_4$	C$_7$D$_8$	1.22 (d)	4.62 (septet)	146a

[1]Chemical shifts relative to internal TMS ($\delta = 0$).
*a*M.H. Chisholm, D.L. Clark, K. Folting, J.C. Huffman, and M. Hampden-Smith, *J. Am. Chem. Soc.*, **109**, 7750 (1987).
[2]Relative to Me$_3$SiOSiMe$_3$ ($\delta = 0$).

based on the observation that the downfield chemical shifts relative to trimethyl silicone (TMS) follow the trend: μ_3–OR > μ_2–OR > terminal.

Apart from homometallic alkoxides involving only one type of metal atom, ^{1}H NMR spectra have also proved of great significance in elucidation of structures of heterometallic alkoxides (Section 3) involving more than one different metal atom. This aspect of investigations will be presented in Chapter 3, whilst in this section a brief account is given of the NMR (mainly ^{1}H and ^{13}C) spectra of homo- and heteroleptic metal alkoxides arranged group-wise.

3.4.2   Alkoxides of Group 1 Metals

The ^{1}H NMR spectrum of lithium *tert*-butoxide in carbon tetrachloride solution,[438] showed a singlet at δ 1.53 due to methyl protons of the *tert*-butoxo group, indicative of either equivalent OBu^t groups or fast exchange between different environments of the associated molecules. The corresponding peak in *tert*-butyl alcohol appears at δ 1.23 and this shifting of the Bu^t signal to lower field in LiOBu^t led Bains to assume a higher covalency of the Li–O bond in alkoxide than of the O–H bond in alcohol. Variable temperature ^{1}H NMR studies on *tert*-butoxides of heavier alkali metals (K, Rb, Cs) by Weiss *et al.*[443] have shown that the Bu^t group rotates freely above −20°C but below this temperature the rotation is restricted. Variable temperature ^{1}H and ^{13}C NMR studies[28] have shown that the derivatives [LiOBu^t]$_n$ and [LiOCMe$_2$Ph]$_6$ exhibit a single peak for the methyl groups at room temperature, which do not change on cooling to −53°C.

In benzene-d$_6$ and toluene-d$_8$ the complexes [MOBu^t.Bu^{t}OH]$_\infty$ (M = K, Rb, Cs) exhibit only a single ^{1}H NMR peak due to *t*-butoxo ligand environment[29] at room temperature. The ^{1}H NMR resonances gradually broadened upon cooling to $-53°$C, but the solution could not be cooled to sufficiently low temperature to observe coalescence. Furthermore, the alcohol adducts undergo a rapid alkoxide ligand exchange process on the ^{1}H NMR time scale at room temperature.

The corresponding cubane derivatives [MOBu^t]$_4$ (M = K, Rb, Cs) in benzene and toluene solution exhibit a single,[29] sharp peak for the *tert*-butoxo group environment. The broadening of the line could not be observed even cooling to $-53°$C.

The ^{7}Li NMR of [LiOBu^t]$_n$ in toluene-d$_8$ at 25°C shows only one resonance[29] at δ -1.73, consistent with the observed solid-state structure[28] in which all the lithium environments are equivalent. The behaviour of [LiOBu^t]$_n$ and [LiOCMe$_2$Ph]$_6$ in THF-d$_8$ was markedly different. For example, these derivatives exhibit two peaks in a ratio of 4:1 and 9:1 at $-80°$C, respectively, favouring an equilibrium in solution of the type shown below:

$$\frac{1}{n}[\text{LiOR}]_n \underset{\text{toluene}}{\overset{\text{THF}}{\rightleftharpoons}} \frac{1}{4}[\text{LiOR.thf}]_4 \tag{2.181}$$

Undoubtedly, the more sterically demanding CMe$_2$Ph groups favour the formation of a reduced amount of [LiOCMe$_2$Ph.THF]$_n$ in solution.

3.4.3 Alkoxides of Group 2 and 12 Metals

The ^{1}H NMR spectra of normal alkoxides of heavier alkaline earth metals (Ca, Sr, Ba), beryllium as well as magnesium, zinc, cadmium and mercury do not appear to have been studied, probably owing to their highly polymeric nature and insolubility in common organic solvents.

The ^{1}H NMR spectrum of beryllium di-*tert*-butoxide in benzene showed a singlet ($\delta = 1.44$), but in perdeuteriomethylcyclohexane, two lines were observed[385] at δ 1.40 and 1.25 in 2:1 ratio and the spectrum remained unchanged from 33–100°C. In carbon tetrachloride solution a similar spectrum (δ 1.38, 1.22; 2:1) was observed consistent with structure (2-IV), involving two three-coordinated berylliums at the ends and a tetrahedrally coordinated beryllium in the middle.

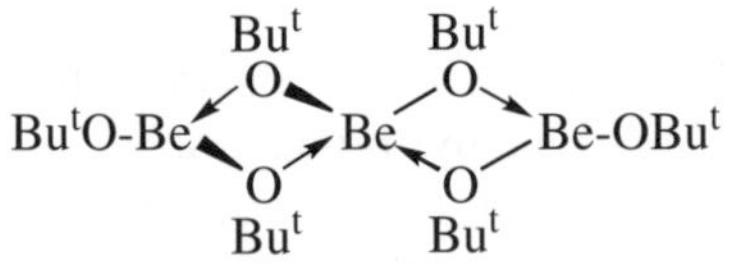

(2-IV)

The ^{1}H NMR spectrum of the dimeric more bulky triethylmethoxide derivative of beryllium Be$_2$(OCEt$_3$)$_4$ showed only one type of OCEt$_3$ groups indicating rapid exchange between terminal and bridging alkoxide groups in contrast to the behaviour of Be$_3$(OBu^t)$_6$ (2-IV).

The ^{1}H NMR spectra of only a few heavier alkaline earth metal alkoxide derivatives that are soluble in suitable organic solvents have been studied so far. The complex Ba$_2$(OCPh$_3$)$_4$(THF)$_3$ shows only a single line for aryl ring protons down to $-43°$C, and the ^{13}C NMR spectrum also shows only one aryl environment.[48]

In toluene-d_8 the soluble barium *tert*-butoxide complex $[Ba(OBu^t)_2(HOBu^t)_2]_4$ shows ^1H NMR signals[549] at δ 0.67 (OH), and 1.18 (Bu^t), whereas ^{13}C NMR peaks appear at δ 36.2 and 68.5.

The ^1H NMR spectra of some soluble alkylberyllium-, alkylmagnesium-, and alkylzinc-*tert*-butoxides have also been studied by Coates and co-workers.[390] The tetrameric methylberyllium *tert*-butoxide exhibits two peaks at δ 0.46 and 1.47 due to CH_3Be and OBu^t groups, respectively, in the intensity ratio 1:3. However, the spectrum in perdeuteriomethylcyclohexane (C_7D_{14}) at 33°C shows the splitting of CH_3Be (δ −0.94, −0.71) and OBu^t (δ 1.37, 1.48) protons, and the spectral pattern remained the same at 33 and 100°C.

Interestingly, ^1H NMR spectrum of the tetrameric methylberyllium methoxide $[CH_3BeOCH_3]_4$ in benzene-d_6[385] shows a single peak at δ 0.61 due to the CH_3Be group and two resonances at δ 3.50 and 3.24 due to methoxo group protons in the intensity ratio 1:1. It is noteworthy that a symmetrical structure (2-V) would give rise to two singlets only.

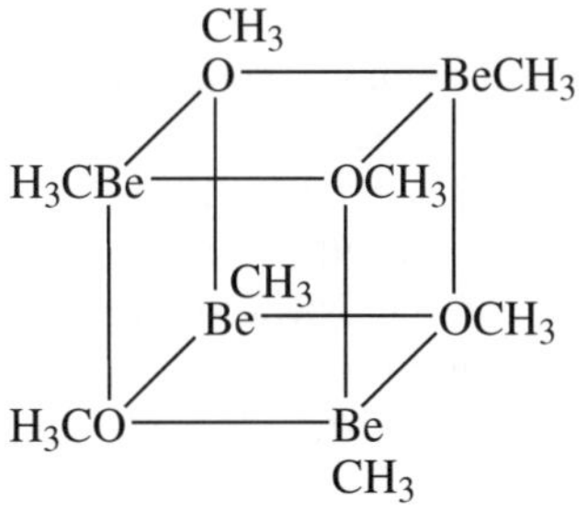

(2-V)

The ^1H NMR spectrum (in benzene-d_6) of tetrameric methylmagnesium *tert*-butoxide shows two broad peaks at δ 1.94 and 4.06 due to CH_3Mg and OBu^t groups, respectively.[390] On increasing the temperature to 100°C, the peak due to OBu^t protons splits into two unequal signals and CH_3Mg group protons appear as a very sharp, single line.

Although ^{25}Mg NMR spectroscopy has been used for characterization of organomagnesium compounds.[550] such studies appear not to have been carried out on magnesium alkoxides.

The ^1H NMR spectrum (in benzene-d_6) of dimeric *tert*-butylzinc *tert*-butoxide[551] exhibits two singlets of equal intensity at δ 1.28 (Bu^tZn) and 1.34 (OBu^t), consistent with the dimeric structure (2-VI):

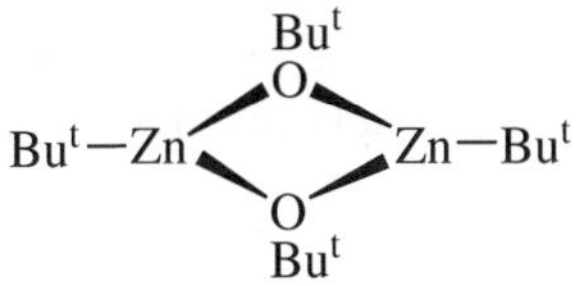

(2-VI)

3.4.4 Alkoxides of Yttrium, Lanthanides, and Actinides

3.4.4.1 Yttrium and lanthanum alkoxides

A number of yttrium and lanthanum alkoxides have been investigated by ^{1}H and ^{13}C spectral studies[18,21] by the research groups of Evans[160,161,164,346,552] and Bradley et al.[345,349-349c] in an attempt to provide preliminary evidence for the structures of these derivatives. Although the spectral patterns are generally more complex, the data are interpretable in terms of X-ray crystallographically established structures of typical compounds of each series. For example, the ^{1}H NMR spectrum (in C_6D_6) of $Y_3(\mu_3$-$OBu^t)(\mu_3$-$Cl)(\mu_2$-$OBu^t)_3(OBu^t)_4(THF)_2$ exhibits several broad overlapping resonances in the δ 1.1–2.1 region attributable to the t-butoxide ligands.[160] In THF-d_8, these peaks split into six sharp well-resolved signals with relative intensities of 1:2:1:2:1:1. The ^{13}C {^{1}H} NMR spectrum of the complex in THF-d_8 also showed six chemically nonequivalent OBu^t groups, consistent with the solid-state structure[160] of the derivative.

The ^{1}H NMR spectrum of $La_3(OBu^t)_9(thf)_2$ in aromatic solvents shows broad resonances characteristic of fluxional behaviour.[160] In THF-d_8, the resonance due to OBu^t groups sharpens. Using the structure–shift correlations developed for $Y_3(OBu^t)_8Cl(thf)_2$, the lanthanum complex contains two equivalent μ_3-OBu^t groups, three μ_2-OBu^t groups in a ratio of 1:2, and four terminal OBu^t groups. Even at 500 MHz, the terminal OBu^t peaks are not fully resolved, so the number of distinct environments is indeterminate. The ^{13}C NMR pattern is consistent with ^{1}H NMR data.

At room temperature, the ^{1}H and ^{13}C NMR studies on tris-$tert$-alkoxides of yttrium and lanthanum[345] reveal that the complexes are fluxional and require low temperatures to slow down the exchange between distinguishable alkoxo ligands. Even at around $-43°$C ^{1}H NMR spectra of the trinuclear complexes $[M_3(OBu^t)_9(HOBu^t)_2]$ (M = Y, La) exhibit a complex pattern, each having five major peaks with intensities in the ratio 2:2:2:1:4, appearing from lower to higher fields. For interpretation of such complex spectral data, the knowledge gained from the X-ray crystallographically characterized lanthanum complex[345] proved invaluable. The ^{1}H NMR spectral data of these and some alkoxide complexes of Y and La are listed in Table 2.20.

The ^{1}H and ^{13}C NMR spectra[346] of $[(Ph_3CO)_2La(\mu_2$-$OCPh_3)_2La(OCPh_3)_2]$ show two types of phenyl resonances, consistent with terminal and bridging groups in the dimer.

3.4.4.2 Thorium and uranium alkoxides

At room temperature the ^{1}H NMR spectrum of $[Th(OBu^t)_4(py)_2]$ reveals that the molecule is fluxional[406] and low-temperature ($-75°$C) spectra in toluene-d_8 failed to freeze out a limiting structure. In benzene-d_6 the room temperature ^{1}H NMR spectrum[406] of $[Th_2(OBu^t)_8(HOBu^t)]$ shows only one broad signal due to OBu^t groups at δ 1.54 and a broad peak at δ 3.22 for alcoholic OH protons in the intensity ratio 80:1, consistent with these assignments.

In an interesting ^{1}H NMR experiment the reaction of $Th_2(OBu^t)_8(HOBu^t)$ with pyridine has been studied[406] which supports the stepwise changes as illustrated in scheme 2.4:

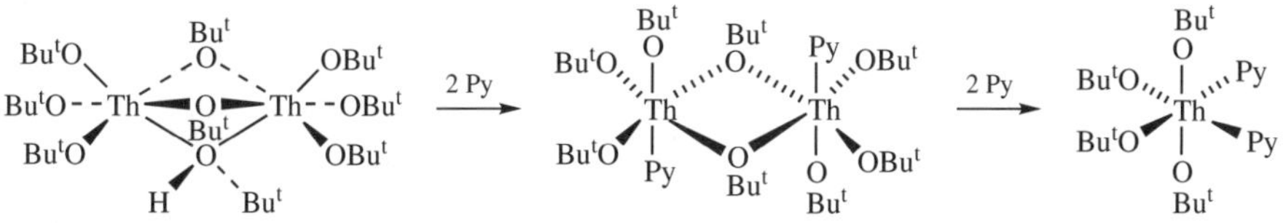

Scheme 2.4

It seems clear that octahedral coordination is the preferred coordination environment around thorium for the t-butoxo ligand.

^{1}H NMR spectral studies[407] of $Th_2(OCHPr^i_2)_8$ in noncoordinating solvents (C_6D_6 or C_7D_7) reveal the presence of two different alkoxide environments (broadened methine resonances at δ 2.0 and 3.72 and a smaller sharp set of resonances at δ 1.75 and 3.38; the two overlapping sets of diastereotopic methyl group resonances are seen in the region δ 1.0–1.2) consistent with dimer/monomer equilibrium in solution at room temperature (Eq. 2.181a):

$$Th_2(OCHPr^i_2)_8 \rightleftharpoons 2Th(OCHPr^i_2)_4 \qquad (2.181a)$$

A detailed ^{1}H NMR investigation of the above system has been used to determine both kinetic and thermodynamic parameters for the equilibrium. Thermodynamic parameters for the equilibrium process are $\Delta H^\circ = 17\,kcal\,mol^{-1}$, $\Delta G^\circ = 5\,kcal\,mol^{-1}$, $\Delta S^\circ = 40\,cal\,deg^{-1}\,mol^{-1}$.

Karraker *et al.*[553] recorded the ^{1}H NMR spectrum of dimeric uranium pentaethoxide in $CFCl_3$ and $CDCl_3$ at $-65°C$ and observed a large signal at δ 2.26 and a small one at δ 1.06 due to methyl protons and a large very broad signal at δ 20.2 and a smaller very broad signal at δ 16.1 due to the methylene protons. Below $-65°C$, the spectrum is very complex and on the basis of the quoted results it was assumed that above $-65°C$, the product is dimeric whereas below this temperature it is oligomeric. The ^{13}C NMR data are consistent with these structural characteristics.[554]

The ^{1}H NMR spectrum of uranium pentaisopropoxide at 13°C shows a doublet at δ 1.26, a much larger broad signal at δ 1.78 and a very broad signal, which is just detectable above the unavoidable noise, at δ -15. These results have been interpreted in terms of the dimer/monomer behaviour of the isoproxide. At $-65°C$, the high-field signal splits into two equal signals and the dimeric species predominates below this temperature.[553]

In benzene-d_6 the ^{1}H NMR spectrum[330] of $U_2(OBu^t)_8(HOBu^t)$ shows only one broad Bu^t signal at δ 1.5 and a broad signal at δ 13.0 due to the OH proton; the intensity ratio of these two signals is $\sim$80:1. Variable temperature ^{1}H NMR studies in toluene-d_8 in the range $+25$ to $-90°C$ have been incapable of "freezing out" a static structure.

Although ^{1}H NMR peaks in benzene-d_6 at 22°C of X-ray crystallographically characterized complex $U_2I_4(OPr^i)_4(HOPr^i)_2$ are quite broad and paramagnetically shifted[109] as is typical of U(IV) systems,[555,556] the above formulation is consistent with the observed ^{1}H NMR data: δ 32(br, CHMe$_2$), -5 (vv br, CHMe$_2$).

Variable temperature ^1H and $^{31}P\{^1H\}$ spectral studies[557] of $UO_2(OBu^t)_2(Ph_3PO)_2$ reveal that both *cis* and *trans* isomers exist in solution.

In addition to the above, ^1H NMR spectroscopy has been used to investigate[558] a number of monocyclopentadienyl uranium alkoxides.

3.4.5   Alkoxides of Group 4 Metals

Bradley and Holloway[548,559] measured the ^1H NMR spectra of straight as well as branched chain alkoxides of titanium. The trimeric straight chain alkoxides like ethoxide, *n*-propoxide, and *n*-butoxide show a constant chemical shift over a wide range of temperature up to 160°C. Even at very low concentration, the dissociation up to this high temperature does not appear to be complete. At a low temperature (below $-20°C$) a new sharp peak appears in the ethoxide spectrum which raises the possibility of the presence of the tetrameric species. The ^1H NMR spectra of titanium tetraisopropoxide or isobutoxide show temperature- and concentration-dependent behaviour. The spectra of these derivatives do not show any splitting even below $-50°$ indicating that the exchange of terminal and bridging alkoxo groups is very fast.

On the basis of four broad bands observed in the spectrum of titanium tetramethoxide in the ratio 1:2:3:2, Weingarten and Van Wazer[560] assigned a tetrameric structure, for the product in solution, which was later confirmed by X-ray crystallography.[436] At room temperature tetrameric titanium tetraethoxide $[Ti(OEt)_4]_4$, shows only one set of CH_2CH_3 signals. Although the rapid bridge–terminal exchange was slowed down at low temperature, no simple assignment could be made.[548]

The ^1H NMR spectra of tetra-*t*-butoxides of Ti, Zr, and Hf show only one Bu^t resonance, consistent with their monomeric behaviour[548] in solution. A solution of titanium *tert*-butoxide in *tert*-butyl alcohol at room temperature gave two signals for *tert*-butyl protons thereby demonstrating that the exchange of *tert*-butoxo group between $Ti(OBu^t)_4$ and Bu^tOH is slow on the NMR time scale probably owing to steric hindrance. However, ^1H NMR spectra of mixtures of $M(OBu^t)_4$ (M = Zr or Hf) and Bu^tOH provide evidence for rapid intermolecular exchange of *t*-butoxide ligands.[548]

^1H NMR spectra of freshly distilled trimeric samples of the isopropoxides of Zr and Hf in $CDCl_3$ exhibit only one type of isopropoxo group, while aged samples indicate the presence of a number of environments for OPr^i ligands. This observation could be interpreted in terms of transformation of trimer to tetramer on ageing, which is consistent with molecular weight measurements.[548]

^1H and ^{13}C NMR spectra of $M_2(OPr^i)_8(Pr^iOH)_2$ (M = Zr, Hf) at 25°C are so simple as to be structurally uninformative, as a result of rapid fluxionality.[460] At $-80°C$, the NMR in toluene-d_8 are too complex to be accounted for by a single edge-shared bioctahedral structure as established by X-ray crystallography.[460]

The room temperature ^1H and ^{19}F NMR spectra of both $Ti\{OCH(CF_3)_2\}_4(NCMe)_2$ and $Ti\{OCH(CF_3)_2\}_4(thf)_2$[272] show only one type of nondiastereotopic $OCH(CF_3)_2$ ligand. The appearance of broad NMR signals suggests a dynamic exchange process, which could be due to *cis* to *trans* isomerization by dissociation of a ligand to form a five-coordinate species.

Heteroleptic isopropoxides of titanium $[TiCl_2(OPr^i)(Pr^iOH)(\mu\text{-}Cl)]_2$ and $[TiCl_2(OPr^i)(HOPr^i)(\mu\text{-}OPr^i)]_2$ are dynamic in solution[561] with the appearance of one set of ^1H NMR signals for the isopropyl groups at ambient temperatures. Variable temperature

^{1}H NMR studies show that they are static and remain in the dimeric forms in solution at low temperatures.

3.4.6   Alkoxides of Group 5 Metals

The ^{1}H NMR spectrum of dimeric tantalum pentamethoxide in *n*-octane solution at 100°C gave a sharp signal at δ 4.22 which indicated the rapid intramolecular exchange of terminal and bridging methoxo groups.[562] The low-temperature ^{1}H NMR spectra of $M_2(OMe)_{10}$ (M = Nb, Ta) show three signals of intensity ratio 2:2:1, consistent with a dimeric bioctahedral structure with two bridging methoxo groups. Interestingly the temperature dependence of their spectra (see for example, Fig. 2.4) indicates that terminal exchange of methoxides occurs faster than bridge–terminal exchange.[563] Similar behaviour was also observed for the other normal alkoxides of niobium and tantalum.

The ^{1}H NMR spectra of the isopropoxides[562] of niobium and tantalum were found to be both temperature (Fig. 2.5) and concentration (Fig. 2.6) dependent, consistent with the equilibrium: $M_2(OPr^i) \rightleftharpoons 2M(OPr^i)_5$. For example the pentaisopropoxide of tantalum in cyclohexane solution gave three septets in the range δ 4.7–5.0. The intensity of the high-field signals increased on raising the temperature at the expense of the

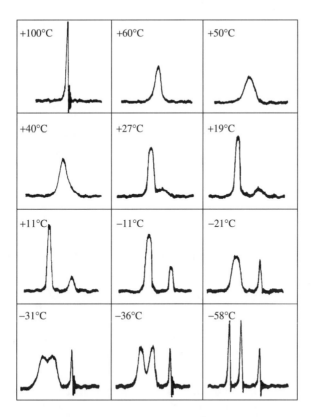

Figure 2.4   Variable temperature ^{1}H NMR spectra of $[Ta(OMe)_5]_2$.

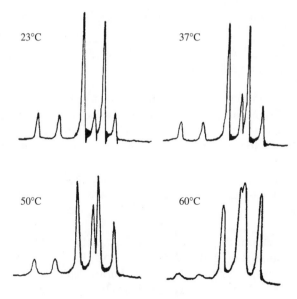

Figure 2.5   Variable temperature ^{1}H NMR spectra of $Ta_2(OPr^i)_{10} \rightleftharpoons Ta(OPr^i)_5$.

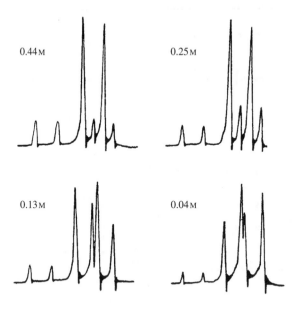

Figure 2.6 ^{1}H NMR spectra of $Ta_2(OPr^i)_{10} \rightleftharpoons 2Ta(OPr^i)_5$.

other two signals whereas decreasing the temperature had the reverse effect on the spectrum.[562] On adding isopropyl alcohol to the pentaisopropoxide, the high field septet was shifted to still higher field by coalescence with the isopropyl alcohol signal and the intensities of the two low-field septets was diminished. On this basis the high-field septet was assumed to be due to monomeric $Ta(OPr^i)_5$ and the low-field septets due to resolved terminal and bridging isopropoxo groups of dimeric $Ta_2(OPr^i)_{10}$. The added alcohol rapidly exchanged with the isopropoxy protons of the monomer but not with the protons of the dimer, although the intensities of the latter were diminished. The existence of a monomer/dimer behaviour in tantalum isopropoxide was supported by the appearance of the spectrum in benzene solution which gave three doublets in the range δ 1.23–1.28. The low-field doublets were in the intensity ratio 4:1 (due to dimeric species) and their decrease in intensity with increase in temperature or decrease in concentration with the simultaneous increase in intensity of the upfield doublet (Figs 2.5 and 2.6) accorded with the conversion of dimer to monomer species. The measurements of equilibrium constants at various temperatures indicate the equilibrium constants were twice the values in benzene than in carbon tetrachloride.

The ^1H NMR spectrum of niobium pentaisopropoxide at room temperature indicates a broad resonance and an average doublet which at low temperatures was resolved into peaks of the dimer (at δ 1.25 and 4.90) and monomer (at δ 1.19 and 4.68). Although the monomer/dimer equilibrium for niobium pentaisopropoxide also existed in benzene solution, the value of the equilibrium constant at a given temperature was about 100 times greater for niobium than for tantalum isopropoxide, thereby demonstrating the greater degree of dissociation of niobium isopropoxide into monomeric species.

The ^1H NMR spectra of niobium and tantalum *tert*-butoxides[562] gave single peaks at δ 1.37 and 1.36, respectively which did not change with temperature or concentration. The addition of *tert*-butyl alcohol gave a single coalesced signal indicating rapid alcohol interchange, a behaviour which is in contrast to that observed in case of titanium tetra-*tert*-butoxide.[548] This abnormal behaviour was not clearly understood on steric grounds, which should prevent the exchange between *tert*-butoxo groups and *tert*-butyl alcohol much more in niobium or tantalum penta-alkoxides than in titanium tetra-alkoxide.

Riess and Hubert-Pfalzgraf[564,565] measured the temperature and concentration dependence of the ^1H NMR spectra of niobium and tantalum pentamethoxides in nonpolar solvents like toluene, carbon disulphide, and octane at ambient temperature and at low temperatures (−11 to −74°C) and at different concentrations and confirmed that at low temperatures and higher concentrations methyl protons having peak intensity ratio 2:2:1 are obtained. This is consistent with the X-ray structure of $[(MeO)_4Nb(\mu\text{-}OMe)_2Nb(OMe)_4]$.[566] The coalescence of these peaks occurs at higher temperature (10°C) and low concentration (0.01 M).

On the basis of variable temperature ^1H NMR studies Holloway[563] reasserted that in dimeric niobium and tantalum alkoxides, the terminal–terminal alkoxide exchange occurs at a faster rate than terminal–bridging exchange.[562]

Oxovanadium alkoxides, $VO(OR)_3$ (R = Me, Et, Pr^i, Bu^s, Bu^t, CH_2CH_2F, CH_2CH_2Cl, CH_2CCl_3), have been investigated by ^{51}V NMR studies,[567] which indicate that the limiting (low concentration) $\delta(^{51}V)$ values depend on the bulk of R (highest ^{51}V shielding for Bu^t). Shielding decreases with increasing concentration (more pronounced for smaller R groups), owing to the formation of oligomers involving alkoxo bridging. Similar observations were also reported by Lachowicz and Thiele[568] in 1977 on ^{51}V NMR data for oxovanadium alkoxides. For example, the chemical shift for oxovanadium trimethoxide changes more than 40 ppm when the concentration increase from 1 mM to more than 50 mM.

Interestingly, the ^{51}V chemical shift of oxovanadium trialkoxides of bulky and chiral alcohols, $VO(OR')_3$ (where $R' = $ 1-admantanyl, *endo*-borneyl, and *exo*-norborneyl)[569] did not change in toluene, hexanes, and chloroform solutions from 1.0 to 500 mM; this is possibly due to the tendency of the bulky alkoxo ligands to inhibit association. The ^{1}H NMR chemical shifts for the protons on the carbon α to the oxygen shift from 1.3 to 1.5 ppm upon vanadylation. Protons on remote sites show a shift ranging from 0 to 0.4 ppm. The ^{13}C NMR spectra exhibit the carbon α to the oxygen shift from 19 to 25 ppm downfield. All the other carbons shift less than 2 ppm. The broadening of the signal for the α-carbon indicated the possibility of the occurrence of exchange process at ambient temperature. Furthermore, the hydrolysis of the oxovanadium trialkoxides have also been investigated in organic solvents using ^{51}V NMR spectroscopy.[569]

3.4.7 Alkoxides of Group 6 Metals

A number of chromyl(VI) alkoxides have been investigated by NMR spectroscopy[233] and the observed data have been interpreted either in terms of a static dimeric structure in solution or due to rapid exchange (on NMR time scale) between terminal and bridging alkoxo ligands.

$Mo_2(OBu^t)_6$ shows only a methyl signal in the ^{1}H NMR spectrum at 1.42 ppm downfield from hexamethyldisiloxane.[360] Variable temperature ^{1}H and ^{13}C NMR spectroscopic studies[361] on $Mo_2(OR)_6$ (R $= PhMe_2C$, Me_3C, Me_3CCH_2) tend to support the well-established formulation containing Mo≡Mo bonds for these alkoxides.

The diamagnetic tetrameric ethoxide $Mo_4(OEt)_{12}$ obtained by the reaction of $Mo_2(NMe_2)_6$ with ethanol ($\geq$6 equiv.) shows a very complex ^{1}H NMR spectrum due to overlapping of several ethyl resonances.[361] In benzene-d_6 the ^{13}C{^{1}H} spectrum at 40°C shows eight methylene carbon signals in the intensity ratio 1:1:1:1:2:2:2:2, consistent with the tetrameric, $Mo_4(OEt)_{12}$ structure.

The diamagnetic $Mo_2(OPr^i)_8$ gives sharp NMR spectra[371] at room temperature. The low-temperature limiting spectra (^{1}H and ^{13}C) are consistent with the adoption of a structure having a pair of bridging alkoxo ligands: $[Pr^iO_3Mo(\mu\text{-}OPr^i)]_2$ (see also Chisholm *et al.*[372]).

The ^{1}H NMR spectrum of the diamagnetic $W_4(OMe)_{16}$ shows rigid behaviour on the NMR time scale by depicting eight singlets,[367] consistent with the maintenance of the centrosymmetric structure found in the solid state. It is noteworthy that the tetrameric titanium ethoxide $Ti_4(OEt)_{16}$ dissociates in solution. The ^{1}H NMR spectrum of $W_4(OEt)_{16}$ shows many overlapping methylene and methyl resonances.

Hexaalkoxides $M(OR)_6$ (R $=$ Me, Et, Pr^n, Pr^i, $CH_2CH{=}CH_2$) are only known for tungsten (M $=$ W) and ^{1}H and ^{13}C NMR spectra of these compounds[570] in the temperature range +90 to −90°C exhibit in all cases only one kind of alkyl group.

Room temperature ^{1}H NMR spectra (in C_6D_6) for $W_2(OMe)_{10}$ and $Re_2(OMe)_{10}$ exhibit three sharp singlets in a 2:2:1 ratio consistent with an edge-shared bioctahedral geometry.[429]

3.4.8 Alkoxides of Group 13 Metals

The observations of Mehrotra[293] regarding the existence of aluminium alkoxides in different oligomeric forms such as dimers, trimers, and tetramers depending upon the experimental conditions and the ramification of the alkyl groups, attracted the interest of a number of workers such as Bains[571] Shiner et al.,[489] Kleinschmidt,[572] Oliver and Worrall,[492,573] Fieggen,[574] Akitt and Duncan,[575] and Kriz et al.[576] to investigate by NMR the structures of aluminium alkoxides in solution, and considerable light has been thrown on the same in the past three decades. A preliminary ^{1}H NMR study on dimeric aluminium *tert*-butoxide $Al_2(OBu^t)_6$ was made by Bains[571] who observed two types of Bu^t signals in both benzene and carbon tetrachloride solutions due to terminal and bridging *t*-butoxo groups. No evidence for dissociation of the dimer at higher temperatures or in basic solvents (dioxane, *tert*-butyl alcohol) has been obtained. The above observation was confirmed by the presence of two signals of intensity 1:2 in the ^{1}H NMR spectra[481] for $[M(OBu^t)_3]_2$ (M = Al, Ga) in conformity with a rigid dimeric structure (2-VII) in solution with two bridging and four terminal OBu^t ligands. The dimeric structure proposed earlier by Mehrotra[293] in 1953 on the basis of molecular weight measurements was later confirmed for $[Al(OBu^t)_3]_2$ by X-ray crystallography.[338]

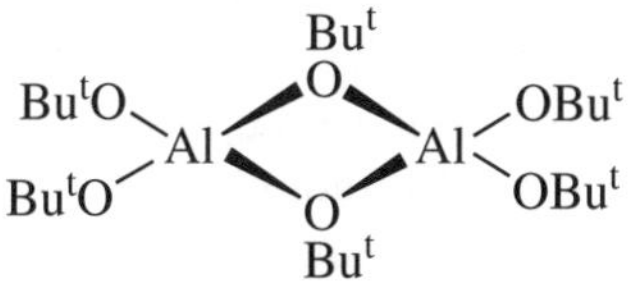

(2-VII)

The mixed alkoxide $Al(OPr^i)(OBu^t)_2$ is also dimeric[571] and its ^{1}H NMR spectrum[492] shows four resonances. The complex spectrum could not be resolved but lent support for an interesting unsymmetrical structure (2-VIII):

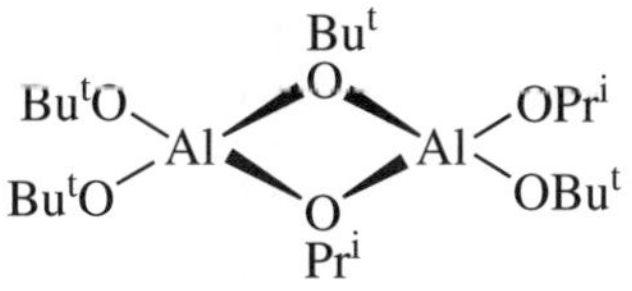

(2-VIII)

In order to explain the "ageing" phenomenon of aluminium isopropoxides observed by Mehrotra, it was suggested by Bradley,[577] that the tetramer might involve a unique structure (2-IX) with a central octahedral aluminium and three peripheral tetrahedral aluminium centres:

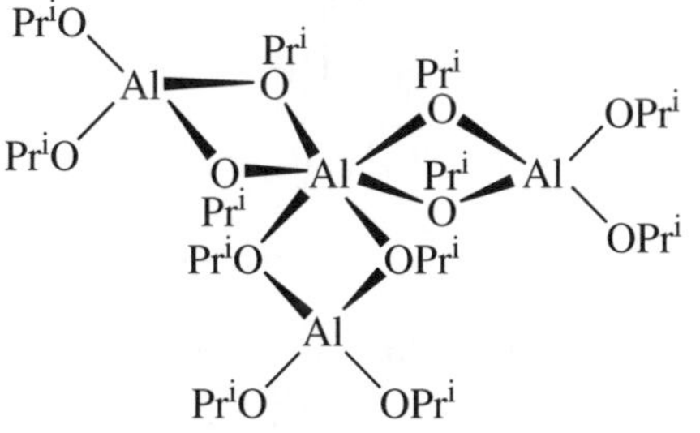

(2-IX)

^{1}H NMR studies on solution of tetrameric [Al(OPr^i)$_3$]$_4$ by Shiner *et al.*[489] and Mehrotra and Mehrotra[578] show three doublets for the CH$_3$ protons in a 1:1:2 ratio (Fig. 2.7), as the methyl groups in the bridging isopropoxo ligands are nonequivalent. These observations are consistent with structure (2-IX) which has been confirmed by X-ray crystallography[579,580] (when R = Pr^i).

In order to study the structure of aluminium isopropoxide in detail, Worrall *et al.*[490,492,581] measured the ^{1}H NMR spectra of aluminium isopropoxide under different experimental conditions. The freshly distilled aluminium isopropoxide on crystallization at −20°C was dissolved in benzene and the ^{1}H NMR spectrum of this solution was almost identical to that obtained by Shiner *et al.*[489] consisting of three sets of doublets, thus confirming yet again the proposed tetrameric structure for aluminium isopropoxide. However, these workers assumed the nonequivalence of methyl protons to be due to the asymmetric nature of the molecule which has D$_3$ symmetry and in principle is optically active.

The results of Shiner *et al.*[489] and those of Mehrotra and Mehrotra[578] are presented in Table 2.21. In the case of the 100 MHz spectrum, two distinct sets of septets with equal intensity ratio were observed; these peaks could be assigned to the methine protons of the terminal and bridging isopropoxo groups. Following Worrall *et al.*[490,492] the explanation of the splitting of methyl protons in the spectra of tetrameric aluminium isopropoxide could be sought in terms of magnetic nonequivalence of the two methyls on the same isopropoxide group due to asymmetry, but then the corresponding methine protons on the isopropoxide groups in similar environment should be identical. This general conclusion was confirmed by the appearance of only two sets of methine proton

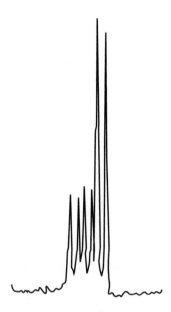

Figure 2.7 ^{1}H NMR spectrum of aluminium isopropoxide in CDCl$_3$ at 60 MHz.

Table 2.21 ^{1}H NMR spectra of $Al_4(OPr^i)_{12}$

| Solvent | Methyl protons chemical shift (δ) | | | Reference |
	Terminal	Bridging		
Benzene	1.27	1.37	1.65	596
	1.30	1.39	1.66	439
Cyclohexane	1.16	1.34	1.54	578a
Carbon tetrachloride	1.10	1.28	1.47	596
	1.24^1	1.30^1	1.48^1	596
	1.12	1.28	1.47	489
Chloroform-d	1.13	1.32	1.50	578
Carbon disulphide	1.09	1.26	1.43	578a

^{1}At 100 MHz; all other data at 60 MHz.

peaks of equal intensity ratio. According to Shiner et al.,[489] the spectrum ought to have shown three sets of peaks for methine protons.

^{1}H NMR studies by Oliver and Worrall[573] have convincingly demonstrated that the asymmetric central aluminium in tetrameric $Al_4(OCH_2R)_{12}$ ($R = C_6H_5$, $4\text{-}ClC_6H_4$) causes nonequivalence of the methylene protons with consequent appearance of an AB quartet ($J_{AB} = 11$ Hz). In the spectra of both the compounds the bridging CH_2 groups appear as well-defined quartets ($\delta_{AB} = 22.5$ and 21.3 Hz, respectively) at 60 MHz, but terminal methylenes gave unresolved singlets. However, at 220 MHz AB quartets ($\delta_{AB} = 0.5$ and 6 Hz converted to 60 MHz equivalents) have been observed even for terminal methylene protons. The larger δ_{AB} values for the bridging CH_2 groups could be ascribed to their closer proximity to the asymmetric aluminium centre.

Ayres et al.[582] measured the ^{1}H NMR spectrum of trimeric aluminium tribenzyloxide in carbon tetrachloride solution at 40°C which showed a singlet due to terminal aryloxo groups and a four line multiplet with intensity ratio 2:1 downfield relative to the TMS due to bridging benzyloxo groups. No solvent or internal reference was suitable for the measurement of temperature-dependent ^{1}H NMR spectra in the wide range of temperature. However, benzene was used up to 70°C and biphenyl in the temperature range 70–170°C. The spectra showed a broad peak below 70°C, probably due to a mixture of trimeric and dimeric species; the dimeric species appeared to dominate in the range 85–140°C. It was assumed that monomeric species may also exist above 170°C, but the ^{1}H NMR measurement was rather impracticable. Huckerby et al.[491] observed a singlet and a quartet for aluminium tribenzyloxide, but the intensity ratio was found to be 1:1 rather than 2:1 as observed by Ayres et al.[582] in their ^{1}H NMR spectrum. On this basis Huckerby et al.[491] assumed the presence of equal numbers of terminal and bridging benzyloxide groups which is a requirement of the tetrameric species.

Shiner et al.[489] examined the ^{1}H NMR spectra of super-cooled trimeric liquid aluminium isopropoxide in various solvents, which showed the presence of a single isopropoxide species with a chemical shift close to that observed for the terminal isopropoxo groups of the tetramer. At low temperatures the trimer signals broadened and, in some cases, split into two (approximately in 1:2 intensity ratio), consistent with the requirements of a cyclic trimer (2-X).

Kleinschmidt[572] proposed an alternative structure for the trimer (2-XI) involving a central five-coordinated aluminium bridged to two four-coordinated aluminium atoms.

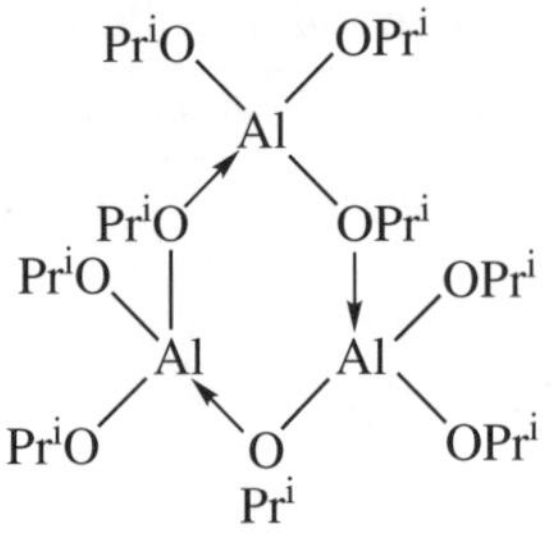

(2-X)

Fieggen[574] found evidence in support of this structure from the ^{1}H NMR spectrum of the trimer in chloroform solution.

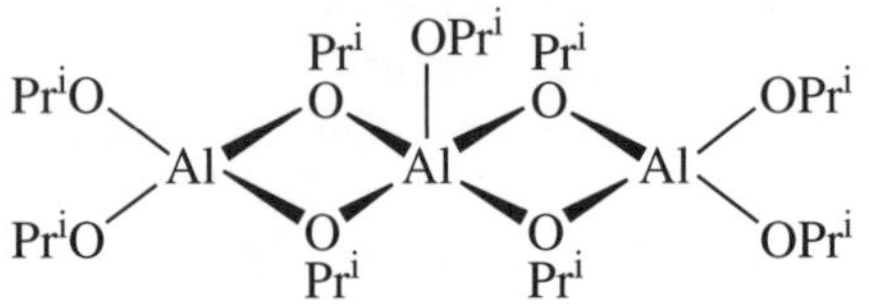

(2-XI)

The ^{27}Al NMR spectrum of the tetrameric aluminium isopropoxide investigated by Kleinschmidt[572] showed a broad and a sharper peak in the ratio 3:1, in agreement with the proposed structure (2-IX).

Akitt and Duncan[583] have provided further ^{27}Al NMR evidence for the structure of the tetramer in solution. The spectrum consisted of a very broad peak due to tetrahedral aluminium and a sharper peak due to the central octahedral aluminium in the required 3:1 ratio of intensities. Owing to the quadrupole moment of the ^{27}Al nucleus, the line width of the NMR signal depends on the symmetry of the electric field and gives independent evidence of the coordination of the aluminium. Furthermore, the structures of a number of other aluminium alkoxides were elucidated in 1984 by ^{27}Al NMR spectroscopy,[576] which supported the earlier proposed structures of a dimer (2-VII) and linear trimer (2-XI). It may be mentioned that a tetrameric structure for aluminium isopropoxide [Al(OPr^i)$_3$]$_4$[579,580] and a dimeric structure for aluminium t-butoxide [Al(OBu^t)$_3$]$_2$[338] have been confirmed by X-ray crystallography.

Attempts to elucidate the structure of gallium isopropoxide have also been made by Oliver and Worrall[495] as well as by Mehrotra and Mehrotra.[578] These workers observed that the ^{1}H NMR spectrum of a freshly prepared benzene solution of gallium isopropoxide shows three doublets due to methyl protons in the approximate intensity ratio expected for a tetrameric structure. A study of the ^{1}H NMR spectrum of an aged sample of gallium isopropoxide shows evidence for dissociation of the tetrameric form into the dimeric form, possibly due to the higher tendency of gallium to attain the stable four-coordination state as compared to the six-coordination state which is less stable.

^{205}Tl NMR studies on tetrameric thallium(I) alkoxides (TlOR)$_4$, support the structure (Fig. 2.1a) containing a Tl$_4$O$_4$ cube.[584]

3.5   Electronic Absorption, Reflectance, and Electron Spin Resonance Spectra

Electronic absorptions[585] may be broadly classified into charge-transfer and ligand-field (d–d) bands. The latter are more helpful in providing information regarding the probable geometries of metal alkoxide complexes. To interpret the spectra of metal complexes in general in which the metal ions have more than one but less than nine (d^2–d^8) electrons, one is required to employ an energy level diagram based upon the Russell-Saunders states of the relevant d^n configuration in the free (uncomplexed) ion. When d–d spectra are observed and assigned, it is possible to derive information on the magnitude of the ligand field splitting in the derivatives, the spin–orbit coupling constants and the Racah parameters, the latter providing a measure of interelectronic repulsions.

The main group metal alkoxides are diamagnetic and do not show ESR spectra. The paramagnetic transition metal complexes studied, however, have yielded information concerning the delocalization of the unpaired electrons.

Titanium trimethoxide[219] was expected to have an octahedral environment around Ti(III), but the reflectance spectrum gave a d–d transition at $10\,000\,\mathrm{cm}^{-1}$ and the compound was diamagnetic indicating a strong metal–metal interaction. The methanol-solvated trimethoxide $Ti(OMe)_3(MeOH)_6$ shows a band at $25\,200\,\mathrm{cm}^{-1}$[586] which is considerably higher than that observed for octahedral $Ti(H_2O)_6^{3+}$ at $20\,300\,\mathrm{cm}^{-1}$.[587]

For confirming the predicted stereochemistry of vanadium alkoxides, Bradley *et al.*[588] measured the ESR spectra of vanadium tetramethoxide in solid as well as solution forms (using solvents like methylcyclohexane or benzene) at room temperature and at $-150°C$; broad signals were obtained at $g = 1.955 \pm 0.005$ and 1.94 ± 0.01 in the solid and the solution, respectively, without hyperfine structure. The spectra were in accordance with the distorted octahedral geometry for trimeric species with $^2T_{2g}$ state split by $>1000\,\mathrm{cm}^{-1}$ giving rise to an orbital singlet ground state. Solid vanadium tetraethoxide also gave a very strong broad signal at room temperature ($g = 1.945$). On the other hand the ESR spectra of the tetraethoxide in benzene, methylcyclo-hexane, and carbon disulphide gave narrower signals in the range $g = 1.951$–1.953 with partial resolution of eight line ^{51}V interaction $A_0 = 78$–79.1 gauss.[588] However, the hyperfine structure disappeared at lower temperatures ($-170°C$). These results were in good agreement with the dimeric five-coordinated (trigonal-bipyramidal) vanadium tetraethoxide involving ethoxo bridges. The electronic spectra of solid and solutions of vanadium tetraethoxide were found to be quite similar.[589] The spectra show two d–d transition bands of rather weak intensities at $14\,200$ and $6000\,\mathrm{cm}^{-1}$ and an intense charge-transfer band at $25\,000\,\mathrm{cm}^{-1}$. The band observed at $14\,200\,\mathrm{cm}^{-1}$ ruled out the possibility of involving a tetrahedral or octahedral vanadium(IV) atom, because it was too high in energy and weak in intensity for a tetrahedral species and very low in energy for an octahedral species. If the vanadium atom is five-coordinated, then the two ligand field transitions are expected $\nu_1(^2E' \leftarrow {}^2E'')$ at $6000\,\mathrm{cm}^{-1}$ and $\nu_2(^2A_1' \leftarrow {}^2E'')$ at $14\,200\,\mathrm{cm}^{-1}$ for two ligand field parameters $D_s = 314\,\mathrm{cm}^{-1}$ and $D_t = 1398\,\mathrm{cm}^{-1}$. These values are plausible for the trigonal bipyramidal structure expected for dimeric five-coordinated vanadium tetraethoxide. However, although the values of D_s and D_t are consistent with the D_{3h} point group, the bridging ethoxide oxygen atoms should in principle produce a different electric field from that of the terminal oxygen atoms, and this should cause a lowering of symmetry from D_{3h} to the C_{2v} local symmetry point group with splitting of the $^2E'$ and $^2E''$ states.

The electronic spectra[589] of pure liquid and benzene solution of vanadium tetra-*tert*-butoxide were quite similar and the intense absorption band observed at $28\,000\,cm^{-1}$ was assumed to be due to charge-transfer whereas the sharp weak band at $5920\,cm^{-1}$ was ascribed to an infrared overtone (C–H stretching). The spectrum also contained a broad asymmetric band which was resolved by Gaussian analysis into two d–d transitions at $13\,900$ and $10\,930\,cm^{-1}$ for a distorted tetrahedral species. The ESR spectrum of liquid vanadium tetra-*tert*-butoxide at $30°C$ shows signals, $\langle g \rangle = 1.964 \pm 0.005\,cm^{-1}$ and $\langle a \rangle = 0.006\,4 \pm 0.000\,2\,cm^{-1}$ with eight ^{51}V ($I = 7/2$) hyperfine splitting.[590] It was observed that below $-5°C$, the ESR spectrum of 1–2% $V(OBu^t)_4$ in $Ti(OR)_4$ was anisotropic which did not change appreciably up to $-120°C$, but above $-120°C$ the spectrum becomes isotropic. The ESR spectrum of frozen solid vanadium tetra-*tert*-butoxide at $-196°C$ showed more resolution and the values $g_\parallel = 1.940 \pm 0.005$, $g_\perp = 1.984 \pm 0.005$, $A_\parallel = 0.0125 \pm 0.005$ and $A_\perp = 0.0036 \pm 0.004\,cm^{-1}$ were obtained. The application of the molecular orbital treatment gave a low value of the spin–orbit coupling constant $\lambda = 156\,cm^{-1}$ and the covalency due to involvement of the d orbitals in bond formation.[590] Bradley *et al.*[588] also studied the ESR spectrum of vanadium tetra-*tert*-butoxide and found a similar value of $\langle g \rangle = 1.962$, as had been observed earlier by Kokoszka *et al.*,[590] for the distorted tetrahedral configuration (D_2 symmetry) with a $^2B_1(d_{x^2-y^2}$ orbital) ground state. On this basis, the d–d transitions were assigned to $^2B_2 \leftarrow {}^2B_1$ and $^2E \leftarrow {}^2B_1$ corresponding to the values $10\,930$ and $13\,900\,cm^{-1}$, respectively. The distortion of the tetrahedral structure for the d^1 system (^2E ground state) of vanadium tetra-*tert*-butoxide might be ascribed to the Jahn–Teller effect, but it is not necessarily so because the d^2 Cr(IV) system also shows distortion from regular tetrahedral and the result was explained by Bradley and Chisholm[591] on the basis of covalent bonding in the tetra-*tert*-butoxide.

The electronic spectrum of the blue d^2 chromium tetra-*tert*-butoxide was measured[471] in the region 5000–$40\,000\,cm^{-1}$ to show bands at 9100, $15\,200$, $25\,000$ and $41\,000\,cm^{-1}$ which were assigned to d–d transitions (first three bands) $\nu_1 = {}^2T_2(F) \leftarrow {}^2A_2(F)$; $\nu_2 = {}^3T_1(F) \leftarrow {}^3A_2(F)$; $\nu_3 = {}^3T_1(P) \leftarrow {}^3A_2(F)$ (with $10\,Dq = 9430\,cm^{-1}$ and $B = 795\,cm^{-1}$) and charge-transfer transitions (latter band), respectively assuming that it was tetrahedral. The blue colour of chromium tetra-*tert*-butoxide is due to the presence of the band around $15\,000\,cm^{-1}$. The ν_1 and ν_2 transitions were split into two doublets at 8700, 9500 and $13\,700$, $15\,750\,cm^{-1}$ and this was attributed to distortion from T_d to D_{2d} symmetry. During the study of the ESR spectrum for chromium tetra-*tert*-butoxide in toluene, Alyea *et al.*[471] did not observe any signal below $-175°C$. However, in frozen toluene at $-263°C$, the signals were observed over a range of $120\,000\,G$. The broad absorption band at $g \sim 4$ and a sharp band at $g = 1.962$ were consistent with the distorted tetrahedral symmetry having an orbital singlet ground state ($^3B_1$ in D_{2d} symmetry) with a zero-field splitting. The reflectance spectrum of polymeric chromium dimethoxide shows transitions at $18\,200$ and $22\,200\,cm^{-1}$ which suggested a tetragonally distorted octahedral geometry[219] for this compound.

Dubicki *et al.*[592] measured the diffuse reflectance spectra of chromium dichloride monomethoxide monomethanolate and dimethanolate which gave bands at around $15\,000\,cm^{-1}$ (ν_1), $21\,000\,cm^{-1}$ (ν_2) and $37\,500\,cm^{-1}$ (ν_3), which were assigned to $^4T_{2g} \leftarrow {}^4A_{2g}$, $^4T_{1g} \leftarrow {}^4A_{2g}(F)$ and $^4T_{1g}(P) \leftarrow {}^4A_{2g}$ transitions, respectively. The ligand field parameter $(10\,Dq = 15\,000\,cm^{-1})$ and the above spectral data accorded with an octahedral configuration for the chloride alkoxides.

The electronic reflectance spectra[592a] of the chloride alkoxides of d^2 vanadium(III): $VCl(OMe)_2$, $VCl(OMe)_2MeOH$, $VCl_2(OMe).2MeOH$ exhibited signals around 16 000, 25 000 and 35 700 cm^{-1} which were assigned to $^3T_{2g} \leftarrow {}^3T_{1g}(F)$, $^3T_{1g}(P) \leftarrow {}^3T_{1g}(F)$ and $^3A_{2g} \leftarrow {}^3T_{1g}(F)$ transitions, respectively with the Racah parameter, $B = 745\,cm^{-1}$ and the ligand field parameter $10\,Dq = 17\,400\,cm^{-1}$. These spectra also showed a very weak shoulder at $10\,000\,cm^{-1}$ which was assigned to the spin-forbidden transitions, $^1T_{2g}, {}^1E_g \leftarrow {}^3T_{1g}(F)$. The above observations were well in accordance with the octahedral geometry observed for various vanadium(III) complexes.[593,594]

The reflectance spectra of d^3 chromium(III) methoxide and ethoxide[219,547] agreed with ligand field predictions for an octahedral d^3 ion with $10\,Dq \sim 17\,000\,cm^{-1}$ (Table 2.22).

The reflectance spectrum (transitions at 18 200 and 22 200 cm^{-1}) of d^4 $Cr(OMe)_2$ showed evidence for a tetragonally distorted octahedral configuration[219] for chromium(II).

The reflectance spectrum of d^5 manganese dimethoxide[219] showed spin-forbidden transitions at 18 500, 24 400, 27 070, 29 000, and 31 300 cm^{-1} consistent with an octahedral geometry around Mn(II). Thus a polymeric methoxo-bridged octahedral structure was envisaged.

The high-spin d^6 iron dimethoxide gave a band at around $10\,000\,cm^{-1}$ in its reflectance spectrum which was assumed to arise from the $^5E_g \leftarrow {}^5T_{2g}$ transition required for an octahedral iron(II) ion,[219] and it probably attains a polymeric methoxo-bridged edge-sharing octahedral structure. The diffuse reflectance spectrum of solid d^5 iron trimethoxide[474] showed two sharp bands at 5800 and 7200 cm^{-1} and a broad band with a maximum at 11 000 cm^{-1}. The bands observed at 5800 and 7200 cm^{-1} have been assumed to be a mixture of the overtones of the C–H stretching and bending modes, whereas that at 11 000 cm^{-1} was assigned to a spin-forbidden transition of the tetrahedrally coordinated iron(III) ion.

The electronic spectrum of high-spin d^7 cobalt dimethoxide exhibited bands at 9500, 12 000, 17 900, and 21 000 cm^{-1}. The band at 9500 cm^{-1} was assigned to $^4T_{2g}(F) \leftarrow {}^4T_{1g}(F)$ transitions, that at 12 000 cm^{-1} to $^2E_g(G) \leftarrow {}^4T_{1g}$, and $^4A_{2g}(F) \leftarrow {}^4T_{1g}$ transitions whereas the latter two bands corresponded to $^4A_{2g}(F) \leftarrow {}^4T_{1g}$ and $^4T_{1g}(P) \leftarrow {}^4T_{1g}$ transitions, respectively.[219] The spectrum was

Table 2.22 Electronic spectral data (cm^{-1}) for chromium(III) alkoxides prepared by different routes[7]

Compound	$^4T_{2g} \leftarrow {}^4A_{2g}$ (10 Dq)	$^4T_{1g} \leftarrow {}^4A_{2g}$	B
$Cr(OMe)_3$[1]	17 610	24 150	612
$Cr(OEt)_3$[1]	17 000	23 470	600
$Cr(OBu^t)_3$[1]	17 060	23 700	610
$Cr(OMe)_3$[2]	17 240	23 800	609
$Cr(OEt)_3$[2]	16 370	23 470	654
$Cr(OPr^i)_3$[2]	15 880	22 940	720
$Cr(OMe)_3$[3]	17 180	24 150	613
$Cr(OPr^i)_3$[3]	16 180	23 280	728

[1]Prepared by lithium alkoxide method; [2]Prepared from $Cr(OC_6H_5)_3$; [3]Prepared by alcoholysis of $Cr(OBu^t)_4$.

thus interpreted in terms of an octahedrally coordinated cobalt(II) configuration in cobalt dimethoxide. The diffuse reflectance spectrum of cobalt chloride methoxide[595] showed bands at $7300\,cm^{-1}$ assigned to the $^4T_1(F) \leftarrow {}^4A_{2g}$ transition and at 15 200, 17 600, and 18 900, all being assigned to split $^4T_1(P) \leftarrow {}^4A_2$ transitions. Taking the value of $17\,200\,cm^{-1}$, the average energy in this complex band gave the calculated values of $10\,Dq = 4250\,cm^{-1}$ and the Racah parameter $B = 785\,cm^{-1}$ which are comparable with the values observed for the $CoCl_4^{2-}$ ion $(10\,Dq = 3200\,cm^{-1}$ and $B = 715\,cm^{-1})$.[596] The higher values obtained for the chloride methoxide as compared with the $CoCl_4^{2-}$ ion may be attributed to the positions of methoxide ion and chloride ion in the spectrochemical series since the methoxide is at higher energy than the chloride. The above spectral data were consistent with the tetrahedral coordination of the Co(II) ion. The spectrum of cobalt bromide methoxide was also consistent with the tetrahedral configuration; however, on solvation with methanol to form $CoBr(OMe).2MeOH$ a change from intense blue to pink with corresponding changes in the reflectance spectrum revealed a change in structure from tetrahedral to octahedral.[595]

The diffuse reflectance spectrum of cobalt iodide methoxide was rather complex and showed a large charge transfer band in the visible region. However, some discontinuity was observed around 16 000 and $18\,000\,cm^{-1}$ which was assigned to a $4T_1(P) \leftarrow {}^4A_2$ transition superimposed on the charge transfer edge.

The reflectance spectrum of the d^8 nickel dimethoxide showed bands at 8700, 14 500, and $25\,000\,cm^{-1}$ which were assigned to $^2T_{2g}(F) \leftarrow {}^2A_{2g}(F)$, $^3T_{1g}(F) \leftarrow {}^3A_{2g}$, and $^3T_{1g}(P) \leftarrow {}^3A_{2g}$ transitions, respectively.[219] The bands are in good agreement with the octahedral nickel(II) involving MO_6 configuration. Nickel monochloride monomethoxide also exhibits three bands at 7900, 12 900 and $22\,100\,cm^{-1}$ with the ligand field parameter $Dq = 790$ and Racah parameter $B = 870$ in its reflectance spectrum which are consistent with octahedral geometry for the nickel(II) ion corresponding to $^3T_{2g} \leftarrow {}^3A_{2g}$, $^3T_{1g}(F) \leftarrow {}^3A_{2g}$, and $^3T_{1g}(P) \leftarrow {}^3A_{2g}$ transitions, respectively.[541] The reflectance spectra of various halide alkoxides have also been measured, and show three bands at about the above positions. It was observed that in the series $NiCl(OMe)$, $NiCl(OMe).MeOH$, $Ni_3Cl_2(OMe)_4$, and $Ni_3Cl(OMe)_5$ the $^3T_{2g} \leftarrow {}^3A_{2g}$ transition bands appear at 7900, 8100, 8200, and $8300\,cm^{-1}$, respectively which are consistent with the values 7200 and $8500\,cm^{-1}$ observed for nickel dichloride and nickel dimethoxide, respectively. These observations indicate that the $10\,Dq$ values increased steadily as chloride was replaced by methoxide in the above series.

The diffuse reflectance spectrum of d^9 copper dimethoxide showed broad asymmetric bands with a maximum in the range $13\,800–16\,000\,cm^{-1}$ and these bands were assumed to be due to the $^2T_{2g} \leftarrow {}^2E_g$ transition.[542]

Karraker[597] measured the electronic absorption spectrum of dimeric uranium penta-ethoxide and the weak narrow bands observed at 5405, 5680, 6622, 6934, 10 200, 11 690, and $14\,490\,cm^{-1}$ have been assigned to f–f transitions for a distorted octahedral f^1 system with a spin–orbit coupling constant of $1905\,cm^{-1}$.

3.6 Magnetic Susceptibility Measurements

The magnetic properties of metal alkoxides have been studied in some detail for the first row transition metal alkoxides. Basi and Bradley[334] in 1963 measured the magnetic susceptibility of chromium tetra-*tert*-butoxide which was shown to be paramagnetic in behaviour having a magnetic moment of 2.8 ± 0.3 B.M. as required for a d^2 metal ion. It was shown that the magnetic susceptibility followed the Curie–Weiss law and was independent of temperature.

A detailed study of the magnetic susceptibilities of the methoxides of first row divalent and trivalent transition metals has been published by Winter and co-workers[219,220,474,541,542,592,595,598] in a series of papers and these results along with reflectance spectral data have been used to predict the stereochemistry of these methoxides. The results are summarized in Table 2.23.

Vanadium trimethoxide[589] is also a paramagnetic compound having magnetic moments 1.70–1.79 B.M. in the temperature range -150 to $16°C$. The data appear to indicate a slight variation of the magnetic moment with temperature, and this magnetic behaviour is well in accordance with the distorted octahedral configuration for the trimeric vanadium trimethoxide. Similarly vanadium monochloride dimethoxide shows temperature-dependent magnetic moments varying in the range 1.56–2.09 B.M. over the -176 to $19°C$ temperature range. These low values of magnetic moment have been interpreted in terms of spin interaction between neighbouring metal ions within the polymeric cluster.

Unlike d^2 vanadium trialkoxides, the d^1 tetra-alkoxides exhibit temperature independent magnetic behaviour, *e.g.* dimeric tetramethoxide and monomeric tetra-*tert*-butoxide follow the Curie law and both these compounds have magnetic moments of 1.69 B.M. For the tetraethoxide, the magnetic and electronic spectrum measurements correspond to five-coordinated vanadium having a trigonal bipyramidal structure involving C_{2v} symmetry in which dimerization occurs through an edge-sharing alkoxide

Table 2.23 Magnetic moment, reflectance spectral data, and proposed stereochemistry of first row transition metal methoxides and ethoxides

Metal alkoxide	Colour	Magnetic moment μ_{eff} (B.M.)[1]	Reflectance spectra 10^{-3} (cm^{-1}) tentative assignment	Proposed stereochemistry[2]
Cr(OMe)$_2$	Purple	5.16 ($\theta = 160°$)	18.2 ($d_{xy} \rightarrow d_{x^2-y^2}$); 22.2 ($d_{z^2}, d_{xy}, d_{xy} \rightarrow d_{x^2-y^2}$)	CrO$_4$ (Sp)
Mn(OMe)$_2$	Pale pink	5.96 ($\theta = 35°$)	18.5 ($^4T_{1g}$, G); 24.4 (^4E_g, G); 20.07 ($^4T_{2g}$, D); 29.0 (^4E_g, D); 31.3 ($^4T_{1g}$, P)	MnO$_6$(O$_h$)
Fe(OMe)$_2$	Dark green	5.14 ($\theta = 0$)	10.0 (^5E_g, D)	FeO$_6$(O$_h$)
Co(OMe)$_2$	Pale purple	5.46 ($\theta = 15°$)	9.5 ($^4T_{2g}$, F); 12.0 (^2E_g); 17.9 ($^4A_{2g}$, F); 21.0 ($^4T_{1g}$, P)	CoO$_6$(O$_h$)
Ni(OMe)$_2$	Pale green	3.38 ($\theta = 0$)	8.7 ($^3T_{2g}$); 14.5 ($^3T_{1g}$, F); 25.0 ($^3T_{1g}$, P)	NiO$_6$(O$_h$)
Cu(OMe)$_2$	Royal blue	1.24 (22°C, $\theta = 0$)	15.6 ($d_{xy} \rightarrow d_{x^2-y^2}$); 18.5 ($d_{z^2}, d_{xy}, d_{yz} \rightarrow d_{x^2-y^2}$)	CuO$_4$(Sp)
Ti(OMe)$_3$	Yellow green	Diamagnetic	10.0	—
Cr(OEt)$_3$	Pale green	3.56 ($\theta = 270°$)	16.4 ($^4T_{2g}$); 25.0 ($^4T_{1g}$); 38.1 ($^4T_{1g}$)	CrO$_6$(O$_h$)
Fe(OMe)$_3$	Yellow	5.86 ($\theta = 200°$)	11.1	FeO$_4$(T$_d$)

[1] $\mu_{eff} = 2.84(T + \theta)^{1/2}$; [2] Sp = square planar, O$_h$ = octahedral, T$_d$ = tetrahedral.

bridging structure.[589] On the other hand the monomeric tetra-*tert*-butoxide possesses a distorted tetrahedral configuration having D_{2d} symmetry which is well in accordance with the requirements of the magnetic and electronic spectral data.

Adams *et al.*[219] determined the magnetic moment of chromium triethoxide and the value was $\mu_{eff} = 3.56$ B.M. with $\theta = 270°$. Brown *et al.*[547] found the same value of magnetic moment and also observed the temperature-dependent behaviour of magnetic moments for chromium trimethoxide and triethoxide. These magnetic moment values are less than the spin only value of 3.88 B.M. Assuming this low value to be due to antiferromagnetic behaviour, they extrapolated the curve obtained by plotting χ^{-1} *vs* T from which a high θ value between 270 and 320° was obtained, leading to a value of 3.88 B.M. for the magnetic moment of triethoxide; this value is again close to the requirement of d^3 spin only value. Chromium dimethoxide shows[219] temperature-dependent paramagnetic behaviour having a magnetic moment of 5.16 B.M. with $\theta = 160°$ and it belongs to the d^4 system. The above data for chromium dimethoxide appears to indicate that it exhibits strong antiferromagnetic interaction. Dubicki *et al.*[592] observed the temperature-dependent magnetic behaviour of chromium dichloride monomethoxide monomethanolate; the magnetic moments of this product in the temperature range -177 to $20°C$ was of the order of 3.32–3.66 B.M. These values are less than the spin only value of 3.88 B.M. required for a d^3 ion. The magnetic moment measurements in solvents like acetonitrile, dioxane, and acetone also showed low values and these low values have been assumed to be due to the spin interaction of neighbouring Cr^{3+} ions in solution. The magnetic susceptibility measurements corresponded to its dimeric behaviour and this was supported by the observed molecular weight of the compound. The tetrameric structure shown earlier for this dimeric product has been assumed on the basis that two dimeric molecules will combine to form such a tetrameric structure, thus satisfying the stoichiometric and geometric requirements.

Manganese dimethoxide also shows a temperature-independent magnetic moment of value 5.96 B.M. Thus a comparison of the results indicate that the dimethoxides of divalent metal (manganese, iron, nickel, and cobalt) show temperature-independent magnetic moments between -193 and $77°C$ and thus follow the Curie–Weiss law. The compounds are weakly antiferromagnetic and the crystal lattices involve MO_6 octahedral geometry. The dimethoxides of divalent chromium and copper on the other hand exhibit marked temperature-dependent magnetic moments and are strongly anti-ferromagnetic.

Ferric alkoxides exhibit significant temperature-dependent magnetic behaviour. The magnetic moments for solid trimethoxide and triethoxide, and liquid tri-*n*-butoxide at ambient temperature are found to be 4.51, 4.37, and 4.35 B.M., respectively (against 5.90 B.M. required for a high-spin d^5 system), but these reduce considerably at lower temperatures. For example at $-181°C$, these alkoxides show moments of the order of 3.24, 3.25, and 3.44 B.M., respectively.[474] It has also been noted that these alkoxides follow the Curie–Weiss law with θ values between $-170°$ and $-200°C$ as shown by the curve of the variation of magnetic susceptibility and magnetic moments with temperature (Fig. 2.8a). From the curve it may be expected that at higher temperature, the higher spin levels become densely populated and the slope would correspond to $(\mu_{eff})^2 = 105$ (B.M.)^2 whilst at sufficiently low temperatures the magnetic moment would decrease to the value corresponding to $(\mu_{eff})^2 = 1.0$ (B.M.)^2 per Fe_3 core.

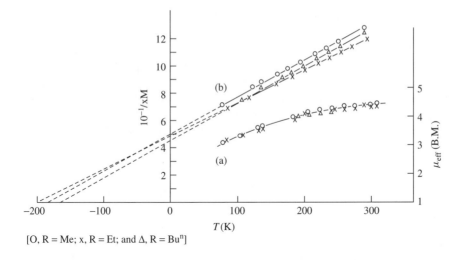

[O, R = Me; x, R = Et; and Δ, R = Bu^n]

Figure 2.8   Variation of magnetic susceptibility of [Fe(OR)$_3$]$_3$ with temperature.

It has been observed that the benzene solutions of d^5 iron trimethoxide and triethoxide do not affect the magnetic moment at the room temperature and the value remains around 4.40 B.M. which is independent of the concentration of the solute. Thus the anomalous magnetic behaviour exhibited by iron(III) alkoxides is intramolecular in its origin. In a detailed study on iron(III) alkoxides Adams et al.[599] interpreted the anomalous temperature dependence of the magnetic susceptibility (Fig. 2.8b) of the trimeric methoxide, ethoxide, and n-butoxide in terms of the spin–spin coupling between three iron atoms building the Fe$_3$ core of a model involving corners sharing FeO$_4$ tetrahedra to give the symmetry point group D$_{3h}$. The paramagnetic behaviour exhibited by iron trialkoxides is consistent with a spin quantum number of $S = 5/2$ for each iron atom (d^5 high spin) and the magnetic behaviour corresponds to the J value in the spin–spin Hamiltonian ($= -2JS_iS_i$ being -15 K).

Similar temperature-dependent magnetic behaviour has also been observed in the case of iron(III) halide alkoxides.[220] For example, iron monochloride dimethoxide shows magnetic moments of 3.67 B.M. and 2.63 B.M. at 19°C and -176°C respectively. Similarly, the magnetic moments per iron atom observed for Fe$_4$Cl$_3$(OMe)$_9$ and Fe$_4$Br$_3$(OMe)$_9$ at room temperature are 4.16 and 4.28 B.M., respectively, which are less than the expected value of 5.90 B.M. for the spin only value for high-spin iron(III) compounds. The magnitude of the variations of susceptibility with temperature is consistent with the magnitude of the interaction $J = -15$ K and -13 K for chloride and bromide alkoxides, respectively, which are comparable with the value $J = -15$ K observed for trialkoxides,[599] and thus exhibits antiferromagnetism. A critical examination of the above values of J indicates that the value is lower for the bromide alkoxide than for the chloride alkoxide. It is well known that bromide bridges provide a more efficient superexchange mechanism than chloride and in the above halide-alkoxides, if the polymerization occurs through halide bridges, on this basis the J value should be more for bromide alkoxide than for chloride alkoxide. However, the reverse order was observed for the above derivatives, thereby indicating that the association occurs through methoxo bridges with the halide groups occupying terminal positions.

The d^7 cobalt(II) methoxide shows variation of magnetic moment with temperature involving an octahedral CoO_6 model ($\mu_{\text{eff}} = 5.46$ B.M., $\theta = 15°$).[219] The halide methoxides also follow the Curie–Weiss law and thus chloride methoxide exhibits a room temperature magnetic moment of 4.89 ± 0.01 B.M.[595] The expected spin only magnetic moment for a high-spin d^7 system is 3.88 B.M. and spin–orbit coupling depends on the symmetry (octahedral or tetrahedral). The magnetic moments were calculated by Figgis and Lewis[600] using a spin–orbit coupling constant $\lambda = -170\,\text{cm}^{-1}$ which gave the values 4.70 and 5.2 B.M., for tetrahedral and octahedral Co(II) and in actual practice these values vary in the range 4.10–4.90 B.M. and 4.70–5.20 B.M., respectively. Applying the magnetic moment and ligand field parameter relationship[560,561] and using $\lambda = -170\,\text{cm}^{-1}$ for the tetrahedral chloride methoxide, the magnetic moment would be expected to be of the order of 4.50 B.M. which is less than the observed value of 4.90 B.M. and, therefore, it may be assumed to involve octahedral geometry. On the other hand, the observed[595] magnetic moment for cobalt bromide methoxide, $\mu_{\text{eff}} = 4.69 \pm 0.01$ B.M., is comparable with that needed for the tetrahedral Co(II) ion, and thus a tetrahedral structure is expected in the latter compound. However, the magnetic moment increases abruptly from 4.69 to 5.10 B.M. in methanol, which indicates that a change in geometry from tetrahedral to octahedal occurs in solution, probably due to solvation. The room temperature magnetic moment of cobalt iodide methoxide was found to be 4.82 ± 0.01 B.M. consistent with a tetrahedral structure for the product. However, the observed value is slightly higher than the 4.69 B.M. observed for the bromide methoxide which may be attributed to the relative positions of bromide and iodide in the spectrochemical series.

The d^8 nickel(II) methoxide follows the Curie–Weiss law as indicated by its magnetic susceptibility measurements in the range -193 to $77°C$ with a magnetic moment of 3.38 B.M. It appears to involve a structure containing NiO_6 octahedra with weak antiferromagnetic interactions.[219] The nickel chloride alkoxides NiCl(OMe), $Ni_3Cl_2(OMe)_4$, and $Ni_3Cl(OMe)_5$ exhibit[541] magnetic moments $\mu_{\text{eff}} \sim 3.2$–3.7 B.M. which are higher than the spin only value of 2.84 B.M. Furthermore, the magnetic susceptibility is temperature dependent and the magnetic moment increases with decreasing temperature. The magnetic moment is not affected by dissolving the chloride methoxide in methanol. On the basis of this anomalous magnetic behaviour shown by nickel chloride methoxide, it appears to involve the cubane structure, having $Ni_4(OMe)_4$ clusters in which the nickel atom is octahedrally surrounded by methoxo bridges and terminal methoxide and chloride groups involving ferromagnetic interactions. On the other hand $NiCl(OPr^i)$ and $Ni_3Cl(OMe)_5$ show temperature-independent magnetic moments which is expected for magnetically dilute nickel(II) ion. Nickel bromide methoxide exhibits a low magnetic moment (3.00 ± 0.03 B.M.) and this anomaly may be ascribed to weak antiferromagnetic interactions. $NiCl(OPr^i)$ and $Ni_3Br(OMe)_5$ on the other hand increase magnetic moments to 3.21 and 3.64 B.M. in methanol and acetone, respectively. The significantly high value in acetone implies a tetrahedral nickel(II) ion, which requires magnetic moments of the order of 3.54–4.20 B.M.

Similar to the d^4 chromium dimethoxide, the d^9 copper(II) alkoxides also exhibit temperature-dependent magnetic susceptibility.[219,542] The magnetic moments of copper dimethoxide and diethoxide at room temperature were reported[221] to be in the range 1.07–1.23 B.M. which were less than the spin only value of 1.73 B.M. This abnormal behaviour was attributed to be due to the highly polymeric structure of copper(II) alkoxides involving tetragonally distorted octahedral copper. On the other hand, the magnetic susceptibility measurements of copper dimethoxide over the wide range of temperature (-193 to $77°C$) indicated a maximum around $-13°C$ and exhibits anti-ferromagnetic behaviour.[542] On the basis of the above observations, Adams et al.[542] concluded that copper(II) methoxide does not possess a distorted octahedral structure as proposed by Brubaker and Wicholas;[221] instead it possesses a linear chain type polymer involving methoxo bridges. A similar type of structure was earlier suggested for copper halides,[601] which also showed magnetic susceptibility maxima similar to that observed in copper(II) methoxide. As compared to dialkoxides, the chloride-alkoxides of copper show higher magnetic moments. Thus monochloride monomethoxide and monochloride monoethoxide have room temperature magnetic moments in the range 1.67–1.68 and 1.39–1.49 B.M., respectively.[542] The higher value of magnetic susceptibility observed for the chloride-methoxide follows the Curie law with a moment $\mu_{\mathrm{eff}} = 2.30$ B.M. per copper atom. This differs from the value deduced from ESR measurements ($g = 2.01$). These workers have interpreted the results in terms of the interaction of each unpaired electron from pairs of ions to give a triplet ground state. The bromide methoxide of copper on the other hand is almost diamagnetic,[600] and therefore it must involve strong magnetic interactions between Cu^{2+} ions (*i.e.* a singlet ground state). It may be mentioned here that copper dibromide possesses stronger magnetic interactions than the dichloride[601] and this behaviour is significantly enhanced on the replacement of a bromide atom with alkoxide.

3.7 Mass Spectral Studies

Mass spectrometry has been a useful technique for elucidating the molecular complexity of metal alkoxides, except for those cases which have extremely high molecular weights or low volatility. There are three features of the mass spectrum that are important in characterizing a metal alkoxide. First, the parent peak can give information regarding the molecular weight of the compound. In cases where high-resolution mass spectral data are obtained this can be an authentic piece of evidence for a specific molecular formulation. However, care must be taken, to ensure that the observed peak in the mass spectrum is not some fragment of larger species. Secondly, an analysis of the isotopic distributions of the parent ion can give information as to the number of metal ions present. Finally, the fragmentation pattern can provide valuable information as to the type and number of ligands bound to the metal centres, particularly in cases of mixed ligand-alkoxide derivatives. Thus mass spectrometry can potentially yield information other than the molecular weight and/or molecular composition of the compound.

For alkali metal alkoxides the mass spectroscopic study is so far limited to lithium *tert*-butoxide.[440] There were controversies[438,439,602] concerning the molecular complexity of lithium *tert*-butoxide, but the appearance of fragment ions $Li_6(OBu^t)_5(OCMe_2)^+$ and $Li_6(OBu^t)_5^+$ in its mass spectrum (at $130°C$) has finally established a hexa-meric structure.[440] The parent molecular ion $Li_6(OBu^t)_5(OCMe_3)^+$ (an odd electron species) on losing one methyl group would give rise to the even-electron species $Li_6(OBu^t)_5(O=CMe_2)^+$ possessing an energetically favoured carbon–oxygen double bond. Furthermore, mass spectroscopic studies by Chisholm et al.[28] in 1991 have also confirmed that hexameric lithium *tert*-butoxide is extremely volatile, exhibiting not only a molecular ion (M^+ 480), but also fragments of the type $[LiOBu^t]_n$ ($n = 1$–5). These observations are in accord with those of Schleyer and co-workers[603] and Hartwell and Brown,[440] who had earlier demonstrated the hexameric form of LiOBu^t in the vapour phase. The derivative [LiOCMe$_2$Ph]$_6$ also exhibits a molecular ion in the vapour phase.[28]

Although the derivative Ca[OC(Bu^t)(CH$_2$OPr^i)$_2$]$_2$ is composed of dimeric units in the solid state,[53] mass spectroscopic studies reveal that it is monomeric in the gas phase.

The suggested tetrameric form[155,158] of some lanthanide isopropoxides [Ln(OPr^i)$_3$]$_4$ (Ln = Nd, Tb, Er, Lu) has been confirmed by their mass spectroscopic studies.[55] Although mass spectra[245,349–352] of none of the X-ray crystallographically character-ized[345,349–352] mononuclear tertiary alkoxides (including fluorinated analogues) gave parent molecular ions, they were interesting in showing the presence of some important di- and or trinuclear species.

Mass spectral studies by Mazdiyasni et al.[604,605] of some tetravalent metal(loid)-tert-amyloxides and hexafluoroisopropoxides such as M(OAm^t)$_4$ (M = Ti, Zr, Hf) and M[OCH(CF$_3$)$_2$]$_4$ (M = Ti, Zr, Hf, Si, Ge) have revealed the monomeric nature of these compounds in the gaseous state, by the detection of parent molecular ions. The appearance of weak parent molecular ion peaks in the mass spectra of tert-butoxides appear to indicate: (i) the general instability of the molecular ion peaks resulting from the successive losses of CH$_3$ and C$_2$H$_5$ groups, and (ii) the thermal instability of these products as is indicated by the presence of strong peaks in the spectra characteristic of OH and H$_2$O fragments at 17 and 18, respectively. The first peak observed for metal hexafluoroisopropoxides is due to the loss of one fluorine atom from the parent species M[OCH(CF$_3$)$_2$]$_4$. A similar behaviour was observed earlier in fluorinated β-diketonates.[606] The m/z values observed for higher mass numbers appear to indicate the general instability of the molecular ions as a result of the successive loss of CF$_3$, CF$_2$, and C$_2$F$_3$ groups and the next observed peak is due to the loss of one OCH(CF$_3$)$_2$ group.

The recent mass spectral findings of Turevskaya et al.[607] on tetra-alkoxides of zirconium and hafnium are usually in good agreement with their molecular complexity.[113,298,432,548] However, the existence of a high intensity of M$_3$O(OR)$_9^+$ or M$_4$O(OR)$_{13}^+$ ions in the mass spectra of [M(OR)$_4$]$_n$ tends to indicate the complex nature of such species. The X-ray crystal structure determination of [M(OR)$_4$]$_n$ is called for to solve such an ambiguity.

The fragmentation pattern for tantalum penta-alkoxides Ta(OR)$_5$ (R = Me, Et, Pr^i, CH$_2$CH$_2$OMe), in their mass spectra[81] is similar to that for alkoxides of zirconium and hafnium,[607] niobium,[608] molybdenum,[609] and tungsten.[610] For example, at first the loss of OR groups occurs, followed by elimination of the molecules of unsaturated hydrocarbons (usually, with the same number of carbon atoms as R). However, the decomposition of Ta(OMe)$_5$ differs from that of other homologues by the loss of HCHO molecules and H atoms. Elimination of ethers, R$_2$O, which is accompanied by the formation of the metal oxoalkoxide ions of the above mentioned alkoxides is typical. In the absence of X-ray crystallographic data, considerable effort had been directed earlier to throw light on the molecular complexities of aluminium trialkoxides in different (solid, liquid, or vapour) states.

The tetrameric behaviour of aluminium triisopropoxide was confirmed from mass spectral studies of [Al(OCHMe$_2$)$_3$]$_4$ and [Al(OCDMe$_2$)$_3$]$_4$ by Fieggen et al.[611] The highest mass fragment was ascribed to Al$_4$(OCHMe$_2$)$_{11}$(OCHMe)$^+$, in addition to very intense peaks due to Al$_4$(OCHMe$_2$)$_{11}^+$, and Al$_4$O(OCHMe$_2$)$_9^+$. Besides these fragments, peaks due to Al_2 or Al_3 fragments were also observed which could be due either to the presence of dimeric and trimeric species or to the fragmentation of tetrameric species. The mass spectrum also showed the presence of metastable peaks arising from the loss of CH_3CHO, $CH_3CH=CH_2$, and $(CH_3)CHOCH(CH_3)_2$, and thus a plausible fragmentation pattern was deduced.

The mass spectrum of dimeric mixed isopropoxide-*tert*-butoxide $(Bu^tO)_2Al(\mu\text{-}OPr^i)_2Al(OBu^t)_2$ was investigated by Oliver and Worrall.[492] On the basis of the mass fragmentation pattern observed, it appears that bridging alkoxo groups break down in two possible ways: one in which an isopropoxo group is lost initially and the other in which a *tert*-butoxo group is lost and the spectrum indicates the loss of either 42 or 56 units, due to C_3H_6 or C_4H_8, respectively, which should proceed by a four-centered mechanism (Scheme 2.5).

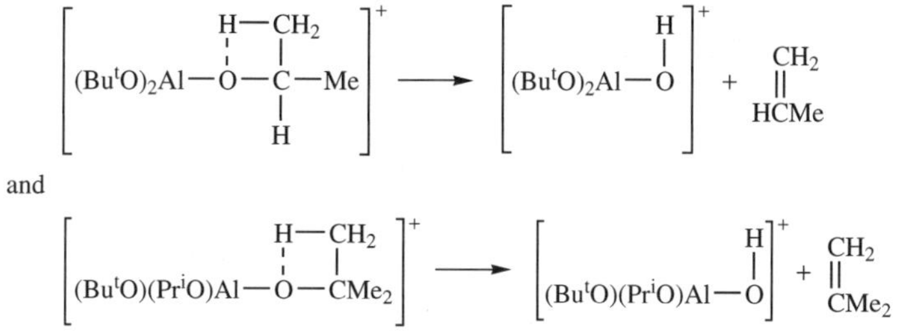

$$\left[(Bu^tO)_2Al-O-\overset{\overset{\displaystyle H-CH_2}{|}}{\underset{\underset{\displaystyle H}{|}}{C}}-Me \right]^+ \longrightarrow \left[(Bu^tO)_2Al-O-\overset{\displaystyle H}{|} \right]^+ + \begin{matrix} CH_2 \\ \| \\ HCMe \end{matrix}$$

and

$$\left[(Bu^tO)(Pr^iO)Al-O-\overset{\overset{\displaystyle H-CH_2}{|}}{CMe_2} \right]^+ \longrightarrow \left[(Bu^tO)(Pr^iO)Al-O-\overset{\displaystyle H}{|} \right]^+ + \begin{matrix} CH_2 \\ \| \\ CMe_2 \end{matrix}$$

Scheme 2.5

Thus the mass fragmentation due to the loss of a bridging alkoxo group could be consistent with an unsymmetrical structure $(Bu^tO)(Pr^iO)Al(\mu\text{-}OBu^t)(\mu\text{-}OPr^i)Al(OBu^t)_2$ in which isopropoxo and *tert*-butoxo groups form bridges for dimerization.

Oliver and Worrall[612] also investigated the mass spectrum of aluminium ethoxide and the proposed mass fragmentation pattern appears to indicate that in addition to tetrameric and other lower species, some pentameric species are also present and the fragmentation scheme of Fig. 2.9 can be deduced for different mass ion peaks:

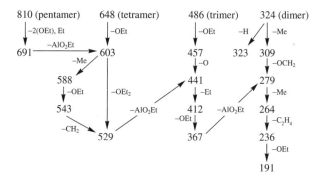

Figure 2.9 Proposed mass fragmentation pattern of $[Al(OEt)_3]_n$.

The mass spectra of gallium ethoxide and isopropoxide have also been investigated[612] and it has been shown that the mass fragmentation schemes (Figs 2.10 and 2.11) follow similar patterns to those observed for aluminium analogues. The spectrum of gallium isopropoxide is easy to interpret as there are $n + 1$ mass ion peaks for n gallium atoms. However, no parent molecular ion peaks were observed and the small peaks at 925 and 679 m/z arise from the loss of OPr^i groups from tetramer and trimer molecules, respectively. Figure 2.10 appears to indicate that the largest number of mass ion peaks is present in the dimer region. Gallium isopropoxide has been found to be dimeric in boiling benzene[184] and ^{1}H NMR spectrum also indicated a dominating dimeric structure.[495]

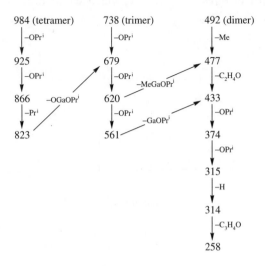

Figure 2.10   Proposed mass fragmentation pattern of [Ga(OPr^i)$_3$]$_4$.

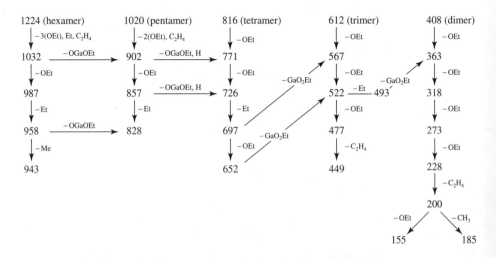

Figure 2.11   Proposed mass fragmentation pattern of [Ga(OEt)$_3$]$_n$.

The mass fragmentation patterns of gallium ethoxide (Fig. 2.11) appear to indicate the presence of higher molecular weight species, pentamers as well as hexamers, in addition to other lower species but in this case also dimeric species predominate over the others.

Lead di-*tert*-butoxide which was X-ray crystallographically characterized to be a trimer[339a] has been shown by mass spectrometry[194] to be dimeric in the gas phase.

3.8 EXAFS and XANES Spectral Studies

There are several attractive features of the Extended X-ray Absorption Fine Structure (EXAFS) technique[613–615] which make it a powerful structural tool; notably (i) it is extremely fast, (ii) both sample preparation and data collection are relatively easy, without the requirement of single crystals, (iii) being sensitive to short-range order in atomic arrangements, it can focus on the local environment of specific absorbing atoms, and (iv) the technique is useful for a wide variety of materials such as amorphous solids, liquids, solutions, gases, polymers, and surfaces.

The usefulness of the EXAFS technique in providing structural information (bonding sites and bond lengths) for metal alkoxides was demonstrated in 1988 by the work of Sanchez and co-workers[616] on titanium tetra-alkoxides, $[Ti(OR)_4]_n$ (R = Et, Pr^i, Bu^n, Am^t). Titanium tetra-*tert*-amyloxide, $[Ti(OAm^t)_4]$ and isopropoxide, $[Ti(OPr^i)_4]$ were shown to be tetrahedral monomeric ($n = 1$) molecules with Ti–O bond lengths of 1.81 and 1.76 Å, respectively, whereas the ethoxide, and n-butoxide, were shown to be trimers ($n = 3$) with average Ti–O bond lengths of 1.80 (terminal) and 2.05 Å (bridging), as well as Ti–Ti interactions (observed for the first time) of 3.1 Å. This confirms the trimeric behaviour previously shown by molecular weight studies.[455]

The X-ray Absorption Near Edge Structure (XANES)[615] spectral features have been used as an additional method for elucidating the stereochemistry as well as the oxidation state of the metal.[616] For example, based on the information available from XANES studies on titanium tetra-alkoxides,[616] the most probable coordination state of titanium in trimeric ethoxide and butoxide $[Ti(OR)_4]_3$ (R = Et, Bu^n) has been suggested to be five.

Although the most direct structural tool is single-crystal X-ray crystallography, yet in some cases (particularly those for which X-ray structures cannot be available) useful complementary information can be obtained by techniques like EXAFS and XANES in addition to those discussed in earlier sections.

4 CHEMICAL PROPERTIES

4.1 Introduction

As stated earlier the metal–oxygen bonds in metal and metalloid alkoxides $[M(OR)_x]_n$ are polarized in the direction, $M^{\delta+}$—$O^{\delta-}$—C owing to higher electronegativity (3.5 on Pauling scale) of oxygen. Consequently the metal atoms are easily susceptible to nucleophilic attack, not only on account of this positive charge, but also owing to the presence of energetically suitable vacant orbitals which can accommodate electrons from nucleophiles. In addition, the oxygen atoms are also susceptible to electrophilic attack. The combined effect of these factors makes metal alkoxides in general a highly reactive class of metallo-organic compounds. Metal alkoxides are, for example, extremely susceptible to hydrolysis even by atmospheric moisture, and require careful handling under stringently anhydrous conditions.

Metal alkoxides react readily with a wide variety of hydroxylic reagents such as water,[3,21,617–619] alcohol,[3–8,21,617] alkanolamines,[4,21,620–627,632] carboxylic acids,[628] β-diketones and β-ketoesters,[629–631] Schiff bases,[632,633] oximes, hydroxylamines,[634,635] and glycols,[2,4,632] to afford homo- and heteroleptic derivatives involving metals in novel coordination states and geometries. The reactions of metal alkoxides with mono-functional reagents (LOH) can be illustrated by Eq. (2.182):

$$M(OR)_x + yLOH \longrightarrow M(OR)_{x-y}(OL)_y + yROH \uparrow \qquad (2.182)$$

These reactions can be carried out quantitatively in the case of metal isopropoxides (or ethoxides) by the azeotropic fractionation of the liberated alcohol with a suitable solvent (e.g. benzene), yielding mixed alkoxides or alkoxide-ligand (carboxylate, β-diketonate, aminoalkoxide) derivatives. Many of these mixed ligand derivatives can be recrystallized and appear to be stable to heat (volatilizing without decomposition under reduced pressure).

A large number of such derivatives synthesized between 1960 and 1980 in the research group of Mehrotra[3,4,7] have been assigned interesting structures mainly on the basis of NMR studies and colligative properties and await X-ray crystallographic elucidation.

In addition to the general procedure illustrated by Eq. (2.182), another route has been employed for the synthesis of metal alkoxides (and particularly analogous siloxides) by the reaction of metal alkoxides with esters (and silyl esters) (Section 4.3). Further, the reactions of metal alkoxides with acyl halides (or hydrogen halides) and unsaturated substrates such as CO, CO_2, CS_2, organic-isocyanates and isothiocyanates, chloral, etc. have been utilized for the preparation of halide-alkoxides (Section 4.11) and interesting insertion products (Section 4.12), respectively.

Metal alkoxides are also sometimes reactive towards other substrates having reactive protons of $-NH_2$ or $-SH$ groups. In these cases, however, the reactions are controlled mainly by thermodynamic factors and are governed by the comparative stability and strength of M–O, M–N, and M–S bonds. For example, the $-NH_2$ groups of alkanol-amines and even substituted analogues are reactive with tin(IV) alkoxides[4,632] but not with analogues of titanium(IV).[4,632]

It is worth mentioning that a number of new classes of metal-organic compounds have become conveniently accessible only by the use of metal alkoxides as synthons. For example, the synthesis of pure aluminium tricarboxylates, $Al(O_2CR)_3$ (even the existence of which was seriously doubted in 1949 by Gray and Alexander),[636] after numerous reported failures over the previous two decades was successfully achieved in 1953 by the preparation of $Al(O_2CR)_3$ (R = $C_{15}H_{31}$, $C_{17}H_{35}$) through the reactions of aluminium alkoxides[637] with carboxylic acids:

$$[Al(OPr^i)_3]_4 + 16RCOOH \xrightarrow{\text{facile}} 4(RCO_2)_2Al(\mu\text{-}OPr^i)_2Al(O_2CR)_2 + 8Pr^iOH$$

$$(2.183)$$

$$[\text{Al}(\text{OPr}^i)(\text{O}_2\text{CR})_2]_2 + 2\text{RCOOH} \xrightarrow{\text{slow}} 2\text{Al}(\text{O}_2\text{CR})_3 + 2\text{Pr}^i\text{OH} \uparrow \qquad (2.184)$$

Similarly, preparation of anhydrous and/or unsolvated lanthanide tris-acetylacetonates remained a challenge till 1965, as the efforts to remove ligated water from $\text{Ln}(\beta\text{-dik})_3.2\text{H}_2\text{O}$ (obtained from aqueous media) led to hydrolyzed products. However, pure tris-β-diketonates of lanthanum,[638] praseodymium, and neodymium[635] could be easily synthesized by the reactions of their alkoxides (methoxides/isopropoxides) with 3 equivalents of acetylacetone (acacH):

$$\text{La}(\text{OMe})_3 + 3\text{acacH} \longrightarrow \text{La}(\text{acac})_3 + 3\text{MeOH} \uparrow \qquad (2.185)$$

$$\text{Ln}(\text{OPr}^i)_3 + 3\text{acacH} \longrightarrow \text{Ln}(\text{acac})_3 + 3\text{Pr}^i\text{OH} \uparrow \qquad (2.186)$$

where Ln = Pr, Nd.

In spite of the fact that a wide variety of structurally interesting mixed ligand-alkoxide derivatives of a large number of metals and metalloids are known, X-ray structural data are available only in a very few cases such as $\text{Ce}_2(\text{OPr}^i)_6(\mu\text{-OC}_2\text{H}_4\text{NMeC}_2\text{H}_4\text{NMe}_2)_2$,[640] $[\text{Ti}\{\text{OCH}(\text{CF}_3)_2\}_2(\text{OEt})_2(\text{HOEt})]_2$,[272] $[\text{Fe}(\text{OMe})_2(\text{OCH}_2\text{Cl})]_{10}$,[641,642] $[\text{Al}(\text{OSiMe}_3)_2(\text{acac})]_2$,[643] $[\text{Al}(\text{OPr}^i)_2(\text{acac})]_3$,[643] $[\text{Al}(\text{OPr}^i)(\text{Et}_2\text{acac})_2]_2$,[643] and $[\text{Sn}(\text{OBu}^t)_3(\text{OAc})(\text{py})]$.[644]

The reactions of metal alkoxides appear to be generally influenced by steric factors. For example, reactions with highly branched alcohols (Bu^tOH, Am^tOH) are generally slower and may be even sterically hindered in some cases as is illustrated[293] by Eqs (2.187)–(2.189):

$$\text{Al}(\text{OEt})_3 + 2\text{Bu}^t\text{OH} \xrightarrow{\text{facile}} \text{Al}(\text{OEt})(\text{OBu}^t)_2 + 2\text{EtOH} \uparrow \qquad (2.187)$$

$$2\text{Al}(\text{OEt})(\text{OBu}^t)_2 + \text{Bu}^t\text{OH} \xrightarrow{\text{slow}} \text{Al}_2(\text{OEt})(\text{OBu}^t)_5 + \text{EtOH} \uparrow \qquad (2.188)$$

$$\text{Al}_2(\text{OEt})(\text{OBu}^t)_5 + \text{Bu}^t\text{OH} \xrightarrow{\text{negligible}} \text{Al}_2(\text{OBu}^t)_6 + \text{EtOH} \qquad (2.189)$$

In the following pages the general reactions of metal alkoxides with a wide variety of reagents are discussed under the following headings, with special reference to the synthesis and salient features of the new products isolated:

(1) Hydrolysis reactions (Section 4.2)
(2) Reactions with alcohols, phenols, and silanols (as well as organic and silyl esters) (Section 4.3)
(3) Reactions with alkanolamines (Section 4.4)
(4) Reactions with β-diketones and β-ketoesters (Section 4.5)
(5) Reactions with carboxylic acids and acid anhydrides (Section 4.6)
(6) Reactions with glycols (including some special types of polyhydroxy reagents) (Section 4.7)
(7) Reactions with oximes and N,N-diethylhydroxylamine (Section 4.8)
(8) Reactions with Schiff bases and β-ketoamines (Section 4.9)
(9) Reactions with thiols (Section 4.10)
(10) Reactions with halogens, hydrogen halides, acyl halides, and metal halides (Section 4.11)

(11) Reactions with unsaturated substrates (Section 4.12)

(12) Reactions with coordinating ligands (Section 4.13)

(13) Reactions with main group organometallic compounds (Section 4.14)

(14) Tischtchenko, Meerwein–Ponndorf–Verley, and Oppenauer oxidation reactions (Section 4.15)

(15) Interactions between two different metal alkoxides (Section 4.16)

(16) Special reactions of metal–metal bonded alkoxides (Section 4.17).

The main aim of this section is to highlight the potential of metal alkoxides as unique synthons for a variety of novel metallo-organic compounds with interesting properties, structures, and applications.

4.2 Hydrolysis Reactions

The high reactivity of metal alkoxides towards water leads to a complex hydrolysis and polymerization chemistry.[3,21,617–619,645] During hydrolysis, the alkoxo ($^-$OR) groups are replaced either by hydroxo (OH^-) or oxo (O^{2-}) ligands. This reaction is influenced by a number of factors such as: (i) the nature of the alkyl (R) group, (ii) the nature of the solvent, (iii) the concentration of the species present in solution, (iv) the water to alkoxide molar ratio, and (v) the temperature. It appears reasonable to suggest that the initial step should involve the coordination of a water molecule through its oxygen to the metal in a facile nucleophilic process as represented by Eq. (2.190), which represents the initial stage of a continued hydrolysis process:

Hydrolysis

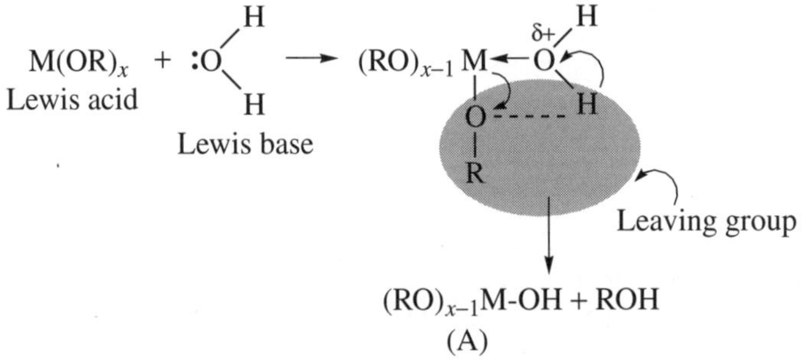

$$(2.190)$$

$$(RO)_{x-1}M\text{-}OH + ROH$$
$$(A)$$

The hydroxo-metal alkoxides (A) tend to react further in two ways (Eqs 2.191 and 2.192) to form the metal oxo-alkoxide derivatives.

Dealcoholation

$$(RO)_{x-1}MO\boxed{H + RO}M(OR)_{x-1} \longrightarrow (RO)_{x-1}MOM(OR)_{x-1} + ROH$$
$$(2.191)$$

Dehydration

$$(RO)_{x-1}MO\boxed{H + HO}M(OR)_{x-1} \longrightarrow (RO)_{x-1}MOM(OR)_{x-1} + H_2O$$
$$(2.192)$$

The above reactions generally proceed further to yield finally the hydroxides (hydrated oxides) and sometimes hydroxy-alkoxides.

Under carefully controlled conditions, however, intermediate metal oxo-alkoxides, $MO_y(OR)_{x-2y}$ may be isolated.[21,617-619,645,646] (See also Chapter 5.)

A detailed mechanistic study of hydrolysis reactions was made by Bradley[617,645] who had isolated a number of metal oxo-alkoxides and assigned plausible structures for such products. There has been a renewal of interest[21,646] in the chemistry of metal oxo-alkoxides (Chapter 5), some of which have been isolated under anhydrous conditions also. Further, these hydrolysis reactions have assumed unprecedented technological significance in view of their importance in the preparation of ceramic materials by the sol–gel process (Chapter 7) which involves controlled hydrolysis of metal alkoxides (or some other precursors).

4.3   Reactions with Alcohols, Phenols, and Silanols (as well as Organic and Silyl Esters)

4.3.1   Alcohol Exchange Reactions

Metal alkoxides (of early transition, lanthanide, actinide, and main group metals, except silicon) react with a variety of primary, secondary, and tertiary alcohols (R′OH) to set up an equilibrium rapidly of the type shown in Eq. (2.193):

$$M(OR)_x + yR'OH \rightleftharpoons M(OR)_{x-y}(OR')_y + yROH \qquad (2.193)$$

There appear to be a number of important factors which influence the facility and extent of substitution in alcoholysis reactions (Eq. 2.193): (i) the solubility of the reactant and product metal alkoxides, (ii) the steric demands of the ligands and the alcohol, (iii) the relative ΔH^0 of ionization of the reactant and product alcohols, (iv) the relative bond strengths of the alkoxides, (v) the presence of strongly bridged and more easily replaceable terminal alkoxo groups, and (vi) the electron-donating and withdrawing nature of the groups attached to oxygen.

Interestingly, the reaction represented by Eq. (2.193) can be pushed to yield homoleptic alkoxides $M(OR')_x$ by fractionating out the liberated alcohol, provided it is more volatile. In some cases a solvent like benzene which forms an azeotrope with the liberated alcohol (EtOH or Pr^iOH) not only facilitates fractionation of the alcohol produced, but also makes it possible to carry out reactions in different stoichiometric ratios of the reactants to obtain the desired heteroleptic alkoxides $M(OR)_{x-y}(OR')_y$. Another advantage of this procedure is the possibility of monitoring the progress of the reaction by estimating the liberated isopropyl alcohol (or ethyl alcohol) by a convenient oxidimetric method.[205,647] The alcohol-interchange reactions have, therefore, been extensively used as synthetic strategies (Section 2.7) for the formation of a variety of new homo- and heteroleptic metal alkoxides.

An interesting observation was reported[61] in the replaceability of only three methoxide groups of insoluble $[Zr(OMe)_4]_n$ with excess tertiary butyl alcohol resulting in the formation of the kinetically and thermodynamically favoured derivative $[(Bu^tO)_3Zr(\mu\text{-}OMe)_2Zr(OBu^t)_3]$ which does not appear to react further with tertiary butyl alcohol (Eq. 2.194)

$$2[Zr(OMe)_4]_n + 6nBu^tOH \longrightarrow n[(Bu^tO)_3Zr(\mu\text{-}OMe)]_2 + 6nMeOH \uparrow \qquad (2.194)$$

B

The product (B) is also obtained in an exothermic equimolar reaction between $Zr(OBu^t)_4$ and MeOH.

For alcohols with highly electron-withdrawing groups such as $R' = CH(CF_3)_2$, only partial substitution[272] is observed when reactions for example, of $Ti(OR)_4$ ($R = Et$ or Pr^i) with $(CF_3)_2CHOH$ were carried out

$$Ti(OR)_4 + 2HOCH(CF_3)_2 \xrightarrow{\text{toluene}} Ti\{OCH(CF_3)_2\}_2(OR)_2(HOR) + ROH \quad (2.195)$$
$$\text{(excess)}$$

In a comparative study of the alcoholysis reactions of titanium alkoxides,[270] the facility of replacement reactions appears to follow the trend: $OBu^t > OPr^i > OEt > OMe$. This type of comparative trend appears to be particularly marked in the alkoxides of some later 3d transition metals.[7,225] For example, primary alkoxides of d^8 nickel(II) do not appear to undergo alcoholysis reactions with secondary or tertiary and other primary alcohols, whereas the reactions of nickel(II) tert-butoxide with primary alcohols are highly facile.

The compound $Mo_2(OPr^i)_6$ shows no tendency to associate but upon partial alcoholysis with MeOH the tetranuclear crystalline compound $Mo_4(OPr^i)_{10}(OMe)_2$ was isolated.[648]

The reactions of $M_2(OBu^t)_6$ ($M = Mo, W$) with less sterically demanding alcohols RCH_2OH ($R = c\text{-}C_4H_7$, $c\text{-}C_5H_9$ $c\text{-}C_6H_{11}$, Pr^i) afford a series of structurally related homoleptic alkoxides $M_4(OCH_2R)_{12}$ according to Eq. (2.196):[649]

$$2M_2(OBu^t)_6 + 12RCH_2OH \xrightarrow{\text{hexane}} M_4(OCH_2R)_{12} + 12Bu^tOH \quad (2.196)$$

Investigations on alcohol-interchange reactions of copper alkoxides with various fluorinated alcohols have produced exciting and diverse results. For example, it has been observed that approximately 50% substitution of tert-butoxo groups occurred in reactions between copper(I) tert-butoxide and a wide variety of fluoroalcohols.[650] Similarly copper(II) methoxide reacts with excess $(CF_3)_2CHOH$ in refluxing benzene to produce an insoluble solid of approximate stoichiometry $[Cu\{OCH(CF_3)_2\}(OMe)]_n$.[175] However, reactions of copper(II) methoxide with fluoroalcohols in diethyl ether in the presence of a variety of amines (pyridine (py), N,N,N',N'-tetraethylethylenediamine (teeda), N,N,N',N-tetramethylethylenediamine (tmeda), $2,2'$-bipyridine (bipy)) yield the corresponding solvated copper(II) fluoroalkoxides (Eq. 2.197):[175]

$$\frac{1}{n}[Cu(OMe)_2]_n + 2R_fOH + L \longrightarrow Cu(OR_f)_2(L) + 2MeOH \quad (2.197)$$

$$(R_f = CH(CF_3)_2; L = (py)_2, bipy, teeda, tmeda.$$
$$R_f = CMe(CF_3)_2; L = (py)_2, bipy, tmeda)$$

Interestingly, reaction of $Cu(OBu^t)_2$ with a moderate excess of $(CF_3)_3COH$ in hydrocarbon solvents affords two primary products: $[Cu_4(\mu\text{-}OBu^t)_6\{OC(CF_3)_3\}_2]$ and $[Cu_3(\mu\text{-}OBu^t)_4\{OC(CF_3)_3\}_2]$, with their relative amounts determined by the stoichiometry of the reactants.[651]

4.3.2 Phenol–Alcohol Exchange Reactions

The characteristic ability of metal alkoxides to undergo facile exchange reactions with hydroxylic reagents has been extensively utilized for preparing homo- and heteroleptic phenoxides of a large number of elements[652] by the general reaction (Eq. 2.198):

$$M(OR)_x + y PhOH \longrightarrow M(OR)_{x-y}(OPh)_y + y ROH \uparrow \qquad (2.198)$$

(M = Y,[155] Pr, Nd,[153] Sm,[155] Gd, and Er;[157] $y = 1, 2, 3$; R = Pr^i; $x = 3$. M = U;[653] $y = 1, 2, 3, 4$; R = Et; $x = 5$. M = Ti,[654] Zr;[655] $y = 4$; R = Bu^t, $x = 4$. M = Ti;[276] $y = 1, 2, 3, 4$; R = Et, Pr^i; $x = 4$. M = VO;[656] $y = 3$; R = Et; $x = 3$. M = Al;[657] $y = 1, 2, 3$; R = Bu^t; $x = 3$. M = Si.[658,659] Ge;[660] $y = 1, 2, 3, 4$; R = Et; $x = 4$)

This method tends to work well with simple phenol or phenols substituted with electron-withdrawing groups (F, Cl, NO_2), owing to the increased acidity of the phenols and their lower volatility relative to alcohols.

In some cases, a control of stoichiometry allows the isolation of mixed alkoxide-phenoxide derivatives.[153,155,157,276,652] In the case of uranium(v) ethoxide, only partial substitution can be achieved even by reacting with an excess of phenol.[653]

Reaction of $Mo_2(OPr^i)_6$ with excess 2,6-dimethylphenol yields only partially substituted product[661] $Mo_2(OPr^i)_2(OC_6H_3Me_2$-2,6$)_4$ for steric reasons.

The important role of steric congestion around the central atom in determining the composition of the final product has also been demonstrated[662–664] during the reactions of other metal alkoxides [Ti(OPr^i)$_4$, Zr(OPr^i)$_4$.Pr^{i}OH, M(OPr^i)$_5$ with M = Nb or Ta, Al(OPr^i)$_3$, Sb(OPr^i)$_3$) with sterically hindered phenols (HOC$_6$H$_3$Me$_2$-2,6; HOC$_6$H$_3$Pr$^i_2$-2,6; HOC$_6$H$_3$Bu$^t_2$-2,6].

It is noteworthy that calix[n]arenes, which are a family of macrocyclic reagents with $n = 4, 5, 6, 7$, or 8 phenolic groups in the ring (2-XII) can also find the metal ions via the deprotonated phenolate ligand.[665,666] For example, treating two equivalents of Ti(OR)$_4$ (R = Pr^i, Bu^t) or Zr(OPr^i)$_4$ with p-$tert$-butylcalix[8]arene (LH$_8$ when $n = 8$) in the presence of a base such as alkali metal (M′ = Li, Na, K) alkoxides or amines (R′NH$_2$) has led to the synthesis[665] of derivatives of the type [M′ or R′NH$_3$]$^+$ [(MOR)$_2$(p-$tert$-butylcalix[8]arene)]$^-$ (M = Ti, Zr). Chisholm et al.[667] have shown that the reaction of $Mo_2(OBu^t)_6$, with monomethylated calix[4]arene ligand (H$_3$L′) in benzene at room temperature affords an interesting derivative $Mo_2[\eta^4$-L′$)_2$.

$$
\begin{bmatrix}
\text{Bu}^t \\
\text{[ring structure]} \\
\text{CH}_2 \\
\text{OH}
\end{bmatrix}_n \quad (n = 4, 5, 6, 7, 8)
$$

(2-XII)

4.3.3 Silanol–Alcohol Exchange Reactions

Like alcohols, silanols have been found to react with metal alkoxides to form the corresponding homo- and heteroleptic metal siloxides (Eq. 2.199).

$$M(OR)_x + y R'_3 SiOH \rightleftharpoons M(OR)_{x-y}(OSiR'_3)_y + y ROH \uparrow \qquad (2.199)$$

In these cases also, the forward reaction could be pushed further by fractionating out the more volatile alcohol (ROH) liberated during the reaction.

The silanolysis reactions appear to be subject to less steric hindrance than the analogous alcoholysis reactions. For example, the final products in the reactions of $Al(OPr^i)_3$ with Me_3COH and Me_3SiOH[668] have been reported to be $Al_2(OPr^i)(OCMe_3)_5$ and $Al_2(OSiMe_3)_6$, respectively. However, the reaction of $Al(OPr^i)_3$ with $HOSi(OBu^t)_3$ led to the formation of $Al_2(OPr^i)_2[OSi(OBu^t)_2]_4$ only.[669]

A complicating factor affecting the reactions of metal alkoxides with silanols is the condensation tendency of silanols to yield hexa-alkyldisiloxanes and water (Eq. 2.200); this may be to some extent avoided by adding the silanol slowly to the reaction mixture.

$$R'_3Si-O\boxed{H+HO}-SiR'_3 \longrightarrow R'_3Si-O-SiR'_3 + H_2O \qquad (2.200)$$

In spite of this difficulty the reactions of a wide range of metal(loid) alkoxides with silanols (Eq. 2.199) under controlled conditions have been successfully employed for the synthesis of a range of metal siloxides such as $Ti(OR)_3\{OSi(OBu^t)_3\}$ (R = Pr^n,[670] Bu^t),[671] $Ti(OPr^i)\{OSi(OBu^t)_3\}_3$,[672] $Ti(OPr^i)_2\{OSi(OBu^t)_3\}_2$.[672]

Attempts to prepare niobium pentakis(trimethylsiloxide) by the reaction of $Nb(OEt)_5$ with Me_3SiOH did not yield the expected product;[6] instead the sublimed (in vacuo) material corresponded in analysis to $[Nb(OSiMe_3)_4]_2O$. As compared to nonvolatile titanium tetrakis(triphenysiloxide), the trimethylsiloxide derivatives of titanium and zirconium as well as tantalum pentakis(trimethylsiloxide) could be purified in vacuo by either distillation or sublimation.[678]

The metal tetrasiloxides differ from the analogous alkoxides in their degree of polymerization and volatility. This is primarily due to the smaller steric effect of trialkylsiloxo compared with tertiary alkoxo groups. Furthermore, trialkylsilanols are more acidic than alcohols owing to the presence of (p-d)π bonding in the Si–O bond, and consequently the trialkylsiloxide derivatives may experience a difference in degree of covalency compared with tertiary alkoxides. However, steric factors appear to offer the most plausible explanation for higher association in metal siloxides.

Metal siloxides are more resistant to hydrolysis than their alkoxide analogues; this salient difference in hydrolytic stability of metal siloxides may be ascribed to the water-repelling property of trialkylsiloxo groups.

As discussed already (Section 2.8), metal alkoxides undergo transesterification reactions with organic as well as silyl esters as illustrated by Eqs (2.201), (2.202), and (2.203):

$$M(OR)_x + yCH_3COOR' \rightleftharpoons M(OR)_{x-y}(OR')_y + yCH_3COOR \qquad (2.201)$$

$$M(OR)_x + yCH_3COOSiR'_3 \rightleftharpoons M(OR)_{x-y}(OSiR'_3)_y + yCH_3COOPr^i \qquad (2.202)$$

where M = a wide range of s-, p-, d-, and f-block metals.

$$M(OPr^i)_5 + 5CH_3COOC_6H_5 \longrightarrow M(OC_6H_5)_5 + 5CH_3COOPr^i \qquad (2.203)$$

where M = Nb, Ta.

4.4 Reactions with Alkanolamines

Alkanolamines represent another class of alcohols derived from ammonia (NH_3) by the substitution of one, two, or three hydrogen atoms by $-(CRR')_nOH$ groups, where R, R' = H or an alkyl and n is generally two. Depending upon the number of hydrogen atom(s) of NH_3 replaced by $-(CRR')_nOH$ group(s), the derivatives are known as mono-, di-, and tri-alkanolamines.

Monoalkanolamines such as $HOCH_2CH_2NH_2$, $HOCH_2CH_2NHMe$, and $HOCH_2CH(Me)NH_2$ with active hydrogen atoms at both the nitrogen and oxygen ends of the molecule are potential bidentate chelating ligands.[680–687] Nitrogen could be expected to form a dative bond more readily, but the higher acidity of the hydroxyl hydrogen tends to make it more reactive towards the formation of an O-bonded substitution product. The aminoalkoxides that result from deprotonation show an increased tendency both to chelate (Fig. 2.12f) and to form additional bonds (Fig. 2.12b–e). On chelation (Fig. 2.12f), the amino protons tend to become more reactive and may form derivatives of the type $MOCH_2CH_2NR'R$ (R', R = H, H; R, alkyl or aryl) under suitable experimental conditions.

The hydrogen atom(s) on nitrogen of alkanolamines can be activated to form a cyclic product (Fig. 2.12g) particularly with metals which have a high affinity for nitrogen.

An interesting feature of primary and secondary aminoalkoxides of the type $[M(OCH(R)CH_2NMe_2)_x]_n$ is their low molecularity and higher volatility compared to those of 2-methoxyethoxide analogues. For example, $Cu(OCHRCH_2NMe_2)_2$ (R = H, Me) is both monomeric and volatile, whereas copper(II) alkoxyalkoxides are oligomeric or even polymeric.[285] The same observation is valid for zinc derivatives.[688] This difference could be ascribed to the higher donating chelating tendency of NR_2 compared to OR substituents on the alcohol.

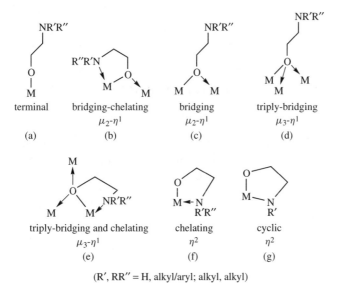

$(R', RR'' = H, alkyl/aryl; alkyl, alkyl)$

Figure 2.12 Ligating possibilities of monoalkanolamines.

4.4.1 Monoalkanolamines

A comparative study of the reactivity of metal alkoxides towards ethanolamine, $HOCH_2CH_2NH_2$, indicated distinct differences between the reactivity of the hydroxyl and amino groups. For example, reactions of a large number of metal (Sm, Gd, Er, Yb, Y, Ti, V, Nb, U, Fe, Ni, B, Si, Sb) alkoxides with ethanolamines proceed with the metallation of the hydroxyl proton only as shown by Eq. (2.204):

$$M(OR)_x + yHO(CH_2)_mNH_2 \xrightarrow{benzene} M(OR)_{x-y}[O(CH_2)_mNH_2]_y + yROH \uparrow$$

(2.204)

(M = Sm,[689] Gd, Er, Yb, Y;[148,156] $x = 3$; $y = 1, 2$ or 3; $m = 2$; R = Pr^i.
M = Ti,[623,624] $x = 4$, $y = 1, 2, 3$ or 4; $m = 2$; R = Et, Bu^n, Pr^i. M = OV;[691]
$x = 3$; $y = 1, 2, 3$; $m = 2$, R = Et. M = Nb,[625] Ta;[626] $x = 5$; $y = 1$; $m = 2$;
R = Et. M = U;[627] $x = 5$; $y = 1, 2, 3, 4$ or 5; $m = 2$; R = Et. M = Fe;[283]
$x = 3$; $y = 1, 2$ or 3; $m = 2$, R = Pr^i. M = Ni;[692] $x = 2$; $y = 1$ or 2; $m = 2$;
R = Me, Pr^i. M = B,[693] Al,[694] Ga;[185] $x = 3$, $y = 1, 2$ or 3; $m = 2$; R = Pr^i.
M = B,[695] $x = 3$; $y = 3$; $m = 2$ or 3; R = Me. M = Si (in the presence of Na);[696]
$x = 4$; $y = 4$; $m = 2$; R = Et. M = Sb;[627] $x = 3$; $y = 1, 2$, or 3; $m = 2$; R = Et)

Similar reactions with different types of N-alkyl- and N,N-dialkyl-aminoalcohols (R'OH) have yielded derivatives which are generally soluble (in organic solvents) and volatile. Some typical reactions are illustrated by Eqs (2.205):

$$M(OR)_x + yR'OH \xrightarrow{benzene} M(OR)_{x-y}(OR')_y + yROH \uparrow$$

(2.205)

(M = Ti,[623,624] Zr;[697] $x = 4$; $y = 1, 2, 3$, or 4; R' = CH_2CH_2NHMe, CH_2CH_2NHEt,
$CH_2CH_2NMe_2$, $CH_2CH_2NEt_2$, $CHMeNMe_2$; R = Pr^i. M = Cu;[698] $x = 2$; $y = 2$;
R' = $CH_2CH_2NMe_2$, $CHMeCH_2NMe_2$,[698,699] $CH_2CH_2NMeCH_2CH_2NMe_2$;[698]
R = Me. M = B;[695] $x = 3$; $y = 3$; R' = CH_2CH_2NHMe, $CH_2CH_2NMe_2$,
$CH_2CH_2NEt_2$, $(CH_2)_3NMe_2$, $(CH_2)_3NEt_2$; R = Me. M = Al;[700] $x = 3$; $y = 1, 2$, or 3;
R' = $(CH_2)_2NHMe$, $(CH_2)_2NHEt$, $(CH_2)_2NEt_2$, $(CH_2)_2NMe_2$, $CHMeCH_2NEt_2$;
R = Pr^i)

By contrast, with alkoxides of niobium,[625] tantalum,[626] germanium,[701] tin,[621] and antimony[627] both the hydroxylic and the amino group protons of ethanolamine undergo replacement reactions to form cyclic derivatives (Eq. 2.206):

$$M(OR)_x + yHOCH_2CH_2NH_2 \xrightarrow{benzene} (RO)_{x-2y}M\left(\begin{array}{c} O-CH_2 \\ | \\ N-CH_2 \\ | \\ H \end{array}\right)_y + 2yROH \uparrow$$

(2.206)

(M = Nb, Ta; $x = 5$; $y = 2$; R = Et (soluble). M = Ge; $x = 4$; $y = 1$;

$R = Pr^i$ (soluble). $M = Sn; x = 4; y = 1$ (soluble), 2 (insoluble);
$R = Pr^i$. $M = Sb; x = 3; y = 1$ (soluble); $R = Et, Pr^i$)

Interestingly, when the reaction of $Ge(OPr^i)_4$ with $HOCH_2CH_2NH_2$ was carried out in 1:2 molar ratio, the product was not $\overline{Ge(OCH_2CH_2NH)_2}$; instead, $Ge(OCH_2CH_2NH)(OCH_2CH_2NH_2)(OPr^i)$ was obtained, and if the reactants were in 1:3 molar ratio, $\overline{Ge(OCH_2CH_2NH)}(OCH_2CH_2NH_2)_2$ was the main product.

In these cases, the increasing reactivity of the amino group in aminoalkoxides may be due to intramolecular coordination of the type:

$$\overset{\diagup}{\underset{\diagdown}{=}}M\overset{O}{\underset{\underset{H_2}{N}}{\diagup}}C_2H_4, \text{ rendering the}$$

amino hydrogens more reactive.

Tin alkoxides form cyclic derivatives with alkanolamines in equimolar ratio and the same is the case with germanium alkoxides but silicon alkoxides show reactivity restricted to the hydroxyl group only. The poor reactivity of silicon alkoxides may be attributed to the fact that silicon prefers to involve its d-orbitals by π-bonding whereas Sn and Ge prefer σ-bonding.[702] This may be reflected in the order of estimated M–O bond energies of Si–O, Ge–O, and Sn–O, which are 112, 85, and 82 kcal mol^{-1}, respectively.[472,702,703]

A comparative study of the reactions of $M(OR)_4$ ($M = Si, Ge, Sn$) with alkanolamines indicated the following order of reactivity: $Sn > Ge \gg Si$.

The non-involvement of the $-NH_2$ group in the formation of intramolecular $N \to Si$ bonding has been substantiated by infrared[516,696,704] and proton magnetic resonance[696,705] studies.

4.4.2 Dialkanolamines

Reactions of metal alkoxides with dialkanolamines afford a variety of structurally interesting products depending on the stoichiometric ratios of the reactants as illustrated by Eqs (2.207)–(2.217):

$$M(OPr^i)_2 + 2(HOC_2H_4)_2NH \xrightarrow{\text{benzene}} M\{OC_2H_4)NH(C_2H_4OH)\}_2 \downarrow + 2Pr^iOH$$
$$\mathbf{C}$$

(2.207)

where $M = Mg, Ca, Sr, Ba$.[706,707]

$$M(OPr^i)_3 + (HOC_2H_4)_2NH \xrightarrow{\text{benzene}} (Pr^iO)M\{OC_2H_4)_2NH\} + 2Pr^iOH \quad (2.208)$$

where $M = Al,$[694] $Ga,$[185] $Sb,$[708] $Fe,$[283] $Y, Gd, Er, Yb.$[148,156]

$$M(OPr^i)_3 + 2(HOC_2H_4)_2NH$$

$$\xrightarrow{\text{benzene}} M[\{(OC_2H_4)_2NH\}\{(OC_2H_4)NH(C_2H_4OH)\}] \downarrow + 3Pr^iOH \quad (2.209)$$
$$\mathbf{D}$$

where M = Al, Sb.[708]

$$2M(OPr^i)_3 + 3(HOC_2H_4)_2NH \xrightarrow{\text{benzene}} M_2[(OC_2H_4)_2NH]_3 + 6Pr^iOH \qquad (2.210)$$

where M = Al.[694]

$$M(OR)_4 \cdot xROH + (HOC_2H_4)_2NH \xrightarrow{\text{benzene}} (RO)_2M[(OC_2H_4)_2NH] + (2+x)ROH$$
$$(2.211)$$

$$(M = Ti,^{709,710} R = Bu^n, x = 0. \; M = Sn;^{621} R = Pr^i, x = 1.$$
$$M = Ge;^{711} R = Et, Pr^i; x = 0)$$

$$M(OR)_4 \cdot xROH + 2(HOC_2H_4)_2NH \xrightarrow{\text{benzene}} M[(OC_2H_4)_2NH]_2 + (4+x)ROH$$
$$(2.212)$$

$$(M = Ge;^{711} R = Et, Pr^i; x = 0. \; M = Sn;^{621} R = Pr^i; x = 1)$$

$$M(OR)_4 \cdot xROH + 3(HOC_2H_4)_2NH$$

$$\xrightarrow{\text{benzene}} M\{(OC_2H_4)_2N(C_2H_4OH)\}\{(OC_2H_4)NH(C_2H_4OH)_2\} + (4+x)ROH$$
$$\mathbf{E}$$
$$(2.213)$$

$$(M = Ti,^{708} R = Pr^i, x = 0, M = Zr, Sn,^{708} R = Pr^i, x = 1)$$

$$R'_2Ge(OR)_2 + R''N(CH_2CH_2OH)_2 \longrightarrow R'_2Ge(OC_2H_4)_2NR'' + 2ROH \qquad (2.214)$$

where R′ = Me, Et; R″ = Me, Pr^n.[712]

$$M(OEt)_5 + (HOC_2H_4)_2NH \xrightarrow{\text{benzene}} (EtO)_3M\{(OC_2H_4)_2NH\} + 2EtOH \qquad (2.215)$$

where M = Nb,[625] Ta.[626]

$$M_2(OEt)_{10} + 4(HOC_2H_4)_2NH$$

$$\xrightarrow{\text{benzene}} M_2(OEt)\{(OC_2H_4)_2NH\}_3\{(OC_2H_4)_2N\} + 9EtOH \qquad (2.216)$$

where M = Nb,[625] Ta.[626]

$$M(OPr^i)_5 + 3(HOC_2H_4)_2NH$$

$$\xrightarrow{\text{benzene}} M\{(OC_2H_4)_2NH\}_2\{(OC_2H_4)NH(C_2H_4OH)\} + 3Pr^iOH \qquad (2.217)$$
$$\mathbf{F}$$

where M = Nb, Ta[713]

The products **C, D, E**, and **F** have been shown (Chapter 3, Section 2.5) to be useful precursors for the synthesis of interesting types of heterometallic alkoxide systems.

4.4.3   Trialkanolamines

The tetradentate trialkanolamines have been shown to react with a variety of s-, p-, d-, and f-block metals to produce thermally robust derivatives (Scheme 2.6), some

$xN(CH_2CHR'OH)_3$

$\xrightarrow{La(OPr^i)_3}$ $La[\{(OC_2H_4)_2N(C_2H_4OH)\}\{(OC_2H_4)N(C_2H_4OH)_2\}] + 3Pr^iOH$
$R' = H;[716]$ $x = 2$

$\xrightarrow{Ti(OPr^i)_4}$ $(Pr^iO)Ti[(OCHR'CH_2)_3N]_x + 3Pr^iOH$
$R' = H,[717,718]$ Me;[715] $x = 1$

$\xrightarrow{Ti(OPr^i)_4}$ $Ti[(OC_2H_4)_2N(C_2H_4OH)]_2 + 4Pr^iOH$
$R' = H;[713]$ $x = 2$

$\xrightarrow{M(OPr^i)_4.Pr^iOH}$ $(Pr^iO)M[(OC_2H_4)_3N]_2 + 4Pr^iOH$
$M = Zr,[713]$ Sn;[621,708] $R' = H; x = 1$

$\xrightarrow{M(OPr^i)_4.Pr^iOH}$ $M[(OC_2H_4)_2N(C_2H_4OH)]_2 + 5Pr^iOH$
$M = Zr,[713]$ Sn;[621,708] $R' = H; x = 2$

$\xrightarrow{M(OEt)_5}$ $(EtO)_2M[(OCH_2CHR')_3N] + 3EtOH$
$M = Nb; R' = H,[625]$ Me;[715] $x = 1$
$M = Ta; R' = H,[626]$ Me;[715] $x = 1$

$\xrightarrow{Al(OR)_3}$ $Al[(OC_2H_4)_3N] + 3ROH$
$R = Pr^i, Bu^t;[694,719,720]$ $x = 1$

$\xrightarrow{R''Si(OEt)_3}$ $R''Si[(OC_2H_4)_3N] + 3EtOH$
$R'' = OEt, Ph;[72]$ $x = 1$

$\xrightarrow{Ge(OR)_4}$ $(RO)Ge[(OC_2H_4)_3N] + 3ROH$
$R = Et, Pr^i;[711]$ $x = 1$

$\xrightarrow{Ge(OR)_4}$ $Ge[(OC_2H_4)_2N(C_2H_4OH)]_2 + 4ROH$
$R = Et, Pr^i;[711]$ $x = 2$

$\xrightarrow{OV(OPr^n)_3}$ $OV[(OCH_2CHR')_3N] + 3Pr^nOH$
$R' = Me, Bu^t, Ph, c\text{-hexyl};[715]$ $x = 1$

$\xrightarrow{RSn(OEt)_3}$ $RSn[(OC_2H_4)_3N] + 3EtOH$
$R = Me; Et, Bu^n, Ph;[722]$ $x = 1$

$R''Ge(OR)_3$
$\xrightarrow{}$ $R''Ge[(OC_2H_4)_3N] + 3ROH$
$x = 1; R = Et$ or $Pr^i; R' = H; R'' = Me; Bu^n, Ph,$ etc.[712]

Scheme 2.6

of which are biologically active[714] and/or useful in a variety of synthetic (Chapter 3, Section 2.6) and catalytic systems.[715]

4.5 Reactions of β-Diketones and β-Ketoesters

The enolic form, $R'C(O)CH=C(OH)R''$ of a β-diketone (R' and R'' represent an alkyl or aryl group or their fluoroanalogue) or β-ketoester ($R' =$ alkyl or aryl, and

$R'' =$ alkoxy or aryloxy group) contains a reactive hydroxy group, which is prone to react readily with metal alkoxides to yield alcohol(s) and a variety of interesting (in terms of compositions, structures, volatility, reactivity, and applicability) homo- and heteroleptic derivatives of a wide variety of metals and metalloids.[629,723,724] Typical general reactions are shown in Eqs (2.218) and (2.219):

$$M(OR)_x + yR'COCH_2COR'' \longrightarrow M(OR)_{x-y}(R'COCHCOR'')_y + yROH \uparrow$$
(2.218)

$$M(OR)_x + yR'COCH_2COOR'' \longrightarrow M(OR)_{x-y}(R'COCHCOOR'')_y + yROH \uparrow$$
(2.219)

The importance of this synthetic route has been shown in the preparation of anhydrous tris-β-diketonates of lanthanides, the hydrated forms of which obtained from aqueous solutions tend to decompose into hydroxy-derivatives. By using isopropoxides or ethoxides as starting materials and removing the liberated alcohol by fractionation as a lower boiling azeotrope with a solvent such as benzene, volatile stable intermediate products may be obtained. The alkoxide groups in these mixed derivatives are much more reactive than the β-diketonate ligands, as illustrated by Eqs (2.220)–(2.223):[725–727]

$$Al(OR)_3 + xR'COCH_2COR'' \xrightarrow{\text{benzene}} (RO)_{3-x}Al(OC(R')CHCOR'')_x + xROH \uparrow$$
(2.220)

where $R = Et$, Pr^i and R', $R'' = CH_3$; or $R' = CH_3$, $R'' = C_6H_5$ or $R' = CH_3$, $R'' = OC_2H_5$; $x = 1$–3.

$$Al(OPr^i)_{3-x}(\beta\text{-dik})_x + (3-x)Bu^tOH$$

$$\xrightarrow{\text{benzene}} Al(OBu^t)_{3-x}(\beta\text{-dik})_x + (3-x)Pr^iOH \uparrow \qquad (2.221)$$

$$Al(OPr^i)_{3-x}(\beta\text{-dik})_x + (3-x)R_3SiOH/R_3SiOOCCH_3$$

$$\xrightarrow{\text{benzene}} Al(OSiR_3)_{3-x}(\beta\text{-dik})_x + (3-x)Pr^iOH \uparrow /CH_3COOPr^i \uparrow \qquad (2.222)$$

where $R = CH_3$, C_2H_5; β-dik = acetylacetonate; $x = 1, 2$.

$$Al(OPr^i)_{3-x}(\beta\text{-dik})_x + (3-x)LH \longrightarrow Al(\beta\text{-dik})_x(L)_{3-x} + (3-x)Pr^iOH \uparrow$$
(2.223)

where LH is another β-diketone, β-ketoester or a ligand such as 8-hydroxyquinoline.

Aluminium tris-β-diketonates are monomeric in nature and have[728] an octahedral geometry. The crystal structure for the [(Et$_2$acac)$_2$Al(μ-OPr^i)$_2$Al(Et$_2$acac)$_2$](Et$_2$acac = 3,5-heptanedione) dimer shows[726] two octahedral aluminium atoms with isopropoxy bridging. The crystal structure of trimeric {Al(OPr^i)$_2$(acac)}$_3$ can be represented[729] by [(acac)$_2$Al(μ-OPr^i)$_2$Al(acac)(μ-OPr^i)$_2$Al(OPr^i)$_2$] with an interesting distribution of the ligand moieties on the three aluminium atoms. It has been shown[729] that the analogous {Al(OSiMe$_3$)$_2$(acac)}$_2$ is dimeric and instead of a symmetrical {(acac)(OSiMe$_3$)Al(μ-OSiMe$_3$)$_2$Al(OSiMe$_3$)(acac)}-type structure with both aluminium atoms in the pentacoordinated state, it contains an unsymmetrical structure of the type [(acac)$_2$Al(μ-OSiMe$_3$)$_2$Al(OSi-Me$_3$)$_2$] in which one of the aluminium atoms is hexacoordinated and the other aluminium atom is tetracoordinated with four (Me$_3$SiO) groups. In conformity with the earlier conclusion(s) on the effect of steric factors on the molecular association of analogous derivatives, $[Al(OSiPh_3)_2(acac)]$ becomes monomeric with a tetrahedral structure.[729]

Dhammani et al.[730-732] have carried out the reactions shown in Eqs (2.224)–(2.226), characterizing similar products by colligative, IR and NMR (^{1}H, ^{13}C and ^{27}Al) measurements:

$$\{Al(CH_3COCHCOR)(OPr^i)_2\}_2 + Ph_3SiOH$$

$$\xrightarrow{-Pr^iOH} [(CH_3COCHCOR)_2Al(\mu\text{-}OPr^i)_2Al(OSiPh_3)(OPr^i)] \qquad (2.224)$$

$$\{Al(CH_3COCHCOR)(OPr^i)_2\}_2 + 2Ph_3SiOH$$

$$\xrightarrow{-2Pr^iOH} [(CH_3COCHCOR)_2Al(\mu\text{-}OPr^i)_2Al(OSiPh_3)_2] \qquad (2.225)$$

$$\{Al(CH_3COCHCOR)(OPr^i)_2\}_2 + HOSiPh_2OH$$

$$\xrightarrow{-2Pr^iOH} [(CH_3COCHCOR)_2Al(\mu\text{-}OPr^i)_2Al(OSiPh_2O)] \qquad (2.226)$$

where R = CH_3, OC_2H_5, C_6H_5.

Stepwise reactions of isopropoxides of other tervalent metals (lanthanides,[733-735] gallium,[736] and antimony[737]) have also been carried out by similar procedures and a number of isopropoxide-β-diketonates as well as tris-β-diketonates have been characterized by physico-chemical techniques.

By contrast, homoleptic tetrakis-β-diketonates of titanium and tin(IV) could not be prepared by this route; reactions of $Ti(OR)_4$ and $Sn(OR)_4$ (with R = Et, Pr^i) with excess of β-diketones/β-ketoesters yield bisalkoxide bis-β-diketonate derivatives only.

The reactions of titanium alkoxides with β-diketones and β-ketoesters in 1:1 and 1:2 molar ratios have been investigated by several workers;[738-740] the 1:2 products were characterized as monomeric derivatives, but there has been a difference of opinion about the dimeric or monomeric nature of 1:1 products which awaits further investigation.

In another investigation,[741] $Ti(OEt)_2(acac)_2$ and $Ti(OPr^i)_2(acac)_2$ were shown to be monomeric volatile products, in which the alkoxide component(s) have been shown to be highly reactive. In addition to facile replacement by hydroxylic reagents, $Ti(OEt)_2(acac)_2$ was found to react with 2 mol HCl, to yield a monomeric volatile product $TiCl_2(acac)_2$ which could be converted into $Ti(OEt)_2(acac)_2$ by reaction with EtOH in the presence of anhydrous NH_3; this led to a correction[742-745] of the well-quoted trimeric nature of the covalent $TiCl_2(acac)_2$ as $\{Ti(acac)_3\}_2TiCl_6$[746] on the basis of its reaction with $FeCl_3$ to yield a product of the formula $\{Ti(acac)_3\}\{FeCl_4\}$, a species the actual nature of which awaits further elucidation by more refined physico-chemical investigations (crystal structure or more refined NMR investigations).

In reactions of titanium alkoxides with benzoylacetone,[747] methylacetoacetate[748] and ethylacetoacetate, bis-β-diketonate or ketoester derivatives were the final products even when excess of these ligands was used. The nonreplaceability of the third or fourth alkoxy groups with β-diketones or ketoesters may most probably be due to the preferred coordination number of 6 for titanium in the bis-derivatives. The trialkoxide monomethylacetoacetate disproportionated to the bis-derivative and tetra-alkoxide, when heated under reduced pressure, whereas bis-derivatives distilled out under similar conditions with slight decomposition.[748] These alkoxide β-diketonates and β-ketoester derivatives undergo alcoholysis reactions to yield new alkoxide β-diketonate or ketoesterate derivatives:

$$(RO)_{(4-x)}Ti(L)_x + (4-x)ROH \longrightarrow (R'O)_{(4-x)}Ti(L)_x + (4-x)ROH \quad (2.227)$$

Earlier workers[749,750] assigned the *trans*-configuration for hexacoordinated bisacetyl-acetonate derivatives. However, on the basis of proton magnetic resonance studies on these derivatives, Bradley and Holloway[751,752] ruled out the possibility of the *trans*-configuration in favour of the *cis*-configuration. These authors found that all of the species Ti(acac)$_2$(OR)$_2$ (R = Me, Et, Pr^i, Bu^t, etc.) existed only in the *cis* (optically active) form over a wide temperature range. Activation energies for intramolecular exchange of ligands in these fluxional molecules showed that steric hindrance of the alkoxo group increased the energy of activation but did not promote the *trans*-form.

Bharara *et al.*[753,754] investigated the reactions of titanium and zirconium alkoxides with fluoro-β-diketones (hexafluoroacetylacetone, benzoyltrifluoroacetone and thenoyl-trifluoroacetone) and observed that in the case of titanium the replacement is limited to disubstitution only whereas zirconium alkoxides form tetrakis-fluoro-β-diketonates. This difference in behaviour of titanium and zirconium alkoxides may be due to the greater atomic radius of zirconium than titanium (Zr = 1.45 Å, Ti = 1.32 Å) as well as to the greater tendency of zirconium to exhibit coordination numbers larger than six.

The hexafluoroacetylacetonate derivatives of titanium[753] are volatile products where-as the zirconium analogues[754] are nonvolatile and therefore could be crystallized from hot benzene. These mixed derivatives interchange their alkoxo groups on treatment with excess of alcohols and show monomeric behaviour in refluxing benzene.

Contrary to earlier observations,[755] Saxena *et al.*[756] have reported the successful synthesis of even the tetrakis derivatives in the step-wise reactions of zirconium isopropoxide with acetylacetone, benzoylacetone, dibenzoylmethane and ethylaceto-acetate. Ebulliometric molecular weight determinations in benzene revealed the mono-meric behaviour of all zirconium β-diketonate derivatives except the triisopropoxide mono-β-diketonate derivatives, which show dimeric behaviour. The bidentate char-acter of the β-diketonate ring in zirconium compounds has been adduced from infrared spectra which show the carbonyl bands in the lower region (1580–1601 cm^{-1}), and the hexacoordinated dimeric structure: (β-dik)(OPr^i)$_2$Zr(μ-OPr^i)$_2$Zr(OPr^i)$_2$(β-dik), with isopropoxide bridging has been suggested for [{Zr(OPr^i)$_3$(β-dik)}$_2$] derivatives.

The above suggested structure was finally confirmed[757] for [Zr$_2$(OPr^i)$_6$(tmhd)$_2$] and [Hf$_2$(OPr^i)$_6$(tmhd)$_2$] in 1999 by X-ray crystallography (where tmhd is 2,2,6,6-tetramethylheptane-3,5-dione).

Similar to the observations in the case of titanium alkoxides, Mehrotra and Gupta[759] have found that tin tetraisopropoxide reacts with acetylacetone, benzoylacetone, and dibenzoylmethane in 1:1 and 1:2 molar ratios to form mono- and bis-β-diketonates, respectively, and further replacement was not observed even when the β-diketones were taken in higher molar ratios. Gaur *et al.*[760] on the other hand observed that alkyltin triisopropoxides reacted with acetylacetone in 1:3 molar ratio to form alkyltin tris-acetylacetonates. The dibenzoylmethane and benzoylacetone could yield only monoiso-propoxide bis-β-diketonates as the final products and the non-replaceability of the third isopropoxo group has been ascribed to steric factors. Monoalkyltin diisopropoxide monoacetylacetonate or monobenzoylacetonate gave a degree of polymerization of about 1.5 in boiling benzene, which suggests the presence of dimeric species. The structure: $(\beta\text{-dik})(OPr^i)BuSn(\mu\text{-}OPr^i)_2SnBu(OPr^i)(\beta\text{-dik})$ has been suggested for the dimeric form, in which tin tends to attain the hexacoordinate state through isopropoxide bridging.

The synthesis of a number of tin(II) β-diketonates has been carried out[761] by the reactions of tin(II) dimethoxide with the β-diketones:

$$Sn(OMe)_2 + 2(R'COCH_2COR'') \longrightarrow Sn(OCR'\!=\!CHCOR'')_2 + 2MeOH \uparrow \quad (2.228)$$

where R', R'' = Me; R' = Me, R'' = CF$_3$; R', R'' = CF$_3$.

Whitley[762] showed that Ta$_2$(OEt)$_{10}$ reacted with excess benzoylacetone (bzac) to give Ta(OEt)$_2$(bzac)$_3$ which may contain eight-coordinated tantalum. Mehrotra and co-workers[763–767] have also investigated the reactions of niobium and tantalum pentaethoxides with acetylacetone, benzoylacetone, dibenzoylmethane, methyl and ethyl acetoacetates and ethylbenzoylacetate and only observed partial replacement in these cases. It has been shown that the above reactions proceeded quite smoothly up to the formation of the bis-derivative, and slowed down considerably beyond that stage, but could be forced to trisubstitution for some β-diketone or ketoester molecules; similar results have been reported[768,769] in the reactions of U(OMe)$_5$ with β-diketones:

$$M(OEt)_5 + x\,acacH \longrightarrow (EtO)_{(5-x)}M(acac)_x + x\,EtOH \quad (2.229)$$

$$M(OEt)_5 + x\,CH_3COCH_2COOR \longrightarrow (EtO)_{(5-x)}M(OCCH_3CHCOOR)_x + x\,EtOH \quad (2.230)$$

where M = Nb, Ta, U; R = Me, Et; x = 1–3.

These derivatives readily interchange their ethoxo groups with higher acohols. The non-replaceability of the last two ethoxo groups by β-diketonate ligands may be due either to steric factors or to the inability of niobium and tantalum to increase their coordination number beyond eight in these derivatives. These derivatives are generally monomeric in boiling benzene and could be distilled unchanged *in vacuo*. Holloway[563] has studied the variable temperature proton magnetic resonance of Ta(OR)$_4$(acac) derivatives which have been shown to be fluxional molecules.

4.6   Reactions with Carboxylic Acids and Acid Anhydrides

The reactions of metal alkoxides with carboxylic acids and acid anhydrides[628] can be represented by Eqs (2.231) and (2.232):

$$M(OR)_x + y\,R'COOH \longrightarrow M(OR)_{x-y}(OOCR')_y + y\,ROH \quad (2.231)$$

$$M(OR)_x + y\,(R'CO)_2O \longrightarrow M(OR)_{x-y}(OOCR')_y + y\,R'COOR \quad (2.232)$$

The reaction of silicon ethoxide with acetic anhydride was investigated as early as 1866 by Friedel and Crafts[770] who reported the preparation of triethoxysilicon mono-acetate. The reaction has been reinvestigated by Post and Hofrichter[771] as well as by Narain and Mehrotra.[772]

In view of the difficulties[628] in the preparation of aluminium tricarboxylates from aqueous systems, the first attempt to prepare a metal carboxylate from its alkoxide appears to have been made in 1932 by McBain and McLatchie[773] from the reaction of aluminium *sec*-butoxide with palmitic acid. However, the end product obtained was only dipalmitate mono-butoxide leading Gray and Alexander[636] to doubt the very existence of tricarboxylates of aluminium. In view of his experience in the reactions of aluminium alkoxides with ramified alcohols,[293] Mehrotra was able to demonstrate[637] that the reaction can be slowly driven to completion by continuous fractionation of isopropyl alcohol formed during the third stage of reaction azeotropically with benzene:

$$Al(OPr^i)_3 + 2RCOOH \xrightarrow[\text{fast}]{\text{benzene}} \tfrac{1}{2}[RCOO)_2Al(\mu\text{-}OPr^i)_2Al(OOCR)_2 + 2Pr^iOH \uparrow$$

$$+ RCOOH \Big| slow$$

$$Al(OOCR)_3 + Pr^iOH \uparrow \qquad (2.233)$$

Rai and Mehrotra[774] later showed that the reaction of $Al(OBu^t)_3$ with carboxylic acids yields the tricarboxylates more readily, as in this case the intermediate product $Al(OBu^t)(OOCR)_2$ is monomeric and continues to react in a facile manner with the third mole of carboxylic acids. In fact the reactions of metal alkoxides with carboxylic acids are facilitated thermodynamically since the carboxylate moiety tends to bind the metal in a bidentate mode, thus:

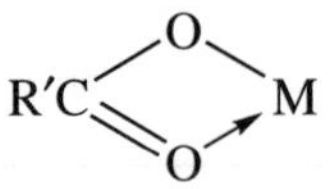

In view of Mehrotra's unique success,[637] the reactions of alkoxides of a number of other metals and metalloids, Ti,[775–779] Zr,[780–783] Si,[628,770–772,784–786] Ge,[787] Ga,[185] Ln,[148,156,788] Fe(III),[283,641,642] V,[789] Nb,[790] Ta,[790] and Sb[791] have been studied extensively. In most cases, however, homoleptic carboxylates could not be obtained, as the intermediate heteroleptic alkoxide carboxylates $M(OR)_{x-y}(OOCR')_y$ tend to decompose with increasing value of y yielding oxo-carboxylates and alkyl esters by intramolecular coupled with intermolecular reactions as illustrated below in a few typical cases.

The reactions of titanium alkoxides with carboxylic acids can be represented by Eqs (2.234)–(2.239):

$$Ti(OR)_4 + R'COOH \longrightarrow Ti(OR)_3(OOCR') + ROH \qquad (2.234)$$

$$Ti(OR)_4 + 2R'COOH \longrightarrow Ti(OR)_2(OOCR')_2 + 2ROH \qquad (2.235)$$

$$Ti(OR)_4 + 3R'COOH \longrightarrow Ti(OR)(OOCR')_3 + 3ROH \qquad (2.236)$$

$$O{=}Ti(OOCR')_2 + R'COOR \qquad (2.237)$$

$$Ti(OR)(OOCR')_3 + Ti(OR)_2(OOCR')_2 \longrightarrow Ti_2O(OR)_2(OOCR')_2 + R'COOR \qquad (2.238)$$

$$Ti_2O(OR)_2(OOCR')_4 + 2R'COOH \longrightarrow Ti_2O(OOCR')_6 + 2ROH \qquad (2.239)$$

where $R = C_2H_5$, $(CH_3)_2CH$; $R' = CH_3$, $C_4H_9^n$, $C_{17}H_{35}$, $C_{21}H_{43}$.

The products up to titanium dialkoxide dicarboxylates are quite stable. Higher carboxylate products tend to decompose in a variety of ways depending on the relative concentrations, solvent, temperature, stirring, and other conditions. Interestingly, a crystalline hexameric product has been isolated in the 1:2 molar reaction of $Ti(OBu^n)_4$ and CH_3COOH and has been X-ray crystallographically characterized[792] as $Ti_6(\mu_2\text{-}O)_2(\mu_3\text{-}O)_2(\mu_2\text{-}OBu^n)_2(OBu^n)_6(OOCCH_3)_8$:

$$2Ti_3(OBu^n)_{12} + 12CH_3COOH$$

$$\longrightarrow Ti_6(OOCCH_3)_8(OBu^n)_8O_4 + 12Bu^nOH + 4CH_3COOBu^n \quad (2.240)$$

The reactions in 1:2 molar ratio of titanium ethoxide or isopropoxide with acetic anhydride are exothermic and yield crystalline $Ti(OR)_2(OOCCH_3)_2$. However, owing to side-reactions of the type indicated above (Eq. 2.240), the final end-product is always the basic acetate, $(CH_3COO)_6Ti_2O$ irrespective of the excess of acid anhydride employed. In fact, titanium tetra-acetate has not been isolated so far even in the reactions of titanium tetrachloride with acetic acid or anhydride.

The reactions of zirconium isopropoxide with fatty acids (caproic, lauric, palmitic, and stearic) were found to be similar to those of titanium isopropoxide, except that the tris-product, $Zr(OPr^i)(OOCR)_3$ was more stable than the titanium analogue and could be isolated at lower temperatures in 1:3 molar reactions. However, it reacted very slowly with a further mole of the acid, and the tetracarboxylate so formed reacted readily with $Zr(OPr^i)(OOCR)_3$ yielding the basic carboxylate $(RCOO)_6Zr_2O$:

$$Zr(OPr^i)(OOCR)_3 + RCOOH \longrightarrow Zr(OOCR)_4 + Pr^iOH \quad (2.241)$$

$$Zr(OOCR)_4 + (Pr^iO)Zr(OOCR)_3 \longrightarrow (RCOO)_3Zr\text{-}O\text{-}Zr(OOCR)_3 + RCOOPr^i$$
$$(2.242)$$

The product $Zr(OOCR)_4$, which could be obtained in the reaction of $ZrCl_4$ with carboxylic acid, was also shown to react with $(Pr^iO)Zr(OOCR)_3$ as shown above.

The reactions of gadolinium, erbium, yttrium, and ytterbium isopropoxides with acids or acid anhydrides yield[148,156,788] mono-, di-, and tri-carboxylates depending upon the stoichiometric ratios of the reactants employed. Similarly, the reactions of ferric isopropoxide with carboxylic acids were found[283] to be straightforward. Interestingly, the preparation of a decameric product $[Fe(OMe)_2(O_2CCH_2Cl)]_{10}$ with a fascinating molecular ferric wheel[641] type structure has opened the possibility[642] of the synthesis of other interesting structures by the reactions of ferric alkoxides with appropriate carboxylic acids.

4.7 Reactions with Glycols

Glycols are dihydroxy alcohols (with $pK_a \sim 2$) and have been found to be highly reactive towards metal alkoxides to form homoleptic and mixed alkoxide-glycolates of metals.[61,301,793–817] The recent interest in glycolate derivatives stems from their ability both to chelate and bridge centres[21,706,708,799–802,804] as well as to the reduced steric demand of a glycolate group in comparison to two alkoxo groups. The combined effect of all these factors allows a metal to increase its coordination number and leads to the formation of higher aggregates in comparison to the parent metal alkoxide. An interesting feature of metal glycolates is the variety in their structural types (Fig. 2.13), which is greatly influenced by (i) the nature of the metals and their oxidation states, (ii) the type of glycolate moiety, and (iii) the stoichiometric ratios of reactants. In the context of the above possibilities, a study of the glycolate (particularly the mixed alkoxide-glycolate) derivatives assumes special interest. Out of the different types of derivatives those with chelated hydroxyl group(s) (Figs 2.13b, d, h, i, m, n) offer the possibility of functioning as precursors for heterometallic alkoxides by their reactivity towards alkoxides of other metals (Section 2.4).

Reeves and Mazzeno[793] as early as 1954 studied cryoscopically the reactions of $Ti(OBu^t)_4$ with various glycols in tertiary butyl alcohol and observed that the products were not simple monomers but were associated species having the titanium to glycol ratio 2:3, 4:6, 2:2, and 3:6 $(Ti_2(OBu^t)_2(glycolate)_6)$. Yamamoto and Kambara[794] carried out the reactions of titanium tetra-alkoxides with excess of 2-methylpentane-2,4-diol and isolated a product of the type $Ti(OCHMeCH_2CMe_2O)_2.(HOCHMeCH_2CMe_2OH)$.

Extensive work on the reactions of alkoxides of different metals and metalloids such as alkaline earth metals,[706] lanthanides,[148–156] titanium,[61,708,793–802] zirconium,[61,708,796–802] uranium,[803] vanadium,[804] niobium[805,819] tantalum,[806] iron,[807] boron,[808] aluminium,[809] silicon,[810–813] germanium,[814] tin,[815] antimony,[301] selenium,[816] and tellurium[817] with a wide variety of glycols has been carried out by Mehrotra et al.

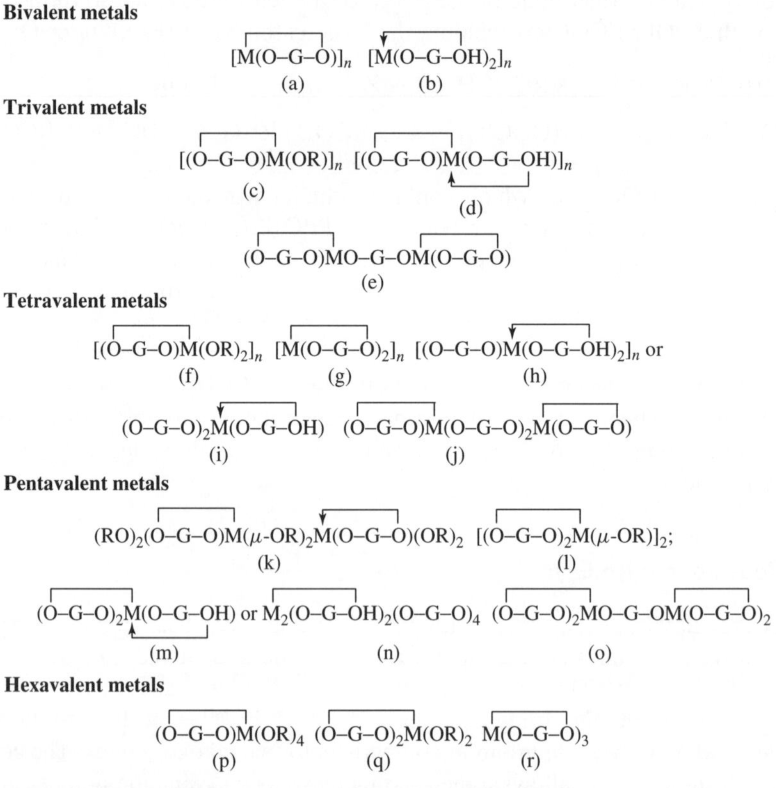

Bivalent metals

$$[M(O\text{–}G\text{–}O)]_n \quad [M(O\text{–}G\text{–}OH)_2]_n$$
$$\text{(a)} \qquad\qquad \text{(b)}$$

Trivalent metals

$$[(O\text{–}G\text{–}O)M(OR)]_n \quad [(O\text{–}G\text{–}O)M(O\text{–}G\text{–}OH)]_n$$
$$\text{(c)} \qquad\qquad\qquad \text{(d)}$$

$$(O\text{–}G\text{–}O)MO\text{–}G\text{–}OM(O\text{–}G\text{–}O)$$
$$\text{(e)}$$

Tetravalent metals

$$[(O\text{–}G\text{–}O)M(OR)_2]_n \quad [M(O\text{–}G\text{–}O)_2]_n \quad [(O\text{–}G\text{–}O)M(O\text{–}G\text{–}OH)_2]_n \text{ or}$$
$$\text{(f)} \qquad\qquad \text{(g)} \qquad\qquad \text{(h)}$$

$$(O\text{–}G\text{–}O)_2M(O\text{–}G\text{–}OH) \quad (O\text{–}G\text{–}O)M(O\text{–}G\text{–}O)_2M(O\text{–}G\text{–}O)$$
$$\text{(i)} \qquad\qquad\qquad\qquad \text{(j)}$$

Pentavalent metals

$$(RO)_2(O\text{–}G\text{–}O)M(\mu\text{-}OR)_2M(O\text{–}G\text{–}O)(OR)_2 \quad [(O\text{–}G\text{–}O)_2M(\mu\text{-}OR)]_2;$$
$$\text{(k)} \qquad\qquad\qquad\qquad \text{(l)}$$

$$(O\text{–}G\text{–}O)_2M(O\text{–}G\text{–}OH) \text{ or } M_2(O\text{–}G\text{–}OH)_2(O\text{–}G\text{–}O)_4 \quad (O\text{–}G\text{–}O)_2MO\text{–}G\text{–}OM(O\text{–}G\text{–}O)_2$$
$$\text{(m)} \qquad\qquad\qquad\qquad \text{(n)} \qquad\qquad\qquad \text{(o)}$$

Hexavalent metals

$$(O\text{–}G\text{–}O)M(OR)_4 \quad (O\text{–}G\text{–}O)_2M(OR)_2 \quad M(O\text{–}G\text{–}O)_3$$
$$\text{(p)} \qquad\qquad \text{(q)} \qquad\qquad \text{(r)}$$

(where O–G–O = a diolate group such as OC_2H_4O, $OCHMeCHMeO$, $OCMe_2CMe_2O$, $OCHMeCH_2CMe_2O$, $OCMe_2(CH_2)_2CMe_2O$)

Figure 2.13 Structural variety in binary and mixed alkoxide-glycolate derivatives of metals.

and other workers according to the general reaction shown in Eq. (2.243):

$$M(OR)_x + (y+z)HO\text{—}G\text{—}OH$$

$$\xrightarrow{\text{benzene}} M(OR)_{x-2y-z}(O\text{—}G\text{—}O)_y(O\text{—}G\text{—}OH)_z + (2y+z)ROH \quad (2.243)$$

where x is the valency of the metal, y and z are integers and their values are dependent on n: if $n=3$ then $y=1$ and $z=0$ or 1.

Plausible structures, suggested mainly from molecular weight measurements, are beginning to be elucidated[21,801,802] by X-ray crystallography.

In addition to the structures displayed in Fig. 2.13, other possibilities also exist as depicted by Crans et al.[804] in the case of oxovanadium 1,2-diolates. It has also been pointed out[804] that the structural differences of the glycol ligands result in very different geometries and demonstrate subtle factors affecting the coordination chemistry of such systems. These studies substantiate the very fine balance between coordination geometries and the respective glycolate ligands in their metal complexes.

Reactions in 1:2 molar ratios of trihydric alcohols such as 1,1,1-tris(hydroxymethyl)-ethane (THME-H_3) and 1,1,1-tris(hydroxymethyl)propane (THMP-H_3) with group 4 metal (Ti, Zr) triisopropoxides in toluene/THF solvent afford X-ray crystallographically characterized tetranuclear products:[818] (THME)$_2$M$_4$(OPr^i)$_{10}$ and (THMP)$_2$M$_4$(OPr^i)$_{10}$; these illustrate the propensity for forming a variety of bridging and chelating modes.

More recently,[819] an X-ray crystallographically characterized niobium(V) alkoxo species derived from THME-H_3 namely [(μ-THME)Nb(OEt)$_2$]$_2$ has been reported to be formed by the reaction of Nb(OEt)$_5$ with THME-H_3.

4.8   Reactions of Oximes and *N,N*-Diethylhydroxylamine

The oxime group ($>$C$=$N—OH) which may be considered to be derived from oxy-imine is amphiprotic with slightly basic nitrogen and a mildly acidic hydroxyl group. Diethylhydroxylamine (Et$_2$N—OH) is a weaker base than NH$_3$ or NEt$_3$, owing to the electron-withdrawing effect of its hydroxyl group.

Both oximes[634,635,820,821] and *N,N*-diethylhdroxylamine[634,635] are ambidentate with the possibility of coordination through oxygen alone or through both nitrogen and oxygen, illustrated in Fig. 2.14.

The reactions of metal alkoxides with a wide range of oximes (Eq. 2.244) provide a convenient and versatile route for the synthesis of a variety of interesting homo- and heteroleptic oximate derivatives:[634,635]

$$M(OR)_x + yR'R''C=NOH \xrightarrow{\text{benzene}} M(OR)_{x-y}(ON=CR'R'')_y + yROH \quad (2.244)$$

where M = titanium,[822-825] zirconium,[826] niobium,[827,828] tantalum,[827,829] boron,[822] aluminum,[830] silicon,[822,831] germanium,[822] tin,[708] arsenic,[832] and antimony.[833] R = Et or Pr^i; x is the valency of the metal; $y=1,2,\ldots x$; R$'$ = H, alkyl, or aryl; R$''$ = alkyl or aryl.

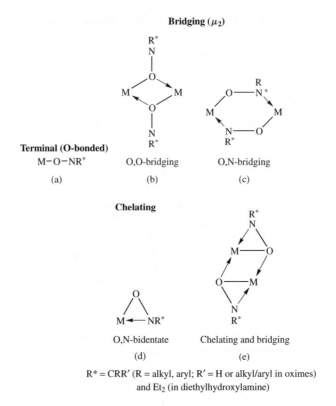

Figure 2.14 Ligating modes of simple oximes and *N,N*-diethylhydroxylamines.

Reactions of *N,N*-diethylhydroxylamine with alkoxides of metals and metalloids as shown in Eq. (2.245) have also been cleaner and quite general.

$$M(OR)_x + yEt_2NOH \xrightarrow{\text{benzene}} M(OR)_{x-y}(ONEt_2)_y + yROH \qquad (2.245)$$

where R = Et or Pr^i.

This reaction (Eq. 2.245) has been employed to prepare binary and mixed alkoxide-*N,N*-diethylaminooxide derivatives of titanium,[823] niobium,[829] tantalum,[829] boron,[834] aluminium,[834] arsenic,[832] and antimony.[833]

The molecular complexity of ketoximate derivatives (similar to the alkoxide analogues) is influenced by the chain length and ramification of the alkyl groups attached to the imino carbon atom.[635] It may be pointed out that generally the same alkyl groups when attached to the imino carbon atom appear to offer more steric hindrance than when attached to the carbinol carbon atom. By contrast, the aldoximate derivatives show a tendency towards higher molecular association depending upon the nature of the alkyl or aryl group attached to the imino carbon atom.

Homo- and heteroleptic *N,N*-diethylaminooxides of aluminium and titanium exhibit molecular complexities of the order of 2.0–2.5 and 1–1.5, respectively. Similar derivatives of niobium, tantalum, boron, arsenic, and antimony are monomeric in boiling benzene.

Interesting structures for oxime and *N,N*-diethylhydroxylamine derivatives of metals and metalloids[634,635] have been suggested, mainly on the basis of molecular weight measurements and spectroscopic studies, but none of these have been so far elucidated by X-ray crystallography.

For example, MONMON six-membered ring structures have been suggested[835] for $Me_2MON=CMe_2$ (M = Al, Ga, In, Tl). However, mass spectral studies of trimethyltin cyclohexanone oximate gave evidence for the dimeric units having four-membered cyclic bridges[836] Sn〈O,O〉Sn rather than the six-membered ring Sn〈O—N, N—O〉Sn In view of the variety in the ligating modes of oximes and hydroxylamines as shown in Fig. 2.12, X-ray crystallographic studies of homoleptic derivatives and mixed ligand-alkoxide compounds assume considerable importance and may be expected to reveal some exciting structural features, some of which have already been established for the products:

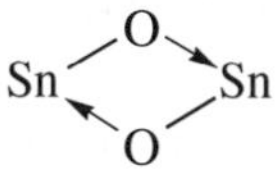

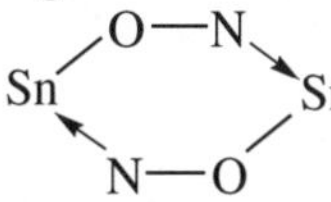

$[\{\eta^5\text{-}C_5H_5)_2Gd(\mu\text{-}\eta^2\text{-}ON=CMe_2)\}_2]$,[837] $(\eta^5\text{-}C_5H_5)(OC)_2Mo(\eta^2\text{-}ON=CR_2)$,[838]   and $(OC)_4Fe(\mu\text{-}NO)(\mu\text{-}NR)Fe(CO)_4$ (where R represents $=CMe_2$).[839]

N-Methylhydroxylamine (MeNHOH) acts as a bidentate ligand through both N and O atoms to form three-membered chelate rings (Fig. 2.12d), which has been demonstrated by X-ray crystallography.[840,841] *N,N*-Diethylhydroxylamine and *N,N*-di-*t*-butylhydroxylamine have also been shown to bind molybdenum(IV)[842] and palladium(II),[843] respectively, in this manner.

4.9   Reactions with Schiff Bases and β-Ketoamines

Similar to reactions of metal alkoxides with mono- and bidentate hydroxylic reagents described in earlier sections, their reactions with Schiff bases and β-ketoamines offer a variety of synthetic and structural possibilities, as illustrated in the following brief account.

Schiff bases involve essentially the azomethine group in addition to some functional groups such as —OH, —SH, —COOH, present in the *ortho* position of the phenyl group bonded either to azomethine carbon or nitrogen or to both.[843] Schiff bases derived from β-diketones and related ligands are known as β-ketoamines. The Schiff bases (including β-ketoamines) are obtained by the condensation of reagents such as salicylaldehyde, 2-hydroxy-1-naphthaldehyde, 2-hydroxy acetophenone, 1-hydroxy-2-acetonaphthone, or β-diketones and related ligands with primary amines, diamines, hydrazines, monoalkanolamines, or 2-aminopyridine. The resulting ligands bind the metals through the available NO, NNO, ONNO, ONS donor sites in the ligand and depending upon the number of functional groups and coordinating centres, these may be classified as (i) monofunctional bidentate, (ii) bifunctional tridentate, and (iii) bifunctional tetradentate ligands (Fig. 2.15), and so on.

Although many routes for the synthesis of metal complexes of Schiff bases and β-ketoamines are available,[844–846] the facile reactivity of metal alkoxides has been utilized for the synthesis of homo- and heteroleptic Schiff base and β-ketoamine derivatives of metals and metalloids with advantage. The most interesting results concern the Schiff bases and β-ketoamines shown in Fig. 2.15, and other ligands obtained by suitably modifying them.

The general reaction of metal alkoxides with monofunctional bidentate (Figs 2.15a–c, designated as SBH) and bifunctional tri- and tetradentate (Figs 2.15d–f, designated as SBH$_2$) ligands can be represented by Eqs (2.246) and (2.247).

$$M(OR)_x + ySBH \xrightarrow{\text{benzene}} M(OR)_{x-y}(SB)_y + yROH \qquad (2.246)$$

$$M(OR)_x + ySBH_2 \xrightarrow{\text{benzene}} M(OR)_{x-2y}(SB)_y + 2yROH \qquad (2.247)$$

where R is generally Et or Pr^i, and x is the valency of the metal or metalloid.

This synthetic procedure has been employed to prepare a wide variety of Schiff base and β-ketoamine derivatives of lanthanides: La,[847–851] Pr,[847–851] Nd,[847,849–851] Sm,[852] Er,[847] Yb;[850] transition metals: Ti,[853–863] Zr,[853,858,862–867] Hf,[868] Nb,[863,869–874] Ta;[870,872,873,875,876] and main group metal(loid)s: B,[877–879] Al,[859,863,878–886] Ga,[847,862,887,888] As,[879] and Sb.[879,889]

It is pertinent to mention the salient features of the reactions represented by Eqs (2.246) and (2.247) and their products. These reactions offer attractive possibilities for preparing derivatives with the metal in interesting coordination environments. For example, reactions of titanium alkoxides with bifunctional tridentate Schiff base ligands (SBH$_2$)[855–858] like o-HOC$_6$H$_4$CH=NCH$_2$(CH$_2$)$_n$OH ($n = 1$ or 2) afford dialkoxide derivatives of the type Ti(OR)$_2$(SB), which exhibit a dimeric octahedral structure involving bridging when R is sterically compact, but monomeric with sterically bulky R such as Bu^t.[858]

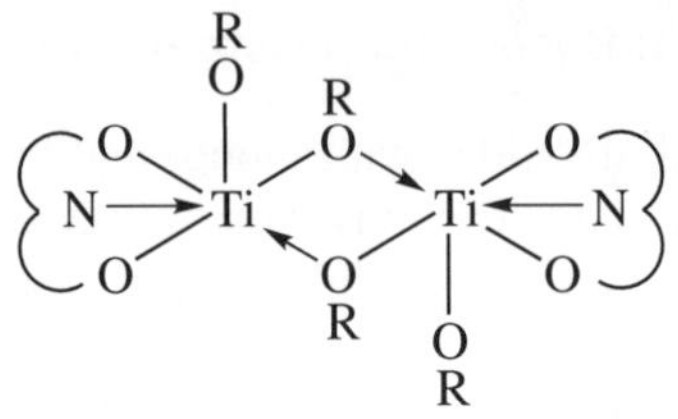

Equimolar reaction of Ti(OPr^i)$_4$ with a bifunctional tetradentate Schiff base ligand like salenH$_2$ (N,N'-ethylenebis(salicylideneimine)) affords[858] a six-coordinate monomeric derivative:

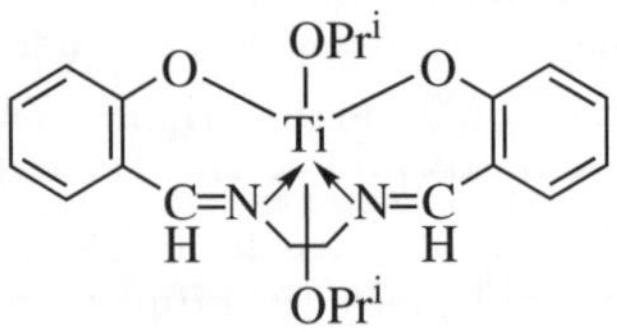

Similar reactions of Al(OPr^i)$_3$ with bifunctional tri- and tetradentate β-ketoamines[847] afford dimeric and monomeric species, respectively in which aluminium atoms are in a five-coordinated environment.

In general, Schiff base and β-ketoamine derivatives of metals and metalloids exhibit variations in the degree of molecular association depending upon their empirical formulae, the valency and covalent radius of the metal, as well as the ligand type.

Monofunctional bidentate

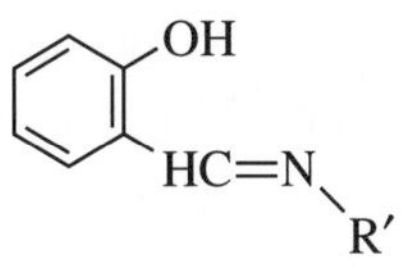

(R′ = alkyl or aryl)
HsalNR′

(a)

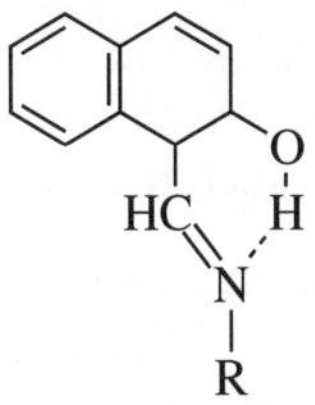

(b)

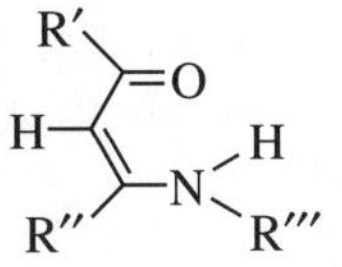

(R′ = R″ = Me, Bu^t, Ph, CF$_3$
R′, R″ = Me, CF$_3$; Me, Ph;
R‴ = alkyl or aryl)

(c)

Bifunctional tridentate

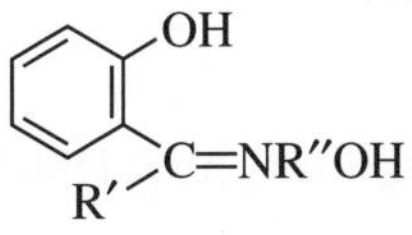

(R′ = H or Me; R″ = C$_2$H$_4$, CH$_2$CHMe)

(d)

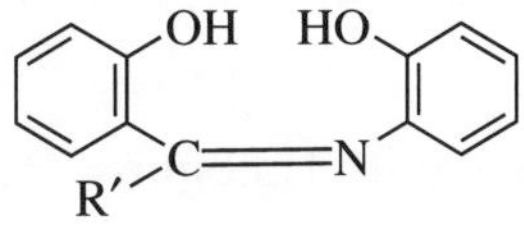

(R′ = H or Me)

(e)

Bifunctional tetradentate

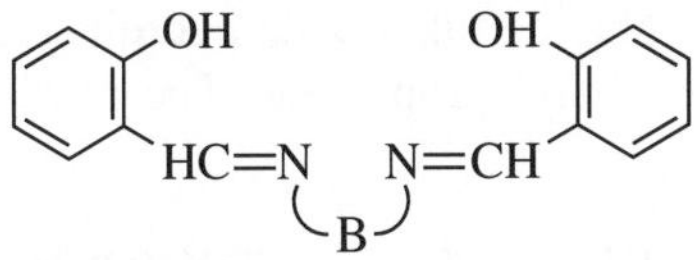

(B = CH$_2$CH$_2$(salen H$_2$), CH$_2$CHMe
(salpnH$_2$), o-C$_6$H$_4$ (salphen H$_2$))

(f)

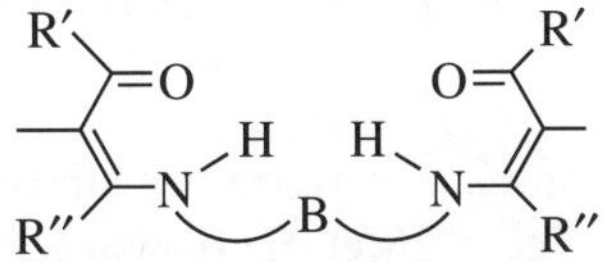

(R′ = R″ = Me, Bu^t, Ph, CF$_3$;
R′, R″ = Me, CF$_3$; Me, Ph;
B = CH$_2$CH$_2$, CH$_2$CHMe, C$_6$H$_4$)

(g)

Figure 2.15 Some of the types of mono- and bi-functional bi- and tri-dentate Schiff bases and β-ketoamines whose reactions with metal alkoxides have been investigated.

Another noteworthy feature is that homoleptic derivatives of a wide variety of mono-functional bidentate Schiff bases and β-ketoamines with a number of metals (Nb, Ta, Al, Sb) may be prepared by the alkoxide route. However, attempted preparations of analogous derivatives of Ti and Zr have not been successful.

4.10　Reactions with Thiols

Although the reactions (Eq. 2.248) of thiols are quite facile with alkoxides of germanium,[890] tin,[891] and antimony,[892] the alkoxides of titanium,[893] zirconium,[893] aluminium,[893] and silicon[893] do not undergo such reactions:

$$M(OR)_x + xR'SH \rightleftharpoons M(SR')_x + xROH \tag{2.248}$$

Qualitatively this difference might be understood on the basis of Pearson's concept of hard and soft acids and bases.

The convenient preparation of thiolates of titanium[894] and uranium[213] from their respective dialkylamides, but not from alkoxide analogues, demonstrates the oxophilic nature of these metals in their higher oxidation states.

4.11　Reactions with Halogens, Hydrogen Halides, Acyl Halides, and Metal Halides

4.11.1　Reactions with Halogens

The reactions of metal alkoxides with halogens were investigated by Nesmeyanov et al.[895] who observed that the reaction between titanium tetra-n-butoxide and chlorine or bromine yielded the dihalide dibutoxide coordinated with one mole of alcohol as $TiX_2(OBu)_2.BuOH$ and n-butyl propionate. The reaction may be represented by Eqs (2.249)–(2.252):

$$Ti(OCH_2R)_4 + X_2 \longrightarrow TiX_2(OCH_2R)_2 + 2RCH_2O \tag{2.249}$$

$$2RCH_2O \longrightarrow RCH_2OH + RCHO \tag{2.250}$$

$$TiX_2(OCH_2R)_2 + RCH_2OH \longrightarrow TiX_2(OCH_2R)_2.RCH_2OH \tag{2.251}$$

$$2RCHO \longrightarrow RCH_2COOR \tag{2.252}$$

where $R = C_3H_7$.

It was observed that the treatment of ferric trimethoxide with excess of chlorine at ambient temperature yielded $FeCl_3.2MeOH$. However, when the above reaction was carried out at $-72°C$, only partial replacement occurred, yielding ferric monochloride dimethoxide.[210]

Jones et al.[379] observed that uranium tetraethoxide reacted with bromine to yield uranium monobromide tetraethoxide which was then used for the preparation of the pentaethoxide (Section 2.10.1)

$$2U(OEt)_4 + Br_2 \longrightarrow 2U(OEt)_4Br \tag{2.253}$$

4.11.2　Reactions with Hydrogen Halides

In view of an interesting gradation observed (Section 2.4) in the reactivity of tetrahalides of silicon, titanium, zirconium, and thorium with ethyl and isopropyl alcohols, the reverse reactions of alkoxides of these elements with hydrogen chloride (bromide) were investigated by Mehrotra[896,897] who demonstrated that the reaction products of metal alkoxides and hydrogen halides were essentially the same as those obtained in the reactions involving the metal chloride and an alcohol:

$$M(OR)_x + yHX \xrightarrow{\text{benzene}} M(OR)_{x-y}(X)_y + yROH \tag{2.254}$$

$$Ti(OR)_4 + 2HX \longrightarrow Ti(OR)_2X_2.ROH + ROH \tag{2.255}$$

where R = Et, Pr^i; X = Cl, Br.

$$2Zr(OR)_4 + 5HX \longrightarrow Zr(OR)X_3.2ROH + Zr(OR)_2X_2.2ROH + ROH \tag{2.256}$$

$$Th(OR)_4 + 4HX \longrightarrow ThX_4.4ROH \tag{2.257}$$

where R = Et, Pr^n, Pr^i; X = Cl, Br.

Gilman and co-workers[898] have also shown that interaction of uranium pentaethoxide and hydrogen chloride in appropriate molar ratios of reactants in ether afforded the products: $UCl(OEt)_4$, $UCl_2(OEt)_3$, and $UCl_3(OEt)_2$.

In contrast to the non-reactivity of $Si(OR)_4$ towards hydrogen halides, germanium[295] and tin tetra-alkoxides[899] show high reactivity on interaction with HCl as shown by Eqs (2.258) and (2.259):

$$Ge(OR)_4 + 4HCl \xrightarrow{\text{benzene}} GeCl_4 + 4ROH \tag{2.258}$$

$$2Sn(OR)_4 + 7HCl \longrightarrow SnCl_3(OR).ROH + SnCl_4.2ROH + 4ROH \tag{2.259}$$

An equimolar reaction of $As(OMe)_5$ and HCl or HF gives $AsX(OMe)_4$ (X = F, Cl);[900] the fluoro derivative was stable *in vacuo* below 30°C whereas the chloro analogue appeared to be unstable and decomposed ($AsCl(OMe)_4 \rightarrow AsO(OMe)_3 + MeCl$) to give arsenyl trimethoxide and methyl chloride.

In contrast to the partial reactions illustrated by Eq. (2.254), complete substitution has been shown to occur to give metal halides that are known to form only alcohol adducts when treated with alcohols:

$$Th(OPr^i)_4 + 4HCl \xrightarrow{\text{benzene}} ThCl_4.4Pr^iOH \tag{2.260}$$

$$Ln(OPr^i)_3 + 3HCl \xrightarrow{\text{benzene}} LnCl_3.3Pr^iOH \tag{2.261}$$

where Ln = a lanthanide.[102]

$$Fe(OPr^i)_3 + 3HCl \longrightarrow FeCl_3.2Pr^iOH + Pr^iOH^{[210]} \tag{2.262}$$

$$Te(OEt)_4 + 4HCl \longrightarrow TeCl_4.3EtOH + EtOH^{[901]} \tag{2.263}$$

4.11.3 Reactions with Acyl Halides

The reactions of metal alkoxides with acyl halides may be illustrated by Eq. (2.264):

$$M(OR)_x + yCH_3COX \xrightarrow{\text{benzene}} M(OR)_{x-y}X_y.zCH_3COOR + (y-z)CH_3COOR \tag{2.264}$$

where R = Et, Pr^i; X = Cl, Br; $y = 1, 2, 3, 4 \ldots x$; $z = 0, 0.5, 1$.

This type of facile reactivity was first observed in 1936 by Jennings *et al.*[902] who synthesized titanium monochloride triethoxide and dichloride diethoxide by the reactions of titanium tetraethoxide and acetyl chloride in the appropriate molar ratios.

Reactions similar to those represented by Eq. (2.264) have been observed for primary and secondary alkoxides of beryllium,[903] yttrium and lanthanides,[148,156,904,905] titanium and zirconium,[647,670,906-911] vanadium,[912] niobium,[913] tantalum,[914] uranium,[915] iron,[916] aluminium,[917] gallium,[918] silicon,[919] germanium,[920] as illustrated by the following equations in a few typical cases:

$$Be(OEt)_2 + 2CH_3COX \longrightarrow BeX_2.2CH_3COOEt \qquad (2.265)$$

$$Zr(OPr^i)_4.Pr^iOH + CH_3COX \longrightarrow ZrCl(OPr^i)_3.Pr^iOH + CH_3COOPr^i \quad (2.266)$$

$$Zr(OPr^i)_4 + xCH_3COX$$
$$\longrightarrow Zr(OPr^i)_{4-x}X_x.yCH_3COOPr^i + (x - y)CH_3COOPr^i \qquad (2.267)$$

where $y = 0, 0.5, 1, 2$ respectively when $x = 1, 2, 3, 4$.

$$Al(OEt)_3 + xCH_3COX \longrightarrow Al(OEt)_{3-x}X_x.yCH_3COOEt + (x - y)CH_3COOEt \qquad (2.268)$$

where $y = 0, 0.5, 1.5$ respectively when $x = 1, 2, 3$.

$$HSi(OPr^i)_3 + xCH_3COCl \longrightarrow HSi(OPr^i)_{3-x}Cl_x + xCH_3COOPr^i \qquad (2.269)$$

All the above reactions have been found to be quite facile and have been mostly carried out with metal isopropoxides and also with ethoxides in some cases. The 1:1 molar reaction of the alcoholate, $Zr(OPr^i)_4.Pr^iOH$ with CH_3COCl showed that the zirconium isopropoxide moiety is more reactive than the coordinated isopropyl alcohol molecule.

In an early paper,[913] the formation of products of lower chloride:zirconium ratios, as compared to the moles of acetyl chloride employed with zirconium tetra-*tert*-butoxide, was ascribed to steric effects. However, in a careful re-examination, the reactions between metal *tert*-butoxides and acetyl halides have been found[923-926] to follow an entirely different course. For example, the reaction between zirconium tertiary butoxide and acetyl chloride was slow, and complete replacement of even one mole of tertiary butoxo group was not achieved; the equimolar reaction product had the average composition, $ZrCl_{0.7}(OBu^t)_{3.3}$. On the basis of further reaction of zirconium chloride tertiary butoxide with acetyl chloride it appeared that a side reaction occurred and instead of higher chloride alkoxides, mixed-alkoxide acetates were formed:

$$Zr(OBu^t)_4 + CH_3COCl \xrightarrow{slow} ZrCl(OBu^t)_3 + CH_3CO_2Bu^t \qquad (2.270)$$

$$ZrCl(OBu^t)_3 + xCH_3COCl \longrightarrow ZrCl_{(1+x)}(OBu^t)_{(3-x)} + xCH_3CO_2Bu^t \qquad (2.271)$$

$$ZrCl(OBu^t)_3 + xCH_3COCl \longrightarrow ZrCl(OCOCH_3)_x(OBu^t)_{(3-x)} + xBu^tCl \qquad (2.272)$$

The above reaction in 1:6 molar ratio afforded zirconium monochloride triacetate $ZrCl(OAc)_3$. The formation of acetate derivatives of the tertiary butoxide was confirmed by the reaction of zirconium tetrachloride with excess of tertiary butylacetate, which yielded zirconium tetraacetate containing a small proportion of chloride as impurity $[ZrCl_{0.2}(OAc)_{3.8}]$.

Contrary to the facile straightforward reactions of ethoxides and isopropoxides of different metals according to Eq. (2.268), an entirely different course of reaction, resulting in the final formation of metal acetate, can be further illustrated by the reaction of aluminium tertiary butoxide with acetyl chloride. In this case, the first stage of the reaction is fast, but the aluminium monochloride di-*tert*-butoxide formed initially reacts[926] with tertiary butylacetate also formed during the reaction, to produce the corresponding mixed alkoxide-acetate:

$$Al(OBu^t)_3 + xCH_3COCl \longrightarrow Al(OBu^t)_{3-x}Cl_x + xCH_3COOBu^t \quad (2.273)$$

$$AlCl_x(OBu^t)_{3-x} + yCH_3COOBu^t \longrightarrow Al(OBu^t)_{3-x}(OCOCH_3)_yCl_{x-y} + yBu^tCl$$
$$(2.274)$$

where $x = 1-3$; $y = 0$ or $< x$.

This assumption was confirmed by treating aluminium trichloride with tertiary butylacetate, whereby aluminium triacetate was finally obtained:

$$AlCl_3 + 3CH_3COOBu^t \longrightarrow Al(OCOCH_3)_3 + 3Bu^tCl \quad (2.275)$$

4.11.4 Reactions with Metal Halides

The reactions between titanium tetra-alkoxides and titanium tetrachloride (excess) at low temperatures lead to the deposition of less soluble titanium trichloride mono-alkoxides:[909,927]

$$Ti(OR)_4 + \underset{(excess)}{3TiCl_4} \longrightarrow 4TiCl_3(OR) \quad (2.276)$$

In the case of zirconium, crystalline chloride-isopropoxide complexes[913] have been obtained according to the reactions illustrated by Eqs (2.277) and (2.278):

$$Zr(OPr^i)_4.Pr^iOH + ZrCl_4.2MeCO_2Pr^i$$
$$\longrightarrow ZrCl_2(OPr^i)_2.Pr^iOH + ZrCl_2(OPr^i)_2.MeCO_2Pr^i + MeCO_2Pr^i \quad (2.277)$$
$$Zr(OPr^i)_4Pr^iOH + ZrCl_2(OPr^i)_2.Pr^iOH \longrightarrow 2ZrCl(OPr^i)_3.Pr^iOH \quad (2.278)$$

These reactions are not confined to metal chloride/metal alkoxide systems, but are also applicable to organometal chloride/organometal alkoxide combinations as demonstrated by Eqs (2.279)–(2.282):

$$xMeTiCl_3 + yMeTi(OR)_3 \xrightarrow{CH_2Cl_2} (x+y)MeTiCl_x(OR)_y{}^{928} \quad (2.279)$$

where $R = Et$, Pr^i; $x = 2$, $y = 1$ or $x = 1$, $y = 2$; the dichloro derivatives are less stable.

$$2RSnCl_3 + RSn(OPr^i)_3 \longrightarrow 3RSnCl_2(OPr^i)^{210} \quad (2.280)$$

$$RSnCl_3 + 2RSn(OPr^i)_3 \longrightarrow 3RSnCl(OPr^i)_2{}^{210} \quad (2.281)$$

$$R_2SnCl_2 + R_2Sn(OMe)_3 \longrightarrow 2R_2Sn(OMe)Cl^{210} \quad (2.282)$$

4.12   Reactions with Unsaturated Substrates

4.12.1   Insertion Reactions into M—OR

4.12.1.1   With OCO, R'NCO, or R'NCNR'

An interesting feature of metal alkoxides is the tendency to react stoichiometrically with some isoelectronic unsaturated molecules like carbon dioxide, alkyl or aryl isocyanates/isothiocyanates, and carbodiimides (X=C=Y) to form metal derivatives as shown in Eq. (2.283):

$$M(OR)_x + y \; X{=}C{=}Y \longrightarrow M(OR)_{x-y} \; (X{-}\overset{\overset{\displaystyle Y}{\|}}{C}OR)_x$$

$$\updownarrow$$

$$M(OR)_{x-y} \left(\begin{array}{c} Y \\ \diagdown \\ X \end{array} \hspace{-1em} \begin{array}{c} \\ COR \\ \end{array} \right)_x \tag{2.283}$$

$$(X,Y = O \text{ or } NR)$$

The bidentate ligating mode of the resulting ligand appears to provide significant driving force for these types of reactions, some of which have been shown to be reversible.

Alkoxides of many metals react readily with CO_2 to give carbamates and alkylcarbonates(bicarbonates). The reversible absorption of approximately one mole of carbon dioxide by solutions (benzene) of some early transition metal alkoxides[931] has been reported to proceed according to Eq. (2.284):

$$M(OR)_x + CO_2 \rightleftharpoons RO\overset{\overset{\displaystyle O}{\|}}{C}OM(OR)_{x-1} \tag{2.284}$$

M = Ti, Zr; R = Et, Bu^n; $x = 4$. M = Nb; R = Et; $x = 5$. M = Fe; R = Et; $x = 3$.

The final products could not be isolated in a pure state and their formulation was inferred from infrared spectroscopic data, the observed stoichiometry of the reaction, and their reaction with ethyl iodide which gave diethylcarbonate.[931]

Molybdenum alkoxides, $Mo_2(OR)_6$, react in hydrocarbon solvents reversibly with two moles of CO_2 to yield the corresponding bis (alkyl carbonates) (Eq. 2.285).[932,933]

$$Mo_2(OR)_6 + CO_2 \underset{\Delta\,(90°C)}{\rightleftharpoons} \tag{2.285}$$

The mechanism proposed for this reaction in solution involves catalysis by adventitious free alcohol and the appropriate monoalkyl carbonic ester as the effective reagent, as shown in Eq. (2.286):

$$CO_2 + ROH \rightleftharpoons HO\overset{\overset{\displaystyle O}{\|}}{C}OR \xrightarrow{Mo_2(OEt)_6} (RO\overset{\overset{\displaystyle O}{\|}}{C}O)_2Mo_2(OR)_4 \qquad (2.286)$$

Nickel and cobalt methoxides, $M(OMe)_2$ (M = Ni, Co), are reported to be unreactive towards carbon dioxide. However, it has been demonstrated that $Cu(OMe)_2$ in the presence of pyridine reacts with two moles of CO_2 to yield the corresponding carbonate (Eq. 2.287):[934]

$$Cu(OMe)_2 + 2CO_2 \xrightleftharpoons{py/25°C} Cu(O\overset{\overset{\displaystyle O}{\|}}{C}OMe)_2(py) \qquad (2.287)$$

Further the dimeric copper(II) acetylacetonate-methoxide was also found to take up two moles of carbon dioxide in the presence of pyridine (Eq. 2.288):[934]

$$(acac)Cu\overset{\overset{\displaystyle Me}{\overset{\displaystyle O}{\diagdown\diagup}}}{\underset{\overset{\displaystyle O}{\underset{\displaystyle Me}{\diagup\diagdown}}}{}}Cu(acac) \xrightarrow{py/CO_2} 2(acac)Cu(O\overset{\overset{\displaystyle O}{\|}}{C}OMe)(py) \qquad (2.288)$$

Tsuda et al.[935] also found that copper(I) t-butoxide in the presence of t-butyl isocyanide or certain other ligands (PEt_3, PPh_3) reacted reversibly with carbon dioxide to give the t-butyl carbonate complex (Eq. 2.289):

$$(Bu^tO)Cu(CNBu^t) + 2CNBu^t + CO_2 \xrightleftharpoons[\text{benzene/reflux}]{\text{benzene/20°C}} (Bu^tO\overset{\overset{\displaystyle O}{\|}}{C}O)Cu(CNBu^t)_3$$
$$(2.289)$$

Titanium and zirconium tetra-alkoxides undergo insertion reactions of both organic isocyanates[936,937] and carbodiimides[938] into M−O bonds, as shown by Eqs (2.290) and (2.291):

$$M(OR)_4 + xR'NCO \longrightarrow M(OR)_{4-x}\{NR'C(O)OR\}_x \qquad (2.290)$$

(M = Ti;[936,937] R = Et, Pr^i, Bu^t; R' = Me, Et, Ph, α-naphthyl.

M = Zr;[938] R = Pr^i; R' = Ph)

$$Ti(OPr^i)_4 + 2R'N=C=NR' \rightleftharpoons (Pr^iO_2)Ti\{\overset{\overset{\displaystyle R'}{|}}{\underset{\underset{\displaystyle OPr^i}{|}}{N}}-C=NR'\}_2 \qquad (2.291)$$

where R' = p-tolyl.[938]

Both types of reaction are reversible in the case of $Ti(OR)_4$.[937]

With niobium[939] and tantalum[940] penta-alkoxides, the degree of insertion of phenyl isocyanate may be controlled, resulting in the formation of mono-, di-, tri-, tetra-, and penta-carbamate derivatives (Eq. 2.292):

$$M(OR)_5 + xPhNCO \longrightarrow M(OR)_{5-x}\{NPhC(O)OR\}_x \qquad (2.292)$$

where M = Nb, Ta; R = Et, Pr^i, Bu^t.

Similar to the earlier transition metal (Ti, Zr, Nb, Ta) alkoxides, methoxide and ethoxide of chromium(III) also react exothermically with organic isocyanates[941] in stoichiometric ratios in benzene, to afford the insertion product as shown by equation (2.293):

$$Cr(OR)_3 + xR'NCO \longrightarrow (RO)_{3-x}Cr\{NR'C(O)OR\}_x \qquad (2.293)$$

$$(R = Me, Et; R' = Ph, 1\text{-naphthyl}; n = 1, 2, \text{ or } 3)$$

Insertion of phenyl isocyanate into the metal-oxygen bonds of dimolybdenum or ditungsten hexa-alkoxides[942,943] affords dinuclear derivatives containing two bridging carbamate ligands, as shown in Eq. (2.294):

$$(2.294)$$

where M = Mo, W; R = Pr^i, Bu^t.

Although nickel dimethoxide does not undergo alcoholysis reactions, it reacts exothermically with phenyl and naphthyl isocyanates to give the corresponding carbamates[944] (Eq. 2.295):

$$Ni(OR)_2 + xR'NCO \longrightarrow (RO)_{2-x}Ni\{NR'C(O)OR\}_x \qquad (2.295)$$

where R = Me, Pr^i; R' = Ph, naphthyl; x = 1, 2.

An interesting example is the formation of alkylzinc carbamates, RZn{NR'C(O)OR} by the interaction of RZnOR with R'NCO, revealing that the Zn−O bond is reactive toward isocyanates, in contrast to the Zn−C bond.[945]

Mercury dimethoxide and phenylmercury methoxide also react with organic isocyanates and isothiocyanates to form carbamate and thiocarbamate derivatives, respectively.[946,947]

Phenyl isocyanate also inserts across the M–OR bonds of arsenic and antimony trialkoxides to form the corresponding carbamates.[948,949]

Infrared spectroscopy has played an important role in the investigation of the reactions of metal alkoxides with heterocumulenes. For example, the formation of pure carbamates according to Eqs (2.290)–(2.295) was indicated by the absence of free ligand band due to ν(N=C=O) around 2275 cm^{-1}. However, in a number of cases on standing for some time, the bond due to ν(N=C=O) again appeared, thus indicating the reversibility of the above reactions. Similarly, pure carbodiimide derivatives obtained from the reaction shown in Eq. (2.291) do not show the presence of the cumene band of the free ligand (R'N=C=N—R') in the range 2500–2000 cm^{-1}. In these cases also the reaction was found by infrared spectral studies to be reversible in solution, particularly at higher temperatures.

4.12.1.2   With ketenes

Ketenes readily insert into Ti–OR[949] and Zr–OR[950] bonds of their tetra-alkoxides as illustrated in Eq. (2.296):[949]

$$M(OR)_4 + x\,Ph_2C{=}C{=}O \longrightarrow (RO)_{4-x}M\{OC(OR){:}CPh_2\}_x$$

$$(2.296)$$

where M = Ti, Zr; R = Et, Pr^i; x = 1, 2.

Ketones and aldehydes undergo subsequent insertion into metal–carbon bonds generated from ketene insertion into Ti–OR bonds (Eq. 2.297):[951]

$$Ti(OR)_4 + 3Ph_2C{=}C{=}O + Me_2CO \longrightarrow (RO)Ti\{OCMe_2Ph_2C\overset{\displaystyle O}{\overset{\|}{C}}OR\}_3 \quad (2.297)$$

The hydrolysis of the above products affords a novel β-hydroxyester, HOCMe$_2$CPh$_2$COR.

Mercury dimethoxide[952] reacts with excess of ketene as illustrated by Eq. (2.298):

$$Hg(OMe)_2 + 2CH_2{=}C{=}O \longrightarrow Hg(CH_2\overset{\displaystyle O}{\overset{\|}{C}}OMe)_2 \quad (2.298)$$

4.12.1.3   With alkenes

The reaction of (dppe)PtMe(OMe) (dppe = bis(1,2-diphenylphosphino)ethane) with F$_2$C=CF$_2$ has been shown to give (dppe)PtMe(CF$_2$CF$_2$OMe), providing the first example of an alkene insertion into an M–OR bond.[953] Interestingly, the platinum alkyl-alkoxide complex does not react with ethene or pentene but similar to 4-tetrafluoroethylene other activated alkenes such as acrylonitrile and methylacrylate also undergo insertion reactions. It is noteworthy that the reaction does not involve nucleophilic attack on a coordinated alkene by free methoxide ion. The transformation of a M–OR bond into a M–R bond appears to be thermodynamically favourable in this instance.

4.12.1.4   With carbon monoxide

The platinum dimethoxide complex, (dppe)Pt(OMe)$_2$, reacts with carbon monoxide to yield (dppe)Pt(CO$_2$Me)$_2$,[954] which provides the first example of a CO migratory insertion into a M–OR bond, rather than the more common attack by RO^- on a carbonyl ligand in the formation of metallocarboxylates.

4.13   Formation of Coordination Compounds

Homoleptic metal alkoxides do not generally form stable molecular adducts with conventional neutral donor ligands,[3] owing to the preferential intermolecular coordination of alkoxo oxygen leading to coordination expansion of the metal and formation of coordinatively saturated oligomeric or polymeric molecules. The latter effect is more pronounced with sterically less demanding alkoxo ligands such as ^-OMe, ^-OEt, or $^-OPr^n$. In some cases, however, the coordination expansion of the metal may also occur alternatively with the lone pair electrons from added donor ligands. Obviously, the added ligand (L) will have to compete thermodynamically with the oligomerization process (Eq. 2.299):[3,8]

$$(RO)_{x-1}M \underset{\underset{R}{O}}{\overset{\overset{R}{O}}{\diamond}} M(OR)_{x-1} + 2L \; \rightleftharpoons \; 2M(OR)_x(L) \qquad (2.299)$$

The reaction (Eq. 2.299) in the forward direction will be facilitated by the higher electrophilicity of the metal atom (M) in the metal alkoxides $[M(OR)_x]_n$ with a weaker $M(\mu\text{-}OR)_2$ bridging system. This could be brought about by the following changes:[3,4,8,21,25] (a) the replacement of R by electron-withdrawing groups (such as CF_3, CH_2CH_2Cl, CH_2CCl_3, $CH(CF_3)_2$, $CMe_2(CF_3)$, $CMe(CF_3)_2$, etc.) to make the alkoxo oxygen less nucleophilic and a weaker bridging agent, (b) an increase in the steric demand of the group R (Bu^t, $CHPr^i_2$, $CHBu^t_2$, CBu^t_3, CPh_3, $CMe(CF_3)_2$, $C(CF_3)_3$) which is more likely to produce a coordinatively unsaturated molecule, and thus provide an opportunity particularly for compact hard oxygen or nitrogen donor ligands to coordinate the metal, and (c) the attachment of electron-attracting groups (such as chloride) to the metal.

These tendencies have been successfully exploited in the synthesis of molecular addition compounds of metal alkoxides. For example, Gilman et al.[898] observed that $U(OCH_2CF_3)_5$ formed addition complexes of moderate stability with ammonia or aliphatic amines, and Bradley et al.[3,146] found that $Zr(OCH_2CCl_3)_4$ formed a stable complex $Zr[OCH_2CCl_3]_4 \cdot (2Me_2CO)$ with acetone, while derivatives $M[OCMe(CCl_3)CH_3]_4$ (M = Ce or Th) give molecular adducts with pyridine. Interestingly, metal alkoxides of the lower alcohols can be readily recrystallized from the parent alcohol as molecular adducts, viz.: $Ti(OBu^i)_4 \cdot Bu^iOH$; $M(OPr^i)_4 \cdot Pr^iOH$ (M = Zr, Hf, Ce, Th, Sn).[3] In addition to these, a few examples of addition compounds of the types $M(OPr^i)_4(NC_5H_5)$ (M = Zr, Ce);[3] $Ti(OPr^i)_4(en)$, $Ti_2(OR)_8(en)$ (R = Et, Pr^i),[955] $TiCl_x(OCH_2CF_3)_{4-x}(NCR)$ (R = Me, Et; $x = 1$, 2);[956] $Ta_2(OPr^i)_{10}(en)$;[955] $Al(OPr^i)_3 \cdot N_2H_4$ and $Al_2(OPr^i)_6 \cdot N_2H_4$,[957] as well as $Al(OCH_2CCl_3)_3(py)$[403] were reported and the possibilities of potential structures discussed.

The 1990s witnessed noteworthy developments[21] in the coordination chemistry of homometallic alkoxides, which resulted in the emergence of fascinating types of molecular adducts such as $[LiOCBut_3(thf)]_2,^{230}$ $Mg(OMe)_2.3.5MeOH,^{38}$ $Ca_9(OC_2H_4OMe)_{18}$ $(HOC_2H_4OMe)_2,^{50}$ $[Ba(OBu^t)_2(HOBu^t)_2]_4,^{549}$ $Cd_9(OC_2H_4OMe)_{18}(HOC_2H_4OMe)_2,^{958}$ $Y_3(OBu^t)_7Cl_2(thf)_2,^{161}$ $Y_3(OBu^t)_8Cl(thf)_2,^{160}$ $La_3(\mu_3\text{-}OBu^t)_2(\mu\text{-}OBu^t)_3(OBu^t)_4$ $(thf)_2,^{160}$ $[Nd(OCBu_3^t)_2(\mu\text{-}Cl)(thf)]_2,^{959}$ $Nd(OCBu_3^t)_3(thf),^{959}$ $Nd_2(OCHPr_2^i)_6L_2$ (L = thf, py; $L_2 = (\mu\text{-}dme)),^{348}$ $U(OCBu_3^t)_xCl_{4-x}(thf)_y$ ($x = 1, y$ is variable; $x = 2, y = 2),^{960}$ $Th_4(OPr^i)_{16}(py)_2,^{165}$ $[TiCl_2(OPr^i)(HOPr^i)(\mu\text{-}Cl)]_2,^{108}$ $[TiCl_2(OPr^i)(HOPr^i)(\mu\text{-}Cl)]_2,^{108}$ and $[TiCl_2(OC_2H_4X)_2(HOC_2H_4X)]_2$ (X = Cl, Br, and I).^{108a}

In addition to the examples of simple addition products already described, dimolybdenum and ditungsten hexa-alkoxides, $M_2(OR)_6(M\equiv M)$, which are coordinatively unsaturated, react reversibly with donor ligands (L) such as amines[365,961] and phosphines[962] to form molecules of the type $M_2(OR)_6L_2$, by simple addition of 2L to $M_2(OR)_6$ without replacement of any of the alkoxo ligands, according to Eq. (2.300):

$$
\begin{array}{c}
\text{RO} \quad \quad \quad \quad \text{OR} \\
\text{RO}\diagdown\!\!\!\!\diagup\text{M}\!\!\equiv\!\!\text{M}\diagup^{\!\!\!\!\text{OR}}_{\diagdown\text{OR}} + 2L \rightleftharpoons \text{RO}
\end{array}
\qquad (2.300)
$$

The state of the above equilibrium is dependent on the nature of M, R, and L, as well as on the temperature. Each metal atom is in a roughly square-planar coordination environment and the metal–metal triple bonding is retained; one of these adducts, $W_2(OPr^i)_6(py)_2$, has been structurally characterized by X-ray crystallography.^{963}

4.14 Reactions with Main Group Organometallic Compounds

There are a number of reactions known involving metal alkoxides and organometallic compounds of main group elements (Li, Mg, Al, Sn, or Zn).[21] The fundamental reaction generally involves elimination of a main group metal alkoxide. The composition of the reaction products is considerably influenced by: (i) the nature and chemistry of the reactants as well as their stoichiometry, (ii) the reaction conditions, and (iii) the presence of other reactive substrates in the reaction medium. Some typical reactions are illustrated by Eqs (2.301)–(2.319).

The group 2 metal alkoxides and their alkyls undergo an interesting redistribution reaction (Eq. 2.301):

$$MR_2 + M(OR')_2 \longrightarrow \tfrac{1}{2}[RMOR']_4 \qquad (2.301)$$

(M = Be;^{385} R = Me; R' = Bu^t. M = Mg;^{389} R = Me; R' = CEt$_2$Me.

M = Mg;^{964} R = C$_5$H$_5$; R' = Et)

$$Er(OBu^t)_3 + 3LiBu^t \longrightarrow ErBu_3^t.3LiOBu^{t^{965}} \qquad (2.302)$$

$$Er(OBu^t)_3 + 3LiBu^t \longrightarrow ErBu_3^t.(tmeda)_2 + 3LiOBu^{t^{965}} \qquad (2.303)$$

$$Ln(OBu^t)_3 + 4LiBu^t + 2TMEDA \xrightarrow{\text{pentane}} [Li(tmeda)_2][LnBu_4^t] + 3LiOBu^{t965,966}$$

(2.304)

where Ln = Er, Lu.

$$Ln(OBu^t)_3 + 3AlMe_3 \longrightarrow [Ln\{(\mu\text{-}OBu^t)(\mu\text{-}Me)AlMe_2\}_3]$$ (2.305)

where Ln = Y, Pr, Nd.[966a]

Out of the above, the lutetium derivative $[Li(tmeda)_2][LuBu_4^t]$ has been characterized X-ray crystallographically.

The following two reactions represent an efficient synthetic route to organometallic compounds of highly oxidizing Ce(IV).

$$Ce(OPr^i)_4 + \tfrac{3}{2}Mg(C_5H_5)_2 \longrightarrow (C_5H_5)_3Ce(OPr^i) + \tfrac{3}{2}Mg(OPr^i)_2^{967}$$ (2.306)

$$Ce(OPr^i)_4 + 3Me_3SnC_5H_5 \longrightarrow (C_5H_5)_3Ce(OPr^i) + 3Me_3SnOPr^{i968}$$ (2.307)

A number of interesting organometallic compounds of cerium have also been reported as arising from the reactions of $Ce(OPr^i)_4.Pr^iOH$ or $Ce(OPr^i)_4$ with cyclooctatetraene (C_8H_8) in the presence of $AlEt_3$. The composition of the final product depends[967] on the reaction conditions (Eqs 2.308–2.310):

$$Ce(OPr^i)_4Pr^iOH$$

$$\xrightarrow[\text{toluene},140°C]{+AlEt_3/3C_8H_8} Ce_2(C_8H_8)_3 + 10AlEt_2(OPr^i) + 2C_2H_6$$

(2.308)

$$\xrightarrow[\text{toluene},100°C]{+8AlEt_3/2C_8H_8} 2(C_8H_8)Ce(\mu\text{-}OPr^i)_2AlEt_2$$ (2.309)

$$\xrightarrow[110°C]{\substack{+10AlEt_3\,(\text{excess})/4C_8H_8 \\ (\text{solvent})}} 2Ce(C_8H_8)_2 + 10AlEt_2(OPr^i) + 2C_2H_6 + 8[Et]$$ (2.310)

where 8[Et] = eight ethyl radicals.

Methyllithium reacts with $U(OCHBu_2^t)_4$ in hexane to afford heterobimetallic lithium uranium methyl alkoxide,[229] the structure of which is based upon two-coordinate lithium and five-coordinate uranium with the geometry of uranium being near that of a square pyramid with the methyl group occupying the apical site:

$$U(OCHBu_2^t)_4 + LiMe \xrightarrow{\text{hexane}} (Bu_2^tCHO)_2MeU(\mu\text{-}OCHBu_2^t)_2Li$$ (2.311)

The type of reactions illustrated in Eqs (2.302)–(2.305), (2.309), and (2.311) assume special significance for their role in the development of novel organoheterometallic alkoxide chemistry (Chapter 3, Section 2.1.5.1).

The interactions of early 3d transition metal alkoxides with lithium alkyls, Grignard reagents, or dimethylzinc yield interesting homo- and heteroleptic alkyls as shown in

Eqs (2.312)–(2.315):

$$Ti(OPr^i)_4 + 4LiR \xrightarrow[-60°C]{petroleum} TiR_4 + 4LiOPr^i \qquad (2.312)$$

where R = 1-adamantylmethyl.[970]

$$VO(OPr^i)_3 + ZnMe_2 \longrightarrow MeVO(OPr^i)_2 + \tfrac{1}{4}[MeZnOPr^i]_4{}^{971} \qquad (2.313)$$

$$M(OPr^i)_5 + 3LiR \longrightarrow MR_3(OPr^i)_2 + 3LiOPr^i \qquad (2.314)$$

where M = Nb, Ta; R = Me, CH_2Bu^t.[972]

$$Ta(OMe)_5 + 4LiCH_2SiMe_3 \longrightarrow Ta(OMe)(CH_2SiMe_3)_4 + 4LiOMe^{972} \qquad (2.315)$$

The alkylation of alkoxides of niobium and tantalum rarely lead to total substitution. Typically the reaction is free from any complication and the products are pure up to the trisubstitution stage.

In the reactions with $Cr(OBu^t)_4$, the solvent plays a determining role. For example, reactions with methyllithium or Grignard reagents in the presence of donor solvents (such as THF or diethyl ether) yield the corresponding *n*-bonded tris(organo)chromium complex.

$$Cr(OBu^t)_4 + 4LiR \xrightarrow[-78°C]{petroleum} CrR_4 + 4LiOBu^t \qquad (2.316)$$

where R = Me, Pr^i, Bu^n, Bu^i, CH_2Bu^t, Bu^t,[973,974] 1-adamantylmethyl.[970]

$$2Cr(OBu^t)_4 \begin{cases} \xrightarrow[THF]{+8PhMgBr} 2CrPh_3(thf)_3 + Ph-Ph + 8Mg(OBu^t)Br^{975} & (2.317) \\[2ex] \xrightarrow[THF]{+8EtMgBr} 2CrEt_3(thf)_x + 8Mg(OBu^t)Br + 2C_2H_6 + C_2H_4{}^{975} & (2.318) \\[2ex] \xrightarrow[THF]{+8MeMgBr} 2CrMe_3(thf)_x + 8Mg(OBu^t)Br + 2CH_4 & (2.319) \end{cases}$$

4.15 Tischtchenko, Meerwein–Ponndorf–Verley (M–P–V) and Oppenauer Oxidation Reactions

4.15.1 Introduction

Metal alkoxides are known to act as potential catalysts (Chapter 7, Section 4) and this property has been utilized successfully in organic syntheses. Before dealing with the Tischtchenko, Meerwein–Ponndorf–Verley, and Oppenauer oxidation reactions, it is worth briefly mentioning the Cannizzaro reaction which appears to be the main basis of the above three types of reactions. The reaction in which aldehydes that have no α-hydrogen for enolization disproportionate in the presence of a strong base, to give equal amounts of the corresponding alcohol and a salt of carboxylic acid (Eqs 2.320–2.322) is known as Cannizzaro's reaction (an oxidation–reduction reaction involving aldehydes and a strong base). One molecule of the aldehyde acts as a hydride ($H:^-$) donor and another functions as an acceptor.

$$2HCHO \xrightarrow{\text{50\% NaOH}} CH_3OH + HCOO^-Na^+ \tag{2.320}$$

$$2Bu^tCHO \xrightarrow{\text{Conc NaOH}} Bu^tCH_2OH + Bu^tCOO^-Na^+ \tag{2.321}$$

$$2C_6H_5CHO \xrightarrow{\text{Conc NaOH}} C_6H_5CH_2OH + C_6H_5COO^-Na^+ \tag{2.322}$$

However, an aldehyde containing an α-hydrogen atom undergoes an aldol condensation in the presence of a dilute base or acid. For example, two moles of acetaldehyde combine to form β-hydroxy butyraldehyde (Eq. 2.323):

$$CH_3CHO + CH_3CHO \xrightarrow{\text{HO}^-} CH_3CH(OH)CH_2CHO \tag{2.323}$$

4.15.2   Tischtchenko Reaction

Claisen[976] in 1887 modified the Cannizzaro reaction by using sodium ethoxide in place of sodium hydroxide, and found that benzaldeyde was converted to benzyl benzoate. In 1906 Tischtchenko[977] observed that sodium alkoxides can be used for the formation of esters from both aliphatic and aromatic aldehydes. Nord et al.[978–980] found that all aldehydes can be made to undergo the Cannizzaro reaction by treating with Al(OEt)$_3$; under these conditions the acid and alcohol are combined as the ester, and the reaction is known as the Tischtchenko reaction (Eqs 2.324–2.327):

$$2CH_3CHO \xrightarrow{\text{Al(OEt)}_3} CH_3COOCH_2CH_3 \tag{2.324}$$

$$2CH_3CH_2CHO \longrightarrow CH_3CH_2COOCH_2CH_2CH_3 \tag{2.325}$$

$$2CH_2=CHCHO \longrightarrow CH_2=CHCOOCH_2CH_2=CH_2 \tag{2.326}$$

$$2C_6H_5CHO \longrightarrow C_6H_5COOCH_2C_6H_5 \tag{2.327}$$

In all these reactions Al(OEt)$_3$ helps in the condensation of only two moles of aldehydes to form simple esters. However, the heterobimetallic ethoxide Mg[Al(OEt)$_4$]$_2$ catalyses the condensation of three moles of aldehydes to form trimeric glycol ester (Eq. 2.328):

$$2RCH_2CHO \xrightarrow{\text{Mg[Al(OEt)}_4]_2} RCH_2CH(OH)CHRCHO$$

$$\xrightarrow{\text{+RCH}_2\text{CHO}} RCH_2CH(OH)CHRCH_2COOCH_2R \tag{2.328}$$

On the basis of the above observations, Villani and Nord[980] concluded that the catalytic activity of these metal alkoxides in the condensation of aldehydes depends upon the nature of the metal alkoxides. Thus, when a strongly basic alkoxide like sodium ethoxide is used, the aldol type of condensation occurs whereas with an acidic alkoxide such as aluminium ethoxide, the simple esters are formed. However, weakly basic alkoxides [Mg[Al(OEt)$_4$]$_2$, Ca[Al(OEt)$_4$]$_2$, or Na[Al(OEt)$_4$]], which are intermediate between sodium ethoxide and aluminium ethoxide as regards their acidic or basic character, cause the formation of glycol esters. The bifunctional activity of the complex catalyst Mg[Al(OEt)$_4$]$_2$ or Ca[Al(OEt)$_4$]$_2$, may be due to the intermediate basicity of the complex which is formed from strongly basic Mg(OR)$_2$ and acidic Al(OR)$_3$. However, it has also been observed that aldehydes other than those containing an α-CH$_2$ group do not form a glycol ester with mildly basic alkoxides; the condensation of α-ethylbutyraldehyde results in the formation of a glycol ester only in the presence of strongly basic sodium ethoxide.

The mechanism of the Tischtchenko reaction was discussed by Lin and Day[981] who proposed that in the initial stages, the aluminium atom of aluminium alkoxide coordinates with the carbonyl group of the aldehyde (Eq. 2.329),

$$
\underset{\overset{|}{R-C=O}}{\overset{H}{|}} + Al(OR)_3 \longrightarrow \underset{\overset{|}{R-\underset{\oplus}{C}=O} \rightarrow Al(OR)_3}{\overset{H}{|}} \qquad (2.329)
$$

Consequently a positive charge is induced on the carbonyl carbon atom which facilitates the reaction to proceed further (Eq. 2.330):

$$
\begin{array}{c}
\overset{H}{\underset{|}{R-\underset{\oplus}{C}}} + \overset{H}{\underset{|}{O=CR}} \\
\underset{\oplus \ominus}{|} \\
O-\bar{A}l(OR)_3
\end{array}
\longrightarrow
\begin{array}{c}
\overset{H \frown H}{\underset{|}{R-C-O-C-R}} \\
| \\
O-Al(OR)_3 \\
\downarrow
\end{array}
\qquad (2.330)
$$

$$
RCOOCH_2R + Al(OR)_3
$$

by oxidation process (loss of hydride ion) coupled with simultaneous acceptance of hydride ion (reduction) by another aldehyde molecule finally yielding the organic ester.

Mixed Tischtchenko reactions were also studied by Nord.[982–984] In these reactions he isolated benzyl isovalerate and isoamyl benzoate from benzaldehyde and isovaleraldehyde as well as benzyl acetate, ethyl acetate, and benzyl benzoate from benzaldehyde and acetaldehyde. Orloff[985] also reported the synthesis of benzyl isobutyrate and isobutyl benzoate from benzaldehyde and isobutyraldehyde. However, the mechanism of mixed Tischtchenko reactions is rather complex. For example, in the reaction of benzaldehyde and n-butyraldehyde, either of the carbonyl groups may coordinate with the aluminium alkoxide. In fact the carbonium ion formed in butyraldehyde is more active than that in benzaldehyde and further reaction may proceed with the carbonium ion of the former type. The overall mechanism may be explained by the following equations (2.331)–(2.334):

$$
\underset{\overset{|}{C_3H_7-C=O}}{\overset{H}{|}} + Al(OR)_3 \longrightarrow \underset{\overset{|}{C_3H_7-\underset{\oplus}{C}-O-\underset{\ominus}{Al}(OR)_3}}{\overset{H}{|}} \qquad (2.331)
$$

$$\text{C}_6\text{H}_5{-}\overset{\overset{\text{H}}{|}}{\text{C}}{=}\text{O} + \text{Al(OR)}_3 \longrightarrow \text{C}_6\text{H}_5{-}\overset{\overset{\text{H}}{|}}{\overset{\oplus}{\text{C}}}{-}\text{O}{-}\overset{\ominus}{\text{Al}}\text{(OR)}_3 \qquad (2.332)$$

$$\text{C}_3\text{H}_7{-}\overset{\overset{\text{H}}{|}}{\overset{\oplus}{\text{C}}}{-}\text{O}{-}\overset{\ominus}{\text{Al}}\text{(OR)}_3 + \text{C}_6\text{H}_5{-}\overset{\overset{\text{H}}{|}}{\text{C}}{=}\text{O} \longrightarrow \text{C}_3\text{H}_7{-}\overset{\overset{\text{H}}{|}}{\underset{\underset{\ominus}{\text{O}{-}\text{Al(OR)}_3}}{\text{C}}}{-}\text{O}{-}\overset{\oplus}{\text{C}}\text{H}{-}\text{C}_6\text{H}_5$$

(i)

$$(2.333)$$

$$\text{C}_3\text{H}_7{-}\overset{\overset{\text{H}}{|}}{\overset{\oplus}{\text{C}}}{-}\text{O}{-}\overset{\ominus}{\text{Al}}\text{(OR)}_3 + \text{C}_3\text{H}_7{-}\overset{\oplus}{\text{C}}{=}\text{O} \longrightarrow \text{C}_3\text{H}_7{-}\overset{\overset{\text{H}}{|}}{\underset{\underset{\ominus}{\text{O}{-}\text{Al(OR)}_3}}{\text{C}}}{-}\text{O}{-}\overset{\oplus}{\text{C}}\text{HC}_3\text{H}_7$$

(ii)

$$(2.334)$$

The relative amounts of the species (i) and (ii) would depend upon the rate of addition of butyraldehyde. It may be assumed that hydride ion transfer should proceed as shown in the carbonium ion (iii),

$$\text{C}_6\text{H}_5{-}\overset{\overset{\text{H}}{|}}{\underset{\underset{\ominus}{\text{O}{-}\text{Al(OR)}_3}}{\text{C}}}{-}\text{O}{-}\overset{\oplus}{\text{C}}\text{HC}_3\text{H}_7$$

(iii)

However, in actual practice the yield of *n*-butyl benzoate was appreciably low. Thus on the basis of experimental data, it appears that the aldehyde which readily undergoes Tischtchenko reaction, forms the ester in larger amounts. It may be due to the fact that either of the mixed esters predominates over the other and thus eliminates the possibility of ester exchange between two simple esters. It was also observed that the yields of mixed ester could be increased by maintaining a lower concentration of the faster reacting aldehyde. However, with the slower reacting aldehyde such as an α,β-unsaturated aldehyde, if the faster reacting aldehyde is added at a considerably slower rate, the polymerization of unsaturated aldehyde predominates in the reaction. Lin and Day,[981] however, also observed that the formation of mixed ester was more effective in the presence of aluminium isopropoxide catalyst.

In view of the earlier observation that the aldehydes react with nitroparaffins to form nitroalcohols in the presence of catalysts such as NaOH, NaHCO$_3$, KOH, K$_2$CO$_3$, and Na, Villani and Nord[986] observed that the reaction could also be catalysed with the weakly acidic aluminium ethoxide or the weakly basic Mg[Al(OEt)$_4$]$_2$.

$$\text{R}{-}\text{CHO} + \text{H}{-}\overset{\overset{\text{H}}{|}}{\underset{\underset{\text{R}'}{|}}{\text{C}}}{-}\text{NO}_2 \longrightarrow \text{R}{-}\overset{\overset{\text{H}}{|}}{\underset{\underset{\text{OH}}{|}}{\text{C}}}{-}\overset{\overset{\text{H}}{|}}{\underset{\underset{\text{R}'}{|}}{\text{C}}}{-}\text{NO}_2 \qquad (2.335)$$

The formation of esters by the Tischtchenko reaction was also demonstrated by Nakai[987] who showed that magnesium dimethoxide or aluminium triethoxide were effective catalysts.

4.15.3 Meerwein–Ponndorf–Verley Reaction

Another type of reaction involving a metal alkoxide and a carbonyl compound was noticed as early as 1925 by Verley[988] and Meerwein and Schmidt,[989] who observed that alkoxides of magnesium, calcium, and particularly aluminium could catalyse reduction of aldehydes in the presence of excess ethyl alcohol as shown by Eq. (2.336):

$$RCHO + CH_3CH_2OH \overset{Al(OEt)_3}{\rightleftharpoons} RCH_2OH + CH_3CHO \qquad (2.336)$$

Removal of the more volatile acetaldehyde from the reaction medium is easily achieved with a stream of dry nitrogen or hydrogen to drive the reaction to the right.

In 1926, Ponndorf devised a method in which both aldehydes and ketones could be reduced to alcohols by adding excess alcohol and aluminium triisopropoxide[990] (Eqs 2.337–2.339).

$$R_2C=O + Me_2CHOH \overset{Al(OPr^i)_3}{\rightleftharpoons} R_2CHOH + Me_2C=O \qquad (2.337)$$
$$\text{(distil)}$$

$$MeCOCH_2CH_2CH_2Br + Me_2CHOH \overset{Al(OPr^i)_3}{\longrightarrow} MeCH(OH)CH_2CH_2CH_2Br + Me_2C=O$$
$$(2.338)$$

$$Me_2CHOH + MeCH=CHCHO \overset{Al(OPr^i)_3}{\longrightarrow} MeCH=CHCH_2OH + Me_2C=O \quad (2.339)$$

The process of reducing carbonyl compounds (aldehydes or ketones) to alcohols is therefore known as the Meerwein–Ponndorf–Verley reaction. Although alkoxides of a number of metal(loid)s such as sodium, magnesium, titanium, zirconium, iron, boron, aluminium, tin, and antimony have been used for these reactions, those of aluminium are by far preferred, since they tend to give the minimum degree of side reactions. The use of aluminium isopropoxide over other alkoxides was also preferred by Young et al.[991] as well as by Adkins and Cox.[71]

Meerwein et al.[992] originally suggested a mechanism for the reaction (Eq. 2.340) involving coordination of the carbonyl oxygen to the aluminium alkoxide which thereby functioned as a catalyst in the redox reaction.[993]

$$R'R''CO + Al(OCHR_2)_3 \rightleftharpoons R'R''CO \longrightarrow Al(OCHR_2)_3$$
$$\rightleftharpoons (R'R''CHO)(R_2CHO)_2Al \leftarrow OCR_2$$
$$\rightleftharpoons (R'R''CHO)(R_2CHO)_2Al + R_2CO \qquad (2.340)$$

As a result of mechanistic studies by a number of chemists such as Woodward et al.,[994] Jackman et al.,[995–997] McGowan,[998] Rekasheva and Miklykhin,[999] Williams et al.,[1000] and Doering et al.,[1001,1002] it appears that the reaction involves a six-membered cyclic transition state[1003,1004] in which hydrogen transfer occurs from the β-C–H bond of an alkoxide to a coordinated ketone or aldehyde:

As a consequence of the coordination of the carbonyl oxygen with aluminium isopropoxide (R = Me), a partial positive charge will be developed on the carbon atom; this could be the rate-determining step in this reaction mechanism. This view was supported experimentally by McGowan[998] who varied the electron density on the carbonyl carbon atom by substituting an electronegative group X in the *para*-substituted acetophenones $XC_6H_4COCH_3$. However, this did not constitute a conclusive proof of the mechanism. Furthermore, the postulated mechanism has the expected stereochemical consequences, and this aspect has been demonstrated by Jackman *et al.*[995-997] and Doering *et al.*[1001-1002] However, kinetic measurements[996,1005,1005a] are complicated by the oligomeric nature of the aluminium alkoxides.

It is noteworthy that in some cases reduction with metal alkoxides, including aluminium isopropoxide, involves free-radical intermediates (single electron transfer (SET) mechanism).[1006]

The potentialities of the Meerwein–Ponndorf–Verley reaction as an alternative method for preparation of metal alkoxides have also been studied (Eqs 2.341–2.345):[146,1007,1008]

$$Ti(OPr^i)_4 + 4Cl_3CCHO \longrightarrow Ti(OCH_2CCl_3)_4 + 4Me_2C{=}O^{146} \quad (2.341)$$

$$Zr(OPr^i)_4 + 4Cl_3CCHO \longrightarrow Zr(OCH_2CCl_3)_4.2Me_2C{=}O + 2Me_2C{=}O^{146} \quad (2.342)$$

2-Hydroxybenzaldehyde (salicylaldehyde) reacts with isopropoxides of titanium[1007,1008] and zirconium[1008] with displacement of isopropoxo group(s) as isopropyl alcohol to give mixed derivatives (Eqs 2.343–2.345), which on heating undergo reduction of the aldehyde function with elimination of acetone.

$$Ti(OR)_4 + xHOC_6H_4CHO \xrightarrow[\text{room temp}]{\text{benzene}} (RO)_{4-x}Ti(OC_6H_4CHO)_x + xROH \quad (2.343)$$

where R = Et, Pr^i, Bu^t;[1007] $x = 1, 2$.

$$(RO)_{4-x}Ti(OC_6H_4CHO)_x \xrightarrow[\text{reflux}]{\text{benzene}} (RO)_{4-2x}\overline{Ti(OC_6H_4CH_2O)}_x + xR'CHO \quad (2.344)$$

where R = Et, Pr^n;[1008] R' = Me, Et; $x = 1, 2$.

The reaction of $Ti(OPr^i)_4$ with salicylaldehyde in 1:2 molar ratio initially liberated only one mole of acetone and an insoluble product (Eq. 2.345):

$$Ti(OPr^i)_4 + 2HOC_6H_4CHO$$

$$\xrightarrow[\text{reflux}]{\text{toluene}} Ti(OC_6H_4CH_2O)(OC_6H_4CHO)(OPr^i) + (CH_3)_2C{=}O \quad (2.345)$$

On further refluxing, the insoluble product $Ti(OC_6H_4CH_2O)(OC_6H_4CHO)(OPr^i)$ in benzene, again 0.5 mole of acetone was liberated and the soluble product thus obtained corresponded to $Ti_2(OPr^i)(OC_6H_4CH_2O)(OC_6H_4CHO)$.

Similar results were obtained by the reactions of zirconium alkoxide with salicylalde-hyde[1008] except that the rate of these reactions was faster and, unlike the titanium derivative, the product of the 1:2 molar reaction of $Zr(OPr^i)_4 \cdot Pr^iOH$ with HOC_6H_4CHO (salicylaldehyde) did not precipitate; instead the product $Zr_2(OPr^i)(OC_6H_4CH_2O)_3$ (OC_6H_4CHO) was directly obtained.

4.15.4 Oppenauer Oxidation Reaction

The reverse of the Meerwein–Ponndorf–Verley reduction (Section 4.15.3), occurs when a ketone (as a hydride acceptor) in the presence of base is used as the oxidizing agent. It is reduced to a secondary alcohol (Scheme 2.7), and the reaction is known as Oppenauer Oxidation.[1009]

It involves heating the alcohol to be oxidized with an aluminium alkoxide in the presence of a carbonyl compounds, which acts as the hydrogen acceptor. The reaction is an equilibrium process and proceeds through a cyclic transition state.

$$R_2CHOH + Al(OCHMe_2)_3 \longrightarrow R_2CHOAl(OCHMe_2)_2 + Me_2CHOH$$

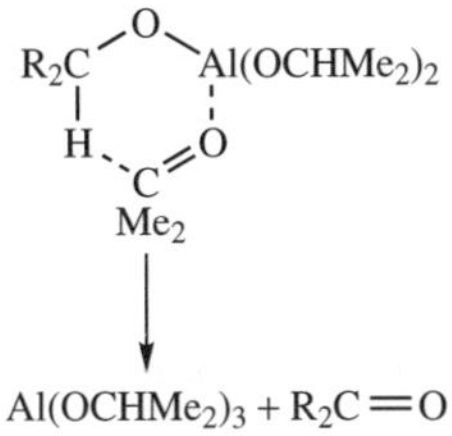

Scheme 2.7

The ketones most commonly used are acetone and butanone. Aluminium *tert*-butoxide is usually employed as the base catalyst (Eq. 2.346).

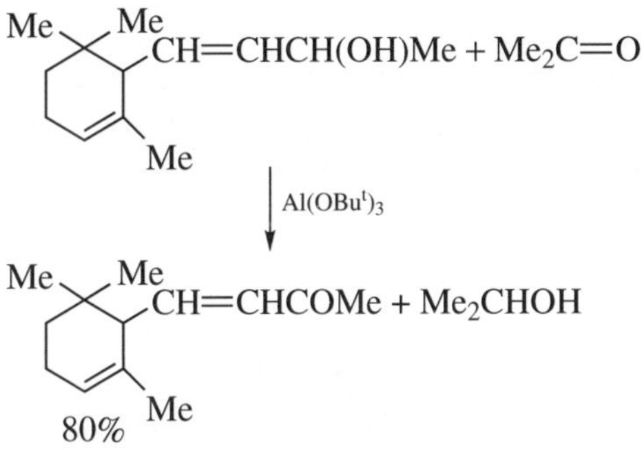

$$(2.346)$$

Cyclohexanone, benzoquinone, and fluorenone have also been used as strong hydrogen acceptors in order to allow reactions to be carried out at temperatures above the boiling point of acetone (Eq. 2.347):

$$(2.347)$$

Since the reaction conditions are nonacidic, this method can be valuable for substances that would not tolerate acidic conditions or the presence of transition metal ions.[1010]

$$(2.348)$$

The lanthanide (especially samarium) alkoxides serve as highly effective catalysts[1011–1013] for Oppenauer-type oxidation of alcohols to aldehydes and ketones (Eq. 2.349):

$$(2.349)$$

The main advantage of Oppenauer oxidation (Eqs 2.346–2.349) is its high selectivity. For example, the reaction takes place under very mild conditions and is highly specific for aldehydes and ketones, so that C=C bonds (including those conjugated with C=O bonds) and many other polyfunctional molecules containing sensitive groups that are destroyed by the conditions of many other oxidations and reductions may be present without themselves being reduced.

In conclusion, Oppenauer oxidation may also be regarded as an elimination of hydride (H:$^-$) ion. The reverse reaction is therefore hydride addition to carbonyl, as in the Meerwein–Ponndorf–Verley reduction (Section 4.15.3):

4.16   Interactions Between Two Different Metal Alkoxides

The reactions which are based on the basic and acidic character of the reactant metal alkoxides have been utilized extensively (Chapter 3, Section 2.1.1) and the topic is, therefore, only mentioned here as a heading, and illustrated by two typical examples[3,4] chosen at random:

$$KOPr^i + 2Zr(OPr^i)_4 \longrightarrow [KZr_2(OPr^i)_9] \tag{2.350}$$

$$Ln(OPr^i)_3 + 3Al(OPr^i)_3 \longrightarrow [Ln\{Al(OPr^i)_4\}_3] \tag{2.351}$$

4.17   Special Reactions of Multiple Metal–Metal Bonded Alkoxides

4.17.1   Reactions of $M_2(OR)_6 (M \equiv M)$

Dimolybdenum and ditungsten hexa-alkoxides with the formula $M_2(OR)_6$ (where R is a bulky group: Pr^i, CH_2Bu^t, Bu^t) have been found to undergo[1014–1020] a wide variety of reactions, generally in a distinct and sometimes unexpected manner. Furthermore, a rich chemistry has been developed in which alkoxo groups function as stabilizing ligands for 12-electron clusters.[1021–1024] The alkoxo oxygen has two filled p orbitals capable of donating π electron density to the metal centres. As these p orbitals are ligand centred, the derivatives are looked upon as coordinatively unsaturated and containing formal metal–metal triple bonds ($\sigma^2\pi^4$).

It is interesting to note that the $M \equiv M$ bonds have some similarity to carbon–carbon triple bonds and provide both a source and a returning site for electrons in oxidative addition[1025,1026] and reductive elimination[374,1027] reactions, wherein the M–M bond order is changed in a stepwise manner downward: $(M \equiv M)^{6+} \to (M = M)^{8+} \to (M-M)^{10+}$ as well as upward: $(M \equiv M)^{6+} \to (M \overline{\equiv} M)^{4+}$, respectively.

4.17.1.1   Oxidative addition reactions
As already mentioned, compounds with M–M multiple bonds have a source of electrons which may be utilized[1025] for oxidative processes (Eqs 2.352 and 2.353):

$$(2.352)$$

For steric reasons $Mo_2(OPr^i)_6$ and $Mo_2(OBu^t)_6$ do not react with Bu^tOOBu^t.[1025]

$$Mo_2(OPr^i)_6 + 2X_2 \xrightarrow[n\text{-}C_6H_{14}]{CCl_4 \text{ or}} X_2(Pr^iO)_2Mo = Mo(Pr^iO)_2 X_2 \tag{2.353}$$

where X = Cl, Br, I.

Attempts to prepare $Mo_2(OPr^i)_6X_2$ (M=M) derivatives by careful addition of one equivalent of halogen were unsuccessful, as such compounds are unstable with respect to disproportionation into $Mo_2(OPr^i)_6$ and $Mo_2(OPr^i)_6X_4$.

$Mo_2(OPr^i)_6$ and similar alkoxides, on the other hand, provide a good source of electrons to ligands that are capable of being reduced upon coordination. For example, Ph_2CN_2 reacts with $Mo_2(OPr^i)_6$ in the presence of pyridine to afford the product $Mo_2(OPr^i)_6(N_2CPh)_2(py)_2$,[1028] in which, considering the diphenyldiazomethane as a two electron (2e) ligand, $Mo=N—N=CPh_2$, the molybdenums may be viewed as $(Mo-Mo)^{10+}$ bonded.

4.17.1.2   Reactions with carbon monoxide

The reactions with hydrocarbon solutions of $M_2(OR)_6$ (M = Mo, W) compounds with CO are complex, and exhibit some remarkable differences depending on the nature of M and R as well as on the reactions conditions. Novel products have been isolated in the carbonylation of triply bonded dimolybdenum and ditungsten hexa-alkoxides and their adducts as is shown by the reactions described below (Eqs 2.354–2.360). For example, the reactions of *tert*-butoxides $M_2(OBu^t)_6$ with one equivalent of CO in hydrocarbon solvent proceed smoothly at room temperature and atmospheric pressure in two stages[1029–1031] as shown by Eqs (2.354) and (2.355):

$$M_2(OBu^t)_6 + CO \rightleftharpoons \qquad\qquad (2.354)$$

$$2Mo_2(OBu^t)_6(\mu\text{-}CO) + 4CO \longrightarrow Mo(CO)_6 + 3Mo(OBu^t)_4 \qquad (2.355)$$

The overall reaction is given by Eq. (2.356).

$$2Mo_2(OBu^t)_6 + 6CO \longrightarrow Mo(CO)_6 + 3Mo(OBu^t)_4 \qquad (2.356)$$

By contrast, the room temperature reaction of $Mo_2(OBu^t)_6$ with CO (1 atm pressure) in hexane–pyridine solvent mixtures yields $Mo(OBu^t)_2(py)_2(CO)_2$, the structure of which has been resolved by single-crystal X-ray study:[1032]

$$Mo_2(OBu^t)_6 + 4CO + 4py \xrightarrow{\text{pyridine–hexane}} 2 \qquad\qquad (2.357)$$

Reactions of isopropoxides, $M_2(OPr^i)_6$, with excess CO afford, in addition to $M(CO)_6$, carbonyl adducts of higher valent metal alkoxides (Eqs 2.358[1030] and 2.359[1033]).

$$4Mo_2(OPr^i)_6 + 18CO \longrightarrow 2Mo(CO)_6 + 3 \quad \text{[structure]} \qquad (2.358)$$
$$\text{(excess)}$$

$$4W_2(OPr^i)_6 + 24CO \longrightarrow 2W(CO)_6 + 3 \quad \text{[structure]} \qquad (2.359)$$

Treatment of $W_2(OPr^i)_6(py)_2$ with two molar equivalents of CO in toluene led to the isolation of two products: $W_2(OPr^i)_6(CO)_4$[1033] and $[(Pr^iO)_3W(\mu\text{-CO})(\mu\text{-}OPr^i)W(OPr^i)_2py]_2$,[1035] which have been characterized by X-ray crystallography.

Reactions of CO with $M_2(OR)_6$ (M = Mo, W; R = Pr^i, CH_2Bu^t) compounds in hydrocarbon solvents at 25°C and 1 atm pressure in the presence of donor ligands ($C_5H_5N(py)$, $HNMe_2$), lead to the formation of $M_2(OR)_6(L)_2(\mu\text{-CO})$ products, which have been structurally characterized[1036] for M = Mo, R = Pr^i and L = pyridine:

$$M_2(OR)_6 + CO + 2L \xrightarrow[\text{25 °C, 1 atm}]{\text{hydrocarbon}} \quad \text{[structure]} \qquad (2.360)$$

4.17.1.3 Reactions with nitric oxide

Hydrocarbon solutions of $Mo_2(OR)_6$ compounds react readily, irreversibly, and apparently quantitatively with NO at room temperature to yield $Mo_2(OR)_6(NO)_2$ according to Eq. (2.361):[1037,1038]

$$Mo_2(OR)_6 + 2NO \longrightarrow \quad \text{[structure]}$$

$$\downarrow + 2Me_2NH \qquad (2.361)$$

$$\text{[structure]}$$

where R = CH_2Bu^t, $CHMe_2$, CMe_3, Pr^i.

In the above reaction of $Mo_2(OPr^i)_6(NO)_2$ with two equivalents of Me_2NH in hydrocarbon solvent, the dimer did not split into a five-coordinate monomer $Mo(OPr^i)_3(NO)(NHMe_2)$; instead, the dinuclear structure was retained and each metal atom became six-coordinate.[1038]

By contrast, an X-ray crystallographically characterized five-coordinate mononuclear tungsten derivative has been obtained by the reaction between $W_2(OBu^t)_6$ and NO in hydrocarbon solutions in the presence of pyridine:[1039]

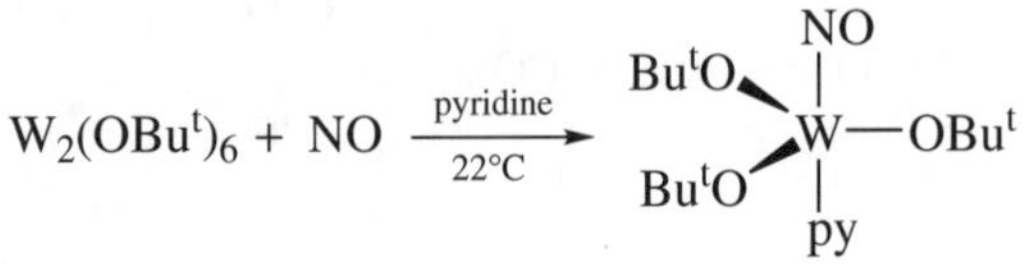

$$W_2(OBu^t)_6 + NO \xrightarrow[22°C]{pyridine} \begin{matrix} & NO \\ Bu^tO & | \\ & W-OBu^t \\ Bu^tO & | \\ & py \end{matrix} \qquad (2.362)$$

4.17.1.4   Reactions with alkenes or alkynes

Reactions of alkenes[1040–1042] and alkynes with $M_2(OR)_6$ compounds lead to a variety of interesting reactions and products depending upon the system, as shown by Eqs. (2.361)–(2.363):

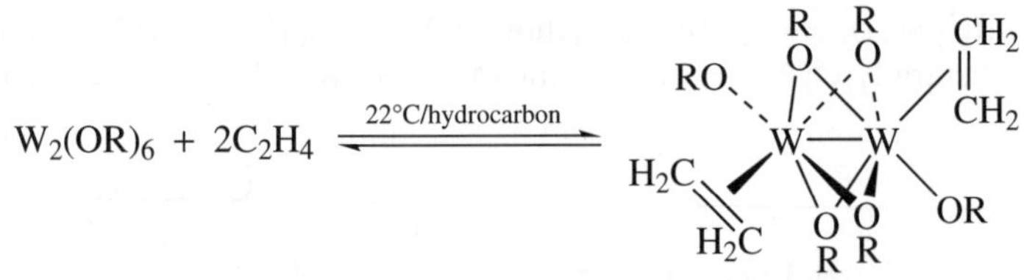

$$W_2(OR)_6 + 2C_2H_4 \rightleftharpoons^{22°C/hydrocarbon} \qquad (2.363)$$

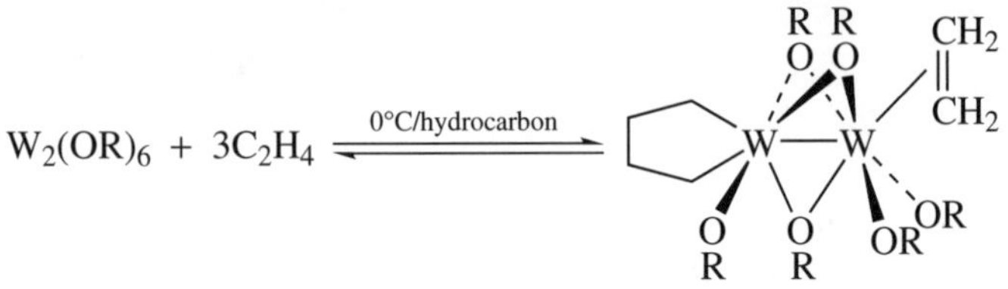

$$W_2(OR)_6 + 3C_2H_4 \rightleftharpoons^{0°C/hydrocarbon} \qquad (2.364)$$

$$W_2(OR)_6 + 3C_2H_4 \rightleftharpoons^{22°C/hydrocarbon} \qquad (2.365)$$

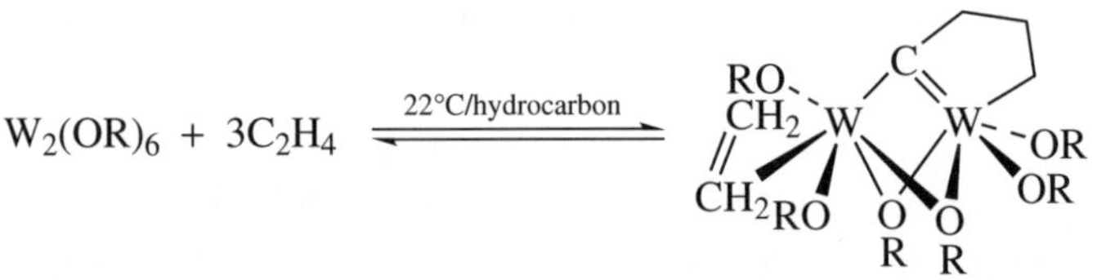

where $R = c$-hex, c-pent, Pr^i, CH_2Bu^t.

By contrast, the sterically encumbered $W_2(OBu^t)_6$ fails to react with $H_2C{=}CH_2$ in hydrocarbon solvents at room temperature.

Reactions between ditungsten hexa-alkoxides and alkynes have led to a remarkable variety of products, depending on the nature of the alkoxide, the alkyne, and the reaction conditions,[1043–1047] as illustrated by Eqs (2.366), (2.368)–(2.370).

The reactions between $(RO)_3M{\equiv}M(OR)_3$ and $R'C{\equiv}CR''$ or $R'C{\equiv}N$[1043] may be viewed as complementary redox reactions. The dinuclear metal centre as well as the alkyne and nitrile are potential six-electron (6e) reducing agents. It has been possible to prepare $(Bu^tO)_3W{\equiv}CR'$ ($R' = Me$, Et, Ph) or $(Bu^tO)_3W{\equiv}N$ by the reaction of $W_2(OBu^t)_6$ with $R'C{\equiv}CR'$ or $R'C{\equiv}N$, respectively (Eqs 2.366–2.368) in which isolobality of CR′ and W(OR)$_3$ is apparent:

$$W_2(OBu^t)_6 + RC{\equiv}CR \xrightarrow{\text{hexane}} 2(Bu^tO)_3W{\equiv}CR \qquad (2.366)$$

where R = Me, Et, Pr^n.

$$W_2(OBu^t)_6 + 2EtC{\equiv}CR \longrightarrow 2(Bu^tO)_3W{\equiv}CR + EtC{\equiv}CEt \qquad (2.367)$$

where R = Ph, SiMe$_3$, —CH=CH$_2$

$$W_2(OR)_6 + 2py + R'C{\equiv}CR' \longrightarrow W_2(OR)_6(\mu\text{-}C_2R_2')(py)_x \qquad (2.368)$$

(R = Bu^t; R′ = H; x = 1. R = Pr^i; R′ = H; x = 2. R = CH$_2$Bu^t; R′ = Me; x = 2)

$$W_2(OR)_6 + 3R'C{\equiv}CR' \longrightarrow W_2(OR)_6(\mu\text{-}C_4R_4')(C_2R_2') \qquad (2.369)$$

where R = Pr^i, CH$_2$Bu^t; R′ = H, Me.[1046,1047]

The last reaction (Eq. 2.370) is favoured for less sterically demanding combinations of ligands.

Although W$_2$(OBu^t)$_6$ does not react with PhC≡CPh[1043] at room temperature, its reaction in molar ratios between 1:1 and 3:1 at about 70°C in toluene produces W$_2$(OBu^t)$_4$(μ-CPh)$_2$ and W$_2$(OBu^t)$_4$(PhC≡CPh)$_2$ in moderate yields.[1044]

Furthermore, the reaction of W$_2$(OBu^t)$_6$ with EtC≡CEt at 75–80°C for about three days affords moderate yields of a remarkable hexanuclear product, [W$_3$(OBu^t)$_5$(μ-O)(μ-CEt)O]$_2$, which has been characterized by X-ray crystallography.[1045]

Reaction of Mo$_2$(OBu^t)$_6$ with phenylacetylene in pentane produces the alkylidyne complex Mo(≡CC$_6$H$_5$)(OBu^t)$_3$ in 60% yield[1049] (Eq. 2.370):

$$Mo_2(OBu^t)_6 + C_6H_5C{\equiv}CH \longrightarrow (Bu^tO)_3Mo{\equiv}CC_6H_5 \qquad (2.370)$$

In the presence of quinuclidine (quin), both Mo(≡CC$_6$H$_5$)(OBu^t)$_3$(quin) and Mo(≡CH)(OBu^t)$_3$(quin) have been identified.

$$W_2(OBu^t)_6 + RC{\equiv}N \longrightarrow (Bu^tO)_3W{\equiv}N + (Bu^tO)_3W{\equiv}CR \qquad (2.371)$$

where R = Me, Ph.

In contrast, W$_2$(OCMe$_2$(CF$_3$))$_6$ reacts reversibly with acetonitrile[1048] to form addition complex W$_2$(OCMe$_2$(CF$_3$))$_6$(NCCH$_3$).

4.17.1.5 Reactions with ketones or aldehydes

W$_2$(OCH$_2$Bu^t)$_6$(py)$_2$ reacts with ketones or aldehydes in hydrocarbon solvents at room temperature to afford alkenes by a reductive coupling and deoxygenation of the C–O bonds[1050] according to Eq. (2.372):

$$W_2(OCH_2Bu^t)_6(py)_2 + 2R_2C{=}O \xrightarrow[\text{hexane}]{22°C} W_2O_2(OCH_2Bu^t)_6 + R_2C{=}CR_2 + 2py$$

$$(2.372)$$

4.17.1.6   Reaction with P_4

Reaction of $W_2(OCH_2Bu^t)_6(HNMe_2)_2$ with P_4 in hot (76°C) toluene affords two crystallographically characterized products, $(Me_2NH)(Bu^tCH_2O)_3W(\eta^3\text{-}P_3)$[1051] and $W_3(\mu_3\text{-}P)(OCH_2Bu^t)_9$.[1052]

4.17.1.7   Reactions with MeCOX or Me_3SiX ($X = Cl, Br, I$) or PF_3

Reactions between MeCOX or Me_3SiX and $Mo_2(OPr^i)_6$ in 1.5:1 molar ratio in hydrocarbon solvents yield $Mo_4X_3(OPr^i)_9$ and $MeCOPr^i$ or Me_3SiOPr^i, respectively (Eq. 2.373):[1053]

$$2Mo_2(OPr^i)_6 + 3R'X \longrightarrow Mo_4X_3(OPr^i)_9 + 3R'OPr^i \qquad (2.373)$$

where $R' = MeCO$ or Me_3Si.

This reaction (Eq. 2.373) when carried out with 2 equivalents of MeCOX or Me_3SiX affords $Mo_4X_4(OPr^i)_8$, but the iodo derivative was labile to elimination of 2 Pr^iI.[1053]

The reaction of $Mo_2(OBu^t)_6$ with PF_3 in 1:2 molar ratio in a hydrocarbon solvent affords the X-ray crystallographically characterized tetranuclear complex $Mo_4(\mu\text{-}F)_4(OBu^t)_8$.[1054]

In an extension of this procedure, Chisholm *et al.*[1053] investigated the reaction of $Mo_2(OPr^i)_6$ with PF_3 and isolated a rectangular cluster $Mo_2(\mu\text{-}F)_2(\mu\text{-}OPr^i)_2(OPr^i)_8$.

4.17.2   Reactions of $W_2(OCH_2Bu^t)_8$ ($W{=}W$)

The reactions of $W_2(OR)_8$-type derivatives are less developed compared to those of $M_2(OR)_6$. However, a report on the reactions of $W_2(OCH_2Bu^t)_8$ tends to indicate that this could become a rich area of chemistry,[1055] as shown in Fig. 2.16.

4.17.3   Reactions of $M_4(OR)_{12}$ with Carbon Monoxide, Isocyanides, Nitriles, Nitric Oxide, and Alkynes

When tetranuclear alkoxides $W_4(OR)_{12}$ ($R = Pr^i$, CH_2Bu^t) were allowed to react with CO in hexane at 0°C, carbido-oxo alkoxides, $W_4(\mu_4\text{-}C)O(OR)_{12}$ were isolated by cleavage of the carbonyl C–O bond[1021, 1022, 1056] and characterized by X-ray crystallography.[1056]

The reactions between $W_4(OR)_{12}$ ($R = CH_2Pr^i$, $CH_2\text{-}c\text{-}Pent$) and the isonitriles RNC ($R = Bu^t$, CH_2Ph, $C_6H_2Me_3\text{-}2,4,6$) in hydrocarbon solvents yielded μ_4-carbido clusters $W_4(\mu_4\text{-}C)(OR)_{14}$ by cleavage of the C≡N bond.[1024] By contrast, under similar conditions, the tetranuclear alkoxides fail to react with nitriles (MeC≡N and NC(CH_2)_5CN) and internal alkynes (MeC≡CMe, EtC≡CEt).

Nitric oxide reacts with $W_4(OR)_{12}$ in hydrocarbon solvents at 22°C in the presence of pyridine[1024] to yield $[W(OR)_3(NO)(Py)]_2$ (Eq. 2.374). The compound (with $R = CH_2Pr^i$) has been characterized by X-ray crystallography.

$$W_4(OR)_{12} + 4NO + 4Py \xrightarrow[22°C]{\text{hydrocarbon}} 2 \qquad (2.374)$$

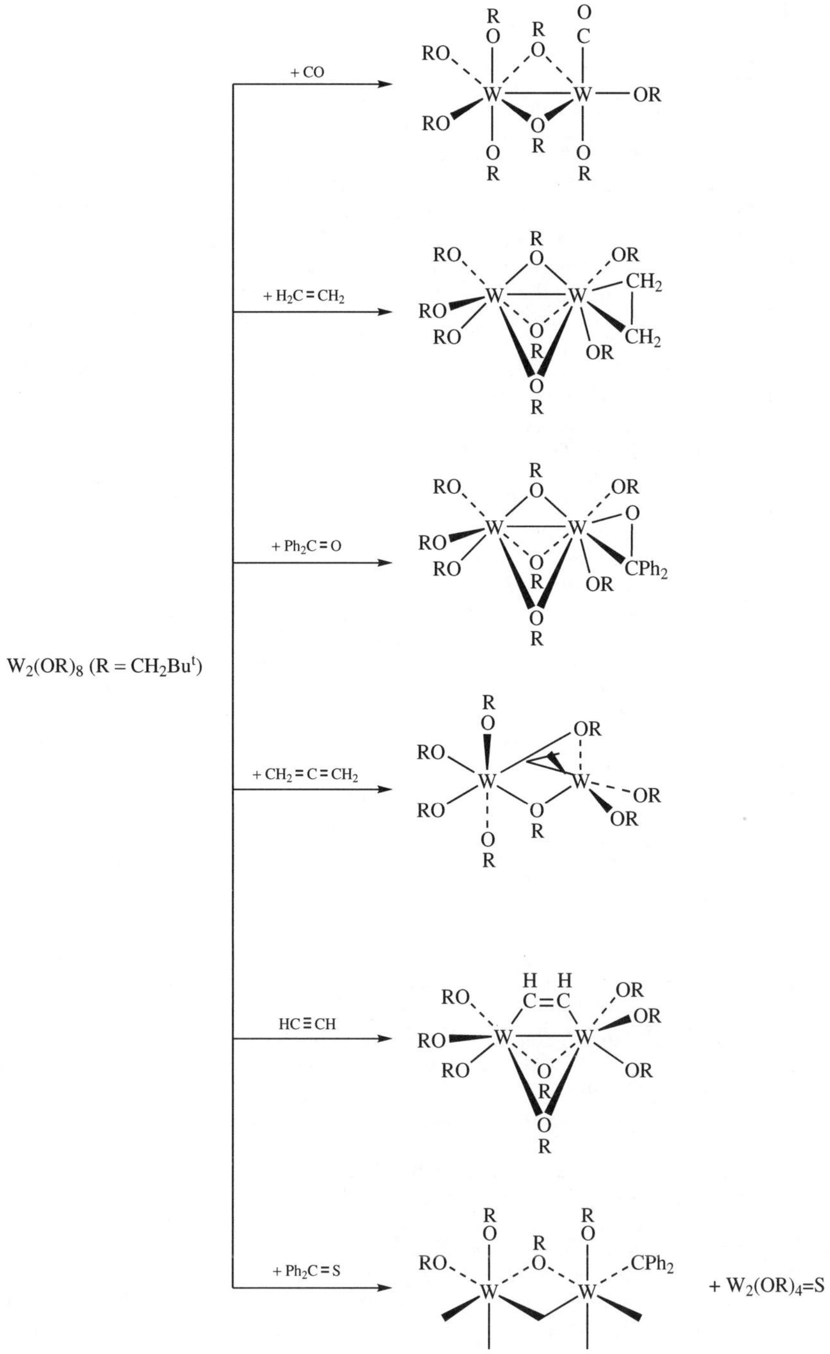

Figure 2.16 Some reactions of $W_2(OCH_2Bu^t)_8$.[1055]

REFERENCES

1. J.J. Ebelman, *Ann.*, **57**, 331 (1846).
2. J.J. Ebelman and M. Bouquet, *Ann. Chim. Phys.*, **17**, 54 (1846).
3. D.C. Bradley, *Prog. Inorg. Chem.*, **2**, 303 (1960).
4. R.C. Mehrotra, *Inorg. Chim. Acta Rev.*, **1**, 99 (1967).
5. D.C. Bradley and K.J. Fisher, in *M.T.P. International Reviews of Science* (H.J. Emeleus and D.W.A. Sharp, eds), Vol. 5, Part I, 65–91, Butterworths, London (1972).
6. D.C. Bradley, *Adv. Inorg. Chem. Radiochem.*, **15**, 259 (1972).
7. R.C. Mehrotra, *Adv. Inorg. Chem. Radiochem.*, **26**, 269 (1983).
8. M.H. Chisholm and I.P. Rothwell, in *Comprehensive Coordination Chemistry* (G. Wilkinson, R.D. Gillard and J.A. McCleverty, eds), Vol. 2, 335, Pergamon, London (1987).
9. R.C. Mehrotra, S.K. Agarwal, and Y.P. Singh, *Coord. Chem. Rev.*, **68**, 101 (1985).
10. H.E. Bryndza and W. Tam, *Chem. Rev.*, **88**, 1163 (1988).
11. R.C. Mehrotra, *J. Non-Cryst. Solids*, **100**, 1 (1988).
12. D.C. Bradley, *Chem., Rev.*, **89**, 1317 (1989).
13. R.C. Mehrotra, in *Structure and Bonding* (R. Relsfeld and C.K. Jorgensen, eds), Vol. 77, 1, Springer-Verlag, Berlin (1992).
14. L.G. Hubert-Pfalzgraf, *Appl. Organometal. Chem.*, **6**, 627 (1992).
15. S. Sakka, *J. Sol–Gel Sci. Technol.*, **3**, 69 (1994).
16. D.C. Bradley, *Polyhedron*, **13**, 1111 (1994).
17. W.G. Van der Sluys and A.P. Sattelberger, *Chem. Rev.*, **90**, 1027 (1990).
18. R.C. Mehrotra, A. Singh, and U.M. Tripathi, *Chem. Rev.*, **91**, 1287 (1991).
19. R.C. Mehrotra, A. Singh, and S. Sogani, *Chem. Rev.*, **94**, 1643 (1994).
20. R.C. Mehrotra, A. Singh, and S. Sogani, *Chem. Soc. Rev.*, **23**, 215 (1994).
21. R.C. Mehrotra and A. Singh, *Prog. Inorg. Chem.*, **46**, 239 (1997) and references therein.
22. W.A. Herrmann, N.W. Huber, and O. Runte, *Angew. Chem., Int. Ed. Engl.*, **34**, 2187 (1995).
23. L.G. Hubert-Pfalzgraf, *Coord. Chem. Rev.*, **178–180**, 967 (1998).
24. C.J. Willis, *Coord. Chem. Rev.*, **88**, 133 (1988).
25. D.C. Bradley, *Phil. Trans. Roy. Soc. London*, **A330**, 167 (1990).
26. D.C. Bradley, in *Preparative Inorganic Reactions*, (W. Jolly, ed.) Vol. 2, 169, Wiley, New York (1965).
27. See reference 240 of reference 21.
28. M.H. Chisholm, S.R. Drake, A.A. Naiini, and W.E. Streib, *Polyhedron,* **10**, 805 (1991).
28a. P.G. Williard and G.J. MacEwan, *J. Am. Chem. Soc.*, **111**, 7671 (1989); P.J. Wheatley, *J. Chem. Soc.*, 4270 (1960).
29. M.H. Chisholm, S.R. Drake, A.A. Naiini, and W.E. Streib, *Polyhedron*, **10**, 337 (1991).
30. N.Ya. Turova, A.V. Novoselova, and K.N. Semenenko, *Zhur. Neorg. Khim. (Engl. Trans.)*, **4**, 453 (1959).
31. N.Ya. Turova and A.V. Novoselova, *Usp. Khim.*, **34**, 385 (1965); *Chem. Abstr.*, **62**, 14213b (1965).
32. H. Lund and J. Bjerrum, *Chem. Ber., B*, **64**, 210 (1931).
33. C.A. Cohen, *U.S. Patent*, 2,287,088 (1943); *Chem. Abstr.*, **37**, 141 (1943).
34. A.G. Rheinpreussen, *Ger. Patent*, 1,230,004 (1966); *Chem. Abstr.*, **66**, 37426 (1967).
35. E. Szarwasy, *Chem. Ber.*, **30**, 806, 1836 (1897).
36. M. Quinet, *Bull. Soc. Chim. Fr.*, **5**, 1201 (1935).
37. H. Thoms, M. Epple, H. Viebrock, and A. Reller, *J. Mater. Chem.*, **5**, 589 (1995).
38. Z.A. Starikova, A.I. Yanovsky, E.P. Turevskaya, and N.Ya. Turova, *Polyhedron*, **16**, 967 (1997).
39. H.D. Lutz, *Z. Anorg. Allg. Chem.*, **339**, 308 (1965).
40. H.D. Lutz, *Z. Anorg. Allg. Chem.*, **353**, 207 (1967); **356**, 132 (1968).
41. N.Ya. Turova, B.A. Popovkin, and A.V. Novoselova, *Z. Anorg. Allg. Chem.*, **365**, 100 (1969).
42. K.S. Mazdiyasni, R.T. Dolloff, and J.S. Smith II, *J. Am. Ceram. Soc.*, **52**, 523 (1969).

43. J.S. Smith II, R.T. Dolloff, and K.S. Mazdiyasni, *J. Am. Ceram. Soc.*, **53**, 91 (1970).
44. H. Staeglich and E. Weiss, *Chem. Ber.*, **14**, 901 (1978).
45. J.G. Bednorz and K.A. Müller, *Z. Physik, B*, **64**, 189 (1986).
46. K.A. Muller and J.G. Bednorz, *Science*, **217**, 1133 (1987).
47. A.P. Purdy, C.F. George, and J.H. Callahan, *Inorg. Chem.*, **30**, 2812 (1991).
48. S.R. Drake, W.E. Streib, K. Folting, M.H. Chisholm, and K.G. Caulton, *Inorg. Chem.*, **31**, 3205 (1992).
49. K.G. Caulton, M.H. Chisholm, S.R. Drake, and K. Folting, *J. Chem. Soc., Chem. Commun.*, 1349 (1990).
50. S.C. Goel, M.A. Matchett, M.Y. Chiang, and W.E. Buhro, *J. Am. Chem. Soc.*, **113**, 1844 (1991).
51. K.G. Caulton, M.H. Chisholm, S.R. Drake, and J.C. Huffman, *J. Chem. Soc., Chem. Commun.*, 1498 (1990).
52. W.S. Rees Jr and D.A. Moreno, *J. Chem. Soc., Chem. Commun.*, 1759 (1991).
53. W.A. Herrmann, N.W. Huber, and T. Priermeier, *Angew. Chem., Int. Ed. Engl.*, **33**, 105 (1994).
54. K.S. Mazdiyasni, C.T. Lynch, and J.S. Smith, *Inorg. Chem.*, **5**, 342 (1966).
55. L.M. Brown and K.S. Mazdiyasni, *Inorg. Chem.*, **9**, 2783 (1970).
56. O. Poncelet and L.G. Hubert-Pfalzgraf, *Polyhedron*, **8**, 2183 (1989).
57. O. Poncelet, L.G. Hubert-Pfalzgraf, J.C. Daran, and R. Astier, *J. Chem. Soc., Chem. Commun.*, 1846 (1989).
58. D.C. Bradley, H. Chudzynska, D.M. Frigo, M.E. Hammond, M.B. Hursthouse, and M.A. Mazid, *Polyhedron*, **9**, 719 (1990).
59. O. Poncelet, W.J. Sartain, L.G. Hubert-Pfalzgraf, K. Folting, and K.G. Caulton, *Inorg. Chem.*, **28**, 263 (1989).
60. G. Helgesson, S. Jagner, O. Poncelet, and L.G. Hubert-Pfalzgraf, *Polyhedron*, **10**, 1559 (1991).
61. N.Ya. Turova, E.P. Tureuskaya, V.G. Kessler, A.I. Yanovsky, and Y.T. Struchkov, *J. Chem. Soc., Chem. Commun.*, 21 (1993).
62. N.Ya. Turova, E.P. Turevskaya, M.I. Turevskaya, A.I. Yanovsky, V.G. Kessler, and D.E. Tacheboukov, *Polyhedron*, **17**, 899 (1998).
63. N.Ya. Turova, V.G. Kessler, and S.I. Kuchciko, *Polyhedron*, **10**, 2617 (1991).
64. D.C. Bradley, B.N. Chakravarti, and A.K. Chatterjee, *J. Chem. Soc.*, 99 (1958).
65. J.H. Gladstone and A. Tribe, *J. Chem. Soc.*, 394 (1881).
66. H. Wislicenus and O. Kaufman, *Chem. Ber.*, **28**, 1323 (1893).
67. H.W. Hillyer, *Am. Chem. J.*, **19**, 37 (1897).
68. W.E. Tischtschenko, *J. Russ. Chem. Ges.*, **31**, 694, 784 (1899).
69. W.C. Child and H. Adkins, *J. Am. Chem. Soc.*, **45**, 3013 (1923).
70. W.G. Young, W.H. Harting, and F.S. Crossley, *J. Am. Chem. Soc.*, **58**, 100 (1936).
71. H. Adkins and F.W. Cox, *J. Am. Chem. Soc.*, **60**, 1151 (1938).
72. S.J. Teichner, *Compt. Rend.*, **237**, 810 (1953).
73. S. Coffey and V. Boyd, *J. Chem. Soc.*, 2468 (1954).
74. G.C. Whitaker, *Metal Organic Compounds*, Advances in Chem. Series, **23**, 184 (1959).
75. A. Lamy, *Ann. Chim.*, **3**, 373 (1864).
76. B. Szilard, *Z. Electrochem.*, **12**, 393 (1906).
77. *Gerr. Offen. Pat.* 2,005,835 (1970); *Fr. Pat.* 2,091,229 (1972).
78. H. Lehmkuhl and W. Eisenbach, *Ann. Chem.*, 672 (1975).
79. V.A. Shreider, E.P. Turevskaya, N.I. Kozlova, and N.Ya. Turova, *Inorg. Chim. Acta*, **13**, L73 (1981).
80. N.I. Kozlova, N.Ya. Turova, and E. Turevskaya, *Sov. J. Coord. Chem.*, **8**, 339 (1982).
81. N.Ya. Turova, A.V. Korolev, D.E. Tchebukov, and A.I. Belokon, *Polyhedron*, **15**, 3869 (1996).
82. S.I. Kucheiko, N.Ya. Turova, and V.A. Schreider, *Russ. J. Gen. Chem.*, **55**, 2396 (1985).
83. N.Ya. Turova and V.G. Kessler, *Russ. J. Gen. Chem.*, **60**, 113 (1990).
84. N.Ya. Turova, V.G. Kessler, and S.I. Kucheiko, *Polyhedron*, **10**, 2617 (1991).

85. V.G. Kessler, A.V. Mironov, N.Ya. Turova, A.I. Yanovsky, and Yu.T. Struchkov, *Polyhedron*, **12**, 1573 (1993).
86. V.G. Kessler, E.P. Turevskaya, and S.I. Kucheiko, *Mater. Res. Soc. Symp. Proc.*, **346**, 3 (1994).
87. N.Ya. Turova, E.P. Turevskaya, V.G. Kessler, N.I. Kozlova, and A.I. Belokon, *Russ. J. Inorg. Chem.*, **37**, 26 (1992).
88. S.I. Kucheiko, V.G. Kessler, and N.Ya. Turova, *Sov. J. Coord. Chem.*, **13**, 586 (1987).
89. T.V. Rogova, unpublished results.
90. T.V. Rogova and N.Ya. Turova, *Sov. J. Coord. Chem.*, **11**, 448 (1985).
91. E.P. Turevskaya, N.I. Kozlova, N.Ya. Turova, B.A. Popovkin, and M.I. Yanovskaya, *Phys. Chem. Technol.*, **2**, 925 (1989).
92. V.G. Kessler, A.V. Shevel'kov, and G.V. Khvorykh, *Russ. J. Inorg. Chem.*, **40**, 1424 (1995).
93. J.S. Banait and P.K. Pahil, *Synth. React. Inorg. Met.-Org. Chem.*, **16**, 1217 (1986); *Polyhedron*, **5**, 1865 (1986).
94. J.S. Banait, S.K. Deol, and B. Singh, *Synth. React. Inorg. Met.-Org. Chem.*, **20**, 1331 (1990).
95. V.G. Kessler, A.N. Panov, N.Ya. Turova, Z.A. Starikova, A.I. Yanovsky, F.M. Dolgushin, A.P. Pisarevsky, and Yu.T. Struchkov, *J. Chem. Soc., Dalton Trans.*, 21 (1998).
96. E.P. Kovsman, S.I. Andruseva, L.I. Solovjeva, V.I. Fedyaev, M.N. Adamova, and T.V. Rogova, *Sol. Gel. Sci. Technol.*, **2**, 61 (1994).
97. G. Wilkinson, F.G.A. Stone, and E.W. Abel (eds), *Comprehensive Organometallic Chemistry I*, Vols 1–9, Pergamon Press, Oxford (1982); E.W. Abel, F.G.A. Stone, and G. Wilkinson (eds), *Comprehensive Organometallic Chemistry II*, Vols 1–14, Pergamon Press, Oxford (1995).
98. W.J. Power and G.A. Ozin, *Adv. Inorg. Chem. Radiochem.*, **23**, 140 (1982).
99. P.B. Hitchcock, M.F. Lappert, G.A. Lawless, and B. Royo, *J. Chem. Soc., Chem. Commun.*, 141 (1990).
100. S. Halut-Desportes and M. Philoche-Levisalies, *Acta Crystallogr., B*, **34**, 432 (1978).
101. R.C. Mehrotra, T.N. Misra, and S.N. Misra, *J. Indian Chem. Soc.*, **42**, 351 (1965).
102. S.N. Misra, T.N. Misra, and R.C. Mehrotra, *J. Inorg. Nucl. Chem.*, **27**, 105 (1965).
103. J.M. Batwara, U.D. Tripathi, and R.C. Mehrotra, *J. Chem. Soc., A*, 991 (1967).
104. A. Rosenheim, V. Samter, and J. Davidsohn, *Z. Anorg. Allg. Chem.*, **35**, 447 (1903).
105. G. Jantsch and W. Urbach, *Helv. Chim. Acta*, **2**, 490 (1919).
106. R.C. Mehrotra, K.N. Mahendra, and M. Aggarwal, *Proc. Indian Acad. Sci. (Chem. Sci.)*, **93**, 719 (1984).
107. D.C. Bradley, M.A. Saad, and W. Wardlaw, *J. Chem. Soc.*, 2002 (1954).
108. Y.T. Wu, Y.C. Ho, C.C. Lin, and H.M. Gau, *Inorg. Chem.*, **35**, 5948 (1996).
108a. C.H. Winter, P.H. Sheridon, and M.J. Heeg, *Inorg. Chem.*, **30**, 1962 (1991).
109. W.G. Van der Sluys, J.C. Huffman, D.S. Ehler, and N.N. Sauer, *Inorg. Chem.*, **31**, 1316 (1992).
110. T. Colclough, W. Gerrard, and M.F. Lappert, *J. Chem. Soc.*, 3006 (1996).
111. M.F. Lappert, *Chem. Rev.*, **56**, 959 (1956).
112. D.E. Young, L.R. Anderson, and W.C. Cox, *J. Chem. Soc., C*, 736 (1971).
113. D.C. Bradley, R.C. Mehrotra, and W. Wardlaw, *J. Chem. Soc.*, 5020 (1952).
114. L. Sheng-Lich and W. Hsiuchin, *J. Chinese Chem. Soc.*, **13**, 188 (1966); *Chem. Abstr.*, **68**, 2947 (1968).
115. W. Rothe, *Ger. Patent*, 1,298,972; *Chem. Abstr.*, **71**, 90804 (1969).
116. M. Wojnowska and W. Wojnowski, *Rocz. Chem.*, **44**, 1019 (1970); *Chem. Abstr.*, **73**, 130577 (1970).
117. F. Augst and F. Sufraga, *Ger. Offen Patent*, 2,033,373 (1971); *Chem. Abstr.*, **74**, 125839 (1971).
118. J.R. Van Wazer, *Phosphorus and its Compounds*, Vols 1 and 2, Interscience, New York (1958, 1961).
119. G.O. Doak and L.D. Freedman, *Chem. Rev.*, **61**, 31 (1961).
120. G. Tokar and I. Simonyi, *Magyar Kem Golyoirat*, **63**, 172 (1957).
121. R.K. Mehrotra, *Ph.D. Thesis*, University of Gorakhpur, India (1962).

122. P.N. Kapoor, *Ph.D. Thesis*, University of Rajasthan, Jaipur, India (1965).
123. D. Brown in *Comprehensive Inorganic Chemistry* (J.C. Bailar Jr, H.J. Emeleus, R.S. Nyholm, and A.F. Trotman-Dickenson, eds), Vol. 3, Pergamon Press, Oxford (1973).
124. D.L. Tabern, W.R. Orndorff, and L.M. Dennis, *J. Am. Chem. Soc.*, **47**, 2039 (1925).
125. D.C. Bradley, L. Kay, and W. Wardlaw, *Chem. Ind. (London)*, 746 (1953).
126. C. Miner Jr, L.A. Bryan, R.P. Holysz, and G.W. Pedlow Jr, *Ind. Eng. Chem., Ind. Ed.*, **39**, 1368 (1947).
127. D. Ridge and M. Todd, *J. Chem. Soc.*, 2637 (1949).
128. D.B. Denny, D.Z. Denny, and K.S. Tseng, *Phosphorus Sulfur*, **22**, 33 (1985).
129. A. Fischer, *Z. Anorg. Allg. Chem.*, **81**, 170 (1913).
130. O. Klejnot, *Inorg. Chem.*, **4**, 1668 (1965).
131. H. Funk and H. Naumann, *Z. Anorg. Allg. Chem.*, **343**, 294 (1966).
132. D.P. Rillema and C.H. Brubaker Jr, *Inorg. Chem.*, **8**, 1645 (1969).
132a. D.P. Rillema, W.J. Reagen, and C.H. Brubaker Jr, *Inorg. Chem.*, **8**, 587 (1969).
133. L.B. Anderson, F.A. Cotton, D. De Marco, A. Fang, W.H. Isley, B.W.S. Kolthammer, and R.A. Walton, *J. Am. Chem. Soc.*, **103**, 5078 (1981).
134. J. Nelles, *Brit. Patent*, 512,452 (1939); *Chem. Abstr.*, **34**, 3764 (1940).
135. D.C. Bradley and W. Wardlaw, *Nature (London)*, **165**, 75 (1950).
136. H. Meerwein, *U.S. Patent*, 1689356 (1929); *Chem. Abstr.*, **23**, 156 (1929).
137. W.G. Bartley and W. Wardlaw, *J. Chem. Soc.*, 421 (1958).
138. M. Merbach and U.P. Carrard, *Helv. Chim. Acta.*, **54**, 2771 (1971).
139. K.S. Mazdiyasni and B.J. Schaper, *J. Less-Common Met.*, **30**, 105 (1973).
140. R.N.P. Sinha, *M.Sc. Thesis*, University of London (1957).
141. D.C. Bradley, M.A. Saad, and W. Wardlaw, *J. Chem. Soc.*, 1091 (1954).
141a. D.C. Bradley, M.A. Saad, and W. Wardlaw, *J. Chem. Soc.*, 3488 (1954).
142. D.C. Bradley, B.N. Chakravarti, and A.K. Chatterjee, *J. Inorg. Nucl. Chem.*, **3**, 367 (1957).
143. D.C. Bradley, A.K. Chatterjee, and W. Wardlaw, *J. Chem. Soc.*, 2260 (1956).
144. D.C. Bradley, B. Harder, and F. Hudswell, *J. Chem. Soc.*, 3318 (1957).
145. D.C. Bradley, F.M. Abd El-Halim, E.A. Sadek, and W. Wardlaw, *J. Chem. Soc.*, 2032 (1952).
145a. D.C. Bradley and W. Wardlaw, *J. Chem. Soc.*, 280 (1951).
146. D.C. Bradley, R.P.N. Sinha, and W. Wardlaw, *J. Chem. Soc.*, 4651 (1958).
146a. D.B. Denny, D.Z. Denny, P.J. Hammond, and Y.F. Hsu, *J. Am. Chem. Soc.*, **103**, 2340 (1981).
147. K.F. Tesh, T.P. Hanusa, J.C. Huffman, and C.J. Huffman, *Inorg. Chem.*, **31**, 5572 (1992).
148. U.D. Tripathi, *Ph.D. Thesis*, University of Rajasthan, Jaipur, India (1970).
149. J.M. Batwara, U.D. Tripathi, R.K. Mehrotra, and R.C. Mehrotra, *Chem. Ind. (London)*, 1379 (1966).
150. S.N. Misra, T.N. Misra, R.N. Kapoor, and R.C. Mehrotra, *Chem. Ind. (London)*, 120 (1963).
151. T.N. Misra, *Ph.D. Thesis*, University of Rajasthan, Jaipur, India (1963).
152. S.N. Misra, *Ph.D. Thesis*, University of Rajasthan, Jaipur, India (1965).
153. S.N. Misra, T.N. Misra, and R.C. Mehrotra, *Aust. J. Chem.*, **21**, 797 (1968).
154. B.S. Sankhla, S.N. Misra, and R.N. Kapoor, *Chem. Ind. (London)*, 382 (1965).
155. B.S. Sankhla, S.N. Misra, and R.N. Kapoor, *Aust. J. Chem.*, **20**, 2013 (1967).
156. J.M. Batwara, *Ph.D. Thesis*, University of Rajasthan, Jaipur, India (1969).
157. R.C. Mehrotra and J.M. Batwara, *Inorg. Chem.*, **9**, 2505 (1970).
158. A. Mehrotra and R.C. Mehrotra, *Indian J. Chem.*, **10**, 532 (1972).
159. R.A. Andersen, D.H. Templeton, and A. Zalkin, *Inorg. Chem.*, **17**, 1962 (1978).
160. W.J. Evans, M.S. Sollberger, and T.P. Hanusa, *J. Am. Chem. Soc.*, **110**, 1841 (1988).
161. W.J. Evans and M.S. Sollberger, *Inorg. Chem.*, **27**, 4417 (1988).
162. P.S. Gradeff, F.G. Schreiber, K.C. Brooks, and R.E. Sievers, *Inorg. Chem.*, **24**, 1110 (1985).

163. P.S. Gradeff, F.G. Schreiber, and H. Mauermann, *J. Less-Common Met.*, **126**, 335 (1986).
164. W.J. Evans, T.J. Deming, J.M. Olofson, and J.W. Ziller, *Inorg. Chem.*, **28**, 4027 (1989).
165. D.M. Barnhart, D.L. Clark, J.C. Gordon, J.C. Huffman, and J.G. Watkin, *Inorg. Chem.*, **33**, 3939 (1994).
166. F. Bischoff and H. Adkins, *J. Am. Chem. Soc.*, **46**, 256 (1924).
167. N. Brnicevic, F. Mustovic, and R.E. McCarley, *Inorg. Chem.*, **27**, 4532 (1988).
168. H.L. Krauss and G. Munster, *Z. Anorg. Allg. Chem.*, **352**, 24 (1967).
169. M. Bochmann, G. Wilkinson, G.B. Young, M.B. Hursthouse, and K.M.A. Malik, *J. Chem. Soc., Dalton Trans.*, 901 (1980).
170. T.M. Gilbert, A.M. Landes, and R.D. Rogers, *Inorg. Chem.*, **31**, 3438 (1992).
171. H. Rothfuss, J.C. Huffman, and K.G. Caulton, *Inorg. Chem.*, **33**, 187 (1994).
172. A.C.C. Wong, G. Wilkinson, B. Hussain, M. Motevalli, and M.B. Hursthouse, *Polyhedron*, **7**, 1363 (1988).
173. D.M. Hoffman, D. Lappas, and D.A. Wierda, *J. Am. Chem. Soc.*, **111**, 1531 (1989).
174. D.M. Hoffman, D. Lappas, and D.A. Wierda, *J. Am. Chem. Soc.*, **115**, 10538 (1993).
175. P.M. Jeffries, S.R. Wilson, and G.S. Girolami, *Inorg. Chem.*, **31**, 4503 (1992).
176. T. Blackmore, M.I. Bruce, P.J. Davidson, M.Z. Iqbal, and F.G.A. Stone, *J. Chem. Soc. A*, 3153 (1970).
177. H.E. Bryndza, J.C. Calabrese, M. Marsi, D.C. Roc, W. Tam, and J.E. Bercaw, *J. Am. Chem. Soc.*, **108**, 4805 (1986).
177a. H.E. Bryndza, S.A. Kretchmar, and T.H. Tulip, *J. Chem. Soc. Chem. Commun.*, 977 (1985).
178. N.W. Alcock, A.W.G. Platt, and P.G. Pringle, *J. Chem. Soc., Dalton Trans.*, 139 (1989).
178a. N.W. Alcock, A.W.G. Platt, and P.G. Pringle, *Inorg. Chim. Acta*, **128**, 215 (1987).
178b. N.W. Alcock, A.W.G. Platt, and P.G. Pringle, *J. Chem. Soc., Dalton Trans.*, 2273 (1987).
179. H. Meerwein and T. Bersin, *Ann.*, **455**, 23 (1927); **476**, 113 (1929).
180. D.C. Bradley, E.V. Caldwell, and W. Wardlaw, *J. Chem. Soc.*, 4775 (1957).
181. R.C. Mehrotra and R.K. Mehrotra, *Curr. Sci. (India)*, 241 (1964).
182. R. Reinmann and A. Tanner, *Z. Naturforsch.*, **20(B)**, 524 (1965).
183. L. Moegele, *Z. Anorg. Allg. Chem.*, **363**, 166 (1968).
184. S.R. Bindal, V.K. Mathur, and R.C. Mehrotra, *J. Chem. Soc., A*, 863 (1969).
185. S.R. Bindal, *Ph.D. Thesis*, University of Rajasthan, Jaipur, India (1973).
186. S. Chatterjee, S.R. Bindal, and R.C. Mehrotra, *J. Indian Chem. Soc.*, **53**, 867 (1976).
187. A. Mehrotra and R.C. Mehrotra, *Inorg. Chem.*, **11**, 2170 (1970).
188. N.V. Sidgwick and A.W. Laubengayer, *J. Am. Chem. Soc.*, **54**, 948 (1932).
189. T. Athar, R. Bohra, and R.C. Mehrotra, *Synth. React. Inorg. Met.-Org. Chem.*, **19**, 195 (1989).
190. M.J. Hampden-Smith, T.A. Wark, A. Rheingold, and J.C. Huffman, *Can. J. Chem.*, **69**, 121 (1991).
191. H. Reuter and M. Kremser, *Z. Anorg. Allg. Chem.*, **259**, 598 (1991).
192. R.C. Mehrotra, A.K. Rai, and A. Jain, *Polyhedron*, **10**, 1103 (1991).
193. R. Papiernik, L.G. Hubert-Pfalzgraf, and M.C. Massiani, *Inorg. Chim. Acta*, **165**, 1 (1989).
194. R. Papiernik, L.G. Hubert-Pfalzgraf, and M.C. Massiani, *Polyhedron*, **10**, 1657 (1991).
195. J.D. Smith, in *Comprehensive Inorganic Chemistry* (J.C. Bailar Jr, H.J. Emeleus, R.S. Nyholm, and A.F. Trotman-Dickenson, eds), Vol. 2, Ch. 21, 547–683, Pergamon Press, Oxford (1973).
196. W. Herrmann, in *Methoden der Organischen chemie*, 4th edn, Vol. 6, Part 2, 363. G. Thieme Verlag, Stuttgart (1963).
197. K. Moedritzer, *Inorg. Synth.*, **11**, 181 (1968).
198. J.M. Crafts, *Bull. Soc. Chim.*, **14**, 102 (1870).
199. C. Russias, F. Damm, A. Deluzarche, and A. Maillard, *Bull. Soc. Chim. France*, 2275 (1966).
200. O.D. Dubrovina, *Uchenye Zapiski Kazan Gosudarst Univ. im. V.I. Ulyanova-Lenina*, **116**, 3 (1956); *Chem. Abstr.*, **51**, 6534i (1957).

201. A. Killard, A. Deluzarche, J.C. Marie, and L. Havas, *Bull. Soc. Chim. France*, 2962 (1965).
202. R.C. Mehrotra and A.K. Rai, *Indian J. Chem.*, **4**, 537 (1966).
203. W.J. Evans, J.H. Hains Jr, and J.W. Ziller, *J. Chem. Soc., Chem. Commun.*, 1628 (1989).
204. M.C. Massiani, R. Papiernik, L.G. Hubert-Pfalzgraf, and J.C. Daran, *J. Chem. Soc., Chem. Commun.*, 301 (1990).
205. M.C. Massiani, R. Papiernik, L.G. Hubert-Pfaslzgraf, and J.C. Daran, *Polyhedron*, **10**, 437 (1991).
206. K.H. Whitmire, C.M. Jones, M.D. Burkart, J.C. Hutchison, and A.L. McKnight, *Mat. Res. Soc. Symp. Proc.*, **271**, 149 (1992).
206a. C.M. Jones, M.D. Burkart, R.E. Bachman, D.L. Serra, S.J. Hwu, and K.H. Whitmire, *Inorg. Chem.*, **32**, 5136 (1993).
207. G. Dupuy, *Compt. Rend.*, **240**, 2238 (1955).
208. R.C. Mehrotra and S.N. Mathur, *J. Indian Chem. Soc.*, **42**, 1 (1965).
209. M. Veith, P. Hobein, and R. Rosler, *Z. Naturforsch.* **44b**, 1067 (1989).
210. D.P. Gaur, G. Srivastava, and R.C. Mehrotra, *J. Organomet. Chem.*, **63**, 221 (1973).
211. D.P. Gaur, *Ph.D. Thesis*, University of Rajasthan, Jaipur, India (1973).
212. A.G. Davies, L. Smith, and P.S. Smith, *J. Organomet. Chem.*, **39**, 279 (1972).
213. R.G. Jones, G. Karmas, G.A. Martin Jr, and H. Gilman, *J. Am. Chem. Soc.*, **78**, 4285 (1956).
214. M. Arora and R.C. Mehrotra, *Indian J. Chem.*, **7**, 399 (1969).
215. D.C. Bradley and M.M. Faktor, *Chem. Ind. (London)*, 1332 (1958).
216. D.C. Bradley, B.N. Chakravarti, and A.K. Chatterjee, *J. Inorg. Nucl. Chem.*, **12**, 71 (1959).
217. E.T. Samulski and D.G. Karraker, *J. Inorg. Nucl. Chem.*, **29**, 993 (1967).
218. D.C. Bradley and M.L. Mehta, *Can. J. Chem.*, **40**, 1710 (1962).
219. R.W. Adams, E. Bishop, R.L. Martin, and G. Winter, *Aust. J. Chem.*, **19**, 207 (1966).
220. G.A. Kakos and G. Winter, *Aust. J. Chem.*, **22**, 97 (1969).
221. C.H. Brubaker Jr, and M. Wicholas, *J. Inorg. Nucl. Chem.*, **27**, 59 (1965).
222. R.C. Mehrotra and M. Arora, *Z. Anorg. Allg. Chem.*, **370**, 300 (1969).
223. E.P. Turevskaya, N.Ya. Turova, and A.V. Novoselova, *Izv. Akad. Nauk. SSSR, Ser. Khim.*, 1667 (1968); *Chem. Abstr.*, **69**, 102603 (1968).
224. T.V. Talalaeva, G.V. Zenina, and K.A. Kocheshkov, *Dokl. Akad. Nauk. SSSR.*, **171**, 122 (1966).
225. R.C. Mehrotra, *Coord. Chem. (IUPAC)*, **21**, 113 (1981).
226. M. Bochmann, G. Wilkinson, G.B. Young, M.B. Hursthouse, and K.M.A. Malik, *J. Chem. Soc., Dalton Trans.*, 1863 (1980).
227. P.B. Hitchcock, M.F. Lappert, and I.A. MacKinnon, *J. Chem. Soc., Chem. Commun*, 1557 (1988).
227a. P.B. Hitchcock, M.F. Lappert, and I.A. MacKinnon, *J. Chem. Soc., Chem. Commun.*, 1015 (1993).
228. J.L. Stewart and R.A. Andersen, *J. Chem. Soc., Chem. Commun*, 1846 (1987).
229. T.V. Lubben, P.T. Wolczanski, and G.D. Van Duyne, *Organometallics*, **3**, 977 (1984).
230. J. Hoslef, H. Hope, B.D. Murray, and P.P. Power, *J. Chem. Soc., Chem. Commun.*, 1438 (1983).
231. M.M. Olmstead, P.P. Power, and G. Sigel, *Inorg. Chem.*, **25**, 1027 (1986).
232. S.L. Chadha and K. Uppal, *Bull. Soc. Chim. France*, **3**, 431 (1987).
233. S.L. Chadha, V. Sharma, and A. Sharma, *J. Chem. Soc., Dalton Trans.*, 1253 (1987).
234. S.L. Chadha, *Inorg. Chim. Acta.*, **156**, 173 (1989).
235. R. Choukroun, A. Dia, and D. Gervais, *Inorg. Chim. Acta*, **34**, 211 (1979).
236. G.E.M. Jones and O.L. Hughes, *J. Chem. Soc.*, 1197 (1934).
237. E.F. Caldin and G. Long, *J. Chem. Soc.*, 3737 (1954).
238. H. Steinberg, *Organoboron Chemistry*, Vol. 1, Interscience, New York (1964).
239. G.W.A. Kahlbaum, K. Roth, and P. Seidler, *Z. Anorg. Allg. Chem.*, **29**, 223 (1902).
240. J.S. Skelcy, J.E. Rumminger, and K.O. Groves, *U.S. Patent*, 3,494,946 (1970); *Chem. Abstr.*, **72**, 110787 (1970).

241. M.F. Shostakovskii, I.A. Shikhiev, and D.A. Kochkin, *Izv. Akad. Nauk, SSSR, Otd. Khim. Nauk.*, 941 (1953); *Chem. Abstr.*, **49**, 1541 (1955).

242. M.F. Shostakovskii, D.A. Kochkin, and V.M. Rogov, *Izv. Akad. Nauk, SSSR, Otd. Khim. Nauk.*, 953 (1955).

243. R.M. Meals, *U.S. Patent*, 2,826,599 (1958); *Chem. Abstr.*, **52**, 10638 (1958).

244. P. Pascal and A. Dupire, *C.R. Acad. Sci., Paris*, **195**, 14 (1932); *Chem. Abstr.*, **26**, 5064 (1932).

245. B.D. Chernokalsknu, V.S. Gamoyurova, and G.K. Kamai, *Vysshikh Uchebn Zavedenu Khim.i-Khim Tekhnol*, **8**, 959 (1965).

246. R.D. Gigauri, G.K. Kamai, and M.M. Ugulava, *Zh. Obshch. Khim.*, **41**, 336 (1971); *Chem. Abstr.*, **75**, 35064 (1971).

247. W. Prandtl and L. Hess, *Z. Anorg. Chem.*, **82**, 103 (1913).

248. F. Cartan and C.N. Caughlan, *J. Phys. Chem.*, **64**, 175 (1960).

249. R.C. Mehrotra and R.K. Mittal, *Z. Anorg. Allg. Chem.*, **327**, 111 (1964).

250. C.W. Decock and L.V. McAffee, *Inorg. Chem.*, **24**, 4293 (1985).

251. G.B. Goodwin and M.E. Kenney, *Inorg. Chem.*, **29**, 1216 (1990).

252. M. Akiyama, E. Suzuki, and Y. Ono, *Inorg. Chim. Acta.*, **207**, 259 (1993).

253. R.C. Mehrotra, V.D. Gupta, and G. Srivastara, *Rev. Silicon, Germanium, Tin, Lead Compd*, **1**, 299 (1975)

254. R.C. Mehrotra and S. Mathur, *J. Organomet. Chem.*, **6**, 11 (1966).

255. R.C. Mehrotra and S. Mathur, *J. Organomet. Chem.*, **7**, 233 (1967).

256. G.S. Sasin, *J. Org. Chem.*, **18**, 1142 (1953).

257. D.L. Alleston, A.G. Davies, M. Hancock, and R.F.M. White, *J. Chem. Soc.*, 5469 (1963).

258. W.J. Considine, J.J. Ventura, A.J. Gibbons, and A. Ross, *Can. J. Chem.*, **41**, 1239 (1963).

259. A. Rieche and J. Dahlmann, *Ann.*, **675**, 19 (1964).

260. B.R. Laliberte, W. Davidson, and M.C. Henry, *J. Organomet. Chem.*, **5**, 526 (1966).

261. A.G. Davies, D.C. Kleinschmidt, P.R. Palan, and S.C. Vasishta, *J. Chem. Soc.*, 3972 (1971).

262. A.J. Bloodworth and A.G. Davies in *Organotin Compounds* (A.K. Sawyer, ed.), Vol. 1, Marcel Dekker, New York (1971).

263. R.K. Ingham, S.D. Rosenberg, and H. Gilman, *Chem. Rev.*, **66**, 459 (1969).

264. W.P. Newmann, *The Organic Chemistry of Tin*, J. Wiley & Sons, London (1970).

265. R.C. Poller, *The Chemistry of Organotin Compounds*, Logos Press, London (1970).

266. M.G. Voronkov and J. Romadans, *Zh. Obshch. Khim.*, **39**, 2785 (1969); *Chem. Abstr.* **72**, 111581 (1970).

267. M.F. Shostakovskii, N.V. Komarov, and T.I. Ermolova, *Izv. Akad. Nauk. SSSR, Ser. Khim.*, 1170 (1969); *Chem. Abstr.*, 71, 50152 (1969).

268. M.F. Shostakovskii, N.V. Komarov, and T.I. Ermolova, *Dokl. Akad. Nauk. SSSR*, **184**, 1117 (1969); *Chem. Abstr.*, **70**, 115282 (1969).

269. A.J. Bloodworth, *J. Chem. Soc. C*, 205 (1970).

270. I.D. Verma and R.C. Mehrotra, *J. Chem. Soc.*, 2966 (1960).

271. I. Shiihava, W.T. Schwartz Jr, and H.W. Post, *Chem. Rev.*, **61**, 1 (1961).

272. C. Campbell, S.G. Bott, R. Larsen, and W.G. Van der Sluys, *Inorg. Chem.*, **33**, 4950 (1994).

273. D.C. Bradley, R.C. Mehrotra, and W. Wardlaw, *J. Chem. Soc.*, 2027 (1952).

274. D.C. Bradley, R.C. Mehrotra, and W. Wardlaw, *J. Chem. Soc.*, 4204 (1952).

275. D.C. Bradley and M.M. Faktor, *Nature (London)*, **184**, 55 (1959).

276. R.C. Mehrotra and I.D. Verma., *J. Indian Chem. Soc.*, **38**, 147 (1961).

277. D.C. Bradley, R.C. Mehrotra, and W. Wardlaw, *J. Chem. Soc.*, 1634 (1953).

278. R.K. Mittal, *Ph.D. Thesis*, University of Rajasthan, Jaipur, India (1963).

279. D.C. Bradley, B.N. Chakravarti, and W. Wardlaw, *J. Chem. Soc.*, 4439 (1956); ibid., 2381 (1956).

280. D.C. Bradley, W. Wardlaw, and A. Whitley, *J. Chem. Soc.*, 1139 (1956); ibid., 726 (1955).

281. R.C. Mehrotra and P.N. Kapoor, *J. Less-Common Met.*, **10**, 354 (1966); ibid., 726 (1955).
282. D.C. Bradley, R.K. Multani, and W. Wardlaw, *J. Chem. Soc.*, 126 (1958).
283. P.P. Sharma, *Ph.D. Thesis*, University of Rajasthan, Jaipur, India (1966).
284. J.V. Singh, *Ph.D. Thesis*, Delhi University (1980).
285. S.C. Goel, K.S. Kramer, P.C. Gibbons, and W.E. Buhro, *Inorg. Chem.*, **28**, 3619 (1989).
286. D.C. Bradley, A.K. Chatterjee, and W. Wardlaw, *J. Chem. Soc.*, 2600 (1957).
287. D.C. Bradley, A.K. Chatterjee, and W. Wardlaw, *J. Chem. Soc.*, 3469 (1956).
288. M.A. Saad, *Ph.D. Thesis*, University of London (1954).
289. R.G. Jones, E. Bindschadler, G. Karmas, G.A. Martin Jr, J.R. Thirtle, F.A. Yoeman, and H. Gilman, *J. Am. Chem. Soc.*, **78**, 4289 (1956).
290. D.C. Bradley, R.N. Kapoor, and B.C. Smith, *J. Inorg. Nucl. Chem.*, **24**, 863 (1962).
291. D.C. Bradley, R.N. Kapoor, and B.C. Smith, *J. Chem. Soc.*, 1034 (1963).
292. R.C. Mehrotra and G. Srivastava, *J. Indian Chem. Soc.*, **38**, 1 (1961).
293. R.C. Mehrotra, *J. Indian Chem. Soc.*, **30**, 585 (1953).
294. R.C. Mehrotra, *J. Indian Chem. Soc.*, **31**, 85 (1954).
295. G. Chandra, *Ph.D. Thesis*, University of Rajasthan, Jaipur, India (1965).
296. G.S. Sasin and R. Sadin, *J. Org. Chem.*, **20**, 770 (1955).
297. G.P. Mack and E. Parker, *U.S. Patent*, 2,727,917 (1956); *Chem. Abstr.*, **50**, 10761b (1956).
298. J.C. Marie, *Ann. Chem.*, **6**, 969 (1961).
299. R.C. Mehrotra and V.D. Gupta, *J. Indian Chem. Soc.*, **41**, 537 (1964).
300. M.F. Shostakovaskii, R.G. Mirskov, and V.M. Vleasov, *Khim. Atsetilena*, 171 (1968); *Chem. Abstr.*, **71**, 81487 (1969).
301. R.C. Mehrotra and D.D. Bhatnagar, *J. Indian Chem. Soc.*, **42**, 327 (1965).
302. S.N. Mathur, *Ph.D. Thesis*, University of Rajasthan, Jaipur, India (1966).
303. W.L. Jolly, *Modern Inorganic Chemistry*, 201–207, McGraw-Hill, New York (1991).
304. S. Patai, *The Chemistry of Functional Groups*, Interscience, New York (1971).
305. R.C. Mehrotra, *J. Indian Chem. Soc.*, **31**, 904 (1954).
306. D.C. Bradley and M.J. Hillyer, *Trans. Faraday Soc.* **62**, 2374 (1966).
307. D.C. Bradley and M.J. Hillyer, *Trans. Faraday Soc.*, **62**, 2367 (1966).
308. R.C. Mehrotra and D.P. Gaur, unpublished results.
309. R.C. Mehrotra, *Alkoxides and Alkylalkoxides of Metals and Metalloids*, Presidential Address, 54th Indian Science Congress, Hyderabad, India (1967).
310. R.C. Mehrotra, R.K. Mittal, and A.K. Rai, *Alkoxides of Metals and Metalloids*, Rajasthan University Studies, Science Section Chemistry (1964–1965).
311. V.D. Gupta, *Ph.D. Thesis*, University of Rajasthan, Jaipur, India (1965).
312. D.C. Bradley, B.N. Chakravarti, A.K. Chatterjee, W. Wardlaw, and A. Whitley, *J. Chem. Soc.*, 99 (1958).
313. R.C. Mehrotra, *Pure Appl. Chem.*, **13**, 111 (1966).
314. I.A. Arnautova, G.Ya. Zhigalin, E.Yu. Merkel, and M.V. Sobolev, *Khim. Prom.*, **44**, 827 (1968); *Chem. Abstr.*, **70**, 48970 (1969).
315. D.C. Bradley, L. Kay, and W. Wardlaw, *J. Chem. Soc.*, 4916 (1956).
316. C. Eaborn, *Organosilicon Compounds*, Butterworth, London (1960).
317. S. Mathur, G. Chandra, A.K. Rai, and R.C. Mehrotra, *J. Organomet. Chem.*, **4**, 371 (1965).
318. S. Mathur, *Ph.D. Thesis*, University of Rajasthan, Jaipur, India (1967).
319. R.C. Mehrotra and S.N. Mathur, *J. Indian Chem. Soc.*, **41**, 11 (1964).
320. D.C. Bradely, E.V. Caldwell and W. Wardlow, *J. Chem. Soc.*, **79**, 4775 (1957).
321. R.C. Mehrotra, *J. Am. Chem. Soc.*, **76**, 2266 (1954); R.H. Baker, *J. Am. Chem. Soc.*, **60**, 2673 (1938).
322. R.K. Mittal and R.C. Mehrotra, *Z. Anorg. Allg. Chem.*, **331**, 89 (1964).
323. R.C. Mehrotra and P.N. Kapoor, *J. Less-Common Met.*, **7**, 98 (1964).
324. P.N. Kapoor and R.C. Mehrotra, *J. Less-Common Met.*, **10**, 66 (1966).
325. D.C. Bradley and I.M. Thomas, *Chem. Ind. (London)*, 1231 (1958).
326. D.C. Bradley and I.M. Thomas, *J. Chem. Soc.*, 3404 (1959).
327. R.C. Mehrotra and G. Srivastava, unpublished results.

328. M.F. Lappert, *J. Chem. Soc.*, 817 (1961).
329. M.F. Lappert, *J. Chem. Soc.*, 542 (1962).
330. W.G. Van der Sluys, A.P. Sattelberger, and M. McElfresh, *Polyhedron*, **9**, 1843 (1990).
331. I.M. Thomas, *Can. J. Chem.*, **39**, 1386 (1961).
332. D.C. Bradley and I.M. Thomas, *J. Chem. Soc.*, 3857 (1960); *Can. J. Chem.*, **40**, 449, 1355 (1962).
333. D.C. Bradley and M.L. Mehta, *Can. J. Chem.*, **40**, 1183 (1962).
334. J.S. Basi and D.C. Bradley, *Proc. Chem. Soc.*, 305 (1963).
335. J.S. Basi, D.C. Bradley, and M.H. Chisholm, *J. Chem. Soc. A*, 1433 (1971).
336. D.C. Bradley and M.H. Chisholm, *Acc. Chem. Res.*, **9**, 273 (1976).
337. M.H. Chisholm, V.F. Distasi, and W.E. Streib, *Polyhedron*, **9**, 253 (1990).
338. R.H. Cayton, M.H. Chisholm, D.R. Davidson, V.F. Distasi, P. Du, and J.C. Huffman, *Inorg. Chem.*, **30**, 1020 (1991).
339. M.A. Matchett, M.Y. Chiang, and W.E. Buhro, *Inorg. Chem.*, **29**, 358 (1990).
339a. S.C. Goel. M.Y. Chiang, and W.E. Buhro, *Inorg. Chem.*, **29**, 4640 (1990)
339b. S.C. Goel, M.Y. Chiang, and W.E. Buhro, *Inorg. Chem.*, **29**, 4646 (1990).
339c. S. Boulmaaz, R. Papiernik, L.G. Hubert-Pfalzgraf, J. Vaissermann, and J.C. Daran, *Polyhedron*, **11**, 1331 (1992).
340. N.W. Huber, *Dissertation*, Technical University Munich, Germany (1994) and references therein.
341. R. Anwander, *Dissertation*, Technical University Munich, Germany (1993) and references therein.
342. R. Anwander, *Top. Curr. Chem.*, **179**, 149 (1996).
343. H.A. Stecher, A. Sen, and A. Rheingold, *Inorg. Chem.*, **27**, 1130 (1988).
344. H.A. Stecher, A. Sen, and A. Rheingold, *Inorg. Chem.*, **28**, 3280 (1989).
345. D.C. Bradley, H. Chudzynska, M.B. Hursthouse, and M. Motevalli, *Polyhedron*, **10**, 1049 (1991).
346. W.J. Evans, R.E. Golden, and J.W. Ziller, *Inorg. Chem.*, **30**, 4963 (1991).
347. W.A. Herrmann, R. Anwander, M. Kleine, and W. Scherer, *Chem. Ber.*, **125**, 1971 (1992).
348. D.M. Barnhart, D.L. Clark, J.C. Huffman, R.L. Vincent, and J.G. Watkin, *Inorg. Chem.*, **32**, 4077 (1993).
348a. D.M. Barnhart, D.L. Clark, J.C. Gordon, J.C. Huffman, J.G. Watkin, and B.D. Zwick, *J. Am. Chem. Soc.*, **115**, 8461 (1993).
349. D.C. Bradley, H. Chudzynska, M.E. Hammond, M.B. Hursthouse, M. Motevalli, and Wu. Ruowen, *Polyhedron*, **11**, 375 (1992).
349a. D.C. Bradley, H. Chudzynska, M.B. Hursthouse, and M. Motevalli, *Polyhedron*, **12**, 1907 (1993).
349b. D.C. Bradley, H. Chudzynska, M.B. Hursthouse, M. Motevalli, and R. Wu, *Polyhedron*, **13**, 1 (1994).
349c. D.C. Bradley, H. Chudzynska, M.B. Hursthouse, and M. Motevalli, *Polyhedron*, **13**, 7 (1994).
350. B. Horvath and E.G. Horvath, *Z. Anorg. Allg. Chem.*, **457**, 51 (1979).
351. B.D. Murray, H. Hope, and P.P. Power, *J. Am. Chem. Soc.*, **107**, 169 (1985).
351a. G.A. Sigel, R.A. Bartlett, D. Decker, M.M. Olmstead, and P.P. Power, *Inorg. Chem.*, **26**, 1773 (1987).
352. T. Fieldberg, P.B. Hitchcock, M.F. Lappert, S.J. Smith, and A.J. Thorne, *J. Chem. Soc., Chem. Commun.*, 939 (1985).
353. W.A. Herrmann, R. Anwander, F.C. Munck, W. Scherer, V. Dufaud, N.W. Huber, and G.R.J. Artus, *Z. Naturforsch*, **B49**, 1789 (1994).
354. Y.F. Radkov, E.A. Pedorova, S.Y. Khorshev, G.S. Kalinina, M.N. Bochkarev, and G.A. Razuvaev, *J. Gen. Chem. U.S.S.R.*, **55**, 1911 (1985).
355. R. Anwander, F.C. Munck, T. Priermeier, W. Scherer, O. Runte, and W.A. Herrmann, *Inorg. Chem.*, **36**, 3545 (1997).
356. W.A. Herrmann, R. Anwander, and M. Denk, *Chem. Ber.*, **125**, 2399 (1992).
357. W.J. Evans, R. Anwander, U.H. Berlekamp, and J.W. Ziller, *Inorg. Chem.*, **34**, 3583 (1995).

358. L.A. Mlinea, S. Suh, and D.M. Hoffman, *Inorg. Chem.*, **38**, 4447 (1999).
359. F.A. Cotton and R.A. Walton, *Multiple Bonds Between Metal Atoms*, Oxford University Press, Oxford (1993).
360. M.H. Chisholm and W. Reichert, *J. Am. Chem. Soc.*, **96**, 1249 (1974).
361. M.H. Chisholm, F.A. Cotton, C.A. Murillo, and W.W. Reichert, *Inorg. Chem.*, **16**, 1801 (1977).
362. M.H. Chisholm, W.W. Reichert, F.A. Cotton, and C.A. Murillo, *J. Am. Chem. Soc.*, **99**, 1652 (1977).
363. I.P. Parkin, J.C. Huffman, and K. Folting, *J. Chem. Soc., Dalton Trans.*, 2343 (1992).
364. M.H. Chisholm and M. Extine, *J. Am. Chem. Soc.*, **97**, 5625 (1975).
365. A. Akiyama, M.H. Chisholm, F.A. Cotton, M.W. Extine, D.A. Haitko, D. Little, and P.E. Fanwick, *Inorg. Chem.*, **18**, 2266 (1979).
366. M.H. Chisholm, J.C. Huffman, and J. Leonelli, *J. Chem. Soc., Chem. Commun.*, 270 (1981).
367. M.H. Chisholm, J.C. Huffman, C.C. Kirkpatrick, J. Leonelli, and K. Folting, *J. Am. Chem. Soc.*, **103**, 6093 (1981).
368. M.H. Chisholm, K. Folting, M.J. Hampden-Smith, and C.A. Smith, *Polyhedron*, **6**, 1747 (1987).
369. M. Akiyama, D. Little, M.H. Chisholm, D.A. Haitko, F.A. Cotton, and M.W. Extine, *J. Am. Chem. Soc.*, **101**, 2504 (1979).
370. M. Akiyama, M.H. Chisholm, F.A. Cotton, M.W. Extine, D.A. Haitko, J. Leonelli, and D. Little, *J. Am. Chem. Soc.*, **103**, 779 (1981).
371. M.H. Chisholm, W.W. Reichert, and P. Thornton, *J. Am. Chem. Soc.*, **100**, 2744 (1978).
372. M.H. Chisholm, F.A. Cotton, M.W. Extine, and W.W. Reichert, *Inorg. Chem.*, **17**, 2944 (1978).
373. M.H. Chisholm, C.E. Hammond, and J.C. Huffman, *Polyhedron*, **8**, 129 (1989).
374. M.H. Chisholm, K. Folting, J.C. Huffman, and R.J. Tatz, *J. Am. Chem. Soc.*, **106**, 1153 (1984).
375. M.H. Chisholm, K. Folting, J.C. Huffman, E.F. Putilina, W.E. Streib, and R.J. Tatz, *Inorg. Chem.*, **32**, 3771 (1993).
376. R.G. Abbott, F.A. Cotton, and L.R. Falvello, *Polyhedron*, **9**, 1821 (1990).
377. R.G. Abbott, F.A. Cotton, and L.R. Falvello, *Inorg. Chem.*, **29**, 514 (1990).
378. T.W. Bentley, P.von.R. Schleyer, in *Advances in Physical Organic Chemistry* (V. Gold and D. Bethell, eds), Vol. 14, 1, Academic Press, New York (1977).
379. R.G. Jones, E. Bindschadler, G. Karmas, F.A. Yoeman, and H. Gilman, *J. Am. Chem. Soc.*, **78**, 4287 (1956).
380. R.G. Jones, E. Bindschadler, D. Blume, G. Karmas, G.A. Martin Jr, J.R. Thirtle, and H. Gilman, *J. Am. Chem. Soc*, **78**, 6027 (1956).
381. D.A. Brown, D. Cunningham, and W.K. Glass, *Chem. Commun.*, 306 (1966); *J. Chem. Soc. A*, 1563 (1968).
382. N. Hagihara and H. Yamazaki, *J. Am. Chem. Soc.*, **81**, 3160 (1959).
383. A. Sen, H.A. Stecher, and A.L. Rheingold, *Inorg. Chem.*, **31**, 473 (1992).
384. M.A. Giardello, W.A. King, S.P. Nolan, M. Porchia, C. Sishta, and T.J. Marks, in *Energetics of Organometallic Species, NATO ASI Ser.*, **367**, 35 (1992).
385. G.E. Coates and A.H. Fishwick, *J. Chem. Soc. A*, 477 (1968).
385a. R.A. Andersen and G.E. Coates, *J. Chem. Soc., Dalton Trans.*, 1171 (1974).
385b. R.A. Andersen and G.E. Coates, *J. Chem. Soc., Dalton Trans*, 1244 (1975).
386. E.C. Ashby, J. Nackashi, and G.E. Parris, *J. Am. Chem., Soc.*, **97**, 3162 (1975).
386a. J.A. Nackashi and E.C. Ashby, *J. Organomet. Chem.*, **35**, C1 (1972).
387. H.O. House, R.A. Latham, and G.M. Whitesides, *J. Org. Chem.*, **32**, 2481 (1967).
388. G.E. Coates, J.A. Heslop, M.E. Redwood, and D. Ridley, *J. Chem. Soc. A*, 1118 (1968).
388a. G.E. Coates and D. Ridley, *Chem. Commun.*, 560 (1966).
389. B.J. Wakefield, *Adv. Inorg. Chem. Radiochem.*, **11**, 341 (1968).
390. G.E. Coates and D. Ridley, *J. Chem. Soc.*, 1870 (1965).
390a. W.A. Herrmann, S. Bogdanovic, J. Behm, and M. Denk, *J. Organomet. Chem.*, **430**, C33 (1992).
391. J.G. Noltes and J. Boersma, *J. Organomet. Chem.*, **12**, 425 (1968).

392. G.E. Coates and D. Ridley, *J. Chem. Soc. A*, 1064 (1966).
393. R. Haran, C. Jouany, and J.P. Laurent, *Bull. Soc. Chim. France*, 457 (1968).
394. K.S. Chong, S.J. Rettig, A. Storr, and J. Trotter, *Can. J. Chem.*, **57**, 586 (1979).
395. W.M. Cleaver and A.R. Barron, *Organometallics*, **12**, 1001 (1993).
396. G. Beck, P.B. Hitchcock, M.F. Lappert, and I.A. Mackinnon, *J. Chem. Soc., Chem. Commun.*, 1312 (1989).
397. A.W. Duff, P.B. Hitchcock, M.F. Lappert, R.G. Taylor, and J.A. Segal, *J. Organomet. Chem.*, **293**, 271 (1985).
398. J.A. Samuels, K. Folting, J.C. Huffman, and K.G. Caulton, *Chem. Mater.*, **7**, 929 (1995).
398a. J.A. Samuels, E.B. Lobkovsky, W.E. Streib, K. Folting, J.C. Huffman, J.W. Zwanziger, and K.G. Caulton, *J. Am. Chem. Soc.*, **115**, 5093 (1993).
399. R.K. Minhas, J.J.H. Edema, S. Gambarotta, and A. Meetsma, *J. Am. Chem. Soc.*, **115**, 6710 (1993).
400. M.H. Chisholm, F.A. Cotton, M.W. Extine, and D.C. Rideout, *Inorg. Chem.*, **18**, 120 (1979).
401. S.E. Kegley, C.J. Schaverien, J.H. Freudenberger, and R.G. Bergman, *J. Am. Chem. Soc.*, **109**, 6563 (1987).
402. K. Osakada, T. Takizawa, M. Tanaka, and T. Yamamoto, *J. Organomet. Chem.*, **473**, 359 (1994).
403. H.F. Klein and H.H. Karsch, *Chem. Ber.*, **108**, 944 (1975).
404. Y.J. Kim, K. Osakada, K. Sugita, T. Yamamoto, and A. Yamamoto, *Organometallics*, **7**, 2182 (1988).
405. T. Saegusa and T. Ueshima, *Inorg. Chem.*, **6**, 1679 (1967).
406. D.L. Clark and J.G. Watkin, *Inorg. Chem.*, **32**, 1766 (1993).
407. D.M. Barnhart, D.L. Clark, J.C. Gordon, J.C. Huffman, J.G. Watkin, and B.D. Zwick, *Inorg. Chem.*, **34**, 5416 (1995).
408. A.B. Goel and R.C. Mehrotra, *Indian J. Chem.*, **16A**, 428 (1978).
409. R.G. Herman, K. Klier, G.W. Simmons, B.P. Finn, B.J. Bulko, and T.P. Kobylinski, *J. Catal.*, **56**, 407 (1979).
410. S. Mehta, G.W. Simmons, K. Klier, and R.G. Herman, *J. Catal.*, **57**, 339 (1979).
411. E.C. Ashby and A.B. Goel, *Inorg. Chem.*, **20**, 1096 (1981).
412. G.V. Goeden and K.G. Caulton, *J. Am. Chem. Soc.*, **103**, 7354 (1981).
413. T.L. Neils and J.M. Burlitch, *Inorg. Chem.*, **28**, 1607 (1989).
414. H. Nöth and H.J. Suchy, *Z. Anorg. Allg. Chem.*, **358**, 44 (1968).
415. E. Frankland and B.F. Duppa, *J. Chem. Soc.*, **17**, 29 (1864).
415a. E. Frankland and B.F. Duppa, *Ann.*, **135**, 29 (1865).
416. V. Ragosin, *Chem. Ber.*, 26, 380 (1893).
417. M.H. Abraham, *J. Chem. Soc.*, 4130 (1960).
418. M.H. Abraham and A.G. Davies, *J. Chem. Soc.*, 429 (1959).
419. S.I. Bunov, S.I. Frolov, V.G. Kiselev, V.S. Bogdanov, B.M. Mikhailov, *Zh. Obshch Khim.*, **40**, 1311 (1970).
420. A.J. Lundeen and J.E. Yates, *U.S. Patent*, 3,600,417 (1971); *Chem. Abstr.*, **75**, 151349 (1971).
421. T.V. Lubben and P.T. Wolczanski, *J. Am. Chem. Soc.*, **109**, 424 (1987); *J. Am. Chem. Soc.*, **107**, 701 (1985).
422. L.M. Engelhardt, J. Mac, B. Harrowfied, M.F. Lappert, I.A. MacKimmon, B.H. Newton, C.L. Raston, B.W. Skelton, and A.H. White, *J. Chem. Soc., Chem. Commun.*, 846 (1986).
423. H. Schmidbaur and H.F. Klein, *Chem. Ber.*, **100**, 1129 (1967).
424. R.E.A. Dear, *J. Org. Chem.*, **33**, 3959 (1968).
425. I.I. Lapkin, T.N. Povarnitsyana, and I.A. Kostareva, *Zh. Obshch Khim.*, **38**, 1598 (1968); *Chem. Abstr.*, **70**, 11749 (1969).
426. D.C. Bradley and D.A.W. Hill, *J. Chem. Soc.*, 2101 (1963).
427. E.A. Cuellar and T.J. Marks, *Inorg. Chem.*, **20**, 2129 (1981).
428. E. Jacob, *Angew. Chem., Int. Ed. Engl.*, **21**, 142 (1982).
429. J.C. Bryan, D.R. Wheeler, D.L. Clark, J.C. Huffman, and A.P. Sattelberger, *J. Am. Chem. Soc.*, **113**, 3184 (1991).

430. M.R. Duttera, V.W. Day, and T.J. Marks, *J. Am. Chem. Soc.*, **106**, 2907 (1984).
431. J. Caruso, M.J. Handen-Smith, A.L. Rheingold, and G. Yap, *J. Chem. Soc., Chem. Commun.*, 157 (1995).
432. D.C. Bradley, R.C. Mehrotra, J.D. Swanwick, and W. Wardlaw, *J. Chem. Soc.*, 2025 (1953).
433. D.C. Bradley, *Nature (London)*, **182**, 1211 (1958).
434. J.A. Ibers, *Nature (London)*, **197**, 686 (1963).
435. D.C. Bradley, *Ph.D. Thesis*, University of London (1950).
436. D.A. Wright and D.A. Williams, *Acta Crystallogr.* **B24**, 1107 (1968).
437. P. Dunn, *Aust. J. Appl. Sci.*, **10**, 458 (1959).
438. M.S. Bains, *Can. J. Chem.*, **42**, 945 (1964).
439. I.B. Golonanov, A.P. Simonov, A.K. Priskunov, T.V. Talaleava, G.W. Tsareva, and K.A. Kocheshkov, *Dokl. Akad. Nauk. SSSR*, **149**, 835 (1963)
440. G.E. Hartwell and T.L. Brown, *Inorg. Chem.*, **5**, 1257 (1966).
441. R.C. Mehrotra and M.M. Agrawal, *J. Chem. Soc.*, 1026 (1967).
442. R.E.A. Dear, F.W. Fox, R.J. Fredericks, G.E. Gilbert, and D.K. Huggins, *Inorg. Chem.*, **9**, 2590 (1970).
443. E. Weiss, H. Alsdorf, and H. Kühr, *Angew. Chem., Int. Ed. Engl.*, **6**, 801 (1967).
444. M.E. Redwood and C.J. Willis, *Can. J. Chem.*, **43**, 1893 (1965).
445. M.E. Redwood and C.J. Willis, *Can. J. Chem.*, **45**, 389 (1967).
446. A.G. Sharpe, *Inorganic Chemistry*, Longman, London (1981).
447. M. Arora, *Ph.D. Thesis*, University of Rajasthan, Jaipur, India (1966).
448. R.A. Andersen and G.E. Coates, *J. Chem. Soc., Dalton Trans.*, 2153 (1972).
449. H.M.M. Shearer and C.B. Spencer, *Chem. Commun.*, 194 (1966).
450. G.E. Coates and M. Tranah., *J. Chem. Soc. A*, 236 (1967).
451. E.C. Ashby, G.F. Willard, and A.B. Goel, *J. Org. Chem.*, **44**, 1221 (1979).
452. D.M. Barnhart, D.L. Clark, J.C. Gordon, J.C. Huffman, J.G. Watkin, and B.D. Zwick, *J. Am. Chem. Soc.*, **115**, 8461 (1993).
453. R.G. Jones, E. Bindschadler, G.A. Martin Jr, J.R. Thirtle, and H. Gilman, *J. Am. Chem. Soc.*, **79**, 4921 (1957).
454. C.N. Caughlan, H.S. Smith, W. Katz, W. Hodgson, and R.W. Crowe, *J. Am. Chem. Soc.*, **73**, 5652 (1951).
455. D.C. Bradley and C.E. Holloway, *Inorg. Chem.*, **3**, 1163 (1964).
456. R.L. Martin and G. Winter, *J. Chem. Soc.*, 2947 (1961).
457. C.G. Barraclough, R.L. Martin, and G. Winter, *J. Chem. Soc.*, 758 (1964).
458. P.N. Kapoor, R.N. Kapoor, and R.C. Mehrotra, *Chem. Ind. (London)*, 1314 (1968).
459. N.M. Cullinane, S.J. Chard, G.F. Price, B.B. Millward, and G. Langlois, *J. Appl. Chem.*, **1**, 400 (1951).
460. B.A. Vaartstra, J.C. Huffman, P.S. Gradeff, L.G. Hubert-Pfalzgraf, J.C. Daran, S. Parraud, K. Yunlu, and K.G. Caulton, *Inorg. Chem.*, **29**, 3126 (1990).
461. D.C. Bradley, *Nature*, **174**, 323 (1954).
462. D.C. Bradley and J.D. Swanwick, *J. Chem. Soc.*, 3207 (1958).
463. D.C. Bradley and J.D. Swanwick, *J. Chem. Soc.*, 748 (1959).
464. D.C. Bradley and M.M. Faktor, *J. Appl. Chem.*, **9**, 5425 (1959).
465. D.C. Bradley and M.M. Faktor, *Trans. Faraday Soc.*, **55**, 2117 (1959).
466. D.C. Bradley, A.K. Chatterjee, and W. Wardlaw, *J. Chem. Soc.*, 1091 (1954).
467. D.C. Bradley and M.J. Hillyer, *Trans. Faraday Soc.*, **62**, 2382 (1966).
468. D.C. Bradley, B.N. Chakravarti, and W. Wardlaw, *J. Chem. Soc.*, 2381 (1956).
469. D.C. Bradley, W. Wardlaw, and A. Whitley, *J. Chem. Soc.*, 726 (1955).
470. D.C. Bradley and C.E. Holloway, *J. Chem. Soc. A*, 219 (1968).
471. E.C. Alyea, J.S. Basi, D.C. Bradley, and M.H. Chisholm., *J. Chem. Soc. A*, 772 (1971).
472. T.L. Cottrell, *The Strength of Chemical Bonds*, 2nd edn, Butterworth, London (1958).
473. P. Gray and A. Williams, *Chem. Rev.*, **59**, 239 (1959).
474. R.W. Adams, R.L. Martin, and G. Winter, *Aust. J. Chem.*, **19**, 363 (1966).
475. H. Steinberg and D.L. Hunter, *Ind. Eng. Chem.*, **49**, 174 (1957).
476. A. Scattergood, W.H. Miller, and J. Gammon, *J. Am. Chem. Soc.*, **67**, 2150 (1944).
477. L. Kahovec, *Z. Physik. Chem.*, **B43**, 109 (1939).

478. G.J. Baumgarten and C.L. Wilson, *J. Am. Chem. Soc.*, **81**, 2440 (1959).
479. C.H. Dupuy and R.W. King, *Chem. Rev.*, **60**, 431 (1960).
480. T. Charnley, H.A. Skinner, and B.N. Smith, *J. Chem. Soc.*, 2288 (1952).
481. N.V. Sidgwick, *The Chemical Elements and Their Compounds*, Vol. 1, Oxford University Press, London (1950).
482. J.T.F. Fenwick and J.W. Wilson, *J. Chem. Soc., Dalton Trans.*, 1324 (1972).
483. A. Finch and P.J. Gardner, *Prog. Boron Chem.*, **3**, 177 (1970).
484. H. Ulich and W. Nespital, *Z. Physik. Chem. A*, **165**, 294 (1933).
485. R.A. Robinson and D.A. Peak, *J. Phys. Chem.*, **39**, 1125 (1935).
486. R.C. Menzies, *J. Am. Chem. Soc.*, **43**, 2309 (1921).
487. S.M. McElvain and W.R. Davie, *J. Am. Chem. Soc.*, **73**, 1400 (1951).
488. R.C. Wilhoit, *J. Phys. Chem.*, **61**, 114 (1957).
489. V.J. Shiner, D. Whittaker, and V.P. Fernandez, *J. Am. Chem. Soc.*, **85**, 2318 (1963).
490. J.G. Oliver, P.K. Phillips, and I.J. Worrall, *J., Inorg. Nucl. Chem.*, **31**, 1609 (1969).
491. T.N. Huckerby, J.G. Oliver, and I.J. Worrall, *Inorg. Nucl. Chem. Lett.*, **5**, 749 (1969).
492. J.G. Oliver and I.J. Worrall, *J. Chem. Soc. A*, 845 (1970).
493. R.C. Wilhoit, J.R. Burton, F.T. Kuo, S.R. Huang, and A. Viguesnel, *J. Inorg. Nucl. Chem.*, **24**, 851 (1962).
494. J.W. Wilson, *J. Chem. Soc. A*, 981 (1971).
495. J.G. Oliver and I.J. Worrall, *Inorg. Nucl. Chem. Lett.*, **5**, 455 (1969).
496. V. Bazant, V. Chvalovsky, and J. Rathousky, *Organosilicon Compounds*, Vols 1 and 2, Academic Press, New York (1965).
497. A.P. Kreshkov and D.A. Karteev, *Zh. Obsch. Khim.*, **27**, 2715 (1957).
498. T. Asahara and K. Kanabu, *J. Chem. Soc. Japan*, **55**, 589 (1952).
499. R.C. Mehrotra and G. Chandra, *J. Indian Chem. Soc.*, **39**, 235 (1962).
500. Yu.Kh. Saulov, A.Ya. Zhyuryutina, A.S. Petrov, Yu.B. Nadzhafov, and A.A. Oigenblik, *Zh. Fiz. Khim.*, **42**, 1523 (1968); *Chem. Abstr.*, **69**, 105749 (1968).
501. G.W. Heunisch, *U.S. Clearing House Fed. Sci. Tech. Inf.*, 667897 (1967); *Chem. Abstr.*, **69**, 111341 (1968).
502. J.D. Kennedy, W. McFarlane, P.J. Smith, and R.F.M. White, *J. Chem. Soc., Perkin Trans. II*, 1785 (1973).
503. R.C. Mehrotra and B.P. Bachlas, *J. Organomet. Chem.*, **22**, 121 (1970).
504. J. Mendelsohn, J.C. Pommier, and J. Valade, *C.R. Acad. Sci. Paris C*, **263**, 921 (1966).
505. R.C. Mehrotra, V.D. Gupta, and G. Srivastava, *Rev. Silicon, Germanium, Tin, Lead*, **1**, 299 (1975).
506. O.D. Dubrovina, *Uchenye Zapiski Kazan Gosudarst Univ. im. V.I. Ulyanova Lenina*, **116**, 3–70 (1956).
507. G.K. Kamai, *Uchenye Zapiski Kazan Gosudarst, Univ.*, **115**, 43 (1955); D. Hass, *Z. Chem.*, **8**, 150 (1968).
508. W.A. Herrmann, N.W. Huber, R. Anwander, and T. Priemeier, *Chem. Ber*, **126**, 1127 (1993).
509. R. Levaillant, *Ann. Chim.*, **6**, 459 (1936).
510. R.C. Mehrotra and S.N. Mathur, *J. Indian Chem. Soc.*, **41**, 111 (1964).
511. C.G. Barraclough, D.C. Bradley, J. Lewis, and I.M. Thomas, *J. Chem. Soc.*, 2601 (1961).
512. J.V. Bell, J. Heisler, H. Tannenbaum, and J. Goldenson, *Anal. Chem.*, **25**, 1720 (1953).
513. D.L. Guertin, S.E. Wiberley, W.H. Bauer, and J. Goldenson, *J. Phys. Chem.*, **60**, 1018 (1956).
514. L. Maijs, I. Strauss, and I. Vevere, *Latv. PSR Zinat. Akad. Vestis, Khim. Ser.*, 498 (1969); *Chem. Abstr.*, **72**, 27887 (1970).
515. H. Kriegsmann and K. Licht, *Z. Electrochem.*, **62**, 1163 (1958); H. Kriegsmann and K. Licht, *Z. Electrochem.*, **68**, 617 (1964).
516. R. Forneris and E. Funck, *Z. Electrochem.*, **62**, 1130 (1958).
517. A.N. Lazarev, *Opt. J. Spektr.*, 8511 (1960); *Chem. Abstr.*, **56**, 15063 (1962).
518. G. Kesslev and H. Kriegsmann, *Z. Anorg. Allg. Chem.*, **342**, 63 (1963).
519. W.E. Newton and E.G. Rochow, *J. Chem. Soc. A*, 2664 (1970).
520. H. Murata, K. Kawai, and M. Yokoo, *J. Chem. Soc. Japan*, **77**, 893 (1956).

521. I. Simonand and H.O. McMohan, *J. Chem. Phys.*, **20**, 905 (1952).
522. A.P. Kreshkov, Y.Y. Mikhailenko, and G.F. Yakimovich, *Zh. Fiz. Khim.*, **28**, 537 (1954); *Chem. Abstr.*, **48**, 13427 (1954).
523. M.O. Bulanin, B.N. Dolgov, T.A. Speranskaya, and N.P. Kharitonov, *Zh. Fiz. Khim.*, **31**, 1321 (1957); *Chem. Abstr.*, **52**, 2541 (1958).
524. R.N. Hazeldine and R.J. Marklow, *J. Chem. Soc.*, 962 (1956).
525. A.L. Smith, *Spectrochim. Acta*, **16**, 87 (1960).
526. A. Marchand, J. Valade, M.T. Forel, M.L. Josien, and R. Calas, *J. Chim. Phys.*, **59**, 1142 (1962).
527. H. Murata, *J. Chem. Phys.*, **20**, 1184 (1952).
528. A.N. Lazarev, K. Poiker, and T.F. Tenisheva, *Dokl. Akad. Nauk. SSSR*, **175**, 1322 (1967); *Chem. Abstr.*, **68**, 62855 (1968).
529. K.A. Andrianov, A.A. Mamedov, N.A. Chumaevskii, and L.M. Volkova, *Izv. Akad. Nauk. SSSR, Ser. Khim.*, 2151 (1968); *Chem. Abstr.*, **70**, 7674 (1969).
530. A.A.V. Stuart, C. Lalau, and H. Breederveld, *Recl. Trav. Chim.*, **74**, 747 (1955).
531. O.H. Johnson and H.E. Fritz, *J. Am. Chem. Soc.*, **75**, 718 (1953).
532. M. Lebedeff, A. Marchand, and J. Valade, *Compt. Rend.*, **267**, 813 (1968).
533. S. Mathur, R. Ouaki, V.K. Mathur, R.C. Mehrotra, and J.C. Maire, *Indian J. Chem.*, **7**, 284 (1969).
534. F.K. Butcher, W. Gerrard, E.F. Mooney, R.G. Rees, and H.A. Willis, *Spectrochim. Acta*, **20**, 51 (1964).
535. A. Marchand, J. Mendelsohn, and J. Valade, *Compt. Rend.*, **259**, 1737 (1964).
536. R.A. Cummins and J.V. Evans, *Spectrochim. Acta*, **21**, 1016 (1965).
537. J. Mendelsohn, A. Marchand, and J. Valade, *J. Organomet. Chem.*, **6**, 25 (1966).
538. T. Tanaka, *Organomet. Chem. Rev. A*, **5**, 1 (1970).
539. R.C. Poller, *J. Inorg. Nucl. Chem.*, **24**, 593 (1962).
540. J.C. Maire and R. Ouaki, *Helv. Chim. Acta*, **51**, 1151 (1968).
541. A.G. Kruger and G. Winter, *Aust. J. Chem.*, **23**, 1 (1970).
542. R.W. Adams, C.G. Barraclough, R.L. Martin, and G. Winter, *Aust. J. Chem.*, **20**, 2351 (1967).
543. R.W. Adams, R.L. Martin, and G. Winter, *Aust. J. Chem.*, **20**, 773 (1967).
544. C.T. Lynch, K.S. Mazdiyasni, J.S. Smith, and W.J. Crawford, *Anal. Chem.*, **36**, 2332 (1964).
545. D.C. Bradley and A.H. Westlake, *Proc. Symp. Coord. Chem.*, Tihany, Hungary (M. Beck, ed.), 309–315, Publ. House Hung. Acad. Sci., Budapest (1965).
546. W.R. Russo and W.H. Nelson, *J. Am. Chem. Soc.*, **92**, 1521 (1970).
547. D.A. Brown, D. Cunningham, and W.K. Glass, *J. Chem. Soc. A*, 1563 (1968).
548. D.C. Bradley and C.E. Holloway, *J. Chem. Soc. A*, 1316 (1968).
549. B. Borup, J.A. Samuels, W.E. Streib, and K.G. Caulton. *Inorg. Chem.*, **33**, 994 (1994).
550. E.W. Lindsell, in *Comprehensive Organometallic Chemistry II* (E.W. Abel, F.G.A. Stone, and G. Wilkinson, eds), Vol. 1, 77, Pergamon Press, Oxford (1995).
551. G.E. Coates and P.D. Roberts, *J. Chem. Soc. A*, 1233 (1967).
552. W.J. Evans, J.M. Olofson, and J.W. Ziller, *J. Am. Chem. Soc.*, **112**, 2308 (1990).
553. D.G. Karraker, T.H. Siddall, and W.E. Stewart, *J. Inorg. Nucl. Chem.*, **31**, 711 (1969).
554. P.G. Eller and P.J. Vergamini, *Inorg. Chem.*, **22**, 3184 (1983).
555. U(IV) is a f^2 ion having a paramagnetic ground state of $^3H_4$. For a general discussion of the effects of paramagnetic species on NMR spectra, see: E.D. Becker, *High Resolution NMR, Theory and Chemical Applications*, Academic Press, New York (1980).
556. F.A. Cotton, D.O. Marler, and W. Schwotzer, *Inorg Chem.*, **23**, 4211 (1984).
557. C.J. Burns, D.C. Smith, A.P. Sattelberger, and H.B. Gray, *Inorg. Chem.*, **31**, 3724 (1992).
558. B. Delavaux-Nicot and M. Ephritikhine, *J. Organomet. Chem.*, **399**, 77 (1990).
559. D.C. Bradley and C.E. Holloway, *Proc. Paint Res. Inst.*, **37**, 487 (1965).
560. H. Weingarten and J.R. Van Wazer, *J. Am. Chem. Soc.*, **87**, 724 (1965).
561. Y.T. Wu, Y.C. Ho, C.C. Lin, and H.M. Gau, *Inorg. Chem.*, **35**, 5948 (1996).
562. D.C. Bradley and C.E. Holloway, *J. Chem. Soc. A*, 219 (1968).
563. C.E. Holloway, *J. Coord. Chem.*, **1**, 253 (1972).

564. J.G. Riess and L.G. Hubert-Pfalzgraf, *Bull. Soc. Chim. France*, 2401 (1968).
565. L.G. Hubert-Pfalzgraf and J.G. Riess, *Bull. Soc. Chim. France*, 4348 (1968).
566. A.A. Pinkerton, D. Schwarzebac, L.G. Hubert-Pfalzgraf, and J.G. Riess, *Inorg. Chem.*, **15**, 1196 (1976).
567. W. Priebsch and D. Rehder, *Inorg. Chem.*, **29**, 3013 (1990).
568. A. Lachowicz and K.H. Thiele, *Z. Anorg. Allg. Chem.*, **434**, 271 (1977).
569. D.C. Crans, H. Chen, and R.A. Felty, *J. Am. Chem. Soc.*, **114**, 4543 (1992).
570. D.C. Bradley, M.H. Chisholm, M.W. Extine, and M.E. Stage, *Inorg. Chem.*, **16**, 1794 (1977).
571. M.S. Bains, *Can. J. Chem.*, **40**, 381 (1962).
572. D.C. Kleinschmidt, *Ph.D. Thesis*, Indiana University, Bloomington, Indiana (1967).
573. J.G. Oliver and I.J. Worrall, *J. Chem. Soc. A.*, 1389 (1970).
574. W. Fieggen, *Doctoral Thesis*, University of Amsterdam (1970).
575. J.W. Akitt and R.H. Duncan, *J. Magn. Reson.*, **15**, 162 (1974).
576. C.B. Kriz, A. Lycka, J. Fusek, and S. Hermanek, *J. Magn. Reson.*, **60**, 375 (1984).
577. D.C. Bradley, *Advan. Chem. Ser.*, **23**, 10 (1959).
578. A. Mehrotra and R.C. Mehrotra, *Proc. Indian Natl. Sci. Acad.*, **40**, 215 (1974).
578a. D.C. Bradley and C.E. Holloway, unpublished results; C.E. Holloway, *Ph.D. Thesis*, University of Western Ontario, Canada (1966).
579. N.Ya. Turova, V.A. Kuzunov, A.I. Yanovskii, N.G. Borkii, Yu T. Struchkov, and B.L. Tarnopolskii, *J. Inorg. Nucl. Chem.*, **41**, 5 (1979).
580. K. Folting, W.E. Streib, K.G. Caulton, O. Poncelet, and L.G. Hubert-Pfalzgraf, *Polyhedron*, **10**, 1639 (1991).
581. I.J. Worrall, *J. Chem. Educ.*, **461**, 510 (1969).
582. D.C. Ayres, M. Barnard, and M.R. Chambers, *J. Chem. Soc. B*, 1385 (1967).
583. J.W. Akitt and R.H. Duncan, unpublished results.
584. P.J. Burke, R.W. Matthew, and D.G. Gilles, *J. Chem. Soc., Dalton Trans.*, 1439 (1980).
585. A.B.P. Lever, *Inorganic Electronic Spectroscopy*, 2nd edn, Elsevier, Amsterdam/New York (1984).
586. A.N. Nesmeyanov, O.V. Nogina, A.M. Berlin, A.S. Girshovich, and G.V. Shatalov, *Izv. Akad. Nauk SSSR, Otd. Khim. Nauk*, 2146 (1961); *Chem. Abstr.*, **57**, 11221 (1962).
587. H. Hartmann and H.L. Schaefer, *Z. Phys. Chem.*, **197**, 116 (1951).
588. D.C. Bradley, R.H. Moss, and K.D. Sales, *Chem. Commun.*, 1255 (1969).
589. E.C. Alyea and D.C. Bradley, *J. Chem. Soc. A*, 2330 (1969).
590. G.F. Kokoszka, H.C. Allen Jr, and G. Gordon, *Inorg. Chem.*, **5**, 91 (1966).
591. D.C. Bradley and M.H. Chisholm, *J. Chem. Soc. A*, 2741 (1971).
592. L. Dubicki, G.A. Kakos, and G. Winter, *Aust. J. Chem.*, **21**, 1461 (1968).
592a. L. Dubicki, G.A. Kakos, and G. Winter, *Aust. J. Chem.*, **23**, 15 (1970).
593. R.J.H. Clark and M.L. Greenfield, *J. Chem. Soc. A*, 409 (1967).
594. D.J. Machin and K.S. Murray, *J. Chem. Soc. A*, 1948 (1967).
595. G.A. Kakos and G. Winter, *Aust. J. Chem.*, **20**, 2343 (1967).
596. F.A. Cotton, D.M.L. Goodgame, and M.J. Goodgame, *J. Am. Chem. Soc.*, **83**, 4690 (1961).
597. D.G. Karraker, *Inorg. Chem.*, **3**, 1618 (1964).
598. G. Winter, *Inorg. Nucl. Chem. Lett.*, **2**, 161 (1966).
599. R.W. Adams, C.G. Barraclough, R.L. Martin, and G. Winter, *Inorg. Chem.*, **5**, 346 (1966).
600. B.N. Figgis and J. Lewis, *Prog. Inorg. Chem.*, **6**, 110 (1964).
601. C.G. Barraclough and C.F. Ng, *Trans. Faraday Soc.*, **60**, 836 (1965).
602. A.P. Simonov, D.N. Shigorin, T.V. Talalaeva, and K.A. Kocheshkov, *Izv. Akad. Nauk. SSSR. Otd. Khim. Nauk*, 1126 (1962).
603. J.D. Kahn, A. Haag, and P. von, R. Schleyer, *J. Phys. Chem.*, **92**, 212 (1988).
604. K.S. Mazdiyasni, B.J. Spacher, and L.M. Brown, *Inorg. Chem.*, **10**, 889 (1971).
605. L.M. Brown and K.S. Mazdiyasni, *Anal. Chem.*, **10**, 1243 (1969).
606. C. Chattorj, C.T. Lynch, and K.S. Mazdiyasni, *Inorg. Chem.*, **1**, 2501 (1968).
607. E.P. Turevskaya, N.I. Kozlova, N.Ya. Turova, A.I. Belokon, D.V. Berdyev, V.G. Kessler, and Yu.K. Grishin, *Russ. Chem. Bull.*, **44**, 734 (1995); E.P. Turevskaya,

N.I. Kozlova, N.Ya. Turova, A.I. Belokon, D.V. Berdyev, and V.G. Kessler, *Bull. Acad. Sci. Russ., Div. Chem. Sci.*, 752 (1995).

608. A.I. Yanovsky, E.P. Turevskaya, N.Ya. Turova, F.M. Dolgushin, A.P. Pisarevsky, A.S. Batsanov, and Yu.T. Struchkov, *Russ. J. Inorg. Chem.*, **39**, 1246 (1994).

609. N.I. Kozlova, V.G. Kessler, N.Ya. Turova, and A.I. Belokon, *Soviet. J. Coord. Chem.*, **15**, 867 (1989).

610. S.I. Kucheiko, N.Ya. Turova, and N.I. Kozlova, *Soviet J. Coord. Chem.*, **11**, 1656 (1985).

611. W. Fieggen, H. Gerding, and N.M.M. Nibbering, *Rec. Trav. Chem.*, **87**, 377 (1968).

612. J.G. Oliver and I.J. Worrall, *J. Chem. Soc. A*, 2347 (1970).

613. B.K. Teo, *Acc. Chem. Res.*, **13**, 412 (1980).

614. B.K. Teo, in *EXAFS: Basic Principles and Data Analysis*, Springer-Verlag, Berlin (1986).

615. D.E. Sayers and B.A. Bunker, in *X-ray Absorption: Principles, Applications, Techniques of EXAFS, SEXAFS, and XANES* (D.C. Konigsberger and R. Prins, eds), 211–253, Wiley-Interscience, New York (1988).

616. F. Babonneau, S. Doeuff, A. Leaustic, C. Sanchez, C. Cartier, and M. Verdaguer, *Inorg. Chem.*, **27**, 3166 (1988).

617. D.C. Bradley, in *Inorganic Polymers* (F.G.A. Stone and W.A.G. Graham, eds), Ch. 7, 410, Academic Press, London (1962).

618. J. Livage, M. Henry, and C. Sanchez, *Prog. Solid State Chem.*, **18**, 259 (1988).

619. C. Sanchez and J. Livage, *New J. Chem.*, **14**, 513 (1990).

620. D.P. Gaur, G. Srivastava, and R.C. Mehrotra, *Z. Anorg. Allg. Chem.*, **398**, 72 (1973).

621. R.C. Mehrotra and V.D. Gupta, *Indian J. Chem.*, **5**, 643 (1967).

622. D.P. Gaur, G. Srivastava, and R.C. Mehrotra, *J. Organomet. Chem.*, **65**, 195 (1974).

623. P.C. Bharara, V.D. Gupta, and R.C. Mehrotra, *Z. Anorg. Allg. Chem.*, **403**, 337 (1973).

624. P.C. Bharara, V.D. Gupta, and R.C. Mehrotra, *J. Indian Chem. Soc.*, **51**, 849 (1974).

625. R.C. Mehrotra and P.N. Kapoor, *J. Indian Chem. Soc.*, **44**, 468 (1967).

626. R.C. Mehrotra and P.N. Kapoor, *Indian J. Chem.*, **5**, 505 (1967).

627. K. Kumar, S. Dubey, S.N. Misra, and R.N. Kapoor, *Egyptian J. Chem.*, **15**, 267 (1972); *Chem. Abstr.*, **79**, 86904 (1973).

628. R.C. Mehrotra and R. Bohra, *Metal Carboxylates*, Academic Press, London (1983).

629. R.C. Mehrotra, R. Bohra, and D.P. Gaur, *Metal β-Diketonates and Allied Derivatives*, Academic Press, London (1978).

630. A.R. Siedle, in *Comprehensive Coordination Chemistry* (G. Wilkinson, R.D. Gillard, and J.A. McCeverty, eds), Vol. 2, Ch. 15.4, 365, Pergamon Press, Oxford (1987).

631. R.C. Mehrotra, *Pure Appl. Chem.*, **60**, 1349 (1988).

632. D.C. Bradley, R.C. Mehrotra, and D.P. Gaur, *Metal Alkoxides*, Academic Press, London (1978).

633. J.P. Tandon, *Newer Trends in the Chemistry of Schiff Base and Allied Derivatives*, Presidential Address, 74th Indian Science Congress, Bangalore (1987).

634. A. Singh, V.D. Gupta, G. Srivastava, and R.C. Mehrotra, *J. Organomet. Chem.*, **64**, 145 (1974).

635. R.C. Mehrotra, A.K. Rai, A. Singh, and R. Bohra, *Inorg. Chim. Acta. Rev.*, **13**, 91 (1975).

636. V.R. Gray and A.E. Alexander, *J. Phys. Colloid Chem.*, **53**, 23 (1949).

637. R.C. Mehrotra, *Nature (London)*, **172**, 74 (1953).

638. M.M. Faktor and D.C. Bradley, unpublished results.

639. R.C. Mehrotra, T.N. Misra, and S.N. Misra, *Indian J. Chem.*, **3**, 525 (1965).

640. L.G. Hubert-Pfalzgraf, N.E. Khokh, and J.C. Daran, *Polyhedron*, **11**, 59 (1992).

641. K.L. Taft, C.D. Delfs, G.C. Papaeffthymiou, S. Foner, D. Gatteschi, and S.J. Lippard, *J. Am. Chem. Soc.*, **116**, 823 (1994).

642. R.C. Mehrotra and A. Singh, *Chemtracts*, **5**, 346 (1993).

643. J.H. Wengrovius, M.F. Garbauskas, E.A. Williams, R.C. Going, P.E. Donahue, and J.E. Smith, *J. Am. Chem. Soc.*, **108**, 982 (1986).

644. J. Caruso, C. Roger, F. Schwertfeger, M.J. Hampden-Smith, A.L. Rheingold, and G. Yap, *Inorg. Chem.*, **34**, 449 (1995).

645. D.C. Bradley, *Coord. Chem. Rev.*, **2**, 299 (1967).
646. R.C. Mehrotra and A. Singh, *Chem. Soc. Rev.*, **26**, 1 (1996).
647. D.C. Bradley, F.M.A. Halim, and W. Wardlaw, *J. Chem. Soc.*, 3450 (1950).
648. M.H. Chisholm, C.E. Hammond, M. Hampden-Smith, J.C. Huffman, and W.G. Van der Sluys, *Angew. Chem., Int. Ed. Engl.*, **26**, 904 (1987).
649. M.H. Chisholm, K. Folting, C.E. Hammond, M.J. Hampden-Smith, and K.G. Moodley, *J. Am. Chem. Soc.*, **111**, 5300 (1989).
650. M.E. Gross, *J. Electrochem. Soc.*, **138**, 2422 (1991).
651. A.P. Purdy, C.F. George, and G.A. Brewer, *Inorg. Chem.*, **31**, 2633 (1992).
652. K.C. Malhotra and R.L. Martin, *J. Organomet. Chem.*, **239**, 159 (1982).
653. K.W. Bagnall, A.K. Bhandari, and D. Brown, *J. Inorg. Nucl. Chem.*, **37**, 1815 (1975).
654. G. Pfeifer, T. Flora, and Magyar Kem, *Polyoirat.*, **70**, 375 (1964).
655. R.N. Kappor and R.C.Mehrotra, *J. Am. Chem. Soc.*, **82**, 3495 (1960).
656. W.Fiegger, H. Gerding, and N.M.M. Nibbering, *Rec. Trans. Chem.*, **87**, 377 (1968).
657. J.G. Oliver and I.J. Worrall, *J. Chem. Soc. (A)*, 2347 (1970).
658. R. De Forchand, *Compt. Rend.*, **176**, 20 (1923).
659. W. Rodziewiez and T. Jasunski, *Roczniki Chem.*, **27**, 332 (1953).
660. B. Smith, *Acta Chem. Scand.*, **9**, 1337 (1955).
661. R.C. Mehrotra and G. Chandra, *J. Indian Chem. Soc.*, **39**, 235 (1962).
662. T.W. Coffindaffer, I.P. Rothwell and J.C. Huffman, *Inorg. Chem.*, **22**, 2906 (1983).
663. A. Shah, A. Singh and R.C. Mehrotra, unpublished results; A. Shah, *Ph. D. Thesis*, University of Rajasthan, Jaipur, India (1987).
664. T. Athar, R. Bohra, and R.C. Mehrotra, *Indiana J. Chem.*, **28A**, 492 (1989).
665. G.E. Hofmeister, E. Alvarado, J.A. Leary, D.I. Yoon, and S.F. Pedersen, *J. Am. Chem. Soc.*, **112**, 8843 (1990).
666. D. Max Roundhill, *Prog. Inorg. Chem.*, **43**, 533 (1995).
667. M.H. Chisholm, K. Folting, W.E. Streib, and De-Dong Wu, *Inorg. Chem.*, **38**, 5219 (1999).
668. K. Folting, W.E. Streib, K.G. Caulton, O. Poncelet, and L.G. Hubert-Pfalzgraf, *Polyhedron*, **10**, 1639 (1991).
669. I. Kijima, T. Yamamoto, and Y. Abe, *Bull. Chem. Soc. Japan*, **44**, 3193 (1971).
670. I.D. Verma and R.C. Mehrotra, *J. Prakt. Chem.*, **8**, 235 (1959).
671. A.N. Nesmeyanov and D.V. Nogina, *Dokl. Akad. Nauk. SSSR*, **117**, 249 (1957).
672. Y. Abe and I. Kijima, *Bull. Chem. Soc. Japan*, **43**, 466 (1970).
673. D. Danforth, *J. Am. Chem. Soc.*, **80**, 2585 (1958).
674. D.C. Bradley and I.M. Thomas, *Chem. Ind. (London)*, 17 (1958).
675. V.A. Zeitler and C.A. Brown, *J. Am. Chem. Soc.*, **79**, 4616 (1957).
676. Y. Abe, K. Hayama, and I. Kijima, *Bull. Chem. Soc. Japan*, **45**, 1258 (1972).
677. D.C. Bradley, R.N. Kapoor, and B.C. Smith, *J. Chem. Soc.*, 204 (1963).
678. D.C. Bradley and I.M. Thomas, *J. Chem. Soc.*, 3404 (1959).
679. Y. Abe and I. Kijima, *Bull. Chem. Soc. Japan*, **42**, 1148 (1969).
680. D.W. Meek and C.S. Springer Jr, *Inorg. Chem.*, **5,** 445 (1966).
681. J.A. Bertrand and P.G. Eller, *Prog. Inorg. Chem.*, **21**, 29 (1976).
682. F.A. Cotton and G.N. Mott, *Inorg. Chem.*, **22**, 1136 (1983).
683. T. Lindgren, R. Sillanpaa, T. Nortia, and K. Pihlaja, *Inorg. Chim. Acta*, **82**, 1 (1984).
684. H. Muhonen, A. Pajunen, and R. Hamalainen, *Acta Crystallogr., Sect. B*, **36**, 2790 (1980).
685. Y. Takahashi, H. Hayashi, and Y. Ohya, *Mater. Res. Soc. Proc.*, **271**, 401 (1992).
686. Y. Ohya, T. Tanaka, and Y. Takahashi, *Jap. J. Appl. Phys.*, **32**, 4163 (1993).
687. H. Tanaka, K. Tadanga, N. Tohge, and T. Minami, *Jap. J. Appl. Phys.*, **34**, L1155 (1995).
688. S.C. Goel, M.Y. Chiang, and W.E. Buhro, *Inorg. Chem.*, **29**, 4646 (1990).
689. B.S. Sankhla and R.N. Kapoor, *Bull. Chem. Soc., Japan*, **40**, 1381 (1967).
690. D.M. Puvi and R.C. Mehrotra, *J. Indian Chem. Soc.*, **39**, 447 (1962).
691. R.K. Mittal, *Z. Anorg. Allg. Chem.*, **351**, 309 (1969).
692. B.P. Baranwal, P.C. Bharara, and R.C. Mehrotra, *Transition Met. Chem.*, **2**, 204 (1977).
693. R.C. Mehrotra and G. Srivastava, *J. Indian Chem.* Soc., **39**, 521 (1962).

694. R.C. Mehrotra and R.K. Mehrotra, *J. Indian Chem. Soc.*, **39**, 677 (1962).
695. K.D. Edwards, G.H. Pearson, M. Kevin Woodrum, and K. Niedenzu, *Inorg. Chim. Acta*, **194**, 81 (1992).
696. R.C. Mehrotra and P. Bajaj, *J. Organomet. Chem.*, **24**, 611 (1970).
697. P.C. Bharara, V.D. Gupta, and R.C. Mehrotra, *Synth. React. Inorg. Met.-Org. Chem.*, **7**, 537 (1977).
698. S.C. Goel, K.S. Kramer, M.Y. Chiang, and W.E. Buhro, *Polyhedron*, **9**, 611 (1990).
699. S.C. Goel and W.E. Buhro, *Inorg. Synth.*, **31**, 294 (1997).
700. R. Duggal and R.C. Mehrotra, *Inorg. Chim. Acta.*, **98**, 121 (1985).
701. R.C. Mehrotra and G. Chandra, *Indian J. Chem.*, **11**, 497 (1965).
702. A.E. Beezer and C.T. Mortimer, *J. Chem. Soc. A*, 514 (1966).
703. J.L. Bills and F.A. Cotton, *J. Phys. Chem.*, **68**, 802 (1964).
704. T. Tanaka, *Bull. Chem. Soc. Japan*, **33**, 446 (1960).
705. R.C. Mehrotra and P. Bajaj, *J. Organomet. Chem.*, **25**, 359 (1970).
706. M. Bhagat, *Ph.D. Thesis*, University of Rajasthan, Jaipur, India (1998).
707. M. Sharma, A. Singh, and R.C. Mehrotra, *Polyhedron*, **18**, 77 (1999).
708. Jayshree Godhwani, *Ph.D. Thesis*, University of Rajasthan, Jaipur, India (1997); Jayshree Godhwani, A. Singh, and R.C. Mehrotra, unpublished results.
709. E.V. Kuznetsov, E.K. Ignatieva, and L.A. Emikh, *Zh. Obshch. Khim.*, **39**, 1816 (1969); *Chem. Abstr.*, **72**, 2961 (1970).
710. E.V. Kuznetsov, E.K. Ignatieva, and L.A. Emikh, *Zh. Obshch. Khim.*, **39**, 1820 (1969); *Chem. Abstr.*, **72**, 2962 (1970).
711. R.C. Mehrotra and G. Chandra, unpublished results.
712. V.F. Mironov, *Main Group Met. Chem.*, **12**, 355 (1989).
713. Manish K. Sharma, A. Singh, and R.C. Mehrotra, unpublished results.
714. M.G. Voronkov, *Pure Appl. Chem.*, **13**, 35 (1966).
714a. M.G. Voronkov and V.P. Baryshok, *J. Organomet. Chem.*, **239**, 199 (1982).
714b. S.N. Tandura, M.G. Voronkov, and N.V. Alekseev, *Top. Curr. Chem.*, **131**, 99 (1986).
715. W.A. Nugent and R.L. Harlow, *J. Am. Chem. Soc.*, **116**, 6142 (1994).
716. V.G. Kessler, L.G. Hubert-Pfalzgraf, S. Halut, and J.C. Daran, *J. Chem. Soc., Chem. Commun.*, 705 (1994).
717. H.J. Cohen, *J. Organomet. Chem.*, **9**, 177 (1967).
718. W.M.P.B. Menge and J.G. Verkade, *Inorg. Chem.*, **30**, 4628 (1991).
719. F. Hein and P.W. Albert, *Z. Anorg. Allg. Chem.*, **269**, 67 (1952).
720. M.J. Lacy and C.G. McDonald, *Aust. J. Chem.*, **29**, 1119 (1976).
721. A.B. Finestone, *U.S. Patent*, 2,953,545 (1960); *Chem. Abstr.*, **55**, 4045 (1961).
722. M. Zeldin and J. Ochs, *J. Organomet. Chem.*, **86**, 369 (1975).
723. R.C. Mehrotra, *J. Indian Chem. Soc.*, **55**, 1 (1978).
724. R.C. Mehrotra, *Coordination Chemistry of Metal β-Diketonates*, XXVth ICCC Nanjing, *Pure. Appl. Chem.* (IUPAC), **60**, 1349 (1988).
725. R.K. Mehrotra and R.C. Mehrotra, *Can. J. Chem.*, **39**, 795 (1961).
726. J.H. Wengrovius, M.F. Garbauskas, E.A. Williams, R.C. Going, P.E. Donahue, and I.F. Smith, *J. Am. Chem. Soc.*, **108**, 982 (1986).
727. R.C. Mehrotra and A.K. Rai, *Polyhedron*, **10**, 1967 (1991).
728. Y.P. Singh, S. Saxena, and A.K. Rai, *Synth. React. Inorg. Metal.-Org. Chem.*, **14**, 237 (1984).
729. M.F. Garbauskas, J.H. Wengrovius, R.C. Going, and J.S. Kasper, *Acta Cryst. Cryst. Struct. Commun.*, **C40**, 1536 (1984).
730. A. Dhammani, R. Bohra, and R.C. Mehrotra, *Main Group Met. Chem.*, **18**, 687 (1995).
731. A. Dhammani, R. Bohra, and R.C. Mehrotra, *unpublished results*.
732. A. Dhammani, *Ph.D. Thesis*, University of Rajasthan, Jaipur, India (1997).
733. D.C. Bradley and M.M. Faktor, unpublished results.
734. R.C. Mehrotra, T.N. Misra, and S.N. Misra, *Indian J. Chem.*, **3**, 525 (1965); ibid. **5**, 372 (1967).
735. B.S. Sankhla and R.N. Kapoor, *Can. J. Chem.*, **44**, 1369 (1966); *Aust. J. Chem.*, **20**, 685 (1967).
736. R.C. Mehrotra and S.R. Bindal, *Can. J. Chem.*, **47**, 2661 (1969).

737. K. Kumar, S.N. Misra, and R.N. Kapoor, *Indian J. Chem*, **7**, 1249 (1969).
738. F. Schmidt, *Angew. Chem.*, **64**, 536 (1952).
739. R.E. Reeves and L.W. Mazgeno, *J. Am. Chem. Soc.*, **76**, 2533 (1954)
740. A. Yamamoto and S. Kambara, *J. Am. Chem. Soc.*, **79**, 4344 (1957).
741. D.M. Puri, K.C. Pande, and R.C. Mehrotra, *J. Less-Common Met.*, **4**, 393, 481 (1962).
742. K.C. Pande and R.C. Mehrotra, *Chem. Ind.* (*London*), **35**, 1198 (1958).
743. M. Cox, J. Lewis, and R.S. Nyholm, *J. Chem. Soc.*, 6116 (1964).
744. T.J. Pinnavaia and R.C. Fay, *Inorg. Chem.*, **7**, 502 (1968).
745. B.D. Podolesov, V.B. Jardanovska, M.R. Karunoski, and D.N. Toser, *J. Inorg. Nucl. Chem.*, **36**, 1495 (1974).
746. W. Dilthey, *Berichte*, **36**, 1833 (1903); ibid. **37**, 589 (1904).
747. D.M. Puri and R.C. Mehrotra, *J. Indian Chem. Soc.*, **39**, 499 (1962).
748. D.M. Puri and R.C. Mehrotra, *J. Less-Common Met.*, **3**, 253 (1961).
749. R.C. Fay and R.N. Lowry, *Inorg. Chem.*, **6**, 1512 (1967).
750. N. Serpone and R.C. Fay, *Inorg. Chem.*, **6**, 1835 (1967).
751. D.C. Bradley and C.E. Holloway, *Chem. Commun.*, 284 (1965).
752. D.C. Bradley and C.E. Ilolloway, *J. Chem. Soc. A*, 282 (1969).
753. P.C. Bharara, V.D. Gupta, and R.C. Mehrotra, *Synth. React. Inorg. Metal.-Org. Chem.*, **5**, 59 (1975).
754. P.C. Bharara, V.D. Gupta, and R.C. Mehrotra, *Indian J. Chem.*, **13**, 725 (1975).
755. D.M. Puri, *J. Indian Chem. Soc.*, **47**, 535 (1970).
756. U.B. Saxena, A.K. Rai, V.K. Mathur, R.C. Mehrotra, and D. Radford, *J. Chem. Soc. A*, 904 (1970).
757. M. Morstein, *Inorg. Chem.*, **38**, 125 (1999).
758. K.A. Fleeting, P. O'Brien, D.J. Otway, A.J.P. White, D.J. Williams, and A.C. Jones, *Inorg. Chem.*, **38**, 1432 (1997).
759. R.C. Mehrotra and V.D. Gupta, *J. Organomet. Chem.*, **4**, 237 (1965).
760. D.P. Gaur, G. Srivastava, and R.C. Mehrotra, *Indian J. Chem.*, **12**, 399 (1974).
761. P.F.R. Ewings, P.G. Harrison, and D.E. Fenton, *J. Chem. Soc. Dalton Trans.*, 821 (1975).
762. A. Whitley, *Ph.D. Thesis*, University of London (1954).
763. P.N. Kapoor and R.C. Mehrotra, *J. Less-Common Met.*, **8**, 339 (1965).
764. R.N. Kapoor, S. Prakash, and P.N. Kapoor, *Bull. Chem. Soc. Japan*, **40**, 1384 (1967).
765. R.C. Mehrotra and P.N. Kapoor, *J. Less-Common Met.*, **7**, 176 (1964).
766. R.C. Mehrotra and P.N. Kapoor, *J. Less-Common Met.*, **7**, 453 (1964).
767. R.C. Mehrotra, R.N. Kapoor, S. Prakash, and P.N. Kapoor, *Aust. J. Chem.*, **19**, 2079 (1966).
768. A.M. Bhandari and R.N. Kapoor, *Can. J. Chem.*, **44**, 1369 (1966).
769. S. Dubey, S.N. Misra, and R.N. Kapoor, *Z. Naturforsch*, **25b**, 476 (1970).
770. C. Friedel and J. Crafts, *Ann. Chim. Phys.*, **9**, 10 (1866).
771. H.W. Post and C.H. Hofrichter Jr, *J. Org. Chem.*, 443 (1940).
772. R.P. Narain and R.C. Mehrotra, *J. Indian Chem. Soc.*, **41**, 755 (1964).
773. J.W. McBain and W.L. McLatchie, *J. Am. Chem. Soc.*, **54**, 3266 (1932).
774. A.K. Rai and R.C. Mehrotra, unpublished results.
775. K.C. Pande and R.C. Mehrotra, *Z. Anorg. Allg. Chem.*, **290**, 87, 95 (1957); ibid. **291**, 97 (1957).
776. K.C. Pande and R.C. Mehrotra, *J. Prakt. Chem.*, **5**, 101 (1957).
777. K.C. Pande and R.C. Mehrotra, *Chem. Ind. (London)*, 114 (1957).
778. I.D. Verma and R.C. Mehrotra, *J. Prakt. Chem.*, **8**, 235 (1959); ibid. **10**, 247 (1960).
779. E.V. Kusnetsov, E.V. Ignat'eva, and R.M. Kudosova, *Zh. Obshch. Khim.*, **80**, 1797 (1970); *Chem. Abstr.*, **74**, 60, 381 (1971).
780. R.N. Kapoor and R.C. Mehrotra, *Chem. Ind. (London)*, 68 (1958).
781. R.N. Kapoor and R.C. Mehrotra, *J. Am. Chem. Soc.*, **80**, 3569 (1958).
782. R.N. Kapoor and R.C. Mehrotra, *J. Chem. Soc.*, 422 (1959).
783. R.N. Kapoor and R.C. Mehrotra, *J. Am. Chem Soc.*, **82**, 3495 (1960).
784. A.W. Dearing and E.E. Reid, *J. Am. Chem. Soc.*, **50**, 3058 (1928).
785. R.C. Mehrotra and B.C. Pant, *Indian J. Chem.*, **1**, 380 (1963).

786. R.C. Mehrotra and B.C. Pant, *J. Indian Chem. Soc.*, **40**, 623 (1963).
787. S. Mathur and R.C. Mehrotra, *J. Organomet. Chem.*, **7**, 227 (1967).
788. S. Dubey, A.M. Bhandani, S.N. Misra, and R.N. Kapoor, *Indian J. Chem.*, **7**, 701 (1969).
789. F. Preuss, J. Woitschach, and H. Schug, *J. Inorg. Nucl. Chem.*, **35**, 3723 (1973).
790. S. Prakash and R.N. Kapoor, *Inorg. Chim. Acta*, **5**, 372 (1971).
791. Y. Matsmura, M. Shindo, and R. Okawara, *J. Organomet. Chem.*, **27**, 357 (1971).
792. S. Doeuff, Y. Dromzee, F. Taulelle, and C. Sanchez, *Inorg. Chem.*, **28**, 4439 (1989).
793. R.E. Reeves and L.W. Mazzeno, *J. Am. Chem. Soc.*, **76**, 2533 (1954).
794. Y. Yamamoto and S. Kambara, *J. Am. Chem. Soc.*, **81**, 2663 (1959).
795. D.M. Puri and R.C. Mehrotra, *Indian J. Chem.*, **5**, 448 (1967).
796. D.M. Puri, *J. Indian Chem. Soc.*, **47**, 535 (1970).
797. U.B. Saxena, *Ph.D. Thesis*, University of Rajasthan, Jaipur, India (1968).
798. U.B. Saxena, A.K. Rai, and R.C. Mehrotra, *Inorg. Chim. Acta.*, **7**, 681 (1973).
799. R.C. Mehrotra and A. Singh, *Phosphorus, Sulfur, Silicon*, **124**, **125**, 153 (1997).
800. R.C. Mehrotra, A. Singh, M. Bhagat, and J. Godhwani, *J. Sol–Gel. Sci. Technol.*, **13**, 45 (1998).
801. C.A. Zechmann, J.C. Huffman, K. Folting, and K.G. Caulton, *Inorg. Chem.*, **37**, 5856 (1998) and references therein.
802. A. Singh and R.C. Mehrotra, *Chemtracts*, **12**, 607 (1999).
803. S. Dubey, A.M. Bhandari, S.N. Misra, and R.N. Kapoor, *Indian J. Chem.*, **8**, 97 (1970).
804. D.C. Crans, R.A. Felty, H. Chen, H. Eckert, and N. Das, *Inorg. Chem.*, **33**, 2427 (1994).
805. R.C. Mehrotra and P.N. Kapoor, *J. Less-Common Met.*, **8**, 419 (1965).
806. R.N. Kapoor, S. Prakash, and P.N. Kapoor, *Z. Anorg. Allg. Chem.*, **351**, 219 (1967).
807. P.P. Sharma and R.C. Mehrotra, *J. Indian Chem. Soc.*, **46**, 123 (1969).
808. R.C. Mehrotra and G. Srivastava, *J. Chem. Soc.*, 1032, 3819, 4045 (1962); R.C. Mehrotra and G. Srivastava, *J. Indian Chem. Soc.*, **39**, 203 (1962).
809. R.C. Mehrotra and R.K. Mehrotra, *J. Indian Chem. Soc.*, **39**, 635 (1962).
810. R.C. Mehrotra and R.P. Narain, *Indian J. Chem.*, **5**, 444 (1967).
811. R.C. Mehrotra, *Pure Appl. Chem.*, **13**, 111 (1966).
812. M.G. Voronkov, I. Romadane, V.A. Pestunovich, and I. Mazeika, *Khim. Geterotsikl. Soedin*, **6**, 972 (1968); *Chem. Abstr.*, **70**, 68820t (1969).
813. H.G. Emblem and K. Hargreaves, *J. Inorg. Nucl. Chem.*, **30**, 721 (1968).
814. R.C. Mehrotra and G. Chandra, *J. Chem. Soc.*, 2804 (1963).
815. R.C. Mehrotra and V.D. Gupta, *J. Indian Chem. Soc.*, **32**, 727 (1966).
816. R.C. Mehrotra and S.N. Mathur, *J. Indian Chem. Soc.*, **42**, 814 (1965).
817. R.C. Mehrotra and S.N. Mathur, *J. Indian Chem. Soc.*, **42**, 749 (1965).
818. T.J. Boyle, R.W. Schwatz, R.J. Doedens, and J.W. Ziller, *Inorg. Chem.*, **34**, 1110 (1995).
819. T.J. Boyle, T.M. Alam, D. Dimons, G.J. Moore, C.D. Buchheit, H.N. Al-Shareef, E.R. Mechenbeer, R.R. Bear, and J.W. Ziller, *Chem. Mater.*, **9**, 3187 (1997).
820. A. Chakravorty, *Coord. Chem. Rev.*, **13**, 1 (1974) and references therein.
821. R.C. Mehrotra, in *Comprehensive Coordination Chemistry* (G. Wilkinson, R.D. Gillard, and J.A. McCleverty, eds), Vol. 2, Ch. 13.8, 269–291, Pergamon Press, London (1987).
822. A. Singh, *Ph.D. Thesis*, University of Rajasthan, Jaipur, India (1972).
823. A. Singh, C.K. Sharma, A.K. Rai, V.D. Gupta, and R.C. Mehrotra, *J. Chem. Soc. A*, 2440 (1971).
824. A. Singh, A.K. Rai, and R.C. Mehrotra, *Indian J. Chem.*, **12A**, 512 (1974).
825. P. Rupani, A. Singh, A.K. Rai, and R.C. Mehrotra, *Indian J. Chem.*, **19A**, 449 (1988).
826. A. Singh, A.K. Rai, and R.C. Mehrotra, *Inorg. Chim. Acta.*, **7**, 450 (1973).
827. R. Bohra, *Ph.D. Thesis*, University of Rajasthan, Jaipur, India (1974).
828. R. Bohra, A.K. Rai, and R.C. Mehrotra, *Indian J. Chem.*, **12**, 855 (1974).
829. R.C. Mehrotra, A.K. Rai, and R. Bohra, *Z. Anorg. Allg. Chem.*, **399**, 338 (1973).
830. A. Singh, A.K. Rai, and R.C. Mehrotra, *Indian J. Chem.*, **11**, 478 (1974).
831. A. Singh, A.K. Rai, and R.C. Mehrotra *J. Chem. Soc., Dalton Trans.*, 1911 (1972).
832. R.C. Mehrotra, A.K. Rai, and R. Bohra, *Synth. React. Inorg. Met.-Org. Chem.*, **4**, 167 (1974).

833. R.C. Mehrotra, A.K. Rai, and R. Bohra, *J. Indian Chem. Soc.*, **51**, 304 (1974).
834. C.K. Sharma, *Ph.D. Thesis*, University of Rajasthan, Jaipur, India (1970).
835. J.R. Jennings and K. Wade, *J. Chem. Soc. A*, 1333 (1967).
836. P.G. Harrison and J.J. Zuckerman, *Inorg. Chem.*, **9**, 175 (1970).
837. Z. Wu, X. Zhou, W. Zhang, Z. Xu, X. You, and X. Huang, *J. Chem. Soc., Chem. Commun.*, 813 (1994).
838. G.P. Khare and R.J. Doedens, *Inorg. Chem.*, **16**, 907 (1977).
839. G.P. Khare and R.J. Doedens, *Inorg. Chem.*, **15**, 86 (1976).
840. K. Wieghardt, E. Hofer, W. Holzbach, B. Nuber, and J. Weiss, *Inorg. Chem.*, **19**, 2927 (1980).
841. W. Holzbach, K. Wieghardt, and J. Weiss, *Z. Naturforsch.*, **36B**, 289 (1981).
842. S.F. Gheller, T.W. Hambley, P.R. Traill, R.T.C. Brownlee, M.J. O'Connor, M.R. Snow, and A.G. Wedd, *Aust. J. Chem.*, **35**, 2183 (1982).
843. M. Calligaris and L. Randaccio, in *Comprehensive Coordination Chemistry* (G. Wilkinson, R.D. Gillard, and J.A. McCleverty, eds), Vol. 2, Ch. 20.1, 715–738, Pergamon Press, London (1987).
844. R.H. Holm, G.W. Everett Jr, and A. Chakravarty, *Prog. Inorg. Chem.*, **7**, 83 (1966).
845. S. Yamada, *Coord. Chem. Rev.*, **2**, 83 (1967).
846. M.D. Hobday and T.D. Smith, *Coord. Chem. Rev.*, **9**, 311 (1973).
847. J. Dayal (J. Sahai), *Ph.D. Thesis*, University of Rajasthan, Jaipur, India (1972).
848. S.K. Agrawal and J.P. Tandon, *Synth. React. Inorg. Met.-Org. Chem.*, **4**, 387 (1974).
849. S.K. Agrawal and J.P. Tandon, *J. Inorg. Nucl. Chem.*, **37**, 949 (1975).
850. S.K. Agrawal and J.P. Tandon, *J. Inorg. Nucl. Chem.*, **37**, 1994 (1975).
851. S.K. Agrawal and J.P. Tandon, *Monatsh. Chem.*, **110**, 401 (1979).
852. S.P. Mittal, R.V. Singh, and J.P. Tandon, *Synth. React. Inorg. Met.-Org. Chem.*, **12**, 269 (1982).
853. P. Prashar, *Ph.D. Thesis*, University of Rajasthan, Jaipur, India (1969).
854. S.R. Gupta and J.P. Tandon, *Z. Naturforsch.*, **25B**, 1090 (1970).
855. S.R. Gupta and J.P. Tandon, *Z. Naturforsch.*, **25B**, 1231 (1971).
856. P. Prashar and J.P. Tandon, *J. Indian Chem. Soc.*, **47**, 1081 (1970).
857. P. Prashar and J.P. Tandon, *Z. Anorg. Allg. Chem.*, **383**, 81 (1971).
858. S.R. Gupta, *Ph.D. Thesis*, University of Rajasthan, Jaipur, India (1971).
859. J. Uttamchandani, S.K. Mehrotra, A.M. Bhandari, and R.N. Kapoor, *Synth. React. Inorg. Met.-Org. Chem.*, **8**, 439 (1978).
860. E.C. Alyea, A. Malik, and P.H. Merrell, *Transition Met. Chem.*, **4**, 172 (1979).
861. R.K. Sharma, R.V. Singh, and J.P. Tandon, *Synth. React. Inorg. Met.-Org. Chem.*, **9**, 519 (1979).
862. S. Mishra and A. Singh, unpublished results.
863. J. Uttamchandani and R.N. Kapoor, *Indian J. Chem.*, **24A**, 242 (1985).
864. P. Prashar and J.P. Tandon, *J. Less-Common Met.*, **15**, 219 (1968).
865. J. Uttamchandani and R.N. Kapoor, *Monatsh. Chem.*, **110**, 841 (1979).
866. M. Singh, B.L. Mathur, K.S. Gharia, and R. Mehta, *Inorg. Nucl. Chem. Lett.*, **17**, 291 (1981).
867. R.V. Singh, R.K. Sharma, and J.P. Tandon, *Synth. React. Inorg. Met.-Org. Chem.*, **11**, 139 (1981).
868. V. Verma, S. Kher, and R.N. Kapoor, *Synth. React. Inorg. Met.-Org. Chem.*, **13**, 943 (1983).
869. S.R. Gupta and J.P. Tandon, *Ann. Soc. Bruxe.*, **88**, 251 (1974).
870. P. Prashar and J.P. Tandon, *Z. Naturforsch.*, **25B**, 31 (1970).
871. P. Prashar and J.P. Tandon, *Z. Naturforsch.*, **26B**, 11 (1971).
872. A.K. Narula, B. Singh, S. Gupta, and R.N. Kapoor, *Transition Met. Chem.*, **9**, 379 (1984).
873. S. Gupta, B. Singh, A.K. Narula, and R.N. Kapoor, *Synth. React. Inorg. Met.-Org. Chem.*, **15**, 723 (1985).
874. J. Uttamchandani and R.N. Kapoor, *Indian J. Chem.*, **24A**, 242 (1985).
875. P. Prashar and J.P. Tandon, *Bull. Chem. Soc. Japan*, **43**, 1244 (1970).

876. J. Uttamchandani, S.K. Mehrotra, A.M. Bhandari, and R.N. Kapoor, *Transition Met. Chem.*, **1**, 249 (1976).
877. J.P. Tandon, *J. Indian Chem. Soc.*, **63**, 451 (1986) and references therein.
878. M. Goyal, *Ph.D. Thesis*, University of Rajasthan, Jaipur, India (1996).
879. S. Goyal and A. Singh, unpublished results.
880. J. Dayal and R.C. Mehrotra, *Synth. React. Inorg. Met.-Org. Chem.*, **1**, 287 (1971).
881. J. Dayal and R.C. Mehrotra, *Indian J. Chem.*, **10**, 435 (1972).
882. R.N. Prasad and J.P. Tandon, *Z. Naturforsch.*, **28B**, 153 (1973).
883. J. Dayal and R.C. Mehrotra, *Z. Naturforsch.*, **27B**, 25 (1972).
884. J.P. Tandon and R.N. Prasad, *Z. Naturforsch.*, **28B**, 63 (1973).
885. J.P. Tandon and R.N. Prasad, *Monatsch. Chem.*, **104**, 1064 (1973).
886. R.N. Prasad, *Ph.D. Thesis*, University of Rajasthan, Jaipur, India (1973).
887. R.N. Prasad and J.P. Tandon, *J. Less-Common Met.*, **37**, 141 (1974).
888. R.N. Prasad and J.P. Tandon, *J. Inorg. Nucl. Chem.*, **37**, 35 (1975).
889. A. Saxena, J.P. Tandon, T. Birchall, B. Ducourant, and G. Mascherpa, *Polyhedron*, **3**, 661 (1984).
890. R.C. Mehrotra, V.D. Gupta, and D. Sukhani, *J. Inorg. Nucl. Chem.*, **29**, 83 (1967).
891. R.C. Mehrotra, V.D. Gupta, and D. Sukhani, *J. Inorg. Nucl. Chem.*, **29**, 1577 (1967).
892. R.C. Mehrotra, V.D. Gupta, and S. Chatterjee, *Aust. J. Chem.*, **21**, 2929 (1968).
893. R.C. Mehrotra, V.D. Gupta, and D. Sukhani, *Inorg. Chim. Acta*, **2**, 111 (1968).
894. D.C. Bradley and P.A. Hammersley, *J. Chem. Soc. A*, 1894 (1967).
895. A.N. Nesmeyanov, R.K. Reidlina, and O.V. Nogina, *Iz. Akad. Nauk. SSSR, Otdel. Khim. Nauk*, 518 (1951).
896. R.C. Mehrotra, *J. Indian Chem. Soc.*, **30**, 731 (1953).
897. R.C. Mehrotra, *J. Indian Chem. Soc.*, **32**, 759 (1955).
898. R.G. Jones, E. Bindschadler, D. Blume, G.A. Martin, J.R. Thirtle, and H. Gilman, *J. Am. Chem. Soc.*, **78**, 6027 (1956).
899. R.C. Mehrotra and V.D. Gupta, *J. Indian Chem. Soc.*, **43**, 155 (1966).
900. D. Hass and I. Cech, *Z. Chem.*, **9**, 384 (1969); *Chem. Abstr.*, **72**, 18060 (1970).
901. R.C. Mehrotra and S.N. Mathur, *Indian J. Chem.*, **5**, 206 (1967).
902. J.S. Jennings, W. Wardlaw, and W.J.R. Way, *J. Chem. Soc.*, 637 (1936).
903. R.C. Mehrotra and M. Arora, *J. Less-Common Met.*, **17**, 181 (1969).
904. R.C. Mehrotra and R.A. Misra, *Can. J. Chem.*, **42**, 717 (1964).
905. R.C. Mehrotra and R.A. Misra, *J. Chem. Soc.*, 43 (1965).
906. R.C. Mehrotra and R.A. Misra, *J. Indian Chem. Soc.*, **45**, 656 (1968).
907. S.N. Misra, T.N. Misra, and R.C. Mehrotra, *Indian J. Chem.*, **5**, 439 (1967).
908. R.N. Kapoor and B.S. Sankhla, *Can. J. Chem.*, **44**, 2131 (1966).
909. D.C. Bradley, D.C. Hancock, and W. Wardlaw, *J. Chem. Soc.*, 2773 (1952).
910. N.M. Cullinane, S.J. Chard, G.E. Price, and B.B. Millward, *J. Appl. Chem.*, **2**, 250 (1952).
911. D.C. Bradley, F.M. El-Halim, and W. Wardlaw, *J. Chem. Soc.*, 3450 (1950).
912. I.D. Verma and R.C. Mehrotra, *J. Less-Common Met*, **1**, 263 (1960).
913. D.C. Bradley, F.M. El-Halim, R.C. Mehrotra, and W. Wardlaw, *J. Chem. Soc.*, 4609 (1952).
914. R.C. Mehrotra and R.N. Kapoor, *J. Less-Common Met.*, **3**, 188 (1961).
915. R.C. Mehrotra and R.A. Misra, *Indian J. Chem.*, **6**, 669 (1968).
916. R.K. Mittal and R.C. Mehrotra, *Z. Anorg. Allg. Chem.*, **332**, 189 (1964).
917. R.C. Mehrotra and P.N. Kapoor, *J. Less-Common Met.*, **10**, 348 (1966).
918. S. Prakash, P.N. Kapoor, and R.N. Kapoor, *J. Prakt. Chem.*, **36**, 24 (1967).
919. R.N. Kapoor and A.M. Bhandari, *J. Prakt. Chem.*, **35**, 284 (1967).
920. R.C. Mehrotra and P.P. Sharma, *J. Indian Chem. Soc.*, **44**, 74 (1967).
921. R.C. Mehrotra and R.K. Mehrotra, *J. Indian Chem. Soc.*, **39**, 1 (1962); *J. Prakt. Chem.*, **16**, 251 (1962).
922. S.R. Bindal, P.N. Kapoor, and R.C. Mehrotra, *Inorg. Chem.*, **7**, 384 (1968).
923. A.F. Reilly and H.W. Post, *J. Org. Chem.*, **16**, 383 (1951).
924. R.C. Mehrotra and G. Chandra, *Recl. Trav. Chem.*, **82**, 683 (1963).
925. R.C. Mehrotra and V.D. Gupta, *J. Indian Chem. Soc.*, **43**, 155 (1966).

926. R.C. Mehrotra and S.N. Mathur, *Indian J. Chem.*, **5**, 206 (1967).
927. A.N. Nesmeyanov, E.M. Brainina, and R.Kh. Freidlina, *Dokl. Akad. Nauk. SSSR*, **94**, 249 (1954).
928. K. Clauss, *Justus Liebigs Ann. Chem.*, **711**, 19 (1968); *Chem. Abstr.*, **68**, 95911 (1968).
929. I.P. Goldshtein, N.N. Zemlyanski, O.P. Shamagina, E.N. Guryanova, E.M. Panov, N.A. Slovokhotova, and K.A. Kocheshkov, *Proc. Acad. Sci. USSR*, **163**, 715 (1965).
930. A.G. Davies and P.G. Harrison, *J. Chem. Soc. C*, 298 (1967).
931. M. Hidai, T. Hikita, and Y. Uchida, *Chem. Lett. Japan.*, 521 (1972).
932. M.H. Chisholm, F.A. Cotton, M.W. Extine, and R.L. Kelly, *J. Am. Chem. Soc.*, **100**, 5764 (1978).
933. M.H. Chisholm, F.A. Cotton, M.W. Extine, and W.W. Reichert, *J. Am. Chem. Soc.*, **100**, 1727 (1978).
934. T. Tsuda and T. Saegusa, *Inorg. Chem.*, **11**, 2561 (1972).
935. T. Tsuda, S. Sanada, and T. Saegusa, *J. Organomet. Chem.*, **116**, C10 (1976).
936. H. Burger, *Monatsch. Chem.*, **95**, 671 (1964).
937. O. Meth-Cohn, D. Thorpe, and H.J. Twitchett, *J. Chem. Soc.* (C), 132 (1970).
938. P.C. Bharara, V.D. Gupta, and R.C. Mehrotra, *Indian J. Chem.*, **13**, 156 (1975).
939. R.C. Mehrotra, A.K. Rai, and R. Bohra, *J. Inorg. Nucl. Chem.*, **36**, 1887 (1974).
940. R.C. Mehrotra, A.K. Rai, and R. Bohra, unpublished results.
941. K.N. Mahendra, P.C. Bharara, and R.C. Mehrotra, *Inorg. Chim. Acta.*, **25**, 15 (1977).
942. M.H. Chisholm, F.A. Cotton, K. Folting, J.C. Huffman, A.L. Rattermann, and E.S. Shamshoum, *Inorg. Chem.*, **23**, 4423 (1984).
943. F.A. Cotton and E.S. Shamshoum, *J. Am. Chem. Soc.*, **107**, 4662 (1985).
944. R.C. Mehrotra, P.C. Bharara, and B.P. Baranwal, *Indian J. Chem.*, **15A**, 458 (1977).
945. J. Boersma and J.G. Noltes, *Organozinc Coordination Chemistry*, 61, International Lead and Zinc Research Organization, Inc., New York (1968); J.G. Noltes, *Recl. Trav. Chim. Pays-Bas*, **84**, 126 (1965).
946. A.G. Davies and G.J.D. Peddle, *Chem. Commun.*, 96 (1965).
947. A.J. Bloodworth and J. Serlin, *J. Chem. Soc., Perkin Trans. I*, 261 (1973).
948. V.L. Foss, E.A. Besolova, and I.F. Lutsenko, *Zh. Obshch. Khim.*, **35**, 759 (1965).
949. C. Blandy and D. Gervais, *Inorg. Chim. Acta*, **47**, 197 (1981).
950. C. Blandy and M. Hiliwa, *C.R. Seances Acad. Soc.*, Ser. 2, **296**, 51 (1983).
951. L. Vuitel and A. Jacot-Guillarmod, *Synthesis*, 608 (1972).
952. I.F. Lutsenko, V.L. Foss, and A.L. Ivanova, *Proc. Acad. Sci. USSR, Chem. Sect.*, **141**, 1270 (1971); *Chem. Abstr.*, **56**, 12920 (1962).
953. H.E. Bryndza, J.C. Calabrese, and S.S. Wreford, *Organometallics*, **3**, 1603 (1984).
954. H.E. Bryndza and T.H. Tulip, *Organometallics*, **4**, (1985).
955. M.S. Bains and D.C. Bradley, *Can. J. Chem.*, **40**, 2218 (1962).
956. M.S. Bains and D.C. Bradley, *Can. J. Chem.*, **40**, 1350 (1962).
957. M. Basso-Bert and D. Gervais, *Inorg. Chim. Acta*, **34**, 191 (1979).
958. S. Boulmaaz, R. Papiernik, L.G. Hubert-Pfalzgraf, J. Vaissermann, and J.C. Daran, *Polyhedron*, **11**, 1331 (1992).
959. M. Wedler, J.W. Gilje, U. Pieper, D. Stalke, M. Noltemeyer, and F.T. Edelmann, *Chem. Ber.*, **124**, 1163 (1991).
960. C. Baudin, D. Baudry, M. Ephritikhine, M. Lance, A. Navaza, M. Nierlick, and J. Vigner, *J. Organomet. Chem.*, **415**, 59 (1991).
961. M.H. Chisholm, F.A. Cotton, M.W. Extine, and W.W. Reichert, *J. Am. Chem. Soc.*, **100**, 153 (1978).
962. M.J. Chetcuti, M.H. Chisholm, J.C. Huffman, and J. Leonelli, *J. Am. Chem. Soc.*, **105**, 292 (1983).
963. M.H. Chisholm, F.A. Cotton, M.W. Extine, M. Millar, and B.R. Stults, *Inorg. Chem.*, **16**, 2407 (1977).
964. H. Lehmkuhl, K. Mehler, R. Benn, A. Rufinska, and C. Kruger, *Chem. Ber.*, **119**, 1054 (1986).
965. H. Schumann, in *Organometallics of the f-Elements* (T.J. Marks and R.D. Fischer, eds), 81, D. Reidel, Dordrecht (1978); H. Schumann, *Angew. Chem., Int. Ed. Engl.* **23**, 474 (1984).

966.   H. Schumann, W. Genthe, E. Hahn, J. Pickardt, H. Schwarz, and K. Eckart, *J. Organomet. Chem.*, **306**, 215 (1986).

966a.   P. Biagini, G. Lugli, L. Abis, and R. Millini, *J. Organomet. Chem.*, **474**, C16 (1994).

967.   A. Greco, S. Cesca, and G. Bertolini, *J. Organomet. Chem.*, **113**, 321 (1976).

968.   A. Gulino, M. Casarin, V.P. Conticello, J.G. Gaudiello, H. Mauermann, F.I. Fragala, and T.J. Marks, *Organometallics*, **7**, 2360 (1988).

969.   A. Greco, G. Bertolini, and S. Cesca, *Inorg. Chim. Acta.*, **21**, 245 (1977).

970.   M. Bochmann, G. Wilkinson, and G.B. Young, *J. Chem. Soc., Dalton Trans.*, 1879 (1980).

971.   A. Lachowicz and K.H. Thiele, *Z. Anorg. Allg. Chem.* **431**, 88 (1977).

972.   L. Chamberlain, J. Keddinton, and I.P. Rothwell, *Organometallics*, **1**, 1098 (1982).

973.   W. Kruse, *J. Organomet. Chem.*, **42**, C39 (1972).

974.   W. Mowat, A.J. Shortland, N.J. Hill, and G. Wilkinson, *J. Chem. Soc., Chem. Commun.*, 770 (1973).

975.   T. Tsuda and J.K. Kochi, *Bull. Chem. Soc. Japan*, **45**, 648 (1972).

976.   L. Claisen, *Chem. Ber.*, **20**, 646 (1887).

977.   W. Tischtchenko, *Chem. Zentr.*, **77**, 1309, 1556, 1558 (1906).

978.   M.S. Kulpinski and F.F. Nord, *Nature*, **151**, 363 (1943).

979.   M.S. Kulpinski and F.F. Nord, *J. Org. Chem.*, **8**, 256 (1943).

980.   F.J. Villani and F.F. Nord, *J. Am. Chem. Soc.*, **69**, 2605 (1947).

981.   I. Lin and A.R. Day. *J. Am. Chem. Soc.*, **74**, 5133 (1952).

982.   F.F. Nord, *Biochem. Z.*, **106**, 275 (1920).

983.   F.F. Nord, *Beitrage Physiologie*, **2**, 301 (1924).

984.   F.F. Nord, *Ger. Patent*, 434, 728 (1926).

985.   N.A. Orloff, *Bull. Chem. Soc. France*, **35**, 360 (1924).

986.   F.J. Villani and F.F. Nord, *J. Am. Chem. Soc.*, **69**, 2608 (1947).

987.   R. Nakai, *Biochem. Z.*, **152**, 258 (1925); *Chem. Abstr.*, **19**, 2807 (1947).

988.   A. Verley, *Bull. Soc. Chim. France*, **37**, 537 (1925); *Chem. Abstr.*, **19**, 2635 (1925).

989.   H. Meerwein and R. Schmidt, *Justus Liebigs Ann. Chem.*, **444**, 221 (1925).

990.   W. Ponndorf, *Z. Angew. Chem.*, **39**, 138 (1926); *Chem. Abstr.*, **20**, 1611 (1926).

991.   W.G. Young, W.H. Hartung, and F.S. Crossley, *J. Am. Chem. Soc.*, **58**, 100 (1936).

992.   H. Meerwein, B. Von Bock, B. Kirschnick, W. Lenz, and A. Migge, *J. Prakt. Chem.*, **147**, 211 (1936).

993.   L.P. Hammett, *Physical Organic Chemistry*, 352, McGraw-Hill, New York (1940).

994.   R.B. Woodward, N.L. Wendler, and F.J. Brutschky, *J. Am. Chem. Soc.*, **67**, 1426 (1945).

995.   L.M. Jackman and J.A. Mills, *Nature*, **164**, 789 (1949).

996.   L.M. Jackman, A.K. Macbeth, and J.A. Mills, *J. Chem. Soc.*, 2641 (1949).

997.   L.M. Jackman, J.A. Mills, and J.S. Shannon, *J. Am. Chem. Soc.*, **72**, 4814 (1950).

998.   J.C. McGowan, *Chem. Ind. (London)*, 601 (1951).

999.   A.F. Rekasheva and G.P. Miklykhin, *Dokl. Akad. Nauk. SSSR*, **78**, 283 (1951).

1000.   E.D. Williams, K.A. Kreiger, and A.R. Day, *J. Am. Chem. Soc.*, **75**, 2404 (1953).

1001.   W. von E. Doering, and R.W. Young, *J. Am. Chem. Soc*, **72**, 631 (1950).

1002.   W. von E. Doering, and T.C. Aschner, *J. Am. Chem. Soc.*, **75**, 393 (1953).

1003.   V.J. Shiner and D. Whittaker, *J. Am. Chem. Soc.*, **85**, 2337 (1963).

1004.   E.W. Warnhoff, P. Reynolds-Warnhoff, and M.Y.H. Wong, *J. Am. Chem. Soc.*, **102**, 5956 (1980).

1005.   M.S. Bains, *Ph.D. Thesis*, University of London (1959).

1005a. M.S. Bains and D.C. Bradley, *Chem. Ind. (London)*, 1032 (1962).

1006.   E.C. Ashby, A.B. Goel, and J.N. Argyropoulos, *Tetrahedron Lett.*, **23**, 2273 (1982).

1007.   R.C. Mehrotra and I.D. Verma, *J. Less-Common Met.*, **3**, 321 (1961).

1008.   R.C. Mehrotra, V.D. Gupta, and P.C. Bharara, *Indian J. Chem.*, **11**, 814 (1973).

1009.   For a review, see: C. Djerassi, *Org. React.*, **6**, 207 (1951).

1010.   P.D. Bartlett and W.P. Giddings, *J. Am. Chem. Soc.*, **82**, 1240 (1960).

1011.   J.L. Namy, J. Souppe, J. Collin, and H.B. Kagan, *J. Org. Chem.*, **49**, 2045 (1984).

1012.   J.L. Namy and H.B. Kagan, *Nouv. J. Chem.*, **10**, 229 (1986).

1013.   G.A. Molander, *Chem. Rev.*, **92**, 29 (1992).

1014.   M.H. Chisholm, *Polyhedron*, **2**, 681 (1983).

1015. M.H. Chisholm, D.M. Hoffman, and J.C. Daran, *Chem. Soc. Rev.*, **14**, 69 (1985).
1016. M.H. Chisholm, *Angew. Chem., Int. Ed. Engl.*, **25**, 21 (1986); M.H. Chisholm, *J. Organomet. Chem.*, **334**, 77 (1987).
1017. M.H. Chisholm, *J. Organomet. Chem.*, **400**, 235 (1990).
1018. T.A. Budzichowski, M.H. Chisholm, and K. Folting, *Chem. Eur. J.*, **2**, 110 (1996).
1019. R.C. Mehrotra and A. Singh, *Chemtracts*, **10**, 596 (1997).
1020. M.H. Chisholm, D.M. Hoffman, J. McCandless Northius, and J.C. Hoffman, *Polyhedron*, **16**, 839 (1997).
1021. M.H. Chisholm, K. Folting, M.J. Hampden-Smith, and C.E. Hammond, *J. Am. Chem. Soc.*, **111**, 7283 (1989).
1022. M.H. Chisholm, C.E. Hammond, J.C. Huffman, and V.J. Johnston, *J. Organomet. Chem.*, **394**, C16 (1990).
1023. M.H. Chisholm, K. Folting, V.J. Johnston, and C.E. Hammond, *J. Organomet. Chem.*, **394**, 265 (1990).
1024. M.H. Chisholm, V.J. Johnston, and W.E. Streib, *Inorg. Chem.*, **31**, 4081 (1992).
1025. M.H. Chisholm, C.C. Kirkpatrick, and J.C. Huffman, *Inorg. Chem.*, **20**, 871 (1981).
1026. M.H. Chisholm, J.C. Huffman, and A.L. Ratermann, *Inorg. Chem.*, **22**, 4100 (1983).
1027. M.H. Chisholm, J.C. Huffman, and R.J. Tatz, *J. Am. Chem. Soc.*, **105**, 2075 (1983).
1028. M.H. Chisholm and R.L. Kelly, *Inorg. Chem.*, **18**, 2321 (1979).
1029. M.H. Chisholm, F.A. Cotton, M.W. Extine, and R.L. Kelly, *J. Am. Chem. Soc.*, **100**, 2256 (1978).
1030. M.H. Chisholm, F.A. Cotton, M.W. Extine, and R.L. Kelly, *J. Am. Chem. Soc.*, **101**, 7645 (1979).
1031. M.H. Chisholm, D.M. Hoffman, and J.C. Huffman, *Organometallics*, **4**, 986 (1985).
1032. M.H. Chisholm, J.C. Huffman, and R.L. Kelly, *J. Am. Chem. Soc.*, **101**, 7615 (1979).
1033. F.A. Cotton and W. Schwotzer, *J. Am. Chem. Soc.*, **105**, 5639 (1983).
1034. M.H. Chisholm, unpublished results.
1035. F.A. Cotton and W. Schwotzer, *J. Am. Chem. Soc.*, **105**, 4955 (1983).
1036. M.H. Chisholm, J.C. Huffman, J. Leonelli, and I.P. Rothwell, *J. Am. Chem. Soc.*, **104**, 7030 (1982).
1037. M.H. Chisholm, F.A. Cotton, M.W. Extine, and R.L. Kelly, *J. Am. Chem. Soc.*, **100**, 3354 (1978).
1038. M.H. Chisholm, J.C. Huffman, and R.L. Kelly, *Inorg. Chem.*, **19**, 2762 (1980).
1039. M.H. Chisholm, F.A. Cotton, M.W. Extine, and R.L. Kelly, *Inorg. Chem.*, **18**, 116 (1978).
1040. M.H. Chisholm, J.C. Huffman, and M.J. Hampden-Smith, *J. Am. Chem. Soc.*, **111**, 5284 (1989).
1041. M.H. Chisholm and M.J. Hampden-Smith, *J. Am. Chem. Soc.*, **109**, 5871 (1987).
1042. M.H. Chisholm and M.J. Hampden-Smith, *Angew. Chem., Int. Ed. Engl.*, **26**, 903 (1987).
1043. R.R. Schrock, M.L. Listemann, and L.G. Sturgeof, *J. Am. Chem. Soc.*, **104**, 4291 (1982).
1044. F.A. Cotton, W. Schwotzer, and E.S. Shamshoum, *Organometallics*, **2**, 1167 (1983).
1045. F.A. Cotton, W. Schwotzer, and E.S. Shamshoum, *Organometallics*, **2**, 1340 (1983).
1046. M.H. Chisholm, K. Folting, D.M. Hoffman, J.C. Huffman, and J. Leonelli, *J. Chem. Soc., Chem. Commun.*, 589 (1983).
1047. M.H. Chisholm, D.M. Hoffman, and J.C. Huffman, *J. Am. Chem. Soc.*, **106**, 6806 (1984).
1048. M.H. Chisholm *J. Chem. Soc., Dalton Trans.*, 1781 (1996).
1049. H. Strutz and R.R. Schrock, *Organometallics*, **3**, 1600 (1984).
1050. M.H. Chisholm and J.A. Klang, *J. Am. Chem. Soc.*, **111**, 2324 (1989); M.H. Chisholm, K. Folting, and J.A. Klang, *Organometallics*, **9**, 602 (1990); M.H. Chisholm, K. Folting, and J.A. Klang, *Organometallics*, **9**, 607 (1990).
1051. M.H. Chisholm, J.C. Huffman, and J.W. Pasterczyk, *Inorg. Chim. Acta.*, **133**, 17 (1987).
1052. M.H. Chisholm, K. Folting, and J.W. Pasterczyk, *Inorg. Chem.*, **27**, 3057 (1988).
1053. M.H. Chisholm, D.L. Clark, R.J. Errington, K. Folting, and J.C. Huffman, *Inorg. Chem.*, **27**, 2071 (1988).

1054. M.H. Chisholm, J.C. Huffman, and R.L. Kelly, *J. Am. Chem. Soc.*, **101**, 7100 (1979);
       M.H. Chisholm, D.L. Clark, and J.C. Huffman, *Polyhedron*, **4**, 1203 (1985).
1055. M.H. Chisholm, *Angew. Chem., Int. Ed. Engl.*, **36**, 52 (1997).
1056. M.H. Chisholm, C.E. Hammond, V.J. Johnston, W.E. Streib, and J.C. Huffman, *J. Am. Chem. Soc.*, **114**, 7056 (1992).

3

Heterometallic Alkoxides

1 INTRODUCTION

Heterometal alkoxides represent a rapidly growing category of novel molecular species in which alkoxo derivatives of two or more different metals are held together, generally by alkoxo, chloro, or oxo bridging ligands.

   As some rather diverse names are still in use for this interesting class of compounds, it is appropriate to recapitulate some of them, particularly as the genesis of the current term "heterometal alkoxides" itself has a historical perspective reflecting the gradual elucidation of their chemical behaviour. In 1929, Meerwein and Bersin[1] reported the formation of the so called "alkoxo salts", *e.g.* $K^+\{Al(OR)_4\}^-$ in the titrations of basic alkali alkoxides with alkoxy derivatives of less electropositive metals in neutral solvents like nitrobenzene. In 1958, Bartley and Wardlaw[2] showed that the alkali metal (M) zirconium alkoxides actually had the formula $NaZr_2(OR)_9$ instead of $[NaHZr(OEt)_6]$ reported by Meerwein and Bersin. These $NaZr_2(OR)_9$ derivatives were volatile covalent compounds. A large number of similar derivatives were synthesized amongst others by Mehrotra and co-workers, and a review article[3] entitled "Chemistry of Double Alkoxides of Various Elements" appeared in 1971, followed by a general survey[4] in 1978 and another account[5] in 1983 dealing with "bimetallic alkoxides" of transition metals. The name "mixed alkoxides" was also coined for these compounds by Chisholm and Rothwell[6] in 1988. Compounds incorporating three different metals: $[(Pr^iO)_2Al(\mu\text{-}OPr^i)_2Be(\mu\text{-}OPr^i)_2Zr(OPr^i)_3]$ and $[(Pr^iO)_2Al(\mu\text{-}OPr^i)_2Be(\mu\text{-}OPr^i)_2Nb(OPr^i)_4]$ were first reported[7] in 1985 and these were called "trimetallic alkoxides". This was soon followed in 1988 by the work on tri- and tetrametallic alkoxides of copper[8] and other later 3d transition metals.[9]

   In 1987 Mehrotra[10] drew attention to potential applications of these easily purifiable and soluble "bimetallic" and higher alkoxides as precursors for the preparation of glasses and ceramic materials, which immediately brought forth the disclosure of an earlier conjecture by Dislich.[11] All this finally led to the first publication entitled "Polymetallic Alkoxides — Precursors for Ceramics",[12] in which a variety of possible applications of such species as precursors were spelled out by Mehrotra in 1988. In view of the possibility of the term "polymetallic alkoxides" being confused with polymerized forms of homo-metal systems, such as $[M(OR)_x]_n$, the name was almost immediately changed to "Heterometal alkoxides", which has been gradually adopted since then in a number of review articles.[13-29] In the present account, the term "heterometal alkoxides" has been employed to represent these compounds in general, with bi-, tri-, and tetraheterometallic alkoxides denoting molecular species containing two, three, or four different metals, respectively.

More than two decades after the preparation of a large number of the so-called "alkoxo salts" by Meerwein and Bersin[1] in 1929, $U\{Al(OPr^i)_4\}_4$ was synthesized in 1952 by Albers et al.[30] and evidence for the formation of $M\{U(OEt)_6\}_n$ (M = Na, Ca, Al) was obtained by Jones et al.[31] in 1956. Similarly, formation of a number of anionic methoxide species was indicated in the potentiometric titrations of chlorides of metals (B, Al, Ti, Nb, Ta) with lithium methoxide in methanol by Gut[32] in 1964. Ludman and Waddington[33] studied the conductometric titrations of a wide variety of Lewis acids with basic metal methoxides and reported the formation of "alkoxo salts" of the type $KB(OMe)_4$ and $K_3Fe(OMe)_6$. Schloder and Protzer[34] also synthesized a number of bimetallic alkoxides of aluminium with the formulae, $MAl(OMe)_4$ and $M'\{Al(OMe)_4\}_2$ where M and M' are alkali and alkaline earth metals respectively.

The synthesis of volatile (under reduced pressure, indicating stability to heat) and soluble (in organic solvents) derivatives of alkali metals and zirconium, with rather unexpected compositions, e.g. $NaZr_2(OEt)_9$ by Bartley and Wardlaw[2] gave a new orientation to the field. Attention was drawn to the apparently covalent characteristics of such derivatives, in spite of the strongly electropositive alkali and alkaline earth metal component, in review articles by Mehrotra and Kapoor,[35,36] in whose group a large number of such derivatives of alkali and alkaline earth metals were synthesized, with confirmation[3] of the covalent characteristics of the derivatives by conductivity measurements. Such covalent heterometallic alkoxides have also been given the classification "molecular" in a review by Caulton and Hubert-Pfalzgraf.[16]

The coordination models of the following types suggested for tetraisopropoxoaluminate derivatives:

(M = Be, Mg, Ca, Sr, Ba, Zn, Cd, Cr, Mn, Fe, Co, Ni, Cu, In, Th, etc.), as well as

have been confirmed in a number of recent publications by the X-ray structural elucidation of molecular species such as: $[(HOPr^i)_2Mg]\{(\mu\text{-}OPr^i)_2Al(OPr^i)_2\}_2$,[37] $[(Pr^iOH)_2K(\mu\text{-}OPr^i)_2Al(OPr^i)_2]_n$,[38] $[\{Pr\{(\mu\text{-}OPr^i)_2Al(OPr^i)_2\}_2(Pr^iOH)(\mu\text{-}Cl)\}_2]$,[39] and $[Er\{(\mu\text{-}OPr^i)_2Al(OPr^i)_2\}_3]$.[40]

It is interesting to note that in the penultimate example, two $[\{Al(OPr^i)_4\}_2(Pr^iOH)$ $(\mu\text{-}Cl)\}_2]$ groups are held together by chloride bridges. With the rapidly increasing applications of heterometal alkoxides for mixed metal–oxide ceramic materials, evidence is accumulating for the formation of a large number of oxo-heterometal alkoxides (Chapter 5), which may well be intermediate stages in the hydrolytic sol–gel process. X-ray structural studies on a number of such oxo-derivatives have shown that in these, alkoxo moieties of two or three elements are linked together via oxo-ligands of various bridging modes in addition to μ_2- or μ_3-bridging alkoxide groups. The formation of oxo-derivatives under stringently anhydrous conditions indicates the possibility of their formation by some alternative pathway, such as elimination of ether or alkene.

Chelation of a central metal atom by ligands (represented in general by L) like $\{Al(OR)_4\}^-$, $\{Nb(OR)_6\}^-$, $\{Ta(OR)_6\}^-$, and $\{Zr_2(OR)_9\}^-$ gives rise to a large number of homoleptic bimetallic alkoxides, which can be represented by ML_x (where x is the principal valency and $2x$ is the coordination number of M); this coordination number would be higher ($3x$ or $4x$) when the alkoxometallate ligand is tri- or tetradentate: e.g. (before six-coordinate) nickel in $Ni\{(\mu\text{-}OMe)_3Al(OMe)\}_2$ (Section 3.3.1).

In addition to the homoleptic bimetallic derivatives like $[M\{Al(OPr^i)_4\}_x]$ hetero-leptic tri- and tetrametallic derivatives with the general formula, $[M\{Al(OPr^i)_4\}_a\{Nb(OPr^i)_6\}_b\{Zr_2(OPr^i)_9\}_cX_{x-a-b-c}]$ (where X is a ligand of the type Cl, OR, acac, etc.) have also been synthesized in the laboratories of Mehrotra[7,14,15,17-25] since 1985. Many of these have been shown to volatilize unchanged under reduced pressure. These "stable" derivatives have thus added a novel dimension to heterometallic coordination systems.[23] The prominent effect of steric factors on the stability and structural features of some bimetallic species has been observed in a comparison of similar pairs: (i) $[NaZr_2(OPr^i)_9]$ and $[\{NaZr(OBu^t)_5\}_2]$;[41] (ii) $[Ni\{(\mu\text{-}OMe)_3Al(OMe)\}_2]$ and $[Ni\{(\mu\text{-}OPr^i)(\mu\text{-}OBu^t)Al(OBu^t)_2\}_2]$.[42] Some other novel types of heterometallic alkoxide/glycolate/aminoalkoxide derivatives have also been synthesized.[43]

In view of their volatility and solubility in organic solvents, these heterometallic alkoxides have become attractive precursors for ceramic materials by the MOCVD[26] and sol–gel[27] processes (Chapter 6). Interesting observations that the framework of these heterometallic alkoxides, for example $Mg\{Al(OEt)_4\}_2$[44] and $Ba\{Zr_2(OPr^i)_9\}_2$,[45] remains unaltered at least during the early stage(s) of hydrolysis, has further underlined the stability of these systems. The above far-reaching observations and the variety of routes now available for various heterometal alkoxide systems have led workers to anticipate the synthesis of "single source" precursors,[22-25] with a composition corresponding to the targeted ceramic material, although the possibilities of alteration(s) in composition during the sol–gel treatment have to be monitored carefully.

In the following periodic table, the underlined elements are those of which mixed metal–oxide ceramic materials have been prepared,[46] whilst the encircled ones indicate the elements whose heterometallic alkoxide chemistry has been investigated:

H   He

(Li) (Be)   (B) C N O F Ne

(Na) (Mg)   (Al) (Si) P S Cl Ar

(K) (Ca) (Sc) (Ti) (V) (Cr) (Mn) (Fe) (Co) (Ni) (Cu) (Zn) (Ga) (Ge) (As) Se Br Kr

(Rb) (Sr) (Y) (Zr) (Nb) (Mo) Tc Ru Rh Pd Ag (Cd) (In) (Sn) (Sb) Te I Xe

(Cs) (Ba) (Ln*) (Hf) (Ta) (W) Re Os Ir Pt Au (Hg) (Tl) (Pb) (Bi) Po At Rn

Fr Ra An**

*Ln = La Ce Pr Nd Pr Pm Sm Eu Gd Tb Dy Ho Er Tm Yb Lu

**An = Ac Th Pa U

2 SYNTHESIS

As discussed in several review articles[16–19,21,24,28,29,35,47] the procedures employed for the synthesis of heterobimetallic alkoxides are in general similar to those utilized for the homometallic species. Some of the heterobimetallic derivatives with reactive group(s) (generally chloride) on the central metal atom can be conveniently converted to tri- or higher metallic systems by treating these with different alkali alkoxometallates.

2.1 Preparative Routes to Homoleptic Heterobimetallic Alkoxides

These can be broadly sub-divided into several categories.

2.1.1 Reactions Between Component Metal Alkoxides

The formation of alkali alkoxometallates in the reactions of alkali alkoxides with those of less basic alkoxides continues to be the most common route, although reactions of a few other alkoxides have also been utilized for this purpose:

2.1.1.1 Preparation of alkali alkoxometallates (A-1)

Following the procedure of Meerwein and Bersin[1] and others,[30,31,33,34] this method has been extended to the bimetallic alkoxides of alkali metals (Lewis bases) with those of less basic metals and metalloids, beryllium,[48,49] zinc,[48,50] boron,[51] aluminium,[1,34,52,53] gallium,[52] tin(II),[54–56] tin(IV),[57] antimony(III),[58,59] bismuth,[59] titanium,[41,60] niobium(IV),[61] zirconium,[2,41,62–64] thorium,[52] niobium(V),[52,65–67] tantalum(V)[52,65] and copper.[68] Equations (3.1)–(3.3) reflect a few typical reactions used in the synthesis of bimetallic alkoxides involving alkali metals (M):

$$MOR + M'(OR)_x \longrightarrow MM'(OR)_{x+1} \tag{3.1}$$

$x = 1; M' = Cu(I); M = Na, R = Bu^t.$

$x = 2; M' = Sn(II); M = Li, Na, K, Rb, Cs; R = Bu^t.$

$x = 3; M' = Al, Sb(III), Bi(III); M = Li, Na, K, Rb, Cs. R = Et, Pr^i, Bu^t.$

$x = 4; M' = Ti; M = Li, Na, K; R = Pr^i.$

$M' = Zr, Sn(IV); M = K, Rb, Cs; R = Bu^t.$

$M' = Nb(IV); M = Na; R = Pr^i.$

$x = 5; M' = Nb, Ta; M = Li, Na, K, Rb, Cs;$

$R = Me, Et, Pr^i, Bu^t, CH_2CMe_3, CH_2SiMe_3.$

$$MOR + 2M'(OR)_4 \longrightarrow MM'_2(OR)_9 \tag{3.2}$$

$M' = Ti, Zr, Hf, Sn(IV), Th; M = Li, Na, KTl(I); R = Et, Pr^i.$

$M' = Nb(IV); M = Na; R = Me.$

$$2MOPr^i + 3Zr(OPr^i)_4 \longrightarrow M_2Zr_3(OPr^i)_{14} \tag{3.3}$$

$[M = Na, K]$

2.1.1.2 Preparation of bimetallic alkoxides of some other metals (A-2)

In addition to the formation of alkali alkoxometallates by the reactions of alkali alkoxides (strong bases) with alkoxides of a variety of metals and metalloids (Lewis acids), formation of heterometal alkoxides has been shown to occur even between alkoxides of such similar metals as aluminium and gallium[69] as well as niobium and tantalum.[70] However, the formation constant of the latter derivative has been found to be statistical, which precludes the isolation of this bimetallic alkoxide in view of the equilibrium:

$$(MeO)_4Nb(\mu\text{-}OMe)_2Ta(OMe)_4 \rightleftharpoons Nb(OMe)_5 + Ta(OMe)_5 \qquad (3.4)$$

An early conjecture of Dislich[11] in this direction is noteworthy. The extraordinary homogeneity of the final ceramic product obtained by the sol–gel hydrolysis of a mixture of alkoxides of various metals in alcoholic solution, led him to suggest that in addition to the advantage of much more efficient mixing at the molecular level in the initial precursor solution, some sort of complexation reactions must be occurring amongst the alkoxides of various metals in the initial stages:

$$mSi(OR)_4 + nB(OR)_3 + pAl(OR)_3 + q(NaOR)$$

$$\xrightarrow{\text{Complexation in alcoholic medium}} (Si_mB_nAl_pNa_q)(OR)_{4m+3n+3p+q} \qquad (3.5)$$

$$\xrightarrow[\text{–ROH}]{\text{Hydrolysis by added water}} (Si_mB_nAl_pNa_q)(OH)_{4m+3n+3p+q} \qquad (3.6)$$

Sol changing to gel by partial

dehydration or dealcoholation
reactions | followed by
  | sintering

$$(Si_mB_nAl_pNa_q)(O)_{(4m+3n+3p+q)/2}$$

Ultrahomogeneous
oxide ceramic or glassy material

Although the equilibria involved in this type of complexation may be rather labile and require further detailed investigations, yet examples of the isolation of a few well-defined heterometallic alkoxides from mixtures of component alkoxides may be mentioned here. For example, bimetallic isopropoxides of lanthanides with aluminium can be conveniently volatilized out of a reaction mixture of $Ln(OPr^i)_3$ and $Al(OPr^i)_3$ (Eq. 3.7):

$$\frac{1}{n}\{Ln(OPr^i)_3\}_n + \frac{3}{4}\{Al(OPr^i)_3\}_4 \longrightarrow [Ln\{Al(OPr^i)_4\}_3] \qquad (3.7)$$

In fact, with the use of excess $Al(OPr^i)_3$ in the initial reaction mixture, the more volatile $Al(OPr^i)_3$ distils out first followed by $Ln\{Al(OPr^i)_4\}_3$ in the temperature range 160–180°C/1 mm (the volatility increasing with lanthanide contraction). However, if the much less volatile $\{Ln(OPr^i)_3\}_n$ is in excess, the bimetallic isopropoxide can be distilled out first, leaving the excess (>1/3 mole) of $Ln(OPr^i)_3$ behind.[71,72] Formation of a similar product, $In\{Al(OPr^i)_4\}_3$ has been reported[69,73] in the reaction of $In(OPr^i)_3$ with $Al(OPr^i)_3$ in 1:3 molar ratio.

The reaction of thorium isopropoxide with excess aluminium isopropoxide yields[74] a product of the composition, $Th\{Al(OPr^i)_4\}_4$. However, similar reactions of

$Zr(OPr^i)_4.Pr^iOH$,[52] $Nb(OPr^i)_5$ and $Ta(OPr^i)_5$[75] with excess $Al(OPr^i)_3$ yield 1:2 products only on volatilization,[52,74] irrespective of the excess of $Al(OPr^i)_3$ taken (Eqs 3.8 and 3.9).

$$Zr(OPr^i)_4.Pr^iOH + 2Al(OPr^i)_3 \xrightarrow{Pr^iOH} (Pr^iO)_2Al(\mu\text{-}OPr^i)_2Zr(OPr^i)_2(\mu\text{-}OPr^i)_2Al(OPr^i)_2$$

$$(3.8)$$

$$M(OPr^i)_5 + 2Al(OPr^i)_3 \xrightarrow{Pr^iOH} (Pr^iO)_2Al(\mu\text{-}OPr^i)_2M(OPr^i)_3(\mu\text{-}OPr^i)_2Al(OPr^i)_2$$

$$(3.9)$$

where M = Nb or Ta.

The hafnium analogue of zirconium has been prepared[76] in the 1:2 molar reaction between $Hf(OPr^i)_4.Pr^iOH$ and $Al(OPr^i)_3$, with its structure recently elucidated by X-ray crystallography.[77]

The products, $[(Pr^iO)_2Zr\{Al(OPr^i)_4\}_2]$ and $[(Pr^iO)_3M\{Al(Pr^iO)_4\}_2]$ are volatile and monomeric in organic solvents. However, dimeric volatile products are obtained in the 1:1 molar reactions of $Al(OPr^i)_3$ with $Zr(OPr^i)_4.Pr^iOH$, $Nb(OPr^i)_5$ and $Ta(OPr^i)_5$ (Eqs 3.10 and 3.11):

$$2Zr(OPr^i)_4.Pr^iOH + 2Al(OPr^i)_3 \xrightarrow{Pr^iOH} \{(OPr^i)_2Al(\mu\text{-}OPr^i)_2Zr(OPr^i)_2(\mu\text{-}OPr^i)\}_2$$

$$(3.10)$$

$$2M(OPr^i)_5 + 2Al(OPr^i)_3 \longrightarrow \{(Pr^iO)_2Al(\mu\text{-}OPr^i)_2M(OPr^i)_3(\mu\text{-}OPr^i)\}_2$$

$$(3.11)$$

where M = Nb or Ta.

A few more illustrative examples of the synthesis of heterometal alkoxides by the interactions of component alkoxides are represented by Eqs (3.12)–(3.23).

$$M(OBu^t)_2 + M'(OBu^t) \longrightarrow M(OBu^t)_3M' \qquad (3.12)$$

where M = Ge(II), Sn(II), Pb(II); M' = In, Tl.[54]

$$2Sn(OBu^t)_2 + M(OBu^t)_2 \longrightarrow M\{Sn(OBu^t)_3\}_2 \qquad (3.13)$$

where M = Sr, Ba.[78]

$$2Tl(OR) + M(OR)_4 \longrightarrow Tl_2M(OR)_6 \qquad (3.14)$$

M = Sn(IV); R = Et.[55] M = Zr; R = $CH(CF_3)_2$.[79]

$$Al(OPr^i)_3 + 3Ga(OPr^i)_3 \longrightarrow Al\{Ga(OPr^i)_4\}_3 \text{[73]} \qquad (3.15)$$

$$Ga(OPr^i)_3 + 3Al(OPr^i)_3 \longrightarrow Ga\{Al(OPr^i)_4\}_3 \text{[73]} \qquad (3.16)$$

$$Mg(OEt)_2 + 2Sb(OEt)_3 \longrightarrow \tfrac{1}{2}Mg_2Sb_4(OEt)_{16} \text{[80]} \qquad (3.17)$$

$$\text{Ba(OPr}^i)_2 \quad \begin{cases} \xrightarrow{\text{Ti(OPr}^i)_4 \ 1:2} \frac{1}{2}[\text{BaTi}_2(\text{OPr}^i)_{10}]_2{}^{81} & (3.18) \\[2mm] \xrightarrow{\text{Ti(OPr}^i)_4 \ 1:3} [\text{Ti(OPr}^i)_5\text{Ba}\{\text{Ti}_2(\text{OPr}^i)_9\}]^{81} & (3.19) \\[2mm] \xrightarrow{\text{Ti(OPr}^i)_4 \ 1:4} [\text{Ba}\{\text{Ti}_2(\text{OPr}^i)_9\}_2]^{81} & (3.20) \end{cases}$$

$$M(\text{OBu}^t)_2 + Zr(\text{OBu}^t)_4 \xrightarrow{\text{ether}} MZr(\text{OBu}^t)_6 \tag{3.21}$$

where M = Sn(II), Pb(II).[82]

$$Ba(\text{OBu}^t)_2 + 2Zr(\text{OBu}^t)_4 \longrightarrow BaZr_2(\text{OBu}^t)_{10}{}^{83} \tag{3.22}$$

$$2Ba(\text{OCEt}_3)_2 + 4Cu(\text{OCEt}_3) \longrightarrow Ba_2Cu_4(\text{OCEt}_3)_8{}^{84} \tag{3.23}$$

2.1.2 Reactions of Magnesium and Alkaline Earth Metals with Alcohols in the Presence of Other Metal (Al, Zr, Nb, etc.) Alkoxides (B)

Reactions of alkaline earth metals and magnesium in alcohols is rather slow even in the presence of catalysts such as $HgCl_2$ and/or I_2; this may be due to the low solubility of their bisalkoxides, $\{M(OR)_2\}_n$ in alcohols, resulting from their highly associated nature (large value of n). The dissolution of these bivalent metals appears to be considerably facilitated[35,36] by the presence of metal alkoxides like $Al(OR)_3$, $Zr(OR)_4$, $Nb(OR)_5$, or $Ta(OR)_5$ (Eqs 3.24–3.27):

$$M + 2ROH \xrightarrow{\text{slow}} \frac{1}{n}[M(OR)_2]_n + H_2 \uparrow \atop \text{(low solubility)} \tag{3.24}$$

$$M + 2ROH + 2M'(OR)_3 \xrightarrow{\text{facile}} M\{M'(OR)_4\}_2 + H_2 \uparrow^{85} \tag{3.25}$$

$$M + 2ROH + 4M''(OR)_4 \xrightarrow{\text{facile}} M\{M''_2(OR)_9\}_2 + H_2 \uparrow^{86} \tag{3.26}$$

$$M + 2ROH + 2M'''(OR)_5 \xrightarrow{\text{facile}} M\{M'''(OR)_6\}_2 + H_2 \uparrow^{87-89} \tag{3.27}$$

where M = Mg, Ca, Sr, Ba; R = Et, Pr; M' = Al, Ga; M'' = Zr, Hf; M''' = Nb, Ta.

It has been suggested[16] that the reactivity of the alcohols may be enhanced by the formation of adducts with metal alkoxides, $\begin{smallmatrix} R \\ \diagdown \\ O \longrightarrow M'(OR)_3 \\ \diagup \\ H \end{smallmatrix}$, as a result of which the alcoholic proton is rendered more labile and hence, more reactive as a result of the electron drift shown above.

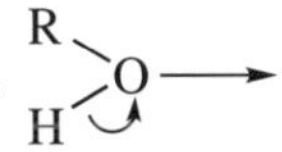

In the case of magnesium, the liquid product $Mg\{Al(OPr^i)_4\}_2$ showed[90] a tendency to yield an X-ray characterizable crystalline product, $Mg_2Al_3(OPr^i)_{13}$ on standing for a few weeks (Eq. 3.28):

$$[Mg\{Al(OPr^i)_4\}_2]_n \longrightarrow \frac{n}{2}Mg_2Al_3(OPr^i)_{13} + \frac{n}{2}Al(OPr^i)_3 \tag{3.28}$$

$Mg_2Al_3(OPr^i)_{13}$ can also be prepared in good yield by the reaction of Mg and $Al(OPr^i)_3$ in 2:3 molar ratio in isopropanol (Eq. 3.29):

$$2Mg + 3Al(OPr^i)_3 + 4Pr^iOH \longrightarrow Mg_2Al_3(OPr^i)_{13} \tag{3.29}$$

Recently, the reactions of Ba and Ca with $Al(OPr^i)_3$ in 1:3 and 2:3 reactions have been found to yield[91] corresponding products as represented by Eqs (3.30) and (3.31).

$$M + 3Al(OPr^i)_3 + 2Pr^iOH \longrightarrow MAl_3(OPr^i)_{11} + H_2 \uparrow \tag{3.30}$$

$$2M + 3Al(OPr^i)_3 + 4Pr^iOH \longrightarrow M_2Al_3(OPr^i)_{13} + 2H_2 \uparrow \tag{3.31}$$

2.1.3   Reactions of Metal Halides (Nitrate) with Alkali Alkoxometallates (C)

Albers *et al.*[30] investigated this route for the preparation of uranium(IV) tetraisopropoxoaluminate by reacting uranium tetrachloride with a gel of sodium tetraisopropoxoaluminate in isopropanol (Eq. 3.32):

$$UCl_4 + 4Na\{Al(OPr^i)_4\} \longrightarrow U\{Al(OPr^i)_4\}_4 + 4NaCl \downarrow \tag{3.32}$$

The greater solubility of $KAl(OPr^i)_4/KGa(OPr^i)_4$[92] in isopropanol as compared to their sodium analogues was used to advantage by Mehrotra and co-workers for the synthesis of a variety of heterometal isopropoxides of Al (and Ga) with lanthanons,[71,72] In,[69,73] Th,[74] Sn(IV),[93] Be, Zn, Cd, Hg(II),[94] and Sn(II)[95] (Eqs 3.33–3.35):

$$MCl_3 + 3KAl(OPr^i)_4 \longrightarrow M\{Al(OPr^i)_4\}_3 + 3KCl \downarrow \tag{3.33}$$

where M = La, Pr, Nd, Sm, Gd, Ho, Er, Yb, Lu, Sc, Y, In.

$$ThCl_4 + 4KAl(OPr^i)_4 \longrightarrow Th\{Al(OPr^i)_4\}_4 + 4KCl \downarrow \tag{3.34}$$

$$MCl_2 + 2KAl(OPr^i)_4 \longrightarrow M\{Al(OPr^i)_4\}_2 + 2KCl \downarrow \tag{3.35}$$

where M = Be, Zn, Cd, Hg(II), Sn(II).

Although $Th\{Al(OPr^i)_4\}_4$ could be isolated by the 1:4 molar reaction of $ThCl_4$ with $KAl(OPr^i)_4$, a similar reaction between $CeCl_4$ and $KAl(OPr^i)_4$ was precluded by the instability of $CeCl_4$. It might be worth investigating the possible synthesis of $Ce\{Al(OPr^i)_4\}_4$ by the reaction of $(NH_4)_2Ce(NO_3)_6$ or $(C_5H_5NH)_2CeCl_6$ with $KAl(OPr^i)_4$ in 1:4 molar ratio (*cf.* the preparation[96] of $Ce(OBu^t)_4$ and $NaCe_2(OBu^t)_9$ by the reaction between $(NH_4)_2Ce(NO_3)_6$ and $NaOBu^t$ in appropriate molar ratios (Eqs 3.36–3.38)):

$$(NH_4)_2Ce(NO_3)_6 + 6NaOBu^t$$

$$\xrightarrow{\text{THF}} Ce(OBu^t)_4 \cdot (thf)_2 + 2NH_3 + 6NaNO_3 + 2Bu^tOH \tag{3.36}$$

$$(NH_4)_2Ce(NO_3)_6 + 8NaOBu^t$$

$$\xrightarrow{\text{THF}} Na_2Ce(OBu^t)_6(thf)_4 + 2NH_3 + 6NaNO_3 + 2Bu^tOH \tag{3.37}$$

$$2(NH_4)_2Ce(NO_3)_6 + 13NaOBu^t$$

$$\xrightarrow{THF} NaCe_2(OBu^t)_9 + 4NH_3 + 12NaNO_3 + 4Bu^tOH \qquad (3.38)$$

The reaction of $ZrCl_4$ with $KAl(OPr^i)_4$ in 1:2 molar ratio appears to proceed to the formation of $Cl_2Zr\{Al(OPr^i)_4\}_2$, which on being treated with $KOPr^i$ (2 moles) yields $(OPr^i)_2Zr\{Al(OPr^i)_4\}_2$ (Eq. 3.39) which is also the end product of the reaction between zirconium isopropoxide and excess of aluminium isopropoxide (Eq. 3.8).

$$Cl_2Zr(OPr^i)_2 + 2KAl(OPr^i)_4 \longrightarrow (OPr^i)_2Zr\{Al(OPr^i)_4\}_2 + 2KCl \downarrow \qquad (3.39)$$

Alkoxides of later 3d metals, Mn, Fe, Co, Ni, Cu, are generally polymeric insoluble nonvolatile derivatives. However, a new dimension appears to have been added to the alkoxide chemistry of these metals by the synthesis[97–100] of their monomeric soluble tetraisopropoxoaluminates in reactions of the types shown in Eqs (3.40) and (3.41):

$$MCl_2 + 2KAl(OPr^i)_4 \xrightarrow[C_6H_6]{Pr^iOH} M\{Al(OPr^i)_4\}_2 + 2KCl \downarrow \qquad (3.40)$$

where M = Mn,[101] Fe,[97] Co,[102] Ni,[42] Ga,[103] Cu,[104] Zn,[94] and

$$M'Cl_3 + 3KAl(OPr^i)_4 \longrightarrow M'\{Al(OPr^i)_4\}_3 + 3KCl \downarrow \qquad (3.41)$$

where M' = Cr,[105] Fe.[106]

Stumpp and Hillebrand[107] also described the preparation of a number of other tetra-alkoxoaluminates of Co, Ni and Cu, $M\{Al(OR)_4\}_2$ (R = Me, Et, Pr^i), by a similar procedure, and their spectroscopic properties.

Some $M\{Al(OBu^t)_4\}_n$ derivatives have been synthesized by the reactions of metal chlorides with $KAl(OBu^t)_4$ (prepared by treating $Al(OBu^t)_3$ with 1 mole of $KOBu^t$) (Eq. 3.42):

$$MCl_2 + 2KAl(OBu^t)_4 \longrightarrow M\{Al(OBu^t)_4\}_2 + 2KCl \downarrow \qquad (3.42)$$

where M = Zn,[108] Fe,[109] Co, Ni, Cu,[110] Mn.[101]

Transition metal antimony ethoxides of the type $M_2Sb_4(OEt)_{16}$ (M = Mn, Ni) have been prepared[111,112] from the corresponding metal chloride and $NaSb(OEt)_4$ in 1:2 molar ratio (Eq. 3.43) and their structures have been confirmed[112] by X-ray crystallography.

$$MCl_2 + 2NaSb(OEt)_4 \longrightarrow \tfrac{1}{2}M_2Sb_4(OEt)_{16} + 2NaCl \downarrow \qquad (3.43)$$

Although X-ray structural confirmation has become available only since 1992,[39,53]

the chelating behaviour of the moiety $\begin{smallmatrix} Pr^iO \\ Pr^iO \end{smallmatrix}\!\!>\!Al\!<\!\!\begin{smallmatrix} OPr^i \\ OPr^i \end{smallmatrix}$ had usually been inferred from chemical and other evidence for almost two decades[3] and the structural features of the unit $KAl(OPr^i)_4.2Pr^iOH$, as elucidated[53] by X-ray crystallography in 1993, had been correctly predicted as early as 1971.[3,5]

In addition, the synthesis of a large number of similarly stable (as depicted by volatility) and apparently covalent (as indicated by solubility with low conductivity in organic solvents) derivatives of strongly electropositive alkali metals (M = Na, K), $MZr_2(OPr^i)_9,^{60,62-64}$ $MNb(OPr^i)_6$ and $MTa(OPr^i)_6^{65}$ clearly indicated the chelating behaviour of other moieties such as $Zr_2(OPr^i)_9^-$, $Nb(OPr^i)_6^-$ and $Ta(OPr^i)_6^-$ also.

Following the successful synthesis of tetraisopropoxoaluminates of later 3d transition metals, the chemistry of their niobate, tantalate and zirconate derivatives has also been extensively investigated since 1985: niobate/tantalate, $M(OPr^i)_6^-$ derivatives of Mn(II),[101] chromium,[113] iron,[114] cobalt,[115] nickel,[116] copper,[117] and zirconate, $Zr_2(OPr^i)_9^-$ derivatives of iron,[118,119] cobalt,[120] nickel[121] and copper[122] (Eqs 3.44 and 3.45):

$$MCl_x + x\,KNb(OPr^i)_6 \longrightarrow M\{Nb(OPr^i)_6\}_x + x\,KCl \downarrow \qquad (3.44)$$
$$\text{or} \qquad\qquad\qquad \text{or}$$
$$KTa(OPr^i)_6 \qquad\quad M\{Ta(OPr^i)_6\}_x$$

where M = Cr(III), Fe(III), Fe(II), Co(II), Ni(II), Cu(II).

$$MCl_x + x\,KZr_2(OPr^i)_9 \longrightarrow M\{Zr_2(OPr^i)_9\}_x + x\,KCl \downarrow \qquad (3.45)$$

where M = Fe(II), Fe(III), Co(II), Ni(II), Cu(II).

In addition, examples may also be cited of a few other novel types of bimetallic alkoxometallates, which have been described since 1990: $MTi(OPr^i)_5^{60}$ (M = Li, Na, K); $BaCu_6(OCEt_3)_8;^{123}$ $Na_2Cu\{OCH(CF_3)_2\}_4;^{124}$ $Ba\{Cu(OCMe(CF_3)_2)_3\}_2;^{125}$ $Cu(I)_2Zr_2(OPr^i)_{10};^{126}$ $MZn(OBu^t)_3^{127}$ (M = Na, K); and $M[Sn_2(OPr^i)_9]_2^{128}$ (M = Mg, Ba, Zn, Cd).

A bimetallic hexaisopropoxoniobate of barium, $Ba\{Nb(OPr^i)_6\}_2$ which was earlier prepared by the reaction of metallic barium with isopropanol in the presence of 2 moles of $Nb(OPr^i)_5$, was also synthesized by the alternative route[129] of Eq. (3.46), which could probably be extended to other systems:

$$BaI_2 + 2K\{Nb(OPr^i)_6\} \longrightarrow Ba\{Nb(OPr^i)_6\}_2 + 2KI \downarrow \qquad (3.46)$$

2.1.4   Reactions of Two Metal Halides with Alkali Alkoxides (D)

This is really a modification of the earlier procedures in which the alkoxides of two metals formed *in situ* by the reactions of the respective metal halides with alkali alkoxides interact together to yield the bimetallic alkoxide; the method can be illustrated by the "one-pot" synthesis[72] of $Ln\{Al(OPr^i)_4\}_3$ by the reaction shown in Eq. (3.47):

$$LnCl_3 + 3AlCl_3 + 12K + 12Pr^iOH \longrightarrow Ln\{Al(OPr^i)_4\}_3 + 12KCl + 6H_2 \uparrow \quad (3.47)$$

where Ln = Gd, Ho, Er.

2.1.5   Miscellaneous Methods for the Synthesis of Heterobimetallic Alkoxides

In addition to the procedures described above, substitution/addition reactions of metal hydrides and alkyls with alkoxides of other metals may be utilized for the synthesis of various heterometal alkoxide derivatives.

2.1.5.1 From hydrides (E-1)

This route can be illustrated by Eqs (3.48)–(3.50):

$$NaH + Ti(OPr^i)_4 + Pr^iOH \longrightarrow NaTi(OPr^i)_5 + H_2 \uparrow^{60} \tag{3.48}$$

$$Zr_2(OPr^i)_8.2Pr^iOH + KH \longrightarrow KZr_2(OPr^i)_9 + Pr^iOH + H_2 \uparrow^{64} \tag{3.49}$$

$$\begin{array}{c} \xrightarrow{+2B(OPr^i)_3} Mg\{(\mu\text{-}OPr^i)_2B(H)(OPr^i)\}_2{}^{130} \\ MgH_2 \\ \xrightarrow{+2Al(OPr^i)_3} Mg\{(\mu\text{-}OPr^i)(\mu\text{-}H)Al(OPr^i)_2\}_2{}^{130} \end{array} \tag{3.50}$$

2.1.5.2 From amides and bis(trimethylsilyl) amides (E-2)

This route, which has been extensively exploited for the synthesis of homo-metal alkoxides, can be illustrated by Eqs (3.51) and (3.52):[131]

$$2Zr(OPr^i)_4.Pr^iOH + LiNMe_2 \longrightarrow LiZr_2(OPr^i)_9.Pr^iOH + HNMe_2 \uparrow \tag{3.51}$$

$$Ba\{N(SiMe_3)_2\}_2 + 4Zr(OPr^i)_4.Pr^iOH \longrightarrow Ba\{Zr_2(OPr^i)_9\}_2$$
$$+ 2Pr^iOH + 2HN(SiMe_3)_2 \uparrow \tag{3.52}$$

2.1.5.3 From alkyls (E-3)

Instead of the expected elimination reactions involving small molecules like methane or amines, addition reactions of some metal alkoxides with alkyls and amides of other metals have been reported, *e.g.*:

$$U(OPr^i)_6 + 3LiMe \longrightarrow (MeLi)_3U(OPr^i)_6{}^{132} \tag{3.53}$$

$$U(OPr^i)_6 + 3MgR_2 \longrightarrow (R_2Mg)_3U(OPr^i)_6{}^{133} \tag{3.54}$$

Interestingly, the reaction of $LiC(SiMe_3)_3$ with $B(OMe)_3$ yielded[134] $\{(Me_3Si)_3C\}B(OMe)_2$ and $LiB(OMe)_4$; the latter compound on crystallization from methanol gave the heterometallic methoxide $(MeOH)_2Li(\mu\text{-}OMe)_2B(OMe)_2$.

Similar to the reaction of $Ti(OPr^i)_4$ with NaH, the reaction of $Ti(OPr^i)_4$ with Bu^nLi in isopropanol also yielded[60] $LiTi(OPr^i)_5$ (Eq. 3.55):

$$Bu^nLi + Ti(OPr^i)_4 \xrightarrow{Pr^iOH} LiTi(OPr^i)_5 + Bu^nH \tag{3.55}$$

2.2 Synthesis of Heterometallic Chloride Alkoxides (F)

Although homoleptic heterometallic isopropoxometallates, ML_4, of a number of quadrivalent metallic elements like M = Sn(IV), Ce(IV), Th(IV) and U(IV) have been synthesized by the reactions of MCl_4 with 4 moles of reagents like $K\{Al(OPr^i)_4\}$, yet only bis products could be obtained[4,5] in the reactions of chlorides of smaller elements like Zr(IV) and Nb(V):

$$ZrCl_4 + 2K\{Al(OPr^i)_4\} \longrightarrow ZrCl_2\{Al(OPr^i)_4\}_2 + 2KCl \downarrow \tag{3.56}$$

$$NbCl_5 + 2K\{Al(OPr^i)_4\} \longrightarrow NbCl_3\{Al(OPr^i)_4\}_2 + 2KCl \downarrow \tag{3.57}$$

This appears to be due to steric factors, as the remaining chloride groups on the products remain reactive and these can be replaced by –OR groups by reactions with potassium alkoxide (KOR).

Partially substituted products, $MCl_{x-y}L_y$, have been reported by the reactions of anhydrous chlorides of various metals with alkali alkoxometallates in suitable stoichiometric ratios (Eqs 3.58–3.62):

$$MCl_2 + KAl(OPr^i)_4 \xrightarrow[\text{isopropanol}]{\text{benzene}} ClM\{Al(OPr^i)_4\} + KCl \downarrow \qquad (3.58)$$

where M = Be, Mg, Zn, Cd.[94]

$$CdCl_2 + KZr_2(OPr^i)_9 \xrightarrow[-KCl]{\text{benzene}} \tfrac{1}{2}[\{Cd\{Zr_2(OPr^i)_9\}(\mu\text{-}Cl)\}_2]^{135} \qquad (3.59)$$

$$CdI_2 + KZr_2(OPr^i)_9 \longrightarrow ICd\{Zr_2(OPr^i)_9\} + KI \downarrow {}^{136} \qquad (3.60)$$

$$PrCl_3.3Pr^iOH + 2KAl(OPr^i)_4 \xrightarrow[-2KCl]{\text{benzene}} [Pr\{Al(OPr^i)_4\}_2Pr^iOH(\mu\text{-}Cl)]_2{}^{39} \quad (3.61)$$

In addition to the above structurally characterized halide isopropoxometallate derivatives, a large number of derivatives of other metals have been isolated by similar reactions: $ClCo\{Al(OPr^i)_4\}$;[137] $ClCo\{Nb(OPr^i)_6\}$;[137] $ClCo\{Zr_2(OPr^i)_9\}$;[137] $ClFe\{Al(OPr^i)_4\}$;[138] $ClCu\{Al(OPr^i)_4\}$;[139] $ClCu\{Ta(OPr^i)_6\}$;[8] $ClCu\{Zr_2(OPr^i)_9\}$;[8,126] $ClCu\{Ti_2(OPr^i)_9\}$;[81] $Cl_xSn\{Zr_2(OPr^i)_9\}_{4-x}$;[140] $ClSn\{Zr_2(OPr^i)_9\}$;[141,142] $Cl_2Y\{Ti_2(OPr^i)_9\}$;[81] $Cl_2La\{Zr_2(OPr^i)_9\}$[143] and $ClLa\{Zr_2(OPr^i)_9\}_2$.[143] Thus the general method of synthesizing chloride alkoxometallate derivatives of different metals (M) can be easily described by the following general reaction (Eq. 3.62) with $y < x$:

$$MCl_x + yKL \xrightarrow[-yKCl]{\text{benzene}} MCl_{x-y}(L)_y \qquad (3.62)$$

where L = $Al(OPr^i)_4$, $Nb(OPr^i)_6$, $Ta(OPr^i)_6$, $Zr_2(OPr^i)_9$, $Ti_2(OPr^i)_9$, etc.

A variety of heterometallic chloride alkoxides have been synthesized by a reaction[144] of the type shown in Eq. (3.63):

$$3LiMe + 3Bu^t_3COH + CrCl_3 \xrightarrow{\text{ether–THF}} [Cr(OCBu^t)_3.LiCl.2thf] + 2LiCl + 3CH_4 \uparrow$$
$$(3.63)$$

Similarly, an interaction of LiCl with $Nd(OCBu^t_3)_3$ in THF (Eq. 3.64) yields[145] a derivative $[(thf)Li(\mu\text{-}Cl)Nd(OCBu^t_3)_3]$. which has been characterized by X-ray crystallography:

$$LiCl + Nd(OCBu^t_3)_3 \xrightarrow{\text{THF}} [(thf)Li(\mu\text{-}Cl)Nd(OCBu^t_3)_3]. \qquad (3.64)$$

Chlorides of bivalent metals like Mg, Zn, Sn(II) are insoluble in nonpolar solvents like n-heptane/benzene, but their dissolution in solutions of tetraalkoxides of Ti and Zr in such solvents readily yields heterometallic chloride alkoxide derivatives (Eqs 3.65 and 3.66):

$$MgCl_2 + 2Ti(OEt)_4 \xrightarrow{n\text{-heptane}} [Mg\{Ti_2(OEt)_8Cl\}(\mu\text{-}Cl)]_2{}^{146} \qquad (3.65)$$

$$MCl_2 + 2Zr(OPr^i)_4 \xrightarrow{toluene} [M\{Zr_2(OPr^i)_8Cl\}(\mu\text{-}Cl)]_2 \qquad (3.66)$$

where M = Be, Mg, Zn, Sn(II).[147]

All the derivatives obtained in the above reactions are white crystalline solids, dimeric in solution (benzene) except the beryllium derivative which also dimerizes on ageing. The beryllium product was distilled at ~190°C under 0.05 mm pressure whereas under these conditions the magnesium derivative yielded a sublimate corresponding in analysis to $Mg_4Zr_6Cl_3(OPr^i)_{23}O_3$. All these products appear to be chloride bridged, which is corroborated by the crystal structure of the Mg–Ti product.[146]

Chloride alkoxides of Be, Mg, Zn and Sn(II) with Nb(v) have also been synthesized by a similar series of reactions (Eqs 3.67 and 3.68):[148]

$$MCl_2 + Nb(OPr^i)_5 \xrightarrow[68\,hr(stir)]{benzene} \tfrac{1}{2}[\{M(Nb(OPr^i)_5Cl)\}(\mu\text{-}Cl)]_2 \qquad (3.67)$$

$$MCl_2 + 2Nb(OPr^i)_5 \xrightarrow[68\,hr(stir)]{} [M\{Nb(OPr^i)_6Cl\}_2] \qquad (3.68)$$

2.3 Synthesis of Higher Heterometallic Alkoxides and Allied Derivatives

Starting with bimetallic chloride (iodide) alkoxides synthesized by the general reaction of Eq. (3.69):

$$MCl_x + yKL \longrightarrow MCl_{x-y}L_y + yKCl \downarrow \qquad (3.69)$$

where $y < x$, a variety of heterotrimetallic and -tetrametallic alkoxides of Be,[7] lanthanons;[17,149] Zn and Cd;[150,151] Sn(II) and (IV);[152,153] Mn(II);[101] Fe(II);[109] Fe(III);[9,154,155] Co(II);[156] Ni(II);[157] Cu(II)[8,158] and Mg[159] can be synthesized as illustrated in a few representative cases by Eqs (3.70)–(3.79):

$$\{Al(OPr^i)_4\}BeCl + K\{Zr_2(OPr^i)_9\} \xrightarrow{-KCl} \{Al(OPr^i)_4\}Be\{Zr_2(OPr^i)_9\} \qquad (3.70)$$

$$\{Zr_2(OPr^i)_9\}MCl + K\{Al(OPr^i)_4\} \xrightarrow{-KCl} \{Zr_2(OPr^i)_9\}M\{Al(OPr^i)_4\} \qquad (3.71)$$

where M = Mn(II), Fe(II), Co(II), Ni(II).

$$\{Zr_2(OPr^i)_9\}NiCl + KGa(OPr^i)_4 \xrightarrow{-KCl} \{Zr_2(OPr^i)_9\}Ni\{Ga(OPr^i)_4\} \qquad (3.72)$$

$$\{Zr_2(OPr^i)_9\}MCl + KTa(OPr^i)_6 \xrightarrow{-KCl} \{Zr_2(OPr^i)_9\}M\{Ta(OPr^i)_6\} \qquad (3.73)$$

where M = Fe(II), Co(II).

$$\{Nb(OPr^i)_6\}CoCl + KAl(OPr^i)_4 \xrightarrow{-KCl} \{Nb(OPr^i)_6\}Co\{Al(OPr^i)_4\} \qquad (3.74)$$

$$\{Al(OPr^i)_4\}MCl + KGa(OPr^i)_4 \xrightarrow{-KCl} \{Al(OPr^i)_4\}M\{Ga(OPr^i)_4\} \qquad (3.75)$$

where M = Co(II), Ni(II).

$$\{Zr_2(OPr^i)_9\}_2FeCl + KAl(OPr^i)_4 \xrightarrow{-KCl} \{Zr_2(OPr^i)_9\}_2Fe\{Al(OPr^i)_4\} \qquad (3.76)$$

$$\{Zr_2(OPr^i)_9\}FeCl_2 + 2KAl(OPr^i)_4 \xrightarrow{-KCl} \{Zr_2(OPr^i)_9\}Fe\{Al(OPr^i)_4\}_2 \qquad (3.77)$$

$$\{Zr_2(OPr^i)_9\}FeCl_2 + KAl(OPr^i)_4 \xrightarrow{-KCl} \{Zr_2(OPr^i)_9\}Fe\{Al(OPr^i)_4\}Cl \qquad (3.78)$$

$$\{Zr_2(OPr^i)_9\}Fe\{Al(OPr^i)_4\}Cl + K\{Ta(OPr^i)_6\}$$
$$\xrightarrow{-KCl} \{Zr_2(OPr^i)_9\}Fe\{Al(OPr^i)_4\}\{Ta(OPr^i)_6\} \qquad (3.79)$$

All the heterotri- and tetrametallic derivatives described above are moisture sensitive, hydrocarbon soluble, monomeric solids. Almost all these can be volatilized unchanged under reduced pressure, except some of the niobate and tantalate products, which tend to dissociate yielding $Nb(OPr^i)_5/Ta(OPr^i)_5$.

The first crystallographically characterized heterotrimetallic isopropoxide was synthesized[136,160] in 1996 by the reactions shown in Eqs (3.80) and (3.81):

$$CdI_2 + KM_2(OPr^i)_9 \xrightarrow{toluene} ICd\{M_2(OPr^i)_9\} + KI \downarrow \qquad (3.80)$$

$$ICd\{M_2(OPr^i)_9\} + KBa(OPr^i)_3 \longrightarrow \tfrac{1}{2}[\{Cd(OPr^i)_3\}Ba\{M_2(OPr^i)_9\}]_2 + KI \downarrow \qquad (3.81)$$

where M = Zr, Hf, Ti.

A distinctive feature hitherto unnoticed in heterometallic alkoxide chemistry is the exchange of two chelating ligands around the central metal atom Ba; this rearrangement is probably favoured by greater oxophilicity of barium and its tendency to attain higher coordination states. Similar syntheses have been reported[141,160–163] as illustrated by Eqs (3.82)–(3.86):

$$SnCl_2 + KM_x(OPr^i)_y \longrightarrow ClSn\{M_x(OPr^i)_y\} + KCl \downarrow \qquad (3.82)$$

(M = Al; $x = 1$; $y = 4$. M = Nb/Ta; $x = 1$, $y = 6$. M = Zr; $x = 2$; $y = 9$)

$$MI_2 + K\{M'_2(OPr^i)_9\} \longrightarrow [IM\{M'_2(OPr^i)_9\}] + KI \downarrow \qquad (3.83)$$

(M = Sn; M' = Zr, Ti. M = Pb; M' = Zr, Ti. M = Cd, M' = Sn(IV))

$$ClSn\{Zr_2(OPr^i)_9\} + KAl(OPr^i)_4 \longrightarrow \{Al(OPr^i)_4\}Sn\{Zr_2(OPr^i)_9\} + KCl \downarrow \qquad (3.84)$$

$$CuCl_2 + KHf_2(OPr^i)_9 \longrightarrow ClCu\{Hf_2(OPr^i)_9\} + KCl \downarrow \qquad (3.85)$$

$$ClCu\{Zr_2(OPr^i)_9\} + KM(OPr^i)_4 \longrightarrow [\{M(OPr^i)_4\}Cu\{Zr_2(OPr^i)_9\}] + KCl \downarrow \qquad (3.86)$$

where M = Al, Ga.[8]

In addition to hetero-bimetallic and -trimetallic alkoxides, some heterotetrametallic alkoxides have also been reported[9] as illustrated already by Eqs (3.76) and (3.77).

In spite of considerable initial scepticism, the synthesis of an increasing number of well-characterized heterometallic alkoxides is now well established and attempts are being directed towards the synthesis of "single-source" precursors for preparation of mixed metal–oxide ceramic materials by the sol–gel or MOCVD processes.[12,20,22,24–27,43] Before proceeding to a brief account of some novel types of heterometal glycolate and aminoalkoxide derivatives, it is appropriate to mention four review articles[47,163–165] summarizing the state of knowledge about the conventional heterometallic alkoxide systems at the end of the twentieth century.

2.4   Synthesis of Heterometallic Glycolate Alkoxide Derivatives

The facile dissolution of alkaline earth metals in alcohols in the presence of alkoxides of metals like aluminium (Section 2.1.2) has been ascribed to the formation of alcohol adducts of the metal alkoxides $\begin{smallmatrix} R \\ H \end{smallmatrix}\!>\!O \longrightarrow Al(OR)_3$ rendering the proton of the adduct alcohol more labile. A similar type of lability of protons of glycol molecules coordinated in derivatives like titanium trisglycolate[166] has been utilized[43,167,168] for synthesis of bi- and triheterometallic glycolate alkoxides by reactions of the types shown in Eqs (3.87)–(3.89):

$$M(OPr^i)_4 + 3G\!\!\begin{smallmatrix} OH \\ OH \end{smallmatrix} \xrightarrow{\text{benzene}} G\!\left\{\!\begin{smallmatrix} O \\ O \end{smallmatrix}\!\right\}_2\!\!M\!\!\begin{smallmatrix} O \\ O \end{smallmatrix}\!\!G + 4Pr^iOH\uparrow \quad (3.87)$$

$$\mathbf{(1)}\ \text{Soluble}$$

where $M = Ti, Zr, Hf, Sn(IV);\ G\!\!\begin{smallmatrix} OH \\ OH \end{smallmatrix} = \text{2-methylpentane-2,4-diol.}$

$$\mathbf{(1)} + 2Al(OPr^i)_3 \longrightarrow [\{Al(OPr^i)_2\}_2 M\!\!\left(\!\begin{smallmatrix} O \\ O \end{smallmatrix}\!G\right)_3] + 2Pr^iOH\uparrow \quad (3.88)$$

$$\mathbf{(2)}\ \text{Soluble}$$

$$\mathbf{(1)} + 2Nb(OPr^i)_5 \longrightarrow [\{Nb(OPr^i)_4\}_2 M\!\!\left(\!\begin{smallmatrix} O \\ O \end{smallmatrix}\!G\right)_3] + 2Pr^iOH\uparrow \quad (3.89)$$

These two reactions can also be carried out in two consecutive stages of 1:1 molar reactions with $Al(OPr^i)_3$ and $Nb(OPr^i)_5$, yielding a heterotrimetallic derivative

$$[\{Al(OPr^i)_2\}\{Nb(OPr^i)_4\}\{M\!\!\left(\!\begin{smallmatrix} O \\ O \end{smallmatrix}\!G\right)_3\}].$$

Homometallic glycolates of alkaline earth metals (M) have been prepared[169] by the routes shown in Eqs (3.90)–(3.92):

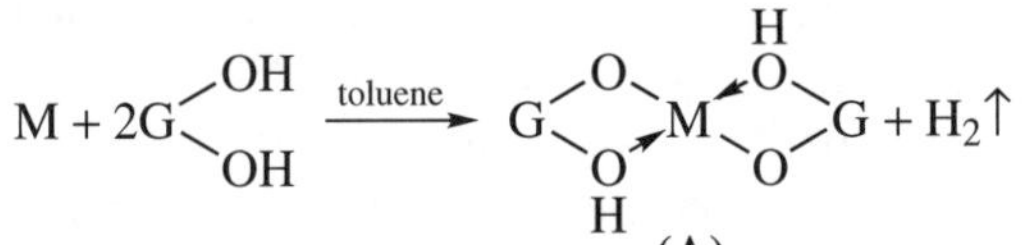

$$M + 2G\begin{smallmatrix}OH\\OH\end{smallmatrix} \xrightarrow{\text{toluene}} G\underset{\underset{H}{O}}{\overset{O}{\underset{}{<}}}M\underset{O}{\overset{\overset{H}{O}}{>}}G + H_2\uparrow$$

(3.90)

(A)

Colourless, soluble, monomeric solid

$$M(OPr^i)_2 + 2G\begin{smallmatrix}OH\\OH\end{smallmatrix} \xrightarrow{\text{benzene}} G\underset{\underset{H}{O}}{\overset{O}{<}}M\underset{O}{\overset{\overset{H}{O}}{>}}G + 2Pr^iOH\uparrow$$

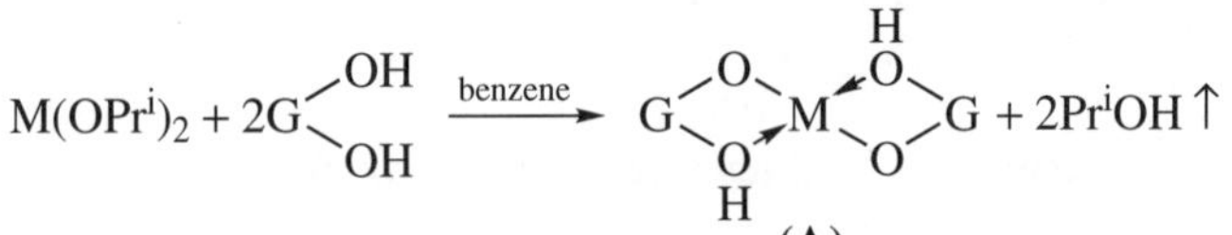

(3.91)

(A)

$$(A) + 2Al(OPr^i)_3 \longrightarrow [\{Al(OPr^i)_2\}_2M(OGO)_2] + 2Pr^iOH\uparrow$$

(3.92)

(B)

Colourless, soluble, monomeric solid

Heterobimetallic glycolate alkoxides of alkaline earth metals (M) can also be prepared by the reactions shown in Eqs (3.93) and (3.94):

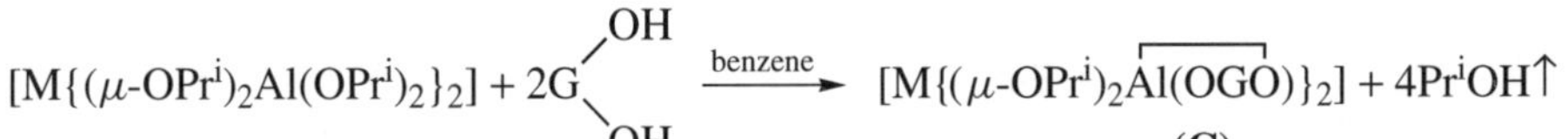

$$[M\{(\mu\text{-}OPr^i)_2Al(OPr^i)_2\}_2] + 2G\begin{smallmatrix}OH\\OH\end{smallmatrix} \xrightarrow{\text{benzene}} [M\{(\mu\text{-}OPr^i)_2\overline{Al(OGO)}\}_2] + 4Pr^iOH\uparrow$$

(C)

(3.93)

$$M\{Al(OPr^i)_4\}_2 + 4G\begin{smallmatrix}OH\\OH\end{smallmatrix} \xrightarrow{\text{benzene}} M\{\overline{Al(OGO)_2}\}_2 + 8Pr^iOH\uparrow$$

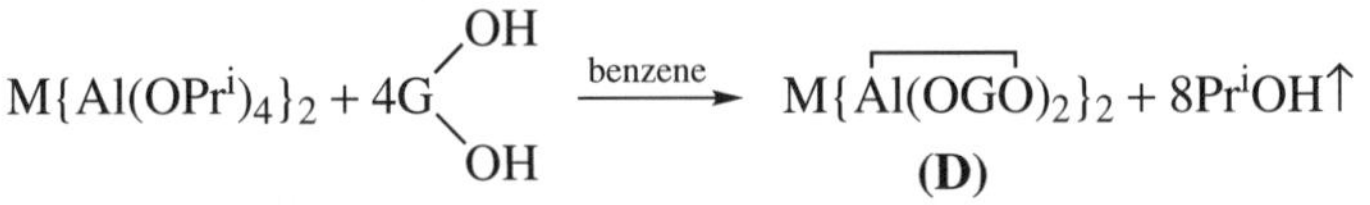

(3.94)

(D)

All the products (**A**), (**B**), (**C**), and (**D**) are colourless, monomeric soluble, nonvolatile solids which do not yield by disproportionation any $Al(OPr^i)_3$, when (**B**) and (**C**) are heated under reduced pressure.

The reactions of $Zr_2(OCMe_2CMe_2O)_2(OCMe_2CMe_2OH)_4$ with $Ti(OPr^i)_4$ in 1:1 and 1:2 molar reactions have been shown[170] to yield $Ti:Zr_2$ and $Ti:Zr$ heterobimetallic crystalline products (characterized by X-ray crystallography) according to Eqs (3.95) and (3.96):

$$Zr_2(OCMe_2CMe_2O)_2(OCMe_2CMe_2OH)_4 + Ti(OPr^i)_4$$

$$\xrightarrow[-2Pr^iOH]{\text{THF}} [TiZr_2(OCMe_2CMe_2O)_4(OCMe_2CMe_2OH)_2(OPr^i)_2] \quad (3.95)$$

$$Zr_2(OCMe_2CMe_2O)_2(OCMe_2CMe_2OH)_4 + 2Ti(OPr^i)_4$$

$$\xrightarrow[-4Pr^iOH]{\text{THF}} [Ti_2Zr_2(OCMe_2CMe_2O)_6(OPr^i)_4] \quad (3.96)$$

2.5   Synthesis of Heterometallic Diethanolaminate Alkoxide Derivatives

Similar to homometallic glycolates of alkaline earth metals ($M = Mg$, Ca, Sr, Ba), diethanolaminates also can be prepared by dissolution of the metals or by reactions of their isopropoxides with diethanolamine (Eqs 3.97 and 3.98):

$$M + 2HN(C_2H_4OH)_2 \xrightarrow{\text{toluene}} M\{(OC_2H_4)(HOC_2H_4)NH\}_2 \downarrow + H_2 \uparrow \qquad (3.97)$$

$$M(OPr^i)_2 + 2HN(C_2H_4OH)_2 \longrightarrow M\{(OC_2H_4)(HOC_2H_4)NH\}_2 \downarrow + 2Pr^iOH \uparrow \quad (3.98)$$
$$(\textbf{A}') \text{ Colourless, insoluble solid}$$

The colourless solids ($\textbf{A}'$) differ from ($\textbf{A}$) (Section 2.4) in being insoluble in organic solvents. However, on being reacted with $Al(OPr^i)_3$, ($\textbf{A}'$) also yields[171] a hetero-bimetallic diethanolaminate product ($\textbf{B}'$) which is a colourless, soluble, monomeric non-volatile solid (Eq. 3.99).

$$\textbf{A}' + 2Al(OPr^i)_3 \xrightarrow{\text{benzene}} \{Al(OPr^i)_2\}_2M\{(OC_2H_4)_2NH\}_2 + 2Pr^iOH \uparrow \qquad (3.99)$$
$$(\textbf{B}')$$

Other colourless, soluble monomeric solids ($\textbf{C}'$) and ($\textbf{D}'$), with the compositions $M\{Al(OPr^i)_2\}\{(OC_2H_4)_2NH\}_2$ and $M\{Al(OC_2H_4NHC_2H_4O)_2\}_2$, respectively, can also be synthesized by reactions similar to those of Eqs (3.93) and (3.94).

Heterobimetallic diethanolaminate derivatives of a number of other metals have also been synthesized[172] by similar reactions (Eqs 3.100–3.102):

$$M(OPr^i)_3 + 2HN(C_2H_4OH)_2$$

$$\xrightarrow{\text{benzene}} M\{(OC_2H_4)(HOC_2H_4)NH\}\{(OC_2H_4)_2NH\} \downarrow + 3Pr^iOH \uparrow \qquad (3.100)$$
$$(\textbf{E}), (\textbf{F})$$

where $M = Al$, Sb.

$$Ti(OPr^i)_4 + 3HN(C_2H_4OH)_2$$

$$\xrightarrow{\text{benzene}} Ti\{(OC_2H_4)_2NH\}\{(HOC_2H_4)(OC_2H_4)NH\}_2 \downarrow + 4Pr^iOH \uparrow \qquad (3.101)$$
$$(\textbf{G})$$

$$M'(OPr^i)_4 \cdot Pr^iOH + 3HN(C_2H_4OH)_2$$

$$\xrightarrow{\text{benzene}} M'\{(OC_2H_4)_2NH\}\{(HOC_2H_4)(OC_2H_4)NH\}_2 \downarrow + 5Pr^iOH \uparrow \qquad (3.102)$$
$$(\textbf{H}), (\textbf{I})$$

where $M' = Zr$, Sn(IV).

The products ($\textbf{E}$), ($\textbf{F}$), ($\textbf{G}$), ($\textbf{H}$), and ($\textbf{I}$) are colourless, insoluble nonvolatile solids, which can be converted into heterobimetallic soluble monomeric products by reactions represented Eqs (3.103)–(3.105):

$$(\textbf{E}) + Nb(OPr^i)_5 \xrightarrow{\text{benzene}} \{Nb(OPr^i)_4\}\{Al[(OC_2H_4)_2NH]_2\} + Pr^iOH \uparrow \qquad (3.103)$$

$$(\mathbf{F}) + Al(OPr^i)_3 \xrightarrow{\text{benzene}} \{Al(OPr^i)_2\}\{Sb[(OC_2H_4)_2NH]_2\} + Pr^iOH \uparrow \qquad (3.104)$$

$$(\mathbf{G}), (\mathbf{H}), (\mathbf{I}) + 2Al(OPr^i)_3 \xrightarrow{\text{benzene}} \{Al(OPr^i)_2\}_2\{M[(OC_2H_4)_2NH]_3\} + 2Pr^iOH \uparrow \quad (3.105)$$
$$(\mathbf{G'}), (\mathbf{H'}), (\mathbf{I'})$$

In contrast to insoluble products $(\mathbf{E})$ to $(\mathbf{I})$, the heterobimetallic products $(\mathbf{E'})$ to $(\mathbf{I'})$ are soluble colourless monomeric products.[173]

2.6　Synthesis of Heterometallic Triethanolaminate (tea) Alkoxide Derivatives

The reactions of isopropoxides of alkaline earth metals $(M = Ca, Sr, Ba)$ with $(HOC_2H_4)_3N(H_3tea)$ in 1:2 molar ratio yield colourless insoluble nonvolatile solids $(\mathbf{J})$ with the composition, $M(H_2tea)_2$, which on treatment with 4 moles of $Al(OPr^i)_3$ afford[173] soluble monomeric heterobimetallic triethanolaminate products $(\mathbf{K})$ according to Eqs (3.106) and (3.107):

$$M(OPr^i)_2 + 2H_3tea \xrightarrow{\text{benzene}} M(H_2tea)_2 \downarrow + 2Pr^iOH \uparrow \qquad (3.106)$$
$$(\mathbf{J})$$

$$(\mathbf{J}) + 4Al(OPr^i)_3 \xrightarrow{\text{benzene}} [\{Al(OPr^i)_2\}_4\{M(tea)_2\}] + 4Pr^iOH \uparrow \qquad (3.107)$$
$$(\mathbf{K})$$

Triethanolaminates of Al, Sb(III) and Sn(IV) have also been prepared[174] by the reactions shown in Eqs (3.108)–(3.112):

$$M'(OPr^i)_3 + 2H_3tea \xrightarrow{\text{benzene}} M'(H_2tea)(Htea) \downarrow + 3Pr^iOH \uparrow \qquad (3.108)$$
$$(\mathbf{O}), (\mathbf{P}) \text{ Colourless insoluble solid}$$

where $M' = Al, Sb$.

$$Sn(OPr^i)_4 \cdot Pr^iOH + 2H_3tea \xrightarrow{\text{benzene}} Sn(Htea)_2 + 5Pr^iOH \uparrow \qquad (3.109)$$
$$(\mathbf{Q}) \text{ Insoluble solid}$$

$$(\mathbf{O}) + 3Nb(OPr^i)_5 \xrightarrow{\text{benzene}} [\{Nb(OPr^i)_4\}_3\{Al(tea)_2\}] + 3Pr^iOH \uparrow \quad (3.110)$$
$$\text{Soluble, monomeric solid}$$

$$(\mathbf{P}) + 3Al(OPr^i)_3 \xrightarrow{\text{benzene}} [\{Al(OPr^i)_2\}_3\{Sb(tea)_2\}] + 3Pr^iOH \uparrow \quad (3.111)$$
$$\text{Soluble, monomeric solid}$$

$$(\mathbf{Q}) + 2Al(OPr^i)_3 \longrightarrow [\{Al(OPr^i)_2\}_2\{Sn(tea)_2\}] + 2Pr^iOH \uparrow \qquad (3.112)$$
$$\text{Soluble, monomeric solid}$$

A number of bimetallic titanium and zirconium (M') derivatives with aluminium and niobium have been synthesized[43,165,167,168] by the stepwise reactions shown in Eqs (3.113)–(3.115):

$$M'(OPr^i)_4 + 2H_3tea \longrightarrow M'(Htea)_2 + 4Pr^iOH \uparrow \qquad (3.113)$$
$$(\mathbf{T}), (\mathbf{U}) \text{ Insoluble solids}$$

$$(\mathbf{T}), (\mathbf{U}) + 2Al(OPr^i)_3 \xrightarrow{\text{benzene}} [\{Al(OPr^i)_2\}_2\{M'(tea)_2\}] + 2Pr^iOH \uparrow \quad (3.114)$$
$$(\mathbf{V}), (\mathbf{W}) \text{ Soluble, monomeric solids}$$

$$(\mathbf{T}), (\mathbf{U}) + 2Nb(OPr^i)_5 \xrightarrow{\text{benzene}} [\{Nb(OPr^i)_4\}_2\{M'(tea)_2\}] + 2Pr^iOH \uparrow \quad (3.115)$$
$$(\mathbf{X}), (\mathbf{Y}) \text{ Soluble, monomeric solids}$$

All the products (**V**), (**W**), (**X**), and (**Y**) are white crystalline solids, soluble in hydrocarbon solvents, out of which the titanium–niobium product $\{Nb(OPr^i)_4\}_2\{Ti(tea)_2\}$ could be distilled[175] unchanged at 205°C/0.05 mm.

A survey of the literature has revealed[176] that the triethanolaminate lanthanum derivative, $La(H_2tea)(Htea)$ reacts with 3 moles of $Nb(OPr^i)_5$ to give a heterobimetallic derivative (Eqs 3.116 and 3.117), which has been characterized by X-ray crystallography:

$$La(OPr^i)_3 + 2H_3tea \xrightarrow{\text{room temp.}} La(H_2tea)(Htea) + 3Pr^iOH \uparrow \quad (3.116)$$
$$\text{Insoluble}$$

$$La(H_2tea)(Htea) + 3Nb(OPr^i)_5 \xrightarrow[\text{r.t.}]{\text{dissolution}} \{La(tea)_2\}\{Nb(OPr^i)_4\}_3 + 3Pr^iOH \uparrow \quad (3.117)$$
$$\text{Soluble}$$

The heterobimetallic La–Nb product is readily soluble in hydrocarbons and sublimes at 180–250°C under 0.1 mm pressure.

3 PROPERTIES OF HETEROMETALLIC ALKOXIDES

3.1 Introduction

A close parallelism in the properties of homo- and heterometal alkoxides has been elucidated by Mehrotra and Singh.[47] The role of mainly steric (coupled with inductive) factors was established in the homometal species since the early 1950s by a systematic study of the oligomerization and volatility of the alkoxides of metals (groupwise) in a number of publications from the research school of Bradley.[177] Two interesting publications emphasizing the similarity of the neopentyloxides of titanium and zirconium[178] to their secondary amyloxides, followed by a similar conclusion for aluminium[179] tended to indicate the more significant role of steric factors. Bradley[180] made a highly significant contribution in 1958 by his conclusion that metals in homometal alkoxides generally tend to attain the higher (preferred) coordination number by the lowest possible degree of oligomerization through alkoxo(μ_2-/μ_3-) bridges between similar metals. A similar generalization was arrived at by Caulton et al.[181] that the ratio of $m/(a+b)$ in closed polyhedral structures of heterobimetallic alkoxides, $[M_aM'_b(OR)_m]$ tends to be typically low. All the M and M' metal atoms are enclosed in a closed polyhedron by μ_2- and μ_3-OR-type bridging to yield a compact structural unit; the heterometal atoms try to fit themselves in convenient geometries: the metal atoms in $[KZr_2(OPr^i)_9]^{64}$ and $[BaZr_2(OPr^i)_{10}]^{131}$ arrange themselves in a triangular pattern whereas the metal atoms of $[K_2Zr_2(OPr^i)_{10}]^{131}$ and $[K_4Zr_2O(OPr^i)_{10}]^{64}$ are arranged respectively in tetrahedral and octahedral geometries. The last example, $[K_4Zr_2O(OPr^i)_{10}]$, depicts the formation of an oxo-derivative (Chapter 5) with the same objective of the formation of a compact unit.

The covalent behaviour[2,3,35,36] of alkali derivatives of the $\{Zr_2(OPr^i)_9\}^-$ ligand in fact initiated an interest in the chemistry of such heterometallic alkoxides, at a time when such covalent complexes as $[Cs\{Y(CF_3COCHCOCF_3)_4\}]^{[183]}$ and $[K\{Sc(CF_3COCHCOCF_3)_4\}]^{[184]}$ were quoted as rare examples of covalent alkali complexes in an article by Nyholm $et\ al.^{[185]}$

The structure of $\{Zr_2(OPr^i)_9\}^-$ has been correlated with the structure of dimeric $\{Zr(OPr^i)_4.Pr^iOH\}_2,^{[186]}$ which was shown to be an edge-sharing bi-octahedron, $(Pr^iO)_3(Pr^iOH)Zr(\mu\text{-}OPr^i)_2Zr(Pr^iOH)(OPr^i)_3$.

$$\left[(Pr^iO)_3Zr \underset{\substack{O \\ | \\ Pr^i \\ / \quad \backslash \\ Pr^i \quad H}}{\overset{\substack{Pr^i \quad H^- \\ \backslash \quad / \\ O \\ Pr^i \\ | \\ O}}{<}} Zr(OPr^i)_3 \right]$$

(3-I)

The face-sharing bi-octahedral anion (3-II) can obviously be assumed[187] to be derived from the edge-sharing $Zr_2(OPr^i)_8.2Pr^iOH^{[186]}$ by the deprotonation of one adduct Pr^iOH molecule coupled with the loss of the second ligated Pr^iOH molecule. The formation of such a face-sharing anion ligand was suggested[3] as early as 1971 and its polydentate ligating tendencies were confirmed by (i) alcoholysis reactions with ramified alcohols showing that the terminal isopropoxide groups were replaced with greater facility than the bridging isopropoxides, (ii) the nonconducting behaviour of $[NaZr_2(OPr^i)_9]$ and $[KZr_2(OPr^i)_9]$ in isopropanol,[41] and (iii) the ^1H NMR spectrum[76] of the hafnium analogue, which showed the presence of six terminal (nonbridging) and three bridging isopropoxide groups. The different bonding modes of alkoxo-metallate(IV) ligands (Ti, Zr, Hf, and Sn) and the chemistry of their various metal derivatives have been summarized in a number of review articles.[164,165,188]

$$\left\{ (Pr^iO)_3Zr \underset{\substack{OPr^i \\ OPr^i}}{\overset{OPr^i}{<}} Zr(OPr^i)_3 \right\}^-$$

(3-II)

Further, the chemistry of formation of a variety of heterometallic alkoxides with the above and other coordinating ligands has been described in another review article[163] entitled "New perspectives in the tailoring of the heterometallic alkoxide derivatives".

The chemistry of heterometallic alkoxides involving a common chelating ligand, $\{Al(OPr^i)_4\}^-$ finds a close parallelism in an early suggestion by Bradley[189] that tetrameric aluminium isopropoxide $\{Al(OPr^i)_3\}_4$ may be represented by $Al\{(\mu\text{-}OPr^i)_2Al(OPr^i)_2\}_3$ in which the central aluminium is hexacoordinated by being ligated with three bidentate $\{Al(OPr^i)_4\}^-$ ligands. This unusual structure of

$\{Al(OPr^i)_3\}_4$ has since been confirmed by X-ray crystallography.[190,191] It also provided a clue for the suggestion in 1968 by Mehrotra and Agrawal[192] of a similar structure, $Ln\{(\mu\text{-}OPr^i)_2Al(OPr^i)_2\}_3$ for volatile tetraisopropoxoaluminates of lanthanons (and other trivalent elements); this has also been confirmed crystallographically for $Er\{Al(OPr^i)_4\}_3$ in 1996 by Wijk *et al*.[40]

In addition to the chemistry of heterometallic derivatives of alkoxometallate(IV),[188] and tetraalkoxoaluminate[92] ligands, the properties of hexa alkoxoniobates and -tantalates[87-89,113-116] of various metals have also been extensively studied. An account of these different alkoxometallates is presented in Section 3.3, after presentation of their general properties in Section 3.2.

3.2 General Properties

As mentioned earlier, the heterometal alkoxides tend to form compact units, which are volatile and generally monomeric in organic solvents. In view of some inherent difficuties[193] in the X-ray structural elucidation of metal alkoxide systems, most of the earlier work (till the early 1980s) on identification and characterization of heterometal alkoxides was based on chemical analyses, colligative properties, volatility (indicating stability to heat and ease of purification), and physicochemical investigations like UV-Vis, IR, NMR (^1H, ^{13}C, ^{27}Al), and mass spectroscopy coupled with magnetic measurements (particularly of paramagnetic later 3d systems). An account of these has already been published in review articles[3,4,97] and these are, therefore, only referred to in specific cases. During the 1990s, NMR studies based on a few other metals (^{119}Sn, ^{110}Cd, ^{89}Y, etc.)[16,17,28,29,47,103] have also played an important role in throwing light on the coordination geometries of such metals in heterometal alkoxide systems.

Mention may also be made of the extensive phase-rule type of studies based on the solubility isotherms of $M(OR)_m$–$M'(OR)_n$ systems in the research school of Turova,[194] as illustrated by Gibbs–Roseboom triangular plots of $NaOMe$–$Fe(OMe)_3$–$MeOH$, $Ba(OBu^n)_2$–$Ti(OBu^n)_4$–Bu^nOH, $Bi(OEt)_3$–$WO(OEt)_4$–$EtOH$, $Ba(OMe)_2$–$Ta(OMe)_5$–$MeOH$, $NaOMe$–$Al(OMe)_3$–$MeOH$, $Ca(OEt)_2$–$Ta(OEt)_4$–C_6H_6, and $Al(OPr)_3$–$Hf(OPr)_4$–Pr^iOH systems.

The chemical properties of heterometallic alkoxides are in general similar to their homometal counterparts: (i) hydrolysis, (ii) alcoholysis, (iii) trans-esterification reactions, (iv) reactivity with carboxylic acids[195] and enolic forms of chelating ligands such as β-diketones/β-ketoesters.[196-198] The hydrolytic reactions are now widely employed for the preparation of homogeneous mixed metal–oxide ceramic materials and the rest have found wide applications for the synthesis of a variety of novel metallo-organic derivatives (sometimes unique), which are not often available through any other synthetic route.[197]

Just as in the case of homometal alkoxides, the alcoholysis reactions have also provided a convenient techique for throwing light on the structural features of heterometallic alkoxides. Whereas alcoholysis reactions are quite facile with simple primary alcohols, resulting in the replacement of all the alkoxide groups of the heterometal alkoxides, the reactions with sterically demanding alcohols enables the distinction between terminal (nonbridged) and bridged alkoxide groups, as the intermediate product surrounded by sterically demanding alkoxide groups (obtained by initial replacement of terminal and possibly some of the bridged alkoxide groups) does not permit the additional ramified alcohol molecule to come close enough to the reacting metal atom for the initiation of further alcoholysis reactivity. For example all the isopropoxide groups in $M\{(\mu\text{-}OPr^i)_2Al(OPr^i)_2\}_n$ (where M is a bi-, tri- or tetravalent metal atom) are easily replaced by *n*-butoxy groups when treated with excess *n*-butanol. However, with tertiary butyl alcohol, only 3 out of 4 isopropoxide groups around each aluminium atom can be replaced, as the product $M\{(OPr^i)(OBu^t)Al(OBu^t)_2\}_n$ appears to become immune to further reactivity with the next tertiary butyl alcohol molecule. For example, a derivative $BaAl_3(OPr^i)_{11}$ synthesized[91] by dissolving barium in isopropanol in the presence of 3 moles of $Al(OPr^i)_3$ (Eq. 3.30) has been assigned the plausible structure (3-III) on the basis of NMR and alcoholysis studies:

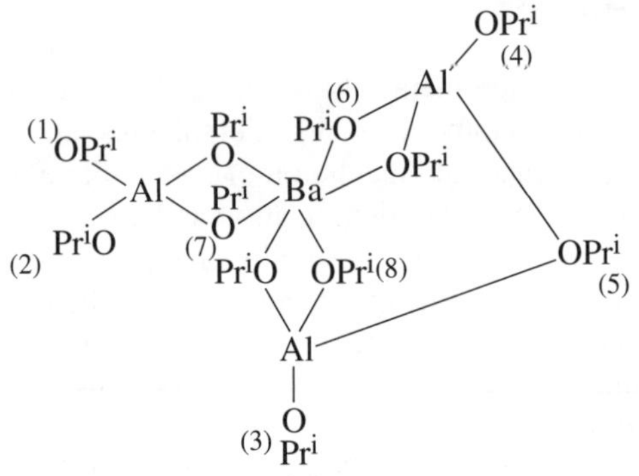

(3-III)

The proposed structure has been supported by the observation that only five isopropoxide groups [(1) to (5)] are replaced by treatment with excess Et_3COH, whereas on treatment with the less sterically demanding species Me_3COH, eight isopropoxide groups, (1) to (5) initially, then (6), (7), and (8), are finally replaced, when each of the three aluminium atoms is left bonded with three tertiary butoxy and one isopropoxy group.

The gradual stepwise methanolysis and very mild hydrolysis reactions[197] represented by Eqs (3.118)–(3.120) are highly interesting and confirm the suggested structures:

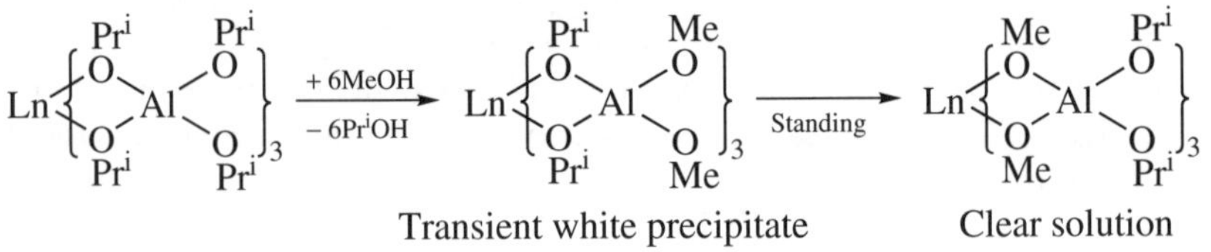
+ 6MeOH
– 6PriOH
Standing
Transient white precipitate
Clear solution

(3.118)

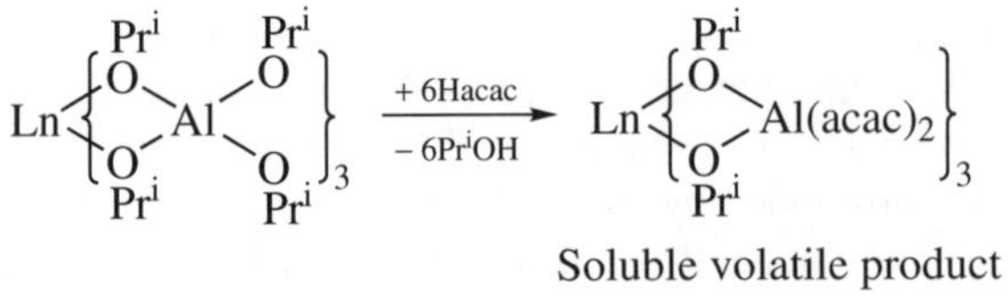
+ 6Hacac
– 6PriOH
Soluble volatile product

(3.119)

$$Ln\left\{\begin{matrix} \overset{Pr^i}{O} & \overset{Pr^i}{O} \\ \diagdown & \diagup \\ O & Al & O \\ \diagup & & \diagdown \\ \underset{Pr^i}{} & \underset{Pr^i}{} \end{matrix}\right\}_3 \xrightarrow[-6Pr^iOH]{+6HOH} Ln\left\{\begin{matrix} \overset{Pr^i}{O} & OH \\ \diagdown & \diagup \\ O & Al \\ \diagup & \diagdown \\ \underset{Pr^i}{} & OH \end{matrix}\right\}_3 \xrightarrow{Standing} Ln\left\{\begin{matrix} \overset{H}{O} & \overset{Pr^i}{O} \\ \diagdown & \diagup \\ O & Al & O \\ \diagup & & \diagdown \\ \underset{H}{} & \underset{Pr^i}{} \end{matrix}\right\}_3$$

Transient white precipitate   Clear solution

Heating $\downarrow$ $-Pr^iOH, -H_2O$

$$LnAlO_3$$

(3.120)

The mechanism suggested above is confirmed by ^1H NMR studies, indicating that as expected the terminal isopropoxide groups are replaced initially as indicated by the insolubility of the intermediate products in Eqs (3.118) and (3.120), but these methoxide/hydroxide groups are preferentially transferred to the bridging positions (bringing about higher thermodynamic stability); the external environment of the final products thus becomes similar to the original reactant, $Ln\{(\mu\text{-}OPr^i)_2Al(OPr^i)_2\}_3$ which brings about the observed solubility.

An interesting observation was made by Jones et al.[44] that in the sol–gel process for the preparation $MgAl_2O_4$ by the hydrolysis of $Mg\{Al(OEt)_4\}_2$ in the presence of triethanolamine, (Eq. 3.121) the original framework of the precursor bimetallic ethoxide remains almost unaltered (as revealed by ^1H and ^{27}Al NMR studies), indicating the stability of the coordinated structure:

$$Mg\left\{\begin{matrix} \overset{Et}{O} & OEt \\ \diagdown & \diagup \\ O & Al \\ \diagup & \diagdown \\ \underset{Et}{} & OEt \end{matrix}\right\}_2 \longrightarrow Mg\left\{\begin{matrix} \overset{H}{O} & OEt \\ \diagdown & \diagup \\ O & Al \\ \diagup & \diagdown \\ \underset{H}{} & OEt \end{matrix}\right\}_2 \longrightarrow Mg\left\{\begin{matrix} \overset{H}{O} & OH \\ \diagdown & \diagup \\ O & Al \\ \diagup & \diagdown \\ \underset{H}{} & OH \end{matrix}\right\}_2 \longrightarrow MgAl_2O_4$$

(3.121)

A similar observation about the retention of the original framework in the initial steps of hydrolysis of $Mg\{Nb(OC_2H_4OMe)_6\}_2$,[199] $[LiZr(OPr^i)_5]_5$ and $[Ba\{Zr_2(OPr^i)_9\}_2]$[45] also points to the stability of the original structure of the bimetallic alkoxide precursor. Detailed investigations on the mechanism of the formation of $BaTiO_3$ from barium–titanium-ethoxide systems by Turova et al.[200] tend to point to similar conclusions.

In view of their importance in the sol–gel preparation of spinel type materials, detailed stepwise hydrolytic reaction studies of $[Ca\{Al(OPr^i)_4\}_2]$[201,202] and $[Mg\{Al(OPr^i)_4\}_2]$[203] have been carried out with similar results. With the synthesis of heteropolymetallic alkoxides as precursors, the importance of extending hydrolytic studies to such systems is obvious.[20,24]

As illustrated by a number of examples in Section 1, the X-ray structural elucidation of a rapidly increasing number of heterometallic alkoxides (Chapter 5) has in general confirmed their coordination models[165,204] with chelating ligands like $\{M(OR)_{n+1}\}^-$ (M = Al, Ga, Nb, Ta) and $\{M_2(OR)_{2n+1}\}^-$ (M = Zr, Hf, Sn(IV)) of metals with valency n, as suggested since 1971.

Heterometallic alkoxides, therefore, constitute a novel class of heterometal coordination systems, stabilized by alkoxide bridges without the support of any auxiliary ligands (like CO) or metal–metal bonds. The factors responsible for this extraordinary stability of heterometallic alkoxides are not yet fully understood, but the formation of a thermodynamically favoured heterometal system appears to be governed by parameters like size and electronegativity of the component metals as well as the ramification of the alkyl group and the presence of electronegative (*e.g.* fluorine) or coordinating substituents (*e.g.* alkoxy, amino, etc.) on the alkyl group(s) involved.

Since the basic information available from physico-chemical techniques (*e.g.* conductivity, spectroscopy, etc.) is essentially similar to that available from the corresponding homometal systems, it will not be repeated here (see Chapter 2, Section 3), but applications are illustrated for a few typical systems in Section 3.3. Further, although more definitive information is available from crystal structure elucidations, this will not be dealt with in this chapter, in view of the separate treatment in Chapter 4.

3.3   Properties of Tetra-alkoxoaluminate and Hexa alkoxoniobate (-tantalate) Derivatives

As the above three ligands generally function in a bidentate manner, their heterometallic coordination alkoxides resemble one another in forming derivatives of the type

$$
\underset{R}{\overset{R}{\underset{O}{\diagup}}} M \underset{O}{\overset{O}{\diagup}} Al(OR)_2 \quad \text{and} \quad M \underset{O}{\overset{O}{\diagup}} Nb(OR)_4 \; (Ta)
$$

the main difference being that the ligated aluminium atoms are tetrahedral whereas the niobium (tantalum) atoms are octahedral. These are, therefore, being dealt with together.

3.3.1   Tetraalkoxoaluminate Derivatives

Of the tetra-alkoxoaluminates of various alkali metals (Li, Na, K, Rb, Cs), the potassium derivative was found in 1976[205] to show the maximum solubility and volatility. This solubility behaviour was ascribed to the suitability of the size of potassium enabling the formation of a di-isopropanolate adduct,

$$
\begin{bmatrix} Pr^iO & & Pr^i & Pr^i \\ & Al & O & O-H \\ Pr^iO & & K & \\ & & O & O-H \\ & & Pr^i & Pr^i \end{bmatrix}
$$

on the lines of the well-known suggestion by Sidgwick and Brewer in 1925 about the formation of an adduct

$$
\begin{bmatrix} Ph & C-O & OH_2 \\ HC & & K & \\ Me & C-O & OH_2 \end{bmatrix}
$$

with similar covalent characteristics, as confirmed by Nyholm *et al.*[185] in 1970. This suggestion of $(Pr^iOH)_2K(\mu\text{-}OPr^i)_2Al(OPr^i)_2$ adduct formation received confirmation in 1993 from the X-ray structural elucidation of this low-melting polymeric unit by Gilje *et al.*[38] In fact, these workers have shown that the unsolvated $KAl(OPr^i)_4$ is a white solid (m.p. 380°C) insoluble in noncoordinating solvents, whereas the adduct $KAl(OPr^i)_4.2Pr^iOH$ (m.p. 31°C) is soluble in aromatic solvents. The most intense peak in the mass spectrum of this material corresponds to $[K_2Al_2(OPr^i)_7(HOPr^i)]^+$ and a major fragment contains three K and two Al atoms, which might correspond to a product of the composition $K_3Al_2(OPr^i)_9$, isolated[35] in a 3:2 molar reaction of $KOPr^i$ and $Al(OPr^i)_3$. Compared to $KAl(OPr^i)_4$, which dissolves in isopropanol to give a homogeneous solution, $NaOPr^i$ yields only a thick suspension of white powder in isopropanol.

The coordination sphere of K in $[(Pr^iO)_2Al(\mu\text{-}OPr^i)_2K.2Pr^iOH]$ was found to be similar to that of magnesium[37] in $[\{(Pr^iO)_2Al(\mu\text{-}OPr^i)_2\}_2Mg.2Pr^iOH]_2$. This latter product along with a series of volatile soluble products with the formula, $M\{M'(OR)_4\}_2$ ($M = Mg, Ca, Sr, Ba; M' = Al, Ga; R = Et, Pr^i$) were isolated (Eq. 3.25) by the dissolution of metal M in alcohol in the presence of 2 moles of $M'(OR)_3$. This comparatively facile dissolution of bivalent metal in alcohol has been ascribed[16] to the coordination of alcohol to the $M(OR)_3$, rendering the proton more labile and reactive. The mass and NMR spectra of $Mg\{Al(OBu^s)_4\}_2$ liquid, and the X-ray structure[206] of $Mg(\mu\text{-}OPh)_2Al(OPh)_2$, confirm a similar coordination model. Table 3.1 lists some representative tetra-alkoxoaluminate derivatives.

Compared to the polymeric insoluble and nonvolatile nature of homo-alkoxides of bivalent metals $[M(OR)_2]_n$ in general (where M = (i) Mg, Ca, Sr, Ba (Eq. 3.25), (ii) Be, Zn, Cd, Hg(II), Sn(II) (Eq. 3.35), and (iii) Mn(II), Fe(II), Co(II), Ni(II), Cu(II) (Eq. 3.40)), their tetraalkoxoaluminate derivatives $M\{Al(OR)_4\}_2$ are volatile and soluble in organic solvents, in which they exhibit monomeric (and in a few cases slightly associated) behaviour.

Synthesized by the reactions (Eqs 3.35 and 3.40) of their anhydrous chlorides with 2 moles of $KAl(OPr^i)_4$, their framework $M\{Al(OPr^i)_4\}_2$ (M = Mg, Zn, Cd, Hg) appears to be so stable that even in 1:1 molar reaction of MCl_2 with $KAl(OPr^i)_4$, only half the amount of $M\{Al(OPr^i)_4\}_2$ is obtained leaving the remaining MCl_2 unreacted. However, a product $ClBe\{Al(OPr^i)_4\}$ could be obtained in such a 1:1 molar reaction; this was converted by reaction with $KOPr^i$ into volatile monomeric $(Pr^iO)Be\{Al(OPr^i)_4\}$ which could be distilled at $\sim157°C/0.3$ mm pressure. However, this monomeric product tended to dimerize on standing for a few days as shown by its molecular weight and ^1H NMR spectrum. This change could be ascribed to the tendency of beryllium to attain the four-coordination state by assuming a structure of the type: $[(Pr^iO)_2Al(\mu\text{-}OPr^i)_2Be(\mu\text{-}OPr^i)_2Be(\mu\text{-}OPr^i)_2Al(OPr^i)_2].$[207] This observation immediately led to the synthesis of the first heterotrimetallic alkoxides[7] $[(Pr^iO)_2Al(\mu\text{-}OPr^i)_2Be(\mu\text{-}OPr^i)_2Zr(OPr^i)_3]$ and $[(Pr^iO)_2Al(\mu\text{-}OPr^i)_2Be(\mu\text{-}OPr^i)_2Nb(OPr^i)_4]$ by the reactions of $(Pr^iO)_2Al(\mu\text{-}OPr^i)_2Be(OPr^i)$ with $Zr(OPr^i)_4$ and $Nb(OPr^i)_5$, respectively.

The structural features of derivatives (iii) of later 3d transition metals, $M\{Al(OPr^i)_4\}_2$[97,101–104] were elucidated by their UV-Vis spectra and paramagnetic characteristics, techniques commonly employed in the conventional coordination derivatives dealing with such $3d^5-3d^9$ systems and explaining the observations with the help of ligand field theory. For example, the selective exchange ability of branched

Table 3.1 Preparation and characterization of some representative tetraalkoxo-metallate derivatives

Compound	Synthetic route[1]	Characterization techniques	References
$KAl(OPr^i)_4.2Pr^iOH$	A-1	IR, NMR, MS, X-ray	1, 35, 38, 52, 53, 76, 204
$Mg\{Al(OPr^i)_4\}_2.2Pr^iOH$	B	NMR, X-ray	35, 36, 37
$Mg\{Al(OBu^s)_4\}_2$	F	NMR, MS	206
$M\{Al(OPr^i)_4\}_2$ M = Mg, Ca, Sr, Ba	B	IR, NMR, volatility, solubility, MW, alcoholysis	35, 36, 85
$Mg_2Al_3(OPr^i)_{13}$	B	IR, NMR, X-ray	90
$BaAl_3(OPr^i)_{11}$, $Ba_2Al_3(OPr^i)_{13}$	B	IR, NMR, MW, alcoholysis	91
$M\{Al(OPr^i)_4\}_2$, $ClM\{Al(OPr^i)_4\}$ M = Be, Zn, Cd, Hg(II)	C	IR, NMR, MW, solubility, volatility, alcoholysis	94
$\{Al(OPr^i)_4BeOPr^i\}_2$	C	NMR, MW alcoholysis	207
$[M\{Al(OPr^i)_4\}_2]$ M = Mn(II), Fe(II), Co(II), Ni(II), Cu(II)	C		42, 97, 100–104
$M\{Al(OBu^t)_4\}_2$ M = Mn, Fe, Co, Ni, Cu, Zn	C	UV-Vis, paramagnetism, solubility, volatility, alcoholysis	101, 109, 110, 108
$[M\{Al(OPr^i)_4\}_3]$ M = Cr(III), Fe(III), In(III)	C		105, 106, 69, 77
$Ln\{Al(OPr^i)_4\}_3$	A-2, C	NMR, MS, X-ray of $[Er\{Al(OPr^i)_4\}_3]$, volatility, solubility	80 40
$[(Pr^iO)_2Zr\{Al(OPr^i)_4\}_2]$	A-2, C		209–211
$[(Pr^iO)_2Hf\{Al(OPr^i)_4\}_2]$	A-2	X-ray	77
$[(Pr^iO)_3M\{Al(OPr^i)_4\}_2]$ M = Nb, Ta	A-2	NMR, volatility, MW	75
$[\{Pr\{(\mu\text{-}OPr^i)_2Al(OPr^i)_2\}_2(Pr^iOH)(\mu\text{-}Cl)\}_2]$	C	X-ray	39
$[I_2Sn\{Al(OPr^i)_4\}_2]$	C	X-ray	236
$[Th\{Al(OPr^i)_4\}_4]$	C	NMR, volatility, MW	52, 74

[1]See Section 2 for details of routes listed.

alkoxides of Co and Ni with primary alcohols has been explained[97] on the basis of ligand field effects. Alkoxoaluminates of these transition metals are again volatile and monomeric, whereas their simple alkoxides are polymerized insoluble entities.

These alkoxoaluminates of later 3d metals undergo facile exchange reactions with other alcohols or β-diketones. The properties of a few of these bimetallic alkoxide products are illustrated in Table 3.2.

Table 3.2 **Properties of tetraalkoxoaluminates and allied derivatives of a few later 3d metals**[97]

Compound	Nature	Volatility (°C/mm)	MW found (Calc.)	μ_{eff} (B.M./K)	Dq (cm^{-1})
Cr[Al(OPr^i)$_4$]$_3$	Green viscous liquid	190/0.6	838 (842)	3.90/100 3.92/297	1527
Mn[Al(OPr^i)$_4$]$_2$	Brown viscous liquid	145/0.6	607 (582)	5.95/88 5.90/297	–
Fe[Al(OPr^i)$_4$]$_3$	Brown viscous liquid	130/0.8	1003 (846)	5.06/89 5.39/296	–
Fe[Al(OPr^i)$_4$]$_2$	Brown semi-solid	125/0.8	605 (582)	4.53/87 5.06/297	–
Co[Al(OPr^i)$_4$]$_2$	Purple liquid	140/0.6	629 (586)	4.62/92 4.39/297	873
Ni[Al(OPr^i)$_4$]$_2$	Pink liquid	120/0.6	570 (585)	3.29/90 3.39/294	823 (film) 895 (Pr^{i}OH) 912 (py)
Cu[Al(OPr^i)$_4$]$_2$	Greenish-blue liquid	135/0.6	604 (593)	1.22/87 1.64/297	1380
Cr[Al(OPr^i)$_3$(OMe)]$_3$	Light-green solid	215/0.8	–	3.84/300	1572
Cr[Al(OPr^i)$_2$(OMe)$_2$]$_3$	Green solid	255/0.8*	–	3.86/300	1582
Cr[Al(OEt)$_4$]$_3$	Green solid	235/0.8*	–	3.86/300	1577
Cr[Al(OCH$_2$CF$_3$)$_4$]$_3$	Brown viscous liquid	120/0.8*	1353 (1321)	3.86/300	1520
Cr[Al(OPr^i)$_3$(acac))]$_3$	Green pasty solid	140/0.8	967 (962)	3.82/300	1675
Cr[Al(OPr^i)$_2$(acac)$_2$]$_3$	Green solid	180/0.8*	1233 (1082)	3.83/300 –	1692 –
Fe[Al(OPr^i)$_2$(OAm^t)$_2$]$_3$	Brown solid	270/0.6*	976 (1014)	– –	–
Ni[Al(OPr^i)(OBu^t)$_3$]$_2$	Purple solid	140/0.6*	691 (670)	3.14/88 3.47/300	736
Ni[Al(OPr^i)$_2$(acac)$_2$]$_2$	Brown pinkish solid	240/0.6*	708 (745)	3.28/300	769
Cu[Al(OCH$_2$CF$_3$)$_4$]$_4$	Blue semi-solid	115/0.5	–	1.72/300	1602
Cu[Al(OPr^i)(OBu^t)$_3$]$_2$	Green solid	150/0.5*	–	1.74/300	1230 (benzene) 1490 (Nujol)

*Indicates sublimation.
[1]See Section 2 for details of routes listed.

An interesting trend in the ligating behaviour of {Al(OR)$_4$}$^-$ can be illustrated by nickel derivatives.[42] The generally bidentate nature of {Al(OR)$_4$}$^-$, ligands, should be reflected in the tetrahedral or square planar nature of nickel in Ni{Al(OPr^i)$_4$}$_2$. However, its spectral and magneto-chemical properties appear to reveal a tendency for the nickel atom to achieve octahedral geometry, possibly by adopting a structure of the type Ni{(μ-OPr^i)$_3$Al(OPr^i)}$_2$ in which {Al(OPr^i)$_4$}$^-$ behaves as a tridentate ligand. Its electronic spectra (Fig. 3.1) show ν_2 and ν_3 bands characteristic of octahedral geometry at about 13 000 and 24 000 cm^{-1}. In addition, the appearance of a doublet in the region 17 000–19 000 cm^{-1} indicates the presence of some tetrahedral species also in equilibrium with the octahedral one; the octahedral form becomes predominant in isopropyl

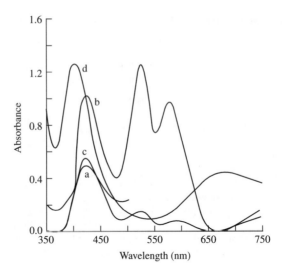

Figure 3.1 Spectra of [Ni{Al(OPr^i)$_4$}$_2$] in (a) film, (b) C$_6$H$_6$, (c) Pr^{i}OH, (d) C$_5$H$_5$N.

alcohol or pyridine solutions owing to the formation of bis-adducts. This change in geometry is also reflected in a change of colour (pink in benzene solution or as thin film *versus* yellowish green in isopropanol/pyridine solution). The octahedral form would obviously be predominant in Ni{Al(OMe)$_4$}$_2$, but the tetrahedral form should be the predominant species in Ni{Al(OPr^i)(OBu^t)$_3$}$_2$ and Ni{Al(OBu^t)$_4$}$_2$.[208]

As mentioned at the end of Section 3.1, the tetra-alkoxoaluminates of some trivalent metals (Sc, Y, Ln,[40,80,192] Cr(III),[105] and In[69,73]) exhibit structures similar to that of Al{(μ-OPr^i)$_2$Al(OPr^i)$_2$}$_3$, in which the central aluminium atom can be replaced by other trivalent metals. These are again all monomeric and volatile, which suggests that they are more stable than [Al(OPr^i)$_3$]$_4$, which distils with dissociation as a dimer.

Except for the tetraisopropoxoaluminates of cerium(IV) and thorium with aluminium, diisopropoxozirconium bis-tetraisopropoxoaluminate derivatives of the type [(Pr^{i}O)$_2$Al(μ-OPr^i)$_2$Zr(OPr^i)$_2$(μ-OPr^i)$_2$Al(OPr^i)$_2$] can only be obtained in the case of smaller zirconium atoms. The structure of this derivative, those of ZrAl$_2$(OPr^i)$_{10-x}$L$_x$ (where L = acetylacetone, benzoylactone, dibenzoyl methane, methylacetoacetate, or ethylacetoacetate), have been elucidated mainly by ^{1}H NMR studies and can be represented[209,210] by

$$\underset{L}{\overset{L}{>}}\text{Al}(\mu\text{-OPr}^i)_2\overset{\overset{L}{|}}{\underset{\underset{L}{|}}{\text{Zr}}}(\mu\text{-OPr}^i)_2\text{Al}\underset{L}{\overset{L}{<}}$$

in which the terminal isopropoxo groups of the original derivative [(Pr^{i}O)$_2$Al(μ-OPr^i)$_2$Zr(OPr^i)$_2$(μ-OPr^i)$_2$Al(OPr^i)$_2$] have been replaced by bidentate β-diketone or β-ketoester ligands (L). It is interesting to record that the X-ray structure of the analogous [HfAl$_2$(OPr^i)$_{10}$] shown in 1997[77] to conform in general to the structures suggested above.

3.3.2 Hexaalkoxoniobate and -tantalate Derivatives

Compared to the tetra-alkoxoaluminates, the corresponding hexa-alkoxoniobate and -tantalate derivatives of most of the metals are less stable. Of these two, the niobate products tend to disproportionate more readily; this is reflected in the conductivity curves obtained[211] in the titrations of niobium and tantalum pentaisopropoxide with alkali (Na, K) isopropoxides in isopropyl alcohol (see Fig. 3.2).

All the MM'(OR)$_6$ derivatives (M = Li, Na, K; M' = Nb, Ta; R = Me, Et, Pr^i, Bu^t)[35,36,212,213] are white solids soluble in the parent alcohols (solubility decreasing from Li to K). If attempts are made to purify them by distillation/sublimation under reduced pressure, the niobium derivatives tend to disproportionate more readily, yielding the corresponding volatile niobium alkoxide and alkali alkoxide residues. By contrast, MTa(OR)$_6$ derivatives (M = Na, K; R = Me, Et) are stable up to $\sim$ 320°C/ 0.1 mm (the boiling points of Ta(OMe)$_5$ and Ta(OEt)$_5$ are 180°C/10 mm and 202°C/ 10 mm, respectively).

Hexaisopropoxotantalates of Li, Na, K, and Cs tend to sublime around 220°C under 0.1 mm pressure and all the corresponding tertiary butoxy analogues, M{M'(OBu^t)$_6$} (M = Li, Na, K, Cs; M' = Nb, Ta) can be volatilized at lower temperatures around 110–120°C/0.1 mm. The thermal stability of the MTa(OR)$_6$ derivatives appears to follow the orders: Li > Na > K and OBu^t > OPr^i > OEt $\approx$ OMe. Molecular weight studies also appear to indicate the same pattern.

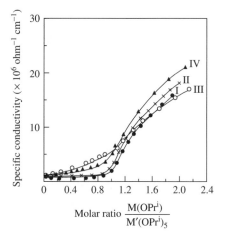

Figure 3.2 Titration between M'(OPr^i)$_5$ and M(OPr^i). Curve I: ($\bullet$) NaOPr^i M/4.77 *vs* Ta(OPr^i)$_5$ M/93.74; Curve II: ($\times$) KOPr^i M/4.96 *vs* Ta(OPr^i)$_5$ M/93.74; Curve III: (o) NaOPr^i M/4.77 *vs* Nb(OPr^i)$_5$ M/94.24; Curve IV: ($\blacktriangle$) KOPr^i M/4.98 *vs* Nb(OPr^i)$_5$ M/94.91.

The corresponding alkaline earth derivatives[35,36,88,89,212,213] (Eq. 3.27) are soluble in organic solvents and exhibit monomeric behaviour in refluxing benzene. Almost all of these can be volatilized in fairly high yields under reduced pressure. They also undergo complete alcoholysis with primary alcohols (say n-butanol), but yield[36] only partially substituted tertiary butoxy products of the formula $[M\{M'(OPr^i)_3(OBu^t)_3\}_2]$ (M = Mg, Ca, Sr, Ba; M' = Nb, Ta).

A few derivatives of later 3d transition metals and samarium, $M\{M'(OPr^i)_6\}_2$ and $M\{M'(OPr^i)_6\}_3$, (M = Cr(III),[113] Mn(II),[101] Fe(II, III),[114] Co(II),[115] Ni(II),[116] Cu(II),[117]) have also been synthesized (Eq. 3.44) by the reactions of their anhydrous chlorides with stoichiometric quantities of $K\{M'(OPr^i)_6\}$ in benzene–isopropanol medium. Some of these are listed in Table 3.3.

As emphasized in the opening paragraph of this section, heteroalkoxoniobates and -tantalates are less stable than those of aluminates and zirconates; this aspect is also reflected in the chemistry of their homo-alkoxides. The chemistry of hetero-metal alkoxides of those two metals is dominated in most cases (particularly with primary alcohols) by their tendencies for disproportionation as well as formation of oxo-derivatives. Two review articles[212,213] on these heterometal alkoxides reflect these aspects prominently, even questioning the validity of some earlier findings.

Table 3.3   Preparation and characterization of a few hexaalkoxoniobate and -tantalate derivatives

Compound	Synthetic route[1]	Characterization techniques	References
$M\{M'(OR)_6\}$ M = Li, Na, K, Cs; M' = Nb, Ta; R = Me, Et, Pr^i, Bu^t, CH_2SiMe_3	A-1	Conductivity, volatility, MW, X-ray structures of $\{LiNb(OEt)_6\}_n$ and $LiNb(OCH_2SiMe_3)_6$	35, 36 211, 212 213 67
$M\{M'(OR)_6\}_2$ M = Mg, Ca, Sr, Ba; M' = Nb, Ta; R = Et, Pr^i	B	Volatilization, MW, alcoholysis, X-ray structure of $Ba\{Nb(OPr^i)_6\}_2.2Pr^iOH$	35, 36, 88, 89, 129 213, 214
$Cr\{M'(OPr^i)_6\}_3$ M' = Nb, Ta	C	Alcoholysis, electronic spectra, magnetic susceptibility, ESR	113
$Mn\{Nb(OPr^i)_6\}_2$	C	Alcoholysis, electronic spectra, magnetic susceptibility, ESR	101
$Fe\{M(OPr^i)_6\}_2$	C	Alcoholysis, electronic spectra, magnetic susceptibility, ESR	114
$Fe\{M(OPr^i)_6\}_3$ M = Nb, Ta	C	Alcoholysis, electronic spectra, magnetic susceptibility, ESR	114
$M\{M'(OPr^i)_6\}_2$ M = Co(II), Ni(II), Cu(II); M' = Nb, Ta	C	Alcoholysis, electronic spectra, magnetic susceptibility, ESR	115 116 117, 214
$[(Pr^iO)La\{(\mu-OPr^i)_2Nb(OPr^i)_4\}$ $\{(\mu-OPr^i)_3Nb(OPr^i)_3\}]$	C	Solubility, X-ray	214

[1]See Section 2 for details of routes listed.

The X-ray structural study of LiNb(OEt)$_6$[212,213] and [(Pr^{i}OH)$_2$Ba{Nb(OPr^i)$_6$}$_2$][214] reveals familiar patterns. However, LaNb$_2$(OPr^i)$_{13}$ reveals[214] the bonding of the central La atom by a bidentate and another tridentate {Nb(OPr^i)$_6$} group in addition to coordination with a free isopropanol molecule. This manner of novel double (bi- as well as tri-) types of bonding indicates the need for further detailed investigations in such systems.

3.4   Heterometallic Alkoxides Involving Alkoxometallate(IV) Ligands

3.4.1   Introduction

The synthesis of volatile and soluble (in organic solvents) covalent[2,3,35,36,183–185] nona-alkoxodizirconates of alkali metals such as KZr$_2$(OR)$_9$ (R = Et, Pr^i), by Bartley and Wardlaw[2] in 1958 initiated an entirely new dimension in the chemistry of heterometallic alkoxides. Historically, a nona-alkoxostannate derivative, NaSn$_2$(OEt)$_9$ had been described by Bradley et al.,[216] in 1957 as an intermediate during their efforts to synthesize tin tetra-alkoxides by the reactions of SnCl$_4$ with NaOEt.

It was Mehrotra[41] who drew attention to the uniqueness of [M{Zr$_2$(OPr^i)$_9$}] (M = Li, Na, K) derivatives based on their nonconducting behaviour in isopropanol. These workers also synthesized similar derivatives of titanium and tin(IV) by reacting them with alkali isopropoxides in 2:1 molar ratio; both of these could be recrystallized without any change in composition. However, whereas the zirconium derivatives could be distilled unchanged under reduced pressure, their titanium analogues tended to disproportionate on heating yielding volatile Ti(OPr^i)$_4$; conductometric titrations (Fig. 3.3) also did not give clear inflexions at 2:1 ratio of the reactants except in the case of zirconium.

However, a derivative with the composition [M$_2$Zr$_3$(OPr^i)$_{14}$] was obtained when the molar ratio of MOPr^i and Zr(OPr^i)$_4$.Pr^{i}OH was 1:1 or >1:1; this product could be recrystallized without change in its composition and could be volatilized under reduced pressure.

The 1:1 molar reaction of alkali and titanium isopropoxide also yielded a crystallizable derivative of composition M{Ti(OPr^i)$_5$}. Interestingly, a similar reaction of MOBu^t and Zr(OBu^t)$_4$ also resulted in a crystallizable volatile dimeric product [{KZr(OBu^t)$_5$}$_2$]. The basic information on the more dominant was thus set out in nona-alkoxodimetallate {M$_2$(OR)$_9$}$^-$, penta-alkoxometallate {M(OR)$_5$}$^-$, and hexa-alkoxometallate {M(OR)$_6$}$^{2-}$ derivatives were thus laid in this early publication.

For the highly interesting and novel {Zr$_2$(OPr^i)$_9$}$^-$ ligand, a plausible structure (Fig. 3.4) involving two face-sharing octahedra was suggested in 1971,[3] which was capable of encapsulating the alkali ion. Some evidence for this was furnished by alcoholysis with ramified alcohols and ^{1}H NMR spectroscopy[76] in 1972.

This type of structure in a simple anionic form is depicted[61] in a few niobium(IV) solvent-separated ion-pair derivatives like [Na(MeOH)][Nb$_2$(OMe)$_9$]. The relationship of the above structure of {Zr$_2$(OPr^i)$_9$}$^-$ anion to the edge-sharing structure[186] of Zr$_2$(OPr^i)$_8$(Pr^{i}OH)$_2$ was depicted by Evans et al.[187] in 1997. The flexibility of the above type of {M$_2$(OR)$_9$}$^-$ (M = Zr, Hf, Th, Ce(IV), U(IV), Sn(IV), Ti(IV), Nb(IV), etc.) ligands in binding the central heterometal atoms in the tetra-, tri-, or bi-dentate

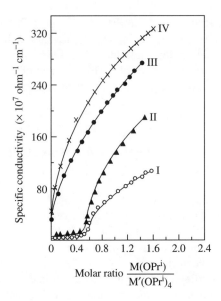

Figure 3.3 Titration between $M'(OPr^i)_4$ and $M(OPr^i)$. Curve I: (○) $NaOPr^i$ M/4.77 vs $Zr(OPr^i)_4.Pr^iOH$ M/80.49; Curve II: (▲) $KOPr^i$ M/9.97 vs $Zr(OPr^i)_4.Pr^iOH$ M/43.011; Curve III: (●) $NaOPr^i$ M/9.94 vs $Ti(OPr^i)_4$ M/37.7; Curve IV: (×) $KOPr^i$ M/9.94 vs $Ti(OPr^i)_4$ M/37.7.

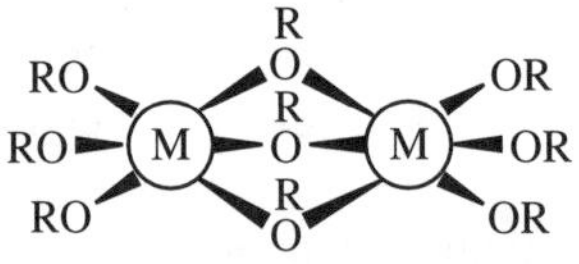

Figure 3.4 Schematic representation of a nona-alkoxodimetallate(IV) ligand.

manner[188] that suits their preferred coordination states is remarkable. The possibilities of the alkoxometallate(IV) ligands to function as simpler $\{M(OR)_5\}^-$ and $\{M(OR)_6\}^{2-}$ also enhance their versatility in a remarkable manner, as revealed by their X-ray crystal structural studies at an ever-increasing pace since 1984;[217] these have in general confirmed the conclusions about their plausible structures on the basis of simpler colligative, spectroscopic, and other physico-chemical studies. A number of representative examples of these types of tetra-alkoxometallate(IV) (including a few heterotermetallic) derivatives are listed in Table 3.4. and an attempt is made in the following paragraphs to rationalize their structural features (see also Chapter 4).

Table 3.4   Preparation and characterization of some heterometallic alkoxides involving $\{M_2(OR)_9\}^-$, $\{M(OR)_5\}^-$ and $\{M(OR)_6\}^{2-}$ ligands

Compound	Synthetic route[1]	Characterization techniques	References
Nona-alkoxodimetallate derivatives			
Li$\{Zr_2(OR)_9\}$	A-1	MW, volatility	2
R = Et, Pr^i, Pr^n, Bu^n			
M$'\{M_2(OPr^i)_9\}$	A-1	IR, MW, volatility,	41
M$'$ = Li, Na, K, Cs;	A-1	Alcoholysis	
M = Ti, Zr, Hf, Sn			
Li$\{Zr_2(OPr^i)_9\}(Pr^iOH)$	E-2	^{1}H, ^{13}C NMR, X-ray	131
[Na(MeOH)][Nb$_2$(OMe)$_9$]	A-2	X-ray	61
K$\{Zr_2(OBu^t)_9\}$	A-1	NMR	41, 182
K$\{U_2(OBu^t)_9\}$	C	NMR, X-ray	217
Na$\{Ce_2(OBu^t)_9\}$	C	IR, NMR, X-ray	218
Na$\{Th_2(OBu^t)_9\}$	E-2	IR, NMR, X-ray	219
Mg$\{Zr_2(OPr^i)_9\}Cl$	C	IR, NMR, MW	221
Ba$\{Zr_2(OPr^i)_9\}_2$	C, E-2	IR, NMR, X-ray	131
Ba$\{Zr_2(OPr^i)_{10}\}$	E-2	NMR, X-ray	131
M$'\{M_2(OPr^i)_9\}_2$			
M$'$ = Mg, Ca, Sr, Ba;	B	} IR, NMR, MW	{ 36, 86
M = Zr, Hf.			
M$'$ = Zn, Cd; M = Zr	C		224
BaZr$_3$(OPr^i)$_{14}$	B	MW, crystallization, alcoholysis	86
Sr$\{Ti_2(OEt)_9\}_2$	A-1	IR, NMR, MS, X-ray, solubility	223
[Cd$\{Zr_2(OPr^i)_9\}Cl]_2$	C	NMR, MW, X-ray	135
[Sn$\{Zr_2(OPr^i)_9\}Cl]_2$	C	NMR, X-ray	141, 142
[(C$_5$H$_5$)Sn$\{Zr_2(OPr^i)_9\}]$	C	NMR, X-ray	142
[Sn$\{Hf_2(OPr^i)_9Cl]_2$	C	NMR, X-ray	142
[ICd(M$_2$(OPr^i)$_9$)]	C	^{113}Cd NMR, X-ray	160, 161
M = Ti, Hf, Sn			
[ISn$\{Zr_2(OPr^i)_9\}]$		^{119}Sn NMR, X-ray	161
ClM$'\{Sn_2(OPr^i)_9\}$	C	IR, ^{1}H, ^{13}C, ^{119}Sn NMR	221
M$'$ = Zn, Cd			
[(C$_5$H$_5$)Sn$\{M_2(OPr^i)_9\}]$	C	^{119}Sn NMR, X-ray	142
M = Zr, Hf			
M$'\{Sn_2(OPr^i)_9\}_2$	C	^{119}Sn NMR, X-ray	221
M$'$ = Zn, Cd			
M$\{Zr_2(OPr^i)_9\}_2$	C	Spectra, MW, magnetic moments, solubility, volatility	118–122
M = Fe, Co, Ni, Cu			
(ClCu$\{Zr_2(OPr^i)_9\}$	C	X-ray	126
LaCl$_x\{(Zr_2OPr^i)_9\}_{3-x}$	C	IR, NMR	143
SmCl$_x\{Zr_2(OPr^i)_9\}_{3-x}$	C	IR, NMR	220
Cl$_2$Y$\{Ti_2(OPr^i)_9\}$	C	IR, variable-temperature NMR, X-ray, MW	{ 81
ClCu$\{Ti_2(OPr^i)_9\}$	C	}	{ 81
[$\{Zr_2(OPr^i)_9Eu(\mu\text{-}I)\}_2$]	C	} X-ray	187
[$\{Zr_2(OPr^i)_9NdCl(\mu\text{-}Cl)\}_2$]	C	X-ray	81
[$\{Ti(OPr^i)_5\}Ba\{Ti_2(OPr^i)_9\}]$	B	X-ray	81
Cl$_x$Sn$\{Zr_2(OPr^i)_9\}_{4-x}$	C	IR, NMR	140

(*continued overleaf*)

Table 3.4 (*Continued*)

Compound	Synthetic route[1]	Characterization techniques	References
Pentaalkoxometallate derivatives			
M′{Ti(OPr^i)$_5$}$_n$	A-1	MW, IR, NMR, X-ray	41, 60, 225
M′ = Li; $n = 2$.			
M′ = Na, K; $n = \infty$			
[M′{M(OBu^t)$_5$}]$_2$	A-1		
M′ = Li, Na; M = Ti, Zr		MW, volatility,	41, 60
M′ = K; M = Zr		NMR, X-ray	41, 182
M′ = K, M = Sn(IV)		NMR, X-ray	56
{Ti(OPr^i)$_5$}Ba{Ti$_2$(OPr^i)$_9$}	A-2	IR, NMR, X-ray	81
[I$_2$Sn{Ti(OPr^i)$_5$}$_2$]	C	IR, NMR, X-ray	236
[I$_3$Sn{Zr(OPr^i)$_5$}.Pr^{i}OH]	C	IR, NMR, X-ray	236
Hexaalkoxometallate derivatives			
SnZr(OPr^i)$_6$	A-2	^{119}Sn NMR	82
M′Zr(OBu^t)$_6$	A-2	^{119}Sn NMR	82
M′ = Sn(II), Pb		^{207}Pb NMR	82
Sn^{II}Sn^{IV}(OPr^i)$_6$	C	^{1}H, ^{119}Sn NMR	226
[Na$_2${Zr(OCH(CF$_3$)$_2$)$_6$}.C$_6$H$_6$]	A-1, C	NMR, X-ray	227
[{(COD)Rh}$_2${Sn(OEt)$_6$}]	C	NMR, X-ray	229
[{(COD)Rh}$_2${Ti(OPr^n)$_6$}]	C	NMR, X-ray	230
Tl$_2$Sn(OEt)$_6$	A-1	NMR, X-ray	55
[(dme)$_2$Na$_2$Ce(OBu^t)$_6$]	C	NMR, X-ray	228
[BaZr(OBu^t)$_6$(thf)$_2$]	A-1	IR, NMR	83
[BaZr(OH)(OPr^i)$_5$(Pr^{i}OH)]$_2$	A-1	IR, NMR	83
Heterotrimetallic alkoxides			
[Cd(OPr^i)$_3$BaM$_2$(OPr^i)$_9$]$_2$	A-1, C	NMR, X-ray	136, 160, 162
M = Ti, Zr, Hf			
[Cd(OPr^i)$_3$Sr{Zr$_2$(OPr^i)$_9$}]$_2$	A-1, C	NMR, X-ray	235
[{Zr$_2$(OPr^i)$_9$}M{Al(OPr^i)$_4$}]	C	IR, NMR, MW, UV-Vis	8, 9, 101, 109
M = Mn, Fe, Co, Ni			156–159
{Al(OPr^i)$_4$}Be{Zr(OPr^i)$_5$}	C	IR, NMR, MW	7
[{Zr$_2$(OPr^i)$_9$}Fe{Ta(OPr^i)$_6$}]	C	MW, volatility	109
[{Al(OPr^i)$_4$}Fe{Zr$_2$(OPr^i)$_4$}$_2$]	C	MW, volatility	9
[Y{Al(OPr^i)$_4$}$_x${Zr$_2$(OPr^i)$_9$}$_{3-x}$]	C	^{1}H, ^{13}C, ^{89}Y NMR	222
$x = 1, 2$			

[1]See Section 2 for details of routes listed.

3.4.2 *Different Modes of Bonding of Nona-alkoxodimetallate(IV) ligands*

3.4.2.1 *Tetradentate mode of coordination*

A critical survey of a large number of derivatives with {M$_2$(OR)$_9$}$^-$ ligands suggests an idealized core structure (Fig. 3.5) in which all the metals attain a six-coordinate state.

Factors such as size of M, size, valency, and preferred coordination state of M′, and bulk of the alkoxo group(s) determine the requirement for additional L...L ligands (as a bidentate neutral ligand) or zero/one/two monovalent unidentate ligand(s) on M′.

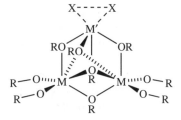

Figure 3.5 Schematic representation of the structural pattern of M'{M₂(OR)₉} derivatives (where M is a tetravalent metal and M' is a mono- or divalent metal).

In view of the availability of actual X-ray structure elucidations (Chapter 4), this brief description will be focussed mainly on the information available from other physico-chemical characteristics. For example, the ^{1}H and ^{13}C NMR spectra of KZr$_2$(OPr^i)$_9$ (Fig. 3.5, with R = Pr^i; M = Zr and M' = K) in C$_6$D$_6$ at 25°C are in complete accord[64] with the C$_{2v}$ symmetry of the suggested structure: ^{1}H: δ 1.2–1.6 (doublets, intensity ratio = 1:2:2:2:2); ^{13}C{^{1}H}: δ 68–71 (CH, intensity ratio = 4:1:2:2) and δ 27–28.5 (CH$_3$, intensity ratio = 2:2:2:1:2).

In fact, this type of structure for KZr$_2$(OPr^i)$_9$ was first[41] arrived at by the replaceability of a maximum of 5 out of 9 OPr^i groups by bulkier OBu^t groups when KZr$_2$(OPr^i)$_9$ was treated with excess of tertiary butanol, fractionating out the liberated isopropanol azeotropically with benzene:

$$\text{KZr}_2(\text{OPr}^i)_9 + 5\text{Bu}^t\text{OH} \longrightarrow \text{KZr}_2(\text{OPr}^i)_4(\text{OBu}^t)_5 + 5\text{Pr}^i\text{OH} \uparrow \qquad (3.122)$$
$$\textbf{(A)}$$

The formation of the final product **(A)** (characterized by careful analysis of metal, isopropoxide, and tertiary butoxide contents separately) was explained on the basis that the replacement of four terminal isopropoxide groups on two zirconium atoms and a bridging OBu^t group between them sterically hindered further coordination with another tertiary butyl alcohol molecule. In view of the repeated confirmation of the plausible structure(s) in a large number of such cases (Section 1), a brief account of the revealing parameters will be included particularly in cases for which X-ray structural data are not yet available.

In fact, the formation of the final product KZr$_2$(OPr^i)$_4$(OBu^t)$_5$ has been further confirmed by the instability of KZr$_2$(OBu^t)$_9$, which could be isolated[41,182] by the interaction of KOBu^t and Zr(OBu^t)$_4$ in a 1:2 molar reaction; the reaction did result in a quantitative yield of recrystallizable white powder corresponding in analysis to KZr$_2$(OBu^t)$_9$, which was, however, found to disproportionate (Eq. 3.123) into the more volatile Zr(OBu^t)$_4$ (confirmed[182] by thermogravimetric analysis) leaving behind crystalline [KZr(OBu^t)$_5$]$_n$:

$$\text{KZr}_2(\text{OBu}^t)_9 \xrightarrow[\text{vacuum}]{\Delta} \frac{1}{n}[\text{KZr}(\text{OBu}^t)_5]_n + \text{Zr}(\text{OBu}^t)_4 \uparrow \qquad (3.123)$$

The product $[KZr(OBu^t)_5]_n$ was actually isolated[41,182] in 1:1 molar reaction of $KOBu^t$ and $Zr(OBu^t)_4$ and characterized[182] by X-ray crystallography.

In view of the larger size of the central metal atoms, the stability of $[KU_2(OBu^t)_9]$,[217] $[NaCe_2(OBu^t)_9]$,[218] and $[NaTh_2(OBu^t)_9]$[219] can be easily understood.

In contrast to the stability of $KZr_2(OPr^i)_9$,[41] the comparative instability of similarly synthesized $KTi_2(OPr^i)_9$ was explained on the basis of the difficulty of the smaller titanium (0.64 Å) atom to accommodate six OPr^i groups around itself like zirconium (0.80 Å). However, Veith[164] has been able to characterize X-ray crystallographically the identity of products such as $KTi_2(OPr^i)_9$ (low temperature technique) and of $BaTi_3(OPr^i)_{14}$ as $\{Ti(OPr^i)_5\}Ba\{Ti_2(OPr^i)_9\}$;[140] the latter finding is of special interest as it might finally throw light on the nature of products like $K_2Zr_3(OPr^i)_{14}$[41] and may lead to the isolation of similar derivatives, $MTi_3(OPr^i)_{14}$, of other divalent metals like Ca, Sr, Zn, Cd, Ni, Co, and Cu.

The stability of monomeric $[ClCu\{Zr_2(OPr^i)_9\}]$[176] in contrast to the dimeric $[Cd\{Zr_2(OPr^i)_9(\mu\text{-}Cl)\}]_2$[135] again illustrates the effect of the size of central metal atom. Further, the monomeric nature of isostructural $Cd\{M_2(OPr^i)_9\}I$ derivatives (M = Zr, Hf, Ti, Sn(IV)) can also be ascribed to the large size of I[160-163] compared to that of Cl.[135] Although $[ClSn\{Zr_2(OPr^i)_9\}]_2$ is also dimeric like $[ClCd\{Zr_2(OPr^i)_9\}]_2$, the $\{Zr_2(OPr^i)_9\}^-$ ligand binds Sn(II) in a bidentate manner[141,142] in place of the usual tetradentate ligating mode[188] in the cadmium analogue; this difference has been ascribed[141] to the presence of a lone pair of electrons on tin(II). The replacement of Cl in $[ClSnM_2(OPr^i)_9]_2$ by C_5H_5 (cyclopentadienyl) by interaction with NaC_5H_5 led[142] to monomeric derivatives $[(C_5H_5)SnM_2(OPr^i)_9]$, as confirmed by ^{119}Sn NMR spectra and X-ray crystallography.

An eight-coordinated barium derivative, $[Ba\{Zr_2(OPr^i)_9\}_2]$[131] has been crystallographically characterized; the structure consists of a "bow-tie" or "spiro" Zr_2BaZr_2 unit wherein barium is eight-coordinated by two face-shared bi-octahedral $\{Zr_2(OPr^i)_9\}$ units.

3.4.2.2 Tri- and bidentate modes of coordination

The small lithium ion (0.78 Å) interacts with only three isopropoxo groups[131] of a $\{Zr_2(OPr^i)_9\}^-$ unit in contrast to the four utilized by its larger congeners like Na^+ (0.98 Å) and K^+ (1.33 Å). Lithium appears to be too small to span the distance between isopropoxo groups on both zirconium centres (see Chapter 4). The fourth coordination site on lithium is occupied by an isopropanol molecule, which is hydrogen bonded to an oxygen of a terminal isopropoxo group on zirconium.

The structure of $[Sn\{Zr_2(OPr^i)_9(\mu\text{-}I)\}_2]$[161] also shows a $\{Zr_2(OPr^i)_9\}^-$ unit interacting with Sn(II) in a tridentate fashion, which appears to result from the presence of the stereochemically active lone pair of electrons on tin(II).

The electronic spectra of $[Co\{Zr_2(OPr^i)_9\}_2]$[115] and $[Cu\{Zr_2(OPr^i)_9\}_2]$[122] show the central transition metals Co and Cu to be hexacoordinate, indicating a tridentate bonding mode of the $\{Zr_2(OPr^i)_9\}^-$ ligand.

Although the size of lead(II) is larger (1.32 Å) than that of tin(II) (0.93 Å), $\{Zr_2(OPr^i)_9\}^-$ appears to bind the former in a bidentate (η^2) manner, as revealed by the X-ray structure of $[Pb\{Zr_2(OPr^i)_9\}(\mu\text{-}OPr^i)]_2$ with a "serpentine" rather than "close" pattern[181] as exhibited by the $[Sn(\mu\text{-}OPr^i)_3Zr(OPr^i)_3]$ derivative in its crystal structure. In fact, Caulton et al.[82] have shown that the reactions of $Zr(OBu^t)_4$ and $M(OBu^t)_2$ (M = Sn, Pb) yielded

$$M \overset{OBu^t}{\underset{OBu^t}{\overset{|}{\leftarrow}}} OBu^t \Rightarrow Zr \overset{OBu^t}{\underset{OBu^t}{\overset{|}{\leftarrow}}} OBu^t$$

in a straightforward manner similar to $Sn(\mu_3\text{-}OPr^i)_3Zr(OPr^i)_3$, but the corresponding reaction of $Pb(OPr^i)_2$ with $Zr(OPr^i)_4$ (Eq. 3.124) yields an equimolar mixture of $Pb_4Zr_2(OPr^i)_{16}$ and $Pb_2Zr_4(OPr^i)_{20}$, which have been shown by X-ray crystallography (cf. Chapter 4) to adopt "serpentine" structures:

$$6Pb(OPr^i)_2 + 6Zr(OPr^i)_4 \longrightarrow Pb_4Zr_2(OPr^i)_{16} + Pb_2Zr_4(OPr^i)_{20} \qquad (3.124)$$

3.4.3 Pentaalkoxometallate Derivatives

The role of steric effects in determining the stoichiometry and structure of a heterometallic alkoxide can be further illustrated by the comparative stability of $M'M_2(OR)_9$ and $M'M(OR)_5$ systems.

For isopropoxo ligands, a number of smaller tetravalent metals: Sn^{4+} $(0.74\,\text{Å})$,[221] Hf^{4+} $(0.84\,\text{Å})$,[76] and Zr^{4+} $(0.87\,\text{Å})$[2,41,86,126,224] form stable $M'\{M_2(OPr^i)_9\}$, derivatives, whereas with $R = Bu^t$, stable products are obtained with Ce^{4+} $(0.94\,\text{Å})$,[218] U^{4+} $(0.97\,\text{Å})$,[217] and Th^{4+} $(0.99\,\text{Å})$.[219] Titanium $(0.64\,\text{Å})$ was earlier considered to be too small to accommodate six isopropoxy groups around itself, as required for a structure of $Ti_2(OPr^i)_9$ type, but Veith[164] and Baxter et al.[223] have synthesized a number of derivatives of this type for titanium also.

Amongst the alkali metal derivatives of the type $MM'(OR)_5$, the formation of a stable volatile dimeric $\{KZr(OBu^t)_5\}_2$ derivative has been confirmed.[182] In the case of titanium, $\{LiTi(OPr^i)_5\}_2$ has been shown[60] to be dimeric in solution as well as in the solid state, whereas the analogous Na and K derivatives are polymeric, with linear and nonlinear chains respectively as revealed[225] by their X-ray structures.

3.4.4 Hexaalkoxometallate(IV) Derivatives

Although it has not been possible to confirm the structures of $Pb(\mu\text{-}OBu^t)_3Zr(OBu^t)_3$ and $Sn(\mu\text{-}OR)_3Zr(OR)_3$ $(R = Bu^t, Pr^i)$ with hexa-alkoxozirconium ligands by X-ray crystallography, the structure is clearly indicated by ^1H, ^{13}C, ^{207}Pb, and ^{119}Sn NMR data and other physico-chemical characteristics.[82] An interesting derivative of this type, $Sn^{II}Sn^{IV}(OPr^i)_6$ has been obtained[226] by the reaction of $Sn\{N(SiMe_3)_2\}_2$ with $Sn(OPr^i)_4.Pr^iOH$ in isopropyl alcohol; its ^1H spectrum at $-20°C$ exhibits two equal-intensity peaks in both the methyl and the methine regions, as expected from the structure: $Sn^{II}(\mu\text{-}OPr^i)_3Sn^{IV}(OPr^i)_3$. Further, its $^{119}Sn\{^1H\}$ NMR spectrum in $C_6D_5CD_3$ shows two peaks at δ, -344 and -584, consistent with two tin atoms in different coordination environments.

The hexaalkoxometallate type of bonding has also been confirmed by actual crystal structures in a number of other derivatives: $[Tl_2Zr\{OCH(CF_3)_2\}_6]$,[79] $[Na_2Zr\{OCH(CF_3)_2\}_6.C_6H_6]$,[227] $[(DME)_2Na_2Ce(OBu^t)_6]$,[228] $[\{(COD)Rh\}_2\{Sn(OEt)_6\}]$,[229] $[\{(COD)Rh\}_2\{Ti(OPr^n)_6\}]$,[230] and $[Tl_2Sn(OEt)_6]$.[55]

The structure of $[Tl_2Zr\{OCH(CF_3)_2\}_6]$ can be represented by $Tl(\mu\text{-}OR_f)_3Zr(\mu_2\text{-}OR_f)_3Tl$ $(R_f = HC(CF_3)_2)$. The derivative $[NaZr\{OCH(CF_3)_2\}_6.C_6H_6]$ is structurally similar, with the difference that lone pair sites on sodium are occupied by benzene ligands along with the presence of four instead of six doubly bridged fluoroalkoxo groups. These crystallographic studies have also indicated evidence of secondary heterometal (Tl or Na)...F interactions.[227] In view of the highly interesting findings with $\{Zr_2(OPr^i)_9\}^-$ ligands, it would be worthwhile to investigate the ligating behaviour of the fluorinated analogue, $\{Zr_2(OR_f)_9\}^-$ further.

The derivatives $[\{(COD)Rh\}_2\{Sn(OEt)_6\}]^{229}$ and $[\{(COD)Ph\}_2\{Ti(OPr^n)_6\}]^{230}$ are also similar, with the difference that the two terminally bonded M–O–C groups on titanium are almost linear (160°), whereas they are more bent (125.5°) in the case of tin; the acute bond angle appears to be a general feature in main-group metal chemistry.

The structure of $[\{(dme)Na\}_2Ce(\mu_3\text{-}OBu^t)(\mu\text{-}OBu^t)_3(OBu^t)_2]^{228}$ is interesting in that $(dme)_2Na^+$ electrophiles occupy adjacent rather than *trans*-faces, which would have given a more linear Na/Ce/Na geometry. By contrast, $Tl_2Sn(OEt)_6^{55}$ exists as a one-dimensional polymer in the solid state, owing to the lone pair electrons on thallium.

3.5 Heterometallic Alkoxides with Other Chelating Ligands

Previous sections have dealt with heterobimetallic alkoxides with the four most commonly used ligands: $\{Al(OR)_4\}^-$, $\{Nb(OR)_6\}^-$, $\{Ta(OR)_6\}^-$, and various types of alkoxometallates(IV). A brief description is presented in this section of a few alkoxometallate(II) and other ligands.

A detailed study has been made in the research school of Veith[54] on a variety of derivatives with $\{M(OBu^t)_3\}^-$ type ligands (M = a divalent metal like Ge, Sn, and Pb), involving reactions of the following types.

$$M'OBu^t + M(OBu^t)_2 \longrightarrow \frac{1}{n}\{M'M(OBu^t)_3\}_n \qquad (3.125)$$

where M = Ge, Sn, Pb; M' = Li, Na, K, Rb, Cs.

$$2M'Br + Na_2M_2(OBu^t)_6 \longrightarrow 2[M'M(OBu^t)_3] + 2NaBr \qquad (3.126)$$

where M = Ge, Sn, Pb; M' = In, Tl.

$$M'Cl_2 + Na_2M_2(OBu^t)_6 \longrightarrow MM'_2(OBu^t)_6 + 2NaCl \qquad (3.127)$$

where M' = Mg, Mn, Zn and M = Ge or M' = Mn, Zn and M = Pb.

These are interesting "cage" compounds.

Another type of compound (3-IV) has also been obtained,[54,163,231] with the formula $[M'_2M_2(OBu^t)_8]$ (M = Ge, Sn, Pb; M' = Mg, Ni, Cr, Co, Mn).

(3-IV)

Heterometallic alkoxides of subvalent main group metals (ns^2 configuration) located at the apices of one or two (fused) trigonal bipyramids are able to function as excellent electron donors to various transition metal carbonyls as illustrated[163] in two typical examples (3-V) and (3-VI)

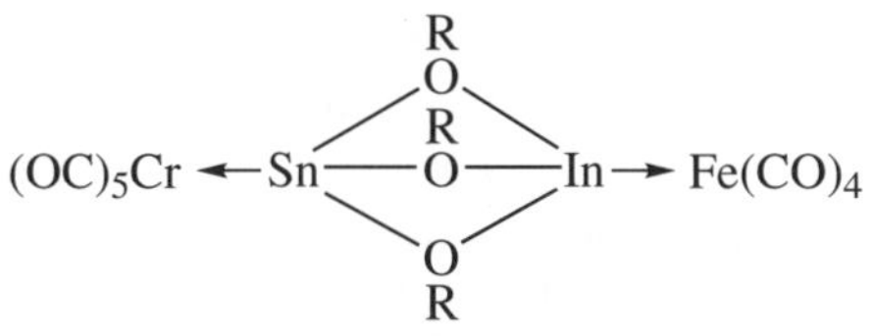

(3-V)

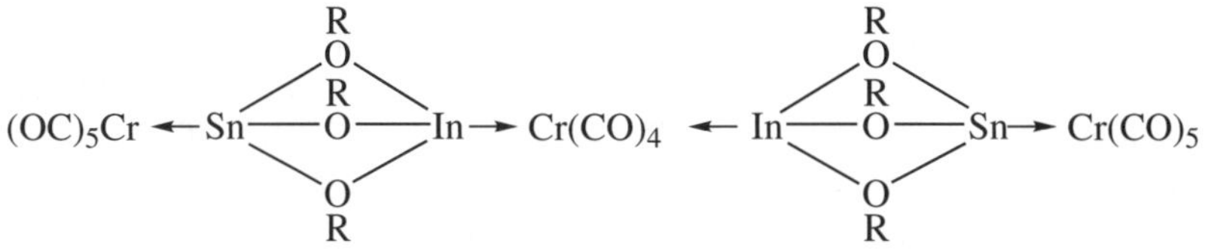

(3-VI)

In these carbonyl adducts the reactivity sequences are $In^I \gg Ge^{II} > Sn^{II} > Tl^I \sim Pb^{II}$ and $Ni(CO)_4 > Fe(CO)_5 > Mo(CO)_6 > Cr(CO)_6 > W(CO)_6$.

Purdy and George[127] obtained two interesting alkali (Na, K) tritertiarybutoxozincates by the 1:3 molar reaction of $ZnCl_2$ with alkali tertiary butoxide:

$$ZnCl_2 + 3M(OCMe_3) \longrightarrow \tfrac{1}{2}[\{MZn(OCMe_3)_3\}_2] + 2MCl \qquad (3.128)$$

Both these zincates are soluble in hydrocarbon solvents and are volatile (Na and K derivatives start to sublime at 50°C and 90°C, respectively, under vacuum).Their structures are centrosymmetric, with alkali metal atoms positioned above and below the Zn_2O_2 plane.

For other chelating ligands, a few examples are mentioned below by the formulae of the derivatives: $KSb(OBu^t)_4$, $\{KBi(OBu^t)_4\}_n$, $K_2Sb(OBu^t)_5 \cdot C_4H_8O_2,$[59] $M_2Sb_4(OEt)_{16}$ (M = Mn, Ni),[111,112] $MSb(OR)_6$ (M = Li, Na, K; R = Me, Pr^n, Pr^i, Bu^n, Bu^t),[232] and $YBa_2(OBu^t)_7(Bu^tOH)$.[233]

3.6   Higher Heterometal Alkoxides and Single-source Precursors

Synthesis (by metathetic reactions) of a large number of hetero-tri and tetrametallic alkoxides of a number of metals has been reported since 1985 (Section 2.3 and Table 3.4(c)) mainly from the research school of Mehrotra[7−9,17,101,149−159] and the potentiality of their applications as "single-source" precursors[12,20,22,24,25,43] for high-purity heterometallic-oxide ceramic materials has been emphasized since 1988.

However, the lingering scepticism about these novel and stable (as depicted by their volatility) coordination systems has been allayed only since 1996 when the crystal structure of the first heterotrimetallic derivative $[\{Cd(OPr^i)_3\}Ba\{Zr_2(OPr^i)_9\}]$

was determined,[136] which not only confirmed its existence, but also revealed a spontaneous change in its composition from the expected (on the basis of the synthetic route) $\{Ba(OPr^i)_3\}Cd\{Zr_2(OPr^i)_9\}$ to the thermodyamically more stable form $\{Cd(OPr_3^i)\}Ba\{Zr_2(OPr^i)_9\}$. Mehrotra[234] immediately drew attention to this achievement in the field of heterometal alkoxides. This has been followed by crystal structure characterization of several other similar derivatives, such as $[\{Cd(OPr^i)_3\}Ba\{M_2(OPr^i)_9\}]$ (M = Ti, Hf)[160] and $[\{Cd(OPr^i)_3\}Sr\{Zr_2(OPr^i)_9\}]$.[235]

As heteroleptic halide heterometal alkoxides remain as starting materials for the higher heterometallic derivatives as discussed in Section 2.3, these are receiving renewed attention: e.g. by the synthesis of $[I_2Sn\{Al(OPr^i)_4\}_2]$, $[I_2Sn\{Ti(OPr^i)_5\}_2]$, $[I_3Sn\{Zr(OPr^i)_5\}(Pr^iOH)]$, and $[I_2Sn\{Mo(C_5H_5)(CO)_3\}_2]$.[236] Work on the isolation of this last interesting cyclopentadienyl framework, had been preceded by the synthesis[142] of $[(C_5H_5)Sn\{Zr_2(OPr^i)_9\}]$ from $[ClSn\{Zr_2(OPr^i)_9\}]_2$, which had been reported earlier.[141]

It may be relevant to recall that Veith et al.[83] have evinced interest in attempting to tune metal stoichiometry in heterometal alkoxide systems and have been successful in the synthesis of $[BaZr(OH)(OPr^i)_5(Pr^iOH)_3]$ and $[BaZr(OBu^t)_6(thf)_2]$, which they have described as the first structurally characterized molecular precursor leading to the important ceramic material $BaZrO_3$.

It might be appropriate to conclude this brief account of heterometallic alkoxide systems by mentioning $[(Bu^tO)_8Li_4K_4]$, prepared by Clegg et al.[237] in their continuing efforts to develop hetero-s-block metal chemistry.

4   FUTURE OUTLOOK

Although it is gratifying that the X-ray structural elucidation of several heterometallic alkoxides has in general confirmed the plausible frameworks suggested for them on the basis of simpler physico-chemical techniques, yet the emergence of an interesting variety of actual structural patterns is playing a vital role in a much better understanding of the factors responsible for the extraordinary stability of this fascinating group of heterometallic coordination systems.

In addition to these academic developments, the synthesis of single-source heteropolymetallic alkoxide precursors for high-purity mixed metal–oxide ceramic materials, including superconductors, is becoming a possibility,–for which more penetrating investigations on novel alkoxometallate systems and their hydrolytic and pyrolytic behaviour are required.

REFERENCES

1. H. Meerwein and T. Bersin, *Ann.*, **476**, 113 (1929).
2. W.G. Bartley and W. Wardlaw, *J. Chem. Soc.*, 421 (1958); *see also* W.G. Bartley, *Ph.D. Thesis* London University (1953).
3. R.C. Mehrotra and A. Mehrotra, *Inorg. Chim. Acta. Rev.*, **5**, 127 (1971).
4. D.C. Bradley, R.C. Mehrotra, and D.P. Gaur, *Metal Alkoxides*, Academic Press, London (1978).

5. R.C. Mehrotra, *Adv. Inorg. Chem. Radiochem.*, **26**, 269 (1983).
6. M.H. Chisholm and I.P. Rothwell, in *Comprehensive Coordination Chemistry* (G. Wilkinson, J.A. McCleverty, and R.D. Gillard, eds), Vol. 2, 335–364, Pergamon Press, London (1988).
7. R.C. Mehrotra and M. Aggarwal, *Polyhedron*, **4**, 845, 1141 (1985).
8. R.K. Dubey, A. Singh, and R.C. Mehrotra, *J. Organometal. Chem.*, (Eaborn Issue), **341**, 869 (1988).
9. R.K. Dubey, A. Shah, A. Singh, and R.C. Mehrotra, *Recl. Trav. Chim. Pays-Bas* (Van der Kirk Issue), **107**, 237 (1988).
10. R.C. Mehrotra, 'Invited Lectures at Kyoto (1987 & 1990)', *J. Non-cryst. Solids*, **100**, 1 (1988); **121**, 1 (1990).
11. H. Dislich, *Angew. Chem., Int. Ed. Engl.*, **10**, 367 (1971).
12. R.C. Mehrotra, *Mat. Res. Soc. Symp. Proc.*, **121**, 81 (1988).
13. D.C. Bradley, *Chem. Rev.*, **89**, 1317 (1989).
14. R.C. Mehrotra and A. Singh, *Proc. Ind. Nat. Sci. Acad.*, **55A**, 347 (1989).
15. R.C. Mehrotra, *Chemtracts*, **2**, 389 (1990).
16. K.G. Caulton and L.G. Hubert-Pfalzgraf, *Chem. Rev.*, **90**, 969 (1990).
17. R.C. Mehrotra, A. Singh, and U.M. Tripathi, *Chem. Rev.*, **91**, 1287 (1991).
18. R.C. Mehrotra, *Nat. Acad. Sci. Letters*, **14**, 153 (1991).
19. R.C. Mehrotra and A.K. Rai, *Polyhedron*, **10**, 1967 (1991).
20. R.C. Mehrotra, in *Chemistry, Spectroscopy and Applications of Sol–Gel Glasses* (R. Reisfeld and C.K. Jorgensen, eds), Structure and Bonding, Vol. 77, 1–36, Springer-Verlag, Berlin (1992).
21. R.C. Mehrotra, *J. Non-cryst. Solids*, **145**, 1 (1992).
22. R.C. Mehrotra, *Indian. J. Chem.*, **31A**, 492 (1992).
23. R.C. Mehrotra, *Nat. Acad. Sci. Letters*, **16**, 41 (1993).
24. R.C. Mehrotra, *J. Sol–Gel Sci. Technol.*, **2**, 1 (1994).
25. R.C. Mehrotra, in *Sol–Gel Processing and Applications* (Y. Attia, ed.), 41–60, Plenum Publications, New York (1994); R.C. Mehrotra and A. Singh, in *Sol–Gel Processing of Advanced Ceramics* (F.D. Gyanam, ed.), 11–36, Oxford & IBH Publishing, Delhi (1996).
26. D.C. Bradley, *Polyhedron*, **13**, 1111 (1994).
27. L.G. Hubert-Pfalzgraf, *Polyhedron*, **13**, 1181 (1994).
28. R.C. Mehrotra, A. Singh, and S. Sogani, *Chem. Soc. Rev.*, **23**, 215 (1994).
29. R.C. Mehrotra, A. Singh, and S. Sogani, *Chem. Rev.*, **94**, 1643 (1994).
30. H. Albers, M. Deutsch, W. Krastinat, and H. Von Asten, *Chem. Ber.*, **85**, 267 (1952).
31. R.G. Jones, E. Bindschadler, D. Blume, G. Karmas, G.A. Martin Jr, J.R. Thirtle, and H. Gilman, *J. Am. Chem. Soc.*, **78**, 6027 (1956).
32. R. Gut, *Helv. Chim. Acta*, **47**, 2262 (1964).
33. C.J. Ludman and T.C. Waddington, *J. Chem. Soc. A*, 1816 (1966).
34. R. Von Schloder and H. Protzer, *Z. Anorg. Allgem. Chem.*, **340**, 23 (1965).
35. R.C. Mehrotra and P.N. Kapoor, *Coord. Chem. Rev.*, **14**, 1 (1974).
36. R.C. Mehrotra, *Proc. Ind. Nat. Sci. Acad.*, **42**, 1 (1976).
37. J. Sassmannshausen, R. Riedel, K.B. Pflanz, and H. Chimiel, *Z. Naturforsch*, **48B**, 7 (1993).
38. J.A. Meese-Marktscheffel, R. Weimann, H. Schumann, and J.W. Gilje, *Inorg. Chem.*, **32**, 5894 (1993).
39. U.M. Tripathi, A. Singh, R.C. Mehrotra, S.C. Goel, M.Y. Chiang, and W.E. Buhro, *J. Chem. Soc., Chem. Commun.*, 152 (1992).
40. M. Wijk, R. Norrestam, M. Nygren, and G. Westin, *Inorg. Chem.*, **35**, 1077 (1996).
41. R.C. Mehrotra and M.M. Agarwal, *J. Chem. Soc.*, 1026 (1967).
42. R.C. Mehrotra and J.V. Singh, *Can. J. Chem.*, **62**, 1003 (1984).
43. R.C. Mehrotra, A. Singh, M. Bhagat, and J. Godhwani, *J. Sol–Gel Sci. Technol.*, **13**, 45 (1998).
44. K. Jones, T.J. Davies, H.G. Emblem, and P. Parkes, *Mat. Res. Soc. Symp. Proc.*, **73**, 111 (1986).
45. R. Kuhlman, B.A. Vaartstra, W.E. Streib, J.C. Huffman, and K.G. Caulton, *Inorg. Chem.*, **32**, 1272 (1993).

46. H. Dislich, *J. Non-cryst. Solids*, **73**, 599 (1985).
47. R.C. Mehrotra and A. Singh, *Prog. Inorg. Chem.*, **46**, 239 (1997).
48. M. Arora, *Ph.D. Thesis*, University of Rajasthan, Jaipur, India (1968).
49. M. Arora and R.C. Mehrotra, *Indian J. Chem.*, **7A**, 399 (1969).
50. R.C. Mehrotra and M. Arora, *Z. Anorg. Allgem. Chem.*, **370**, 300 (1969).
51. A.S. Vavilkin and Z.T. Dmitrieva, *Zh. Obshch. Khim.*, **59**, 1970 (1989).
52. M.M. Agrawal, *Ph.D. Thesis*, University of Rajasthan, Jaipur, India (1968).
53. N.C. Jain, A.K. Rai, and R.C. Mehrotra, *Proc. Indian Acad. Sci.*, **84A**, 98 (1976).
54. M. Veith, *Chem. Rev.*, **90**, 3 (1990).
55. M.J. Hampden-Smith, D.E. Smith, and E.N. Duesler, *Inorg. Chem.*, **28**, 3399 (1989).
56. M. Veith and R. Roesler, *Z. Naturforsch.*, **41B**, 1071 (1986).
57. M. Veith and M. Reimers, *Chem. Ber.*, **123**, 1941 (1990).
58. T. Athar, R. Bohra, and R.C. Mehrotra, *Main Group Met. Chem.*, **10**, 399 (1987).
59. M. Veith, E.-Chul Yu, and V. Huch, *Chem. Eur. J.*, **1**, 26 (1995).
60. M.J. Hampden-Smith, D.S. Williams, and A.L. Rheinhold, *Inorg. Chem.*, **29**, 4076 (1990).
61 F.A. Cotton, M.P. Diebold, and W.J. Roth, *Inorg. Chem.*, **27**, 3596 (1988).
62. C.K. Sharma, S. Goel, and R.C. Mehrotra, *Indian J. Chem.*, **14A**, 878 (1976).
63. B.A. Vaartstra, J.C. Huffman, P.S. Gradeff, L.G. Hubert-Pfalzgraf, J.C. Daran, S. Parraud, K. Yunlu, and K.G. Caulton, *Inorg. Chem.*, **29**, 3126 (1990).
64. B.A. Vaartstra, W.E. Streib, and K.G. Caulton, *J. Am. Chem. Soc.*, **112**, 8593 (1990).
65. R.C. Mehrotra, M.M. Agrawal, and P.N. Kapoor, *J. Chem. Soc. A*, 2673 (1968).
66. A. Nazeri-Eshgi, A.X. Kuang, and J.D. Mackenzie, *J. Mater. Sci.*, **25**, 3333 (1990).
67. S.C. Goel, J.A. Hollingsworth, A.M. Beatty, K.D. Robinson, and W.E. Buhro, *Polyhedron*, **17**, 781 (1998).
68. A.P. Purdy and C.F. George, *Inorg. Chem.*, **33**, 761 (1994).
69. A. Mehrotra and R.C. Mehrotra, *Inorg. Chem.*, **11**, 2170 (1972).
70. J.G. Reiss and L.G. Hubert-Pfalzgraf, *Inorg. Chem.*, **14**, 2854 (1975).
71. R.C. Mehrotra, M.M. Agrawal, and A. Mehrotra, *Synth. React. Inorg. Met.-Org. Chem.*, **3**, 181, 407 (1973).
72. R.C. Mehrotra, J.M. Batwara, and P.N. Kapoor, *Coord. Chem. Rev.*, **31**, 67 (1980).
73. A. Mehrotra, *Ph.D. Thesis*, University of Rajasthan, Jaipur, India (1971).
74. R.C. Mehrotra and M.M. Agrawal, unpublished results (1968).
75. S. Govil, P.N. Kapoor, and R.C. Mehrotra, *Inorg. Chim. Acta*, **15**, 43 (1975).
76. R.C. Mehrotra and A. Mehrotra, *J. Chem. Soc., Dalton Trans.*, 1203 (1972).
77. E.P. Turevskaya, D.V. Berdyev, N.Ya. Turova, Z.A. Starikova, A.I. Yanovsky, Yu.T. Struchkov, and A.I. Belokon, *Polyhedron*, **16**, 663 (1997).
78. M. Veith, D. Kafer, and V. Huch, *Angew. Chem., Int. Ed. Engl.*, **25**, 375 (1986).
79. J.A. Samuels, J.W. Zwanziger, E.B. Lobkovsky, and K.G. Caulton, *Inorg. Chem.*, **31**, 4046 (1992).
80. U. Benm, K. Lashgari. R. Norrestam, M. Nygren, and G. Westin, *Solid State Chem.*, **103**, 366 (1993).
81. M. Veith, S. Mathur, and V. Huch, *Inorg. Chem.*, **36**, 2391 (1997).
82. D.J. Teff, J.C. Huffman, and K.G. Caulton, *Inorg. Chem.*, **35**, 2981 (1996); (*cf.* R.C. Mehrotra, *Chemtracts*, **7**, 184 (1995)).
83. M. Veith, S. Mathur, V. Huch, and T. Decker, *Eur. J. Inorg. Chem.*, 1327 (1998).
84. A.P. Purdy and C.F. George, *Polyhedron*, **14**, 761 (1995).
85. R.C. Mehrotra, S. Goel, R.B. King, and K.C. Nainan, *Inorg. Chim. Acta*, **29**, 131 (1978).
86. S. Govil and R.C. Mehrotra, *Aust. J. Chem.*, **28**, 2125 (1975).
87. E.V. Turevskaya, N.Ya. Turova, A.V. Korolev, A.I. Yanovsky, and Yu.T. Struchkov, *Polyhedron*, **14**, 1531 (1995).
88. S. Govil, P.N. Kapoor, and R.C. Mehrotra, *J. Inorg. Nucl. Chem.*, **38**, 172 (1976).
89. S. Goel, A.B. Goel, and R.C. Mehrotra, *Synth. React. Inorg. Met.-Org. Chem.*, **6**, 251 (1976).
90. J.A. Meese-Marketscheffel, K. Fukuchi, M. Kido, G. Tachibana, C.M. Jensen, and J.W. Gilje, *Chem. Mater.*, **5**, 755 (1993).
91. M. Bhagat, A. Singh, and R.C. Mehrotra, *Main Group Met. Chem.*, **20**, 89 (1997) and unpublished results (1999).

92. R.C. Mehrotra and J. Singh, *J. Ind. Chem. Soc.*, **54**, 109 (1977).
93. R.C. Mehrotra, A.K. Rai, and N.C. Jain, *J. Inorg. Nucl. Chem.*, **40**, 349 (1977).
94. M. Aggrawal, C.K. Sharma, and R.C. Mehrotra, *Synth. React. Met.-Org. Chem.*, **13**, 571 (1983).
95. R.C. Mehrotra, N.C. Jain, and R.R. Goel, *Synth. React. Inorg. Met.-Org. Chem.*, **11**, 345 (1981).
96. W.J. Evans, T.J. Deming, J.M. Olofson, and J.W. Ziller, *Inorg. Chem.*, **28**, 4027 (1989).
97. R.C. Mehrotra, *Coord. Chem.*, (IUPAC), **21**, 113 (1981).
98. R.C. Mehrotra, *J. Ind. Chem. Soc.*, **59**, 715 (1982).
99. R.C. Mehrotra, *Proc. Ind. Nat. Sci. Acad.*, **52A**, 954 (1986).
100. J. Singh and R.C. Mehrotra, *Trans. Met. Chem.*, **9**, 148 (1984).
101. R.K. Dubey, A. Singh, and R.C. Mehrotra, *Indian J. Chem.*, **31A**, 156 (1992).
102. J. Singh and R.C. Mehrotra, *J. Coord. Chem.*, **13**, 273 (1984).
103. R.C. Mehrotra and A. Singh, *J. Ind. Chem. Soc.*, **70**, 885 (1993).
104. J. Singh and R.C. Mehrotra, *Z. Anorg. Allgem. Chem.*, **522**, 211 (1984).
105. J. Singh and R.C. Mehrotra, *Inorg. Chem.*, **23**, 1046 (1984).
106. A. Shah, J. Singh, A. Singh, and R.C. Mehrotra, *Indian J. Chem.*, **30A**, 1018 (1991).
107. E. Stumpp and U. Hillebrand, *Z. Naturforsch.*, **34B**, 262 (1979).
108. S. Sogani, A. Singh, and R.C. Mehrotra, *Main Group Met. Chem.*, **15**, 197 (1992).
109. R. Gupta, A. Singh, and R.C. Mehrotra, *Indian J. Chem.*, **30A**, 592 (1991).
110. G. Garg, R.K. Dubey, A. Singh, and R.C. Mehrotra, *Polyhedron*, **10**, 1733 (1991).
111. G. Westin and M. Nygren, *J. Mater. Chem.*, **4**, 1275 (1994).
112. U. Bemm, R. Norrestam, M. Nygren, and G. Westin, *Acta Cryst.*, **C51**, 1260 (1995).
113. S.K. Agarwal and R.C. Mehrotra, *Inorg. Chim. Acta*, **112**, 117 (1986).
114. A. Shah, A. Singh, and R.C. Mehrotra, *Indian J. Chem.*, **28A**, 392 (1989); **30A**, 1018 (1991).
115. R.K. Dubey, A. Singh, and R.C. Mehrotra, *Bull. Chem. Soc. Jpn.*, **61**, 983 (1988).
116. R. Jain, A.K. Rai, and R.C. Mehrotra, *Z. Naturforsch*, **40B**, 1371 (1985).
117. R.K. Dubey, A. Singh, and R.C. Mehrotra, *Trans. Met. Chem.*, **10**, 473 (1985).
118. A. Shah, A. Singh, and R.C. Mehrotra, *Indian J. Chem.*, **26A**, 485 (1987).
119. A. Shah, A. Singh, and R.C. Mehrotra, *Inorg. Chim. Acta*, **118**, 151 (1986).
120. R.K. Dubey, A. Singh, and R.C. Mehrotra, *Inorg. Chim. Acta*, **143**, 169 (1988).
121. R. Jain, A.K. Rai, and R.C. Mehrotra, *J. Inorg. Chem. (China)*, **3**, 96 (1987).
122. R.K. Dubey, A. Singh, and R.C. Mehrotra, *Polyhedron*, **6**, 427 (1987).
123. A.P. Purdy and C.F. George, *Polyhedron*, **17**, 4041 (1998).
124. A.P. Purdy, C.F. George, and J.H. Callahan, *Inorg. Chem.*, **30**, 2812 (1991).
125. A.P. Purdy and C.F. George, *Inorg. Chem.*, **30**, 1970 (1991).
126. B.A. Vaarstra, J.A. Samuels, E.H. Barash, J.D. Martin, W.E. Streib, C. Gasser and K.G. Caulton, *J. Organometal. Chem.*, **449**, 191 (1993).
127. A.P. Purdy and C.F. George, *Polyhedron*, **13**, 709 (1994).
128. S. Sogani, A. Singh, and R.C. Mehrotra, *Polyhedron*, **13**, 709 (1994).
129. S. Boulmaz, R. Papiernik, L.G. Hubert-Pfalzgraf, and J.C. Daran, *Eur. J. Solid State Inorg. Chem.*, **30**, 583 (1993).
130. A.B. Goel, E.C. Ashby, and R.C. Mehrotra, *Inorg. Chim. Acta*, **62**, 161 (1982).
131. B.A. Vaartstra, J.C. Huffman, W.E. Streib, and K.G. Caulton, *J. Chem. Soc., Chem. Commun.*, 1750 (1990); *Inorg. Chem.*, **30**, 3068 (1991).
132. E.R. Sigurdson and G. Wilkinson, *J. Chem. Soc., Dalton Trans.*, 812 (1977).
133. J.L. Stewart and R.A. Andersen, *J. Chem. Soc., Chem. Commun.*, 1846 (1987).
134. S.S. Al-Juaid, C. Eaborn, M.N.A. El-Kheli, P.B. Hitchcock, P.D. Lickiss, M.E. Molla, J.D. Smith, and J.A. Zora, *J. Chem. Soc., Dalton Trans.*, 447 (1989).
135. S. Sogani, A. Singh, R. Bohra, R.C. Mehrotra, and M. Noltemeyer, *J. Chem. Soc., Chem. Commun.*, 738 (1991).
136. M. Veith, S. Mathur, V. Huch, *J. Am. Chem. Soc.*, **118**, 903 (1996).
137. R.K. Dubey, A. Singh, and R.C. Mehrotra, *Inorg. Chim. Acta*, **143**, 169 (1988).
138. R. Gupta, *Ph.D. Thesis*, University of Rajasthan, Jaipur, India (1991).
139. R.C. Chhipa, A. Singh, and R.C. Mehrotra, *Indian J. Chem.*, **20**, 396 (1989).
140. S. Mathur and V. Huch, *Inorg. Chem.*, **36**, 2394 (1997).

141. S. Mathur, A. Singh, and R.C. Mehrotra, *Polyhedron*, **12**, 1073 (1993).
142. M. Veith, C. Mathur, S. Mathur, and V. Huch, *Organometallics*, **16**, 1292 (1997).
143. U.M. Tripathi, A. Singh, and R.C. Mehrotra, *Polyhedron*, **10**, 949 (1991).
144. J. Hvoslef, H. Hope, B.D. Murray, and P.P. Power, *J. Chem. Soc., Chem. Commun.*, 1438 (1983).
145. F.T. Edelmann, A. Steiner, D. Stalke, J.W. Gilje, S. Jagner, and M. Hakansson, *Polyhedron*, **13**, 539 (1991).
146. L. Malpezzi, U. Zucchini, L.T. Dall' Occo, *Inorg. Chim. Acta*, **180**, 245 (1991).
147. M. Bhagat, A. Singh, and R.C. Mehrotra, *Synth. React. Inorg. Met.-Org. Chem*, **28**, 997 (1998).
148. M. Bhagat, A. Singh and R.C. Mehrotra, *Indian J. Chem.*, **37A**, 820 (1998).
149. U.M. Tripathi, A. Singh, and R.C. Mehrotra, unpublished results.
150. R.C. Mehrotra, A. Singh, and S. Sogani, *Chem. Soc. Rev.*, **23**, 215 (1994).
151. S. Sogani, A. Singh, and R.C. Mehrotra, *Main Group Met. Chem.*, **15**, 197 (1992); *Indian J. Chem.*, **32A**, 345 (1993).
152. S. Mathur, A. Singh, and R.C. Mehrotra, *Polyhedron*, **11**, 341 (1992).
153. S. Mathur, *Ph.D. Thesis*, University of Rajasthan, Jaipur, India (1992).
154. R. Gupta, A. Singh, and R.C. Mehrotra, *New J. Chem.*, **15**, 65 (1991).
155. R. Gupta, A. Singh, and R.C. Mehrotra, *Indian J. Chem.*, **32A**, 310 (1993).
156. G. Garg, A. Singh, and R.C. Mehrotra, *Indian J. Chem.*, **30A**, 688 (1991).
157. G. Garg, A. Singh, and R.C. Mehrotra, *Indian J. Chem.*, **30A**, 866 (1991).
158. R.C. Chhipa, A. Singh, and R.C. Mehrotra, *Synth. React. Inorg. Met.-Org. Chem.*, **20**, 989 (1990).
159. S. Sogani, A. Singh, and R.C. Mehrotra, *Indian J. Chem.*, **34A**, 449 (1995).
160. M. Veith, S. Mathur, and V. Huch, *Inorg. Chem.*, **35**, 7295 (1996).
161. M. Veith, S. Mathur, and V. Huch, *J. Chem. Soc., Dalton Trans.*, 2485 (1996).
162. M. Veith, S. Mathur, C. Mathur, and V. Huch, *J. Chem. Soc., Dalton Trans.*, 2101 (1997).
163. M. Veith, S. Mathur, and C. Mathur, *Polyhedron*, **17**, 1005 (1998).
164. M. Veith, in *Advances in Metallo-organic Chemistry*, (R. Bohra, ed.), 1–18, RBSA Publishers, Jaipur, India (1999).
165. R.C. Mehrotra, A. Singh, M. Sharma, and J. Godhawani, in *Advances in Metallo-organic Chemistry* (R. Bohra, ed.), 238–252, RBSA Publishers, Jaipur, India (1999).
166. G.J. Gainsford, T. Kemmitt, C. Lensik, and N.B. Milestone, *Inorg. Chem.*, **34**, 746 (1995).
167. R.C. Mehrotra and A. Singh, *Phosphorus, Sulphur, Silicon*, **124–125**, 153 (1997).
168. J. Godhwani, A. Singh, and R.C. Mehrotra, *Synth. React. Inorg. Met.-Org. Chem.* (1999) in press.
169. M. Sharma, A. Singh, and R.C. Mehrotra, *Indian J. Chem.*, **38A**, 1209 (1999).
170. C.A. Zechmann, J.C. Huffman, K. Folting, and K.G. Caulton, *Inorg. Chem.*, **37**, 5886 (1998).
171. M. Sharma, A. Singh, and R.C. Mehrotra, *Polyhedron*, **19**, 27 (2000).
172. M. Sharma, *Ph.D. Thesis*, University of Rajasthan, Jaipur, India (1999).
173. M. Sharma, A. Singh, and R.C. Mehrotra, *Synth. React. Inorg. Met.-Org. Chem., Ph.D. Thesis*, University of Rajasthan, Jaipur (2000).
174. J. Godhwani, *Ph.D. Thesis*, University of Rajasthan, Jaipur, India (1999).
175. M.K. Sharma, A. Singh, and R.C. Mehrotra, *Synth. React. Inorg. Met.-Org. Chem.*, (1999).
176. V.G. Kessler, L.G. Hubert-Pfalzgraf, S. Halut, and J.C. Daran, *J. Chem. Soc., Chem. Commun.*, 705 (1994).
177. D.C. Bradley, *Prog. Inorg. Chem.*, **2**, 303 (1960).
178. D.C. Bradley, R.C. Mehrotra, and W. Wardlaw, *J. Chem. Soc.*, 2027 (1952).
179. R.C. Mehrotra, *J. Ind. Chem. Soc.*, **31**, 85 (1954).
180. D.C. Bradley, *Nature*, **182**, 1211 (1958).
181. D.J. Teff, J.C. Huffman, and K.G. Caulton, *Inorg. Chem.*, **34**, 2491 (1995).
182. D.J. Teff, J.C. Huffman, and K.G. Caulton, *Inorg. Chem.*, **33**, 6289 (1994).
183. S.J. Lippard, *J. Am. Chem. Soc.*, **88**, 4300 (1966).
184. M.Z. Gurevich, B.D. Stepin, L.N. Komissavova, N.E. Labedeva, and T.M. Sas, *Zh. Neorg. Khim.*, **16**, 93 (1971).

185. D. Bright, A.J. Kolombos, G.H.W. Milburn, R.S. Nyholm, and M.R. Truter, *J. Chem. Soc., Chem. Commun.*, 49 (1970).
186. B.A. Vaarstra, J.C. Huffman, P.S. Gradeff, L.G. Hubert-Pfalzgraf, J.C. Daran, S. Parraud, K. Yunlu, and K.G. Caulton, *Inorg. Chem.*, **29**, 3126 (1990).
187. W.J. Evans, M.A. Greci, M.A. Ansari, and J.W. Ziller, *J. Chem. Soc., Dalton Trans.*, 4503 (1997).
188. R.C. Mehrotra and A. Singh, *Polyhedron*, **17**, 689 (1998).
189. D.C. Bradley, *Adv. Chem. Ser.*, **23**, 10 (1959).
190. N.Ya. Turova, V.A. Kuzunov, A.I. Yanovsky, N.G. Bokii, Yu.T. Struchkov, and B.L. Turnopolsky, *J. Inorg. Nucl. Chem.*, **41**, 5 (1979).
191. K. Folting, W.E. Streib, K.G. Caulton, O. Poncelet, and L.G. Hubert-Pfalzgraf, *Polyhedron*, **10**, 1639 (1991).
192. R.C. Mehrotra and M.M. Agrawal, *J. Chem. Soc., Chem. Commun.*, 469 (1968).
193. D.C. Bradley, *Phil. Trans. Roy. Soc., London*, **A330**, 167 (1990).
194. N.Ya. Turova, E.P. Turevskaya, M.I. Yanovskaya, A.I. Yanovsky, V.G. Kessler, and D.E. Tcheboukov, *Polyhedron*, **17**, 899 (1998).
195. R.C. Mehrotra and R. Bohra, *Metal Carboxylates*, Academic Press, London (1983).
196. R.C. Mehrotra, D.P. Gaur, and R. Bohra, *Metal β-Diketonates and Allied Derivatives*, Academic Press, London (1978).
197. R.C. Mehrotra, in *Recent Trends in Inorganic Chemistry* (A. Chakravorty ed.), 256–275, Indian National Science Academy Publications (1986).
198. R.C. Mehrotra, in *Sol–Gel Science & Technology* (M.A. Aegerter, M. Jafelicci Jr, D.F. Souza, and E.D. Zanotto, eds), 17–60, World Scientific, Singapore (1989).
199. L.C. Francis, V.J. Ola, and D.A. Payne, *J. Mater. Sci.*, **25**, 5007 (1990).
200. E.P. Turevskaya, M.I. Yanovskaya, V.K. Lymar, and N.Ya. Turova, *Russ. J. Inorg. Chem.*, **38**, 563 (1993).
201. J. Rai and R.C. Mehrotra, *J. Non-Cryst. Solids*, **134**, 23 (1991).
202. J. Rai and R.C. Mehrotra, *Main Group Met. Chem.*, **15**, 209 (1992).
203. J. Rai and R.C. Mehrotra, *J. Non-Cryst. Solids*, **152**, 119 (1993).
204. R.C. Mehrotra, in *Facets of Coordination Chemistry* (B.V. Agarwal and K.N. Munshi, eds), World Scientific, Singapore (1993).
205. N.C. Jain, A.K. Rai, and R.C. Mehrotra, *Indian J. Chem.*, **14A**, 256 (1976).
206. J.A. Meese-Markscheffel, R.E. Cramer, and J.W. Gilje, *Polyhedron*, **13**, 1045 (1994).
207. R.C. Mehrotra and M. Aggrawal, *Polyhedron*, 4, 1141 (1985).
208. G. Garg, R.K. Dubey, A. Singh, and R.C. Mehrotra, *Polyhedron*, **10**, 1733 (1991).
209. R. Jain, A.K. Rai, and R.C. Mehrotra, *Polyhedron*, **5**, 1017 (1986).
210. R. Jain, A.K. Rai, and R.C. Mehrotra, *Inorg. Chim. Acta*, **126**, 99 (1987).
211. R.C. Mehrotra, M.M. Agrawal, and P.N. Kapoor, *J. Chem. Soc., A*, 2673 (1968).
212. E.P. Turevskaya, N.Ya. Turova, A.V. Koraler, A.I. Yanovsky, and Yu.T. Struchkov, *Polyhedron*, **14**, 1531 (1995).
213. D. Eichorst, D.A. Payne, S.R. Wilson, and K.E. Hasord, *Inorg. Chem.*, **29**, 1458 (1990).
214. D.E. Tchebukov, N.Ya. Turova, A.V. Korolev, and A.I. Belokon, *Russ. J. Inorg. Chem.*, **42**, 10 (1997).
215. S. Mathur, A. Singh, and R.C. Mehrotra, *Polyhedron*, **12**, 1073 (1993).
216. D.C. Bradley, E.V. Caldwell, and W. Wardlaw, *J. Chem Soc. A*, 4775 (1957).
217. F.A. Cotton, D.O. Marler, and W. Schwotzer, *Inorg. Chem.*, **23**, 4211 (1984).
218. W.J. Evans, T.J. Deming, J.M. Olofson, and Z.W. Ziller, *Inorg. Chem.*, **28**, 4027 (1989).
219. D.L. Clark and J.G. Watkin, *Inorg. Chem.*, **32**, 1766 (1993).
220. G. Garg, A. Singh, and R.C. Mehrotra, *Polyhedron*, **12**, 1399 (1993).
221. S. Sogani, A. Singh, and R.C. Mehrotra, *Polyhedron*, **14**, 621 (1995).
222. U.M. Tripathi, A. Singh, and R.C. Mehrotra, *Polyhedron*, **12**, 1947 (1993).
223. I. Baxter, S.R. Drake, M.B. Hursthouse, K.M. Abdul Malik, D.M.P. Mingos, J.C. Plakatouras, and D.J. Otway, *Polyhedron*, **17**, 625 (1998).
224. S. Sogani, A. Singh, and R.C. Mehrotra, *Main Group Met. Chem.*, **13**, 375 (1990).
225. T.J. Boyle, D.C. Bradley, M.J. Hampden-Smith, A. Patel, and J.W. Ziller, *Inorg. Chem.*, **34**, 5893 (1995).
226. D.J. Teff, C.D. Miner, D.V. Baxter, and K.G. Caulton, *Inorg. Chem.*, **37**, 2547 (1998).

227. J.A. Samuels, B. Labkorsky, W.E. Streib, K. Folting, J.C. Huffman, J.W. Zwanziger, and K.G. Caulton, *J. Am. Chem. Soc.*, **115**, 5093 (1993).
228. S.N. Putlin, E.V. Antipov, O. Chaissen, and M. Mariezo, *Nature*, **363**, 56 (1993).
229. T.A. Wark, E.A. Gulliver, and M.J. Hampden-Smith, *Inorg. Chem.*, **29**, 4360 (1990).
230. V.W. Day, T.A. Eberspacher, J. Haq, W.G. Klemperer, and B. Zhong, *Inorg. Chem.*, **34**, 3549 (1995).
231. M. Veith, *Phosphorus, Sulphur, Silicon*, **11**, 195 (1989).
232. T. Athar, R. Bohra, and R.C. Mehrotra, *Indian J. Chem.*, **28A**, 302 (1989).
233. B. Barup, W.E. Streib, and K.G. Caulton, *Chem. Ber.*, **129**, 1003 (1996).
234. R.C. Mehrotra, *Chemtracts*, **7**, 182 (1995).
235. M. Veith, S. Mathur, and V. Huch, *Phosphorus, Sulphur, Silicon*, **124–125**, 493 (1997).
236. M. Veith, S. Mathur, C. Mathur, and V. Huch, *Organometallics*, **17**, 1044 (1998).
237. W. Clegg, A.M. Drummond, S.T. Liddle, R.E. Mulvey, and A. Robertson, *J. Chem. Soc., Chem. Commun.*, 1559 (1999).

4

X-Ray Crystal Structures of Alkoxo Metal Compounds

1 INTRODUCTION

Since 1980 an increasing number of single-crystal X-ray structure determinations has been carried out and it is not feasible to describe more than a representative number in detail. A number of reviews have been published.[1-5] In fact only a few homoleptic metal alkoxides have been structurally characterized and the majority of structures determined involve metal complexes containing other ligands besides the alkoxo group. Therefore most of the structural data in this chapter are reported in tabular form.

The uninegative alkoxide ion RO⁻ contains a donor oxygen atom with three pairs of electrons available for covalent bonding to a metal ion.

In the nonbridging mode the angle RÔM should reflect the degree of π-donation in the bonding (4-I–4-III).

(4-I)	(4-II)	(4-III)

The extent of π-bonding depends on the availability of vacant π-type orbitals on the metal ion.

Alkoxide bridges are a characterisic feature of metal alkoxide structures and in principle are also affected by π-bonding supplementing the σ-bonds (4-IV–4-VII).

(4-IVa)	(4-IVb)	(4-V)
(4-VIa)	(4-VIb)	(4-VII)

Thus, whereas modes (4-IV) and (4-VI) would imply symmetrical bridges, (4-V) should be unsymmetrical. In addition to the all-σ-bonding μ_3-mode (4-VII) involving two-electron two-centred localized bonds, there is also the possibility of μ_4-bridging, etc. involving delocalized bonding.

A distinguishing feature of the alkoxo group as a ligand is its steric effect. By increasing the size and chain branching of the R-group the steric effect of RO$^-$ may be increased sufficiently to prevent alkoxo-bridging and thus result in a mononuclear metal alkoxide. This effect is obviously greater in the alkoxides of metals in higher valencies, e.g. M(OR)$_6$, M(OR)$_5$, M(OR)$_4$, owing to their greater intramolecular congestion. The electronic effect of the alkoxo ligand also plays a part in structure determination. An electronegative substituent X (e.g. CF$_3$) in XCH$_2$O– will reduce the electron density on the donor oxygen thereby weakening the bridging modes (4-IV–4-VII). At the same time the electrophilic nature of the metal ion will be enhanced, leading to the acquisition of a supplementary neutral ligand L (e.g. THF, py, etc.) and the formation of a mononuclear adduct [M(OCH$_2$X)$_x$L$_y$]. This is well illustrated by lanthanum which forms the trinuclear tertiary butoxide [La$_3$(μ_3-OBu^t)$_2$(μ-OBu^t)$_3$(OBu^t)$_4$(Bu^{t}OH)$_2$][6] but gives the mononuclear [La{OCMe(CF$_3$)$_2$}$_3$(thf)$_3$][7] with hexafluoro-tertiary butoxide, although the metal is six-coordinated in both molecules.

In an early attempt to rationalize the known structures of homoleptic metal alkoxides [M(OR)$_x$]$_n$ it was noted by Bradley[8] that the alkoxides which were soluble in common organic solvents usually exhibited low degrees of oligomerization (e.g. $n = 2, 3,$ or 4). It was proposed that in the less sterically demanding alkoxo groups (e.g. MeO, EtO, Pr^{n}O, etc.) the metal M utilized the bridging propensity of the alkoxo group to achieve the smallest oligomer compatible with all the metal atoms attaining their preferred coordination number. To minimize the complexity of a polynuclear metal alkoxide it is necessary to maximize the number of bridging alkoxo groups between adjacent pairs of metal atoms. For tetrahedral or octahedral metal coordination this means face-sharing rather than edge-sharing polyhedra. Thus, although a trivalent metal can achieve four-coordination (distorted tetrahedral) by forming an edge-bridged dimer [M$_2$(μ-OR)$_2$(OR)$_4$] (Fig. 4.1) and a pentavalent metal can achieve six-coordination (distorted octahedral) by also forming an edge-bridged dimer [M$_2$(μ-OR)$_2$(OR)$_8$] (Fig. 4.2), a tetravalent metal would require face-bridging octahedra in order to form the trimer [M$_3$(μ-OR)$_6$(OR)$_6$] (Fig. 4.3), the minimum sized oligomer for M(OR)$_4$. Molecular weight measurements on super-cooled titanium tetraethoxide in benzene solution indicated that a trimer was the predominant species, but the X-ray crystal structures of [Ti$_4$(OMe)$_{16}$], [Ti$_4$(OEt)$_{16}$], and [Ti$_4$(OEt)$_4$(OMe)$_{12}$] all revealed the edge-sharing tetranuclear structure [Ti$_4$(μ_3-OR)$_2$(μ-OR)$_4$(OR)$_{10}$][9] (Fig. 4.4). It was observed by Chisholm[10] that this particular structure is favoured by a number of species [M$_4$(OR)$_x$X$_y$L$_z$] (where X = anionic ligand; L = neutral ligand; $x + y + z = 16$). It thus appears that edge-sharing is favoured although

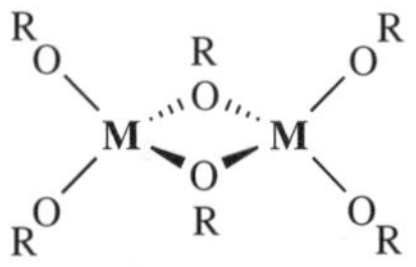

Figure 4.1 Structure of [M$_2$(μ-OR)$_2$(OR)$_4$].

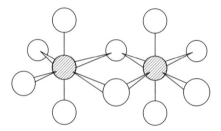

Figure 4.2 M_2O_{10} core structure of $[M_2(\mu\text{-}OR)_2(OR)_8]$.

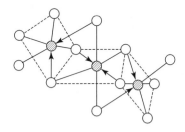

Figure 4.3 M_3O_{12} core structure of $[M_3(\mu\text{-}OR)_6(OR)_6]$.

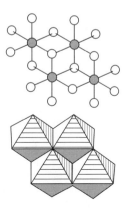

Figure 4.4 M_4O_{16} core structure of $[M_4(\mu_3\text{-}OR)_2(\mu\text{-}OR)_4(OR)_{10}]$.

Cotton *et al.*[11] found that face-sharing octahedra are involved in the mixed valency compound $[U_2(\mu\text{-}OBu^t)_3(OBu^t)_6]$ in contrast to uranium pentaisopropoxide which has edge-sharing octahedra $[U_2(\mu\text{-}OPr^i)_2(OPr^i)_8]$. The face-sharing bi-octahedral unit was also found in $[KU_2(OBu^t)_9]$,[11] $[NaCe_2(OBu^t)_9]$,[12] $[NaTh_2(OBu^t)_9]$[13] and $[KZr_2(OBu^t)_9]$.[14]

Another structural motif is present in the trinuclear species $[La_3(\mu_3\text{-}OBu^t)_2(\mu\text{-}OBu^t)_3(OBu^t)_4(Bu^tOH)_2]$[6] and $[Y_3(\mu_3\text{-}Cl)(\mu_3\text{-}OBu^t)(\mu\text{-}OBu^t)_3(OBu^t)_4(thf)_2]$[15] (Fig. 4.5). The same framework is present in $[W_3(\mu_3\text{-}X)(\mu_3\text{-}OR)(\mu\text{-}OR)_3(OR)_6]$ (where X = O or NH).[16]

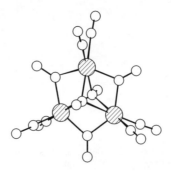

Figure 4.5 $M_3(\mu_3\text{-OC})_2(\mu\text{-OC})_3(OC)_6$ core structure of $[M_3(\mu_3\text{-OR})_2(\mu\text{-OR})_3(OR)_4(ROH)_2]$.

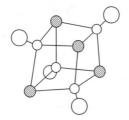

Figure 4.6 $M_4(\mu_3\text{-OC})_4$ core structure of $[M_4(\mu_3\text{-OR})_4]$.

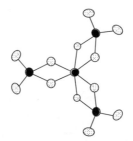

Figure 4.7 M_4O_{12} core structure of $[M_4(\mu\text{-OR})_6(OR)_6]$.

The cubane structure $[M_4(\mu_3\text{-OR})_4]$ (Fig. 4.6) is favoured by the alkali metal tertiary butoxides (M = K, Rb, Cs)[17] and thallium(I) alkoxides,[18] and also related species such as $[Me_4Zn_4(\mu\text{-OMe})_4]$.[19] Another common tetranuclear structure $[M_4(\mu\text{-OR})_6(OR)_6]$ (Fig. 4.7) involves one central octahedral metal bridged to three tetrahedral metals in a chiral configuration (D_3). Crystalline aluminium isopropoxide was the first example established to have this structural type.[20] Interestingly in the heterometallic complex $[ErAl_3(OPr^i)_{12}]$ the central position is occupied by the erbium ion in a distorted trigonal prismatic configuration.[21]

Finally we draw attention to the unique dimeric alkoxides of molybdenum and tungsten $M_2(OR)_6$ which are held together by metal–metal triple bonds without the

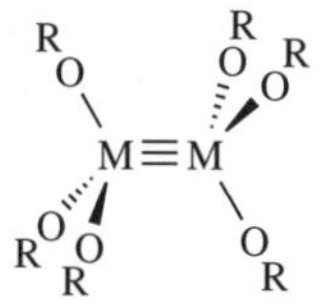

Figure 4.8 Structure of $[M_2(OR)_6]$.

support of bridging alkoxo ligands[22] (Fig. 4.8). Chisholm and co-workers have established a fascinating branch of alkoxide chemistry based on the chemical reactivity of these molecules.

Notwithstanding attempts to rationalize the many known structures of metal alkoxides it remains problematical at best to predict any detailed structures. It is not clear, for example, why tetranuclear titanium tetra-alkoxides adopt the structure shown in Fig. 4.4 in preference to the alternative cubane structure $[M_4(\mu_3\text{-}OR)_4(OR)_{12}]$ in which the facial configuration of three nonbridging alkoxo groups on each metal enjoys the enhanced π-bonding by virtue of being *trans* to the non-π-bonding μ_3-alkoxo groups.

2 STRUCTURES OF ALKOXO COMPOUNDS OF METALS

2.1 Structures of Alkali Metal Alkoxides

2.1.1 *Lithium*

Lithium methoxide forms a sheet polymer $[LiOMe]_n$ in which the lithiums are four-coordinated by μ_4-OMe groups[23] but with more sterically hindered groups a hexanuclear species $[LiOCMe_2Ph]_6$ (Fig. 4.9) is formed involving 3-coordinated Li with μ_3-alkoxo groups.[24] With excessively bulky alkoxo groups, dimeric molecules $[Li(\mu\text{-}OCBu_3^t)]_2$ are

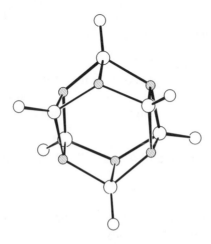

Figure 4.9 $Li_6(\mu_3\text{-}OC)_6$ core structure of $[Li_6(\mu_3\text{-}OR)_6]$.

formed.[25] Bond length data on these and other alkoxo lithium complexes are presented in Table 4.1. Also included are some structurally characterized complexes involving $LiOBR_2$ and $LiOSiPh_3$ molecules.

2.1.2 Sodium

Sodium methoxide is isostructural with lithium methoxide[33] whereas the *tert*-butoxide has a novel structure containing both hexameric and nonameric molecules.[34] The triphenylsiloxide $[Na_4(OSiPh_3)_4(H_2O)_3]$ has a cubane Na_4O_4 framework but three of the four Na atoms are terminally bonded by water molecules.[35]

2.1.3 Potassium, Rubidium, and Caesium

The methoxides of potassium, rubidium, and caesium are isostructural, with a layer structure different from the lithium and sodium analogues.[36] The K, Rb, and Cs atoms are five-coordinated in a tetragonal pyramidal configuration. The *tert*-butoxides of K, Rb, and Cs all form tetranuclear molecules with the cubane structure[37] as do the trimethyl siloxides $[MOSiMe_3]_4$.[38] However, it was shown that the solvates $[MOBu^t.Bu^tOH]$ (M = K, Rb) crystallize in one-dimensional chain structures (Fig. 4.10) in which the metals are in distorted tetrahedral configurations and strong hydrogen bonds link the asymmetric $MOBu^t.Bu^tOH$ units.[39] Some phenylsiloxo complexes of potassium have also been structurally characterized. In $[K(18$-crown-$6)(OSiPh_3)]$ there are dimers due to bridging involving the ether oxygens giving the potassiums an eight-coordination.[30] With dimethoxyethane (DME) as supplementary ligand the novel complex $[K_8(OSiPh_3)_8(MeOC_2H_4OMe)_3]$ was obtained in which two K_4O_4 cubane units are linked by a DME molecule (Fig. 4.11).[30] The other DME molecules each chelate one potassium whilst one Ph_3SiO group in each K_4O_4 unit is also involved in *ipso*-C....K bonding (C–K = 3.078 Å) with one of its phenyl groups. Each potassium is in a different coordination environment with coordination numbers of 3, 4, and 5. In the complex $[KOSiPh_2(OSiPh_2)O.SiPh_2OH]_2$ the dimeric molecule is formed by the silanolate bridges and the potassiums each achieve three-coordination by bonding with the terminal SiOH oxygen of the hexaphenyl trisiloxane ligand which thereby forms an eight-membered ring.[40] Interestingly the shortest *ipso*-C....K bonds present are 3.157(5) Å and 3.185(5) Å.

The directional nature of the M–O bonds in the alkali metal alkoxides together with the solubility of these compounds in organic solvents and the absence of electrical conductivity suggests that they are significantly covalent in nature.

2.2 Beryllium, Magnesium, and Alkaline Earth Alkoxides

2.2.1 Beryllium

Beryllium dialkoxides are generally insoluble nonvolatile compounds, but the use of bulky alkoxo groups imparts solubility and volatility. Thus the *tert*-butoxide is trimeric[42] and the triethylcarbinol derivative $[Be(OCEt_3)_2]$ is dimeric,[3] but to date no crystal structure has been reported.

Table 4.1 Alkali metal alkoxides

Compound	Metal coordination	M–O Bond lengths (Å)		Bond angles (°) $\widehat{MOC}$	Reference
		Terminal	Bridging		
[LiOMe]$_\infty$	4	—	1.95	114.5	23
[Li(μ_3-OCMe$_2$Ph)]$_6$	3	—	1.874–1.979 (3) (av. 1.915)	115.9–139.9 (av. 124.0)	24
[Li(μ-OCBu$^t_3$)]$_2$	2	—	—	—	25
[Li{μ-OB[CH(SiMe$_3$)$_2$]$_2$}$_2$]$_2$	2	—	1.75 (8)	—	25
[Li{OB[CH(SiMe$_3$)$_2$]$_2$}{Me$_2$NC$_2$H$_4$NMe$_2$}]	3	1.667 (15)	1.78 (2)	—	25
[Li(μ-OCBu^{t}CH$_2$PMe$_2$)]$_2$	3	1.78 (1)	—	—	26
[Li(μ-OCBu^{t}CH$_2$PPh$_2$)]$_2$	3	1.792 (9)	—	—	26
[Li(μ-OCBu^t)$_3$(thf)]$_2$	3	—	1.84 (1)	—	27
[Li{μ-OB(Mes)$_2$}(thf)]$_2$	3	—	1.849 (7)	—	28
[Li(μ_3-N-methylpseudoephedrate)]$_2$	3	—	1.870–1.985 (13) (av. 1.929)	—	29
[Li(μ-OSiPh$_3$)(MeOC$_2$H$_4$OMe)]$_2$	3	—	1.881–1.910 (9) (av. 1.895)	133.0–143.5 (av. 136.9)	30
[{(Ph$_2$SiOLi)$_2$O(thf)}$_2$(DABCO)$_2$]	3	—	1.803, 1.877 (10)	—	31
[{(Ph$_2$SiOLi)$_2$O(thf)}$_2$(DABCO)$_2$]	4	—	1.889, 2.014 (10)	—	31
[{(Ph$_2$SiOLi)$_2$O(thf)}$_2$(4,4'-bipy)$_2$]	4	—	1.983 (8)	—	31
[LiOCH(C$_6$H$_4$OMe)NMeC$_2$H$_4$NMe$_2$]$_4$	4	—	1.852–1.909 (8)	—	32
[NaOMe]$_\infty$	4	—	2.32 (1)	110.6	33
(NaOBu^t)$_6$	4	—	2.14–2.32 (3) (av. 2.24)	—	34

(continued overleaf)

235

Table 4.1 (Continued)

Compound	Metal coordination	M–O Bond lengths (Å)		Bond angles (°) $\widehat{MOC}$	Reference
		Terminal	Bridging		
$(NaOBu^t)_9$	4	–	2.15–2.39 (3) (av. 2.25)	–	34
$[Na_4(OSiPh_3)_4(H_2O)_3]$	3 and 4	–	2.256–2.311 (9) (av. 2.284)	–	35
$[KOMe]_\infty$	5	–	2.60, 2.80	–	36
$[RbOMe]_\infty$	5	–	2.72, 2.95	–	36
$[CsOMe]_\infty$	5	–	2.93, 3.14	–	36
$[KOBu^t]_4$	3	–	2.56 (2)	–	37
$[KOBu^t]_4$	3	–	2.623 (1)	125.4	39
$[RbOBu^t]_4$	3	–	2.757 (4)	124.6	39
$[KOSiMe_3]_4$	3	–	2.61 (4)	123	38
$[RbOSiMe_3]_4$	3	–	2.74 (6)	123	38
$[KOBu^t.Bu^tOH]$	4	–	2.620, 2.737 (2)	–	39
$[RbOBu^t.Bu^tOH]$	4	–	2.767 (3), 2.869 (3)	–	39
$[K(18\text{-crown-}6)(OSiPh_3)]_2$	8	–	2.625 (2)	–	30
$[K_8(OSiPh_3)_8(MeOC_2H_4OMe)_3]$	3, 4, 5	–	2.592 (3)–2.757 (3) (av. 2.662)	–	30
$[\overline{KOSiPh_2(OSiPh_2)OSiPh_2OH}]_2$	3	–	2.681 (4), 2.812 (4)	134.4, 118.1	40

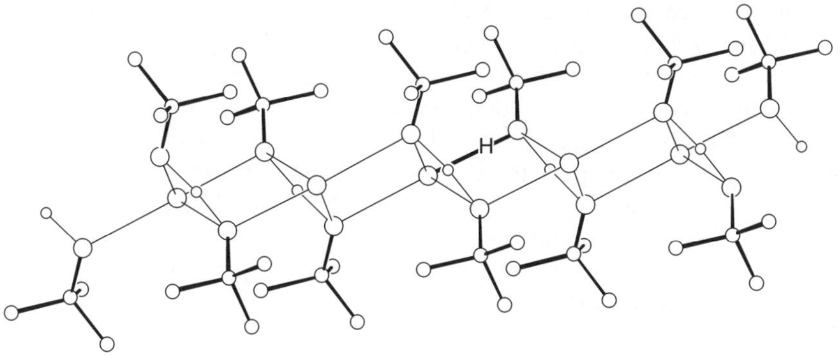

Figure 4.10   Chain structure of [KOBu^t, Bu^{t}OH]$_\infty$.

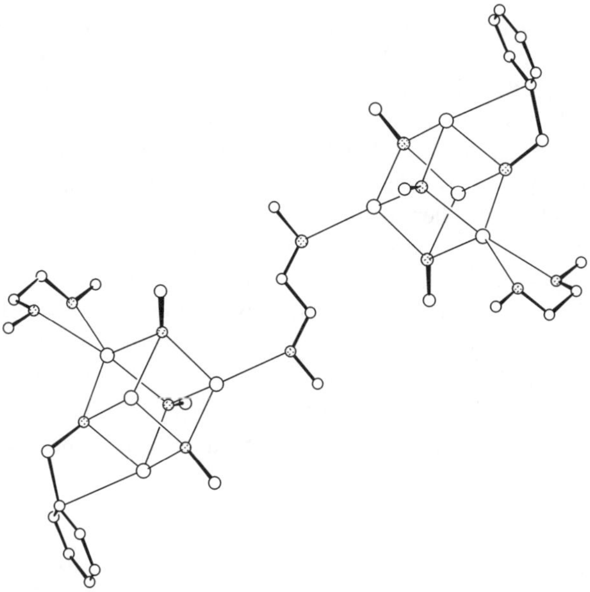

Figure 4.11   Structure of [K$_8$(OSiPh$_3$)$_8$(dme)$_3$] (Ph groups omitted).

2.2.2 *Magnesium*

It has long been known that the sparingly soluble magnesium methoxide slowly deposits massive crystals from saturated methanolic solutions but attempts to characterize them have been thwarted by their facile loss of coordinated methanol leading to disintegration of the crystals. However Turova and co-workers have recently succeeded in determining the crystal structure of the solvate Mg(OMe)$_2$.3.5MeOH.[41] This remarkable structure contains neutral molecules [Mg$_4$(μ_3-OMe)$_4$(OMe)$_4$(MeOH)$_8$] having a cubane Mg$_4$O$_4$ structure, dipositive cations [Mg$_4$(μ_3-OMe)$_4$(OMe)$_2$(MeOH)$_{10}$]$^{2+}$ also having a cubane Mg$_4$O$_4$ structure, and hydrogen-bridged [MeO..H..OMe]$^-$ anions, together with lattice molecules of MeOH. Thus the compound is properly formulated as $\{[Mg_4(OMe)_6(MeOH)_{10}]^{2+}[(MeO)_2H]_2^-[Mg_4(OMe)_8(MeOH)_8]8MeOH\}$. In the neutral cubane structure each Mg is octahedrally coordinated by three μ_3-bridging methoxides and three terminal ligands comprising one methoxo group and two methanol molecules. In the di-cation each Mg also has three bridging methoxo groups but there are only two methoxo groups with ten methanol molecules in the terminal positions. From the Mg–O bond distances given in Table 4.2 (p. 241) it is evident that there is little difference between bridging and terminal bond lengths although there is a large variation in individual values. There is also extensive hydrogen bonding involving both coordinated and lattice methanol molecules.

2.2.3 Calcium

Calcium, strontium, and barium dialkoxides containing less sterically demanding alkoxo ligands are also insoluble and nonvolatile, and very bulky or functionalized alkoxo groups are required to solubilize these derivatives. Only three crystal structures of calcium alkoxides have been reported. The complex $[Ca_9(OC_2H_4OMe)_{18}(HOC_2H_4OMe)_2]^{43}$ involves four different bonding modes for the alkoxo ligand with 6 seven-coordinated and 3 six-coordinated calciums (Fig. 4.12). By using the very bulky ligand OR $(R = CPh_2(CH_2C_6H_4Cl))$ the dimeric molecule $[Ca(\mu-OR)(OR)(thf)]_2$ was obtained with calcium exhibiting a distorted tetrahedral configuration.[44] With the triphenylsiloxide ligand and ammonia as a supporting ligand the dimeric molecule $[Ca_2(\mu-OSiPh_3)_3(OSiPh_3)(NH_3)_4]$ is formed.[45] The structure is unsymmetrical (Fig. 4.13) with one octahedrally coordinated Ca having one siloxide and two NH_3 ligands in terminal positions with three facially bridging siloxides connected to a five-coordinated Ca containing two terminal NH_3 molecules. The terminal bonded siloxide ligand has the shortest Ca–O bond length and the widest $Ca\widehat{O}Si$ bond angle whilst the five-coordinated Ca has shorter Ca–O bond lengths than the six-coordinated Ca and smaller $Ca\widehat{O}Si$ angles than the six-coordinated Ca (Table 4.2). See also $[Ca_2\{\mu-OC(CH_2Pr^i)_2Bu^t\}_4]$.[444]

2.2.4 Strontium

The only strontium alkoxo crystal structures reported so far all involve siloxy ligands. For example the triphenylsiloxide of strontium forms a mononuclear complex with THF and 15-crown-5 as supporting ligands in $[Sr(OSiPh_3)_2(15\text{-crown-}5)(thf)]$.[46] The Sr atom is eight-coordinated to five oxygens from the 15-crown-5 ligand, two from the siloxides and one from the THF ligand. Barium forms an analogous complex. In the mixed ligand complex $[Sr_3(\mu_3\text{-}OSiPh_3)_2(\mu\text{-}OSiPh_3)(tmhd)_3]$ (tmhd = 2,2,6,6-tetramethylheptane-3,5-dionate) a trinuclear molecule is formed.[47] One Sr atom is in a distorted octahedral configuration and the other two are five-coordinated in distorted trigonal-bipyramidal (TBP) configurations. The three metals are in a triangular arrangement capped by two μ_3-$OSiPh_3$ ligands and bridged by one μ-$OSiPh_3$ and two μ-tmhd ligands in the metal plane. One tmhd is chelating and the other two are chelating and bridging. The two five-coordinated strontiums are involved in agostic interactions with neighbouring carbon atoms. Another trinuclear species was obtained with the tetraphenyl disiloxanediolate ligand in conjunction with tetraglyme

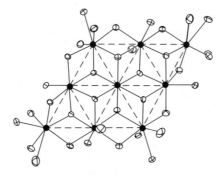

Figure 4.12 Ca_9O_{36} core structure of
$[Ca_9(OC_2H_4OMe)_{18}(HOC_2H_4OMe)_2]$.

Figure 4.13 $Ca_2(\mu\text{-}O)_3ON_4$
core   structure   of $[Ca_2(\mu\text{-}$
$OSiPh_3)_3(OSiPh_3)(NH_3)_4]$ $(SiPh_3$
groups omitted).

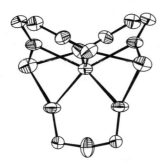

Figure 4.14 $Sr_3(OSiOSiO)_3$ core structure of
$[Sr_3(OSiPh_2OSiPh_2O)_3(tetraglyme)_2]$ (Ph groups
omitted).

$[Sr_3\{O(SiPh_2O)_2\}_3(tetraglyme)_2]$.[47] The three Sr atoms are in a nearly linear array
with the central one being six-coordinated (distorted trigonal prismatic) and flanked
by two seven-coordinated ones which are connected by the bridging oxygens of the
disiloxanediolate ligands (Fig. 4.14).

2.2.5   Barium

A few structures of complex barium alkoxides have been determined. With the bulky
triphenylsiloxide ligand the trimer $[Ba_3(\mu_3\text{-}OSiPh_3)_2(\mu\text{-}OSiPh_3)_3(OSiPh_3)(thf)]$ was

characterized.[48] The structure consists of a triangle of Ba atoms capped above and below the plane by two μ_3-OSiPh$_3$ ligands and equatorially bridged by three μ-OSiPh$_3$ ligands. One barium is four-coordinated by two μ_3-OSiPh$_3$ and two μ-OSiPh$_3$ ligands whilst the others are five-coordinated, one having the terminal OSiPh$_3$ ligand and the other having the THF molecule. In view of the unusual coordination of the four-coordinated barium the possibility of agostic C....Ba interactions was investigated but the closest approach to a phenyl carbon was 3.31 Å. The first barium alkoxide to be structurally characterized was the mononuclear [Ba{OC$_2$H$_4$N(C$_2$H$_4$OH)$_2$}$_2$]EtOH which is chelated by two quadridentate (O$_3$N) ligands making the barium eight-coordinated.[49] The lattice ethanol molecules are involved in extensive hydrogen bonding. Bulky tertiary alkoxides and siloxides give rise to dimeric species. Thus the triphenylcarbinolate has the formula [Ba$_2$(μ-OCPh$_3$)$_3$(OCPh$_3$)(thf)$_3$].[50] The molecular structure is unsymmetrical with one five-coordinated barium having one terminal OCPh$_3$ ligand and one THF whilst the other barium has two terminal THF molecules. The dimer is sustained by three μ-OCPh$_3$ face-sharing ligands making each barium five-coordinated (square pyramidal). With the very bulky tris-*tert*-butylsiloxide another unsymmetrical dimer [Ba$_2$(μ-OSiBu$_3^t$)$_3$(OSiBu$_3^t$)(thf)] is formed but in this molecule the barium atoms are four-coordinated.[50] With the less sterically demanding hexafluoroisopropoxo ligand (hfip) the pentanuclear complex [Ba$_5$(μ_5-OH)(μ_3-hfip)$_4$(μ-hfip)$_4$(hfip)(thf)$_4$(H$_2$O)].thf was obtained[51] (Fig. 4.15) having a distorted square pyramidal array of barium ions encapsulating the μ_5-OH ligand. The apical barium has a coordination number of 6 as have three of the basal bariums with the fourth having six-coordination owing to the additional molecule of coordinated water. Agostic CF....Ba interactions are also featured. The mononuclear [Ba(OSiPh$_3$)$_2$(15-crown-5)(thf)][46] is isostructural with the strontium analogue.[46]

Finally it is interesting to note that the tetranuclear complex [Ba$_4$(μ_3-OBu^t)$_4$(OBu^t)$_4$(Bu^{t}OH)$_8$] has a structure based on the cubane unit Ba$_4$(μ_3-OBu^t)$_4$.[52] Each barium ion is six-coordinated by virtue of terminal bonding to one *tert*-butoxo ligand and two *tert*-butanol molecules with intramolecular hydrogen bonding.

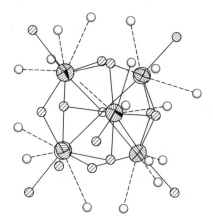

Figure 4.15 Ba$_5$(μ_5-O)(μ_3-O)$_4$(μ-O)$_4$O$_6$ core structure of [Ba$_5$(μ_5-OH)(μ_3-hfip)$_4$(μ-hfip)$_4$(hfip)(thf)$_4$(H$_2$O)], (hfip = OCH(CF$_3$)$_2$; C and H atoms omitted).

Table 4.2 Magnesium and alkaline earth Alkoxides

Compound	Metal coordination	M–O Bond lengths (Å)		Bond angles (°) MOC	Reference
		Terminal	Bridging		
$[Mg_4(\mu_3\text{-}OMe)_4(OMe)_4(MeOH)_8]$	6	2.04–2.13 (1) (av. 2.09)	2.05–2.12 (1) (av. 2.09)	–	41
$[Mg_4(\mu_3\text{-}OMe)_4(OMe)_2(MeOH)_{10}]^{2+}$	6	2.01–2.15 (1) (av. 2.09)	2.05–2.14 (1) (av. 2.09)	–	41
$[Ca_9(\mu_3\text{-}OR)_8(\mu\text{-}OR)_8(OR)_8(HOR)_2]$ $R = C_2H_4OMe$	6 and 7	2.313 (9)	av. μ, 2.291 (8) av. μ_3, 2.390 (8)	–	43
$[Ca(\mu\text{-}OR)(OR)(thf)]_2$ $R = OCPh_2CH_2C_6H_4Cl$	4	2.105 (2)	2.268 (2)	167.4 (t) 132.2, 123.9 (b)	44
$[Ca_2(\mu\text{-}OSiPh_3)_3(OSiPh_3)(NH_3)_4]$	5 and 6	2.179 (5)	av. Ca (6) 2.374 (5) av. Ca (5) 2.265 (5)	162.9 (t) 141.5 (b, av.) 132.5 (b, av.)	45
$[Ca_2(\mu,\eta^3\text{-}L)(\mu,\eta^2\text{-}L)_2(\eta^2\text{-}L)]$ $L = Bu^t(Pr^iOCH_2)_2CO$	6	alkoxo, 2.181 (4) ether, 2.531 (8)	2.269, 2.362 (8)	–	444
$[Sr(OSiPh_3)_2(15\text{-crown-5})(thf)]$	8	2.363, 2.322 (6)	–	168.2, 167.3	46
$[Sr_3(\mu\text{-}OSiPh_3)(\mu_3\text{-}OSiPh_3)(tmhd)_3]$ $tmhd = Bu^tCO(CH)COBu^t$	5 and 6	–	(μ) 2.436, 2.440 (3) (μ_3) av. 2.524 and 2.549 (3)	(μ) 144.3, 121.2 (μ_3) 118.9–133.8	47
$[Sr_3(O(SiPh_2O)_2]_3(tetraglyme)_2]$	6 and 7	–	2.414–2.608 (6) av. 2.493	–	47
$[Ba_3(\mu_3\text{-}OSiPh_3)_2(\mu\text{-}OSiPh_3)_3(OSiPh_3)(thf)]$	4 and 5	2.472 (4)	av. μ_3, 2.69 (9) av. μ, 2.64 (4)	–	48
$[Ba\{OC_2H_4N(C_2H_4OH)_2\}_2]$	8	2.73–2.78 (1) (av. 2.75)	–	124.3–126.6 (av. 125.2)	49
$[Ba_2(\mu\text{-}OCPh_3)_3(OCPh_3)(thf)_3]$	5	2.409 (4)	2.544–2.650 and 2.835	163.5 (t)	50
$[Ba_2(\mu\text{-}OSiBu^t_3)_3(OSiBu^t_3)(thf)]$	4	2.413 (16)	2.461–2.832 (21)	174.7 (t)	50
$[Ba_5(\mu_5\text{-}OH)(\mu_3\text{-}hfip)_4(\mu\text{-}hfip)_4(hfip)(thf)_4(H_2O)]$ $hfip = OCH(CF_3)_2$	6 and 7	2.64 (2)	av. μ = 2.63 (3) av. μ_3 = 2.76 (3)	129.2–135.1 (b) 126.8 (t)	51
$[Ba(OSiPh_3)_2(15\text{-crown-5})(thf)]$	8	2.466, 2.482 (4)	–	166.9, 160.5	46
$[Ba_4(\mu_3\text{-}OBu^t)_4(OBu^t)_4(Bu^tOH)_8]$	6	2.673 (8)	μ_3, 2.675–2.728 (6) av. 2.702	148.6–170.2 (t) 115.3–118.1 (b)	52

b = bridging, t = terminal.

2.3　Alkoxides of p-Block Elements: Al, Ga, In, Tl

2.3.1　Aluminium

The structural chemistry of aluminium alkoxides has posed many problems, owing to the tendency of aluminium to attain coordination numbers of 4, 5, or 6 depending on the steric requirements of the ligands.

The first alkoxide to be structurally characterized was the crystalline tetranuclear isopropoxide $[Al_4(\mu\text{-}OPr^i)_6(OPr^i)_6]$ (Fig. 4.7)[53] which appears to be the thermodynamically stable form. This molecule contains a central octahedral aluminium "chelated" by three $Al(\mu\text{-}OPr^i)_2(OPr^i)_2$ groups containing distorted tetrahedral aluminiums. This structure had been postulated twenty years earlier in order to distinguish it from the trimeric form which was then believed to involve a ring structure with tetrahedral aluminium.[54] The crystal structure of the tetramer has been redetermined at low temperature confirming the original structure with some improvement in the refinement.[55] The trinuclear molecule $[ClAl\{(\mu\text{-}OPr^i)_2AlCl_2\}_2]$[56] has an interesting structure containing one five-coordinated Al (trigonal bipyramid) bonded to one terminal Cl and bridged by isopropoxo ligands to two tetrahedrally coordinated Al atoms which are each bonded to two terminal chlorides. The analogous structure $[\{Pr^iO\}\{Al\{(\mu\text{-}OPr^i)_2Al(OPr^i)_2\}_2\}]$ was proposed for the aluminium isopropoxide trimer based on NMR and mass spectral data.[57] As expected the structure of the tetraisopropoxo aluminate anion $[Al(OPr^i)_4]^-$ was shown to be tetrahedral with some distortion due to hydrogen bonding with the cyclohexylammonium counter-ions.[58] Owing to steric hindrance of the *tert*-butoxo ligands aluminium *tert*-butoxide gives a dimeric species $[Al_2(\mu\text{-}OBu^t)_2(OBu^t)_4]$ containing pseudo-tetrahedral Al.[59]

Another dinuclear species $[Al_2(\mu\text{-}NMe_2)(\mu\text{-}OBu^t)_4]$ was isolated as an intermediate in the reaction of $[Al_2(NMe_2)_6]$ with Bu^tOH which gave $[Al(OBu^t)_3.Me_2NH]$ as the kinetically stable product. Interestingly the thermodynamically stable dimeric $[Al_2(OBu^t)_6]$ obtained from the dimethylamine adduct does not react with the amine to reform the adduct.[60] The structures of the dimeric dialkylaluminium derivatives of 1-menthol and 1-borneol have been reported[61] (Table 4.3 pp. 244–245). The *tert*-butoxo aluminium hydrides are worthy of comment. For $AlH_2(OBu^t)$ a tetranuclear cubane structure $[Al_4(\mu_3\text{-}OBu^t)_4H_8]$ with five-coordinated Al might have been anticipated, but it was shown to be dimeric $[Al_2(\mu\text{-}OBu^t)_2H_4]$ as was the monohydride $[Al_2(\mu\text{-}OBu^t)_2(OBu^t)_2H_2]$.[62] The unsymmetrical dinuclear $[(Bu^tO)(Bu^tOO)Al(\mu\text{-}OBu^t)_2Al(mesal)_2]$ (where mesal = methylsalicylate) contains pseudo-tetrahedral and distorted octahedral aluminiums bridged by *tert*-butoxo ligands with a terminal *tert*-butylperoxo ligand attached to the four-coordinated metal.[63] The liquid dimeric $[Al_2(OPr^i)_4(acac)_2]$ slowly rearranges to a stable crystalline trimer, which has an unsymmetrical structure containing two octahedral and one tetrahedral aluminium $[(acac)_2Al(\mu\text{-}OPr^i)_2Al(\mu\text{-}OPr^i)_2Al(OPr^i)_2]$.[64] Some trimethylsiloxo compounds have been structurally characterized. In $[(Me_2NH_2)^+\{Al(OSiMe_3)_4\}^-]$ hydrogen bonding between the counter-ion N–H protons and the siloxo oxygens causes some distortion in the tetrahedral aluminium whilst the dimeric $[Al_2(\mu\text{-}OSiMe_3)_2(OSiMe_3)_4]$ has a similar Al_2O_6 core structure to the *tert*-butoxide.[65]

In the dimer $[(acac)_2Al(\mu\text{-}OSiMe_3)_2Al(OSiMe_3)_2]$[64] one Al is pseudo-tetrahedral and the other pseudo-octahedral in preference to a more symmetrical structure $[Al(\mu\text{-}OSiMe_3)_2(acac)_2]$ containing equivalent five-coordinated metals. With the more bulky Ph_3SiO ligands a mononuclear pseudo-tetrahedral complex $[Al(OSiPh_3)_2(acac)]$ is formed.[64]

2.3.2 Gallium

Although no structure of a homoleptic alkoxide of gallium has yet been reported some five-coordinated species involving functionalized alkoxo ligands are known.[66-68] The hydrido complex $[GaH\{(OC_2H_4)_2NMe\}]_2$[66] contains the chelating ligand N-methyl diethanolaminate and is dimerized through bridging by one of the oxygens, giving a distorted TBP configuration for each gallium with the hydrides each occupying an equatorial site. Similarly in the bridging N,N-dimethylethanolaminates $[GaH_2(\mu\text{-}OC_2H_4NMe_2)]_2$ and $[GaMe_2(\mu\text{-}OC_2H_4NMe_2)]_2$ the galliums are in distorted TBP configurations.[67] The pyridylmethoxo ligand also bridges in $[Me_2Ga\{(\mu\text{-}OCH_2)C_5H_4N)\}]_2$ to form a five-coordinated dimer.[68] All of these complexes contain the Ga_2O_4 four-membered ring (data are presented in Table 4.3). In $[Bu^t_2Ga(\mu\text{-}OBu^t)]_2$ the bridging tert-butoxo groups from a centrosymmetrical dimer with distorted tetrahedral galliums.[69] Replacing the tert-butoxo ligand with the bulkier triphenylmethoxo ligand gave rise to the mononuclear three-coordinated compound $[Bu^t_2Ga(OCPh_3)]$.[70] Similar dimeric species $[Ga(\mu\text{-}OCH_2SiMe_3)(CH_2SiMe_3)_2]_2$[71] and $[Ga(\mu\text{-}OBu^t)H_2]_2$[62] have also been structurally characterized.

2.3.3 Indium and Thallium

To date two structures of organoindium alkoxides have been reported, $[In(\mu\text{-}OEt)Bu^t_2]_2$[72] and $[In(\mu\text{-}OCH_2SiMe_3)(CH_2SiMe_3)_2]_2$.[71] Even less structural data is available on thallium alkoxides. The tetranuclear $[Tl_4(\mu_3\text{-}OMe)_4]$ was shown to have a distorted cubane Tl_4O_4 framework (Fig. 4.6) although the Tl–O bond distances were not determined. Recently the remarkable chain copolymer $[WCl_2(PMe_2Ph)_4, (TlOCH_2CF_3)_4]_n$ has been shown to crystallize as two polymorphs.[74] Weak Tl....Cl interactions and charge-transfer are believed to support the structure in which the cubane Tl_4O_4 unit is present.

2.4 Alkoxides of p-Block Elements: Ge, Sn, Pb

2.4.1 Germanium

Most of the structures of quadrivalent germanium alkoxo compounds reported to date involve the triethanolaminate ligand $[RGe\{(OC_2H_4)_3N\}]$ (R = Et, naphthyl, $\frac{1}{2}CH_2$, Br, Bu^t, NCS).[75-77] In each case the germatrane structure (Fig. 4.16) contains a five-coordinated Ge in a TBP configuration with the three oxygen atoms in the equatorial plane (data in Table 4.4, pp. 248–249). Other Ge(IV) compounds studied are the six-coordinated $[Ge(OMe)_2(porphyrinate)]$[78] and the four-coordinated $[Ge(OEt)H\{CH(SiMe_3)_2\}_2]$.[79]

The first homoleptic Ge(II) alkoxide to be structurally determined was the mononuclear $[Ge(OCBu^t_3)_2]$ containing two-coordinated Ge owing to the extreme bulkiness of the ligands. A noteworthy feature is the acute $\widehat{OGeO}$ angle of 85.9°. With less demanding ligands as in $[Ge_2(\mu\text{-}OBu^t)_2Cl_2]$ a dimer is formed with trigonal

Table 4.3 Alkoxides of aluminium, gallium, indium, and thallium

Compound	Metal coordination	M–O Bond lengths (Å)			Bond angles (°) MOC	Reference
		Terminal	Bridging			
[Al$_4$(μ-OPr^i)$_6$(OPr^i)$_6$]	4 and 6	av. 1.71 (2)	1.94 (1) (oct) 1.80 (2) (tet)		146 (t, av.) 129 (b, av.)	53
[Al$_4$(μ-OPr^i)$_6$(OPr^i)$_6$]	4 and 6	av. 1.700 (5)	1.925 (5) (oct) 1.802 (5) (tet)		140 (t, av.) 129 (b, av.)	55
[Al$_3$Cl$_5$(μ-OPr^i)$_4$]	4 and 5	–	1.917, 1.804 (7) (tbp) 1.801, 1.756 (5) (tet)		128 (b)	56
[(C$_6$H$_{11}$NH$_3$)$^+$ {Al(OPr^i)$_4$}$^-$]	4	av. 1.74 (1)	–		av. 126	58
[Al$_2$(μ-OBu^t)$_2$(OBu^t)$_4$]	4	1.698–1.681 (3) av. 1.688	1.831–1.824 (3) av. 1.828		144 (t, av.) 130 (b, av.)	59
[Al$_2$(μ-NMe$_2$)(μ-OBu^t)(OBu^t)$_4$]	4	1.671–1.697 (4) av. 1.683	1.819 and 1.823 (3) av. 1.821		–	60
[Al(OBu^t)$_3$ (Me$_2$NH)]	4	1.699–1.744 (2) av. 1.718	–		–	60
[Al$_2$(μ-l-mentholate)$_2$Me$_4$]	4	–	1.835–1.847 (3) av. 1.841		–	61
[Al$_2$(μ-l-mentholate)$_2$Bu$^i_4$]	4	–	1.833–1.852 (9) av. 1.841		–	61
[Al$_2$(μ-l-borneolate)$_2$Me$_4$]	4	–	1.829–1.861 (5) av. 1.841		–	61
[Al$_2$(μ-OBu^t)$_2$H$_4$]	4	–	1.810 and 1.815 (3) av. 1.812		129.6 and 131.4	62
[Al$_2$(μ-OBu^t)$_2$(OBu^t)$_2$H$_2$]	4	1.675 (3)	1.817 (3)		144.3 (t) 130.1 (b, av.)	62
[(Bu^{t}O$_2$)(Bu^{t}O)Al(μ-OBu^t)$_2$Al(mesal)$_2$] mesal = methylsalicylate	4 and 6	1.69 (1)	Not reported		Not reported	63
[Al$_3$(μ-OPr^i)$_4$(OPr^i)$_2$(acac)$_3$]	4 and 6	1.662, 1.716 (7)	av. 1.909 (7) (oct) 1.766, 1.794 (6) (tet)		140 (t, av.) 129 (b, av.)	64

	C.N.				Ref.
$[(Me_2NH_2)^+\{Al(OSiMe_3)_4\}^-]$	4	1.733–1.780 (5) av. 1.752	—	133.8–143.0 av. 137	65
$[Al_2(\mu\text{-}OSiMe_3)_2(OSiMe_3)_4]$	4	1.667, 1.678 (5) av. 1.672	1.813 (4)	174.3, 159.4 (t)	65
$[(acac)_2Al(\mu\text{-}OSiMe_3)_2Al(OSiMe_3)_2]$	4 and 6	1.700 (8)	1.788 (6)		64
$[Al(OSiPh_3)_2(acac)]$	4	1.680, 1.700 (4)	—	169.2 (t) 129.6, 132.8 (t)	64
$[GaH\{(OC_2H_4)_2NMe\}]_2$	5	av. 1.847 (2)	av. 1.960 (8) and 2.018 (2)	165.2, 143.3 (t) 115.9 (t, av.) 117.9 and 125.2 (b, av.)	66
$[Ga_2H_4(\mu\text{-}OC_2H_4NMe_2)_2]$	5	—	1.911 (3) and 2.053 (3)	123.0 and 131.3	67
$[Ga_2Me_4(\mu\text{-}OC_2H_4NMe_2)_2]$	5	—	1.913 (3) and 2.078 (3)	124.8 and 129.8	67
$[Ga_2Me_4\{\mu\text{-}OCH_2(C_5NH_4)\}_2]$	5	—	1.925 (3) and 2.073 (3)	125.0 and 128.5	68
$[Ga_2Bu_4^t(\mu\text{-}OBu^t)_2]$	4	—	1.990 (2) and 1.989 (2)	127.6 and 128.3	69
$[Ga(\mu\text{-}OCPh_3)Bu_2^t]$	3	1.831 (4)	—	127.5	70
$[Ga_2(\mu\text{-}OCH_2SiMe_3)_2(CH_2SiMe_3)_4]$	4	—	1.953–1.976 (3) av. 1.967	av. 118.8 and 122.4	71
$[Ga_2(\mu\text{-}OBu^t)_2H_4]$	4	—	1.902 and 1.908 (9) av. 1.905	129.2	62
$[Ga_2(\mu\text{-}OBu^t)_2(OBu^t)_2H_2]$	4	—	1.904 and 1.907 (4) av. 1.905	128.4 and 129.2	62
$[In_2(\mu\text{-}OEt)_2Bu_4^t]$	4	—	2.147, 2.165 (5) av. 2.156	131.0	72
$[In_2(\mu\text{-}OCH_2SiMe_3)_2(CH_2SiMe_3)_4]$	4	—	2.163–2.182 (2) av. 2.173	av. 125.6 and 131.4	71
$[Tl_4(\mu_3\text{-}OCH_2CF_3)_4]$	3	—	2.464–2.567 (15) av. 2.518	106.0–128.8 av. 116.2	72

tbp = trigonal bipyramidal, tet = tetrahedral, oct = octahedral; b = bridging, t = terminal.

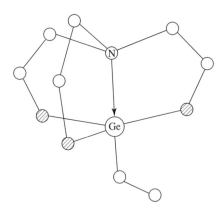

Figure 4.16 Structure
of [EtGe(OCH$_2$CH$_2$)$_3$N]
(H atoms omitted).

pyramidal three-coordinated Ge.[81] The dimeric structure is also maintained in [Ge$_2$(μ-OBu^t)$_2$(OSiPh$_3$)$_2$] in which the *tert*-butoxo group is preferred as the bridging ligand.[82]

2.4.2 Tin

Numerous structures of Sn(II) and Sn(IV) alkoxo compounds are known. In tin(II) ethyleneglycollate the ligand behaves as a chelating and a bridging group giving rise to a three-dimensional network with three-coordinated Sn.[83] The volatile [Sn$_2$(μ-OBu^t)$_2$(OBu^t)$_2$] was shown by electron diffraction analysis to have the dimeric structure in the gas phase with three-coordinated Sn(II) atoms and the terminal *tert*-butoxo in the *trans* configuration.[80] This structure was confirmed by single crystal X-ray diffraction along with the structure of [Sn$_2$(μ-OBu^t)$_2$Cl$_2$].[84] A previous structure determination of [Sn$_2$(μ-OBu^t)$_2$(OBu^t)$_2$] had shown that the dimers are linked into a chain structure by bridging weakly bonding THF molecules.[84] An interesting feature of [M$_2$(μ-OBu^t)$_2$Cl$_2$] (M = Ge, Sn) molecules is that in the germanium compound the chlorines are disposed *syn* with respect to the Ge$_2$O$_2$ ring whilst in the tin compound they are *anti*. In the dimer [Sn$_2$(μ-OBu^t)$_2${N(SiMe$_3$)$_2$}$_2$][85] the bulky terminal silylamido ligands also adopt the *trans* configuration. The di-anionic N-donor ligand Me$_2$Si(NBu^t)$_2$$^{2-}$ acts as a chelate to one Sn and bridges the other in the novel dinuclear molecule [Me$_2$Si(NBu^t)$_2$Sn$_2$(μ-OBu^t)Cl] which also contains a bridging *tert*-butoxo group.[86] Finally it is interesting to note that [Sn$_2$(μ-OBu^t)$_2$(OSiPh$_3$)$_2$] which is isostructural with the germanium analogue also forms a heteronuclear complex with Mo(CO)$_5$ through one tin atom acting as a donor to the molybdenum in [{Sn$_2$(μ-OBu^t)$_2$(OSiPh$_3$)$_2$}{Mo(CO)$_5$}].[82]

A few tin(IV) alkoxo compounds have been structurally characterized in recent years. It was found that the quadridentate ligand triethanolamine does not form a tin analogue of the germatranes since only two of the three OH protons are replaced giving an octahedrally coordinated tin atom in [Sn{(OC$_2$H$_4$)$_2$NC$_2$H$_4$OH}$_2$].[86] Octahedral tin was also found in the mixed dimer [Sn$_2$(μ-OPr^i)$_2$(acac)$_2$].[87] The steric effects of the bulky *tert*-butoxo ligand produced the mononuclear homoleptic alkoxide Sn(OBut)$_4$ with distorted tetrahedral tin.[88] The tetraisopropoxide crystallized with coordinated isopropanol giving the dimeric $[Sn_2(\mu\text{-}OPr^i)_2(OPr^i)_6(Pr^iOH)_2]$ species involving edge-sharing octahedra (*cf.* Fig. 4.2) distorted by the presence of asymmetric hydrogen bonding between axially coordinated alcohol ligands and isopropoxo groups.[88] Thus the terminal isopropoxides *trans* to the bridging ligands have significantly shorter Sn–O bonds (1.943 Å) than those involved in hydrogen bonding (2.098 Å and 2.142 Å) which are longer than the bridging ligands (2.085 Å). Although the OH-proton was not located, it is probably located nearer to the isopropoxo group with the longest Sn–O bond (2.142 Å). A similar structure was found for the isobutoxo complex $[Sn_2(\mu\text{-}OBu^i)_2(OBu^i)_6(Bu^iOH)_2]$ (data in Table 4.4).[89] In $[Sn_2(\mu\text{-}OCH_2Ph)_2(OCH_2Ph)_6(Me_2NH)_2]$ the terminal alcohol ligand was replaced by dimethylamine but the bi-octahedral structure was the same.[90] The tetrakis-2-phenylpropoxide $Sn(OCMe_2Ph)_4$ on thermolysis (110°C) undergoes cyclo-metallation to form the dimeric $[Sn_2(\mu\text{-}OCMe_2C_6H_4)_2(OCMe_2C_6H_4)_2]$ containing both bridging and chelating ligands with five-coordinated (distorted TBP) tin atoms with the Sn–C(aryl) bonds occupying equatorial positions[90] (data in Table 4.4). Other octahedral tin alkoxo compounds have been reported. The dimer $[Sn_2(\mu\text{-}OBu^i)_2(OBu^i)_2(Bu^iOH)_2Cl_2]$ has the familiar bi-octahedral structure[91] whereas the triphenylsiloxo complex with pyridine $[Sn(OSiPh_3)_4(C_5H_5N)_2]$ is a mononuclear octahedral complex[92] with the pyridine ligands occupying *trans* positions.

2.4.3   Lead

The first lead alkoxide to be structurally determined was the Pb(II) derivative of the 2-methoxyethoxo ligand $[Pb(\mu\text{-}OC_2H_4OMe)_2]_n$ which forms chains of apical–equatorial edge-shared trigonal bipyramids with one equatorial position occupied by the stereochemically active lone pair of electrons. The ligands are monodentate and form unsymmetrical alkoxo-bridges (Fig. 4.17).[93] With the bulky *tert*-butoxo lligand the trinuclear compound $[Pb(\mu\text{-}OBu^t)_3Pb(\mu\text{-}OBu^t)_3Pb]$ was obtained.[94] The molecule has a linear system of three lead atoms with the central one being octahedrally coordinated by the six bridging *tert*-butoxo groups and the outer Pb atoms being three-coordinated (data in Table 4.4). This molecule appears to be the prototype for a large number of heterotrinuclear species $[M(\mu\text{-}OBu^t)_3M'(\mu\text{-}OBu^t)_3M]$ where M = Ge and M' = Mg, Ca, Cd, Eu, Pb; M = Sn and M' = Ca, Sr, Cd, Pb; and M = Pb and M' = Sr, Ba.[94]

2.5   Alkoxides of p-Block Elements; Sb, Bi

Antimony pentamethoxide forms a dinuclear molecule $[Sb_2(\mu\text{-}OMe)_2(OMe)_8]$ with the familiar edge-sharing bi-octahedral structure (Fig. 4.2).[95] The *p*-tolyl antimony-bis-tetramethylethyleneglycolate forms a mononuclear five-coordinated Sb(v) com-

Figure 4.17   Chain structure of $[Pb(\mu\text{-}OC_2H_4OMe)_2]_n$ (C and H atoms omitted).

Table 4.4 Alkoxides of germanium, tin, and lead

Compound	Metal coordination	M–O Bond lengths (Å)		Bond angles (°) $\widehat{MOC}$	Reference
		Terminal	Bridging		
[Ge((OC$_2$H$_4$)$_3$N]Et]	5	1.69–1.78 av. 1.74	–	118–124 av. 121	75
[Ge((OC$_2$H$_4$)$_3$N](C$_{10}$H$_7$)]	5	1.77–1.78 av. 1.77	–	118–123 av. 120	75
[[Ge(OC$_2$H$_4$)$_3$N]$_2$CH$_2$]	5	1.77–1.81 av. 1.79	–	118–122 av. 121	76
[Ge((OC$_2$H$_4$)$_3$N]Br]	5	1.726–1.808 (15) av. 1.780	–	114.5–118.3 av. 115.8	76
[Ge[(OC$_2$H$_4$)$_3$N]Bu^t]	5	1.807 (3)	–	120.8	76
[Ge[(OC$_2$H$_4$)$_3$N](NCS)]	5	See Cambridge Crystallographic Data Base[1]			77
[Ge(OMe)$_2$(porphyrinate)]	6	1.822 (3)	–	124.0	78
[Ge(OEt)H{CH(SiMe$_3$)$_2$}$_2$]	4	1.797 (5)	–	124.4	79
[Ge(OCBu$^t_3$)$_2$]	2	1.832 and 1.896 (6) av. 1.87 (4)	–	131.9 and 134.0	80
[Ge$_2$(μ-OBu^t)$_2$Cl$_2$]	3	–	1.986–2.028 (6) av. 2.009	118.8–122.8 av. 120.7	81
[Ge$_2$(μ-OBu^t)$_2$(OSiPh$_3$)$_2$]	3	1.814 (2)	1.966 and 1.967 (2)	150.2 (t)	82
[Sn(μ-OC$_2$H$_4$O-μ)]$_\infty$	3	–	2.07, 211 (2)[2] 2.31, 2.34 (2)[3]	–	83
[Sn$_2$(μ-OBu^t)$_2$(OBu^t)$_2$] (gas phase electron diffraction)	3	1.97 (2)	2.16 (1)	132 (t) 127 (b)	80
[Sn$_2$(μ-OBu^t)$_2$(OBu^t)$_2$]	3	2.009 (4)	2.128 and 2.165 (4)	129.8 (t) 123.8 and 127.4 (b)	81
[Sn$_2$(μ-OBu^t)$_2$Cl$_2$]	3	–	2.139 (4)	125.1, 127.1 (b)	81
[Sn$_2$(μ-OBu^t)$_2$Cl$_2$]	3	–	2.153, 2.156 (2)	123.8, 127.8 (b)	84
[Sn$_2$(μ-OBu^t)$_2$[N(SiMe$_3$)$_2$]]	3	–	2.168, 2.173 (4)	124.4, 127.6 (b)	85
[Me$_2$Si(NBu^t)$_2$Sn$_2$(μ-OBu^t)Cl]	3	–	2.125, 2.151 (2)	127.0, 133.2 (b)	86
[Sn$_2$(μ-OBu^t)$_2$(OSiPh$_3$)$_2$]	3	1.949 (5)	2.079, 2.099 (4)	149.4 (t)	82

[{Sn$_2$(μ-OBu^t)$_2$(OSiPh$_3$)$_2$}{Mo(CO)$_5$}]	3 and 4	1.968, 2.000 (3)	2.083–2.171 (3) av. 2.126	149.4, 155.2 (t)	82
[Sn{(OC$_2$H$_4$)$_2$NC$_2$H$_4$OH}$_2$]	6	2.01, 2.04	–	–	86
[Sn$_2$(μ-OPr^i)$_2$(OPr^i)$_4$(acac)$_2$]	6	1.957, 1.974 (5)	2.079, 2.117 (4)	124.4, 122.0 (t) 128.1 (b)	87
[Sn(OBu^t)$_4$]	4	1.946, 1.949 (2)	–	124.1, 125.0	88
[Sn$_2$(μ-OPr^i)$_2$(OPr^i)$_6$(Pr^{i}OH)$_2$]	6	1.923, 1.964 (7); 2.098, 2.142 (6)	2.080, 2.091 (4)	160, 139 (t) 126 (b)	88
[Sn$_2$(μ-OBu^i)$_2$(OBu^i)$_6$(Bu^{i}OH)$_2$]	6	1.932, 1.943 (9); 2.108, 2.114 (9)	2.069, 2.094 (7)	–	89
[Sn$_2$(μ-OCH$_2$Ph)$_2$(OCH$_2$Ph)$_6$(Me$_2$NH)$_2$]	6	1.978–2.010 (3) av. 1.988	2.113–2.133 (3) av. 2.122	123 (t, av.) 124 (b, av.)	90
[Sn$_2$(μ-OCMe$_2$C$_6$H$_4$)$_2$(OCMe$_2$C$_6$H$_4$)$_2$]	5	2.007 (3)	2.056 and 2.258 (2)	127.6 (r) 114.2 (b)	90
[Sn(OSiPh$_3$)$_4$(C$_5$H$_5$N)$_2$]	6	1.976–2.004 (7) av. 1.987	–	150.6–159.2 av. 154.8	92
Pb(μ-OC$_2$H$_4$OMe)$_2$]$_\infty$	4	–	2.23, 2.25; 2.38, 2.51	–	93
Pb(μ-OBu^t)$_3$Pb(μ-OBu^t)$_3$Pb]	3 and 6	–	2.16 and 2.55 (2)	–	94

b = bridging, r = ring, t = terminal.
[2]Intrachelating. [3]Interchelating.

pound with the TBP configuration. Each chelating glycolate ligand bonds at one axial and one equatorial position and the *p*-tolyl group occupies the remaining equatorial position.[96] Antimony trimethoxide forms a layer structure $[Sb(\mu\text{-OMe})_3]_n$ in which each Sb(III) is linked by unsymmetrical methoxo bridges to three others giving a coordination number of 6.[97] The first bismuth alkoxide structure was determined simultaneously by two independent groups.[98] The compound $[Bi_2(\mu\text{-}OC_2H_4OMe)_4)_4(OC_2H_4OMe)_2]_n$ has a chain structure comprising dimeric units involving five-coordinated (square pyramidal) bismuth atoms bonded to one terminal (apical) and four bridging (shared with two neighbours) 2-methoxyethoxo ligands. The edge-sharing square pyramids have unsymmetrical alkoxo bridges and it is possible that the vacant octahedral site on the Bi atom is occupied by the lone electron pair. None of the ligands is chelating (Fig. 4.18).

The monomeric mixed ligand complex $[Bi(OSiPh_3)_3(thf)_3]$ has a distorted octahedral configuration owing to the presence of the bismuth lone pair and the two sets of ligands are arranged meridionally.[99] In the dimeric $[Bi(\mu\text{-hfip})_2(hfip)_4(thf)_2]$ (hfip = $OCH(CF_3)_2$) the bismuth adopts the square pyramidal five-coordination with lone pairs occupying the vacant octahedral sites. The edge-sharing bridges are markedly unsymmetrical with the THF ligands bonded *trans* to the shorter Bi–O bridge bond (Fig. 4.19).[100] Similar structures were reported for the neopentyloxides $[Bi_2(\mu\text{-ONp})_2(ONp_4(NpOH)_2]$ $(Np = CH_2Bu_3^t)$ and $[Bi_2(\mu\text{-ONp})_4(C_5H_5N)_2]$ (data in Table 4.5).[101]

2.6   Alkoxides of Scandium and Yttrium

2.6.1   Scandium

Owing to its relative scarcity and lack of any rich mineral sources the chemistry of scandium had been neglected until recently.

The first crystal structure determined of a scandium alkoxide was that of the dimeric *tert*-butoxide $[Sc_2(\mu\text{-OBu}^t)_2(OBu^t)_4]$ containing four-coordinated Sc in edge-bridged distorted tetrahedra as in Fig. 4.1.[102] With the less sterically demanding but more acidic

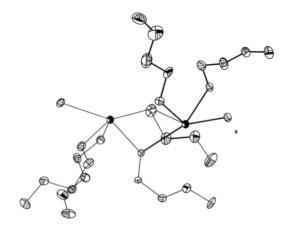

Figure 4.18   Chain structure of $[Bi_2(\mu\text{-}OC_2H_4OMe)_4(OC_2H_4OMe)_2]_\infty$

Table 4.5 Alkoxides of antimony and bismuth

Compound	Metal coordination	M–O Bond lengths (Å)		Bond angles (°) $\widehat{\text{MOC}}$	Reference
		Terminal	Bridging		
[Sb$_2$(μ-OMe)$_2$(OMe)$_8$]	6	1.937–1.942 (2) av. 1.939	2.095, 2.108 (2)	122.8 (t, av.) 120.0 (b, av.)	95
[(tolyl)Sb(O$_2$C$_2$Me$_4$)$_2$]	5	1.972 (1) (ax) 1.941 (2) (eq)	–	–	96
[Sb(μ-OMe)$_3$]$_\infty$	6	–	1.986–2.012 (8) av. 2.001 2.564–3.003 (7) av. 2.818	122.4–124.9 av. 123.5	97
[Bi$_2$(μ-OC$_2$H$_4$OMe)$_4$(OC$_2$H$_4$OMe)$_2$]$_\infty$	5	2.07, 2.11 (2)	2.20–2.21 (1) av. 2.205 2.53–2.58 (1) av. 2.55	–	98a
[Bi$_2$(μ-OC$_2$H$_4$OMe)$_4$(OC$_2$H$_4$OMe)$_2$]$_\infty$	5	2.071, 2.114 (6)	2.200–2.230 (6) av. 2.216 2.513–2.573 (6) av. 2.531	121.8	98b
[Bi(OSiPh$_3$)$_3$(thf)$_3$]	6	2.04 (1)	–	164.8	99
[Bi$_2$(μ-hfip)$_2$(hfip)$_4$(thf)$_2$] hfip = OCH(CF$_3$)$_2$	5	2.064, 2.116 (7)	2.188, 2.688 (7)	–	100
[Bi$_2$(μ-ONp)$_2$(ONp)$_2$(NpOH)$_2$] Np = CH$_2$Bu^t	5	2.098, 2.247 (8) (2.579 coord. NpOH)	2.072–2.546 (6)	–	101
[Bi$_2$(μ-ONp)$_2$(ONp)$_4$(C$_5$H$_5$N)]	5	2.058, 2.109 (4)	2.170, 2.606 (4)	–	101

b = bridging, t = terminal; ax = axial, eq = equatorial.

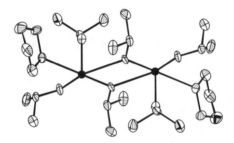

Figure 4.19 Structure of $[Bi_2(\mu\text{-}hfip)_2(hfip)_4(thf)_2]$ (hfip $= OCH(CF_3)_2$; H and F atoms omitted).

hfip ligand an ammine was isolated $[Sc_2(\mu\text{-}hfip)_2(hfip)_4(NH_3)_2]$ containing octahedral Sc in an edge-sharing dimer with *cis* terminal alkoxo groups *trans* to the bridging alkoxo groups.[103] The ammonia ligands adopt the *trans* configuration (Fig. 4.20).

2.6.2   Yttrium

The first structurally characterized organoyttrium containing an alkoxo ligand was the dimeric bis-cyclopentadienyl enolate $[(C_5H_4CH_3)_4Y_2(\mu\text{-}OCH{=}CH_2)_2]$.[104] The first monomeric, homoleptic yttrium alkoxo compound involved a bulky functionalized chelating ligand as in $[Y(OCBu^t_2CH_2PMe_2)_3]$.[105] The first trinuclear alkoxo yttrium compound reported was the mixed ligand species $[Y_3(\mu_3\text{-}OBu^t)(\mu_3\text{-}Cl)(\mu\text{-}OBu^t)_3(OBu^t)_4(thf)_2]$ which has a triangular array of distorted octahedral yttrium atoms capped by a $\mu_3\text{-}OBu^t$ on one side and a $\mu_3\text{-}Cl$ on the other. The three $\mu\text{-}OBu^t$ ligands are in the triangular plane. Each metal has one terminal OBu^t group, one has a second *tert*-butoxide and the other two have terminal THF ligands.[106] This structure is related to that shown in Fig. 4.5. The same framework structure was found in the related complex $[Y_3(\mu_3\text{-}OBu^t(\mu_3\text{-}Cl)(\mu\text{-}OBu^t)_3Cl(OBu^t)(thf)_2]$ in which one of the terminal *tert*-butoxo groups is replaced by a chloride,[107] and also in $[Y_3(\mu_3\text{-}OBu^t)_2(\mu\text{-}OBu^t)_3(OBu^t)_4(Bu^tOH)_2]$.[6]

A remarkable decameric compound was formed by the 2-methoxyethoxo ligand in which each Y atom is seven-coordinated (pentagonal bipyramid) $[Y_{10}(OC_2H_4OMe)_{30}]$ (Fig. 4.21).[108] Each yttrium atom is linked to two neighbouring yttrium atoms by the bridging alkoxo moiety of two ligands which also chelate through the OMe function. One ligand acts as a monodentate group on each metal. The overall molecular structure is a unique ten-membered ring. Replacing some $MeOC_2H_4O$ groups by acac gave the trinuclear species $[Y_3(\mu_3\text{-}OC_2H_4OMe)_2(\mu\text{-}OC_2H_4OMe)_3(acac)_4]$ which has a triangle of eight-coordinated yttrium capped above and below the Y_3-plane by $\mu_3\text{-}OC_2H_4OMe$ groups which also chelate through the OMe function. Three $\mu\text{-}OC_2H_4OMe$ groups span the three yttriums, with two chelating and the third not. Each yttrium is also chelated by an acac ligand (Fig. 4.22).[109] This molecule illustrates the versatility and various bonding modes of the 2-methoxyethoxo ligand.

A novel trinuclear mono-cation has also been structurally characterized as $[Y_3(\mu_3\text{-}OBu^t)(\mu_3\text{-}Cl)(\mu\text{-}OBu^t)_3(OBu^t)_3(thf)_3]^+$. It exhibits the familiar triangular Y_3 unit held by $\mu_3\text{-}OBu^t$ and $\mu_3\text{-}Cl$ capping ligands together with three in-plane $\mu_2\text{-}OBu^t$ ligands.

Figure 4.20 $Sc_2(\mu\text{-}O)_2O_4N_4$ core structure of $[Sc_2(\mu\text{-}hfip)_2(hfip)_4(NH_3)_4]$ (hfip = $OC(CF_3)_2$; ⊘ = N).

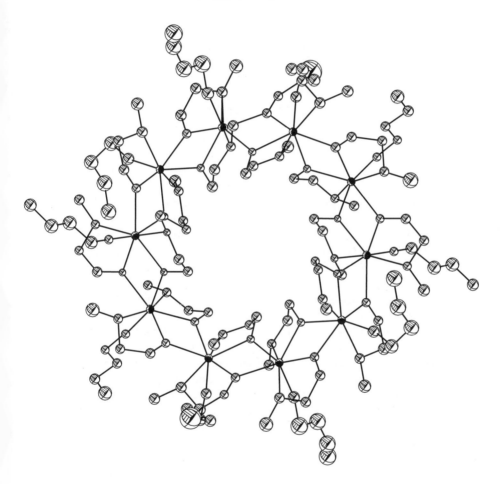

Figure 4.21   Structure of $[Y_{10}(OC_2H_4OMe)_{30}]$ (H atoms omitted).

Each metal has a terminal OBu^t group on the same side as the $\mu_3\text{-}OBu^t$ group and a THF ligand on the $\mu_3\text{-}Cl$ side completing the octahedral coordination.[110] A dinuclear mono-cation, $[Y_2(\mu\text{-}OBu^t)_2(\mu\text{-}Cl)(OBu^t)_2(thf)_4]^+$, was also isolated.[110] The two octahedral yttrium atoms are bridged by two OBu^t groups and one chloride in a confacial bi-octahedral structure. Each yttrium is terminally bonded to one OBu^t group and two THF ligands. In the mononuclear mono-cation $[Y(OBu^t)Cl(thf)_5]^+$

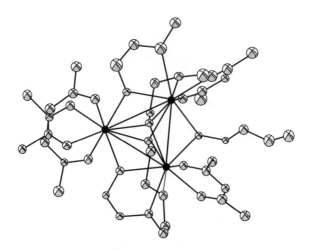

Figure 4.22   Structure of [Y$_3$(μ_3-OC$_2$H$_4$OMe)$_2$(OC$_2$H$_4$OMe)$_3$(acac)] (acac = acetylacetonate; H atoms omitted).

the seven-coordinated yttrium has a OBu^t group and a chloride in the axial positions with the five THF ligands occupying the equatorial positions.[110] Some structures have also been determined on mono- and bis-cyclopentadienyl yttrium alkoxo compounds. The dimeric species [(C$_5$H$_4$SiMe$_3$)$_4$Y$_2$(μ-OMe)$_2$],[111] [(C$_5$Me$_5$)$_2$Y$_2$(μ-OBu^t)$_2$(OBu^t)$_2$], [(C$_5$H$_5$)$_2$Y$_2$(μ-OBu^t)$_2$], [(C$_5$H$_4$SiMe$_3$)$_2$Y$_2$(μ-OBu^t)$_2$(OBu^t)$_2$] and [(C$_9$H$_7$)$_2$Y$_2$(μ-OBu^t)$_2$(OBu^t)$_2$] all contained Y(μ-OR)$_2$Y bridges (data in Table 4.6, pp. 256–257).[112]

The bridging propensity of the alkoxo ligand can be reduced both sterically by using bulky alkyl groups and electronically by using electronegative alkyl groups, which reduce the electron density on the alkoxo oxygen donor atom. Both factors are operating in the fluorinated tertiary alkoxo groups, and the mononuclear yttrium compounds [Y{OCMe(CF$_3$)$_2$}$_3$(thf)$_3$] and [Y{OCMe(CF$_3$)$_2$}$_3$(diglyme)] (Fig. 4.23) were obtained using the hexafluoro-*tert*-butoxo (hftb) ligand.[113] Both complexes are octahedral with the facial configuration. The tris-THF complex exhibits the same structure as the previously reported tris-triphenylsiloxo complex [Y(OSiPh$_3$)$_3$(thf)$_3$] (Fig. 4.24).[114] The homoleptic tris-triphenylsiloxide forms an unsymmetrical dimer [Y$_2$(μ-OSiPh$_3$)$_2$(OSiPh$_3$)$_4$] with distorted tetrahedral yttrium atoms.[115] Interestingly a mononuclear five-coordinated adduct [Y(OSiPh$_3$)$_3$(OPBu$^n_3$)$_2$] was formed with tri-*n*-butylphosphine oxide ligands occupying the axial positions in a TBP configuration.[116] In the salt [K(dme)$_4$]$^+$[Y(OSiPh$_3$)$_4$(dme)]$^-$ the yttrium mono-anion adopts a distorted octahedral configuration.[116] The dimethyl *tert*-butylsiloxo ligand gave rise to the mononuclear adduct [Y(OSiMe$_2$Bu^t)$_3$(thf)$_3$] in the presence of THF but in the presence of excess silanol the novel unsymmetrical dinuclear species [(Me$_2$Bu^{t}SiOH)(Me$_2$Bu^{t}SiO)$_2$Y(μ-OSiMe$_2$Bu^t)$_2$Y(OSiMe$_2$Bu^t)$_2$] was formed.[116] One yttrium is in a distorted tetrahedral configuration while the other one is approximately trigonal bipyramidal, and this is reflected in the bond lengths (Table 4.6).

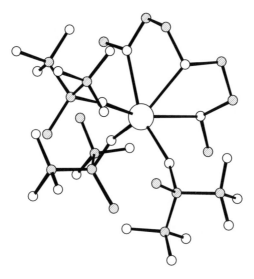

Figure 4.23 Structure of [Y(hftb)$_3$(diglyme)] (hftb = OCCH$_3$(CF$_3$)$_2$; diglyme = MeOC$_2$H$_4$OC$_2$H$_4$OMe; ⊘ = carbon; H and F atoms omitted).

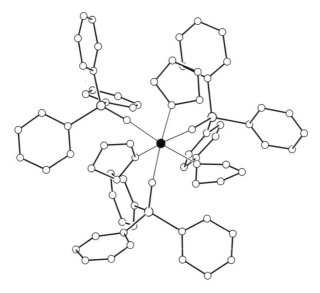

Figure 4.24 Structure of [Y(OSiPh$_3$)$_3$(thf)$_3$ (H atoms omitted).

Table 4.6 Alkoxides of scandium and yttrium

Compound	Metal coordination	M–O Bond lengths (Å)		Bond angles (°)	Reference
		Terminal	Bridging	$\widehat{MOC}$	
[Sc₂(μ-OBuᵗ)₂(μ-OBuᵗ)₄]	4	1.855 (6)	2.052, 2.069 (8)	174.1 (t)	102
[Sc₂(μ-hfip)₂(hfip)₄(NH₃)₂]	6	1.921, 1.964 (5)	2.172, 2.208 (4)	177, 146 (t)	103
hfip = OCH(CF₃)₂					
[(CH₃C₅H₄)₄Y₂(μ-OCH=CH₂)₂]	4¹	–	2.275, 2.290 (3)	127.3	104
[Y(OCBuᵗ₂CH₂PMe₂)₃]	6	2.090 (4)	–	–	105
[Y₃(μ₃-OBuᵗ)(μ₃-Cl)(μ-OBuᵗ)₃(OBuᵗ)₄(thf)₂]	6	2.037–2.073 (15)	av. μ₃, 2.39 (12); av. μ₂, 2.30 (5)	173.8 (t, av.); 126.6 (b, av.)	106
[Y₃(μ₃-OBuᵗ)(μ₃-Cl)(μ-OBuᵗ)₃Cl(OBuᵗ)₃(thf)₂]	6	1.97–2.01 (2)	av. μ₃, 2.36–2.45 (3); av. μ₂, 2.25–2.32 (2)	–	107
[Y₁₀(μ-OC₂H₄OMe)₂₀(OC₂H₄OMe)₁₀]	7	av. 2.09 (2)	av. 2.29 (2)	167.4 (t)	108
[Y₃(μ₃-OC₂H₄OMe)₂(μ-OC₂H₄OMe)₃(acac)₄]	8	–	av. μ₃, 2.47; av μ₂, 2.26	–	109
[Y₃(μ₃-OBuᵗ)(μ₃-Cl)(μ-OBuᵗ)₃(OBuᵗ)₃(thf)₃]⁺	6	2.00–2.02 (1); av. 2.01	av. μ₃, 2.37 (1); av. μ₂, 2.28 (1)	178 (t, av.)	110
[Y₂(μ-OBuᵗ)₂(μ-Cl)(OBuᵗ)₂(thf)₄]⁺	6	1.97 (2)	2.24, 2.26 (2)	176 (t)	110
[Y(OBuᵗ)Cl(thf)₅]⁺	7	2.026 (4)	–	179.5	110
[(C₅H₄SiMe₃)₄Y₂(μ-OMe)₂]	4¹	–	2.217, 2.233 (3)	122.5, 131.0	111
[(C₅Me₅)₂Y₂(μ-OBuᵗ)₂(OBuᵗ)₂]	4¹	1.995, 2.018 (9)	2.229–2.282 (9); av. 2.256	–	112

Compound	Coordination				Ref.
[(C$_5$H$_5$)$_2$Y$_2$(μ-OBu^t)$_2$(OBu^t)$_2$]	4[1]	2.001 (12)	2.230 (7)	–	112
[(C$_5$H$_4$SiMe$_3$)$_2$Y$_2$(μ-OBu^t)$_2$(OBu^t)$_2$]	4[1]	2.015, 2.023 (6)	2.231–2.242 (6) av. 2.235	–	112
[(C$_9$H$_7$)$_2$Y$_2$(μ-OBu^t)$_2$(OBu^t)$_2$]	4[1]	2.005, 2.017 (7)	2.203–2.247 (7) av. 2.224	–	112
[Y{OCMe(CF$_3$)$_2$}$_3$(thf)$_3$]	6	2.079–2.123 (11) av. 2.099	–	159.2–173.5 av. 166.8	113
[Y{OCMe(CF$_3$)$_2$}$_3$(diglyme) diglyme = MeO(C$_2$H$_4$O)$_2$Me	6	2.059–2.084 (16) av. 2.072	–	159.1–171.0 av. 163.5	113
[Y(OSiPh$_3$)$_3$(thf)$_3$]	6	2.118–2.138 (18) av. 2.131	–	168.5–174.4	114
[Y$_2$(μ-OSiPh$_3$)$_2$(OSiPh$_3$)$_4$]	4	2.058, 2.062 (5)	2.211, 2.288 (5)	162.3, 177.8 (t) 111.1, 142.5 (b)	115
[Y(OSiPh$_3$)$_3$(OPBu$^n_3$)$_2$]	5	2.118–2.129 (3) av. 2.123	–	158.1–174.6 av. 163.5	116
[Y(OSiPh$_3$)$_4$(dme)]$^-$	6	2.143, 2.147 (5) (trans) 2.192, 2.196 (5) (cis)	–	163.7–175.1 av. 168.7	116
[Y(OSiMe$_2$Bu^t)$_2$(HOSiMe$_2$Bu^t)(μ-OSiMe$_2$Bu^t)$_2$Y(OSiMe$_2$Bu^t)$_2$]	4 and 5	2.046, 2.060 (18) (tet) 2.072, 2.104, 2.487 (18) (tbp)	2.239, 2.242 (18) (tet) 2.271, 2.280 (17) (tbp)	175 (tet, t) 161 (tbp, t)	116

b = bridging, t = terminal; tbp = trigonal bipyramidal, tet = tetrahedral.
[1]Coordination doubtful.

257

2.7   Alkoxides of the Lanthanides

2.7.1   Lanthanum

Lanthanum alkoxides have been the subject of considerable research in the 1990s. The first chloride-free alkoxo compound to be structurally characterized by X-ray crystallography was $[La_3(\mu_3-OBu^t)_2(\mu-OBu^t)_3(OBu^t)_4(Bu^tOH)_2]$ having a cluster of octahedra (Fig. 4.25).[6]

Considerations of bond lengths and bond angles was required to locate the positions of the coordinated Bu^tOH molecules because the –OH hydrogen could not be identified. From Table 4.7 (pp. 264–265) it can be seen that the order of La–O bond lengths is terminal $< \mu$-bridging $< \mu_3$-bridging as expected while the LaÔC angles are in the order terminal $> \mu$-bridging $> \mu_3$-bridging with the terminal ligands (excluding coordinated Bu^tOH) being close to linear. There is no doubt that the compound $[La_3(\mu_3-OBu^t)_2(\mu-OBu^t)_3(OBu^t)_4(thf)_2]$ has the analogous structure with the coordinated Bu^tOH molecules replaced by THF molecules.[106] With the very bulky triphenylcarbinolate ligand, a dimeric compound with distorted tetrahedral lanthanums was obtained $[La_2(\mu-OCPh_3)_2(OCPh_3)_4]$[117] in which the terminal La–O bond lengths (av. 2.179 Å) were significantly shorter than in the octahedrally coordinated $tert$-butoxo compound (av. 2.280 Å). With the neopentyloxo ligand a tetranuclear molecule was formed involving a square of lanthanum atoms bonded by double $\mu-OCH_2Bu^t$ bridges to each neighbour. The terminal ligand occupies the apical position in a distorted square pyramidal five-coordinated La (Fig. 4.26).[118] The close approach of some neopentyloxo methylene groups to the metal atom indicated agostic La....H–C interactions.

As mentioned in the introduction to this chapter the use of the bulky electronegative hexafluoro tertiary butoxo group (hftb) gave rise to the mononuclear octahedral complex $[La(hftb)_3(thf)_3]$ (Fig. 4.27)[7] which like the analogous yttrium compound[113] adopted the facial configuration. With the triphenylsiloxo ligand the octahedral complex $[La(OSiPh_3)_3(thf)_3]$[116] was obtained which was isomorphous with the yttrium analogue (Fig. 4.24).[114]

2.7.2   Cerium

Cerium can adopt either the tervalent or quadrivalent states and a few crystal structures of alkoxo compounds have been reported in recent years. The first authentic

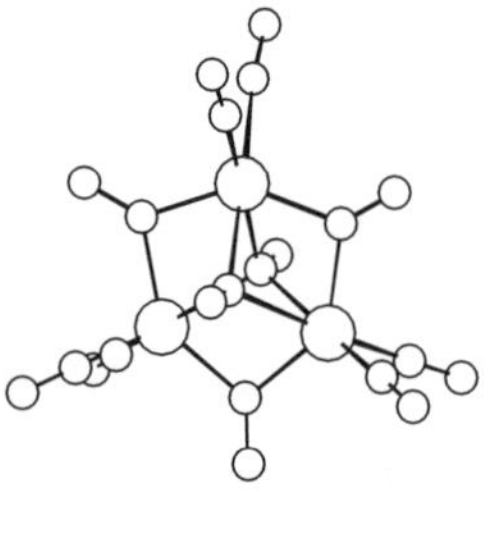

Figure 4.25 $La_3(\mu_3-OC)_2(\mu-OC)_3(OC)_6$ core structure of $[La_3(\mu_3-OBu^t)_2(\mu-OBu^t)_3(OBu^t)_4(Bu^tOH)_2]$ (CH$_3$ groups omitted).

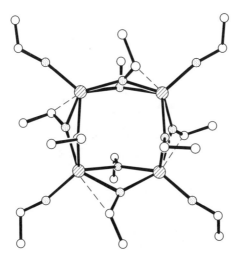

Figure 4.26 Structure of [La$_4$(μ-OCH$_2$Bu^t)$_8$(OCH$_2$Bu^t)$_4$ (⊘ = La; ○ = O; H atoms and CH$_3$ groups omitted).

Figure 4.27 Structure of [La(hftb)$_3$(thf)$_3$]   (hftb = OCCH$_3$(CF$_3$)$_2$; ○ = O; H and F atoms omitted).

Ce(IV) organometallic alkoxide was the mononuclear tris-cyclopentadienyl cerium *tert*-butoxide [(C$_5$H$_5$)$_3$Ce(OBu^t)] which has a pseudo-tetrahedral configuration (data in Table 4.7).[119] The first homoleptic alkoxo cerium(III) compound characterized structurally was the dimeric [Ce$_2$(μ-OCHBu$^t_2$)$_2$(OCHBu$^t_2$)$_4$] obtained by thermolysis of the monomeric [Ce(OCBu$^t_3$)$_3$].[120] The dimer exhibits the familiar pseudo-tetrahedral structure (Fig. 4.1). The remarkable mixed ligand Ce(IV) complex [Ce(OBu^t)$_2$(NO$_3$)$_2$(Bu^{t}OH)$_2$] has bidentate nitrato ligands giving a formal eight-coordination configuration to the cerium (Fig. 4.28).[121] The coordinated Bu^{t}OH ligands are clearly distinguishable from the Ce–OBu^t groups by their longer Ce–O bond lengths (av. 2.525 Å *vs* av. 2.025 Å) and more acute CeÔC angles (av. 138.2° *vs* av. 169.5°). The dimeric [Ce$_2$(μ-OPr^i)$_2$(OPr^i)$_6$(Pr^{i}OH)$_2$] has the expected structure based on the familiar edge-sharing bi-octahedral configuration (Fig. 4.2).[122] The coordinated Pr^{i}OH ligands were identified by longer Ce–O bonds

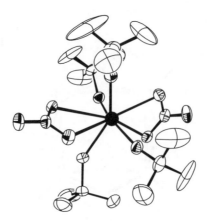

Figure 4.28 Structure of
$[Ce(OBu^t)_2(NO_3)_2(Bu^tOH)_2]$
(H atoms omitted).

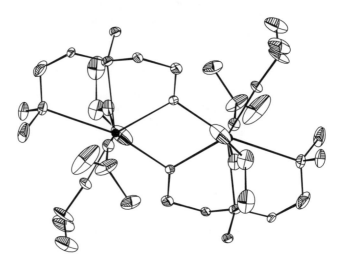

Figure 4.29 Structure of $[Ce_2\{\mu\text{-}OC_2H_4N(Me)C_2H_4NMe_2\}_2(OPr^i)_6]$ (H atoms omitted).

(av. 2.322 Å *vs* 2.041 Å) and more acute Ce$\widehat{O}$C angles (av. 140° *vs* 173.5°). Although $[Ce(OBu^t)_2(OCBu^t_3)_2]$ was unstable the Ce(III) compound $[Ce_2(\mu\text{-}OBu^t)_2(OCBu^t_3)_4]$ was isolated and characterized.[123] The molecule has the familiar pseudo-tetrahedral bridged structure with the less bulky *tert*-butoxo ligands taking the bridging positions (data in Table 4.7). The use of the aminoalcohol $Me_2NC_2H_4N(Me)C_2H_4OH$ gave rise to the centrosymmetric dimer $[Ce_2\{\mu\text{-}OC_2H_4N(Me)C_2H_4NMe_2\}_2(OPr^i)_6]$ in which all the isopropoxo ligands occupy terminal positions and the tridentate functionalized alkoxo group acted as a bridging chelating ligand giving seven-coordinated Ce(IV) (Fig. 4.29).[124] The novel eight-coordinated Ce(IV)

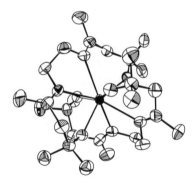

Figure 4.30   Structure of $[Ce\{OC(CF_3)_2CH_2C(Me)=NC_2H_4N=C(Me)CH_2C(CF_3)_2O\}_2]$ (H and F atoms omitted).

complex $[Ce\{OC(CF_3)_2CH_2C(Me)=NC_2H_4N=C(Me)CH_2C(CF_3)_2O\}_2]$ (Fig. 4.30) was obtained using the tetradentate functionalized diolate ligand obtained by template condensation of ethylene diamine and 5,5,5-trifluoro-4-hydroxy-4-trifluoromethyl-2-pentanone.[125] The structure involves a distorted square antiprism with the two diolate ligands occupying meridional configurations. The triphenylsiloxo ligand gave rise to the mononuclear $[Ce(OSiPh_3)_3(thf)_3]$[126] isostructural with the yttrium and lanthanum analogues (Table 4.7). The unsolvated compound $[Ce_2(\mu\text{-}OSiPh_3)_2(OSiPh_3)_4]$ has the familiar dimeric structure with a distorted tetrahedral structure. There appears to be interaction between the *ipso*-carbon atom of one phenyl group of the bridging $OSiPh_3$ group with a cerium atom (Ce . . . . C $= 2.982(9)$ Å).[117]

2.7.3   Praseodymium

Some structures of praseodymium(III) complexes with fluorinated tertiary alkoxo ligands have been reported. The tris-trifluoro-*tert*-butoxide adopts the trinuclear structure $[Pr_3(\mu_3\text{-}tftb)_2(\mu\text{-}tftb)_3(tftb)_2]$($tftb = OCMe_2(CF_3))$[127] with the three metals in an isosceles triangle capped above and below by two $\mu_3\text{-}OCMe_2(CF_3)$ ligands. The μ-ligands span the edges of the triangle, and one Pr atom, which has two terminal alkoxo groups, is in a distorted octahedral coordination while the other two, with one terminal ligand apiece, are five-coordinated. However, the five-coordinated metals are close enough to fluorine atoms of CF_3 groups (Pr . . . .F, 2.756, 2.774 Å) to suggest distorted six- and seven-coordination (Fig. 4.31). This trinuclear structure is clearly related to the other $[M_3(\mu_3\text{-}OR)_2(\mu\text{-}OR)_3(OR)_4L_2]$ structures (Fig. 4.5) but without the neutral donor ligands L. With the even more acidic hftb group the ammonia produced in a side reaction was captured by the praseodymiums to give the dimeric ammine $[Pr_2(\mu\text{-}hftb)_2(hftb)_4(NH_3)_4]$. The centrosymmetric molecule has the typical edge-shared octahedral structure (*cf.* Fig. 4.20) but with one hftb and one ammonia in the Pr_2O_2 plane and the other hftb at right angles to the Pr_2O_2 plane and *trans* to the second ammonia on each praseodymium.[128] This is a different arrangement from that found in $[Sc_2(\mu\text{-}hfip)_2(hfip)_4(NH_3)_4]$.[103]

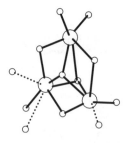

Figure 4.31 $Pr_3(\mu_3\text{-O})_2(\mu\text{-O})_3O_4$ core structure of $[Pr_3(\mu_3\text{-tftb})_2(\mu\text{-tftb})_3(\text{tftb})_4]$ (tftb = $OC(CH_3)_2CF_3$; $C(CH_3)_2CF_3$ groups omitted).

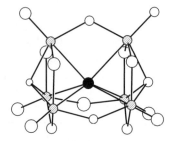

Figure 4.32 $Nd_6(\mu_6\text{-Cl})(\mu_3\text{-O})_2(\mu\text{-O})_9O_6$ core structure of $[Nd_6(\mu_6\text{-Cl})(\mu_3\text{-OPr}^i)_2(\mu\text{-OPr}^i)_9(\text{OPr}^i)_6]$ ($\oslash$ = Nd, $\bullet$ = Cl).

2.7.4 Neodymium

The first structurally characterized neodymium alkoxo compound reported was the μ_6-Cl centred complex $[Nd_6(\mu_6\text{-Cl})(\mu_3\text{-Pr}^i)_2(\mu\text{-OPr}^i)_9(\text{OPr}^i)_6]$ (Fig. 4.32).[129] The central chloride ion is surrounded by a trigonal prism of neodymium ions. Each Nd has one terminal isopropoxo ligand *trans* to the chloride, with four bridging isopropoxides making a distorted octahedral coordination. It was noted by Evans and Sollberger[107] that this structure was related to the trinuclear $[Y_3(\mu_3\text{-OBu}^t)(\mu_3\text{-Cl})(\mu\text{-OBu}^t)_3(\text{OBu}^t)_4(\text{thf})_2]$, which by the replacement of the THF molecules and addition of a trinuclear $[Y_3(\mu\text{-OBu}^t)(\mu\text{-OBu}^t)_3(\text{OBu}^t)_5]$ moiety would give the $[M_6(\mu_6(\text{Cl})(\mu_3\text{-OBu}^t)_2(\mu\text{-OBu}^t)_9(\text{OPr}^i)_6]$ structure. The functionalized chelating ligand $OCMe_2CH_2PMe_2$ formed a mononuclear six-coordinated Nd(III) complex $[Nd(OCMe_2CH_2PMe_2)_3]$ with the facial configuration isostructural with the yttrium analogue.[105] The mixed ligand chloride-bridged dimer $[Nd_2(\mu\text{-Cl})_2(\text{OCBu}_3^t)_4(\text{thf})_2]$ exhibits a five-coordinated neodymium (data in Table 4.7).[130] With acetonitrile as the supplementary ligand the mononuclear complex $[Nd(\text{OCBu}_3^t)_3(\text{MeCN})_2]$ was obtained with a trigonal bipyramidal Nd with the "tritox" ligands in the equatorial plane.[131] The homoleptic complex $[Nd_4(\mu\text{-ONp})_8(\text{ONp})_4]$ forms a tetranuclear species with five-coordinate (square pyramidal) Nd which is isostructural with the La analogue (Fig. 4.26).[118] With the "ditox" ligand $OCHBu_2^t$ the unsymmetrical dinuclear complex $[(ditox)_2Nd(\mu\text{-}ditox)_2Nd(ditox)_2(CH_3CN)]$ was formed having 1 five-coordinated and 1 four-coordinated metal in the molecule.[132]

Dinuclear complexes with five-coordinated Nd (distorted trigonal bipyramids) were obtained using the di-isopropylcarbinolate $[Nd_2(\mu\text{-}OCHPr_2^i)(OCHPr_2^i)_4(thf)_2]$ and $[Nd_2(\mu\text{-}OCHPr_2^i)_2(OCHPr_2^i)_4(C_5H_5N)_2]$. When the THF ligands were replaced by dimethoxyethane (DME) a chain polymer was formed with the DME ligands linking μ-dinuclear units in $[Nd_2(\mu\text{-}OCHPr_2^i)_2(OCHPr_2^i)_4(MeOCH_2CH_2OMe)]_n$.[133] The mixed ligand mono-anion $[Nd(OSiMe_3)\{N(SiMe_3)_2\}_3]^-$ has a distorted tetrahedral Nd.[134]

2.7.5 Samarium

Several organosamarium alkoxides have been structurally characterized. In the dinuclear molecule $[(C_5Me_5)_2Sm(\mu\text{-}O_2C_{16}H_{10})Sm(C_5Me_5)_2]$ each bis-pentamethylcyclopentadienyl samarium moiety is bonded to an oxygen of the dihydroindenoindene diolate bridging ligand giving rise to three- or seven-coordination depending on whether the C_5Me_5 ligand is counted as monodentate or tridentate.

Similarly the THF adduct $[(thf)(C_5Me_5)_2Sm(\mu\text{-}O_2C_{16}H_{10})Sm(C_5Me_5)_2(thf)]$ is either four- or eight-coordinated.[135] Similar geometry is found in the mononuclear complex $[(C_5Me_5)_2Sm(OC_4H_8C_5Me_5)(thf)]$.[136]

In $[(thf)(C_5Me_5)_2Sm(\mu\text{-}OSiMe_2OSiMe_2O)Sm(C_5Me_5)_2(thf)]$ the two $(C_5Me_5)_2Sm(thf)$ moieties are linked by the tetramethyldisiloxane diolate bridging ligand.[137] Using the potentially tridentate (ON_2) ligand $L = OC(Bu^t)(2\text{-}CH_2NC_5H_3Me\text{-}6)_2$ the novel five-coordinated complex SmL_2 was obtained with two ligands acting as bidentate, with *trans* oxygen and nitrogen donor atoms in the basal positions of a distorted square pyramid, with the third ligand's oxygen occupying the axial position as a monodentate ligand.[138]

2.7.6 Europium and Dysprosium

The only alkoxo europium complex reported to date is the dimeric $[(C_5Me_5)_2Eu_2(\mu\text{-}OBu^t)_2(OBu^t)_2]$[139] which was found to be isostructural with the fully characterized yttrium analogue.[112]

A five-coordinated (TBP) mononuclear dysprosium complex $[Dy(OCHBu_2^t)_3(CH_3CN)_2]$[132] has also been structurally characterized (Table 4.7).

2.7.7 Ytterbium

Complexes of both Yb(II) and Yb(III) have been structurally characterized. The mixed ligand Yb(II) complex $[Yb_2(\mu\text{-}OCBu_3^t)_2\{N(SiMe_3)_2\}\{N(SiMe_3)_2\}_2]$ is dimeric with bridging alkoxo ligands giving the familiar pseudo-tetrahedral structure.[140] The pseudo-tetrahedral configuration is also present in the mononuclear Yb(II) complex $[Yb(OCBu_3^t)_2(thf)_2]$.[140] The dinuclear organoytterbium complex has bridging ethoxo groups and coordinated diethylether ligands.[141]

The mixed ligand complex $[Yb\{OC(Bu^t)(2\text{-}CH_2NC_5H_3Me\text{-}6)\}_2\{N(SiMe_3)_2\}]$ involves a five-coordinated Yb(III) ion with two bidentate functionalized alkoxo ligands and a terminal (axial) bis-trimethylsilylamide.[138] Another five-coordinated Yb(II) dimer

Table 4.7 Alkoxides of lanthanides

Compound	Metal coordination	M–O Bond lengths (Å)		Bond angles (°) $\widehat{MOC}$	Reference
		Terminal	Bridging		
[La₃(μ_3-OBuᵗ)₂(μ-OBuᵗ)₃(OBuᵗ)₄(BuᵗOH)₂]	6	2.195–2.315 (16) av. 2.280	2.408–2.460 (12) (μ_2) av. 2.440; 2.505–2.622 (14) (μ_3) av. 2.571	177.4 (t, av.); 129.5 (μ_2, av.); 122.3 (μ_3, av.)	6
[La₂(μ-OCPh₃)₂(OCPh₃)₄]	4	2.175, 2.184 (2)	2.389, 2.483 (2)	160.8, 174.1 (t); 120.9 (b)	117
[La₄(μ-ONp)₈(ONp)₄] Np = CH₂Buᵗ	5	2.157, 2.169 (7)	2.376–2.445 (9) av. 2.405	162.4 (t); 126.5 (b, av.)	118
[La(hftb)₃(thf)₃] hftb = OCMe(CF₃)₂	6	2.222–2.237 (15) av. 2.229	—	160.9–176.1	7
[La(OSiPh₃)₃(thf)₃]	6	2.203–2.246 (7) av. 2.226	—	av. 167.6; 167.9, 173.1	116
[(C₅H₅)₃Ce(OBuᵗ)]	4ˡ	2.045 (6)	—	176.3	119
[Ce₂(μ-OCHBuᵗ₂)₂(OCHBuᵗ₂)₄]	4	2.142, 2.152 (3)	2.363 (3)	—	120
[Ce(OBuᵗ)₂(NO₃)₂(BuᵗOH)₂]	8	2.026, 2.023 (5)	—	168.9, 170.2	121
[Ce₂(μ-OPrⁱ)₂(OPrⁱ)₆(PrⁱOH)₂]	6	2.037, 2.046 (9)	2.310, 2.320 (6)	173, 174 (t); 127 (b)	122
[Ce₂(μ-OBuᵗ)₂(OCBuᵗ₃)₄]	4	2.157, 2.162 (4)	2.422, 2.430	168.8, 174.3 (t)	123
[Ce₂[μ-OC₂H₄N(Me)C₂H₄NMe₂]₂(OPrⁱ)₆]	7	2.122–2.137 (5) av. 2.130	2.328, 2.411 (4)	151.1–174.8 (t); av. 161.3; 120.9 (b, av.)	124
[CeL₂] L = {OC(CF₃)₂CH₂C(Me)=NCH₂}₂	8	2.196–2.230 (2) av. 2.211	—	—	125
[Ce(OSiPh₃)₃(thf)₃]	6	2.208–2.234 (4) av. 2.222	—	169.9–177.0 av. 174.4	126
[Ce₂(μ-OSiPh₃)₂(OSiPh₃)₄]	4	2.141–2.185 (7)	2.345–2.583 (5)	161.4, 166.5 (t); 109.8 (b)	117
[Pr₃(μ_3-tftb)₂(μ-tftb)₃(tftb)₄] tftb = OCMe₂(CF₃)	5 and 6	2.112–2.168 (14) av. 2.144	2.407–2.834 (11) (μ_3) av. 2.583; 2.371–2.459 (12) (μ_2) av. 2.417	—	127
[Pr₂(μ-hftb)₂(hftb)₄(NH₃)₄] hftb = OCMe(CF₃)₂	6	2.200, 2.222 (5)	2.462 (5)	—	128
[Nd(OCMe₂CH₂PMe₂)₃]	6	2.174 (2)	—	—	105

Compound	Coordination no.				Ref.
[Nd₂(μ-Cl)₂(OCBu^t₃)₄(thf)₂]	5	2.090–2.114 (3)	—	—	130
[Nd(OCBu^t₃)₃(CH₃CN)₂]	5	2.149–2.171 (5) av. 2.162	—	167.5–171.8 av. 170.3	131
[Nd₄(μ-ONp)₈(ONp)₄] Np = CH₂Bu^t	5	2.138 (8)	2.320–2.381 (12) av. 2.341	163.9 (t) 126.6 (b, av.)	118
[Nd₂(μ-OCHBu^t₂)₂(OCHBu^t₂)₄(CH₃CN)]	4 and 5	2.132–2.151 (4) av. 2.143	2.359–2.394 (4) av. 2.374	166.4 (t, av.) 127.0 (b, av.)	132
[Nd₆(μ₆-Cl)(μ₃-OPr^i)₂(μ-OPr^i)₉(OPr^i)₆]	6	2.05 (2)	2.36 (4) (μ, av.) 2.45 (5) (μ₃, av.)	—	129
[Nd₂(μ-Cl)₂(OCBu^t₃)₄(thf)₂]	5	2.108–2.114 av. 2.105	—	163.8–174.4 av. 168	130
[Nd₂(μ-OCHPr^i₂)₂(OCHPr^i₂)₄(thf)₂]	5	2.146, 2.160 (4)	2.368, 2.394 (4)	159.6, 170.2 (t) 117.9, 130.6 (b)	133
[Nd₂(μ-OCHPr^i₂)₂(OCHPr^i₂)₄(C₅H₅N)₂]	5	2.133, 2.158 (4)	2.380, 2.383 (4)	173.6, 176.1 (t) 120.8, 130.8 (b)	133
[Nd₂(μ-OCHPr^i₂)(OCHPr^i₂)₄(μ-dme)]∞ dme = MeOC₂H₄OMe	5	2.144–2.145 (6)	2.359–2.391 (5)	169.3, 174.3 (t) 117.7, 133.0 (b)	133
[Li(thf)₄]⁺[Nd(OSiMe₃)₃]⁻	4	2.173 (11)	—	177.1	134
[(C₅Me₅)₂Sm(μ-O₂C₁₆H₁₀)Sm(C₅Me₅)₂]	3¹	2.08 (2)	—	173.2	135
[(thf)(C₅Me₅)₂Sm(μ-O₂C₁₆H₁₀)Sm(C₅Me₅)₂(thf)]	4¹	2.099 (9)	—	166.3	135
[C₅Me₅)₂Sm(OC₄H₈C₅Me₅)(thf)]	4¹	2.08 (1)	—	165.2	136
[(thf)(C₅Me₅)₂Sm(μ-OSiMe₂OSiMe₂O)Sm(C₅Me₅)₂)(thf)]	4¹	2.157 (5)	—	173.6	137
[Sm[OC(Bu^t)(2-CH₂NC₅H₃Me-6)]₃]	5	2.116 (2) (t) 2.142, 2.172 (3) (c)	—	168.6 (t) 137.8, 147.8 (c)	138
[Dy(OCHBu^t₃)₃(CH₃CN)₂]	5	2.057, 2.064, 2.063 (5)	—	177.0, 177.2, 177.7	132
[Yb₂(μ-OCBu^t₃)₂[N(SiMe₃)₂]₂]	3	—	2.294–2.320 (6) av. 2.307	129.6–136.2 av. 133.0	140
[Yb(OCBu^t₃)₂(thf)₂]	4	2.07, 2.09 (2)	—	155, 175	140
[{(Me₃Si)₃C}₂Yb₂(μ-OEt)₂(Et₂O)₂]	4	—	2.267, 2.276 (10)	120.9, 121.0	141
[Yb{OC(Bu^t)(2-CH₂NC₅H₃Me-6)}₂[N(SiMe₃)₂]]	5	—	2.057, 2.072 (3)	142.3, 143.0 (c)	138
[Yb₂(μ-OSiMe₂Bu^t)₂(OSiMe₂Bu^t)₂(dme)₂]	5	2.163 (11)	2.269, 2.311 (8)	172.9 (t) 119.1, 141.5 (b)	142
[Lu₂(μ-OCMe₂CH₂OMe)₃(OCMe₂CH₂OMe)₃]	6 and 7	2.053, 2.058 (3) (t) 2.123 (c)	2.196–2.373 (3) av. 2.260	166.4, 174.8 (t) 126.6 (c) 129.7 (b, av.)	143

b = bridging, c = chelating, t = terminal.
¹Coordination doubtful.

265

was formed using the dimethyl-*tert*-butyl siloxo ligand with DME as supporting ligand in $[Yb_2(\mu\text{-}OCMe_2Bu^t)_2(OCMe_2Bu^t)_2(dme)_2]$.[142]

2.7.8 Lutetium

Lutetium(III) forms a novel unsymmetrical dinuclear molecule $[Lu_2(\mu\text{-}OCMe_2CH_2OMe)_3(OCMe_2CH_2OMe)_3]$ in which one Lu is seven-coordinated and the other six-coordinated (Fig. 4.33).[143] The six-coordinated Lu is bonded to three bridging ligands (one of which is chelating) and two terminals whilst the seven-coordinated metal is bonded to three bridging (two of which are chelating) and one chelating ligand.

2.8   Alkoxides of the Actinides

2.8.1   Thorium

Although thorium(IV) alkoxides have been known for many years it was only in 1992 that the X-ray crystal structure of a homoleptic thorium compound was reported. Using the bulky di-isopropylcarbinolate ligand the dimeric compound $[Th_2(\mu\text{-}OCHPr^i_2)_2(OCHPr^i_2)_6]$ was obtained in which each Th is in a distorted TBP coordination.[144] The monomeric *tert*-butoxide $[Th(OBu^t)_4(C_5H_5N)_2]$ has a distorted octahedral configuration with pyridine ligands occupying *cis* positions.[145] With the less sterically demanding Et_2CHO ligand an edge-shared bi-octahedral complex $[Th_2(\mu\text{-}OCHEt_2)_2(OCHEt_2)_6(C_5H_5N)_2]$ was obtained. Each thorium is coordinated to one pyridine ligand located *cis* to the bridging alkoxo groups in a centrosymmetric molecule.[146] With the isopropoxo ligand the remarkable tetranuclear complex $[Th_4(\mu\text{-}OPr^i)_6(OPr^i)_{10}(C_5H_5N)_2]$ was obtained (Fig. 4.34).[146] The thorium ions are all octahedrally coordinated forming a zigzag chain through *cis* pairs of bridging isopropoxides. The two inner thoriums each have *cis* pairs of terminal isopropoxo groups whilst the end thoriums each have a pyridine ligand acting as chain blocker. The series of edge-sharing bi-octohedral dimers give rise to a centrosymmetrical molecule. The edge-sharing bi-octahedral structure was also found in the dimer $[Th_2(\mu\text{-}OPr^i)_2(OPr^i)_2I_4(Pr^iOH)_2]$ where the bridging isopropoxo group is unsymmetrical with a short Th–O (2.327 Å) *trans* to an iodide and the longer Th–O

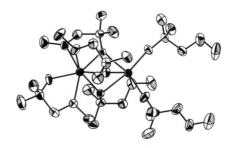

Figure 4.33 Structure of $[Lu_2(\mu\text{-}OCMe_2CH_2OMe)_3(OCMe_2CH_2OMe)_3]$ (H atoms omitted).

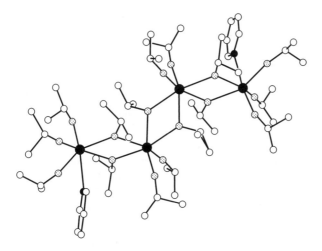

Figure 4.34 Structure of [Th$_4$(μ-OPr^i)$_6$(OPr^i)$_{10}$(py)$_2$] ($\bullet$ = N; $\bigcirc$ = O; $\bullet$ = Th; H atoms omitted).

(2.399 Å) is *trans* to a terminal isopropoxo group, which has a very short Th–O (2.056 Å) bond. The coordinated isopropanol was *trans* to an iodide and appeared to be hydrogen bonded to the iodide on the adjacent thorium ion.[147] Addition of the strong Lewis base quinuclidine to the dimeric [Th$_2$(OCHPr$_2^i$)$_8$] produced the mononuclear adduct [Th(OCHPr$_2^i$)$_4$(quinuc)] with five-coordinated (TBP) thorium with the quinuclidine occupying an axial site.[148] In [Th(OCHPr$_2^i$)$_3$I(C$_5$H$_5$N)$_2$] the Th is octahedrally coordinated with the alkoxo ligands in the facial configuration and the pyridines in the *cis* configuration.[148] A most remarkable organometallic thorium alkoxide is the hexanuclear complex [(C$_5$H$_5$)Th$_2$(μ-OPr^i)$_3$(OPr^i)$_4$]$_3$ which has a ring structure in which the pentahapto-cyclopentadienyls act as bridging ligands between pairs of thoriums which in turn are linked facially by three μ-OPr^i ligands.[149]

2.8.2   *Uranium*

Only a few structures of homoleptic uranium alkoxides have been reported. The hexamethoxide [U(OMe)$_6$] is a mononuclear molecule with an octahedral UO$_6$ framework.[150] The mixed valency [U(IV), U(V)] complex [U$_2$(μ-OBu^t)$_3$(OBu^t)$_6$] has a face-sharing bi-octahedral structure. The U....U separation (3.549 Å) precludes any metal–metal bonding and the fact that the U–O bond distances for one metal are significantly shorter than for the other suggests that this molecule contains isolated U(IV) and U(V) atoms.[151] The uranium pentaisopropoxide is also a dimer [U$_2$(μ-OPr^i)$_2$(OPr^i)$_8$] with the edge-shared bi-octahedral structure (Fig. 4.2) (data in Table 4.8, pp. 269–270).[151] The iodo compound [U$_2$(μ-OPr^i)$_2$(OPr^i)$_2$I$_4$(Pr^{i}OH)$_2$] also has the edge-shared bi-octahedral structure with the coordinated isopropanol ligands *trans* to iodides and hydrogen bonded as in the analogous thorium compound.[152] The mononuclear bis-benzopinacolate complex with THF [U(OC$_2$Ph$_4$O)$_2$(thf)$_2$] has a distorted octahedral structure with the THF molecules *cis* to one another.[153] Structures of [U(OCBu$_3^t$)(BH$_4$)$_3$(thf)] and [U(OCBu$_3^t$)$_3$(BH$_4$)] have been determined.[154] The first compound has a five-coordinated (TBP) uranium with the "tritox" ligand in an axial position and the tetrahydroborato ligands in equatorial positions. The second compound has a distorted tetrahedral structure (Table 4.8). Octahedral complexes $[U(OCHPh_2)(BH_4)_3(thf)_2]$ and $[U(OCHPh_2)_2(BH_4)_2(thf)_2]$ have also been structurally characterized. In the former compound the three BH_4 ligands occupy one face with the two THF ligands *cis* to one another. The latter compound adopts a *trans–trans–trans* configuration.[155] Finally we mention some organo uranium alkoxides. The bis-allyl bis-isopropoxide is dimeric with isopropoxo bridges $[U_2(\mu\text{-}OPr^i)(OPr^i)_2(C_3H_5)_4]$.[156] Assuming the uranium coordination to be pseudo-trigonal bipyramidal, the bridges span axial–equatorial positions with the terminal isopropoxo groups in the other axial positions. With the cyclooctatetraenyl ligand (COT) the dinuclear complexes $[(cot)_2U_2(\mu\text{-}OEt)_2(BH_4)_2]$ and $[(cot)_2U_2(\mu\text{-}OPr^i)_2(OPr^i)_2]$ have been characterized. Assuming that the COT ligand occupies one coordination site then the uraniums are in distorted tetrahedral configurations.[157]

2.9   Alkoxides of Titanium, Zirconium, and Hafnium

2.9.1   *Titanium*

The first X-ray crystal structure determination of a tetraalkoxide was of the centrosymmetric tetranuclear ethoxo compound $[Ti_4(\mu_3\text{-}OEt)_2(\mu\text{-}OEt)_4(OEt)_{10}]$.[158] Each titanium atom is in a distorted octahedral coordination as indicated in Fig. 4.35. Since solution molecular weight measurements showed that the trimer $[Ti_3(OEt)_{12}]$ is present, the crystal structure determination clearly emphasized that it is unwise to attempt to predict crystal structures from solution data. Various attempts have been made to deduce the structure of the trimer[9] and a recent X-ray absorption study (XANES and EXAFS) inclines to the view that five-coordinated (TBP) titanium is present in a symmetrical molecule $[Ti_3(\mu\text{-}OEt)_3(OEt)_9]$.[159] The tetranuclear structure in the crystalline state was reinforced by the partial structure of the mixed alkoxide $[Ti_4(OMe)_4(OEt)_{12}]$ which had the same Ti_4O_{16} framework as in Fig. 4.35.[160] The same structure was also found for the tetramethoxide $[Ti_4(\mu_3\text{-}OMe)_2(\mu\text{-}OMe)_4(OMe)_{10}]$.[161]

With the tetradentate nitrilotriethoxo ligand titanium forms a dimeric complex $[Ti_2\{\mu\text{-}OC_2H_4)N(C_2H_4O)_2\}_2(OPr^i)_2]$, in which one arm of the chelating ligand forms

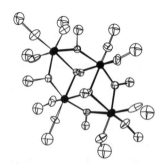

Figure 4.35   Structure of $[Ti_4(\mu_3\text{-}OMe)_2(\mu\text{-}OMe)_4(OMe)_{10}]$ (H atoms omitted).

Table 4.8 Alkoxides of the actinides

Compound	Metal coordination	M–O Bond lengths (Å)		Bond angles (°) $\widehat{MOC}$	Reference
		Terminal	Bridging		
[Th$_2$(μ-OCHPr$^i_2$)$_2$(OCHPr$^i_2$)$_6$]	5	2.141–2.161 (11) av. 2.154	2.408, 2.463 (11)	–	144
[Th(OBu^t)$_4$(C$_5$H$_5$N)$_2$]	6	2.161, 2.204 (6)		160.8, 171.6	145
[Th$_2$(μ-OCHEt$_2$)$_2$(OCHEt$_2$)$_6$(C$_5$H$_5$N)$_2$]	6	2.147–2.183 (6) av. 2.170	2.416, 2.438 (5)	159.6–171.5 (t) av. 166.5 129.5 (b)	146
[Th$_4$(μ-OPr^i)$_6$(OPr^i)$_{10}$(C$_5$H$_5$N)$_2$]	6	2.150–2.199 (10) av. 2.172	2.389–2.446 (10) av. 2.420	160.1–176.2 (t) av. 167.2	146
[Th$_2$(μ-OPr^i)$_2$(OPr^i)$_2$L$_4$(Pr^{i}OH)$_2$]	6	2.056 (7)	2.327, 2.399 (6)	173.4 (t) 135.1 (b)	147
[Th(OCHPr$^i_2$)$_4$(quinuc)] quinuc = quinuclidine	5	2.152 (12) (ax) 2.169 (14) (eq)	–	162.8–176.6 av. 168.5	148
[Th(OCHPr$^i_2$)$_3$I(C$_5$H$_5$N)$_2$]	6	2.132, 2.137, 2.139 (8)	–	169.9–175.1 (t) av. 172.5	148
[(C$_5$H$_5$)Th$_2$(μ-OPr^i)$_3$(OPr^i)$_4$]$_3$	6^1	2.09–2.18 (2) av. 2.14	2.37–2.46 (2) av. 2.42	170.8 (t, av.) 129.7 (b, av.)	149
[U$_2$(μ-OBu^t)$_3$(OBu^t)$_6$]	6	2.062, 2.073 (7) (U[1]) 2.106, 2.107 (7) (U[2])	2.251, 2.271 (9) (U[1]) 2.457, 2.502 (9) (U[2])	163.9–176.0 (t) av. 169.3 121.9–135.5 (b) av. 130.0	151

(continued overleaf)

269

Table 4.8 (*Continued*)

Compound	Metal coordination	M–O Bond lengths (Å)		Bond angles (°) $\widehat{\text{MOC}}$	Reference
		Terminal	Bridging		
$[U_2(\mu\text{-}OPr^i)_2(OPr^i)_8]$	6	2.02–2.05 (1) av. 2.03	2.28, 2.29 (1)	160–165 (t) av. 163 128 (b)	151
$[U_2(\mu\text{-}OPr^i)_2(OPr^i)_2L_4(Pr^iOH)_2]$	6	2.02 (1)	2.27, 2.32 (1)	166.7 (t)	152
$[U(OC_2Ph_4O)_2(thf)_2]$	6	2.131–2.162 (4) av. 2.152	–	–	153
$[U(OCBu^t_3)(BH_4)_3(thf)]$	5	1.97 (1)	–	178.6	154
$[U(OCBu^t_3)_3(BH_4)]$	4	2.073–2.077 (7) av. 2.075	–	167.9–172.8 av. 170.6	154
$[U(OCHPh_2)(BH_4)_3(thf)_2]$	6	2.037 (3)	–	166.7	155
$[U(OCHPh_2)_2(BH_4)_2(thf)_2]$	6	2.070 (4)	–	161.0	155
$[U_2(\mu\text{-}OPr^i)_2(OPr^i)_2(C_3H_5)_4]$	5[1]	2.056 (13)	2.271, 2.413 (10)	178 (t) 120, 129 (b)	156
$[(cot)_2U_2(\mu\text{-}OEt)_2(BH_4)_2]$ cot $= C_8H_8\text{-}cyclo$	5[1]	–	2.296 (5)	117.8, 130.3	157
$[(cot)_2U_2(\mu\text{-}OPr^i)_2(OPr^i)_2]$	5[1]	2.027, 2.070 (8)	2.297–2.317 (6) av. 2.308	161.7, 168.3 (t) 112.2–134.2 (b) av. 123.7	157

b = bridging, t = terminal; ax = axial, eq = equatorial.
[1]Coordination doubtful.

an unsymmetrical bridge giving each metal a distorted octahedral coordination (data in Table 4.9, pp. 276–279).[161] Use of the bulky Ph_3SiO ligand in place of the isopropoxide gave rise to the monomeric five-coordinated (TBP) "titanatrane" $[Ph_3SiOTi(OC_2H_4)_3N]$, whereas the acetato derivative $[Ti_2\{(\mu\text{-}OC_2H_4O)N(C_2H_4O)_2\}_2(OAc)_2]$ was dimeric through the bridging chelate ligand but the bidentate acetato group gave the titanium the coordination number of seven.[163] The dimethylamido derivative is also dimeric with distorted octahedral Ti, with one arm of the chelate ligand bridging $[Ti_2\{(\mu\text{-}OC_2H_4)N(C_2H_4O)_2\}_2(NMe_2)_2]$. This compound differs from the isopropoxo derivative in having the dimethylamido-nitrogen *trans* to one of the bridging oxygens whereas the isopropoxo ligand is *trans* to a nitrogen atom.[164] The structure of the isopropylthiolate $[Ti_2\{(\mu\text{-}OC_2H_4)N(C_2H_4O)_2\}_2(SPr^i)_2]$ corresponds to that of the dimethylamido derivative.[165] Interestingly the dimer formed with the pinacolato ligand $[Ti\{(\mu\text{-}OC_2H_4)N(C_2H_4O)_2\}]_2$ involves a diolate bridge linking the two "titanatrane" moieties giving rise to five-coordinated (TBP) Ti with the diolate oxygens *trans* to the nitrogen of the chelating ligand.[165] Homochiral trialkanolamines $N(CH_2CHROH)_3$ have been used in metalatrane synthesis. Thus with titanium the monomeric chlorotitanatrane $[ClTi(OCHPr^iCH_2)_2N]$ has been characterized as a five-coordinate (TBP) complex with chloride and chelate ligand nitrogen in the axial positions.[166] Special interest was generated by the structure of the dinuclear *tert*-butylperoxo titanatrane $[Ti_2\{(\mu\text{-}OC_2H_4)N(C_2H_4O)_2\}_2(O_2Bu^t)_2]$ in which the η^2-peroxo ligand gave rise to seven-coordinated titanium.[167] This molecule has significance as a model for an intermediate in the oxidation of various organic substrates by *tert*-butyl hydroperoxide. Sharpless et al.[168] had proposed the bonding of an η^2-*tert*-butylperoxo group to the dinuclear titanium tartrate catalysts involved in the asymmetric epoxidation of allylic alcohols. The structure of the diisopropoxo titanium-N,N'-dibenzyltartramide dimer $[(Pr^iO)_4Ti_2\{\mu\text{-}OCH(CONHCH_2Ph)CH(CONHCH_2Ph)O\}_2]$ revealed distorted octahedral titanium atoms each coordinated facially by a tartramide ligand through two diolate oxygens and one of the carbonyl groups with one diolate oxygen bridging to the adjacent metal atom (Fig. 4.36). One isopropoxo group is *trans* to the bridging diolate oxygen and the other is *trans* to the carbonyl oxygen. Exchange of the two isopropoxo groups and dissociation of the carbonyl oxygen exposes a meridional set of coordination sites to accommodate the allylic alkoxo group and the η^2-butylperoxo ligand.[168] Later work using the diisopropyltartrato ligand produced dinuclear, trinuclear, and tetranuclear complexes which were structurally characterized.[169] A centrosymmetrical dimer was formed by the mixed ligand complex $[Ti_2(\mu\text{-}OPr^i)_2(OPr^i)_2(OPr^i)_4\{HC(SO_2CF_3)_2\}_2]$ which contains titanium in distorted octahedral coordination (Table 4.9).[170] In the bis-chelated complex $[Ti(OEt)_2(diket)_2]$ (diket = 4,4,4-trifluoro-1-phenyl-1,3-butanedionate) the terminal ethoxo groups, as expected, are in *cis* positions in the octahedral structure.[171]

Although titanium dichloride dialkoxide alcoholates have been known for many years[9] it was only recently that an X-ray structure was reported.[172] The dimeric complex $[Ti_2(\mu\text{-}OC_2H_4Cl)_2(OC_2H_4Cl)_2Cl_4(ClC_2H_4OH)_2]$ forms an edge-shared bi-octahedral molecule with bridging alkoxo groups. The terminal alkoxo group is *trans* to the bridging alkoxide and the alcohol ligand is *trans* to a chloride on one Ti atom and appears to be hydrogen bonded to the chloride on the adjacent Ti (Table 4.9). The 3,3'-disubstituted-1,1'-bi-2-naphtholates R_2BINO (R = Me, Me_2Bu^tSi) gave rise to the mononuclear $[Ti(OPr^i)_2\{(Bu^tMe_2Si)_2bino\}]$ four-coordinated complex and

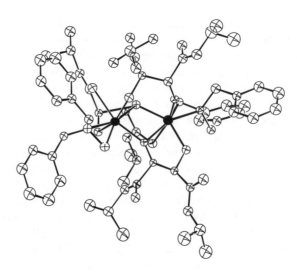

Figure 4.36 Structure of [Ti$_2$(dipt)$_3$\{ON(CH$_2$Ph)$_2$\}$_2$] (dipt = diisopropyltartrate; H atoms omitted).

the binuclear [Ti$_2$(μ-Me$_2$bino)$_2$(OPr^i)$_4$] five-coordinated complex. Another four-coordinated complex, [(Pr^{i}O)$_3$Ti\{(Bu^{t}Me$_2$Si)$_2$bino\}Ti(OPr^i)$_3$], was also characterized (Table 4.9).[173]

The hexadentate ligand *N,N,N',N',N'*-tetrakis-(2-hydroxy propyl) ethylene diamine (H4THPED) gave rise to the dimeric complex [Ti(thped)]$_2$ in which the Ti is seven-coordinated distorted monocapped trigonal prism or distorted pentagonal biprism.[174] The tetraneopentyloxide was found to be a dimer [Ti$_2$(μ-OCH$_2$Bu^t)$_2$(OCH$_2$Bu$_t$)$_6$] with the titanium atoms exhibiting distorted TBP five-coordination.[201]

It is interesting to note that the mixed alkoxide [Ti$_2$(μ-OPr^i)$_2$(OPr^i)$_2$(hfip)$_4$] is dimeric in contrast to the monomeric Ti(OPr^i)$_4$.[175] Doubtless the electronegative hfip groups enhance the electrophilicity of the titanium thus strengthening the isopropoxo bridges and countering the steric effects of the ligands. Nevertheless the bridge bonding is very unsymmetrical (1.900 Å, 2.121 Å) and the compound is monomeric in solution and in the vapour phase. The Ti atoms are five-coordinated with the TBP configuration. Another mixed ligand dimer [Ti$_2$(μ-OPr^i)$_2$(OPr^i)$_2$Cl$_4$] also exhibits five-coordinated titanium.[176] The *cis*-octahedral configuration for the dialkoxo-bis-acetylacetonates Ti(OR)$_2$(acac)$_2$ proposed on ^{1}H NMR evidence in 1965[177] was confirmed by the X-ray crystal structure of [Ti(OPr^i)$_2$(acac)$_2$].[176] However, a series of trialkoxo-monoacetylacetonates [Ti$_2$(μ-OR)$_2$(OR)$_4$(acac)$_2$] (R = Me, Et, Pr^i) and mono-2,2,6,6-tetramethylheptane-3,5-dionates (tmhd) [Ti$_2$(μ-OR)$_2$(OR)$_4$(thmd)$_2$] (R = Me, Pr^n, Pr^i) were shown to be dimeric with octahedral Ti.[178] The terminal alkoxo groups adopted the *cis* configuration in all of these compounds. Some interesting tetranuclear complexes [Ti$_4$(OPr^i)$_{10}$L$_2$] were obtained using the tridentate ligands L = tris(hydroxymethylethane) (THME) and tris(hydroxymethylpropane) (THMP). In both compounds the Ti$_4$O$_{16}$ framework contains two types of distorted octahedral titanium atoms, one type having three terminal isopropoxo groups in the *fac* configuration and the other type having two *cis* terminal isopropoxo groups analogous to the terminal ligands in $Ti_4(OR)_{16}$. The two ligands L each have two μ-OCH_2 groups and one μ_3-OCH_2 group, *e.g.* $(\mu$-$OCH_2)_2CMe(CH_2O$-$\mu_3)$.[179]

From a reaction involving $Ti(OPr^i)_4$ and 2-hydroxy-ethylmethacrylate the remarkable pentanuclear mixed ligand molecule $[Ti_5(\mu_4$-$OC_2H_4O)(\mu_3$-$OC_2H_4O)(\mu$-$OPr^i)(OPr^i)_9]$ containing ethane-1,2-diolate groups was obtained.[180] This asymmetric structure (Fig. 4.37) contains 1 seven-coordinate, 3 six-coordinate and 1 five-coordinate titanium atom whilst the chelating–bridging diolate ligands exhibit four types of bonding mode.

The isolated $[Ti(OC_2H_4O)_3]_2^-$ octahedral tris-chelated diolato complex has been structurally characterized in the salts $Na_2[Ti(OC_2H_4O)_3](HOC_2H_4OH)_4$ and $K_2[Ti(OC_2H_4O)_3](HOC_2H_4OH)_{2.5}$.[181] The titanium tris-glycolate is also present in the barium salt $[Ba(HOC_2H_4OH)_4(H_2O)]_2^+[Ti(OC_2H_4O)_3]_2^-$. The cation and anion are linked by hydrogen bonds from the coordinated glycol hydroxyls in the cation interacting with diolate oxygens in the anion.[182] In the mixed ligand complex $[Ti_2(\mu$-$OEt)_2(OEt)_4(O_2CCH_2NH_2)_2]$ in which the glycinate ligand is chelating, the terminal ethoxo groups are in *cis* positions. The glycinate nitrogen is *trans* to one of the terminal ethoxo groups whilst the carboxylate oxygen is *trans* to a bridging ethoxo group.

Some interesting titanosilsesquioxanes have been structurally characterized. The compounds were obtained from the reactions of the cubic trisilanols $[R_7Si_7O_9(OH)_3]$ ($R = c$-pentyl, c-hexyl) with $Ti(OPr^i)_4$. The crystalline dimer $[Ti_2(\mu$-$OMe)_2\{O_{12}Si_7(C_6H_{11})_7\}(MeOH)_2]$ exhibited octahedral titaniums[184] whereas the mononuclear $[Ti\{O_{12}Si_7(C_6H_{11})_7\}(OSiMe_3)]$ trimethylsilanolate contains tetrahedral titanium.[185] These compounds are of special interest as soluble titanium-centred catalysts for the epoxidation of alkenes by *tert*-butylhydroperoxide (see Chapter 7).

A few organotitanium alkoxides have been structurally determined. In bis-benzyl titanium diethoxide a centrosymmetric dimer $[(C_6H_5CH_2)_4Ti_2(\mu$-$OEt)_2(OEt)_2]$ is formed involving five-coordinated Ti in a distorted TBP configuration.[186] In bis-methyl titanium tritox methoxide (tritox = tri-*tert*-butyl carbinolate, Bu_3^tCO) the methoxo

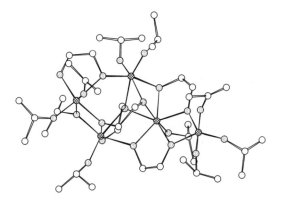

Figure 4.37 Structure of $[Ti_5(\mu_4$-$OC_2H_4O)(\mu_3$-$OC_2H_4O)_3(\mu$-$OC_2H_4O)(\mu$-$OPr^i)(OPr^i)_9]$ ($\circledast$ = Ti; $\bigcirc$ = O; H atoms omitted).

bridged dimer [Me$_4$Ti$_2$(μ-OMe)$_2$(tritox)$_2$] has titanium in the square pyramidal configuration.[187] The bulky tritox ligands occupy the apical positions with the Ti atoms located 0.71 Å above the basal planes comprising two *cis* methyl groups and two bridging methoxides with the two tritox groups being *trans* across the dimer. The 1,3-bis-(trimethylsilyl)cyclopentadienyl titanium *cis*-1,3,5-cyclohexan-triolate forms a monomer [{Me$_3$Si}$_2$C$_5$H$_3$}Ti(O$_3$C$_6$H$_9$)].[188]

The bis-cyclopentadienyl titanium(III) alkoxides are dimeric [(C$_5$H$_5$)$_4$Ti$_2$(μ-OR)$_2$] (R = Me, Et) and paramagnetic, exhibiting antiferromagnetic behaviour.[189] A novel mononuclear Ti(III) alkoxide was obtained using the functionalized alcohol Me$_2$PCH$_2$CBu$_2^t$(OH). The complex [TiCl(OCBu$_2^t$CH$_2$PMe$_2$)$_2$] involves titanium in a TBP configuration with the P atoms of the chelating alkoxo ligands occupying the axial positions and the oxygens and the chlorine in the equatorial positions.[190]

2.9.2 Zirconium

Relatively few structures of alkoxo zirconium compounds have been reported. Although a dimeric structure [Zr$_2$(μ-OPr^i)$_2$(OPr^i)$_6$(Pr^{i}OH)$_2$] had been proposed for the crystalline zirconium isopropoxide isopropanolate many years ago[191] the edge-shared bi-octahedral structure was not properly determined by X-ray crystallography until 1990.[192] An interesting feature of this structure is the location of the coordinated isopropanol molecules which are each linked by a hydrogen bond to the terminal isopropoxo ligand on the adjacent zirconium atom Pr^{i}O–H....OPr^i. This causes a marked distortion due to bending of the Zr–OPr^i bonds which are drawn together by the hydrogen bond. A similar situation occurs in the Hf,[170,200] Ce(IV)[122,192] and Sn(IV) analogues.[88,89] The bulky tris(trimethylsilyl)silyl ligand formed the monomeric four-coordinated zirconium *tert*-butoxo compound [(Me$_3$Si)$_3$SiZr(OBu^t)$_3$] which contains a Zr–Si bond.[193] The bidentate ligand *trans*-1,2-cyclohexanediolate bridges two ZrCl$_3$ moieties in the binuclear compound [(thf)$_2$Cl$_3$Zr(O$_2$C$_6$H$_{10}$)ZrCl$_3$(thf)$_2$].[194] The octahedral Zr atoms are each bonded to three chlorines in a meridional configuration *cis* to the Zr–O bonds. The remarkable pentanuclear amido imidonitride cluster Zr$_5$(μ_5-N)(μ_3-NH)$_4$(μ-NH$_2$)$_4$(OCBu$_3^t$)$_5$] is stabilized by the terminal tritox ligands.[195] The zirconiums are all octahedral and the structure bears a striking resemblance to the pentanuclear oxo isopropoxo clusters [M$_5$(μ_5-O)(μ_3-OPr^i)$_4$(μ-OPr^i)$_5$] (M = Sc, Y, Ln). The structure of monomeric Zr(hftb)$_4$ (hftb = OCMe(CF$_3$)$_2$) revealed an essentially tetrahedral configuration about the Zr atom.[196] Another mononuclear mixed ligand complex is [(oep)Zr(OBu^t)$_2$] (oep = octaethylporphyrin) where the *tert*-butoxo groups occupy *cis* positions above the plane of the porphyrin ring.[197] The Zr–O–C bonds are practically linear.

The tridentate ligand tris(hydroxymethyl)ethane forms a tetranuclear complex [Zr$_4${μ-OCH$_2$)$_3$CMe}$_2$(μ-OPr^i)$_2$(OPr^i)$_8$], which exhibits a different structure from the titanium analogue.[169]

2.9.3 Hafnium

Only a few structures of hafnium alkoxo compounds have been reported. As expected, hafnium isopropoxide [Hf$_2$(μ-OPr^i)$_2$(OPr^i)$_6$(Pr^{i}OH)$_2$] has the edge-shared bi-octahedral structure like that of the analogous zirconium compound.[200] The coordinated isopropanol in $[Hf_2(OPr^i)_8(Pr^iOH)_2]$ may be replaced by pyridine and the structure of $[Hf_2(\mu\text{-}OPr^i)_2(OPr^i)_6(Pr^iOH)(py)]$ shows that the edge-shared bi-octahedral configuration is retained. The coordinated Pr^iOH molecule is clearly hydrogen bonded to the terminal isopropoxo group on the adjacent hafnium atom.[200] In the mixed ligand complex $[(Bu^tO)_3Hf(\mu\text{-}Cl)(\mu\text{-}OBu^t)(\mu\text{-}NHPMe_3)Hf(OBu^t)_3]$ the structure comprises a confacial bi-octahedral configuration with three different bridging ligands.[198] A mononuclear octahedral hafnium compound $[Hf(OSiPh_2OSiPh_2OSiPh_2O)_2(py)_2]$ was obtained using the chelating hexaphenyl trisiloxane diolate ligands supplemented by pyridine donors in the *cis* configuration. Eight-membered hafnasiloxane rings are present in a distorted octahedral structure.[199] The Hf–O bonds *trans* to the pyridines are shorter (1.961 Å) than those *cis* (2.014 Å).

2.10   Alkoxides of Vanadium, Niobium, and Tantalum

2.10.1   *Vanadium*

The structures of many alkoxo vanadium compounds have been reported but they all contain oxo ligands and are accordingly dealt with in the chapter on metal oxo alkoxides.

2.10.2   *Niobium*

The Nb(v) complex $[Nb(OPr^i)_3(diket)(NCS)]$ (diket = PhCO(CH)COPh) has a *fac*-octahedral structure.[202] The first homoleptic niobium alkoxide structure determined was the dimeric pentamethoxide $[Nb_2(\mu\text{-}OMe)_2(OMe)_8]^{203}$ and this confirmed the edge-shared bi-octahedral configuration (Fig. 4.2) originally proposed on the basis of variable-temperature ^1H NMR spectroscopy.[204] Two independent centrosymmetrical molecules are present in the unit cell (data in Table 4.10, p. 281). The niobium(IV) dichloride dimethoxide complexes $[Nb_2(\mu\text{-}OMe)_2(OMe)_2Cl_4(L)_2]$ (L = MeOH, MeCN) also exhibit the edge-shared bi-octahedral structure.[205] In the methanolate the four chlorines occupy the same plane as the $Nb(\mu\text{-}OMe)_2Nb$ double bridge. Interestingly the methanol hydroxyls form intermolecular hydrogen bonds to neighbouring chlorides. In the acetonitrile complex the MeCN ligands are in the $ClNb(\mu\text{-}OMe)_2NbCl$ plane with methoxo groups *trans* to chlorines in the axial positions. These diamagnetic molecules have d^1–d^1, Nb–Nb single bonds (Nb–Nb, 2.781 Å, 2.768 Å) in contrast to the $d^0[Nb_2(OMe)_{10}]$ (Nb–Nb, 3.5 Å). A similar structure is shown by the Nb(III) complex $[Nb_2(\mu\text{-}OPr^i)(\mu\text{-}Cl)(Pr^iOH)_4Cl_4]$ which has a $Nb(\mu\text{-}OPr^i)(\mu\text{-}Cl)Nb$ double bridge.[206] The Pr^iOH ligands *trans* to chlorine (Nb–O, 2.09 Å, 2.13 Å) are significantly shorter than those *trans* to the bridging isopropoxo group (Nb–O, 2.22 Å, 2.27 Å) and there is intramolecular H....Cl hydrogen bonding. The formal double bond (Nb=Nb) character is reflected in the short niobium–niobium distance (2.611 Å).

The $[Nb_2(\mu\text{-}OMe)_3(OMe)_6]^-$ anion has the confacial bi-octahedral structure in three related salts $[Mg(MeOH)_6]_2[Nb_2(OMe)_9]Cl_3$, $[Mg(MeOH)_6][Nb_2(OMe)_9]I.2MeOH$ and $[Na(MeOH)_6][Nb_2(OMe)_9]$.[207] The niobium–niobium bond distance is quite short for a formally d^1–d^1 single bond (Nb–Nb, 2.632–2.652, av. 2.640 Å) whilst the

Table 4.9 Alkoxides of titanium, zirconium, and hafnium

Compound	Metal coordination	Bond lengths (Å)		Bond angles (°) $\widehat{MOC}$	Reference
		Terminal	Bridging		
[Ti₄(μ₃-OEt)₂(μ-OEt)₄(OEt)₁₀]	6	1.77 (3)	2.23 (3) (μ₃) / 2.03 (3) (μ₂)	–	158
[Ti₄(μ₃-OMe)₂(μ-OMe)₄(OMe)₁₀]	6	1.780–2.078 (12) / av. 1.90	2.134–2.196 (9) (μ₃) / av. 2.16 / 1.959–2.084 (9) (μ₂) / av. 2.01	140.0–160.6 (t) / av. 150 / 117.1–125.7 (μ₂) / 114.6–118.3 (μ₃)	161
[Ti₂{(μ-OC₂H₄N(C₂H₄O)₂}₂(OPrⁱ)₂]	6	1.833 (1) (OPrⁱ) / 1.856, 1.872 (1) (L)	1.998, 2.108 (1)	13.6 (t, PrⁱO) / 125.8, 126.8 (t, L) / 117.2, 123.7 (b)	162
[Ti{(OC₂H₄)₃N}(OSiPh₃)]	5	1.834			163
[Ti₂{(μ-OC₂H₄)N(C₂H₄O)₂}₂(NMe₂)₂]	6	1.873, 1.877 (1)	1.939, 2.160 (1)	–	164
[Ti₂{(μ-OC₂H₄)N(C₂H₄O)₂}₂(SPrⁱ)₂]	6	1.845 (2)	1.941, 2.117 (2)	–	165
[Ti{(μ-OC₂H₄)N(C₂H₄O)₂}₂(OCMe₂)₂]	5	1.776 (2) (diolate) / 1.830–1.841 (3) (L) / av. 1.836	–	159.5 (diolate)	165
[ClTi(OCHPrⁱCH₂)₃N]	5	1.800–1.818 (2) / av. 1.811	–	–	166
[Ti₂{(μ-OC₂H₄)N(C₂H₄O)₂}₂(O₂Buᵗ)₂]	7	1.913, 2.269 (2) (peroxo) / 1.850, 1.882 (3) (L)	1.994, 2.048 (2) (L)ˣ	–	167
[(PrⁱO)₄Ti₂{(μ-OCH(CONHCH₂Ph)CH(CONHCH₂Ph)O)₂]	6	1.805 (9)	1.973, 2.160 (9)	159.1	168
[Ti₂(μ-OPrⁱ)₂(OPrⁱ)₄{HC(SO₂CF₃)₂}₂]	6	1.735, 1.780 (4)	1.927, 2.108 (3)	–	170
[Ti(OEt)₂(PhCOCHCOCF₃)₂]	6	1.760 (2)	–	139.4	171
[Ti₂(μ-OC₂H₄Cl)₂(OC₂H₄Cl)₂Cl₄(ClC₂H₄OH)₂]	6	1.746 (5)	1.956, 2.087 (5)	154.2 (t) / 127.3 (b)	172
[Ti(OPrⁱ)₂{(BuᵗMe₂Si)₂bino}] bino = binaphtholate	4	1.740, 1.769 (2)	–	149, 160	173
[Ti₂(μ-Me₂bino)₂(OPrⁱ)₄]	5	1.741, 1.771 (9)	–	–	173

Compound					
$[(Pr^iO)_3Ti\{(Bu^tMe_2Si)_2bino\}Ti(OPr^i)_3]$	4	1.730–1.756 (5) av. 1.739	—	156.1–168.0 av. 163	173
$[Ti(thped)]_2$ thped = $[\{OCH(CH_3)CH_2\}_2NCH_2]_2$	7	1.880–1.905 (3) av. 1.892	2.025, 2.078 (3)	—	174
$[Ti_2(\mu\text{-}OCH_2Bu^t)_2(OCH_2Bu^t)_6]$	5	1.775–1.811 (5) av. 1.791	1.960–2.116 (4) av. 2.036	—	201
$[Ti_2(\mu\text{-}OPr^i)_2(OPr^i)_2(hfip)_4]$	5	1.752 (2) (Pr^iO) 1.835, 1.849 (2) (hfip)	1.900, 2.121 (2) (Pr^iO)	175.3 (t, Pr^iO) 136.7, 141.5 (t, hfip)	175
$[Ti_2(\mu\text{-}OPr^i)_2(OPr^i)_2Cl_4]$	5	1.7259 (11)	1.8686, 2.1543 (11)	165.9 (t) 121.8, 132.2 (b)	176
$[Ti(OPr^i)_2(acac)_2]$	6	1.7877, 1.8041 (14) (Pr^iO) 1.9920, 2.0892 (15) (acac) av. 2.034	—	139.3, 159.6 (Pr^iO) 129.7, 132.4 (acac)	176
$[Ti_2(\mu\text{-}OMe)_2(OMe)_4(acac)_2]$	6	1.8004 (14), 1.8038 (13) (MeO) 2.0168 (13), 2.0789 (13) (acac)	1.9996, 2.0562 (13)	134.3, 142.5 (t, MeO) 119.4, 124.9 (b, MeO) 130.8, 130.6 (acac)	178
$[Ti_2(\mu\text{-}OEt)_2(OEt)_4(acac)_2]$	6	1.796, 1.805 (1) (EtO) 2.020, 2.084 (1) (acac)	1.991, 2.064 (1)	137.1, 142.9 (t, EtO) 120.5, 127.0 (b, EtO) 132.0, 133.1 (acac)	178
$[Ti_2(\mu\text{-}OPr^i)_2(OPr^i)_4(acac)_2]$	6	1.782, 1.784 (4) (Pr^iO) 2.030, 2.073 (4) (acac)	1.969, 2.101 (3)	150.3–160.8 (t, Pr^iO) 125.2, 129.7 (b, Pr^iO) 132.0, 133.0 (acac)	178
$[Ti_2(\mu\text{-}OMe)_2(OMe)_4(tmhd)_2]$ tmhd = $Bu^tCO(CH)COBu^t$	6	1.795, 1.818 (1) (MeO) 2.009, 2.079 (1) (tmhd)	1.971, 2.065 (1)	129.2, 149.7 (t, MeO) 122.5, 123.1 (b, MeO) 134.2, 135.1 (tmhd)	178

(continued overleaf)

Table 4.9 *(Continued)*

Compound	Metal coordination	Bond lengths (Å)		Bond angles (°) $\widehat{MOC}$	Reference
		Terminal	Bridging		
[Ti$_2$(μ-OPr^n)$_2$(OPr^n)$_4$(tmhd)$_2$]	6	1.783, 1.792 (7) (Pr^{n}O); 1.998, 2.055 (2) (tmhd)	1.940, 2.069 (6)	137.3, 145.8 (t, Pr^{n}O); 112, 137 (b, Pr^{n}O); 134.2, 135.7 (tmhd)	178
[Ti$_2$(μ-OPr^i)$_2$(OPr^i)$_4$(thmd)$_2$]	6	1.802, 1.816 (1) (Pr^{i}O); 2.043, 2.075 (1) (tmhd)	1.968, 2.094 (1)	129.8, 151.0 (t, Pr^{i}O); 123.2, 129.8 (b, Pr^{i}O); 134.5, 135.3 (tmhd)	178
[Ti$_4$(μ_3-OCH$_2$)CMe(CH$_2$O-μ)$_2$]$_2$(OPr^i)$_{10}$]	6	1.787–1.839 (1) (Pr^{i}O)	2.04 (μ, L)^l; 2.20 (μ_3, L)^l	120 (μ, L, av.); 114.3, 115.9, 121.8 (μ_3, L)	179
[Ti$_4${(μ_3-OCH$_2$)CEt(CH$_2$O-μ)$_2$}$_2$(OPr^i)$_{10}$]	6	1.787–1.839 (1) (Pr^{i}O)	2.04 (μ, L)^l; 2.20 (μ_3, L)^l	119 (μ, L, av.); 114.2, 116.2, 121.2 (μ_3, L)	179
[Ti$_5$(μ_4-OC$_2$H$_4$O)(μ_3-OC$_2$H$_4$O)$_3$(μ-OC$_2$H$_4$O)(μ-OPr^i)(OPr^i)$_9$]	7, 6, 5	1.72–1.81 (Pr^{i}O); av. 1.77 (1); 1.85 (1) (OC$_2$H$_4$, av.)	1.93, 2.11 (1) (Pr^{i}O); 2.00, 2.26 (1) (OC$_2$H$_4$)	152.8 (t, Pr^{i}O)	180
Na$_2$[Ti(OC$_2$H$_4$O)$_3$](HOC$_2$H$_4$OH)$_4$	6	1.915–1.969 (4); av. 1.953	–	112.1–117.3; av. 115.0	181
K$_2$[Ti(OC$_2$H$_4$O)$_3$](HOC$_2$H$_4$OH)$_{2.5}$	6	1.899–1.986 (2); av. 1.943	–	110.9–118.3; av. 115.0	181
[Ba(HOC$_2$H$_4$OH)$_4$(H$_2$O)$_9$]$^{2+}$[Ti(OC$_2$H$_4$O)$_3$]$^{2-}$	6	av. 1.93 (2)	–	–	182
[Ti$_2$(μ-OEt)$_2$(OEt)$_4$(O$_2$CCH$_2$NH$_2$)$_2$]	6	1.765, 1.813 (6) (EtO)	1.955, 2.087 (5)	–	183
[Ti$_2$(μ-OMe)$_2${O$_{12}$Si$_7$(C$_6$H$_{11}$)$_7$}(MeOH)]	6	1.981 (6) (glycinate); 1.837 (7) (silsesq); 2.210 (7) (MeOH)	2.004, 2.022 (7) (μ, MeO)	–	184
[Ti{O$_{12}$Si$_7$(C$_6$H$_{11}$)$_7$}(OSiMe$_3$)]	4	1.84 (2) (Me$_3$SiO)	–	180 (Me$_3$SiO); 176.0 (t)	185
[(C$_6$H$_5$CH$_2$)$_4$Ti$_2$(μ-OEt)$_2$(OEt)$_2$]	5	1.837 (8)	1.865, 2.104 (7)	122.1, 130.6 (b)	186

$[Me_4Ti_2(\mu\text{-}OMe)_2(tritox)_2]$ tritox = Bu^t_3CO	5	1.752 (6) (tritox)	2.015 (4) (MeO)	174.4 (t) 125.3 (b)	187
$[\{(Me_3Si)_2C_5H_3\}Ti(O_3C_6H_9)]$	4[1]	1.824, 1.832, 1.833 (2)	–	114.9, 115.0, 115.1	188
$[(C_5H_5)_4Ti_2(\mu\text{-}OMe)_2]$	4[1]	–	2.065 (2)	125.6	189
$[(C_5H_5)_4Ti_2(\mu\text{-}OEt)_2]$	4[1]	–	2.076 (3)	132.1	189
$[Ti(OCBu^t_3CH_2PMe_2)_2Cl]$	5	1.841, 1.843 (3)	–	–	190
$[Zr_2(\mu\text{-}OPr^i)_2(OPr^i)_6(Pr^iOH)_2]^2$	6	1.942, 1.945 (5) (Pr^{i}O) 2.066, 2.271 (5) (Pr^{i}OH)	2.161, 2.179 (4)	Term. 169.2, 174.0 (t) 130.7, 132.1 (Pr^{i}OH) 123.7, 129.0 (b)	192
$[(Me_3Si)_3SiZr(OBu^t)_3]$	4	1.90, 1.88 (1)	–	165, 167	193
$[(thf)_2Cl_3Zr(O_2\text{-}C_6H_{10})ZrCl_3(thf)_2]$	6	1.870 (5)	–	176.0	194
$[Zr_5N(NH)_4(NH_2)_4(OCBu^t_3)_5]$	6	av. 1.927 (19)	–	av. 177.7	195
$[Zr\{OCMe(CF_3)_2\}_4]$	4	1.902–1.930 (12) av. 1.916	–	153.2–174.6 av. 163.8	196
$[(oep)Zr(OBu^t)_2]$ oep = octaethylporphyrin	6	1.947, 1.948 (2)	–	175.6, 178.1	197
$[Zr_4\{(\mu\text{-}OCH_2)_3CMe\}_2(\mu\text{-}OPr^i)_2(OPr^i)_8]$	6	av. 1.94 (1)	2.171 (6) (Pr^{i}O) 2.16 (2) (L)	–	179
$[Hf_2(\mu\text{-}OPr^i)_2(OPr^i)_6(Pr^iOH)_2]$	6	1.932, 1.938, 1.911, 1.944 (8) (Pr^{i}O) 2.143, 2.185, 2.082, 2.243 (7) (Pr^{i}OH)	2.148, 2.160, 2.155, 2.156 (6)	167.4 (t, av.)	200
$[(Bu^tO)_3Hf(\mu\text{-}Cl)(\mu\text{-}OBu^t)(\mu\text{-}NHPMe_3)Hf(OBu^t)_3]$	6	av. 1.938 (7)	2.202, 2.209 (7)	–	198
$[Hf(OSiPh_2OSiPh_2OSiPh_2O)_2(py)_2]$	6	1.961, 2.014 (5)	–	148.8, 158.9	199
$[Hf_2(\mu\text{-}OPr^i)_2(OPr^i)_6(Pr^iOH)(py)]$	6	1.929, 1.944 (7) (Pr^{i}O) 2.154 (6) (Pr^{i}OH)	2.175, 2.178 (5)	–	200

b = bridging, t = terminal; L = $N(C_2H_4O)_3$.

[1]Coordination doubtful; [2]Average of data from four independent molecules.

terminal Nb–OMe bonds are longer than normal owing to interionic hydrogen bonding with the methanol molecules of neighbouring complex cations.

The steric effect of the bulky $N(SiMe_3)_2$ ligand is evident in the structure of $[Nb_2(\mu\text{-}OMe)_2Cl_2(NSiMe_3)_2\{N(SiMe_3)_2\}_2]$ which is constrained to five-coordination for the Nb(v) atoms.[208]

2.10.3 Tantalum

The organotantalum alkoxo compound $[Ta(OBu^t)_2(CHBu^t)\{(Me_2N)_2C_6H_3\}]$ is mononuclear with a distorted trigonal pyramidal configuration.[209] In the seven-coordinated cation $[Ta\{O(C_2H_4O)_3C_2H_4O\}Cl_2]^{2+}$ all five oxygens of the tetraethyleneglycolate are in the pentagonal plane.[210] In the mixed ligand complex $[Ta_2(\mu\text{-}OMe)_2(OMe)_2(NSiMe_3)_2\{N(SiMe_3)_2\}_2]$ the binuclear five-coordinated Ta(v) molecule involves methoxo bridges.[254]

2.11 Alkoxides of Chromium, Molybdenum, and Tungsten

2.11.1 Chromium

The structures of a few alkoxo compounds of chromium have been determined. The first homoleptic molecule to be structurally characterized was the monomeric Cr(IV) tetrakis-(2,2,4,4-tetramethyl-3-pentanolate) $[Cr(OCHBu^t_2)_4]$ which exhibited a slightly distorted tetrahedral configuration[211] (data in Table 4.11, pp. 300–319). With the functionalized alkoxo group $OCMe_2CH_2OMe$ the tris-chelated compound $[Cr(OCMe_2CH_2OMe)_3]$ was characterized as the *mer* octahedral Cr(III) complex.[212] A novel trigonal planar three-coordinated Cr(II) complex was obtained as the dimer $[Cr_2(\mu\text{-}OCHBu^t_2)_2(OCBu^t_3)_2]$ using bulky alkoxo ligands.[213] The terminal Cr–O–C bonds were almost linear and the short Cr–O bond lengths suggested that ligand-to-metal π-bonding was operative. A dimeric structure was also found in the cyclopentadienyl chromium-*tert*-butoxide $[(C_5H_5)_2Cr_2(\mu\text{-}OBu^t)_2]$.[214] Magnetic susceptibility measurements revealed antiferromagnetism in this Cr(II) (d^4) dimer and the short Cr–Cr distance (2.65 Å) suggests that metal–metal bonding is present.

2.11.2 Molybdenum

The first reported structure of a mononuclear alkoxo molybdenum compound featured the octahedral Mo(IV) trialkylsiloxy complex $[trans\text{-}Mo(OSiMe_3)_4(Me_2NH)_2]$ which had nearly linear Mo–O–Si bond angles (Table 4.11).[215] The remarkable carbonyl complex $[Mo(OBu^t)_2(CO)_2(py)_2]$, which is formally Mo(II), has *trans-tert*-butoxo ligands with *cis* carbonyls and *cis* pyridines.[216] The bulky 1-adamantolato ligand gave a mononuclear five-coordinated (distorted TBP) complex $[Mo(1\text{-}ado)_4(Me_2NH)]$ (1-ado = 1 adamantolato) with the dimethylamine ligand occupying an axial position.[217] The difference in bond lengths (0.066 Å) between the axial Mo–O (1.963 Å) and the average equatorial Mo–O (1.897 Å) is noteworthy. See also $[Mo(OBu^t)_4(NNCPh_2)]$.[305] The tridentate 3,5-dimethylpyrazolylato ligand forms octahedral nitrosyl alkoxo complexes $[Mo(OR)(OR')(NO)(Me_2pz)]$

Table 4.10 Alkoxides of vanadium, niobium, and tantalum

Compound	Metal coordination	M–O Bond lengths (Å)		Bond angles (°) $\widehat{\text{MOC}}$	Reference
		Terminal	Bridging		
[Nb(OPr^i)$_3${PhCO(CH)COPh}(NCS)]	6	1.828, 1.834, 1.845 (5)	–	148.5, 153.5, 159.7	202
[Nb$_2$(μ-OCH$_2$SiMe$_3$)$_2$(OCH$_2$SiMe$_3$)$_8$]	6	1.896–1.919 (4)	2.134, 2.174 (2)	133.0–151.3 (t) 120.6, 121.1 (b)	458
[Nb$_2$(μ-OMe)$_2$(OMe)$_2$Cl$_4$(MeOH)$_2$]	6	1.811 (4) (MeO) 2.147 (4) (MeOH)	2.046, 2.048 (3)	159.6 (t, MeO) 129.5 (t, MeOH) 137.0, 137.4 (b)	205
[Nb$_2$(μ-OMe)$_2$(OMe)$_2$Cl$_4$(MeCN)$_2$]	6	1.814, 1.831 (9) (MeO)	2.016–2.043 (10) av. 2.033	152.1, 152.4 (t)	205
[Nb$_2$(μ-OPr^i)(μ-Cl)(Pr^{i}OH)$_4$Cl$_4$]	6	2.02, 2.09 (2) (Pr^{i}OH, ax) 2.22, 2.27 (2) (Pr^{i}OH, eq)	2.02, 2.03 (2)	136, 143 (b) Pr^{i}OH, (t, Pr^{i}OH ax) (t, Pr^{i}OH, eq)	206
[Nb$_2$(μ-OMe)$_3$(OMe)$_6$]$^{-1}$	6	1.943–1.980 av. 1.966 (5)	2.019–2.076 av. 2.069 (2)	130.7–132.1 (t) av. 131.4 139.6–140.6 (b) av. 140.3	207
[Nb$_2$(μ-OMe)$_2$Cl$_2$(NSiMe$_3$)$_2${N(SiMe$_3$)$_2$}$_2$]	5	–	2.104, 2.157 (2)	118.7 (b)	208
[Nb$_2$(μ-OMe)$_2$(OMe)$_8$]^2	6	1.896 (7) (*trans*) 1.898 (8) (*cis*)	2.134 (7)	135.2–160.6 (t) av. 145.3 121.2–123.6 (b) av. 122.5	203
[Ta(OBu^t)$_2$(CHBu^t){(Me$_2$N)$_2$C$_6$H$_3$}]	5	1.872, 1.914 (3)	–	141.6, 154.9	209
[Ta{O(C$_2$H$_4$O)$_3$C$_2$H$_4$O)Cl$_2$]$^{2+}$[TaCl$_6$]$^{2-}$	7	1.89 (1) (olate) 2.25 (2) (ether, av.)	–	–	210
[Ta$_2$(μ-OMe)$_2$(OMe)$_2$(NSiMe$_3$)$_2${N(SiMe$_3$)$_2$}$_2$]	5	1.954 (15)	2.079, 2.130 (2)	–	254

b = bridging, t = terminal; ax = axial, eq = equatorial.
[1] For counter-ions, see text; [2] Average of data from two independent molecules.

(Me$_2$pz = 3,5-dimethylpyrazolylato; R = R' = Pr^i and R = Et, R' = Pr^i) containing linear Mo–N–O angles.[218] A similar structure was found in the compound [Mo(OPr^i)$_2$(NO){HB(impz)$_3$}] [HB(impz)$_3$ = isopropylmethylpyrazol-1-yl].[219] It is noteworthy in these formally Mo(II) complexes that the Mo–O bond distances of the alkoxo–Mo groups are relatively short. The formally Mo(VI) nitrido complexes [Mo(μ-N)(OR)$_3$]$_n$ (R = Pr^i, Bu^t) have infinitely linear structures due to bridging involving the Mo≡N → Mo system. The long dative N → Mo bond (2.88 Å, cf. 1.66 Å for Mo≡N) suggests a weak interaction and the complex is fully dissociated in benzene solution.[220] Interestingly, methylation of the *tert*-butoxo complex in the presence of 4-Bu^{t}pyridine gave rise to the tetranuclear nitrido-bridged species [Mo$_4$(μ-N)$_4$(OBu^t)$_4$Me$_8$(NC$_5$H$_4$-4-Bu^t)$_2$].[221] The eight-membered Mo$_4$(μ-N)$_4$ ring has alternating short (1.664 Å, 1.689 Å) and long (2.147 Å, 2.209 Å) Mo–N bonds with two five-coordinated and two six-coordinated Mo atoms. Reaction of MoN(OBu^t)$_3$ with ethyleneglycol and pinacol gave the tris-chelated [Mo(OC$_2$H$_4$O)$_3$] and [Mo(OCMe$_2$CMe$_2$O)$_3$] complexes with Mo(VI) in distorted octahedral coordination.[222]

The metal–metal triple bonded diamagnetic dimers Mo$_2$(OR)$_6$ are unique in having no bridging alkoxo ligands (Fig. 4.8). The structure of Mo$_2$(OCH$_2$Bu^t)$_6$ showed that the Mo$_2$O$_6$ system had virtual D$_{3d}$ symmetry with ethane-like geometry.[223] With less sterically demanding groups (OMe, OEt) polymeric compounds [Mo(OR)$_3$]$_n$ are formed whereas the bulkier ligands (OPr^i, OBu^t, OSiMe$_3$, OSiEt$_3$) all give dinuclear Mo≡Mo species Mo$_2$(OR)$_6$. On the other hand the Mo(IV) tetraisopropoxide forms [Mo$_2$(μ-OPr^i)$_2$(OPr^i)$_6$] with unsymmetrical isopropoxo bridges and five-coordinated molybdenum.[224] The Mo$_2$O$_8$ system is composed of two MoO$_5$ trigonal bipyramids joined along a common axial–equatorial edge. The two terminal ligands in equatorial sites have shorter Mo–O bond lengths (1.872 Å, 1.884 Å) than the one in the axial position (1.976 Å).

The dinuclear species Mo$_2$(OR)$_6$ form adducts with a variety of Lewis bases [Mo$_2$(OR)$_6$L$_2$] (L = NH$_3$, MeNH$_2$, Me$_2$NH, Me$_3$N, Me$_2$PhP) and the structure of [Mo$_2$(OSiMe$_3$)$_6$(Me$_2$NH)$_2$] has been determined.[225] Reversible CO$_2$ insertion also occurs with Mo$_2$(OR)$_6$ and the structure of [Mo$_2$(O$_2$COBu^t)$_2$(OBu^t)$_4$] shows that the O$_2$COBu^t groups bridge across the two Mo atoms, and the *tert*-butoxo groups are all in terminal positions (Fig. 4.38).[226]

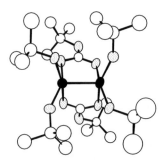

Figure 4.38 Structure of [Mo$_2$(μ-O$_2$COBu^t)$_2$(OBu^t)$_4$] (○ = O; ● = Mo; H atoms omitted).

Although $Mo_2(OBu^t)_6$ reacts readily with CO to form $Mo(CO)_6$ and $Mo(OBu^t)_4$ (Eq. 4.1),

$$2Mo_2(OBu^t)_6 + 6CO \longrightarrow Mo(CO)_6 + 3Mo(OBu^t)_4 \qquad (4.1)$$

the dinuclear carbonyl complex $[Mo_2(\mu\text{-}OBu^t)_2(\mu\text{-}CO)(OBu^t)_4]$ was isolated as an unstable intermediate.[227] The molybdenum atoms are in distorted square pyramidal five-coordination sharing a common triangular face with the carbonyl carbon atom at the common apex. This diamagnetic compound appears to involve Mo≡Mo (Mo–Mo, 2.498 Å).

Addition of nitric oxide to $Mo_2(OPr^i)_6$ gave the diamagnetic dimer $[Mo_2(\mu\text{-}OPr^i)_2(OPr^i)_4(NO)_2]$ comprising edge-bridged trigonal bipyramids. The near linear (Mo$\widehat{N}$O, 178°) nitrosyls occupy terminal axial positions and the two bridging isopropoxo groups form short bonds in the equatorial positions and long bonds in the axial positions. The Mo–Mo distance (3.335 Å) precludes metal–metal bonding.[228] The exceptionally long axial bridging Mo–O bond (2.195 Å vs 1.951 Å equatorial) reflects the high trans influence of the NO ligand. Addition of dimethylamine to $[Mo_2(\mu\text{-}OPr^i)_2(OPr^i)_4(NO)_2]$ gave the unstable $[Mo_2(\mu\text{-}OPr^i)_2(OPr^i)_4(Me_2NH)_2]$ which has the edge-shared isopropoxo-bridged bi-octahedral structure.[229] The linear NO ligands are trans to the unsymmetrical bridging isopropoxo groups which again reveal the trans influence of the nitrosyl groups. The terminal Pr^{i}O group trans to a bridging Pr^{i}O has a significantly shorter Mo–O bond length (Mo–O, 1.909 Å) than the terminal Pr^{i}O group trans to the Me_2NH ligand (Mo–O, 1.935 Å).

Addition of halogen X_2 (X = Cl, Br) to $Mo_2(OPr^i)_6$ gave the unstable halide alkoxides $[Mo_2(\mu\text{-}OPr^i)_2(OPr^i)_4X_4]$.[230] Each compound has the edge-shared isopropoxo-bridged bi-octahedral structure with cis terminal halides and trans terminal Pr^{i}O ligands. The Mo–Mo distances (X = Cl, 2.731 Å; X = Br, 2.739 Å) are consistent with the presence of single metal–metal bonds in these Mo(v) complexes. The very short Mo–O bond distances for the terminal Pr^{i}O groups suggest that π-donation for Pr^{i}O oxygen to molybdenum is very pronounced.

Addition of arylazides ArN_3 to $Mo_2(OBu^t)_6$ gave rise to the dinuclear bis-tert-butoxo-bis-arylimido-molybdenum(VI) complexes $[Mo_2(\mu\text{-}NAr)_2(NAr)_2(OBu^t)_4]$.[231] In the p-tolyl derivative the dimer consists of two equatorial–axial edge-sharing trigonal bipyramids with terminal tolylimido ligands (near linear Mo$\widehat{N}$C) The Bu^{t}O ligands occupy equatorial positions. Addition of diphenyldiazo methane to $Mo_2(OR)_6$ in the presence of pyridine gave the dinuclear $[Mo_2(\mu\text{-}OR)_3(OR)_3(N_2CPh_2)_2(py)](R = Pr^i, CH_2Bu^t)$.[232] The unsymmetrical molecule (Fig. 4.39) has a confacial Pr^{i}O-bridged bioctahedral structure with a terminal $NNCPh_2$ ligand attached to each Mo. One Mo has two terminal Pr^{i}O ligands and the other has one Pr^{i}O and a pyridine ligand in terminal positions. The Mo–Mo distance (2.662 Å) is consistent with the presence of a single metal–metal bond.

Addition of alkynes RC_2R' to $Mo_2(OR)_6$ (R'' = Pr^i, Bu^t, CH_2Bu^t) in the presence of pyridine gave rise to alkyne-bridged alkoxo-bridged complexes $[Mo_2(\mu\text{-}RC_2R')(\mu\text{-}OR)_2(OR)_4(py)_2]$.[233] The structure of $[Mo_2(\mu\text{-}C_2H_2)(\mu\text{-}OPr^i)_2(OPr^i)_4(py)_2]$ revealed that the acetylene molecule bridges the two molybdenum atoms in a crosswise manner (Fig. 4.40) with two bridging Pr^{i}O groups in a pseudo-confacial bi-octahedron configuration and the pyridines trans to the $\mu\text{-}C_2H_2$ group. The Mo–Mo distance (2.554 Å) is suggestive of a metal–metal double bond. In the case of the neopentoxide, addition

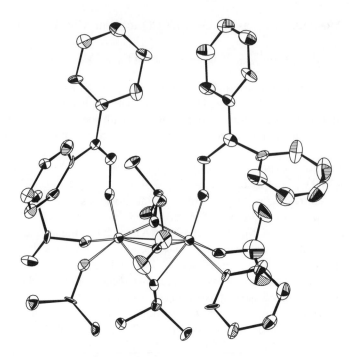

Figure 4.39   Structure   of $[Mo_2(\mu\text{-}OPr^i)_3(OPr^i)_2(N_2CPh_2)(py)]$ (H atoms omitted).

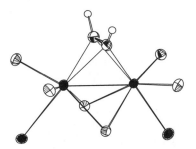

Figure 4.40 $Mo_2(\mu\text{-}C_2H_2)(\mu\text{-}O)_2O_4N_2$ core structure of $[Mo_2(\mu\text{-}C_2H_2)(\mu\text{-}OPr^i)_2(OPr^i)_4(py)_2]$.

of excess acetylene produced $[Mo_2(\mu\text{-}C_4H_4)(\mu\text{-}OCH_2Bu^t)(OCH_2Bu^t)_5(py)]$ in which one Mo is part of a molybdacyclopentadiene moiety and is further bonded to a bridging Bu^tCH_2O, two terminal Bu^tCH_2O ligands and a pyridine whilst the other Mo is bonded to a bridging Bu^tCH_2O ligands and π-bonds to the metallocycle.

Further research on the carbonylation of $Mo_2(OR)_6$ in the presence of Lewis bases L gave the dimer $[Mo_2(\mu\text{-}CO)(\mu\text{-}OPr^i)_2(OPr^i)_4(py)_2]$ which has the confacial bi-octahedral structure with μ-CO and two μ-OPr^i bridges.[234] The pyridine ligands are *trans* to the carbonyl ligand. The structure of the mixed ligand (Mo≡Mo) dimer

1,2-$Mo_2(OBu^t)_2(CH_2SiMe_3)_4$ is ethane-like (C_{2h} symmetry).[235] Addition of dialkyl cyanamides R_2NCN to $Mo_2(OR)_6$ compounds produced $[Mo_2(OR)_6(\mu\text{-}NCNR_2)]$ complexes and the crystal structure $[Mo_2(OR)_6(\mu\text{-}OCH_2Bu^t)(\mu\text{-}NCNMe_2)(OCH_2Bu^t)_5]$ was determined. One Mo atom is bonded to two terminal Bu^tCH_2O ligands, the η^2-CN portion of the cyanamide and one bridging Bu^tCH_2O ligand. The second Mo is bonded to three terminal Bu^tCH_2O ligands, the bridging nitrogen of the cyanamide and the bridging Bu^tCH_2O ligand. The terminal Bu^tCH_2O *trans* to the bridging cyanamide nitrogen has a longer Mo–O bond length (1.954 Å) than the other terminal neopentoxide groups (av. 1.87 Å) whilst the alkoxo bridge is unsymmetrical (Mo–O, 1.999, 2.146 Å).[236]

In the dimer $[Mo_2(\mu\text{-}\eta^1,\eta^1\text{-}O_2CPh)_2(OBu^t)_4]$ the Mo≡Mo unit (d, 2.236 Å) is spanned by a pair of *cis* bridging benzoato ligands with terminal Bu^tO ligands *trans* to each of the benzoate oxygens.[237] By contrast in the acetylacetonate $[Mo_2(OBu^t)_4(acac)_2]$ (Mo≡Mo, 2.237 Å) the diketonato groups are chelating in preference to bridging.[238] Addition of *tert*-butyl isonitrile gave the edge-bridged bi-octahedral molecule $[Mo_2(\mu\text{-}OCH_2Bu^t)_2(OCH_2Bu^t)_2(CNBu^t)_2(acac)_2]$.[239] Alkoxo-ligands in $Mo_2(OR)_6$ can be replaced by thiolato groups as in $[Mo_2(OPr^i)_2(SAr)_4]$ (Ar = 2,4,6-trimethylphenyl) where the M≡M bond distance (2.230 Å) is typical for this class of compound.[240] The arylselenato complex $[Mo_2(OPr^i)_2(SeAr)_4]$ (Mo≡Mo, 2.219 Å) is similar.[299] In the novel bromo complex $[Mo_2(\mu\text{-}OCH_2Bu^t)_2(\mu\text{-}Br)(OCH_2Bu^t)_4Br(py)]$ each Mo is in a distorted octahedral environment and the Mo–Mo distance (2.534 Å) implies metal–metal double bonding.[241]

Reaction of $Mo_2(OBu^t)_6$ with diphenyldiazomethane gave the mononuclear five-coordinated (TBP) $[Mo(OBu^t)_4(NNCPh_2)]$ where the equatorial Mo–O bond lengths (av. 1.89 Å) are shorter than the axial Mo–O bond (1.944 Å). With $Mo_2(OPr^i)_6$ in the presence of pyridine the binuclear confacial bi-octahedral complex $[Mo_2(\mu\text{-}OPr^i)_3(NNCPh_2)_2(py)]$ with Mo–Mo = 2.66 Å indicative of a single metal–metal bond.[242] One Mo atom is coordinated to an $NNCPh_2$ ligand, two terminal Pr^iO ligands and three bridging Pr^iO ligands whilst the other Mo is coordinated to one $NNCPh_2$ ligand, one terminal Pr^iO, one pyridine, and three bridging Pr^iO ligands. The bridges are distinctly unsymmetrical. The binuclear phenylisocyanate insertion product $[Mo_2(OPr^i)_4\{NPhC(O)(OPr^i)\}_2]$ has two *cis* bridging bidentate ligands and four terminal isopropoxo ligands with a Mo≡Mo (2.221 Å).[243]

The quadruply metal–metal bonded dimers $[Mo_2(OPr^i)_4(py)_4]$, $[Mo_2(OCH_2Bu^t)_4(PMe_3)_4]$, $[Mo_2(OCH_2Bu^t)_4(Me_2NH)_4]$, and $[Mo_2(OPr^i)_4(Pr^iOH)_4]$ are of special interest (Mo≡Mo; 2.195, 2.209, 2.133, 2.110 Å, respectively) in having no bridging ligands.[244] Generally in the $[Mo_2(OR)_4L_4]$ complexes the OR groups are *cis* to the L ligands and the two halves of the molecule are in the eclipsed conformation (Fig. 4.41). When L = Pr^iOH or Me_2NH there is hydrogen bonding RO....HOR($HNMe_2$) across the metal–metal bond which is thereby shortened. The same situation occurs in $[Mo_2(OC_5H_9\text{-}c)_4(c\text{-}C_5H_9OH)_4]$ although it was not possible to distinguish c-C_5H_9O ligands from c-C_5H_9OH in this structure.[245]

The unsymmetrical Mo≡Mo (2.235 Å) complex $[Mo_2(OPr^i)_4(CH_2Ph)_2(PMe_3)]$ has one molybdenum bonded to benzyl groups and one PMe_3 ligand.[246,247] With the bidentate phosphine $(Me_2P)_2CH_2$ (dmpm) the symmetrical dimer

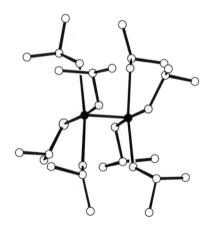

Figure 4.41 Structure of [$Mo_2(OPr^i)_4(Pr^iOH)_4$] ($\bigcirc$ = O; $\bullet$ = Mo; H atoms omitted).

[$Mo_2(OPr^i)_4(CH_2Ph)_2(dmpm)$] was formed with the dmpm ligand bridging across the Mo≡Mo bond (2.253 Å).[247]

Interestingly in the dimer [$Mo_2(OPr^i)_4(dmpe)_2$] the dmpe ligands chelate a single Mo atom giving an unsymmetrical Mo(IV)-Mo(0) complex with a Mo≡Mo triple bond (2.236 Å).[248]

From the reaction of 2-butyne with [$Mo_2(OCH_2Bu^t)_6(NCNEt_2)$] the novel complexes [$Mo_2(\mu\text{-}OCH_2Bu^t)(\mu\text{-}C_4Me_4(OCH_2Bu^t)_5(NCNEt_2)$] and [$Mo_2(\mu\text{-}OCH_2Bu^t)$] {$\mu\text{-}C_4Me_3CH_2(NEt_2)NH\}(OCH_2Bu^t)_5$] were obtained.[249] Addition of quinones to $Mo_2(OPr^i)_6$ gave rise to edge-shared isopropoxo-bridged bi-octahedral complexes [$Mo_2(\mu\text{-}OPr^i)_2(OPr^i)_4(O_2C_6Cl_4)_2$] and [$Mo_2(\mu\text{-}OPr^i)_2(OPr^i)_2(O_2C_{14}H_8)_3$] in which the diolate ligands are chelating.[250] The adduct of 1,4-diisopropyl-1,4-diazabutadiene with $Mo_2(OPr^i)_6$ is the unsymmetrical binuclear complex [$Mo_2(\mu\text{-}OPr^i)_2(OPr^i)_4(Pr^iNCHCHNPr^i)$] in which one Mo is bonded to three terminal Pr^iO ligands and two isopropoxo bridges whilst the other is chelated by the nitrogen-donor ligand, one terminal Pr^iO ligand and the two isopropoxo bridges. The Mo coordination is distorted TBP in each case.[251] In the dimer [$Mo_2(\mu\text{-}NMe_2)_2(NMe_2)_2(OC_6H_{10}\text{-}4Me)_4$] which also has five-coordinated Mo (TBP) the c-4-MeC$_6$H$_{10}$O ligands are all terminal.[252] Treatment of $Mo_2(NMe_2)_6$ with perfluoro-*tert*-butanol gave the Mo≡Mo (2.216 Å) bonded centrosymmetric dimer [$Mo_2\{OC(CF_3)_3\}_4(NMe_2)_2$] having an anti, staggered rotational conformation and a planar C_2NMo≡$MoNC_2$ unit.[253] A similar conformation was exhibited by [$Mo_2(OCPh_3)_2(NMe_2)_4$].[300]

With the hexafluoro-*tert*-butoxo ligand (hftb) the familiar non-bridged dinuclear molecule [$Mo_2(hftb)_6$] was obtained.[255] In the arylimido-bridged complex [$Mo_2(\mu\text{-}NAr)_2(OBu^t)_2$] (Ar = 2,6-diisopropylphenyl), terminal *tert*-butoxo ligands give a distorted tetrahedral structure for this Mo(IV) compound (Mo-Mo, 2.654 Å).[256] In the metal–metal triply bonded (Mo≡Mo, 2.254 Å) dimer [$Mo_2(OBu^t)_4(NHPh)_2(NH_2Ph)_2$] the eclipsed geometry is due to hydrogen bonding across the metal–metal bond.[257]

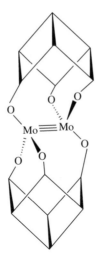

Figure 4.42 Structure of $[Mo_2\{O_{12}Si_7(C_6H_{11}\text{-}c)_7\}_2]$ (cyclohexyl groups omitted).

Using the tridentate trisilanolate ligand $(c\text{-}C_6H_{11})_7Si_7O_{12}^{3-}$ the interesting dimer $[Mo_2\{O_{12}Si_7(C_6H_{11}\text{-}c)_7\}_2]$ was obtained (Fig. 4.42) in which the $Mo\equiv Mo$ bond (2.215 Å) is retained.[258]

Interestingly the discrete anion in $[K\text{-}(18\text{-crown-6})]^+[Mo_2(\mu\text{-}OCH_2Bu^t)(OCH_2Bu^t)_6]^-$ has a $Mo\equiv Mo$ triple bond (2.218 Å) buttressed by a single bridging alkoxo group.[259]

The reaction of Ph_2CS with $Mo_2(OCH_2Bu^t)_6$ in the presence of PMe_3 produced the unsymmetrical binuclear $[(Bu^tCH_2O)_4Mo(\mu\text{-}OCH_2Bu^t)(\mu\text{-}S)Mo(OCH_2Bu^t)(CPh_2)(PMe_3)]$ in which the six-coordinated Mo is bonded to four terminal Bu^tCH_2O ligands and edge bridged through μ-S and $\mu\text{-}OCH_2Bu^t$ to the five-coordinated Mo which is also bonded to one terminal Bu^tCH_2O ligand, the carbene CPh_2 and the PMe_3 ligand.[260]

Addition of PF_3 to $Mo_2(OBu^t)_6$ caused exchange of Bu^tO for F with the formation of the fluorine-bridged tetranuclear complex $[Mo_4(\mu\text{-}F)_4(OBu^t)_8]$ containing a bisphenoid of molybdenum atoms with two short $Mo\equiv Mo$ (2.26 Å) and four long nonbonding Mo....Mo (3.75 Å) (Fig. 4.43). The complex $[Mo_4(\mu\text{-}F)_3(\mu\text{-}NMe_2)(OBu^t)_8]$ has a similar structure.[261] In the reaction with PMe_3 the tetranuclear molecule is converted into the typical dinuclear species $[Mo_2(OBu^t)_4F_2(PMe_3)_2]$ with an unbridged $Mo\equiv M$ (2.27 Å).[262]

Interestingly the tetranuclear chloride isopropoxide $[Mo_4(\mu\text{-}OPr^i)_8Cl_4]$ has a square Mo_4 configuration (delocalized Mo....Mo, 2.378 Å) with bridging Pr^iO ligands and terminal chlorides giving square pyramidal coordination for Mo(III). In contrast the bromide analogue adopts the "butterfly" or open tetrahedron structure with five short (av. 2.50 Å) and one long Mo–Mo (3.287 Å). This complex contains terminal and bridging (μ and μ_3) isopropoxo ligands and terminal bromines $[Mo_4(\mu_3\text{-}OPr^i)_2(\mu\text{-}OPr^i)_4(OPr^i)_2Br_4]$ (Fig. 4.44).[263] In the related molecule $[Mo_4(\mu_3\text{-}OPr^i)_2(\mu\text{-}OPr^i)_4(OPr^i)_3Br_3]$ one of the bromines attached to a backbone Mo is replaced by a terminal isopropoxo ligand.[264] In $[Mo_4(\mu\text{-}OPr^i)_2(OPr^i)_8F_2]$ there is a rectangle of Mo atoms with two short $Mo\equiv Mo$ (2.23 Å) and two long nonbonding

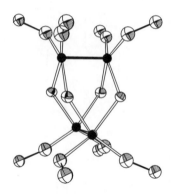

Figure 4.43 $Mo_4(\mu\text{-}F)_4(OC)_8$ core structure of $[Mo_4(\mu\text{-}F)_4(OBu^t)_8]$ (CH_3 groups omitted).

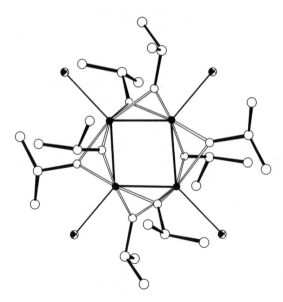

Figure 4.44 Structure of $[Mo_4(\mu\text{-}OPr^i)_8Cl_4]$ (H atoms omitted).

Mo–Mo (3.41 Å). The two halves of this centrosymmetric molecule are joined by an isopropoxo and a fluoride bridge.[264]

In the novel tetranuclear nitrido complex $[Mo_4(\mu_3\text{-}N)_2(\mu\text{-}OPr^i)_2(OPr^i)_{10}]$ there is a zig-zag chain of four Mo atoms all in one plane which includes the two bridging Pr^iO ligands and two terminal isopropoxides.[265] The mixed alkoxide $[Mo_4(\mu\text{-}OMe)_2(\mu\text{-}OPr^i)_2(OPr^i)_8]$ has two nonbridged $Mo\equiv Mo$ (2.238 Å) units joined by methoxo and isopropoxo bridges in a centrosymmetric molecule.[266] Another "butterfly" configuration is exhibited in $[Mo_4(\mu_3\text{-}OCH_2Bu\text{-}c)(\mu\text{-}OCH_2Bu\text{-}c)_4(OCH_2Bu\text{-}c)_7(c\text{-}BuCH_2OH)]$. One triangular face is capped by the μ_3-OR group and each edge is also bridged by μ-OR groups. The alcohol ligand is coordinated to one of the backbone Mo atoms.[267]

The novel hydrido alkoxide [Mo$_4$(μ_3-OBu^t)(μ_3-H)(μ-OBu^t)$_2$(μ-H)$_2$(OBu^t)$_4$(Me$_2$NH)] also has a "butterfly" structure with the two backbone Mo atoms having the same environment but the wingtip Mo atoms are different, with one bonded to the μ_3-H, two μ_2-H, a terminal Bu^{t}O and the coordinated Me$_2$NH whilst the other is bonded to the μ_3-OBu^t, two μ_2-OBu^t and one terminal *tert*-butoxo ligand.[268] Addition of potassium hydride to Mo$_2$(OCH$_2$Bu^t)$_6$ in the presence of 18-crown-6 gave the tetranuclear anionic complex in [K(18-crown-6)]$^+$[Mo$_4$(μ_4-H)(μ_3-OCH$_2$Bu^t)$_2$(μ-OCH$_2$Bu^t)$_4$(OCH$_2$Bu^t)$_6$]. The presence of the unique μ_4-H (Fig. 4.45) causes a widening of the wingtip-to-wingtip Mo....Mo distance (3.741 Å).[269]

The structures of some hexanuclear clusters have been determined: (C$_{18}$H$_{36}$N$_2$O$_6$Na)$_2^{2+}$[Mo$_6$(μ_3-Cl)$_8$(OMe)$_6$]$^{2-}$,[270]   [Na$_2$(MeOH)$_{10}$]$^{2+}$[Mo$_6$(μ_3-Cl)$_8$(OMe)$_6$]$^{2-}$ and [Na$_2$(MeOH)$_{10}$]$^{2+}$[Mo$_6$(μ_3-OMe)$_8$(OMe)$_6$]$^{2-}$.[271]

2.11.3   Tungsten

Many alkoxotungsten compounds have been structurally characterized in various oxidation states and coordination environments. The dichlorotetramethoxo tungsten(VI) complex [W(OMe)$_4$Cl$_2$] has the *cis* octahedral structure[272] whereas [W(OBu^t)$_4$I$_2$] adopts the *trans* configuration.[273] In the W(IV) complex [W(OCH$_2$CF$_3$)$_2$Cl$_2$(PMe$_2$Ph)$_2$] the trifluoroethoxo groups are *trans* but the chlorines and phosphines are *cis*.[274] The pseudo-octahedral [W(OCH$_2$Bu^t)$_4$(py)(OCPh$_2$)] has the η^2-benzophenone ligand *cis* to the pyridine.[275] The trisilanolate complex [W{O$_{12}$Si$_7$(C$_6$H$_{11}$-c)$_7$}(NMe$_2$)$_3$] has the facial configuration imposed by the tridentate polysiloxanolate ligand.[276] Mononuclear octahedral W(VI) compounds involving organoimido groups have also been characterized. In [W(OBu^t)$_3$(NPh)Cl(Bu^{t}NH$_2$)] the *tert*-butoxo groups are meridionally bonded with the phenylimido group *trans* to the *tert*-butylamine.[277] A similar structure was found in the *tert*-butylimido complex [W(OBu^t)$_3$(NBu^t)Cl(Bu^{t}NH$_2$)],[278] whilst in [W(OPr^i)$_4$(NPh)(py)] the phenylimido ligand is *trans* to the pyridine ligand.[279]

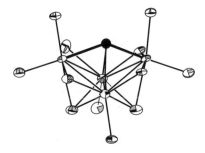

Figure 4.45   Mo$_4$(μ_4-H)(μ_3-O)$_2$(μ-O)$_4$O$_6$ core structure of [Mo$_4$(μ_4-H)(μ_3-OR)$_2$(μ-OR)$_4$(OR)$_6$]   (R = CH$_2$Bu^t; ● = μ_4-H; H and C atoms omitted).

The homoleptic tris-1,2-ethane-diolate W(VI) complex $[W(OC_2H_4O)_3]$ was one of the first tungsten alkoxo compounds to be structurally determined.[280] A similar structure was found in $[W(OC_2Me_4O)_3]$.[281] In the anion of the salt $Li[W\{OC_2(CF_3)_4O\}_2(NPh)Cl]$ the chlorine is *cis* to the phenylimido group in an octahedral structure.[282] Other diolato complexes studied are *cis*-$[W(OC_2Me_4O)_2Cl_2]$,[283] $[W(OC_2H_4O)(OC_2Me_4O)_2]$, $[W(OC_2H_4O)(OC_2Me_4O)_2, (PhOH)_2]$, $[W(OC_2H_4O)(OC_2Me_4O)_2, (PhNH_2)_2]$,[282] $[W(OC_2Me_4O)_2(OPh)_2]$,[285] and $[W(OC_2Me_4O)_2\{OCH(CH_2Cl)CH_2O\}]$.[286] Two interesting five-coordinated (TBP) W(VI) diolato complexes have been reported $[W\{OC_2(CF_3)_4O\}(NBu^t)(H_2NBu^t)]$ and $[W\{OC_2(Ph)_2O\}(NBu^t)(NHBu^t)_2]$ but no W–O bond length data was given.[287]

One of the first five-coordinated alkoxo tungsten compounds structurally characterized was the nitrosyl complex $[W(OBu^t)_3(NO)(py)]$ in which the nitrosyl and pyridine ligands occupy the axial positions.[288] The thiotungsten tetra-*tert*-butoxide compound $[W(OBu^t)_4S]$ exhibit the square pyramidal configuration.[289] The pseudo-trigonal pyramidal complex $[W(OCH_2Bu^t)_3(P_3)(Me_2NH)]$ has the amine and the novel P_3 ligand occupying the axial positions.[290]

Some interesting organotungsten alkoxides have also been reported. In the benzylidyne complex $[W(OBu^t)_3(CPh)]$ a mononuclear tetrahedrally coordinated tungsten was found.[291] Two hexafluoroisopropoxo tungstenacyclobutadiene complexes $[W(hfip)_3(C_3Et_3)]$[292] and $[W(hfip)_3\{CBu^t(CH)CBu^t\}]$[293] have been characterized. The C_3R_3 moiety chelates the W atom in the equatorial plane of the trigonal bipyramid with *tert*-butoxo groups in the axial and one equatorial position.

Like molybdenum, tungsten forms the dinuclear d^3–d^3, M≡M molecules $W_2(OR)_6$ in which the metal–metal triple bond is unsupported by bridging groups. In $[W_2(OPr^i)_6(py)_2]$ the triple bond (W≡W, 2.332 Å) links two $W(OPr^i)_3(py)$ units in a slightly staggered conformation with the pyridine ligands adjacent to each other.[294] See also $[W_2(OBu^t)_6(4$-Me-py$)_2]$, $[W_2(tftb)_6(py)_2]$, and $[W_2(tftb)_6(NCxylyl)_2]$ (Table 4.11).[322] In the alkyltungsten alkoxide $[1,2$-$Bu_2^iW_2(OPr^i)_4]$ (W–W, 2.309 Å) the short W–O bond lengths (av. 1.88 Å) and wide WÔC angles (av. 138°) suggest the presence of oxygen-to-tungsten π-donation.[295] In the unsymmetrical complex $[(p$-tolyl$)(Me_2NH)(OPr^i)_2W≡W(OPr^i)_2(p$-tolyl$)]$ (W≡W, 2.317 Å) the weakly bound dimethylamine ligand is *trans* to a *p*-tolyl ligand.[295] The parent molecule $[W_2(OPr^i)_6]$ (W≡W, 2.315 Å) exists in the staggered ethane-like conformation and dimerizes to the tetranuclear species $[W_4(OPr^i)_{12}]$.[296]

The reaction between $W_2(NMe_2)_6$ and $(CF_3)_2CO$ produced the symmetrical complex $[W_2\{OC(CF_3)_2(NMe_2)\}_2(NMe_2)_4]$ (W≡W, 2.301 Å).[297] In $[W_2(OC_6H_{11}$-$c)_6]$ (W≡W, 2.340 Å) the cyclohexanolate ligands adopt the staggered conformation but in the pinacolate $[W_2(OCMe_2CMe_2O)_3]$ (W≡W, 2.274 Å) the diolate ligands bridge across the metal–metal bond in a near eclipsed conformation.[298]

In the symmetrical arylselenato derivative $[W_2(OPr^i)_2(SeAr)_4]$ (Ar = mesityl; W≡W, 2.308 Å) the ligands adopt the ethane-like conformation with the mesityl groups distal and isopropyl groups proximal to the W≡W bond.[299] In $[W_2(OBu^t)_4(HNPh)_2(PhNH_2)_2]$ (W≡W, 2.322 Å), as in the analogous molybdenum complex, the ligands are held in the eclipsed conformation by the girdle of N–H....O hydrogen bonds around the metal–metal bond.[257] The two *tert*-butoxo ligands on each tungsten are in the *trans* configuration. In $[W_2(OCPh_3)_2(NMe_2)_4]$ (W≡W, 2.307 Å) The ethane-like conformation is also found with ligands adopting the gauche configuration whereas in $[W_2(OSiPh_3)_2(NMe_2)_4]$ (W≡W, 2.295 Å) the *anti* configuration occurs.[300] The structure of optically active (+) $[W_2(OC_{10}H_{19})_6]$ (W≡W, 2.338 Å) containing mentholato ligands has also been determined.[301] In the mixed ligand symmetrical dimer $[W_2(OBu^t)_4(O_3SCF_3)_2(PMe_3)_2]$ (W≡W, 2.421 Å) the eclipsed conformation is found with the *tert*-butoxo ligands *trans* to one another.[302] The symmetrical dimer $[W_2(hftb)_4(NMe_2)_2]$ has the staggered *anti* conformation.[321] The benzyltungsten isopropoxide diketonate $[(PhCH_2)(tmhd)(Pr^iO)W≡W(OPr^i)(tmhd)(CH_2Ph)]$ (W≡W, 2.328 Å) also has a nonbridged metal–metal triple bond with the ligands in a staggered conformation.[303] In the symmetrical carbonyl complex $[W_2(tftb)_6(CO)_2]$ (W≡W, 2.450 Å) the two halves of the molecule are eclipsed and the carbonyls are in the *syn* configuration.[304] Studies on the chemical reactivity of the ditungsten species $[(RO)_3W≡W(OR)_3]$ have produced a multitude of interesting new dinuclear species containing a variety of different bridging ligands.

In $[W_2(\mu\text{-}OMe)_2(OMe)_8]$ the edge-shared bi-octahedral structure (Fig. 4.2) was found.[306] The diamagnetism and W–W distance (2.790 Å) indicated the presence of a single W–W bond. A similar structure occurs in the W(v) chloride ethoxide $[W_2(\mu\text{-}OEt)_2(OEt)_4Cl_4]$ (W–W, 2.715 Å) in which the chlorines are *cis* and the terminal ethoxo ligands *trans* to one another.[307] Interestingly the W(IV) complexes $[W_2(\mu\text{-}OR)_2(OR)_2(ROH)_2Cl_4]$ (R = Me, Et; W=W, 2.481, 2.483 Å) have nearly the same structure with an ROH ligand replacing a terminal alkoxo group on each tungsten. Hydrogen bonding ROH....OR occurs between *syn*-alcohol and alkoxo ligands across the metal–metal bond.[308] Exchange reactions with other alcohols R′OH produced a series of mixed ligand complexes $[W_2(\mu\text{-}OR)_2(OR')_2(R'OH)_2Cl_4]$ (R = Et, R′ = Pr^i; R = Et, R′ = Et_2CH) and $[W_2(\mu\text{-}OPr^i)_2(OPr^i)_2(Pr^iOH)_2Cl_4]$ all having the same overall edge-shared bi-octahedral structure.[309]

Reaction of $[W_2(OEt)_4Cl_4(EtOH)_2]$ with ketones R_2CO gave rise to novel W(v) dinuclear complexes $[W_2(\mu\text{-}OEt)_2(OCR_2CR_2O)_2Cl_4]$ containing the bridging diolate ligands in place of the terminal ethoxo and ethanol ligands (Fig. 4.46).[310] On the other hand α-diketones R′CO.COR′ reacted with $(RO)_3WW(OR)_3$ to give centrosymmetrical dinuclear molecules $[W_2(\mu\text{-}OR)_2(OR)_4(OCR'=CR'O)_2]$ (R = Pr^i; R′ = Me, *p*-tolyl; W–W, 2.745, 2.750 Å) in which the alkylene diolate ligands are chelated to tungsten(v) (Fig. 4.47).[311] In the W(VI) complex $[W_2(\mu\text{-}OBu^t)_2(OBu^t)_4(CNMe_2)_2]$ the metal is five-coordinated (TBP) with a carbyne $CNMe_2$ and a bridging Bu^tO occupying axial positions. The bridging Bu^tO in the equatorial position is shorter (W–O, 1.95 Å) than

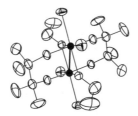

Figure 4.46 Structure of $[W_2(\mu\text{-}OEt)_2(OC_2Me_4O)_2Cl_4]$ (H atoms omitted).

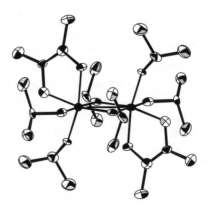

Figure 4.47 Structure of $[W_2(\mu\text{-}OPr^i)_2(OPr^i)_4(OCMe=CMeO)_2]$ (H atoms omitted).

the one in the axial position (2.42 Å) owing to the *trans* influence of the carbyne (W≡C, 1.77 Å).[312] See also $[W_2(\mu\text{-}OBu^t)_2(OBu^t)_4(CMe)_2]$ (Table 4.11).[323]

The unsymmetrical chloride methoxide $[Cl_2(MeO)_2W(\mu\text{-}OMe)_2W(OMe)_3Cl]$ also has the edge-shared bi-octahedral structure.[313] The remarkable compound $[W_2(\mu\text{-}OCH_2Bu^t)(\mu\text{-}H)(\mu\text{-}I)(OCH_2Bu^t)_5(Me_2NH)]$ (W=W, 2.456 Å) has the confacial bi-octahedral structure with three different bridging groups.[314] A similar hydrido-bridged confacial bi-octahedral structure was also found in $[W_2(\mu\text{-}OC_5H_9\text{-}c)_2(\mu\text{-}H)(OC_5H_9\text{-}c)_5(Me_2NH)]$ (W=W, 2.438 Å).[315] Reaction of the latter compound gave rise to the novel alkyl, alkene complex $[Et(C_2H_4)(c\text{-}C_5H_9O)W(\mu\text{-}OC_5H_9\text{-}c)_3W(OC_5H_9\text{-}c)_3]$ (W–W, 2.668 Å) in which the confacial bridging is sustained by three alkoxo ligands. Other related molecules are $[W_2(\mu\text{-}OC_5H_9\text{-}c)_2(\mu\text{-}CH_2CCH_2)(OC_5H_9\text{-}c)_6]$, $[W_2(\mu\text{-}OC_5H_9\text{-}c)_3(OC_5H_9\text{-}c)_5\{\eta^2\text{-}O\overline{C(CH_2)_3}CH_2\}]$, $[W_2(\mu\text{-}OPr^i)_3(OPr^i)_4H(\eta^2\text{-}OCPh_2)]$, and $[W_2(\mu\text{-}OPr^i)_2(\mu\text{-}\eta^2\text{-}HNCPh=CPhN]$ in which the nitrogen-containing ligand is both chelating and bridging.[316]

Reaction of xylylisonitrile ($2,6\text{-}Me_2C_6H_3NC$) with $W_2(OBu^t)_6$ gave the dinuclear $[W_2(\mu\text{-}OBu^t)_2(\mu\text{-}CNC_6H_3Me_2)(OBu^t)_4]$ (W–W, 2.525 Å) in which the five-coordinated metals are in a face-sharing bi-square pyramidal configuration with the carbon atom of the bridging isonitrile occupying the apical position.[316] See also $[W_2(tftb)_6(CNC_6H_3Me_2)_2]$ (Table 4.11).[322] With $W_2(OPr^i)_6$ the unsymmetrical $[W_2(\mu\text{-}OPr^i)_2(\mu\text{-}CNC_6H_3Me_2)(OPr^i)_4(py)]$ species was formed. One metal is in distorted octahedral and the other is in square pyramidal coordination in a confacial bridged structure.[316]

Reaction of the tetranuclear $W_4(OBu^i)_{12}$ with nitric oxide gave the centrosymmetrical dinuclear $[W_2(\mu\text{-}OBu^i)_2(OBu^i)_4(NO)_2(py)_2]$ with the edge-shared isobutoxo-bridged bi-octahedral structure.[317] The isobutoxo bridges are unsymmetrical with the W–O bond *trans* to the linear NO ligand being longer (2.204 Å) than the one *trans* to the terminal isobutoxo ligand (2.054 Å).

A number of binuclear organoimido alkoxo complexes have been studied. The edge-shared bi-octahedral structure was found for $[W_2(\mu\text{-}OMe)_2(OMe)_6(NPh)_2]$. The practically linear (WN̂C, 174°) imido group is *trans* to a bridging methoxo ligand.[277] A similar structure was exhibited by $[W_2(\mu\text{-OEt})_2(OEt)_4Cl_2(NPh)_2]$ with the phenylimido ligand *trans* to one bridging ethoxo (W–O, 2.213 Å) and a terminal ethoxo *trans* to the other (W–O, 2.003 Å).[318] Other similar structures with unsymmetrical alkoxo bridges were found in $[W_2(\mu\text{-OMe})_2(OMe)_6(NC_6H_4Me)_2]$ and $[W_2(\mu\text{-OPr}^i)_2(OPr^i)_6(NC_6H_4Me)_2]$ but in $(NBu_4^n)^+[W_2(\mu\text{-OMe})(\mu\text{-Cl})_2Cl_4(NPh)_2]^-$ the anion has a distorted confacial bi-octahedral structure in which one of the bridging chlorides is *trans* to both terminal phenylimido groups.[319] The typical edge-shared bi-octahedral structure was found for $[W_2(\mu\text{-OPr}^i)_2(OPr^i)_6(NPh)_2]$.[279]

An unusual dinuclear complex was formed by pinacolate (2,3-dimethyl butane-2,3-diolate) ligands with the phenylamido-tungsten (W-$\widehat{N}$-C, 134°) moiety, $[(\text{pinacolato})_2(PhNH)W(\mu\text{-OCMe}_2CMe_2O)W(\text{pinacolato})_2(NHPh)]$. Each tungsten is in a distorted octahedral coordination with two chelating pinacolates, a phenylamido group, and the bridging pinacolate.[320]

In the binuclear complex $[W_2(\mu\text{-NHMeC}_2H_4NHMe)(OEt)_6]$ (W≡W, 2.296 Å) the dimethylethylenediamine ligand straddles the metal–metal bond giving an eclipsed conformation to the ethoxo groups.[324]

The linking of alkyne and nitrile functions at ditungsten centres has led to the formation of several interesting dinuclear compounds. An example is the unsymmetrical diene–diimido complex $[W_2(\mu\text{-OCH}_2Bu^t)_2(\mu\text{-NCMe}=CMeCMe=CMeN)(OCH_2Bu^t)_4(py)]$ (W–W, 2.617 Å) in which the diimido ligand chelates one tungsten and bridges the other in a confacial bi-octahedral structure.[325] A similar configuration is found in the related molecule $[W_2(\mu\text{-OPr}^i)_2(\mu\text{-NCMeC}_2H_2CMeN)(OPr^i)_5]$ (W–W, 2.576 Å).[325] In $[W_2(\mu\text{-OBu}^t)(\mu\text{-NCPhCHCH})(OBu^t)_5]$ (W–W, 2.674 Å) the alkylidene imido ligand chelates one tungsten through C and N donors but bridges the other through N.[325] The confacial bi-octahedral configuration is found in $[W_2(\mu\text{-OPr}^i)_2(\mu\text{-NCPhCHCH}_2)(OPr^i)_5]$ (W–W, 2.584 Å).[325] Addition of bis-*p*-tolyldiazomethane $(MeC_6H_4)_2CN_2$ to $W_2(OBu^t)_6$ gave the binuclear $[W_2\{\mu\text{-NNC}(C_6H_4Me)_2\}_2(OBu^t)_6]$ (W–W, 2.675 Å) in which each W atom is in a distorted (TBP) configuration with the NNC $(C_6H_4Me)_2$ ligands bridging a common axial–equatorial edge.[305]

The compound $[W_2\{\mu\text{-NPhCO}(OBu^t)\}_2(OBu^t)_4]$ (W≡W, 2.290 Å) formed by insertion of PhNCO into the W-OBu^t bonds of $W_2(OBu^t)_6$, is a centrosymmetric molecule with two *trans* bridging bidentate ligands.[243] The precursor molecule $[W_2(\mu\text{-OCNPh})(OBu^t)_6]$ (W–W, 2.488 Å) was converted by addition of PMe$_3$ into the insertion product $[W_2\{\mu\text{-NPhCO}(OBu^t)\}(OBu^t)_5(PMe_3)]$ (W–W, 2.361 Å).[326] The bis-*p*-tolylcarbodiimide adduct $[W_2\{\mu\text{-N}(C_6H_4Me)CN(C_6H_4Me)\}(OBu^t)_6]$ (W–W, 2.482 Å) has a similar structure to the isocyanate adduct.[327] The bis-isopropyl and bis-cyclohexyl carbodiimides gave analogous structures (data in Table 4.11).[328] In the CO_2 insertion product $[W_2(\mu\text{-O}_2COBu^t)_2(OBu^t)_4]$ (W≡W, 2.315 Å) the bidentate bridging ligands adopt the cisoid conformation.[329] The pivalato ligand $Bu^tCO_2^-$ also spans the W≡W in the symmetrical molecule $[W_2(\mu\text{-O}_2CBu^t)_2(OBu^t)_4]$ (W≡W, 2.312 Å) in which the two carboxylates adopt a cisoid configuration which is also found in the related molecule $[W_2(\mu\text{-O}_2CBu^t)_2(OBu^t)_2(NPr^i)_2]$ (W≡W, 2.336 Å).[330]

Addition of α,β-unsaturated ketones and aldehydes to $[W_2(OCH_2Bu^t)_6(py)_2]$ gave rise to some novel complexes. With acrolein a bis-adduct was formed $[W_2(\mu\text{-OCH}_2Bu^t)_2(\mu\text{-OCH}_2CH_2CH_2)(\eta^2\text{-OCHCH:CH}_2)(OCH_2Bu^t)_4]$ in which one ligand is chelating (O,C) and bridging whilst the other is a terminally bonded η^2-OC.[331] With methylvinylketone a symmetrical molecule $[W_2(\mu\text{-OCHMeCH}_2CH_2)_2(OCH_2Bu^t)_6]$ is formed whilst crotonaldehyde coordinates in η^2-OC mode in [W$_2$(μ-OCH$_2$Bu^t)$_2$(OCH$_2$Bu^t)$_4$(η^2-OCHCH:CHCH$_3$)$_2$].[331] In the diolate complex [W$_2$(μ-L)$_3$], where L = 2,5-dimethylhexane-2,5-diolate, the molecule retains the ethane-like staggered conformation with 3 eight-membered diolate rings spanning the metal–metal triple bond (W≡W, 2.363 Å) (Fig. 4.48).[332] In the dimethylamine adduct [W$_2$(μ-L)(η^2-L)$_2$(Me$_2$NH)$_2$] there is one bridging diolate ligand and the other two each chelate one of the metal atoms whilst the two amine ligands are *anti* to one another about the metal–metal bond (W≡W, 2.320 Å).[332]

Reactions of W$_2$(OR)$_6$ with CO have produced some interesting products. With the isopropoxide the dinuclear complex [W$_2$(μ-OPr^i)$_2$(μ-CO)(OPr^i)$_4$(py)$_2$] (W–W, 2.499 Å) was obtained. This molecule has the same confacial bi-octahedral structure as the isomorphous molybdenum complex.[234] With W$_2$(OBu^t)$_6$ the compound formed [W$_2$(μ-OBu^t)$_2$(μ-CO)(OBu$^t_4$)][333] which is isomorphous and isostructural with the molybdenum analogue.[227] Also formed in the reaction of [W$_2$(OPr^i)$_6$(py)$_2$] with CO is the interesting complex [W$_2$(μ-OPr^i)$_2$(OPr^i)$_4$(CO)$_4$] in which an octahedral W(OPr^i)$_6$ moiety acts as a bidentate bridge to the W(CO)$_4$ moiety.[334] In this edge-sharing bi-octahedral complex the bridges are very unsymmetrical with W–O bonds in the W(OPr^i)$_6$ moiety being shorter (1.96 Å) than those in the W(μ-OPr^i)$_2$(CO)$_4$ moiety (2.25 Å).

Reaction of the alkylidyne complex [W$_2$(OBu^t)$_6$(CMe)$_2$] with CO gave the unsymmetrical complex [W$_2$(μ-OBu^t)(μ,η^4-C$_2$Me$_2$)(OBu^t)$_5$(CO)] composed of two trigonal bipyramids joined by axial (μ-OBu^t) and equatorial (μ-C$_2$Me$_2$) ligands. One W atom is bonded to three terminal Bu^{t}O ligands and the other to two terminal Bu^{t}O and one CO ligand.[335] Addition of CO to [W$_2$(OBu^t)$_6$(CNMe$_2$)$_2$] producing bridging η^2-ketenyl ligands in the symmetrical dimer [W$_2$(μ,η^2-OCCNMe$_2$)$_2$(OBu^t)$_6$].[335] With [W(OPr^i)$_3$(CNEt$_2$)(py)$_2$] carbonylation produced the symmetrical [W$_2$(μ,η^2-

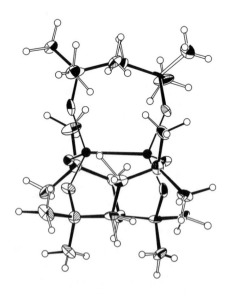

Figure 4.48 Structure of [W$_2$(OCMe$_2$C$_2$H$_4$CMe$_2$O)$_3$].

OCCNet$_2$)$_2$(OPr^i)$_6$(py)$_2$] also containing the bridging η^2-ketenyl ligands but the analogous reaction with [W(OPr^i)$_3$(CNMe$_2$)(py)$_2$] gave the unsymmetrical molecule [W$_2$(μ-OPr^i)$_3$(OPr^i)$_3${η^2-C$_2$(NMe$_2$)$_2$}(CO)$_2$]. The latter has a confacial bi-octahedral structure with one W atom terminally bonded to one Pr^{i}O and two CO ligands whilst the other is bonded terminally to two Pr^{i}O ligands and η^2-bonding alkyne.[335] Carbonylation of [W$_2$(hftb)$_4$(NMe$_2$)$_2$] gave rise to the unsymmetrical dinuclear complex [W$_2$(μ-CO)$_2$(μ-NMe$_2$)(hftb)$_4$(CO)(NMe$_2$)] (hftb = OCMe(CF$_3$)$_2$) which has a confacial bi-octahedral structure with one bridging NMe$_2$ ligand and two semi-bridging carbonyls. One tungsten is bonded terminally to two Pr^{i}O and one CO ligand and the other tungsten is terminally bonded to two Pr^{i}O and one NMe$_2$ ligand.[304]

Reactions of W$_2$(OR)$_6$ with alkynes have led to the isolation of several interesting complexes. From W$_2$(OBu^t)$_6$ and PhC$_2$Ph two compounds were obtained: [W$_2$(μ-CPh)$_2$(OBu^t)$_4$] (W–W, 2.665 Å) involving two tetrahedra sharing a common edge and [W$_2$(μ-C$_2$Ph$_2$)$_2$(OBu^t)$_4$] (W–W, 2.677 Å).[336] With other tungsten alkoxides and other alkynes the following binuclear complexes were characterized: [W$_2$(μ-OPr^i)$_2$(μ-C$_2$H$_2$)(OPr^i)$_4$(py)$_2$] (W–W, 2.567 Å), [W$_2$(μ-OCH$_2$Bu^t)(μ-C$_2$Me$_2$)(OCH$_2$Bu^t)$_5$(py)$_2$] (W–W, 2.602 Å), and [W$_2$(μ-OPr^i)(μ-C$_4$Me$_4$)(OPr^i)$_5$(C$_2$Me$_2$)] (W–W, 2.852 Å);[337–339] [W$_2$(μ-OPr^i)(μ-C$_4$H$_4$)(OPr^i)$_5$(C$_2$H$_2$)],[339] [W$_2$(μ-OBu^t)(μ-C$_2$H$_2$)(OBu^t)$_5$(py)].[338]

The binuclear complex [W$_2$(μ-CSiMe$_3$)$_2$(OPr^i)$_4$][340] reacted with acetylene giving [W$_2$(μ-CSiMe$_3$)(μ-CHCHCSiMe$_3$)(OPr^i)$_4$][340] and with 2,6-Me$_2$C$_6$H$_3$NC giving [W$_2$(μ-CSiMe$_3$)(μ-CCSiMe$_3$)(OPr^i)$_4$(NC$_6$H$_3$Me$_2$)].[341]

Other organo-tungsten isopropoxo compounds characterized are [W$_2$(μ-OPr^i)$_2$(OPr^i)$_2$(CH$_2$Ph)$_2$(η^2-C$_2$Me$_2$)$_2$] and [W$_2$(μ-C$_4$Me$_4$)(μ-CH$_2$Ph)(OPr^i)$_4$].[342]

Interesting alkyne [W$_2$(μ-OBu^t)(μ-C$_2$Me$_2$)(OBu^t)$_5$(CO)][343] and allene complexes [W$_2$(μ-C$_3$H$_4$)(OBu^t)$_6$], [W$_2$(μ-OBu^t)$_2$(μ-C$_3$H$_4$)(OBu^t)$_4$(η^2-C$_3$H$_4$)], and [W$_2$(μ-OBu^t)$_2$(μ-C$_3$H$_4$)(OBu^t)$_4$(CO)$_2$] have been prepared.[344] Alkyne adducts have also been prepared from reactions involving the neopentoxide [W$_2$(OCH$_2$Bu^t)$_6$(py)$_2$], e.g. [W$_2$(μ-OCH$_2$Bu^t)(μ-C$_2$H$_2$)(OCH$_2$Bu^t)$_5$(py)$_2$] and [W$_2$(μ-OCH$_2$Bu^t(μ-C$_2$Et$_2$)(OCH$_2$Bu^t)$_5$(py)].[345] Addition of ethylene gave the interesting binuclear bis-adduct [W$_2$(μ-OCH$_2$Bu^t)$_4$(OCH$_2$Bu^t)$_2$(η^2-C$_2$H$_4$)] (W–W, 2.533 Å) with four unsymmetrical neopentoxo bridges between the two tungsten atoms. The longer bridge bonds (av. 2.31 Å, cf. 2.00 Å) are trans to the terminal alkoxo or ethylene ligands.[346] This type of complex appears to be a precursor to other species formed in the reaction of ethylene with W$_2$(OR)$_6$, e.g. [W$_2$(μ-OPr^i)$_3$(OPr^i)$_3$(η^2-CH$_2$CH$_2$CH$_2$CH$_2$)(η^2-C$_2$H$_4$)] (W–W, 2.643 Å) in which the (CH$_2$)$_4$ ligand is chelating one tungsten atom to form a tungstacyclopentane ring.[347] In [W$_2$(μ-OPr^i)(OPr^i)$_5$(μ,η^2-CCH$_2$CH$_2$CH$_2$)(py)] (W–W, 2.674 Å) the hydrocarbyl ligand bridges two tungstens, chelating one of them.[347]

The 1,2-R$_2$W$_2$(OR)$_4$ compounds also react with alkynes producing a variety of products, e.g. [W$_2$(μ-OPr^i)$_2$(OPr^i)$_2$(C$_3$H$_7$-n)$_2$(η^2-C$_2$Me$_2$)$_2$] (W–W, 2.681 Å),[348] [W$_2$(μ-OPr^i)$_2$(OPr^i)$_2$(CH$_2$Ph)$_2$(η^2-C$_2$Me$_2$)$_2$] (W–W, 2.668 Å),[349] [W$_2$(μ-OPr^i)$_2$(OPr^i)$_2$(Ph)$_2$(η^2-C$_2$Me$_2$)$_2$] (W–W, 2.658 Å),[349] [W$_2$(μ-OPr^i)$_2$(OPr^i)$_2$(CH$_2$Bu^t)$_2$(CEt)$_2$],[349] and [Me$_2$(Bu^{t}O)W(μ-OBu^t)(μ-C$_2$Me$_2$)W(OBu^t)$_2$(py)] (W–W, 2.622 Å).[349] Related molecules are [W$_2$(μ-CPh)(μ,η^2-C$_4$Me$_4$)H(OPr^i)$_4$] (W–W, 2.755 Å),[350] [W$_2$(μ-CSiMe$_3$)(μ,η^2-C$_4$Me$_4$)(OPr^i)$_4$(CH$_2$SiMe$_3$)] (W–W, 2.780 Å),[350] [W$_2$(μ-C$_2$Me$_2$)$_2$(OPr^i)$_4$] (W–W, 2.611 Å),[351] [W$_2$(μ-OCH$_2$Bu^t)$_2${μ,η^2-CH$_2$(CH)$_2$CH$_2$}(OCH$_2$Bu^t)$_4$(py)] (W–W, 2.471 Å),[352] and [W$_2$(μ-OCH$_2$Bu^t)$_4$(OCH$_2$Bu^t)$_2$(η^2-C$_2$H$_4$)$_2$] (W–W, 2.533 Å).[353] In

$[(Bu^tO)_3W \equiv C-C \equiv W(OBu^t)_3]$ the WC_2W system is linear and the two $W(OBu^t)_3$ moieties are staggered.[354]

Another binuclear complex in which the W atoms are well separated by the bridging ligand is $[W_2\{\mu\text{-}CCCNH(C_6H_3Me_2\text{-}2,6)(OSiMe_2Bu^t)_5\}\{CN(C_6H_3Me_2)\}_4]$ obtained from the reaction of $[W_2(\mu\text{-}CCH)(OSiMe_2Bu^t)_5]$ with xylylisocyanide. The bridging $\mu\text{-}CCCNH(C_6H_3Me_2)$ ligand may be viewed as an η^2-aminoalkyne to one tungsten (six-coordinated) and an alkylidyne to the other (five-coordinated).[355] Bridging alkylidyne ligands are featured in $[W_2(\mu\text{-}CCHCHMe)(\mu\text{-}OSiMe_2Bu^t)(OSiMe_2Bu^t)_5H]$ (W–W, 2.658 Å) and $[W_2\{\mu\text{-}CC(CH_2)(CHCH_2)\}(\mu\text{-}OSiMe_2Bu^t)(OSiMe_2Bu^t)_5H]$ (W–W, 2.561 Å).[356]

In $[W_2(\mu,\eta^2\text{-}CCH_2CHCHS)(\mu\text{-}OBu^t)(OBu^t)_4(\eta^1\text{-}\overline{CCHCHCHS})]$ (W–W, 2.663 Å) the butadiene thiolate ligand is chelating one W atom and bridging to the other.[357] A similar situation is present in $[W_2(\mu\text{-}OBu^t)(\mu,\eta^2\text{-}CCH_2CHCHO)(OBu^t)_4(\eta^1\text{-}\overline{CCHCHCHO})]$ (W–W, 2.649 Å).[357] In $[W_2(\mu\text{-}OBu^t)(\mu,\eta^2\text{-}C_4H_4)(OBu^t)_5(CO)]$ (W–W, 2.901 Å) the Bu^tO bridge is asymmetric with the long W–O bond (2.224 Å, cf. 2.000 Å) being trans to the terminal carbonyl ligand in a confacial pseudo-octahedral structure.[358]

A few examples are known with phosphorus or phosphine bridges. In $[W_2(\mu\text{-}OPr^i)(\mu\text{-}P_2)(OPr^i)_5(py)]$ (W–W, 2.69 Å) a pseudo-tetrahedral W_2P_2 unit forms the core of a confacial pseudo-octahedral structure in which the shorter W–O bridging bond (1.989 Å) is trans to the pyridine ligand.[359] In the diphenylphosphido bridged complex $[W_2(\mu\text{-}NPh_2)_2(OBu^t)_4]$ the W_2P_2 unit is a puckered ring and each W atom is in a distorted tetrahedral configuration, probably due to the presence of a bent metal–metal bond (W–W, 2.59 Å).[360] In the bis-cyclohexylphosphido complex $[W_2\{\mu\text{-}P(C_6H_{11}\text{-}c)_2\}_3\{PH(C_6H_{11}\text{-}c)_2\}(OCH_2Bu^t)_3]$ (W–W, 2.618 Å) there is a triple phosphido bridge with five-coordinated tungstens.[361] The diphosphine ligand dmpm (dimethylphosphino methane) straddles the two W atoms in $[W_2(\mu\text{-}dmpm)(OPr^i)_4(CH_2Ph)_2]$ (W–W, 2.347 Å) and in $[W_2(\mu\text{-}dmpm)(OPr^i)_4Bu^i_2]$ (W–W, 2.349 Å).[362]

The tertiary phosphine PMe_3 acts as a terminal ligand in $[W_2(\mu\text{-}H)(\mu\text{-}CPh)(OPr^i)_4(PMe_3)_3]$ (W–W, 2.54 Å) and $[W_2(\mu\text{-}H)(\mu\text{-}CPr^i)(OPr^i)_4(PMe_3)_3]$ (W–W, 2.50 Å). Each molecule has one pseudo-octahedral W bonded to three PMe_3 ligands, a terminal Pr^iO, the bridging hydride, and alkylidyne ligands whilst the other W is five-coordinated by three terminal Pr^iO ligands and the two bridging ligands.[362] By contrast the dimethylphosphinoethane (dmpe) ligand acts as a chelate in $[(dmpe)_2W(\mu\text{-}H)_2W(OPr^i)_4]$ (W–W, 2.496 Å) giving an edge-shared bi-octahedral structure in this unsymmetrical molecule.[363]

In contrast to the abundance of binuclear alkoxo tungsten compounds there are relatively few trinuclear species. In $[W_3(\mu_3\text{-}CMe)(\mu\text{-}OPr^i)_3(OPr^i)_6]$ the triangle of W atoms is capped by the μ_3-ethylidyne ligand and bridged by three Pr^iO ligands in the W_3 plane. Each metal has two terminal Pr^iO ligands giving five-coordination (square pyramidal) (Fig. 4.49).[364] The same general framework of atoms is found in $[W_3(\mu_3\text{-}P)(\mu\text{-}OCH_2Bu^t)_3(OCH_2Bu^t)_6]$.[365] In the imido complex $[W_3(\mu_3\text{-}NH)(\mu_3\text{-}OPr^i)(\mu\text{-}OPr^i)_3(OPr^i)_6]$ there are capping ligands above and below the W_3 plane analogous to the oxo-complex $[W_3(\mu_3\text{-}O)(OPr^i)_{10}]$.[366] A less symmetrical structure obtains in $[W_3(\mu_3\text{-}CMe)(\mu\text{-}OPr^i)_3(OPr^i)_3Cl_2]$ in which W[1]–$\widehat{W[2]}$–W[3] = 81° (cf. 60° for an equilateral triangle). One tungsten W[1] is in a pseudo-trigonal

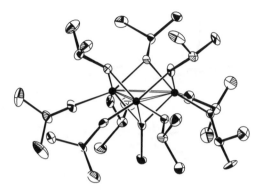

Figure 4.49   Structure   of $[W_3(\mu_3\text{-CMe})(\mu\text{-}OPr^i)_3(OPr^i)_6]$ (H atoms omitted).

bipyramidal configuration while W[2] and W[3] are pseudo-octahedral.[367] The trimeric molecule $[W_3(\mu\text{-N})_3(\text{tftb})_9]$ has a completely different structure with a W_3N_3 ring with alternating long and short W–N bonds (1.72, 2.14 Å). The three alkoxo ligands in the W_3N_3 plane are *trans* to the long W–N bond while the other six are at right angles above and below the plane giving each W a TBP coordination.[368] This behaviour is in contrast to that of $[W(\mu\text{-N})(OBu^t)_3]_n$ which forms infinite linear polymers with alternating short (1.740 Å) and long (2.661 Å) tungsten–nitrogen bonds.[323]

Carbonylation of $[W_3(\mu_3\text{-CMe})(OPr^i)_9]$ gave a novel trinuclear species $[(Pr^iO)_3W(\mu\text{-CMe})(\mu\text{-}OPr^i)W(OPr^i)(\mu\text{-}OPr^i)_3W(OPr^i)(CO)_2]$ involving a bent $W[1]\text{–}\widehat{W}[2]\text{–}W[3]$ (150°). Two of the tungstens have TBP configurations and the third is octahedral.[369] Another trinuclear complex with a triangular (isosceles) W_3 system is $[W_3(\mu_3\text{-H})(\mu\text{-}OPr^i)_2(\mu\text{-NPh})(OPr^i)_7(\text{py})]$ which contains 1 five-coordinated and 2 six-coordinated W atoms. The capping hydride is above the plane of the three metal atoms whilst the bridging donor atoms are all below it.[279]

A few tetranuclear alkoxo tungsten compounds have been characterized. The homoleptic alkoxides $[W_4(\mu_3\text{-OR})_2(\mu\text{-OR}_4)(OR)_{10}]$ (R = Me, Et) have the edge-sharing distorted octahedral structure (Fig. 4.50) similar to that of $[Ti_4(OR)_{16}]$ (Fig. 4.35) but the presence of metal–metal bonding involving W(IV) (d^2) ions produces a rigid structure which is maintained in solution.[370]

In the tungsten(IV) hydride isopropoxide $[W_4(\mu\text{-H})_2(OPr^i)_6(OPr^i)_8]$ a zig-zag line of four tungsten atoms is present with evidence of W=W bonding localized between the outer and inner W atoms (W–W, 2.446 Å) but not between the inner W atoms (W–W, 3.407 Å).[371] The outer and inner W atoms share a hydride and two isopropoxo ligands in a confacial bi-octahedral system whereas the two inner W atoms are edge-sharing with two isopropoxo ligands.

Interestingly the tetranuclear W(III) triisopropoxide $[W_4(\mu\text{-}OPr^i)_4(OPr^i)_8]$ contains a distorted rhombus of tungsten atoms with two short W–W distances (2.50 Å) and two longer (2.73 Å) together with a longer diagonal W–W distance (2.81 Å). The four unsymmetrical edge-sharing Pr^iO groups are in the W_4-plane whilst the terminal ligands (two per W atom) are four above and four below this plane (Fig. 4.51).[296]

Another remarkable tetranuclear complex is $[W_4(\mu\text{-}OPr^i)_2(\mu_3,\eta^2\text{-CO})_2(OPr^i)_{10}(\text{py})_2]$ which contains the unusual μ_3,η^2-CO ligand. Two of the tungstens are five-coordinated

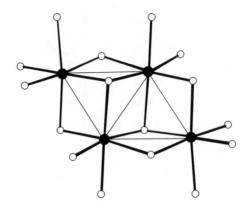

Figure 4.50 $W_4(\mu_3\text{-O})_2(\mu\text{-O})_4O_{10}$ core structure of $[W_4(\mu_3\text{-OR})_2(\mu\text{-OR})_4(OR)_{10}]$ ($R = Me$, Et; R groups omitted).

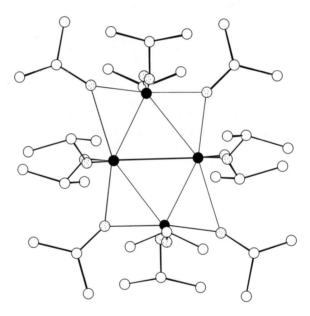

Figure 4.51   Structure of $[W_4(\mu\text{-OPr}^i)_4(OPr^i)_8]$ ($\bigcirc = O$; H atoms omitted).

(distorted TBP) and two (bonded to pyridine) are distorted octahedral.[372] Without the coordinated pyridines all 4 W atoms are five-coordinated (TBP) in $[W_4(\mu_4\text{-OPr}^i)_2(\mu_3,\eta^2\text{-CO})_2(OPr^i)_{10}]$.[333]

In $[W_4(\mu\text{-OEt})_6(\mu\text{-CSiMe}_3)_2(OEt)_8]$ there is again a zig-zag chain of tungsten atoms with alternating short (2.51 Å) and long (3.51 Å) W–W distances. The outer two metal atoms are in confacial bi-octahedral units which are edge bridged by the μ-CSiMe$_3$ ligands.[364]

The interesting carbido methylimido isopropoxo complex $[W_4(\mu_4\text{-C})(\mu\text{-OPr}^i)_4(\mu\text{-NMe})(\text{OPr}^i)_8]$ has a "butterfly" W_4 core of atoms with the μ_4-carbide atom closer to the wingtip tungstens (W–C, 1.914 Å, 1.956 Å) than to the backbone tungstens (W–C, 2.241 Å, 2.251 Å). The backbone W atoms are octahedrally coordinated whereas the wingtip tungstens are five-coordinated.[373]

Yet another W_4 configuration is found in the carbonyl complex $[W_4(\mu_4,\eta^2\text{-CO})(\mu\text{-OBu}^i)_5(\text{OBu}^i)_7(\text{CO})_2]$ (Fig. 4.52). In this spiked-triangular arrangement three of the W atoms are octahedrally coordinated and the fourth is in a pseudo-square pyramidal configuration. The two terminal carbonyls are bonded to one tungsten in the *cis* configuration.[374] Another W_4 chain structure with alternating short (2.721 Å) and long (3.504 Å) W–W distances is found in $[W_4(\mu\text{-CEt})_2(\mu\text{-C}_2\text{Me}_2)(\mu\text{-OPr}^i)_2(\text{OPr}^i)_4(\eta^2\text{-C}_2\text{Me}_2)_2]$ which contains both bridging and chelating alkyne ligands. The two halves of this symmetrical molecule are held together by weak isopropoxo bridges.[351]

Another tetranuclear "butterfly" configuration is found in $[W_4(\mu_4\text{-C})(\mu\text{-NMe})(\mu\text{-OCH}_2\text{Bu}^t)_4(\text{OCH}_2\text{Bu}^t)_7(\text{H})]$ in which the hydride is terminally bonded to one of the backbone tungsten atoms.[375]

A most remarkable structure is exhibited by the hydrido complex $[W_4(\mu\text{-H})_4(\mu\text{-OPr}^i)(\mu\text{-dmpm})_3(\text{OPr}^i)_7]$ or $[(\text{Pr}^i\text{O})_3W(\mu\text{-H})(\mu\text{-OPr}^i)W(\mu\text{-dmpm})_3(\mu\text{-H})W(\mu\text{-H})_2W(\text{OPr}^i)_4]$ in which each W atom has a different coordination. The outer W atoms are metal–metal bonded (W–W, 2.451, 2.499 Å) to their neighbours but the inner two metals, which are linked by three bridging dmpm ligands and one hydride, are not metal–metal bonded.[363]

One hexanuclear structure $[W_6(\mu\text{-H})_4(\mu\text{-OPr}^i)_7(\mu\text{-CPr}^i)(\text{OPr}^i)_5(\text{H})]$ has been reported.[363] This is based on the familiar octahedral W_6 cluster configuration with 12 bridging (4 hydride, 7 isopropoxide, 1 alkylidyne) ligands and 6 terminal (5 isopropoxide, 1 hydride) ligands.

Finally it is noteworthy that extended-chain metal–alkylidyne polymers $[W\{\mu\text{-C}(4)C_6H_2N\text{-}3,5\text{-Me}_2\}(\text{OBu}^t)_3]_n$ and $[W\{\mu\text{-C}(3)C_6H_4N\}(\text{OBu}^t)_3]_n$ have been structurally characterized. In the 4-alkynyl-3,5-dimethylpyridyl complex the tungstens are in the square pyramidal five-coordination but in the 3-alkynylpyridyl complex the tungstens adopt the TBP configuration.[376]

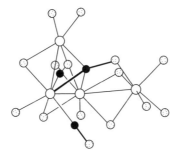

Figure 4.52 $W_4(\mu_4\text{-CO})(\mu\text{-O})_5O_7(\text{CO})_2$ core structure of $[W_4(\eta^2,\mu_4\text{-CO})(\mu\text{-OR})_5(\text{OR})_7(\text{CO})_2]$ (R = CH$_2$Pr^i; ● = C of CO; ○ = O; R groups omitted).

Table 4.11 Alkoxides of chromium, molybdenum, and tungsten

Compound	Metal coordination	M–O Bond lengths (Å)		Bond angles (°) $\widehat{MOC}$	Reference
		Terminal	Bridging		
[Cr(OCHBu$^t_2$)$_4$]	4	1.771, 1.774 (3)	–	140.5, 141.1	211
[Cr(OCMe$_2$CH$_2$OMe)$_3$]	6	1.892, 1.923, 1.931 (2) (alkoxo) 2.051, 2.060, 2.096 (2) (ether)	–	–	212
[Cr$_2$(μ-OCHBu^t)$_2$(OCBu$^t_3$)$_2$]	3	1.838 (3)	1.964, 1.986 (4)	171.0, 174.9 (t) 133.1 (b)	213
[(C$_5$H$_5$)$_2$Cr$_2$(μ-OBu^t)$_2$]	3^1	–	1.91–1.98 (1) av. 1.96	129–139 av. 133	214
[Mo(OSiMe$_3$)$_4$(Me$_2$NH)$_2$]	6	1.950, 1.951 (4)	–	170.8, 174.2	215
[Mo(OBu^t)$_2$(CO)$_2$(py)$_2$]	6	1.94, 1.95 (1)	–	140, 141	216
[Mo(1-ado)$_4$(Me$_2$NH)] 1-ado = 1-adamantolato	5	1.963 (2) (ax) 1.888, 1.897, 1.916 (2) (eq)	–	140.3 (ax) 136.8, 137.8, 141.7 (eq)	217
[Mo(OBu^t)$_4$(NNCPh$_2$)]	5	1.944 (3) (ax) 1.889 (3) (eq. av.)		141.4 (ax) 137.5 (eq, av.)	305
[Mo(OPr^i)$_2$(NO)(Me$_2$pz)] Me$_2$pz = 3,5-dimethylpyrazolylato	6	1.900, 1.908 (4)	–	133.4	218
[Mo(OEt)(OPr^i)(NO)(Me$_2$pz)]	6	1.900, 1.901 (4)	–	129.5, 129.8	218
[Mo(OPr^i)$_2$(NO){HB(impz)$_3$}] HB(impz)$_3$ = isopropylmethylpyrazol-1-yl	6	1.88 (1)	–	130.5	219
[Mo(μ-N)(OBu^t)$_3$]$_\infty$	5	1.882 (4)	–	135.1	220
[Mo(μ-N)(OPr^i)$_3$]$_\infty$	5	1.894 (17)	–	128.6	220
[Mo$_4$(μ-N)$_4$(OBu^t)$_4$Me$_8$(NC$_5$H$_4$Bu^t)$_2$]	5 and 6	1.882 (2)	–	–	221
[Mo(OC$_2$H$_4$O)$_3$]	6	1.890–1.918 (5) av. 1.903	–	116.0–119.3 av. 118.4	222
[Mo(OCMe$_2$CMe$_2$O)$_3$]	6	1.902–1.911 (5) av. 1.907	–	118.7–120.5 av. 119.8	222
[Mo$_2$(OCH$_2$Bu^t)$_6$]	4^1	1.855–1.905 (6) av. 1.876	–	114.5, 134.2, 134.2	223

Compound		M–O (br)	M–O (t)	¹³C NMR	Ref.
[Mo₂(μ-OPrⁱ)₂(OPrⁱ)₆]	5	1.872–1.976 (3) av. 1.911	1.958, 2.111 (3)	123.3–134.7 (t) av. 129.2	224
[Mo₂(OSiMe₃)₆(Me₂NH)₂]	5	1.925–1.981 (5) av. 1.949	–	131.1, 137.4 (b) 134.1–175.6 av. 157.2	225
[Mo₂(O₂COBuᵗ)₂(OBuᵗ)₄]	5	1.860–1.903 (6) (BuᵗO) av. 1.885	2.111–2.142 (6) (BuᵗOCO₂) av. 2.126	123.4, 154.6 (t, BuᵗO) av. 139.3	226
[Mo₂(μ-OBuᵗ)₂(μ-CO)(OBuᵗ)₄]	5	1.876, 1.888 (7)	2.074, 2.093 (8)	152.6, 153.3 (t) 135.0, 141.6 (b)	227
[Mo₂(μ-OPrⁱ)₂(OPrⁱ)₄(NO)₂]	5	1.850, 1.861 (7)	1.951, 2.195 (6)	125.7, 129.9 (t) 118.3, 134.8 (b)	228
[Mo₂(μ-OPrⁱ)₂(OPrⁱ)₄(NO)₂(Me₂NH)₂]	6	1.909, 1.935 (5)	2.062, 2.196 (5)	129.9, 131.2 (t) 120.6, 134.0 (b)	229
[Mo₂(μ-OPrⁱ)₂(OPrⁱ)₄Cl₄]	6	1.808–1.819 (3) av. 1.814	2.013–2.020 (4) av. 2.016	143.8–146.0 (t) av. 145.1 135.7–139.0 (b) av. 137.3	230
[Mo₂(μ-OPrⁱ)₂(OPrⁱ)₄Br₄]	6	1.805, 1.818 (5)	2.012, 2.014 (4)	147.7, 148.4 (t) 134.0, 140.2 (b)	230
[Mo₂(μ-NC₇H₇)₂(NC₇H₇)₂(OPrⁱ)₄]	5	1.882–1.900 (7) av. 1.892	–	138.0–149.6 av. 141.8	231
[Mo₂(μ-OPrⁱ)₃(OPrⁱ)₂(N₂CPh₂)₂(py)]	6	1.96, 1.96, 1.98 (1) av. 2.12	2.11–2.24 (1) av. 2.12	–	232
[Mo₂(μ-C₂H₂)(μ-OPrⁱ)₂(OPrⁱ)₄(py)₂]	6¹	1.923–1.964 (3) av. 1.941	2.106–2.187 (3) av. 2.146	133.9–138.9 (t) av. 135.5 123.5–126.9 (b) av. 125.3	233

(continued overleaf)

Table 4.11 (*Continued*)

Compound	Metal coordination	M–O Bond lengths (Å)		Bond angles (°) $\widehat{\text{MOC}}$	Reference
		Terminal	Bridging		
[Mo₂(μ-C₄H₄)(μ-OPrⁱ)₂(OPrⁱ)₅(py)]	6¹	1.88, 1.94, 1.94 (2)	2.06–2.17 (2)	134.1, 154.3, 172.1 (t) 137.0, 144.1 (b)	233
[Mo₂(μ-CO)(μ-OPrⁱ)₂(OPrⁱ)₄(py)₂]	6	1.909, 1.920 (5)	2.098, 2.113 (6)	135.2, 135.3 (t) 137.3, 146.9 (b)	234
[Mo₂(OBuᵗ)₂(CH₂SiMe₃)₄]	4	1.865 (8)	–	158.4	235
[Mo₂(OCH₂Buᵗ)(μ-η′,η²-NCNMe₂)(OCH₂Buᵗ)₅]	5¹	1.870–1.954 (2) av. 1.904	1.999, 2.146 (2)	117.0–135.8 (t) av. 127.6 128.6, 135.2 (b)	236
[Mo₂(μ-O₂CPh)₂(OBuᵗ)₄]	5¹	1.868, 1.903 (6)	2.125, 2.158 (6) (benzoate)	124.6, 154.7 (BuᵗO)	237
[Mo₂(OBuᵗ)₄(acac)₂]	5¹	1.888–1.919 (3) (BuᵗO) av. 1.900 2.064–2.117 (3) (acac) av. 2.092	–	117.8–137.8 (BuᵗO) av. 131.8 133.6–135.9 (acac) av. 134.6	238
[Mo₂(μ-OCH₂Buᵗ)₂(OCH₂Buᵗ)₂(CNBuᵗ)₂(acac)₂]	6	1.998, 2.020 (8) (neop) 2.098, 2.105 (7) (acac)	2.042, 2.054 (6)	125.6 (t, neop) 129.6 (b neop) 127.7, 128.3 (acac)	239
[Mo₂(OPrⁱ)₂(SAr)₄] Ar = 2,4,6-trimethylphenyl	4	1.878 (2)	–	138.4	240
[Mo₂(OPrⁱ)₂(SeAr)₄]	4	1.873 (6)	–	140.0	299
[Mo₂(μ-OCH₂Buᵗ)₂(μ-Br)(OCH₂Buᵗ)₄Br(py)]	6	1.872–1.905 (5) av. 1.887	2.020–2.118 (5) av. 2.061	128.8–135.3 (t) av. 130.6 123.6–136.5 (b) av. 130.5	241

Compound	Coord. no.	M–O (Å)	M–O (bridging) (Å)	O–M–O (°)	Ref.
[Mo(OBu^t)_4(NNCPh_2)]	5	1.944 (3) (ax) 1.876, 1.892, 1.899 (3) (eq)	–	141.4 (ax) 135.9, 137.7, 138.8 (eq)	242
[Mo_2(μ-OPr^i)_3(OPr^i)_3(NNCPh_2)_2(py)]	6	1.962, 1.966 (7) (Mo[1]) 1.989 (8) (Mo[2])	2.114, 2.127, 2.239 (7) (Mo[1]) 2.046, 2.040, 2.155 (7) (Mo[2])	125.2 (t, Mo[1]) 130.3, 139.0, 141.4 (b, Mo[1]) 136.1 (t, Mo[1]) 120.9, 129.1, 131.1 (b, Mo[2])	242
[Mo_2(OPr^i)_4[NPhC(O)(OPr^i)]_2]	5^1	1.88, 1.91 (2) Lig.† 2.18 (2)	–	119, 139 Lig.† 116	243
[Mo_2(OPr^i)_4(py)_4]	5^1	2.026–2.033 (3) av. 2.03	–	125.0–131.7 av. 128.8	244, 245
[Mo_2(OCH_2Bu^t)_4(PMe_3)_4]	5^1	1.980–2.088 (25) av. 2.03	–	124.8–134.1 av. 128.8	244, 245
[Mo_2(OCH_2Bu^t)_4(Me_2NH)_4]	5^1	2.073–2.085 (12) av. 2.08	–	121.0–127.0 av. 124.2	244, 245
[Mo_2(OPr^i)_4(Pr^iOH)_4]	5^1	2.092 (9) (Pr^iO) 2.170 (7) (Pr^iOH)	–	117.9	244, 245
[Mo_2(OC_5H_9-c)_4(c-C_5H_9OH)_4]	5^1	2.082–2.150 (16) av. 2.123	–	–	245
[Mo_2(OPr^i)_4(CH_2Ph)_2(PMe_3)]	4^1 and 5^1	1.920 (4) (Mo[1]) 1.874, 1.881, 1.911 (4) (Mo[2])	–	137.5 (Mo[1]) 118.1, 145.1, 148.5 (Mo[2])	246, 247
[Mo_2(OPr^i)_4(CH_2Ph)_2(dmpm)] dmpm = bis-dimethyl phosphinomethane	5^1	1.910–1.982 (6) av. 1.939	–	121.5–145.8 av. 134.5	247
[Mo_2(OPr^i)_4(dmpe)_2]	5^1	2.00 (1)	–	128	248
[Mo_2(μ-OCH_2Bu^t)(μ-C_4Me_4)(OCH_2Bu^t)_5(NCNEt_2)]	6^1	1.918–2.015 (5) av. 1.951	2.073, 2.150 (4)	126.3–140.4 (t) av. 131.3 121.5, 128.1 (b)	249

(continued overleaf)

303

Table 4.11 (*Continued*)

Compound	Metal coordination	M–O Bond lengths (Å)		Bond angles (°) $\widehat{MOC}$	Reference
		Terminal	Bridging		
[Mo₂(μ-OCH₂Buᵗ)(μ-C₄Me₃CH₂(NEt₂)NH)(OCH₂Buᵗ)₅]	6[1]	1.890–1.944 (5) av. 1.928	2.127, 2.135 (5)	128.8–143.9 (t) av. 135.9 125.2, 125.9 (b)	249
[Mo₂(μ-OPrⁱ)₂(OPrⁱ)₄(O₂C₆Cl₄)₂]	6	1.817, 1.844 (5) (PrⁱO) 2.014, 2.016 (5) (Diol)	2.020, 2.026 (5)	140.1, 165.1 (t) 134.2, 139.5 (b)	250
[Mo₂(μ-OPrⁱ)₂(OPrⁱ)₂(O₂C₁₄H₈)₃]	6	1.840, 1.845 (4) (PrⁱO) 1.941–2.030 (4) (Diol) av. 1.984	1.974, 1.979 (4) (Mo[1]) 2.045, 2.061 (4) (Mo[2])	151.7, 154.3 (t) 131.9–142.0 (b) av. 136.9	250
[Mo₂(μ-OPrⁱ)₂(OPrⁱ)₄(PrⁱNC₂H₂NPrⁱ)]	5	1.887, 1.894, 1.919 (3)	1.981, 2.097 (4) (Mo[1]) 1.959, 2.147 (4) (Mo[2])	–	251
[Mo₂(μ-NMe₂)₂(NMe₂)₂(OC₆H₁₀-4-Me)₄]	5	1.990, 1.966 (2)	–	–	252
[Mo₂{OC(CF₃)₃}₄(NMe₂)₂]	4[1]	1.909, 1.923 (10)	–	147.2, 164.2	253
[Mo₂{OCMe(CF₃)₂}₆]	4[1]	1.88 (1)	–	149	255
Mo₂(NAr)₂(OBuᵗ)₄] Ar = 2,6-diisopropylphenyl	4	1.887, 1.892 (3)	–	–	256
[Mo₂(OBuᵗ)₄(NHPh)₂(NH₂Ph)₂]	5[1]	1.926–1.979 (4) av. 1.954	–	132.8–144.3 av. 140.5	257
[Mo₂{O₁₂Si₇(C₆H₁₁-c)₇}₂]	4[1]	av. 1.90 (1)	–	–	258
[Mo₂(μ-OCH₂Buᵗ)(OCH₂Buᵗ)₆]	5[1]	1.921–1.975 (6) av. 1.951	2.143–2.159 (6) av. 2.150	–	259
[(BuᵗCH₂O)₄Mo(μ-OCH₂Buᵗ)(μ-S)Mo(OCH₂Buᵗ)(CPh₂)(PMe₃)]	5 and 6	1.861–1.945 (6) av. 1.913	2.116, 2.119 (6)	–	260
[Mo₄(μ-F)₄(OBuᵗ)₈]	5[1]	1.850–1.992(6) av. 1.925	–	114.3–165.7 av. 139.5	261

304

Compound		M–O (terminal)	M–O (bridging)	M–O–M (°)	Ref.
$[Mo_4(\mu\text{-}F)_3(\mu\text{-}NMe_2)(OBu^t)_8]$	5^1	1.867–1.946 (5) av. 1.903	—	121.8–156.9 av. 139.7	261
$[Mo_2(OBu^t)_4F_2(PMe_3)_2]$	5^1	1.863–1.938 (6) av. 1.902	—	124.8–167.5 av. 144.2	262
$[Mo_4(\mu\text{-}OPr^i)_8Cl_4]$	5	—	1.981, 2.078 (4)	136.8, 138.7	263
$[Mo_4(\mu_3\text{-}OPr^i)_2(\mu\text{-}OPr^i)_4(OPr^i)_2Br_4]$	5^1	1.843 (4)	2.140, 2.150, 2.158 (4) (μ_3) 2.020–2.046 (4) (μ) av. 2.029	—	263
$[Mo_4(\mu_3\text{-}OPr^i)_2(\mu\text{-}OPr^i)_4(OPr^i)_3Br_3]$	5^1	1.860, 1.862, 1.938 (9)	2.090–2.162 (7) (μ_3) av. 2.133 2.008–2.067 (7) (μ_2) av. 2.040	—	264
$[Mo_4(\mu\text{-}OPr^i)_2(\mu\text{-}F)_2(OPr^i)_8]$	5^1	1.875–1.913 (3) av. 1.894	2.113, 2.117 (3)	—	264
$[Mo_4(\mu_3\text{-}N)(\mu\text{-}OPr^i)_2(OPr^i)_{10}]$	5	av. 1.92 (1)	2.058, 2.139 (2)	—	265
$[Mo_4(\mu\text{-}OMe)_2(\mu\text{-}OPr^i)_2(OPr^i)_8]$	5^1	av. 1.91 (2)	2.11 (1)	—	266
$[Mo_4(\mu_3\text{-}OCH_2Bu\text{-}c)(\mu\text{-}OCH_2Bu\text{-}c)_4(OCH_2Bu\text{-}c)_7(c\text{-}BuCH_2OH)]$	5 and 6	1.876–2.011 (9) av. 1.940 2.342 (10) (c-BuCH$_2$OH)	2.078–2.163 (9) (μ_3) av. 2.117 2.026–2.105 (9) (μ) av. 2.061	—	267
$[Mo_4(\mu_3\text{-}OBu^t)(\mu_3\text{-}H)(\mu\text{-}OBu^t)_2(\mu\text{-}H)_2(OBu^t)_4(Me_2NH)]$	5 and 6	1.93–1.95 (1)	2.08, 2.33, 2.35 (μ_3) 2.03–2.05 (μ_2)	—	268
$[Mo_4(\mu_4\text{-}H)(\mu_3\text{-}OCH_2Bu^t)_2(\mu\text{-}OCH_2Bu^t)_4(OCH_2Bu^t)_6]$	6	1.933–2.032 (7) av. 1.983	2.135–2.180 (7) (μ_3) av. 2.157 2.092–2.172 (7) (μ_2) av. 2.133	—	269
$[Mo_6(\mu_3\text{-}Cl)_8(OMe)_6]^{2-}$	5^1	2.046 (4)	—	—	270

(continued overleaf)

Table 4.11 (*Continued*)

Compound	Metal coordination	M–O Bond lengths (Å)		Bond angles (°) $\widehat{MOC}$	Reference
		Terminal	Bridging		
[Mo$_6$(μ_3-Cl)$_8$(OMe)$_6$]$^{2-}$	5^1	2.032, 2.038, 2.080 (3) av. 2.051	–	126	271
[Mo$_6$(μ_3-OMe)$_8$(OMe)$_6$]$^{2-}$	5^1	2.092, 2.149, 2.150 (6) av. 2.130	av. 2.173	119.9	271
[W(OMe)$_4$Cl$_2$-*cis*]	6	1.82–1.87 (4) av. 1.84	–	131–143 av. 138	272
[W(OBu^t)$_4$I$_2$-*trans*]	6	1.840 (4)	–	160.5	273
[W(OCH$_2$CF$_3$)$_2$Cl$_2$(PMe$_2$Ph)$_2$]	6	1.844, 1.852 (5)	–	142.2, 144.8	274
[W(OCH$_2$Bu^t)$_4$(py)(OCPh$_2$)]	6^1	1.897–1.967 (6) av. 1.924	–	128.4–136.8 av. 131.7	275
[W{O$_{12}$Si$_7$(C$_6$H$_{11}$-*c*)$_7$(NMe$_2$)$_3$]	6	1.947 (4)	–	162.4	276
[W(OBu^t)$_3$(NPh)Cl(Bu^{t}NH$_2$)]	6	1.85, 1.86, 1.91 (1)	–	143–153 av. 148	277
[W(OBu^t)$_3$(NBu^t)Cl(Bu^{t}NH$_2$)]	6	1.864–1.897 (7) av. 1.884	–	–	278
[W(OPr^i)$_4$(NPh)(py)]	6	1.912–1.977 (16) av. 1.951	–	123.9–132.5 av. 130.0	279
[W(OC$_2$H$_4$O)$_3$]	6	1.880–1.932 (15) av. 1.907	–	117–121 av. 120	280
[W(OC$_2$Me$_4$O)$_3$]	6	1.914–1.920 (3) av. 1.916	–	119.2–120.6 av. 119.7	281
Li[W(OC$_2$(CF$_3$)$_2$O)$_2$(NPh)Cl]	6	1.929–2.053 (16) av. 1.987	–	124.2, 125.1	282
Cis-[W(OC$_2$Me$_4$O)$_2$Cl$_2$]	6	1.874–1.892 (8) av. 1.882	–	–	283
[W(OC$_2$H$_4$O)(OC$_2$Me$_4$O)$_2$]	6	1.889–1.924 av. 1.905	–	118.3–120.6 av. 119.5	284
[W(OC$_2$H$_4$O)(OC$_2$Me$_4$O)$_2$(PhOH)$_2$]	6	1.884–1.936 (6) av. 1.903	–	118.0–121.5 av. 119.7	284

Compound	C.N.	W–O (Å)		Angle (°)	Ref.
[W(OC$_2$H$_4$O)(OC$_2$MeO$_4$)$_2$](PhNH$_2$)$_2$	6	1.886–1.901 (4) av. 1.894	—	117.4–121.3 av. 119.8	284
[W(OC$_2$Me$_4$O)$_2$(OPh)$_2$]	6	1.889–1.920 (5) (diolate) av. 1.904 1.884, 1.916 (6) (PhO)	—	119.5–120.5 (diolate) av. 120 138.6, 144.5 (PhO)	285
[W(OC$_2$Me$_4$O)$_2$\{OCH(CH$_2$Cl)CH$_2$O\}]	6	1.890–1.930 (7) av. 1.907	—	117.1–121.6 av. 119.7	286
[W(OBu^t)$_3$(NO)(py)]	5	1.876, 1.887, 1.898 (6)	—	134.3–136.0 av. 135.4	288
[W(OBu^t)$_4$(S)]	5	av. 1.886 (3)	—	av. 143.5	289
[W(OCH$_2$Bu^t)$_3$(P$_3$)(Me$_2$NH)]	5	av. 1.908 (5)	—	av. 136.6	290
[W(OBu^t)$_3$(CPh)]	4	av. 1.865 (4)	—	av. 141	291
[W(hfip)$_3$(C$_3$Et$_3$)] hfip = OCH(CF$_3$)$_2$	5	1.932–1.982 (11) av. 1.958	—	129.4–138.6 av. 133.2	292
[W(hfip)$_3$\{CBu^t(CH)CBu^t\}]	5	1.954–1.959 (7) av. 1.957	—	135.3–138.7 av. 137.3	293
[W$_2$(OPr^i)$_6$(py)$_2$]	5^1	1.86–2.04 (2) av. 1.94	—	119–126 (*trans* Pr^{i}O) av. 121 136, 138 (*trans* to py)	294
[W$_2$(OBu^t)$_6$(4-Me-py)$_2$]	5^1	1.930–1.941 (4) av. 1.937	—	132.2–152.5 av. 140.5	322
[W$_2$(tftb)$_6$(py)$_2$]	5^1	1.923–1.958 (5) av. 1.945	—	127.2–148.3 av. 137.2	322
[W$_2$(tftb)$_6$(NCxylyl)$_2$]	5^1	1.933–1.931 (6) av. 1.935	—	131.2–148.1 av. 139.3	322

(continued overleaf)

Table 4.11 *(Continued)*

Compound	Metal coordination	M–O Bond lengths (Å)		Bond angles (°) $\widehat{MOC}$	Reference
		Terminal	Bridging		
[1,2-Bu$_2^i$W$_2$(OPr^i)$_4$]	4^1	1.878, 1.880 (6)	–	137.6, 138.5	295
[1,2-(p-tolyl)$_2$W$_2$(OPr^i)$_4$(Me$_2$NH)]	4^1 and 5^1	1.848–1.942 (17) av. 1.896	–	120.3–144.1 av. 133.4	295
[W$_2$(OPr^i)$_6$]	4^1	1.861, 1.873, 1.880 (16)	–	136.8–170 av. 150	296
[W$_2$\{(OC(CF$_3$)$_2$(NMe$_2$)\}$_2$(NMe$_2$)$_4$]	4^1	2.007 (3)	–	127.1	297
[W$_2$(OC$_6$H$_{11}$-c)$_6$]	4^1	1.856–1.880 (4) av. 1.872	–	141.0–144.9 av. 142.3	298
[W$_2$(OC$_2$Me$_4$O)$_3$]	4^1	1.885–1.931 (7) av. 1.903	–	133.3–139.1 av. 136.5	298
[W$_2$(OPr^i)$_2$(SeC$_6$H$_2$Me$_3$)$_4$]	4^1	1.858 (6)	–	140.5	299
[W$_2$(OBu^t)$_4$(NHPh)$_2$(PhNH$_2$)$_2$]	5^1	1.930–1.974 (7) av. 1.950	–	132.1–144.5 av. 140.0	257
[W$_2$(OCPh$_3$)$_2$(NMe$_2$)$_4$]	4^1	1.963 (9)	–	140.2	300
[W$_2$(OSiPh$_3$)$_2$(NMe$_2$)$_4$]	4^1	1.925 (4)	–	153.8	300
[W$_2$(OC$_{10}$H$_{19}$)$_6$]	4^1	1.840–1.933 (8) av. 1.890	–	114.0–143.4 av. 132.7	301
[W$_2$(OBu^t)$_4$(O$_3$SCF$_3$)$_2$(PMe$_3$)$_2$]	5^1	1.836–1.908 (15) (Bu^{t}O) av. 1.876 2.185, 2.209 (14) (OTf)	–	136.5–147.4 (Bu^{t}O) av. 142.2 153.5, 156.2 (OTf)	302
[W$_2$(hftb)$_4$(NMe$_2$)$_2$] hftb = OCMe(CF$_3$)$_2$	5^1	1.907, 1.959 (3)	–	–	321
[W$_2$(OPr^i)$_2$(CH$_2$Ph)$_2$(tmhd)$_2$] tmhd = Bu^{t}CO(CH)COBu^t	5^1	1.872 (4) (Pr^{i}O) 2.077, 2.091 (4) (tmhd)	–	140.7 (Pr^{i}O) 134.5, 136.3 (tmhd)	303

$[W_2(tftb)_6(CO)_2]$ $tftb = OCMe_2(CF_3)$	5^1	1.896–1.925 (11) av. 1.912	–	–	304
$[W_2(\mu\text{-}OMe)_2(OMe)_8]$	6	1.887, 1.963 (6)	2.028 (6)	135.1 (t)	306
$[W_2(\mu\text{-}OPr^i)_2(OPr^i)_6Cl_2]$	6	1.837, 1.913 (8)	2.013, 2.076 (7)	136.5, 146.5 (t) 138.7 (b)	381
$[W_2(\mu\text{-}OEt)_2(OEt)_4Cl_4]$	6	1.820–1.828 (4)	2.011, 2.013 (4)	144.0, 145.1 (t) 136.5, 137.1 (b)	307
$[W_2(\mu\text{-}OMe)_2(OMe)_2(MeOH)_2Cl_4]$	6	1.950 (5) (MeO) 2.036 (6) (MeOH)	2.032, 2.036 (6)	131.3, 130.1 (t) 130.1, 132.5 (b)	308
$[W_2(\mu\text{-}OEt)_2(OEt)_2(EtOH)Cl_4]$	6	1.955, 1.968 (7) (EtO) 2.016, 2.023 (5) (EtOH)	2.015–2.030 (6) av. 2.022	130.4 (t, av.) 132.8 (b, av.)	308
$[W_2(\mu\text{-}OPr^i)_2(Pr^iO)_2(Pr^iOH)_2Cl_4]$	6	1.936, 1.967 (7) (Pr^{i}O) 1.999, 2.004 (7) (Pr^{i}OH)	2.014–2.038 (7) av. 2.024	132.6 (t, av.) 141.3 (b, av.)	309
$[W_2(\mu\text{-}OEt)_2(OPr^i)_2(Pr^iOH)_2Cl_4]$	6	1.987, 1.995 (8) (Pr^{i}O)	2.023, 2.034 (9) (EtO)	131.4, 132.2 (t) 133.5, 141.2 (b)	309
$[W_2(\mu\text{-}OEt)_2(OCHEt)_2(Et_2CHOH)_2Cl_4]$	6	1.973, 1.983 (6) (Et$_2$CHO) 2.008, 2.010 (4) (Et$_2$CHOH)	2.020–2.033 (EtO) av. 2.026	130.2 (t, av.) 135.9 (b, av.)	309
$[W_2(\mu\text{-}OEt)_2(OC_2Me_4O)_2Cl_4]$	6	1.815, 1.827 (5) (diolate)	2.036, 2.062 (5) (EtO)	155.9, 160.4 (diol) 132.1, 133.1 (EtO)	310
$[W_2(\mu\text{-}OEt)_2\{(OC(MeEt)C(MeEt)O\}_2Cl_4]$	6	1.812, 1.822 (7) (diolate)	2.034, 2.048 (7) (EtO)	154.4, 161.7 (diol) 131.6, 132.2 (EtO)	310
$[W_2(\mu\text{-}OPr^i)_2(OPr^i)_4(OCMe{=}CMeO)_2]$	6	1.841, 1.925 (5) (Pr^{i}O) 1.925, 2.020 (6) (diolate)	1.979, 2.097 (6) (Pr^{i}O)	115.9, 119.1 (diol)	311

(continued overleaf)

Table 4.11 (*Continued*)

Compound	Metal coordination	M–O Bond lengths (Å) Terminal	M–O Bond lengths (Å) Bridging	Bond angles (°) $\widehat{\text{MOC}}$	Reference
[W₂(μ-OPrⁱ)₂(OPrⁱ)₄(OCAr=CArO₂)] Ar = p-tolyl	6	1.856, 1.928 (9) (PrⁱO) 1.926, 2.020 (9) (diolate)	1.985, 2.091 (10) (PrⁱO)	127.7, 144.2 (PrⁱO) 114.7, 122.4 (diol) 135.0 (PrⁱO)	311
[W₂(μ-OBuᵗ)₂(OBuᵗ)₄(CNMe₂)₂]	5	1.88, 1.89 (1)	1.94, 1.95 (1) (eq) 2.42, 2.43 (1) (ax)	—	312
[W₂(μ-OBuᵗ)₂(OBuᵗ)₄(CMe)₂I₂]	5	1.886, 1.897 (4)	1.934 (eq) 2.484 (ax)	137.0, 138.5 (t) 129.9 (b)	323
[W₂(μ-OMe)₂(OMe)₅Cl₃]	6	1.804–1.978 (7) av. 1.852	1.986–2.055 (7) av. 2.026	136.5–148.7 (t) av. 144.6 138.1, 139.6 (b)	313
[W₂(μ-OCH₂Buᵗ)(μ-H)(μ-I)(OCH₂Buᵗ)₅(Me₂NH)]	6	1.868–1.966 (5) av. 1.905	2.052, 2.053 (5)	120.9–134.2 (t) av. 129.2 133.0, 142.5 (b)	314
[W₂(μ-OC₅H₉-c)₂(μ-H)(OC₅H₉-c)₅(Me₂NH)]	6	1.855–2.013 (13) av. 1.921	2.056–2.132 (16) av. 2.089	121.3–137.1 (t) av. 131.0 134.9–143.8 (b) av. 140.6	315
[W₂(μ-OBuᵗ)₂(μ-NC₅H₃Me₂)(OBuᵗ)₄]	5	1.858–1.910 (15) av. 1.886	2.056–2.101 (15) av. 2.079	138.4–156.5 (t) av. 146.3 134.6–141.3 (b) av. 137.3	316
[W₂(μ-OPrⁱ)₂(μ-CNC₆H₃Me₂)(OPrⁱ)₄(py)]	5 and 6	1.875, 1.913 (13) (5-coord) 1.928, 1.941 (13) (6-coord)	2.063, 2.092 (14) (5-coord) 2.074, 2.087 (13) (6-coord)	128.7–143.0 (t) av. 135.0 134.8–146.3 (b) av. 138.8	316
[W₂(μ-OBuⁱ)₂(OBuᵗ)₄(NO)₂(py)₂]	6	1.900–1.915 (6)	2.054, 2.204 (6)	125.9, 128.6 (t) 118.5, 130.0 (b)	317

310

Compound					Ref.
[W$_2$(μ-OMe)$_2$(OMe)$_6$(NPh)$_2$]	6	1.89–1.92 (3) av. 1.91	2.05, 2.16 (2)	131–151 (t) av. 139 121, 128 (b)	277
[W$_2$(μ-OEt)$_2$(OEt)$_4$Cl$_2$(NPh)$_2$]	6	1.835–1.865 (9)	2.003, 2.213 (8)	129.7, 139.0 (t) 118.6, 126.8 (b)	318
[W$_2$(μ-OMe)$_2$(OMe)$_4$(NC$_6$H$_4$Me)$_2$]	6	1.886–1.922 (6) av. 1.902	2.072, 2.181 (6)	126.8–138.0 (t) av. 131.6 120.1, 125.1 (b)	139
[W$_2$(μ-OPr^i)$_2$(OPr^i)$_4$(NC$_6$H$_4$Me)$_2$]	6	1.905–1.927 (4) av. 1.915	2.029, 2.243 (3)	137.4, 138.3 (t) 130.0, 132.1 (b)	319
(NBu$_4^n$)$^+$[W$_2$(μ-OMe)(μ-Cl)$_2$Cl$_4$(NPh)$_2$]$^-$	6	—	2.002, 2.024 (5)	134.3, 135.2	319
[W$_2$(μ-OPr^i)$_2$(OPr^i)$_6$(NPh)$_2$]	6	1.877, 1.932 (4) av. 1.908	2.039, 2.227 (4)	127.2–169.2 (t) av. 141.1 121.1, 124.1 (b)	279
[W$_2$(μ-OCMe$_2$CMe$_2$O)(O$_2$C$_2$Me$_4$)$_4$(NHPh)$_2$]	6	1.85–2.00 (2) (c) av. 1.93 1.80 (2) (b)	—	118–148 (b)	320
[W$_2$(μ-NHMeC$_2$H$_4$NHMe)(OEt)$_6$]	5^1	1.88–1.97 (2) av. 1.92	—	—	324
[W$_2$(μ-OCH$_2$Bu^t)$_2$(μ-NC$_4$Me$_4$N)(OCH$_2$Bu^t)$_4$(py)]	6	1.870–1.967 (8) av. 1.926	2.037–2.226 (8) av. 2.119	127.2–132.6 (t) av. 129.8 124.9–135.6 (b) av. 130.6	325
[W$_2$(μ-OPr^i)$_2$(μ-NMeC$_2$H$_2$CMeN)(OPr^i)$_5$]	6	1.875–1.962 (7) av. 1.915	2.055–2.131 (7) av. 2.086	127.2–142.0 (t) av. 133.3 128.6–146.4 (b) av. 138.8	325
[W$_2$(μ-OBu^t)(μ-NCPhCHCH)(OBu^t)$_5$]	5	1.878–1.913 (4) av. 1.892	2.058–2.107 (4)	133.3–151.0 (t) av. 144.9 137.6, 138.3 (b)	325

(continued overleaf)

Table 4.11 (*Continued*)

Compound	Metal coordination	M–O Bond lengths (Å)		Bond angles (°) $\widehat{MOC}$	Reference
		Terminal	Bridging		
$[W_2(\mu\text{-}OPr^i)_2(\mu\text{-}NCPhCHCH_2)(OPr^i)_5]$	6	1.851–1.953 (5) av. 1.908	2.025–2.184 (5) av. 2.096	133.3–141.2 (t) av. 135.9 133.0–145.9 (b) av. 138.6	325
$[W_2\{\mu\text{-}NNC(C_6H_4Me)_2\}_2(OBu^t)_6]$	5¹	1.858–1.944 (9) av. 1.892	–	138.9–146.0 av. 143.6	305
$[W_2\{\mu\text{-}NPhCO(OBu^t)\}_2(OBu^t)_4]$	5¹	1.895, 1.942 (5)	–	127.1, 154.3	243
$[W_2(\mu\text{-}OCNPh)(OBu^t)_6]$	5¹	1.851–1.932 (9) av. 1.887	–	124.0–155.6 av. 143.8	326
$[W_2\{\mu\text{-}PhCO(OBu^t)\}(OBu^t)_5(PMe)_3]$	5¹	1.85–2.07 (3) av. 1.967	–	–	326
$[W_2\{\mu\text{-}N(C_6H_4Me)CN(C_6H_4Me)\}(OBu^t)_6]$	5¹	1.848–1.950 (10) av. 1.903	–	127.0–155.0 av. 143.5	327
$[W_2\{\mu\text{-}N(C_6H_{11}\text{-}c)CN(C_6H_{11}\text{-}c)\}(OBu^t)_6]$	5¹	1.860–1.959 (6) av. 1.905	–	129.0–154.4 av. 144.9	328
$[W_2(\mu\text{-}NPr^iCNPr^i)(OBu^t)_6]$	5¹	1.839–1.946 (13) av. 1.897	–	129.0–155.0 av. 144.5	328
$[W_2(\mu\text{-}O_2COBu^t)_2(OBu^t)_4]$	5¹	1.885–1.926 (14) av. 1.903	–	125–159 av. 140	329
$[W_2(\mu\text{-}O_2CBu^t)_2(OBu^t)_4]$	5¹	1.883–1.909 (12) (Bu^{t}O) av. 1.890 2.091–2.177 (12) (carboxylate) av. 2.134	–	124.5, 150.3 (Bu^{t}O) 117.1, 119.7 (carboxylate)	330
$[W_2(\mu\text{-}O_2CBu^t)_2(OBu^t)_2(NPr^i_2)_2]$	5¹	1.78, 1.92 (3) (Bu^{t}O) 2.007–2.264 (26) (carboxylate) av. 2.148	–	160, 163 (Bu^{t}O)	330

[W₂(μ-OCH₂Buᵗ)₂(μ-OCH₂CH₂CH₂CH₂)(η²-OCHCH:CH₂)(OCH₂Buᵗ)₄]	6¹	1.88 (2) (BuᵗCH₂O, av.)	2.025–2.167 (6) (BuᵗCH₂O) av. 2.088	—	331
[W₂(μ-OCHMeCH₂CH₂)₂(OCH₂Buᵗ)₆]	6	1.840–1.913 (8) (BuᵗCH₂O) av. 1.872	—	—	331
[W₂(μ-OCH₂Buᵗ)₂(OCH₂Buᵗ)₄(OCHCH:CHMe)₂]	6¹	1.85 (2) (BuᵗCH₂O)	2.07 (3) (BuᵗCH₂O, av.)	—	331
[W₂(μ-OCMe₂C₂H₄CMe₂O)₃]	4¹	1.870–1.879 (8) av. 1.874	—	148.3–149.8 av. 149.2	332
[W₂(μ-OCMe₂C₂H₄CMe₂O)(η²-OCMe₂C₂H₄CMe₂O)₂(Me₂NH)₂]	5¹	1.925–1.976 (9) av. 1.954	—	126.7–148.8 av. 136.2	332
[W₂(μ-OPrⁱ)₂(μ-CO)(OPrⁱ)₄(py)₂]	6	1.91, 1.92 (2)	2.12, 2.15 (2)	132.3, 133.9 (t); 134.7, 148.7 (b)	234
[W₂(μ-OBuᵗ)₂(μ-CO)(OBuᵗ)₄]	5	1.852, 1.890 (10)	2.082, 2.099 (11)	147.2, 149.3 (t); 134.3, 142.0 (b)	333
[W₂(μ-OPrⁱ)₂(OPrⁱ)₄(CO)₄]	6	1.826–1.876 (12) av. 1.857	1.955, 1.967 (12) (W[1]); 2.239, 2.258 (12) (W[2])	107.4, 108.6 (W[1])	334
[W₂(μ-OBuᵗ)(μ-C₂Me₂)(OBuᵗ)₃(CO)]	5	1.876–1.923 (10) av. 1.916	2.064, 2.065 (10)	134.0–158.2 (t) av. 145.8; 139.7, 140.5 (b)	335
[W₂(μ-OCCNMe₂)₂(OBuᵗ)₆]	5¹	1.883–1.931 (6) av. 1.911	—	129.0–150.7 av. 140.6	335
[W₂(μ-OCCNEt₂)₂(OPrⁱ)₆(py)₂]	6¹	1.923–1.952 (8) av. 1.936	—	133.0–137.7 av. 136.0	335

(continued overleaf)

Table 4.11 *(Continued)*

Compound	Metal coordination	M–O Bond lengths (Å)		Bond angles (°) $\widehat{MOC}$	Reference
		Terminal	Bridging		
[W₂(μ-OPrⁱ)₃(OPrⁱ)₃{C₂(NMe₂)}(CO)₂]	6	1.861–1.903 (16) av. 1.878	2.006–2.212 (16) av. 2.138	132.9–146.5 (t) av. 141.2 122.4–137.3 (b) av. 128.4	335
[W₂(μ-CO)₂(μ-NMe₂){OCMe(CF₃)₂}₄(CO)(NMe₂)]	6	2.026–2.056 (10) (W[1]) av. 2.043 1.900–1.931 (10) (W[2]) av. 1.914	–	–	304
[W₂(μ-CPh)₂(OBuᵗ)₄]	4¹	1.831–1.855	–	147.8, 156.2	336
[W₂(μ-C₂Ph₂)₂(OBuᵗ)₄]	4¹	1.844–1.882 (9) av. 1.860	–	144.2–175.0 av. 159.4	336
[W₂(μ-OPrⁱ)₂(μ-C₂H₂)(OPrⁱ)₄(py)₂]	6¹	1.927–1.950 (9) av. 1.941	2.099–2.183 (9) av. 2.142	133.2–134.4 (t) av. 134.6 122.9–126.9 (b) av. 124.9	337, 338
[W₂(μ-OCH₂Buᵗ)(μ-C₂Me₄)(OCH₂Buᵗ)₅(py)₂]	6¹	1.936–1.973 (7) av. 1.953	2.025, 2.123 (7)	134.0–141.6 (t) av. 136.7 126.3, 130.3 (b)	337, 338
[W₂(μ-OPrⁱ)(μ-C₄H₄)(OPrⁱ)₅(C₂H₂)]	6¹	1.901–1.931 (7) av. 1.914	2.004, 2.152 (7)	125.6–137.5 (t) av. 133.5 132.9, 137.9 (b)	339
[W₂(μ-OPrⁱ)(μ-C₄Me₄)(OPrⁱ)₅(C₂Me₂)]	6¹	1.911–1.945 (11) av. 1.930	1.995, 2.112 (11)	131.5–141.0 (t) av. 134.0 133.1, 138.3 (b)	337, 339
[W₂(μ-OBuᵗ)(μ-C₂H₂)(OBuᵗ)₆(py)]	5¹	1.914–1.958 (6) av. 1.930	1.999, 2.083 (6)	144.8–152.9 (t) av. 150.4 138.9, 139.6 (b)	338
[W₂(μ-CSiMe₃)₂(OPrⁱ)₄]	4	1.84–1.87 (1) av. 1.855	–	144.8–149.3 av. 148.0	340

[W₂(μ-CSiMe₃)(μ-CHCHCSiMe₃)(OPrⁱ)₄]	4¹	1.851–1.894 (8) av. 1.873	–	140.8–169.4 av. 152.9	340
[W₂(μ-CSiMe₃)(μ-CCSiMe₃)(OPrⁱ)₄(NC₆H₃Me₂)]	4 and 5¹	1.851–1.921 (5) av. 1.897	–	–	341
[W₂(μ-OPrⁱ)₂(OPrⁱ)₂(CH₂Ph)₂(C₂Me₂)₂]	5¹	av. 1.85 (1)	av. 2.15	–	342
[W₂(μ-C₄Me₄)(μ-CH₂Ph)(OPrⁱ)₄]	4¹	av. 1.89 (2)	–	–	342
[W₂(μ-OBuᵗ)(μ-C₂Me₂)(OBuᵗ)₅(CO)]	5¹	av. 1.90(2)	av. 2.065 (10)	–	343
[W₂(μ-C₃H₄)(OBuᵗ)₆]	4¹	1.866–1.933 (5) av. 1.893	–	140.0–150.8 av. 144.7	344
[W₂(μ-OBuᵗ)₂(μ-C₃H₄)(OBuᵗ)₄(η²-C₃H₄)]	6¹	1.849–1.961 (7) av. 1.898	1.985–2.306 (7) av. 2.119	144.5–158.4 (t) av. 151.9 132.8–141.1 (b) av. 136.9	344
[W₂(μ-OBuᵗ)₂(μ-C₃H₄)(OBuᵗ)₄(CO)₂]	6¹	1.857–1.96 (3) av. 1.889	1.967–2.315 (26) av. 2.143	140.0–168.2 (t) av. 155.3 127.9–141.5 (b) av. 134.6	344
[W₂(μ-OCH₂Buᵗ)(μ-C₂H₂)(OCH₂Buᵗ)₅(py)₂]	6¹	1.911–1.979 (17) av. 1.947	2.012, 2.070 (12)	129.8–139.6 (t) av. 134.7 132.7, 133.9 (b)	345
[W₂(OCH₂Buᵗ)(μ-C₂Et₂)(OCH₂Buᵗ)₅(py)]	6¹	1.848–1.989 (12) av. 1.930	2.055, 2.120 (6)	115.7–144.8 (t) av. 133.9 125.6, 128.4 (b)	345
[W₂(μ-OCH₂Buᵗ)₄(OCH₂Buᵗ)₂(η²-C₂H₄)₂]	6¹	1.88, 1.89 (1)	2.31 (1) (long, av.) 2.00 (1) (short, av.)	–	346
[W₂(μ-OPrⁱ)₃(OPrⁱ)₃{η²-(CH₂)₄}(η²-C₂H₄)]	6¹	1.806–1.938 (8) av. 1.889	2.146 (8) (long, av.) 2.056 (8) (short, av.)	132.4–164.7 (t) 136.1–145.4 (b)	347

(continued overleaf)

Table 4.11 (*Continued*)

Compound	Metal coordination	M–O Bond lengths (Å)		Bond angles (°) $\widehat{MOC}$	Reference
		Terminal	Bridging		
[W₂(μ-OPrⁱ)(OPrⁱ)₅{μ,η²-C(CH₂)₃}(py)]	5 and 6	1.887–1.929 (5) av. 1.901	2.028, 2.156 (5)	133.7–144.7 (t) 131.3, 144.1 (b)	347
[W₂(μ-OPrⁱ)₂(OPrⁱ)₂(C₃H₇)₂(η²-C₂Me₂)₂]	5[1]	1.88 (1)	2.05, 2.15 (1)	138.9 (b)	348
[W₂(μ-OPrⁱ)₂(OPrⁱ)₂(CH₂Ph)₂(η²-C₂Me₂)₂]	5[1]	1.856, 1.857 (4)	2.153 (4) (long, av.) 2.072 (4) (short, av.)	137.8, 145.3 (t)	349
[W₂(μ-OPrⁱ)₂(OPrⁱ)₂(Ph)₂(η²-C₂Me₂)₂]	5[1]	1.850, 1.866 (6)	2.170 (5) (long, av.) 2.088 (5) (short, av.)	139.2, 151.0 (t)	349
[W₂(μ-OPrⁱ)₂(OPrⁱ)₂(CEt)₂(CH₂Buᵗ)₂]	5	1.882 (5)	2.365 (5) (long) 1.961 (5) (short)	134.6 (t)	349
[Me₂(BuᵗO)W(μ-OBuᵗ)(μ-C₂Me₂)W(OBuᵗ)₂(py)]	6[1]	1.873–1.917 (9) av. 1.897	2.046, 2.054 (9)	146.4–158.7 (t) 138.1, 142.7 (b)	349
[W₂(μ-CPh)(μ.η²-C₄Me₄)H(OPrⁱ)₄]	6[1]	1.870–1.925 (18) av. 1.896	–	130.6–155.0	350
[W₂(μ-CSiMe₃)(μ,η²-C₄Me₄)(OPrⁱ)₄(CH₂SiMe₃)]	5 and 6	1.883–1.934 (5) av. 1.904	–	140.7–150.0	350
[W₂(μ-C₂Me₂)₂(OPrⁱ)₄]	6[1]	1.871, 1.920 (14)	–	142.4, 144.9	351
[W₂(μ-OCH₂Buᵗ)₂(μ,η²-C₄H₆)(OCH₂Buᵗ)₄(py)]	6[1]	1.909–2.049 (7) av. 1.966	2.078–2.146 (7) av. 2.100	–	352
[W₂(μ-OCH₂Buᵗ)₄(OCH₂Buᵗ)₂(η²-C₂H₄)₂]	6[1]	1.885, 1.892 (7)	1.993 (7) (short, av.) 2.308 (7) (long, av.)	130.5, 181.7 (t) 127.7–136.5 (b)	353
[(BuᵗO)₃W(μ-C₂)W(OBuᵗ)₃]	4[1]	1.859–1.887 (10) av. 1.875	–	137.4–146.4	354
[W₂{μ-CCCNH(C₆H₃Me₂)}(OSiMe₂Buᵗ)₅{CN(C₆H₃Me₂)}₄]	5, 6	1.93–2.07 (2) av. 1.98	–	129.1–157.5	335
[W₂(μ-CCHCHMe)(μ-OSiMe₂Buᵗ)(OSiMe₂Buᵗ)₅H]	5	1.87–1.94 (1) av. 1.91	2.05, 2.10 (1)	153.7–162.9 (t) 137.0, 143.3 (b)	356
[W₂{μ-CC(CH₂)(CHCH₂)}(μ-OSiMe₂Buᵗ)(OSiMe₂Buᵗ)₅H]	5	1.84–1.94 (1) av. 1.89	2.01, 2.11 (1)	155.5–164.9 (t) 139.8 (b)	356

[W$_2$(μ-OBu^t)(μ,η^2-C$_4$H$_4$S)(OBu^t)$_4$(η'-C$_4$H$_3$S)]	5	1.858, 1.875 (6)	2.047, 2.083 (6)	–	357
[W$_2$(μ-OBu^t)(μ,η^2-C$_4$H$_4$O)(OBu^t)$_4$(η'-C$_4$H$_3$O)]	5	1.875–1.973 (7) av. 1.908	2.044, 2.085 (6)	130.7 (t)	357
[W$_2$(μ-OBu^t)(μ,η^2-C$_4$H$_4$)(OBu^t)$_5$(CO)]	6^1	1.863–1.921 (6) av. 1.904	2.000, 2.224 (6)	–	358
[W$_2$(μ-OPr^i)(μ-P$_2$)(OPr^i)$_5$(py)]	6^1	1.892–1.932 (7) av. 1.909	1.989, 2.074 (7)	–	359
[W$_2$(μ-PPh$_2$)$_2$(OBu^t)$_4$]	4^1	1.95 (1) (long) 1.85 (1) (short)	–	–	360
[W$_2$\{μ-P(C$_6$H$_{11}$-c)$_2$\}$_3$\{PH(C$_6$H$_{11}$-c)$_2$\}(OBu^t)$_3$]	5	1.97, 1.99, 2.00 (1)	–	–	361
[W$_2$(μ-dmpm)(OPr^i)$_4$(CH$_2$Ph)$_2$] dmpm = Me$_2$PCH$_2$PMe$_2$	5^1	1.869–1.977 (15) av. 1.921	–	120.5–142.9	362
[W$_2$(μ-dmpm)(OPr^i)$_4$Bu$_2^i$] dmpm = Me$_2$PCH$_2$PMe$_2$	5^1	1.900–1.975 (7) av. 1.939	–	121.0–141.7	362
[W$_2$(μ-H)(μ-CPh)(OPr^i)$_4$(PMe$_3$)$_3$]	5 and 6	1.903–2.074 (14) av. 1.962	–	126.4–144.7	362
[W$_2$(μ-H)(μ-CPr^i)(OPr^i)$_4$(PMe$_3$)$_3$]	5 and 6	1.872–2.090 (16) av. 1.968	–	122.8, 137.4	362
[W$_2$(μ-H)(μ-CPh)(OPr^i)$_4$(quin)$_2$]	5	1.921–1.969 (7) av. 1.938	–	130.3–139.1	362
[W$_2$(μ-H)$_2$(OPr^i)$_4$(dmpe)$_2$] dmpe = Me$_2$PC$_2$H$_4$PMe$_2$	6	1.935–1.982 (5) av. 1.960	–	123.8–134.9	363
[W$_3$(μ_3-CMe)(μ-OPr^i)$_3$(OPr^i)$_6$]	5	1.900–1.929 (8) av. 1.916	2.023–2.063 (7) av. 2.039	135.8–142.9 (t) 128.6–133.6 (b)	364
[W$_3$(μ_3-P)(μ-OCH$_2$Bu^t)$_3$(OCH$_2$Bu^t)$_6$]	5	av. 1.92 (1)	av. 2.05 (2)	–	365
[W$_3$(μ_3-CMe)(μ-OPr^i)$_3$(OPr^i)$_3$Cl$_2$]	5 and 6	1.857, 1.909 (17) av. 1.887	1.979–2.093 (16) av. 2.047	–	367

(continued overleaf)

Table 4.11 (*Continued*)

Compound	Metal coordination	M–O Bond lengths (Å)		Bond angles (°) $\widehat{MOC}$	Reference
		Terminal	Bridging		
[W$_3$(μ-N)$_3$(tftb)$_9$] tftb = OCMe$_2$(CF$_3$)	5	1.860–1.909 (*trans* to N) av. 1.891 1.904–1.913 (*trans* to tftb) av. 1.907	–	–	368
[W(μ-N)(OBu^t)$_3$]$_{3\infty}$	5	1.872 (7)	–	136.6	323
[W$_3$(μ-OPr^i)$_4$(μ-CMe)(OPr^i)$_5$(CO)$_2$]	5 and 6	1.840–1.884 (18) av. 1.860	1.943–2.231 (17) av. 2.093	129.7–149.8 (t) 125.7–140.8 (b)	369
[W$_3$(μ_3-H)(μ-OPr^i)$_2$(μ-NPh)(OPr^i)$_7$(py)]	5 and 6	1.894–1.983 (10) av. 1.941	1.988–2.104 (9) av. 2.047	128.3–138.1 (t) 134.4–141.1 (b)	279
[W$_4$(μ_3-OEt)$_2$(μ-OEt)$_4$(OEt)$_{10}$]	6	1.96	2.18 (μ_3) 2.02 (μ_2)	–	370
[W$_4$(μ-H)$_2$(μ-OPr^i)$_6$(OPr^i)$_8$]	6	1.892–1.959 (7) av. 1.913	2.005–2.187 (6) av. 2.091	125.9–134.3 (t) 121.1–145.0 (b)	371
[W$_4$(μ-OPr^i)$_4$(OPr^i)$_8$]	4^1	1.885–1.987 (17) av. 1.919	2.184, 2.59 (4) (long) 1.872, 1.976 (14) (short)	142.2, 144.3 (t) 130.0–140.1 (b)	296
[W$_4$(μ-OPr^i)$_2$(μ_3,η^2-CO)$_2$(OPr^i)$_{10}$(py)$_2$]	5 and 6	1.905–1.921 (14) av. 1.915	2.012, 2.075 (14)	–	372
[W$_4$(μ-OPr^i)$_2$(μ_3,η^2-CO)$_2$(OPr^i)$_{10}$]	5	1.876–1.919 (8) av. 1.891	2.009, 2.079 (7)	133.0–141.1 (t) 132.2, 134.9 (b)	333

$[W_4(\mu\text{-OEt})_6(\mu\text{-CSiMe}_3)_2(\text{OEt})_8]$	6	1.893–1.999 (5) av. 1.925	2.061–2.081 (5) av. 2.093	128.5–134.2 (t) 127.7–143.0 (b)	364
$[W_4(\mu_4\text{-C})(\mu\text{-OPr}^i)_4(\mu\text{-NMe})(\text{OPr}^i)_8]$	5 and 6	1.857–2.056 (12) av. 1.946	2.007–2.127 (9) av. 2.053	126.2–149.4 (t) 128.3–138.8 (b)	373
$[W_4(\mu_4,\eta^2\text{-CO})(\mu\text{-OBu}^i)_5(\text{OBu}^i)_7(\text{CO})_2]$	5, 6	1.844–2.073 (25) av. 1.922	1.997–2.171 (25) av. 2.080	129.1–150.6 (t)	374
$[W_4(\mu\text{-CEt})_2(\mu\text{-C}_2\text{Me}_2)_2(\mu\text{-OPr}^i)_2(\text{OPr}^i)_4(\eta^2\text{-C}_2\text{Me}_2)_2]$	5^1 and 6^1	1.885, 1.919 (8)	1.998, 2.291 (7)	132.0, 142.1 (t) 126.0 (b)	351
$[W_4(\mu_4\text{-C})(\mu\text{-NMe})(\mu\text{-OCH}_2\text{Bu}^t)_4(\text{OCH}_2\text{Bu}^t)_7(\text{H})]$	5 and 6	1.889–2.031 (18) av. 1.940	2.020–2.112 (19) av. 2.058	121.8–138.6 (t) 119.1–145.1 (b)	375
$[W_4(\mu\text{-H})_4(\mu\text{-dmpm})_3(\mu\text{-OPr}^i)(\text{OPr}^i)_7]$ dmpm = Me$_2$PCH$_2$PMe$_2$	5 and 6	1.89–1.97 (1) av. 1.933	2.03, 2.28 (1)	121–138 (t) 132, 140 (b)	363
$[W_6(\mu\text{-H})_4(\mu\text{-CPr}^i)(\mu\text{-OPr}^i)_7(\text{OPr}^i)_5(\text{H})]$	5^1	1.95–2.02 (1) av. 1.976	1.99–2.21 (1) av. 2.06	–	363
$[W(\mu\text{-CC}_6\text{H}_2\text{N-Me}_2)(\text{OBu}^t)_3]_\infty$	5	1.886–1.940 (10) av. 1.907	–	–	376
$[W(\mu\text{-CC}_6\text{H}_4\text{N})(\text{OBu}^t)_3]_\infty$	5	1.874–1.889 (6) av. 1.883	–	–	376

b = bridging, c = chelating, t = terminal; ax = axial, eq = equatorial.
^{1}Coordination doubtful.
†Liq. = {NPhC(O)(OPr^i)}$_2$, ¶OTf = O$_3$SCF$_3$.

2.12   Alkoxides of Manganese, Technetium, and Rhenium

Only a few structures have been determined of alkoxo compounds of manganese and rhenium.

2.12.1   Manganese and Technetium

The bulky triphenylmethoxo ligand gives rise to the mononuclear Mn(II) complex [Mn(OCPh$_3$)$_2$(py)$_2$] in which the manganese is in an irregular four-coordinated configuration (O$\widehat{\text{Mn}}$O, 140.4°; N$\widehat{\text{Mn}}$N, 95.6°).[377]

An octahedrally coordinated Mn(IV) complex has been obtained using the tridentate ligand 3-(5-chlorosalicylidene amino)propanolate (5-Cl-salahp) [Mn(5-Cl-salahp)$_2$]. The chelating ligands adopt the meridional configuration.[378] The mononuclear Mn(IV) cation [MnL]$^+$ (L = N,N',N''-tris((2R)-2-olato-3-methylbutyl)-1,4,7-triazacyclononane) was obtained as the PF$_6$$^-$ salt using the hexadentate pendant-arm macrocyclic ligand L. The steric hindrance of this bulky ligand gave rise to a distorted octahedral configuration.[379]

In the tetranuclear complexes [Mn$_4$(μ_3-OMe)$_4$(MeOH)$_4$L$_4$] (L = 2,6-tetramethyl-heptane-3,5-dionate; 1,3-diphenyl-propane-1,3-dionate) the Mn$_4$(μ_3-OMe)$_4$ core has a cubane-like configuration (Fig. 4.6) with each metal having one chelating diketonate and a coordinated methanol to give an octahedral configuration.[380]

Although some oxo alkoxides of technetium have been characterized there are no homoleptic alkoxide structures available.

2.12.2   Rhenium

Several rhenium alkoxo structures have recently been reported. The homoleptic pentamethoxide has the edge-bridged dimeric structure [Re$_2$(μ-OMe)$_2$(OMe)$_8$] (Re≡Re, 2.532 Å) with a rhenium–rhenium double bond (Table 4.12, p. 322).[306] In the [Re$_2$(μ-Cl)$_2$(μ-dppm)$_2$Cl$_3$(OEt)] (dppm = diphenylphosphinomethane) (Re–Re, 2.667 Å) the ethoxo group is in a terminal position.[381]

Some dinuclear organorhenium alkoxo complexes have also been characterized. The nonbridged dimer [Re$_2$(OBu^t)$_4$(CBu^t)$_2$] (Re≡Re, 2.396 Å) has the ethane-like configuration with the alkylidyne ligands *trans* across the metal–metal double bond. A similar structure is present in [Re$_2${OCMe(CF$_3$)$_2$}$_4$(CBu^t)$_2$] (Table 4.12).[382] In the interesting alkylidyne alkylidene alkoxo complex *syn*-[Re{OCMe(CF$_3$)$_2$}$_2$(CBu^t)(CHBu^t)(thf)] the five-coordinated Re(VII) is in a face-capped tetrahedral or distorted TBP configuration, owing to the weakly bound THF ligand being *trans* to the alkylidyne.[382] In *anti*-[Re{OCMe(CF$_3$)$_2$}$_2$(CBu^t)(CHC$_5$H$_4$FeC$_5$H$_5$)] the Re atom is in a distorted tetrahedral coordination.[383]

The trinuclear cluster [Re$_3$(μ-Cl)$_3$(OBu^t)$_6$] (Re≡Re, 2.438 Å) has an interesting structure with the triangular plane of Re$_3$(μ-Cl)$_3$ and the terminal *tert*-butoxo groups placed three above and three below the plane. There are three short Re–O bonds (av. 1.82 Å) on one side of the plane and three long bonds (av. 1.99 Å) on the other side. The inner triangular Re$_3$ core is held together by metal–metal double bonds.[385] In [Re$_3$(μ-Cl)(μ-mentholate)$_2$(mentholate)$_5$Cl] (Re≡Re, 2.389 Å) the isosceles triangle of Re$_3$ atoms is bridged by one chloride and two alkoxo ligands. The terminal chloride is bonded to the Re atom which is bridged by alkoxo ligands.[385] The homoleptic *triangulo*-$[Re_3(\mu\text{-}OPr^i)_3(OPr^i)_6]$[386] is in equilibrium with the hydride $[Re_3(\mu\text{-}OPr^i)_3(OPr^i)_5H]$ and acetone. The hydride has the Re_3 triangle (Re=Re, 2.368 Å) with the three bridging Pr^iO groups in the same plane. The terminal isopropoxide on the Re bonded to the hydride is bent towards the $Re_3(\mu\text{-}O)_3$ plane.[387] Four other Re cluster alkoxo complexes $[Re_3(\mu\text{-}OPr^i)_3(H)(Cl)(OPr^i)_4(PMe_3)]$ (Re–Re, 2.376 Å), $[\{Re_3(\mu\text{-}OPr^i)_3Cl(OPr^i)_4\}(\mu\text{-}Cl)_2]$ (Re–Re, 2.384 Å), $[Re_3(\mu\text{-}OCH_2Bu^t)_3(OCH_2Bu^t)_6(PMe_3)]$ (Re–Re, 2.387 Å), and $[Re_3(\mu\text{-}OCH_2Bu^t)_2(\mu\text{-}Cl)(OCH_2Bu^t)_4Cl(PMe_3)_2]$ (Re–Re, 2.449 Å) have also been reported.[533]

2.13   Alkoxides of Iron and Osmium

Relatively few structures of the iron group alkoxo compounds have been reported. With the bulky triphenylmethoxo ligand the mononuclear Fe(II) complex $[Fe(OCPh_3)_2(thf)_2]$ was obtained. It has an irregular four-coordinate geometry with a very wide $\widehat{OFeO}$ (alkoxo) angle (154.4°) and an acute $\widehat{OFeO}$ (THF) angle (93.6°).[377]

The Os(IV) Schiff base alkoxo complex $[Os(OPr^i)_2(salen)]$ (salen = ethylene-bis-salicylidineimine) has the *trans* octahedral configuration (Table 4.13).[388] In the Os(VI) nitrido chlorodiolato anions in the compounds $(PPh_4)[OsNCl_2(O_2C_2H_4)]$ and $(PPh_4)[OsNCl_2(O_2C_2Me_4)]$ the nitride atom occupies the apical position in a distorted square pyramid with *cis* chlorines and the chelating diolate forming the base.[389] The structures of numerous oxo alkoxo complexes of iron and osmium are detailed in Chapter 5.

2.14   Alkoxides of Cobalt, Rhodium, and Iridium

2.14.1   Cobalt

The structure of the Co(II) derivative of glycerol $[Co\{OCH_2CH(O)CH_2OH\}]_n$ revealed a chain polymeric structure with the dinegative glycero-late anion acting as a chelating and bridging ligand giving the cobalt ions a TBP five-coordination.[390] Another mononuclear octahedral Co(II) complex $[Co\{OC(CF_3)_2CH_2C(Me)=N(C_2H_4O)_2C_2H_4N=C(Me)CH_2C(CF_3)_2O\}]$ was formed by the template condensation of hexafluoro diacetone alcohol with 1,2-bis(2-aminoethoxyethane). The hexadentate ligand gave a distorted octahedral configuration with alkoxo oxygens *cis*, imine nitrogen *trans*, and ether oxygens *cis* to one another.[391]

The tridentate fluorinated sulphur-containing diol $HOC(CF_3)_2CH_2SCH_2C(CF_3)_2OH$ gave the five-coordinated (TBP) Co(II) complex $[Co\{OC(CF_3)_2CH_2SCH_2C(CF_3)_2O\}(py)_2]$ with pyridine as supplementary ligands. The sulphur donor and one pyridine ligand occupy the axial positions.[392]

By using very bulky ligands the anionic complex $[Co(OCBu_3^t)_2\{N(SiMe_3)_2\}]^-$ was obtained with Co(II) in a planar three-coordinated configuration.[393] With the triphenylmethoxo ligand the mononuclear four-coordinated $[Co(OCPh_3)_2(thf)_2]$ and the dinuclear three-coordinated $[Co_2(\mu\text{-}OCPh_3)_2(OCPh_3)_2]$ neutral complexes were characterized. The binuclear complexes $[Co_2\{\mu\text{-}OC(C_6H_{11}\text{-}c)_3\}_2\{OC(C_6H_{11}\text{-}c)_3\}_2]$ and $[Co_2(\mu\text{-}OSiPh_3)_2(OSiPh_3)_2(thf)_2]$ exhibited three-coordinated and four-coordinated Co(II) respectively.[394]

Table 4.12 Alkoxides of manganese, technetium, and rhenium

Compound	Metal coordination	M–O Bond lengths (Å) Terminal	M–O Bond lengths (Å) Bridging	Bond angles (°) $\widehat{\text{MOC}}$	Reference
$[Mn^{II}(OCPh_3)_2(py)_2]$	4	1.956 (4)	–	131.2	377
$[Mn^{IV}(5\text{-Cl-salahp})_2]$ 5-Cl-salahp = $OC_2H_6NCH_2C_6H_3Cl(O)$	6	1.842, 1.854 (2) (alkoxo)	–	–	378
$[Mn^{IV}L]^+$ L = N,N',N''-tris(2-olato-3-methylbutyl)-1,4,7-triazacyclononane	6	1.892, 1.905 (2) (aryloxo); 1.819, 1.825, 1.831 (3)	–	–	379
$[Mn_4(\mu_3\text{-OMe})_4(MeOH)_4L_4]$ L = $Me_3CCO(CH)COCMe_3$	6	2.172 (9) (MeO); 2.273 (1) (MeOH)	–	–	380
$[Mn_4(\mu_3\text{-OMe})_4(MeOH)_4L_4]$ L = $PhCO(CH)COPh$	6	2.12 (4) (diket); 2.174 (1) (MeO); 2.253 (9) (MeOH); 2.13 (3) (diket)	–	–	380
$[Re_2(\mu\text{-OMe})_2(OMe)_8]$	6	1.907, 1.943 (4)	2.036 (4)	124.9, 125.6 (t)	306
$[Re_2(\mu\text{-Cl})_2(dppm)_2Cl_3(OEt)]$ dppm = $Ph_2PCH_2PPh_2$	6	2.085 (14)	–	129	381
$[Re_2(OBu^t)_4(CBu^t)_2]$	4^1	1.901, 1.909 (5)	–	135.9	382
$[Re_2OCMe(CF_3)_2]_4(CBu^t)_2]$	4^1	1.925, 1.932 (5)	–	136.9, 140.5	382
$[Re\{OCMe(CF_3)_2\}_2(CBu^t)(CHBu^t)(thf)]$	5	1.954 (7) alkoxo, 2.398 (8) thf,	–	av. 142.8	383
$[Re\{OCMe(CF_3)_2\}_2(CBu^t)(CHC_5H_4FeC_5H_5)]$	4	1.916, 1.928 (6)	–	140.3, 141.8	384
$[Re(\mu\text{-Cl})_3(OBu^t)_6]$	4^1	1.982, 2.001, 2.009 (16) (long); 1.794, 1.824, 1.844 (14) (short)	–	141.1–145.4	385
$[Re_3(\mu\text{-Cl})(\mu\text{-mentholate})_2(\text{mentholate})_5Cl]$	4^1	1.877–1.889 (15) av. 1.880	2.053–2.103 (14) av. 2.084	132.8 (t); 121.0, 129.1 (b)	385
$[Re_3(\mu\text{-OPr}^i)_3(OPr^i)_5H]$	4^1	av. 1.90 (1)	av. 2.11 (1)	–	387
$[Re_3(\mu\text{-OPr}^i)_3(H)Cl(OPr^i)_4(PMe_3)]$	5^1 and 4^1	1.887–1.913 (6); av. 1.900	2.073–2.125 (5); av. 2.100	–	533
$[[Re(\mu\text{-OPr}^i)_3Cl(OPr^i)_4]_2(\mu\text{-Cl})_2]$	5^1 and 4^1	1.860–1.911 (11); av. 1.885	2.083–2.123 (10); av. 2.105	–	533
$[Re_3(\mu\text{-OCH}_2Bu^t)_3(OCH_2Bu^t)_6(PMe_3)]$	5^1 and 4^1	1.903–2.047 (8); av. 1.941	2.088–2.127 (7); av. 2.113	–	533
$[Re_3(\mu\text{-OCH}_2Bu^t)_2(\mu\text{-Cl})(OCH_2Bu^t)_4Cl(PMe_3)_2]$	5^1 and 4^1	1.900, 1.915 (8)	2.073, 2.118 (7)	–	533

Table 4.13 Alkoxides of iron, ruthenium, osmium, cobalt, rhodium and iridium

Compound	Metal coordination	M–O Bond lengths (Å)		Bond angles (°) $\widehat{MOC}$	Reference
		Terminal	Bridging		
[Fe(OCPh₃)₂(thf)₂]	4	1.883 (1) (alkoxide), 2.134 (2) (thf)	–	130.9 (alkoxide)	377
[Osᴵⱽ(OPrⁱ)₂(salen)}] salen = C₂H₄(NCHC₆H₄O)₂	6	1.920 (3) (PrⁱO), 2.023 (2) salen-O	–	123.1 (PrⁱO)	388
(PPh₄)[OsNCl₂(O₂C₂H₄)]	5	1.930, 1.946 (4)	–	109.7, 113.4	389
(PPh₄)[OsNCl₂(O₂C₂Me₄)]	5	1.934, 1.938 (6)	–	112.0, 116.5	389
[Co{OCH₂CH(O)CH₂CH₂OH}]∞	5	1.97 (2)	2.07 (2)	103.7, 113.7	390
[Co{OC(CF₃)₂CH₂C(Me)=N(C₂H₄O)₂ C₂H₄N=C(Me)CH₂C(CF₃)₂O}]	6	1.933, 1.952 (4) (alkoxide), 2.317, 2.339 (4) (ether)	–	124.8–127.9	391
[Co{OC(CF₃)₂CH₂SCH₂C(CF₃)₂O}(py)₂]	5	1.911 (4)	–	126.9, 126.9	392
[Co(OCBuᵗ₃)₂[N(SiMe₃)₂]]⁻	3	1.849, 1.851 (7)	–	156.7, 157.3	393
[Co(OCPh₃)₂(thf)₂]	4	1.859, 1.880 (9) (alkoxo), 2.047, 2.076 (11) (thf)	–	126.9, 132.0	394
[Co₂(µ-OCPh₃)₂(OCPh₃)₂]	3	1.811, 1.814 (4)	1.963–1.982 (4) av. 1.969	129.0, 137.1	394
[Co₂{µ-OC(C₆H₁₁-c)₃}₂{OC(C₆H₁₁-c)₃}₂]	3	1.763, 1.798 (4)	1.944–1.966 (5) av. 1.953	145.2, 158.1	394
[Co₂(µ-OSiPh₃)₂(OSiPh₃)₂(thf)₂]	4	1.845, 1.858 (3) (siloxo), 2.020, 2.036 (3) (thf)	1.977–1.993 (3) av. 1.988	161.3, 170.7	394
[Co₈(µ-OMe)₈(µ-O₂CMe)₁₆]	6	–	1.878–1.918 (5)	–	396
[Ir(H)(OMe)(PMe₃)₄]⁺	6	2.118 (8)	–	119.4	397
[Ir(OCH₂CF₃)(H)₂{P(C₆H₁₁-c)₃}₂]	5	2.032 (10)	–	138.0	398
[Ir(OCH₂CF₃)(H)₂{P(C₆H₁₁-c)₃}₂(CO)]	6	2.169 (7)	–	118.4	398

323

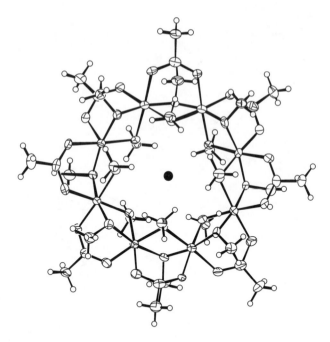

Figure 4.53   Structure of $[\{Co_8(\mu\text{-OMe})_{16}(O_2CMe)_8\}(NH_4^+)]$ ($\bullet$ = N of guest NH_4^+).

The tetranuclear species $[Co_4(\mu_3\text{-OMe})_4(\eta^2\text{-chp})(chp)_3(MeOH)_7]$ (hchp = 6-chloro-2-pyridone) has been obtained. The molecule contains a cubane-like $Co_4(\mu_3\text{-OMe})_4$ unit.[395] The remarkable octanuclear molecule $[Co_8(\mu\text{-OMe})_{16}(\mu,\eta^2\text{-O}_2CMe)_8]$ has a symmetrical toroidal shape containing the eight Co atoms in a plane octagon formed from four $[Co_2(\mu\text{-OMe})_2(\mu\text{-O}_2CMe)]$ units, forming eight edge-shared distorted CoO_6 octahedra (Fig. 4.53). An NH_4^+ ion is sequestered into the centre of the toroidal molecule and is counterbalanced by a PF_6^- anion in the crystal.[396]

2.14.2   Iridium and Rhodium

In the salt $[IrH(OMe)(PMe_3)_4]^+(PF_6)^-$ the Ir(III) is octahedrally coordinated with hydride and methoxide *cis* to each other.[397] Other Ir(III) complexes characterized are the five-coordinated $[Ir(OCH_2CF_3)(H)_2\{P(C_6H_{11}\text{-}c)_3\}_2]$ and the six-coordinated $[Ir(OCH_2CF_3)(H)_2\{P(C_6H_{11}\text{-}c)_3\}_2(CO)]$ (Table 4.13, p. 323).[398] The Rh(I) alkoxide $[Rh(OCH_2CF_3)(PMe_3)_3]$ has the expected square planar configuration.[399]

2.15   Alkoxides of Nickel, Palladium, and Platinum

2.15.1   Nickel

Several alkoxo complexes of nickel have been structurally characterized.

One of the first structures reported was that of the tetranuclear $[Ni_4(\mu\text{-OMe})_4(EtOH)_4(OC_6H_4CHO)_4]$ which has a cubane-like $Ni_4(\mu\text{-OMe})_4$ core. Each

Table 4.14   Alkoxides of nickel, palladium, and platinum

Compound	Metal coordination	M–O Bond lengths (Å)		Bond angles (°) $\widehat{MOC}$	Reference
		Terminal	Bridging		
[Ni₄(μ-OMe)₄(MeOH)₄(OC₆H₄CHO)₄]	6	2.099 (9) (MeOH)	2.019, 2.043, 2.054 (8)	117.2, 120.7, 120.8 (b)	400
[Ni₂{OC(CF₃)₂CH₂CMe=N(CH₂)₆N=CMeCH₂C(CF₃)₂O}₂]	4	1.841, 1.843 (2)	–	–	401
[Ni{O₂C₂(CF₃)₄}₂]²⁻	4	1.852 (3)	–	111.8	402
[Ni₄(μ-OMe)₄(MeOH)₄(OC₆H₂Cl₃)₄]	6	2.06–2.09 (MeOH) av. 2.08	2.02–2.08 av. 2.04	118–123	403
[Ni{OC(CF₃)₂CH₂C(Me)=NC₂H₄C₆H₄N}₂]	6	2.086 (4)	–	126.6	404
[Ni{OC(CF₃)₂CH₂PPh₂}₂]	4	1.837–1.844 (7) av. 1.839	–	121.1–123.9	405
[Ni{O₂C(CF₃)₂}L] L = cyclo-N=C(Me)CH₂C(Me)₂NH(C₃H₆)NH(C₃H₆)	5	2.037, 2.039 (4)	–	91.1, 91.4	406
[Ni{OC(CF₃)₂CH₂C(O)=CHC(CF₃)₂OH}(Me₂NC₂H₄NMe₂)]	4	1.825, 1.897	–	119.4, 123.1	407
[Ni₄(μ-OMe)₄(MeOH)₄{PhCO(CH)COPh}₄]	6	–	2.015–2.072 (9) av. 2.045	–	408
[Pd{OC(CF₃)₂CH₂PPh₂}Cl(PMe₂Ph)]	4	2.046 (2)	–	122.7	409
[Pd{OC(CF₃)₂C₆H₄SMe}Cl(PMePh₂)]	4	2.053 (3)	–	125.0	410
[Pt(OMe)₂(Ph₂PC₂H₄PPh₂)]	4	2.037, 2.041 (7)	–	117.0, 121.0	411
[Pt(OMe)Me(Ph₂PC₂H₄PPh₂)]	4	1.990 (10)	–	119	411
[Pt{OC(CF₃)₂CH₂SMe}(PPh₃)₂](BF₄)	4	2.041 (5)	–	121.2	412
[Pt{OC(CF₃)₂CH₂SMe}(PMePh₂)](BF₄)	4	2.058 (6)	–	117.3	412
[Pt{OC(CF₃)₂CH₂PPh₂}₂]	4	2.030–2.034 (6)	–	–	413
[Pt{O₂C(CF₃)₂}(PPh₃)₂]	4	2.040, 2.056 (5)	–	91.8, 92.2	414
[Pt(OCH₂CH₂PPh₂)₂]	4	2.039 (5)	–	114.8	416
[Pt{OCH(CF₃)₂}Me(PMe₃)₂]	4	2.07 (1)	–	125.3	417
[Pt{OCH(CF₃)₂}{HOCH(CF₃)₂}Me(PMe₃)₂]	4	2.07 (3)	–	123	418
[Pt(D-mannitolate)(DPPP)]	4	2.033, 2.042 (9)	–	–	419

b = bridging.

325

Ni(II) atom is chelated by a salicylaldehydato ligand and with one coordinated methanol achieves an octahedral configuration (Table 4.14, p. 325).[400] The tetradentate iminoalkoxo ligand formed from the template reaction of $(CF_3)_2C(OH)CH_2COMe$ with the diamine $NH_2(CH_2)_6NH_2$ gave a dimeric Ni(II) complex [$Ni_2\{\mu$-$OC(CF_3)_2CH_2CMe{=}N(CH_2)_6N{=}CMeCH_2C(CF_3)_2)_2O\}_2$] in which the two square planar (*trans*-O_2N_2) nickel units are linked by $N(CH_2)_6N$ moieties.[401] A mononuclear square planar complex was also found in the anion of $K_2[Ni\{O_2C_2(CF_3)_4\}_2]$. $4H_2O$ in which the perfluoropinacolato ligands are chelating.[402] Another tetranuclear complex with the cubane-like $Ni_4(\mu$-$OMe)_4$ core was established in [$Ni_4(\mu$-$OMe)_4(MeOH)(OC_6H_2Cl_3$-$2,4,6)$], where the 2,4,6-trichlorophenoxo group acts as an O, Cl chelating ligand to complete the octahedral coordination of the nickel.[403]

Template condensation of $MeC(O)CH_2C(CF_3)_2OH$ with $NC_5H_4CH_2CH_2NH_2$ gave rise to the octahedral complex [$Ni\{OC(CF_3)_2CH_2C(Me){=}NCH_2CH_2C_6H_4N\}_2$]. The tridentate chelating ligands occupy facial configurations.[404]

The phosphino functionalized fluorinated alkoxo ligand $PPh_2CH_2C(CF_3)_2O^-$ gave the mononuclear *trans*-square planar chelated Ni(II) complex [$Ni\{OC(CF_3)_2CH_2PPh_2\}_2$].[405]

A five-coordinated complex was formed using the chelating fluorinated diol $(CF_3)_2CO_2^{2-}$ and the tridentate triaza-macrocyle 2,2,4-trimethyl-1,5,9-triaza-cyclododec-1-ene ligand L in [$Ni\{O_2C(CF_3)_2)\}L$]. A distorted square pyramidal configuration was found and the four-membered NiOCO chelate ring led to acute Ni$\widehat{O}$C bond angles (91°).[406] The potentially tridentate fluorinated ketodiol $(CF_3)_2C(OH)CH_2C(O)CH_2C(CF_3)_2OH$ acts as a bidentate chelating ligand in the square planar complex [$Ni\{OC(CF_3)_2CH_2C(O){=}CHC(CF_3)_2OH\}(tmed)$](tmed = $Me_2NC_2H_4NMe_2$).[407] Another tetranuclear cubane-like structure was found in [$Ni_4(\mu$-$OMe)_4(MeOH)_4(dbm)_4$] (dbm = PhCO(CH)COPh).[408]

2.15.2   Palladium and Platinum

Some square planar Pd(II) complexes involving functionalized fluorinated alkoxo ligands have also been characterized: *e.g.* [$Pd\{OC(CF_3)_2CH_2PPh_2\}Cl(PMe_2Ph)$],[409] [$Pd\{OC(CF_3)_2C_6H_4SMe\}(PMePh_2)Cl$].[410]

Several square planar Pt(II) complexes have been structurally characterized such as the bis-phosphine-stabilized methoxo compounds [$Pt(OMe)_2(Ph_2PC_2H_4PPh_2)$] and [$Pt(OMe)Me(Ph_2PC_2H_4PPh_2)$].[411] The functionalized fluorinated alkoxo ligands are also well represented in [$Pt\{OC(CF_3)_2CH_2SMe\}(PPh_3)_2$]$^+$(BF$_4$)$^-$, [$Pt\{OC(CF_3)_2CH_2SMe\}(PMePh_2)_2$]$^+$(BF$_4$)$^-$,[412] [*cis*-$Pt\{OC(CF_3)_2CH_2PPh_2\}_2$][413] and [$Pt\{O_2C(CF_3)_2\}(PPh_3)_2$].[414] Complexes of nonfluorinated phosphino alkoxo ligands have also revealed square planar Pt(II) as in [$Pt\{OCMe_2CH_2PPh_2\}_2$][415] and [$Pt(OCH_2CH_2PPh_2)_2$].[416] The stability of chelate-alkoxo complexes was deemed to be kinetic in origin. Interestingly the *cis*-complex [$Pt\{OCH(CF_3)_2\}Me(PMe_3)_2$][417] readily forms the hydrogen-bonded adduct [$Pt\{OCH(CF_3)_2\}\{HOCH(CF_3)_2\}Me(PMe_3)_2$][418] but the Pt–O bond length (2.07 Å) is the same in each molecule. In the structure of [Pt(D-mannitolate)(dppp)] (dppp = $Ph_2P(CH_2)_3PPh_2$) the mannitolate ligand is bonded to Pt by the oxygens on C(3) and C(4) giving a 2,5-dioxaplatinacyclopentane chelate ring.[419]

Table 4.15 Alkoxides of copper, silver, and gold

Compound	Metal coordination	M–O Bond lengths (Å)		Bond angles (°) $\overline{\text{MOC}}$	Reference
		Terminal	Bridging		
$[Cu_4(\mu\text{-}OBu^t)_4]$	2	–	1.813–1.882 (10) av. 1.852	119.8–127.3	420
$[Cu_2(\mu\text{-}OBu^t)_2(PPh_3)_2]$	3	–	1.960, 1.996 (4)	126.9, 132.1	421
$[Cu_4(\mu\text{-}OSiPh_3)_4]$	2	–	1.833–1.856 (4) av. 1.844	117.8–134.7	422
$[Cu_2(\mu\text{-}OSiPh_3)_2(PMe_2Ph)_2]$	3	–	1.946, 2.046 (2)	123.2, 133.5	422
$[Cu_4\{\mu\text{-}OSiPh(OBu^t)_2\}_4]$	2	–	1.827–1.844 (5) av. 1.835	131.9, 132.1	423
$[Cu_2\{\mu\text{-}O(CH_2)_4N=C(Me)CH_2C(CF_3)_2O\}_2]$	4	–	1.892–1.924 (2) av. 1.905	121.2, 127.2	424
$[Cu_2\{\mu\text{-}O(CH_2)_3N=C(Me)CH_2C(CF_3)_2O\}_2]$	4	–	1.876–1.926 (2) av. 1.899	122.5, 128.7	424
$[Cu_2\{\mu\text{-}OC(CF_3)_2CH_2C(Me)=N$ $(CH_2)_5N=C(Me)CH_2C(CF_3)_2O\}_2]$	4	1.857–1.876 (7) av. 1.864	–	123.9–126.2	425
$[Cu\{OC(CF_3)_2CH_2CHNHR\}_2]$ R = menthyl	4	1.906, 1.934 (8)	–	118.8, 120.2	426
$[Cu\{OC(CF_3)_2(OH)\}_4]^{2-}$	4	av. 1.933 (6)	–	–	427
$[Cu\{O_2C(CF_3)_2\}L]$ L = $cyclo\text{-}N=C(Me)CH_2C(Me)_2NH(C_3H_6)NH(C_3H_6)$	5	1.977, 1.999 (4)	–	90.9, 91.4	406
$[Cu\{OC(CF_3)_2OC(CF_3)_2O\}(Me_2NC_2H_4NMe_2)]$	4	1.890–1.895 (4) av. 1.893	–	120.8–121.7	428

(continued overleaf)

Table 4.15 *(Continued)*

Compound	Metal coordination	M–O Bond lengths (Å)		Bond angles (°) MOC	Reference
		Terminal	Bridging		
[Cu{OC(CF$_3$)$_2$CH$_2$C(O)=CHC(CF$_3$)$_2$OH}(Me$_2$NC$_2$H$_4$NMe$_2$)]	4	1.874, 1.881 (8)	–	121.0, 124.0	407
[Cu{OCH$_2$C(Me)$_2$NH$_2$}$_2$]	4	1.916 (3)	–	112.5	429
[Cu{OCH(Me)CH$_2$NMe$_2$}$_2$]	4	1.865 (3)	–	–	430
[Cu{OCH$_2$CH$_2$N(Me)CH$_2$CH$_2$NMe$_2$}$_2$]	5	1.887, 1.892 (17)	–	–	430
[{(CF$_3$)$_3$CO}{Cu(μ-OBu^t)$_2$Cu}$_3${OC(CF$_3$)$_3$}]	3 and 4	1.797 (5)	1.86 short 1.95 long	120.7–138.5	431
[Cu{OCH(CF$_3$)$_2$}$_2$(Me$_2$NC$_2$H$_4$NMe$_2$)]	4	1.895, 1.910 (3)	–	116.1, 120.9	432
[Cu{OCMe(CF$_3$)$_2$}$_2$(Me$_2$NC$_2$H$_4$NMe$_2$)]	4	1.884 (3)	–	125.5	432
[Cu$_4$(μ_3,η'-OC$_2$H$_4$OPr^i)$_4$(acac)$_4$]	5	1.901–1.921 (acac) av. 1.910	1.939–1.953 (3) (alkoxo) av. 1.948	–	433
[Cu$_4$(μ-η^4-L)$_2$(prz)$_4$(MeOH)$_2$]$^{2+}$ L = MeOC(C$_5$H$_4$N)$_2$O; prz = pyrazolate	4 and 5	2.396 (4) (MeOH)	1.934, 1.964 (4)	115.8, 117.4	434
[Cu$_3$(μ-OCH(CH$_2$NMe$_2$)$_2$)$_4$(MeOH)Cl$_2$]	5	1.948 (7) (alkoxo) 2.06 (1) (MeOH)	1.948, 1.984 (6)	115.0 (t) 111.6, 112.5 (b)	435
[Cu$_4$(μ-OMe)$_4$(μ.η^2-O$_2$CMe)$_4$]	4	1.933–1.966 (3) (acetate) av. 1.948	1.914–1.920 (3) (MeO) av. 1.917	–	436

b = bridging, t = terminal.

2.16   Alkoxides of Copper, Silver, and Gold

2.16.1   *Copper*

The structures of various Cu(I) and Cu(II) alkoxo complexes have been reported (Table 4.15).

The first homoleptic Cu(I) alkoxide structure determined was the tetranuclear *tert*-butoxide $[Cu_4(\mu\text{-}OBu^t)_4]$ which has a planar Cu_4O_4 ring (Fig. 4.54) with linear two-coordinated copper rather than the cubane-like structure which might have been anticipated.[420] Addition of PPh$_3$ produced the dimer $[Cu_2(\mu\text{-}OBu^t)_2(PPh_3)_2]$ with three-coordinated copper. The $Cu_2P_2O_2$ core is not coplanar owing to the dihedral angle of 143.4° about the O....O axis.[421] Likewise the triphenylsiloxo group gave the tetranuclear $[Cu_4(\mu\text{-}OSiPh_3)_4]$ with linear two-coordinated Cu(I) and the dinuclear $[Cu_2(\mu\text{-}OSiPh_3)_2(PMe_2Ph)_2]$.[422] Even the very bulky $(Bu^tO)_2(Ph)SiO$ ligand produces the tetranuclear $[Cu_4\{\mu\text{-}OSi(OBut)_2Ph\}_4]$ structure with the planar Cu_4O_4 core and linear two-coordinated Cu(I).[423]

Numerous structures of alkoxo copper(II) complexes have been determined. The tridentate diolate ligands $[O(CH_2)_4N\text{=}C(Me)CH_2C(CF_3)_2O]^{2-}$ and $[O(CH_2)_3N\text{=}C(Me)CH_2C(CF_3)_2O]^{2-}$ gave binuclear complexes $[Cu_2\{\mu\text{-}O(CH_2)_nN\text{=}C(Me)CH_2C(CF_3)_2O\}_2]$ ($n = 4, 3$). Each molecule has square planar Cu(II) atoms with an essentially near planar $ONCu(\mu\text{-}O)_2CuNO$ framework.[424]

In the dinuclear fluorinated iminoalkoxo complex $[Cu_2\{\mu\text{-}OC(CF_3)_2CH_2C(Me)\text{=}N(CH_2)_5N\text{=}C(Me)CH_2C(CF_3)_2O\}_2]$, in which the tetradentate ligands bridge the two copper(II) centres, the coordination around the metal is distorted planar towards tetrahedral.[425] Other related Cu(II) complexes are $[Cu\{OC(CF_3)_2CH_2CHNHR\}_2]$ (R = menthyl),[426] $(PPh_4)_2^+[Cu\{OC(CF_3)_2(OH)\}_4]^{2-}$,[427] $[Cu\{O_2C(CF_3)_2\}L]$ (L = triazacyclododecene),[406] $[Cu\{OC(CF_3)_2OC(CF_3)_2O\}(Me_2NC_2H_4NMe_2)]$,[428] and $[Cu\{OC(CF_3)_2CH_2C(O)\text{=}CHC(CF_3)_2OH\}(Me_2NC_2NMe_2)]$.[407] Bis-chelated complexes have been obtained using aminoalkoxo ligands, *e.g.* $[Cu\{OCH_2C(Me)_2NH_2\}_2]$,[429] $[Cu\{OCH(Me)CH_2NMe_2\}_2]$, and $[Cu\{OCH_2CH_2N(Me)CH_2CH_2NMe_2\}_2]$.[430] In the latter compound the Cu(II) is five-coordinated (distorted square pyramidal) because one ligand is tridentate and the other one bidentate. The mixed ligand complex $[\{(CF_3)_3CO\}Cu(\mu\text{-}OBu^t)_2Cu(\mu\text{-}OBu^t)_2Cu(\mu\text{-}OBu^t)_2Cu\{OC(CF_3)_3\}]$ has a novel linear structure with the inner copper atoms

Figure 4.54   Structure of $[Cu_4(\mu\text{-}OBu^t)_4]$ (H atoms omitted).

being four-coordinated (compressed tetrahedral) and the outer ones three-coordinated (trigonal planar). The trinuclear species $[Cu_3(OBu^t)_4\{OC(CF_3)_3\}_2]$ probably has a linear structure too.[431] In the tetranuclear compound the bridging Cu–O bonds to the three-coordinated Cu are shorter (1.86 Å) than those to the four-coordinated copper (av. 1.95 Å). The mononuclear complexes $[Cu(hfip)_2(tmed)]$ and $[Cu(hftb)_2(tmed)]$ (hfip = $OCH(CF_3)_2$, hftb = $OC(Me)(CF_3)_2$, tmed = $Me_2NC_2H_4NMe_2$) have distorted square planar Cu(II) and are appreciably volatile.[432]

The mixed ligand tetranuclear species $[Cu_4(\mu_3,\eta^1\text{-}OC_2H_4OPr^i)_4(acac)_4]$ has a cubane-like Cu_4O_4 core and with the chelating acac ligand each Cu(II) is five-coordinated (distorted square pyramidal).[433]

The tridentate ligand 1,1-bis-(2-pyridyl)-1-methoxy methoxide reacted with Cu(II) in methanol in the presence of pyrazolate ions to form the tetranuclear complex $[Cu_4L_2(prz)_4(MeOH)_2]^{2+}(ClO_4)_2^-$ (L = $MeOC(C_5H_4N)_2O$; prz = $C_3H_3N_2$) in which two coppers are square planar and two are square pyramidal (Fig. 4.55). The di-cation is centrosymmetric, consisting of a dinuclear unit with two coppers bridged by the bridging bis-chelating alkoxo ligand and one bridging pyrazolate ion, linked to another dinuclear unit by two interdimer bridging pyrazolates. The five-coordinated Cu has a coordinated methanol ligand.[434] The tridentate aminoalkoxo ligand $(Me_2NCH_2)_2CHO^-$ which also acts as a bis-chelating bridging ligand forms the trinuclear complex $[Cu_3\{\mu\text{-}OCH(CH_2NMe_2)_2\}_4(MeOH)Cl_2]$ in which the 3 five-coordinated Cu(II) atoms are in a nonlinear array.[435]

A novel tetranuclear structure is exhibited by the mixed ligand complex $[Cu_4(\mu\text{-}OMe)_4(\mu,\eta^2\text{-}O_2CMe)_4]$. The four copper atoms in a plane are alternately double bridged by μ-OMe and μ,η^2-O_2CMe ligands, and the molecule exhibits strong antiferromagnetic behaviour.[436]

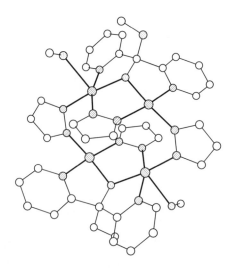

Figure 4.55 Structure of $[Cu_4\{\mu\text{-}OC(C_5H_4N)_2(OMe)\}_2(\mu\text{-}N_2C_3H_3)_4(MeOH)_2]^{2+}$ ($\oslash$ = Cu, $\bigcirc$ = O; $\oslash$ = N; H atoms omitted).

2.16.2   Silver and Gold

Although several alkoxo complexes of Ag(I) and Au(I) have been prepared, usually stabilized by organophosphine supplementary ligands, no X-ray structures have yet been reported.

2.17   Alkoxides of Zinc, Cadmium, and Mercury

2.17.1   Zinc

The first organozinc alkoxo compound structurally characterized was the tetrameric $[Me_4Zn_4(\mu_3\text{-OMe})_4]$ (Fig. 4.56) which has the cubane-like Zn_4O_4 core with tetrahedrally coordinated zinc (Table 4.16, p. 332).[437] The *tert*-butoxide has been shown to adopt a similar structure in $[Me_4Zn_4(\mu_3\text{-OBu}^t)_4]$.[438]

In the heptanuclear molecule $[Zn(\mu_3\text{-OMe})_6(ZnMe)_6]$, which is centrosymmetric, two Zn_4O_4 cubanes are fused through one octahedrally coordinated zinc giving a dicubane structure.[439] The steric effect of bulky ligands led to the formation of the dimer $[Zn_2(\mu\text{-OCEt}_3)_2\{N(SiMe_3)_2\}_2]$ containing three-coordinated (trigonal planar) zinc.[440] Using the tridentate functionalized alkoxo ligand the mononuclear octahedral complex $[Zn(1,4,7\text{-}\eta^3\text{-OCH=CNMeCH}_2CH_2NMe_2)_2]$ was obtained with the donor oxygen atoms *trans* to one another.[440]

The dimeric complex $[EtZn\{\mu,\eta^2\text{-OC(Me)=CHNEtBu}^t\}_2ZnEt]$ has four-coordinated zinc.[441] The asymmetric dimer $[\{\eta^2\text{-}(Bu^tO)_3SiO\}Zn\{\mu\text{-OSi(OBu}^t)_3\}\{\mu,\eta^2\text{-}OSi(OBu^t)_3\}Zn\{OSi(OBu^t)_3\}]$ is of interest in exhibiting four different coordinating modes for the tri-*tert*-butoxy siloxo ligand (Fig. 4.57).[442]

2.17.2   Cadmium

Two alkoxo cadmium structures have been reported. The bifunctional methoxyethoxo ligand gave rise to the nonanuclear complex $[Cd_9(OC_2H_4OMe)_{18}](MeOC_2H_4OH)_2$ containing six-coordinated metal atoms.[443] The ligand is present in five different coordination modes as in $[Cd_9(\mu_3,\eta^2\text{-OR})_6(\mu_3,\eta^1\text{-OR})_2(\mu,\eta^2\text{-OR})_6(\mu,\eta^1\text{-OR})_2(OR)_2]$ $(OR = OC_2H_4OMe)$ with the hydrogen of two lattice alcohol molecules bonded to the terminal alkoxo groups (Fig. 4.58). By using the bulky potentially

Figure 4.56   Structure of $[MeZn(\mu_3\text{-}OMe)]_4$ (H atoms omitted).

Table 4.16 Alkoxides of zinc, cadmium, and mercury

Compound	Metal coordination	M–O Bond lengths (Å)		Bond angles (°) $\widehat{MOC}$	Reference
		Terminal	Bridging		
[Me$_4$Zn$_4$(μ_3-OMe)$_4$]	4	–	2.047–2.114 (16) av. 2.078	118.4–123.5 av. 121.0	437
[Me$_4$Zn$_4$(μ-OBu^t)$_4$]	4	–	av. 2.06 (2)	–	438
[Zn(μ_3-OMe)$_6$(ZnMe)$_4$(μ_3-OMe)$_2$(ZnMe)$_2$]	4 and 6	–	2.105 (Zn, oct, av.) 2.069 (Zn, tet, av.)	–	439
[Zn$_2$(μ-OCEt$_3$)$_2${N(SiMe$_3$)$_2$}$_2$]	3	–	1.930, 1.936 (3)	128.1, 133.3	440
[Zn(1,4,7,-η^3-OCH=CN(Me)C$_2$H$_4$NMe$_2$)$_2$]	6	1.991, 1.997 (3)	–	–	440
[EtZn(μ, η^2-OC(Me)=CHNEtBu^t)$_2$ZnEt]	4	–	2.02, 2.12 (1)	–	441
[(η^2-L)Zn(μ-L)($\mu.\eta^2$-L)Zn(L)] L = (Bu^{t}O)$_3$SiO	4	1.798, 1.870 (4)	1.940–2.141 (4) av. 1.994	158.2 (t)	442
[Cd$_9$(OC$_2$H$_4$OMe)$_{18}$]	6	2.129 (6) (alkoxo) 2.58 (ether, av.)	2.256–2.380 (6) (μ_3) av. 2.314 2.165–2.239 (6) (μ_2) av. 2.195	96.8, 128.1 (b)	443
[Cd$_2$(μ,η^2-OR)$_2$(η^2-OR)$_2$] R = Bu^t(Pr^{i}OCH$_2$)$_2$C	5	2.062 (11) (alkoxo) 2.518 (15) (ether)	2.182 (13)	–	444

oct = octahedral, tet = tetrahedral; b = bridging, t = terminal.

332

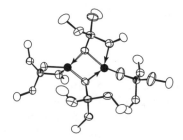

Figure 4.57 Structure of $[Zn_2\{\mu,\eta^2\text{-}OSi(OBu^t)_3\}\{\mu\text{-}OSi(OBu^t)_3\}\{\eta^2\text{-}OSi(OBu^t)_3\}\{OSi(OBu^t)_3\}]$ ($\bullet$ = Zn; CH_3 groups omitted).

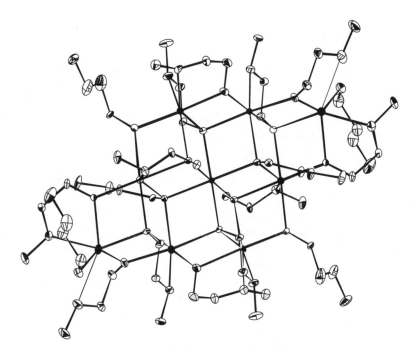

Figure 4.58 Structure   of $[Cd_9(\mu_3,\eta^2\text{-}OR)_6(\mu_3\text{-}OR)_2(\mu,\eta^2\text{-}OR)_6(\mu\text{-}OR)_2(OR)_2(ROH)_2]$ (R = CH_2CH_2OMe; H atoms omitted).

tridentate functionalized ligand $[Bu^t(Pr^iOCH_2)_2CO]$ the binuclear cadmium complex $[Cd_2(\mu,\eta^2\text{-}OR)_2(\eta^2\text{-}OR)_2]$ with five-coordinated (distorted TBP) Cd was obtained.[444]

3 STRUCTURES OF HETEROMETAL ALKOXIDES

The structures of many heterometal alkoxides have been reported in recent years involving heterodimetallic and heterotrimetallic compounds.[445]

3.1   Alkali Metal Dimetallic Alkoxides

3.1.1   Lithium

Complexes of general formula $[Li_4K_4(OSiMe_3)_{8-x}(OBu^t)_x]$ have been prepared and the structure of $[Li_4K_4(OSiMe_3)_{6.8}(OBu^t)_{1.2}]$ has been determined. The array of metals comprises two orthogonal trigonal prisms fused at their base which consists of a near square plane of potassium ions with two lithiums above and two below the plane. The trimethylsiloxo or *tert*-butoxo ligands cap each of the four triangular and four trapezoidal faces.[446] The lithiums are three-coordinated and the potassiums four-coordinated. The lithium-bridged spirocyclic magnesiasiloxane complex $[\{(py)_2Li\}_2\mu-\{OSi(Ph)_2OSi(Ph)_2OSi(Ph)_2O\}\mu-\{OSi(Ph)_2OSi(Ph)_2O\}Mg]$ has four-coordinated lithium and four-coordinated (distorted tetrahedral) magnesium[447] (Table 4.17, pp. 339–345).

Four-coordinated Li and pseudo-four-coordinated yttrium are found in $[Li(thf)_2(\mu-OBu^t)_2Y\{C_5H_4SiMe_3\}_2]$.[448] In the trinuclear complex $[(thf)_2Li(\mu-OBu^t)(\mu,\eta^2-CH_2SiMe_2)Y(CH_2SiMe_3)(\mu-OBu^t)Li(thf)_2]$ the four-coordinated lithiums are bridged by Bu^tO and $Me_2Si(CH_2)_2$ ligands, giving the Y a distorted square pyramidal configuration.[449] In the remarkable salt $[\{(thf)Li\}_4(\mu_4-OBu^t)_4(\mu_4-Cl)Y(CH_2SiMe_3)_x(OBu^t)_{1-x}]^+[Y(CH_2SiMe_3)_4]^-$ salt the pentanuclear cation has a five-coordinated Y and four-coordinated Li.[450] Interestingly, the chloride ion is incorporated into the same structure as the lithiums. The dinuclear complex $[(thf)Li(\mu-OBu^t)_2Y\{CH(SiMe_3)_2\}_2]$ has three-coordinated (trigonal planar) Li and four-coordinated (pseudo-tetrahedral) Y.[451] The $[(thf)_3Li(\mu-Cl)Y(OCBu^t_3)_3]$ complex also has four-coordinated metals.[452]

With titanium(IV) the tetranuclear compound $[Li_2(\mu_3-OPr^i)_2(\mu-OPr^i)_4Ti_2(OPr^i)_4]$ has distorted tetrahedral Li and distorted TBP titanium.[453] In the tetranuclear mixed ligand complex $[\{(thf)Li\}_2(\mu_3-OMe)_2(\mu-OPr^i)_4\{Ti(OPr^i)_2\}_2]$ the lithium is four-coordinated and the Ti(IV) is octahedrally coordinated.[176]

The bulky ligand Bu^t_3CO gave rise to the dinuclear complex $[(Et_2O)_2Li(\mu-Cl)ZrCl(OCBu^t_3)_2]$ with four-coordinated Li and the pseudo-trigonal bipyramidal zirconium(IV).[454] In the heterometallasiloxane $[(Me_5C_5)_2Zr(\mu,\eta^2-OSiPh_2O)(\mu-OH)Li(\mu-OH)Li(\mu,\eta^2-OSiPh_2O)Zr(C_5Me_5)_2]$ another tetranuclear structure involves three-coordinated Li and pseudo-four-coordinated zirconium.[455]

In the hafnium complex $[Li_2(\mu_3-OPr^i)_2(\mu-OPr^i)_4Hf_2(OPr^i)_4]$ the hafniums are in an edge-bridged bi-octahedral structure $Hf_2(OPr^i)_{10}$ which is bridged by two lithiums in pyramidal three-coordinated configurations.[200] The corresponding Nb(v) complex $[(EtOLi)_2(\mu_3-OEt)_2(\mu-OEt)_4Nb_2(OEt)_2(OH)_2]$ has octahedral niobium and four-coordinated lithium.[456] However, in $[Li(\mu-OEt)_4Nb(OEt)_2]_n$ an infinite helical polymer occurs owing to the *cis* configuration of the terminal Nb-OEt groups.[457] In the structure of the tetranuclear $[(Me_3SiCH_2O)_3Nb(\mu_3-OCH_2SiMe_3)(\mu-OCH_2SiMe_3)_2Li_2(\mu_3-OCH_2SiMe_3)(\mu-OCH_2SiMe_3)_2Nb(OCH_2SiMe_3)_3]$, the two niobium octahedra are bridged by the two lithium atoms which are each bonded to four alkoxo groups.

In the lithiatantalasiloxane complex $[(py)_2Li\{\mu,\eta^2-O(SiPh_2O)_2\}_2\{\eta^2-O(SiPh_2O)_2Ta\}]$ the three tetraphenyl siloxanediolate ligands chelate the octahedral Ta(v) and two of them bridge the lithium.[459]

In the binuclear Cr(II) complex the bulky $Bu_3^t CO$ ligand restricts the chromium to three-coordination,[27] whilst in $[Li(\mu-OCBu_3^t)_2Mn\{N(SiMe_3)_2\}]$ the lithium is two-coordinated and the Mn(II) is three-coordinated.[460] In the trinuclear bromo complex $[(thf)_2Li(\mu-Br)_2Mn(\mu-OCBu_3^t)_2Li]$ one Li is two-coordinated and the other is four-coordinated like the Mn(II).[460]

In the Co(II) complex $[(thf)_3Li(\mu-Cl)Co(OCBu_3^t)_2]$ the Li is four-coordinated and the Co three-coordinated (trigonal planar), but in $[Li(\mu-OCBu_3^t)_2Co\{N(SiMe_3)_2\}]$ with the very bulky silylamide ligand the Li is two-coordinated and the Co three-coordinated.[393] Two trinuclear cobaltodisiloxanes have been reported. In $[(tmeda)Li\{\mu-O(SiPh_2O)_2\}Co-\mu-Li(tmeda)]$ the Co is in a distorted (flattened) tetrahedral configuration and the lithiums are pseudo-tetrahedral, whilst in $[(py)_2Li\{\mu-O(SiPh_2O)_2\}_2(py)Co\mu-CoCl(py)]$ one Co is five-coordinated (distorted square pyramidal) with a pyridine ligand occupying the apical position.[461]

A centrosymmetrical molecule is given by $[(Pr^iOH)Li\{\mu,\eta^2-OCH(Me)CH_2NMe_2\}_2NiCl]_2$ in which the lithiums are in distorted tetrahedral configurations and the Ni(II) ions are five-coordinated (distorted square pyramidal) by two chelating dimethylamino propanolate ligands and a terminal chloride.[462]

The Cu(II) ion in $[(py)_2Li\{\mu,\eta^2-O(SiPh_2)_2\}Cu\mu-Li(py)_2]$ is in a distorted square planar configuration and the Li is pseudo-tetrahedral.[463] Square planar Cu(II) is also found in the diolato complex $[(H_2O)_2Li\{\mu,\eta^2-O_2(CH)_2(CH_2)_2O\}_2Cu\mu-Li(H_2O)_2]$ with pseudo-tetrahedral lithium.[464] The preference of Cu(I) for linear two-coordination is evident in $[Li_4Cu_4(\mu_3-OBu^t)_4(\mu-OBu^t)_8]$ which has three-coordinated Li in a large structure (Fig. 4.59).[465]

Lithium also forms heterometal alkoxides with nontransition metals such as Al or Sn. In $[(Et_2O)Li\{\mu-OC(Me)Ph_2\}(\mu-OC_6H_2Bu_2^tMe-2,4,6)AlMe(OC_6H_2Bu_2^tMe-2,4,6)]$ the Al is pseudo-tetrahedral and the Li three-coordinated,[466] whereas in $[(thf)_2Li(\mu-OCEt_3)_2Al(OCEt_3)Cl]$ the lithium is pseudo-tetrahedral.[467]

In the Sn(II) complex $[Li_2(\mu_3-OBu^t)_2(\mu-OBu^t)_4Sn_2]$ the lithiums are four-coordinated but the tin atoms are three-coordinated (pyramidal) with a $Li_2O_6Sn_2$ cage built from two seco-norcubane $Sn_2Li_2O_3$ units sharing a Li_2O_2 four-membered ring.[468] In the Sn(IV)

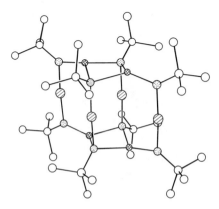

Figure 4.59   Structure of $[Li_4Cu_4(\mu_3-OBu^t)_4(\mu-OBu^t)_4]$ ($\oslash = Cu$, $\circledast = Li$, $\bigcirc = O$; H atoms omitted).

complex [{(thf)$_2$Li}$_2${μ-O(SiPh$_2$O)$_2$}$_2$SnCl$_2$] the tin is in a distorted *trans*-octahedral configuration and the lithium is pseudo-tetrahedral.[469]

3.1.2   Sodium

Numerous sodium metal alkoxides have been characterized including the remarkable complexes [Na$_8$M(μ_9-Cl)(μ_4-OBu^t)(μ_3-OBu^t)$_8$(OBu^t)] (M = Y, Eu) in which a chloride ion is encapsulated in the cage of a square antiprism of sodium ions capped on one square face by the six-coordinated yttrium or europium atom (Fig. 4.60). The lone terminal alkoxo ligand is bonded to the tervalent metal and there are two sets of four μ_3-bridging alkoxides and one μ_4-bridging alkoxide on the face opposite the tervalent metal.[470] In the chloride-free complex [Na$_3$Y{μ-OCH(CF$_3$)$_2$}$_6$(thf)$_3$] the Y is in a distorted octahedral coordination and each Na is bonded to two alkoxo ligands and one terminal THF in a trigonal planar configuration.[471] The yttrium can only accommodate five of the bulkier (F$_3$C)$_2$MeCO ligands and the six-coordination is completed by a terminal THF in [Na$_2$Y{μ-OCMe(CF$_3$)$_2$}$_4${OCMe(CF$_3$)$_2$}(thf)$_3$]. Each Na is bonded to two alkoxo ligands and a terminal THF.[471] However, there are Na....F agostic interactions implying a higher coordination for the Na ion.

In the sodium cerium(IV) *tert*-butoxide [(dme)$_2$Na$_2$(μ_3-OBu^t)$_2$(μ-OBu^t)$_2$Ce(OBu^t)$_2$] the cerium is in a distorted octahedral configuration with two *cis*-terminal, two *cis*-μ-bridging and two *cis*-μ_3-bridging *tert*-butoxo groups. Each sodium is five-coordinated by two μ_3-bridging and one μ-bridging Bu^{t}O groups and a chelating DME ligand.[472] In the Sm(III) compound [Na(μ,η^2-OR)$_2$Ce(η^2-OR')$_2$] (R = C(CMe$_3$)(2-CH$_2$NC$_5$H$_3$Me-6)$_2$; R' = C(CMe$_3$):2-CHNC$_5$H$_3$Me-6)) the Sm is in distorted octahedral coordination and the Na distorted tetrahedral.[138]

A distorted confacial bi-octahedral unit is found in the sodium thorium(IV) complex [NaTh$_2$(μ_3-OBu^t)$_2$(μ-OBu^t)$_3$(OBu^t)$_4$] in which the two μ_3-ligands cap the NaTh$_2$ triangular unit.[145]

In sodium titanium(IV) isopropoxide [Na(μ-OPr^i)$_4$Ti(OPr^i)]$_n$ there is a linear chain comprising alternating distorted trigonal bipyramidal Ti moieties with very distorted bridging tetrahedral Na atoms.[473]

The anion of the salt ([(Ph$_3$P)$_2$N]$^+$)$_3$[Na{Mo$_3$(CO)$_6$(NO)$_3$(μ-OMe)$_3$(μ_3-O)}$_2$]$^{3-}$ has the sodium ion sandwiched in an octahedral coordination by the six bridging MeO groups of two trinuclear units of octahedrally coordinated triangles of Mo atoms each capped by a μ_3-O atom.[474]

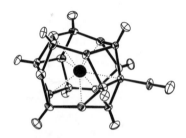

Figure 4.60   YNa$_8$(μ_9-Cl)(μ_4-OC)(μ_3-OC)$_8$(OC) core structure of [YNa$_8$(μ_9-Cl)(μ_4-OBu^t)(μ_3-OBu^t)$_8$(OBu^t)] ($\bullet$ = Cl; CH$_3$ groups omitted).

In the Cr(II) complex $[Na_4Cr_2(\mu_3\text{-}OPr^i)_8(thf)_4]$ the two chromiums are in axial positions and the four sodiums in the equatorial positions of an octahedron with triple-bridging Pr^iO ligands capping the eight faces. Each Cr atom is above the plane of its four bridging Pr^iO ligands whilst each Na achieves five-coordination by virtue of the terminally bound THF ligand.[475] In the sodium chromium(II) bis-disiloxanediolate $[(thf)_4Na_2\{\mu\text{-}O(SiPh_2O)_2\}_2Cr]$ the Na ions are four-coordinated and the chromiums square planar.[476]

In the hydrido complex $[(diglyme)Na(\mu\text{-}H)(\mu\text{-}OPr^i)_4W_2(OPr^i)_4]$ (W–W, 2.431 Å) there is a confacial bi-octahedral $W_2(\mu\text{-}H)(\mu\text{-}OPr^i)_2$ core with the five-coordinated Na bridging two apices of this unit.[313]

A cubane-like $NaTc_3O_4$ core is present in $[Na(\mu_3\text{-}OMe)_4\{Tc(CO)_3\}_3]$ where the octahedral coordination of each Tc atom is completed by three terminal carbonyl ligands.[477]

In the Cu(II) complex $[Na_2\{\mu\text{-}OCH(CF_3)_2\}_4Cu]$ the sodium ions are edge bridged with a near square planar CuO_4 unit but five fluorines are involved in Na....F agostic interactions.[478] The octanuclear compound $[Na_4Cu_4(\mu_3\text{-}OCEt_3)_4(\mu\text{-}OCEt_3)_4]$ has a structure similar to that of $[Li_4Cu_4(OBu^t)_8]$ (Fig. 4.59) with linear two-coordinated Cu(I).[465]

The sodium tin(II) *tert*-butoxide $[Na_2(\mu_3\text{-}OBu^t)_2(\mu\text{-}OBu^t)_4Sn]$, like its lithium analogue, has the $Sn_2O_6M_2$ cage comprising two *seco*-norcubane $Sn_2M_2O_3$ units sharing a Na_2O_2 four-membered ring.[468] By contrast the bridging-chelating ligand $(Bu^tO)_3SiO$ gave rise to the dinuclear species $[Na\{\mu,\eta^2\text{-}OSi(OBu^t)_3\}_3Pb]$ and $[K\{\mu,\eta^2\text{-}OSi(OBu^t)_3\}_3Sn]$ in which the alkali metal is six-coordinated and the heterometal is three-coordinated (pyramidal).[484]

3.1.3 Potassium

Various potassium metal alkoxides have been structurally characterized. One of the first examples reported was the homoleptic complex $[K(\mu_3\text{-}OBu^t)_2(\mu\text{-}OBu^t)_3U_2(OBu^t)_4]$ which has the confacial bioctahedral U_2O_9 framework with the K ion occupying a pocket between the two uranium(IV) atoms (U–U, 3.631 Å).[151]

In $[K(\mu_3\text{-}OPr^i)(\mu\text{-}OPr^i)_3Ti(OPr^i)]_n$ both metals are five-coordinated and alternate in infinite zig-zag chains significantly different in structure from the analogous $[NaTi(OPr^i)_5]_n$.[473]

A novel structure is exhibited by $[(py)_2K_2(\mu\text{-}OCH_2Bu^t)_8M_2]$ (M = Mo, W). The bimetallic core (Mo–Mo, 2.257 Å; W–W, 2.335 Å) is encapsulated in a pseudo-O_8 cube of alkoxo oxygens which are bridged to the two potassiums on opposite faces.[259] By contrast in $[M_2(\mu_3\text{-}OBu^t)_2(\mu\text{-}OBu^t)_4Zn_2]$ (M = Na, K) the three-coordinated alkali metals are located above and below the Zn_2O_2 plane.[479]

In potassium aluminium isopropoxide isopropanolate $[(Pr^iOH)_2K(\mu\text{-}OPr^i)_4Al]_n$ the alternating K and Al atoms are joined by double isopropoxo bridges in a zig-zag linear chain. The two coordinated isopropanols give the potassium a distorted octahedral configuration whilst the Al is distorted tetrahedral.[480]

In the Sn(II) complex $[K(\mu_3\text{-}OBu^t)_2(\mu\text{-}OBu^t)Sn]_n$ both metals are in distorted five-coordinated configurations in a polymeric structure of alternating potassium and tin atoms linked by bridging alkoxo groups in a structure different from the Li and Na analogues.[468] The binuclear potassium tin(II) triphenylsiloxide exists in two forms both containing the $K(\mu\text{-}OSiPh_3)_3Sn$ unit with three-coordinated tin. In one form (monoclinic) the potassium is further coordinated by two chelating DME ($CH_3OC_2H_4OMe$) ligands and in the other (triclinic) the K ion is bonded to one bidentate and one monodentate DME.[481] With Sn(IV) a helical polymer chain is formed of alternating four-coordinated K and five-coordinated (TBP) Sn linked by double alkoxo bridges $[K(\mu\text{-}OBu^t)_4Sn(OBu^t)]_n$.[482]

With the tervalent non-transition metals Sb and Bi the chain polymers $[K(\mu\text{-}OBu^t)_4M']_n$ (M' = Sb, Bi) and formed; the KO_4 units have square planar configurations and the $M'O_4$ units display Ψ-trigonal bipyramidal structures.[483]

In $[(\mu,\eta^2\text{-dioxane})K_2(\mu_3\text{-}OBu^t)_2(\mu\text{-}OBu^t)_3Sb]$ the K_2SbO_5 unit involves a trigonal pyramid of oxygens with the metal ions inserted in the three equatorial edges giving four-coordinated Sb(II) atoms. The dioxane ligands bridge across the K_2SbO_5 units to generate a polymer and give the potassium atoms five-coordination.[483]

3.2   Magnesium, Strontium, or Barium Bimetallic Alkoxides

3.2.1   *Magnesium*

Several magnesium heterometal alkoxides have been reported. The complex $[(\eta^2\text{-dioxane})_2Mg(\mu\text{-}OMe)_4Al_2Me_4]_n$ (Table 4.18, pp. 348–352) is an infinite polymer in which the trinuclear $MgAl_2O_4$ units are crosslinked by bridging dioxane ligands giving the Mg atom an octahedral configuration and the Al atoms a pseudo-tetrahedral coordination.[485]

In the hexanuclear complex $[Mg_2(\mu_3\text{-}OEt)_4(\mu\text{-}OEt)_4(\mu\text{-}Cl)_4Ti_4(OEt)_8]$ all of the metals are in distorted octahedral configurations with the two halves of the molecule held together by an edge-sharing $Mg_2(\mu\text{-}Cl)_2$ bridge.[486] A large family of trinuclear molecules $[M'(\mu\text{-}OBu^t)_3M(\mu\text{-}OBu^t)_3M']$ (M = Mg, Ca, Sr, Ba; M' = Ge, Sn, Pb) has been reported. The central metal, as in $[Ge(OBu^t)_3Mg(OBu^t)_3Ge]$, is octahedral and the outer metals are three-coordinated (pyramidal).[487] An interesting variation on this theme is the linear tetranuclear bimetallic homoleptic complexes $[Bu^tO)M'(\mu\text{-}OBu^t)_2M(\mu\text{-}OBu^t)_2M(\mu\text{-}OBu^t)_2M'(OBu^t)]$, (M = Mg, M' = Sn; M = Cr, M' = Sn; M = Mn, M' = Sn; M = Co, M' = Ge, Sn; M = Ni, M' = Ge, Sn) in which M is in a distorted tetrahedral configuration and M' is three-coordinated (pyramidal).[523] Also of special interest are the pentanuclear trimetallic mixed ligand alkoxo complexes $[M\{(\mu\text{-}OBu^t)_3M'M''(CO)_x\}_2]$ (M = Sr, M' = Sn, M'' = Fe, x = 4; M = Ba, M' = Sn, M'' = Cr, x = 5) in which the central metal M is octahedrally coordinated whilst M' and M'' are joined by a metal–metal bond.[524]

The homoleptic hexanuclear magnesium antimony(III) complex $[Mg_2(\mu_3\text{-}OEt)_2(\mu\text{-}OEt)_8Sb_4(OEt)_6]$ has a core of two edge-sharing MgO_6 octahedra with edge-bridging $(EtO)Sb(\mu\text{-}OEt)_2$ units chelating each Mg and tridentate $EtOSb(\mu_3\text{-}OEt)(\mu\text{-}OEt)_2$ units bridging the two Mg atoms.[488] The "bent" (AlMgAl = 141.6°) trinuclear complex $[(Pr^iOH)_2Mg(\mu\text{-}OPr^i)_4Al_2(OPr^i)_4]$ has *cis*-octahedral Mg and tetrahedral Al.[489] One of the terminal Pr^iO groups on each Al is hydrogen bonded to a Pr^iOH which is coordinated to the magnesium.

3.2.2   *Strontium*

Strontium, cadmium, and hafnium are present in the remarkable trimetallic octanuclear complex $[Sr_2Cd_2Hf_4(OPr^i)_{24}]$ (Fig. 4.61).[200] This centrosymmetric molecule has

Table 4.17 Alkali metal bimetallic alkoxides

Compound	Coordination no. Alkali metal (M)	Coordination no. Other metal (M')	M–O Bond lengths (Å) Terminal	M–O Bond lengths (Å) Bridging	Bond angles (°) MOC	Reference
[{(py)$_2$Li}μ-{O(SiPh$_2$O)$_3$}μ-{O-(SiPh$_2$O)$_2$Mg]	4	4	–	LiO, 1.925–2.000 (9) av. 1.963 MgO, 1.923–1.954 (3) av. 1.936	–	447
[(thf)$_2$Li(μ-OBu^t)$_2$Y{C$_5$H$_4$(SiMe$_3$)}$_2$]	4	4^1	–	LiO, 1.950, 1.968 (28) YO, 2.148, 2.157 (11)	–	448
[{(thf)$_2$Li(μ-OBu^t)}$_2$(μ_2-CH$_2$SiMe$_2$)Y(CH$_2$SiMe$_3$)$_2$]	4	5	–	LiO, 1.913, 1.953 (12) YO, 2.174, 2.192 (4)	–	449
[{(thf)Li}$_4$(μ_4-OBu^t)$_4$(μ_4-Cl)$_4$Y(CH$_2$SiMe$_3$)$_x$(OBu^t)$_{1-x}$]$^+$	4	5	–	LiO, 1.982–2.025 (14) av. 1.996 YO, 2.261–2.276 (5) av. 2.269	–	450
[(thf)Li(μ-OBu^t)$_2$Y{CH(SiMe$_3$)$_2$}$_2$]	3	4	–	LiO, 1.92 (2) YO, 2.117, 2.125 (8)	LiO, 116.9, 126.9 (b) YO, 140.8, 151.6 (b)	451
[(thf)$_3$Li(μ-Cl)Y(OCBu$^t_3$)$_3$]	4	4	–	YO, 2.150, 2.158, 2.171 (3)	YO, 170.2 (t)	452
[Li$_2$(μ_3-OPr^i)$_2$(μ-OPr^i)$_4$Ti(OPr^i)$_4$]	4	5	TiO, 1.782, 1.829 (9)	LiO, 1.928, 1.930 (23) (μ) 1.982, 1.984 (23) (μ_3) TiO, 1.886, 1.900 (9) (μ) 1.988 (8) (μ_3)	LiO, 123.3 (μ_3, b) 132.7 (μ, b) TiO, 126.6 (μ_3, b) 128.6, 129.3 (μ, b) TiO, 140.3, 170.2 (t)	453
[{(thf)Li}$_2$(μ_3-OMe)$_2$(μ-OPr^i)$_4${Ti(OPr^i)$_2$}$_2$]	4	6	TiO, 1.815, 1.833 (7)	LiO, 1.997 (15) (μ_3) 1.853 (16) (μ) TiO, 2.123 (6) (μ_3) 1.957 (7) (μ)	LiO, 123.1 (μ_3, b) 132.0 (μ, b) TiO, 118.1 (μ_3, b) 127.7 (μ, b)	176
[Et$_2$O)$_2$Li(μ-Cl)$_2$ZrCl(OCBu$^t_3$)$_2$]	4	5	ZrO, 1.889, 1.900 (8)	–	TiO, 141.1, 167.0 (t) ZrO, 166.0, 171.9 (t)	454

(continued overleaf)

Table 4.17 *(Continued)*

Compound	Coordination no. Alkali metal (M)	Other metal (M')	M–O Bond lengths (Å) Terminal	Bridging	Bond angles (°) $\widehat{MOC}$	Reference
[(C₅Me₅)₂Zr{OSiPh₂OLiOH}]₂	3	4[1]	ZrO, 1.994 (3)	LiO, 1.793, 1.843 (9); ZrO, 2.049 (4)	–	455
[Li₂(μ₃-OPrⁱ)₂(μ-OPrⁱ)₄Hf₂(OPrⁱ)₄]	3	6	HfO, 1.937 (8)	LiO, 2.06 (3) (μ₃); 1.86 (3) (μ); HfO, 2.229 (8) (μ₃); 2.073 (9) (μ)	–	200
[(EtOLi)₂(μ₃-OEt)₂(μ-OEt)₄Nb₂(OEt)₂(OH)₂]	4	6	LiO, 1.90 (2); NbO, 1.867 (6)	LiO, 2.06 (2) (μ₃); 1.88 (2) (μ); NbO, 2.218 (5) (μ₃); 1.995 (6) (μ)	LiO, 123 (t); NbO, 151 (t); LiO, 102.4 (μ₃, b); 124, 129 (μ, b); NbO, 123.3 (μ₃, b); 125 (μ, b)	456
[Li(μ-OEt)₄Nb(OEt)₂]∞	4	6	NbO, 1.88 (2)	LiO, 1.94 (3); NbO, 1.98 (1)	–	457
[(RO)₃Nb(μ₃-OR)(μ-OR)₂Li₂]₂ R = CH₂SiMe₃	4	6	LiO, 1.974, 1.980 (12); NbO, 1.881–1.902 (5); av. 1.893	LiO, 1.976, 2.163 (13) (μ₃); 1.974, 1.980 (12) (μ); NbO, 2.054 (4) (μ₃); 2.045 (4) (μ)	–	458
[(py)₂Li{μ.η²-O(SiPh₂O)}₂Ta{O(SiPh₂O₂)}]	4	6	TaO, 1.918–1.974 (3); av. 1.947	LiO, 1.998 (10); TaO, 2.032 (4)	TaO, 136.9–141.1 (t); LiO, 122.7, 129.0 (b); TaO, 132.7, 133.9 (b)	459
[(thf)₂Li(μ-OCBuᵗ₃)(μ-Cl)Cr(OCBuᵗ₃)]	4	3	CrO, 1.881 (4)	LiO, 1.920 (11); CrO, 1.991 (3)	CrO, 158.0 (t)	27
[Li(μ-OCBuᵗ₃)₂Mn[N(SiMe₃)₂}]	2	3	–	LiO, 1.808, 1.813 (14); MnO, 1.980, 2.019 (4)	MnO, 150.7	460
[(thf)₃Li(μ-Br)₂Mn(μ-OCBuᵗ₃)₂Li]	2 and 4	4	–	LiO, 1.816, 1.871 (20); MnO, 2.016 (7)	MnO, 151.0	460

340

Compound						Ref.
[(thf)$_3$Li(μ-Cl)Co(OCBu$^t_3$)$_2$]	4	3	1.838, 1.840 (5)	–	138.0, 139.3	393
[Li(μ-OCBu$^t_3$)$_2$Co{N(SiMe$_3$)$_2$}]	2	3	–	LiO, 1.844, 1.845 (19); CoO, 1.925, 1.939 (6)	CoO, 144.4, 147.2 (b)	393
[{(tmeda)Li}μ-[O(SiPh$_2$O)$_2$]$_2$Co]	4	4	–	LiO, 1.940 (9); CoO, 1.974 (4)	–	461
[(py)$_2$Liμ,η^2-{O(SiPh$_2$O)$_2$}$_2$(py)Coμ-CoCl(py)]	4	4 and 5	–	LiO, 1.863 (19); CoO, 1.982–2.198 (7); av. 2.065	–	461
[Pr^{i}OH)Liμ,η^2-{OCH(Me)CH$_2$NMe$_2$}$_2$NiCl]$_2$	4	5	–	LiO, 1.89, 1.99, 2.03 (1); NiO, 1.937, 2.017 (5)	–	462
[(py)$_2$Liμ,η^2-{O(SiPh$_2$O)$_2$}$_2$Cuμ-Li(py)$_2$]	4	4	–	CuO, 1.896–1.944 (8); av. 1.919	SiOCu, av. 123.7	463
[(H$_2$O)$_2$Liμ,η^2-{O$_2$(CH$_2$)(CH$_2$)O}$_2$Cuμ-Li(H$_2$O)$_2$]	4	4	–	LiO, 1.956, 1.962 (4); CuO, 1.918, 1.944 (2)	–	464
[Li$_4$Cu$_4$(μ_3-OBu^t)$_4$(μ-OBu^t)$_4$]	3	2	–	LiO, 1.79–1.94 (1); av. 1.89; CuO, 1.823–1.855 (5); av. 1.840	–	465
[(Et$_2$O)Liμ-[OC(Me)Ph]$_2${μ-OC$_6$H$_2$Bu$^t_2$Me)AlMe(OC$_6$H$_2$Bu$^t_2$Me)]	3	4	AlO, 1.736 (2)	LiO, 1.910, 1.922 (7); AlO, 1.805, 1.808 (2)	LiO, 118.8, 121.0; AlO, 165.6 (t); 134.0, 146.3 (b)	466
[(thf)$_2$Li(μ-OCEt$_3$)$_2$Al(OCEt$_3$)Cl]	4	4	AlO, 1.696 (4)	LiO, 1.965, 1.968 (8); AlO, 1.761, 1.766 (4)	AlO, 153.0 (t); LiO, 128.9, 130.0 (b); AlO, 135.3, 137.3 (b)	467

(continued overleaf)

341

Table 4.17 *(Continued)*

Compound	Coordination no.		M–O Bond lengths (Å)		Bond angles (°) $\widehat{MOC}$	Reference
	Alkali metal (M)	Other metal (M')	Terminal	Bridging		
[Li₂(μ₃-OBuᵗ)₂(μ-OBuᵗ)₄Sn₂]	4	3	—	LiO, 2.098, 2.133 (10) (μ₃); 1.926, 1.933 (8) (μ); SnO, 2.099 (3) (μ₃); 2.084, 2.095 (3) (μ)	LiO, 132.1–133.7; SnO, 117.6–122.1	468
[{(thf)₂Li}₂{μ-O(SiPh₂O)₂}₂SnCl₂]	4	6	—	LiO, 1.980, 1.998 (13); SnO, 2.041, 2.053 (4)	—	469
[Na₈Eu(μ₉-Cl)(μ₄-OBuᵗ)(μ₃-OBuᵗ)₈(OBuᵗ)]	4¹	6	Eu, 2.090 (8)	NaO, 2.379 (μ₄, av.) 2.305 (μ₃, av.); EuO, 2.335 (μ₃, av.)	EuO, 178.8 (t)	470
[Na₈Y(μ₉-OH)(μ₄-OBuᵗ)(μ₃-OBuᵗ)₈(OBuᵗ)]	4¹	6	YO, 2.022 (16)	NaO, 2.364 (μ₄, av.) 2.264 (μ₃, av.); YO, 2.77 (μ₃, av.)	YO, 176.3 (t)	470
[Na₈Y(μ₉-Cl)(μ₄-OBuᵗ)(μ₃-OBuᵗ)₈(OBuᵗ)]	4¹	6	YO, 2.044 (7)	NaO, 2.364 (μ₄, av.) 2.284 (μ₃, av.); YO, 2.283 (μ₃, av.)	Term. YO, 177.4 (t)	470
[Na₃Y{μ-OCH(CF₃)₂}₆(thf)₃]	3	6	NaO(thf), 2.258 av.	NaO, av. 2.300; YO, av. 2.235	NaO, 116.6–123.5 (b); YO, 134.1–142.9 (b)	471
[Na₂Y{μ-OCMe(CF₃)₂}₄{OCMe(CF₃)₂}(thf)₃]	3¹	6	NaO(thf), 2.338 av.; YO (alkoxo), 2.150 (4); YO(thf) 2.425 (4)	NaO, av. 2.598; YO, av. 2.234	YO, 169.6 (t); (thf), av. 127.8; NaO, 108.5–115.1 (b); YO, 152.1–163.2 (b)	471
[(dme)₂Na₂(μ₃-OBuᵗ)₂(μ-OBuᵗ)₂Ce(OBuᵗ)₂]	5	6	NaO(dme), 2.582 av.; CeO, 2.141 av.	NaO, 2.384 (μ₃, av.) 2.354 (μ, av.); CeO, 2.373 (μ₃, av.) 2.230 (μ, av.)	CeO, 172.6 (t); NaO, 123.9 (μ₃, b) 119.3 (μ, b); CeO, 134.1 (μ₃, b) 149.3 (μ, b)	472

Compound						Ref.
[Na(μ,η²-OR)₂Sm(η²-OR')] R = C(CMe₃)(2-CH₂NC₅H₃Me-6)₂ R' = C(CMe₃); (2-2CHNC₅H₃Me-6)	4	6	SmO, 2.239 (4) (c)	NaO, 2.322 (5) SmO, 2.222 (5)	NaO, 119.3 (b) SmO, 143.3–146.4 (b)	138
[NaTh₂(μ₃-OBuᵗ)₂(μ-OBuᵗ)₃(OBuᵗ)₂]	4	6	ThO, 2.152, 2.157 (8)	NaO, 2.400 (μ₃, av.) 2.607 (10) (μ) ThO, 2.491 (μ₃, av.) 2.372 (μ, av.)	ThO, 167.0, 168.1 (t) NaO, 114.5, 120.3 (b) ThO, 125.8–158.2 (b)	145
[Na(μ-OPrⁱ)₄Ti(OPrⁱ)]∞	4	5	1.811 (10)	NaO, 2.288, 2.320 (9) TiO, 1.865–1.936 (8) av. 1.899	–	473
[Na{Mo₃(CO)₆(NO)₃(μ-OMe)₃(μ₃-O)}₂]	6	6	–	NaO, 2.41, 2.41, 2.44 (1) MoO, 2.05 (μ₃, av.) 2.19 (μ, av.)	MoO, 116.6–119.9 (b)	474
[Na₄Cr₂(μ₃-OPrⁱ)₈(thf)₄]	5	4	–	NaO, av. 2.429 CrO, av. 1.989	–	475
[(thf)₄Na₂{μ-O(SiPh₂O)₂}₂Cr]	4	4	NaO(thf), av. 2.271	NaO, av. 2.226 CrO, av. 1.986	CrOSi, av. 131.3	476
[(diglyme)Na(μ-H)(μ-OPrⁱ)₄W₂(OPrⁱ)₄]	5	6	NaO(diglyme) av. 2.457 WO, av. 1.965	NaO, av. 2.324 WO, av. 2.058	WO, 123.2–137.4 (t) NaO, 127.5, 128.9 (b) WO, 127.1–146.8 (b)	313
[Na(μ₃-OMe)₄{Tc(CO)₃}₃]	3	3	–	NaO, av. 2.388 TcO, av. 2.167	–	477
[Na₂{μ-OCH(CF₃)₂}₄Cu]	7¹	4	–	NaO, 2.281, 2.318 (6) CuO, 1.916, 1.968 (5)	–	478
[Na₄Cu₄(μ₃-OCEt₃)₄(μ-OCEt₃)₄]	3	2	–	NaO, av. 2.271 CuO, av. 1.831	–	465

(continued overleaf)

Table 4.17 (*Continued*)

Compound	Coordination no. Alkali metal (M)	Other metal (M')	M–O Bond lengths (Å) Terminal	Bridging	Bond angles (°) M–O–C	Reference
[Na₂(μ₃-OBuᵗ)₂(μ-OBuᵗ)₄Zn₂]	3	4	–	NaO, 2.427 (5) (μ_3) 2.226 (3) (μ) ZnO, 2.050 (2) (μ_3) 1.892 (4) (μ)	ZnO, 126.9 (μ_3, b) 130.8, 136.1 (μ, b)	479
[Na₂(μ₃-OBuᵗ)₂(μ-OBuᵗ)₄Sn]	4	3	–	NaO, 2.271–2.414 SnO, 2.095–2.111	NaO, 126.7–130.3 SnO, 118.8–123.1	468
[K(μ₃-OBuᵗ)₂(μ-OBuᵗ)₃U₂(OBuᵗ)₄]	4	6	UO, 2.10–2.14 (2) av. 2.122	KO, 2.93, 2.94 (2) (μ_3) 2.74, 2.76 (2) (μ_2) UO, 2.43 (μ_3, av.) 2.26 (μ, av.)	UO, 168–175 (t) UO, 126–128 (μ_3, b) 127–153 (μ, b)	151
[K(μ₃-OPrⁱ)(μ-OPrⁱ)₃Ti(OPrⁱ)]	5	5	1.814 (1)	KO, 2.784 (1) (μ_3) 2.820 (μ, av.) TiO, 1.979 (1) (μ_3) 1.876 (μ, av.)	–	473
[(py)₂K₂(μ-OCH₂Buᵗ)₈Mo₂]	5	5¹	–	KO, av. 2.745 MoO, av. 1.997	–	259
[(py)₂K₂(μ-OCH₂Buᵗ)₈W₂]	5	5¹	–	KO, av. 2.770 WO, av. 1.998	–	259
[K₂(μ₃-OBuᵗ)₂(μ-OBuᵗ)₄Zn₂]	3	4	–	KO, 2.771 (5) (μ_3) 2.556 (4) (μ) ZnO, 2.038 (3) (μ_3) 1.899 (5) (μ)	ZnO, 127.0 (μ_3, b) 134.4, 1.38.4 (μ, b)	479
[(PrⁱOH)₂K(μ-OPrⁱ)₄Al]∞	6	4	(PrⁱOH)K, 2.668, 2.725 (8)	KO, 2.695–3.136 (6) av. 2.898 AlO, 1.713–1.776 (6) av. 1.743	–	480

[K(μ_3-OBu^t)$_2$(μ-OBu^t)Sn]$_\infty$	5	5	—	KO, 2.805–3.188 (6) (μ_3) 2.564 (5) (μ) SnO, 2.067, 2.074 (5) (μ_3) 2.061 (5) (μ)	KO, 114–134 (μ_3, b) 129.9 (μ, b) SnO, 121.7, 123.1 (μ_3, b) 127.6 (μ, b)	468
[(η^2-dme)$_2$K(μ-OSiPh$_3$)$_3$Sn] monoclinic	7	3	KO(dme) 2.744–2.824 (3)	KO, 2.783–2.904 (3) SnO, 2.057–2.092 (2)	SnOSi; 124.9–138.3	481
[(η'-dme)(η^2-dme)K(μ-OSiPh$_3$)$_3$Sn]	6	3	KO(dme) 2.734–2.831 (11)	KO, 2.718–2.891 (11) SnO, 2.049–2.091 (9)	SnOSi; 127.0–133.2	481
[K(μ-OBu^t)$_4$Sn(OBu^t)]	4	5	SnO, 2.005 (8)	KO, 2.643–2.803 (8) SnO, 1.979–2.057 (7)	SnO, 141.7 (t) KO, 113.9–33.8 (b) SnO, 126.5–136.6 (b)	482
[K(μ-OBu^t)$_4$Sb]$_\infty$	4	4	—	KO, 2.650–2.967 (2) SbO, 1.971–2.181 (2)	KO, 117.1, 117.8 (b) SbO, 123.5, 126.5 (b)	483
[(μ,η^2-dioxane)K$_2$(μ_3-OBu^t)$_2$(μ-OBu^t)$_3$Sb]$_\infty$	5	4	—	KO, 2.777, 2.778 (3) (μ_3) 2.592, 2.717 (3) (μ) SbO, 1.979, 1.985 (3) (μ_3) 2.144 (3) (μ)	KO, 121.2, 122.0 (μ_3, b) 126.9, 134.0 (μ, b) SbO, 127.1, 128.1 (μ_3, b) 131.8 (μ, b)	483
[Na{μ,η^2-OSi(OBu^t)$_3$}$_3$Pb]	6	3	NaO 2.455–2.858 (23) (c)	NaO, 2.420–2.619 (15) PbO, 2.119–2.152 (15)	NaOSi; 96.6–109.8 (b) PbOSi; 147–155 (b)	484
[K{μ,η^2-OSi(OBu^t)$_3$}$_3$Sn]	6	3	KO 2.704–2.730 (8) (c)	KO, 2.797–2.865 (6) SnO, 2.038–2.058 (6)	KOSi; 98.3–100.1 (b) SnOSi; 158.4–159.7 (b)	484

b = bridging, c = chelating, t = terminal.
[1]Coordination doubtful.

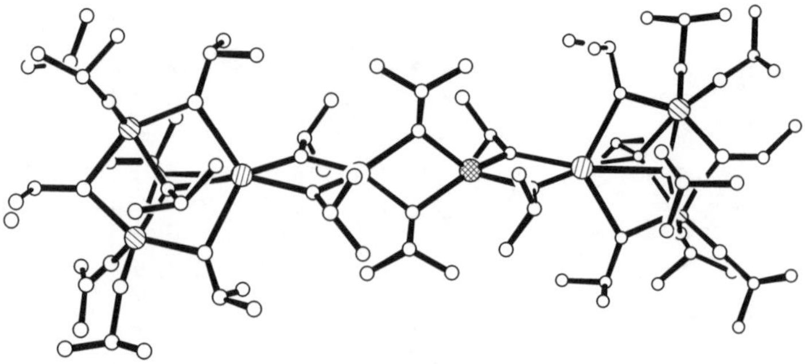

Figure 4.61   Structure of $[Sr_2Cd_2Hf_4(\mu_3\text{-}OPr^i)_4(\mu\text{-}OPr^i)_{12}(OPr^i)_8]$ (⊛ = Cd; ⊘ = Sr, ◍ = Hf; H atoms omitted).

a linear central core of $[Sr(\mu\text{-}OPr^i)_2Cd(\mu\text{-}OPr^i)_2Cd(\mu\text{-}OPr^i)_2Sr]$ capped at each end by octahedrally coordinated confacially bridged $Hf_2(OPr^i)_9$ moieties acting as η^4-chelating units. The $M_2(OR)_9$ moiety is also featured in the structure of $[M(\mu_3\text{-}OEt)_4(\mu\text{-}OEt)_4\{Ti_2(\mu\text{-}OEt)(OEt)_4\}_2]$ (M = Ca, Sr, Ba) in which the central bivalent metal ion is eight-coordinated (distorted square antiprism).[490,491] With the more bulky isopropoxo ligands the coordination of strontium is limited to six in the trinuclear complex $[(Pr^iOH)_3(Pr^iO)Sr_2(\mu_3\text{-}OPr^i)_2(\mu\text{-}OPr^i)_3Ti(OPr^i)_2]$ which has an isosceles triangle of metal atoms capped above and below the triangle by $\mu_3\text{-}OPr^i$ ligands and bound by three bridging isopropoxides in the plane. This molecule has the familiar M_3O_{11} framework found in other metal alkoxide structures (Figs 4.5 and 4.25).[490]

3.2.3   Barium

In the trinuclear complex $[Ba\{\mu\text{-}OCH(CF_3)_2\}_4Y_2(thd)_4]\{thd = Bu^tC(O)CHC(O)Bu^t\}$ the three metal atoms are in a non-linear chain (YB̂aY, 129.3°) and the central Ba ion is joined to the Y ions by double fluoroalkoxo bridges and interacts with eight fluorine atoms from neighbouring CF_3 groups. The yttrium ions are six-coordinated (trigonal prismatic).[492]

   With the isopropoxo ligand a novel tetranuclear species $[(Pr^iO)_3Ti(\mu\text{-}OPr^i)_2Ba(\mu_3\text{-}OPr^i)_2(\mu\text{-}OPr^i)_2\{Ti_2(\mu_3\text{-}OPr^i)_2(\mu\text{-}OPr^i)(OPr^i)_4\}]$ is formed, with the central six-coordinated (trigonal prismatic) Ba ion being bridged by bidentate $Ti(OPr^i)_5$ and quadridentate $Ti_2(OPr^i)_9$ units.[493]

   With zirconium the hexanuclear molecule $[(Pr^iO)_4(\mu_3\text{-}OPr^i)_2(\mu\text{-}OPr^i)Zr_2(\mu\text{-}OPr^i)_2Ba(\mu\text{-}OPr^i)_2Ba(\mu\text{-}OPr^i)_2Zr_2(\mu_3\text{-}OPr^i)_2(\mu\text{-}OPr^i)(OPr^i)_4]$ was obtained, with six-coordinated metals. The confacial bi-octahedral $Zr_2(OPr^i)_9$ units act as tetradentate bridging groups to each barium.[494] The pentanuclear species $[BaZr_4(OPr^i)_{18}]$ has a similar structure to that of $[BaTi_4(OEt)_{18}]$.

   The novel trimetallic octanuclear homoleptic complexes $[Ba_2Cd_2M_4(\mu_3\text{-}OPr^i)_4(\mu\text{-}OPr^i)_{12}(OPr^i)_8]$ (M = Ti, Zr, Hf) were the first of this class of compound to be structurally characterized.[495] These structures, which are analogous to that of $[Sr_2Cd_2Hf_4(OPr^i)_{24}]$ (Fig. 4.61) already mentioned,[200] contain a linear array of Ba(μ-OPr^i)$_2$Cd(μ-OPr^i)$_2$Cd(μ-OPr^i)$_2$Ba, containing four-coordinated (distorted tetrahedral) Cd with each Ba capped with a quadridentate chelating M$_2$(OPr^i)$_9$ moiety. In the trinuclear complex [(Pr^{i}O)$_4$Nb(μ-OPr^i)$_2$Ba(HOPr^i)$_2$(μ-OPr^i)$_2$Nb(OPr^i)$_4$] the three octahedrally coordinated metals are nonlinear (NbBaNb, 137°) owing to the *cis* configuration of the terminally coordinated PriOH ligands on the central barium ion.[496,497] Hydrogen bonding involving the isopropanol OH proton and a terminal Nb–OPr^i oxygen causes a significant lengthening of the Nb–O bond.

A number of barium copper alkoxo compounds have been characterized. The first homoleptic complex reported was the volatile species [Ba{μ-OCMe(CF$_3$)$_2$}$_4${CuOCMe(CF$_3$)$_2$}$_2$] in which the central Ba ion is coordinated by four oxygens and eight fluorines. The Cu(II) is three-coordinated (trigonal planar).[498] In the hexanuclear barium copper(I) complex [Ba$_2$(μ-OR)$_2${μ,η^2-OR(Cu(μ-OR)Cu)OR}$_2$] (R = CEt$_3$) the two Ba ions are bridged by two alkoxo groups and chelated by two (RO)Cu$_2$(OR)$_2$ moieties containing linear two-coordinated copper(I) atoms.[465] The Ba$_2$O$_2$ ring is near normal to the BaOCuOCuOBaOCuOCuO plane.

Some mixed ligand complexes involving functionalized alkoxo ligands have been reported. In [Ba$_2$Cu$_2$(μ_3,η^2-OC$_2$H$_4$OMe)$_4$(μ,η^2-acac)$_2$(acac)$_2$(MeOC$_2$H$_4$OH)$_2$] the metals are arranged in a rhomboid structure with four μ_3-alkoxo ligands in a plane with the two CuO$_2$(acac) units. The Cu(II) atoms are five-coordinated (square pyramidal) whilst the Ba ions are nine-coordinated (distorted capped rectangular antiprism). In the pentanuclear [BaCu$_4$(μ_3-OC$_2$H$_4$OMe)$_2$(μ-OC$_2$H$_4$OMe)$_2$(OC$_2$H$_4$OMe)$_2$(Bu^{t}COCHCOBu^t)$_4$] the Ba ion is ten-coordinated and the binuclear Cu$_2$O$_9$ units are confacial bi-octahedra which act as η^4-chelating groups to the barium.[500] In the binuclear species [η^2-HOC$_2$H$_4$OH)$_3$Ba(μ,η^2-O$_2$C$_2$H$_4$)$_2$Cu] the barium ion is eight-coordinated (distorted cubic) and the copper four-coordinated (square planar).[501]

The trinuclear mixed ligand complex [Ba$_2$Cu(μ_3,η^2-OCNMeCH$_2$NMe$_2$)$_2$(μ,η^2-Bu^{t}COCHCOBu^t)$_2$(Bu^{t}COCHCOBu^t)$_2$(Pr^{i}OH)$_2$] is unsymmetrical in structure. One eight-coordinated barium is bonded to two terminal Pr^{i}OH ligands, the oxygens of two μ_3,η^2-OCHMeCH$_2$NMe$_2$ ligands and two diketonates whilst the other is seven-coordinated to one chelating diketonate, two bridging diketonates, one chelating aminoalkoxide, and the μ_3-O of another chelating aminoalkoxide. The Cu(II) atom is five-coordinated (square pyramidal) to two μ_3-O atoms of aminoalkoxides, one nitrogen, and a chelating diketonate.[502]

In the barium dialuminium tetrakis-1,2-ethanediolate [Ba(μ,η^2-O$_2$C$_2$H$_4$)$_2$Al$_2$(O$_2$C$_2$H$_4$)$_2$] a two-dimensional layer polymer is formed by bi-trigonal bipyramidal Al$_2$(μ,η^2-O$_2$C$_2$H$_4$)$_2$ dimers linked *via* four diolate ligands to four neighbouring dimers with the barium ions surrounded by four dimer units in an eight-coordinated structure.[503]

3.3   Yttrium and Lanthanide Bimetallic Alkoxides

3.3.1   *Yttrium*

Some mixed ligand yttrium heterometal alkoxides have been characterized. In the trinuclear complex [YCu$_2$(μ_3,η^2-OC$_2$H$_4$OMe)(μ,η^2-OC$_2$H$_4$OMe)$_2$(hfd)$_2$(thd)$_2$] (hfd = CF$_3$COCHCOCF$_3$; thd = Bu^{t}COCHCOBu^t) the Y is bonded to eight oxygens (av. 2.39 Å) and weakly to a ninth (2.79 Å) whilst one Cu is four-coordinated (square planar) and the

Table 4.18 Bimetallic alkoxides of magnesium, calcium, strontium, and barium

Compound	Coordination no. Mg, Ca, Sr, Ba	Coordination no. Other metal	M–O Bond lengths (Å) Terminal	M–O Bond lengths (Å) Bridging	Bond angles (°) $\widehat{MOC}$	Reference
[L₂Mg(μ-OMe)₄Al₂Me₄] L = η^2-dioxane	6	4	–	MgO, 2.05, 2.06 (1); AlO, 1.80, 1.86 (1)	MgO, 139.9, 141.5 (b); AlO, 121.2, 121.3 (b)	485
[(μ-Cl)Mg(μ_3-OEt)₂(μ-OEt)(μ-Cl)Ti₂(μ-OEt)(OEt)₄]₂	6	6	TiO, 1.753–1.770 (5) av. 1.760	MgO, 2.088, 2.109 (4) (μ_3); 2.041 (μ); TiO, 2.100–2.177, (μ_3) av. 2.134; 1.904–2.044 (4) (μ) av. 1.974	–	486
[Ge(μ-OBuᵗ)₃Mg(μ-OBuᵗ)₃Ge]	6	3	–	MgO, 2.119 (4); GeO, 1.899 (3)	–	487
[Ge(μ-OBuᵗ)₃Ca(μ-OBuᵗ)₃Ge]	6	3	–	CaO, 2.36 (1); GeO, 1.903 (10)	–	487
[Sn(μ-OBuᵗ)₃Ca(μ-OBuᵗ)₃Sn]	6	3	–	CaO, 2.37 (1); SnO, 2.087 (9)	–	487
[Sn(μ-OBuᵗ)₃Sr(μ-OBuᵗ)₃Sn]	6	3	–	SrO, 2.523 (3); SnO, 2.078 (3)	SrOC, 142.8; SnOC, 124.9	487
[Pb(μ-OBuᵗ)₃Sr(μ-OBuᵗ)₃Pb]	6	3	–	SrO, 2.47 (2); PbO, 2.21 (2)	–	487
[Pb(μ-OBuᵗ)₃Ba(μ-OBuᵗ)₃Pb]	6	3	–	BaO, 2.73 (3); PbO, 2.22 (3)	–	487
[(BuᵗO)Sn(μ-OBuᵗ)₂Mg(μ-OBuᵗ)]₂	4	3	SnO, 2.009 (4)	MgO, 1.958–1.969 (3) av. 1.964; SnO, 2.125, 2.129 (3)	–	523
[Sr{(μ-OBuᵗ)₃SnFe(CO)₄}₂]	6	4 and 5	–	SrO, 2.550 (6); SnO, 2.027 (6)	–	524
[Ba{(μ-OBuᵗ)₃SnCr(CO)₅}₂]	6	4 and 6	–	BaO, 2.705 (6); SnO, 2.034 (5)	–	524

Compound						Ref.
[Mg$_2$(μ_3-OEt)$_2$(μ-OEt)$_8$Sb$_4$(OEt)$_6$]	6	4	Sb-O, 2.003, 2.040 (5)	MgO, 2.136, 2.139 (5) (μ_3) 2.007–2.120 (5) (μ) av. 2.079 SbO, 2.323 (5) (μ_3) 1.990–2.208 (5) (μ) av. 2.057	–	488
[(Pr^{i}OH)$_2$Mg(μ-OPr^i)$_4$Al$_2$(OPr^i)$_4$]	6	4	MgO(Pr^{i}OH), 2.127, 2.173 (7) AlO, H-bonded, 1.726, 1.728 (6) AlO, 1.679, 1.689 (6)	MgO, 2.063–2.108 av. 2.081 AlO, 1.769–1.779 (5) av. 1.772	–	489
[Sr$_2$Cd$_2$Hf$_4$(μ_3-OPr$^i_4$(μ-OPr^i)$_{12}$(OPr^i)$_8$]	6	4 and 6	HfO, 1.896–1.934 (10) av. 1.916	SrO, 2.642, 2.650 (8) (μ_3) 2.406–2.61 (10) (μ) av. 2.504 CdO, 2.152–2.172 (9) (μ) av. 2.163 HfO, 2.208–2.254 (8) (μ_3) 2.234 2.030–2.187 (10) (μ) av. 2.104	–	200
[Sr(μ_3-OEt)$_4$(μ-OEt)$_4${Ti$_2$(μ-OEt)(OEt)$_4$}$_2$]	8	6	TiO, 1.764–1.814 (5) av. 1.787	SrO, 2.578–2.604 (4) (μ_3) av. 2.590 2.585–2.620 (5) (μ) av. 2.605 TiO, 2.085–2.149 (5) (μ_3) av. 2.111 1.878–2.077 (6) (μ) av. 1.981	–	491

(continued overleaf)

349

Table 4.18 *(Continued)*

Compound	Coordination no.		M–O Bond lengths (Å)		Bond angles (°)	Reference
	Mg, Ca, Sr, Ba	Other metal	Terminal	Bridging	$\widehat{MOC}$	
[(Pr^{i}OH)$_3$(Pr^{i}O)Sr$_2$(μ_3-OPr^i)$_2$(μ-OPr^i)$_3$Ti(OPr^i)$_2$]	6	6	Pr^{i}O/Pr^{i}OH, Sr, 2.520, 2.535 (4) TiO, 1.816 (4)	SrO, 2.520, 2.528 (4) (μ_3) 2.399–2.524 (6) (μ) av. 2.460 TiO, 2.116 (4) (μ_3) 1.991, 2.017 (5) (μ)	—	491
[Ca(μ_3-OEt)$_4$(μ-OEt)$_4${Ti$_2$(μ-OEt)(OEt)$_4$}$_2$]	8	6	TiO, 1.770–1.803 (12)	CaO, 2.458–2.518 (9) (μ_3) 1.902–1.935 (8) (μ) TiO, 2.079–2.148 (10) (μ_3) 2.023, 2.060 (13) (μ)	—	490
[Ba(μ_3-OEt)$_4$(μ-OEt)$_4${Ti$_2$(μ-OEt)(OEt)$_4$}$_2$]	8	6	TiO, 1.70–1.90 (3)	BaO, 2.74–2.84 (2) (μ_3) 1.86–1.94 (2) (μ) TiO, 2.07–2.16 (2) (μ_3) 2.03–2.10 (3) (μ)	—	490
[Ba{μ-OCH(CF$_3$)$_2$}$_4$Y$_2${thd}$_4$] thd = Bu^{t}C(O)CHC(O)Bu^t	12^1	6	—	BaO, 2.63–2.68 (1) YO, 2.21–2.25 (1)	—	492
[(Pr^{i}O)$_3$Ti(μ-OPr^i)$_2$Ba(μ_3-OPr^i)$_2$(μ-OPr^i)$_2${Ti$_2$(μ_3-OPr^i)$_2$(μ-OPr^i)(OPr^i)$_4$}]	6	5 and 6	TiO, 1.786–1.824 (5) av. 1.799	BaO, 2.724, 2.765 (4) (μ_3) 2.614–2.737 (4) (μ) av. 2.684 TiO, 2.148–2.169 (4) (μ_3) av. 2.157 1.931–2.057 (4) (μ) av. 1.989	—	493

[(Pr^iO)_4(μ_3-OPr^i)_2(μ-OPr^i)_2Zr_2(μ-OPr^i)_2Ba(μ-OPr^i)_2Ba(μ-OPr^i)_2Zr_2(μ_3-OPr^i)_2(μ-OPr^i)_4](OPr^i)_4]	6	6	ZrO, 1.924–1.940 (7) av. 1.930	BaO, 2.873, 2.895 (6) (μ_3) 2.551–2.787 (7) (μ) av. 2.660 ZrO, 2.219–2.252 (6) (μ_3) av. 2.239 2.054–2.202 (6) (μ) av. 2.127	ZrO, 171.6 (t, av.) BaO, 127.5 (b, av.) ZrO, 129.8 (b, av.)	494
[Ba_2Cd_2Ti_4(μ_3-OPr^i)_4(μ-OPr^i)_12(OPr^i)_8]	6	4 and 6	TiO, 1.752–1.979 (7) av. 1.849	BaO, 2.422, 2.802 (6) (μ_3) 2.337–2.702 (6) (μ) av. 2.553 CdO, 1.959–2.280 (6) av. 2.101 TiO, 2.124 (μ_3, av.) 1.969 (μ, av.)	–	495
[Ba_2Cd_2Zr_4(μ_3-OPr^i)_4(μ-OPr^i)_12(OPr^i)_8]	6	4 and 6	ZrO, av. 2.044	BaO, 2.719 (μ_3, av.) 2.521, 2.699 (11) (μ_2) CdO, 1.979, 2.179 (11) ZrO, 2.320 (μ_3, av.) 2.146 av. (μ, av.)	–	495
[Ba_2Cd_2Hf_2(μ_3-OPr^i)_4(μ-OPr^i)_12(OPr^i)_8]	6	4 and 6	HfO, av. 1.906	BaO, 2.831 (μ_3, av.) 2.552, 2.754 (8) (μ) CdO, 2.160, 2.177 (9) HfO, 2.235 (μ_3, av.) 2.018 (μ, av.)	–	495
[(Pr^iOH)_2Ba(μ-OPr^i)_4{Nb(OPr^i)_4}_2]	6	6	Pr^iOH·Ba, 2.724, 2.734 (7) NbO, 1.872–2.046 (8) av. 1.925	BaO, 2.642–2.707 (7) NbO, 2.007–2.046 (7)	NbO, 124.4–135.2 (t) NbO, 125.9–160.4 (b)	496

(continued overleaf)

Table 4.18 (*Continued*)

Compound	Coordination no. Mg, Ca, Sr, Ba	Other metal	M–O Bond lengths (Å) Terminal	Bridging	Bond angles (°) $\widehat{MOC}$	Reference
[(Pr^{i}OH)$_2$Ba(μ-OPr^i)$_4$[Nb(OPr^i)$_4$]$_2$]	6	6	Pr^{i}OH·Ba, 2.725, 2.740 (5); NbO, H-bonded, 2.025, 2.051; NbO, 1.889–1.913	BaO, 2.651–2.729; NbO, 2.005–2.042	NbO, H-bonded, 123.7, 125.3 (t); NbO 143.2–161.9 (t)	497
[Ba{μ-OCMe(CF$_3$)$_2$}$_4${CuOCMe(CF$_3$)$_2$}$_2$]	4^l	3	CuO, 1.781 (7)	BaO, 2.636–2.644; CuO, 1.878, 1.889 (7)	–	498
[Ba$_2$Cu$_4$(μ-OCEt$_3$)$_8$]	4	2	–	BaO, av. 2.59; CuO, av. 1.83	–	465
[Ba$_2$Cu$_2$(μ_3,η^2-OC$_2$H$_4$OMe)$_4$(acac)$_4$(MeOC$_2$H$_4$OH)$_2$]	9	5	Ba-acac, 2.668, 2.727 (3); Cu-acac, 1.938, 1.948 (4)	BaO, 2.749–2.833 (3) (μ_3); CuO, 1.929, 1.933 (3) (μ_3)	–	499
[BaCu$_4$(μ_3-OC$_2$H$_4$OMe)$_2$(μ-OC$_2$H$_4$OMe)$_2$(OC$_2$H$_4$OMe)$_2$(Bu^{t}COCHCOBu^t)$_4$]	10	6	BaO, 2.85, 2.89 (2); CuO, 1.88–2.55 (2)	BaO, 2.68–3.08 (2) (μ_3); 2.73–3.02 (2) (μ); CuO, 1.88–2.45 (2) (μ_3); 1.90–2.81 (2) (μ)	–	500
[(η^2-HOC$_2$H$_4$OH)$_3$Ba(μ,η^2-O$_2$C$_2$H$_4$)$_2$Cu]	8	4	Ba-glycol, 2.743–2.848 (3); Cu-glycolate 1.909, 1.920 (3)	BaO, 2.732, 2.754 (3); CuO, 1.922, 1.931 (3)	–	501
[Ba$_2$Cu(μ_3,η^2-OCHMeCH$_2$NMe$_2$)$_2$(μ,η^2-diket)$_2$(η^2-diket)(Pr^{i}OH)$_2$] diket = Bu^{t}COCHCOBu^t	8 and 7	5	Pr^{i}OH·Ba 2.78, 2.81 (2); Ba-diket 2.59–2.74 (1); Cu-diket 1.92 (1)	BaO, 2.70–2.75 (1) (μ_3); 2.67–2.87 (1) (μ); CuO, 1.92, 1.93 (1) (μ_3); 2.28 (1) (μ)	–	502
[BaAl$_2$(μ,η^2-O$_2$C$_2$H$_4$)$_2$(O$_2$C$_2$H$_4$)$_2$]$_\infty$	8	5	AlO, 1.72–1.88 (1)	BaO, 2.68–2.94 (2); AlO, 1.863, 1.924 (7)	–	503

b = bridging, t = terminal.
^{l} Coordination doubtful.

other five-coordinated (distorted square pyramid) (Table 4.19, pp. 354–355). The yttrium is chelated by two hfd ligands, and the copper(II) atoms are each chelated by one thd ligand.[504] In the tetranuclear complex $[YCu_3(\mu_3,\eta^2\text{-}OC_2H_4OMe)_2(\mu,\eta^2\text{-}OC_2H_4OMe)(\mu\text{-}OC_2H_4OMe)_2(thd)_4]$ each metal is chelated by one thd ligand and the Y atom is eight-coordinated and the Cu atoms five-coordinated (square pyramid).[504] Another tetranuclear species $[Y_2Cu_2(\mu_3,\eta^2\text{-}OC_2H_4OMe)_2(\mu,\eta^2\text{-}OC_2H_4OMe)_2(\mu\text{-}OC_2H_4OMe)_2(hfd)_4]$ has 2 eight-coordinated Y atoms which are each chelated by two hfd ligands, a μ_3,η^2-OC_2H_4OMe ligand, a μ,η^2-OC_2H_4OMe ligand and a μ-OC_2H_4OMe ligand. The two Cu atoms are five-coordinated (square pyramidal) in a centrosymmetrical molecule.[504]

Yttrium titanium chloride isopropoxide $[YTi_2(\mu_3\text{-}OPr^i)_2(\mu\text{-}OPr^i)_3(OPr^i)_4Cl_2]$ has an isosceles triangle of octahedral metals with μ_3-OPr^i ligands capping above and below the YTi_2 plane which is girdled by three μ-OPr^i ligands in the plane. The Y atom has two terminal chlorides and the Ti atoms each have two terminal isopropoxo ligands. The $YTi_2O_9Cl_2$ framework is related to the $M_3(\mu_3\text{-}OR)_2(\mu\text{-}OR)_3(OR)_4L_2$ structures (Figs 4.5 and 4.25).[493]

Some novel yttrium organoaluminium complexes have been reported. In the trinuclear complex $[(thf)(Bu^tO)Y\{(\mu\text{-}OBu^t)(\mu\text{-}CH_3)AlMe_2\}_2]$ the Y atom is cis-octahedrally coordinated and the aluminiums are four-coordinated (distorted tetrahedral).[505] In the tetranuclear $[Y\{(\mu\text{-}OBu^t)(\mu\text{-}CH_3)AlMe_2\}_3]$ the yttrium is tris-chelated by the bridging $(\mu\text{-}OBu^t)(\mu\text{-}CH_3)AlMe_2$ ligands in the facial configuration. A related complex is the trinuclear organo-yttrium compound $[(C_5H_4SiMe_3)Y\{(\mu\text{-}OBu^t)(\mu\text{-}CH_3)AlMe_2\}_2]$.[506]

3.3.2 Lanthanides

The tetranuclear $[La\{(OC_2H_4)_3N\}_2\{Nb(OPr^i)_4\}_3]$ contains eight-coordinated (bicapped trigonal antiprism) La and distorted octahedral niobium (Fig. 4.62).[506] A homoleptic trinuclear lanthanum niobium complex $[(Pr^iO)_3Nb(\mu\text{-}OPr^i)_3La(OPr^i)(\mu\text{-}OPr^i)_2Nb(OPr^i)_4]$ has also been established. The distorted octahedral La is triple-bridged to one Nb and double-bridged to the other, generating a bent (136.2°) $Nb\widehat{La}Nb$ axis.[497]

In the centrosymmetrical hexanuclear complex $[\{(Pr^iO)_2Al(\mu\text{-}OPr^i)_2\}_2Pr(Pr^iOH)(\mu\text{-}Cl)_2Pr(Pr^iOH)\{(\mu\text{-}OPr^i)_2Al(OPr^i)_2\}_2]$ the praseodymium is seven-coordinated (distorted capped trigonal prism) and the aluminiums are in typical distorted tetrahedral coordination.[507] The trinuclear $[Ge(\mu\text{-}OBu^t)_3Eu(\mu\text{-}OBu^t)_3Ge]$ contains six-coordinated Eu(II) and three-coordinated Ge(II).[487]

3.4 Titanium Group Heterometallic Alkoxides

In addition to the previously mentioned heterometallic species formed by alkali metal, alkaline earth, or yttrium and lanthanide ions with titanium, zirconium, or hafnium, a number of other heterometallic complexes of the latter elements with transition and nontransition metals have been reported.

3.4.1 Titanium

A novel example formally involving a Ti≡Rh triple bond (Ti–Rh, 2.214 Å) is $[Ti(\mu,\eta^2\text{-}OCMe_2CH_2PPh_2)_3Rh]$ containing a pseudo-tetrahedral Ti(III) atom (Table 4.20,

Table 4.19 Heterometallic alkoxides of yttrium and lanthanum

Compound	Coordination no.		M–O Bond lengths (Å)		Bond angles (°)	Reference
	Y, ln	Heterometal	Terminal	Bridging	$\widehat{MOC}$	
$[YCu_2(OC_2H_4OMe)_3(hfd)_2(thd)_2]$ $hfd = CF_3COCHCOCF_3$ $thd = Bu^tCOCHCOBu^t$	9^1	5^1	Y(hfd), 2.337–2.389 (5) Cu(thd), 1.888–1.911 (5) Alkoxo ether O YO, 2.687, 2.789 (6) CuO, 2.439 (5)	YO, 2.505 (6) (μ_3) 2.218, 2.258 (5) (μ) CuO, 1.925–2.006 (4) (μ_3) 1.897, 1.925 (5) (μ)	–	504
$[YCu_3(OC_2H_4OMe)_4(thd)_4]$	8	5	Y(thd), 2.305, 2.349 (7) Cu(thd), 1.897–1.947 (9) Alkoxo ether O YO, 2.484, 2.530 (11) CuO, 2.588 (11)	YO, 2.425, 2.464 (7) (μ_3) 2.265–2.315 (8) (μ) CuO, 1.977–2.407 (8) (μ_3) 1.916–2.664 (μ)	–	504
$[Y_2Cu_2(OC_2H_4OMe)_6(hfd)_4]$	8	5	Y(hfd), 2.329–2.415 (10) Alkoxo ether O YO, 2.526 (11) CuO, 2.119 (11)	YO, 2.327 (9) (μ_3) 2.235, 2.243 (8) (μ) CuO, 1.909, 2.008 (8) (μ_3) 1.909, 1.923 (9) (μ)	–	504
$[YTi_2(\mu_3\text{-}OPr^i)_2(\mu\text{-}OPr^i)_3(OPr^i)_4Cl_2]$	6	6	TiO, 1.743–1.784 (9) av. 1.768	YO, 2.308, 2.373 (8) (μ_3) 2.261, 2.346 (9) (μ) TiO, 2.129–2.207 (8) (μ_3) av. 2.181 1.994, 2.057 (8) (μ) av. 2.019	–	493

354

[(thf)(Bu^{t}O)Y{(μ-OBu^t)(μ-CH$_3$)AlMe$_2$}$_2$]	6	4	YO, 2.005 (4)	YO, 2.252, 2.256 (6) / AlO, 1.841 (6)	505
[Y{(μ-OBu^t)(μ-CH$_3$)AlMe$_2$}$_3$]	6	4	Y(thf), 2.388 (7)	YO, 2.209 (14) / AlO, 1.864 (20)	505
[(C$_5$H$_4$SiMe$_3$)Y{(μ-OBu^t)(μ-CH$_3$)AlMe$_2$}$_2$]	5[1]	4	–	YO, 2.280 (4) / AlO, 1.84 (4)	449
[La{OC$_2$H$_4$)$_3$N}$_2$[Nb(OPr^i)$_4$}$_3$]	8	6	NbO, 1.915 (8)	LaO, av. 2.499 (6) / NbO, av. 2.078 (6)	506
[(Pr^{i}O)$_3$Nb(μ-OPr^i)$_3$La(OPr^i)(μ-OPr^i)$_2$Nb(OPr^i)$_4$]	6	6	LaO, 2.164 (7) / Nb[1]O, 1.859–1.880 / Nb[2]O, 1.898–1.921	LaO, 2.455, 2.465 (7) / NbO, 2.009–2.087	497
[{(Pr^{i}O)$_2$Al(μ-OPr^i)$_2$}$_2$Pr(Pr^{i}OH)μ(Cl)]$_2$	7	4	PrO(Pr^{i}OH), 2.525 (8) / AlO, 1.724 (t+b, av.)	PrO, av. 2.397	507
[Ge(μ-OBu^t)$_3$Eu(μ-OBu^t)$_3$Ge]	6	3	–	EuO, 2.496 (6) / GeO, 1.894 (7)	487

b = bridging, t = terminal.
[1]Coordination doubtful.

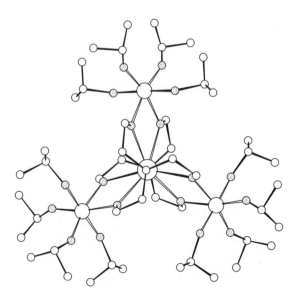

Figure 4.62 Structure of $[La\{\mu,\eta^4\text{-}(OC_2H_4)_3N\}_2\{Nb(OPr^i)_4\}_3]$ ($\bigcirc = O$; H atoms omitted).

pp. 358–360).[508] In the trinuclear complex $[\{(Pr^iO)_2Ti(\mu\text{-}OPr^i)Ti(OPr^i)_2\}(\mu_3\text{-}OPr^i)_2(\mu\text{-}OPr^i)_2CuCl]$ there is an isosceles triangle of metal atoms capped by (μ_3-OPr^i) ligands above and below the metal plane and bridged by three (μ-OPr^i) ligands in the plane. This $[MM'_2(\mu_3\text{-}OR)_2(\mu\text{-}OR)_3]$ core is typically found in heterometallic and homometallic alkoxide complexes (Figs 4.5 and 4.25). In this complex each six-coordinated Ti has *cis* terminal alkoxo ligands whilst the five-coordinated (distorted TBP) Cu(II) has one terminal chloride.[493]

A similar structure was reported for $[ICd\{Ti_2(OPr^i)_9\}]$ with a five-coordinated Cd atom.[495] Another example is found in the cationic complex $[Sn\{Ti_2(OPr^i)_9\}]^+$, in which the Sn(II) atom has no terminal ligands but, assuming its lone pair of electrons occupies an equatorial position, it has a pseudo-trigonal bipyramidal configuration. Interestingly the Sn-μ_3-OR bond distances are shorter than the Sn-μ-OR bonds.[509]

A very remarkable compound is the salt $[Ti_3(OPr^i)_{11}]^+[Sn_2I_6(OPr^i)_3]^-$ which contains discrete homometallic ions. The cation has the familiar $[M_3(\mu_3\text{-}OR)_2(\mu\text{-}OR)_3(OR)_6]$ structure whilst the anion has the confacial bi-octahedral structure $[I_3Sn(\mu\text{-}OR)_3SnI_3]$.[509]

In the pentanuclear complex $[\{Ti(OPr^i)_4\}\{OPb_4(NBu^t)_3\}]$ the cubane-like $[OPb_4(\mu_3\text{-}NBu^t)_3]$ moiety is coordinated to the five-coordinated (TBP) titanium through its oxygen so that strictly speaking it is a heterometallic oxoalkoxide.[510]

3.4.2 Zirconium

The tetranuclear mixed ligand complex $[Zr_2Co_2(\mu_3\text{-}OPr^n)_2(\mu\text{-}OPr^n)_4(OPr^n)_4(acac)_2]$ contains distorted octahedral Zr(IV) and trigonal bipyramidal Co(II). It is a centrosymmetrical molecule containing an edge-shared bi-octahedral $[Zr_2(\mu\text{-}OR)_2(OR)_8]$ unit that incorporates two Co (acac) moieties, which are bridged by one μ_3-OR and two μ-OR groups above and below the $Zr_2(\mu\text{-}OR)_2$ plane.[511]

The trinuclear $[ClCu\{Zr_2(OPr^i)_9\}]$ has the $[CuZr(\mu_3\text{-}OPr^i)_2(\mu\text{-}OPr^i)_3]$ core similar to the analogous titanium complex.[512] The homoleptic Cu(I)Zr(IV) complex has a different structure $[Cu_2Zr_2(OPr^i)_{10}]$ containing the confacial bi-octahedral $Zr_2(OPr^i)_9$ unit into which the $Cu_2(\mu\text{-}OPr^i)$ group is inserted by the copper atoms bridging with terminal isopropoxo ligands on each Zr. The $(\mu\text{-}OPr^i)$ Cu$(\mu\text{-}OPr^i)$ system has a linear O–Cu–O arrangement.[512]

The centrosymmetric dimer $[\{Cd[Zr_2(OPr^i)_9](\mu\text{-}Cl)\}_2]$ is interesting compared to the monomeric $[ClCu\{Zr_2(OPr^i)_9\}]$ because the bridging chlorides give the Cd atoms a distorted octahedral configuration.[513] Interestingly in the dimeric complexes $[\{M_2(OPr^i)_9\}Sn(\mu\text{-}Cl)]_2$ (M = Zr, Hf) the $M_2(OPr^i)_9$ moieties act as bidentate bridging ligands towards the Sn(II) atoms, which are joined by an unsymmetrical $(\mu\text{-}Cl)_2$ bridge.[522] On the other hand in the dimeric $[\{Zr_2(OPr^i)_9\}Sn(\mu\text{-}I)]_2$ the $Zr_2(OPr^i)_9$ unit acts as a tridentate ligand towards the Sn(II) atom which is thus five-coordinated.[525]

The $Zr_2(OPr^i)_9$ unit is also present in the centrosymmetric hexanuclear complex $[\{Zr_2(OPr^i)_9\}Pb(\mu\text{-}OPr^i)_2Pb\{Zr_2(OPr^i)_9\}]$ in which the Pb(II) atoms are 4-coordinated or pseudo-trigonal bipyramidal, assuming the presence of the lone pair of electrons occupying the vacant equatorial site.[514] The centrosymmetric $[(Pr^iO)_3Zr(\mu\text{-}OPr^i)_3Pb(\mu\text{-}OPr^i)Pb(\mu\text{-}OPr^i)_2Pb(\mu\text{-}OPr^i)Pb(\mu\text{-}OPr^i)_3Zr(OPr^i)_3]$ has a zig-zag chain structure with two pyramidal three-coordinated Pb(II) atoms and two pseudo-trigonal bipyramidal five-coordinated.[514]

The trinuclear $[Tl(\mu\text{-}hfip)_3Zr(\mu\text{-}hfip)_3Tl]$ (hfip = $OCH(CF_3)_2$) is also centrosymmetric with octahedral Zr and three-coordinated Tl(I) which also have six Tl....F contacts.[515]

3.4.3 Hafnium

The hafnium cadmium complex $[ICd\{Hf_2(OPr^i)_9\}]$ has a similar trinuclear structure to that of the previously described titanium cadmium analogue,[495] as does the hafnium copper(II) complex $[ClCu\{Hf_2(OPr^i)_9\}]$.[200] The trinuclear $[(Pr^iO)_2Hf(\mu\text{-}OPr^i)_4\{Al(OPr^i)_2\}_2]$ has a cis-octahedral Hf chelated by two bidentate bridging $(\mu\text{-}OPr^i)_2Al(OPr^i)_2$ tetrahedra.[516]

3.5 Vanadium Group Heterometallic Alkoxides

The trinuclear niobium(v) cadmium mixed ligand complex $[(Pr^iO)_3Nb(\mu\text{-}OPr^i)_2(\mu,\eta^2\text{-}O_2CMe)Cd(\mu\text{-}OPr^i)_2(\mu,\eta^2\text{-}O_2CMe)Nb(OPr^i)_3]$ has a central six-coordinated (trigonal prismatic) Cd atom which is bridged to two $\mu\text{-}OPr^i$ ligands and one bridging (η^2) acetate to each octahedral Nb atom (Table 4.21, pp. 362–363).[517] Other heterometallic alkoxides involving niobium or tantalum have been dealt with in earlier sections.

3.6 Chromium Group Heterometallic Alkoxides

In the tetranuclear homoleptic dimeric complex $[(Bu^tO)Sn(\mu\text{-}OBu^t)_2Cr(\mu\text{-}OBu^t)]_2$ the inner Cr(II) atoms are four-coordinated and the outer Sn(II) atoms are three-coordinated (pyramidal) (Table 4.21).[523] In the mixed ligand tetranuclear complex

Table 4.20 Heterometallic alkoxides of titanium, zirconium, and hafmium

Compound	Coordination no.		M–O Bond lengths (Å)		Bond angles (°) $\widehat{MOC}$	Reference
	Ti, Zr, Hf	Hetero-metal	Terminal	Bridging		
[Ti(μ,η^2-OCMe$_2$CH$_2$PPh$_2$)$_3$Rh]	4^1	4^1	TiO, 1.827–1.833 (4) av. 1.830	–	TiO, 137.9–144.1 av. 141.9	508
[ClCu{Ti$_2$(OPr^i)$_9$}]	6	5	TiO, 1.772–1.790 (2) av. 1.780	CuO, 2.149, 2.157 (2) (μ_3) 2.007, 2.008 (2) (μ) TiO, 2.088–2.155 (2) (μ_3) av. 2.124 1.981–2.087 (2) (μ) av. 2.029	–	493
[ICd{Ti$_2$(OPr^i)$_9$}]	6	5	TiO, 1.774 (4)	CdO, 2.281 (5) (μ_3) 2.337, 2.343 (6) (μ) TiO, 2.107–2.167 (5) (μ_3) av. 2.149 1.925–2.047 (7) (μ) av. 2.004	–	495
[Sn{Ti$_2$(OPr^i)$_9$}]$^+$ (SnCl$_3$)$^-$	6	5^1	TiO, 1.755 (4)	SnO, 2.214, 2.164 (4) (μ_3) 2.321, 2.422 (4) (μ) TiO, 2.014, 2.027 (4) (μ)	–	509
[{Ti(OPr^i)$_4$}{OPb$_4$(NBu^t)$_3$}]	5	3	TiO 1.76 (5) (ax) TiO 1.83 (3) (eq)	TiO, 2.09 (3) (μ_4) PbO, 2.23 (3) (μ_4)	–	510
[Zr$_2$Co$_2$(μ_3-OPr^n)$_2$(μ-OPr^n)$_4$(OPr^n)$_4${acac)$_2$]	6	5	ZrO, 1.913, 1.928 (7) Co(acac), 1.924, 2.014 (6)	ZrO, 2.225–2.229 (5) (μ_3) 2.138, 2.144 (5) (μ) CoO, 2.454 (5) (μ_3) 1.930, 1.935 (5) (μ)	ZrO, 168.3, 171.6 (t) ZrO, 127.6 (μ, b) CoO, 120.3	511
[ClCu{Zr$_2$(OPr^i))$_9$]	6	5	ZrO, 1.914–1.936 (6) av. 1.926	ZrO, 2.221–2.257 (6) (μ_3) av. 2.240 2.092–2.208 (6) (μ) av. 2.145 CuO, 2.186, 2.207 (6) (μ_3) 1.998, 1.999 (6) (μ)	ZrO, 172.9–174.4 (t) ZrO, 123.6–133.3 (μ_3, b) 128.5–138.4 (μ, b) CuO, 114.0–116.7 (μ_3, b) 124.4, 132.6 (μ, b)	512

Compound			M–OR	M–O	M–O–M	Ref.
$[Cu_2Zr_2(OPr^i)_{10}]$	6	2^1	ZrO, 1.930–1.942 (5) av. 1.937	ZrO, 2.110–2.225 (5) av. 2.179 CuO, 1.836–1.858 (5) av. 1.849	ZrO, 171.3, 177.3 (t) ZrO, 122.6–139.0 (b) CuO, 118.7–121.1 (b)	512
$[\{Cd[Zr_2(OPr^i)_9](\mu\text{-Cl})\}_2]$	6	6	Not reported	CdO, av. 2.379 (4)	–	513
$[\{Zr_2(OPr^i)_9\}Sn(\mu\text{-Cl})]_2$	6	4	ZrO, 1.905–1.946 (8) av. 1.920	ZrO, 2.328–2.371 (6) (μ_3) av. 2.351 2.085–2.250 (6) (μ) av. 2.185 SnO, 2.259, 2.286 (6) (μ_3) 2.154, 2.161 (7) (μ)	–	522
$[\{Hf_2(OPr^i)_9\}Sn(\mu\text{-Cl})]_2$	6	4	HfO, 1.896–1.953 (13) av. 1.924	HfO, 2.305–2.362 (10) (μ_3) av. 2.334 2.096–2.272 (10) (μ) av. 2.183 SnO, 2.315, 2.332 10) (μ_3) 2.178, 2.191 (11) (μ)	–	522
$[\{Zr_2(OPr^i)_9\}Sn(\mu\text{-I})]_2$	6	5	ZrO, 1.913–1.979 (5) av. 1.938	ZrO, 2.310–2.337 (4) (μ_3) av. 2.327 2.115–2.246 (5) (μ) av. 2.164 SnO, 2.367, 2.404 (4) (μ_3) 2.196 (4) (μ)	–	525
$[Pb_2Zr_4(OPr^i)_{20}]$	6	4	ZrO, 1.886–1.939 (22) av. 1.931	ZrO, 2.158–2.269 (22) av. 2.220 PbO, 2.239, 2.309, 2.425, 2.72	–	514

(continued overleaf)

Table 4.20 (*Continued*)

Compound	Coordination no.		M–O Bond lengths (Å)		Bond angles (°) $\widehat{MOC}$	Reference
	Ti, Zr, Hf	Hetero-metal	Terminal	Bridging		
[Pb₄Zr₂(OPrⁱ)₁₆]	6	3 and 4	ZrO, 1.855–1.995(18) av. 1.927	ZrO, 2.135–2.271 (21) av. 2.221 PbO, 2.232–2.431 (20) av. 2.318	–	514
[Tl{(μ-OCH(CF₃)₂)₃Zr{μ-OCH(CF₃)₂}₃Tl]	6	3[l]	–	ZrO, 2.035–2.090 (7) av. 2.060 TlO, 2.740–2.831 (11) av. 2.795	–	515
[ICd{Hf₂(OPrⁱ)₉}]	6	5	HfO, 1.915–1.923 (8) av. 1.919	HfO, 2.255–2.277 (6) (μ₃) av. 2.263 2.030–2.173 (6) (μ) av. 2.103 CdO, 2.315, 2.329 (6) (μ₃) 2.306, 2.361 (7) (μ)	–	495
[ClCu{Hf₂(OPrⁱ)₉}]	6	5	HfO, 1.921–1.932 (10) av. 1.927	HfO, 2.210–2.233 (8) (μ₃) av. 2.219 2.074–181 (8) (μ) av. 2.120 CuO, 2.183, 2.187 (9) (μ₃) 2.001, 2.008 (10) (μ)	–	200
[(PrⁱO)₂Hf(μ-OPrⁱ)₄{Al(OPrⁱ)₂}₂]	6	4	HfO, 1.913 (4) AlO, 1.667, 1.689 (6)	HfO, 2.140, 2.213 (4) AlO, 1.787, 1.803 (5)	HfO, 177.4 (t) AlO, 138.5, 160.9 (t) HfO, 126.3 127.9 (b) AlO, 128.7, 129.8 (b)	516

b = bridging, t = terminal; ax = axial, eq = equatorial.
[l] Coordination doubtful.

360

$[W_2Fe_2(\mu_3\text{-S})_2(\mu\text{-OPr}^i)_2(OPr^i)_4(CO)_5(py)]$ the four metal atoms form a planar rhomboidal core (Fe–Fe, 2.648 Å; W–W, 2.714 Å; W–Fe, 2.648 Å, 2.699 Å). A μ_3-S ligand caps each W_2Fe and WFe_2 triangle above and below the metal plane. In this unsymmetrical molecule there is one $W(\mu\text{-OPr}^i)W$ bridge and one $W(\mu\text{-OPr}^i)Fe$ bridge. One W has three terminal isopropoxo ligands and its neighbour has only one.[518] The μ-C atom produces a linear W≡C–Ru system in $[(Bu^tO)_3W\equiv C\text{-Ru}(CO)_2(C_5H_5)]$ and the three Bu^tO ligands on tungsten are staggered with respect to the three ligands on ruthenium.[519] The trinuclear complex $[(C_5H_5)CoW_2(\mu\text{-OCH}_2Bu^t)_2(OCH_2Bu^t)_4]$ (W–W, 2.504 Å) has two W atoms bridged by two Bu^tCH_2O ligands and the Co atoms of the cyclopentadienyl cobalt moiety.[520] A pentanuclear complex was formed by bridging the Pt(II) atom in $trans$-$Pt(PMe_2Ph)_2$ to two $W_2(OBu^t)_5$ (W–W, 2.445 Å) units by unsymmetrical μ,μ,η^3-C_2 dicarbide fragments.[521]

3.7 Manganese, Nickel, Copper, and Cobalt Heterometallic Alkoxides

3.7.1 *Manganese*

The tetranuclear homoleptic dimeric species $[(Bu^tO)M'(\mu\text{-OBu}^t)_2M(\mu\text{-OBu}^t)]_2$ (M = Mn, M' = Sn; M = Co, M' = Ge, Sn; M = Ni, M' = Ge, Sn) all have similar linear configurations with the inner two M atoms four-coordinated and the outer two M' atoms three-coordinated (Table 4.21).[523] The homoleptic hexanuclear complexes $[M_2(\mu_3\text{-OEt})_2(\mu\text{-OEt})_8Sb_4(OEt)_6]$ (M = Mn, Ni) have the same structure as the magnesium analogue[488] with a core of two edge-sharing MO_6 octahedra with edge-bridging $(EtO)Sb(\mu\text{-OEt})_2$ units chelating each M atom and tridentate $(EtO)Sb(\mu_3\text{-OEt})(\mu\text{-OEt})$ groups bridging the two M atoms.[522]

3.7.2 *Copper*

The dinuclear complex $[(acac) Cu(\mu\text{-OSiMe}_3)_2Al(OSiMe_3)_2]$ contains an unsymmetrical Cu $(\mu\text{-OSiMe}_3)_2$ Al bridged unit with square planar Cu(II) and distorted tetrahedral aluminium.[433] The mixed ligand complex $[CuPb(\mu_3\text{-OC}_2H_4NMe_2)(\mu\text{-OC}_2H_4NMe_2)(\mu\text{-Cl})_2]_\infty$ is a linear polymer formed by bridging centrosymmetrical dimer units involving five-coordinated (distorted square pyramidal) Cu(II) atoms and distorted octahedral Pb(II) atoms. The iodo analogue $[CuPb(\mu_3\text{-OC}_2H_4NMe_2)_2(\mu\text{-OC}_2H_4NMe_2)(\mu\text{-I})_2]_\infty$ has a similar structure.[526]

3.7.3 *Cobalt*

The hexanuclear dimeric mixed ligand termetallic alkoxo complex $[(\mu\text{-OBu}^t)Co(\mu\text{-OBu}^t)_2Sn(OBu^t)Fe(CO)_4]_2$ is of considerable interest. It has a linear array of 4-coordinated metals $(Bu^tO)Sn(\mu\text{-OBu}^t)_2Co(\mu\text{-OBu}^t)_2Co(OBu^t)_2Sn(OBu^t)$ capped at each end by metal–metal bonded $Fe(CO)_4$ groups.[523]

Table 4.21 Heterometallic alkoxides of niobium, chromium, tungsten, manganese, cobalt, nickel and copper

Compound	Coordination no.		M–O Bond lengths (Å)		Bond angles (°)	Reference
	M	M'	Terminal	Bridging	$\widehat{MOC}$	
[(Pr^{i}O)$_3$Nb(μ-OPr^i)$_2$(μ,η^2-O$_2$CMe)Cd(μ-OPr^i)$_2$(μ,η^2-O$_2$CMe)Nb(OPr^i)$_3$]	6	6	NbO, av. 1.877	NbO(OPr^i), av. 2.00 (OAc), av. 2.167 CdO(OPr^i), av. 2.312 (OAc), av. 2.265	NbO, 151.0	517
[(Bu^{t}O)Sn(μ-OBu^t)$_2$Cr(μ-OBu^t)]$_2$	4	3	SnO, 2.016 (5)	CrO, 2.011–2.052 (5) av. 2.031 SnO, 2.126, 2.128 (5)	–	523
[(Bu^{t}O)$_3$WCRu(CO)$_2$(C$_5$H$_5$)]	4^1	4^1	WO, av. 1.89	–	–	519
[(C$_5$H$_5$)CoW$_2$(μ-OCH$_2$Bu^t)$_2$(OCH$_2$Bu^t)$_4$]	5^1	3^1	WO, av. 1.89	WO, av. 2.06	–	520
[{(Bu^{t}O)$_5$W$_2$(μ,μ,η^3-C$_2$)}Pt(PMe$_2$Ph)$_2$]	5^1	4	WO, 1.883–1.930 (6) av. 1.906	–	WO, 129.2–154.9 (t)	521
[(Bu^{t}O)Sn(μ-OBu^t)$_2$Mn(μ-OBu^t)]$_2$	4	3	SnO, 1.994 (9)	MnO, 2.030–2.068 av. 2.048 SnO, 2.104, 2.122 (8)	–	523
[Mn$_2$(μ_3-OEt)$_2$(μ-OEt)$_8$Sb$_4$(OEt)$_6$]	6	4	SbO, 1.940–2.041 (5) av. 1.996	MnO, 2.233 (4) (μ_3) 2.074–2.229 (4) (μ) av. 2.172 SbO, 2.311 (4) (μ_3) 1.994–2.206 (4) (μ) av. 2.057	–	522
[(Bu^{t}O)Ge(μ-OBu^t)$_2$Co(μ-OBu^t)]$_2$	4	3	GeO, 1.817 (4)	CoO, 1.942–1.987 (5) av. 1.966 GeO, 1.952, 1.963 (5)	–	523
[(Bu^{t}O)Sn(μ-OBu^t)$_2$Co(μ-OBu^t)]$_2$	4	3	SnO, 2.016 (5)	CoO, 1.955–1.985 (4) av. 1.970	–	523

[(Bu^{t}O)Sn(μ-OBu^t)$_2$Ni(μ-OBu^t)]$_2$	4	3	SnO, 2.000 (7)	NiO, 1.931–1.965 (5) av. 1.950 SnO, 2.105, 2.116 (6)	–	523
[Ni$_2$(μ_3-OEt)$_2$(μ-OEt)$_8$Sb$_4$(OEt)$_6$]	6	4	SbO, 1.950–2.033 (5) av. 1.993	NiO, 2.097 (4) (μ_3) 2.010–2.117 (4) (μ) av. 2.069 SbO, 2.313 (4) (μ_3) 1.987–2.182 (5) (μ) av. 2.050	–	522
[(acac)Cu(μ-OSiMe$_3$)$_2$Al(OSiMe$_3$)$_2$]	4	4	Cu(acac), 1.889–1.894 (9) av. 1.892 AlO(OPr^i), 1.678–1.714 (9) av. 1.697	CuO, 1.973–1.997 (8) av. 1.983 AlO, 1.799–1.818 (9) av. 1.808	–	433
[CuPb(μ_3-OC$_2$H$_4$NMe$_2$)(μ-OC$_2$H$_4$NMe$_2$)(μ-Cl)$_2$]$_\infty$	5	6	–	CuO, 2.01 (2) (μ_3) 1.94 (3) (μ) PbO, 2.36, 2.66 (2) (μ_3) 2.36 (2) (μ)	–	526
[CuPb(μ_3-OC$_2$H$_4$NMe$_2$)(μ-OC$_2$H$_4$NMe$_2$)(μ-I)$_2$]$_\infty$	5	6	–	CuO, 1.936 (6) (μ_3) 1.903 (6) (μ) PbO, 2.551, 2.560 (7) (μ_3) 2.339 (6) (μ)	–	526
[(μ-OBu^t)$_2$Co(μ-OBu^t)$_2$Sn(OBu^t)Fe(CO)$_4$]$_2$	4	4Sn 5Fe	SnO, 1.957 (7)	CoO, 1.950–2.007 (5) av. 1.978 SnO, 2.057 (5)	–	523

t = terminal.

3.8   Zinc, Cadmium, Indium, Thallium, Germanium, Tin, and Lead Heterometallic Alkoxides

3.8.1   Zinc

The mixed ligand tetranuclear complex $[Zn_2Pb_2(\mu,\eta^2\text{-}OC_2H_4OMe)_4(\mu,\eta^2\text{-}O_2CMe)_4]$ has a cyclic structure with alternating four-coordinated (tetrahedral) Zn and six-coordinated (distorted octahedral) Pb atoms (Table 4.22, pp. 365–366).[527]

3.8.2   Cadmium

Cadmium and Ge(II) and Sn(II) give the trinuclear homoleptic alkoxides $[M'(\mu\text{-}OBu^t)_3M(\mu\text{-}OBu^t)_3M']$ (M = Cd, M' = Ge, Sn) with the typical structure reported previously with six-coordinated M and three-coordinated M'.[487] Cadmium is also present in the trinuclear species $[ICd(Sn_2OPr^i_9)]$ which has the well-established core structure $MM'_2(\mu_3\text{-}OR)_2(\mu\text{-}OR)_3$ found in numerous other trimetallic complexes.[495,525]

3.8.3   Indium and Thallium

Indium(I) and thallium(I) are also involved in the structures of heterometallic alkoxides. For example in the binuclear complex $[Tl(\mu\text{-}OBu^t)_3Sn]$ both metals are three-coordinated (pyramidal).[528]

Both $[In(\mu\text{-}OBu^t)_3Sn]$ and $[Tl(\mu\text{-}OBu^t)_3Sn]$ moieties can act as Lewis bases by virtue of the electron lone pairs on each metal atom, which coordinate to other metal carbonyls to give trimetallic and even tetrametallic mixed ligand heterometallic alkoxides.[524] Some examples are $[Tl(\mu\text{-}OBu^t)_3Sn.Mo(CO)_5]$, $[(CO)_5Mo.In(\mu\text{-}OBu^t)_3Ge.Mo(CO)_5]$, $[(CO)_4Fe.In(\mu\text{-}OBu^t)_3Sn.Fe(CO)_4]$, $[cis\text{-}(CO)_4Cr\{In(\mu\text{-}OBu^t)_3Ge\}_2]$, $[trans\text{-}(CO)_4Mo\{Sn(\mu\text{-}OBu^t)_3Tl\}_2]$, $[cis\text{-}(CO)_4Cr\{In(\mu\text{-}OBu^t)_3Sn.Cr(CO)_5\}_2]$ and $[cis\text{-}(CO)_4Mo\{In(\mu\text{-}OBu^t)_3Sn.Cr(CO)_5\}_2]$.

3.8.4   Germanium, Tin, and Lead

Several other heterometallic alkoxides involving Ge, Sn, or Pb have been reported. The dimeric $Ge_2(\mu\text{-}OBu^t)_2(OBu^t)_2$ is coordinated to $Ni(CO)_3$ in the tetranuclear $[\{(CO)_3Ni\}(Bu^tO)Ge(\mu\text{-}OBu^t)]_2$. The double $\mu\text{-}OBu^t$ bridge is also present in the dinuclear organometallic alkoxo compound $[(Bu^tO)Ge(\mu\text{-}OBu^t)_2Sn,\eta^1\text{-}(C_5H_5)]$ in which the $\eta^1\text{-}C_5H_5$ group and the terminal Bu^tO ligand are *cis* with respect to the GeO_2Sn ring.[530] In the analogous complex $[(Bu^tO)Sn(\mu\text{-}OBu^t)_2Pb(\eta^5\text{-}C_5H_5)]$ it is noteworthy that the cyclopentadienyl ligand is bonded in pentahapto mode to the Pb atom.[522] The trinuclear homoleptic complexes $[M'(\mu\text{-}OBu^t)_3M(\mu\text{-}OBu^t)_3M']$ (M = Pb, M' = Ge, Sn) have the typical structures characteristic of this class of molecule.[487]

Two novel examples of pentanuclear termetallic alkoxo carbonyl complexes are $[(Bu^tO)_2Ge(\mu\text{-}OBu^t)\{\mu\text{-}Fe(CO)_4\}Pb\{\mu\text{-}Fe(CO)_4\}(\mu\text{-}OBu^t)Ge(OBu^t)_2]$ in which the inner Pb atom is four-coordinated by two bridging $Fe(CO)_4$ and two bridging Bu^tO groups whereas the four-coordinated Ge has two terminal Bu^tO groups. However, in $[Pb\{\mu\text{-}OBu^t)_3Sn.Fe(CO)_4\}_2]$ the central Pb atom is six-coordinated entirely by bridging Bu^tO groups and the Sn atoms are bonded to terminal $Fe(CO)_4$ groups.[531]

Table 4.22 Heterometallic alkoxides of zinc, cadmium, indium, thallium, germanium, tin, and lead

Compound	Coordination no.		M-O Bond lengths (Å)		Bond angles (°)	Reference
	M	M', M''	Terminal	Bridging	$\widehat{MOC}$	
$[Zn_2Pb_2\{\mu,\eta^2\text{-}OC_2H_4OMe\}_4(\mu,\eta^2\text{-}O_2CMe)_4]$	4	6	PbO(acetate), 2.48 (2); PbO(CH$_2$OMe), 2.84 (2)	ZnO, av. 1.97; PbO, av. 2.26	–	527
$[Ge(\mu\text{-}OBu^t)_3Cd(\mu\text{-}OBu^t)_3Ge]$	6	3	–	CdO, 2.33 (1); GeO, 1.893 (9)	–	487
$[Sn(\mu\text{-}OBu^t)_3Cd(\mu\text{-}OBu^t)_3Sn]$	6	3	–	CdO, 2.34 (1); SnO, 2.080 (9)	–	487
$[ICdSn_2(\mu_3\text{-}OPr^i)_2(\mu\text{-}OPr^i)_3(OPr^i)_4]$	5	6	SnO, 1.958, 1.972 (9)	CdO, 2.370 (7) (μ_3); 2.297 (8) (μ); SnO, 2.189, 2.198 (8) (μ_3); 2.048, 2.145 (8) (μ)	–	525
$[In(\mu\text{-}OBu^t)_3Sn]$	3	3	–	InO, 2.413 (4); SnO, 2.032 (4)	–	524
$[Tl(\mu\text{-}OBu^t)_3Sn]$	3	3	–	TlO, 2.595 (7); SnO, 2.023 (9)	TlOC, 123.3; SnOC, 134.8	524, 528
$[Tl(\mu\text{-}OBu^t)_3SnMo(CO)_5]$	3	4^1, 6^1	–	TlO, 2.557 (15); SnO, 2.017 (9)	–	524
$[(CO)_5MoIn(\mu\text{-}OBu^t)_3GeMo(CO)_5]$	4^1	4^1	–	InO, 2.244 (9); GeO, 1.842 (9)	–	524
$[(CO)_4FeIn(\mu\text{-}OBu^t)_3SnFe(CO)_4]$	4^1	5^1	–	InO, 2.187 (20); SnO, 2.034 (20)	–	524
$[cis\text{-}(CO)_4Cr\{In(\mu\text{-}OBu^t)_3Ge\}_2]$	4^1	3	–	InO, 2.236 (9); GeO, 1.872 (8)	–	524
$[trans\text{-}(CO)_4Mo\{Sn(\mu\text{-}OBu^t)_3Tl\}_2]$	3	4^1	–	TlO, 2.568 (50); SnO, 2.036 (30)	–	524
$[cis\text{-}(CO)_4Cr\{In(\mu\text{-}OBu^t)_3SnCr(CO)_5\}]$	4^1	4^1	–	InO, 2.263 (15); SnO, 2.025 (12)	–	524
$[cis\text{-}(CO)_4Mo\{In(\mu\text{-}OBu^t)_3SnCr(CO)_5\}_2]$	4^1	4^1	–	InO, 2.271 (17); SnO, 2003 (6)	–	524

(continued overleaf)

Table 4.22 (*Continued*)

Compound	Coordination no.			M-O Bond lengths (Å)		Bond angles (°)	Reference
	M	M', M''		Terminal	Bridging	$\widehat{MOC}$	
[{(CO)₃Ni}(Bu^tO)Ge(μ-OBu^t)]₂	4[1]	4[1]		GeO, 1.780 (6)	GeO, 1.932 (6)	–	529
[(Bu^tO)Ge(μ-OBu^t)₂Sn(σ,η'-C₅H₅)]	3	3		GeO, 1.839 (10)	GeO, 1.935, 1.960 (9) SnO, 2.212, 2.218 (9)	–	530
[(Bu^tO)Sn(μ-OBu^t)₂Pb(η⁵-C₅H₅)]	3	3[1]		SnO, 1.96 (2)	SnO, 2.042 (11) PbO, 2.362 (11)	–	522
[Ge(μ-OBu^t)₃Pb(μ-OBu^t)₃Ge]	6	3		–	PbO, 2.57 (1) GeO, 1.90 (1)	–	487
[Sn(μ-OBu^t)₃Pb(μ-OBu^t)₃Sn]	6	3		–	PbO, 2.56 (2) SnO, 2.074 (8)	–	487
[Bu^tO)₂Ge(μ-OBu^t){μ-Fe(CO)₄}]₂Pb	4[1]	4	6	GeO, 1.75, 1.78 (1)	PbO, 2.64 (1) GeO, 1.80 (1)	–	531
[Pb{(μ-OBu^t)₃SnFe(CO)₄}₂]	6	4	5	–	PbO, 2.482–2.730 (7) SnO, 1.992–2.031 (7)	–	531
[{(cod)Rh}₂(μ-OEt)₄Sn(OEt)₂] cod = cyclooctadiene	6	4		SnO, 1.979, 1.980 (5)	SnO, 2.075–2.103 (4) av. 2.089 RhO, 2.050–2.082 (4) av. 2.065	SnOC, 125.4, 125.6 (t)	532
[ClSn(μ-OBu^t)₂Al(OBu^t)₂]	3	4		AlO, 1.668, 1.682 (8)	SnO, 2.152, 2.154 (7) AlO, 1.800, 1.815 (7)	–	533

t = terminal.
[1] Coordination doubtful.

366

In the organometallic alkoxo complex $[\{(cod)Rh\}_2(\mu\text{-}OEt)_4Sn(OEt)_2]$ (cod = cyclooctadiene), the distorted octahedral Sn(IV) has *cis*-terminal ethoxo ligands.[532] The dinuclear complex $[ClSn(\mu\text{-}OBu^t)_2Al(OBu^t)_2]$ has three-coordinated (pyramidal) Sn(II) and distorted tetrahedral Al.[533]

REFERENCES

1. M.H. Chisholm and I.P. Rothwell, in Comprehensive Coordination Chemistry (G. Wilkinson, R.D. Gillard, and J.A. McCleverty, eds), Vol. 2, Pergamon, London (1987).
2. K.G. Caulton and L.G. Hubert-Pfalzgraf, *Chem. Rev.*, **90**, 969 (1990).
3. R.C. Mehrotra, A. Singh, and S. Sogani, *Chem. Soc. Rev.*, **23**, 215 (1994).
4. M.H. Chisholm, *Chem. Soc. Rev.*, **24**, 79 (1995).
5. M.I. Khan and J. Zubieta, *Progress in Inorganic Chemistry*, Vol. 43, 1–149, Wiley Interscience, New York (1995).
6. D.C. Bradley, H. Chudzynska, M.B. Hursthouse, and M. Motevalli, *Polyhedron*, **10**, 1049 (1991).
7. D.C. Bradley, H. Chudzynska, M.B. Hursthouse, and M. Motevalli, *Polyhedron*, **13**, 7 (1994).
8. D.C. Bradley, *Nature (London)*, **182**, 1211 (1958).
9. D.C. Bradley, R.C. Mehrotra, and D.P. Gaur, *Metal Alkoxides*, 92–98, Academic Press, London (1978).
10. M.H. Chisholm, *A.C.S. Symp. Ser.*, **211**, 243 (1983).
11. F.A. Cotton, D.O. Marler, and W. Schwotzer, *Inorg. Chem.*, **23**, 4211 (1984).
12. W.J. Evans, T.J. Deming, J.M. Olofson, and J.W. Ziller, *Inorg. Chem.*, **28**, 4027 (1989).
13. T. Fjeldberg, P.B. Hitchcock, M.F. Lappert, S.J. Smith, and A.J. Thorne, *J. Chem. Soc., Chem. Commun.*, 939 (1985).
14. C.K. Sharma, S. Goel, and R.C. Mehrotra, *Indian J. Chem.*, **14A**, 878 (1976).
15. W.J. Evans, M.S. Sollberger, and T.P. Hanusa, *J. Am. Chem. Soc.*, **110**, 1841 (1988).
16. M.H. Chisholm, D.L. Clark, M.J. Hampden-Smith, and D.M. Hoffman, *Angew. Chem.*, **101**, 446 (1989).
17. E. Weiss, H. Alsdorf, and H. Kuhr, *Angew. Chem. Int. Ed. Engl.*, **6**, 801 (1967).
18. L.F. Dahl, G.L. Davies, D.L. Wampler, and R. West, *J. Inorg. Nucl. Chem.*, **24**, 357 (1962).
19. H.M.M. Shearer and C.B. Spencer, *Acta Crystallogr.*, **B36**, 2046 (1980).
20. N.Ya. Turova, V.A. Kuzunov, A.I. Yanovskii, N.G. Borkii, Yu.T. Struchkov, and B.L. Tarnopolskii, *J. Inorg. Nucl. Chem.*, **41**, 5 (1979).
21. M. Wijk, R. Norrestam, M. Nygren, and G. Westin, *Inorg. Chem.*, **35**, 1077 (1996).
22. M.H. Chisholm, F.A. Cotton, C.A. Murillo, and W.W. Reichert, *Inorg. Chem.*, **16**, 1801 (1977).
23. P.J. Wheatley, *J. Chem. Soc.*, 4270 (1961).
24. M.H. Chisholm, S.R. Drake, A.A. Naiini, and W.E. Streib, *Polyhedron*, **10**, 805 (1991).
25. G. Beck, P.B. Hitchcock, M.F. Lappert, and I.A. MacKinnon, *J. Chem. Soc., Chem. Commun.*, 1312 (1989).
26. L.M. Engelhardt, J.M. Harrowfield, M.F. Lappert, I.A. MacKinnon, B.H. Newton, C.L. Raston, B.W. Skelton, and A.H. White, *J. Chem. Soc., Chem. Commun.*, 846 (1986).
27. J. Hvoslef, H. Hope, B.D. Murray, and P.P. Power, *J. Chem. Soc., Chem. Commun.*, 1438 (1983).
28. K.J. Weese, R.A. Bartlett, B.D. Murray, M.M. Olmstead, and P.P. Power, *Inorg. Chem.*, **26**, 2409 (1987).
29. E.M. Arnett, M.A. Nichols, and A.T. McPhail, *J. Am. Chem. Soc.*, **112**, 7059 (1990).
30. M.J. McGeary, K. Folting, W.E. Streib, J.C. Huffman, and K.G. Caulton, *Polyhedron*, **10**, 2699 (1991).
31. M. Motevalli, D. Shah, and A.C. Sullivan, *J. Organomet. Chem.*, **513**, 239 (1996).

32. J.E. Davies, P.R. Raithby, R. Snaith, and A.E.H. Wheatley, *Chem. Commun.*, 1721 (1997).
33. E. Weiss, *Z. Anorg. Allg. Chem.*, **332**, 197 (1964).
34. J.E. Davies, J. Kopf, and E. Weiss, *Acta Crystallogr.*, **B38**, 2251 (1982).
35. A. Mommertz, K. Dehnicke, and J. Magull, *Z. Naturforsch.*, **51b**, 1583 (1996).
36. E. Weiss and H. Alsdorf, *Z. Anorg. Allg. Chem.*, **372**, 206 (1970).
37. E. Weiss, H. Alsdorf, and H. Kuhr, *Angew. Chem. Int. Ed. Engl.*, **6**, 801 (1967); E. Weiss, H. Alsdorf, H. Kuhr, and H.-F. Grützmacher, *Chem. Ber.*, **101**, 3777 (1968).
38. E. Weiss, K. Huffmann, and H.-F. Grützmacher, *Chem. Ber.*, **103**, 1190 (1970).
39. M.H. Chisholm, S.R. Drake, A.A. Naiini, and W.E. Streib, *Polyhedron*, **10**, 337 (1991).
40. B. Laermann, M. Lazell, M. Motevalli, and A.C. Sullivan, *J. Chem. Soc., Dalton Commun.*, 1263 (1997).
41. Z.A. Starikova, A.I. Yanovsky, E.P. Turevskaya, and N.Ya. Turova, *Polyhedron*, **16**, 967 (1997).
42. G.E. Coates and A.H. Fishwick, *J. Chem. Soc. A*, 477 (1968).
43. S.C. Goel, M.A. Matchett, M.Y. Chiang, and W.E. Buhro, *J. Am. Chem. Soc.*, **113**, 1844 (1991).
44. K.F. Tesh, T.M. Hanusa, J.C. Huffman, and C.J. Huffman, *Inorg. Chem.*, **31**, 5572 (1992).
45. J.A. Darr, S.R. Drake, M.B. Hursthouse, and K.M. Abdul Malik, *Inorg. Chem.*, **32**, 5704 (1993).
46. W.A. Wojtczak, M.J. Hampden-Smith, and E.N. Duesler, *Inorg. Chem.*, **35**, 6638 (1996).
47. I. Baxter, J.A. Darr, S.R. Drake, M.B. Hursthouse, K.M. Abdul Malik, and D.M.P. Mingos, *J. Chem. Soc., Dalton Trans.*, 2875 (1997).
48. K.G. Caulton, M.H. Chisholm, S.R. Drake, and W.E. Streib, *Angew. Chem. Int. Ed. Engl.*, **29**, 1483 (1990).
49. O. Poncelet, L.G. Hubert-Pfalzgraf, L. Toupet, and J.-C. Daran, *Polyhedron*, **10**, 2045 (1991).
50. S.R. Drake, W.E. Streib, K. Folting, M.H. Chisholm, and K.G. Caulton, *Inorg. Chem.*, **31**, 3205 (1992).
51. H. Vincent, F. Labrize, and L.G. Hubert-Pfalzgraf, *Polyhedron*, **13**, 3323 (1994).
52. B. Borup, J.A. Samuels, W.E. Streib, and K.G. Caulton, *Inorg. Chem.*, **33**, 994 (1994).
53. N.Ya. Turova, V.A. Kozunov, A.I. Yanovskii, N.G. Bokii, Yu.T. Struchkov and B.L. Tarnopol'skii, *J. Inorg. Nucl. Chem.*, **41**, 5 (1979).
54. D.C. Bradley, *Adv. Chem. Ser.*, **23**, 10 (1959).
55. K. Folting, W.E. Streib, K.G. Caulton, O. Poncelet, and L.G. Hubert-Pfalzgraf, *Polyhedron*, **10**, 1639 (1991).
56. A.I. Yanovskii, V.A. Kozunov, N.Ya. Turova, N.G. Furmanova, and Yu.T. Struchkov, *Dokl. Akad. Nauk. SSSR*, **244**, 119 (1979).
57. W. Fieggen and H. Gerding, *Rec. Trav. Chim. Pays-Bas*, **89**, 175 (1970); **90**, 410 (1971); D.C. Kleinschmidt, V.J. Shiner, and D. Whittaker, *J. Org. Chem.*, **38**, 3334 (1973).
58. B. Beagley, K. Jones, P. Parkes, and R.G. Pritchard, *Synth. React. Inorg. Met.-Org. Chem.*, **18**, 465 (1988).
59. R.H. Cayton, M.H. Chisholm, E.R. Davidson, V.F. DiStasi, Ping Du, and J.C. Huffman, *Inorg. Chem.*, **30**, 1020 (1991).
60. M.H. Chisholm, V.F. DiStasi, and W.E. Streib, *Polyhedron*, **9**, 253 (1990).
61. M.L. Sierra, R. Kumar, V.S.J. de Mel, and J.P. Oliver, *Organometallics*, **11**, 206 (1992).
62. M. Veith, S. Faber, H. Wolfanger, and V. Huch, *Chem. Ber.*, **129**, 381 (1996).
63. J. Lewinski, J. Zachara, and E. Grabska, *J. Am. Chem. Soc.*, **118**, 6794 (1996).
64. M.F. Garbauskus, J.H. Wengrovius, R.C. Going, and J.S. Kasper, *Acta Crystallogr.*, **C40**, 1536 (1984).
65. M.H. Chisholm, J.C. Huffman, and J.L. Wesemann, personal communication.
66. S.R. Rettig, A. Storr, and J. Trotter, *Can. J. Chem.*, **52**, 2206 (1974).
67. S.R. Rettig, A. Storr, and J. Trotter, *Can. J. Chem.*, **53**, 58 (1975).
68. S.R. Rettig, A. Storr, J. Trotter, and K. Uhrich, *Can. J. Chem.*, **62**, 2783 (1984).
69. M.B. Power, W.M. Cleaver, A.W. Apblett, A.R. Barron, and J.W. Ziller, *Polyhedron*, **11**, 477 (1992).
70. W.M. Cleaver and A.R. Barron, *Organometallics*, **12**, 1001 (1993).

71. U. Dembowski, T. Pape, R. Herbst-Irmer, E. Pohl, H.W. Roesky, and G.M. Sheldrick, *Acta Cryst.*, **C49**, 1309 (1993).
72. D.C. Bradley, D.M. Frigo, M.B. Hursthouse, and B. Hussain, *Organometallics*, **7**, 1112 (1988).
73. L.F. Dahl, G.L. Davis, D.L. Wampler, and R. West, *J. Inorg. Nucl. Chem.*, **24**, 357 (1962).
74. H. Rothfuss, K. Folting, and K.G. Caulton, *Inorg. Chim. Acta*, **212**, 165 (1993).
75. L.O. Atovmyan, Ya.Ya. Bleidelis, A.A. Kemme, and R.B. Shibaeva, *Zhur. Strukt. Khim.*, **11**, 318 (1970); **14**, 103 (1973).
76. S.N. Gurkova, A.I. Gusev, N.V. Alekseev, R.I. Segel'man, T.K. Gar, and N.Yu. Khromova, *Zhur. Strukt. Khim.*, **23**, 101 (1982); **24**, 83 (1983); **24**, 162 (1983).
77. S.P. Narula, S. Soni, R. Shankar, and R.K. Chadha, *J. Chem. Soc., Dalton Trans.*, 3055 (1992).
78. A. Mavrides and A. Tulinsky, *Inorg. Chem.*, **15**, 2723 (1976).
79. M.F. Lappert, S.J. Miles, J.L. Atwood, M.J. Zaworotko, and A.J. Carty, *J. Organomet. Chem.*, **212**, C4 (1981).
80. T. Fjeldberg, P.B. Hitchcock, M.F. Lappert, S.J. Smith, and A.J. Thorne, *J. Chem. Soc., Chem. Commun.*, 939 (1985).
81. M. Veith, P. Hobein, and R. Rösler, *Z. Naturforsch.*, **44b**, 1067 (1989).
82. M. Veith, C. Mathur, and V. Huch, *J. Chem. Soc., Dalton Trans.*, 995 (1997).
83. R. Nesper and H.G. von Schnering, *Z. Naturforsch.*, **37b**, 1144 (1982).
84. Chr Zybill and G. Müller, *Z. Naturforsch.*, **43b**, 45 (1988).
85. M.J. McGeary, K. Folting, and K.G. Caulton, *Inorg. Chem.*, **28**, 405 (1989).
86. R. Fiedler and H. Follner, *Monatsh. Chem.*, **108**, 319 (1977).
87. C.D. Chandler, G.D. Fallon, A.J. Koplick, and B.O. West, *Aust. J. Chem.*, **40**, 1427 (1987).
88. M.H. Hampden-Smith, T.A. Wark, A.L. Rheingold, and J.C. Huffman, *Can. J. Chem.*, **69**, 121 (1991).
89. C.D. Chandler, J. Caruso, M.J. Hampden-Smith, and A.L. Rheingold, *Polyhedron*, **14**, 2491 (1995).
90. G.D. Smith, V.M. Visciglio, P.E. Fanwick, and I.P. Rothwell, *Organometallics*, **11**, 1064 (1992).
91. H. Reuter and M. Kremser, *Z. Kristallog.*, **203**, 158 (1993).
92. J. Caruso, T.M. Alam, M.J. Hampden-Smith, A.L. Rheingold, and G.A.P. Yap, *J. Chem. Soc., Dalton Trans.*, 2659 (1996).
93. S.C. Goel, M.Y. Chiang, and W.E. Buhro, *Inorg. Chem.*, **29**, 4640 (1990).
94. M. Veith, J. Hans, L. Stahl, P. May, V. Huch, and A. Sebald, *Z. Naturforsch.*, **46b**, 403 (1991).
95. N. Tempel, W. Schwarz, and J. Weidlein, *Z. Anorg. Allgem. Chem.*, **474**, 157 (1981).
96. R. Holmes, R.O. Day, V. Chandrasekhar, and J.M. Holmes, *Inorg. Chem.*, **26**, 163 (1987).
97. V. Ensinger, W. Schwarz, B. Schrutz, K. Sommer, and A. Schmidt, *Z. Anorg. Allgem. Chem.*, **544**, 181 (1987).
98. M.A. Matchett, M.Y. Chiang, and W.E. Buhro, *Inorg. Chem.*, **29**, 360 (1990); M.-C. Massiani, R. Papiernik, L.G. Hubert-Pfalzgraf, and J-C. Daran, *J. Chem. Soc., Chem. Commun.*, 301 (1990). See also Ref. 99.
99. M-C. Massiani, R. Papiernik, L.G. Hubert-Pfalzgraf, and J.-C. Daran, *Polyhedron*, **10**, 437 (1991).
100. C.M. Jones, M.D. Burkart, R.E. Bachman, D.L. Serra, S.-J. Hwu, and K.H. Whitmire, *Inorg. Chem.*, **32**, 5136 (1993).
101. T.J. Boyle, D.M. Pedrotty, B. Scott, and J.W. Ziller, *J. Amer. Chem. Soc.*, **121**, 12104 (1999).
102. M.E. Hammond, *Ph.D. Thesis*, University of London (1991).
103. D.C. Bradley, H. Chudzynska, M.E. Hammond, M.B. Hursthouse, M. Motevalli, and Ruowen Wu, *Polyhedron*, **11**, 375 (1992).
104. W.J. Evans, R. Dominguez, and T.P. Hanusa, *Organometallics*, **5**, 1291 (1986).
105. P.B. Hitchcock, M.F. Lappert, and I.A. MacKinnon, *J. Chem. Soc., Chem. Commun.*, 1557 (1988).
106. W.J. Evans, M.S. Sollberger, and T.P. Hanusa, *J. Am. Chem. Soc.*, **110**, 1841 (1988).

107. W.J. Evans and M.S. Sollberger, *Inorg. Chem.*, **27**, 4417 (1988).
108. O. Poncelet, L.G. Hubert-Pfalzgraf, J.-C. Daran, and R. Astier, *J. Chem. Soc., Chem. Commun.*, 1846 (1989).
109. O. Poncelet, L.G. Hubert-Pfalzgraf, and J.-C. Daran, *Inorg. Chem.*, **29**, 2882 (1990).
110. W.J. Evans, J.M. Olofson, and J.W. Ziller, *J. Am. Chem. Soc.*, **112**, 2308 (1990).
111. W.J. Evans, M.S. Sollberger, J.L. Shreeve, J.M. Olofson, J.H. Hain Jr, and J.W. Ziller, *Inorg. Chem.*, **31**, 2492 (1992).
112. W.J. Evans, T.J. Boyle, and J.W. Ziller, *Organometallics*, **12**, 3998 (1993).
113. D.C. Bradley, H. Chudzynska, M.B. Hursthouse, and M. Motevalli, *Polyhedron*, **12**, 1907 (1993).
114. M.J. McGeary, P.S. Coan, K. Folting, W.E. Streib, and K.G. Caulton, *Inorg. Chem.*, **28**, 3284 (1989).
115. P.S. Coan, M.J. McGeary, E.B. Lobkovsky, and K.G. Caulton, *Inorg. Chem.*, **30**, 3570 (1991).
116. M.J. McGeary, P.S. Coan, K. Folting, W.E. Streib, and K.G. Caulton, *Inorg. Chem.*, **30**, 1723 (1991).
117. W.J. Evans, R.E. Golden, and J.W. Ziller, *Inorg. Chem.*, **30**, 4963 (1991).
118. D.M. Barnhart, D.L. Clark, J.C. Gordon, J.C. Huffman, J.G. Watkin, and W.D. Zwick, *J. Am. Chem. Soc.*, **115**, 8461 (1993).
119. W.J. Evans, T.J. Deming, and J.W. Ziller, *Organometallics*, **8**, 1581 (1989).
120. H.A. Stecher, A. Sen, and A.L. Rheingold, *Inorg. Chem.*, **28**, 3280 (1989).
121. W.J. Evans, T.J. Deming, J.M. Olofson, and J.W. Ziller, *Inorg. Chem.*, **28**, 4027 (1989).
122. P. Toledano, F. Ribot, and C. Sanchez, *Acta Crystallogr.*, **C46**, 1419 (1990).
123. A. Sen, H.A. Stecher, and A.L. Rheingold, *Inorg. Chem.*, **31**, 473 (1992).
124. L.G. Hubert-Pfalzgraf, N. El Khokh, and J.-C. Daran, *Polyhedron*, **11**, 59 (1992).
125. J.H. Timmons, J.W.L. Martin, A.E. Martell, P. Rudolf, A. Clearfield, J.A. Arner, S.J. Loeb, and C.J. Willis, *Inorg. Chem.*, **19**, 3553 (1980).
126. P.S. Gradeff, K. Yunlu, T.J. Deming, J.M. Olofson, R.J. Doedens, and W.J. Evans, *Inorg. Chem.*, **29**, 420 (1990).
127. D.C. Bradley, H. Chudzynska, M.B. Hursthouse, M. Motevalli, and Ruowen Wu, *Polyhedron*, **12**, 2955 (1993).
128. D.C. Bradley, H. Chudzynska, M.B. Hursthouse, M. Motevalli, and Ruowen Wu, *Polyhedron*, **13**, 1 (1994).
129. R.A. Andersen, D.H. Templeton, and A. Zalkin, *Inorg. Chem.*, **17**, 1962 (1978).
130. M. Wedler, J.W. Gilje, U. Pieper, D. Stalke, M. Noltemeyer, and F.T. Edelmann, *Chem. Ber.*, **124**, 1163 (1991).
131. W.A. Herrmann, R. Anwander, M. Kleine, and W. Scherer, *Chem. Ber.*, **125**, 1971 (1992).
132. W.A. Herrman, R. Anwander, and W. Scherer, *Chem. Ber.*, **126**, 1533 (1993).
133. D.M. Barnhart, D.L. Clark, J.C. Huffman, R.L. Vincent, and J.D. Watkin, *Inorg. Chem.*, **32**, 4077 (1993).
134. F.T. Edelmann, A. Steiner, D. Stalke, J.W. Gilje, S. Jagner, and M. Håkansson, *Polyhedron*, **13**, 539 (1994).
135. W.J. Evans, D.K. Drummond, L.A. Hughes, H. Zhang, and J.L. Atwood, *Polyhedron*, **7**, 1693 (1988).
136. W.J. Evans, T.A. Vlibarri, L.R. Chamberlain, J.W. Ziller, and D. Alvarez Jr, *Organometallics*, **9**, 2124 (1990).
137. W.J. Evans, T.A. Vlibarri, and J.W. Ziller, *Organometallics*, **10**, 134 (1991).
138. W.J. Evans, R. Anwander, U.H. Berlekamp, and J.W. Ziller, *Inorg. Chem.*, **34**, 3583 (1995).
139. W.J. Evans, J.L. Shreeve, and J.W. Ziller, *Organometallics*, **13**, 731 (1994).
140. J.R. van den Hende, P.B. Hitchcock, and M.F. Lappert, *J. Chem. Soc., Chem. Commun.*, 1413 (1994); *J. Chem. Soc., Dalton Trans.*, 2251 (1995).
141. J.R. van den Hende, P.B. Hitchcock, S.A. Holmes, M.F. Lappert, and S. Tian, *J. Chem. Soc., Dalton Trans.*, 3933 (1995).
142. D.J. Duncalf, P.B. Hitchcock, and G.A. Lawless, *Organomet. Chem.*, **506**, 347 (1996).

143. R. Anwander, F.C. Munck, T. Priermeier, W. Scherer, O. Runte, and W.A. Herrmann, *Inorg. Chem.*, **36**, 3545 (1997).
144. D.L. Clark, J.C. Huffman, and J.G. Watkin, *J. Chem. Soc., Chem. Commun.*, 266 (1992).
145. D.L. Clark and J.G. Watkin, *Inorg. Chem.*, **32**, 1766 (1993).
146. D.M. Barnhart, D.L. Clark, J.C. Gordon, J.C. Huffman, and J.G. Watkin, *Inorg. Chem.*, **33**, 3939 (1994).
147. D.M. Barnhart, T.M. Frankcom, P.L. Gordon, N.N. Sauer, J.A. Thompson, and J.G. Watkin, *Inorg. Chem.*, **34**, 4862 (1995).
148. D.M. Barnhart, D.L. Clark, J.C. Gordon, J.C. Huffman, J.G. Watkin, and W.D. Zwick, *Inorg. Chem.*, **34**, 5416 (1995).
149. D.M. Barnhart, R.J. Butcher, D.L. Clark, J.C. Gordon, J.G. Watkin, and W.D. Zwick, *New J. Chem.*, **19**, 503 (1995).
150. E.A. Cuellar, S.S. Miller, T.J. Marks, and E. Weitz, *J. Am. Chem. Soc.*, **105**, 4580 (1983).
151. F.A. Cotton, D.O. Marler, and W. Schwotzer, *Inorg. Chem.*, **23**, 4211 (1984).
152. W.G. Van Der Sluys, J.C. Huffman, D.S. Ehler, and N.N. Sauer, *Inorg. Chem.*, **31**, 1318 (1992).
153. C. Villiers, R. Adam, M. Lance, M. Nierlich, J. Vigner, and M. Ephritikhine, *J. Chem. Soc., Chem. Commun.*, 1144 (1991).
154. C. Baudin, D. Baudry, M. Ephritikhine, M. Lance, A. Navaza, M. Nierlich, and J. Vigner, *J. Organomet. Chem.*, **415**, 59 (1991).
155. R. Adam, C. Villiers, M. Ephritikhine, M. Lance, M. Nierlich, and J. Vigner, *New J. Chem.*, **17**, 455 (1993).
156. M. Brunelli, G. Perego, G. Lugli, A. Mazzei, *J. Chem. Soc., Dalton Trans.*, 861 (1979).
157. T. Arliguie, D. Baudry, M. Ephritikhine, M. Nierlich, M. Lance, and J. Vigner, *J. Chem. Soc., Dalton Trans.*, 1019 (1992).
158. J.A. Ibers, *Nature (London)*, **197**, 686 (1963).
159. F. Babonneau, S. Doeuff, A. Leaustic, C. Sanchez, C. Cartier, and M. Verdaguer, *Inorg. Chem.*, **27**, 3166 (1988).
160. R.D. Witters and C.N. Caughlan, *Nature (London)*, **205**, 1312 (1965).
161. D.A. Wright and D.A. Williams, *Acta Crystallogr.*, **B24**, 1107 (1968).
162. R.L. Harlow, *Acta Crystallogr.*, **C39**, 1344 (1983).
163. W.M.P.B. Menge and J.G. Verkade, *Inorg. Chem.*, **30**, 4628 (1991).
164. A.A. Naiini, W.M.P.B. Menge, and J.G. Verkade, *Inorg. Chem.*, **30**, 5009 (1991).
165. A.A. Naiini, S.L. Ringrose, Y. Su, R.A. Jacobson, and J.G. Verkade, *Inorg. Chem.*, **32**, 1290 (1993).
166. W.A. Nugent and R.L. Harlow, *J. Am. Chem. Soc.*, **116**, 6142 (1994).
167. G. Boche, K. Möbus, K. Harms, and M. Marsch, *J. Am. Chem. Soc.*, **118**, 2770 (1996).
168. I.D. Williams, S.F. Pedersen, K.B. Sharpless, and S.J. Lippard, *J. Am. Chem. Soc.*, **106**, 6430 (1984).
169. S.F. Pedersen, J.C. Dewar, R.R. Eckman, and K.B. Sharpless, *J. Am. Chem Soc.*, **109**, 1279 (1987).
170. A.R. Siedle and J.C. Huffman, *Inorg. Chem.*, **29**, 3131 (1990).
171. Jin Ling Wang, Fang Ming Miao, Xiu Ju Fan, Xiao Feng, and Ji Tao Wang, *Acta Crystallogr.*, **C46**, 1633 (1990).
172. C.H. Winter, P.H. Sheridan, and M.J. Heeg, *Inorg. Chem.*, **30**, 1962 (1991).
173. T.J. Boyle, D.L. Barnes, J.A. Heppert, L. Morales, F. Takusagawa, and J.W. Connolly, *Organometallics*, **11**, 1112 (1992).
174. D.F. Evans, J. Parr, S. Rahman, A.M.Z. Slavin, D.J. Williams, C.Y. Wong, and J.D. Woollins, *Polyhedron*, **12**, 337 (1993).
175. J. Fisher, W.G. Van Der Sluys, J.C. Huffman, and J. Sears, *Synth. React. Inorg. Met.-Org. Chem.*, **23**, 479 (1993).
176. R.J. Errington, personal communication (1998).
177. D.C. Bradley and C.E. Holloway, *J. Chem. Soc., Chem. Commun.*, 284 (1965); *J. Chem. Soc. A*, 282 (1969).
178. R.J. Errington, J. Ridland, W. Clegg, R.A. Coxall, and J.M. Sherwood, *Polyhedron*, **17**, 659 (1998).
179. T.J. Boyle, R.W. Schwartz, R.J. Doedens, and J.W. Ziller, *Inorg. Chem.*, **34**, 1110 (1995).

180. N. Pajot, R. Papiernik, L.G. Hubert-Pfalzgraf, J. Vaissermann, and S. Parraud, *J. Chem. Soc., Chem. Commun.*, 1817 (1995).
181. G.J. Gainsford, T. Kemmitt, C. Lensink, and N.B. Milestone, *Inorg. Chem.*, **34**, 746 (1995).
182. V.W. Day, T.A. Eberspacher, M.H. Frey, W.G. Klemperer, S. Liang, and D.A. Payne, *Chem. Mater.*, **8**, 330 (1996).
183. U. Schubert, S. Tewinkel, and F. Möller, *Inorg. Chem.*, **34**, 995 (1995).
184. T. Maschmeyer, M.C. Klunduk, C.M. Martin, D.S. Shephard, J.M. Thomas, and B.F.G. Johnson, *Chem. Commun.*, 1847 (1997).
185. M. Crocker, R.H.M. Herold, and A.G. Orpen, *Chem. Commun.*, 2411 (1997).
186. H. Stoeckli-Evans, *Helv. Chim. Acta*, **58**, 373 (1975).
187. T.V. Lubben and P.T. Wolczanski, *J. Am. Chem. Soc.*, **109**, 424 (1987).
188. D.M. Chocette, W.E. Buschmann, M.M. Olmstead, and R.P. Planalp, *Inorg. Chem.*, **32**, 1062 (1993).
189. E. Samuel, J.F. Harrod, D. Gourier, Y. Dromzee, F. Robert, and Y. Jeannin, *Inorg. Chem.*, **31**, 3252 (1992).
190. P.B. Hitchcock, M.F. Lappert, and I.A. Mackinnon, *J. Chem. Soc., Chem. Commun.*, 1015 (1993).
191. D.C. Bradley, R.C. Mehrotra, J.D. Swanwick, and W. Wardlaw, *J. Chem. Soc.*, 2025 (1953).
192. B.A. Vaartstra, J.C. Huffman, P.S. Gradeff, L.G. Hubert-Pfalzgraf, J.-C. Daran, S. Parraud, K. Yunlu, and K.G. Caulton, *Inorg. Chem.*, **29**, 3126 (1990).
193. R.H. Heyn and T.D. Tilley, *Inorg. Chem.*, **28**, 1769 (1989).
194. B. Galeffi, M. Simard, and J.D. Wuest, *Inorg. Chem.*, **29**, 955 (1990).
195. M.M. Banaszak Holl and P.T. Wolczanski, *J. Am. Chem. Soc.*, **114**, 3854 (1992).
196. J.A. Samuels, E.B. Lobkovsky, W.E. Streib, K. Folting, J.C. Huffman, J.W. Zwanziger, and K.G. Caulton, *J. Am. Chem. Soc.*, **115**, 5093 (1993).
197. H. Brand and J. Arnold, *Organometallics*, **12**, 3655 (1993).
198. T. Rübenstahl, F. Weller, and K. Dehnicke, *Z. Krist.*, **210**, 537 (1995).
199. M. Lazell, M. Motevalli, S.A.A. Shah, C.K.S. Simon, and A.C. Sullivan, *J. Chem. Soc., Dalton Trans.*, 1449 (1996).
200. M. Veith, S. Mathur, C. Mathur, and V. Huch, *J. Chem. Soc., Dalton Trans.*, 2101 (1997).
201. T.J. Boyle, T.M. Alam, E.R. Mechenbier, B.L. Scott, and J.W. Ziller, *Inorg. Chem.*, **36**, 3293 (1997).
202. F. Dahan, R. Kergoat, M.-C. Senechal-Tocquer and J.E. Guerchais, *J. Chem. Soc., Dalton Trans.*, 2202 (1976).
203. A.A. Pinkerton, D. Schwarzenbach, L.G. Hubert-Pfalzgraf, and J.G. Riess, *Inorg. Chem.*, **15**, 1196 (1976).
204. D.C. Bradley and C.E. Holloway, *J. Chem. Soc. A*, 219 (1968); L.G. Hubert-Pfalzgraf and J.G. Riess, *Bull. Soc. Chim. France*, 2401 (1968); 4348 (1968); 1202 (1973).
205. F.A. Cotton, M.P. Diebold, and W.J. Roth, *Inorg. Chem.*, **26**, 3319 (1987).
206. F.A. Cotton, M.P. Diebold, and W.J. Roth, *Inorg. Chem.*, **26**, 3323 (1987).
207. F.A. Cotton, M.P. Diebold, and W.J. Roth, *Inorg. Chem.*, **27**, 3596 (1988).
208. A. Antinolo, A. Otero, F. Urbanos, S. Garcia-Blanco, S. Martinez-Carrera, and J. Sanz-Aparicio, *J. Organomet. Chem.*, **350**, 25 (1988).
209. H.C.L. Abbenhuis, M.H.P. Rietveld, H.F. Haarman, M.P. Hogerheide, A.L. Spek, and G. van Koten, *Organometallics*, **13**, 3259 (1994).
210. B.M. Bulychev and V.K. Bel'skii, *Russ. J. Inorg. Chem.*, **40**, 1765 (1995).
211. M. Bochmann, G. Wilkinson, G.B. Young, M.B. Hursthouse, and K.M.A. Malik, *J. Chem. Soc., Dalton, Trans.*, 1863 (1980).
212. W.A. Herrmann, N.W. Huber, R. Anwander, and T. Priermeier, *Chem. Ber.*, **126**, 1127 (1993).
213. B.D. Murray, H. Hope, and P.P. Power, *J. Am. Chem. Soc.*, **107**, 169 (1985).
214. M.H. Chisholm, F.A. Cotton, M.W. Extine, and D.C. Rideout, *Inorg. Chem.*, **18**, 120 (1979).
215. M.H. Chisholm, W.W. Reichert, and P. Thornton, *J. Am. Chem. Soc.*, **100**, 2744 (1978).
216. M.H. Chisholm, J.C. Huffman, and R.L. Kelly, *J. Am. Chem Soc.*, **101**, 7615 (1979).

217. M. Bochmann, G. Wilkinson, G.B. Young, M.B. Hursthouse, and K.M.A. Malik, *J. Chem. Soc., Dalton Trans.*, 901 (1980).

218. J.A. McCleverty, A.E. Rae, I. Wolochowicz, N.A. Bailey, and J.M.A. Smith, *J. Chem. Soc., Dalton Trans.*, 951 (1982).

219. M. Cano, J.V. Heras, A. Monge, E. Pinilla, E. Santamaria, H.A. Hinton, C.J. Jones, and J.A. McCleverty, *J. Chem, Soc., Dalton Trans.*, 2281 (1995).

220. D.M.-T. Chan, M.H. Chisholm, K. Folting, J.C. Huffman, and N.S. Marchant, *Inorg. Chem.*, **25**, 4170 (1986).

221. W.A. Herrmann, S. Bogdanovic, J. Behm, and M. Denk, *J. Organomet. Chem.*, **430**, C33 (1992).

222. S. Buth, S. Wocadlo, B. Neumüller, F. Weller, and K. Dehnicke, *Z. Naturforsch.*, **47b**, 706 (1992).

223. M.H. Chisholm, W.W. Reichert, F.A. Cotton, and C.A. Murillo, *J. Am. Chem. Soc.*, **99**, 1652 (1977); M.H. Chisholm, F.A. Cotton, C.A. Murillo and W.W. Reichert, *Inorg. Chem.*, **16**, 1801 (1977).

224. M.H. Chisholm, F.A. Cotton, M.W. Extine, and W.W. Reichert, *Inorg. Chem.*, **17**, 2944 (1978).

225. M.H. Chisholm, F.A. Cotton, M.W. Extine, and W.W. Reichert, *J. Am. Chem. Soc.*, **100**, 153 (1978).

226. M.H. Chisholm, F.A. Cotton, M.W. Extine, and W.W. Reichert, *J. Am. Chem. Soc.*, **100**, 1727 (1978).

227. M.H. Chisholm, R.L. Kelly, F.A. Cotton, and M.W. Extine, *J. Am. Chem. Soc.*, **100**, 2256 (1978); M.H. Chisholm, F.A. Cotton, M.W. Extine and R.L. Kelly, *J. Am. Chem. Soc.*, **101**, 7645 (1979).

228. M.H. Chisholm, F.A. Cotton, M.W. Extine, and R.L. Kelly, *J. Am. Chem. Soc.*, **100**, 3354 (1978).

229. M.H. Chisholm, J.C. Huffman, and R.L. Kelly, *Inorg. Chem.*, **19**, 2762 (1980).

230. M.H. Chisholm, C.C. Kirkpatrick, and J.C. Huffman, *Inorg. Chem.*, **20**, 871 (1981).

231. M.H. Chisholm, K. Folting, J.C. Huffman, C.C. Kirkpatrick, and A.L. Ratermann, *J. Am. Chem. Soc.*, **103**, 1305 (1981); M.H. Chisholm, K. Folting, J.C. Huffman and A.L. Ratermann, *Inorg. Chem.*, **21**, 978 (1982).

232. M.H. Chisholm, K. Folting, J.C. Huffman, and A.L. Ratermann, *J. Chem. Soc., Chem. Commun.*, 1229 (1981).

233. M.H. Chisholm, J.C. Huffman, and I.P. Rothwell, *J. Am. Chem Soc.*, **103**, 4245 (1981); M.H. Chisholm, K. Folting, J.C. Huffman, and I.P. Rothwell, *J. Am. Chem. Soc.*, **104**, 4389 (1982).

234. M.H. Chisholm, J.C. Huffman, J. Leonelli, and I.P. Rothwell, *J. Am. Chem Soc.*, **104**, 7030 (1982). See also F.A. Cotton and W. Schwotzer, *Inorg. Chem.*, **22**, 387 (1983).

235. M.H. Chisholm, K. Folting, J.C. Huffman, and I.P. Rothwell, *Organometallics*, **1**, 251 (1982).

236. M.H. Chisholm, J.C. Huffman, and N.S. Marchant, *J. Am. Chem Soc.*, **105**, 6162 (1983); M.H. Chisholm, J.C. Huffman, and N.S. Marchant, *Organometallics*, **6**, 1073 (1987).

237. M.H. Chisholm, J.C. Huffman, and C.C. Kirkpatrick, *Inorg. Chem.*, **22**, 1704 (1983).

238. M.H. Chisholm, K. Folting, J.C. Huffman, and A.L. Ratermann, *Inorg. Chem.*, **23**, 613 (1984).

239. M.H. Chisholm, J.F. Corning, K. Folting, J.C. Huffman, A.L. Ratermann, I.P. Rothwell, and W.E. Streib, *Inorg. Chem.*, **23**, 1037 (1984).

240. M.H. Chisholm, J.F. Corning, and J.C. Huffman, *Inorg. Chem.*, **23**, 754 (1984).

241. M.H. Chisholm, J.C. Huffman, A.L. Ratermann, and C.A. Smith, *Inorg, Chem.*, **23**, 1596 (1984).

242. M.H. Chisholm, K. Folting, J.C. Huffman, and A.L. Ratermann, *Inorg. Chem.*, **23**, 2303 (1984).

243. M.H. Chisholm, F.A. Cotton, K. Folting, J.C. Huffman, A.L. Ratermann, and E.S. Shamshoum, *Inorg. Chem.*, **23**, A423 (1984).

244. M.H. Chisholm, K. Folting, J.C. Huffman, and R.J. Tatz, *J. Am. Chem. Soc.*, **106**, 1153 (1984).

245. M.H. Chisholm, K. Folting, J.C. Huffman, E.F. Putilina, W.E. Streib, and R.J. Tatz, *Inorg. Chem.*, **32**, 3771 (1993).
246. M.H. Chisholm, J.C. Huffman, and R.J. Tatz, *J. Am. Chem. Soc.*, **106**, 5385 (1984).
247. M.H. Chisholm, K. Folting, J.C. Huffman, K.S. Kramer, and R.J. Tatz, *Organometallics*, **11**, 4029 (1992).
248. M.H. Chisholm, J.C. Huffman, and W.G. Van Der Sluys, *J. Am. Chem. Soc.*, **109**, 2514 (1987).
249. M.H. Chisholm, K. Folting, J.C. Huffman, and N.S. Marchant, *Organometallics*, **5**, 602 (1986); M.H. Chisholm, J.C. Huffman, and N.S. Marchant, *Polyhedron*, **7**, 919 (1988).
250. T.P. Blatchford, M.H. Chisholm, and J.C. Huffman, *Inorg. Chem.*, **27**, 2059 (1988).
251. M.H. Chisholm, K. Folting, J.C. Huffman, and J.J. Koh, *Polyhedron*, **8**, 123 (1989).
252. M.H. Chisholm, C.E. Hammond, and J.C. Huffman, *Polyhedron*, **8**, 129 (1989).
253. R.G. Abbott, F.A. Cotton, and L.R. Falvello, *Polyhedron*, **9**, 1821 (1990).
254. D.C. Bradley, M.B. Hursthouse, A.J. Howes, A.N. de M. Jelfs, J.D. Runnacles, and M. Thornton-Pett, *J. Chem. Soc., Dalton Trans.*, 841 (1991).
255. T.M. Gilbert, A.M. Landes, and R.D. Rogers, *Inorg. Chem.*, **31**, 3438 (1992).
256. J. Robbins, G.C. Bazan, J.S. Murdzek, M.B. O'Regan, and R.R. Schrock, *Organometallics*, **10**, 2902 (1991).
257. M.H. Chisholm, I.P. Parkin, W.E. Streib, and K.S. Folting, *Polyhedron*, **10**, 2309 (1991).
258. T.A. Budzichowski, S.T. Chacon, M.H. Chisholm, F.J. Feher, and W.E. Streib, *J. Am. Chem. Soc.*, **113**, 689 (1991).
259. T.A. Budzichowski, M.H. Chisholm, J.D. Martin, J.C. Huffman, K.G. Moodley, and W.E. Streib, *Polyhedron*, **12**, 343 (1993); T.A. Budzichowski, M.H. Chisholm, K. Folting, J.C. Huffman, and W.E. Streib, *J. Am. Chem. Soc.*, **117**, 7428 (1995).
260. T.A. Budzichowski, M.H. Chisholm, and K. Folting, *Chem. Eur. J.*, **2**, 110 (1996).
261. M.H. Chisholm, J.C. Huffman, and R.L. Kelly, *J. Am. Chem. Soc.*, **101**, 7100 (1979).
262. M.H. Chisholm, D.L. Clark, and J.C. Huffman, *Polyhedron*, **4**, 1203 (1985).
263. M.H. Chisholm, R.J. Errington, K. Folting, and J.C. Huffman, *J. Am. Chem. Soc.*, **104**, 2025 (1982).
264. M.H. Chisholm, D.L. Clark, R.J. Errington, K. Folting, and J.C. Huffman, *Inorg. Chem.*, **27**, 2071 (1988).
265. M.H. Chisholm, K. Folting, J.C. Huffman, J. Leonelli, N.S. Marchant, C.A. Smith, and L.C.E. Taylor, *J. Am. Chem. Soc.*, **107**, 3722 (1985).
266. M.H. Chisholm, C.E. Hammond, M. Hampden-Smith, J.C. Huffman, and W.G. Van Der Sluys, *Angew. Chem. Int. Ed. Engl.*, **26**, 904 (1987).
267. M.H. Chisholm, K. Folting, C.E. Hammond, and M.J. Hampden-Smith, *J. Am. Chem. Soc.*, **110**, 3314 (1988); M.H. Chisholm, K. Folting, C.E. Hammond, M.J. Hampden-Smith, and K.G. Moodley, *J. Am. Chem. Soc.*, **111**, 5300 (1989).
268. M.H. Chisholm, J.C. Huffman, K.S. Kramer, and W.E. Streib, *J. Am. Chem Soc.*, **115**, 9866 (1993).
269. T.A. Budzichowski, M.H. Chisholm, J.C. Huffman, and O. Eisenstein, *Angew. Chem. Int. Ed. Engl.*, **33**, 191 (1994).
270. N. Perchenek and A. Simon, *Zeit. Anorg. Allg. Chem.*, **619**, 103 (1993).
271. M.H. Chisholm, J.A. Heppert, and J.C. Huffman, *Polyhedron*, **3**, 475 (1984).
272. L.B. Handy, *Acta Crystallogr.*, **B31**, 300 (1975).
273. F.A. Cotton, W. Schwotzer, and E.S. Shamshoum, *Inorg. Chem.*, **23**, 4111 (1984).
274. H. Rothfuss, J.C. Huffman, and K.G. Caulton, *Inorg. Chem.*, **33**, 187 (1994).
275. M.H. Chisholm, K. Folting, and J.A. Klang, *Organometallics*, **9**, 607 (1990).
276. M.H. Chisholm, T.A. Budzichowski, F.J. Feher, and J.W. Ziller, *Polyhedron*, **11**, 1575 (1992).
277. A.J. Nielson and J.M. Waters, *Polyhedron*, **1**, 561 (1982); A.J. Nielson, J.M. Waters and D.C. Bradley, *Polyhedron*, **4**, 285 (1985).
278. D.C. Bradley, A.J. Howes, M.B. Hursthouse, and J.D. Runnacles, *Polyhedron*, **10**, 477 (1991).
279. J.T. Barry, M.H. Chisholm, K. Folting, J.C. Huffman, and W.E. Streib, *Polyhedron*, **16**, 2113 (1997).
280. J. Scherle and F.A. Schröder, *Acta Crystallogr.*, **B30**, 2772 (1974).

281. M.H. Chisholm, I.P. Parkin, W.E. Streib, and O. Eisenstein, *Inorg. Chem.*, **33**, 812 (1994).
282. A.A. Danopoulos, C. Redshaw, A. Vaniche, G. Wilkinson, B. Hussain-Bates, and M.B. Hursthouse, *Polyhedron*, **12**, 1061 (1993).
283. A. Lehtonen and R. Sillanpaa, *Polyhedron*, **14**, 455 (1995).
284. A. Lehtonen and R. Sillanpaa, *Polyhedron*, **14**, 1831 (1995).
285. A. Lehtonen and R. Sillanpaa, *J. Chem. Soc., Dalton Trans.*, 2701 (1995).
286. A. Lehtonen and R. Sillanpaa, *Acta Crystallogr.*, **C51**, 1270 (1995).
287. D.M.-T. Chan, W.C. Fultz, W.A. Nugent, D.C. Roe, and T.H. Tulip, *J. Am. Chem. Soc.*, **107**, 251 (1985).
288. M.H. Chisholm, F.A. Cotton, M.W. Extine, and R.L. Kelly, *Inorg. Chem.*, **18**, 116 (1979).
289. M.H. Chisholm, J.C. Huffman, and J.W. Pasterczyk, *Polyhedron*, **6**, 1551 (1987).
290. M.H. Chisholm, J.C. Huffman, and J.W. Pasterczyk, *Inorg. Chem. Acta*, **133**, 17 (1987).
291. F.A. Cotton, W. Schwotzer, and E.S. Shamshoum, *Organometallics*, **3**, 1770 (1984).
292. J.H. Freudenberger, R.R. Schrock, M.R. Churchill, A.L. Rheingold, and J.W. Ziller, *Organometallics*, **3**, 1563 (1984).
293. M.R. Churchill and J.W. Ziller, *J. Organomet. Chem.*, **286**, 27 (1985).
294. K. Akiyama, M.H. Chisholm, F.A. Cotton, M.W. Extine, D.A. Haitko, D. Little, and P. Fanwick, *Inorg. Chem.*, **18**, 2266 (1979).
295. M.H. Chisholm, B.W. Eichhorn, K. Folting, J.C. Huffman, and R.J. Tatz, *Organometallics*, **5**, 1599 (1986).
296. M.H. Chisholm, D.L. Clark, K. Folting, and J.C. Huffman, *Angew. Chem. Int. Ed. Engl.*, **25**, 1014 (1986); M.H. Chisholm, D.L. Clark, K. Folting, J.C. Huffman, and M. Hampden-Smith, *J. Am. Chem. Soc.*, **109**, 7750 (1987).
297. H.W. Roesky, N. Bertel, F. Edelmann, R. Herbst, E. Egert, and G.M. Sheldrick, *Z. Naturforsch.*, **41b**, 1506 (1986).
298. M.H. Chisholm, K. Folting, M. Hampden-Smith, and C.A. Smith, *Polyhedron*, **6**, 1747 (1987).
299. M.H. Chisholm, J.C. Huffman, I.P. Parkin, and W.E. Streib, *Polyhedron*, **9**, 2941 (1990).
300. M.H. Chisholm, I.P. Parkin, J.C. Huffman, E.M. Lobkovsky, and K. Folting, *Polyhedron*, **10**, 2839 (1991).
301. I.P. Parkin and K. Folting, *J. Chem. Soc., Dalton Trans.*, 2343 (1992).
302. M.H. Chisholm, K.S. Kramer, J.D. Martin, J.C. Huffman, E.B. Lobovsky, and W.E. Streib, *Inorg. Chem.*, **31**, 4469 (1992).
303. M.H. Chisholm, E.F. Putilina, K. Folting, and W.E. Streib, *J. Clust. Sci.*, **5**, 67 (1994).
304. T.A. Budzichowski, M.H. Chisholm, D.B. Tiedke, J.C. Huffman, and W.E. Streib, *Organometallics*, **14**, 2318 (1995).
305. M.H. Chisholm, K. Folting, J.C. Huffman, and A.L. Ratermann, *Inorg. Chem.*, **23**, 2303 (1984).
306. J.C. Bryan, D.R. Wheeler, D.L. Clark, J.C. Huffman, and A.P. Sattelberger, *J. Am. Chem. Soc.*, **113**, 3184 (1991).
307. F.A. Cotton, D. DeMarco, B.W.S. Kolthammer, and R.A. Walton, *Inorg. Chem.*, **20**, 3048 (1981).
308. L.B. Anderson, F.A. Cotton, D. DeMarco, A. Fang, W.H. Ilsley, B.W.S. Kolthammer, and R.A. Walton, *J. Am. Chem. Soc.*, **103**, 5078 (1981).
309. F.A. Cotton, L.R. Falvello, M.F. Friedrich, D. DeMarco, and R.A. Walton, *J. Am. Chem. Soc.*, **105**, 3088 (1983); F.A. Cotton, D. DeMarco, L.R. Falvello, and R.A. Walton, *J. Am. Chem. Soc.*, **104**, 7375 (1982); L.B. Anderson, F.A. Cotton, D. DeMarco, L.R. Falvello, S.M. Tetrick, and R.A. Walton, *J. Am. Chem. Soc.*, **106**, 4743 (1984).
310. M.H. Chisholm, J.C. Huffman, and A.L. Ratermann, *Inorg. Chem.*, **22**, 4100 (1983).
311. M.H. Chisholm, J.C. Huffman, and N.S. Marchant, *J. Am. Chem. Soc.*, **105**, 6162 (1983).
312. C. Redshaw, G. Wilkinson, B. Hussain-Bates, and M.B. Hursthouse, *J. Chem. Soc., Dalton Trans.*, 555 (1992).
313. M.H. Chisholm, J.C. Huffman, and C.A. Smith, *J. Am. Chem. Soc.*, **108**, 222 (1986).
314. S.T. Chacon, M.H. Chisholm, K. Folting, M. Hampden-Smith, and J.C. Huffman, *Inorg. Chem.*, **30**, 3122 (1991).
315. J.T. Barry, S.T. Chacon, M.H. Chisholm, J.C. Huffman, and W.E. Streib, *J. Am. Chem. Soc.*, **117**, 1974 (1995).

316. M.H. Chisholm, D.L. Clark, D. Ho, and J.C. Huffman, *Organometallics*, **6**, 1532 (1987).
317. M.H. Chisholm, V.J. Johnston, and W.E. Streib, *Inorg. Chem.*, **31**, 4081 (1992).
318. P.A. Bates, A.J. Nielson, and J.M. Waters, *Polyhedron*, **6**, 163 (1987).
319. W. Clegg, R.J. Errington, and C. Redshaw, *J. Chem. Soc., Dalton Trans.*, 3189 (1992).
320. P.A. Bates, A.J. Nielson, and J.M. Waters, *Polyhedron*, **4**, 999 (1985).
321. T.A. Budzichowski, M.H. Chisholm, D.B. Tiedke, N.E. Gruhn, and D.L. Lichtenberger, *Polyhedron*, **17**, 705 (1998).
322. T.A. Budzichowski, M.H. Chisholm, K. Folting, J.C. Huffman, W.E. Streib, and D.B. Tiedke, *Polyhedron*, **17**, 857 (1998).
323. M.H. Chisholm, D.M. Hoffman, and J.C. Huffman, *Inorg. Chem.*, **22**, 2903 (1983).
324. M.J. Chetcuti, M.H. Chisholm, J.C. Huffman, and J. Leonelli, *J. Am. Chem. Soc.*, **105**, 292 (1983).
325. M.H. Chisholm, D.M. Hoffman, and J.C. Huffman, *J. Am. Chem. Soc.*, **106**, 6815 (1984).
326. F.A. Cotton and E.S. Shamshoum, *J. Am. Chem. Soc.*, **107**, 4662 (1985).
327. F.A. Cotton, W. Schwotzer, and E.S. Shamshoum, *Organometallics*, **3**, 461 (1985).
328. F.A. Cotton and E.S. Shamshoum, *Polyhedron*, **4**, 1727 (1985).
329. R.G. Abbott, F.A. Cotton, and E.S. Shamshoum, *Gazz. Chim. Ital.*, **116**, 91 (1986).
330. T.A. Budzichowski, M.H. Chisholm, K. Folting, M.G. Fromhold, and W.E. Streib, *Inorg. Chim. Acta*, **235**, 339 (1995).
331. M.H. Chisholm, E.A. Lucas, A.C. Sousa, J.C. Huffman, K. Folting, E.B. Lobkovsky, and W.E. Streib, *J. Chem. Soc., Chem. Commun.*, 847 (1991).
332. M.H. Chisholm, I.P. Parkin, K. Folting, E.B. Lobkovsky, and W.E. Streib, *J. Chem. Soc., Chem. Commun.*, 1673 (1991); M.H. Chisholm, I.P. Parkin, K. Folting, and E. Lobkovsky, *Inorg. Chem.*, **36**, 1636 (1997).
333. M.H. Chisholm, D.M. Hoffman, and J.C. Huffman, *Organometallics*, **4**, 986 (1985).
334. F.A. Cotton and W. Schwotzer, *J. Am. Chem. Soc.*, **105**, 5639 (1983).
335. M.H. Chisholm, D. Ho, J.C. Huffman, and N.S. Marchant, *Organometallics*, **8**, 1626 (1989). See also, M.H. Chisholm, J.C. Huffman, and N.S. Marchant, *J. Chem. Soc., Chem. Commun.*, 717 (1986).
336. F.A. Cotton, W. Schwotzer, and E.S. Shamshoum, *Organometallics*, **2**, 1167 (1983).
337. M.H. Chisholm, K. Folting, D.M. Hoffman, J.C. Huffman, and J. Leonelli, *J. Chem. Soc., Chem. Commun.*, 589 (1983).
338. M.H. Chisholm, K. Folting, D.M. Hoffman, and J.C. Huffman, *J. Am. Chem. Soc.*, **106**, 6794 (1984).
339. M.H. Chisholm, D.M. Hoffman, and J.C. Huffman, *J. Am. Chem. Soc.*, **106**, 6806 (1984).
340. M.H. Chisholm, J.C. Huffman, and J.A. Heppert, *J. Am. Chem. Soc.*, **107**, 5116 (1985).
341. M.H. Chisholm, J.A. Heppert, J.C. Huffman, and W.E. Streib, *J. Chem. Soc., Chem. Commun.*, 1771 (1985).
342. M.H. Chisholm, B.W. Eichhorn, and J.C. Huffman, *J. Chem. Soc., Chem. Commun.*, 861 (1985).
343. M.H. Chisholm, B.K. Conroy, J.C. Huffman, and N.S. Marchant, *Angew. Chem. Int. Ed. Engl.*, **25**, 446 (1986).
344. S.T. Chacon, M.H. Chisholm, K. Folting, J.C. Huffman, and M. Hampden-Smith, *Organometallics*, **10**, 3722 (1991).
345. M.H. Chisholm, B.K. Conroy, K. Folting, D.M. Hoffman, and J.C. Huffman, *Organometallics*, **5**, 2457 (1986).
346. R.H. Cayton, S.T. Chacon, M.H. Chisholm, and J.C. Huffman, *Angew. Chem. Int. Ed. Engl.*, **29**, 1026 (1990).
347. M.H. Chisholm, J.C. Huffman, and M. Hampden-Smith, *J. Am. Chem. Soc.*, **111**, 5284 (1989).
348. M.H. Chisholm, B.W. Eichhorn, and J.C. Huffman, *Organometallics*, **6**, 2264 (1987).
349. M.H. Chisholm, B.W. Eichhorn, K. Folting, and J.C. Huffman, *Organometallics*, **8**, 49 (1989).
350. M.H. Chisholm, B.W. Eichhorn, and J.C. Huffman, *Organometallics*, **8**, 67 (1989).
351. M.H. Chisholm and B.W. Eichhorn, *Organometallics*, **8**, 80 (1989).
352. M.H. Chisholm, J.C. Huffman, E.A. Lucas, and E.B. Lubkovsky, *Organometallics*, **10**, 3424 (1991).

353. S.T. Chacon, M.H. Chisholm, O. Eisenstein, and J.C. Huffman, *J. Am. Chem. Soc.*, **144**, 8497 (1992).

354. K.G. Caulton, R.H. Cayton, M.H. Chisholm, J.C. Huffman, E.B. Lobkovsky, and Z. Xue, *Organometallics*, **11**, 321 (1992); T.M. Gilbert and R.D. Rogers, *Acta Crystallogr.*, **C49**, 677 (1993).

355. M.H. Chisholm, C.M. Cook, J.C. Huffman, and J.D. Martin, *Organometallics*, **12**, 2354 (1993).

356. S.T. Chacon, M.H. Chisholm, C.M. Cook, M.J. Hampden-Smith, and W.E. Streib, *Angew. Chem. Int. Ed. Engl.*, **31**, 462 (1992); M.H. Chisholm, C.M. Cook, J.C. Huffman, and W.E. Streib, *Organometallics*, **12**, 2677 (1993).

357. M.H. Chisholm, S.T. Haubrich, J.D. Martin, and W.E. Streib, *J. Chem. Soc., Chem. Commun.*, 683 (1994); *J. Am. Chem. Soc.*, **119**, 1634 (1997).

358. M.H. Chisholm, D.M. Hoffman, J. McC. Northius, and J.C. Huffman, *Polyhedron*, **16**, 839 (1997).

361. T.A. Budzichowski, M.H. Chisholm, J.D. Martin, J.C. Huffman, K.G. Moodley, and W.E. Streib, *Polyhedron*, **12**, 343 (1993).

362. R.J. Blau, M.H. Chisholm, B.W. Eichhorn, J.C. Huffman, K.S. Kramer, E.B. Lobovsky, and W.E. Streib, *Organometallics*, **14**, 1855 (1995).

363. M.H. Chisholm, K. Folting, K.S. Kramer, and W.E. Streib, *J. Am. Chem. Soc.*, **119**, 5528 (1997).

364. M.H. Chisholm, D.M. Hoffman, and J.C. Huffman, *Inorg. Chem.*, **23**, 3683 (1984); M.H. Chisholm, K. Folting, J.A. Heppert, D.M. Hoffman, and J.C. Huffman, *J. Am. Chem. Soc.*, **107**, 1234 (1985).

365. M.H. Chisholm, K. Folting, and J.W. Pasterczyk, *Inorg. Chem.*, **27**, 3057 (1988).

366. M.H. Chisholm, D.M. Hoffman, and J.C. Huffman, *Inorg. Chem.*, **24**, 796 (1985).

367. M.H. Chisholm, K. Folting, J.C. Huffman, and J.A. Klang, *Organometallics*, **7**, 1033 (1988).

368. M.H. Chisholm, K. Folting-Streib, D.B. Tiedtke, F. Lemoigno, and O. Eisenstein, *Angew. Chem. Int. Ed. Engl.*, **34**, 110 (1995).

369. M.H. Chisholm, K. Folting, J.C. Huffman, J.A. Klang, and W.E. Streib, *Organometallics*, **8**, 89 (1989).

370. M.H. Chisholm, J.C. Huffman, and J. Leonelli, *J. Chem. Soc., Chem. Commun.*, 270 (1981).

371. M. Akiyama, D. Little, M.H. Chisholm, D.A. Haitko, F.A. Cotton, and M.W. Extine, *J. Am. Chem. Soc.*, **101**, 2504 (1979); M. Akiyama, M.H. Chisholm, F.A. Cotton, M.W. Extine, D.A. Haitko, J. Leonelli, and D. Little, *J. Am. Chem. Soc.*, **103**, 779 (1981).

372. F.A. Cotton and W. Schwotzer, *J. Am. Chem. Soc.*, **105**, 4955 (1983).

373. M.H. Chisholm, D.L. Clark, J.C. Huffman, and C.A. Smith, *Organometallics*, **6**, 1280 (1987). First reported in ref. 265.

374. M.H. Chisholm, K. Folting, M.J. Hampden-Smith, and C.E. Hammond, *J. Am. Chem. Soc.*, **111**, 7283 (1989); M.H. Chisholm, K. Folting, V.J. Johnston, and C.E. Hammond, *J. Organomet. Chem.*, **394**, 265 (1990).

375. M.H. Chisholm, V. Johnston, J.D. Martin, E. Lobkovsky, and J.C. Huffman, *J. Clust. Sci.*, **4**, 105 (1993).

376. T.P. Pollagi, S.J. Geib, and M.D. Hopkins, *J. Am. Chem. Soc.*, **116**, 6051 (1994).

377. R.A. Bartlett, J.J. Ellison, P.P. Power, and S.C. Shoner, *Inorg. Chem.*, **30**, 2888 (1991).

378. X. Li, M.S. Lah, and V.L. Pecoraro, *Acta Crystallogr.*, **C45**, 1517 (1989).

379. I.A. Fallis, L.J. Farrugia, N.M. Macdonald, and R.D. Peacock, *J. Chem. Soc., Dalton Trans.*, 2759 (1993).

380. L.E. Pence, A. Caneschi, and S.J. Lippard, *Inorg. Chem.*, **35**, 3069 (1996).

381. T.J. Barder, F.A. Cotton, D. Lewis, W. Schwotzer, S.M. Tetrick, and R.A. Walton, *J. Am. Chem. Soc.*, **106**, 2882 (1984).

382. R. Toreki, R.R. Schrock, and M.G. Vale, *J. Am. Chem. Soc.*, **113**, 3610 (1991).

383. R. Toreki, R.R. Schrock, and W.M. Davis, *J. Am. Chem. Soc.*, **114**, 3367 (1992).

384. R. Toreki, G.A. Vaughan, R.R. Schrock, and W.M. Davis, *J. Am. Chem. Soc.*, **115**, 127 (1993).

385. A.C.C. Wong, G. Wilkinson, B. Hussain, M. Motevalli, and M.B. Hursthouse, *Polyhedron*, **7**, 1363 (1988).
386. D.M. Hoffman, D. Lappas, and D.A. Wierda, *J. Am. Chem. Soc.*, **111**, 1531 (1989).
387. D.M. Hoffman, D. Lappas, and D.A. Wierda, *J. Am. Chem. Soc.*, **115**, 10538 (1993).
388. W.-K. Cheng, K.-Y. Wong, W.-F. Tong, T.-F. Lai, and C.-M. Che, *J. Chem. Soc., Dalton Trans.*, 91 (1992).
389. S. Buth, K. Harms, P. König, S. Wocadlo, F. Weller, and K. Dehnicke, *Z. Anorg. Allg. Chem.*, **619**, 853 (1993).
390. E.W. Radoslovic, M. Raupach, P.G. Slade, and R.M. Taylor, *Aust. J. Chem.*, **23**, 1963 (1970); P.G. Slade, E.W. Radoslovic, and M. Raupach, *Acta Crystallogr.*, **B27**, 2432 (1971).
391. E. Konefal, S.J. Loeb, D.W. Stephan, and C.J. Willis, *Inorg. Chem.*, **23**, 538 (1984).
392. R.T. Boere, W.M. Brown, D.W. Stephan, and C.J. Willis, *Inorg. Chem.*, **24**, 593 (1985).
393. M.M. Olmstead, P.P. Power, and G. Sigel, *Inorg. Chem.*, **25**, 1027 (1986).
394. G.A. Sigel, R.A. Bartlett, M.M. Olmstead, and P.P. Power, *Inorg. Chem.*, **26**, 1773 (1987).
395. E.K. Brechin, S.G. Harris, S. Parsons, and R.E.P. Winpenny, *Chem. Commun.*, 1439 (1996).
396. J.K. Beattie, T.W. Hambley, J.A. Klepetko, A.F. Masters, and P. Turner, *Chem. Commun.*, 45 (1998).
397. D. Milstein, J.C. Calabrese, and I.D. Williams, *J. Am. Chem. Soc.*, **108**, 6387 (1986).
398. D.M. Lunder, E.B. Lobkovsky, W.E. Streib, and K.G. Caulton, *J. Am. Chem. Soc.*, **113**, 1837 (1991).
399. S.E. Kegley, C.J. Schaverien, J.H. Freudenberger, R.G. Bergman, S.P. Nolan, and C.D. Hoff, *J. Am. Chem. Soc.*, **109**, 6563 (1987).
400. J.E. Andrew and A.B. Blake, *J. Chem. Soc. A*, 1456 (1969).
401. J.W.L. Martin, N.C. Payne, and C.J. Willis, *Inorg. Chem.*, **17**, 3478 (1978).
402. D.M. Barnhart and E.C. Lingafelter, *Cryst. Struct. Comm.*, **11**, 733 (1982).
403. Y.A. Simonov, A.A. Dvorkin, G.S. Matuzenko, M.A. Yampolskaya, T.S. Gifeisman, N.V. Gerbeleyu and T.I. Malinovski, *Koord. Khim.*, **10**, 1247 (1984).
404. S.J. Loeb, D.W. Stephan, and C.J. Willis, *Inorg. Chem.*, **23**, 1509 (1984).
405. R.T. Boere, C.D. Montgomery, N.C. Payne, and C.J. Willis, *Inorg. Chem.*, **24**, 3680 (1985).
406. P. Bradford, R.C. Hynes, N.C. Payne, and C.J. Willis, *J. Am. Chem. Soc.*, **112**, 2647 (1990).
407. J.J. Vittal and C.J. Willis, *Can. J. Chem.*, **71**, 1051 (1993).
408. M.S. El Fallah, E. Rentschler, A. Caneschi, and D. Gatteschi, *Inorg. Chim. Acta*, **247**, 23 (1996).
409. C.D. Montgomery, N.C. Payne, and C.J. Willis, *Inorg. Chim. Acta*, **117**, 103 (1986).
410. R.T. Boere, D.E. Esser, C.J. Willis, D.W. Stephan, and T.W. Obal, *Can. J. Chem.*, **65**, 798 (1987).
411. H.E. Bryndza, J.C. Calabrese, M. Marsi, D.C. Roe, W. Tam, and J.W. Bercaw, *J. Am. Chem. Soc.*, **108**, 4805 (1986).
412. R.T. Boere, N.C. Payne, and C.J. Willis, *Can. J. Chem.*, **64**, 1474 (1986).
413. C.D. Montgomery, N.C. Payne, and C.J. Willis, *Inorg. Chem.*, **26**, 519 (1987).
414. R.C. Hynes, C.J. Willis, and N.C. Payne, *Acta Crystallogr.*, **C48**, 42 (1992).
415. N.W. Alcock, A.W.G. Platt, and P.G. Pringle, *J. Chem. Soc., Dalton Trans.*, 2273 (1987).
416. N.W. Alcock, A.W.G. Platt, and P.G. Pringle, *J. Chem. Soc., Dalton Trans.*, 139 (1989).
417. K. Osakada, Y.-J. Kim, and A. Yamamoto, *J. Organomet. Chem.*, **382**, 303 (1990).
418. K. Osakado, Y.-J. Kim, M. Tanaka, S. Ishiguro, and A. Yamamoto, *Inorg. Chem.*, **30**, 197 (1991).
419. M.A. Andrews, E.J. Voss, G.L. Gould, W. Klooster, and T.F. Koetzle, *J. Am. Chem Soc.*, **116**, 5730 (1994).
420. T. Greiser and E. Weiss, *Chem. Ber.*, **109**, 3142 (1976).
421. T.H. Lemmen, G.V. Goeden, J.C. Huffman, R.L. Geerts, and K.G. Caulton, *Inorg. Chem.*, **29**, 3680 (1990).

422. M.J. McGeary, R.C. Wedlich, P.S. Coan, K. Folting, and K.G. Caulton, *Polyhedron*, **11**, 2459 (1992).
423. K.W. Terry, C.G. Lugmair, P.K. Gantzel, and T.D. Tilley, *Chem. Mater.*, **8**, 274 (1996).
424. J.H. Timmons, J.W.L. Martin, A.E. Martell, P. Rudolf, A. Clearfield, S.J. Loeb, and C.J. Willis, *Inorg. Chem.*, **20**, 181 (1981).
425. D.L. Barber, S.J. Loeb, J.W.L. Martin, N.C. Payne, and C.J. Willis, *Inorg. Chem.*, **20**, 272 (1981).
426. S.J. Loeb, J.F. Richardson, and C.J. Willis, *Inorg. Chem.*, **22**, 2736 (1983).
427. R.C. Hynes, N.C. Payne, and C.J. Willis, *J. Chem. Soc., Chem. Commun.*, 744 (1990).
428. R.C. Hynes, C.J. Willis, and N.C. Payne, *Acta Crystallogr.*, **C52**, 2173 (1996).
429. H. Muhonen, *Acta Crystallogr.*, **B37**, 951 (1981).
430. S.C. Goel, K.S. Kramer, M.Y. Chiang, and W.E. Buhro, *Polyhedron*, **9**, 611 (1990).
431. A.P. Purdy, C.F. George, and G.A. Brewer, *Inorg. Chem.*, **31**, 2633 (1992).
432. P.M. Jeffries, S.R. Wilson, and G.S. Girolami, *Inorg. Chem.*, **31**, 4503 (1992).
433. C. Sirio, O. Poncelet, H.G. Hubert-Pfalzgraf, J.C. Daran, and J. Vaissermann, *Polyhedron*, **11**, 177 (1992).
434. J. Manzur, A.M. Garcia, M.T. Garland, V. Acuna, O. Gonzalez, O. Pena, A.M. Atria, and E. Spodine, *Polyhedron*, **15**, 821 (1996).
435. S. Wang, *Acta Crystallogr.*, **C52**, 41 (1996).
436. L.P. Wu, T. Kuroda-Sowa, M. Maekawa, Y. Suenaga, and M. Munakata, *J. Chem. Soc., Dalton Trans.*, 2179 (1996).
437. H.M.M. Shearer and C.B. Spenser, *Chem. Commun.*, 194 (1966).
438. W.A. Hermann, S. Bogdanovic, J. Behm, and M. Denk, *J. Organomet. Chem.*, **430**, C33 (1992).
439. M.L. Ziegler and J. Weiss, *Angew. Chem. Int. Ed. Engl.*, **9**, 905 (1970).
440. S.C. Goel, M.Y. Chiang, and W.E. Buhro, *Inorg. Chem.*, **29**, 4646 (1990).
441. M.R.P. van Vliet, G. van Koten, P. Buysingh, J.T.B.H. Jastrzebski, and A.L. Spek, *Organometallics*, **6**, 537 (1987); J.T.B.H. Jastrzebski, J. Boersma, G. van Koten, W.J.J. Smeets, and A.L. Spek, *Rec. Trav. Chem. Pays-Bas*, **107**, 263 (1988).
442. K. Su, T.D. Tilley and M.J. Sailor, *J. Am. Chem. Soc.*, **118**, 3459 (1996).
443. S. Boulmaaz, R. Papiernik, L.G. Hubert-Pfalzgraf, J. Vaissermann, and J-C. Daran, *Polyhedron*, **11**, 1331 (1992).
444. W.A. Herrmann, N.W. Huber and T. Priermeier, *Angew. Chem. Int. Ed. Engl.*, **33**, 105 (1994).
445. M. Veith, S. Mathur, and C. Mathur, *Polyhedron*, **17**, 1005 (1998).
446. K.B. Renkema, R.J. Matthews, T.L. Bush, S.K. Hendges, R.N. Redding, F.W. Vance, M.E. Silver, S.A. Snow, and J.C. Huffman, *Inorg. Chim. Acta*, **244**, 185 (1996).
447. M. Motevalli, D. Shah, and A.C. Sullivan, *J. Chem. Soc., Chem. Commun.*, 2427 (1994).
448. W.J. Evans, T.J. Boyle, and J.W. Ziller, *Organometalllics*, **12**, 3998 (1993).
449. W.J. Evans, T.J. Boyle, and J.W. Ziller, *J. Organomet. Chem.*, **462**, 141 (1993).
450. W.J. Evans, J.L. Shreeve, R.N.R. Broomhall-Dillard, and J.W. Ziller, *J. Organomet. Chem.*, **50**, 7 (1995).
451. W.J. Evans, R.N.R. Broomhall-Dillard, and J.W. Ziller, *Organometallics*, **15**, 1351 (1996).
452. F.T. Edelmann, A. Steiner, J.W. Gilje, S. Jagner, and M. Håkansson, *Polyhedron*, **13**, 539 (1994).
453. M.J. Hampden-Smith, D.S. Williams, and A.L. Rheingold, *Inorg. Chem.*, **29**, 4076 (1990).
454. T.V. Lubben, P.T. Wolczanski, and G.D. Van Duyne, *Organometallics*, **3**, 977 (1984).
455. I. Abrahams, C. Simon, M. Motevalli, S.A.A. Shah, and A.C. Sullivan, *J. Organomet. Chem.*, **521**, 301 (1996).
456. A.I. Yanovskii, E.P. Turevskaya, N.Ya. Turova, and Yu.T. Struchkov, *Koord. Khim.*, **11**, 110 (1985).
457. D.J. Eichorst, D.A. Payne, S.R. Wilson, and K.E. Howard, *Inorg. Chem.*, **29**, 1458 (1990).
458. S.C. Goel, J.A. Hollingsworth, A.M. Beatty, K.D. Robinson, and W.E. Buhro, *Polyhedron*, **17**, 781 (1998).
459. M. Lazell, M. Motevalli, S.A.A. Shah, and A.C. Sullivan, *J. Chem. Soc., Dalton Trans.*, 3363 (1997).

460. B.D. Murray and P.P. Power, *J. Am. Chem. Soc.*, **106**, 7011 (1984).
461. M.B. Hursthouse, M.A. Mazid, M. Motevalli, M. Sanganee, and A.C. Sullivan, *J. Organomet. Chem.*, **381**, C43 (1990).
462. L.G. Hubert-Pfalzgraf, V.G. Kessler, and J. Vaissermann, *Polyhedron*, **16**, 4197 (1997).
463. M.B. Hursthouse, M. Motevalli, M. Sanganee, and A.C. Sullivan, *J. Chem. Soc., Chem. Commun.*, 1709 (1991).
464. M. Klaassen and P. Klüfers, *Z. Anorg. Allg. Chem.*, **619**, 661 (1993).
465. A.P. Purdy and C.F. George, *Polyhedron*, **14**, 761 (1995).
466. M.B. Power, S.G. Bott, J.L. Atwood, and A.R. Barron, *J. Am. Chem. Soc.*, **112**, 3446 (1990).
467. W.J. Evans, T.J. Boyle, and J.W. Ziller, *Polyhedron*, **11**, 1093 (1992).
468. M. Veith and R. Rösler, *Z. Naturforsch.*, **41b**, 1071 (1986).
469. I. Abrahams, M. Motevalli, S.A.A. Shah, and A.C. Sullivan, *J. Organomet. Chem.*, **492**, 99 (1995).
470. W.J. Evans, M.S. Sollberger, and J.W. Ziller, *J. Am. Chem. Soc.*, **115**, 4120 (1993).
471. F. Laurent, J.C. Huffman, K. Folting, and K.G. Caulton, *Inorg. Chem.*, **34**, 3980 (1995).
472. W.J. Evans, T.J. Deming, J.M. Olofson, and J.W. Ziller, *Inorg. Chem.*, **28**, 4027 (1989).
473. T.J. Boyle, D.C. Bradley, M.J. Hampden-Smith, A. Patel, and J.W. Ziller, *Inorg. Chem.*, **34**, 5893 (1995).
474. S.W. Kirtley, J.P. Chantan, R.A. Love, D.L. Tipton, T.N. Sorrell, and R. Bau, *J. Am. Chem. Soc.*, **102**, 3451 (1980).
475. J.J.H. Edema, S. Gambarotta, W.J.J. Smeets, and A.L. Spek, *Inorg. Chem.*, **30**, 1380 (1991).
476. M. Motevalli, M. Sanganee, P.D. Savage, S.A.A. Shah, and A.C. Sullivan, *J. Chem. Soc., Chem. Commun.*, 1132 (1993).
477. W.A. Herrmann, R. Alberto, J.C. Bryan, and A.P. Sattelberger, *Chem. Ber.*, **124**, 1107 (1991).
478. A.P. Purdy, C.F. George, and J.H. Callahan, *Inorg. Chem.*, **30**, 2812 (1991).
479. A.P. Purdy and C.F. George, *Polyhedron*, **13**, 709 (1994).
480. J.A. Meese-Marktscheffel, R. Weiman, H. Schumann, and J.W. Gilje, *Inorg. Chem.*, **32**, 5894 (1993).
481. M.J. McGeary, R.H. Cayton, K. Folting, J.C. Huffman, and K.G. Caulton, *Polyhedron*, **11**, 1369 (1992).
482. M. Veith and M. Reimers, *Chem. Ber.*, **123**, 1941 (1990).
483. M. Veith, E.-C. Yu, and V. Huch, *Chem. Eur. J.*, **1**, 26 (1995).
484. K.W. Terry, K. Su, T.D. Tilley, and A.L. Rheingold, *Polyhedron*, **17**, 891 (1998).
485. J.L. Atwood and G.D. Stucky, *J. Organomet. Chem.*, **13**, 53 (1968).
486. L. Malpezzi, U. Zucchini, and T. Dall'Occo, *Inorg. Chim. Acta*, **180**, 245 (1991).
487. M. Veith, J. Hans, L. Stahl, P. May, V. Huch, and A. Sebald, *Z. Naturforsch.*, **46b**, 403 (1991).
488. U. Bemm, K. Lashgari, R. Norrestam, M. Nygren, and G. Westin, *J. Solid State Chem.*, **103**, 366 (1993).
489. J. Sassmannshausen, R. Riedel, K.B. Pflanz, and H. Chmiel, *Z. Naturforsch.*, **48b**, 7 (1993).
490. E.P. Turevskaya, V.G. Kessler, N. Ya. Turova, A.P. Pisarevsky, A.I. Yanovsky, and Y.T. Struchkov, *J. Chem. Soc., Chem. Commun.*, 2303 (1994).
491. I. Baxter, S.R. Drake, M.B. Hursthouse, K.M.A. Malik, D.M.P. Mingos, J.C. Plakatouras, and D.J. Otway, *Polyhedron*, **17**, 625 (1998).
492. F. Labrize, L.G. Hubert-Pfalzgraf, J.-C. Daran, and S. Halut, *J. Chem. Soc., Chem. Commun.*, 1556 (1993); F. Labrize, L.G. Hubert-Pfalzgraf, J-C. Daran, S. Halut and P. Tobaly, *Polyhedron*, **15**, 2707 (1996).
493. M. Veith, S. Mathur, and V. Huch, *Inorg. Chem.*, **36**, 2391 (1997).
494. B.A. Vaarstra, J.C. Huffman, W.E. Streib, and K.G. Caulton, *J. Chem. Soc., Chem. Commun.*, 1750 (1990); *Inorg. Chem.*, **30**, 3068 (1991). See also E.P. Turevskaya,

D.V. Berdyev, N.Ya. Turova, and I.M. Yanovskaya, *Russ. J. Inorg. Chem.*, **40**, 1527 (1995).

495. M. Veith, S. Mathur, and V. Huch, *J. Am. Chem. Soc.*, **118**, 903 (1996); *Inorg. Chem.*, **35**, 7295 (1996).

496. S. Boulmaaz, R. Papiernik, L.G. Hubert-Pfalzgraf, and J.-C. Daran, *Eur. J. Solid State Inorg. Chem.*, **30**, 583 (1993).

497. E.P. Turevskaya, N.Ya. Turova, A.V. Korolev, A.I. Yanovsky, and Yu.T. Struchkov, *Polyhedron*, **14**, 1531 (1995).

498. A.P. Purdy and C.F. George, *Inorg. Chem.*, **30**, 1969 (1991).

499. N.N. Sauer, E. Garcia, K.V. Salazar, R.R. Ryan, and J.A. Martin, *J. Am. Chem. Soc.*, **112**, 1524 (1990).

500. W. Bidell, V. Shklover, and H. Berke, *Inorg. Chem.*, **31**, 5561 (1992).

501. C.P. Love, C.C. Torardi, and C.J. Page, *Inorg. Chem.*, **31**, 1784 (1992).

502. F. Labrize, L.G. Hubert-Pfalzgraf, J. Vaissermann, and C.B. Knobler, *Polyhedron*, **15**, 577 (1996).

503. M.C. Cruickshank and L.S. Dent Glasser, *Acta Crystallogr.*, **C41**, 1014 (1985).

504. W. Bidell, J. Döring, H.W. Bosch, H.-U. Hund, E. Plappert, and H. Berke, *Inorg. Chem.*, **32**, 502 (1993).

505. W.J. Evans, T.J. Boyle, and J.W. Ziller, *J. Am. Chem. Soc.*, **115**, 5084 (1993).

506. V.G. Kessler, L.G. Hubert-Pfalzgraf, S. Halut and J.-C. Daran, *J. Chem. Soc., Chem. Commun.*, 705 (1994).

507. U.M. Tripathi, A. Singh, R.C. Mehrotra, S.C. Goel, M.Y. Chiang, and W.E. Buhro, *J. Chem. Soc., Chem. Commun.*, 152 (1992).

508. L.M. Slaughter and P.T. Wolczanski, *Chem. Commun.*, 2109 (1997).

509. M. Veith, S. Mathur, and V. Huch, *Chem. Commun.*, 2197 (1997).

510. R. Papiernik, L.G. Hubert-Pfalzgraf, M. Veith, and V. Huch, *Chem. Ber./Recueil*, **130**, 1361 (1997).

511. R. Schmid, A. Mosset, and J. Galy, *Acta Crystallogr.*, **C47**, 750 (1991).

512. B.A. Vaarstra, J.A. Samuels, E.H. Barash, J.D. Martin, W.E. Streib, C. Gasser, and K.G. Caulton, *J. Organomet. Chem.*, **449**, 191 (1993).

513. S. Sogani, A. Singh, R. Bohra, R.C. Mehrotra, and M. Nottemeyer, *J. Chem. Soc., Chem. Commun.*, 738 (1991).

514. D.J. Teff, J.C. Huffman, and K.G. Caulton, *Inorg. Chem.*, **35**, 2981 (1996).

515. J.A. Samuels, J.W. Zwanziger, E.B. Lobkovsky, and K.G. Caulton, *Inorg. Chem.*, **31**, 4046 (1992).

516. E.P. Turevskaya, D.V. Berdyev, N.Ya. Turova, Z.A. Starikova, A.I. Yanovsky, Yu.T. Struchkov, and A.I. Belokon, *Polyhedron*, **16**, 663 (1997).

517. S. Boulmaaz, R. Papiernik, and L.G. Hubert-Pfalzgraf, *Chem. Mater.*, **3**, 779 (1991).

518. M.H. Chisholm, J.C. Huffman, and J.J. Koh, *Polyhedron*, **8**, 127 (1989).

519. S.L. Latesky and J.P. Selegue, *J. Am. Chem. Soc.*, **109**, 4731 (1987).

520. M.H. Chisholm, V.J. Johnston, O. Eisenstein, and W.E. Streib, *Angew. Chem. Int. Ed. Engl.*, **31**, 896 (1992).

521. R.J. Blau, M.H. Chisholm, K. Folting, and R.J. Wang, *J. Chem. Soc., Chem. Commun.*, 1582 (1985); *J. Am. Chem. Soc.*, **109**, 4552 (1987).

522. M. Veith, C. Mathur, S. Mathur, and V. Huch, *Organometallics*, **16**, 1292 (1997).

523. M. Veith, D. Käfer, J. Koch, P. May, L. Stahl, and V. Huch, *Chem. Ber.*, **125**, 1033 (1992).

524. M. Veith, S. Weidner, K. Kunze, D. Käfer, J. Hans, and V. Huch, *Coord. Chem. Rev.*, **137**, 297 (1994).

525. M. Veith, S. Mathur, and V. Huch, *J. Chem. Soc., Dalton Trans.*, 2485 (1996).

526. O.Yu. Vassilyeva, L.A. Kovbasyuk, V.N. Kokozay, B.W. Skelton, and W. Linert, *Polyhedron*, **17**, 85 (1998).

527. L.F. Francis, D.A. Payne, and S.R. Wilson, *Chem. Mater.*, **2**, 645 (1990).

528. M. Veith and R. Rösler, *Angew. Chem. Int. Ed. Engl.*, **21**, 858 (1982).

529. M. Grenz, E. Hahn, W.-W. du Mont, and J. Pickardt, *Angew. Chem. Int. Ed. Engl.*, **23**, 61 (1984).
530. M. Veith, C. Mathur, and V. Huch, *Organometallics*, **15**, 2858 (1996).
531. M. Veith and J. Hans, *Angew. Chem. Int. Ed. Engl.*, **30**, 878 (1991).
532. T.A. Wark, E.A. Gulliver, M.J. Hampden-Smith, and A.L. Rheingold, *Inorg. Chem.*, **29**, 4362 (1990).
533. W.-W. Zhuang, S.G. Bott, A. Schmitz, and D.M. Hoffman, *Polyhedron*, **17**, 879 (1998).

5

Metal Oxo-alkoxides

1 INTRODUCTION

Metal oxo-alkoxides $[MO_x(OR)_{(y-2x)}]_z$ may be regarded as intermediates between the oligomeric metal alkoxides $[M(OR)_y]_n$ $(x = 0)$ and the three-dimensional macromolecular metal oxides $[MO_{y/2}]$ $(x = y/2;\ y = $ oxidation state of M). Indeed most metal alkoxides are readily converted by hydrolysis to the corresponding metal oxide (Eq. 5.1).

$$M(OR)_y + \frac{y}{2}H_2O \longrightarrow MO_{y/2} + yROH \qquad (5.1)$$

For many years this moisture sensitivity of metal alkoxides caused problems in the synthesis and characterization of pure metal alkoxides. Special techniques were developed for the elimination of water from reactants, solvents, and apparatus, thereby enabling not only the synthesis of metal alkoxides but also their characterization by infrared spectra, NMR spectra, mass spectra, and single-crystal X-ray diffraction analysis.

Apart from the pioneering researches of Bradley and co-workers,[1] who studied the controlled hydrolysis of the alkoxides of some early transition metals, there was little interest in the metal oxo-alkoxides until the sol–gel technique was developed for the conversion of metal alkoxide precursors into metal oxides and heterometal oxides (Chapter 7).[2,3] A concise review of the chemistry of metal oxo-alkoxides published by Mehrotra and Singh[4] reveals the rapid growth of interest in this field. The most notable feature has been the dramatic increase in the number of crystal structure determinations in recent years. However, our knowledge of the mechanism of formation of metal oxo-alkoxides leaves much to be desired.

2 SYNTHETIC METHODS

Although the hydrolysis of a metal alkoxide is the obvious route to the formation of a metal oxo-alkoxide other methods have been identified and these will also be mentioned.

2.1 Hydrolysis

It is presumed that the first stage in the hydrolysis of a metal alkoxide is the formation of a hydroxo derivative which then condenses to form the oxo ligand (Eqs 5.2 and 5.3).

$$M(OR)_y + H_2O \longrightarrow M(OH)(OR)_{y-1} + ROH \qquad (5.2)$$

$$M(OH)(OR)_{y-1} + M(OR)_y \longrightarrow M_2O(OR)_{2y-2} + ROH \qquad (5.3)$$

Studies on the hydrolysis of titanium tetra-alkoxides revealed that the reactions represented by both Eqs (5.2) and (5.3) are extremely rapid, and no hydroxo compounds were isolated.[5] Similar results were found with the hydrolysis of the alkoxides of zirconium,[6] tantalum(v),[7] tin(iv),[8] and uranium(v).[8] In these ebulliometric studies of the hydrolysis of metal alkoxides in their progenitive alcohols it was found that the degrees of polymerization of metal oxo-alkoxides formed were surprisingly low for a given degree of hydrolysis, and structures were proposed based on certain models. However, just as titanium tetraethoxide, which was essentially trimeric in solution, proved to be a tetramer in the crystalline state so the oxo-ethoxide $[Ti_6O_4(OEt)_{16}]$ proposed in solution was shown to be $[Ti_7O_4(OEt)_{20}]$ by X-ray crystallography, thus emphasizing the difficulty of assigning structures to species in solution. Klemperer and co-workers have used ^{17}O enriched water to great effect in studying the hydrolysis of titanium alkoxides.[9] They were able to assign chemical shifts to the oxo-ligand in different environments (e.g. μ-O, 650–850; μ_3-O, 450–650; μ_4-O, 250–450 ppm) by reference to the known structures $[Ti_7O_4(OEt)_{20}]$, $[Ti_8O_6(OCH_2Ph)_{20}]$, and $[Ti_{10}O_8(OEt)_{24}]$. The ^{17}O NMR spectrum of partially hydrolysed $Ti(OEt)_4$ showed the presence of $[Ti_7O_4(OEt)_{20}]$ and $[Ti_8O_6(OEt)_{20}]$ but not $[Ti_{10}O_8(OEt)_{24}]$. Using magic angle spinning (MAS) ^{17}O NMR they demonstrated the presence of mainly μ_3-O with some μ_4-O in the xerogels formed by hydrolysis of $Ti(OEt)_4$ and the formation of first anatase (δ_{17_O}, 561 ppm) and then rutile (δ_{17_O}, 591 ppm) (TiO_2) on firing the xerogel above 600°C. Although $[Ti_3(\mu_3\text{-}O)(\mu_3\text{-}OPr^i)(\mu\text{-}OPr^i)_3(OPr^i)_6]$ was insufficiently stable to be isolated it was identified by ^{17}O NMR in the initial stages of hydrolysis of $Ti(OPr^i)_4$ in isopropanol with the undecanuclear species $[Ti_{11}O_{13}(OPr^i)_{18}]$ being formed as the more stable species.[10] The latter compound and the dodecanuclear species $[Ti_{12}O_{16}(OPr^i)_{16}]$, which occurs in two isomeric forms, were isolated and characterized structurally.[11] The importance of steric factors was demonstrated by the fact that the mixed ligand trinuclear compound $[Ti_3(\mu_3\text{-}O)(\mu_3\text{-}OMe)(\mu\text{-}OPr^i)_3(OPr^i)_6]$, which was identified in solution by ^{17}O NMR, was sufficiently stable to be crystallized from solution and its X-ray structure was determined.[10] The value of ^{17}O and ^{13}C NMR in identifying complex polyoxo-alkoxides in solution was illustrated by studies on the hydrolysis of $Ti(OBu^t)_4$ in Bu^tOH. The hydrolysis, which was relatively slow at room temperature, gave initially a species proposed to be $[Ti_3O(OBu^t)_{10}]$ but prolonged heating at 100°C gave rise to the giant molecule $[Ti_{18}O_{27}(OH)(OBu^t)_{17}]$ whose ^{17}O NMR spectrum was in accord with the requirements of the X-ray crystal structure.[12] By using ^{17}O MAS NMR it was shown that the sol–gel produced directly by hydrolysis of $Ti(OBu^t)_4$ was significantly different in constitution from the product obtained by hydrolysis of $[Ti_{18}O_{27}(OH)(OBu^t)_{17}]$, a result which clearly has some bearing on the mechanism of the sol–gel process. The structures of the known titanium oxo-alkoxides and those of zirconium, niobium, tantalum, tin, and aluminium will be discussed in Section 5.

As an alternative to adding water to a metal alkoxide it has been shown that acetone will undergo aldol-type condensation in the presence of a metal alkoxide leading to the formation of an oxo-alkoxide. For example the zinc alkoxide $[Zn(OCEt_3)_2]$ was converted to a gel[65] whilst titanium tetraisopropoxide gave a trinuclear species

$[Ti_3(\mu_3\text{-}O)(\mu\text{-}OPr^i)_4\{Me_2C(O)CHC(O)CH_2C(O)Me_2\}]^{66}$ in which the tridentate ligand formed by condensation of acetone spans the three Ti atoms.

2.2 Formation of Oxo-alkoxides by Nonhydrolytic Reactions of Alkoxides

Metal oxo-alkoxides unintentionally obtained during attempts to prepare metal alkoxides have generally resulted from the presence of adventitious water leading to hydrolysis, but it appears that the formation of some oxo-alkoxides is due to the instability of the sought after alkoxide.

An early example was the failure to obtain penta-tertiary alkoxides of niobium owing to the preferential formation of the oxo compounds $Nb_2O(OR)_8$ and $NbO(OR)_3$ in contrast to tantalum which formed thermally stable $Ta(OBu^t)_5$. This behaviour was explained in terms of the greater tendency of niobium to form the metal-oxo double bond.[13] In the reactions of $W(NMe_2)_6$ with ROH the hexa-alkoxides $W(OR)_6$ (R = Me, Et, Pr^i, allyl) were formed but not the tertiary butoxide which gave $WO(OBu^t)_4$ instead.[14] The tendency of $MoO(OR)_4$ compounds to form $MoO_2(OR)_2$ by ether elimination has also been noted (Eq. 5.4).[15]

$$L_xM(OR)_2 \longrightarrow L_xM{=}O + R_2O \tag{5.4}$$

The formation of multiple metal–oxo bonds by penta or hexa valent transition metals (d^0) is not surprising but the more recent discovery that scandium, yttrium, and the tervalent lanthanides form oxo-isopropoxides rather than triisopropoxides was unexpected.[16] Considerable efforts have been devoted to ensuring that the formation of these pentanuclear oxo-alkoxides $[M_5(\mu_5\text{-}O)(\mu_3\text{-}OPr^i)_4(\mu\text{-}OPr^i)_4(OPr^i)_5]$ was not due to hydrolysis, and there is evidence that elimination of ether is involved. An alternative mode of decomposition would be alkene elimination with concomitant formation of alcohol (Eq. 5.5).

$$L_xM\{OCH(CH_3)_2\}_2 \longrightarrow L_xM{=}O + CH_2{=}CHMe + Me_2CHOH \tag{5.5}$$

This process would be expected to be facilitated by tertiary alkoxo groups but it is significant that the tris-tertiary alkoxides of these metals are readily prepared and are thermally stable. It is also noteworthy that whereas scandium forms pentanuclear oxo-alkoxides $Sc_5O(OR)_{13}$ with ethoxo and isopropoxo ligands, the trifluorethoxo and hexafluoroisopropoxo ligands form stable tris-alkoxides.[17] The thermal desolvation of $[Ce_2(OPr^i)_8(Pr^iOH)_2]$ gave rise to the oxo-species $[Ce_4O(OPr^i)_{14}]$ by elimination of diisopropyl ether.[67]

Heterometallic oxo-alkoxides may also be formed by conversion from heterometallic alkoxides. Thus the insoluble $[Pb(OPr^i)_2]_x$ reacted with $Ti(OPr^i)_4$ to form $[Pb_4Ti_4O_3(OPr^i)_{18}]$ and $[Pb_2Ti_2(\mu_4\text{-}O)(\mu_3\text{-}OPr^i)_2(\mu\text{-}OPr^i)_4(OPr^i)_4]$ presumably by ether elimination.[18] Similarly, from reactions of barium alkoxides and titanium alkoxides the heterometallic oxo-alkoxides $\{[Ba_4Ti_4O_4(OPr^i)_{16}(Pr^iOH)_4][Ba_4Ti_4O_4(OPr^i)_{10}(Pr^iOH)_3]\}^{19}$ and $[Ba_2Ti_4O(OMe)_{18}(MeOH)_7]^{20}$ were obtained as a result of ether elimination, although similar complexes are readily prepared when the barium alkoxide solution is aerobically oxidized.[21]

Refluxing a toluene solution of $Zr(OBu^t)_4$ and $Pb_3(OBu^t)_6$ for 24 hours led to the formation of $[Pb_3ZrO(OBu^t)_8]$ with concomitant formation of alkene and *tert*-butanol (Eq. 5.6).[22]

$$Zr(OBu^t)_4 + Pb_3(OBu^t)_6 \longrightarrow Pb_3ZrO(OBu^t)_8 + Me_2C{=}CH_2 + Bu^tOH \qquad (5.6)$$

It was also shown that $Pb_3(OBu^t)_6$ is stable in refluxing toluene but is converted by catalytic amounts of Bu^tOH, $LiNMe_2$ or $HN(SiMe_3)_2$ into the oxo-alkoxide $[Pb_4O(OBu^t)_6]$ which is readily formed by stoichiometric hydrolysis of lead *tert*-butoxide. The oxo-alkoxide $[Pb_4O(OBu^t)_6]$ reacted readily with $Zr(OBu^t)_4$ to form $[Pb_3ZrO(OBu^t)_8]$ whereas the heterometallic alkoxide $[PbZr(OBu^t)_6]$ did not react with $[Pb_3(OBu^t)_6]$. The mechanism of oxo-formation in these reactions has been discussed.[22]

The reaction of $Zn(OPr^i)_2$ with $Ta(OPr^i)_5$ gave the oxo-alkoxide $[ZnTa_2O_2(OPr^i)_8]$[23] whilst reactions of MCl_2 (M = Ni, Mn) with $[NaSb(OEt)_4]$ gave $[Ni_5Sb_3O_2(OEt)_{15}(EtOH)_4]$[24] and $[Mn_8Sb_4O_4(OEt)_{20}]$.[25]

Another mode of decomposition of the metal-bound *tert*-butoxo ligand is the formation of a hydrido oxo-*tert*-butoxo complex by alkene elimination from the anion $[W_2(\mu\text{-}OBu^t)(OBu^t)_7]^-$ (Eq. 5.7).[26]

$$[W_2(\mu\text{-}OBu^t)(OBu^t)_6]^- \longrightarrow [W_2(\mu\text{-}O)(\mu\text{-}H)(OBu^t)_6]^- + Me_2C{=}CH_2 \qquad (5.7)$$

In the reaction of $MoCl_5$ with ethanol it was shown that the presence of the oxo ligands in the product $[Cl_2(O)Mo(\mu\text{-}OEt)_2(\mu\text{-}HOEt)Mo(O)Cl_2]$ was due to elimination of ethyl chloride.[27] However, hydrolysis by water produced by reaction of ethanol with HCl cannot be ruled out (Eq. 5.8).

$$2MoCl_5 + 5EtOH \longrightarrow [Mo_2(\mu\text{-}OEt)_2(\mu\text{-}HOEt)(O)_2Cl_4] + 4HCl + 2EtCl \qquad (5.8)$$

Another process leading to oxo-alkoxide formation is ester elimination involving reaction between a metal alkoxo group and a metal carboxylate group (Eq. 5.9).

$$MOR + MO_2CR' \longrightarrow MOM + R'CO_2R \qquad (5.9)$$

Thus the reaction of $Ti(OBu^n)_4$ (1 mol) with acetic acid (2 mol) did not give $Ti(OBu^n)_2(OAc)_2$ but instead produced $Ti_6O_4(OBu^n)_8(OAc)_8$ (Eq. 5.10).[28]

$$6Ti(OBu^n)_4 + 12HOAc \longrightarrow Ti_6O_4(OBu^n)_8(OAc)_8 + 4BuOAc + 12BuOH \qquad (5.10)$$

The reaction of $Sn(OBu^t)_4$ with $Sn(OAc)_4$ in refluxing toluene led to the formation of $[Sn_6O_6(OBu^t)_6(OAc)_6]$ with elimination of *tert*-butylacetate (Eq. 5.11).

$$3Sn(OBu^t)_4 + 3Sn(OAc)_4 \longrightarrow Sn_6O_6(OBu^t)_6(OAc)_6 + 6Bu^tOAc \qquad (5.11)$$

However, in the presence of pyridine the ester elimination was prevented and ligand exchange occurred instead (Eq. 5.12).[29]

$$3Sn(OBu^t)_4 + 3Sn(OAc)_4$$
$$\longrightarrow 2Sn(OBu^t)_3(OAc) + 2Sn(OBu^t)_2(OAc)_2 + 2Sn(OBu^t)(OAc)_3 \qquad (5.12)$$

Similarly the reaction of $Sn(OBu^t)_4$ with Me_3SiOAc caused ester exchange (Eq. 5.13):

$$Sn(OBu^t)_4 + Me_3SiOAc \longrightarrow Sn(OBu^t)_3(OSiMe_3) + Bu^tOAc \qquad (5.13)$$

In the presence of pyridine the reaction took a different course (Eq. 5.14):

$$Sn(OBu^t)_4 + Me_3SiOAc + py \longrightarrow Sn(OBu^t)_3(OAc)py + Me_3SiOBu^t \qquad (5.14)$$

It was suggested that in these reactions the pyridine coordinates to the tin atom thus competing with the carbonyl oxygen of the acetate ligand and preventing ester elimination.[30]

Ester elimination has been employed successfully in the synthesis of heterometallic oxo-alkoxides (Eq. 5.15).[31,32]

$$Ca(OAc)_2 + 2Al(OPr^i)_3 \longrightarrow CaAl_2O_2(OPr^i)_4 + 2Pr^iOAc \qquad (5.15)$$

Even termetallic species may be synthesized using the appropriate sequence of reactions (Eqs 5.16 and 5.17).[33]

$$Ca(OAc)_2 + Al(OPr^i)_3 \longrightarrow CaAlO(OPr^i)_2(OAc) + Pr^iOAc \qquad (5.16)$$

$$CaAlO(OPr^i)_2(OAc) + Ti(OPr^i)_4 \longrightarrow CaAlTiO_2(OPr^i)_5 + Pr^iOAc \qquad (5.17)$$

Another method of introducing the oxo ligand is by direct oxygenation with O_2 or by oxygen abstraction from an oxygen-containing substrate.

Chisholm and co-workers have shown that reactions involving $[Mo_2(OR)_6]$ or $[Mo(OR)_4]_x$ molecules and O_2 are affected by the steric effect of the R group and afford a plethora of novel molybdenum oxo-alkoxides (Eq. 5.18).[34]

$$Mo_2(OBu^t)_6 + 2O_2 \longrightarrow 2MoO_2(OBu^t)_2 + 2Bu^tO \qquad (5.18)$$

No intermediates were found in this reaction and the butoxo radicals abstract hydrogen from solvent molecules. Reaction of $[Mo_2(OPr^i)_6]$ with O_2 proceeds through the intermediates $[Mo_3O(OPr^i)_{10}]$ and $[Mo_6O_{10}(OPr^i)_{12}]$ before finally producing the unstable $MoO_2(OPr^i)_2$ which was stabilized by coordination to 2,2'-bipyridyl $[MoO_2(OPr^i)_2(bpy)]$.

Oxygenation in the presence of pyridine gave $[MoO(OPr^i)_4(py)]$, $[Mo_4O_8(OPr^i)_4(py)_4]$, and $[MoO_2(OPr^i)_2(py)_2]$. With the neopentyloxide $[Mo_2(OCH_2Bu^t)_6]$ the main product was $[MoO_2(OCH_2Bu^t)_2]$, isolated as the adduct with bpy, and a small percentage of $[MoO(OCH_2Bu^t)_4]$ with $[Mo_3O(OCH_2Bu^t)_{10}]$ as a transient intermediate. With the mononuclear Mo(IV) alkoxide $Mo(OBu^t)_4$ concentrated solutions gave rise to $[MoO(OBu^t)_4]$ whereas dilute solutions produced $[MoO_2(OBu^t)_2]$. With the binuclear compound $[Mo_2(OPr^i)_8]$ oxygenation gave $[MoO(OPr^i)_4]$. In pyridine, which forms $[Mo(OPr^i)_4(py)_2]$, the product was $[MoO_2(OPr^i)_2(py)_2]$. The tetrakis-neopentyloxide gave rise to $[MoO(OCH_2Bu^t)_4]$. Some of these oxo-alkoxides were also obtained in high yields using alkoxide/oxo-alkoxide reactions (Eqs 5.19–5.21):

$$Mo_2(OPr^i)_6 + MoO(OPr^i)_4 \longrightarrow Mo_3O(OPr^i)_{10} \qquad (5.19)$$

$$2Mo_3O(OPr^i)_{10} + 16MoO_2(OPr^i)_2 \longrightarrow 3Mo_6O_{10}(OPr^i)_{12} + 4MoO(OPr^i)_4 \qquad (5.20)$$

$$Mo_2(OPr^i)_8 + 6MoO_2(OPr^i)_2 \longrightarrow Mo_6O_6(OPr^i)_{12} + 2MoO(OPr^i)_4 \qquad (5.21)$$

The structures of several of these molybdenum oxo-alkoxides have been determined and are discussed in Section 5.

Tungsten oxo-alkoxides have been obtained from the reaction of ketones with the dinuclear $[W_2(OR)_6(py)_2]$ compounds (Eq. 5.22).[35]

$$2W_2(OPr^i)_6(py)_2 + 2Me_2CO \longrightarrow W_4O_2(OPr^i)_{12} + C_2Me_4 + 4py \qquad (5.22)$$

With the neopentyloxide a stable adduct is formed containing oxo and μ-propylidene ligands by reductive cleavage of the ketonic carbonyl group (Eq. 5.23).[36]

$$W_2(OCH_2Bu^t)_6(py)_2 + Me_2CO \longrightarrow W_2(\mu\text{-}CMe_2)(\mu\text{-}OCH_2Bu^t)(O)(OCH_2Bu^t)_5(py)$$
$$(5.23)$$

This adduct reacted further with acetone to produce tetramethylethylene and the oxo-complex $[W_2O_2(OCH_2Bu^t)_6]$. An analogous adduct formed from $[W_2(OCH_2Bu^t)_6(py)_2]$ and cyclopropane carboxaldehyde (C_3H_5CHO) retained the cyclopropane ring in the bridging cyclopropylmethylidene complex $[(Bu^tCH_2O)_3W(\mu\text{-}CHC_3H_5)(\mu\text{-}OCH_2Bu^t)_2W(O)(OCH_2Bu^t)(py)]$ in which the oxo ligand is bound terminally to one tungsten.[37]

The tungsten trialkoxides $[W(OR)_3]_n$ $(n = 2$ or $4)$ undergo various reactions with carbon monoxide including cleavage of CO to form carbido- and oxo-alkoxides. The tetranuclear complex $[W_4(\mu_4\text{-}C)(\mu\text{-}O)(OR)_{12}]$ $(R = Pr^i$ or $CH_2Bu^t)$ were thereby isolated and characterized.[38]

The Ta(III) siloxide $[Ta(OSiBu_3^t)_3]$ also abstracted an oxo ligand from CO (Eq. 5.24).[39]

$$4Ta(OSiBu_3^t)_3 + 2CO \longrightarrow 2[(O)Ta(OSiBu_3^t)_3] + [(Bu_3^tSiO)_3TaC_2Ta(OSiBu_3^t)_3]$$
$$(5.24)$$

Oxygen transfer from nitrosobenzene to $[W_2(OBu^t)_6]$ gave rise to the μ-oxo complex $[W_2(\mu\text{-}O)(\mu\text{-}OBu^t)_2(NPh)_2(OBu^t)_4]$[40] whilst the siloxo complex $[W_2(OSiMe_2Bu^t)_6]$ reacted with nitric oxide forming the oxo-complexes $trans$-$[W(O)(OSiMe_2Bu^t)_4(py)]$ and $[W_2(\mu\text{-}O)_2(OSiMe_2Bu^t)_4(O)_2(py)_2]$.[41]

Finally it is noteworthy that the tetranuclear mixed valency cerium oxo-isopropoxide $[Ce_4(\mu_4\text{-}O)(\mu_3\text{-}OPr^i)_2(\mu\text{-}OPr^i)_4(OPr^i)_7(Pr^iOH)]$ was obtained by photolysis of $[Ce_2(\mu\text{-}OPr^i)_2(OPr^i)_6(Pr^iOH)_2]$.[42]

2.3   Reactions of Metal Oxochlorides with Alkali Alkoxides

An alternative approach to the synthesis of metal oxo-alkoxides involves the reaction of the metal oxo-chloride with the requisite amount of sodium alkoxide (Eq. 5.25):

$$MO_xCl_y + yNaOR \longrightarrow MO_x(OR)_y + yNaCl \qquad (5.25)$$

An early example of this method was its use in the preparation of uranyl dialkoxides $[UO_2(OR)_2(ROH)_x]$ (Eq. 5.26):[43]

$$UO_2Cl_2 + 2LiOMe + MeOH \longrightarrow UO_2(OMe)_2, MeOH + 2LiCl \qquad (5.26)$$

Lithium methoxide (Eq. 5.26) was used instead of sodium methoxide because uranyl methoxide which was insoluble in methanol could thus be separated from lithium chloride which was soluble. The ethoxide $[UO_2(OEt)_2(EtOH)_2]$ was also prepared using lithium ethoxide but the isopropoxide $[UO_2(OPr^i)_2(Pr^iOH)]$, being appreciably soluble in isopropanol, was prepared using sodium isopropoxide. The triphenyl phosphine oxide adduct of uranyl *tert*-butoxide $[UO_2(OBu^t)_2(Ph_3PO)_2]$ was prepared similarly from $[UO_2Cl_2(Ph_3PO)_2]$ and $KOBu^t$.[44] The vanadium(v) oxo-trialkoxides $[VO(OR)_3]_n$ were readily prepared from $VOCl_3$,[45-47] but the reaction of $[Li_2O_2C_2(CF_3)_4]$ with CrO_2Cl_2 led to reduction to Cr(v) in the form of $[LiCr(O)_2\{O_2C_2(CF_3)_4\}]$.[48] Several W(vi) oxo-tetra-alkoxides have been prepared from $WOCl_4$ by reactions with NaOR or alcohol and ammonia.[49]

2.4 Preparation of Metal Oxo-alkoxides from Metal Oxides

An alternative approach to the synthesis of metal oxo-alkoxides is the reaction of the metal oxide with an alcohol (Eq. 5.26).

$$MO_x + 2ROH \longrightarrow MO_{x-1}(OR)_2 + H_2O \qquad (5.26)$$

This is in effect a reversal of the hydrolysis of a metal alkoxide (Eqs 5.2 and 5.3), although there is little evidence for the reversibility of the hydrolysis process. Nevertheless, the reaction of some organometallic oxo-compounds with alcohols to form alkoxo derivatives is well documented (*e.g.* Eq. 5.27).[1]

$$R_3GeOGeR_3 + 2R'OH \longrightarrow 2R_3GeOR' + H_2O \qquad (5.27)$$

Several oxo-alkoxides of vanadium and molybdenum have been prepared from reactions of the metal oxide or oxometallate anion with alcohols. Thus the reaction of MoO_3 with ethyleneglycol gave $[MoO_2(OC_2H_4OH)_2]$[50] and with 2,2'-oxodiethanol gave $[MoO_2(OC_2H_4OC_2H_4O)]$.[51] Refluxing $MoO_3.2H_2O$ with methanol in the presence of molecular sieve (4A) gave mainly $[Mo_2O_5(OMe)_2]$ with some $[Na_4\{Mo_8O_{24}(OMe)_4\}].8MeOH$.[52]

The vanadium(v) oxo-trialkoxides, $[VO(OR)_3]$ (R = Et, Pr^n, Pr^i, Bu^s, Bu^t, C_2H_4Cl, C_2H_4F, CH_2CCl_3) were prepared by refluxing finely divided V_2O_5 with alcohol and benzene and removing the water produced (Eq. 5.28) by azeotropic distillation.[46]

$$V_2O_5 + 6ROH \longrightarrow 2VO(OR)_3 + 3H_2O \qquad (5.28)$$

Zubieta *et al.* have obtained the interesting polyoxovanadate anions involving V(IV): $[V_{10}O_{16}\{EtC(CH_2O)_3\}_4]^{4-}$, $[V_{10}O_{13}\{EtC(CH_2O)_3\}_5]^-$,[53] $[V_{10}O_{14}(OH)_2\{(OCH_2)_3CCH_2OH\}_4]^{2-}$,[54] $[Ba\{V_6O_7(OH_3)\}\{(OCH_2)_3CMe\}_3].3H_2O$ and $[Na_2\{V_6O_7\}\{(OCH_2)_3CEt\}_4]$[55] by hydrothermal reactions of a mixture of vanadium oxides [V(III) and V(v)]. Some mixed valency [V(IV), V(v)] species $[V_{10}O_{16}\{(OCH_2)_3CR\}_4]^{2-}$ (R = Et, Me)[54] and $(Me_3NH)[V_6O_7(OH)_3\{(OCH_2)_3CMe\}_3]$[55] were similarly obtained. Other syntheses have utilized the solubility of quaternary ammonium salts of polyvanadates to obtain the polyoxo-alkoxo vanadate ions $[V_6O_{13}\{(OCH_2)_3CR\}_2]^{3-}$ (R = Me,

OH, NO_2) (Eq. 5.29):[56]

$$2H_3V_{10}O_{28}^{3-} + 6RC(CH_2OH)_3 \longrightarrow 3V_6O_{13}\{(OCH_2)_3CR\}^{2-} + 12H_2O + V_2O_5$$

$$(5.29)$$

The methoxo polyvanadate $[V_6O_{12}(OMe)_7]^-$ has been prepared by a similar method.[57]

Using organic-soluble tetrabutyl ammonium salts of $[Mo_2O_7]^{2-}$ and $[Mo_8O_{26}]^{4-}$ the mixed alkoxo polyoxomolybdate species $[Mo_8O_{20}(OMe)_4\{(OCH_2)_3CMe\}_2]^{2-}$[58] and $[Mo_3O_6(OMe)\{(OCH_2)_3CMe\}_2]^-$[59] were obtained. The interesting heteropolyalkoxo metallate anion $[V_2Mo_2O_8(OMe)_2\{(OCH_2)_3C(CH_2OH)\}_2]^{2-}$ was prepared by the reaction of $[V_2O_2Cl_2\{(OCH_2)_2C(CH_2OH)_2\}_2]$ with $[Mo_2O_7]^{2-}$.[60] The oxo vanadatrane $[OV(OC_2H_4)_3N]$ was prepared by treating ammonium metavanadate with triethanolamine.[61]

3   CHEMICAL REACTIVITY OF OXO-ALKOXIDES

The further hydrolysis of metal oxo-alkoxides to produce metal oxides is of considerable importance in the sol−gel process.

Klemperer and co-workers[9-12] have shown that with increasing oxo content titanium oxo-alkoxides become increasingly resistant to hydrolysis in alcoholic solutions, although gel formation occurs readily by hydrolysis in nonalcoholic solvents.[11]

Alcohol exchange of metal oxo-alkoxides has also been studied. Thus the molybdenum oxo-alkoxides $[MoO_2(OR)_2]$ ($R = Pr^i$, CH_2Bu^t) were obtained from the tertiary butoxo compound (Eq. 5.30).[34]

$$MoO_2(OBu^t)_2 + 2ROH \longrightarrow MoO_2(OR)_2 + 2Bu^tOH \qquad (5.30)$$

Of considerable interest is the selective alkoxo exchange shown by the more hydrolysed titanium oxo-alkoxides. The two isomeric forms of $[Ti_{12}O_{16}(OPr^i)_{16}]$ both reacted with ethanol (Eq. 5.31) to form isomers of $[Ti_{12}O_{16}(OEt)_6(OPr^i)_{10}]$ by exchange of terminal isopropoxo groups bonded to the 6 five-coordinated Ti atoms in each molecule.[11]

$$Ti_{12}O_{16}(OPr^i)_{16} + 6EtOH \longrightarrow Ti_{12}O_{16}(OEt)_6(OPr^i)_{10} + 6Pr^iOH \qquad (5.31)$$

Similarly $[Ti_{11}O_{13}(OPr^i)_{18}]$ was converted to $[Ti_{11}O_{13}(OEt)_5(OPr^i)_{13}]$[11] and the giant oxo-tert-butoxide $[Ti_{18}O_{28}H(OBu^t)_{17}]$ also exchanged the 5 terminal Bu^tO groups bonded to five-coordinated titanium with tertiary amyl alcohol.[12] These reactions demonstrated the robust nature of the metal−oxo core structure in these compounds.

Alcohol exchange of the pentanuclear oxo-alkoxides $[M_5O(OPr^i)_{13}]$ (M = Gd, Pr) led to reorganization of the M_5O_{14} framework and formation of hexanuclear $[Gd_6(\mu_4\text{-}O)(OC_2H_4OMe)_{16}]$[62] and octanuclear $[Pr_8(\mu_4\text{-}O)_4(OC_2H_4OMe)_{16}(Me_3PO)_2]$[63] species. Some surprising results were obtained in alcohol exchange reactions involving the sparingly soluble $[UO_2(OMe)_2.(MeOH)]$.[43] With isopropanol a disproportionation occurred with formation of insoluble $[U_2O_5(OPr^i)_2(Pr^iOH)_2]$ and soluble $[UO(OPr^i)_4(Pr^iOH)]$. The latter compound disproportionated on heating in vacuo producing the volatile hexa-alkoxide $U(OPr^i)_6$. Similar results were obtained using tert-butanol and the hexa-alkoxide $U(OBu^t)_6$ was remarkably air stable. Other work[64] showed that in the reaction of UO_2Cl_2 with $KOBu^t$ in THF the soluble trinuclear complex

$[(Bu^tO)_4U(\mu\text{-}O)(\mu\text{-}OBu^t)U(O)_2(\mu\text{-}O)(\mu\text{-}OBu^t)U(OBu^t)_4]$ was formed in 30% yield with the remainder being present as an insoluble compound. It thus appears that in nonaqueous solutions the UO_2^{2+} moiety is susceptible to reaction with certain alkoxo ligands.

4 PHYSICAL PROPERTIES

In general the physical properties of the metal oxo-alkoxides $[M_xO_y(OR)_z]$ depend on the degree of hydrolysis (y/z) and the nature of the alkyl group R. For a given R-group, solubility and volatility tend to decrease with increase in the degree of hydrolysis and increase in oligomerization (x). Lengthening of the alkyl chain or chain branching enhances solubility. With low degrees of hydrolysis there is a tendency for dispro-portionation to occur on heating owing to the greater volatility of the parent metal alkoxide.

For example in the hydrolysis of titanium ethoxide it was found that precipitation occurred at $y/x \sim 1.75$ giving a solid having the composition $[Ti_2O_3(OEt)_2]$. More-over, heating $[TiO(OEt)_2]_n$ in vacuo until no more $Ti(OEt)_4$ was recovered led to a nonvolatile residue with the same composition (Eq. 5.32).[5]

$$3TiO(OEt)_2 \longrightarrow Ti_2O_3(OEt)_2 + Ti(OEt)_4 \qquad (5.32)$$

Replacing the ethoxo ligand by *tert*-butoxo gave on hydrolysis $[Ti_{18}O_{28}H(OBu^t)_{17}]$, which is soluble in hydrocarbon solvents although its degree of hydrolysis is similar to that of the insoluble $[Ti_2O_3(OEt)_2]_n$.[12] Some oxo-alkoxides can be volatilized unchanged and give fragment ions in their mass spectra in which the core M_xO_y structure remains intact.

4.1 Vibrational Spectra

Most publications on metal oxo-alkoxides list infrared spectra and occasionally Raman spectra. As with the parent metal alkoxides, assigning the spectral frequencies to specific vibrational modes is fraught with difficulties due to coupling of different vibra-tions in polyatomic molecules. However, it is usually possible to distinguish terminal metal–oxygen vibrations from bridging (μ, μ_3, μ_4, μ_5, and μ_6) metal–oxygen vibra-tions. Some data on terminal M=O stretching vibrations are given in Table 5.1 which includes bridging frequencies where available. The polyoxo-alkoxide clusters of vana-dium are prominently featured and Zubieta et al.[54] have noted that in the decavanadates containing the tridentate ligands $(OCH_2)_3CR$ the $V(IV)O_t$ vibrations (940–960 cm^{-1}) may be distinguished from those due to $V(V)O_t$ (970–980 cm^{-1}). The terminal M=O frequencies for a number of $Mo_2(OR)_2$ (M = Cr, Mo) and $MO(OR)_4$ (M = Mo, W) compounds have also been assigned.[15,34,69,70]

4.2 NMR Spectra

NMR spectroscopy has played an increasingly important role in characterizing metal oxo-alkoxides since the 1970s. In principle the environments of the alkyl groups may

Table 5.1 Metal oxygen terminal frequencies

Compound	$\nu(M\text{–}O)^1$ (cm^{-1})	$\nu(MOM)$ (cm^{-1})	Reference
$(Bu_4N)_2[V_6O_{13}\{(OCH_2)_3CNO_2\}_2]$	960, 944	800, 719	54
$(Bu_4N)_2[V_6O_{13}\{(OCH_2)_3CCH_2OH\}_2]$	950	810, 715	54
$(Bu_4N)_2[V_6O_{13}\{(OCH_2)_3CMe\}_2]$	960, 940	816, 710	54
$(Bu_4N)_2[V_6O_{13}\{(OCH_2)_3CNHCOCHCH_2\}_2]$	956	809, 723	54
$(py.H)_2[V_6O_{13}\{(OCH_2)_3CMe\}_2]$	958	789, 704	54
$Na_2[V_6O_7\{(OCH_2)_3CEt\}_2]$	950	–	54
$(NH_4)_4[V_{10}O_{16}\{(OCH_2)_3CEt\}_4]$	972, 942	840, 773	54
$(Et_4N)[V_{10}O_{16}\{(OCH_2)_3CEt\}_5]$	979, 941	836, 778	54
$(Me_3NH)_2[V_{10}O_{14}(OH)_2\{(OCH_2)_3CCH_2OH\}_4]$	974	838	54
$Na_2[V_{10}O_{16}\{(OCH_2)_3CEt\}_4]$	998, 987, 977, 964, 940	847, 775	54
$K_2[V_{10}O_{16}\{(OCH_2)_3CEt\}_4].2H_2O$	995, 979, 970, 961, 943	850, 780	54
$(Bu_4N)_2[V_{10}O_{16}\{(OCH_2)_3CMe\}_4]$	983, 970, 950	838, 780	54
$[V_2O_2Cl_2\{(OCH_2)_2CMe(CH_2OH)\}_2]$	957	–	68
$(Bu_4N)_2[V_4O_4\{(OCH_2)_3CEt\}_2(SO_4)_2(H_2O)_2]$	966, 945	–	68
$[CrO_2Cl(OCPh_3)]$	981, 968	–	69
$[CrO_2(OCPh_3)_2]$	980, 963	–	69
$[CrO_2Cl(OSiPh_3)]$	989, 981	–	69
$[CrO_2(OSiPh_3)_2]$	986, 974	–	69
$[MoO_2\{(OCH_2)_2C_5H_3N\}]_\infty$	909	850	70
$[MoO(OPr^i)_4]$	915	–	34
$[MoO(OBu^t)_4]$	967	–	34
$[MoO_2(OBu^t)_2]$	968, 930 (920, 887)	–	34
$[MoO_2(OBu^t)_2(bpy)]$	912, 888 (863, 843)	–	34
$[MoO_2(OPr^i)_2(bpy)]$	899, 880	–	34
$[MoO_2(OCH_2Bu^t)_2(bpy)]$	918, 893 (872, 851)	–	34
$[Mo_4O_8(OPr^i)_4(py)_4]$	951, 918 (904, 873)	–	34
$[Mo_2(O)_2(OMe)_8]$	950, 920	–	15
$(NEt_4)[WO_2(OCMe_2CMe_2O)(CH_2Bu^t)]$	930, 885	–	71
$[W_4O(OPr^i)_{10}]$	–	667 (635)	72
$[W_4OCl(OPr^i)_9]$	–	678	72
$[W_2(O)_2O(OCMe_2CMe_2O)_2(OCMe_2CMe_2OH)_2]$	959	787	73
$[W_2(O)_2O\{O\overline{CH(CH_2)_5C}HO\}_2\{O\overline{CH(CH_2)_5C}HOH\}_2]$	964	787	73
$[W_2(O)_2O\{O\overline{CH(CH_2)_6C}HO\}_2\{O\overline{CH(CH_2)_6C}HOH\}_2]$	959	791	73
$(Ph_4As)[Tc(O)(OCH_2CH_2S)_2]$	948	–	74
$[Tc(O)Cl(OC_2H_4O)(phen)]$	952	854	75
$[Re_2(O)_2O(OMe)_6]$	982, 968	756	76
$[ReO(OBu^t)_4]$	992	–	76
$[Li\{ReO(OPr^i)_5\}]$	987	–	76
$[Os_2(O)_2O_2(O_2C_6H_{10})_2(C_7H_{13}N)_2]$	919	879	77

Table 5.1   (*continued*)

Compound	$\nu(\text{M–O})^1$ (cm^{-1})	$\nu(\text{MOM})$ (cm^{-1})	Reference
$[\text{Os}_2(\text{O})_2\text{O}_2(\text{OCMe}_2\text{CH}_2\text{NBu}^t)_2]$	960	660	78
$[\text{U}_3(\text{O})_4(\text{OBu}^t)_{10}]$	931, 898	–	64
$[\text{UO}_2(\text{OBu}^t)_2(\text{Ph}_3\text{PO})_2]$	861 (822)	–	44

^{1}Values in parentheses are $\nu(\text{M}-^{18}\text{O})$ frequencies (cm^{-1}).

be explored using ^{1}H and ^{13}C, the oxo ligands using ^{17}O, and the metal using a range of metal nuclei (^{27}Al, ^{119}Sn, ^{207}Pb, ^{89}Y, ^{47,49}Ti, ^{51}V, ^{93}Nb, and ^{183}W). Fluxional behaviour has been studied using variable-temperature (VT) studies and in many cases the limiting low-temperature spectrum is in accordance with the known X-ray crystal structure. Metals with $I = \frac{1}{2}$ nuclei give useful spectra but the quadrupolar nuclei give broad lines with obvious disadvantages. Although the ^{17}O nucleus is quadrupolar the chemical shift[8] and line width are very sensitive to the oxo environment. Klemperer and co-workers[9-12] have demonstrated the value of ^{17}O NMR in characterizing titanium oxo-alkoxides. Some data on ^{17}O chemical shifts are given in Table 5.2 relating specifically to the oxo ligand. Unambiguous spectra were obtained by carrying out hydrolysis with ^{17}O-enriched water. Thus the Mo=O terminal oxo ligands in $\text{MoO}_2(\text{OR})_2$ and $\text{MoO}(\text{OR})_4$ gave shifts in the 970–862 ppm range[34] and the bridging oxo-ligands (μ, μ_3, μ_4) were identified in a number of titanium oxo-alkoxides.[9-12] The chemical shifts for μ_4- and μ_5-O in several aluminium oxo-alkoxides have also been determined.[79] Heterometal oxo-alkoxides have also been studied using ^{17}O NMR. In the cases of the niobotungstate anions in $(\text{Bu}_4\text{N})_3[\text{Nb}_2\text{W}_4\text{O}_{18}(\text{OMe})]$ and $(\text{Bu}_4\text{N})_3[\text{Nb}_2\text{W}_4\text{O}_{18}(\text{OBu}^t)]$ it was possible to identify to which metal atoms the different oxo ligands were bonded.[80] Similar data have been reported for several more hexanuclear complexes $(\text{NBu}_4)_x[(\text{MeO})\text{MW}_5\text{O}_{18}]^{x-}$ (M = Ti, Zr; $x = 3$. M = Nb, Ta; $x = 2$).[81] In these cases the μ_6- ^{17}O resonances were seen at -56 to -71 ppm (see Table 5.2). In the lead oxo-alkoxides $[\text{Pb}_4(\mu_4\text{-O})(\text{OBu}^t)_6]$ and $[\text{Pb}_6(\mu_3\text{-O})_4(\text{OBu}^t)_4]$ and the zirconium lead heterometal oxo-alkoxide $[\text{Pb}_3\text{Zr}(\mu_4\text{-O})(\text{OBu}^t)_8]$ the relatively sharp ^{17}O resonances show the expected coupling to the ^{207}Pb (22%) isotope.[22] The ^{17}O resonances for μ_3-O(Ba, Ti) and μ_5-O(Ba$_2$Ti$_3$) were in accordance with the X-ray crystal structure of the hetero metal oxo-alkoxide $[\text{Ba}_4\text{Ti}_{13}(\mu_5\text{-O})_6(\mu_3\text{-O})_{12}(\text{OC}_2\text{H}_4\text{OMe})_{24}]$.[82]

4.3 Mass Spectra

Mass spectral studies on metal oxo-alkoxides have sometimes given valuable analytical data. As with metal alkoxides the parent molecular ion $P = [\text{M}_x\text{O}_y(\text{OR})_z]^+$ is usually of very low intensity and the strongest high mass fragment ion is often $P - \text{Me}$, $P - \text{R}$, or $P - \text{OR}$.

The appearance of $P - \text{Me}$ or $P - \text{R}$ is then a confirmation that the oxo-alkoxide has retained the integrity of its M_xO_y core in spite of the disruptive effects of 70 eV electrons. For example, although $[\text{Sc}_5\text{O}(\text{OEt})_{13}]$ gave a parent molecular ion[17] the pentanuclear isopropoxo species $[\text{M}_5(\mu_5\text{-O})(\text{OPr}^i)_{13}]$ did not and were thus characterized as $(P - \text{OPr}^i)$ (M = Sc, Y) or $(P - 3\text{OPr}^i)$ (M = Yb)[16a]. Indium oxo-isopropoxide was

Table 5.2   Chemical shifts from ^{17}O NMR

Compound	(M=O)[1]	(μ_2-O)	(μ_3-O)	(μ_4-O)	(μ_5-O)	Reference
[Mo(O)(OPr^i)$_4$]	894	—	—	—	—	34
[Mo(O)(OBu^t)$_4$]	970	—	—	—	—	34
[Mo(O)$_2$(OBu^t)$_2$]	862	—	—	—	—	34
[Mo(O)$_2$(OBu^t)$_2$(py)$_2$]	885	—	—	—	—	34
[Mo(O)$_2$(OPr^i)$_2$(py)$_2$]	878	—	—	—	—	34
[Mo(O)$_2$(OCH$_2$Bu^t)$_2$(py)$_2$]	872	—	—	—	—	34
[Ti$_3$(μ_3-O)(OMe)(OPr^i)$_9$]	—	—	554	—	—	10
[Ti$_7$(μ_4-O)$_2$(μ_3-O)$_2$(OEt)$_{20}$]	—	—	537	364	—	9
[Ti$_8$(μ_3-O)$_4$(μ-O)$_2$(OCH$_2$Ph)$_{20}$]	—	726	533	—	—	9
[Ti$_{10}$(μ_4-O)$_4$(μ_3-O)$_2$(μ-O)$_2$(OEt)$_{24}$]	—	782	525	367, 360	—	9
[Ti$_{18}$(μ_4-O)$_4$(μ_3-O)$_{20}$(μ-O)$_4$(H)(OBu^t)$_{17}$]	—	760, 699	580, 560, 530	440, 439, 407	—	12
[Al$_4$(μ_4-O)(OBu^i)$_{10}$(Bu^{i}OH)]	—	—	—	82.1	—	79
[Al$_4$(μ_4-O)(OBu^i)$_5$(OCH$_2$CF$_3$)$_5$(Bu^{t}OH)]	—	—	—	80.6	—	79
[Al$_4$(μ_4-O)(OCH$_2$CF$_3$)$_{10}$(CF$_3$CH$_2$OH)]	—	—	—	85.9	—	79
[Al$_5$(μ_5-O)(OBu^i)$_{13}$]	—	—	—	—	91.7	79
[Al$_5$(μ_5-O)(OBu^i)$_{12}$(OBu^n)]	—	—	—	—	93.6	79
(Bu$_4$N)$_3$[Nb$_2$W$_4$O$_{18}$(OMe)]	813 (Nb) 732 (W) 721 (W)	458, 449 (Nb, W) 393, 381 (W$_2$)	—	—	—	80
(Bu$_4$N)$_3$[Nb$_2$W$_4$O$_{18}$(OBu^t)]	784 (Nb) 715 (W)	527 (Nb$_2$) 450, 448, 444 (Nb, W) 385, 382 (W$_2$)	—	—	—	80

(Bu$_4$N)$_2$[NbW$_5$O$_{18}$(OMe)]	758, 748 (W)	472 (Nb, W) 405, 402 (W$_2$)	–	–	–71, μ_6-O	81
(Bu$_4$N)$_2$[TaW$_5$O$_{18}$(OMe)]	756, 744 (W)	425 (Ta, W) 404, 403 (W$_2$)	–	–	–68, μ_6-O	81
(Bu$_4$N)$_3$[TiW$_5$O$_{18}$(OMe)]	721, 713 (W)	525 (Ti, W) 390, 380 (W$_2$)	–	–	–58, μ_6-O	81
(Bu$_4$N)$_3$[ZrW$_5$O$_{18}$(OMe)]	727, 715 (W)	530 (Zr, W) 387, 375 (W$_2$)	–	–	–56, μ_6-O	81
[Pb$_4$(μ_4-O)(OBu^t)$_6$]	–	–	–	198	–	22
[Pb$_6$(μ_3-O)$_4$(OBu^t)$_4$]	–	–	228	–	–	22
[Pb$_3$Zr(μ_4-O)(OBu^t)$_8$]	–	–	–	197	–	22
[Ba$_4$Ti$_{13}$(μ_5-O)$_6$(μ_3-O)$_{12}$(OC$_2$H$_4$OMe)$_{24}$]	949, 917, 913, 907	651, 633, 577, 575, 571, 555, 539	651 (Ba, Ti$_2$)	–	487 (Ba$_2$, Ti$_3$)	82
(NBu$_4$)$_2$[Mo(O)$_5$(μ_6-O)(μ-O)$_{11}$(μ-OMe)(NO)]	–	–	–	–	–38 (μ_6)	194

^{1}^{7}O chemical shifts are downfield from H$_2$^{17}O

unstable. The mononuclear oxo vanadium triethanolaminate $[OV\{(OCH_2)_3N\}]$ gave a parent molecule ion[83] whereas the oxoniobium trialkoxides $[ONb(OR)_3]$ (R = Me, Et, Pr^i, Bu^n) gave evidence for disproportionation to $Nb(OR)_5$ and $NbO_2(OR)$ in the mass spectrometer.[84] On the other hand $[Ta_2O(OPr^i)_8(Pr^iOH)]$ gave evidence of dinuclear fragment ions[85] and both $[Me(C_2Me_4O_2)Re(O)(\mu\text{-}O)Re(O)(O_2C_2Me_4)Me]$ and $[Me_3Re(O)(\mu\text{-}O)Re(O)Me_3]$ gave parent ions.[86] Interestingly the oxo-bridged Fe(III) complex $[(L)FeOFe(L)]$ containing the pentadentate ligand L = $[MeC\{CH_2N\colon CMe.CH_2C(CF_3)_2\}O_2(CH_2NH_2)]$ gave $(L)Fe^+$ and $(L)FeO^+$ fragment ions due to unequal splitting of the dinuclear species.[87]

The tetranuclear lead oxo-*tert*-butoxide $[Pb_4O(OBu^t)_6]$ gave a substantial parent molecular ion but the most intense lead-containing species was $[Pb_4O(OBu^t)_5]^+$ showing that the Pb_4O_5 core was stable.[22]

4.4   Magnetic Properties

Some of the transition metal oxo-alkoxide cluster molecules have shown interesting magnetic properties in view of the current theory of magnetic exchange mechanisms. Some examples of systems subjected to detailed magnetic studies are: $(NEt_4)_2[Mn_{10}O_2Cl_8\{(OCH_2)_3CMe\}_6]$,[88] $[Mn_{13}O_8(OEt)_6(O_2CPh)_{12}]$,[89] $[\{Fe_6O_2(OMe)_{12}(tren)_2\}(O_3SCF_3)_2]$ (tren = $\beta\beta'\beta''$-triaminotriethyl-amine)$[N(CH_2CH_2NH_2)_3]$,[90] $[NMe_4)_2[Fe_6O(OMe)_3Cl_6\{(OCH_2)_3CMe\}_3]$,[91] $[Fe_{12}O_2(OMe)_{18}Cl_{0.7}(O_2CCH_2Cl)_{5.3}(CH_3OH)_4]$,[92] $Na[Fe_4O(OH)\{OCH[CH_2N(CH_2CO_2)_2]_2\}_2\{O_2CCH(Me)NH_3\}_2]$,[93] and $[Zr_4Cu_4O_3(OPr^i)_{18}]$.[129]

5   X-RAY CRYSTAL STRUCTURES

In the 1978 book *Metal Alkoxides*[1] only three authentic X-ray crystal structures of metal oxo-alkoxides were described, namely $[V_2(O)_2(OMe)_6]$, $[Ti_7O_4(OEt)_{20}]$, and $[Nb_8O_{10}(OEt)_{20}]$. Since then there has been an explosive growth in the number of structure determinations with examples of μ_6-O, μ_5-O, μ_4-O, μ_3-O, μ-O, and O_t oxo-environments.[4]

In the metal (and heterometal) oxo-alkoxides the metals adopt coordination numbers that are common in the parent metal alkoxides, but the presence of the oxo ligand, with its ability to bond to six metals (μ_6) in the centre of a molecule, is often structure determining. Since the emphasis in this section is on the behaviour of the oxo ligand rather than the metal the structures of heterometal oxo-alkoxides will be considered together with those of the metal oxo-alkoxides. Owing to the very large number of structures known, only a few will be considered in detail and the data on the majority will be tabulated.

5.1   Alkali-metal Containing Oxo-alkoxides (Including Heterometallic Compounds)

Although no alkali metal oxo-alkoxide has yet been structurally characterized the structure of the remarkable heteroalkali metal complex $[Li_8K_2(O)(OBu^t)_8(Me_2NC_2H_4NMe_2)_2]$

has been reported.[94] In fact this molecule has a μ_8-oxo ligand at the core of a 16-vertex $Li_8O_8(Bu^t)_8$ cage which is bridged on opposite sides by two $K(Me_2NC_2H_4NMe_2)$ moieties (Fig. 5.1).

An octanuclear lithium titanium oxo-isopropoxide $[Li_4Ti_4(\mu_5\text{-}O)_2(\mu\text{-}O)_2(OPr^i)_{12}]$ has also been reported.[95] The lithiums are four-coordinated, whilst the μ_5-O ligand has trigonal bipyramidal coordination and the μ-O gives a nonlinear bridge. The hexanuclear $[Na_2Mo_4(O)_6(\mu\text{-}O)_2(OPr^i)_{10}(Pr^iOH)_2]$ consists of two $NaMo_2$ triangles (capped by μ_3-OPr^i groups) joined by NaOMo-bridges. Both Mo atoms (one with two terminal oxo-ligands and the other with one) are in distorted octahedral configurations and the sodium atoms (bonded to one Pr^iOH ligand) are five-coordinated.[95] However, the tetranuclear $[Na_2W_2(O)_2(\mu_3\text{-}OEt)_2(\mu\text{-}OEt)_4(OEt)_8]$ consists of four edge-sharing octahedra, reminiscent of the $[Ti_4(\mu_3\text{-}OR)_2(\mu\text{-}OR)_4(OR)_{10}]$ structure.[97] The sodium zirconium oxo-ethoxide $[Na_4Zr_6(\mu_5\text{-}O)_2(OEt)_{24}]$ has two μ_5-oxo ligands in a symmetrical molecule (Fig. 5.2) containing 4 five-coordinated sodiums and 6 octahedrally coordinated zirconiums.[98]

In the hexanuclear compound $[K_4Sb_2(\mu_6\text{-}O)(OBu^t)_8(thf)_4]$ the central oxo ligand is in a distorted octahedron formed by the four potassiums and two (cis) antimony atoms.[99] In the highly symmetrical molecule $[K_4Sb_4(\mu_4\text{-}O)_2(OPr^i)_{12}]$ the μ_4-oxo ligands are

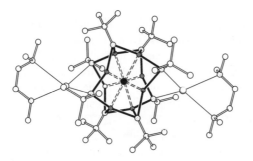

Figure 5.1   Structure of $[Li_8K_2(\mu_8\text{-}O)(\mu_3\text{-}OBu^t)_8(Me_2NC_2H_4NMe_2)_2]$ ($\bullet = \mu_8$-oxo, $\oslash = $ Li, $\bigcirc = $ oxygen of OBu^t, $\oslash = $ N in TMEDA; H atoms omitted).

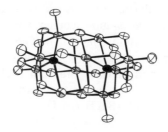

Figure 5.2 $Na_4Zr_6(\mu_5\text{-}O)_2(\mu_3\text{-}O)_4(\mu\text{-}O)_{14}O_6$ core structure of $[Na_4Zr_6(\mu_5\text{-}O)_2(\mu_3\text{-}OEt)_4(\mu\text{-}OEt)_{14}(OEt)_6]$ ($\bullet = \mu_5$-oxo; Et groups omitted).

tetrahedrally bonded in K_2Sb_2O units which have significantly shorter metal–oxo bond lengths than in the μ_6-O complex.[99]

5.2   Magnesium and Alkaline Earth-containing Oxo-alkoxides (Including Heterometallic Compounds)

5.2.1   Magnesium and Calcium

In the magnesium molybdenum(v) oxo-methoxide $[Mg_2Mo_2(O)_2(OMe)_{10}(MeOH_4)]$ compound the oxo-ligands are terminally bonded to the molybdenums.[100] Both metals are octahedrally coordinated, with an $Mg_2Mo_2O_{16}$ structure similar to that of the $Ti_4(OR)_{16}$ molecules. The structure is preserved in the Mo(vi) derivative $[Mg_2Mo_2(O)_4(OMe)_8(MeOH)_4]$ in which the terminally bonded oxo-ligands are *cis* to each other.[100]

In the hexanuclear calcium oxo-ethoxide $[Ca_6(\mu_4\text{-}O)_2(\mu_3\text{-}OEt)_4(OEt)_4(EtOH)_6]\cdot8EtOH$ the μ_4-oxo ligands are bonded to octahedrally coordinated calciums in a structure comprising two Ca_4O_4 cubane units sharing a common Ca_2O_2 face.[101] Some hexanuclear barium oxo-alkoxides have also been characterized.

5.2.2   Barium

The *tert*-butoxide $[H_3Ba_6(\mu_5\text{-}O)(OBu^t)_{11}(OCEt_2CH_2O)(thf)_3]$ has a square-based pyramid of six-coordinated bariums with a μ_5-oxo (or μ_5-OH) and the sixth barium is attached in a five-coordinated mode.[102] In $[H_4Ba_6(\mu_6\text{-}O)(OC_2H_4OMe)_{14}]$ the μ_6-oxo ligand is encapsulated within the octahedron of barium atoms, which are all eight-coordinated with the methoxymethoxo ligands exhibiting three different modes of bonding.[103] Again in the pentanuclear species $[Ba_5(\mu_5\text{-}O)\{\mu_3\text{-}OCH(CF_3)_2\}_4\{\mu\text{-}OCH(CF_3)_2\}_4\{(CF_3)_2CHOH\}(thf)_4(H_2O)]$ there is a square pyramid of Ba atoms bonded to either a μ_5-oxo or μ_5-hydroxo ligand.[104] Four of the bariums are six-coordinated and the fifth is seven-coordinated.

The centrosymmetrical heterometallic oxo-alkoxide-diketonate molecule $[Ba_2Y_4(\mu_6\text{-}O)(\mu_3\text{-}OEt)_8(Bu^tCOCHCOBu^t)_6]$ also contains an octahedrally encapsulated μ_6-oxo-ligand.[105]

The structures of several barium titanium oxo-alkoxides have been determined due to their interest as precursors for the production of $BaTiO_3$. The methoxo complex $[Ba_2Ti_4(\mu_3\text{-}O)(\mu_3\text{-}OMe)_6(\mu\text{-}OMe)_2(OMe)_{10}(MeOH)_7]$ has a remarkable structure (Fig. 5.3) containing nine-coordinated Ba ions and octahedral Ti atoms.[20] Even more remarkable is the structure of the barium titanium oxo-isopropoxide which contains two slightly different molecules $[Ba_4Ti_4(\mu_4\text{-}O)_4(\mu_3\text{-}OPr^i)_2(\mu\text{-}OPr^i)_8(OPr^i)_6(Pr^iOH)_4]$ and $[Ba_4Ti_4(\mu_4\text{-}O)_4(\mu_3\text{-}OPr^i)_2(\mu\text{-}OPr^i)_9(OPr^i)_5(Pr^iOH)_3]$ in the unit cell.[19,106] Each molecule has a Ba_4O_4 distorted cubane core with each oxo ligand bonded to a titanium atom. All of the bariums are seven-coordinated and the titaniums are five-coordinated. The structure of $[Ba_4Ti_4(\mu_4\text{-}O)_4(OPr^i)_{16}(Pr^iOH)_3]$ has also been determined independently.[107]

The methoxymethoxo ligand stabilized the giant barium titanium oxo-alkoxide $[Ba_4Ti_{13}(\mu_5\text{-}O)_6(\mu_3\text{-}O)_{12}(OC_2H_4OMe)_{24}]$. The central octahedral Ti is bonded to the six trigonal pyramidal μ_5-oxo ligands which are each bonded axially to two bariums and equatorially to two of the external titaniums, which are also bonded to two μ_3-oxo

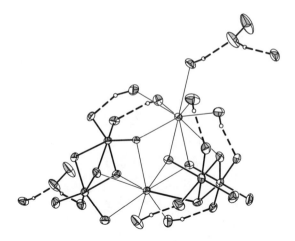

Figure 5.3 $Ba_2Ti_4(\mu_3\text{-}O)(\mu_3\text{-}O)_6(\mu\text{-}O)_2O_{10}(OH)_7$ core structure of $[Ba_2Ti_4(\mu_3\text{-}O)(\mu_3\text{-}OMe)_6(\mu\text{-}OMe)_2(OMe)_{10}(MeOH)_7]$ (Me groups omitted).

ligands.[82] The bridging and chelating $CH_3OC_2H_4O$ ligands with the μ-oxo ligands generate a twelve-coordination for the barium atoms.

Finally it is noteworthy that barium and cerium(IV) form the mixed ligand hexanuclear compound $[Ba_4Ce(\mu_6\text{-}O)(\mu_3\text{-}OPr^i)_8(OPr^i)_2(Bu^tCOCHCOBu^t)_4]$.[125] The μ_6-oxo atom is encapsulated in an octahedron of metals (*trans*-Ce_2).

5.3   Scandium, Yttrium, Lanthanide, and Actinide-containing Oxo-alkoxides (Including Heterometallic Compounds) (Table 5.3)

5.3.1   *Scandium, Yttrium, and Lanthanides*

Evans and co-workers have determined the structure of several interesting yttrium oxo-alkoxides. The cyclopentadienyl yttrium oxide-methoxide $[(C_5H_5)_5Y_5(\mu_5\text{-}O)(\mu_3\text{-}OMe)_4(\mu\text{-}OMe)_4]$[108] has the square pyramidal Y_5O_9 core structure, which was subsequently found in the pentanuclear species $[M_5(\mu_5\text{-}O)(\mu_3\text{-}OR)_4(\mu\text{-}OR)_4(OR)_5]$ (M = Sc, Y, lanthanides) which have the cyclopentadienyl groups replaced by terminal alkoxo ligands. The giant centrosymmetrical molecule $[Y_{14}(\mu_4\text{-}O)_2(\mu_3\text{-}Cl)_2(\mu_3\text{-}OBu^t)_4(\mu\text{-}Cl)_8(\mu\text{-}OBu^t)_{14}(OBu^t)_{10}(thf)_4]$ appears to be constructed of three units A–C. The outer unit A consists of the familiar trinuclear $[Y_3(\mu_3\text{-}Cl)(\mu_3\text{-}OBu^t)(\mu\text{-}OBu^t)_3(OBu^t)_3Cl(thf)_2]$ which is linked to the next unit B $[Y_3(\mu_3\text{-}O)(\mu_3\text{-}OBu^t)(\mu\text{-}OBu^t)_2(\mu\text{-}Cl)(\mu\text{-}OBu^t)_2Cl]$ which in turn is linked to the central mononuclear unit C $[Y(OBu^t)_2Cl]$. A is linked by its terminal chloride to one Y in B which is also linked by its μ_3-O to the Y in C. The whole molecule consists of the sequence ABCCBA with both C groups being chloride bridged to the terminal chlorides in the B units and by chloride bridges to each other.[109] The trinuclear anion $[(C_5H_5)_6Y_3(\mu_3\text{-}O)(\mu\text{-}OMe)_2]^-$ has also been reported.[110] The tetranuclear complex $[Y_4(\mu_4\text{-}O)(\mu_5\text{-}OBu^t)(\mu\text{-}OBu^t)_3(\mu\text{-}Cl)_3(OBu^t)_3(thf)_3]$ which has three-fold symmetry can be considered as comprising the trinuclear anion $[Y_3(\mu_3\text{-}OBu^t)(\mu\text{-}OBu^t)_3(OBu^t)_3Cl_3]^{3-}$ on top of which the cation $[Y(thf)_3]^{3-}$ is clamped by

Table 5.3 Scandium, yttrium, lanthanide, and actinide-containing oxo-alkoxides (including heterometallic compounds)

Compound	Metal coord.	Oxo coord.	M–oxo bond lengths (Å)	M$_x$–oxo bond angles (°)	Reference
[Li$_8$K$_2$(O)(OBu^t)$_8$(tmeda)$_2$] tmeda = Me$_2$NC$_2$H$_4$NMe$_2$	3, 4 (Li) 4 (K)	μ_8	3-coord. Li, 1.895 (av.) 4-coord. Li, 2.195 (av.)	Li$_2$O, 61.2–151.3	94
[Li$_4$Ti$_4$(μ_5-O)$_2$(μ-O)$_2$(OPr^i)$_{12}$]	4 (Li) 5, 6 (Ti)	μ_5, μ	μ_5 2.039 (av.) μ 1.864 (av.)	μ_5, 87.3–174.0 μ_2, 107.0	95
[Na$_2$Mo$_4$(O)$_6$(μ-O)$_2$(OPr^i)$_{12}$]	5 (Na) 6 (Mo)	t, μ	μ, 1.722 (Mo), 2.256 (Na) Mo(O)$_2$, 1.691 (3) MoO, 1.715 (3)	MoONa, 144.5	96
[Na$_2$W$_2$(O)$_2$(OEt)$_{14}$]	6, 6	t	WO, 1.72 (2)	–	97
[Na$_4$Zr$_6$(O)$_2$(OEt)$_{24}$]	5 (Na) 6 (Zr)	μ_5	NaO, 2.666, 3.018 (6) ZrO, 2.169, 2.169, 2.186 (5)	NaOZr, 89.4, 163.1 Zr$_2$O, 104.3 (av.)	98
[K$_4$Sb$_2$(μ_6-O)(OBu^t)$_8$(thf)$_4$]	6 (K) 5 (Sb)	μ_6	SbO, 2.004 (5) KO, 3.047 (8)	Sb$_2$O, 107.1 SbOK, 91.8–97.2, 161.1 K$_2$O, 69.3–80.9, 156.1	99
[K$_4$Sb$_4$(μ_4-O)$_2$(OPr^i)$_{12}$]	6 (K) 5 (Sb)	μ_4	SbO, 1.982 (5) KO, 2.693 (5)	Sb$_2$O, 105.8 SbOK, 104.3–105.2 K$_2$O, 130.1	99
[Mg$_2$Mo$_2$(O)$_2$(OMe)$_{10}$(MeOH)$_4$]	6, 6	t	MoO, 1.685 (4)	–	100
[Ca$_6$(μ_4-O)$_2$(μ_3-OEt)$_4$(OEt)$_4$(EtOH)$_6$]	6	μ_4	CaO, 2.402–2.529 (5)	–	101
[H$_3$Ba$_6$(μ_5-O)(OBu^t)$_{11}$(OCEt$_2$CH$_2$O)(thf)$_3$]	6, 5	μ_5	BaO, 2.74–2.86 (1) av. 2.79	–	102
[H$_4$Ba$_6$(μ_6-O)(OC$_2$H$_4$OMe)$_{14}$]	8	μ_6	BaO, 2.90–2.93	–	103
[Ba$_5$(μ_5-O)(hfip)$_8$(hfipH)(thf)$_4$(H$_2$O)] hfip = OCH(CF$_3$)$_2$	6, 7	μ_5	BaO 2.87 (3)	–	104
[Ba$_2$Y$_4$(μ_6-O)(μ_3-OEt)$_8$(Bu^{t}COCHCOBu^t)$_6$]	7, 7	μ_6	BaO, 3.027 (2) YO, 2.363, 2.378 (3)	–	105
[Ba$_2$Ti$_4$(μ_3-O)(μ_3-OMe)$_6$(μ-OMe)$_2$(OMe)$_{10}$(MeOH)$_7$]	9 (Ba) 6 (Ti)	μ_3	BaO, 2.615, 2.706 (6) TiO, 1.695 (6)	Ba$_2$O, 102.6	22

Compound	Coordination number	μ	M–O distances (Å)	Angles (°)	Reference
$[Ba_4Ti_4(\mu_4\text{-}O)_4(OPr^i)_{16}(Pr^iOH)_4]$	7 (Ba), 5 (Ti)	μ_4	TiO, 1.70–1.73 (1); 1.78–1.81 (1)	Ba_3TiO, 88.6–150.4, 100.4–115.0	19, 106
$[Ba_4Ti_4(\mu_4\text{-}O)_4(OPr^i)_{16}(Pr^iOH)_3]$	7 (Ba), 5 (Ti)	μ_4	TiO, 1.711, 1.714 (2); 1.787, 1.795 (2) BaO, 2.603 (av.), 2.615 (av.), 2.779 (av.), 2.824 (av.)	Ba_3TiO, 90.2–153.6, 103.0–114.4	107
$[Ba_4Ti_{13}(\mu_5\text{-}O)_6(\mu_3\text{-}O)_{12}(OC_2H_4OMe)_{24}]$	12 (Ba), 6 (Ti)	μ_5, μ_3	Ba-μ_5-O, 3.080 (9) Ba-μ_3-O, 2.916 (9) Ti-μ_5-O, 2.020–2.069 (9) Ti-μ_3-O, 1.832 (9)	–	82
$[(C_5H_5)_5Y_5(\mu_5\text{-}O)(OMe)_8]$	6^l	μ_5	YO 2.34–2.41 (2) (bs) (av. 2.38) YO 2.27 (2) (ap) YO, 2.13–2.37 (2)	–	108, 110
$[Y_{14}(\mu_4\text{-}O)_2(\mu_3\text{-}Cl)_2(\mu_3\text{-}OBu^t)_4(\mu\text{-}Cl)_8(\mu\text{-}OBu^t)_{14}(OBu^t)_{10}(thf)_4]$	6	μ_4	–	–	109
$[Y_4(\mu_4\text{-}O)(\mu_3\text{-}OBu^t)(\mu\text{-}OBu^t)_3(\mu\text{-}Cl)_3(OBu^t)_3(thf)_3]$	6, 7	μ_4	YO 2.304 (13) (bs) YO 2.059 (23) (ap)	94.6, 121.9	111
$[Y_5(\mu_5\text{-}O)(\mu_3\text{-}OPr^i)_4(\mu\text{-}OPr^i)_4(OPr^i)_5]$	6	μ_5	YO 2.31–2.39 (4) (bs) YO 2.34 (3) (ap)	88.3–95.4, 170.3	16b
$[Pr_5(\mu_5\text{-}O)(\mu_3\text{-}OPr^i)_4(\mu\text{-}OPr^i)_4(OPr^i)_5]$	6	μ_5	PrO 2.466–2.560 (bs) PrO 2.463 (ap)	88.2–95.8, 170.0	112
$[Yb_5(\mu_5\text{-}O)(\mu_3\text{-}OPr^i)_4(\mu\text{-}OPr^i)_4(OPr^i)_5]$	6	μ_5	YbO 2.300–2.339 (24) (bs) YbO 2.285 (23) (ap)	88.9–95.2, 171.3	16c
$[Y_4Pr(\mu_5\text{-}O)(\mu_3\text{-}OPr^i)_4(\mu\text{-}OPr^i)_4(OPr^i)_5]$	6	μ_5	Y^2O 2.38–2.45 (2) (bs) Y^2O 2.41 (1) (ap)	88.5–95.2, 171.3	63
$[Nd_5(\mu_5\text{-}O)(\mu_3\text{-}OPr^i)_2(\mu\text{-}OPr^i)_6(OPr^i)_5(Pr^iOH)_2]$	6	μ_5	NdO 2.468, 2.355 (13) (eq) NdO 2.719 (1) (ax)	85.9–95.7, 132.1, 170.8	16d

(continued overleaf)

Table 5.3 (*continued*)

Compound	Metal coord.	Oxo coord.	M-oxo bond lengths (Å)	M_x-oxo bond angles (°)	Reference
[Gd$_6$(μ_4-O)(μ_3-OR)$_4$(μ-OR)$_8$(OR)$_4$] R = C$_2$H$_4$OMe	7, 8	μ_4	8-coord. GdO, 2.173, 2.220 (8); 7-coord. GdO, 2.275, 2.305 (8)	102.3–113.0; av. 109.5	62
[Ce$_4$(μ_4-O)(μ_3-OPr^i)$_2$(μ-OPr^i)$_4$(OPr^i)$_7$(Pr^{i}OH)]	6, 7	μ_4	7-coord. CeO, 2.483 (5); 6-coord. CeO, 2.242 (2)	97.4, 99.0, 99.2, 152.3	42
[Ce$_4$(μ_4-O)(μ_3-OPr^i)$_2$(μ-OPr^i)$_4$(OPr^i)$_8$]	6, 7	μ_4	7-coord. CeO, 2.52 (1); 6-coord. CeO, 2.19, 2.22 (1)	97.8, 98.5, 98.5, 98.6, 100.9, 153.6	67
[Pr$_8$(μ_4-O)(μ_3-OR)$_4$(μ-OR)$_{10}$(OR)$_2$(OPMe$_3$)$_2$] R = C$_2$H$_4$OMe	8	μ_4	PrO, 2.276–2.330 (5), 2.488–2.762 (5)	102.5–115.8, 95.6, 161.7	63
[Sm$_4$Ti(μ_5-O)(μ_3-OPr^i)$_2$(μ-OPr^i)$_6$(OPr^i)$_6$]	6 (Sm) 5 (Ti)	μ_5	SmO, 2.58, 2.70 (2); TiO, 1.85 (2)	Sm$_2$O, 84.1, 84.3, 91.0, 163.4; SmOTi, 98.3, 134.5	113
[Li$_8$Y$_8$(μ_4-O)$_2$(μ_3-OBu^t)$_4$(μ-OBu^t)$_{12}$(μ-Cl)$_4$(OBu^t)$_8$]	6 (Y) 2, 3 (Li)	μ_4	YO, 2.40, 2.44, 2.47, 2.52 (2)	Y$_2$O, 96.5, 179.1	114
[Gd$_2$Zr$_6$(μ_4-O)$_2$(μ-OPr^i)$_{10}$(μ-OAc)$_6$(OPr^i)$_{10}$]	6 (Zr) 7 (Gd)	μ_4	GdO, 2.421 (8); ZrO, 2.151, 2.180, 2.195 (8)	Zr$_2$O, 103.5, 105.0, 137.3; ZrOGd, 102.0, 102.2, 102.2	115
[La$_2$Mo$_4$(O)$_4$(μ_4-O)$_4$(μ-OPr^i)$_8$(OPr^i)$_6$]	8 (La) 6 (Mo)	t, μ_4	LaO, 2.728 (3); MoOμ_4, 1.813 (4); M = O, 1.696 (5)	–	116
[U$_3$(μ_3-O)(μ_3-OBu^t)(μ-OBu^t)$_3$(OBu^t)$_6$]	6	μ_3	UO, 2.27 (3)	–	117
[U$_3$(O)$_2$(μ-O)$_2$(μ-OBu^t)$_2$(OBu^t)$_8$]	6	t, μ	UO, 1.753 (6) (t); UO, 1.923, 2.301 (6) (bs)	–	64
[UO$_2$(OBu^t)$_2$(Ph$_3$PO)$_2$]	6	t	UO, 1.789, 1.795 (6) (t)	–	44

ap = apical, ax = axial, bs = basal, eq = equatorial, t = terminal.
[1] Coordination no. doubtful. [2] Randomized Y$_4$Pr atoms.

bridging through the three chlorides and the oxo ligand which becomes μ_4 (Fig. 5.4).[111] The apical Y is thus seven-coordinated whilst the basal Y atoms are six-coordinated.

As already mentioned the tervalent metal isopropoxides were shown to be pentanuclear oxo-isopropoxides $[M_5(\mu_5\text{-O})(\mu_3\text{-OPr}^i)_4(\text{OPr}^i)_5]$ with the square pyramidal structure for the M_5O_{14} core (Fig. 5.5). Structures have been reported for $M = Y,^{16b}$ Pr,[112] Yb,[16c] and In,[16c] and also for the heterometallic compound in which the two metals were distributed randomly over the five sites $[Y_4\text{PrO}(\text{OPr}^i)_{13}]$.[63]

Although single crystals of $[Er_5(\mu_5\text{-O})(\mu_3\text{-OPr}^i)_4(\mu\text{-OPr}^i)_4(\text{OPr}^i)_5]$ could not be obtained, the X-ray powder diffraction data showed that it was isomorphous and isostructural with Pr, Yb, and In compounds which are all monoclinic $P2_1/n$ (cf. the yttrium compound which is orthorhombic).[112] Interestingly neodymium formed the pentanuclear complex $[Nd_5(\mu_5\text{-O})(\mu_3\text{-OPr}^i)_2(\mu\text{-OPr}^i)_6(\text{OPr}^i)_5(\text{Pr}^i\text{OH})_2]$ in which the μ_5-oxo ligand is at the centre of a trigonal bipyramid of Nd atoms.[16d]

Treatment of $[Gd_5(\mu_5\text{-O})(\text{OPr}^i)_{13}]$ with 2-methoxyethanol gave rise to the hexa-nuclear species $[Gd_6(\mu_4\text{-O})(\mu_3\text{-OC}_2\text{H}_4\text{OMe})_4(\mu\text{-OC}_2\text{H}_4\text{OMe})_8(\text{OC}_2\text{H}_4\text{OMe})_4]$.[62] The

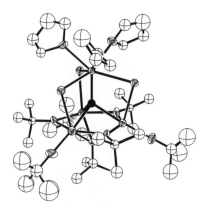

Figure 5.4   Structure of $[Y_4(\mu_4\text{-O})(\mu_3\text{-OBu}^t(\mu\text{-OBu}^t)_3(\mu\text{-Cl})_3(\text{OBu}^t)_3(\text{thf})_3]$ ($\bullet = \mu_4$-oxo; H atoms omitted).

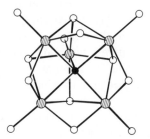

Figure 5.5 $M_5(\mu_5\text{-O})(\mu_3\text{-O})_4(\mu\text{-O})_4\text{O}_5$ core structure of $[M_5(\mu_5\text{-O})(\mu_3\text{-O})_4(\mu\text{-OPr}^i)_4(\text{OPr}^i)_5]$ ($\bullet = \mu_5$-oxo, $\oslash = M$; Pr^i groups omitted).

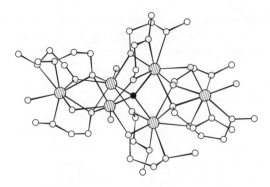

Figure 5.6   Structure of $[Gd_6(\mu_4\text{-}O)(\mu_3,\eta^2\text{-}OR)_4(\mu,\eta^2\text{-}OR)_6(\mu\text{-}OR)_2(OR)_4]$ (R = C_2H_4OMe; $\bullet$ = μ_4-oxo, $\circleddash$ = O, $\oslash$ = Gd; H atoms omitted).

molecule (Fig. 5.6) is composed of a central $Gd_4(\mu_4\text{-}O)(OC_2H_4OMe)_{10}$ tetrahedron containing 2 seven-coordinate and 2 eight-coordinate gadoliniums with two edges capped by $Gd(OC_2H_4OMe)_3$ units containing eight-coordinated Gd. The methoxyethoxo ligands exhibit four different modes of bonding.

The cerium(IV) isopropoxide $[Ce_2(\mu\text{-}OPr^i)_2(OPr^i)_6(Pr^iOH)]$ underwent photoreduction to the Ce(III)Ce$_3$(IV) tetranuclear oxo-isopropoxide $[Ce_4(\mu_4\text{-}O)(\mu_3\text{-}OPr^i)_2(\mu\text{-}OPr^i)_4(OPr^i)_7(Pr^iOH)]$ which has a "butterfly" structure[42] whereas thermal desolvation gave the Ce(IV) complex $[Ce_4(\mu_4\text{-}O)(\mu_3\text{-}OPr^i)_2(\mu\text{-}OPr^i)_4(OPr^i)_8]$ which has a flattened tetrahedral Ce_4O core with 2 six-coordinated and 2 seven-coordinated cerium atoms.[67]

Treatment of $[Pr_5(\mu_5\text{-}O)(OPr^i)_{13}]$ with 2-methoxyethanol in the presence of Me$_3$PO gave rise to the octanuclear species $[Pr_8(\mu_4\text{-}O)_4(\mu_3\text{-}OC_2H_4OMe)_4(\mu\text{-}OC_2H_4OMe)_{10}(OC_2H_4OMe)_2(OPMe_3)_2]$. In this centrosymmetrical molecule all of the metal atoms are eight-coordinated, 2 of the μ_4-oxo atoms are tetrahedrally coordinated and the other 2 are in an irregular four-coordination.[63]

Another interesting pentanuclear heterometallic oxo-alkoxide is $[Sm_4Ti(\mu_5\text{-}O)(\mu_3\text{-}OPr^i)_2(\mu\text{-}OPr^i)_6(OPr^i)_6]$ in which the μ_5-oxo ligand is encapsulated in a trigonal bipyramid of metal atoms with Ti an equatorial position.[113] The giant molecule $[Li_8Y_8(\mu_4\text{-}O)_2(\mu_3\text{-}OBu^t)_4(\mu\text{-}OBu^t)_{12}(\mu\text{-}Cl)_4(OBu^t)_8]$ is composed of two outer $[Y_4(\mu_4\text{-}O)(\mu_3\text{-}OBu^t)_2(\mu\text{-}OBu^t)_4(\mu\text{-}Cl)_2(OBu^t)_4]^{2-}$ anions joined by chloride bridges to a central $[Li_6(\mu_3\text{-}OBu^t)_4]^{2+}$ unit with two additional lithium ions in the lattice. The yttriums are all six-coordinated and the lithiums two- and three-coordinated.[114] It is fascinating to see covalently bonded lithium chloride built into this structure. Since the Y_4 unit has a "butterfly" configuration the μ_4-oxo ligand is linearly bonded to the wing-tip metals and makes an angle of 96.5° with the backbone metals.

In the mixed ligand octanuclear heterometallic compound $[Gd_2Zr_6(\mu_4\text{-}O)_2(\mu\text{-}OPr^i)_{10}(\mu\text{-}OAc)_6(OPr^i)_{10}]$ the molecule may be viewed as two halves $[GdZr_3(\mu_4\text{-}O)(\mu\text{-}OPr^i)_5(\mu\text{-}OAc)_2(OPr^i)_5]$ joined by two acetate bridges at the Gd atoms. The configuration at the μ_4-oxo ligand is like a trigonal pyramid with Gd at the apex; the zirconiums are six-coordinated and the gadoliniums seven-coordinated.[115]

The hexanuclear $[La_2Mo_4(O)_4(\mu_4\text{-O})_4(\mu\text{-OPr}^i)_8(OPr^i)_6]$ has a tetragonal bipyramidal configuration of metal atoms with the lanthanums *trans* to each other. Each Mo atom has one terminal oxo ligand and the μ_4-oxo ligands are each bonded to two La and two Mo atoms. The bridging isopropoxo groups span the LaMo edges of the polyhedron and each metal has one terminal isopropoxo group.[116]

5.3.2 Actinides

Among the actinides a few structures of uranium oxo-alkoxides have been reported. The trinuclear $[U_3(\mu_3\text{-O})(\mu_3\text{-OBu}^t)(\mu\text{-OBu}^t)_3(OBu^t)_6]$ uranium(IV) complex has the familiar M_3O_{11} core structure but there is no U....U bonding in contrast to $[Mo_3O(OCH_2Bu^t)_{10}]$, which exhibits metal–metal bonding.[117] The trinuclear uranium(VI) compound $[U_3(O)_2(\mu\text{-O})_2(\mu\text{-OBu}^t)_2(OBu^t)_8]$ has a central uranium containing a linear UO_2 group bridged by unsymmetrical μ-O oxo groups and μ-OBu^t groups to the outer U atoms.[64] The mononuclear $[UO_2(OBu^t)_2(Ph_3PO)_2]$ has the linear UO_2 group and *cis* phosphine oxide ligands.[44]

5.4 Titanium Sub-group Metal-containing Oxo-alkoxides (Including Heterometallic Compounds) (Table 5.4)

5.4.1 Titanium

Numerous titanium oxo-alkoxides have been structurally characterized including $[Ti_7O_4(OEt)_{20}]$,[9,118] $[Ti_{10}O_8(OEt)_{24}]$[9] and $[Ti_{16}O_{16}(OEt)_{32}]$[118] which with the unhydrolysed crystalline tetrameric ethoxide $[Ti_4(OEt)_{16}]$ conform to a general formula $[Ti_{(4+3x)}O_{4x}(OEt)_{4(x+4)}]$ ($x = 0, 1, 2, 4$) implying an increment of $Ti_3O_4(OEt)_4$ between each member of the series. Interestingly this is very similar to the series proposed by Bradley *et al.*[5] $[Ti_{(3+3x)}O_{4x}(OEt)_{4(x+3)}]$ based on the analysis of the variation of number average degree of polymerization with degree of hydrolysis of the tetraethoxide in boiling ethanolic solution. The actual structures of the oxo-ethoxides obtained to date all involve octahedra. In the heptanuclear $[Ti_7(\mu_4\text{-O})_2(\mu_3\text{-O})_2(\mu\text{-OEt})_8(OEt)_{12}]$ the central TiO_6 octahedron shares six edges through two *cis*-μ_4-oxo ligands, two *trans*-μ_3-oxo ligands, and two *cis*-μ-ethoxo ligands to six other octahedra (Fig. 5.7).[118]

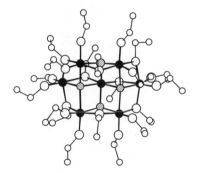

Figure 5.7 Structure of $[Ti_7(\mu_4\text{-}O)_2(\mu_3\text{-O})_2(\mu\text{-OEt})_8(OEt)_{12}]$ (◍ = μ_4-oxo, ◒ = μ_3-oxo, ● = Ti; H atoms omitted).

Table 5.4 Titanium and zirconium-containing oxo-alkoxides (including heterometallic compounds)

Compound	Metal coord.	Oxo coord.	M–Oxo bond lengths (Å)	M_x–Oxo bond angles (°)	Reference
$[Ti_7(\mu_4\text{-}O)_2(\mu_3\text{-}O)_2(\mu\text{-}OEt)_8(OEt)_{12}]$	6	μ_4, μ_3	μ_4, 2.029, 2.037–2.149 (5) av. 2.067 μ_3, 1.877–2.034 (5) av. 1.963	μ_4, 96.3–102.2 and 150.5	9, 118
$[Ti_3(\mu_3\text{-}O)(\mu_3\text{-}OMe)(\mu\text{-}OPr^i)_3(OPr^i)_6]$	6	μ_3	1.954–1.984 (5) av. 1.971	96.0–97.1 av. 96.7	10
$[Ti_6(\mu_3\text{-}O)_2(\mu\text{-}O)_2(\mu\text{-}OPr^i)_2(\mu\text{-}OAc)_8(OPr^i)_6]$	6	μ_3, μ	μ_3, 1.890, 1.916, 2.080 (6) μ, 1.963, 2.079 (6)	–	119
$[Ti_6(\mu_3\text{-}O)_2(\mu\text{-}O)_2(\mu\text{-}OBu^n)_2(\mu\text{-}OAc)_8(OBu^n)_6]$	6	μ_3, μ	μ_3, 1.883, 1.909, 2.104 (4) μ, 1.746, 1.880 (4)	μ_3, 101.8, 128.4, 129.8 av. 120.0 μ, 137.1	120
$[Ti_3(\mu_3\text{-}O)\{(\mu_3\text{-}Me_2C(O)CH{:}C(O)CH_2(O)\}(\mu\text{-}OPr^i)_3(OPr^i)_4]$	6	μ_3	1.969 (4)	–	66
$[Ti_3(\mu_3\text{-}O)(\mu_3\text{-}Cl)(\mu\text{-}OCH_2Bu^t)_3(OCH_2Bu^t)_6]$	6	μ_3	1.929, 1.935, 1.939 (7)	103.6, 103.6, 103.7	121
$[Ti_{18}(\mu_4\text{-}O)_4(\mu_3\text{-}O)_{20}(\mu\text{-}O)_4(H)(OBu^t)_{17}]$	4, 5, 6	μ_4, μ_3, μ	tet. Ti-μ_4-O, 1.769–1.836 (12) av. 1.812 oct. Ti-μ_4-O, 2.218–2.287 (12) av. 2.254 All other bridging TiO between 1.812 and 2.254	Ti_4O, 109.5	12
$[Ti_2Pb_2(\mu_4\text{-}O)(\mu_3\text{-}OPr^i)_2(\mu\text{-}OPr^i)_4(OPr^i)_4]$	6 (Ti) 5(Pb)	μ_4	TiO, 2.01, 2.03 (1) PbO, 2.33, 2.36 (1)	Pb_2O, 99.6 PbOTi, 95.0, 95.4 Ti_2O, 163.0	18
$[TiPb_4(\mu_4\text{-}O)(\mu_3\text{-}NBu^t)_3(OPr^i)_4]$	5 (Ti) 3 (Pb)	μ_4	TiO, 2.09 (3) PbO, 2.22, 2.23 (3)	Pb_2O, 104.1–104.5	122

		t, μ_6, μ			
$(NBu_4)_3[(MeO)TiW_5(O)_5(\mu_6\text{-}O)(\mu\text{-}O)_{12}]$	6, 6		$TiO(\mu_6)$, 2.211 (11); $TiO(\mu)$, 1.931–1.967 (11) av. 1.947; $WO(\mu_6)$, 2.333 av.; $WO(\mu)$, 1.914 av.; WO, 1.719 av. (t)	—	81
$[Zr_{13}(\mu_4\text{-}O)_8(\mu\text{-}OMe)_{24}(\mu\text{-}OMe)_{12}]$	7, 8	μ_4	2.19, 2.23 (1)	—	123
$[Zr_3(\mu_3\text{-}O)(\mu_3\text{-}Cl)(\mu\text{-}OCH_2Bu^t)_3(OCH_2Bu^t)_6]$	6	μ_3	2.082, 2.089, 2.090 (3)	104. 4 av.	98
$[Zr_3(\mu_3\text{-}O)(\mu_3\text{-}OBu^t)(\mu\text{-}OH)(\mu\text{-}OBu^t)_2(OBu^t)_6]$	6	μ_3	2.094, 2.108 (4)	100.1, 101.9	98
$[Zr_3Pb(\mu_4\text{-}O)(\mu_3\text{-}OPr^i)_2(\mu\text{-}OPr^i)_4(\mu\text{-}OAc)_2(OPr^i)_4]$	6 (Zr), 5 (Pb)	μ_4	ZrO, 2.15, 2.22, 2.23 (1); PbO, 2.310 (9)	—	121
$[Zr_4Pb_2(\mu_4\text{-}O)_2(\mu\text{-}OEt)_8(\mu\text{-}OAc)_2(OEt)_4(OAc)_2]$	6, 7 (Zr), 4 (Pb)	μ_4	ZrO, 2.17; PbO, 2.32	—	126
$[ZrPb_3(\mu_4\text{-}O)(\mu_3\text{-}OBu^t)(\mu\text{-}OBu^t)_5(OBu^t)_2]$	6 (Zr), 5, 3 (Pb)	μ_4	ZrO, 2.23 (4); PbO, 2.15, 2.26, 2.26 (3)	ZrOPb, 100.9, 101.0, 127.8; Pb$_2$O, 100.8, 111.2, 111.6	22
$[Zr_3Fe(\mu_4\text{-}O)(\mu\text{-}Pr^n)_6(OPr^n)_4(acac)_3]$	7 (Zr), 5 (Fe)	μ_4	ZrO, 2.106, 2.111, 2.123 (8); FeO, 2.124 (9)	Zr$_2$O, 114.3, 114.5, 114.7; ZrOFe, 104.0, 104.6, 104.8	127
$[Zr_2Co_4(\mu_6\text{-}O)(\mu\text{-}OPr^n)_8(OPr^n)_2(acac)_4]$	6, 6	μ_6	ZrO, 2.207 (1); CoO, 2.363, 2.377 (1)	90, 180	128
$[Zr_4Cu_4(\mu_4\text{-}O)(\mu\text{-}OPr^i)_{10}(OPr^i)_8]$	2 (Cu), 6 (Zr)	μ_4	CuO, 1.821–1.843 (9) av. 1.830	Cu$_2$O, 94.2, 94.4, 102.3, 119.4, 122.8, 125.0	129
$[Zr_4Cu_4(\mu_4\text{-}O)_3(\mu\text{-}OPr^i)_{10}(OPr^i)_8]$	4 (Cu), 6 (Zr)	μ_4	ZrO, 2.214, 2.231 (11); CuO, 1.9655, 1.9679 (25)	Cu$_2$O, 90, 180; Cu$_2$O, 94.9; ZrOCu, 98.1, 98.6, 140.3, 144.3; Zr$_2$O, 92.3	129

oct = octahedral, t = terminal, tet = tetrahedral.

The compound $[Ti_{16}(\mu_4\text{-}O)_4(\mu_3\text{-}O)_8(\mu\text{-}O)_4(\mu\text{-}OEt)_{16}(OEt)_{16}]$ was obtained in two forms, I (stable in air, monoclinic) and II (sensitive to moisture, tetragonal) due to different packings of the hexadecanuclear molecules in the crystals. The $Ti_{16}O_{48}$ core in each form is built up of two orthogonal blocks of eight TiO_6 octahedra. In each block six octahedra form a layer structure (NiAs-type) with the seventh and eighth being on either side (Fig. 5.8). The coordination around the μ_4-oxo atoms is unsymmetrical (Ti_2O angles, 94.4–150.3°) whilst some μ_3-O are pyramidal and others trigonal planar.[118] In the decanuclear molecule $[Ti_{10}(\mu_4\text{-}O)_4(\mu_3\text{-}O)_2(\mu\text{-}O)_2(\mu\text{-}OEt)_{10}(OEt)_{14}]$ (Fig. 5.9) it can be seen how three more octahedra have been attached to the $Ti_7O_4(OEt)_{20}$ structure.[9]

The hydrolysis of $Ti(OCH_2Ph)_4(PhCH_2OH)$ gave rise to the crystalline octanuclear compound $[Ti_8(\mu_3\text{-}O)_4(\mu\text{-}O)_2(\mu\text{-}OCH_2Ph)_8(OCH_2Ph)_{12}]$ (Fig. 5.10)[9] which contains 4 octahedral and 4 five-coordinated titanium atoms presumably due to the greater steric effect of the benzyloxo ligand compared to the ethoxide. In solution there was ^{17}O

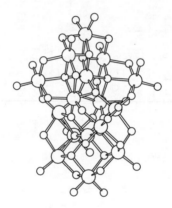

Figure 5.8 $Ti_{16}O_{48}$ core structure of $[Ti_{16}(\mu_4\text{-}O)_4(\mu_3\text{-}O)_8(\mu\text{-}O)_4(\mu\text{-}OEt)_{16}(OEt)_{16}]$.

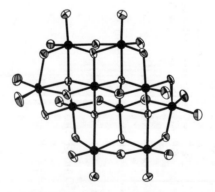

Figure 5.9 $Ti_{10}O_{32}$ core structure of $[Ti_{10}(\mu_4\text{-}O)_4(\mu_3\text{-}O)_2(\mu\text{-}O)_2(\mu\text{-}OEt)_{10}(OEt)_{14}]$ ($\bullet$ = Ti; Et groups omitted).

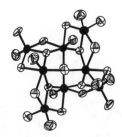

Figure 5.10 Ti_8O_{26} core structure of
$[Ti_8(\mu_3\text{-}O)_4(\mu\text{-}O)_2(\mu\text{-}OCH_2Ph)_8(OCH_2Ph)_{12}]$
($\bullet = Ti$).

NMR evidence for the corresponding ethoxo compound $[Ti_8(\mu_3\text{-}O)_4(\mu\text{-}O)_2(OEt)_{20}]$ but crystals could not be obtained. In view of the exclusively octahedral titanium coordination in $Ti_4(OEt)_{16}$, $Ti_7O_4(OEt)_{20}$, $Ti_{10}O_8(OEt)(OEt)_{24}$, and $Ti_{16}O_{16}(OEt)_{32}$ it seems unlikely that the Ti_8 species of ethoxide would introduce five-coordinated Ti.

Further consequences of steric effects were revealed in the hydrolysis of $Ti(OPr^i)_4$ which is practically monomeric with tetrahedral titanium. Removal of ispropoxo groups by hydrolysis relaxes the steric congestion and the ^{17}O NMR spectrum pointed to the formation of the unstable trinuclear $[Ti_3(\mu_3\text{-}O)(OPr^i)_{10}]$ in the initial stages of hydrolysis, although this compound could not be crystallized out of solution. However, in the presence of methanol the stable crystalline compound $[Ti_3(\mu_3\text{-}OMe)(\mu_3\text{-}OPr^i)_3(OPr^i)_6]$ was obtained.[10] The structure involves edge-sharing of three TiO_6 octahedra as in $[U_3O(OBu^t)_{10}]$ and $[Mo_3O(OPr^i)_{10}]$. Further hydrolysis of the isopropoxide led to the isolation of $[Ti_{11}O_{13}(OPr^i)_{18}]$ and $[Ti_{12}O_{16}(OPr^i)_{16}]$ with the latter compound being resistant to further hydrolysis in alcoholic solution but more susceptible in non-hydroxylic solvents.[11] Treatment of the Ti_{11} species with ethanol gave the crystalline compound $[Ti_{11}O_{13}(OEt)_5(OPr^i)_{13}]$ and its structure (Fig. 5.11) revealed the presence of 6 octahedral and 5 five-coordinated Ti atoms. The Ti_{12} compound exists in two interconvertible isomeric forms each containing equal numbers of octahedral and five-coordinate titaniums (Fig. 5.12).

The structure of the hexanuclear mixed ligand oxo-alkoxide $[Ti_6(\mu_3\text{-}O)_2(\mu\text{-}O)_2(\mu\text{-}OPr^i)_2(\mu\text{-}OAc)_8(OPr^i)_6]$ showed the presence of exclusively octahedral titanium.[119] Similarly the centrosymmetric hexanuclear $[Ti_6(\mu_3\text{-}O)_2(\mu\text{-}O)_2(\mu\text{-}OBu^n)_2(\mu\text{-}OAc)_8(OBu^n)_6]$ also contained octahedral titanium and the μ_3-oxo atoms were in trigonal planar configurations.[120]

Another trinuclear oxo-isopropoxide has been obtained by allowing $Ti(OPr^i)_4$ to hydrolyse through the condensation of acetone to give $[Ti_3(\mu_3\text{-}O)\{\mu_3\text{-}Me_2C(O)CH\text{:}C(O)CH_2CMe_2(O)\}(\mu\text{-}OPr^i)_3(OPr^i)_4]$.[66] The Ti_3O_{11} core framework is apparently stabilized by the tridentate ligand which is chelating and bridging the three Ti atoms. A variation on the Ti_3O_{11} theme was found in $[Ti_3(\mu_3\text{-}O)(\mu_3\text{-}Cl)(\mu\text{-}OCH_2Bu^t)_3(OCH_2Bu^t)_6]$ where a μ_3-oxygen is replaced by a μ_3-chloride.[121]

With the aid of a Siemens Molecular Analysis Research Tool Charge Coupled Device (SMART CCD) area detector the complex structure of the giant molecule $[Ti_{18}O_{28}H(OBu^t)_{17}].Bu^tOH$ was successfully unravelled.[12] The structure of the $Ti_{18}O_{45}$ framework was described as a pentacapped Keggin unit in which a $Ti_{13}O_{40}$ (Keggin)

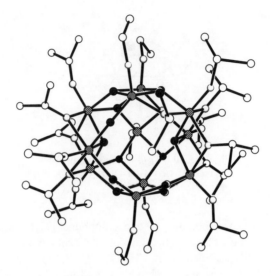

Figure 5.11   Structure of $[Ti_{11}(\mu_3\text{-}O)_{13}(\mu\text{-}OPr^i)_7(OPr^i)_6(OEt)_5]$ ($\bullet$ = μ_3-oxo, ⊛ = Ti; H atoms omitted).

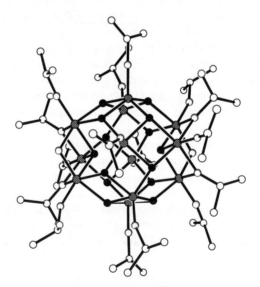

Figure 5.12   Structure of $[Ti_{12}(\mu_3\text{-}O)_{14}(\mu\text{-}O)_2(\mu\text{-}OPr^i)_4(OPr^i)_{12}]$ ($\bullet$ = μ_3-oxo, ⊛ = Ti; H atoms omitted).

structure is capped by five TiO moieties which lie above five of the six square faces of the T_d-distorted cuboctahedron. As shown in Fig. 5.13 all seventeen of the terminal positions are occupied by *tert*-butoxo ligands and the single hydrogen atom is present in the μ-OH group and is hydrogen bonded to the lattice Bu^tOH molecule.

At the centre of the $Ti_{18}O_{45}$ framework is a tetrahedral Ti atom which is linked by four μ_4-oxo ligands to four groups of three TiO_6 octahedra (Keggin structure).

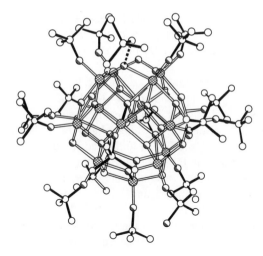

Figure 5.13 Structure of [Ti$_{18}$(μ_4-O)$_4$(μ_3-O)$_{20}$(μ-O)$_4$H(OBu^t)$_{17}$] (⬣ = Ti; H atoms omitted).

The 5 capping five-coordinated titaniums adopt distorted trigonal bipyramidal (TBP) configurations.

The structures of some heterometallic titanium-containing molecules have been reported. In the tetranuclear compound [Ti$_2$Pb$_2$(μ_4-O)(μ_3-OPr^i)$_2$(μ-OPr^i)$_4$(OPr^i)$_4$] the μ_4-oxo ligand has an irregular configuration (Ti$\widehat{O}$Ti, 163°; Pb$\widehat{O}$Pb, 99.6°; Pb$\widehat{O}$Ti, 95.0° and 95.4°), the titaniums are octahedrally coordinated and the lead atoms five-coordinated (distorted square pyramidal with the vacant octahedral site occupied by a lone electron pair).[18]

The presence of μ-acetato ligands in [Ti$_2$Pb$_2$(μ_4-O)(μ-OPr^i)$_5$(μ-OAc)$_2$(OPr^i)$_3$] generates a different structure with one lead atom four-coordinated and the other five-coordinated together with distorted octahedral titaniums.[121] In the novel complex [TiPb$_4$(μ_4-O)(μ_3-NBu^t)$_3$(OPr^i)$_4$] the oxo ligand in the cubane-like Pb$_4$(μ_3-O)(μ_3-NBu^t)$_3$ unit coordinates to a TBP titanium through an axial position.[122]

In the methoxo titanotungstate anion [(MeO)TiW$_5$O$_{18}$]$^{3-}$ the M$_6$O$_{19}$ framework is retained with a central μ_6-oxo, twelve μ-oxo and five terminal oxo ligands with the methoxo ligand bound terminally to the titanium atom.[81]

5.4.2 Zirconium

The first zirconium oxo-alkoxide structure reported was the tridecanuclear [Zr$_{13}$(μ_4-O)$_8$(μ-OMe)$_{24}$(OMe)$_{12}$] which contains a central eight-coordinated (almost cubic) zirconium connected by μ_4-oxo ligands to 12 seven-coordinated metals which are linked by bridging methoxo groups. Each of the exterior zirconiums is bonded to a terminal methoxo ligand.[123] Recently the trinuclear compounds [Zr$_3$(μ_3-O)(μ_3-Cl)(μ-OCH$_2$Bu^t)$_3$(OCH$_2$Bu^t)$_6$]   and   [Zr$_3$(μ_3-O)(μ_3-OBu^t)(μ-OH)

(μ-OBu^t)$_2$(OBu^t)$_6$] were reported.[98] Both molecules are modelled on the familiar M$_3$O$_{11}$ structure involving three ZrO$_6$ octahedra. It is noteworthy that in the *tert*-butoxide the OH-group occupied a μ-position rather than the μ_3-position that might have been expected.

Some zirconium lead heterometallic oxo-alkoxides have been characterized. In the tetranuclear mixed ligand compound [Zr$_3$Pb(μ_4-O)(μ_3-OPr^i)$_2$(μ-OPr^i)$_4$(μ-OAc)$_2$(OPr^i)$_4$] the central oxo ligand is in a distorted (squashed tetrahedral) configuration whilst the zirconiums are distorted octahedral and the lead atom is five-coordinated (tetragonal pyramid with lone electron pair in the vacant octahedral site).[121] The acetato ligands are bidentate bridging. In the hexanuclear complex [Zr$_4$Pb$_2$(μ_4-O)(μ-OEt)$_8$(μ-OAc)$_2$(OEt)$_4$(OAc)$_2$] two acetates are bidentate bridging and two are chelating. Two of the zirconiums are six-coordinated and two are seven-coordinated.[126] The oxo-*tert*-butoxide [ZrPb$_3$(μ_4-O)(μ_3-OBu^t)(μ-OBu^t)$_5$(OBu^t)$_2$] has a μ_4-oxo ligand in a distorted tetrahedral coordination. The zirconium in six-coordinated, 2 lead atoms are five-coordinated but the third is only three-coordinated (pyramidal).[22]

Seven-coordinated Zr is also found in the mixed ligand compound [Zr$_3$Fe(μ_4-O)(μ-OPr^n)$_6$(OPr^n)$_4$(acac)$_3$]. The μ_4-oxo ligand is tetrahedrally coordinated and the Fe(III) atom is five-coordinated (distorted TBP).[127] However, with Co(II) the hexanuclear [Zr$_2$Co$_4$(μ_6-O)(μ-OPr^n)$_8$(OPr^n)$_2$(acac)$_4$] was obtained. This has a μ_6-oxo ligand encapsulated in an octahedron (*trans*-Zr$_2$) of metal atoms in the familiar cubic M$_6$O$_{19}$ core framework with all metals octahedrally coordinated.[128]

Some interesting zirconium copper(I) and zirconium copper(II) oxo-isopropoxides have been characterized.[129] In [Zr$_4$Cu$_4$(μ_4-O)(μ-OPr^i)$_{10}$(OPr^i)$_8$] the central μ_4-oxo ligand is in a distorted tetrahedral Cu$_4$O system linked to two confacial bioctahedral Zr$_2$(OPr^i)$_9$ moieties by bridging isopropoxo ligands. The copper(I) atoms are two-coordinated (linear) and the zirconiums octahedral. The centrosymmetric Cu(II) complex [Zr$_4$Cu$_4$(μ_4-O)$_3$(μ-OPr^i)$_{10}$(OPr^i)$_8$] has a central μ_4-oxo ligand exhibiting a novel square planar configuration in the planar Cu$_4$O(OPr^i)$_2$ fragment which is capped by two confacial bioctahedral Zr$_2$O(OPr^i)$_8$ moieties.

5.5   Vanadium Sub-group Metal-containing Oxo-alkoxides (Including Heterometallic Compounds) (Table 5.5)

5.5.1   *Vanadium*

The chemistry of vanadium is replete with oxo vanadium complexes including a large number of vanadium oxo-alkoxides.

The first structural determination revealed that vanadyl(V) trimethoxide had a polymeric structure built up from weakly bound dimer units [VO(μ-OMe)(OMe)$_2$]$_2$ giving the vanadium a distorted octahedral configuration.[130] The 2-chloroethoxo-derivative gave a weakly bound dimer with distorted TBP coordination[46] as did the cyclopentyloxo-derivative.[134] The distorted TBP configuration was also found in several mononuclear oxovanadatrane molecules [OV(OC$_2$H$_4$)$_3$N],[83,131] [OV(OCHMeCH$_2$)$_3$N],[83] [OV{(OC$_2$H$_4$)$_2$(CHEtCH$_2$O)}],[132] and [OV(OCHBu^{t}CH$_2$)$_3$N].[133] By contrast the binuclear complex [V$_2$(O)$_2$Cl$_2$(OC$_2$H$_4$O)$_2$] had a cyclic structure with distorted tetrahedral vanadium atoms,[135] whilst the complex [V$_2$(O)$_2${(OCH$_2$)$_2$CMe(CH$_2$OH)}$_2$Cl$_2$] had edge-sharing octahedral V(V) atoms.[136]

Table 5.5 Vanadium sub-group metal-containing oxo-alkoxides (including heterometallic compounds)

Compound	Metal coord.	Oxo coord.	M–Oxo bond lengths (Å)	M_x–Oxo bond angles (°)	Reference
[V₂(O)₂(μ-OMe)₂(OMe)₄]	6	t, μ	VO, 1.51, 1.58 (t) VO, 2.25, 2.30 (μ)	—	130
[V₂(O)₂(μ-OC₂H₄Cl)₂(OC₂H₄Cl)₆]	5	t	VO, 1.584 (2) (t)	—	46
[V₂(O)₂(μ-OC₅H₉-c)₂(OC₅H₉-c)₆]	5	t	VO, 1.595 (3) (t)	—	134
[OV(OC₂H₄)₃N]	5	t	VO, 1.633 (6) (t)	—	83
[OV(OC₂H₄)₃N]	5	t	VO, 1.601 (6) (t)	—	131
[OV(OCHMeCH₂)₃N]	5	t	VO, 1.617 (13) (t)	—	83
[OV{(OC₂H₄)₂N(CHEtCH₂O)}]	5	t	VO, 1.614 (2) (t)	—	132
[OV(OCHBuᵗCH₂)₃N]	5	t	VO, 1.612 (2) (t)	—	133
[V₂(O)₂Cl₂(OC₂H₄O)₂]	4	t	VO, 1.578 (2) (t)	—	135
[V₂(O)₂{(OCH₂)₂CMe(CH₂OH)}₂Cl₂]	6	t	VO, 1.592 (4) (t)	—	60, 136
(PPh₄)₂[V₂(O)₂(μ-OMe)₂Cl₄]	5	t	VO, 1.581 (4) (t)	—	60, 136
(PPh₄)₂[V₂(O)(μ-OEt)₂Cl₄]	5	t	VO, 1.592 (6) (t)	—	60, 136
[V₄(O)₄(μ-O)₄(μ₃-OMe)₂(OMe)₂(bpy)₂]	6	t, μ	VO, 1.597, 1.610, (1) (t) VO, 1.84 (av.) (μ)	V₂O, 113.9, 118.0	137
(Bu₄N₂)[V₄(O)₄{(OCH₂)₃CEt}₂(SO₄)₂(H₂O)₂]	6	t	VO, 1.592, 1.609 (6) (t)	—	60
[V₄(O)₄(μ-O)₂(μ-OEt)₂(OEt)₄(phen)₂]	4, 6	t, μ	V(ıᵥ)O, 1.592 (3) (t)	—	138
(NBu₄)₂[V(O)₆(μ₆-O)(μ-O)₆{(OCH₂)₃CNO₂}₂]	6	t, μ, μ₆	VO, 1.602 (av.) (t) VO, 1.819 (av.) (μ) VO, 2.244 (av.) (μ₆)	V₂O, 109.7 (av.) (μ) V₂O, 84.7 and 95.3 (av.) (μ₆)	56
Ba[V(O)₆(μ₆-O)(μ-OH)₃{(OCH₂)₃CMe}₃]	6	t, μ₆	VO, 1.596, 1.667 (24) (t) VO, 2.272, 2.328 (19) (μ₆)	V₂O, 89.2, 90.2, 179.2 (μ₆)	55
Na₂[V(O)₆(μ₆-O){(OCH₂)₃CEt}₄]	6	t, μ₆	VO, 1.593 (7) (t) VO, 2.315 (2) (μ₆)	—	55
(Me₃NH)[V(O)₆(μ₆-O)(μ-OH)₃{(OCH₂)₃CMe}₃]	6	t, μ₆	VO, 1.600, 1.608 (7) (t) VO, 2.236, 2.382 (8) (μ₆)	V₂O, 89.5 (μ₆)	55
Na[V₆(O)₆(μ₃-F)(μ-OH)₃{(OCH₂)₃CMe}₃]	6	t	VO, 1.549, 1.634 (t)	—	55
(NBu₄)[V₆(O)₆(μ₆-O)(μ-O)₅(μ-OMe)₇]	6	t, μ, μ₆	VO, 1.591 (av.) (t)	V₂O, 112.6 (μ)	57

(continued overleaf)

413

Table 5.5 (*continued*)

Compound	Metal coord.	Oxo coord.	M–Oxo bond lengths (Å)	M_x–Oxo bond angles (°)	Reference
$(NH_4)_4[V_{10}(O)_8(\mu_6\text{-}O)_2(\mu_3\text{-}O)_2(\mu\text{-}O)_4\{(OCH_2)_3CEt\}_4]$	6	t, μ, μ_3, μ_6	VO, 1.826 (av.) (μ); VO, 2.212 (av.) (μ_6); VO, 1.586–1.599 (3) (t)	–	53
$(NEt_4)[V_{10}(O)_8(\mu_6\text{-}O)_2(\mu_3\text{-}O)(\mu\text{-}O)_2\{(OCH_2)_3CEt\}_5]$	6	t, μ, μ_3, μ_6	VO, 1.656–2.017 (3) (μ); VO, 1.933–1.981 (3) (μ_3); VO, 2.014–2.447 (3) (μ_6); VO, 1.57–1.61 (1) (t)	–	53
$(Me_3NH)_2[V_{10}(O)_8(\mu_6\text{-}O)_2(\mu_3\text{-}O)_2(\mu\text{-}O)_2(\mu\text{-}OH)_2\{(OCH_2)_3CEt\}_4]$	6	t, μ, μ_3, μ_6	VO, 1.66–2.03 (1) (μ); VO, 1.91–2.00 (1) (μ_3); VO, 2.01–2.54 (1) (μ_6); VO, 1.596 (3) (av.) (t)	–	54
$Na_2[V_{10}(O)_8(\mu_6\text{-}O)_2(\mu_3\text{-}O)_2(\mu\text{-}O)_4\{(OCH_2)_3CEt\}_4]$	6	t, μ, μ_3, μ_6	VO, 1.869 (3) (av.) (μ); VO, 1.929 (3) (μ_3); VO, 2.289 (3) (av.) (μ_6); VO, 1.595 (4) (av.) (t)	–	54
$(NBu_4)_2[V_{10}(O)_8(\mu_6\text{-}O)_2(\mu_3\text{-}O)_2(\mu\text{-}O)_4\{OCH_2CMe\}_4]$	6	t, μ, μ_3, μ_6	VO, 1.795 (4) (av.) (μ); VO, 1.943 (4) (μ_3); VO, 2.275 (4) (av.) (μ_6); VO, 1.593 (5) (av.) (t)	–	54
$(NBu_4)_2[V_2Mo_2(O)_8(OMe)_2\{(OCH_2)_3C(CH_2OH)\}_2]$	6, 6	t	VO, 1.773 (4) (μ); VO, 1.935 (4) (μ_3); VO, 2.259 (4) (av.) (μ_6); VO, 1.614, 1.624 (6) (t); MoO, 1.700, 1.723 (6) (t)	–	60
$[Nb(O)(OEt)Cl_2(bpy)]$	6	t	NbO, 1.71 (t)	–	139
$[Nb_4(\mu\text{-}O)_2(\mu\text{-}OEt)_2(\mu\text{-}Cl)_2(OEt)_2Cl_6(thf)_4]$	6	μ	NbO, 1.816, 2.008 (7) (μ)	Nb_2O, 177.7	140
$[Nb_8(\mu\text{-}O)_2(\mu\text{-}O)_8(\mu\text{-}OEt)_6(OEt)_{14}]$	6	μ, μ_3	NbO, 2.022, 2.102 (4) (μ_3)	106.9–111.6 (μ_3)	84

414

Compound		Bridging	M–O distances	Angles	Ref.
[Nb$_4$(μ-O)$_4$(μ-O$_2$CCMe:CH$_2$)$_4$OPr^i)$_8$]	6	μ	NbO, 1.889–1.936 (3) (μ) av. 1.905	146.2, 150.2	142
[Nb$_4$(μ$_4$-O)$_2${(μ-(OCH$_2$)$_2$CMeCO$_2$}$_2$(OEt)$_{10}$]	6	μ	NbO, 1.911, 1.945 (5) (μ)	114.4	143
(NBu$_4$)$_3$[Nb$_2$W$_4$(O)$_6$(μ$_6$-O)(μ-O)$_{11}$(μ-OMe)]	6	t, μ, μ$_6$	MO, 1.59–1.74 (1) (t), av. 1.66, MO, 1.80–2.04 (1) (μ), av. 1.92, MO, 2.28–2.43 (1) (μ$_6$), av. 2.35	86.2–98.8, av. 90.0; and 172.9–176.7 (μ$_6$)	80
(NBu$_4$)$_3$[Nb$_2$W$_4$(O)$_5$(μ$_6$-O)(μ-O)$_{12}$(OSiMe$_2$Bu^t)]	6	t, μ, μ$_6$	MO, 1.68–1.72 (1) (t), av. 1.70	88.9–90.9, av. 90.0; and 178.2–179.8 (μ$_6$)	80
[Nb$_4$Pb$_6$(μ$_4$-O)$_4$(μ$_3$-OEt)$_4$(μ-OEt)$_{12}$(OEt)$_8$]	6, 6	μ$_4$	MO, 1.87–2.03 (2) (μ), av. 1.93, MO, 2.27–2.37 (2) (μ$_6$), av. 2.33 PbO, 2.38 (av.) (μ$_4$)	–	144
[Ta$_2$(μ-O)(μ-OPr^i)(OPr^i)$_7$(Pr^{i}OH)]	6	μ	TaO, 1.929, 1.930 (9) (μ)	114.3	85
[Ta$_7$(μ$_3$-O)$_3$(μ-O)$_6$(μ-OPr^i)$_4$(OPr^i)$_{13}$]	6	μ, μ$_3$	TaO, 1.977–2.141 (10) (μ$_3$) av. 2.065,	Ta$_3$O, 98.3–145.4 Ta$_2$O, 106.7–140.0	85
[Ta$_8$(μ$_3$-O)$_2$(μ-O)$_8$(μ-OEt)$_6$(OEt)$_{14}$]	6	μ, μ$_3$	TaO, 1.912–2.018 (9) (μ) av. 1.937 TaO, 2.031–2.098 (12) (μ$_3$) av. 2.055	Ta$_3$O, 107.4–111.1	145
[Ta$_5$(μ$_3$-O)$_4$(μ-O)$_3$(μ-OBu^t)(OBu^t)$_{10}$]	6	μ, μ$_3$	TaO, 1.836–2.017 (15) (μ) av. 1.926 TaO, 1.89–2.39 (2), (μ$_3$) av. 2.09	Ta$_2$O, 144.3–145.9 Ta$_3$O, 81.5–143.7	145
[Ta$_4$Zn$_2$(μ$_3$-O)(μ-O)$_2$(μ-OPr^i)$_6$(OPr^i)$_8$I$_2$]	6 (Ta) 4 (Zn)	μ, μ$_3$	TaO, 1.82–2.03 (2), (μ) av. 1.92 TaO, 2.01 (1) av. (μ$_3$) ZnO, 2.10 (1) av. (μ$_3$) TaO, 1.90 (1) av. (μ)	Ta$_2$O, 101.2–111.7 Ta$_2$O, 175.3	23

t = terminal.

415

Some vanadium(IV) oxo-alkoxides have also been studied. In $(PPh_4)_2[V_2(O)_2(\mu\text{-}OR)_2Cl_4]$ (R = Me, Et) the vanadium atoms are in edge-shared square pyramids with the oxo ligands occupying the apical sites.[136] The centrosymmetric tetranuclear V(v) complex $[V_4(O)_4(\mu\text{-}O)(\mu_3\text{-}OMe)_2(OMe)_2(bpy)_2]$ has an interesting structure involving edge-sharing octahedra (Fig. 5.14).[137] The two outer vanadiums each have terminal V=O bonds, and a chelating bipy ligand and are bonded by two μ-O ligands and one μ_3-OMe ligand to the central dimeric $V_2(O)_2(OMe)_2(\mu\text{-}O)_4(\mu_3\text{-}OMe)_2$ moiety.[137] In the vanadium(IV) tetranuclear anion $[V_4(O)_4\{(\mu_3,\mu,\mu\text{-}OCH_2)_3CEt\}_2(\mu\text{-}O_2SO_2)_2(H_2O)_2]^{2-}$ one type of V atom is bonded terminally to an oxo ligand and an aqua ligand and bridged by a sulphato and three oxygens from the organic (alkoxo) ligands. The other type is bonded terminally to an oxo ligand and by bridging to a sulphato and four oxygens from the organic ligands. All of the vanadiums are in distorted octahedral configurations.[60]

A neutral mixed-valence complex $[V_4(O)_4(\mu\text{-}O)_2(\mu\text{-}OEt)_2(OEt)_4(phen)_2]$ was shown to involve two outer $OV(OEt)_2(\mu\text{-}O)$ four-coordinated vanadium(v) moieties bridged to a central dinuclear $V_2(O)_2(\mu\text{-}OEt)_2(phen)_2(\mu\text{-}O)_2$ unit containing octahedral V(IV) atoms.[138]

Some structures of hexanuclear and decanuclear vanadium oxo-alkoxides have been reported. The hexavanadate anion $[V_6O_{13}\{(OCH_2)_3CNO_2\}_2]^{2-}$ has the μ_6-oxo centred cubic V_6O_{19} core with the two tridentate alkoxo ligands clamped on opposite faces of the V_6 octahedron.[56] In the vanadium(IV) complex $[V_6O_7(OH)_3\{(OCH_2)_3CMe\}_3]^{2-}$ there are three symmetrically placed tridentate alkoxide ligands and in $[V_6O_7\{(OCH_2)_3CEt\}_4]^{2-}$ there are four alkoxides completely replacing all twelve μ-oxo atoms of the V_6O_{19} core.[55] The $[V_6O_7(OH)_3\{(OCH_2)_3CMe\}_3]^{-}$ anion contains one V(v) and five V(IV) atoms whilst in $[V_6O_6(\mu_3\text{-}F)(OH)_3\{(OCH_2)_3CMe\}_3]^{-}$ a μ_6-oxo ligand has been replaced by a μ_3-fluoride with consequent distortion of the structure (Fig. 5.15).[55] The μ_3-fluoride also interacts significantly with the other three vanadiums (V-F, 2.245 (6) Å and 2.522 (6) Å). In the V(v) complex anion $[V_6(O)_6(\mu_6\text{-}O)(\mu\text{-}O)_5(\mu\text{-}OMe)_7]^{-}$ the monodentate methoxide ligands occupy seven of the twelve μ-oxo positions.[57]

The tridentate alkoxo ligand $(OCH_2)_3CEt$ has also been attached to the decavanadate core $[V_{10}O_{28}]$ in the decanuclear V(IV) complexes $[V_{10}(O)_8(\mu_6\text{-}O)_2(\mu_3\text{-}O)_2(\mu\text{-}O)_4\{(OCH_2)_3CEt\}_4]^{4-}$ and $[V_{10}(O)_8(\mu_6\text{-}O)_2(\mu_3\text{-}O)(\mu\text{-}O)_2\{(OCH_2)_3CEt\}_5]^{-}$. In addition to the all V(IV) complex $[V_{10}(O)_8(\mu_6\text{-}O)_2(\mu_3\text{-}O)_2(\mu\text{-}O)_2(\mu\text{-}OH)_2\{(OCH_2)_3C(CH_2OH)\}_4]^{2-}$ the mixed valence complexes (VIvVv) $[V_{10}(O)_8(\mu_6\text{-}O)_2(\mu_3\text{-}O)_2(\mu\text{-}O)_4\{(OCH_2)_3CR\}_4]^{2-}$ (R = Me, Et) have also been characterized.[54]

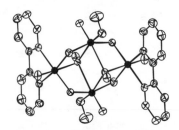

Figure 5.14 Structure of $[V_4(O)_4(\mu\text{-}O)_4(\mu_3\text{-}OMe)_2(OMe)_2(bpy)_2]$ ($\bullet$ = V; H atoms omitted).

Figure 5.15   Structure of $[V_6(O)_6(\mu_3\text{-}$
$F)(\mu\text{-}OH)_3\{(\mu,\mu,\mu,\eta^3\text{-}OCH_2)_3CMe\}_3]^-$
($\bullet = \mu_3\text{-}F$, $\oslash = V$; H atoms omitted).

Since the V(v) sites could be distinguished from the V(iv) sites these complexes were thus class I mixed-valence clusters.

A vanadium(v)-molybdenum(vi) heterometallic oxo-alkoxide $[V_2Mo_2(O)_8$ $(OMe)_2\{(OCH_2)_3C(CH_2OH)\}_2]^{2-}$ has also been characterized. The $V_2Mo_2O_{16}$ core involving octahedral metals is reminiscent of the Ti_4O_{16} core of titanium tetralkoxides.[60]

5.5.2   Niobium

The structures of a few niobium oxo-alkoxides have been reported. The mononuclear $[Nb(O)(OEt)Cl_2(bpy)_2]$ has an octahedral (*trans*-dichloride) structure.[139] In the tetranuclear Nb(iv) compound $[Nb_4(\mu\text{-}O)_2(\mu\text{-}OEt)_2(\mu\text{-}Cl)_2(OEt)_2Cl_6(thf)_4]$ (Nb–Nb, 2.891 Å) the two halves of the centrosymmetric molecule are connected by two linear (μ-O) ligands.[140] The structure of the octanuclear Nb(v) oxo-ethoxide $[Nb_8(\mu_3\text{-}O)_2(\mu\text{-}$ $O)_8(\mu\text{-}OEt)_6(OEt)_{14}]$ (Fig. 5.16) which was originally published in 1968[141] has been redetermined confirming the original structure.[84] The structure may be viewed as two trinuclear $Nb_3(\mu_3\text{-}O)(\mu\text{-}O)_4(\mu\text{-}OEt)_3(OEt)_5$ units containing edge-sharing NbO_6

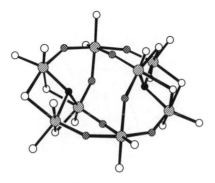

Figure 5.16 Nb_8O_{30} core structure of
$[Nb_8(\mu_3\text{-}O)_2(\mu\text{-}O)_8(\mu\text{-}OEt)_6(OEt)_{14}]$
($\oslash = Nb$, $\bullet = \mu_3\text{-}oxo$, $\circledast = \mu\text{-}oxo$,
$\bigcirc = O$ in OEt).

octahedra joined by two $Nb(\mu\text{-}O)_4(OEt)_2$ octahedra via μ-oxo bridges. In the tetranuclear mixed ligand compound $[Nb_4(\mu\text{-}O)_4(\mu\text{-}O_2CCMe\text{:}CH_2)_4(OPr^i)_8]$ the Nb_4O_4 framework is a planar eight-membered ring. The octahedral niobiums each have two bridging methacrylato ligands and two *cis*-terminal isopropoxo groups.[142] The centrosymmetrical molecule $[Nb_4(\mu\text{-}O)_2\{\mu\text{-}(OCH_2)_2CMeCO_2\}_2(OEt)_{10}]$ consists of two $(EtO)_2Nb(OEt)_3$ moieties linked by two tetradentate bis-methoxo proprionato ligands. Two octahedral niobiums have three (*fac*) terminal ethoxo groups and the other two Nb atoms have two (*cis*) terminal ethoxo groups.[143] Two niobium heterometallic oxo-alkoxide anions $[Nb_2W_4(O)_6(\mu_6\text{-}O)(\mu\text{-}O)_{11}(\mu\text{-}OMe)]^{3-}$ and $[Nb_2W_4(O)_5(\mu_6\text{-}O)(\mu\text{-}O)_{12}(OSiMe_2Bu^t)]^{3-}$ have been characterized. The μ_6-oxo-centred M_6O_{19} core framework has a distorted nonstatistical arrangement of the six metal atoms owing to the presence of several stereoisomers. In the methoxo derivative the methoxo group bridges two niobium atoms whilst the bulkier trialkylsiloxo group occupies a terminal site.[80]

The decanuclear niobium lead oxo-ethoxide $[Nb_4Pb_6(\mu_4\text{-}O)_4(\mu_3\text{-}OEt)_4(\mu\text{-}OEt)_{12}(OEt)_8]$ contains an octahedral framework of Pb atoms four of whose faces are capped by μ_4-oxo ligands which are also each bonded to a $Nb(OEt)_5$ moiety whilst the remaining four faces are capped by μ_3-ethoxo ligands. Each niobium atom is linked to three lead atoms by a μ_4-oxo ligand and three μ-ethoxo ligands. Both metals attain six-coordination.[144]

The structures of tantalum oxo-alkoxides have only been reported since the mid-1990s. Two oxo-isopropoxides have been determined, one being binuclear $[Ta_2(\mu\text{-}O)(\mu\text{-}OPr^i)(OPr^i)_7(Pr^iOH)]$ and the other heptanuclear $[Ta_7(\mu_3\text{-}O)_3(\mu\text{-}O)_6(\mu\text{-}OPr^i)_4(OPr^i)_{13}]$.[85] The binuclear molecule consists of two edge-bridged (μ-O, μ-OPr^i) octahedra distorted by the hydrogen bonding between terminal isopropanol and isopropoxo groups on adjacent metal atoms. The heptanuclear species has a complex structure (Fig. 5.17) involving seven octahedrally coordinated tantalum atoms. Thirty years after the publication of the structure of $[Nb_8O_{10}(OEt)_{20}]$[141] the tantalum analogue $[Ta_8(\mu_3\text{-}O)_2(\mu\text{-}O)_8(\mu\text{-}OEt)_6(OEt)_{14}]$ has now been shown to have the same structure as the niobium compound.[145] The structure of $[Ta_7O_9(OPr^i)_{17}]$ has also been determined independently whilst a novel oxo-*tert*-butoxide was shown to be the pentanuclear $[Ta_5(\mu_3\text{-}O)_4(\mu\text{-}O)_3(\mu\text{-}Bu^t)(OBu^t)_{10}]$ containing highly distorted octahedrally coordinated tantalums.[145]

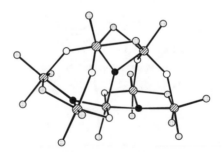

Figure 5.17　Ta_7O_{26} core structure of $[Ta_7(\mu_3\text{-}O)_3(\mu\text{-}O)_6(\mu\text{-}OPr^i)_4(OPr^i)_{13}]$ ($\oslash = Ta$, $\bullet = \mu_3$-oxo, $\bigcirc = \mu$-oxo, $\circledcirc = O$ in OPr^i).

5.5.3   Tantalum

The hexanuclear tantalum heterometallic oxo-isopropoxide [Ta$_4$Zn$_2$(μ_3-O)$_2$(μ-O)$_2$(μ-OPr^i)$_6$(OPr^i)$_8$I$_2$] has a centrosymmetric structure comprising two triangular Ta$_2$Zn(μ_3-O)(μOPr^i)$_3$I units linked by two nearly linear Ta(μ-O)Ta bridges. Tantalum has distorted octahedral and zinc tetrahedral coordination.[23]

5.6   Chromium Sub-group Metal-containing Oxo-alkoxides (Including Heterometallic Compounds) (Table 5.6)

5.6.1   Chromium

A small number of oxo-alkoxides of chromium have been structurally characterized. The mononuclear chromium(VI) compounds [Cr(O)$_2$Cl(OCPh$_3$)] and [Cr(O)$_2$(OCPh$_3$)$_2$] both involve tetrahedrally coordinated chromium.[147] Chromium(V) is five-coordinated (tetragonal pyramidal) with the terminal oxo-ligand in the apical position in the mononuclear [Cr(O){O$_2$C$_2$(CF$_3$)$_4$}$_2$]$^-$ and binuclear [Cr$_2$(O)$_2$(μ-O)$_2${O$_2$C$_2$(CF$_3$)$_4$}$_2$]$^{2-}$ anions.[48]

5.6.2   Molybdenum

There is an abundance of structurally characterized molybdenum oxo-alkoxides. Several mononuclear species have been reported. In the *cis*-dioxo molybdenum(VI) derivatives [Mo(O)$_2$(OC$_2$H$_4$OH)$_2$]^{50} and [Mo(O)$_2$(OCHMeCHMeOH)$_2$]^{148} the metal is in a distorted octahedral configuration. A similar structure was found in [Mo(O)$_2$(OC$_2$H$_4$OMe)$_2$].[149] The tridentate alkoxo ether ligand OC$_2$H$_4$OC$_2$H$_4$O forms a *mer*-complex [Mo(O)$_2$(OC$_2$H$_4$OC$_2$H$_4$O)] in which Mo achieves six-coordination by intermolecular bonding (Mo=O....Mo) giving a linear polymer.[51] A similar situation arises in the pyridine-2,6-dimethoxo complex [Mo(O)$_2${(OCH$_2$)$_2$C$_5$H$_3$N}].[70] In the octahedral *cis*-dioxo complex [Mo(O)$_2$(OPr^i)$_2$(bpy)] the isopropoxo ligands are *trans*.[34] The monoxo molybdenum tetrakis-perfluoro-*tert*-butoxo compound [MoO{OC(CF$_3$)$_3$}$_4$] exhibits five-coordinated Mo (distorted TBP).[150] Alkoxo-bridging leads to binuclear species as in all-*cis*-[Mo$_2$(O)$_4$(μ-OMe)$_2$(OC$_2$Me$_4$OH)$_2$]^{151} and [Mo$_2$(O)$_2$(μ-OMe)$_2$(OMe)$_6$].[15] The novel unsymmetrical binuclear species [(PMe$_3$)$_3$(Bu^{t}CH$_2$O)Mo(μ-O)(μ-OCH$_2$Bu^t)Mo(OCH$_2$Bu^t)$_2$(PMe$_3$)] (Mo–Mo, 2.493 Å) contains octahedral and five-coordinated (TBP) Mo atoms, which are edge-bridged by the oxo and one alkoxo ligand.[152] In the binuclear Mo(V) complex [Mo$_2$(O)$_2$(μ-OEt)$_2$(μ-EtOH)Cl$_4$] (Mo–Mo, 2.683 Å) there is a confacial bi-octahedral structure with the rather unusual bridging ethanol molecule.[27]

The trinuclear molecules [Mo$_3$(μ-O)(μ_3-OR)(μ-OR)$_3$(OR)$_6$] (R = Pr^i, CH$_2$Bu^t) have the familiar triangulo-Mo$_3$ unit capped above and below the metal plane by μ_3-oxo and μ_3-alkoxo ligands. The Mo$_3$O$_{11}$ framework enables each Mo(IV) atom to achieve a distorted octahedral coordination.[34] The electronic structure of these compounds[153] and their relationship to metal oxide structures[154] have been discussed. Two trinuclear anions [Mo$_3$(O)$_7$(hmmp)$_2$]$^{2-}$ and [Mo$_3$(O)$_6$(OMe)(hmmp)$_2$]$^-$ (hmmp = (OCH$_2$)$_3$CMe) have also been determined.[59] In the former complex two metals are *cis*-bonded to two terminal oxo ligands and the third is facially bonded to three. In the

Table 5.6 Chromium sub-group metal-containing oxo-alkoxides (including heterometallic compounds)

Compound	Metal coord.	Oxo coord.	M–Oxo bond lengths (Å)	M_x–Oxo bond angles (°)	Reference
[Cr(O)$_2$Cl(OCPh$_3$)]	4	t	1.576, 1.594 (5)	–	147
[Cr(O)$_2$(OCPh$_3$)$_2$]	4	t	1.578 (2)	–	147
[(py)$_2$H$^+$][Cr(O){O$_2$C$_2$(CF$_3$)$_4$}$_2$]$^-$	5	t	1.533 (4)	–	48
[Li$_2$(H$_2$O)$_2$(py)$_2$][Cr$_2$(O)$_2$(μ-O)$_2${O$_2$C$_2$(CF$_3$)$_4$}$_2$]	5	t, μ	CrO, 1.570 (7) (t); CrO, 1.806, 1.818 (5) (μ)	Cr$_2$O, 92.4 (μ)	48
[MoO$_2$(OC$_2$H$_4$OH)$_2$]	6	t	1.723, 1.729 (10)	–	50
[MoO$_2$(OCHMeCHMeOH)$_2$]	6	t	1.662 (7)	–	148
[MoO$_2$(OC$_2$H$_4$OMe)$_2$]	6	t	1.698, 1.699 (3)	–	149
[MoO$_2$(OC$_2$H$_4$OC$_2$H$_4$O)]$_\infty$	6	t, μ	MoO, 1.63, 1.73 (2) (t); MoO, 2.38 (2) (μ)	Mo$_2$O, 180	51
[MoO$_2${(OCH$_2$)$_2$C$_5$H$_3$N]]	6	t, μ	MoO, 1.710, 1.719 (3) (t); MoO, 2.518 (3) (μ)	Mo$_2$O, 157.1	70
[MoO$_2$(OPr^i)$_2$(bpy)]	6	t	1.689, 1.723 (7)	–	34
[MoO{OC(CF$_3$)$_3$}$_4$]	5	t	1.60 (3)	–	150
[Mo$_2$(O)$_4$(μ-OMe)$_2$(OC$_2$Me$_4$OH)$_2$]	6	t	1.68 (1)	–	151
[Mo$_2$(O)$_2$(μ-OMe)$_2$(OMe)$_6$]	6	t	1.625 (5)	–	15
[Mo$_2$(μ-O)(μ-OCH$_2$Bu^t)(OCH$_2$Bu^t)$_3$(PMe$_3$)$_4$]	5, 6	μ	1.874, 1.996 (4)	80.1	152
[Mo$_2$(O)$_2$(μ-OEt)$_2$(μ-EtOH)Cl$_4$]	6	t	1.638–1.659 (8); av. 1.649	–	27
[Mo$_3$(μ_3-O)(μ_3-OPr^i)(μ-OPr^i)$_3$(OPr^i)$_6$]	6	μ_3	2.055–2.068 (8); av. 2.061	75.7–76.1; av. 75.8	34
[Mo$_3$(μ-O)$_3$(μ_3-OCH$_2$Bu^t)(μ-OCH$_2$Bu^t)$_3$(OCH$_2$Bu^t)$_6$]	6	μ_3	2.026–2.039 (25); av. 2.034	76.5–77.3; av. 76.9	34
(NBu$_4$)$_2$[Mo$_3$(O)$_7$(hmmp)$_2$]; hmmp = (OCH$_2$)$_3$CMe	6	t	1.677–1.754 (3); av. 1.703	–	59
(NBu$_4$)$_2$[Mo$_3$(O)$_6$(OMe)(hmmp)$_2$]	6	t	1.686–1.712 (7); av. 1.697	–	59
[Mo$_4$(O)$_4$(μ_3-O)$_2$(μ-OPr^i)$_4$(OPr^i)$_2$Cl$_4$]	6	t, μ_3	MoO, 1.617, 1.634 (16) (t); MoO, 1.962, 1.993, 2.237 (13) (μ_3)	Mo$_3$O, 84.9, 104.9, 109.4	155

Compound					Ref.
[Mo₄(O)₄(μ₃-O)₂(μ-OPrⁱ)₂(OPrⁱ)₂(py)₄]	6	t, μ, μ₃	MoO, 1.682, 1.697 (3) (t) MoO, 1.937, 1.947 (3) (μ) MoO, 1.978, 2.040, 2.183 (3) (μ₃)	Mo₃O, 80.6, 101.2, 110.6 Mo₂O, 84.1	156
[Mo₄(O)₄(hmmp)₂(OEt)₂]	6	t	1.683–1.689 (5) av. 1.687	–	158
[K(18-crown-6)₂][Mo₄(μ₃-O)(μ-OCH₂Buᵗ)₄(OCH₂Buᵗ)₇]	4, 6	μ₃	2.029, 2.040, 2.046 (4)	73.1, 79.3, 80.1	159
[Mo₆(O)₆(μ-O)₄(μ-OPrⁱ)₆(OPrⁱ)₆]	5, 6	t, μ	MoO, 1.671, 1.673, 1.691 (2) (t) MoO, 1.921–1.939 (2) (μ) av. 1.931	Mo₂O, 83.9, 84.1	34, 160
[Mo₆(O)(μ₆-O)(μ-OEt)₁₂(OEt)₆]	5¹	μ₆	2.1073 (4)	76.5	161
[Mo₆(O)₆(μ₃-O)₂(μ-OEt)₆(μ-Cl)₂Cl₆]	5¹, 6	t, μ₃	MoO, 1.935, 2.001, 2.194 (3) (μ₃) MoO, 1.695–1.759 (2) (t) av. 1.715	Mo₃O, 85.5, 107.2, 164.2	162
[Na(MeOH)₄]₂[Mo₈(O)₁₆(μ₄-O)₂(μ₃-O)₄(μ-O)₂(μ-OMe)₂(OMe)₂]	6	t, μ, μ₃, μ₄	MoO, 1.929, 1.946 (2) (μ) MoO, 1.918–2.260 (2) (μ₃) av. 2.096, MoO 2.251, 2.280 (2) (μ₄)	–	52
[Mo₈(O)₈(μ-O)₈(μ-OMe)₈(PMe₃)₄]	5, 6	t, μ	MoO 1.68 (t) MoO, 1.93 (μ)	Mo₂O, 83.5	163
(NBu₄)₂[Mo₈(O)₁₄(μ-O)₆(μ₃-OMe)₂(OMe)₂[(OCH₂)₃CMe]₂]	6,	t, μ	MoO, 1.697 (7) av. (t) MoO, short, 1.765–1.883 (6) (μ) long, 1.946–2.248 (6) (μ)	–	58
(NBu₄)₄[Ag₂Mo₁₀(O)₈(μ₅-O)₂(μ-O)₁₆(μ-OMe)₈(NO)₂]	4, 6	t, μ, μ₅	MoO, 1.692–1.704 (6) (t) av. 1.698 MoO, 1.716, 1.720 (6) (μ) av. 1.717 AgO, 2.342–2.477 (7) (μ) av. 2.381	–	167a

(continued overleaf)

421

Table 5.6 (continued)

Compound	Metal coord.	Oxo coord.	M–Oxo bond lengths (Å)	M_x–Oxo bond angles (°)	Reference
(NBu$_4$)$_2$[Mo$_5$(O)$_4$(μ_5-O)(μ-O)$_8$(μ-OMe)$_4$(NO)(Na, MeOH)]	5 (Na) 6 (Mo)	t, μ, μ_5	MoO, 1.691–1.699 (7) (t) av. 1.695 Mo$_2$O, 1.905–1.937 (6) (μ) av. 1.919 MoO, 2.122 (5) (μ_5, ap) 2.319–2.363 (5) (μ_5, bs) av. 2.343 MoO(Na), 1.705–1.746 (7) av. 1.723 NaO(Mo), 2.350–2.401 (9) av. 2.379	–	167b
K$_2$[Mo$_5$(O)$_4$(μ_5-O)(μ-O)$_8$(μ-OMe)$_4$(NO)[Na(H$_2$O)(MeOH)]]	7 (Na) 6 (Mo)	t, μ_5, μ	MoO, 1.68–1.70 (1), av. 1.69 (t) Mo$_2$O, 1.900–1.919 (9) (μ) av. 1.91 MoO, 2.154 (8) (μ_5, ap) 2.304–2.363 (9) (μ_5, bs) av. 2.336 MoO(Na), 1.71–1.73 (1) av. 1.72, NaO(Mo), 2.52–2.66 (1) av. 2.59	–	167b
(NMe$_4$)$_2$[Mo$_5$(O)$_4$(μ_5-O)(μ-O)$_8$(μ-OMe)$_4$(NO)(Na, H$_2$O)]	6 (Na) 6 (Mo)	t, μ_5, μ	MoO, 1.691–1.719 (9), av. 1.706 (t) Mo$_2$O, 1.883–1.936 (9) (μ) av. 1.922 MoOμ_5, 2.155 (8) (μ_5, ap) 2.286–2.404 (8) (μ_5, bs) av. 2.345	–	167b

Compound	Coord. no.	Type	Distances (Å)	Angles (°)	Ref.
$(NBu_4)_2[Mo_5(O)_4(\mu_5\text{-}O)(\mu\text{-}O)_8(\mu\text{-}OMe)_4(NO)(Na,dmf)]$	5 (Na), 6 (Mo)	t, μ_5, μ	MoO(Na), 1.717–1.743 (8), av. 1.727, NaO(Mo), 2.39–2.60 (1), av. 2.50; MoO, 1.688–1.705 (7) (t), av. 1.700; Mo_2O, 1.895–1.934 (6) (μ), av. 1.912; MoO, 2.131 (5) (μ_5, ap), 2.312–2.398 (7) (μ_5, bs), av. 2.341; MoO(Na), 1.697–1.723 (7), av. 1.710, NaO(Mo), 2.376–2.407 (9), av. 2.385	—	167b
$(NEt_4)[W(O)_2(OCMe_2CMe_2O)(CH_2Bu^t)]$	5	t	1.712, 1.720 (5)	—	71
$[W(O)(OBu^t)_4(thf)]$	6	t	1.77 (3)	—	164
$[W(O)(OSiMe_2Bu^t)_4(py)]$	6	t	1.68 (1)	—	41
$[W(O)(OBu^t)_2(OC_{10}H_5MeC_{10}H_5MeO)]$	5	t	1.662 (6)	—	165
$[W_2(O)_2(\mu\text{-}O)_2(OSiMe_2Bu^t)_4(py)_2]$	6	t, μ	WO, 1.72 (1) (t); WO, 1.93, 1.95 (1) (μ)	W_2O, 103.2	41
$[W_2(\mu\text{-}O)(\mu\text{-}OBu^t)(OSiMe_2Bu^t)_5(py)_2]$	5, 6	μ	1.89, 1.96 (1)	W_2O, 80.7	41
$[(PhN)(Bu^tO)_2W(\mu\text{-}O)(\mu\text{-}OBu^t)_2W(NPh)(OBu^t)_2]$	6	μ	1.937 (4)	W_2O, 104.3	40
$[(PhN)(HOC_2Me_4O)(OC_2Me_4O)W]_2(\mu\text{-}O)]$	6	μ	1.87, 1.92 (4)	W_2O, 157	166
$[W_2(O)_2(\mu\text{-}OMe)_2(OMe)_6]$	6	t	1.696, 1.702 (7)	—	49
$[W_2(O)_2(\mu\text{-}OC_6H_{11}\text{-}c)_2(OC_6H_{11}\text{-}c)_6]$	6	t	1.691 (4)	—	49
$[(CF_3CO_2)_2(Bu^tCH_2O)_2W(\mu\text{-}OCH_2Bu^t)(\mu\text{-}CMe_2)W(O)(OCH_2Bu^t)(py)]$	5, 6	t	1.694 (4)	—	36
$[(Bu^tCH_2O)_3W(\mu\text{-}CH_2Bu^t)_2(\mu\text{-}CHCH_3H_5\text{-}c)W(O)(OCH_2Bu^t)(py)]$	6	t	1.713 (5)	—	37

(continued overleaf)

Table 5.6 (continued)

Compound	Metal coord.	Oxo coord.	M–Oxo bond lengths (Å)	M_x–Oxo bond angles (°)	Reference
[{(HOC$_2$Me$_4$O)(OC$_2$Me$_4$O)(O)W}$_2$(μ-O)]	6	t, μ	WO, 1.689 (7) (t) WO, 1.900 (2) (μ)	W$_2$O, 161.4	73
[{(HOC$_7$H$_{12}$O)(c-C$_7$H$_{12}$O$_2$)(O)W}$_2$(μ-O)]	6	t, μ	WO, 1.68, 1.71 (2) (t) WO, 1.89, 1.92 (2) (μ)	W$_2$O, 154.3	73
[{(HOC$_8$H$_{14}$O)(c-C$_8$H$_{14}$O$_2$)(O)W}$_2$(μ-O)]	6	t, μ	WO, 1.66, (2) (t) WO, 1.81, 2.00 (2) (μ)	W$_2$O, 154.6	73
[K$_2$ (18-crown-6)$_3$][{(Bu^{t}O)$_3$ W(μ-O)(μ-H)W(OBu^t)$_3$}$_2$]	5	μ	WO, 1.916, 1.932 (12) (μ)	–	26
[W$_3$(μ-O)$_2$(OBu^t)$_8$]	3^1, 4^1, 5^1	μ	WO, 1.81–1.99 (2) (μ) av. 1.89	–	170
[W$_4$(μ-O)$_2$(μ-OPr^i)$_3$(OPr^i)$_9$]	3^1, 4^1, 6^1	μ	WO, 1.845–2.008 (13) (μ) av. 1.913	W$_2$O, 99.8, 101.9	35
[W$_4$(μ-O)(μ-OPr^i)$_2$(OPr^i)$_7$Cl]	2^1, 4^1	μ	WO, 1.921–1.924 (10) (μ)	W$_2$O, 95.6	72
[W$_4$(μ_4-C)(μ-O)(μ-OCH$_2$Bu^t)$_4$(OCH$_2$Bu^t)$_8$]	4^1, 5^1	μ	WO, 1.926–1.933 (11) (μ)	W$_2$O, 89.0	38
[W$_4$(μ_4-C)(μ-O)(μ-OPr^i)$_4$(OPr^i)$_8$]	4^1, 5^1	μ	WO, 1.897, 1.956 (7) (μ)	W$_2$O, 89.4	38
[{(CO)(dmpe)$_2$W·W(O)$_2$}$_2$(μ-O)]	3^1, 5^1	t, μ	WO, 1.73 (1) (t) WO, 1.88 (1) (μ)	W$_2$O, 180	171
[W$_6$(μ-O)$_4$(μ-CEt)$_2$(OBu^t)$_{10}$]	3^1, 4^1	μ	WO, 1.895, 1.896 (8) and 1.755, 2.119 (8) (μ)	W$_2$O, 99.6, 156.0	172
[W$_6$(μ-O)$_4$(μ-CPr^n)$_2$(OBu^t)$_{10}$]	3^1, 4^1	μ	WO, 1.878, 1.906 (9) and 1.774, 2.113 (9) (μ)	W$_2$O, 99.7, 159.6	164
[W$_8$(μ-O)$_4$(μ-OPr^i)$_4$(OPr^i)$_{12}$]	2^1, 4^1	μ	WO, 1.884–1.952 (10) (μ) av. 1.918	W$_2$O, 97.8, 98.4	173

ap = apical, bs = basal, t = terminal.
[1]Coordination number doubtful.

latter complex the third metal is *cis*-bonded to two terminal oxo ligands and a terminal methoxo group.

It is noteworthy that one of the earliest examples of a tetranuclear molybdenum oxo-alkoxide, the mixed ligand species $[Mo_4(O)_4(\mu_3-O)_2(\mu-OPr^n)_4(OPr^n)_2Cl_4]$, has a core framework $Mo_4O_{12}Cl_4$ analogous to the M_4O_{16} core of the titanium tetra-alkoxides. The presence of metal–metal bonding was indicated by Mo–Mo, 2.669 Å in two pairs of metals.[155] A similar structure was found in $[Mo_4(O)_4(\mu_3-O)_2(\mu-O)_2(\mu-OPr^i)_2(py)_4]$ (Mo–Mo, 2.600 Å).[156] However, the *n*-propoxide reported earlier[155] was reformulated as an Mo(v) complex $[Mo_4(O)_4(\mu-O)_2(\mu-OPr^n)_4(Pr^nOH)_2Cl_4]$.[157] The capping nature of the pyrazolylborato ligand (HB(pz)$_3$] generated a tetranuclear species $[Mo_4(O)_4(\mu-O)_4(\mu-OMe)_2(MeOH)_2\{HB(pz)_3\}_2]$ having a zig-zag chain of metal atoms.[157] Using the hmmp ligand the neutral centrosymmetric tetranuclear complex $[Mo_4(O)_4(hmmp)_2(OEt)_2]$ was obtained.[158] By contrast the tetranuclear anion $[Mo_4(\mu_3-O)(\mu-OCH_2Bu^t)_4(OCH_2Bu^t)_7]^-$ has a "butterfly" configuration of the Mo$_4$ unit (Mo–Mo, 2.422 Å, 2.484 Å, 2.608 Å) with one Mo$_3$ triangle being capped by the μ_3-oxo ligand. One wingtip Mo is octahedrally coordinated but the other metals are only four-coordinated to oxygen donor ligands.[159]

The hexanuclear molecule $[Mo_6(O)_6(\mu-O)_4(\mu-OPr^i)_6(OPr^i)_6]$ contains a remarkable serpentine chain of six coplanar Mo atoms linked by bridging oxo or isopropoxo ligands. The structure is centrosymmetrical (Fig. 5.18) with the exterior octahedrally coordinated Mo(vi) atoms edge-sharing $(\mu-OPr^i)_2$ with adjacent five-coordinated (square pyramidal) Mo(v) atoms which in turn are $(\mu-oxo)_2$ bridged and metal–metal bonded (Mo–Mo, 2.585 Å) to the inner Mo(v) atoms.[34, 160]

The interesting μ_6-oxo centred hexanuclear species $[Mo_6(\mu_6-O)(\mu-OEt)_{12}(OEt)_6]$ has a distorted octahedron of Mo atoms joined by μ-OEt bridges along all twelve edges of the octahedron. Each metal also has one terminal ethoxo ligand.[161] The average oxidation state is 3.33 $[2x Mo(IV) + 4x Mo(III)]$ and it is the metal–metal bonding which causes the distortion of the Mo$_6$ core. The structure can be viewed as two trinuclear $[Mo_3(\mu_3-O)(\mu-OEt)_6(OEt)_3]$ clusters each containing eight electrons for Mo–Mo bonding (2.611 Å) and sharing the μ_3-oxo and three μ-OEt ligands with

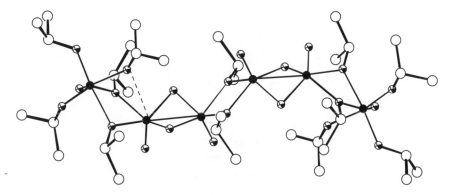

Figure 5.18   Structure of $[Mo_6(O)_6(\mu-O)_4(\mu-OPr^i)_6(OPr^i)_6]$ ($\bullet$ = Mo;   H atoms omitted).

the other cluster giving the central μ_6-oxo ligand. The Mo_6O_{19} core is of course reminiscent of the transition metal isopolyanions.

Another structural motif was exhibited by the structure of the mixed ligand hexanuclear molecule $[Mo_6(O)_6(\mu_3\text{-}O)_2(\mu\text{-}OEt)_6(\mu\text{-}Cl)_2Cl_6]$ with a cyclic array of six Mo(v) atoms in a chair conformation. The six valence electrons are paired off in three Mo–Mo bonds (2.665 Å, 2.672 Å, 2.741 Å) alternating around the ring. The μ_3-oxo ligands are in planar three-coordination (T-shaped).[162]

The octanuclear Mo(vi) complex anion $[Mo_8(O)_{16}(\mu_4\text{-}O)_2(\mu_3\text{-}O)_4(\mu\text{-}O)_2(\mu\text{-}OMe)_2(OMe)_2]^{4-}$ has an Mo_8O_{28} framework similar to that found in various substituted octamolybdates.[52]

Another octanuclear Mo(v) complex $[Mo_8(O)_8(\mu\text{-}O)_8(\mu\text{-}OMe)_8(PMe_3)_4]$ contains a puckered ring of eight Mo atoms. The molecule comprises a tetramer of the $[Mo_2(O)_2(\mu\text{-}O)_2(\mu\text{-}OMe)_2(PMe_3)]$ moiety which involves Mo–Mo bonds (2.567 Å) between the $Mo(\mu\text{-}O)_2Mo$ pairs. Four of the metals which have coordinated PMe_3 ligands are octahedrally coordinated and the other four are five-coordinated (square pyramidal).[163]

In the octanuclear Mo(vi) anion $[Mo_8(O)_{14}(\mu\text{-}O)_6(\mu_3\text{-}OMe)_2(OMe)_2\{(OCH_2)_3CMe\}_2]^{2-}$ the centrosymmetric species may be viewed as two $[Mo_4(O)_7(\mu\text{-}O)_2(\mu_3\text{-}OMe)_3(OMe)\{(OCH_2)_3CMe\}]^-$ units joined by two $(\mu\text{-}O)$ bridges (Fig. 5.19).[58]

A most remarkable mixed ligand heterometallic oxide-alkoxide anion $[Ag_2Mo_{10}(O)_8(\mu_5\text{-}O)_2(\mu\text{-}O)_{16}(\mu\text{-}OMe)_8(NO)_2]^{4-}$ has been characterized.[167a] The centrosymmetrical anions consist of two $Mo_5(O)_4(\mu_5\text{-}O)(\mu\text{-}O)_8(\mu\text{-}OMe)_4(NO)$ units connected by four $Mo(\mu\text{-}O)Ag(\mu\text{-}O)Mo$ bridges involving square planar AgO_4 with a short Ag....Ag distance (2.873 Å). All of the molybdenums are in distorted octahedral configurations.

The pentanuclear $Mo_5O_{13}(OMe)_4(NO)$ moiety was earlier found in the salts of the heterometallic alkali metal complexes $(NBu_4)_2[Mo_5(O)_4(\mu_5\text{-}O)(\mu\text{-}O)_8(\mu\text{-}OMe)_4(NO)(Na, MeOH)]$, $K_2[Mo_5(O)_4(\mu_5\text{-}O)(\mu\text{-}O)_8(\mu\text{-}OMe)_4(NO)\{Na(H_2O)(MeOH)\}]$, $(NMe_4)_2[Mo_5(O)_4(\mu_5\text{-}O)(\mu\text{-}O)_8(\mu\text{-}OMe)_4(NO)(Na,H_2O)]$ and $[NBu_4)_2[Mo_5(O)_4(\mu_5\text{-}O)(\mu\text{-}O)_8(\mu\text{-}OMe)_4(Na,DMF)]$.[167b] Each complex has a square pyramid of Mo atoms with the apical metal bonded to a linear NO ligand *trans* to the μ_5-oxo and the basal

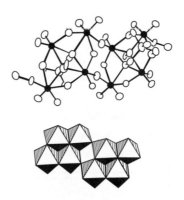

Figure 5.19 Structure of $[Mo_8(O)_{14}(\mu\text{-}O)_6(\mu_3\text{-}OMe)_2(OMe)_2\{(OCH_2)_3CMe\}_2]^{2-}$ ($\bullet$ = Mo; H atoms omitted).

metals bonded to terminal oxo ligands *trans* to the μ_5-oxo ligand. The four μ-OMe ligands bridge the apical Mo to the basal metal atoms which are linked to each other by μ-oxo ligands. The basal Mo atoms also have terminal oxo ligands (*trans* to the bridging μ-OMe oxygens) which are linked to the sodium atom. Each Mo is thus in a distorted octahedral configuration.

5.6.3 Tungsten

Like molybdenum its congener tungsten is represented by many oxo-alkoxides with various nuclearities. Among the mononuclear species the W(VI) anion $[W(O)_2(OCMe_2CMe_2O)(CH_2Bu^t)]^-$ exhibits trigonal pyramidal five-coordination with the terminal oxo ligands occupying equatorial positions with the neopentyl group in an axial site.[71] Two neutral octahedral complexes $[W(O)(OBu^t)_4(thf)]^{164}$ and $[W(O)(OSiMe_2Bu^t)_4(py)]^{41}$ have been reported. In the mixed ligand neutral oxo-alkoxide-aryloxide $[W(O)(OBu^t)_2(OC_{10}H_5Me.C_{10}H_5MeO)]$ the oxo ligand occupies an equatorial site in a distorted TBP structure.[165]

The binuclear W(VI) complex $[W_2(O)_2(\mu\text{-}O)_2(OSiMe_2Bu^t)_4(py)_2]$ has a $(\mu\text{-}O)_2$ edge-shared bioctahedral structure with the terminal oxo-ligands *trans* to the pyridines whereas the W(IV) compound $[W_2(\mu\text{-}O)(\mu\text{-}OBu^t)(OSiMe_2Bu^t)_5(py)_2]$ (W–W, 2.488 Å) has an unsymmetrical structure with octahedral and TBP metals bridged by μ-oxo and μ-*tert*-butoxo ligands.[41] By contrast the W(VI) phenylimido complex $[(PhN)(Bu^tO)_2W(\mu\text{-}O)(\mu\text{-}OBu^t)_2W(NPh)(OBu^t)_2]$ has a confacial bi-octahedral structure with a μ-oxo and two μ-OBu^t bridges[40] and the phenylimido diolato complex $[(PhN)(OC_2Me_4O)(HOC_2Me_4O)W(\mu\text{-}O)W(NPh)(OC_2Me_4O)(OC_2Me_4OH)]$ has an apex sharing bioctahedral structure.[166]

The dimeric molecules $[W_2(O)_2(\mu\text{-}OR)_2(OR)_6]$ (R = Me, c-C_6H_{11}) have terminal oxo ligands *trans* to the bridging alkoxo groups in an edge-shared bioctahedral structure.[49]

An unsymmetrical binuclear W(V) molecule was found in $[(CF_3CO_2)_2(Bu^tCH_2O)_2W(\mu\text{-}OCH_2Bu^t)(\mu\text{-}CMe_2)W(O)(OCH_2Bu^t)(py)]$ which had one octahedral and one TBP metal.[36] In the related molecule $[(Bu^tCH_2O)_3W(\mu\text{-}OCH_2Bu^t)_2(\mu\text{-}CHC_3H_5\text{-}c)\text{-}W(O)(OCH_2Bu^t)(py)]$ (W–W, 2.659Å) both metals achieve octahedral coordination through the triple-bridged confacial bi-octahedral structure.[37]

More examples of μ-oxo apex-sharing bi-octahedral structures are given by the diolato complexes $[(HOC_2Me_4O)(OC_2Me_4O)(O)W(\mu\text{-}O)W(O)(OC_2Me_4O)(OC_2Me_4OH)]$, $[(HOC_7H_{12}O)(O_2C_7H_{12}\text{-}c)(O)W(\mu\text{-}O)W(O)(O_2C_7H_{12}\text{-}c)(OC_7H_{12}OH)]$, and $[(HOC_8H_{14}O)(O_2C_8H_{14}\text{-}c)(O)W(\mu\text{-}O)W(O)(O_2C_8H_{14}\text{-}c)(OC_8H_{14}OH)]$.[73] The novel W(IV) binuclear anion $[(Bu^tO)_3W(\mu\text{-}O)(\mu\text{-}H)W(OBu^t_3)]^-$ (W–W, 2.445 Å) contains five-coordinated tungsten (TBP).[26]

The structure of $[W_3(\mu_3\text{-}O)(\mu_3\text{-}OPr^i)(\mu\text{-}OPr^i)_3(OPr^i)_6]$ has been reported[168] and it is isomorphous with $[Mo_3(\mu_3\text{-}O)(\mu_3\text{-}OPr^i)(\mu\text{-}OPr^i)_3(OPr^i)_6]$.[34] The trinuclear compounds $[W_3(\mu_3\text{-}O)(\mu_3\text{-}OPr^i)(\mu\text{-}OPr^i)_3(OPr^i)_4R_2]$ (R = CH_2Ph, Ph) have related structures.[169] A variation on these trinuclear structures was found in the W(IV) complex $[W_3(\mu\text{-}O)_2(OBu^t)_8]$ which has an isosceles triangle of metal atoms (W–W, 2.45Å, 2.93Å) with *tert*-butoxo bridges across the two longer sides and no capping ligand. Each tungsten atom has a different coordination with respect to oxygen (three, four and five).[170]

A few structures of tetranuclear tungsten oxo-alkoxides have been reported. In the remarkable molecule $[W_4(\mu\text{-O})_2(\mu\text{-OPr}^i)_3(OPr^i)_9]$ where W has the average oxidation state of four there is one triple bonded (W≡W, 2.404 Å) ditungsten $W_2(OPr^i)_5$ unit linked by two μ-oxo ligands to one edge of a confacial bi-octahedral $W_2(\mu_3\text{-OPr}^i)(OPr^i)$ unit (W–W, 2.684 Å). With respect to oxygen the tungstens are three, four, and six-coordinated.[35] Another unusual structure was found for $[W_4(\mu\text{-O})(\mu\text{-OPr}^i)_2(OPr^i)_7Cl]$ which has an apical two-coordinated tungsten $W(OPr^i)Cl$ capping a triangle of four-coordinated basal metals linked by metal–metal bonds. The μ-oxo ligand bridges two of the basal tungsten atoms which are each linked to the third basal tungsten by μ-OPr^i ligands. The oxygen coordination around the basal metals is approximately square planar.[72]

In the tetranuclear μ_4-carbido oxo-alkoxides $[W_4(\mu_4\text{-C})(\mu\text{-O})(\mu\text{-OR})_4(OR)_8]$ (R = CH_2Bu^t, Pr^i) the four metals adopt the "butterfly" configuration. In the neopentyloxide the μ-oxo ligand bridges the backbone tungsten atoms whereas in the isopropoxide it links one backbone W to a wingtip W.[38] A zig-zag array of four W atoms occurs in $[\{(CO)(dmpe)_2W.W(O)_2\}_2(\mu\text{-O})]$ (dmpe = $Me_2PC_2H_4PMe_2$) in which the exterior five-coordinated zero valent tungstens are metal–metal bonded (W–W, 2.648 Å) to three-coordinated pentavalent inner metals, which in turn are linked by the μ-oxo ligand. The diamagnetism is presumed to arise from spin pairing of oxo-bridged interior d^1-d^1 metal centres.[171]

The two centrosymmetrical hexanuclear oxo-alkoxides $[W_6(\mu\text{-O})_4(\mu\text{-CR})_2(OBu^t)_{10}]$ (R = Et, Pr^n) reported[164,172] have the same general structure involving two trinuclear subunits joined by a pair of unsymmetrical μ-oxo bridges (W–O, 1.755 Å, 2.119 Å). Each isosceles W_3 triangle has a μ-oxo and a μ-carbyne ligand bridging the longer sides and each metal has two terminal *tert*-butoxo groups. The only octanuclear tungsten oxo-alkoxide reported to date is the dimeric W_4 unit in $[W_8(\mu\text{-O})_4(\mu\text{-OPr}^i)_4(OPr^i)_{12}]$.[173] This molecule is centrosymmetric and the two W_4 units are connected by a $W(\mu\text{-OPr}^i)_2W$ double bridge. Each W_4 unit consists of a W_3 triangle bridged by two μ-oxo and one μ-OPr^i group, and linked to the fourth tungsten by three tungsten–tungsten bonds.

5.7   Manganese Sub-group Metal-containing Oxo-alkoxides (Including Heterometallic Compounds) (Table 5.7)

5.7.1   Manganese

The use of the tridentate alkoxo ligand $MeC(CH_2O)_3$ gave rise to the decanuclear mixed-valence manganese oxo-alkoxo-chloro anion $[Mn_{10}(\mu_6\text{-O})_2\{(OCH_2)_3CMe\}_6Cl_8]^{2-}$.[88] The $Mn_{10}O_{20}Cl_8$ structure (Fig. 5.20) can be considered as the fusion of two octa-hedral $Mn_6O_{13}Cl_6$ units by eliminating two *cis* manganese moieties. The six alkoxo ligands symmetrically cap the eighteen surface bridging oxo-sites and the mixed valence is accounted for by eight Mn(III) and two Mn(II) atoms.

The remarkable mixed-ligand mixed valence neutral tridecanuclear molecule $[Mn_{13}(\mu_5\text{-O})_6(\mu_3\text{-O})_2(\mu_3\text{-OEt})_6(\mu\text{-O}_2CPh)_{12}]$ has a supercubane structure with a central Mn(IV) atom octahedrally surrounded by six μ_5-oxo ligands which are symmetrically bonded to six Mn(III) and six Mn(II) atoms whose octahedral coordination is completed by the twelve bidentate bridging benzoate ligands.[89]

Table 5.7 Manganese sub-group metal-containing oxo-alkoxides (including heterometallic compounds)

Compound	Metal coord.	Oxo coord.	M–Oxo bond lengths (Å)	M_x–Oxo bond angles (°)	Reference
$(NEt_4)_2[Mn_{10}(\mu_6\text{-}O)_2\{(OCH_2)_3CMe\}_6Cl_8]$	6	μ_6	MnO, 1.95–2.54 (2) (μ_6) av. 2.37	–	88
$[Mn_8Sb_4(\mu_5\text{-}O)_4(\mu_3\text{-}OEt)_2(\mu\text{-}OEt)_{18}]$	5, 6	μ_5	MnO, 2.23–2.38 (μ_5) av. 2.30, SbO, 1.96, 1.97 (μ_5)		25
$(AsPh_4)[Tc(O)(OC_2H_4S)_2]$	5	t	TcO, 1.662 (5) (t)	–	74
$[Tc(O)(OC_2H_4O)Cl(phen)]$ phen = orthophenanthroline	6	t	TcO, 1.661 (4) (t)	–	75
$[Re_2(O)_2(\mu\text{-}O)(\mu\text{-}OMe)_2(OMe)_4]$	6	t, μ	ReO, 1.690, 1.703 (12) (t) ReO, 1.916, 1.917 (13) (μ)	Re$_2$O, 83.8	76
$[\{MeRe(O)(OC_2H_4O)(py)\}_2(\mu\text{-}O)]$	6	t, μ	ReO, 1.691 (8) (t) ReO, 1.859 (1) (μ)	180	86
$[MeRe(O)_2(OC_2Me_4O)]$	5	t	ReO, 1.657, 1.702 (2) (t)	–	86
$[Re(O)\{OCMe(CF_3)_2\}_3(thf)_2]$	6	t	ReO, 1.681 (4) (t)	–	174
$[Re(O)\{OCMe(CF_3)_2\}_3(CHCH{:}CPh_2)(thf)]$	6	t	ReO, 1.674 (7) (t)	–	174

t = terminal.

429

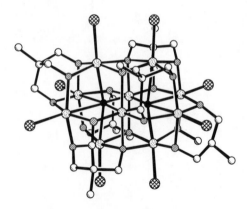

Figure 5.20 Structure of $[Mn_{10}(\mu_6\text{-}O)_2\{(OCH_2)_3CMe\}_6Cl_8]^{2-}$ ($\bigcirc$ = Mn, $\bullet$ = μ_6-oxo, $\oslash$ = O in ligand, $\circledast$ = Cl; H atoms omitted).

The heterometallic oxo-ethoxide $[Mn_8Sb_4(\mu_5\text{-}O)_4(\mu_3\text{-}OEt)_2(\mu\text{-}OEt)_{18}]$ also contains μ_5-oxo ligands with 4 octahedral and 4 five-coordinated Mn atoms and 4 five-coordinated Sb(III) atoms which have the electron lone pair in the vacant octahedral position.[25]

5.7.2 Technetium and Rhenium

Technetium is represented by two oxo-alkoxides, the five-coordinated anion $[Tc(O)(OC_2H_4S)_2]^-$,[74] and the octahedral neutral species $[Tc(O)(OC_2H_4O)Cl(phen)]$.[75] The binuclear rhenium(VI) oxo-methoxide $[Re_2(O)_2(\mu\text{-}O)(\mu\text{-}OMe)_2(OMe)_4]$ (Re–Re, 2.559 Å) has the confacial bi-octahedral structure,[76] whilst in the centrosymmetrical molecule $[(C_5H_5N)(OC_2H_4O)MeRe(O)(\mu\text{-}O)Re(O)Me(OC_2H_4O)(C_5H_5N)]$ the octahedra share a common μ-oxo apex and the mononuclear Re(VII) $[MeRe(O)_2(OC_2Me_4O)]$ has a distorted TBP structure.[86] Examples of distorted octahedral Re(V) are found in the mononuclear species *fac*-$[Re(O)\{OCMe(CF_3)_2\}_3(thf)_2]$ and *syn,mer*-$[Re(O)\{OCMe(CF_3)_2\}_3(CHCH{:}CPh_2)(thf)]$.[174]

5.8 Iron Sub-group Metal-containing Oxo-alkoxides (Including Heterometallic Compounds) (Table 5.8)

5.8.1 Iron

The structures of several iron oxo-alkoxides have been reported, one of the first being the oxo-bridged octahedral Fe(III) complex $[(LFe)_2\mu\text{-}O]$, where L is the pentadentate iminoalkoxo ligand $MeC\{CH_2N{:}CMeCH_2C(CF_3)_2(CH_2NH_2)\}$.[87] The μ_6-oxo ligand is featured in the hexanuclear anions $[Fe_6(\mu_6\text{-}O)\{(OCH_2)_3CMe\}_6]^{2-}$,[175] $[Fe_6(\mu_6\text{-}O)(\mu\text{-}OMe)_{12}(OMe)_6]^{2-}$,[176] and $[Fe_6(\mu_6\text{-}O)(\mu\text{-}OMe)_3\{(OCH_2)_3CMe\}_3Cl_6]^{2-}$.[91] Each complex has either the $[Fe_6O_{19}]^{2-}$ or $[Fe_6O_{13}Cl_6]^{2-}$ framework comparable with other metal oxo-alkoxides or polyoxoanions.

Table 5.8 Iron sub-group metal-containing oxo-alkoxides (including heterometallic compounds)

Compound	Metal coord.	Oxo coord.	M–Oxo bond lengths (Å)	M_x–Oxo bond angles (°)	Reference
$[(LFe)_2(\mu\text{-O})]$ $L = MeC\{CH_2N\text{:}CMeCH_2C(CF_3)_2O\}_2(CH_2NH_2)$	6	μ	1.811 (1)	Fe_2O, 146.6	87
$(NMe_4)_2[Fe_6(\mu_6\text{-O})\{(OCH_2)_3CMe\}_6]$	6	μ_6	FeO, 2.234, 2.259, 2.270 (1) (μ_6)	–	175
$Na_2[Fe_6(\mu_6\text{-O})(\mu\text{-OMe})_{12}(OMe)_6]$	6	μ_6	FeO, 2.26, 2.30, 2.30 (2) (μ_6)	–	176
$(NMe_4)_2[Fe_6(\mu_6\text{-O})(\mu\text{-OMe})_3\{(OCH_2)_3CMe\}_3Cl_6]$	6	μ_6	FeO, 2.24–2.27 (3) (μ_6)	Fe_2O, 89.3–90.7, av. 90.0 178.9	91
$[Fe_{12}(\mu_6\text{-O})_2(OMe)_{18}(OAc)_6(MeOH)_4]$	6	μ_6	FeO, 2.029–2.430 (4) (μ_6)	Fe_2O, 87.0–100.2, av. 89.9 172.0–173.0	177
$[Fe_{12}(\mu_6\text{-O})_2(OMe)_{18}(O_2CCH_2Cl)_{5.3}Cl_{0.7}(MeOH)_4]$	6	μ_6	FeO, 2.009–2.415 (5) (μ_6) av. 2.205	–	92
$[Fe_6(\mu_4\text{-O})_2(OMe)_{12}(tren)_2](O_3SCF_3)_2$	6	μ_4	FeO, 1.911–2.196, (μ_4) av. 2.054	Fe_2O, 96.9–99.9, av. 97.9 157.0	90
$[Os(O)(OC_2H_4O)_2]$	5	t	OsO, 1.670 (t)	–	177
$[Os_2(O)_2(\mu\text{-O})_2(OC_2Me_4O)_2]$	5	t, μ	OsO, 1.675 (7) (t) OsO, 1.92 (av.) (μ)	–	178
$[Os_2(O)_2(\mu\text{-O})_2(O_2C_6H_{10})_2(C_7H_{13}N_2)]$	6	t, μ	OsO, 1.73 (1) (t) OsO, 1.78, 2.22 (1) (μ)	–	77
$[Os_2(O)_2(\mu\text{-O})_2(OCMe_2CH_2Bu^t)_2]$	5	t, μ	OsO, 1.67 (t) OsO, 1.92 (av.) (μ)	–	78
$[Os(O)_2\{OCF(CF_3)C(CF_3)_2O\}(py)_2]$	6	t	OsO, 1.723, 1.732 (5) (t)	–	179

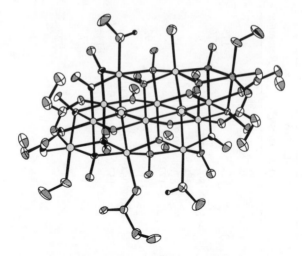

Figure 5.21 Structure of $[Fe_{12}(\mu_6\text{-}O)_2(\mu_3\text{-}OMe)_{10}(\mu\text{-}OMe)_8(O_2CCH_2Cl)_{5.3}Cl_{0.7}(MeOH)_4]$ ($\bigcirc = \mu_6\text{-}oxo$, $\bullet = Fe$; H atoms omitted).

The dodecanuclear mixed ligand mixed-valence iron oxo-alkoxides $[Fe_{12}(\mu_6\text{-}O)_2(OMe)_{18}(OAc)_6(MeOH)_4]^{177}$ and $[Fe_{12}(\mu_6\text{-}O)_2(OMe)_{18}(O_2CCH_2Cl)_{5.3}Cl_{0.7}(MeOH)_4]^{92}$ exhibit the interesting $M_{12}O_{34}$ framework (Fig. 5.21) which is related to the $M_{10}O_{28}$ framework of the decanuclear manganese complex.[88] The two μ_6-oxo ligands link ten of the twelve metal atoms in the dodecanuclear complexes. The mixed-ligand hexanuclear cation $[Fe_6(\mu_4\text{-}O)(\mu\text{-}OMe)_8(OMe)_4(tren)_2]^{2-}$ (tren = $2,2',2''$-triaminotriethylamine) has a centrosymmetric structure with octahedrally coordinated Fe(III).[90]

5.8.2 Osmium

The mononuclear Os(VI) complex $[Os(O)(OC_2H_4O)_2]$ has a square pyramidal structure with the terminal oxo-ligand occupying the apical site. Similarly in the binuclear complex $[Os_2(O)_2(\mu\text{-}O)_2(OC_2Me_4)_2]$ square pyramidal coordination of Os(VI) occurs.[178] Addition of the strong base quinuclidine ($C_7H_{13}N$) produced a dimer exhibiting octahedral Os(VI) $[Os_2(O)_2(\mu\text{-}O)_2(O_2C_6H_{10})_2(C_7H_{13}N)_2]$.[77] The square pyramidal coordination also occurs in the binuclear $[Os_2(O)_2(\mu\text{-}O)_2(OCMe_2CH_2Bu^t)_2]^{78}$, whilst in the mononuclear $[Os(O)_2\{OCF(CF_3)C(CF_3)_2O\}(py)_2]$ the octahedral Os(VI) atom has a linear terminal dioxo osmium unit.[179]

5.9 Oxo-alkoxides of Nickel, Copper, Aluminium, Tin, and Lead (Including Heterometallic Compounds) (Table 5.9)

5.9.1 Nickel and Copper

Although no oxo-alkoxide structures involving cobalt or nickel appear to have been reported there is an interesting octanuclear nickel antimony heterometallic compound $[Ni_5Sb_3(\mu_4\text{-}O)_2(\mu_3\text{-}OEt)_3(\mu\text{-}OEt)_9(OEt)_3(EtOH)_4]$ in which the Ni(II) atoms are octahedrally coordinated, one Sb is four-coordinated (lone pair in the vacant site of a

Table 5.9 Oxo-alkoxides of nickel, copper, aluminium, tin, and lead (including heterometallic compounds)

Compound	Metal coord.	Oxo coord.	M–Oxo bond lengths (Å)	M_x–Oxo bond angles (°)	Reference
[Ni$_5$Sb$_3$(μ_4-O)$_2$(μ_3-OEt)$_3$(μ-OEt)$_9$(OEt)$_3$(EtOH)$_4$]	4^1, 4^1, 6	μ_4	NiO, 2.02–2.15 (μ_4) av. 2.08; SbO, 1.98, 199 (μ_4)	–	24
[Cu$_{18}$(μ_4-O)$_2$(OSiMe$_3$)$_{14}$]	2	μ_4	CuO, 1.809, 1.831, 2.557, 2.580 (9) (μ_4)	Cu$_2$O, 71.8–155.3, av. 99.7	181
[H$_5$Al$_5$(μ_5-O)(μ-OBu^i)$_8$]	5, 6	μ_5	AlO, 1.900 (4) (μ_5, ap) 2.074, 2.092 (3) (μ_5, bs)	Al$_{ap}$-O-Al$_{bs}$, 96.8 (av.) Al$_{bs}$-O-Al$_{bs}$, 89.2 and 166.4 (av.)	182
[(Bu^{i}O)$_5$Al$_5$(μ_5-O)(μ-OBu^i)$_8$]	5, 6	μ_5	AlO, 1.88–1.94 (7) (μ_5, ap) av. 1.90 AlO, 2.071–2.087 (14) (μ_5, bs) av. 2.08	Al$_{ap}$-O-Al$_{bs}$, 89.1–97.1 av. 95.1 Al$_{bs}$-O-Al$_{bs}$, 89.1, 166.0	79
[(EtO)$_8$Al$_{10}$(μ_4-O)$_2$(μ_3-O)$_2$(μ-OEt)$_{14}$]	4, 5, 6	μ_4, μ_3	AlO, 1.834–1.906 (8) (μ_4) av. 1.880 AlO, 1.754–1.828 (8) (μ_3) av. 1.801	Al$_2$O, 96.3–144.7 (μ_4) av. 107.9 Al$_2$O, 97.3–141.5 (μ_3) av. 120	183
[Al$_4$(μ_4-O)(μ-OBu^i)$_5$(OBu^i)$_5$(Bu^{i}OH)]	5	μ_4	AlO, 1.877–1.885 (5) (μ_4) av. 1.880	Al$_2$O, 98.9–101.7 (μ_4) av. 100.4 146.8	184
[Al$_4$(μ_4-O)(μ-OCH$_2$CF$_3$)$_5$(OCH$_2$CF$_3$)$_5$(CF$_3$CH$_2$OH)]	5	μ_4	AlO, 1.871–1.902 (8) (μ_4) av. 1.889	–	185
[Al$_4$(μ_4-O)(μ-OPr^i)$_5$(OPr^i)(Pr^{i}OH)Cl$_4$]	5	μ_4	AlO, 1.82–1.92 (2) (μ_4) av. 1.86	Al$_2$O, 97.9–104.4 (μ_4) av. 101.1 144.4	186
[Al$_8$(μ_4-O)$_2$(μ-OH)$_2$(μ-OBu^i)$_{10}$(OBu^i)$_8$]	5	μ_4 μ-OH	AlO, 1.818–1.917 (3) (μ_4) av. 1.874 AlO(H), 1.786, 1.814 (4)	Al$_2$O, 99.7–145.2 (μ_4) av. 108.1 Al$_2$OH, 100.6	79

(continued overleaf)

Table 5.9 (continued)

Compound	Metal coord.	Oxo coord.	M–Oxo bond lengths (Å)	M_x–Oxo bond angles (°)	Reference
K[Me$_{13}$Al$_7$(μ_3-O)$_3$(μ-OMe)$_3$]	4	μ_3	AlO, 1.792–1.822 (3) (μ_3) av. 1.808	Al$_2$O, 117.0–122.9 (μ_3) av. 119.5	187
[Bu$^t_6$Al$_6$(μ_3-O)$_4$(OCHMeCH$_2$CO$_2$)$_2$]	4, 5	μ_3	AlO, 1.71–1.84 (1) (μ_3) av. 1.81	Al$_2$O, 94.2, 102.1, 155.1 (μ_3)	188
[Sn$_6$(μ_3-O)$_4$(μ_3-OMe)$_4$]	4	μ_3	SnO, 2.05–2.08 (1) (μ_3)	–	189
[Sn$_3$(μ_3-O)(μ-OBu^i)$_3$(OBu^i)$_7$(Bu^{i}OH)$_2$]	6	μ_3	SnO, 2.066, 2.071, 2.074 (4) (μ_3)	Sn$_2$O, 104.4, 104.8, 104.9 (μ_3)	190
[Sn$_6$(μ_3-O)$_6$(μ-OAc)$_6$(OBu^t)$_6$]	6	μ_3	SnO, 2.042–2.055 (5) (μ_3) av. 2.050	Sn$_2$O, 135.0	29
[Pb$_6$(μ_3-O)$_4$(μ-OPr^i)$_4$]	5	μ_3	PbO, 2.12, 2.17, 2.18 (4) (μ_3)	112, 118, 118	193

[1]Coordination number doubtful.

434

trigonal bipyramid) and the other two are five-coordinated (lone pair in the vacant octahedral site).[24]

The only copper oxo-alkoxide structure determined to date is the giant trimethylsiloxo derivative [Cu$_{18}$(μ_4-O)$_2$(OSiMe$_3$)$_{14}$] containing two-coordinated (linear) copper(I).[181]

5.9.2 Aluminium

The first structural report of an aluminium oxo-alkoxide concerned the mixed-ligand complex [H$_5$Al$_5$(μ_5-O)(μ-OBu^i)$_8$] containing the novel Al$_5$O$_9$ framework with a square pyramidal array of aluminium atoms. The apical Al is octahedrally coordinated by one terminal hydride *trans* to the μ_5-oxo ligand and connected to the four basal Al atoms by μ-isobutoxo ligands. Each basal Al is five-coordinated (TBP) by one terminal hydride in an axial position *trans* to the μ_5-oxo ligand and three μ-isobutoxo bridges in the equatorial positions to the apical Al and two adjacent basal aluminiums.[182] Several years later the corresponding oxo-isobutoxide [(Bu^{i}O)$_5$Al$_5$(μ_5-O)(μ-OBu^i)$_8$] (Fig. 5.22) was isolated and characterized and shown to have the same Al$_5$O$_9$ framework.[79] Meanwhile the scandium, yttrium, and lanthanide species [(RO)$_5$M$_5$(μ_5-O)(μ_3-OR)$_4$(μ-OR)$_4$] had been characterized in which all five metals were six-coordinated.[16]

A remarkable decanuclear oxo-ethoxide [Al$_{10}$(μ_4-O)$_2$(μ_3-O)$_2$(μ-OEt)$_{14}$(OEt)$_8$] was reported which contained aluminium in four-, five-, and six-coordination (Fig. 5.23).[183] The average Al–O bond distance for the μ_4-oxo ligand was significantly longer (1.880 Å) than for the μ_3-oxo ligand (1.801 Å). In the tetranuclear compound [Al$_4$(μ_4-O)(μ-OBu^i)$_5$(OBu^i)$_5$(Bu^{i}OH)] the four aluminiums are in a distorted tetrahedral configuration around the μ_4-oxo ligand (Fig. 5.24).[184] Although the OH hydrogen in the isobutanol ligand was not located it was clearly present

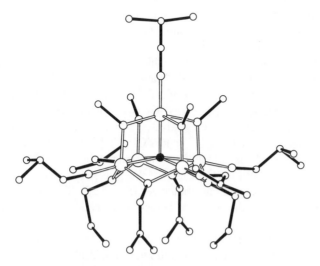

Figure 5.22 Structure of [Al$_5$(μ_5-O)(μ-OBu^i)$_8$(OBu^i)$_5$] ($\bullet$ = μ_5-oxo; H atoms omitted).

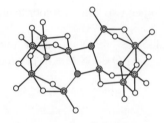

Figure 5.23 $Al_{10}O_{26}$ core structure of $[Al_{10}(\mu_4\text{-}O)_2(\mu_3\text{-}O)_2(\mu\text{-}OEt)_{14}(OEt)_8]$ (⊘ = μ_4-oxo, ⊛ = μ_3-oxo).

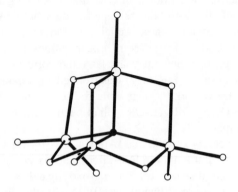

Figure 5.24 Al_4O_{12} core structure of $[Al_4(\mu_4\text{-}O)(\mu\text{-}OBu^i)_5(OBu^i)_5(Bu^iOH)]$ (● = μ_4-oxo).

in a hydrogen bond linking two of the terminal isobutoxo ligands (O....O, 2.43 Å), which had longer Al–O bonds (1.812 Å, 1.819 Å) than the other four terminal Al–O bonds (av. 1.718 Å) and had a wider angle (146.8°) with the μ_4-oxygen than the other aluminiums which had an average Al–μ_4-O–Al angle of 100.4°. Each Al exhibits five-coordination (distorted TBP) with the terminal isobutoxo ligand *trans* to the μ_4-oxo ligand. The same structure was found for the trifluoroethoxo compound $[Al_4(\mu_4\text{-}O)(\mu\text{-}OCH_2CF_3)_5(OCH_2CF_3)_5(CF_3CH_2OH)]$.[185] Earlier a similar metal–oxygen Al_4O_8 framework was revealed in the structure of $[Al_4(\mu_4\text{-}O)(\mu\text{-}OPr^i)_5(OPr^i)(Pr^iOH)Cl_4]$ in which the four chlorine atoms occupy terminal positions.[186] The octanuclear oxo-hydroxo-isobutoxide $[Al_8(\mu_4\text{-}O)(\mu\text{-}OH)_2(\mu\text{-}OBu^i)_{10}(OBu^i)_8]$ has been isolated and structurally characterized.[79] This centrosymmetrical molecule (Fig. 5.25) contains two tetranuclear units linked by a double Al(μ-OH)$_2$Al bridge with five-coordinated (distorted TBP) Al atoms.

The heptanuclear methyl aluminium oxo-methoxide anion $[Me_{13}Al_7(\mu_3\text{-}O)_3(\mu\text{-}OMe)_3]^-$ has a ring of six dimethylaluminiums linked alternately by μ-oxo and μ-methoxo bridges and capped by a monomethylaluminium, which is linked to the oxo-ligands that are μ_3-bonded.[187] All of the aluminiums are in distorted tetrahedral configurations.

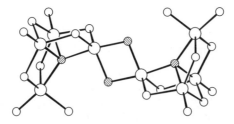

Figure 5.25 Al_8O_{22} core structure of
$[Al_8(\mu_4\text{-O})_2(\mu\text{-OH})_2(\mu\text{-OBu}^i)_{10}(OBu^i)_8]$
($\oslash = \mu_4\text{-oxo}$, $\ominus = \mu\text{-OH}$).

The neutral hexanuclear molecule $[Bu^t_6Al_6(\mu_3\text{-O})_4(OCHMeCH_2CO_2)_2]$ has a
curved Al_6O_4 ladder capped by the tridentate alkoxocarboxylate ligands with 2 five-
coordinate (TBP) and four distorted tetrahedral aluminiums.[188]

5.9.3 Tin

The hexanuclear tin(II) oxo-methoxide $[Sn_6(\mu_3\text{-O})_4(\mu_3\text{-OMe})_4]$ has an adamantane
Sn_6O_4 framework consisting of an octahedral array of tin atoms with μ_3-oxo and μ_3-
methoxo ligands bridging opposite faces of the Sn_6 octahedron in D_{2d} symmetry.[189]
The tin atoms have distorted pseudo-trigonal bipyramidal coordination with the electron lone-pair occupying the vacant equatorial position. In the trinuclear Sn(IV) oxo-
isobutoxide $[Sn_3(\mu_3\text{-O})(\mu\text{-OBu}^i)_3(OBu^i)_7(Bu^iOH)_2]$ the Sn_3 triangle is capped by the
μ_3-oxo ligand and bridged by three μ-isobutoxo groups. The tin atoms are all octahedrally coordinated[190] and the two terminal isobutanol ligands are hydrogen bonded to
terminal isobutoxo ligands on adjacent tin atoms. In the mixed ligand Sn(IV) oxo-*tert*-
butoxide acetate $[Sn_6(\mu_3\text{-O})_6(\mu\text{-OAc})_6(OBu^t)_6]$ the hexanuclear molecule has μ_3-oxo
ligands capping six of the eight faces of the Sn_6 octahedron. Each tin atom is octahedrally coordinated.[29]

The octanuclear mixed ligand heterometallic tin(IV) cadmium oxo-neopentyloxide-
acetate molecule $[Sn_4Cd_4(\mu_4\text{-O})_2(\mu\text{-OAc})_6(\mu\text{-OCH}_2Bu^t)_4(OCH_2Bu^t)_6]$ has a planar
array of four Cd atoms linked alternately by μ-oxo and $\mu\text{-OCH}_2Bu^t$ ligands.
The oxo-ligands are each linked to two Sn atoms in edge-shared bi-octahedral
$Sn_2(\mu\text{-O})(\mu\text{-OCH}_2Bu^t)(\mu\text{-OAc})$ units thus becoming μ_4-oxo groups. The Sn–Sn axes
are parallel to each other and at right angles to the Cd_4O_4 plane with 2 six-coordinate
and 2 seven-coordinate Cd atoms.[191]

In the trinuclear heterometallic mixed-ligand compound $[Sn_2Pb(\mu_3\text{-O})(\mu\text{-OBu}^t)_2(\mu\text{-}$
$OAc)_4(OBu^t)_2]$ each Sn atom is linked to its neighbour by two bridging acetato ligands
and to the Pb atom by one μ-acetato and one $\mu\text{-OBu}^t$ ligand. The octahedral tin
coordination is completed by a terminal *tert*-butoxo group. The Pb atom is in a five-
coordinated pseudo-octahedral configuration with an electron lone-pair occupying the
vacant site.[192]

The hexanuclear Pb(II) oxo-isopropoxide $[Pb_6(\mu_3\text{-O})_4(\mu_3\text{-OPr}^i)_4]$ has the adamantane Pb_6O_4 structure with an octahedron of lead atoms. The μ_3-oxo and μ_3-isopropoxo
ligands occupy opposite pairs of Pb_3 faces in a symmetrical array with five-coordinate
pseudo-octahedral Pb atoms.[193]

REFERENCES

1. D.C. Bradley, R.C. Mehrotra, and D.P. Gaur, *Metal Alkoxides*, 150–167, Academic Press, London (1978).
2. L.C. Klein, *Sol–Gel Technology for Thin Films, Fibers, Preforms, Electronics and Speciality Shapes*, Noyes, Park Ridge, N.J. (1988).
3. C.J. Brinker and G.W. Sherer, *Sol–Gel Science: The Physics and Chemistry of Sol–Gel Processing*, Academic Press, New York (1990).
4. R.C. Mehrotra and Anirudh Singh, *Chem. Soc. Rev.*, **25**, 1 (1996); N.Ya. Turova, E.P. Turevskaya, V.G. Kessler, and M.I. Yanovskaya, *J. Sol–Gel Sci. Technol.*, **2**, 17 (1994).
5. D.C. Bradley, R. Gaze, and W. Wardlaw, *J. Chem. Soc.*, 721 (1955); 3977 (1995); 469 (1957).
6. D.C. Bradley and D.G. Carter, *Can. J. Chem.*, **39**, 1434 (1961); **40**, 15 (1962).
7. D.C. Bradley and H. Holloway, *Can. J. Chem.*, **39**, 1818 (1961); **40**, 62 (1962).
8. D.C. Bradley and H. Holloway, *Can. J. Chem.*, **40**, 1176 (1962).
9. V.W. Day, T.A. Eberspacher, W.G. Klemperer, C.W. Park, and F.S. Rosenberg, in *Chemical Processing of Advanced Materials* (L. Hench and J. West eds), Wiley, New York (1991); V.W. Day, T.A. Eberspacher, W.G. Klemperer, C.W. Park, and F.S. Rosenberg, *J. Am. Chem. Soc.*, **113**, 8190 (1991); Y. Chen, J. Hao, W.G. Klemperer, C.W. Park, and F.S. Rosenberg, *Polym. Preprint, Am. Chem. Soc. Div. Polym. Chem.*, **34**, 250 (1993).
10. V.W. Day, T.A. Eberspacher, Y. Chen, J. Hao, and W.G. Klemperer, *Inorg. Chim. Acta*, **229**, 391 (1995).
11. V.W. Day, T.A. Eberspacher, W.G. Klemperer, and C.W. Park, *J. Am. Chem. Soc.*, **115**, 8469 (1993); see also Y.W. Chen, W.G. Klemperer, and C.W. Park, *Mat. Res. Symp. Proc.*, **271**, 57 (1992).
12. C.F. Campana, Y. Chen, V.W. Day, W.G. Klemperer, and R.A. Sparks, *J. Chem. Soc., Dalton Trans.*, 691 (1996).
13. D.C. Bradley, B.N. Chakravarti, and W. Wardlaw, *J. Chem. Soc.*, 2381 (1956); 4439 (1956).
14. D.C. Bradley, M.H. Chisholm, M.W. Extine, and M.E. Stager, *Inorg. Chem.*, **16**, 1794 (1977).
15. N.Ya. Turova, V.G. Kessler, and S.I. Kucheiko, *Polyhedron*, **10**, 2617 (1991); V.G. Kessler, A.V. Mironov, N.Ya. Turova, A.I. Yanovsky, and Y.T. Struchkov, *Polyhedron*, **12**, 1573 (1993); V.G. Kessler, N.Ya. Turova, and A. Panov, *Polyhedron*, **15**, 335 (1996); V.G. Kessler, A.V. Shevelkov, and L.A. Bengtsson-Kloo, *Polyhedron*, **17**, 965 (1998).
16a. D.C. Bradley, H. Chudzynska, D.M. Frigo, M.B. Hursthouse, and M.A. Mazid, *J. Chem. Soc., Chem. Commun.*, 1258 (1988).
16b. O. Poncelet, W.J. Sartain, L.G. Hubert-Pfalzgraf, K. Folting, and K.G. Caulton, *Inorg. Chem.*, **28**, 263 (1989).
16c. D.C. Bradley, H. Chudzynska, D.M. Frigo, M.E. Hammond, M.B. Hursthouse, and M.A. Mazid, *Polyhedron*, **9**, 719 (1990).
16d. O. Helgesson, S. Jagner, O. Poncelet, and L.G. Hubert-Pfalzgraf, *Polyhedron*, **10**, 1559 (1991).
17. D.C. Bradley and M.E. Hammond, unpublished work; M.E. Hammond, *Ph.D. Thesis*, University of London (1991).
18. S. Danielle, R. Papiernik, L.G. Hubert-Pfalzgraf, S. Jagner, and M. Hakansson, *Inorg. Chem.*, **34**, 628 (1995); R. Papiernik, L.G. Hubert-Pfalzgraf, and F. Chaput, *J. Non-Cryst. Solids*, **147**, 36 (1992).
19. A.I. Yanovsky, M.I. Yanovskaya, V.K. Limar, V.G. Kessler, N.Ya. Turova, and Y.T. Struchkov, *J. Chem. Soc., Chem. Commun.*, 1605 (1991).
20. Z.A. Starikova, A.I. Yanovsky, N.M. Kotova, M.I. Yanovskaya, N.Ya. Turova, and D. Benlian, *Polyhedron*, **16**, 4347 (1997).

21. N.Ya. Turova, E.P. Turevskaya, and M.I. Yanovskaya, *Russ. J. Inorg. Chem.*, **38**, 1137 (1993).
22. D.J. Teff, J.C. Huffman, and K.G. Caulton, *J. Am. Chem. Soc.*, **118**, 4030 (1996); see also K.G. Caulton and L.G. Hubert-Pfalzgraf, *Chem. Rev.*, **90**, 969 (1990).
23. S. Boulmaaz, L.G. Hubert-Pfalzgraf, S. Salut, and J.-C. Daran, *J. Chem. Soc., Chem. Commun.*, 601 (1994).
24. U. Bemm, R. Norrestam, M. Nygren, and G. Westin, *Inorg. Chem.*, **31**, 2050 (1992).
25. U. Bemm, R. Norrestam, M. Nygren, and G. Westin, *Inorg. Chem.*, **34**, 2367 (1995).
26. T.A. Budzichovski, M.H. Chisholm, and W.E. Streib, *J. Am. Chem. Soc.*, **116**, 389 (1994).
27. C. Limberg, S. Parsons, A.J. Downs, and D.J. Watkin, *J. Chem. Soc., Dalton Trans.*, 1169 (1994).
28. A. Doeuff, Y. Dromzee, F. Taulelle, and C. Sanchez, *Inorg. Chem.*, **28**, 4439 (1989).
29. J. Caruso, M.J. Hampden-Smith, A.L. Rheingold, and G. Yap, *J. Chem. Soc., Chem. Commun.*, 157 (1995).
30. J. Caruso, C. Roger, F. Schwertfeger, M.J. Hampden-Smith, A.L. Rheingold, and G. Yap, *Inorg. Chem.*, **34**, 449 (1995).
31. J. Rai and R.C. Mehrotra, *J. Non-Cryst. Solids*, **134**, 23 (1991).
32. Sonika, A.K. Narula, O.P. Vermani, and H.K. Sharma, *J. Organomet. Chem.*, **470**, 67 (1994).
33. R.C. Mehrotra, personal communication.
34. M.H. Chisholm, K. Folting, J.C. Huffman, and C.C. Kirkpatrick, *Inorg. Chem.*, **23**, 1021 (1984) and references therein.
35. T.P. Blatchford, M.H. Chisholm, K. Folting, and J.C. Huffman, *J. Chem. Soc., Chem. Commun.*, 1295 (1984).
36. M.H. Chisholm, K. Folting, and J.A. Klang, *Organometallics*, **9**, 602 (1990).
37. M.H. Chisholm, J.C. Huffman, E.A. Lucas, A. Sousa, and W.E. Streib, *J. Am. Chem. Soc.*, **114**, 2710 (1992).
38. M.H. Chisholm, C.E. Hammond, V.J. Johnston, W.E. Streib, and J.C. Huffman, *J. Am. Chem. Soc.*, **114**, 7056 (1992).
39. P.T. Wolczanski, D.R. Neithamer, R.E. LePointe, R.A. Wheeler, D.S. Richeson, and G.D. Van Duyne, *J. Am. Chem. Soc.*, **111**, 9056 (1989).
40. F.A. Cotton and E.S. Shamshoum, *J. Am. Chem. Soc.*, **106**, 3222 (1984).
41. M.H. Chisholm, C.M. Cook, K. Folting, and W.E. Streib, *Inorg. Chim. Acta*, 198–200, 63 (1992).
42. K. Yunlu, P.S. Gradeff, N. Edelstein, W. Kot, G. Shalimoff, W.E. Streib, B.A. Vaastra, and K.G. Caulton, *Inorg. Chem.*, **30**, 2317 (1991).
43. D.C. Bradley, Amar K. Chatterjee, and Amiya K. Chatterjee, *J. Inorg. Nucl. Chem.*, **12**, 71 (1959).
44. C.J. Burns, D.C. Smith, A.P. Sattelberger, and H.B. Gray, *Inorg. Chem.*, **31**, 3724 (1992).
45. C.N. Caughlan, H.M. Smith, and K. Watenpaugh, *Inorg. Chem.*, **5**, 2131 (1966).
46. W. Priebsch and D. Rehder, *Inorg. Chem.*, **29**, 3013 (1990).
47. D.C. Crans, H. Chen, and R.A. Felty, *J. Am. Chem. Soc.*, **114**, 4543 (1992).
48. H. Nishino and J.K. Kochi, *Inorg. Chim. Acta*, **174**, 93 (1990).
49. W. Clegg, R.J. Errington, P. Kraxner, and C. Redshaw, *J. Chem. Soc., Dalton Trans.*, 1431 (1992).
50. F.A. Schröder, J. Scherle, and R.G. Hazell, *Acta Crystallog.*, **B31**, 531 (1975).
51. A.J. Wilson, B.R. Penfold, and C.J. Wilkins, *Acta Crystallog.*, **C39**, 329 (1983).
52. E.M. McCarron III and R.L. Harlow, *J. Am. Chem. Soc.*, **105**, 6179 (1983).
53. M.I. Khan, Q. Chen, D.P. Goshorn, H. Hope, S. Parkin, and J. Zubieta, *J. Am. Chem. Soc.*, **114**, 3341 (1992).
54. M.I. Khan, Q. Chen, D.P. Goshorn, and J. Zubieta, *Inorg. Chem.*, **32**, 672 (1993).
55. M.I. Khan, Q. Chen, H. Hope, S. Parkin, C.J. O'Connor, and J. Zubieta, *Inorg. Chem.*, **32**, 2929 (1993).
56. Q. Chen and J. Zubieta, *Inorg. Chem.*, **29**, 1458 (1990).
57. D. Hou, G.-S. Kim, K.S. Hagen, and C.L. Hill, *Inorg. Chim. Acta*, **211**, 127 (1993).

58. L. Ma, S. Liu, and J. Zubieta, *J. Chem. Soc., Chem. Commun.*, 440 (1989).
59. L. Ma, S. Liu, and J. Zubieta, *Inorg. Chem.*, **28**, 175 (1989).
60. Y. Chang, Q. Chen, M.I. Khan, J. Salta, and J. Zubieta, *J. Chem. Soc., Chem. Commun.*, 1872 (1993).
61. Z. Y-Zhuang, J. X-Lin, and L. S-Cheng, *Jiegou Huaxue (J. Struct. Chem.)*, **12**, 48 (1993).
62. S. Daniele, L.G. Hubert-Pfalzgraf, and J.-C. Daran, *Polyhedron*, **15**, 1063 (1996).
63. L.G. Hubert-Pfalzgraf, S. Daniele, A. Bennaceur, J.-C. Daran, and J. Vaissermann, *Polyhedron*, **16**, 1223 (1997).
64. C.J. Burns and A.P. Sattelberger, *Inorg. Chem.*, **21**, 3692 (1988).
65. S.C. Goel, M.Y. Chiang, P.C. Gibbons, and W.E. Buhro, *Mater. Res. Soc. Symp. Proc.*, **271**, 3 (1992).
66. J.V. Barkley, J.C. Cannadine, I. Hannaford, M.M. Harding, A. Steiner, J. Tallon, and R. Whyman, *Chem. Commun.*, 1653 (1997).
67. C. Sirio, L.G. Hubert-Pfalzgraf, and C. Bots, *Polyhedron*, **16**, 1129 (1997).
68. Y. Chang, Q. Chen, M.I. Khan, J. Salta, and J. Zubieta, *J. Chem. Soc., Chem. Commun.*, 1872 (1993).
69. P. Stavropoulos, N. Bryson, M.-T. Youinou, and J.A. Osborn, *Inorg. Chem.*, **29**, 1807 (1990).
70. J.M. Berg and R.H. Holm, *Inorg. Chem.*, **22**, 1768 (1983).
71. I. Feinstein-Jaffe, J.C. Dewan, and R.R. Schrock, *Organometallics*, **4**, 1189 (1985).
72. M.H. Chisholm, C.E. Hammond, J.C. Huffman, and J.D. Martin, *Polyhedron*, **9**, 1829 (1990).
73. A. Lehtonen and R. Sillanpaa, *J. Chem, Soc., Dalton Trans.*, 2119 (1994).
74. A.G. Jones, B.V. DePamphilis, and A. Davison, *Inorg. Chem.*, **20**, 1617 (1981).
75. R.M. Pearlstein, C.J.L. Lock, R. Faggiani, C.E. Costello, C.-H. Zeng, A.G. Jones, and A. Davison, *Inorg. Chem.*, **27**, 2409 (1988).
76. P.G. Edwards, G. Wilkinson, M.B. Hursthouse, and K.M.A. Malik, *J. Chem. Soc., Dalton Trans.*, 2467 (1980).
77. B.A. Cartwright, W.P. Griffith, M. Schröder, and A.C. Skapski, *J. Chem. Soc., Chem. Comm.*, 853 (1978).
78. W.P. Griffith, N.T. McManus, and A.C. Skapski, *Inorg. Chim. Acta*, **103**, L5 (1985).
79. I. Abrahams, D.C. Bradley, H. Chudzynska, M. Motevalli, and R. Sinclair, *J. Chem. Soc., Dalton Trans.*, to be published.
80. V.W. Day, W.G. Klemperer, and C. Schwartz, *J. Am. Chem. Soc.*, **109**, 6030 (1987).
81. W. Clegg, M.R.J. Elsegood, R.J. Errington, and J. Havelock, *J. Chem. Soc., Dalton Trans.*, 681 (1996).
82. J.-F. Campion, D.A. Payne, H.K. Chae, J.K. Maurin, and S.R. Wilson, *Inorg. Chem.*, **30**, 3244 (1991).
83. D.C. Crans, H. Chen, O.P. Anderson, and M.M. Miller, *J. Am. Chem. Soc.*, **115**, 6769 (1993).
84. V.G. Kessler, N. Ya. Turova, A.I. Yanovskii, A.I. Belokon, and Yu T. Struchkov, *Russ. J. Inorg. Chem.*, **36**, 938 (1991).
85. A.I. Yanovsky, N. Ya. Turova, A.V. Korolev, D.E. Chebukov, A.P. Pisarevsky, and Yu. T. Struchkov, *Russ. Chem. Bull.*, **45**, 115 (1996).
86. W.A. Herrmann, P. Watzlowik, and P. Kiprof, *Chem. Ber.*, **124**, 1101 (1991).
87. E. Konefal, S.J. Loeb, and C.J. Willis, *Inorg. Chem. Acta*, **115**, 147 (1986).
88. M. Cavaluzzo, Q. Chen, and J. Zubieta, *J. Chem. Soc., Chem. Commun.*, 131 (1993).
89. Z. Sun, P.K. Gantzel, and D.N. Hendrickson, *Inorg. Chem.*, **35**, 6640 (1996).
90. V.S. Nair and K.S. Hagen, *Inorg. Chem.*, **31**, 4048 (1992).
91. A. Cornia, D. Gatteschi, K. Hegetschweiler, L. Hausherr-Primo, and V. Gramlich, *Inorg. Chem.*, **35**, 4414 (1996).
92. A. Caneschi, A. Cornia, S.J. Lippard, G.C. Papaefthymiou, and R. Sessoli, *Inorg. Chim. Acta*, **243**, 295 (1996).
93. S. Yano, T. Inagaki, Y. Yamada, M. Kato, M. Yamasaki, K. Sakai, T. Tsubomura, M. Sato, W. Mori, K. Yamaguchi, and I. Kinoshita, *Chem. Lett.*, 61 (1996).

94. F.M. Mackenzie, R.E. Mulvey, W. Clegg, and L. Horsburgh, *Polyhedron*, **17**, 993 (1998).
95. R. Kuhlman, B.A. Vaarstra, W.E. Streib, J.C. Huffman, and K.G. Caulton, *Inorg. Chem.*, **32**, 1272 (1993).
96. V.G. Kessler, N. Ya Turova and A.I. Yanovsky, *Zh. Obshch. Khim.*, **60**, 2769 (1990); see also ref. 15.
97. A.I. Yanovsky, V.G. Kessler, N. Ya Turova, and Yu T. Struchkov, *Koord. Khim.*, **17**, 54 (1991).
98. W.J. Evans, M.A. Ansari, and J.W. Ziller, *Polyhedron*, **17**, 869 (1998).
99. M. Veith, E.-C. Yu, and V. Huch, *Chem. Eur. J.*, **1**, 26 (1995).
100. M. Antipin, Yu T. Struchkov, A. Shilov, and A. Shilova, *Gazz. Chim. Ital.*, **123**, 265 (1993).
101. N.Ya. Turova, E.P. Turevskaya, V.G. Kessler, A.I. Yanovsky, and Yu.T. Struchkov, *J. Chem. Soc., Chem. Commun.*, 21 (1993).
102. K.G. Caulton, M.H. Chisholm, S.R. Drake, and K. Folting, *J. Chem. Soc., Chem. Commun.*, 1349 (1990).
103. K.G. Caulton, M.H. Chisholm, S.R. Drake, and J.C. Huffman, *J. Chem. Soc., Chem. Commun.*, 1498 (1990).
104. H. Vincent, F. Labrize, and L.G. Hubert-Pfalzgraf, *Polyhedron*, **13**, 3323 (1994).
105. P. Miele, J.D. Foulon, N. Hovnanian, and L. Cot, *J. Chem. Soc. Chem. Commun.*, 29 (1993).
106. A.I. Yanovskii, E.P. Turevskaya, M.I. Yanovskaya, V.G. Kessler, N.Ya. Turova, A.P. Pisarevskii, and Yu.T. Struchkov, *Russ. J. Inorg. Chem.*, **40**, 339 (1995).
107. B. Gaskins, J.L. Lannutti, D.C. Finnen, and A.A. Pinkerton, *Acta Crystallog.*, **C50**, 1387 (1994).
108. W.J. Evans and M.S. Sollberger, *J. Am. Chem. Soc.*, **108**, 6095 (1986).
109. W.J. Evans and M.S. Sollberger, *Inorg. Chem.*, **27**, 4417 (1988).
110. W.J. Evans, M.S. Sollberger, J.L. Shreeve, J.M. Olofson, J.H. Hain, and J.W. Ziller, *Inorg. Chem.*, **31**, 2492 (1992).
111. W.J. Evans, T.J. Boyle, and J.W. Ziller, *J. Am. Chem. Soc.*, **115**, 5084 (1993).
112. D.C. Bradley, M.A. Mazid, and Ruo-wen Wu, unpublished results; Ruo-wen Wu, *Ph.D. Thesis*, University of London (1991).
113. S. Daniele, L.G. Hubert-Pfalzgraf, J.-C. Daran, and S. Halut, *Polyhedron*, **13**, 927 (1994).
114. W.J. Evans, M.S. Sollberger, and T.P. Hanusa, *J. Am. Chem. Soc.*, **110**, 1841 (1988).
115. S. Daniele, L.G. Hubert-Pfalzgraf, J.-C. Daran, and R.A. Toscano, *Polyhedron*, **12**, 2091 (1993).
116. V.G. Kessler, N.Ya. Turova, A.N. Panov, A.I. Yanovsky, A.P. Pisarevsky, and Yu.T. Struchkov, *Polyhedron*, **15**, 335 (1996).
117. F.A. Cotton, D.O. Marler, and W. Schwotzer, *Inorg. Chim. Acta*, **95**, 207 (1984).
118. R. Schmid, A. Mosset, and J. Galy, *J. Chem. Soc., Dalton Trans.*, 1999 (1991).
119. P.A. Laaziz, A. Larbot, C. Guizard, J. Durand, L. Cot, and J. Jofre, *Acta Crystallog.*, **C46**, 2332 (1990).
120. S. Doeff, Y. Dromzee, F. Taulelle, and C. Sanchez, *Inorg. Chem.*, **28**, 4439 (1989).
121. L.G. Hubert-Pfalzgraf, S. Daniele, R. Papiernik, M.-C. Massiani, B. Septe, J. Vaissermann, and J.-C Daran, *J. Mater. Chem.*, **7**, 753 (1997).
122. R. Papiernik, L.G. Hubert-Pfalzgraf, M. Veith, and V. Huch, *Chem. Ber./Recueil*, **130**, 1361 (1997).
123. B. Morosin, *Acta Crystallog.*, **B33**, 303 (1977).
124. W.J. Evans, M.A. Ansari, and J.W. Ziller, *Polyhedron*, **17**, 869 (1998).
125. L.G. Hubert-Pfalzgraf, C. Sirio, and C. Bois, *Polyhedron*, **17**, 821 (1998).
126. L. Ma and D.A. Payne, *Chem. Mater.*, **6**, 875 (1994).
127. R. Schmid, H. Ahamdane, and A. Mosset, *Inorg. Chim. Acta*, **190**, 237 (1991).
128. R. Schmid, A. Mosset, and J. Galy, *Inorg. Chim. Acta*, **179**, 167 (1991).
129. J.A. Samuels, B.A. Vaarstra, J.C. Huffman, K.L. Trojan, W.E. Hatfield, and K.G. Caulton, *J. Am. Chem. Soc.*, **112**, 9623 (1990); J.A. Samuels, W.-C. Chiang, J.C. Huffman, K.L. Trojan, W.E. Hatfield, D.V. Baxter, and K.G. Caulton, *Inorg. Chem.*, **33**, 2167 (1994).

130. C.N. Caughlan, H.M. Smith, and K. Watenpaugh, *Inorg. Chem.*, **5**, 2131 (1966).
131. Zhou Yin-Zhuang, Jin Xiang-Lin, Liu Shun-Cheng, *J. Struct. Chem.*, (*China*), **12**, 48 (1993).
132. Yu.E. Ovchinnikov, Yu.T. Struchkov, V.P. Baryshok, and M.G. Voronkov, *Izvest. Akad. Nauk. SSSR. Ser. Khim.*, 1477 (1993).
133. W.A. Nugent and R.L. Harlow, *J. Am. Chem. Soc.*, **116**, 6142 (1994).
134. F. Hillerns, F. Olbrich, U. Behrens, and D. Rehder, *Angew. Chem. Int. Ed. Engl.*, **31**, 447 (1992).
135. D.C. Crans, R.A. Felty, O.P. Anderson, and M.M. Miller, *Inorg. Chem.*, **32**, 247 (1993).
136. J. Salta and J. Zubieta, *Inorg. Chim. Acta*, **257**, 83 (1997).
137. I. Cavaco, J. Costa Pessoa, M.T. Duarte, P.M. Matias, and R.T. Henriques, *Polyhedron*, **12**, 1231 (1993).
138. H. Kumagai, M. Endo, S. Kawata, and S. Kitagawa, *Acta Crystallog.*, **C52**, 1943 (1996).
139. B. Kramenar and C.K. Prout, *J. Chem. Soc. A*, 2379 (1970).
140. F.A. Cotton, M.P. Diebold, and W.J. Roth, *Inorg. Chem.*, **24**, 3509 (1985).
141. D.C. Bradley, M.B. Hursthouse, and P.F. Rodesiler, *Chem. Commun.*, 1112 (1968).
142. L.G. Hubert-Pfalzgraf, V. Abada, S. Halut, and J. Roziere, *Polyhedron*, **16**, 581 (1997).
143. T.J. Boyle, T.M. Alam, D. Dimos, G.J. Moore, C.D. Buchheit, H.N. Al-Shareef, E. Mechenbier, B.R. Bear, and J.W. Ziller, *Chem. Mater.*, **9**, 3187 (1997).
144. R. Papiernik, L.G. Hubert-Pfalzgraf, J.-C. Daran, and Y. Jeannin, *J. Chem. Soc., Chem. Commun.*, 695 (1990).
145. I. Abrahams, D.C. Bradley, H. Chudzynska, M. Motevalli, and P. O'Shaughnessy, *J. Chem. Soc., Dalton Trans.*, 2685 (2000).
146. S. Boulmaaz, L.G. Hubert-Pfalzgraf, S. Halut, and J.-C. Daran, *J. Chem. Soc., Chem. Commun.*, 601 (1994).
147. P. Stavropoulos, N. Bryson, M.-T. Youinou, and J.A. Osborn, *Inorg. Chem.*, **29**, 1807 (1990).
148. R.J. Butcher, B.R. Penfold, and E. Sinn, *J. Chem. Soc., Dalton Trans.*, 668 (1979).
149. V.G. Kessler, N.Ya. Turova, A.V. Korolev, A.I. Yanovskii, and Yu.T. Struchkov, *Mendeleev Commun.*, 89 (1991).
150. D.A. Johnson, J.C. Taylor, and A.B. Waugh, *J. Inorg. Nucl. Chem.*, **42**, 1271 (1980).
151. C.B. Knobler, B.R. Penfold, W.T. Robinson, and C.J. Wilkins, *Acta Crystallog.*, **B37**, 942 (1981).
152. T.A. Budzichowski, M.H. Chisholm, K. Folting, and K.S. Kramer, *Polyhedron*, **15**, 3085 (1996).
153. M.H. Chisholm, F.A. Cotton, A. Fang, and E.M. Kober, *Inorg. Chem.*, **23**, 749 (1984).
154. M.H. Chisholm, *J. Solid State Chem.*, **57**, 120 (1985).
155. J.A. Beaver and M.G.B. Drew, *J. Chem. Soc., Dalton Trans.*, 1376 (1973).
156. M.H. Chisholm, J.C. Huffman, C.C. Kirkpatrick, J. Leonelli, and K. Folting, *J. Am. Chem. Soc.*, **103**, 6093 (1981).
157. S.A. Koch and S. Lincoln, *Inorg. Chem.*, **21**, 2904 (1982).
158. A.J. Wilson, W.T. Robinson, and C.J. Wilkins, *Acta Crystallog.*, **C39**, 54 (1983).
159. T.A. Budzichowski, M.H. Chisholm, K. Folting, W.E. Streib, and M. Scheer, *Inorg. Chem.*, **35**, 3659 (1996).
160. M.H. Chisholm, K. Folting, J.C. Huffman, and C.C. Kirkpatrick, *J. Chem. Soc., Chem. Commun.*, 189 (1982); see also ref. 34.
161. J.A. Hollingshead and R.E. McCarley, *J. Am. Chem. Soc.*, **112**, 7402 (1990).
162. C. Limbert, S. Parsons, and A.J. Downs, *J. Chem. Soc., Chem. Commun.*, 497 (1994).
163. D.J. Darensbourg, R.L. Gray, and T. Delord, *Inorg. Chim. Acta*, **98**, L39 (1985).
164. F.A. Cotton, W. Schwotzer, and E.S. Shamshoum, *J. Organomet. Chem.*, **296**, 55 (1985).
165. J.A. Heppert, S.D. Dietz, N.W. Eilerts, R.W. Henning, M.D. Morton, F. Takusagawa, and F.A. Kaul, *Organometallics*, **12**, 2565 (1993).
166. P.A. Bates, A.J. Nielson, and J.M. Waters, *Polyhedron*, **4**, 999 (1985).
167a. R. Villanneau, A. Proust, F. Robert, and P. Gouzerh, *Chem. Commun.*, 1491 (1998);
167b. A. Proust, P. Gouzerh, and F. Robert, *Inorg. Chem.*, **32**, 5291 (1993).
168. M.H. Chisholm, K. Folting, J.C. Huffman, and E.M. Kober, *Inorg. Chem.*, **24**, 241 (1985).

169. M.H. Chisholm, K. Folting, B.W. Eichhorn, and J.C. Huffman, *J. Am. Chem. Soc.*, **109**, 3146 (1987).
170. M.H. Chisholm, C.M. Cook, and K. Folting, *J. Am. Chem. Soc.*, **114**, 2721 (1992).
171. M.H. Chisholm, K.S. Kramer, and W.E. Streib, *Angew. Chem. Int. Ed. Engl.*, **34**, 891 (1995).
172. F.A. Cotton, W. Schwotzer, and E.S. Shamshoum, *Organometallics*, **2**, 1340 (1983).
173. M.H. Chisholm, K.S. Kramer, and W.E. Streib, *J. Cluster Sci.*, **6**, 135 (1995).
174. B.T. Flatt, R. Grubbs, R.L. Blanski, J.C. Calabrese, and J. Feldman, *Organometallics*, **13**, 2728 (1994).
175. K. Hegetschweiler, H.W. Schmalle, H.M. Streit, and W. Schneider, *Inorg. Chem.*, **29**, 3265 (1990).
176. K. Hegetschweiler, H.W. Schmalle, H.M. Streit, V. Gramlich, H.-U. Hund, and I. Erni, *Inorg. Chem.*, **31**, 1299 (1992).
177. K.L. Taft, G.C. Papaefthymiou, and S.J. Lippard, *Inorg. Chem.*, **33**, 1510 (1994).
178. F.L. Phillips and A.C. Skapski, *J. Chem. Soc., Dalton Trans.*, 2586 (1975).
179. W.A. Herrmann, S.J. Eder, and W. Scherer, *Chem. Ber.*, **126**, 39 (1993).
180. F.L. Phillips and A.C. Skapski, *Acta Crystallog.*, **B31**, 1814 (1975).
181. T. Greiser, O. Jarchow, K.-H. Klaska, and E. Weiss, *Chem. Ber.*, **111**, 3360 (1978).
182. M. Cesari, *Gazz. Chem. Ital.*, **110**, 365 (1980).
183. A.I. Yanovskii, N.Ya. Turova, N.I. Kozlova, and Yu. T. Struchkov, *Koord. Khim.*, **13**, 242 (1987).
184. R.A. Sinclair, W.B. Gleason, R.A. Newmark, J.R. Hill, S. Hunt, P. Lyon, and J. Stevens, *Chemical Processing of Advanced Materials*, (L.L. Hench and J.K. West eds), 207, John Wiley, New York (1992).
185. S.A. Sangokoya, W.T. Pennington, J. Byers-Hill, G.H. Robinson, and R.D. Rogers, *Organometallics*, **12**, 2429 (1993).
186. N. Ya. Turova, A.I. Yanovskii, V.G. Kessler, N.I. Kozlova, and Yu. T. Struchkov, *Russ. J. Inorg. Chem.*, **36**, 1404 (1991).
187. J.L. Atwood, D.C. Hrncir, R.D. Priester, and R.D. Rogers, *Organometallics*, **2**, 985 (1983).
188. C.J. Harlan, S.G. Bott, B. Wu, R.W. Lenz, and A.R. Barron, *Chem. Commun.*, 2183 (1997).
189. P.G. Harrison, B.J. Haylett, and T.J. King, *J. Chem. Soc., Chem. Commun.*, 112 (1978).
190. H. Reuter and M. Kremer, *Z. Anorg. Allgem. Chemie*, **615**, 137 (1992).
191. C. Chandler, G.D. Fallon, and B.O. West, *J. Chem. Soc., Chem. Commun.*, 1063 (1990).
192. J. Caruso, M.J. Hampden-Smith, and E. Duesler, *J. Chem. Soc., Chem. Commun.*, 1041 (1995).
193. A.I. Yanovskii, N. Ya. Turova, E.P. Turevskaya, and Yu. T. Struchkov, *Russ. J. Coord. Chem.*, **8**, 76 (1982).
194. A. Proust, R. Thouvenot, F. Robert, and P. Gouzerh, *Inorg. Chem.*, **32**, 5299 (1993).

6

Metal Aryloxides

1 INTRODUCTION

The metal–aryloxide bond is not only a component of a rapidly increasing number of inorganic and organometallic compounds,[1] but also occurs in nature in numerous metalloproteins.[2] The amino acid residue tyrosine has been shown to bond to metals through the phenoxide oxygen in the transferrins and other proteins while a key component of many siderophores is the catecholate function that strongly binds iron. The inorganic chemistry of phenolic reagents has a long history. Besides the reaction of simple phenols with metals the phenoxide group is a critical function in many complexing agents as exemplified by 8-hydroxyquinoline (oxine), which was one of the earliest analytical reagents.[3] This and related ligands are useful for the colorimetric and gravimetric determination of metal ions, as well as for their extraction from aqueous solution.[3] The phenoxide group is also a constituent of many bi, tri, and polydentate ligands.[4]

As with metal alkoxide chemistry, the utilization of non-protic solvent systems has expanded the variety of metal aryloxide derivatives that can be isolated, including organometallic derivatives, and the isolation and characterization of simple aryloxide compounds of even the most oxophilic metals. In this regard mention should be made of the work of Funk *et al.* who, in a series of studies dating back to 1937,[5] isolated many transition metal complexes of phenol and its simple substituted derivatives. More recently there has been a growing interest in the organometallic chemistry associated with metal aryloxide compounds. Research has focussed both on the reactivity of the ligands themselves, *e.g.* the insertion chemistry of the M–OAr bonds, and on the organometallic reactivity that can be supported by aryloxide ligation.

2 TYPES OF ARYLOXIDE LIGAND

There are a plethora of ligand types that contain at least one phenoxide nucleus for coordination to metal centres. In this chapter we will initially survey the ligands depending upon the number of phenoxide units present. The chemistry of some of these ligand types, *e.g.* catechols, calix[*n*]arenes, and macrocyclic ligands, especially those containing the salicylaldimine unit, will not be exhaustively reviewed. Instead reference will be given to existing key works.

2.1   Mono-aryloxides

2.1.1   Simple Mono-phenoxides

The conventional substitution chemistry of phenol as well as the importance of substituted phenols as commercial anti-oxidants and monomers in the production of phenylene-oxide polymers make available to the inorganic and organometallic chemist a selection of typically inexpensive ligands. The introduction of alkyl substituents onto the phenoxide ring, usually by Friedel–Crafts and related electrophilic substitution reactions[6] can readily be achieved. The strong *ortho*-metallation directing effect of either the parent or protected phenolic group can be used to build substituted phenols *via* the lithiated nucleus (Scheme 6.1), e.g. by reaction with ketones.[7] Aryl groups can be introduced into the *ortho* position by Suzuki coupling[8] or by arylation with bismuth[9] or lead[10] reagents. Alternative strategies include construction of the phenoxide nucleus via initial condensation reactions followed by aromatization (Scheme 6.1). Alkyl substituents in the *para* position exert little steric influence within the coordination sphere but can help control solubility and modify the spectroscopic properties of metal derivatives. In contrast, the choice of suitable alkyl substituents in either one

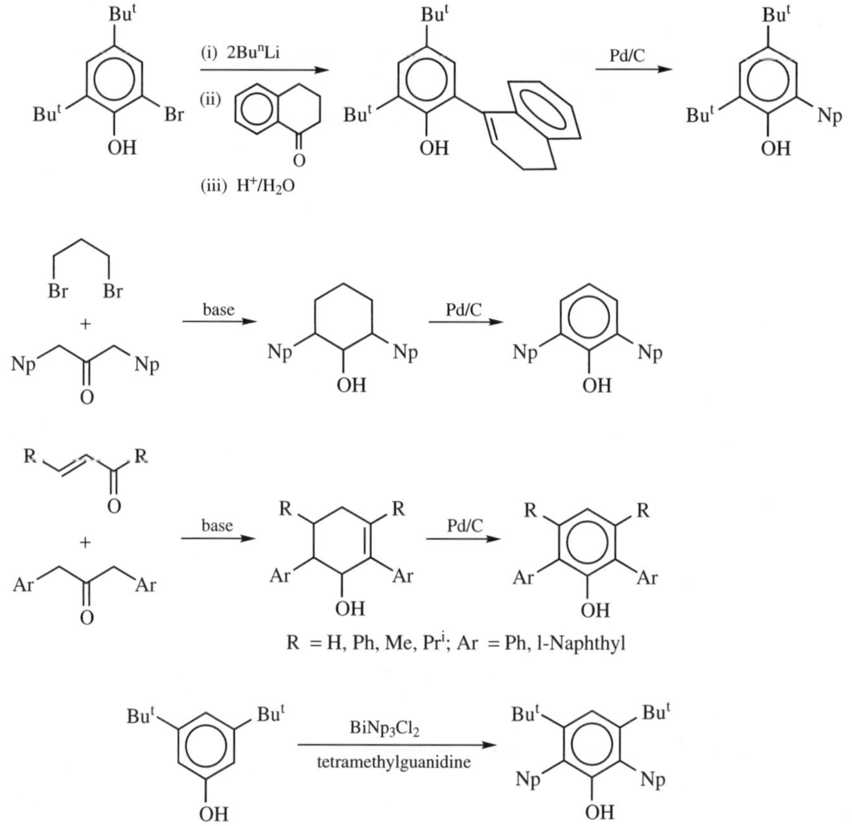

$R = H, Ph, Me, Pr^i; Ar = Ph, 1$-Naphthyl

Scheme 6.1

or both of the *ortho* positions of the phenoxide nucleus allows significant control of the steric requirements of the ligand. This can have a dramatic effect on the structure and reactivity of metal derivatives. Interestingly it has recently been observed that the presence of *meta* substituents can affect the steric properties of ligands by restricting the conformational flexibility of adjacent *ortho*-phenyl substituents.[11] The early resolution of 2-*sec*-butylphenol and some other simple chiral phenols was achieved[12] although their inorganic chemistry has not been developed to any appreciable extent. More recently, chiral 2,6-di-arylphenols that have C_2 symmetry have been obtained via a number of routes.[8,7,13] In some cases resolution of the enantiomers has been achieved and metal derivatives generated.

The use of electronically more imposing substituents can dramatically affect the nature of the phenolic reagents. Electron-withdrawing substituents such as halides and nitro groups are commonly used, and generate more acidic phenols, *e.g.* 2,4,6-trichloro-, pentafluoro-, and 2,4,6-trinitro-phenol.

Phenol can be considered a special example of an enol with the 1,3-cyclohexadienone form being considerably less stable than the hydroxybenzene form; the equilibrium constant for Eq. (6.1) is $\sim 10^{-14}$.[14]

$$\text{(6.1)}$$

The energetics of this tautomerism is affected by substituents on and especially within the arene ring. An important class of ligands, the 2-hydroxy pyridines and quinolines, exemplify this (Eq. 6.2).

$$\text{(6.2)}$$

Here a ground state pyridone and quinolone structure is present in the parent ligand as well as being an important resonance form in many metal complexes.[15] It has also been shown possible to stabilize the 1,3-cyclohexadienone form by attachment to a triosmium cluster framework.[16]

2.1.2 Chelating Mono-phenoxides

Donor atom-containing *ortho* substituents of phenoxides may undergo chelation to the metal centre. The formation of five-membered rings can occur for 2-chloro,[17] 2-amino,[18] 2-methoxy,[19,20] 2-thio[21] and related phenoxides. The coordination chemistry of 2-(dialkylphosphino)phenoxides has been extensively investigated. Chelation of 2-nitro-phenoxides is commonly observed in the solid state. In select systems this coordination can allow reduction of the nitro group, eventually leading to 2-imidophenoxides.[22] The phenoxide group is also an integral part of many important classical bidentate ligands, *e.g.* 8-hydroxyquinoline and derivatives, and many synthetic strategies have been developed to synthesize and study chelating mono-phenoxides. The utilization of phenoxides containing 2-pyridyl, bipyridyl, or phenanthroline groups in the *ortho* position has been explored. One development has been the synthesis of 2-hydroxyphenyl-bis(pyrazolyl)methane ligands. These mono-anionic groups, termed "heteroscorpionate" ligands, have some of the coordination characteristics of traditional tris(pyrazolyl)borates but contain a metal–aryloxide bond.[23-26]

2.1.2.1   *Cyclometallated aryloxides and ligands chelated through carbon*

The simple coordination of aryloxides to metal centres is sometimes followed by carbon–hydrogen bond activation, which leads to new oxa-metallacycle rings such as those shown in the top part of Scheme 6.2. The cyclometallation of simple phenoxide as well as various 2-alkyl- and aryl-phenoxides leads to 4-, 5-, and 6-membered metallacycles.[27,28] It is also possible for *o*-aryl substituents to chelate to metal centres via π-arene interactions.[29] The intramolecular dehydrogenation of alkyl groups[30] or hydrogenation of aryl rings can lead to aryloxides chelated to the metal through π-bonding interactions with the resulting unsaturated olefin or diene functional group.[31] Ligands such as 2-allylphenol can bind simply through oxygen, chelate via an η^2-olefin interaction,[32] or be converted to an η^3-allyl chelate.[33]

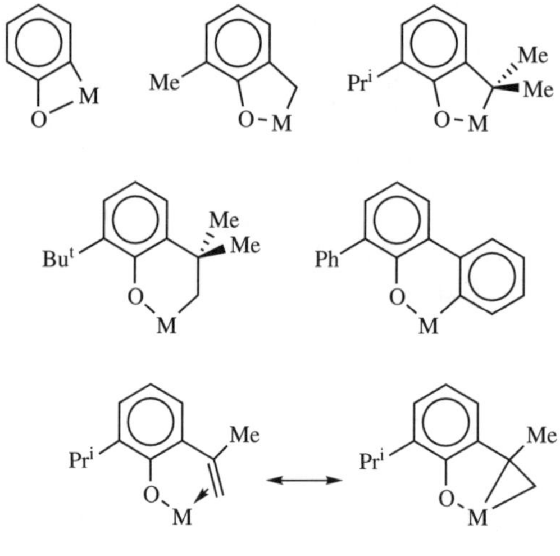

Scheme 6.2

2.1.2.2   *Mono-salicylaldimines*

By far the largest group of chelating mono-aryloxides are those derived from 2-formyl phenols and their condensation with primary amines (Scheme 6.3).[34] The resulting ligands chelate through the phenoxide and the imine functions to generate five-membered chelate rings. The ligands are highly amenable to tuning of their steric and electronic properties through variation of the substituents on nitrogen, at the α-carbon or within the phenoxide ring. The condensation of 2-formyl-phenols with more complex amines has been exploited to yield mono-aryloxide ligands that can chelate through a multitude of neutral and anionic functional groups.

2.1.3   *Binucleating Mono-aryloxides*

Important classes of mono-aryloxides are those containing chelating groups on either side of a central phenoxide function. The geometrical constraints of such ligands

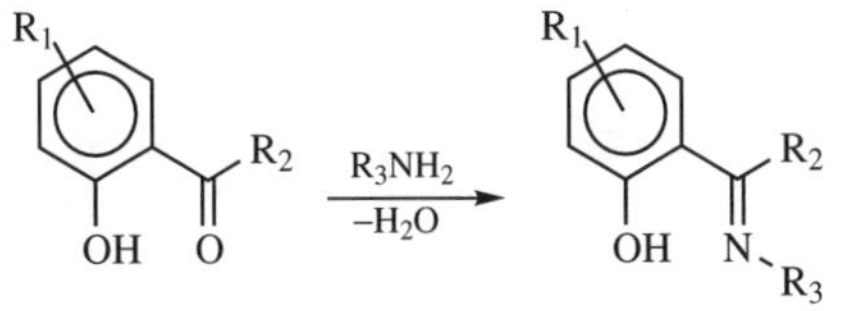

$$\text{R}_1 \underset{\text{OH} \quad \text{O}}{\overset{}{\bigcirc}}\text{R}_2 \quad \xrightarrow[-\text{H}_2\text{O}]{\text{R}_3\text{NH}_2} \quad \text{R}_1 \underset{\text{OH} \quad \text{N}_{\text{R}_3}}{\overset{}{\bigcirc}}\text{R}_2$$

Scheme 6.3

typically preclude chelation of all donor groups to a single metal centre. Instead these ligands form highly stable dinuclear complexes in which two metal centres are bridged by the phenoxide oxygen atom and, typically, other groups.[35] Two common synthetic pathways to ligands of this type proceed via either 2,6-diformyl- or 2,6-di(chloromethyl)phenols (Eq. 6.3) as intermediates.[36]

$$\underset{\text{Cl} \quad \text{OH} \quad \text{Cl}}{\overset{\text{CH}_3}{\bigcirc}} + 2\text{HN(CH}_2\text{py})_2 \quad \xrightarrow[-\ 2[\text{HNEt}_3][\text{Cl}]]{+\ 2\text{NEt}_3} \quad \underset{(\text{pyCH}_2)_2\text{N} \quad \text{OH} \quad \text{N(CH}_2\text{py})_2}{\overset{\text{CH}_3}{\bigcirc}} \qquad (6.3)$$

Non-symmetrical ligands have also been designed such that they impose differing coordination environments about each metal centre.[37]

2.2 Bis-aryloxides

2.2.1 Catechols and Related Ligands

The study of metal derivatives of 1,2-dihydroxybenzene (catechol) is an important area of inorganic chemistry.[38] This importance is emphasized not only by the presence of catecholate binding sites in metalloproteins (*e.g.* enterobactin, which mediates iron uptake by *Escherichia coli* and related bacteria),[39] but also by the existence of enzymes such as catechol dioxygenases, which are involved in oxidative degradation of aromatic hydrocarbons.[40,41] It is also sometimes possible to generate identical metal complexes using *ortho*quinones as the ligand precursor. Hence the relative importance of the dioxy, quinone, and intermediate semi-quinone forms tends to dominate discussions of the spectroscopy, structure, electrochemistry, and reactivity of metal complexes of these ligands.[38] The chemistry of catechols containing alkyl and halide substituents as well as derivatives of other aromatic hydrocarbons such as 9,10-dihydroxyphenanthrene has been extensively investigated.[38]

A similar situation can occur for hydroquinone (1,4-dihydroxybenzene). However, in this case the ligand can coordinate to two metal centres.[42] In one study this ability was utilized to produce interesting covalent three-dimensional networks.[43]

Although not technically bis-aryloxides, 2-thio- and 2-amidophenoxides have the aryloxide bonds considerably perturbed by the *ortho* heteroatom in a fashion similar to that in catediolates.

2.2.2 Biphenoxides and Bis(phenol)methanes

Both biphenol (from oxidative coupling of phenols) and bis(phenol)methane (formaldehyde/phenol condensation) as well as their substituted derivatives have been used as ligands (Scheme 6.4).[44,45] The binaphthol unit (typically generated by oxidative coupling of 2-naphthol) has also been extensively used to generate chiral aryloxides bound through both aryloxide oxygen atoms.[46,47] The low solubility and small steric size of parent binaphthol has led to the development of synthetic routes to substituted derivatives (Scheme 6.4).[48-50] The chiral ligand 6,6'-dimethyl-3,3',5,5'-tetra-*tert*-butyl-1,1'-biphenyl-2,2'-diol has been synthesized, resolved, and applied to Mo-catalysed metathesis reactions.[51,52] A variety of chelating, diaryloxo ligands have been applied to metal-catalysed olefin polymerization studies. Bridging groups include methylene,[53] ethylene,[54] sulfur,[55] tellurium,[56] and sulfinyl units.[57] The metal coordination chemistry of a large number of other bis-phenolic reagents has also been investigated, *e.g.* those derived from salicylic acid and diammines (Scheme 6.5).[58] In some of these ligands the phenoxy groups are chemically non-equivalent.

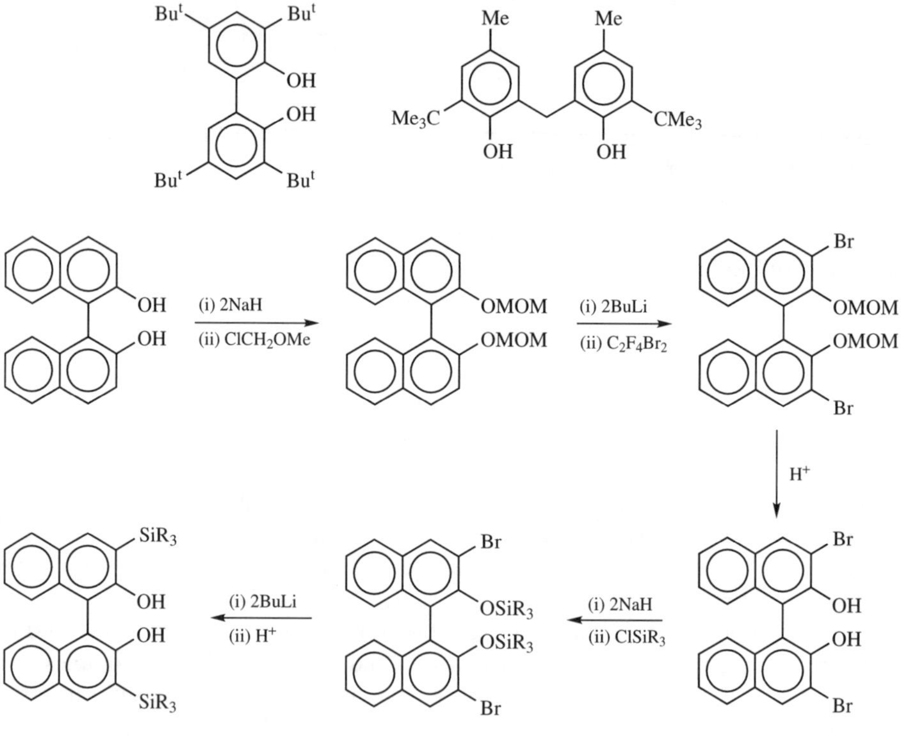

Scheme 6.4

2.2.3 Bis-salicylaldimines

The bis-aryloxide ligands obtained by condensation of two equivalents of salicylaldehyde with diammines, exemplified by bis(salicylaldehyde)ethylenediimine (salen), constitute one of the most important classes of ligands in coordination chemistry.[59]

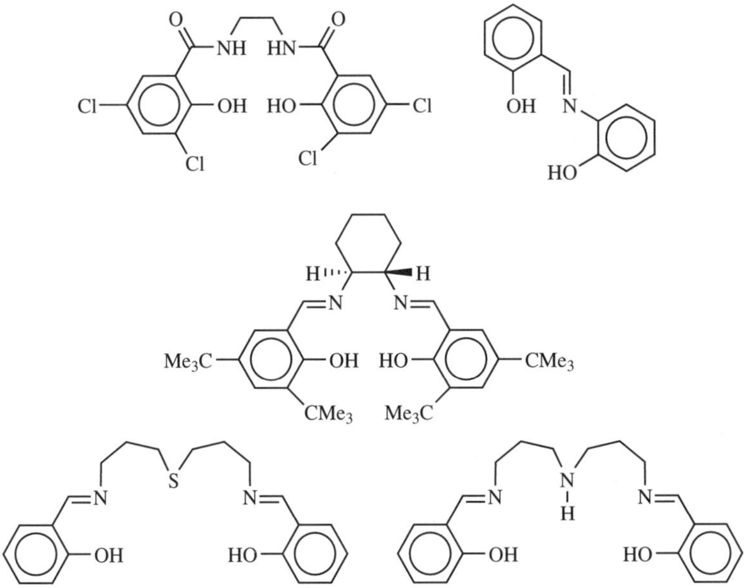

Scheme 6.5

Changing the diammine backbone and/or introducing substituents at the α-carbon or within the phenoxy ring can easily modify these tetradentate ligands. Ligands derived from resolved, chiral diammines have been employed in the synthesis of catalysts for asymmetric epoxidation, and are finding utility in other areas (Scheme 6.5).[60-62] The presence of one or more donor groups in the diammine backbone leads to penta- and hexadentate ligands, some of which favour the formation of dinuclear complexes (Scheme 6.5).

2.2.4   Macrocyclic Bis-phenoxides

The pioneering work of Robson and Pilkington showed the metal-mediated template condensation of 2,6-diformyl-4-methylphenol with 1,3-diaminopropane to produce the dimetal derivatives of a new macrocyclic bis-phenoxide ligand.[63] These ligands favour formation of binuclear species with the metal centres bridged by the two phenoxide oxygen atoms.[64] The parent ligand (Scheme 6.6) has also been synthesized directly from the organic substrates in the absence of a metal template.[65] This initial work was followed by the rapid development of related compartmentalized ligands, *e.g.* partially and totally saturated analogues,[66] ligands with differing size chelate bridges,[67] nonsymmetric coordination environments for the phenoxide bridged metals,[68] and the introduction of extra donor groups into the chelate backbone.[69-72] Much of the work in this area and the related studies utilizing binucleating mono-phenoxides (Section 2.1.3) has been stimulated by the expectation of mimicking the active sites in various metalloenzymes.[73]

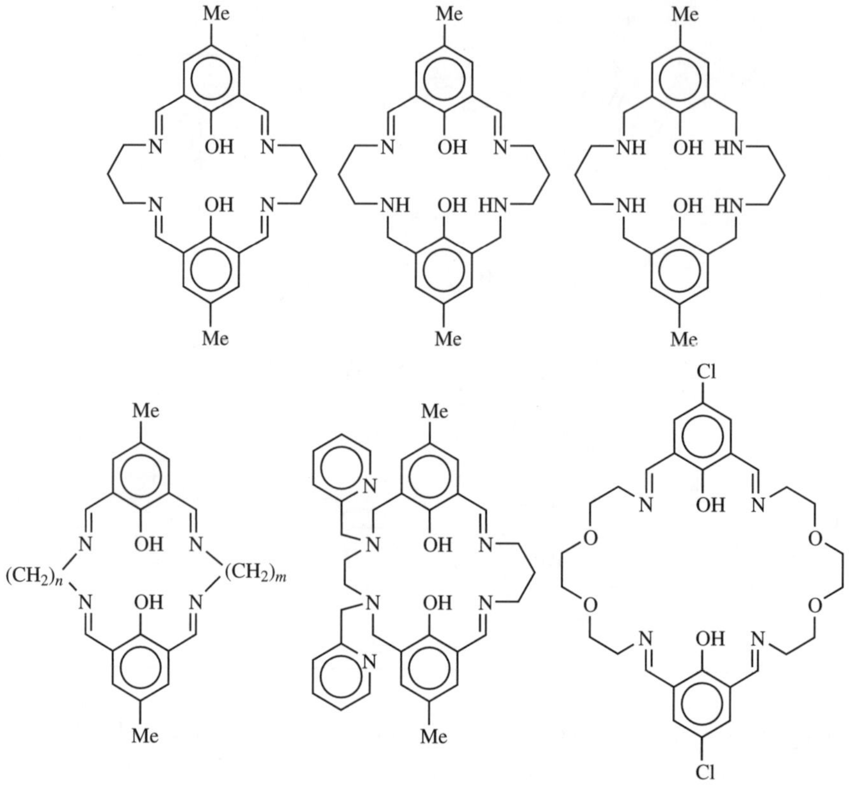

Scheme 6.6

2.3 Poly-aryloxides and Calixarenes

The elaboration of routine synthetic methodologies such as the condensation reaction of 2-formyl- or 2-acyl-phenols with primary amines, sometimes followed by reduction, can be utilized in the synthesis of poly-aryloxide ligands including macrocyclic examples. Some types of tris-aryloxides obtained in this way are shown in Scheme 6.7 (**A**) and (**B**).[74,75] Examples of a nonmacrocyclic tetra-aryloxide (**C**),[76] and a macrocyclic tetra-aryloxide (**D**)[77,78] are also shown. The connection of multiple catecholate nuclei can also generate poly-aryloxides (see Section 2.2.1). An important class of poly-phenoxides is the calix[*n*]arenes (*n* represents the number of phenol units (Scheme 6.7) (**E**)), which can be readily obtained by the condensation of *para*-substituted phenols with formaldehyde.[79] The chemistry of the calixarene ligand has been exhaustively reviewed.[80] An important aspect of the chemistry of these ligands centres upon the various conformers that can be adopted and/or stabilized. In the case of calix[4]arene there are four conformations: "cone" (all oxygen atoms on the same side of the ring), "partial cone" (three up, one down), "1,2-alternate", and "1,3-alternate". By blocking or linking phenoxide oxygens it is possible to generate unusual tris- and bis-aryloxides.[81,82] There are eight possible conformers for calix[6]arene.[83]

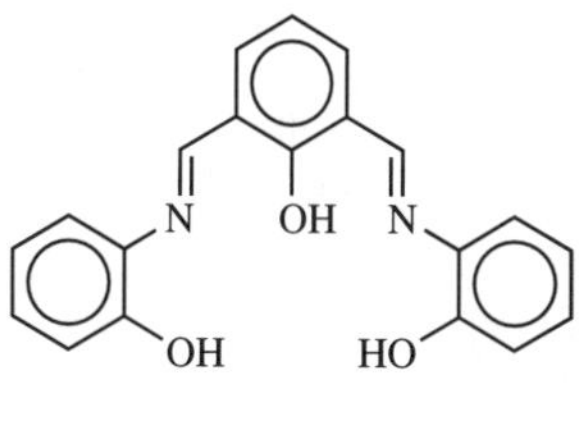

(A)

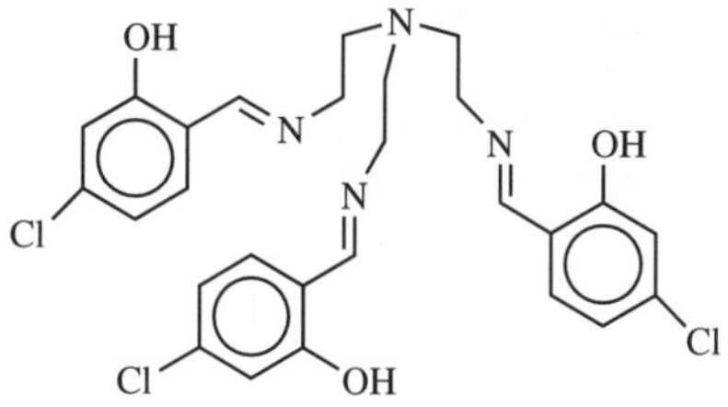

(B)

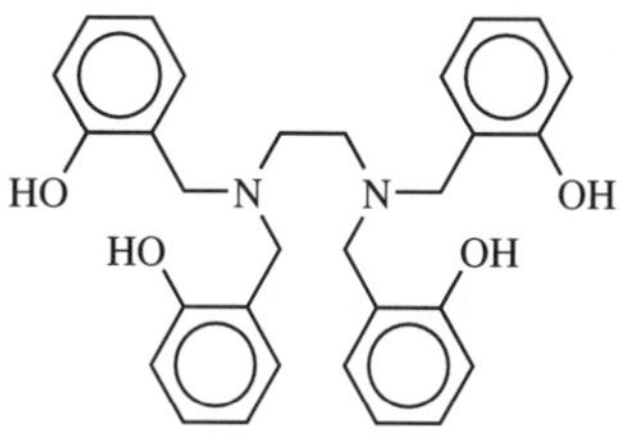

(C)

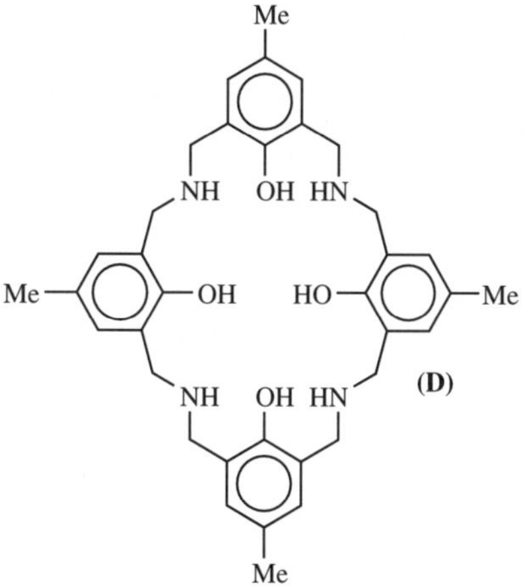

(D)

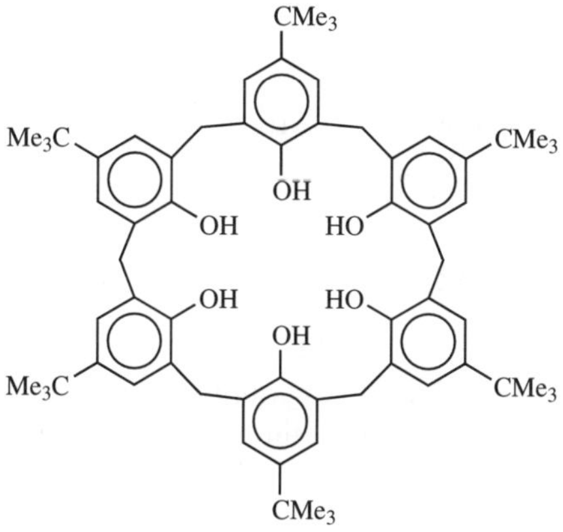

(E)

Scheme 6.7

3   SYNTHESIS OF METAL ARYLOXIDES

There are a large number of synthetic routes to metal aryloxides. The synthetic strategy that is adopted depends largely on the available metal precursors as well as the nature of the ligand that is being used. Many of the synthetic methods used for simple aryloxide derivatives in nonprotic solvents are identical to those used for the corresponding metal alkoxides.

3.1   From the Metal

The reaction of the most electropositive metals with phenols can sometimes lead to direct formation of aryloxide derivatives along with evolved hydrogen (Eq. 6.4[84]).

$$Na + HOAr \xrightarrow{\text{THF}} [NaOAr]_n + \tfrac{1}{2}H_2 \tag{6.4}$$

where $HOAr = HO-\langle\bigcirc\rangle-CH_3$.

Although it is possible to carry out the reaction with the metal in neat liquid or molten phenol, refluxing in a suitable solvent is normally desirable. This method typically produces adducts, coordinated phenol and/or solvent, of the aryloxide complex (*e.g.* Eqs 6.5–6.7).[85,86]

$$6Ba(s) + 12PhOH \xrightarrow{\text{HMPA}} [Ba_6(OPh)_{12}(HMPA)_4] + 6H_2 \tag{6.5}$$

$$3Sr(s) + 6PhOH \xrightarrow{\text{HMPA}} [Sr_3(OPh)_6(HMPA)_5] + 3H_2 \tag{6.6}$$

$$4Sr(s) + 10PhOH \xrightarrow{\text{THF}} [Sr_4(OPh)_8(HOPh)_2(THF)_6] + 4H_2 \tag{6.7}$$

An innovative method involves the use of metal atom vapour procedures to generate a highly activated dispersion of metal in a hydrocarbon solvent for reaction with phenolic or other reagents (Eq. 6.8[87]).

$$M + 2HOAr \xrightarrow{\text{THF}} [M(OAr)_2(THF)_n] + H_2 \tag{6.8}$$

where $M = Ca, n = 3$; $Ba, n = 4$; $HOAr = HO-\langle\bigcirc\rangle$ with Me_3C substituents.

In the case of the reaction of metallic barium with phenol in either THF or HMPA the products isolated were shown to contain both hydride and oxo ligands (Eqs 6.9[88] and 6.10[89]).

$$Ba(s) \xrightarrow[\text{reflux}]{\text{PhOH/THF}} [HBa_5(O)(OPh)_9(THF)_8] \tag{6.9}$$

$$Ba(s) \xrightarrow[\text{HMPA}]{\text{PhOH}} [H_2Ba_8(\mu_5\text{-}O)_2(OPh)_{14}(HMPA)_6] \tag{6.10}$$

In some instances the reaction is accelerated by either amalgamation or the addition of a catalyst (*e.g.* iodine or a mercury(II) halide, Eq. 6.11[90]) while the stoichiometric reaction of mercury and thallium compounds with lanthanide metals has been shown to be a viable route to aryloxide derivatives (Eqs 6.12[91] and 6.13[92]).

$$Y(s) + 3HOPh \xrightarrow{\text{HgCl}_2\,(\text{cat})} [Y(OPh)_3]_n + \tfrac{3}{2}H_2 \tag{6.11}$$

$$2Ln + 3Hg(C_6F_5)_2 + 6HOAr \longrightarrow 2[Ln(OAr)_3] + 3Hg + 6C_6H_5F \quad (6.12)$$

where $Ln = Nd, Er, Sm, Yb, Lu;$ $HOAr =$

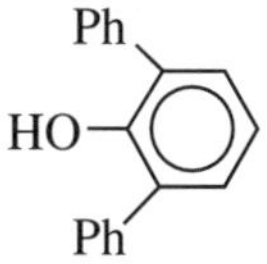

$$Yb + 2TlOAr \xrightarrow{\text{THF}} [Yb(OAr)_2(THF)_3] + 2Tl \qquad (6.13)$$

where $ArO =$

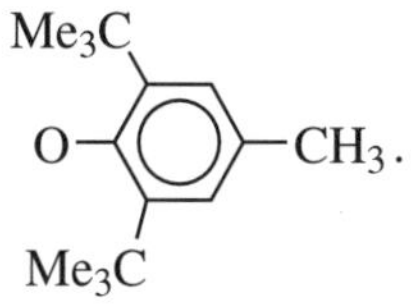

3.2   From Aqueous or Protic Solutions of Salts

The synthesis of many metal derivatives of complexing agents such as 8-hydroxyquinoline and salicylaldehyde oxime begins with aqueous solutions of metal ions (*e.g.* sulfates, nitrates, etc.). Addition of the ligand and suitable control of pH allows either precipitation of the coordination compound or its extraction into nonprotic media for either qualitative determination or work up (Eqs 6.14[93] and 6.15[94]).

$$Al^{3+}(aq) + 3HOAr \xrightarrow{\text{Base}} [Al(OAr)_3] \qquad (6.14)$$

where $HOAr =$

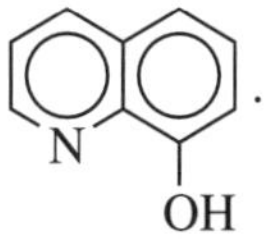

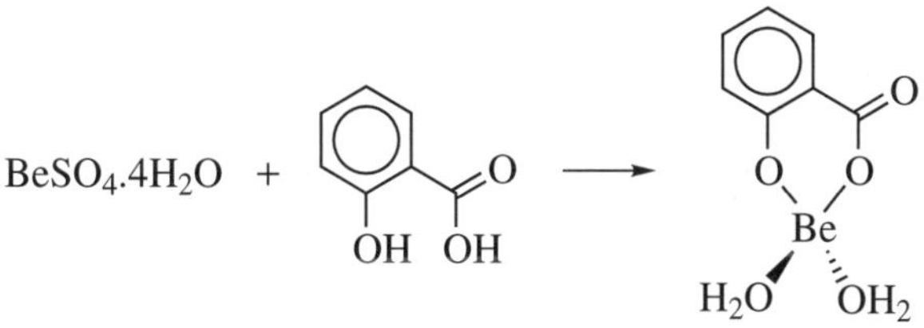

$$BeSO_4.4H_2O \; + \qquad \longrightarrow \qquad (6.15)$$

The synthesis of metal derivatives of strongly acidic phenols such as 2,4,6-trinitrophenol (picric acid) and 2,4-dinitrophenol (Eq. 6.16[95]) is typically carried out in the presence of water or other protic solvents such as acetone/alcohol mixtures.

$$Ca^{2+}(aq) + 2HOAr \xrightarrow{\text{NaOH}} Ca(OAr)_2.7H_2O \qquad (6.16)$$

where HOAr = $HO-\!\!\bigcirc\!\!-NO_2$ (with O_2N substituent). This refers to 6.16 on the previous page.

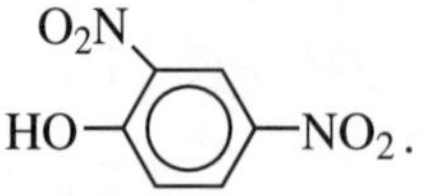

The template synthesis of a number of complexes of macrocyclic ligands is carried out in protic media (Eq. 6.17[96]).

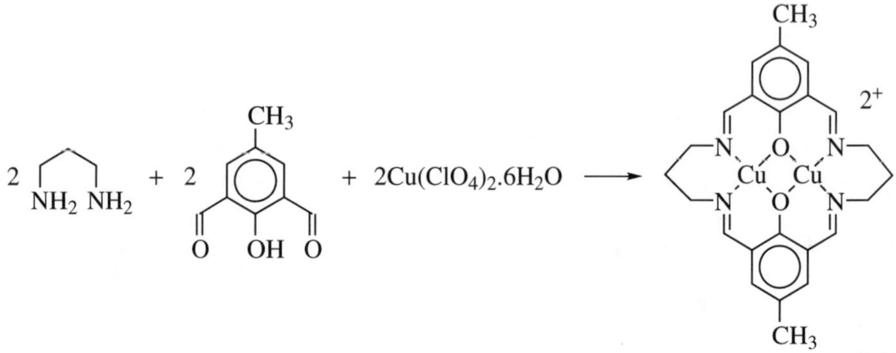

$$2\;H_2N{-}NH_2 \;+\; 2\;(\text{CH}_3\text{-aryl, O OH O}) \;+\; 2Cu(ClO_4)_2{\cdot}6H_2O \;\longrightarrow\; [\text{Cu}_2\text{ macrocycle}]^{2+}$$

(6.17)

3.3 From Metal Halides

The metal chlorides, bromides, and to a lesser extent iodides remain the single most important class of starting materials in inorganic chemistry. Many of them are commercially available in anhydrous form. The formation of aryloxide derivatives by simple addition of phenols to the corresponding anhydrous metal halide is sometimes a viable synthetic route that has been used for the formation of a variety of d-block metal compounds (Eq. 6.18) as well as some aryloxide complexes of the p-block metals.

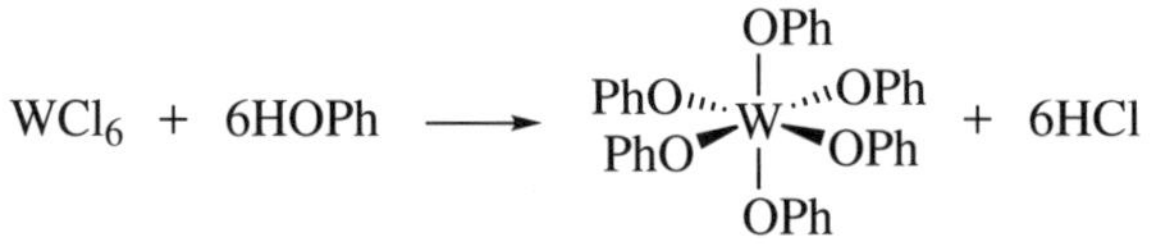

$$WCl_6 \;+\; 6HOPh \;\longrightarrow\; W(OPh)_6 \;+\; 6HCl$$

(6.18)

The HCl that is eliminated can sometimes lead to a secondary reaction that complicates the synthesis, e.g. the dealkylation of tert-butyl phenoxide ligands,[97] and is sometimes removed by addition of a strong base which may aid in completing the reaction (Eqs 6.19[98] and 6.20[99]).

$$GeCl_4 + 4HOPh \;\xrightarrow[-4NH_4Cl]{4NH_3}\; [Ge(OPh)_4]$$

(6.19)

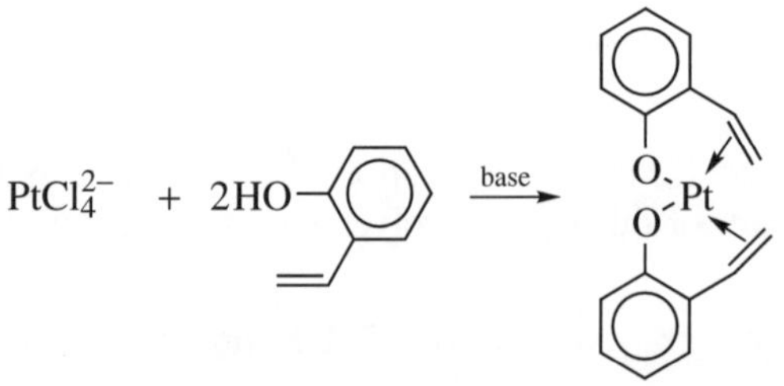

$$PtCl_4^{2-} \;+\; 2HO{-}\bigcirc \;\xrightarrow{\text{base}}\; [Pt(O\text{-aryl})_2]$$

(6.20)

Mixed halo, aryloxide compounds are sometimes obtained either by control of the stoichiometry and reaction conditions or by the use of sterically demanding aryloxide ligands (Eqs 6.21[100] and 6.22[101]).

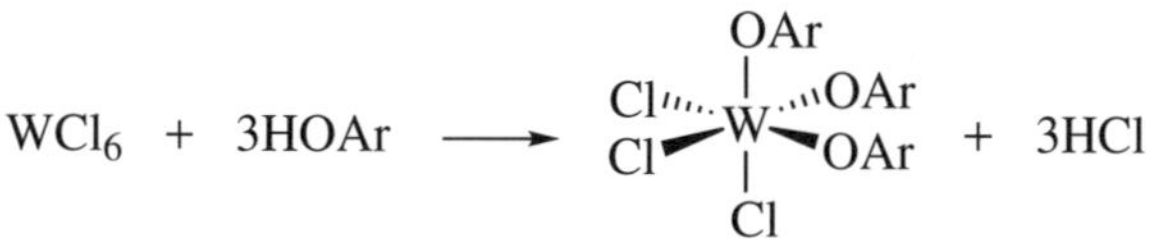

$$WCl_6 \ + \ 3HOAr \ \longrightarrow \ \underset{Cl}{\overset{OAr}{\underset{|}{\overset{|}{Cl^{\prime\prime\prime\prime}\cdots W\cdots OAr}}}} \ + \ 3HCl \qquad (6.21)$$

where HOAr =

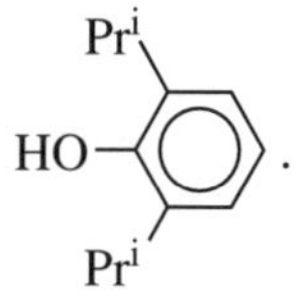

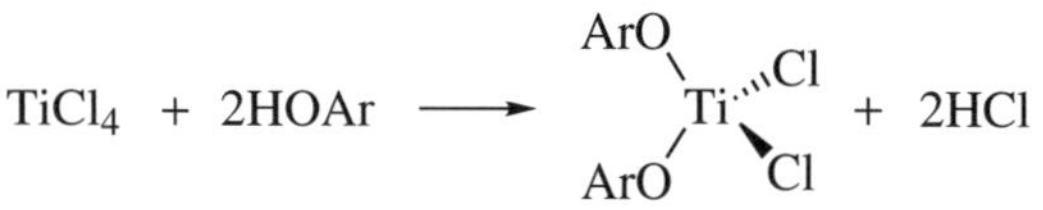

$$TiCl_4 \ + \ 2HOAr \ \longrightarrow \ \underset{ArO}{\overset{ArO}{Ti^{\cdots}}}\overset{Cl}{\underset{Cl}{}} \ + \ 2HCl \qquad (6.22)$$

where ArO =

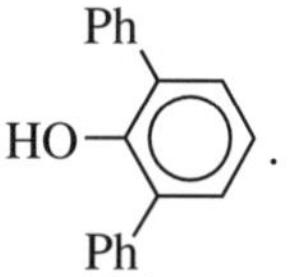

A much more widely used synthetic method entails the metathetical exchange reaction between alkali metal aryloxides and the metal halide. This procedure has been applied to the synthesis of lanthanide, actinide, and d-block metal aryloxides as well as derivatives of the main group metals (Eqs 6.23,[102] 6.24,[103] 6.25,[104] 6.26,[105] 6.27,[106] 6.28,[107] 6.29, and 6.30[108]).

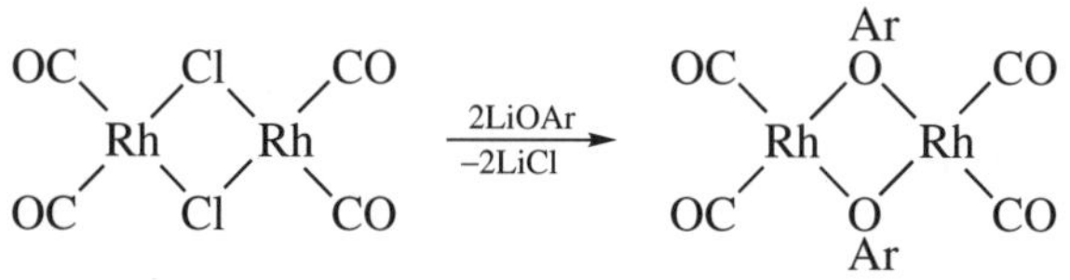

$$\underset{OC}{\overset{OC}{}}\underset{Cl}{\overset{Cl}{}}\underset{CO}{\overset{CO}{Rh\quad Rh}} \xrightarrow[-2LiCl]{2LiOAr} \underset{OC}{\overset{OC}{}}\underset{O}{\overset{O}{}}\underset{CO}{\overset{CO}{Rh\quad Rh}} \qquad (6.23)$$

where OAr =

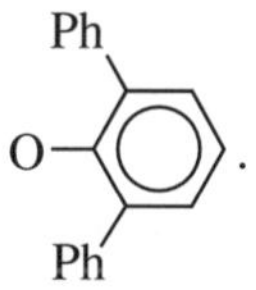

$$2[CuCl_2(en)] + 4NaOPh/2HOPh \longrightarrow [(en)(PhO)Cu(\mu\text{-}OPh)_2Cu(OPh)(en)].2PhOH$$
$$+ \ 4NaCl \qquad (6.24)$$

where en = $H_2NCH_2CH_2NH_2$.

$$YCl_3 + 3NaOAr \xrightarrow{\text{THF}} fac\text{-}[Y(OAr)_3(THF)_3] + 3NaCl \qquad (6.25)$$

$$GeCl_4 + 3LiOAr \longrightarrow \underset{ArO}{\overset{Cl}{\underset{}{Ge}}} \cdots OAr + 3LiCl \qquad (6.26)$$

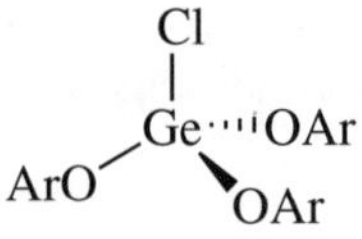

where ArO =

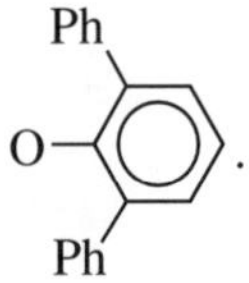

$$TiCl_3(NMe_3)_2 + 3LiOAr \longrightarrow ArO-Ti\overset{\cdots OAr}{\underset{OAr}{}} + 3LiCl + 2HNMe_2 \quad (6.27)$$

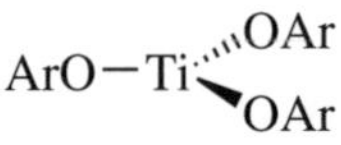

where OAr =

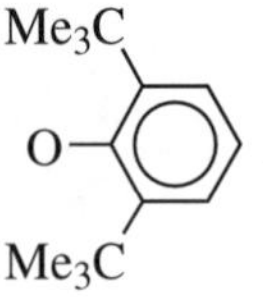

$$WCl_4(SEt_2)_2 + 4LiOAr \longrightarrow \underset{ArO}{\overset{ArO_{\cdots}}{\underset{}{W}}} \overset{\cdots OAr}{\underset{OAr}{}} + 4LiCl + 2Et_2S \quad (6.28)$$

where R = Me, Pr^i; OAr =

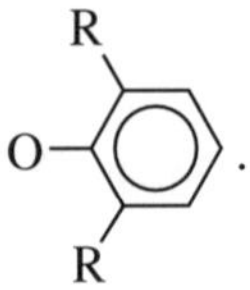

$$MnCl_2(NCCH_3)_2 + 2NaOAr \longrightarrow \underset{ArO}{\overset{ArO}{\underset{}{Mn}}}\overset{\cdots NCCH_3}{\underset{NCCH_3}{}} + 2NaCl \quad (6.29)$$

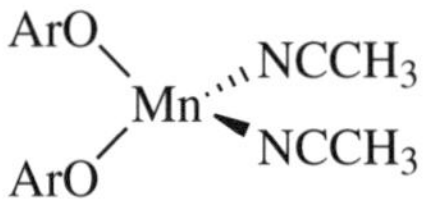

where OAr =

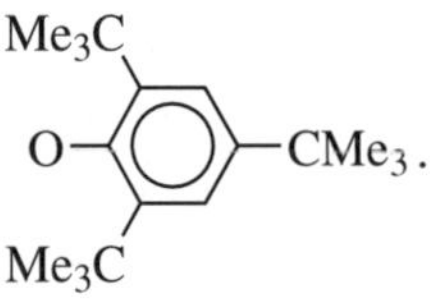

$$CrCl_3(THF)_3 + 3NaOAr \xrightarrow[\text{2. py}]{\text{1. THF}} \underset{py}{\overset{ArO_{\cdots}}{\underset{}{Cr}}}\overset{py}{\underset{OAr}{}} + 3NaCl \quad (6.30)$$

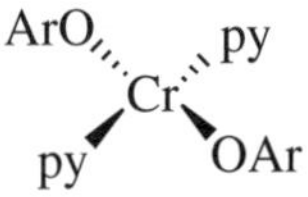

where OAr =

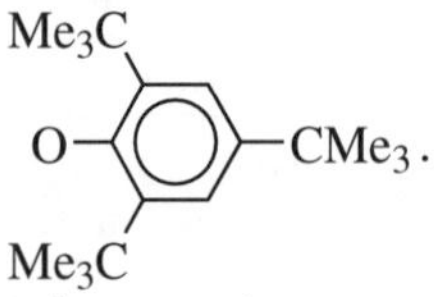

This reaction is sometimes complicated by the formation of anionic products (Eq. 6.31[109]) or double salts involving strong association of the alkali metal with the product metal aryloxide[1e,1f] (Eqs 6.32[110] and 6.33[111]).

$$FeCl_3 + 4LiOAr \xrightarrow[-LiBr]{P(Ph)_4^+Br^-} P(Ph)_4^+[Fe(OAr)_4]^- + 3LiCl \qquad (6.31)$$

where OAr =

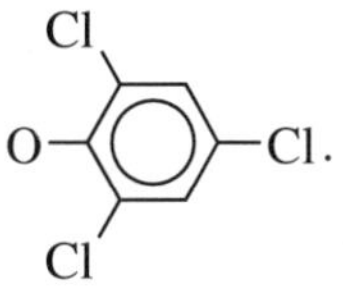

$$SnCl_2 + 3LiOAr \longrightarrow Li \cdots Sn + 2LiCl \qquad (6.32)$$

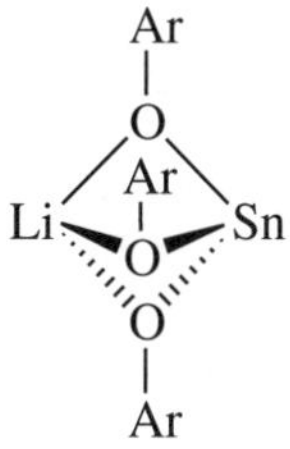

where ArO =

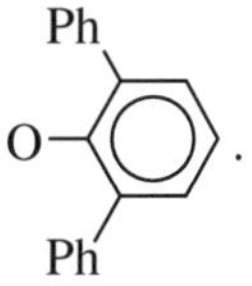

$$2CrCl_2(THF)_2 + 10LiOPh \xrightarrow{THF} [Li_6Cr_2(OPh)_{10}(THF)_6] + 4LiCl \qquad (6.33)$$

The alkali metal is typically bound to the aryloxide oxygen atoms, although in a number of cases it has been shown that interactions between sodium and the phenoxy ring of aryloxide ligand are an important aspect of the bonding in the mixed metal cluster (Eq. 6.34[112]).

$$NdCl_3 + 4KOAr \longrightarrow [KNd(OAr)_4]_n + 3KCl \qquad (6.34)$$

where ArO =

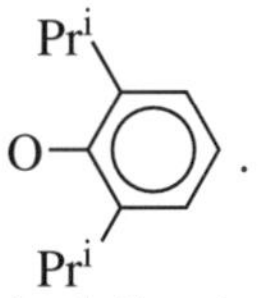

Another method for the replacement of chloride by aryloxide at transition metal centres is utilizing aryl-ethers. In this case the formation of the metal aryloxide is accompanied by the elimination of either alkyl or trialkylsilyl chloride (Eq. 6.35[113]).

$$NbCl_5 + Me_3SiOAr \xrightarrow{THF} Nb(OAr)Cl_4(THF) + Me_3SiCl \qquad (6.35)$$

where ArO =

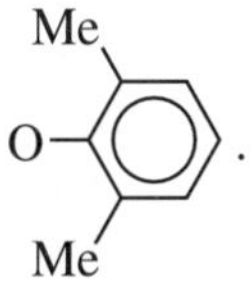

This process can also lead to the formation of metal–aryloxide bonds via activation of pendant methoxy-aryl groups. For example the treatment of [Cp(η^5-C$_5$H$_4$CEt$_2$C$_6$H$_4$OMe-2)TiCl$_2$] with LiBr was found to lead to the chelating phenoxide [Cp(η^5-C$_5$H$_4$CEt$_2$C$_6$H$_4$O-2)TiCl].[114] The use of 2,6-dimethoxyphenyl substituents can lead to bulky phosphine ligands. Early work by Shaw *et al.* showed that ligands such as PBu$^t_2$\{C$_6$H$_3$(OMe)$_2$-2,6\} can generate aryloxy-phosphines with elimination of MeCl at iridium metal centres.[115] This type of reaction is now been extended to a variety of metal systems.[116,117] In some situations the activation by metal centres of the stronger aryl-O bond can occur within aryl-ethers.[118]

3.4   From Metal Dialkylamides

The homoleptic metal dialkylamides are an important class of compounds in inorganic chemistry.[119] They are typically synthesized by treatment of the corresponding halide with lithium or sodium dialkylamide. Although involving an extra synthetic step, there are numerous examples where metal dialkylamide intermediates are useful in the synthesis of metal aryloxide compounds. The reaction normally involves the simple addition of the parent phenol to the metal dialkylamide in a nonprotic, typically hydrocarbon, solvent (Eqs 6.36,[120] 6.37,[121] and 6.38[122]).

$$\text{Ce[N(SiMe}_3)_2]_3 + 3\text{HOAr} \longrightarrow \text{Ce(OAr)}_3 + 3\text{HN(SiMe}_3)_2 \qquad (6.36)$$

where HOAr =

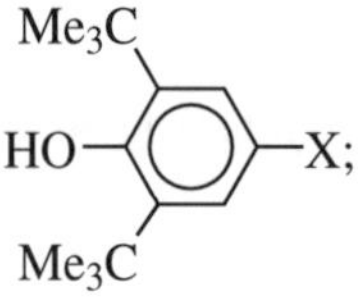

HO—⟨ring⟩—X; X = H, Me, CMe$_3$. (with Me$_3$C substituents)

$$+ 6\text{HNMe}_2 \qquad (6.37)$$

where ArOH =

HO—⟨ring⟩ (with H$_3$C substituents)

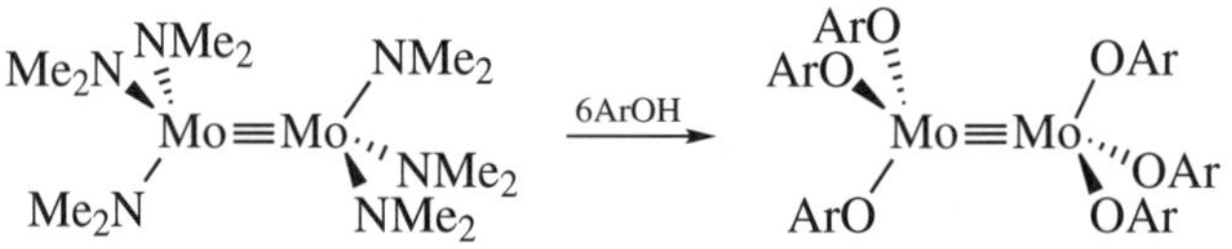

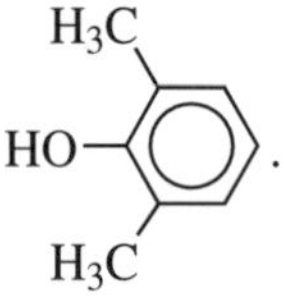

$$[\{\text{Yb(NR}_2)(\mu\text{-NR}_2)\}_2] \xrightarrow[-4\text{HNR}_2]{+4\text{HOAr}} [\{\text{Yb(OAr)}(\mu\text{-OAr})\}_2] \qquad (6.38)$$

where R = SiMe$_3$; OAr =

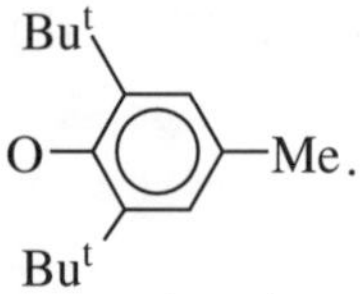

O—⟨ring⟩—Me. (with Bu^t substituents)

The dialkylamine that is produced can be readily removed, leaving the desired product. In some instances, either incomplete substitution can occur (bulky aryloxides,

Eqs 6.39 and 6.40[123]) or the generated dialkylamine can remain in the metal coordination sphere (Eq. 6.41[124]) or even deprotonate acidic phenols (Eq. 6.42[125]). Removal of the bound dialkylamine from adducts of this type can sometimes be difficult.

$$M(NMe_2)_4 + 2HOAr \longrightarrow M(OAr)_2(NMe_2)_2 + 2HNMe_2 \qquad (6.39)$$

where M = Ti, Ge, Sn; ArOH = .

where HOAr = .

$$Mo(NMe_2)_4 + 4HOAr \longrightarrow trans\text{-}[Mo(OAr)_4(HNMe_2)_2] + 2HNMe_2 \qquad (6.41)$$

where ArOH = .

(6.42)

where ArOH = .

3.5 From Metal Alkoxides

The combination of the greater acidity and the typically lower volatility of phenols over alcohols allows metal alkoxides to be used as precursor for the synthesis of corresponding aryloxide derivatives (Eq. 6.43[126,127]).

$$Ge(OR)_4 + 4HOPh \longrightarrow Ge(OPh)_4 + 4HOR \qquad (6.43)$$

In many cases only partial substitution occurs, yielding mixed aryloxide, alkoxide compounds (Eq. 6.44[128]).

$$(6.44)$$

Recent mechanistic work on the formation of rhenium aryloxides by treatment of the corresponding methoxide, ethoxide, or isopropoxide with phenolic reagents, shows the intermediacy of a hydrogen-bonded adduct as well as the exchange pathway shown in Eq. 6.45.[129]

$$(6.45)$$

where X = H, CH$_3$, OCH$_3$, CF$_3$, Cl, NMe$_2$;   L$_2$ = ;   ArO = .

3.6   From Metal Alkyls and Aryls

3.6.1   By Protonolysis Reactions

Metal–alkyl bonds that contain a high degree of carbanion character will react with phenols to produce the corresponding phenoxide and alkane as a readily removed by-product (Eqs 6.46,[130] 6.47,[131] and 6.48[132]).

$$(6.46)$$

$$MgBu_2^n + 4ArOH \xrightarrow{-4Bu^nH} ArO-Mg \underset{O}{\overset{O}{\diamondsuit}} Mg-OAr \qquad (6.47)$$

where HOAr = HO—⟨Me₃C / CH₃ ring⟩.

$$2CpIn + 2 \text{ (ArOH)} \xrightarrow{-2CpH} \text{(In complex)} \qquad (6.48)$$

This method has been used to produce a variety of homoleptic aryloxide complexes and has also been applied to the synthesis of mixed alkyl, aryloxide derivatives of metals. Partial substitution is typically achieved by simple control of the stoichiometry or by using sterically demanding aryloxides (Eqs 6.49,[133,134] 6.50,[135] and 6.51[136]).

$$M(CH_2Ph)_4 + 2ArOH \xrightarrow{-2PhCH_3} \underset{ArO}{\overset{ArO}{\diagdown}} M \underset{CH_2Ph}{\overset{\cdots CH_2Ph}{\diagup}} \qquad (6.49)$$

where M = Ti, Zr, Hf; ArOH = HO—⟨Me₃C ring⟩.

$$2Ga(CH_3)_3 + 2 \text{ (quinolinol)} \xrightarrow{-2CH_4} \text{(Ga complex)} \qquad (6.50)$$

$$\text{(Ta complex with R)} \xrightarrow[-4RH]{4ArOH} \text{(Ta complex with OAr)} \qquad (6.51)$$

where R = CH$_2$SiMe$_3$; ArOH = HO—⟨Ph⟩(Ph) this refers to 6.51 on the previous page.

Kinetic studies have shown a primary kinetic isotope effect upon deuteration of the phenolic proton.[7] In some cases ligand exchange reactions restrict the stoichiometry that can be obtained. One of the most thoroughly studied reactions of this type is the attempted synthesis of the mono-aryloxide derivative by reaction of trimethylaluminum [Al$_2$Me$_6$] with 2,6-di-*tert*-butyl-4-methylphenol (Eq. 6.52)[137] which leads to the bis(aryloxide) owing to the two disproportionation reactions shown in Eq. 6.53.[138]

$$[\text{Al}_2\text{Me}_6] + 2\text{ArOH} \xrightarrow{-2\text{MeH}} 2[(\text{ArO})\text{AlMe}_2] \qquad (6.52)$$

where ArOH = HO—⟨Me$_3$C / Me$_3$C⟩—CH$_3$.

$$3[(\text{ArO})\text{AlMe}_2] \rightleftharpoons [\text{Al}_2\text{Me}_5(\text{OAr})] + [(\text{ArO})_2\text{AlMe}]$$

$$[\text{Al}_2\text{Me}_5(\text{OAr})] \rightleftharpoons \tfrac{1}{2}[\text{Al}_2\text{Me}_6] + [(\text{ArO})\text{AlMe}_2] \qquad (6.53)$$

where ArOH = HO—⟨Me$_3$C / Me$_3$C⟩—CH$_3$.

Changing the aluminium alkyl substrate to [AlBu$_3$] allows isolation of the mono-aryloxide derivative.[89,139]

3.6.2 By Oxygenation of Metal–Aryl Compounds

Under certain circumstances it is possible to insert oxygen into metal–aryl bonds leading to the corresponding aryloxide. This reactivity is a subset of the much larger study of the oxidation of organometallic species in general.[140,141] Mechanistic studies show that for transition metal-aryls, an intermediate peroxo (possibly η^2-bound) can precede formation of the metal aryloxide function.[142] The oxygenation of cyclopalladated compounds has been achieved, leading to chelated aryloxides of palladium.[143] Similarly the treatment of nickel metallacycles with N$_2$O can lead to essentially cyclometallated aryloxides.[144] The addition of electron-deficient olefins to nickel benzyne complexes can lead to metallacycles which are then oxygenated to chelating aryloxides.[145] Treatment of the complex [PtMe(PPh$_2$C$_6$F$_5$)$_2$(THF)] with LiOH was found to produce the phenoxide [PtMe(PPh$_2$C$_6$F$_5$)$_2$(PPh$_2$C$_6$F$_4$O)].[146] The reaction is believed to involve attack of coordinated hydroxide at the electrophilic *ortho* carbon of a C$_6$F$_5$ group. Recently the insertion of O$_2$ into a chromium–phenyl bond has been investigated.[147] The migration of aryl groups to metal–oxo functions can lead to metal aryloxides.[148,149] More aspects of this reactivity are discussed in Section 6.2.8.

3.7 From Metal Hydrides

The elimination of H_2 by addition of phenolic reagents to metal hydrides is an excellent method for the synthesis of alkali metal aryloxide compounds (Eqs 6.54,[150] 6.55,[151] and 6.56[152]).

$$2LiH + 2 \quad \xrightarrow{\text{THF}} \quad + H_2 \quad (6.54)$$

$$2NaH + 2 \quad \xrightarrow{\text{THF}} \quad + H_2 \quad (6.55)$$

$$4NaH + 2 \quad \xrightarrow{\text{dme}} \quad (6.56)$$

The method can also be applied to the synthesis of transition metal aryloxides, with mixed hydrido, aryloxides sometimes being observed and isolated[153] (Eqs 6.57[154] and 6.58[155]).

$$[MoH_4(PMe_2Ph)_4] + 4HOAr \longrightarrow [Mo(OAr)_4(PMe_2Ph)_2] + 4H_2 + 2PMe_2Ph$$
$$(6.57)$$

where ArO = $O-\langle\bigcirc\rangle-Me$.

$$Cp_2^*Th(H)_2 + HOAr \longrightarrow Cp_2^*Th(H)(OAr) + H_2 \qquad (6.58)$$

where ArO = HO $-\langle\bigcirc\rangle$, with Me₃C substituents.

4   BONDING OF METAL ARYLOXIDES

4.1   Bonding Modes

Bradley delineated the important factors that affect the solution and solid-state structures adopted by metal alkoxide compounds many years ago.[156] A similar situation exists for metal aryloxides. The aryloxide ligand can adopt a variety of bonding modes. Quadruply (rare) and triply bridging aryloxides are typically restricted to compounds containing small phenoxides. The commonest examples are contained in cluster compounds of the group 1 and group 2 metals and mixed metal derivatives of the lanthanides. For the group 1 metals the archetypal structures are hexagonal and cubic clusters, *e.g.* $[Li_6(\mu_3\text{-}OPh)_6(THF)_6]$[157] and $[Na_4(\mu_3\text{-}OAr)(DME)_4]$[158] $(OAr = OC_6H_4\text{-}4Me)$. Triply bridging phenoxides are also common in tri- and hexanuclear clusters of Ca, Ba, and Sr, *e.g.* $[Ca_3(OPh)_5(HMPA)_6]$,[159] $[Ba_6(OPh)_{12}(TMEDA)_4]$[159] and $[Sr_3(OPh)_6(HMPA)_5]$[159]. Other examples containing triply bridging aryloxides are mixed metal clusters such as $[(py)_4Na_4Cr_2(OPh)_8]$[160] and $[(THF)_4Na_4Cr_2(OPh)_8]$.[161] A few mixed metal clusters of the lanthanides have been shown to contain quadruply bridging aryloxides, *e.g.* $[Me_4N][La_2Na_2(\mu_4\text{-}OAr)(\mu_3\text{-}OAr)_2(\mu_2\text{-}OAr)_4(OAr)_2(THF)_5]$ $(OAr = 4\text{-methylphenoxide})$.[162] As expected the M–OAr distance increases progressively on moving from terminal to doubly and triply bridging bonding mode. This is highlighted by the compound $[Ba_5(OPh)_9(H)(O)(THF)_8]$[163] where the Ba–OAr distances are 2.54 (1) Å (terminal), 2.65 (3) Å (av., μ_2) and 2.75 (3) Å (av., μ_3).[163]

A large number of compounds contain doubly bridging aryloxides. Some of these compounds contain aryloxides bridging between identical metals, but a large number involve M–O–M′ bridges where one of the metals is a group 1 element. In the majority of cases involving identical elements the two M–O distances and M–O–Ar angles are very similar. There are, however, some examples in which there is a pronounced asymmetry to the aryloxide bridge. As an example, in the compound $[Li_3(\mu_2\text{-}OAr)_3]$ $(OAr = OC_6HBu_2^t\text{-}3,5\text{-}Ph_2\text{-}2,6)$ which contains two coordinate lithium atoms the Li–O distances are 1.78 (1) Å and 1.840 (7) Å while the Li–O–Ar angles are 113.8 (3)° and 134.9 (3)°.[164]

By far the commonest bonding mode for aryloxide ligands is a terminal one in which the phenoxide oxygen atom is bound to only one metal atom. In some cases the phenoxide function may also be a part of a chelate ring and this may constrain the M–O–Ar angle (see below). Chelation may occur *via* donor heteroatoms or *via* metallated or π-bound organic substituents (Section 2.1.2).

Two other important binding modes for aryloxides involve metal interactions with the phenoxide ring. In one situation the aryloxide ligand binds to one metal centre through the oxygen atom while the phenoxide ring is π-bound to another metal centre.[165–167] This type of bonding situation is often encountered for aryloxide derivatives of electropositive metals where there is also a lack of Lewis bases to provide needed electron density to the metal centres. The second situation involves the π-binding of phenols or phenoxides to later transition metal centres.[168–171] In some cases this can lead to η^5-bound cyclohexadienonyl ligands.[102,172–174]

4.2   Bonding for Terminal Metal Aryloxides

One of the most widely discussed aspects of the bonding of alkoxide, aryloxide, and related oxygen donor ligands focusses on the presence and extent of oxygen-p to metal π-bonding as well as the possible importance of an ionic bonding model. A simplistic analysis would conclude that changes in the oxygen atom hybridization from sp^3 to sp^2 and sp allows the oxygen atom to interact with one, two, or three orbitals on the metal centre. Filling these orbitals with 2, 4, or 6 electrons would lead to formal M–O single (σ^2), double (σ^2, π^2) and triple (σ^2, π^4) bonds respectively. The π-components of the multiple bonds can therefore be thought of as arising from rehybridization of oxygen lone pair electron density so that it can be donated to the metal centre. In organic chemistry the π-donation of oxygen electron density from hydroxy and alkoxy substituents is routinely used to rationalize both structure and reactivity. Particularly informative in this regard are aryl alcohols (phenols) themselves as well as related aryl ethers. The increased acidity of phenol over simple alkyl alcohols as well as the relative ease of electrophilic attack on the phenoxide nucleus reflect the delocalization (π-donation) of oxygen electron density into the aromatic ring. This argument can also be used to account for the structural parameters of aryl ethers. The O–Ar bond is consistently shorter than is found for alkyl ethers and the R–O–Ar angle is close to 120° (sp^2) with the alkoxy group lying within the arene plane (maximum π-orbital overlap).[175] In considering metal aryloxide bonding it is important, therefore, to recognize that the oxygen atom is bonded to two potential π-acceptor groups, the phenoxy ring as well as the metal centre. A variety of structural and other studies indicate that aryloxides are weaker π-donor ligands than simple alkoxides (see below).[176]

$$\mathrm{M-O}\diagup^{\mathrm{Ar}} \qquad \mathrm{M{=}O}\diagup^{\mathrm{Ar}} \qquad \mathrm{M{\equiv}O-Ar}$$

The existence of aryloxide oxygen to metal π-bonding should be manifested in a variety of ways. The extent of π-donation would depend on a large number of interrelated factors (formal metal oxidation state, molecular symmetry, coordination number, nature of ancillary ligands, etc.), which ultimately control the electron deficiency of the metal centre and the availability of suitable empty π-acceptor orbitals. How some of these factors influence various parameters is discussed below.

4.2.1 Terminal M–OAr Bond Distances

The arguments presented above imply that as the amount of oxygen to metal π-donation increases so the M–O distance should decrease. The question therefore arises as to what the metal–oxygen distance should be for an aryloxide ligand that is undergoing no π-bonding with the metal centre. Two approaches have been taken to answer this question. The first method tries to estimate what a metal–aryloxide single bond distance should be using structural parameters for ligands that cannot themselves π-bond to metal centres. This approach was originally applied to metal dialkylamido ligands by Chisholm *et al*.[177] Hence on the basis of known metal–alkyl bonds and the difference in covalent radii for carbon and oxygen it is possible to estimate what a particular metal aryloxide bond should be in the absence of π-bonding. This approach has been successfully applied in the literature to metal aryloxide derivatives of both p- and d-block metals. From covalent radii obtained from organic structures it appears that (element)E–O(aryloxide, alkoxide) bonds are approximately 0.10–0.15 Å shorter than corresponding element–alkyl bonds. The parameter $\Delta_{O,C}$ can be defined as

$$\Delta_{O,C} = d(M-O) - d(M-C)$$

and used to estimate the extent of any π-bonding of aryloxide ligands. Applying this analysis to four-coordinate aryloxide derivatives of the group 14 metals showed values of $\Delta_{O,C}$ of -0.15 Å (Sn) and -0.17 Å (Ge) implying little or no π-bonding.[105] In contrast, values of $\Delta_{O,C}$ for identical (to ensure constant ligand electronic and steric factors) derivatives of Ti and Zr were found to be -0.28 Å and -0.29 Å, showing the presence of considerable oxygen-p to metal-d π-bonding. In the case of the group 5 metals Nb and in particular Ta there are now many structurally characterized organometallic compounds of the type $[M(OAr)_x(R)_{5-x}]$. The predominant structural type is trigonal bipyramidal with a few examples of square-pyramidal geometry. The M–O and M–C distances for some of these compounds are presented in Table 6.1 along with calculated values of $\Delta_{O,C}$ for these ligands bound to the same metal centre. It can be seen that the values of $\Delta_{O,C}$ for these compounds are lower than the -0.1 to -0.15 Å predicted for purely σ bonding.

In the case of the group 5 metals Nb and Ta and the group 6 metal W there is a second way to measure the shortening of the M–OAr bond due to π-bonding. For these metals there exist formally saturated (18-electron) derivatives containing aryloxide ligation. For the group 5 metals the compounds $[MeTa(dmpe)_2(CO)_2]$ and $[(ArO)Nb(dmpe)_2(CO)_2]$ have been structurally characterized. The Nb–OAr bond length of 2.181 (4) Å compares with a Ta–CH$_3$ distance of 2.32 (1) Å found for the alkyl. Given the negligible difference found between the M–L bond lengths for derivatives of Nb and Ta we can calculate $\Delta_{O,C} = -0.14$ Å for these two derivatives. We can now observe how the M–OAr bond length varies as the "formal" electron count at the metal (*i.e.* electron count in the absence of π-donation) is decreased. Some of this data is presented in Table 6.2. It can be seen that the M–O distance drops dramatically, as the metal centre becomes more electron deficient. The shortest distance of 1.819 (8) Å is found for the compound *trans*-$[NbCl_4(OC_6H_3Me_2\text{-}2,6)(THF)]$ in which the metal centre is attached to four electronegative chloride ligands which are poorer π-donors than the aryloxide ligand.[178] The decrease of 0.36 Å from the distance found in the di-carbonyl compound is comparable to the decrease of the M–C interatomic

Table 6.1 Metal–ligand bond lengths for mixed alkyl, aryloxides of niobium and tantalum

Compound	Bond length (Å)		$\Delta_{O,C}$ (Å)	Ref.
	M–O	M–C		
$[Ta(OC_6H_3Bu^t_2-2,6)_2Me_3]$	1.930 (6)	2.14 (1)	−0.24	i
	1.945 (6)	2.14 (1)		
		2.25 (1)		
$[Ta(OC_6H_2Bu^t_2-2,6-OMe-4)_2Me_3]$	1.925 (4)	2.151 (8)	−0.25	i
		2.169 (8)		
		2.219 (8)		
$[Nb(OC_6H_3Ph_2-2,6)_2(CH_2C_6H_4Me-4)_3]$	1.915 (2)	2.181 (3)	−0.27	ii
	1.917 (2)	2.181 (3)		
		2.192 (4)		
$[Ta(OC_6H_3Ph_2-2,6)_2(CH_2C_6H_4Me-4)_3]$	1.903 (4)	2.154 (7)	−0.27	ii
	1.906 (4)	2.192 (4)		
		2.181 (3)		
$[Ta(OC_6H_3Me_2-2,6)_2(CH_2C_6H_5)_3]$	1.919 (4)	2.204 (7)	−0.30	iii
	1.887 (4)	2.204 (6)		
$[Ta(OC_6HPh_2-2,6-Me_2-3,5)_2(CH_2SiMe_3)_3]$	1.89 (1)	2.10 (2)	−0.24	iv
	1.90 (1)	2.16 (2)		
		2.13 (1)		
$[Ta(OC_6HPh_2-2,6-Pr^i_2-3,5)_2(CH_2SiMe_3)_3]$	1.922 (3)	2.121 (5)	−0.23	iv
	1.908 (3)	2.130 (5)		
		2.180 (5)		
$[Ta(OC_6H_3Ph_2-2,6)_3(CH_2SiMe_3)_2]$	1.912 (3)	2.090 (6)	−0.19	iv
	1.918 (3)	2.122 (5)		
	1.930 (3)			
$[Ta(OC_6H_3Me_2-2,6)_3(C_6H_5)_2]$	1.848 (5)	2.181 (6)	−0.32	v
	1.879 (4)			
$[Ta(OC_6H_3Pr^i_2-2,6)_3(C_6H_5)_2]$	1.852 (3)	2.206 (5)	−0.34	v
	1.881 (3)	2.209 (5)		
	1.880 (3)			
$[Ta(OC_6H_3Me_2-2,6)_4Me]$	1.899 (4)	2.15 (1)	−0.25	iii

[i]L. Chamberlain, J. Keddington, I.P. Rothwell, and J.C. Huffman, *Organometallics*, **1**, 1538 (1982).
[ii]R.W. Chesnut, G.G. Jacob, J.S. Yu, P.E. Fanwick, and I.P. Rothwell, *Organometallics*, **10**, 321 (1991).
[iii]L.R. Chamberlain, I.P. Rothwell, K. Folting, and J.C. Huffman, *J. Chem. Soc., Dalton Trans.*, 155 (1987).
[iv]J.S. Vilardo, M.A. Lockwood, L.G. Hanson, J.R. Clark, B.C. Parkin, P.E. Fanwick, and I.P. Rothwell, *J. Chem. Soc., Dalton Trans.*, 3353 (1997).
[v]B.D. Steffey, P.E. Fanwick, and I.P. Rothwell, *Polyhedron*, **9**, 963 (1990).

distance found between alkyl and alkylidene derivatives of these metals, *cf.* distances of 2.246 (12) Å and 2.026 (10) Å in $[Cp_2Ta(CH_2)(CH_3)]$.[179]

In the case of tungsten there are a variety of eighteen electron aryloxides to consider. Some examples include the carbonyl anions $[W(OAr)(CO)_5]^-$, the hydrido aryloxide $[W(OPh)(H)_3(PMe_3)_4]$, the cyclometallated derivatives $[W(O-\eta^1-C_6H_4)(H)_2(PMe_3)_4]$ and $[W(O-\eta^1-C_6H_3Me-CH_2)(H)_2(PMe_3)_4]$ and the chelated 2,6-diphenylphenoxides $[W(OC_6H_3Ph-\eta^6-C_6H_5)(H)(PMePh_2)_2]$ and $[W(OC_6H_3Ph-\eta^6-C_6H_5)(Cl)(PMePh_2)_2]$ (Table 6.3). For all these compounds the W–OAr distances fall

Table 6.2   M–OAr distances for various aryloxide derivatives of niobium and tantalum

Compound	Coordination number	Electron count	Bond length (Å) M–OAr	Ref.
[Nb(OC$_6$H$_3$Me$_2$-3,5)(dmpe)$_2$(CO)$_2$]	7	18	2.181 (4)	i
trans-[(Nb(OC$_6$H$_4$Me-4)$_2$(dmpe)$_2$]	6	15	2.022 (3), 2.023 (3)	i
[Ta(OC$_6$H$_3$Pr$^i_2$-2,6)$_2$Cl$_2$(H)(PMe$_2$Ph)$_2$]	7	14	1.902 (3)	ii
[Ta(OC$_6$H$_3$Pr$^i_2$-2,6)$_2$Cl(H)$_2$(PMe$_2$Ph)$_2$]	7	14	1.899 (5), 1.900 (5)	ii
all-trans-[NbCl$_2$(OC$_6$H$_3$Pr$^i_2$-2,6)$_2$(py)$_2$]	6	13	1.903 (2), 1.889 (2)	iii
all-trans-[TaCl$_2$(OC$_6$H$_3$Pr$^i_2$-2,6)$_2$(py)$_2$]	6	13	1.906 (3), 1.896 (3)	iii
all-trans-[NbCl$_2$(OC$_6$H$_3$Pr$^i_2$-2,6)$_2$(PMe$_2$Ph)$_2$]	6	13	1.901 (2)	iii
all-trans-[NbCl$_2$(OC$_6$H$_3$Pr$^i_2$-2,6)$_2$(PPh$_3$)$_2$]	6	13	1.892 (7)	iii
all-trans-[TaCl$_2$(OC$_6$H$_3$Pr$^i_2$-2,6)$_2$(PMe$_3$)$_2$]	6	13	1.904 (2)	iii
cis,mer-[NbCl$_3$(OC$_6$H$_3$Pr$^i_2$-2,6)$_2$(py)]	6	12	1.832 (3), 1.852 (3)	iv
cis,mer-[TaCl$_3$(OC$_6$H$_3$Pr$^i_2$-2,6)$_2$(py)]	6	12	1.841 (4), 1.848 (5)	iv
cis,mer-[NbCl$_3$(OC$_6$H$_3$Pr$^i_2$-2,6)$_2$(PMe$_2$Ph)]	6	12	1.82 (2), 1.83 (2)	iv
trans,mer-[NbCl$_2$(OC$_6$H$_3$Pr$^i_2$-2,6)$_3$(PMe$_2$Ph)]	6	12	1.868 (2), 1.894 (2), 1.913 (2)	iv
trans,mer-[TaCl$_2$(OC$_6$H$_3$Pr$^i_2$-2,6)$_3$(PMe$_2$Ph)]	6	12	1.863 (2), 1.897 (2), 1.913 (2)	iv
trans,mer-[NbCl$_3$(OC$_6$H$_3$Me$_2$-2,6)$_2$(THF)]	6	12	1.829 (6), 1.854 (6)	v
trans-[NbCl$_4$(OC$_6$H$_3$Me$_2$-2,6)(THF)]	6	12	1.819 (8)	v
[Ta(OC$_6$H$_3$Bu$^t_2$-2,6)$_2$Cl(H)$_2$(PMe$_2$Ph)]	6	12	1.896 (3), 1.888 (3)	ii
[NbCl$_3$(OC$_6$HPh$_4$-2,3,5,6)$_2$]	5	10	1.820 (3), 1.870 (3)	vi
[TaCl$_3$(OC$_6$H$_3$Bu$^t_2$-2,6)$_2$]	5	10	1.872 (5), 1.836 (4)	vii
[Nb(OC$_6$H$_4$Me$_2$-2,6)$_5$]	5	10	1.890 (2)–1.910 (2)	i

[i]T.W. Coffindaffer, B.D. Steffy, I.P. Rothwell, K. Folting, J.C. Huffman, and W.E. Streib, *J. Am. Chem. Soc.*, **111**, 4742 (1989).

[ii]B.C. Parkin, J.C. Clark, V.M. Visciglio, P.E. Fanwick, and I.P. Rothwell, *Organometallics*, **14**, 3002 (1995).

[iii]S.W. Schweiger, E.S. Freeman, J.R. Clark, M.C. Potyen, P.E. Fanwick, and I.P. Rothwell, *Inorg. Chim. Acta*, **307**, 63 (2000).

[iv]J.R. Clark, A.L. Pulvirenti, P.E. Fanwick, M. Sigalis, O. Eisenstein, and I.P. Rothwell, *Inorg. Chem.*, **36**, 3623 (1997).

[v]H. Yasuda, Y. Nakayama, K. Takei, A. Nakamura, Y. Kai, and N. Kanehisa, *J. Organomet. Chem.*, **473**, 105 (1994).

[vi]M.A. Lockwood, M.C. Potyen, B.D. Steffey, P.E. Fanwick, and I.P. Rothwell, *Polyhedron*, **14**, 3293 (1995).

[vii]L.R. Chamberlain, I.P. Rothwell, and J.C. Huffman, *Inorg. Chem.*, **23**, 2575 (1984).

Table 6.3 **W–OAr distances for various aryloxide derivatives of tungsten**

Compound	Coordination number	Electron count	Bond length (Å) M–OAr	Ref.
[W(OC$_6$H$_5$)(H)$_3$(PMe$_3$)$_4$]	8	18	2.129 (8)	i
[NEt$_4$][W(OC$_6$H$_3$Ph$_2$-2,6)(CO)$_5$]	6	18	2.165 (9), 2.175 (9)	ii
[NEt$_4$][W(OC$_6$H$_5$)(CO)$_4$(PMePh$_2$)]	6	18	2.191 (6)	ii
[NEt$_4$][W(OC$_6$H$_5$)(CO)$_5$]	6	18	2.182 (19), 2.205 (21)	iii
[W(O-η^1-C$_6$H$_4$)(H)$_2$(PMe$_3$)$_4$]	8	18	2.177 (5)	iv, v
[W(OC$_6$H$_3$Me-CH$_2$)(H)$_2$(PMe$_3$)$_4$]	8	18	2.145 (8)	v
[W(OC$_6$H$_3$Ph-η^6-C$_6$H$_5$)(H)(PMePh$_2$)$_2$]	*	18	2.163 (9)	vi
[W(OC$_6$H$_3$Ph-η^6-C$_6$H$_5$)(Cl)(PMePh$_2$)$_2$]	*	18	2.128 (4)	vi
[(Cp*)W(OAr)(C$_2$Me$_2$)$_2$]	*	16/18	2.072 (7)	vii
[W(OC$_6$H$_3$Ph-η^6-C$_6$H$_5$)(OAr)(PMePh$_2$)]	*	16	2.001 (6), 2.033 (6)	viii, ix
[W(OC$_6$H$_3$Ph-η^6-C$_6$H$_5$)(OAr)(dppm)]	*	16	2.038 (5), 1.986 (5)	viii, ix
all-trans-[W(OC$_6$H$_3$Ph$_2$-2,6)$_2$Cl$_2$(PMe$_2$Ph)$_2$]	6	14	1.966 (4)	x
all-trans-[W(OC$_6$H$_4$Me-4)$_2$Cl$_2$(PMePh$_2$)$_2$]	6	14	1.848 (5), 1.840 (5)	xi
[W(OC$_6$H$_3$Ph-η^1-C$_6$H$_4$)$_2$(PMe$_2$Ph)$_2$]	6	14	1.88 (1), 1.87 (1)	viii, ix
trans-[W(OC$_6$H$_3$Ph$_2$-2,6)$_2$Cl$_3$(THF)]	6	13	1.855 (5), 1.857 (5)	x
[W(OC$_6$H$_2$Cl$_3$-2,4,6)$_4$Cl(OEt$_2$)]	6	13	1.853 (9), 1.964 (10)	xii
cis-[W(OC$_6$H$_3$Ph$_2$-2,6)$_2$Cl$_4$]	6	12	1.851 (3), 1.840 (3)	x
cis-[W(OC$_6$H$_3$Pr$^i_2$-2,6)$_2$Cl$_4$]	6	12	1.814 (8), 1.797 (9)	xiii

*Undefined number.

[i]K.W. Chiu, R.A. Jones, G. Wilkinson, A.M.R. Galas, M.B. Hursthouse, and K.M.A. Malik, *J. Chem. Soc., Dalton Trans.*, 1204 (1981).

[ii]D.J. Darensbourg, B.L. Mueller, C.J. Bischoff, S.S. Chojnacki, and J.H. Reibenspies, *Inorg. Chem.*, **30**, 2418 (1991).

[iii]D.J. Darensbourg, K.M. Sanchez, J.H. Reibenspies, and A.L. Rheingold, *J. Am. Chem. Soc.*, **111**, 7094 (1989).

[iv]D. Rabinovich, R. Zelman, and G. Parkin, *J. Am. Chem. Soc.*, **112**, 9632 (1990).

[v]D. Rabinovich, R. Zelman, and G. Parkin, *J. Am. Chem. Soc.*, **114**, 4611 (1992).

[vi]J.L. Kerschner, E.M. Torres, P.E. Fanwick, I.P. Rothwell, and J.C. Huffman, *Organometallics*, **8**, 1424 (1989).

[vii]M.B. O'Regan, M.G. Vale, J.F. Payack, and R.R. Schrock, *Inorg. Chem.*, **31**, 1112 (1992).

[viii]J.L. Kerschner, I.P. Rothwell, J.C. Huffman, and W.E. Streib, *Organometallics*, **7**, 1871 (1988).

[ix]J.L. Kerschner, P.E. Fanwick, I.P. Rothwell, and J.C. Huffman, *Organometallics*, **8**, 1431 (1989).

[x]J.L. Kerschner, P.E. Fanwick, I.P. Rothwell, and J.C. Huffman, *Inorg. Chem.*, **28**, 780 (1989).

[xi]L.M. Atagi and J.M. Mayer, *Angew. Chem., Int. Ed. Engl.*, **32**, 439 (1993).

[xii]R.M. Kolodziej, R.R. Schrock, and J.C. Dewan, *Inorg. Chem.*, **28**, 1243 (1989).

[xiii]H. Yasuda, Y. Nakayama, K. Takei, A. Nakamura, Y. Kai, and N. Kanehisa, *J. Organomet. Chem.*, **473**, 105 (1994).

in the range 2.13–2.20 Å. As the formal electron count at the tungsten metal centre decreases so there is a dramatic decrease in the W–OAr bond length reaching a value of 1.786 (9) Å for one of the ligands in the compound cis-[W(OC$_6$H$_3$Pr$^i_2$-2,6)$_2$Cl$_4$]. The drop in W–O(aryloxide) distance of >0.3 Å is comparable to that found for the group 5 metals above.

One extremely interesting pair of tungsten compounds are the species *all-trans*-[W(OC$_6$H$_3$Ph$_2$-2,6)$_2$Cl$_2$(PMe$_2$Ph)$_2$][180] and *all-trans*-[W(OC$_6$H$_4$Me-4)$_2$Cl$_2$(PMe$_2$Ph)$_2$].[181] The former compound is paramagnetic (high-spin) with a W–OAr distance of 1.966 (4) Å while the diamagnetic (low-spin) 4-methylphenoxide compound has distances of only 1.848 (5) Å and 1.840 (5) Å. Two factors are possibly at work here. First the steric bulk of the 2,6-diphenylphenoxide ligand may inhibit its approach to the metal centre leading to lower π-bonding and smaller splitting of the d_{xz} and d_{yz} orbitals above the d_{xy} orbital. There is also the possibility that the 2,6-diphenylphenoxide ligand is a poorer π-donor that the 4-methylphenoxide leading to the high-spin situation. In another study it was found possible to finely tune the singlet–triplet energy gap of related d^2-W(IV) systems [W(OC$_6$H$_3$Ph-η^1-C$_6$H$_4$)$_2$(L)$_2$] by changing substituents on the donor pyridine and bipyridine ligands L.[182]

Short M–OAr distances and large M–O–Ar angles are also common for nonchelated, terminal aryloxide derivatives of the electropositive lanthanide metals (Table 6.19). The M–OAr distances for isostructural, monomeric molecules in this case correlate well with the metal ionic radius (lanthanide contraction) and also increase with coordination number.[183]

When one begins to consider structural parameters (and indeed chemistry)[1e] of later d-block metal aryloxides one is confronted with a different situation. This is highlighted by the values of $\Delta_{O,C}$ calculated for later transition metal alkyl, aryloxide compounds (Table 6.4). It can be seen that values range from close to zero to positive numbers, *i.e.* in some compounds the M–OAr bond lengths exceed M–C(alkyl) bond lengths! This is a dramatically different picture from that seen for early transition metal compounds.

Table 6.4   Metal–ligand bond lengths for mixed alkyl, aryloxides of later transition metals

Compound	Bond length (Å)		$\Delta_{O,C}$ (Å)	Ref.
	M–O	M–C		
[Pt(OC$_6$H$_5$)Me(bipy)]	2.001 (5)	2.023 (6)	+0.02	i
[Pt(OC$_6$H$_5$)(I)Me$_2$(bipy)]	2.014 (5)	2.051 (7)	+0.03	i
[(TMEDA)Pd(OC$_6$H$_5$)(Me)]	2.024 (3)	2.010 (5)	+0.01	ii
[(TMEDA)Pd(OC$_6$H$_4$(NO$_2$)-4)(Me)]	2.029 (4)	1.997 (7)	+0.03	ii
trans-[Pd(OC$_6$H$_5$)Me(PMe$_3$)$_2$]	2.106 (3)	2.039 (4)	+0.07	iii
[Pd(PPh$_2$C$_6$H$_4$CH$_2$NMe$_2$)(OC$_6$H$_5$)Me]	2.088 (5)	2.020 (7)	+0.09	iv
cis-[Pd(OC$_6$H$_5$)Me(DMPE)]	2.098 (6)	2.101 (9)	0	v
cis-[Pt(OC$_6$H$_5$)Me(PMe$_3$)$_2$]	2.128 (4)	2.08 (1)	+0.05	vi
cis-[Au(OC$_6$H$_5$)Me$_2$(PPh$_3$)]	2.09 (1)	2.00 (1), 2.03 (3)	+0.07	vii

[i]G.M. Kapteijn, M.D. Meijer, D.M. Grove, N. Veldman, A.L. Spek, and G. van Koten, *Inorg. Chim. Acta*, **264**, 211 (1997).

[ii]G.M. Kapteijn, A. Dervisi, D.M. Grove, H. Kooijman, M.T. Lakin, A.L. Spek, and G. van Koten, *J. Am. Chem. Soc.*, **117**, 10939 (1995).

[iii]Yong-Joo Kim, Kohtaro Osakada, A. Takenaka, and A. Yamamoto, *J. Am. Chem. Soc.*, **112**, 1096 (1990).

[iv]M. Kapteijn, M.P.R. Spee, D.M. Grove, H. Kooijman, A.L. Spek and G. van Koten, *Organometallics*, **15**, 1405 (1996).

[v]A.L. Seligson, R.L. Cowan, and W.C. Trogler, *Inorg. Chem.*, **30**, 3371 (1991).

[vi]K. Osakada, Yong-Joo Kim, and A. Yamamoto, *J. Organomet. Chem.*, **382**, 303 (1990).

[vii]T. Sone, M. Iwata, N. Kasuga, and S. Komiya, *Chem. Lett.*, 1949 (1991).

At first it is troubling to find values of $\Delta_{O,C}$ that exceed the range predicted for pure element–aryloxide bonds on the basis of organic structures. The long M–OAr distances can, however, be rationalized by recognizing the presence of π-antibonding interactions, a theory eloquently propounded by J.M. Mayer.[184] In the later transition metal aryloxides the metal d orbitals with π symmetry are occupied and hence their interaction with the oxygen p-orbitals are a repulsive, filled–filled interaction. As noted by Mayer *"the substantial effect of π antibonding interactions on the stability and reactivity of π donor ligands has not been widely appreciated. This lack of appreciation is due in part to filled–filled interactions being only a subtle and uncommon effect in organic chemistry."* The structural data in Table 6.4 is strong evidence for a π-antibonding situation. There is also evidence that the "pushing back" of the p electron density onto the aryloxide ligands results in a more electron-rich (anionic) oxygen atom. Hence it is common to find these aryloxides undergoing hydrogen bonding upon addition of phenols. Structural studies show that the M–OAr bond length increases only slightly upon formation of the adduct (Table 6.4). It should be noted, however, that despite the presence of this π-antibonding situation there is evidence that the late transition metal aryloxide bond strength is still substantial.[185,186]

An interesting question is why there is no "additional" elongation of the metal–aryloxide bond in the 18-electron compounds of niobium/tantalum and tungsten above (Tables 6.2, 6.3). A possible answer lies in the fact that although electronically saturated the electron density is stabilized in these molecules by the attached π-acceptor carbonyl ligands, hence reducing the π antibonding effect. In the later transition metal compounds the π electron density is essentially metal based with little ligand stabilization.

Turning to the p-block metal aryloxides it was shown above that the M–OAr bond lengths for derivatives of tin and germanium are as expected on the basis of known organic structures, *i.e.* no apparent π-bonding. In these compounds the M–O–Ar angles are all very close to $120°$, again as expected. In contrast, mononuclear, bulky aryloxide derivatives of aluminium and gallium have been shown to possess very short M–OAr bond lengths and large M–O–Ar angles.[187,188] Using the parameter $\Delta_{O,C}$ applied to mixed alkyl, aryloxides Barron *et al.* showed that values ranged from $-0.22\,\text{Å}$ to $-0.28\,\text{Å}$, consistent with a significant π-interaction.[1i] Given the high energy of the 3-d orbitals on aluminium it was proposed that π-donation into the σ^* orbital of adjacent Al–X bonds was occurring. This hypothesis was given strong support by theoretical studies as well as gas-phase photoelectron spectroscopy.[187] There have been suggestions, however, that the bonding in three- and four-coordinate aluminium and gallium aryloxides is best described as ionic in nature. This was argued to account for the short M–OAr distances, lack of observed restricted rotation, and large M–O–Ar angle (see below).[189]

4.2.2 Terminal M–O–Ar Angles

Section 4.2.1 clearly shows a strong correlation between the extent of oxygen-p to metal π-bonding and the M–OAr distance. However, the errors inherent in determining interatomic distances by X-ray diffraction techniques means that only large differences in M–OAr bond lengths can be safely analysed. In contrast, bond angles are typically refined from diffraction data to an accuracy of less than a degree. Given the fact that changing the hybridization of the oxygen atom in a species X–O–Y from sp^3 to sp^2 and sp leads to bond angles of 109°, 120° and 180° implies that this parameter may be an accurate probe of π-bonding in metal aryloxides. The field of metal aryloxide chemistry grew dramatically with the use of sterically bulky phenoxides in order to suppress oligomerization *via* aryloxide bridges and to control stoichiometry. Structural studies showed early transition metal derivatives (group 4 and 5 elements) of these ligands to possess very large (in some cases linear) M–O–Ar angles.[190] The large size of these angles was initially attributed to steric factors both "bending away" the bulky aryl group as well as fostering π-bonding by maintaining a mononuclear environment for the electrophilic metal centre. Even as early as 1966 Watenpaugh and Caughlin had determined the crystal structure of dimeric [Cl$_2$(PhO)Ti(μ-OPh)$_2$Ti(OPh)Cl$_2$] and concluded that the short terminal Ti–OPh distance of 1.74 (1) Å was due to oxygen p to metal d π-bonding.[191] Furthermore the large Ti–O–Ph angle of 166° (clearly not steric in origin) was ascribed to sp hybridization at oxygen with one pair of electrons involved in π-bonding to the phenyl group while the other was donated to the metal centre. As the database of structurally characterized transition metal aryloxides grew it became possible to analyse structural parameters to try and detect any correlation between M–O–Ar angles and the degree of π-donation as measured by the M–OAr distance. For the group 4 and 5 metals Ti, Zr, V, Nb, and Ta studies led to the conclusion that the size of the M–O–Ar angle was a very poor measure of the degree of π-donation.[105,192–194] Plots of M–O distance versus M–O–Ar angle highlight this phenomenon (Fig. 6.1). Included on the plots are unrestrained terminal aryloxides as well as terminal aryloxides that are part of a chelate ring. However, in the case of vanadium the large numbers of

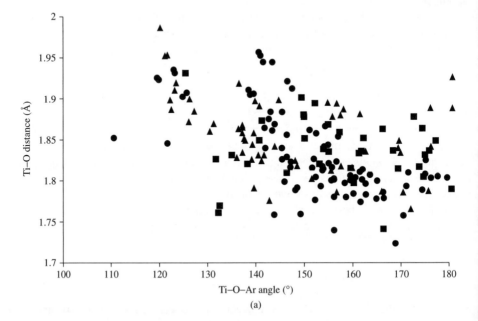

(a)

Figure 6.1   The variation of metal–OAr distance (as a measure of π-donation) with M–O–Ar angle for (a) titanium, (b) zirconium, (c) vanadium, (d) niobium, (e) tantalum. ● = four-coordinate, ■ = five-coordinate, ▲ = six-coordinate, ◆ = seven-coordinate metal.

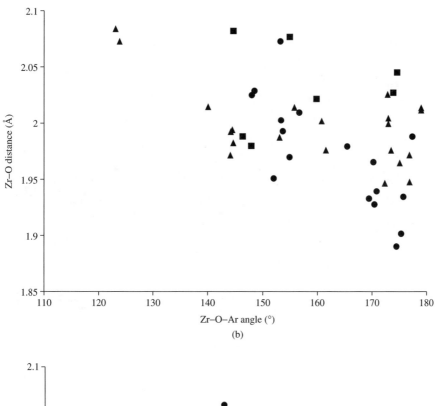

(b)

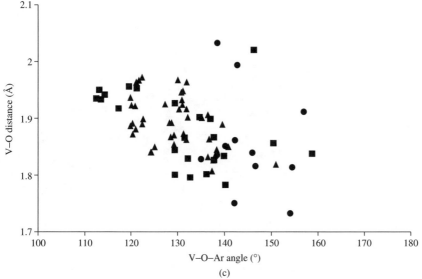

(c)

Figure 6.1   (*Continued*)

salicylaldimine derivatives have been omitted. In some cases the presence of a chelate
ring forces a small angle at oxygen and in some situations this does lead to a longer
M–O bond distance. From the data it can be seen that the shortest metal–aryloxide
bond distances typically occur for the lowest coordination numbers, while the longest

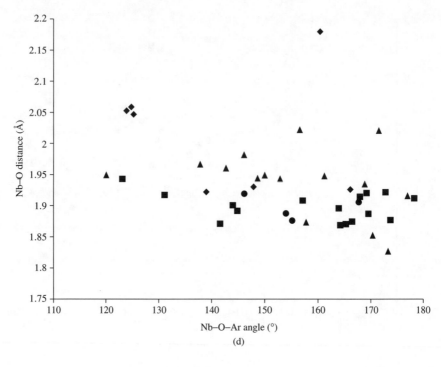

(d)

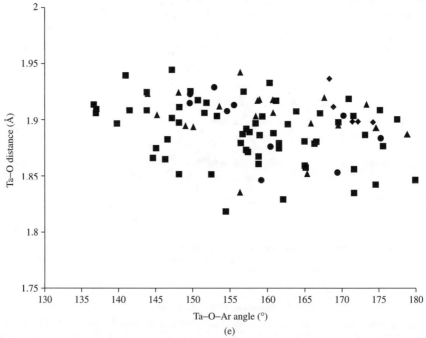

(e)

Figure 6.1   (*Continued*)

as expected are for more electronically saturated compounds. It is difficult, however, to discern any strong trends in these plots. This is particularly true for the group 5 metals Nb and Ta where bond distances of 1.9 Å are associated with angles ranging from 135° to 180°. Interestingly in the case of niobium the longest Nb–OAr distances are for the seven-coordinate compounds [Nb(OC$_6$H$_4$Me-4)(dmpe)$_2$(CO)$_2$] (18-electron) and [Nb(O)(quinol)$_3$] with distances [angles] of 2.181 (4) Å [160°] and 2.055 (av.) Å [125°] respectively. In the latter compound the Nb–O–Ar angle is constrained to 125° (see Section 6.1.1).

Moving across the d-block one predicts that π-bonding effects should decrease except for the highest oxidation state metal centres. In Tables 6.30–6.47 are collected parameters for structurally characterized, non-chelated terminal aryloxide compounds of the group 7–12 metals. These tables show that apart from a few exceptions the M–O–Ar angles lie in the 120–140° range. Notable exceptions are Mn and Fe derivatives of the bulky ligands containing 2,6-di-*tert*-butyl or 2,6-di-(trifluoromethyl) substituents where almost linear angles are observed.

This situation contrasts with that reported for metal alkoxides where there appears to be a reasonable correlation between M–OR distances and M–O–R angles.[195] Furthermore, it has been shown in the compound 1,2-[Mo$_2$(OC$_6$H$_3$Me$_2$-2,6)$_4$(OPr^i)$_2$] that although the M–O(alkoxide) distances are shorter than the M–O(aryloxide) distances, the M–O–Ar angles exceed the M–O–R angles.[196] The consistently large angles found for metal aryloxides may be ascribed to the attachment to the oxygen of two π-accepting substituents which can compete for its electron density. In the case of highly electropositive metal centres, the presence of significant ionic bonding may also be contributing to the high flexibility of the M–O–Ar angle.[183,189]

4.2.3   Electrochemical and Spectrochemical Studies

It is one of the basic tenets of inorganic/organometallic chemistry that π-acceptor ligands such as carbon monoxide stabilize electron-rich metal centres, *i.e.* metals in their lowest formal oxidation states. Conversely it is to be expected that a preponderance of π-donor ligands should lead to a stabilization of high valent metal centres. The π-bonding analysis discussed at length above leads to the prediction that as π-donor ligands are added to a mononuclear d^0-metal centre any empty metal-based orbitals will be pushed up in energy as they take on π^* character. Hence population of these orbitals, *i.e.* reduction of the metal centre, should be made more difficult. This hypothesis was first interrogated by studying the one-electron reduction potentials for the titanium aryloxides [Ti(OAr)$_x$(X)$_{4-x}$] (x = 2, 3, 4) where X is a halide and the aryloxide ligand is bulky enough to maintain mononuclearity. It was shown by electrochemical studies (cyclic voltammetry) that as an aryloxide group replaced a chloride ligand the metal centre became more difficult to reduce by ~400 mV.[197] Much more extensive studies have been carried out on aryloxide derivatives of tungsten.[198–202] Work has focussed on the series of compounds [W(OAr)$_x$Cl$_{6-x}$] and [W(O)(OAr)$_x$Cl$_{4-x}$]. In both cases it was shown that as a chloride was replaced by an aryloxide the reduction potential for formation of the d^1-anion became more negative by 200–300 mV. It was furthermore shown that the reduction potential could be "tuned" by the choice of substituents attached to the phenoxide nucleus, electron-donating substituents making it even more difficult to add an electron to the metal centre.

In the case of the group 5 metals niobium and tantalum the reduction potentials for the mixed alkyl, aryloxides was shown to correlate with absorptions in the electronic spectra.[203] In particular a band(s) assignable to alkyl to metal charge transfer was blue-shifted for compounds that are more difficult to reduce. It was argued that both reduction and the ligand-to-metal charge transfer (LMCT) process are populating a lowest unoccupied molecular orbital (LUMO), that is sensitive to the π-donor aryloxides. In the case of identical derivatives of niobium and tantalum the second row d-block metal compound was more readily reduced (LMCT band red-shifted). Furthermore, the alkoxide $[Nb(OPr^i)_2(CH_2SiMe_3)_3]$ was found to have a more negative (inaccessible) reduction potential compared to the aryloxide $[Nb(OC_6H_3Me_2-2,6)_2(CH_2SiMe_3)_3]$.[203] The much higher energy for the alkyl to metal charge transfer band also indicates that the alkoxide ligand is a stronger π-donor than the aryloxide ligand.

5   ARYLOXIDE LIGAND REACTIVITY

5.1   Cyclometallation of Aryloxide Ligands

The aryloxide ligand is able to undergo cyclometallation at various metal centres *via* a number of mechanistic pathways. The reactivity can involve activation of the *ortho*-CH bond of the phenoxy nucleus itself, as well as aliphatic, benzylic, or aromatic CH bonds of attached substituents. The products of these reactions are typically stable four-, five- or six-membered oxa-metallacycles.[27,28]

The cyclometallation of simple phenoxides can be achieved at mono- and polymetallic centres via oxidative addition pathways. The cluster carbonyl $[Os_3(CO)_{12}]$ initially reacts with phenol to produce the hydrido, phenoxide $[Os_3(\mu_2\text{-}H)(\mu_2\text{-}OPh)(CO)_{11}]$ which on subsequent thermolysis yields the nonacarbonyl whose structural parameters are consistent with a 1,3-cyclohexadienone bonding description (Eq. 6.59).[204]

$$[Os_3(H)(OPh)(CO)_{10}] \xrightarrow{\Delta, -CO}$$

(6.59)

There is also an extensive series of Ru_6 "raft" like clusters that contain ortho-metallated phenoxides bound with not only the oxygen and ortho-metallated carbon bridging two metal centres but also the phenoxy ring π-bound to another metal centre.[205–207]

The addition of phenol to the tungsten complex $[W(\eta^2\text{-}Me_2P\text{-}CH_2)(PMe_3)_4]$ produces the compound $[W(O\text{-}\eta^1\text{-}C_6H_4)(H)_2(PMe_3)_4]$ which contains a four-membered metallacycle ring.[208] The substituted substrates 2,6-dimethylphenol and 2,4,6-trimethylphenol react similarly to yield five-membered metallacycles. The substrates, $HOC_6H_4R\text{-}2$ (R = Et, Pr^i, Bu^t, Ph) undergo metallation at the phenoxide nucleus whereas R = Me produces a five-membered metallacycle. All of the metallated species produce [W(OAr)(H)$_3$(PMe$_3$)$_4$] under H$_2$.

The addition of 4-methylphenol to the ruthenium species [Ru(η^2-Me$_2$P-CH$_2$)(Me)(PMe$_3$)$_4$] produces the cyclometallated aryloxide [Ru(O-η^1-C$_6$H$_4$)(PMe$_3$)$_4$].[209] Reaction with CO and CO$_2$ was found to lead to insertion into the metal–aryl bond to produce five- and six-membered metallacycles respectively.[210]

The overall ortho-metallation of the phenoxide nucleus can also take place with 2-formylphenoxides. The intermediate, chelated acyl formed *via* activation of the formyl CH bond,[211,212] undergoes decarbonylation (typically at Ru or Os metal centres) to produce the four membered metalacycle.[213,214]

A variety of low valent aryloxide derivatives of the early transition metals undergo intramolecular CH bond activation. Attempts to isolate the d^2-species [M(OAr)$_3$] or [M(OAr)$_2$Cl] (OAr = 2,6-di-*tert*-butylphenoxide or 2,6-di-phenylphenoxide; M = Nb, Ta) by reduction of the corresponding d^0-chloride leads instead to bis-cyclometallated compounds (Eq. 6.60).[215]

$$[Ta(OC_6H_3Bu^t_2-2,6)_2Cl_3] \xrightarrow[-H_2]{2Na/Hg, -2NaCl}$$

(6.60)

$$[Ta(OC_6H_3Ph_2-2,6)_3Cl_2] \xrightarrow[-H_2]{2Na/Hg, -2NaCl}$$

The reaction probably proceeds *via* initial oxidative addition of the first CH bond followed by subsequent activation and H$_2$ elimination via σ-bond metathesis (see below). Related reactivity is observed upon reduction (4 Na per W) of the tetrachloride [W(OC$_6$H$_3$Ph$_2$-2,6)$_2$Cl$_4$], although in this case an intermediate [W(OC$_6$H$_3$Ph-η^6-C$_6$H$_5$)(OC$_6$H$_3$Ph$_2$-2,6)(PMePh$_2$)] could be isolated and shown to thermally convert to the bis-cyclometallated product and H$_2$.[29,216]

In the case of the 2,6-diisopropylphenoxide ligand, cyclometallation by low valent metal centres can lead to overall dehydrogenation of the alkyl side chain. The low valent intermediate can be generated by a number of pathways. In the case of niobium, reduction of the dichloride [Nb(OC$_6$H$_3$Pr$^i_2$-2,6)$_3$Cl$_2$] with sodium amalgam (2 Na per Nb) in THF leads to the dehydrogenation product shown in Eq. (6.61).[217,218]

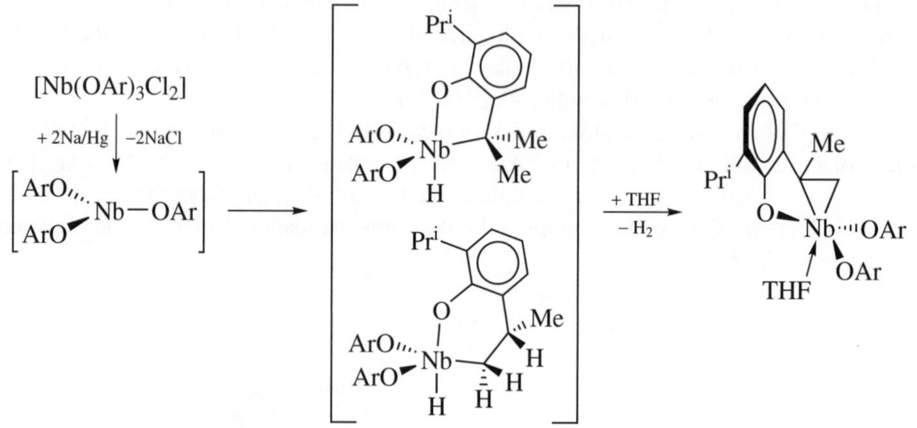

(6.61)

In the case of tantalum, the addition of olefins to the dihydride [Ta(OC$_6$H$_3$Pr$^i_2$-2,6)$_2$(H)$_2$Cl$_2$(L)$_2$] (L = tertiary phosphine) generates one equivalent of alkane along with a dehydrogenation product (Eq. 6.62).[219]

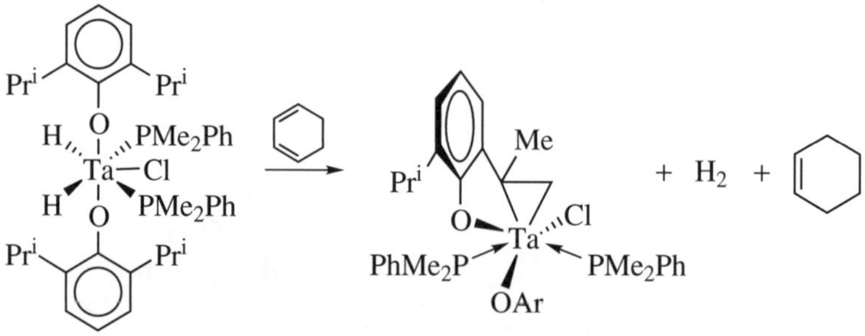

(6.62)

It appears that addition of either the methyne or a methyl CH bond to the low valent metal centre precedes β-hydrogen abstraction and elimination of H$_2$, which can sometimes be detected spectroscopically. The resulting aryloxide contains an ortho-α-methylvinyl group strongly η^2-bound, resulting in a metallacyclopropane ring. This ring can undergo protonation to yield a five-membered metallacycle or undergo ring expansion reactions.[218]

The ligands 2,6-di-*tert*-butylphenoxide and 2,6-diphenylphenoxide undergo cyclometallation at numerous high valent early d-block metal centres. The aliphatic or aromatic CH bonds of these ligands have been activated by metal hydride,[220] alkyl,[221–225] alkylidene (Eq. 6.63),[11,226] alkylidyne (Eq. 6.64)[227,228] and benzyne (o-phenylene)[229] functional groups to produce six-membered oxa-metallacycles.

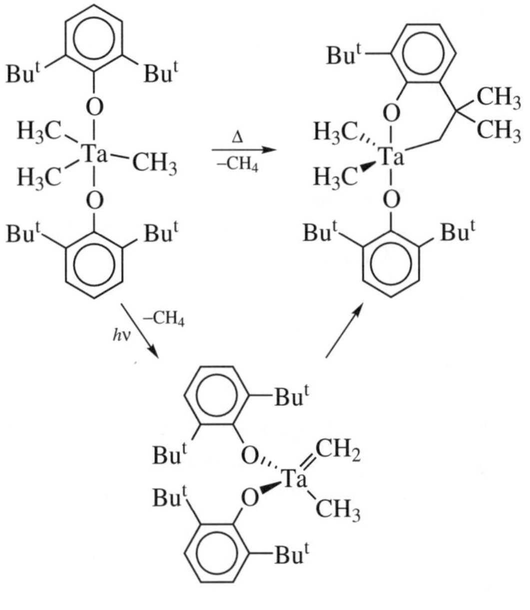

(6.63)

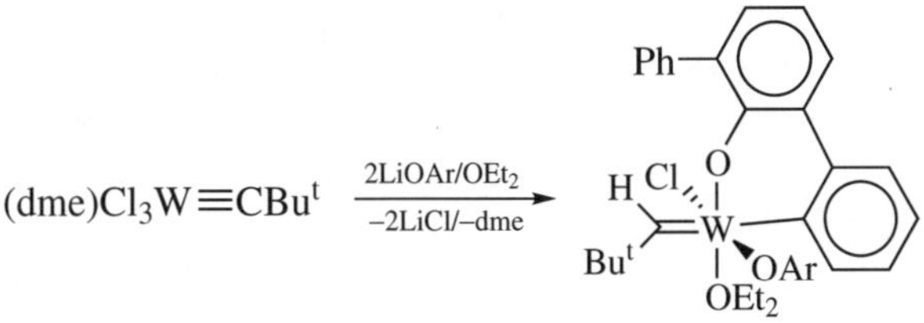

(6.64)

where OAr = $O-$.

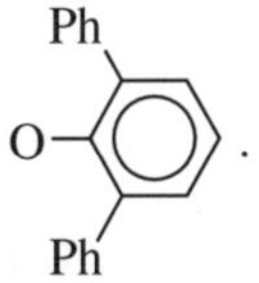

In the case of alkyl leaving groups, mechanistic studies show that σ-bond metathesis pathways are active, proceeding *via* four-centre, four-electron transition states.[230] In the case of unsaturated functional groups, the reaction involves addition of the CH bond across the π-component, so that the substrate hydrogen atom remains (labelling studies) within the product molecule. The cyclometallation of 2,6-dimethylphenoxide by a tantalum alkylidene ligand has also been demonstrated.[231]

The cyclometallation of 2,6-di-*tert*-butylphenoxide ligands at thorium(IV) metal centres can also be facile. Hence thermolysis of the dialkyls [Th(OC$_6$H$_3$Bu$^t_2$-2,6)$_2$(CH$_2$SiMe$_3$)$_2$] and [Cp*Th(OC$_6$H$_3$Bu$^t_2$-2,6)(CH$_2$SiMe$_3$)$_2$] leads to the corresponding mono-cyclometallated species (the latter structurally characterized as the Ph$_3$PO adduct) and one equivalent of TMS.[224,225]

Recent studies have focussed on the generation of metallation-resistant aryloxide ligation. The strategy adopted has involved the introduction of *meta*-substituents onto the 2,6-diphenylphenoxide nucleus. The *meta*-substituents restrict rotation of the *ortho*-phenyl ring into a position for CH-bond activation. In the case of 3,5-di-*tert*-butyl-2,6-diphenylphenoxide, cyclometallation by tantalum alkylidene functional groups is stopped completely (Eq. 6.65).[11]

$$\text{(6.65)}$$

It is also possible for di-aryloxide ligands to undergo cyclometallation. Chisholm *et al*. have shown that addition of 2,2'-methylene-bis(6-*tert*-butyl-4-methylphenol) (di-HOAr) to $[W_2(NMe_2)_6]$ leads to a product $[W_2(\mu_2\text{-}H)(\mu_2\text{-}NMe_2)(\text{di-OAr})(\text{di-OAr-H})(NMe_2)(HNMe_2)]$ in which the central methylene group of one di-aryloxide ligand had been metallated.[232] The reaction involves the addition of the C–H bond to the initial W_2^{6+} core to yield a W_2^{8+} (W≡W) species. Mechanistic studies showed that in fact an equilibrium existed between the metallated compound and $[W_2(\text{di-OAr})_2(NMe_2)_2]$. Addition of pyridine or PMe_3 to this latter compound yielded the corresponding metallated adducts. Remarkably it was shown that the formation of the adduct $[W_2(\text{di-OAr})_2(NMe_2)_2(PMe_3)]$ preceded CH bond activation; *i.e.* the coordination of the donor ligands "turns on" the W_2^{6+} core.[233]

A variety of aryloxide and alkoxide ligands undergo CH bond activation by Sn(IV) metal centres.[234] The reaction appears to be restricted to arene CH bonds and can lead to five- and six-membered stanacycles. Mechanistic studies using substituted *ortho*-arene rings in 2,4,6-triarylphenoxides imply an electrophilic substitution pathway.[235] Mixed alkyl, aryloxides do not undergo cyclometallation but chloro and amido ligands can act as leaving groups to generate HCl and amine respectively. In the case of dimethylamido derivatives of tin the dimethylamine generated by CH bond activation sometimes remains in the coordination sphere of the metal.[234]

5.2   Insertion Chemistry of Metal Aryloxide Bonds

Transition metal aryloxides undergo a number of insertion reactions with small molecules. Some of this reactivity is of fundamental importance, and in a number of cases important mechanistic studies have been carried out.

5.2.1   Insertion of Carbon Monoxide (Carbonylation)

The insertion of carbon monoxide into metal aryloxide bonds appears to be restricted to later transition metal complexes. The initial products of these reactions are aryloxy-carbonyls, which may be stable or undergo further reaction. Three examples of this type of reaction are shown in Eqs (6.66)–(6.68).[236–239]

$$\left[\begin{array}{c} \text{PPh}_2 \\ | \\ \text{Ph}-\text{P}-\text{Pt}-\text{OAr} \\ | \\ \text{PPh}_2 \end{array} \right]^{+} \xrightarrow{\text{CO}} \left[\begin{array}{c} \text{PPh}_2 \\ | \\ \text{Ph}-\text{P}-\text{Pt}-\text{C(O)OAr} \\ | \\ \text{PPh}_2 \end{array} \right]^{+} \qquad (6.66)$$

where $OAr = O-\langle\bigcirc\rangle-R$, R = H, Me, OMe.

$$\text{Ir(CO)(PPh}_3)_2\text{(OPh)} \xrightarrow{\text{CO}} \text{Ir(CO)(PPh}_3)_2\{\text{C(O)OPh}\} \qquad (6.67)$$

$$\text{(dppe)Pt(OPh)}_2 \xrightarrow{\text{CO}} \text{(dppe)Pt}\{\text{C(O)OPh}\}_2 \qquad (6.68)$$

Mechanistic studies show that the first reaction proceeds *via* a migratory insertion reaction. In contrast the second was shown to proceed *via* initial displacement of aryloxide by CO followed by nucleophilic attack at the coordinated carbonyl ligand. In the case of the bis(aryloxycarbonyl), subsequent thermolysis under CO leads to formation of a metal dicarbonyl, CO_2, $ArCO_2Ar$ and deoxygenation of one of the initial aryloxide ligands. A competing reaction is elimination of phenol and formation of a metallolactone.

The addition of CO to the compound *cis*-[Pt(OC$_6$H$_5$)Me(PMe$_3$)$_2$] was found to lead to the isomerized product *trans*-[Pt(OC$_6$H$_5$)Me(PMe$_3$)$_2$] which did not undergo insertion. Interestingly the compound *trans*-[Pt(OC$_6$H$_5$)Me(PEt$_3$)$_2$] was found to insert CO into the metal–methyl bond.[240]

5.2.2   Insertion of Carbon Dioxide and Heterocumulenes

This is a very general reaction for metal aryloxides, and has been observed for both early and late d-block metal centres as well as for the p-block metals.

The reactivity of the carbonyl anions [M(CO)$_5$(OAr)]$^-$ (M = Cr, Mo, W) towards CO_2 and thio analogues has been extensively studied by Darensbourg *et al*. The phenoxides [M(CO)$_5$(OPh)]$^-$ (M = Cr, W) readily insert CO_2 (reversibly), SCO, and CS_2 to produce the corresponding phenyl carbonate, thiocarbonate, and dithiocarbonate respectively (Eq. 6.69).[241]

$$[(\text{OC})_5\text{WOPh}]^- + \text{CO}_2 = [(\text{OC})_5\text{W}\{\text{OC(O)OPh}\}]^-$$

$$[(\text{OC})_5\text{WOPh}]^- + \text{SCO} = [(\text{OC})_5\text{W}\{\text{SC(O)OPh}\}]^-$$

$$[(\text{OC})_5\text{WOPh}]^- + \text{CS}_2 = [(\text{OC})_5\text{W}\{\text{SC(S)OPh}\}]^- \qquad (6.69)$$

The relative rates of the insertion reaction were found to be $CS_2 >$ SCO $> CO_2$ with Cr $>$ W. The substitution compounds [M(CO)$_4$(L)(OAr)]$^-$ were found to undergo insertion of CO_2 much more slowly than the parent carbonyl with L = P(OMe)$_3$ $>$ PMe$_3$ $>$ PPh$_3$ and in this case W $>$ Cr. It was argued there was a steric inhibition of the insertion reaction. It was also found that the more bulky aryloxide compound [W(CO)$_5$(OC$_6$H$_3$Ph$_2$-2,6)]$^-$ did not react with CO_2 but would cleanly form the thio-carbonate (S-bound) with SCO.[242] The mechanism of these reactions is believed to involve attack of the nucleophilic aryloxide oxygen directly on the carbon atom of the electrophilic substrate. This is supported by a lack of retardation of the insertion product (*i.e.* no CO dissociation needed) or formation of $[M(CO)_6]$ (*i.e.* no OAr ionization) when the reaction is carried out under a pressure of CO.

The insertion chemistry of alkyl and hydrido, aryloxy derivatives of the group 10 metals has been studied. Although the phenoxide *trans*-[Pt(H)(OPh)(PEt$_3$)$_2$] undergoes only 10% insertion of CO_2 (1 atm) into the Pt–OPh bond, reaction with PhNCO cleanly produces the corresponding *N*-phenylcarbamato complex.[243] Similarly the nickel compounds *trans*-[Ni(H)(OPh)(L)$_2$] react with PhNCO but in this case there is evidence that the reaction is reversible.[244]

The 4-methylphenoxide complexes *fac*-[Re(OAr)(CO)$_3$L$_2$] (L = PMe$_3$, L$_2$ = diars) show much less tendency to insert electrophiles compared to the methoxide analogues. Neither phenoxide reacts with CO_2 whereas only the PMe$_3$ derivative will insert CS_2. In contrast both methoxides insert both CO_2 and CS_2.[245] Again mechanistic data pointed to a direct attack of metal-bound alkoxide(aryloxide) on the electrophilic carbon atom.

The reactivity of copper aryloxides towards a variety of heterocumulenes has been investigated.[246,247] Insertion of RNCS (R = Me, Ph) into copper(I) aryloxide bonds has been shown to lead to a variety of *N*-alkylamino(aryloxy)methanethiolato complexes depending on the nature of the ancillary ligands. In the absence of any other donor ligation a cluster compound, [Cu{μ_2-SC(=NPh)(OAr)}]$_6$ has been characterized from the insertion of PhNCS into a Cu-4-methylphenoxide bond.[248] Recently the effect of *ortho*-substituents upon the oligomerization of Cu(I) complexes formed by insertion of CS_2 and PhNCS into Cu–OAr bonds has been carried out.[249] Some of the insertion reactions have been shown to be reversible.

The insertion of CO_2, COS, and CS_2 into Zn–OAr and Cd–OAr bonds has been studied. The reactivity is important given the fact that discrete zinc aryloxides will act as catalysts for the copolymerization of epoxides and CO_2.[250,251] The insertion of CO_2 into zinc aryloxides was shown to proceed *via* direct attack by the nucleophilic oxygen. Hence a vacant site at the metal was not needed but small *ortho*-substituents were essential. The complex [Zn(OC$_6$H$_2$Me$_3$-2,4,6)$_2$(py)$_2$] reacted with $^{13}CO_2$ to form a mono(aryl carbonate).[250] In contrast [Cd(OC$_6$H$_3$Ph$_2$-2,6)$_2$(thf)$_2$] failed to react with CO_2 but did undergo insertion with COS and CS_2.[251] The product of CS_2 insertion was crystallized from benzene yielding the dimeric species [(ArOCS$_2$)Cd(μ_2-OAr)$_2$Cd(S$_2$COAr)] (OAr = OC$_6$H$_3$Ph$_2$-2,6). Although highly disordered, the molecular structure was confirmed by an X-ray diffraction study.[251]

The addition of CS_2 to the thallium aryloxides [Tl(OAr)] (Ar = 4-methyl, 4-butyl, 4-*tert*-butyl, 3,5-dimethylphenyl) has been shown to be a good synthetic route to the corresponding [Tl(S$_2$COAr)] salts, which can be used to generate transition metal derivatives.[252]

5.2.3 Insertion of Sulfur Dioxide

The rhodium and iridium complexes [M(ttp)(OPh)] {M = Rh, Ir; ttp = PhP(CH$_2$CH$_2$CH$_2$PPh$_2$)$_2$} has been shown to react with SO$_2$ to produce the corresponding sulfonates, [M(ttp)(SO$_2$OPh).SO$_2$].[253]

Thallium phenoxide undergoes insertion of SO$_2$ to produce [Tl(SO$_2$OPh)].[254]

6 SURVEY OF METAL ARYLOXIDES

6.1 Special Metal Aryloxides

Although technically containing metal–aryloxide functionalities, a number of ligand types can be considered as special examples of aryloxides. In most of these special cases the metal–aryloxide bond is either geometrically constrained or electronically attenuated by other functionalities within the ligand. Detailed discussion of metal catecholates and related 2-thio and 2-aminophenoxides, pyridones, salicylaldimines, and calixarenes is omitted from this chapter. Tables 6.5–6.56 contain details of crystallographically determined structures. These tables focus mainly on simple aryloxide ligands and exclude the extremely large database of structures (mainly later first row d-block metals) in which the aryloxide function is part of a multidentate or macrocyclic ligand.

6.1.1 Metal 8-Hydroxyquinolates

Metal derivatives of 8-hydroxyquinoline are an extensive and important class of aryloxide. Derivatives are known for nearly all metals and are sometimes referred to as oxinate derivatives. Synthetic methods involve straightforward addition of the parent quinol to metal halides, alkyls and amides as well as an extensive aqueous chemistry. A large number of metal "oxinates" have been subjected to single-crystal X-ray diffraction (Table 6.5). The most common bonding mode is terminal, although there are significant examples in which the oxygen atom bridges two metal centres and a few examples of triply bridging quinolates. In nearly all derivatives of 8-hydroxyquinoline the nitrogen atom is also coordinated, leading to a five-membered chelate ring. Structural studies show M–O–Ar angles are constrained to values of 110–120° in such bidentate derivatives. This could possibly restrict the amount of oxygen to metal π-donation that can occur, although the presence of the adjacent pyridine ring will also decrease π-donation. One interesting molecule in this context is [Ti(2Me-8-quin)$_2$(OC$_6$H$_3$Pr$_2^i$-2,6)$_2$] where the unrestrained terminal aryloxide bond, 1.819 (7) Å [157°], is significantly shorter than the quinolate bond length, 1.911 (6) Å [123°].

6.1.2 Metal Salicylaldehydes

A large number of metal derivatives of salicylaldehyde are known. Although the most extensive use has been made of the parent ligand, recent work has also focussed on the use of ligands substituted either within the phenoxy ring or at the α-carbon atom. Structural studies (Table 6.6) show that in nearly all cases chelation via the aldehyde group occurs leading to a constrained M–O–Ar angle of 120–140°. This chelation raises the question of what is the correct bonding description for these molecules. It is possible to draw two distinct resonance structures for the six-membered dioxa-metallacycle ring.

Table 6.5 Metal 8-hydroxyquinolates

Compound	Aryloxide	Bond length (Å) M–O	Bond angle (°) M–O–Ar	Reference
[Be(OAr)₂].2H₂O	2Me-8-quin	1.606 (?) 1.604 (?)	109 110	i
[NH₄][Y{(OAr)₂NBuⁿ}].H₂O	BuⁿN(8-quin)₂	2.289 (?) 2.286 (?)	115 118	ii
[Ti(OAr)₂Cl₂] 6-coord. Ti, *trans* O, *cis* Cl	8-quin	1.888 (11)	122	iii
[CpTi(OAr)₂Cl] *trans* O	8-quin	1.954 (7) 1.987 (6)	121 120	iv
[Ti(OAr')₂(OAr)₂]	8-quin OC₆H₃Pr$^i_2$-2,6	1.953 (6) 1.920 (6) 1.818 (5) 1.813 (6)	121 123 160 155	v
[Ti(OAr)₂(OAr')₂]	2Me-8-quin OC₆H₃Pr$^i_2$-2,6	1.911 (6) 1.819 (7)	123 157	v
[Zr(OAr)₄].3toluene dodecahedral Zr	8-quin	2.106 (6)	120	vi
[V(O)(OAr)₂] tbp, axial N	2Me-8-quin	1.921 (5)	117	vii
[V(O)₂(OAr)(HOAr)].3H₂O tbp, axial N and O of N-protonated quin	2Me-8-quin H-2Me-8-quin	1.957 (3) 1.931 (3)	121 129	viii
[V(OAr)(L)(OAr')].2MeOH octahedral V; V=O, quin and azo-N in plane	8-quin 2-hydroxy-2'-carboxy-5-methylazobenzene	1.841 (4) 1.895 (5)	124 128	ix
[(O)(di-OAr)V(μ-O)V(O)(di-OAr)] tbp V	2-(2'-hydroxyphenyl)-8-quinolinato	1.959 (4) 1.803 (4)	119 129	x
[NBu₄][V(O)₂(OAr)₂].6MeCN *trans* OAr, N *trans* to oxo	8-quin	1.96 (1)	121	xi

Compound	Coordination	Bond length	Ref.	
Na₂[VO₂(OAr)₂]₂.4H₂O.DMF 6-coord V, Na phenoxide bridges	8-quin	1.966 (2) 1.967 (3)	121 121	xii
[NBu₄][V(O)₂(OAr)₂].H₂O 6-coord. V	8-quin	1.98 (1)	122	xii
[(OAr)₂(O)V(μ-O)V(OAr)₂(O)]	8-quin	1.926 (6) 1.892 (6)	120 122	xiii
[V(O)(OAr){PhC(O)=CHCMe=N–N=CPhO}] 6 coord. V, N(quin) *trans* to oxo	8-quin	1.852 (2)	123	xiv
[V(O)(OAr)₂(OPrⁱ)]	8-quin	1.925 (2) 1.902 (2)	121 122	xv
[V(O)(OAr)(azc)].MeOH azc = (2-OC₆H₃Me-4)N=N(2-C₆H₄COO)	8-quin	1.892 (4)	122	xvi

[i] J.C. van Niekerk, H.M.N.H. Irving, and L.R. Nassimbeni, *S. Afr. J. Chem.*, **32**, 85 (1979).

[ii] I.N. Polyakova, Z.A. Starikova, B.V. Parusnikov, and I.A. Krasavin, *Kristallografiya*, **27**, 489 (1982).

[iii] B.F. Studd and A.G. Swallow, *J. Chem. Soc. A*, 1961 (1968).

[iv] J.D. Matthews, N. Singer, and A.G. Swallow, *J. Chem. Soc. A*, 2545 (1970).

[v] P.H. Bird, A.R. Fraser, and C.F. Lau, *Inorg. Chem.*, **12**, 1322 (1973).

[vi] D.F. Lewis and R.C. Fay, *J. Chem. Soc., Chem. Comm.*, 1046 (1974).

[vii] M. Shiro and Q. Fernando, *Anal. Chem.*, **43**, 1222 (1971).

[viii] N. Nakasuka, M. Tanaka, and M. Shiro, *Acta Crystallog., C*, **45**, 1303 (1989).

[ix] J. Chakravarty, S. Dutta, and A. Chakravorty, *J. Chem. Soc., Chem. Comm.*, 1091 (1993).

[x] E. Ludwig, U. Schilde, E. Uhleman, F. Weller, and K. Dehnicke, *Z. Anorg. Allg. Chem.*, **619**, 669 (1993).

[xi] Y. Jeannin, J.P. Launay, and M.A.S. Sedjadi, *J. Coord. Chem.*, **11**, 27 (1981).

[xii] A. Giacomelli, C. Floriani, A.O. de S. Duarte, A.C. Villa, and C. Guastini, *Inorg. Chem.*, **21**, 3310 (1982).

[xiii] S. Yamada, C. Katayama, J. Tanaka, and M. Tanaka, *Inorg. Chem.*, **23**, 253 (1984).

[xiv] J. Chakravarty, S. Dutta, A. Dey, and A. Chakravorty, *J. Chem. Soc., Dalton Trans.*, 557 (1994).

[xv] W.R. Scheidt, *Inorg. Chem.*, **12**, 1758 (1973).

[xvi] J. Chakravarty, S. Dutta, S.K. Chandra, P. Basu, and A. Chakravorty, *Inorg. Chem.*, **32**, 4249 (1993).

(continued overleaf)

Table 6.5 *(Continued)*

Compound	Aryloxide	Bond length (Å) M–O	Bond angle (°) M–O–Ar	Reference
[Nb(O)(OAr)₃]	8-quin	2.047 (3)	125	xvii
		2.055 (3)	124	
		2.055 (3)	124	
		2.058 (3)	125	
[Cr(NBuᵗ)(OAr)₂Cl].0.5Et₂O	8-quin	1.922 (8)	115	xviii
6-coord. Cr, *trans* O, Cl *trans* to NBuᵗ		1.920 (9)	116	
[Mo(O)₂(OAr)₂]	8-quin	1.98 (2)	125	xix, xx, xxi
[Mo(O)(OAr)₂(NNMe₂)]	8-quin	2.019 (8)	118	xxii
cis O(quin), hydrazido *cis* to oxo		2.123 (7)	117	
[H₂quin][Mo(O)(OAr)Cl₃]	8-quin	2.039 (5)	121	xxiii
[W(OAr)₄].C₆H₆	5Br-8-quin	2.070 (9)	121	xxiv
8-coord. W		2.066 (9)	123	
		2.071 (11)	120	
		2.052 (10)	122	
[W(OAr)Cl(CO)₃(PPh₃)]	5,7Cl₂-8-quin	2.120 (5)	117	xxv
7-coord. W				
[W(OAr)₂(CO)₂(PPh₃)].CH₂Cl₂	5,7Cl₂-8-quin	2.096 (7)	118	xxv
7-coord. W		2.106 (7)	118	
mer-[Mn(OAr)₃].MeOH	8-quin	1.905 (7)	116	xxvi
		1.916 (7)	118	
		1.924 (7)	119	
mer-[Mn(OAr)₃].0.5HexOH	8-quin	1.908 (5)	118	xxvii
		1.921 (5)	115	
		1.917 (5)	118	
[(OAr)(O)₂MeRe(μ-O)Re(OAr)(O)₂Me]	8-quin	2.017 (4)	122	xxvii
linear Re–O–Re, Octahedral Re				xxviii
cis-[Tc(O)(OAr)₂Cl]	2Me-8-quin	1.948 (3)	116	xxix
6-coord. Tc		1.994 (3)	120	

Compound	Ligand			Ref
[Tc(O)(OAr)(di-OAr)] O(quin) trans to oxo	8-quin 2-OC$_6$H$_4$CH=N(2-OC$_6$H$_4$)	2.014 (6) 2.008 (7) 2.005 (6)	120 120 119	xxx
[(OAr)$_2$Fe(μ-O)Fe(OAr)$_2$] 5-coord. Fe	2Me-8-quin	1.92 (1) 1.91 (1) 1.94 (1) 1.90 (1)	117 116 115 120	xxxi
cis,cis-[Ru(OAr)Cl(NO)]	2Me-8-quin	1.982 (5) 2.005 (6)	116 112	xxxii
cis,trans-[Ru(OAr)Cl(NO)]	2Me-8-quin	2.006 (6) 1.971 (5)	113 114	xxxii
[NMe$_4$][Ru(OAr)Br$_3$(NO)] octahedral Ru, NO trans to O	2Me-8-quin	1.967 (17)	115	xxxiii

[xvii] S. Garcia-Granda, M.R. Diaz, L. Serra, A. Sanz-Medel, and F. Gomez-Beltran, *Acta Crystallog., C.,* **46**, 753 (1990).

[xviii] Wa-Hung Leung, A.A. Danopoulos, G. Wilkinson, B. Hussain-Bates, and M.B. Hursthouse, *J. Chem. Soc., Dalton Trans.,* 2051 (1991).

[xix] L.O. Atovmyan and Y.A. Sokolova, *Zh. Strukt. Khim.,* **12**, 851 (1971).

[xx] Yu Xiufen, CCSD private communication (1990).

[xxi] S.Z. Haider, K.M.A. Malik, A. Rahman, S. Begum, and M.B. Hursthouse, *J. Bangladesh Acad. Sci.,* **7**, 21 (1983).

[xxii] J. Chatt, B.A.L. Crichton, J.R. Dilworth, P. Dahlstrom, and J.A. Zubieta, *J. Chem. Soc., Dalton Trans.,* 1041 (1982).

[xxiii] K. Yamanouchi, J.T. Huneke, J.H. Enemark, R.D. Taylor, and J.T. Spence, *Acta Crystallogr. B,* **35**, 2326 (1979).

[xxiv] W.D. Bonds Jr, R.D. Archer, and W.C. Hamilton, *Inorg. Chem.,* **10**, 1764 (1971).

[xxv] R.O. Day, W.H. Batschelet, and R.D. Archer, *Inorg. Chem.,* **19**, 2113 (1980).

[xxvi] R. Hems and M.F. Mackay, *J. Cryst. Mol. Struct.,* **5**, 227 (1975).

[xxvii] J. Takacs, P. Kiprof, J.G. Kuchler, and W.A. Herrmann, *J. Organomet. Chem.,* **369**, C1 (1989).

[xxviii] J. Takacs, P. Kiprof, J Riede, and W.A. Herrmann, *Organometallics,* **9**, 782 (1990).

[xxix] B.E. Wilcox, M.J. Heeg, and E. Deutsch, *Inorg. Chem.,* **23**, 2962 (1984).

[xxx] U. Mazzi, F. Refosco, F. Tisato, G. Bandoli, and M. Nicolini, *J. Chem. Soc., Dalton Trans.,* 1623 (1986).

[xxxi] F.E. Mabbs, V.N. McLachlan, D. McFadden, and A.T. McPhail, *J. Chem. Soc., Dalton Trans.,* 2016 (1973).

[xxxii] Y. Kamata, T. Kimura, R. Hirota, E. Miki, K. Mitzumachi, and T. Ishimori, *Bull. Chem. Soc. Jpn.,* **60**, 1343 (1987).

[xxxiii] E. Miki, K. Harada, Y. Kamata, M. Umehara, K. Mizumachi, T. Ishimori, M. Nakahara, M. Tanaka, and T. Nagai, *Polyhedron,* **10**, 583 (1991).

(continued overleaf)

Table 6.5 *(Continued)*

Compound	Aryloxide	Bond length (Å) M–O	Bond angle (°) M–O–Ar	Reference
cis(O,O),*trans*(N,N)-[Ru(OAr)₂Cl(NO)]	8-quin	2.028 (4)	111	xxxiv
[Co(triphos)(OAr)][BPh₄]	8-quin	1.991 (4)	114	xxxv
[Rh(OAr)(COD)]	8-quin	2.051 (6)	113	xxxvi
[Rh(OAr)(CO)(PPh₃)]	8-quin	2.042 (5)	114	xxxvii xxxviii
[Rh(OAr)(Ph₂P-C₆F₄)Br(OH₂)].CH₂Cl₂ dimer due to H-bonding, two molecules	8-quin	2.107 (5) 2.121 (5)	110 109	xxxix
cis-[Rh(OAr)(Ph₂P-C₆F₄)Br(PPh₃)] O *trans* to Br	8-quin	2.03 (1)	114	xl
[Rh(OAr)₂(Ph₂P-C₆F₄)]CH₂Cl₂ distorted octahedron, *trans* O	8-quin	2.034 (4) 2.039 (4)	111 112	xli
[Rh(OAr)(CO)(I)Me(PPh₃)].0.5Me₂CO O *trans* to CO	8-quin	2.036 (4) 2.037 (4)	112 112	xlii
[Ni₄(OAr)₄(OH₂)₄]	μ₂-7-[3,5-dichloro-2-pyridyl)azo]-quin-5-sulfonate	2.072 (7), 2.147 (8) 2.108 (7), 2.072 (7) 2.095 (8), 2.108 (8) 2.147 (7), 2.095 (8)		xliii
[Ni(OAr)₂]	7Prⁱ-8-quin	1.852 (8) 1.979 (5) 1.968 (5)	111 113 115	xliv
[Ni₃(OAr)₆]	8-quin			xlv
	μ₂-8-quin	2.068 (5), 2.086 (5)		
	μ₃-8-quin	2.047 (5), 2.516 (5) 2.024 (5)		

(continued overleaf)

Compound	Ligand	Bond lengths	No.	Ref.
[Ni₃(OAr)₆][ClO₄].EtOH	8-quin	2.046 (6)	113	xlv
		2.018 (6)	115	
	μ₂-quin	2.067 (6), 2.083 (6)		
		2.166 (6), 2.082 (7)		
		2.095 (6), 2.187 (7)		
[Ni(OAr)₂(py)₂] octahedral Ni, *trans* py	5Cl-8-quin	2.042 (6)	110	xlvi
[Ni(OAr)₂(py)₂] octahedral Ni, *cis* py	5,7Cl₂-8-quin	2.048 (5)	111	xlvii
		2.041 (6)	113	
[Ni(OAr)₂(py)₂].py octahedral Ni, *cis* py	5,7Br₂-8-quin	2.049 (6)	112	xlviii
		2.05 (1)	114	
		2.02 (1)	112	
		2.05 (1)	112	
		2.05 (1)	111	
		2.03 (1)	111	
		2.06 (1)	110	
[Ni(OAr)Me(PMe₃)] sq. planar Ni, Me *trans* to O	8-quin	1.924 (3)	112	xlix
[Ni₂(OAr)₃](SCN).H₂O	8-quin	2.058 (1)	112	l
		2.054 (1)	113	

xxxiv Y. Kamata, E. Miki, R. Hirota, K. Mizumachi, and T. Ishimori, *Bull. Chem. Soc. Jpn.*, **61**, 594 (1988).

xxxv M. Fritz, G. Huttner, and W. Imhof, *Chem. Ber.*, **129**, 745 (1996).

xxxvi J.G. Leipoldt and E.C. Grobler, *Inorg. Chim. Acta*, **72**, 17 (1983).

xxxvii L.G. Kuz'mina, Y.S. Varshavskii, N.G. Bokii, Yu. T. Struchkov, and T.G. Cherkasova, *Zh. Strukt. Khim.*, **12**, 653 (1971).

xxxviii J.G. Leipoldt, S.S. Basson, and C.R. Dennis, *Inorg. Chim. Acta*, **50**, 121 (1981).

xxxix F. Barcelo, P. Lahuerta, M.A. Ubeda, C. Foces-Foces, F.H. Cano, and M. Martinez-Ripoll, *J. Organomet. Chem.*, **301**, 375 (1986).

xl F. Barcelo, P. Lahuerta, M.A. Ubeda, A. Cantarero, and F. Sanz, *J. Organomet. Chem.*, **309**, 199 (1986).

xli F.L. Barcelo, J.C. Besteiro, P. Lahuerta, C. Foces-Foces, F.H. Cano, and M. Martinez-Ripoll, *J. Organomet. Chem.*, **270**, 343 (1984).

xlii K.G. van Aswegen, J.G. Leipoldt, I.M. Potgieter, G.J. Lamprecht, A. Roodt, and G.J. van Zyl, *Transition Met. Chem.*, **16**, 369 (1991).

xliii Hu Huang, F. Kai, Y. Asai, M. Hirohata, and M. Nakamura, *Chem. Lett.*, 65 (1991).

xliv A. Friedrich, E. Uhlemann, and F. Weller, *Z. Anorg. Allg. Chem.*, **615**, 39 (1992).

xlv A. Yuchi, H. Murakami, M. Shiro, H. Wada, and G. Nakagawa, *Bull. Chem. Soc. Jpn.*, **65**, 3362 (1992).

xlvi S. Garcia-Granda and F. Gomez-Beltran, *Acta Cryst. C*, **42**, 33 (1986).

xlvii S. Garcia-Granda, P.T. Beurskens, H.J.J. Behm, and F. Gomez-Beltran, *Acta Cryst. C*, **43**, 39 (1987).

xlviii S. Garcia-Granda, C. Jansen, P.T. Beurskens, H.J.J. Behm, and F. Gomez-Beltran, *Acta Cryst. C*, **44**, 176 (1988).

xlix H.-F. Klein, T. Wiemer, M.-J. Menu, M. Dartiguenave, and Y. Dartiguenave, *Inorg. Chim. Acta*, **154**, 21 (1988).

l H. Kiriyama, Y. Yamagata, and K. Suzuki, *Acta Cryst. C*, **42**, 785 (1986).

Table 6.5 (*Continued*)

Compound	Aryloxide	Bond length (Å) M–O	Bond angle (°) M–O–Ar	Reference
[Ni(OAr)₂(quinH)]₂.C₆H₆.H₂O	8-quin	2.064 (5) 2.095 ( ) 2.034 ( ) 2.03 ( )	114 114 111 112	li
[H₂quin][Ni₂(OAr)₃]₂[ClO₄] octahedral Ni, facial O atoms	8-quin	2.02 (1)–2.14 (1)	113	lii
[Ni(OAr)₂(py)₂].py *trans*-O, *cis*-py	5,7Br₂-8-quin	2.049 (6) 2.031 (6)	112 114	liii
[Ni(OAr)₂(OH₂)₂]	2Me-8-quin	2.020 (5) 2.039 (6)	113 112	liv, lv
[Pd{quin-2-CH₂CH₂C(CO₂Et)₂}(py)] sq. planar Pd	Cyclopalladated 2-(ethylmalonate)-8-quin	2.070 (3)	110	lvi
[Pd{quin-2-CH₂C(CO₂Et)₂}(py)] sq. planar Pd	Cyclopalladated 2-(methylmalonate)-8-quin	2.111 (3)	109	lvi
[Pd(OAr)₂].chloroanil sq. planar Pd, plane-to-plane π-complex with chloranil	8-quin	1.98 (2)	110	lvii
[Pd(OAr)₂].C₆H₂(CN)₄	8-quin	2.00 (1)	109	lviii
[Pd(OAr)₂] sq. planar Pd, centrosymmetric	8-quin	2.02 (2)	113	lix
[Pd(OAr)(ArNNC₆H₃OMe)] cyclopalladated 4,4′-dimethoxyazobenzene. N *trans* to O	8-quin	2.018 (4)	113	lx
[Pd(OAr)(NH₂CHMePh)] cyclopalladated diethylmalonate pendant	8-quin-2-{CH₂CH₂C(COOEt)₂}	2.08 (2)	112	lxi

[Pt(OAr)₂](tcnq) sq. planar Pt, O atoms *trans*	8-quin	2.014 (3)	112	lxii
[Pt(μ_2-OAr)Me₃]₂ dimeric structure	μ_2-8-quin	2.22 (3), 2.29 (4)		lxiii
[Cu(OAr)₂]	8-quin	2.07 (1) 2.00 (1)	112 122	lxiv
[Cu(OAr)₂]	8-quin	1.94 (2) 1.93 (2)	110 111	lxv
[Cu(OAr)₂]	8-quin	1.94 (1)	110	lxvi
[Cu(OAr)(OH₂)₂]ₙ sq. planes with apical sulfonic O atoms	5-sulfonic-8-quinolinato	1.936 (3)	111	lxvii
[Cu(OAr)₂]	8-quin	1.917 ( )	112	lxviii
[Cu₂(OAr)₂(OH₂)₂]	μ_2-5-sulfonic-8-quin	1.99 (4)	109	lxix

(continued overleaf)

[lii]A. Yuchi, K. Imai, H. Wada, M. Shiro, and G. Nakagawa, *Bull. Chem. Soc. Jpn.*, **59**, 3847 (1986).

[liii]H. Kiriyama, Y. Yamagata, K. Yonetani, and E. Sekido, *Acta Cryst. C*, **42**, 56 (1986).

[liii]S. Garcia-Granda, C. Jansen, P.T. Beurskens, H.J.J. Behm, and F. Gomez-Beltran, *Acta Cryst. C*, **44**, 176 (1988).

[liv]A. Yuchi, M. Shiro, H. Murakami, H. Wada, and G. Nakagawa, *Bull. Chem. Soc. Jpn.*, **63**, 677 (1990).

[lv]A.T. Rane and V.V. Ravi, *Ind. J. Chem. A*, **21**, 311 (1982).

[lvi]A. Yoneda, G.R. Newkome, Y. Morimoto, Y. Higuchi, and N. Yasuoka, *Acta Cryst. C*, **49**, 476 (1993).

[lvii]B. Kamenar, C.K. Prout, and J.D. Wright, *J. Chem. Soc.*, 4851 (1965).

[lviii]B. Kamenar, C.K. Prout, and J.D. Wright, *J. Chem. Soc. A*, 661 (1966).

[lix]C. K. Prout and A.G. Wheeler, *J. Chem. Soc. A*, 1286 (1966).

[lx]A.M.M. Lanfredi, F. Ugozzoli, M. Ghedini, and S. Licoccia, *Inorg. Chim. Acta*, **86**, 165 (1984).

[lxi]A. Yoneda, T. Hakushi, G.R. Newkome, Y. Morimoto, and N. Yasuoka, *Chem. Lett.*, 175 (1994).

[lxii]P. Bergamini, V. Bertolasi, V. Ferretti, and S. Sostero, *Inorg. Chim. Acta*, **126**, 151 (1987).

[lxiii]J.E. Lydon and M.R. Truter, *J. Chem. Soc.*, 6899 (1965).

[lxiv]F. Kanamaru, K. Ogawa, and I. Nitta, *Bull. Chem. Soc. Jpn.*, **36**, 422 (1963).

[lxv]G.J. Palenik, *Acta Crystallogr.*, **17**, 687 (1964).

[lxvi]R. C. Hoy and R.H. Morriss, *Acta Crystallogr.*, **22**, 476 (1967).

[lxvii]S. Ammor, G. Coquerel, G. Perez, and F. Robert, *Eur. J. Solid State Inorg. Chem.*, **29**, 131 (1992).

[lxviii]S. Ammor, G. Coquerel, G. Perez and F. Robert, *Eur. J. Solid State Inorg. Chem.*, **29**, 445 (1992).

[lxix]S. Petit, G. Coquerel, G. Perez, D. Louer, and M. Louer, *New J. Chem. (Nouv. J. Chim.)*, **17**, 187 (1993).

Table 6.5 *(Continued)*

Compound	Aryloxide	Bond length (Å) M–O	Bond angle (°) M–O–Ar	Reference
[Cu(μ_2-OAr)(CO)]$_4$ tetrahedral Cu held by O bridges	μ_2-2Me-8-quin	2.01 (2)–2.08 (2)		lxx, lxxi
[Cu$_2$(μ_2-OAr)$_2$(dppm)] distorted trigonal pyramid about Cu	μ_2-2Me-8-quin	2.141 (4)–2.336 (6)		lxxi
[Cu(OAr)(PPh$_3$)$_2$] tetrahedral Cu	2Me-8-quin	2.088 (14)	108	lxxi
[Cu(OAr)].3H$_2$O	5-SO$_4$-8-quin	1.952 (3)	110	lxxii
[Au(OAr)(PPh$_3$)].0.5H$_2$O	8-quin	2.373 (?)	112	lxxiii
		2.44 (?)	110	
[Au(OAr)(PPh$_3$)].H$_2$O	8-quin	2.502 (?)	109	lxxiii
[Au(OAr)(Me$_2$NCH$_2$C$_6$H$_4$)][BF$_4$]	8-quin	2.067 (6)	111	lxxiv
[Zn(OAr)$_2$(OH$_2$)$_2$]	8-quin	2.05 (?)	114	lxxv
[Zn(OAr)$_2$(OH$_2$)$_2$]	8-quin	2.066 (?)	111	lxxvi
[Zn$_4$(OAr)$_8$]	8-quin	2.020 (4)	112	lxxvii
		1.965 (4)	107	
[PhHg(μ_2-OAr)$_2$HgPh] loose dimer	μ_2-8-quin	2.176 (4), 2.074 (4)		lxxviii
	μ_2-2Me-8-quin	2.16 (1), 2.79 (1)		
[Hg(OAr)Ph] intermolecular Hg–O = 3.34 (4) (Å)	8-quin	2.19 (4)	120	lxxviii
[Hg(OAr)Ph] second form, intermolecular Hg–O = 3.37 (1) (Å)	8-quin	2.15 (4)	121	lxxviii
		2.06 (2)	121	lxxix
[(OAr)$_2$Al(μ-O)Al(OAr)$_2$] linear Al–O–Al bridge, pseudo-tbp Al, axial N	2Me-8-quin	1.817 (5)	117	lxxx
		1.809 (5)	118	
		1.819 (5)	116	
		1.804 (5)	118	

494

Compound	Ligand	Distances	Angles	Ref.
[Al(OAr)$_3$].MeOH mer O and N	8-quin	1.850 (2) 1.841 (2) 1.881 (2)	118 116 115	lxxxi
[Al(OAr)$_3$].acac	8-quin	1.845 (4) 1.847 (4) 1.857 (4)	117 114 116	lxxxii
[Et$_2$Al(μ_2-OAr)$_2$AlEt$_2$]	μ_2-8-quinolinolato	1.994 (10), 1.868 (9), 2.002 (9), 1.863 (9)		lxxxiii
[Ga(OAr)$_2$Cl] tbp, axial N	2Me-8-quin	1.879 (9) 1.855 (8)	114 114	lxxxiv
[Ga(OAr)$_3$].MeOH mer O and N	8-quin	1.920 (3) 1.964 (3) 1.938 (3)	115 114 114	lxxxi

[lxx] M. Pasquali, P. Fiaschi, C. Floriani, and P.F. Zanazzi, J. Chem. Soc., Chem. Commun., 613 (1983).

[lxxi] C. Floriani, P. Fiaschi, A. Chiesi-Villa, C. Guastini, and P.F. Zanazzi, J. Chem. Soc., Dalton Trans., 1607 (1988).

[lxxii] S. Petit, S. Ammor, G. Coquerel, C. Mayer, G. Perez, and J.-M. Dance, Eur. J. Solid State Inorg. Chem., 30, 497 (1993).

[lxxiii] L.G. Kuz'mina, N.V. Dvortsova, M.A. Porai-Koshits, E.I. Smyslova, K.I. Grandberg, and E.G. Perevalova, Metalloorg. Khim. (Organometallic Chem. in USSR), 2, 1344 (1989).

[lxxiv] J. Vicente, M.T. Chicote, M.D. Bermudez, P.G. Jones, C. Fittshcen, and G.M. Sheldrick, J. Chem. Soc., Dalton Trans., 2361 (1986).

[lxxv] L.L. Merritt Jr, R.T. Cady, and B.W. Mundy, Acta Crystallogr., 7, 473 (1954).

[lxxvi] G.J. Palenik, Acta Crystallogr., 17, 696 (1964).

[lxxvii] Y. Kai, M. Morita, N. Yasuoka, and N. Kasai, Bull. Chem. Soc. Jpn., 58, 1631 (1985).

[lxxviii] C.L. Raston, B.W. Skelton, and A.H. White, Aust. J. Chem., 31, 537 (1978).

[lxxix] L.G. Kuz'mina, M.A. Porai-Koshits, E.I. Smyslova, and K.I. Grandberg, Metalloorg. Khim. (Organometallic Chem. in USSR) 1, 1165 (1988).

[lxxx] Y. Kushi and Q. Fernando, J. Am. Chem. Soc., 92, 91 (1970).

[lxxxi] H. Schmidbaur, J. Lettenbauer, D.L. Wilkinson, G. Muller, and O. Kumberger, Z. Naturforsch. B, 46, 901 (1991).

[lxxxii] M. Ul-Haque, W. Horne, and S.J. Lyle, J. Cryst. Spectrosc., 21, 411 (1991).

[lxxxiii] S.T. Dzugan and V.L. Goedken, Inorg. Chim. Acta, 154, 169 (1988).

[lxxxiv] M. Shiro and Q. Fernando, Anal. Chem., 43, 1222 (1971).

(continued overleaf)

Table 6.5 *(Continued)*

Compound	Aryloxide	Bond length (Å) M–O	Bond angle (°) M–O–Ar	Reference
[Ga(OAr)₂(OAc)] two molecules, 5-coord. Ga	2Me-8-quin	1.879 (2) 1.874 (2) 1.870 (2) 1.885 (2)	114 114 114 116	lxxxv
[Ga(OAr)₂(OAcCN)] tbp, *trans* N	2Me-8-quin	1.877 (2) 1.876 (2) 1.887 (2) 1.872 (2)	115 114 114 114	lxxxv
[(OAr)₂Ga(μ-succinate)₂Ga(OAr)₂] two tbp Ga centers	2Me-8-quin	1.877 (2) 1.883 (3)	114 115	lxxxv
[(c-C₅H₁₀)₂Ga(μ₂-OAr)₂Ga(c-C₅H₁₀)₂]	μ₂-8-quin	1.950 (4), 2.320 (5) 1.932 (4), 2.365 (4)		lxxxvi
[Ga(OAr)Me₂]₂	μ₂-8-quin	1.937 (3), 2.297 (3)		lxxxvii
[Sn(S₂COEt)₂(OAr)₂] distorted six coord. Sn, xanthate mono-S bound	8-quin	2.047 (7)	118	lxxxviii
[(EtOCS₂)Pb(μ₂-OAr)₂Pb(S₂COEt)] non-symmetric Pb–O and Pb–N bridges	μ₂-8-quin	2.296 (5), 2.548 (4)	117	lxxxix
[Sn(OAr)(Et)₂Cl]	8-quin	2.042 (7)	120	xc
cis-[Sn(OAr)₂Cl₂]	8-quin	2.030 (3)	115	xci
[Sn(OAr)₂(Et)(Pr)]	2Me-8-quin	2.065 (6)	125 124	xcii
[Sn(OAr)₂(Bu)₂]	8-quin	2.079 (8) 2.109 (9)	119 119	xciii
[Sn(OAr)₂Me₂]	8-quin	2.106 (7) 2.116 (8)	122 118	xciv
[Sn(OAr)Cl(2-carbomethoxyethyl)] 6-coord. Sn	8-quin	2.040 (3)	120	xcv

[Sb(OAr)₂(S₂COEt)] pentagonal bipyramid, stereochemically active lone pair	8-quin	2.018 (6) 2.105 (5)	120 119	xcvi
[Sb(OAr)(Me)₄] octahedral Sb	8-quin	2.187 (8)	121	xcvii
[Bi(OAr)Cl(Ph)₃] 6-coord. Bi; O *trans* to Cl	2Me-8-quin	2.19 (2)	127	xcviii
[Bi(OAr)Cl(Ph)₃] 6-coord. Bi; O *trans* to Cl	8-quin	2.19 (2)	127	xcix

lxxxv H. Schmidbaur, J. Lettenbauer, O. Kumberger, J. Lachmann, and G. Muller, *Z. Naturforsch. B*, **46**, 1065 (1991).

lxxxvi Wen-Tian Tao, Ying Han, Yao-Zeng Huang, Xiang-Lin Jin, and Jie Sun, *J. Chem. Soc., Dalton Trans.*, 807 (1993).

lxxxvii E.C. Onyiriuka, S.J. Rettig, A. Storr, and J. Trotter, *Can. J. Chem.*, **65**, 782 (1987).

lxxxviii C.L. Raston, A.H. White, and G. Winter, *Aust. J. Chem.*, **31**, 2641 (1978).

lxxxix S.G. Hardin, P.C. Healy, W.G. Mumme, A.H. White, and G. Winter, *Aust. J. Chem.*, **35**, 2423 (1982).

xc Shi Dashuang and Hu Shengzhi, *Jiegou Huaxue* (*J. Struct. Chem.*), **6**, 193 (1987).

xci S.J. Archer, K.R. Koch, and S. Schmidt, *Inorg. Chim. Acta*, **126**, 209 (1987).

xcii V.G.K. Das, Chen Wei, Yap Chee Keong, and E. Sinn, *J. Chem. Soc., Chem. Commun.*, 1418 (1984).

xciii Shi Dashuang and Hu Sheng-zhi, *Jiegou Huaxue* (*J. Struct. Chem.*), **7**, 111 (1988).

xciv E.O. Schlemper, *Inorg. Chem.*, **6**, 2012 (1967).

xcv Seik Weng Ng, Chen Wei, V.G.K. Das, J.-P. Charland, and F.E. Smith, *J. Organomet. Chem.*, **364**, 343 (1989).

xcvi B.F. Hoskins, E.R.T. Tiekink, and G. Winter, *Inorg. Chim. Acta*, **97**, 217 (1985).

xcvii H. Schmidbaur, B. Milewski-Mahrla, and F.E. Wagner, *Z. Naturforsch. B*, **38**, 1477 (1983).

xcviii G. Faraglia, R. Graziani, L. Volponi, and U. Casellato, *J. Organomet. Chem.*, **253**, 317 (1983).

xcix D.H.R. Barton, B. Charpiot, E.T.H. Dau, W.B. Motherwell, C. Pascard, and C. Pichon, *Helv. Chim. Acta*, **67**, 586 (1984).

Table 6.6 Metal salicylaldehyde derivatives

Compound	Aryloxide	M–O Bond length (Å)		Bond angle (°)	Reference
		M–O	M–O=C	M–O–Ar	
[Eu{HB(pz)₃}₂(OAr)] 8-coord. Eu	OC₆H₄CHO-2-OMe-4	2.266 (5)	2.402 (5)	136	i
[Gd(OAr)₂(OH₂)₄](ClO₄)₂(Cl).2.75H₂O square antiprism, py not coordinated	OC₆H₄CHO-2-(HC=NCH₂CH₂py)-6	2.326 (4) 2.340 (4)	2.415 (5) 2.421 (6)	140 138	ii, iii
[Th(OAr)₄]	OC₆H₄CHO-2	2.41 (1) 2.28 (1) 2.26 (1) 2.31 (1)	2.36 (1) 2.67 (1) 2.62 (1) 2.44 (1)	140 145 151 142	iv
[U(O)₂(OAr)(OH₂)] 7-coord. U	OC₆H₂CHO-2-Cl-4-CH(OMe)₂-6	2.82 (2)	2.45 (2)	142	v
[Ti(OAr)₂Cl₂] octahedral Ti, Cl *trans* to one OAr and one formyl	OC₆H₂Me-4-Buᵗ-6-CHO-2	1.84 (1) 1.89 (1)	2.15 (1) 2.098 (9)	138 141	vi
[Zr(OAr)₂Cl₂] octahedral Zr, Cl *trans* to one OAr and one formyl	OC₆H₂Me-4-Buᵗ-6-CHO-2	1.971 (5) 2.011 (5)	2.214 (6) 2.173 (5)	140 141	vi
[Zr(OAr)Cl₃(THF)] octahedral Zr, formyl *trans* to THF	OC₆H₂Me-4-Buᵗ-6-CHO-2	1.953 (6)	2.200 (7)	140	vi
[W(O)(OAr)Cl₃]	OC₆H₄CHO-2	1.872 (?)	2.273 (?)	141	vii
[W(OAr)(CO)(NO)(PMe₃)₂] octahedral W, NO *trans* to OAr, CO *trans* to formyl	OC₆H₄CHO-2	2.111 (7)	2.158 (7)	132	viii
[Mn₂(μ₂-OAr)₂(OAr)₂(HOMe)₂] formyl groups *trans* to μ₂-OAr	OC₆H₄CHO-2 / μ₂-OC₆H₄CHO-2	2.092 (6) 2.102 (7) 2.155 (7), 2.171 (6) 2.179 (6), 2.145 (6)	2.169 (7) 2.153 (8)	131 132	ix
[PPh₄][Tc(OAr)Cl₄] octahedral Tc	OC₆H₄CHO-2	1.98 (2)	2.04 (2)	130	x

Compound	OAr			Ref
[Tc(O)(OAr)(N-salicylidene=D-glucosamine)] distorted octahedral Tc, Tc=O *trans* to aldehyde.	OC_6H_4CHO-2	1.987 (9)	2.360 (9)	124	xi
[Re(OAr)(NC$_6$H$_4$Me-4)Cl$_2$(PPh$_3$)] octahedral Re, Cl mutually *trans*, O *trans* to P	OC_6H_4CHO-2	2.01 (1)	2.08 (1)	133	xii
mer-[Ru(OAr)$_3$]	OC_6H_4CHO-2	1.978 (2) 1.978 (2) 1.986 (2)	2.014 (2) 2.041 (2) 2.039 (2)	122 122 124	xiii
[Ru$_3$(CO)$_8$(μ_2-OAr)$_2$]	μ_2-OC_6H_4CHO-2	2.156 (7), 2.116 (6) 2.178 (7)		124	xiv
[Co(OAr)(di-OAr)] 6-coord. Co, O(OAr) *trans* to O(di-OAr)	OC_5H_4CHO-2-{OC_6H_4CMe=N}$_2$(CH$_2$)$_3$	1.907 (3) 1.894 (3) 1.876 (3)	1.933 (3)	125 117 123	xv
[Co(OAr)$_2$(HOMe)$_2$] *all-trans*	OC_5H_4CHO-2	2.010 (4)	2.076 (4)	127	xvi
[Ni(OAr)$_2$]	OC_5H_4CHO-2-Me-4	1.835 (3)	1.850 (4)	127	xvii
trans-[Ni(OAr)$_2$(OH$_2$)$_2$]	OC_5H_4CHO-2	2.02 (1)	2.02 (1)	126	xviii
[Ni(μ_3-OMe)(OAr)(HOEt)]$_4$ cubic structure	OC_5H_4CHO-2	1.999 (9)	2.016 (9)	127	xix
[Cu(OAr)$_2$] two forms	OC_6H_4CHO-2	1.86 (1) 1.90 (1)	1.98 (1) 1.94 (1)	127 125	xx, xxi
[Cu(OAr)(bipy)][ClO$_4$] some intermolecular Cu–O(OAr) bonding	OC_6H_4CHO-2	1.897 (2)	1.948 (3)	124	xxii
[Cu(OAr)(phen)][ClO$_4$] some intermolecular Cu–O(OAr) bonding	OC_6H_4CHO-2	1.900 (2)	1.941 (2)	124	xxiii
[Cu(OAr)(phen)][NO$_3$]	OC_6H_4CHO-2	1.898 (3)	1.952 (3)	123	xxiv
[Cu(OAr)$_2$(py-4Me)] square pyramid, axial N	OC_6H_4CHO-2	1.91 (1)	2.00 (1)	126	xxv
[Cu(OAr){N(2-py)$_2$}]$_2$[ClO$_4$]$_2$ intermolecular Cu–O(OAr) contacts	OC_6H_4CHO-2	1.932 (2)	1.974 (2)	124	xxvi

(continued overleaf)

Table 6.6 (*Continued*)

Compound	Aryloxide	M–O Bond length (Å)			Bond angle (°) M–O–Ar	Reference
		M–O	M–O=C			
[Cd(OAr)₂(bipy)] 6-coord. Cd	OC₁₀H₆(CMeO)-2	2.207 (4) 2.183 (4)	2.289 (5) 2.260 (4)		131 130	xxvii

[i] R.G. Lawrence, C.J. Jones, and R.A. Kresinski, *J. Chem. Soc., Dalton Trans.*, 501 (1996).

[ii] E. Bakalbassis, O. Kahn, J. Sainton, J.C. Trombe, and J. Galy, *J. Chem. Soc. Chem. Commun.*, 755 (1991).

[iii] M. Andruh, E. Bakalbassis, O. Kahn, J.C. Trombe, and P. Porcher, *Inorg. Chem.*, **32**, 1616 (1993).

[iv] R.J. Hill and C.E.F. Rickard, *J. Inorg. Nucl. Chem.*, **39**, 1593 (1977).

[v] S. Sitran, D. Fregona, U. Casellato, P.A. Vigato, R. Graziano, and G. Faraglia, *Inorg. Chim. Acta*, **121**, 103 (1986).

[vi] L. Matilainen, M. Klinga, and M. Leskela, *J. Chem. Soc., Dalton Trans.*, 219 (1996).

[vii] V.S. Sergienko, A.B. Ilyukhin, Yu. V. Kovalenko, and V.L. Abramenko, *Zh. Neorg. Khim.*, **39**, 1647 (1994).

[viii] A.A.H. van der Zeijden, H.W. Boschi, and H. Berke, *Organometallics*, **11**, 2051 (1992).

[ix] S.-B. Yu, C.-P. Wang, E.P. Day, and R.H. Holm, *Inorg. Chem.*, **30**, 4067 (1991).

[x] U. Mazzi, E. Roncari, G. Bandoli, and D.A. Clemente, *Transition Met. Chem.*, **7**, 163 (1982).

[xi] A. Duatti, A. Marchi, L. Magon, E. Deutsch, V. Bertolasi, and G. Gilli, *Inorg. Chem.*, **26**, 2182 (1987).

[xii] A. Marchi, R. Rossi, A. Duatti, L. Magon, U. Casellato, and R. Graziani, *Transition Met. Chem.*, **9**, 299 (1984).

[xiii] N. Bag, G.K. Lahiri, S. Bhattacharya, L.R. Falvello, and A. Chakravorty, *Inorg. Chem.*, **27**, 4396 (1988).

[xiv] D.S. Bohle, V.F. Breidt, A.K. Powell, and H. Vahrenkamp, *Chem. Ber.*, **125**, 1111 (1992).

[xv] C. Fukuhara, S. Matsuda, K. Katsura, M. Mori, K. Matsumoto, S. Ooi, and Y. Yoshikawa, *Inorg. Chim. Acta*, **142**, 203 (1988).

[xvi] C.E. Pfluger, P.K. Hon, and R.L. Harlow, *J. Cryst. Mol. Struct.*, **4**, 55 (1974).

[xvii] R.D. Mounts and Q. Fernando, *Acta Crystallogr. B*, **30**, 542 (1974).

[xviii] J.M. Stewart, E.C. Lingafelter, and J.D. Breazeale, *Acta Crystallogr.*, **14**, 888 (1961).

[xix] J.E. Andrew and A.B. Blake, *J. Chem. Soc. A*, 1456 (1969).

[xx] A.J. McKinnon, T.N. Waters, and D. Hall, *J. Chem. Soc.*, 3290 (1964).

[xxi] D. Hall, A.J. McKinnon, and T.N. Waters, *J. Chem. Soc.*, 425 (1965).

[xxii] M.T. Garland, J.Y. le Marouille, and E. Spodine, *Acta Cryst. C*, **42**, 1518 (1986).

[xxiii] M.T. Garland, D. Grandjean, and E. Spodine, *Acta Crystallogr. C*, **43**, 1910 (1987).

[xxiv] X. Solans, L. Ruiz-Ramirez, L. Gasque, and J.L. Brianso, *Acta Cryst. C*, **43**, 428 (1987).

[xxv] V.F. Duckworth and N.C. Stephenson, *Acta Crystallogr. B*, **25**, 2245 (1969).

[xxvi] M.T. Garland, J.Y. Le Marouille, and E. Spodine, *Acta Crystallogr. C*, **41**, 855 (1985).

[xxvii] T.A. Annan, C. Peppe, and D.G. Tuck, *Can. J. Chem.*, **68**, 423 (1990).

A contribution from the second resonance form should be indicated structurally by the presence of relatively long M–O–Ar bonds and a concomitant shortening of the M–O ("aldehyde") bond length. For all metals, the M–O distance to aryloxide/alkoxide ligands is considerably shorter than that found for simple donor keto groups. Analysis of the data in Table 6.6 shows that for early d-block metals (*e.g.* Ti, Zr) the distance to the aryloxide oxygen is comparable to that found for simple, terminal aryloxides of these metals. The M–O (aldehyde) distances are 0.2–0.3 Å longer. In contrast it can be seen for the later transition metals that the two parameters are much more comparable. This only partly reflects the decreased amount of aryloxide oxygen π-donation to these metals and there appears to be a significant contribution from the alternative resonance picture for these molecules.

In an interesting reaction, treatment of [AlMe(OC$_6$H$_2$Bu$^t_2$-2,6-Me-4)$_2$] with acetyl chloride was found to yield the complex [AlMe(OC$_6$H$_2$Bu$^t_2$-2,6-Me-4)(OC$_6$H$_2$Bu^t-2-Me-4-CMeO-6)] in which one of the aryloxide ligands was acylated.[255] The structural parameters for this latter compound were discussed in terms of the above two resonance forms. The structural parameters for both α- and β-hydroxy carbonyl ligands bound to aluminium have been discussed.[256]

6.1.3   Metal Salicylates

In nearly all derivatives, this ligand is di-anionic with both the aryloxide and carboxylate oxygen atoms bound to the metal centre (Table 6.7). In one structurally determined example, the carboxylic acid group remains protonated and chelates to the metal through the carbonyl function.[257]

6.1.4   Metal Biphenolates and Binaptholates

The straightforwardly resolved 2,2'-dihydroxy-1,1'-binaphthyl (binol) ligand is one of the most important chiral auxiliaries in chemistry, and its application to organic synthesis has been reviewed.[46,258] The unsubstituted binol has been used in a vast number of asymmetric catalytic applications, although in some of these applications the exact nature (molecularity) of the active species is uncertain.[259] In order to help control solubility and increase the chiral impact and steric size of the ligand, various strategies have been devised to introduce substituents at the 3,3'-positions.[260,261] These bulkier ligands have also been applied to asymmetric syntheses and have been instrumental in allowing the isolation and characterization of discrete molecular species. These bis-aryloxide ligands can adopt a variety of bonding modes (Tables 6.8 and 6.9). Simple chelation to a single metal centre leads to the formation of a seven-membered ring with M–O–Ar angles in the 110–130° range. Alternatively, the ligands can bind in a terminal fashion to two different, nonconnected metal centres. In this case the M–O–Ar angles are less constrained and therefore tend to be larger than for chelating examples. There are also examples of complexes where one of the aryloxides of a bi-phenolate is bridging two metals while the other is terminally bound to one of the metal centres. Much less common are situations in which both oxygen atoms of bi-phenolate or bi-naptholate ligands adopt bridging modes. It is also possible for only one of the aryloxide oxygen atoms to be deprotonated.[262]

Table 6.7   Metal salicylates

Compound	Aryloxide	Bond length (Å) M–O	Bond angle (°) M–O–Ar	Reference
[NMe$_4$]$_2$[Mo(O)$_2$(OAr)$_2$]	OC$_6$H$_4$(COO)-2	1.967 (3)	130	i
cis-dioxo, trans O(OAr)		1.971 (3)	133	
[1-Me-imidazolium]$_2$cis-	OC$_6$H$_4$(COO)-2	1.98 (1)	125	ii
[Mo(O)$_2$(OAr)$_2$]		1.97 (1)	135	
[pyH][Mo(O)$_2$(HOAr)(OAr)]	OC$_6$H$_4$(COOH)-2	1.971 (4)	134	iii
one salicylate protonated and bound through C=O trans to oxo, trans O(OAr)	OC$_6$H$_4$(COO)-2	1.944 (4)	135	
[Mn(OAr)$_2$(bipy)].2MeCN	OC$_6$H$_4$(COO)-2	1.835 (5)	123	iv
6-coord. Mn, cis-O(OAr)		1.823 (5)	126	
[Mn$_2$(μ_2-OMe)$_2$(OAr)$_2$(HOMe)$_4$]	OC$_6$H$_4$(COO)-2	1.889 (3)	127	v
edge-shared bis-octahedron, O(OAr) trans to μ_2-OMe				
trans-[Mn(OAr)(OH$_2$)$_2$(bipy)][ClO$_4$]	OC$_6$H$_4$(COO)-2	1.830 (2)	126	vi
[Fe(OAr)(OAr')(N-Methylimidazole)$_2$]	OC$_6$H$_4$(COO)-2-Me-6	1.930 (4) 1.895 (5)	130	vii
6-coord. Fe, trans OAr	OC$_6$H$_4$ (benzi-midazoyl-methyl)-2			
trans-[Os(O)$_2$(OAr)(py)$_2$]	OC$_6$H$_4$(COO)-2	1.957 (7)	122	viii
6-coord. Os				
[Co(OAr)(triethylenetetramine)]	OC$_6$H$_4$(COO)-2	1.88 (1)	122	ix
6-coord. Co				
[Cu(OAr)$_2$][Cu(2-amino-2-methyl-1-propanol)$_2$]	OC$_6$H$_4$(COO)-2	1.900 (3)	126	x

[i]C.F. Edwards, W.P. Griffith, A.J.P. White, and D.J. Williams, *J. Chem. Soc., Dalton Trans.*, 3813 (1993).
[ii]K. Boutilier, L.R. MacGillivray, S. Subramanian, and M.J. Zaworotko, *J. Crystallogr. Spectrosc. Res.*, **23**, 773 (1993).
[iii]C.F. Edwards, W.P. Griffith, A.J.P. White, and D.J. Williams, *Polyhedron*, **11**, 2711 (1992).
[iv]P.S. Pavacik, J.C. Huffman, and G. Christou, *J. Chem. Soc., Chem. Comm.*, 43 (1986).
[v]Xiang Shi Tan, Wen Xia Tang, and Jie Sun, *Polyhedron*, **15**, 2671 (1996).
[vi]Xiang Shi Tan, Jian Chen, Pei Ju Zheng and Wen Xia Tang, *Acta Crystallogr. C*, **51**, 1268 (1995).
[vii]M.R. McDevitt, A.W. Addison, E. Sinn, and L.K. Thompson, *Inorg. Chem.*, **29**, 3425 (1990).
[viii]C.C. Hinckley, P.A. Kibala, and P.D. Robinson, *Acta Crystallogr. C*, **43**, 842 (1987).
[ix]H. Sekizaki, E. Toyota, and Y. Yamamoto, *Bull. Chem. Soc. Jpn.*, **66**, 1652 (1993).
[x]H. Muhonen, *Acta Crystallogr. B*, **38**, 2041 (1982).

The binol ligand has been used extensively in the synthesis of lanthanoid complexes. These compounds are typically bi-metallic, containing alkali metal ions (M) exemplified by the formula [M$_3$Ln(binaphthoxide)$_3$].[263,264] They can be used as catalysts in a wide variety of important organic transformations.[265]

Heppert and co-workers have extensively investigated the titanium coordination chemistry of binol type ligands.[47] The use of chloride compounds for carrying out enantioselective Diels–Alder reactions has also been evaluated.[47] Schaverien and

Table 6.8 Metal bi-phenolates

Compound	Aryloxide	Bond length (Å) M–O	Bond angle (°) M–O–Ar	Reference
[La(di-OAr){CH(SiMe$_3$)$_2$}(THF)$_3$] 6-coord. La, OAr *trans* to THF	2,2'-(OC$_6$H$_2$Bu$^t_2$-3,5)$_2$	2.271 (9) 2.216 (7)	135 133	i
[Cl$_2$Ti(di-OAr)$_2$TiCl$_2$] bi-phenoxide terminally bound to each 4-coord. Ti	2,2'-(OC$_6$Me$_2$-4,6-Cl$_2$-3,5)$_2$	1.762 (5) 1.757 (5)	168 168	ii
[ClTi(di-OAr)$_3$TiCl] bi-phenoxide terminally bound to each 4-coord. Ti	2,2'-(OC$_6$Me$_2$-4,6-Cl$_2$-3,5)$_2$	1.778 (5) 1.760 (5) 1.769 (5)	152 152 153	ii
[(Pr^{i}O)$_2$Ti(μ_2-di-OAr)$_2$Ti(OPr^i)$_2$] 4-coord. Ti atoms held by bridging OAr	μ_2-2,2'-(OC$_6$Me$_2$-4,6-Cl$_2$-3,5)$_2$	1.858 (5) 1.867 (5)	129 132	ii
[Ti(di-OAr)(CH$_2$Ph)$_2$].0.5Et$_2$O tetrahedral Ti	2,2'-(OC$_6$H$_3$Bu^t-6-OMe-4)$_2$	1.799 (7)	130 110	iii
[Mo(NC$_6$H$_3$Me$_2$-2,6)(=CHCMe$_2$Ph)(di-OAr)]	2,2'-(OC$_6$HMe-6-Bu$^t_2$-3,5)$_2$	1.821 (8) 1.990 (6) 2.003 (6)	97 104	iv
[Mn$_2$(μ_2-di-OAr)$_2$(Hdi-OAr)(bipy)$_2$]	2,2'-(OC$_6$H$_4$(HOC$_6$H$_4$)	2.04 (1)	138	v, vi
5- and 6-coord. Mn, bridging 2,2'-biphenoxide terminally bound to same Mn, monodentate 2,2'-biphenoxide bound to other Mn	μ_2-2,2'-(OC$_6$H$_4$)$_2$	1.87 (1) 1.88 (1) 2.13 (1), 1.954 (1) 2.11 (1), 2.13 (1)	117 122	
[NEt$_3$H]$_2$[Mn(di-OAr)$_2$(Hdi-OAr)] tbp Mn, one monodentate 2,2'-biphenoxide	2,2'-(OC$_6$H$_4$)$_2$ 2,2'-(OC$_6$H$_4$)(HOC$_6$H$_4$)	1.89 (1) 1.984 (9) 1.88 (1)	123 118 116	vi
[Me$_4$N]$_4$[Mn$_{10}$(μ_2-μ_2-di-OAr)$_4$(O)$_4$Cl$_{12}$] Mn$_6$O$_4$ adamantyl framework, 5- and 6-coord. Mn	μ_2-μ_2-2,2'-(OC$_6$H$_4$)$_2$	1.987 (9) 1.91 (2) (av.) 2.14 (1) (av.) 2.18 (2) (av.)	126	vii

(continued overleaf)

Table 6.8 *(Continued)*

Compound	Aryloxide	Bond length (Å) M–O	Bond angle (°) M–O–Ar	Reference
$[Me_4N]_4[Mn_{10}(\mu_2-\mu_2\text{-di-OAr})_4(O)_4Br_{12}]$ Mn_6O_4 adamantyl framework, 5- and 6-coord. Mn	$\mu_2-\mu_2\text{-}2,2'\text{-}(CO_6H_4)_2$	1.91 (3) (av.) 2.14 (1) (av.) 2.16 (3) (av.)		viii
[piperidinium]$_2$[Fe$_2$(μ_2-di-OAr)$_2$(di-OAr)$_2$] 5-coord. Fe	$2,2'\text{-}(OC_6H_4)_2$ $\mu_2\text{-}2,2'\text{-}(OC_6H_4)_2$	1.851 (2) 1.896 (2) 1.933 (2) 2.056 (2), 1.966 (2)	134 119 122	ix
[Et$_4$N]$_2$[(di-OAr)Fe(μ_2-S)$_2$Fe(di-OAr)] tetrahedral Fe	$2,2'\text{-}(OC_6H_4)_2$	1.895 (2) 1.892 (2)	116 126	x, xi

[i] C.J. Schaverien, N. Meijboom, and A.G. Orpen, *J. Chem. Soc., Chem. Commun.*, 124 (1992).

[ii] E.J. Corey, M.A. Letavic, M.C. Noe, and S. Sarshar, *Tetrahedron Lett.*, **35**, 7553 (1994).

[iii] A. van der Linden, C.J. Schaverien, N. Meijboom, C. Ganter, and A.G. Orpen, *J. Am. Chem. Soc.*, **117**, 3008 (1995).

[iv] K.M. Totland, T.J. Boyd, G.G. Lavoie, W.M. Davis, and R.R. Schrock, *Macromolecules*, **29**, 6114 (1996).

[v] J.S. Bashkin, A.R. Schake, J.B. Vincent, H.-R. Chang, Qiaoying Li, J.C. Huffman, G. Christou, and D.N. Hendrickson, *J. Chem. Soc., Chem. Commun.*, 700 (1988).

[vi] A.R. Schake, E.A. Schmitt, A.J. Conti, W.E. Streib, J.C. Huffman, D.N. Hendrickson, and G. Christou, *Inorg. Chem.*, **30**, 3192 (1991).

[vii] D.P. Goldberg, A. Caneschi, and S.J. Lippard, *J. Am. Chem. Soc.*, **115**, 9299 (1993).

[viii] D.P. Goldberg, A. Caneschi, C.D. Delfs, R. Sessoli, and S.J. Lippard. *J. Am. Chem. Soc.*, **117**, 5789 (1995).

[ix] E.W. Ainscough, A.M. Brodie, S.J. McLachlan, and K.L. Brown, *J. Chem. Soc., Dalton Trans.*, 1385 (1983).

[x] D. Coucouvanis, A. Salifoglou, M.G. Kanatzidis, A. Simopoulos, and V. Papaefthymiou, *J. Am. Chem. Soc.*, **106**, 6081 (1984).

[xi] A. Salifoglou, A. Simopoulos, A. Kostikas, R.W. Dunham, M.G. Kanatzidis, and D. Coucouvanis, *Inorg. Chem.*, **27**, 3394 (1988).

Table 6.9 Metal binaphtholates

Compound	Aryloxide	Bond length (Å) M–O	Bond angle (°) M–O–Ar	Reference
[{(THF)$_2$Na}$_3$(μ_2-di-OAr)$_3$Eu]	1,1'-bi-2-naphtholato	2.286 (5) 2.275 (5) 2.336 (5)	123 133 117	i
[{(THF)$_2$Na}$_3$(μ_2-di-OAr)$_3$Nd]	1,1'-bi-2-naphtholato	2.339 (7) 2.242 (7) 2.331 (7)	122 132 118	i
[{(THF)$_2$Na}$_3$(μ_2-di-OAr)$_3$Pr]	1,1'-bi-2-naphtholato	2.365 (7) 2.260 (7) 2.318 (7)	122 133 118	i
[(ansa-Cp)Ti(Hdi-OAr)Cl] tetrahedral Ti	1,1'-bi-2-naphtholato-H	1.967 (2)	155	ii
[(ansa-Cp)Ti(di-OAr)] tetrahedral Ti	1,1'-bi-2-naphtholato	2.010 (2) 2.019 (2)	121 121	ii
[(ansa-Cp)Ti(di-OAr)] tetrahedral Ti	1,1'-bi-2-naphtholato	1.906 (1)	125	iii
[(ansa-Cp)Ti(di-OAr)] tetrahedral Ti	1,1'-bi-2-naphtholato	1.923 (1) 1.908 (1)	119 125	iv
[(ansa-Cp)Ti(di-OAr).n-hexane tetrahedral Ti	1,1'-bi-2-naphtholato	1.926 (2) 1.903 (2)	119 125	iv
[Ti(di-OAr)(OPr^i)$_2$] tetrahedral Ti	3,3'-dimethylbutylsilyl-1,1'-binaphtholato	1.852 (2) 1.846 (2)	110 122	v
[Ti$_2$(μ_2-di-OAr)$_2$(OPr^i)$_4$] distorted tbp, axial OPr^i and μ-O	3,3'-dimethyl-1,1'-binaphtholato	1.832 (6) 1.827 (6) 2.128 (7), 1.969 (7) 1.953 (7), 2.127 (7)	135 132	v

(continued overleaf)

Table 6.9 (*Continued*)

Compound	Aryloxide	Bond length (Å) M–O	Bond angle (°) M–O–Ar	Reference
[Ti$_2$(di-OAr)(OPr^i)$_6$] two tetrahedral Ti linked by a single di-OAr	3,3'-dimethylbutylsilyl-1,1'-binaphtholato	1.825 (3)	156	v
[Cl$_2$Ti(μ_2-di-OAr)$_2$TiCl$_2$] distorted tbp, axial Cl and μ-O	3,3'-dimethyl-1,1'-bi-2-naphtholato	1.762 (2), 1.771 (2) 1.962 (2), 2.064 (2) 1.979 (2), 2.045 (2)		vi
[Ti$_2$(μ_2-di-OAr)(NMe$_2$)$_4$]	3-trimethylsilyl-1,1-binaphtholato	1.875 (4) 2.212 (4) 1.974 (4) 1.850 (4) 2.161 (4) 1.986 (4)	142 108 134 140 113 138	vii
[Ti$_2$(di-OAr)(NMe$_2$)$_6$]	3,3'-trimethylsilyl-1,1-binaphtholato	1.889 (4) 1.879 (4)	162 165	vii
trans-[Ti(di-OAr)(NHMe$_2$)$_2$Cl$_2$]	3,3'-dimethylphenylsilyl-1,1-binaphtholato	1.862 (2) 1.863 (2)	136 134	vii
cis-[Ti(di-OAr)(py)$_2$Cl$_2$]	3,3'-dimethylphenylsilyl-1,1-binaphtholato	1.821 (4) 1.861 (4)	132 125	vii
cis-[Ti(di-OAr)(η^2-Me$_2$NCH$_2$NMe$_2$)Cl$_2$]	3,3'-phenylsilyl-1,1-binaphtholato	1.869 (2) 1.829 (3)	126 133	vii
[(ansa-tetrahydro-indenyl)$_2$Ti(di-OAr)]	1,1'-bi-2-naphtholato	1.935 (7) 1.932 (7)	122 122	viii

Compound	Ligand	Angle	M–O (Å)	Ref.
[TiCl₂(di-OAr)]₂ two tetrahedral Ti linked by terminally bound di-OAr	octahydro-1,1'-bi-2-naphtholato	178 164 173 161	1.726 (5) 1.739 (5) 1.751 (5) 1.750 (5)	ix
[Ti₄(μ₃-O)₂(μ₂-OPrⁱ)₂(di-OAr)₂(OPrⁱ)₆] cluster, di-OAr terminally bound to two Ti	octahydro-1,1'-bi-2-naphtholato	148 157 149 144	1.793 (7) 1.827 (7) 1.813 (8) 1.802 (8)	ix
[Ta(di-OAr)Cl₄][Me₂NH₂]	3,3'-trimethylsilyl-1,1'-binaphtholato	136 135	1.889 (6) 1.893 (7)	vii
[Ta(di-OAr)(NMe₂)₃(NHMe₂)]	3,3'-trimethylsilyl-1,1'-binaphtholato	130 133	2.036 (4) 2.039 (3)	vii
[Ta(di-OAr)(NMe₂)₃(NHMe₂)]	3,3'-methyldiphenylsilyl-1,1'-binaphtholato	134 125	2.043 (3) 2.061 (3)	vii
[Ta(di-OAr)(NMe₂)₃(NHMe₂)]	3,3'-triphenylsilyl-1,1'-binaphtholato	139 124	2.021 (2) 2.075 (2)	vii
[Mo(NC₆H₃Me₂-2,6)(=CHCMe₂Ph)(di-OAr)(thf)] tbp Mo, axial THF and OAr	3,3'-diphenyl-1,1'-bi-2-naphtholato	130	1.988 (5)	x
[W(O)(OBuᵗ)₂(di-OAr)] square pyramidal W	3,3'-dimethyl-1,1'-bi-2-naphtholato	113 125 123	2.012 (5) 1.952 (6) 1.953 (6)	xi

(continued overleaf)

Table 6.9 (*Continued*)

Compound	Aryloxide	Bond length (Å) M–O	Bond angle (°) M–O–Ar	Reference
[W₂(di-OAr)(OBuᵗ)₄] ethane-like, di-OAr terminally bound to each W	3,3'-dimethyl-1,1'-bi-2-naphtholato	1.900 (1) 1.934 (1)	126 130	xii xiii
[W(O)Cl₃(Hdi-OAr)] pseudo-octahedral W with Hdi-OAr OH *trans* to W=O	3,3'-dimethyl-1,1'-bi-2-naphtholato	1.794 (5) 2.262 (6)	135 (HO)	xiv
[(cyclohexanone)(thf)₂Li(μ₂-di-OAr)Al(di-OAr)]	1,1'-bi-2-naphtholato	1.75 (1) (av.) 1.96 (1)	119	xv

[i]H. Sasai, T. Suzuki, N. Itoh, K. Tanaka, T. Date, K. Okamura, and M. Shibasaki, *J. Am. Chem. Soc.*, **115**, 10372 (1993).

[ii]K. Schmidt, A. Reinmuth, U. Rief, J. Diebold, and H.H. Brintzinger, *Organometallics*, **16**, 1724 (1997).

[iii]B.A. Kuntz, R. Ramachandran, N.J. Taylor, J. Guan, and S. Collins, *J. Organomet. Chem.*, **497**, 133 (1995).

[iv]M.S. Erikson, F.R. Fronczek, and M.L. McLaughlin, *J. Organomet. Chem.*, **415**, 75 (1991).

[v]T.J. Boyle, D.L. Barnes, J.A. Heppert, L. Morales, F. Takusagawa, and J.W. Connolly, *Organometallics*, **11**, 1112 (1992).

[vi]T.J. Boyle, N.W. Eilerts, J.A. Heppert, and F. Takusagawa, *Organometallics*, **13**, 2218 (1994).

[vii]M.G. Torn, P.E. Fanwick, and I.P. Rothwell, unpublished results.

[viii]F.R.W.P. Wild, L. Zsolnai, G. Huttner, and H.H. Brintzinger, *J. Organomet. Chem.*, **232**, 233 (1982).

[ix]N.W. Eilerts, J.A. Heppert, M.L. Kennedy, and F. Takusagawa, *Inorg. Chem.*, **33**, 4813 (1994).

[x]K.M. Totland, T.J. Boyd, G.G. Lavoie, W.M. Davis, and R.R. Schrock, *Macromolecules* **29**, 6114 (1996).

[xi]J.A. Heppert, S.D. Dietz, N.W. Eilerts, R.W. Henning, M.D. Morton, and F. Takusagawa, *Organometallics*, **12**, 2565 (1993).

[xii]J.A. Heppert, S.D. Dietz, T.J. Boyle, and F. Takusagawa, *J. Am. Chem. Soc.*, **111**, 1503 (1989).

[xiii]S.D. Dietz, N.W. Eilerts, J.A. Heppert, and D.V. Velde, *Inorg. Chem.*, **32**, 1689 (1993).

[xiv]M.D. Morton, J.A. Heppert, S.D. Dietz, W.H. Huang, D.A. Ellis, T.A. Grant, N.W. Eilerts, D.L. Barnes, F. Takusagawa, and D. Vander Velde, *J. Am. Chem. Soc.*, **115**, 7916 (1993).

[xv]T. Arai, H. Sasai, K. Aoe, K. Okamura, T. Date, and M. Shibasaki, *Angew. Chem. Int. Ed. Engl.*, **35**, 104 (1996).

co-workers have demonstrated that group 4 metal binaphthoxide derivatives can be used to generate isotactic polymers and oligomers from α-olefins.[45]

The development of "Schrock type" metathesis catalysts containing chiral binaphtholate and biphenolate ligands to carry out asymmetric metathesis reactions has been achieved.[50–52,266]

Some recent developments include the synthesis of poly-aryloxides based upon the binol nucleus. Examples include bis(binaphthol)methane[267] and binaphthols containing phenolic substituents in the 3,3′-positions.[268]

6.2 Normal Metal Aryloxides

6.2.1 Group 1 Metal Aryloxides

The aryloxide derivatives of the group 1 metals and particularly lithium and sodium are important starting points for the synthesis of other metal aryloxides. They are generated by addition of the phenol to either the alkyls (lithium, routinely Bu^{n}Li), hydrides (lithium and sodium), or the metal (lithium, sodium, and potassium). As with other electropositive metals, the chemistry of the group 1 elements with 2,4,6-trinitrophenol, metal picrates, is typically carried out with aqueous or other solutions of the corresponding cations. The structural chemistry of group 1 metal compounds continues to attract considerable attention.[269] This interest stems not only from the diverse structures adopted by these compounds but also by the recognition that the degree of aggregation of group 1 metal reagents can strongly influence their reactivity.[270] In the case of the aryloxide derivatives of these metals the degree of oligomerization is sensitive to the steric nature of the phenoxide nucleus as well as the presence of Lewis bases. The degree of association in the solid state has been determined by X-ray diffraction methods for a variety of group 1 metal aryloxides (Tables 6.10–6.13). Also the structures of simple phenoxides [MOPh] (M = Li, Na, K, Rb, and Cs) have been examined by powder diffraction using *ab initio* structure solutions.[271] Many of these studies have been stimulated by the industrially important Kolbe–Schmitt synthesis, which involves the solid-state carboxylation of a group 1 metal phenoxide.[272] Potassium, rubidium, and caesium phenoxide were shown to be isostructural by powder diffraction with two distinct metal environments within infinite chains. Besides a distorted octahedral environment, three-coordinate alkali metals were present with weak phenyl ring interactions.[271] The powder structures of [C$_6$H$_5$OK.xC$_6$H$_5$OH] (x = 2, 3) have also been investigated.[273] In both compounds there are polymeric chains with potassium surrounded by five oxygen atoms and one π-bound phenyl ring.

The lithium derivative of the bulky ligand 2,6-diphenyl-3,5-di-*tert*-butylphenoxide forms a cyclic trimer, [Li(μ_2-OAr)]$_3$ with alternating long/short Li–O distances and two-coordinate Li atoms.[274] A number of polymeric structures of unsolvated sodium phenoxides have been determined (Table 6.11), *e.g.* [NaOC$_6$H$_5$]$_n$[275] and [Na(μ_3-O-η^6-C$_6$H$_4$Me-4)]$_n$.[276] The parent phenoxide has chains of Na$_2$O$_2$ rings stacked together using Na–O and Na–π-arene interactions. In the 4-methyl derivative there are also π-interactions between the phenoxide nucleus and sodium. Recently the structure of [Cs(OC$_6$H$_3$Pr$^i_2$-2,6)] has been determined to consist of infinite chains held together by Cs–O and Cs–π-arene interactions.[277]

Table 6.10 Lithium aryloxides

Compound	Aryloxide	Bond length (Å) $M–O$	Reference
[Li(μ_3-OAr)(THF)]$_6$ two crystalline forms; two chair Li$_3$O$_3$ rings stacked upon one another, 4-coord. Li	μ_3-OC$_6$H$_5$	1.911 (21)–1.990 (22) 1.895 (21)–1.992 (22)	i
[Li(μ_3-OAr)(THF)]$_4$(HOAr) distorted cube with one edge opened up by bridging phenol ligand, 4-coord. Li.	μ_3-OC$_6$H$_5$	1.91 (2)–2.05 (2)	ii
[Li(μ_3-OAr)(nmc)]$_4$ regular cube, 4-coord. Li, N-methyl caprolactam	μ_3-OC$_6$H$_5$	1.88 (2)–2.00 (2)	iii
[Li$_4$(μ_3-OAr)$_3$(NCS)(HMPA)$_4$] cube, 4-coord. Li, bridging HMPA, terminal NCS	μ_3-OC$_6$H$_5$	1.89 (2)–2.02 (2)	iv
[Li$_4$(μ_3-di-OAr)$_2$(HMPA)$_2$] cube, two 4- and two 5-coord. Li	N,N'-ethylene-bis(salicylideneiminato)	2.002 (6) (av.)	v
[Li$_8$(μ_3-di-OAr)$_3$(O)(TMEDA)$_2$] two cubes with one common vertex (oxo group), 4-coord. Li	N,N'-ethylene-bis(salicylideneiminato)	1.985 (9) (mean)	v
[Li$_4$(μ_3-OAr)$_2$(I)$_2$] cube, 4-coord. Li	μ_3-OC$_6$H$_2$(CH$_2$NMe$_2$)$_2$-2,6-Me-4	1.903 (7)–1.934 (9)	vi
Li[μ_4-OArI$_4$ Li$_4$O$_4$ cube, 4-coord. Li, with sodiated methyl groups; Na atoms interact with phenoxide groups	μ_4-OC$_6$H$_2${CH$_2$Na(tmeda)}-2-Me$_2$-4,6	1.97 (2) (av.)	vii
[Li(μ_2-OAr)]$_3$ planar Li$_3$O$_3$ ring, 2-coord. Li	μ_2-OC$_6$HBu$^t_2$-3,5-Ph$_2$-2,6	1.78 (1), 1.840 (7)	viii

[Li(μ₂-OAr)]₃ planar Li₃O₃ ring, 4-coord. Li	$\mu_2\text{-OC}_6\text{H}_2(\text{CH=NPr}^i)_2\text{-2,6-Me-4}$	1.85 (2)–1.90 (3)	ix
[Li(μ₂-OAr)]₃ planar Li₃O₃ ring, 4-coord. Li	$\mu_2\text{-OC}_6\text{H}_2(\text{CH}_2\text{NMe}_2)_2\text{-2,6-Me-4}$	1.846 (9)–1.887 (9)	x
[Li(μ₂-OAr)(OEt₂)]₂ planar Li₂O₂ ring, 3-coord. Li	$\mu_2\text{-OC}_6\text{H}_2\text{Bu}_2^t\text{-2,6-Me-4}$	1.85 (1), 1.87 (1)	xi
[Li(μ₂-OAr)(THF)]₂ planar Li₂O₂ ring, 3-coord. Li	$\mu_2\text{-OC}_6\text{H}_3\text{Bu}_2^t\text{-2,6}$	1.828 (7), 1.867 (7)	xii
[Li(μ₂-OAr)(OEt₂)]₂ planar Li₂O₂ ring, 3-coord. Li	$\mu_2\text{-OC}_6\text{H}_3\text{Bu}_2^t\text{-2,6}$	1.830 (7), 1.880 (7)	xiii
[Li(μ₂-OAr)(DMSO)]₂ planar Li₂O₂ ring, 3-coord. Li	$\mu_2\text{-OC}_6\text{H}_3\text{Bu}_2^t\text{-2,6}$	1.87 (2)	xiv
[(18-crown-6)Li₂(μ₂-OAr)₂] planar Li₂O₂ ring, 5-coord. Li	$\mu_2\text{-OC}_6\text{H}_5$	1.86 (1)–1.90 (1)	xv
[(15-crown-5)₂Li₄(μ₂-OAr)₄] two crystalline forms, 4- and 6-coord. Li	$\mu_2\text{-OC}_6\text{H}_5$	1.836 (9)–1.939 (8)	xvi
[(cyclohexano-15-crown-5)₂Li₄(μ₂-OAr)₄] 4- and 6-coord. Li	$\mu_2\text{-OC}_6\text{H}_5$	1.859 (8)–1.910 (8)	xvii
[Li(μ₂-OAr)(bipy)]₂ planar Li₂O₂ ring, 5-coord. Li	$\mu_2\text{-OC}_6\text{H}_2(\text{NO}_2)_3\text{-2,4,6}$	1.84 (1), 2.16 (2)	xviii
[(dibenzo-36-crown-12)].2[Li(OAr)(OH₂)] 5-coord. Li	$\text{OC}_6\text{H}_2(\text{NO}_2)_3\text{-2,4,6}$	1.936 (5)	xix
[(18-crown-4)Li(OAr)] 5-coord. Li	$\text{OC}_6\text{H}_2(\text{NO}_2)_3\text{-2,4,6}$	1.88 (1)	xx

(continued overleaf)

Table 6.10 (*Continued*)

Compound	Aryloxide	Bond length (Å) M–O	Reference
[(H₂O)Li(μ₂-OAr)₂Li(OH₂)] 5-coord. Li, picrate bound to two different Li through *ortho*-NO₂	μ_2-OC$_6$H$_2$(NO$_2$)$_3$-2,4,6	1.917 (4), 1.940 (3)	xxi

[i]L.M. Jackman, D. Cizmeciyan, P.G. Williard, and M.A. Nichols, *J. Am. Chem. Soc.*, **115**, 6262 (1993).

[ii]M. Pink, G. Zahn, and J. Sieler, *Z. Anorg. Allg. Chem.*, **620**, 749 (1994).

[iii]D. Walther, U. Ritter, S. Gessler, J. Sieler, and M. Kunert, *Z. Anorg. Allg. Chem.*, **620**, 101 (1994).

[iv]P.R. Raithby, D. Reed, R. Snaith, and D.S. Wright, *Angew. Chem., Int. Ed. Engl.*, **30**, 1011 (1991).

[v]S.C. Ball, I. Cragg-Hine, M.G. Davidson, R.P. Davies, M.I. Lopez-Solera, P.R. Raithby, D. Reed, R. Snaith, and E.M. Vogl, *J. Chem. Soc., Chem. Commun.*, 2147 (1995).

[vi]P.A. van der Schaaf, M.P. Hogerheide, D.M. Grove, A.L. Spek, and G. van Koten, *J. Chem. Soc., Chem. Commun.*, 1703 (1992).

[vii]S. Harder and A. Streitwieser, *Angew. Chem., Int. Ed. Engl.*, **32**, 1066 (1993).

[viii]J.S. Vilardo, P.E. Fanwick, and I.P. Rothwell, *Polyhedron*, **17**, 769 (1998).

[ix]M.S. Korobov, V.I. Minkin, L.E. Nivorozhkin, O.E. Kompan, and Yu. T. Struchkov, *Zh. Obshch. Khim.*, **59**, 429 (1989).

[x]P.A. van der Schaaf, J.T.B.H. Jastrzebski, M.P. Hogerheide, W.J.J. Smeets, A.L. Spek, J. Boersma, and G. van Koten, *Inorg. Chem.*, **32**, 4111 (1993).

[xi]B. Cetinkaya, I. Gumrukcu, M.F. Lappert, J.L. Atwood, and R. Shakir, *J. Am. Chem. Soc.*, **102**, 2086 (1980).

[xii]J.C. Huffman, R.L. Geerts, and K.G. Caulton, *J. Cryst. Spectrosc.*, **14**, 541 (1984).

[xiii]G. Kociok-Kohn, J. Pickardt, and H. Schumann, *Acta Cryst. C*, **47**, 2649 (1991).

[xiv]L. Matilainen, M. Klinga, and M. Leskela, *Polyhedron*, **14**, 635 (1995).

[xv]K.A. Watson, S. Fortier, M.P. Murchie, J.W. Bovenkamp, A. Rodrigue, G.W. Buchanan, and C.I. Ratcliffe, *Can. J. Chem.*, **68**, 1201 (1990).

[xvi]M.P. Murchie, J.W. Bovenkamp, A. Rodrigue, K.A. Watson, and S. Fortier, *Can. J. Chem.*, **66**, 2515 (1988).

[xvii]K.A. Watson, S. Fortier, M.P. Murchie, and J.W. Bovenkamp, *Can. J. Chem.*, **69**, 687 (1991).

[xviii]M.S. Hundal, G. Sood, P. Kapoor, and N.S. Poonia. *J. Cryst. Spectrosc.*, **21**, 201 (1991).

[xix]S.M. Doughty, J.F. Stoddart, H.M. Colquhoun, A.M.Z. Slawin, and D.J. Williams, *Polyhedron*, **4**, 567 (1985).

[xx]M. Van Beylen, B. Roland, G.S.D. King, and J. Aerts, *J. Chem. Res.*, **388**, 4201 (1985).

[xxi]J.M. Harrowfield, B.W. Skelton, and A.H. White, *Aust. J. Chem.*, **48**, 1311 (1995).

Table 6.11 Sodium aryloxides

Compound	Aryloxide	Bond length (Å) M–O	Ref.
[(ArO)(DMSO)Na(μ_2-DMSO)$_2$Na(DMSO)(OAr)] unusual structure with bridging DMSO	OC$_6$H$_3$Bu$^t_2$-2,6	2.144 (5)	i
[(Na(OAr)]$_n$ stacked N$_2$O$_2$ rings; held together by intermolecular Na–O and Na-π-arene bonds	OC$_6$H$_5$	2.328 (6)–2.264 (1)	ii
[Na$_6$(μ_4-OAr)$_2$(μ_3-OAr)$_4$(THF)$_6$] two face-fused cubes, 4- and 5-coord. Na	μ_3-OC$_6$H$_5$ μ_4-OC$_6$H$_5$	2.231 (3)–2.365 (2) 2.318 (3)–2.405 (2)	ii
[Na(μ_3-OAr)(tmu)]$_4$ regular cube, 4-coord. Na, tetramethyl urea	μ_3-OC$_6$H$_5$	2.282 (3)–2.315 (4)	ii
[Na(η^6-OAr(OH$_2$)$_3$]$_n$ phenoxide π-bound to Na and hydrogen bonded to H$_2$O ligands. 4-coord. Na	η^6-OC$_6$H$_5$	2.272 (7), 2.343 (7)	iii
[Na(μ_2-OAr)(OH$_2$)]$_n$	μ_2-OC$_6$H$_5$	2.272 (3), 2.340 (4)	iii
[Na(μ_3-O-η^6-Ar)]$_n$ each Na bound to 3 O-atoms and π-bound to a phenoxide ring	μ_3-η^6-OC$_6$H$_4$Me-4	2.231 (4)–2.346 (4)	iv
[Na(μ_3-OAr)(DME)]$_4$ regular cube, 5-coord. Na	μ_3-OC$_6$H$_4$Me-4	2.281 (4)–2.332 (4)	iv
[Na(μ_3-OAr)(nmc)]$_4$ regular cube, 4-coord Na, N-methyl-caprolactam	μ_3-OC$_6$H$_5$	2.294 (8), 2.298 (7)	v
[Na(μ_3-OAr)]$_4$ Na$_4$O$_4$ regular cube, 5-coord. Na	μ_3-OC$_6$H$_2$(CH$_2$NMe$_2$)$_2$-2,6-Me-4	2.325 (4)–2.417 (2)	vi
[Na$_6$(μ_4-OAr)$_2$(μ_3-OAr)$_4$] two face-fused cubes, 4- and 5-coord. Na	μ_3-OC$_6$H$_4$(CH$_2$NMe$_2$)$_2$-2 μ_4-OC$_6$H$_4$(CH$_2$NMe$_2$)$_2$-2	2.234 (5)–2.344 (5) 2.325 (5)–2.478 (5)	vii
[(HOAr)Na(μ_2-OAr)$_2$Na(HOAr)] 5-coord. Na	μ_2-OC$_6$H$_2$(CH$_2$NMe$_2$)$_2$-2,6-Me-4	2.216 (3)–2.289 (3)	vii
[Na$_4$(di-μ_3-OAr)$_2$(DME)$_2$] regular cube, 5-coord. Na	μ_3-OC$_6$H$_4$CH=N-2]$_2$(C$_6$H$_4$)	2.25 (1)–2.39 (1)	viii
[Na(μ_2-OAr)(THF)$_2$]$_2$ planar Na$_2$O$_2$ ring, 4-coord. Na	μ_2-OC$_6$H$_2$(CF$_3$)$_3$-2,4,6	2.306 (6), 2.313 (6)	ix
[(dicyclohexano-18-crown-6)$_2$Na$_4$(μ_2-OAr)$_4$] 4- and 7-coord. Na	μ_2-OC$_6$H$_5$	2.213 (2)–2.323 (2)	x

(continued overleaf)

Table 6.11 (Continued)

Compound	Aryloxide	Bond length (Å) M–O	Ref.
[(dicyclohexano-18-crown-6)Na(OAr)].HOAr 7-coord. Na, OAr hydrogen bonded to HOAr	OC_6H_5	2.363 (4)	xi
[(15-crown-5)$_2$Na$_4$(μ_2-OAr)$_4$] 4- and 6-coord. Na	μ_2-OC_6H_5	2.192 (4)–2.250 (4)	xii
[(cyclohexano-15-crown-5)$_2$Na$_4$(μ_2-OAr)$_4$] 4- and 6-coord. Na	μ_2-OC_6H_5, μ_3-OC_6H_5	2.191 (5)–2.377 (5)	xii
[(dibenzo-24-crown-8)Na$_2$(μ_2-OAr)$_2$] planar Na$_2$O$_2$ ring. 6-coord. Na	μ_2-$OC_6H_4(NO_2)_2$-2	2.296 (4), 2.399 (4)	xiii
[(benzo-15-crown-5)Na(OAr)] 7-coord. Na	$OC_6H_2(NO_2)_3$-2,4,6	2.350 (3)	xiv
[(valinomycin)Na(OAr)(OH$_2$)] [Na(OAr)(OH$_2$)]$_n$ 8-coord. Na	$OC_6H_2(NO_2)_3$-2,4,6 μ_2-$OC_6H_2(NO_2)_3$-2,4,6	2.37 (2) 2.326 (3)	xv xvi
[Na$_2$(μ_2-OAr)$_2$] 5-coord. Na	MeO(CH$_2$)$_2$C$_6$H$_4$NH{(CH$_2$)$_2$OMe}NC$_6$H$_4$O-2	2.21 (1), 2.27 (1)	xvii

[i] L. Matilainen, M. Leskela, and M. Klinga, *J. Chem. Soc., Chem. Commun.*, 421 (1995).

[ii] M. Kunert, E. Dinjus, M. Nauck, and J. Sieler, *Chem. Ber.*, **130**, 1461 (1997).

[iii] J. Sieler, M. Pink, and G. Zahn, *Z. Anorg. Allg. Chem.*, **620**, 743 (1994).

[iv] W.J. Evans, R.E. Golden, and J.W. Ziller, *Inorg. Chem.*, **32**, 3041 (1993).

[v] D. Walther, U. Ritter, S. Gessler, J. Sieler, and M. Kunert, *Z. Anorg. Allg. Chem.*, **620**, 101 (1994).

[vi] P.A. van der Schaaf, J.T.B.H. Jastrzebski, M.P. Hogerheide, W.J.J. Smeets, A.L. Spek, J. Boersma, and G. van Koten, *Inorg. Chem.*, **32**, 4111 (1993).

[vii] M.P. Hogerheide, S.N. Ringelberg, M.D. Janssen, A.L. Spek, J. Boersma, and G. van Koten, *Inorg. Chem.*, **35**, 1195 (1996).

[viii] E. Solari, S. DeAngelis, C. Floriani, A. Chiesi-Villa, and C. Rizzoli, *J. Chem. Soc., Dalton Trans.*, 2471 (1991).

[ix] S. Brooker, F.T. Edelmann, T. Kottke, H.W. Roesky, G.M. Sheldrick, D. Stalke, and K.H. Whitmire, *J. Chem. Soc., Chem. Comm.*, 144 (1991).

[x] M.E. Fraser, S. Fortier, A. Rodrigue, and J.W. Bovenkamp, *Can. J. Chem.*, **64**, 816 (1986).

[xi] M.E. Fraser, S. Fortier, M.K. Markiewicz, A. Rodrigue, and J.W. Bovenkamp, *Can. J. Chem.*, **65**, 2558 (1987).

[xii] K.A. Watson, S. Fortier, M.P. Murchie, and J.W. Bovenkamp, *Can. J. Chem.*, **69**, 687 (1991).

[xiii] D.L. Hughes, *J. Chem. Soc., Dalton Trans.*, 2374 (1975).

[xiv] D.L. Ward, A.I. Popov, and N.S. Poonia, *Acta. Crystallog. C.*, **84**, 238 (1984).

[xv] L.K. Steinrauf, J.A. Hamilton, and M.N. Sabesan, *J. Am. Chem. Soc.*, **104**, 4085 (1982).

[xvi] J.M. Harrowfield, B.W. Skelton, and A.H. White, *Aust. J. Chem.*, **48**, 1311 (1995).

Table 6.12   Potassium aryloxides

Compound	Aryloxide	Bond length (Å) M–O	Ref.
[(dicyclohexano-18-crown-6)K$_2$(μ_2-OAr)$_2$]$_n$ 6-coord. K	μ_2-OC$_6$H$_5$	2.591 (1), 2.611 (2)	i
[(dicyclohexano-18-crown-6)K(OAr)(HOAr)]$_n$ 8-coord. K, OAr hydrogen bonded to HOAr coordinated to next K-atom	OC$_6$H$_5$.HOC$_6$H$_5$	3.103 (4)	ii
[K(μ_2-OAr)(THF)$_3$]$_2$ 6-coord. K, bridging OAr and THF	μ_2-OC$_6$H$_2$(CF$_3$)$_3$-2,4,6	2.752 (3), 2.833 (3)	iii
[K(OAr)]$_n$ 8-coord. K, polymeric structure	μ_2-OC$_6$H$_2$(NO$_2$)$_3$-2,4,6	2.738 (1)	iv
[(18-crown-6)K(OAr)] 8-coord. K	OC$_6$H$_2$(NO$_2$)$_3$-2,4,6	2.741 (3)	v
cis-syn-cis-[(dicyclohexano-18-crown-6)K(OAr)] 7-coord. K	OC$_6$H$_4$(NO$_2$)-4	2.660 (?)	vi
cis-syn-cis-[(dicyclohexane-18-crown-6)K(OAr)]	OC$_6$H$_4$(NO$_2$)-2	2.660 (?)	vii
cis-anti-cis-[(dicyclohexano-18-crown-6)K(OAr)] 7-coord. K	OC$_6$H$_4$(NO$_2$)-2	2.628 (?)	vii

[i]M.E. Fraser, S. Fortier, A. Rodrigue, and J.W. Bovenkamp, *Can. J. Chem.*, **64**, 816 (1986).

[ii]M.E. Fraser, S. Fortier, M.K. Markiewicz, A. Rodrigue, and J.W. Bovenkamp, *Can. J. Chem.*, **65**, 2558 (1987).

[iii]S. Brooker, F.T. Edelmann, T. Kottke, H.W. Roesky, G.M. Sheldrick, D. Stalke, and K.H. Whitmire, *J. Chem. Soc., Chem. Commun.*, 144 (1991).

[iv]J.M. Harrowfield, B.W. Skelton, and A.H. White, *Aust. J. Chem.*, **48**, 1311 (1995).

[v]J.C. Barnes and J. Collard, *Acta Cryst. C*, **44**, 565 (1988).

[vi]L.R. Caswell, J.E. Hardcastle, T.A. Jordan, I. Alam, K.A. McDowell, C.A. Mahan, F.R. Fronczek, and R.D. Gandour, *J. Inclusion Phenomena*, **13**, 37 (1992).

[vii]F.R. Fronczek, R.D. Gandour, L.M.B. Gehrig, L.R. Caswell, K.A. McDowell, and I. Alam, *J. Inclusion Phenomena*, **5**, 379 (1987).

Table 6.13   Rubidium and caesium aryloxides

Compound	Aryloxide	Bond length (Å) M–O	Ref.
[Rb(OAr)]$_n$ 8-coord. Rb, polymeric structure	μ_2-OC$_6$H$_2$(NO$_2$)$_3$-2,4,6	2.927 (3), 3.189 (3)	i
[Cs(OAr)]$_n$ 8-coord. Cs, polymeric structure	μ_2-OC$_6$H$_2$(NO$_2$)$_3$-2,4,6	3.087 (3), 3.372 (3)	i
[Cs(OAr)]$_n$ infinite chains with Cs–O and Cs-η^6-arene interactions	μ_2-OC$_6$H$_3$Pr$^i_2$-2,6	2.875 (4), 2.898 (5)	ii

[i]J.M. Harrowfield, B.W. Skelton, and A.H. White, *Aust. J. Chem.*, **48**, 1311 (1995).

[ii]D.L. Clark, D.R. Click, R.V. Hollis, B.L. Scott, and J.G. Watkin, *Inorg. Chem.*, **37**, 5700 (1998).

In the presence of THF the parent lithium phenoxide has been shown to adopt a hexameric prismatic structure $[Li(\mu_3\text{-OPh})(THF)]_6$ in the solid state.[278] The $[M_4(\mu_3\text{-OAr})_4(L)_x]$ cube appears as a common structural motif for a number of adducts of Li and Na aryloxides (Tables 6.10–6.11). Sometimes the adduct bonds to the metal atom are formed intramolecularly *via* donor groups attached to the phenoxide nucleus, *e.g.* in $[Na\{\mu_3\text{-OC}_6H_2(CH_2NMe_2)_2\text{-2,6-Me-4}\}]_4$.[279] As the bulk of the aryloxide is increased (*ortho*-substituents) the degree of aggregation drops, *e.g.* dinuclear $[Li(\mu_2\text{-OC}_6H_3Bu_2^t\text{-2,6})(THF)]_2$[280] and $[(C_6H_3Bu_2^t\text{-2,6-O})(dmso)Na(\mu_2\text{-dmso})_2Na(dmso)(OC_6H_3Bu_2^t\text{-2,6})]$.[281] The latter compound is unusual in that the aryloxide is terminal with bridging DMSO. Simple monomeric aryloxides can be obtained by the use of multidentate ligands such as crown ethers.

Studies have also been made of the structure of group 1 metal aryloxides in solution. A particularly thorough piece of work is the investigation of the degree of aggregation of lithium aryloxides in various media by ^{7}Li and other NMR techniques by Jackman and co-workers.[278,282–285] Hexamers, tetramers, and dimers have been shown to be present depending on the aryloxide bulk and solvent basicity. In the case of 2-alkylphenoxides $(OC_6H_4R\text{-2})$ as an example in THF, the ratio of dimer (D) to tetramer (T) varies as follows: R = Me, D:T = 0:100; Et, 6:94; Pr^n, 38:62; Pr^i, 40:60; Bu^t, 100:0.[282]

6.2.2 Group 2 Metal Aryloxides

There has been an outburst of research interest in the structures, physical properties, and chemistry of the group 2 metal aryloxides. This is particularly true for the elements strontium and barium where work has been stimulated by the possible use of metal aryloxides as precursors (either via sol–gel or MOCVD processes) for the formation of binary and ternary oxides containing these metals.[286] Synthetic procedures are based on either the halide (Be) alkyl derivatives (Mg, Grignard derivatives, etc.) or the actual metallic element (Ca, Sr, Ba). Structural studies (Tables 6.14–6.18) show for the smaller elements Be, Mg, and Ca that monomeric and dimeric structural motifs dominate, with rarer examples of trinuclear clusters, *e.g.* $[Ca_3(OPh)_5(HMPA)_6][OPh.2HOPh]$.[287] In the case of strontium and barium a more extensive cluster chemistry has been developed for small aryloxide ligands, while monomeric units with terminal aryloxides can be formed with bulky ligands and sufficient additional Lewis bases, *e.g.* $[Ba(OC_6H_2Bu_2^t\text{-2,6-Me-4})_2(THF)_3]$.[288]

Table 6.14   Beryllium aryloxides

Compound	Aryloxide	Bond length (Å) M–O	Bond angle (°) M–O–Ar	Ref.
$[Be(OAr)_2(OEt_2)]$ trigonal planar	$OC_6H_2Bu_3^t\text{-2,4,6}$	1.481 (2)	147	i

^{i}K. Ruhlandt-Senge, R.A. Bartlett, M.M. Olmstead, and P.P. Power, *Inorg. Chem.*, **32**, 1724 (1993).

Table 6.15   Magnesium aryloxides

Compound	Aryloxide	Bond length (Å) M–O	Bond angle (°) M–O–Ar	Ref.
[Mg(μ_2-OAr)$_2$]$_2$	OC$_6$H$_3$Bu$_2^t$-2,6	1.823 (3)	178	i
planar Mg$_2$O$_2$ ring,		1.819 (3)	169	
4-coord. Mg	μ_2-OC$_6$H$_3$Bu$_2^t$-2,6	1.951 (3)–1.971 (3)		
[Mg(OAr)$_2$(THF)$_3$]	OC$_6$H$_2$(CF$_3$)$_3$-2,4,6	1.917 (4)	167	ii
tbp, equatorial OAr		1.923 (4)	179	
[Mg(OAr)$_2$(py)$_2$]	η^2-OC$_6$H$_3$(NO$_2$)$_2$-2,4	1.966 (5)	133	iii
6-coord. Mg		1.940 (5)	136	
[Mg(OAr)$_2$(N-methyl-imidazole)$_2$]	η^2-OC$_6$H$_3$(NO$_2$)$_2$-2,4	1.964 (5)	135	iii
6-coord. Mg		1.962 (5)	133	

[i]J.C. Calabrese, M.A. Cushing Jr, and S.D. Ittel, *Inorg. Chem.*, **27**, 867 (1988).
[ii]H.W. Roesky, M. Scholz, and M. Noltemeyer, *Chem. Ber.*, **123**, 2303 (1990).
[iii]R. Sarma, F. Ramirez, P. Narayanan, B. McKeever, and J.F. Marecek, *J. Am. Chem. Soc.*, **101**, 5015 (1979).

Table 6.16   Calcium aryloxides

Compound	Aryloxide	Bond length (Å) M–O	Bond angle (°) M–O–Ar	Ref.
[Ca$_3$(OAr)$_5$(HMPA)$_6$][OAr.2HOAr]	μ_2-OC$_6$H$_5$	2.33 (1)–2.36 (1)		i
triangular cluster, 6-coord. Ca	μ_3-OC$_6$H$_5$	2.38 (1)–2.44 (1)		
[Ca(OAr)$_2$(THF)$_3$]	OC$_6$H$_2$Bu$_2^t$-2,6-Me-4	2.201 (6)	175	ii
tbp, equatorial OAr		2.210 (6)	177	
[Ca(OAr)$_2$(THF)$_3$]	OC$_6$H$_2$Bu$_2^t$-2,6-Me-4	2.197 (3)	171	iii
tbp, equatorial OAr at −172°C		2.181 (3)	174	
[(glycol-5)Ca(OAr)](OAr)(OH$_2$)	η^2-OC$_6$H$_2$(NO$_2$)$_3$-2,4,6	2.330 (2)		iv
8-coord. Ca				
[Ca$_2$(OAr)$_2$(OH$_2$)$_{10}$](2OAr)(2H$_2$O)	η^2-OC$_6$H$_3$(NO$_2$)$_2$-2,4	2.361 (5)		v
8-coord. Ca				
[Ca(OAr)$_2$(bipy)$_2$]	η^2-OC$_6$H$_2$(NO$_2$)$_3$-2,4,6	2.292 (10)		vi
8-coord. Ca		2.351 (9)		

[i]K.G. Caulton, M.H. Chisholm, S.R. Drake, K. Folting, J.C. Huffman, and W.E. Streib, *Inorg. Chem.*, **32**, 1970 (1993).
[ii]P.B. Hitchcock, M.F. Lappert, G.A. Lawless, and B. Royo, *J. Chem. Soc., Chem. Commun.*, 1141 (1990).
[iii]K.F. Tesh, T.P. Hanusa, J.C. Huffman, and C.J. Huffman, *Inorg. Chem.*, **31**, 5572 (1992).
[iv]T.P. Singh, R. Reinhardt, and N.S. Poonia, *Ind. J. Chem. A*, **23**, 976 (1984).
[v]L.B. Cole and E.M. Holt, *J. Chem. Soc., Perkin Trans.*, 1997 (1986).
[vi]N.S. Poonia, R. Chandra, V.M. Padmanabhan, and V.S. Yadav, *J. Coord. Chem.*, **21**, 167 (1990).

Table 6.17   Strontium aryloxides

Compound	Aryloxide	Bond length (Å) M–O	Bond angle (°) M–O–Ar	Ref.
[Sr$_4$(OAr)$_8$(HOAr)$_2$(THF)$_6$]	OC$_6$H$_5$	2.450 (7)		i
6-coord. Sr	μ_2-OC$_6$H$_5$	2.404 (7)–2.495 (7)		
	μ_3-OC$_6$H$_5$	2.512 (7)–2.558 (6)		
[Sr(OAr)$_2$(THF)$_3$]	OC$_6$H$_2$Bu$^t_3$-2,4,6	2.306 (5)	175	ii
tbp, equatorial OAr		2.323 (5)	176	
[Sr$_3$(OAr)$_6$(HMPA)$_5$]	OC$_6$H$_5$	2.34 (3)	169	iii
	μ_2-OC$_6$H$_5$	2.38 (3)–2.53 (2)		
	μ_3 OC$_6$H$_5$	2.52 (2)–2.65 (2)		
[Ba$_2$Sr$_6$(OAr)$_{14}$(O)$_2$(H)$_2$(HMPA)$_6$]	μ_2-OC$_6$H$_5$	2.52 (3)–2.62 (4)		iv
two fused Sr$_3$Ba$_2$ square	μ_3-OC$_6$H$_5$	2.44 (3)–2.71 (3)		
pyramids with μ_5-O, 6-coord.				
Sr and 6- and 8-coord. Ba				
[Sr(OAr)$_2$].5H$_2$O	μ_2-OC$_6$H$_2$(NO$_2$)$_3$-	2.604 (4), 2.843 (4)		v
8-coord. Sr, polymeric structure	2,4,6			

[i] S.R. Drake, W.E. Streib, M.H. Chisholm, and K.G. Caulton, *Inorg. Chem.*, **29**, 2707 (1990).
[ii] S.R. Drake, D.J. Otway, M.B. Hursthouse, and K.M.A. Malik, *Polyhedron*, **11**, 1995 (1992).
[iii] K.G. Caulton, M.H. Chisholm, S.R. Drake, K. Folting, J.C. Huffman, and W.E. Streib, *Inorg. Chem.*, **32**, 1970 (1993).
[iv] K.G. Caulton, M.H. Chisholm, S.R. Drake, K. Folting, and J.C. Huffman, *Inorg. Chem.*, **32**, 816 (1993).
[v] J.M. Harrowfield, B.W. Skelton, and A.H. White, *Aust. J. Chem.*, **48**, 1311 (1995).

6.2.3   Group 3 Metal and Lanthanide Aryloxides

The synthesis of group 3 metal and lanthanide aryloxides can be achieved utilizing metal halide substrates reacted with alkali metal aryloxides. Homoleptic, base-free metal(III) aryloxides can be formed by this method when the aryloxide is bulky, e.g. [Sc(OC$_6$H$_2$Bu$^t_2$-2,6-Me-4)$_3$],[289] [M(OC$_6$H$_3$Bu$^t_2$-2,6)$_3$] (M = La, Sm),[290] and [Ce(OC$_6$H$_3$Bu$^t_2$-2,6)$_3$][291] (which is also obtained via [Ce{N(SiMe$_3$)$_2$}$_3$].[292] However, in a number of homoleptic derivatives there exists association via π-interaction with the phenoxide nucleus and adjacent metal centres, e.g. [(ArO)$_2$M(μ_2-O-η^6-OAr)$_2$M(OAr)$_2$] (M = La, Sm, Nd; ArO = OC$_6$H$_3$Pr$^i_2$-2,6) (Table 6.19).[293] The reaction of substrates such as [Sm{CH(SiMe$_3$)$_2$}$_3$] (itself prepared from a homoleptic aryloxide) [Cp$_3$Ln] (Ln = Nd, Yb),[91] [Cp$^*_2$Sm(thf)$_2$],[294] and [Sc{N(SiMe$_3$)$_2$}$_2$(THF)$_3$] with phenols has been shown to yield corresponding metal(II) and (III) aryloxides.[295] An important alternative synthetic strategy involves reaction of the lanthanide metal itself with the phenol. This can occur in the presence of aryl-mercury reagents (Eq. 6.12).[296,297] Direct reaction of europium[298] and ytterbium[299] with 2,6-dialkylphenols in liquid ammonia has been reported. Recently N-methylimidazole and acetonitrile have been shown to be suitable solvents for the production of europium aryloxides from the metal.[300] The conversion of ytterbium(III) aryloxides to the corresponding bis-aryloxide has been achieved using Yb metal as the reducing agent.

In the case of 2,6-diphenylphenoxide ancillary ligands, chelation to the electron-deficient metal via π-interactions with *ortho*-phenyl rings has been observed in a large number of cases (Table 6.19). In some compounds only a fraction of the arene ring

Table 6.18 Barium aryloxides

Compound	Aryloxide	Bond length (Å) M–O	Bond angle (°) M–O–Ar	Ref.
[Ba$_6$(OAr)$_{12}$(TMEDA)$_4$] two Ba$_3$ triangles connected by one edge, 6- and 7-coord. Ba	μ_2-OC$_6$H$_5$ μ_3-OC$_6$H$_5$	2.619 (3)–2.676 (3) 2.702 (3)–2.786 (3)		i
[Ba$_5$(OAr)$_9$(H)(O)(THF)$_8$] Ba$_5$ square pyramid with μ_5-O, 7-coord. Ba	OC$_6$H$_5$ μ_2-OC$_6$H$_5$ μ_3-OC$_6$H$_5$	2.54 (1) 2.65 (3) (av.) 2.75 (2) (av.)		ii
[Ba$_5$(OAr)$_9$(OH)(THF)$_5$] Ba$_5$ square pyramid with μ_5-OH, 6- and 7-coord. Ba	OC$_6$H$_5$ μ_2-OC$_6$H$_5$ μ_3-OC$_6$H$_5$	2.551 (23)		iii
[Ba$_4$(OAr)$_6$(O)] Ba$_4$ tetrahedron with μ_4-O, 6- and 7-coord. Ba	μ_2-OC$_6$H$_2$(CH$_2$NMe$_2$)$_3$-2,4,6	2.70 (1), 2.73 (1), 2.74 (1)		iv
[Ba(OAr)$_2$(THF)$_3$] tbp, equatorial OAr	OC$_6$H$_2$Bu$^t_2$-2,6-Me-4	2.38 (1) 2.42 (1)	172 177	v
[Ba$_2$(OAr)$_2$(μ_2-I)$_2$(THF)$_6$] 6-coord. Ba	OC$_6$H$_2$Bu$^t_2$-2,6-Me-4	2.408 (8)	177	vi
[Ba(OAr)$_2$(HMPA)$_2$] distorted tetrahedron	OC$_6$H$_2$Bu$^t_2$-2,6-Me-4	2.414 (8)	178	vii
[Ba(OAr)(OH)(HOAr)(ethanolamine)$_2$]$_n$ 9-coord. Ba	μ_2-OC$_6$H$_3$Bu$^t_2$-3,5	2.676 (11), 2.733 (11)		viii
[Ba$_8$(OAr)$_{14}$(O)$_2$(H)$_2$(HMPA)$_6$] two fused Ba$_5$ square pyramids with μ_5-O, 6- and 8-coord. Ba	μ_2-OC$_6$H$_5$ μ_3-OC$_6$H$_5$	2.581 (12)–2.752 (13) 2.645 (12)–2.858 (13)		ix
[Ba(OAr)$_2$]$_n$ 10-coord. Ba	η^2-OC$_6$H$_4$(NO$_2$)-2	2.711 (3), 2.722 (3)		x
[Ba(OAr)$_2$(1,10-phen)$_3$] 9-coord. Ba	OC$_6$H$_3$(NO$_2$)$_2$-2,4 η^2-OC$_6$H$_3$(NO$_2$)$_2$-2,4	2.602 (4) 2.697 (3)		xi

(continued overleaf)

519

Table 6.18 (*Continued*)

Compound	Aryloxide	Bond length (Å) M–O	Bond angle (°) M–O–Ar	Ref.
[Ba(OAr)$_2$(1,10-phen)$_2$(OCMe$_2$)] 8-coord. Ba	OC$_6$H$_2$(NO$_2$)$_3$-2,4,6	2.702 (3)	143	xii
Ba	η^2-OC$_6$H$_2$(NO$_2$)$_3$-2,4,6	2.728 (3)	128	xiii
[(diaza 21-crown-7)Ba(OAr)$_2$] 11-coord. Ba	η^2-OC$_6$H$_2$(NO$_2$)$_3$-2,4,6	2.62 (1) 2.72 (1)	143 120	viii
[(18-crown-6)Ba(OAr)$_2$](2HOAr)(18-crown-6) 8-coord. Ba	OC$_6$H$_3$Bu^t-3,5	2.570 (8)		viii
[Ba(OAr)$_2$].6H$_2$O 10-coord. Ba, polymeric structure	μ_2-OC$_6$H$_2$(NO$_2$)$_3$-2,4,6	2.771 (3)		xiv

[i] K.G. Caulton, M.H. Chisholm, S.R. Drake, K. Folting, J.C. Huffman, and W.E. Streib, *Inorg. Chem.*, **32**, 1970 (1993).
[ii] K.G. Caulton, M.H. Chisholm, S.R. Drake, and K. Folting, *J. Chem. Soc., Chem. Commun.*, 1349 (1990).
[iii] P. Miele, J.-D. Foulon, N. Hovnanian, and L. Cot, *Polyhedron*, **12**, 267 (1993).
[iv] K.F. Tesh and T.P. Hanusa, *J. Chem. Soc., Chem. Commun.*, 879 (1991).
[v] K.F. Tesh, T.P. Hanusa, J.C. Huffman, and C.J. Huffman, *Inorg. Chem.*, **31**, 5572 (1992).
[vi] K.F. Tesh, D.J. Burkey, and T.P. Hanusa, *J. Am. Chem. Soc.*, **116**, 2409 (1994).
[vii] T.R. Balderrain, J.P. Espinos, A. Fernandez, A.R. Gonzalez-Elipe, D. Leinen, A. Monge, M. Paneque, C. Ruiz, and E. Carmona, *J. Chem. Soc., Dalton Trans.*, 1529 (1995).
[viii] P. Miele, J.-D. Foulon, and N. Hovnanian, *Polyhedron*, **12**, 209 (1993).
[ix] K.G. Caulton, M.H. Chisholm, S.R. Drake, K. Folting, and J.C. Huffman, *Inorg. Chem.*, **32**, 816 (1993).
[x] J.A. Kanters, W.J.J. Smeets, A.J.M. Duisenberg, K. Venkatasubramanian, and N.S. Poonia, *Acta Cryst. C*, **40**, 1699 (1984).
[xi] J.A. Kanters, R. Postma, A.J.M. Duisenberg, K. Venkatasubramanian, and N.S. Poonia, *Acta Cryst. C*, **39**, 1519 (1983).
[xii] R. Postma, J.A. Kanters, A.J.M. Duisenberg, K. Venkatasubramanian, and N.S. Poonia, *Acta Cryst. C*, **39**, 1221 (1983).
[xiii] C.J. Chandler, R.W. Gable, J.M. Gulbis, and M.F. Mackay, *Aust. J. Chem.*, **41**, 799 (1988).
[xiv] J.M. Harrowfield, B.W. Skelton, and A.H. White, *Aust. J. Chem.*, **48**, 1311 (1995).

Table 6.19 Group 3 metal and lanthanide metal aryloxides

Compound	Aryloxide	Bond length (Å) M–O	Bond angle (°) M–O–Ar	Ref.
[Sc(OAr)₃] trigonal planar Sc	OC₆H₂Bu^t₂-2,6-Me-4	1.889 (4) 1.853 (4) 1.864 (4)	173 164 168	i
[Sc(OAr)₂(OH₂)₄](OAr).HOAr	OC₅H₂(NO₂)₃-2,4,6	2.019 (6) 2.046 (6)	148 148	ii
[ClY(μ₂-OAr)₃Y(μ₂-OAr)₃Na]	μ₂-OC₆H₄(CH₂NMe₂)₂-2	Y–O–Y 2.255 (6), 2.360 (6) 2.256 (6), 2.391 (6) 2.236 (7), 2.411 (6) Y–O–Na 2.111 (6), 2.660 (9) 2.096 (7), 2.557 (7) 2.100 (6), 2.661 (8)		iii, iv
[Cp*Y(OAr)₂]	OC₆H₃Bu^t₂-2,6	2.096 (4) 2.059 (3)	129 168	v
[(COT)(THF)Y(μ₂-OAr)₂Y(THF)(COT)]	μ₂-OC₆H₅	2.250 (5)–2.360 (5)		vi
[Y(OAr)(OH₂)₇][OAr)₂·4.5H₂O	OC₆H₂(NO₂)₃-2,4,6	2.244 (5)	145	vii
[(ArO)₂La(μ₂-O-η⁶-OAr)₂La(OAr)₂]	OC₆H₃Pr^i₂-2,6 μ₂-O-η⁶-C₆H₃Pr^i₂-2,6	2.187 (5) 2.198 (5) 2.273 (5)	161 155 165	viii

[i] P.B. Hitchcock, M.F. Lappert, and A. Singh, *J. Chem. Soc., Chem. Commun.*, 1499 (1983).
[ii] J.M. Harrowfield, B.W. Skelton, and A.H. White, *Aust. J. Chem.*, **47**, 397 (1994).
[iii] M.P. Hogerheide, J.T.B.H. Jastrzebski, J. Boersma, W.J.J. Smeets, A.L. Spek, and G. van Koten, *Inorg. Chem.*, **33**, 4431 (1994).
[iv] M.P. Hogerheide, S.N. Ringelberg, D.M. Grove, J.T.B.H. Jastrzebski, J. Boersma, W.J.J. Smeets, A.L. Spek, and G. van Koten, *Inorg. Chem.*, **35**, 1185 (1996).
[v] C.J. Schaverien, J.H.G. Frijns, H.J. Heeres, J. R. van den Hende, J.H. Teuben, and A.L. Spek, *J. Chem. Soc., Chem. Commun.*, 642 (1991).
[vi] H. Schumann, J. Winterfeld, R.D. Kohn, L. Esser, Junquan Sun, and A. Dietrich, *Chem. Ber.*, **126**, 907 (1993).
[vii] J.M. Harrowfield, L. Weimin, B.W. Skelton, and A.H. White, *Aust. J. Chem.*, **47**, 339 (1994).
[viii] R.J. Butcher, D.L. Clark, S.K. Grumbine, R.L. Vincent, R.V. Hollis, B.L. Scott and J.G. Watkin, *Inorg. Chem.*, **34**, 5468 (1995).

(continued overleaf)

Table 6.19 (*Continued*)

Compound	Aryloxide	Bond length (Å) M–O	Bond angle (°) M–O–Ar	Ref.
[La(OAr)$_3$(NH$_3$)$_4$] capped pseudo-octahedral geometry	OC$_6$H$_3$Pr$^i_2$-2,6	2.259 (4)	166	viii
[La(OAr)$_3$(THF)$_2$] tbp, axial THF	OC$_6$H$_3$Pr$^i_2$-2,6	2.233 (8) 2.227 (7) 2.169 (7)	173 152 173	viii
[La(OAr)$_3$(THF)$_2$].2THF distorted tbp, axial OAr and THF	OC$_6$H$_3$Ph$_2$-2,6	2.253 (6) 2.229 (6) 2.239 (6)	171 153 158	ix, x
[La(OAr)$_3$(tetraglyme)] 8-coord. La	OC$_6$H$_3$Me$_2$-2,6	2.28 (2) 2.30 (2) 2.30 (2)	172 173 175	xi
[(THF)Li(μ_2-OAr)$_2$La(OAr)$_2$(THF)]	OC$_6$H$_3$Pr$^i_2$-2,6 μ_2-OC$_6$H$_3$Pr$^i_2$-2,6	2.213 (3), 1.86 (1) 2.204 (3), 1.879 (5)	172 171	xii
[(THF)$_2$Na(μ_2-OAr)$_2$La(OAr)$_2$(THF)$_2$]	OC$_6$H$_3$Pr$^i_2$-2,6 μ_2-OC$_6$H$_3$Pr$^i_2$-2,6	2.224 (6) 2.229 (6) 2.323 (5), 2.362 (6) 2.343 (5), 2.339 (5)	171 172	xii
[Cs(π-Ar-O)$_2$La(OAr)$_2$] two phenoxides π-bound to Cs	OC$_6$H$_3$Pr$^i_2$-2,6 O-π-C$_6$H$_3$Pr$^i_2$-2,6	2.251 (4) 2.231 (5)	150 156	xii
[LaCu(μ_2-OAr)(HOAr)(μ_2-OH)(O$_2$CCF$_3$)$_3$] tbp Cu, 8-coord. La	μ_2-OC$_6$H$_2$(CH$_2$NMe$_2$)$_2$-2,6-Me-4	2.432 (8), 1.963 (8)		xiii
[Cs$_2$La(OAr)$_5$] 5-coord. La, Cs π-bound to Ar rings to form a polymeric structure	OC$_6$H$_3$Pr$^i_2$-2,6	2.32 (av.)		xiv
[La(OAr)$_2$(OH$_2$)$_6$](OAr).6H$_2$O	OC$_6$H$_2$(NO$_2$)$_3$-2,4,6	2.44 (1) 2.440 (9)	144 143	xv

[Me₄N][La₂Na₂(μ₄-OAr)(μ₃-OAr)₂(μ₂-OAr)₄(OAr)₂(THF)₅].THF	OC_6H_4Me-4	2.296 (3) 2.284 (3)	172 172	xvi
	μ_2-OC_6H_4Me-4	2.363 (3)–2.385 (3)		
	μ_3-OC_6H_4Me-4	2.543 (3)–2.558 (3)		
	μ_4-OC_6H_4Me-4	2.549 (3)–2.568 (3)		
[La₂Na₃(μ₄-OAr)₃(μ₂-OAr)₆(dioxane)₅].dioxane	μ_2-OC_6H_4Me-4	2.326 (4)–2.349 (4)		xvi
tbp cluster, axial La, equatorial Na	μ_4-OC_6H_4Me-4	2.527 (4)–2.575 (4)		
[Ce(OAr)₃]	$OC_6H_3Bu^t_2$-2,6	2.140 (5)	166	xvii
pyramidal Ce		2.174 (4)	162	
		2.135 (5)	165	
[Ce(OAr)₃(CNBu^t)₂]	$OC_6H_3Bu^t_2$-2,6	2.239 (5)	176	xvii
tbp, axial C and O		2.225 (7)	175	
		2.231 (7)	153	
[Cp*Ce(OAr)₂]	$OC_6H_3Bu^t_2$-2,6	2.258 (2)	105	xviii, xix
		2.247 (2)	158	
[Ce(OAr)₂(OH₂)₆](OAr)5.75H₂O	$OC_6H_2(NO_2)_3$-2,4,6	2.401 (4)	147	xv
		2.413 (4)	144	

[ix]G.B. Deacon, B.M. Gatehouse, Q. Shen, G.N. Ward, and E.R.T. Tiekink, *Polyhedron*, **12**, 1289 (1993).

[x]G.B. Deacon, Tiecheng Feng, B.W. Skelton, and A.H. White, *Aust. J. Chem.*, **48**, 741 (1995).

[xi]H. C. Aspinall and M. Williams, *Inorg. Chem.*, **35**, 255 (1996).

[xiii]D.L. Clark, R.V. Hollis, B.L. Scott, and J.G. Watkin, *Inorg. Chem.*, **35**, 667 (1996).

[xiii]Liqin Chen, S.R. Breeze, R.J. Rousseau, Suning Wang, and L.K. Thompson, *Inorg. Chem.*, **34**, 454 (1995).

[xiv]D.L. Clark, G.B. Deacon, Tiecheng Feng, R.V. Hollis, B.L. Scott, B.W. Skelton, J.G. Watkin, and A.H. White, *J. Chem. Soc. Chem. Commun.*, 1729 (1996).

[xv]J.M. Harrowfield, Lu Weimin, B.W. Skelton, and A.H. White, *Aust. J. Chem.*, **47**, 321 (1994).

[xvi]W.J. Evans, R.E. Golden, and J.W. Ziller, *Inorg. Chem.*, **32**, 3041 (1993).

[xvii]H.A. Stecher, A. Sen, and A.L. Rheingold, *Inorg. Chem.*, **27**, 1130 (1988).

[xviii]H.J. Heeres, A. Meetsma, and J.H. Teuben, *J. Chem. Soc. Chem. Commun.*, 962 (1988).

[xix]H.J. Heeres, A. Meetsma, J.H. Teuben, and R.D. Rogers, *Organometallics*, **8**, 2637 (1989).

(continued overleaf)

Table 6.19 (*Continued*)

Compound	Aryloxide	Bond length (Å) M–O	Bond angle (°) M–O–Ar	Ref.
[Ce(OAr)₃(Me₄urea)₃] distorted tricapped trigonal prism	OC₆H₂(NO₂)₃-2,4,6	2.442 (5), 2.444 (7) 2.432 (8)		xx
[Ce(OAr)₂(OH₂)₆](OAr).6H₂O two forms, 9-coord. Ce	OC₆H₂(NO₂)₃-2,4,6	I: 2.44 (9), 2.39 (1) II: 2.43 (1), 2.36 (1)		xxi
[Pr(OAr)₂(OH₂)₆](OAr).6H₂O	OC₆H₂(NO₂)₃-2,4,6	2.389 (4) 2.398 (4)	146 146	xv
[Pr(OAr)₂(OH₂)₆](OAr).6H₂O ('aged')	OC₆H₂(NO₂)₃-2,4,6	2.401 (3) 2.418 (3)	145 142	xv
[Pr(OAr)₃(THF)₂] tbp axial THF	OC₆H₃Prⁱ₂-2,6	2.142 (8) 2.158 (9) 2.216 (9)	151 174 173	xxii
[PrCu(μ₂-OAr)(HOAr)(μ₂-OH)(hfacac)₃] tbp Cu, 9-coord. Pr	μ₂-OC₆H₂(CH₂NMe₂)₂-2,6-Me-4	2.424 (8), 1.958 (8)		xxiii
[Pr(OAr)₃(OH₂)₂] distorted pentagonal bipyramid	OC₆H₂(CH₂NMe₂)₃-2,4,6	2.311 (3) 2.280 (3) 2.269 (3)	132 146 163	xxiv
[Nd(OAr)₂(OH₂)₆](OAr).6H₂O]	OC₆H₂(NO₂)₃-2,4,6	2.364 (6) 2.407 (5) 2.407 (5)	149 150 150	xxv
[(ArO)₂Nd(μ₂-O-η⁶-OAr)₂Nd(OAr)₂]	μ₂-O-η⁶-C₆H₃Prⁱ₂-2,6 OC₆H₃Prⁱ₂-2,6	2.124 (9) 2.120 (8) 2.211 (8)	164 164 162	xxii
[Nd(OAr)₃(THF)₂] tbp, axial THF	OC₆H₃Ph₂-2,6			xxii
[Nd(OAr)₃(thf)₂].2THF distorted tbp, OAr and THF axial	OC₆H₃Ph₂-2,6	2.208 (9) 2.16 (1) 2.203 (9)	155 159 170	xxvi
[Nd(OAr)₃(dme)]dme square pyramid, axial OAr, cisoid basal OAr	OC₆H₃Ph₂-2,6	2.198 (5) 2.179 (5) 2.197 (5)	171 154 150	xxvii

Compound / description	Ligand	Distance (Å)	Ref.
[Nd(OC$_6$H$_3$Ph-η^6-C$_6$H$_5$)(OC$_6$H$_3$Ph-η^1-C$_6$H$_5$)(OAr)] pyramidilized Nd due to η^6-C$_6$H$_5$ interaction, also close η^1-C$_6$H$_5$ interaction	OC$_6$H$_3$Ph$_2$-2,6	2.140 (3)	[xxvi]
	OC$_6$H$_3$Ph-η^1-C$_6$H$_5$	2.174 (3)	
	OC$_6$H$_3$Ph-η^6-C$_6$H$_5$	2.193 (5)	
[Nd(OC$_6$H$_3$Ph-η^3-C$_6$H$_5$)(OAr)$_2$(THF)] trigonal pyramid with weaker basal η^3-C$_6$H$_5$ interaction	OC$_6$H$_3$Ph$_2$-2,6	2.186 (2)	[xxvi]
		2.160 (2)	
	OC$_6$H$_3$Ph-η^3-C$_6$H$_5$	2.233 (3)	
[Na(μ_2-OAr)$_3$Nd(OAr)] two independent molecules; in one Na is η^2-, η^2-, and η^1-bound to ortho-Ph groups, in the other it is η^2-, η^2-, and η^2-bound	OC$_6$H$_3$Ph$_2$-2,6	2.208 (4)	[xiv, xxvii]
		2.236 (4)–2.244 (5)	
	μ_2-OC$_6$H$_3$Ph$_2$-2,6	2.181 (6)	
		2.199 (6)–2.278 (4)	
[Na(diglyme)$_2$][Nd(OAr)$_4$] distorted tetrahedral anion	OC$_6$H$_3$Ph$_2$,6	2.175 (4)	[xxvii]
		2.207 (4)	
		2.176 (4)	
		2.191 (4)	
[Cp$_2$Nd(OAr)(THF)$_2$] distorted tbp, axial THF	OC$_6$H$_3$Ph$_2$-2,6	2.239 (8)	[xxix]
[Nd(OAr)$_3$(Me$_4$urea)$_3$] distorted tricapped trigonal prism	OC$_5$H$_2$(NO$_2$)$_3$-2,4,6	2.384 (5), 2.342 (5), 2.382 (5)	[xxx]

(continued overleaf)

[xx] C. Barberato, E.E. Castellano, G. Vicentini, P.C. Isolani, and M.I.R. Lima, *Acta Crystallogr. C*, **50**, 351 (1994).

[xxi] L.F. Delboni, G. Oliva, E.E. Castellano, L.B. Zinner, and S. Braun, *Inorg. Chim. Acta*, **221**, 169 (1994).

[xxii] D.M. Barnhart, D.L. Clark, J.C. Gordon, J.C. Huffman, R.L. Vincent, J.G. Watkin, and B.D. Zwick, *Inorg. Chem.*, **33**, 3487 (1994).

[xxiii] Liqin Chen, S.R. Breeze, R.J. Rousseau, Suning Wang, and L.K. Thompson, *Inorg. Chem.*, **34**, 454 (1995).

[xxiv] S. Daniele, L.G. Hubert-Pfalzgraf, and J. Vaissermann, *Polyhedron*, **14**, 327 (1995).

[xxv] J.M. Harrowfield, Lu Weimin, B.W. Skelton, and A.H. White, *Aust. J. Chem.*, **47**, 321 (1994).

[xxvi] G.B. Deacon, Tiecheng Feng, B.W. Skelton, and A.H. White, *Aust. J. Chem.*, **48**, 741 (1995).

[xxvii] G.B. Deacon, T. Feng, P.C. Junk, B. W. Skelton and A.H. White, *Chem. Ber.*, **130**, 851 (1997).

[xxviii] G.B. Deacon, T. Feng, P.C. Junk, B. W. Skelton and A.H. White, *J. Chem. Soc., Dalton Trans.*, 1181 (1997).

[xxix] G.B. Deacon, S. Nickel and E.R.T. Tiekink, *J Organomet. Chem.*, **409**, C1 (1991).

[xxx] C. Barberato, E.E. Castellano, G. Vicentini, P. C. Isolani and M.I.R. Lima, *Acta Crystallogr. C*, **50**, 351 (1994).

Table 6.19 (*Continued*)

Compound	Aryloxide	Bond length (Å) M–O	Bond angle (°) M–O–Ar	Ref.
[K(μ-Cp)$_2$Nd(O-η^6-Ar)$_2$]$_n$ two-dimensional layered structure, K atom π-bound to 2 Cp and ArO rings	OC$_6$H$_3$Me$_2$-2,6	2.21 (1) 2.19 (1)	147 163	xxxi
[(THF)$_2$(ArO)$_2$Nd(μ_2-OAr)$_2$AlEt$_2$] 6-coord. Nd, 4-coord. Al	OC$_6$H$_3$Me$_2$-2,6 μ_2-OC$_6$H$_3$Me$_2$-2,6	2.153 (9) 2.447 (7), 1.810 (8)	169	xxxii
[K$_3$Nd$_2$(μ_2-OAr)$_6$(μ_4-OAr)$_3$(THF)$_7$] tbp cluster of metal atoms with apical Nd	μ_2-OC$_6$H$_4$Me-4 μ_4-OC$_6$H$_4$Me-4	2.244 (3)–2.305 (3) 2.458 (3)–2.540 (2)		xxxiii
[Nd((μ_2-OAr)$_2$AlMe$_2$]$_4$] octahedral Nd	μ_2-OC$_6$H$_4$Me-4	Nd–O–Al 2.349 (4), 1.845 (7) 2.350 (6), 1.846 (7) 2.368 (4), 1.839 (5)		xxxiii
[Cp*$_2$Sm(OAr)]	OC$_6$HMe$_4$-2,3,5,6	2.13 (1)	172	xxxiv
[Sm(OAr)(OCPhC$_6$H$_6$)$_2$(HMPA)$_2$] tbp, axial HMPA	OC$_6$H$_2$Bu$^t_2$-2,6-Me-4	2.17 (1)	170	xxxv xxxvi
[Sm(OAr)$_2$(OCHPh$_2$)(HMPA)$_2$] tbp, axial HMPA	OC$_6$H$_2$Bu$^t_2$-2,6-Me-4	2.21 (1) 2.22 (1)	170 170	xxxv xxxvi
[Sm(OAr)(OCHPh$_2$)$_2$(HMPA)$_2$] tbp, axial HMPA	OC$_6$H$_3$Bu$^t_2$-2,6	2.202 (4)	171	xxxv xxxvi
[Sm(OAr)$_2$(fluorenoxy)(HMPA)$_2$] tbp, axial HMPA	OC$_6$H$_2$Bu$^t_2$-2,6-Me-4	2.229 (6) 2.243 (5)	174 172	xxxvi
[Sm(OAr)$_2$(OH$_2$)$_6$](OAr).4.75H$_2$O	OC$_6$H$_2$(NO$_2$)$_3$-2,4,6	2.279 (7) 2.370 (6)	161 145	xxxvii
[(ArO)$_2$Sm(μ_2-O-η^6-OAr)$_2$Sm(OAr)$_2$]	OC$_6$H$_3$Pr$^i_2$-2,6 μ_2-O-η^6-C$_6$H$_3$Pr$^i_2$-2,6	2.097 (5) 2.099 (6) 2.195 (5)	164 163 156	xxii
[Sm(OAr)$_3$(THF)$_2$] tbp, axial THF	OC$_6$H$_3$Pr$^i_2$-2,6	2.179 (5) 2.087 (5) 2.166 (5)	175 174 152	xxxviii
fac-[Sm(OAr)$_3$(THF)$_3$] octahedral Sm	OC$_6$H$_3$Pr$^i_2$-2,6	2.158 (2)	164	xxxix

Compound	Ligand	Bond lengths (Å)	Angle (°)	Ref.
[(THF)Li(μ_2-OAr)$_2$Sm(CH$_2$SiMe$_3$)(μ_2-OAr)(μ_2-CH$_2$SiMe$_3$)Li(THF)] square based pyramidal Sm	μ_2-OC$_6$H$_3$Pr$^i_2$-2,6	Sm−O−Li 2.294 (7), 1.89 (1); 2.257 (6), 1.85 (2); 2.246 (6), 1.91 (2)	152	xl
[K(μ_2-OAr)$_2$Sm(O-η^6-Ar)(thf)]$_n$.C$_6$H$_6$ terminal Sm-aryloxide η^6-bound to K of adjacent molecule	O-η^6-C$_6$H$_2$Bu$^t_2$-2,6-Me-4 μ_2-OC$_6$H$_2$Bu$^t_2$-2,6-Me-4	2.336 (7) Sm−O−K 2.362 (6), 2.967 (9); 2.319 (9), 2.778 (6)		xli
[Sm(OAr)$_2$(THF)$_3$].THF tbp, axial THF	OC$_6$H$_2$Bu$^t_2$-2,6-Me-4	2.318 (7) 2.290 (9)	174 171	xlii
[Sm(OAr)$_2$(THF)$_3$] tbp, axial THF	OC$_6$H$_2$Bu$^t_2$-2,6-Me-4	2.331 (11) 2.347 (13)	174 167	xliii, xliv
[Sm(OAr)$_2$(fluorenone ketyl)(THF)$_2$].C$_6$H$_6$ tbp, axial THF	OC$_6$H$_2$Bu$^t_2$-2,6-Me-4	2.165 (4) 2.161 (3)	173 173	xliii, xlv
[Sm(OAr)$_2$(fluorenone ketyl)(HMPA)$_2$].C$_6$H$_6$ tbp, axial HMPA	OC$_6$H$_2$Bu$^t_2$-2,6-Me-4	2.23 (3)	176	xlv

(continued overleaf)

[xxxi]W.J. Evans, M.A. Ansari and S.I. Khan, *Organometallics*, **14**, 558 (1995).

[xxxii]W.J. Evans, M.A. Ansari and J.W. Ziller, *Inorg. Chem.*, **34**, 3079 (1995).

[xxxiii]W.J. Evans, M.A. Ansari and J.W. Ziller, *Polyhedron*, **16**, 3429 (1997).

[xxxiv]W.J. Evans, T.P. Hanusa and K.R. Levan, *Inorg. Chim. Acta*, **110**, 191 (1985).

[xxxv]Zhaomin Hou, T. Yoshimura, and Y. Wakatsuki, *J. Am. Chem. Soc.*, **116**, 11169 (1994).

[xxxvi]T. Yoshimura, Zhaomin Hou and Y. Wakatsuki, *Organometallics*, **14**, 5382 (1995).

[xxxvii]J.N. Harrowfield, B.W. Skelton and A.H. White, *Aust. J. Chem.*, **47**, 359 (1994).

[xxxviii]D.L. Clark, J.C. Gordon, J.G. Watkin, J.C. Huffman, and B.D. Zwick, *Polyhedron*, **15**, 2279 (1996).

[xxxix]Zuowei Xie, Kwoli Chui, Qingchuan Yang, T.C.W. Mak, and Jie Sun, *Organometallics*, **17**, 3937 (1998).

[xl]D.L. Clark, J.C. Gordon, J.C. Huffman, J.G. Watkin, and B.D. Zwick, *Organometallics*, **13**, 4266 (1994).

[xli]W.J. Evans, R. Anwander, M.A. Ansari, and J.W. Ziller, *Inorg. Chem.*, **34**, 5 (1995).

[xlii]GuiZhong Qi, Qi Shen, and YongHua Lin, *Acta Crystallogr. C*, **50**, 1456 (1994).

[xliii]Z. Hou, T. Miyano, H. Yamazaki, and Y. Wakatsuki, *J. Am. Chem. Soc.*, **117**, 4421 (1995).

[xliv]Z. Hou, A. Fujita, T. Yoshimura, A. Jesorka, Y. Zhang, H. Yamazaki, and Y. Wakatsuki, *Inorg. Chem.*, **35**, 7190 (1996).

[xlv]Z. Hou, A. Fujita, Y. Zhang, T. Miyano, H. Yamazaki, and Y. Wakatsuki, *J. Am. Chem. Soc.*, **120**, 754 (1998).

Table 6.19 (*Continued*)

Compound	Aryloxide	Bond length (Å) M–O	Bond angle (°) M–O–Ar	Ref.
[CpSm(OAr)(fluorenone ketyl)(HMPA)] pseudo-tetrahedral Sm	$OC_6H_2Bu^t_2$-2,6-Me-4	2.158 (6)	174	xlv
[{Sm(OAr)$_2$(OEt$_2$)}$_2$(m-fluorenone pinacolate)]	$OC_6H_2Bu^t_2$-2,6-Me-4	2.115 (3), 2.141 (5)	160, 173	xliii, xlv
[(ArO)(thf)$_3$Sm(μ-I)$_2$Sm(OAr)(THF)$_3$] 6-coord. Sm, OAr *trans* to μ-I	$OC_6H_2Bu^t_2$-2,6-Me-4	2.300 (10)	179	xliv
[(ArO)$_2$(thf)Sm(μ-Cl)$_2$Sm(OAr)$_2$(THF)] 5-coord. Sm, square pyramid with axial OAr	$OC_6H_2Bu^t_2$-2,6-Me-4	2.135 (6), 2.110 (7)	165, 167	xliv
[Cp*Sm(OAr)(hmpa)$_2$] psuedo-tetrahedral Sm	$OC_6H_2Bu^t_2$-2,6-Me-4	2.345 (4)	164	xliv
[Sm(OAr)$_2$(I)(thf)$_2$] tbp, axial THF	$OC_6H_2Bu^t_2$-2,6-Me-4	2.153 (7)	164	xliv
[Cp*Sm(μ-OAr)$_2$SmCp*]	μ_2-$OC_6H_2Bu^t_3$-2,4,6	2.425 (5), 2.512 (6)		xlvi
[Cp*$_2$Sm(OAr]	μ_2-$OC_6H_2Bu^t_3$-2,4,6	2.144 (5)	166	xlvi
[(μ,η^5-Cp*)Sm(OAr)(μ-η^5-Cp*)K(thf)$_2$]$_\infty$ infinite chain polymer possible agostic Sm–methyl interaction	μ_2-$OC_6H_3Bu^t_2$-2,6	2.29 (2)	175	xlvi
[(μ,η^5-Cp*)Sm(OAr)(μ-η^5-Cp*)K(thf)$_2$]$_\infty$ possible agostic Sm–methyl interaction	μ_2-$OC_6H_2Bu^t_2$-2,6-Me-4	2.330 (6)	127	xlvi
[Eu(OAr)$_2$(THF)$_3$].THF tbp, axial THF	$OC_6H_2Bu^t_2$-2,6-Me-4	2.321 (5), 2.337 (5)	174, 175	xlvii
[Eu(OAr)$_2$(NCMe$_4$] distorted octahedron	$OC_6H_3Bu^t_2$-2,6	2.313 (12), 2.35 (2)	174, 175	xlviii
[Eu(μ_2-OAr)$_3$Eu(OAr)].toluene Contains three η^1-Ph–Eu and three η^2-Ph–Eu contacts	$OC_6H_3Ph_2$-2,6 μ_2-$OC_6H_3Ph_2$-2,6	2.361 (3) 2.493 (2), 2.463 (3) 2.517 (3), 2.426 (3) 2.428 (2), 2.438 (2)	155	xlix

Compound	OAr		∠(°)	Ref.
[Eu(μ₂-OAr)₄(OAr)₂(μ₃-OH)₂(NCMe)₆] butterfly arrangement of Eu atoms	OC₆H₃Pr$^i_2$-2,6 μ₂-OC₆H₃Pr$^i_2$-2,6	2.284 (6)–2.299 (6) 2.403 (6)–2.543 (6)		xlviii
[(DME)(ArO)Eu(μ₂-OAr)₃Eu(DME)₂]	OC₆H₃Me₂-2,6 μ₂-OC₆H₃Me₂-2,6	2.350 (5) 2.447 (5)–2.597 (5)	167	l
[Eu(OAr)₂(OH₂)₆](OAr).6H₂O	OC₆H₂(NO₂)₃-2,4,6	2.327 (7) 2.372 (7)	151 151	xv
[(L)₃Eu(μ₂-OAr)₃Eu(L)₂(OAr)] L = N-methylimidazole; face-shared bi-octahedron	OC₆H₃Me₂-2,6 μ₂-OC₆H₃Me₂-2,6	2.365 (3) 2.479 (3)–2.632 (3)	158	li
[(L)₂(ArO)Eu(μ₂-OAr)₂(μ₂-L)Eu(L)₂(OAr)] L = NCMe; face-shared bi-octahedron	OC₆H₃Me₂-2,6 μ₂-OC₆H₃Me₂-2,6	2.288 (3) 2.479 (3)–2.632 (3)	178	li
{[Eu(μ₃-η²-OCH₂CH₂OMe)(η²-OCH₂CH₂OMe)(OAr)][H]}₄ tetrahedral Eu	OC₆H₃Me₂-2,6 OC₆H₃Pr$^i_2$-2,6 derivative isostructural but poorly refined	2.420 (8)	154	lii
[Gd(OAr)₃(THF)₂] tbp, axial THF	OC₆H₃Pr$^i_2$-2,6	2.138 (9) 2.156 (11) 2.096 (9)	153 173 176	xxii
[Gd(OAr)₂(OH₂)₆](OAr).6H₂O three forms	OC₆H₂(NO₂)₃-2,4,6	Form I 2.309 (7) 2.363 (7) Form II 2.32 (1) 2.38 (1) Form III 2.34 (2) 2.38 (2)	151 150 148 148 148 147	liii

xlvi Z. Hou, Y. Zhang, T. Yoshimura, and Y. Wakatsuki, *Organometallics*, **16**, 2963 (1997).

xlvii J. R. van den Hende, P.B. Hitchcock, S.A. Holmes, M.F. Lappert, WingPor Leung, T.C.W. Mak, and S. Prashar, *J. Chem. Soc., Dalton Trans.*, 1427 (1995).

xlviii W.J. Evans, M.A. Greci, and J.W. Ziller, *J. Chem. Soc., Dalton Trans.*, 3035 (1997).

xlix G.B. Deacon, C.M. Forsyth, P.C. Junk, B.S. Skelton, and A.H. White, *Chem. Eur. J.*, **5**, 1452 (1999).

l W.J. Evans, W.G. McClelland, M.A. Greci, and J.W. Ziller, *Eur. J. Solid State Inorg. Chem.*, **33**, 145 (1996).

li W.J. Evans, M.A. Greci, and J.W. Ziller, *J. Chem. Soc., Chem. Commun.*, 2367 (1998).

lii W.J. Evans, M.A. Greci, and J.W. Ziller, *Inorg. Chem.*, **37**, 5221 (1998).

liii J.M. Harrowfield, L. Weimin, B.W. Skelton, and A.H. White, *Aust. J. Chem.*, **47**, 349 (1994).

(continued overleaf)

Table 6.19 *(Continued)*

Compound	Aryloxide	Bond length (Å) $M-O$	Bond angle (°) $M-O-Ar$	Ref.
$[Tb(OAr)_2(OH_2)_6](OAr).6H_2O$	$OC_6H_2(NO_2)_3$-2,4,6	2.302 (5) 2.344 (5)	152 150	liii
$[Dy(OAr)(OH_2)_7](OAr)_2.4.5H_2O$	$OC_6H_2(NO_2)_3$-2,4,6	2.250 (5)	145	vii
$[Er(OAr)_3(THF)_2]$ tbp, axial THF	$OC_6H_3Pr^i_2$-2,6	2.072 (10) 2.073 (10) 2.090 (10)	175 173 151	xxii
$[Na(diglyme)_2][Er(OAr)_4]$ distorted tetrahedral anion	$OC_6H_3Ph_2$-2,6	2.068 (6) 2.090 (6) 2.088 (6) 2.080 (6)	162 166 176 171	xxxviii
$[Er(OAr)(OH_2)_7](OAr)_2.4.5H_2O$	$OC_6H_2(NO_2)_3$-2,4,6	2.235 (2)	144	vii
$[Yb(OAr)_3(thf)]$ distorted tetrahedron	$OC_6H_2Bu^t_3$-2,4,6	1.97 (1) 2.02 (1) 2.09 (1)	173 150 148	liv
$[(THF)(ArO)_2Yb(\mu_2\text{-}OH)_2Yb(OAr)_2(THF)].THF$	$OC_6H_2Bu^t_3$-2,4,6	2.047 (7) 2.073 (6)	154 171	liv
$[Yb(OC_6H_3Ph\text{-}\eta\text{-}\eta^6\text{-}C_6H_5)(OC_6H_3Ph\text{-}\eta^1\text{-}C_6H_5)(OAr)]$ trigonal pyramid of O atoms about Yb, possible agostic interaction with CH bond	$OC_6H_3Ph_2$-2,6 $OC_6H_3Ph\text{-}\eta^1\text{-}C_6H_5$ $OC_6H_3Ph\text{-}\eta^6\text{-}C_6H_5$	2.031 (2) 2.061 (4) 2.104 (4)	169 149 135	lv
$[(ArO)Yb(\mu_2\text{-}OAr)_2Yb(OAr)].1.5$toluene two independent molecules; contains η^1- and η^6-Ph interactions to one Yb and η^6-Ph to the other Yb	$OC_6H_3Ph_2$-2,6 $\mu_2\text{-}OC_6H_3Ph_2$-2,6	2.10 (2)–2.17 (2) 2.24 (2)–2.35 (2)		xlix
$[Yb_2(\mu_2\text{-}OAr)_3][Yb(OAr)_4]$.toluene cation contains η^1-, η^1-, and η^6-Ph interactions to one Yb and η^1-, η^2-, and η^6-Ph to the other Yb; tetrahedral anion	$OC_6H_3Ph_2$-2,6 $\mu_2\text{-}OC_6H_3Ph_2$-2,6	2.094 (7) 2.049 (7) 2.073 (8) 2.053 (6) 2.289 (7)–2.327 (7)	155 166 163 164	xlix

Compound	OAr	M–O (Å)	∠	Ref.
[Yb(OAr)₃(THF)₂].THF square pyramid, axial OAr, transoid basal OAr	OC₆H₃Ph₂-2,6	2.038 (7) 2.111 (8) 2.084 (9)	168 168 164	lv, lvi
[Yb(OAr)₃(DME)]0.5DME square pyramid, axial OAr, cisoid basal OAr	OC₆H₃Ph₂-2,6	2.062 (5) 2.051 (5) 2.084 (4)	171 161 148	xxvii
[Cp₂Yb(OAr)(THF)]	1-naphtholato	2.075 (5)	151	lvii
[MeCpYb(OAr)₂(THF)]	OC₆H₂Bu$^t_2$-2,6-Me-4	2.040 (4) 2.078 (4)	168 179	lviii
[Yb(OAr)₂(HMPA)₂] tetrahedral Yb	OC₆H₃Bu$^t_2$-2,6	2.179 (8)	171	lix, lx
[Yb(OAr)(OCPhC₆H₆)₂(HMPA)₂] tbp, axial HMPA	OC₆H₂Bu$^t_2$-2,6-Me-4	2.093 (8)	178	lx
[Yb(OAr)(OCPh₂)₂(HMPA)₂] tbp, axial HMPA	OC₆H₂Bu$^t_2$-2,6-Me-4	2.119 (5)	171	xlv
[Yb(OAr)₂(fluorenone ketyl)(THF)₂] tbp, axial THF	OC₆H₂Bu$^t_2$-2,6-Me-4	2.082 (8) 2.083 (8)	173 172	xlv
[Yb(OAr)₂(THF)₃].THF distorted square pyramid, axial THF	OC₆H₂Bu$^t_3$-2,4,6	2.21 (1) 2.22 (1)	174 177	lxi
[Yb(OAr)₂(OEt₂)₂].OEt₂ tetrahedral Yb	OC₆H₂Bu$^t_2$-2,6-Me-4	2.126 (9) 2.182 (8)	171 167	lxii, lxiii
[Yb(OAr)₂(THF)₂] tetrahedral Yb	OC₆H₂Bu$^t_2$-2,6-Me-4	2.14 (1) 2.135 (9)	173 178	lxii

[liv] G.B. Deacon, Tiecheng Feng, S. Nickel, M.I. Ogden, and A.H. White, Aust. J. Chem., **45**, 671 (1992).

[lv] G.B. Deacon, S. Nickel, P. MacKinnon, and E.R.T. Tiekink, Aust. J. Chem., **43**, 1245 (1990).

[lvi] G.B. Deacon, Tiecheng Feng, S. Nickel, M.I. Ogden, and A.H. White, Aust. J. Chem., **45**, 671 (1992).

[lvii] X.G. Zhou, Z.Z. Wu, and Z.S. Jin, J. Organomet. Chem., **431**, 289 (1992).

[lviii] Yingming Yao, Qi Shen, Jie Sun, and Feng Xue, Acta Crystallogr. C, **54**, 625 (1998).

[lix] Z. Hou, H. Yamazaki, K. Kobayashi, Y. Fujiwara, and H. Taniguchi, J. Chem. Soc., Chem. Commun., 722 (1992).

[lx] T. Yoshimura, Zhaomin Hou, and Y. Wakatsuki, Organometallics, **14**, 5382 (1995).

[lxi] G.B. Deacon, Tiecheng Feng, P. MacKinnon, R.H. Newnham, S. Nickel, B.W. Skelton, and A.H. White, Aust. J. Chem., **46**, 387 (1993).

[lxii] G.B. Deacon, P.B. Hitchcock, S.A. Holmes, M.F. Lappert, P. MacKinnon, and R.H. Newnham. J. Chem. Soc., Chem. Commun., 935 (1989).

(continued overleaf)

Table 6.19 (*Continued*)

Compound	Aryloxide	Bond length (Å) M–O	Bond angle (°) M–O–Ar	Ref.
[Yb(OAr)₂(THF)₃].THF distorted square pyramid, axial THF	OC₆H₂Buᵗ₂-2,6-Me-4	2.22 (1)	165	lxii
[Yb(OAr)₂(THF)₃] tbp, axial OAr	OC₆H₃Ph₂-2,6	2.20 (2) 2.211 (8)	173 171	xxvii
[Yb(OAr)₂(DME)₂] octahedral, *trans* OAr	OC₆H₃Ph₂-2,6	2.202 (9) 2.259 (5) 2.258 (5)	164 150 156	xxvii
[Yb(tri-OAr)] 7-coord. Yb	{(OC₆H₄CH₂-2)₃}-NH(CH₂)₂N(CH₂)₂NH(CH₂)₂NH	2.160 (3) 2.145 (3) 2.170 (3)	134 134 133	lxiv
[(thf)₂(ArO)₂Yb(μ₂-OAr₂AlMe₂] 6-coord. Yb, 4-coord. Al	OC₆H₃Me₂-2,6 μ₂-OC₆H₃Me₂-2,6	2.06 (1) 2.306 (8), 1.85 (1)	165	lxv
[(ArO)Yb(μ₂-OAr)₂Y(OAr)]	OC₆H₂Buᵗ₂-2,6-Me-4 μ₂-OC₆H₂Buᵗ₂-2,6-Me-4	2.10 (2) 2.08 (2) 2.25 (2), 2.37 (2) 2.31 (2), 2.30 (2)	166 170	lxvi, lxvii
[Lu(OAr)(OH)₇][OAr)₂.4.5H₂O [ClLu(μ₂-OAr)₃Na]	OC₆H₂(NO₂)₃-2,4,6 μ₂-OC₆H₂(CH₂NMe₂)₂-2,6-Me-4	2.206 (5) Lu–O–Na 2.143 (3), 2.550 (4) 2.174 (3), 2.391 (4) 2.150 (3), 2.571 (4)	145	vii iii lxviii
[Lu(OAr)₃(THF)₂] tbp, axial THF	OC₆H₃Prⁱ₂-2,6	2.041 (4) 2.048 (4) 2.042 (3)	174 174 153	xxii

lxiii J.R. van den Hende, P.B. Hitchcock, S.A. Holmes, and M.F. Lappert, *J. Chem. Soc., Dalton Trans.,* 1435 (1995).

lxiv LiWei Yang, Shuang Liu, E. Wong, S.J. Rettig, and C. Orvig, *Inorg. Chem.,* **34,** 2164 (1995).

lxv W.J. Evans, M.A. Ansari, and J.W. Ziller, *Inorg. Chem.,* **34,** 3079 (1995).

lxvi J.R. van den Hende, P.B. Hitchcock and M.F. Lappert, *J. Chem. Soc., Chem. Commun.,* 1413 (1994).

lxvii J. R. van den Hende, P.B. Hitchcock, S.A. Holmes, and M.F. Lappert, *J. Chem. Soc., Dalton Trans.,* 1435 (1995).

lxviii M.P. Hogerheide, S.N. Ringelberg, D.M. Grove, J.T.B.H. Jastrzebski, J. Boersma, W.J.J. Smeets, A.L. Spek, and G. van Koten, *Inorg. Chem.,* **35,** 1185 (1996).

carbon atoms approach the metal centre.[301] The π-interactions not only occupy vacant sites around the metal centres but can also control the geometries adopted.[302]

A large number of monomeric, Lewis base (typically O-donor) adducts of the bis and tris aryloxides have now been isolated and studied (Table 6.19). With the bulkiest ligands, only four-coordinate adducts such as [Yb(OC$_6$H$_2$Bu$_3^t$-2,4,6)$_3$(THF)][303] and [Yb(OC$_6$H$_2$Bu$_2^t$-2,6-Me-4)$_2$(thf)$_2$][92] are generated. With intermediate sized aryloxides or donor ligands a large number of five-coordinate adducts have been obtained, e.g. [M(OAr)$_3$(THF)$_2$] (M = La,[304] Pr,[305] Sm,[306] Eu,[307] Gd, Lu[308]) and [Yb(OC$_6$H$_3$Ph$_2$-2,6)$_2$(THF)$_3$].[309] Higher coordination numbers are common for very small or chelating donor ligands, e.g. six-coordinate [Eu(OC$_6$H$_3$Bu$_2^t$-2,6)$_2$(NCMe)$_4$][310] and [Yb(OC$_6$H$_3$Ph$_2$-2,6)$_2$(DME)$_2$][309] as well as seven-coordinate [La(OC$_6$H$_3$Pr$_2^i$-2,6)$_3$(NH$_3$)$_4$][293] and eight-coordinate [La(OC$_6$H$_3$Me$_2$-2,6)$_3$(tetraglyme)].[311]

It is also common that mixed metal derivatives of the lanthanides and group 1 metals are the products of synthesis. This can occur for small aryloxides leading to clusters, e.g. [La$_2$Na$_3$(μ_4-OAr)$_3$(μ_2-OAr)$_6$(dioxane)$_5$] (OAr = OC$_6$H$_4$Me-4),[312] as well as in the generation of more discrete species with bulkier aryloxides, e.g. [(thf)Li(μ_2-OAr)$_2$La(OAr)$_2$(THF)] (OAr = OC$_6$H$_3$Pr$_2^i$-2,6).[313] In some situations the alkali metal interacts with the aryloxide π-nucleus, e.g. [Cs(π-Ar-O)$_2$La(OAr)$_2$] (OAr = OC$_6$H$_3$Pr$_2^i$-2,6).[314] In an unusual reaction, anhydrous [LnCl$_3$] (Ln = Nd or Er) was reacted with [NaOC$_6$H$_3$Ph$_2$-2,6] in 1,3,5-tri-tert-butylbenzene at 300°C to generate [NaLn(OAr)$_4$] which reacts with DME to form the discrete tetra-aryloxide anions.[315]

Aryloxide derivatives can be valuable precursors for the synthesis of lanthanide alkyl compounds. The first homoleptic alkyls, [M{CH(SiMe$_3$)$_2$}$_3$] (M = La, Sm) were prepared by treatment of [M(OC$_6$H$_3$Bu$_2^t$-2,6)$_3$] with [Li{CH(SiMe$_3$)$_2$}].[290] However, only partial substitution of aryloxide ligands occurs in most cases and mixed alkyl, aryloxides can be generated. For example the corresponding tris-aryloxides are a good starting material for the synthesis of [Cp*M(OC$_6$H$_3$Bu$_2^t$-2,6)$_2$] which can be converted to [Cp*CM(OC$_6$H$_3$Bu$_2^t$-2,6){CH(SiMe$_3$)$_2$}] (M = Ce,[291] Y[316,317]). The yttrium compound undergoes hydrogenolysis to the dimeric hydride [Cp*(ArO)Y(μ-H)]$_2$ which has an extensive olefin chemistry, including formation of polymers via [Cp*(ArO)Y(μ-H)(μ-R)Y(OAr)Cp*] intermediates.[317] Neither of the methyl compounds [Cp*(ArO)Y(μ-Me)]$_2$ (ArO = OC$_6$H$_3$Bu$_2^t$-2,6) or [Cp*(Me)Sc(μ-OAr)]$_2$ (ArO = OC$_6$H$_3$Bu$_2^t$-3,5)[318] react with olefins.[317]

The alkylation of precursors [Y(OSiBu^{t}Ar$_2$)(OC$_6$H$_3$Bu$_2^t$-2,6)$_2$] (obtained from the precursor [Y(OSiBu^{t}Ar$_2$){{N(SiMe$_3$)$_2$}$_2$}] with [LiCH$_2$SiMe$_3$] yields the salt complexes [LiY(OSiBu^{t}Ar$_2$)(OC$_6$H$_3$Bu$_2^t$-2,6)(CH$_2$SiMe$_3$)$_2$][319] and [(THF)Li(μ_2-OAr)$_2$Sm(CH$_2$SiMe$_3$)(μ_2-OAr)(μ_2-CH$_2$SiMe$_3$)Li(THF)][320] (ArO = OC$_6$H$_3$Pr$_2^i$-2,6) showing that elimination of LiOAr is sometimes not facile.

6.2.4 Actinide Aryloxides

The metals thorium and uranium (as with the chemistry in general of these elements) dominate actinide aryloxide chemistry. Synthetic strategies normally focus on the halides reacting with group 1 metal aryloxides or reaction of metal amides with phenols.[321,322] An important piece of early work was the demonstrated interconversion of eight-coordinate [UMe$_4$(dmpe)$_2$] and [M(OPh)$_4$(dmpe)$_2$] (M = Th, U: Table 6.20) by addition of phenol or MeLi to each substrate respectively.[323] The reaction of the cyclometallated

Table 6.20 Actinide aryloxides

Compound	Aryloxide	Bond length (Å) M–O	Bond angle (°) M–O–Ar	Ref.
[Th(OAr)$_4$] tetrahedral Th	OC$_6$H$_3$Bu$_2^t$-2,6	2.189 (6)	154	i
[(pyrazolyl)$_2$Th(OAr)$_2$] 8-coord. Th	OC$_6$H$_5$	2.181 (6) 2.172 (6)	173 169	ii
[Th(OAr)$_2$(CH$_2$-py-6Me)$_2$] 6-coord. Th	OC$_6$H$_3$Bu$_2^t$-2,6	2.190 (9)	175	iii
[Th(OAr)$_2$(CH$_2$SiMe$_3$)$_2$]	OC$_6$H$_3$Bu$_2^t$-2,6	2.15 (1) 2.13 (1)	162 177	iv
[Th$_3$(μ_3-H)$_2$(μ_2-H)$_4$(OAr)$_6$] triangular cluster, all terminal OAr	OC$_6$H$_3$Bu$_2^t$-2,6	2.164 (7) 2.126 (7) 2.158 (7) 2.144 (7) 2.139 (7) 2.128 (7)	178 168 179 166 178 171	iv
[Th(OAr)$_4$(py)$_2$].toluene *cis* py ligands	OC$_6$H$_3$Me$_2$-2,6	2.211 (6) 2.188 (6) 2.206 (6) 2.183 (6)	169 176 172 176	i
[(ArO)(H$_2$O)$_6$Th(μ_2-OH$_2$)Th(OAr)(OH$_2$)$_6$](OAr)$_4$.20H$_2$O	OC$_6$H$_2$(NO$_2$)$_3$-2,4,6	2.366 (3)	156	v
[Cp*Th(OAr)$_2$Br] three-legged piano stool	OC$_6$H$_3$Bu$_2^t$-2,6	2.18 (1) 2.139 (8)	162 170	vi

534

Compound	OAr	Distance		
[Cp*Th(OAr)(CH₂SiMe₃)₂] three-legged piano stool	$OC_6H_3Bu^t_2$-2,6,	2.186 (6)	171	vi
[Cp*Th(OC₆H₃Buᵗ-CMe₂CH₂(OAr)(OPPh₂)] distorted tbp, Cp* and cyclometallated O axial	$OC_6H_3Bu^t_2$-2,6 / $OC_6H_3Bu^t$-CMe^2CH_2	2.205 (7), 2.192 (6)	163, 148	vi
[{U₂(OAr)₃(thf)}₂[U(O)₂(thf₂}(μ₂-OAr)₄(μ₃-O)] tetrameric unit, two U(v) and two U(vi) atoms.	OC_6H_5 / μ_2-OC_6H_5	2.03 (2), 2.08 (2), 2.12 (2), 2.46 (2), 2.33 (2), 2.40 (2), 2.42 (2)	174, 164, 158	vii
[U(OAr)₄(dmpe)₂] dodecahedron	OC_6H_5	2.16 (1), 2.17 (1), 2.18 (1), 2.18 (1)	166, 164, 162, 167	viii
[U(OAr)₃(NEt₂)] tetrahedral U	$OC_6H_3Bu^t_2$-2,6	2.140 (4), 2.146 (4), 2.143 (4)	149, 163, 150	ix
[U₂(O-η⁶-Ar)₂(OAr)₄] two 3-coord. U held together by η⁶-phenoxy interactions	$OC_6H_3Pr^i_2$-2,6 / O-η^6-$C_6H_3Pr^i_2$-2,6	2.132 (8), 2.214 (7)	165, 157	x
[U(OAr)₄] tetrahedral U	$OC_6H_3Bu^t_2$-2,6	2.135 (4)	154	i, xi
[U(OAr)₃I] tetrahedral U	$OC_6H_3Bu^t_2$-2,6	2.092 (8), 2.102 (8), 2.114 (11)	170, 160, 170	xii
[U(OAr)₂I₂(THF)] tbp, axial THF and OAr	$OC_6H_3Bu^t_2$-2,6	2.080 (8), 2.073 (9)	173, 166	xii
[(ArO)₃U(μ₂-S)U(OAr)₃].OEt₂ distorted, non tetrahedral geometry about U	$OC_6H_3Bu^t_2$-2,6	2.079 (9), 2.125 (8), 2.119 (6)	172, 174, 172	xii
[Cp*₂U(O)(OAr)]	$OC_6H_3Pr^i_2$-2,6	2.135 (5)	170	xiii
[U(OAr){N(SiMe₃)₂}₃] tetrahedral U	$OC_6H_3Bu^t_2$-2,6	2.145 (8)	159	i
[(thf)₄K][U(OAr)₂Cl₃] tbp U with trans OAr, infinite chain held together by UCl–K–ClU interactions	$OC_6H_3Bu^t_2$-2,6	2.122 (6), 2.123 (6)	155, 153	xiv

(continued overleaf)

Table 6.20 (*Continued*)

Compound	Aryloxide	Bond length (Å) M–O	Bond angle (°) M–O–Ar	Ref.
[U(O)$_2$(OAr)$_2$(py)$_3$]	OC$_6$H$_3$Pr$^i_2$-2,6	2.179 (5)	162	xv
		2.215 (5)	152	
[(thf)$_3$Na(μ_2-OAr)$_2$(μ_2-O)$_2$U(OAr)$_2$]	OC$_6$H$_3$Me$_2$-2,6	2.217 (5)	160	xv
	μ_2-OC$_6$H$_3$Me$_2$-2,6	2.190 (5)	168	
		2.282 (5), 2.436 (4)		
		2.294 (5), 2.445 (5)		
[U{N(CH$_2$CH$_2$NSiMe$_3$)$_3$}(OAr)(μ-OBu^t)Li(THF)] 6-coord. U, terminal OPH	OC$_6$H$_5$	2.188 (8)	178	xvi
[CP$_3$Np(OAr)]	OC$_6$H$_5$	2.136 (7)	160	xvii

[i] J.M. Berg, D.L. Clark, J.C. Huffman, D.E. Morris, A.P. Sattelberger, W.E. Streib, W.G. van der Sluys, and J.G. Watkin, *J. Am. Chem. Soc.*, **114**, 10811 (1992).

[ii] A. Domingos, J. Marcalo, and A. Pires de Matos, *Polyhedron*, **11**, 909 (1992).

[iii] S.M. Beshouri, P.E. Fanwick, I.P. Rothwell, and J.C. Huffman, *Organometallics*, **6**, 2498 (1987).

[iv] D.L. Clark, S.K. Grumbine, B.L. Scott, and J.G. Watkin, *Organometallics*, **15**, 949 (1996).

[v] J.M. Harrowfield, B.J. Peachey, B.W. Skelton, and A.H. White, *Aust. J. Chem.*, **48**, 1349 (1995).

[vi] R.J. Butcher, D.L. Clark, S.K. Grumbine, B.L. Scott, and J.G. Watkin, *Organometallics*, **15**, 1488 (1996).

[vii] A.J. Zozulin, D.C. Moody, and R.R. Ryan, *Inorg. Chem.*, **21**, 3083 (1982).

[viii] P.G. Edwards, R.A. Andersen, and A. Zalkin, *J. Am. Chem. Soc.*, **103**, 7792 (1981).

[ix] P.B. Hitchcock, M.F. Lappert, A. Singh, R.G. Taylor, and D. Brown, *J. Chem. Soc., Chem. Commun.*, 561 (1983).

[x] W.G. van der Sluys, C.J. Burns, J.C. Huffman, and A.P. Sattelberger, *J. Am. Chem. Soc.*, **110**, 5924 (1988).

[xi] W.G. van der Sluys, A.P. Sattelberger, W.E. Streib, and J.C. Huffman, *Polyhedron*, **8**, 1247 (1989).

[xii] L.R. Avens, D.M. Barnhart, C.J. Burns, S.D. McKee, and W.H. Smith, *Inorg. Chem.*, **33**, 4245 (1994).

[xiii] D.S.J. Arney and C.J. Burns, *J. Am. Chem. Soc.*, **115**, 9840 (1993).

[xiv] S.D. McKee, C.J. Burns, and L.R. Avens, *Inorg. Chem.*, **37**, 4040 (1998).

[xv] D.M. Barnhart, C.J. Burns, N.N. Sauer, and J.G. Watkin, *Inorg. Chem.*, **34**, 4079 (1995).

[xvi] P. Roussel, P.B. Hitchcock, N.D. Tinker, and P. Scott, *Inorg. Chem.*, **36**, 5716 (1997).

[xvii] D.J.A. De Ridder, C. Apostolidis, J. Rebizant, B. Kanellakopulos, and R. Maier, *Acta Crystallogr. C*, **52**, 1436 (1996).

amido compound [{(Me$_3$Si)$_2$N}$_2$M{N(SiMe$_3$)-SiMe$_2$CH$_2$}] (M = Th, U) with a variety of phenols has been investigated.[322] Initial protonation of the M–C bond occurs to produce mono-aryloxide, tris-amide compounds [{(Me$_3$Si)$_2$N)}$_3$M(OAr)]. In the case of thorium, stepwise replacement of amide groups occurs with the small 4-*tert*-butylphenol to eventually produce polymeric [Th(OC$_6$H$_4$Bu^t-4)$_4$]$_x$ which can be converted to monomeric [Th(OC$_6$H$_4$Bu^t-4)$_4$(py)$_3$]. With bulkier phenols only intermediate bis- and tris-aryloxides are produced. In contrast the uranium compound underwent protonolysis with even 2,6-di-*tert*-butylphenol to produce homoleptic [U(OC$_6$H$_3$Bu$^t_2$-2,6)$_4$].[324] The thorium analogue was produced via reaction of [ThI$_4$(THF)$_4$] with KOC$_6$H$_3$Bu$^t_2$-2,6 in THF. With smaller aryloxides, KOAr the reaction produces adducts such as [Th(OC$_6$H$_3$Me$_2$-2,6)$_4$(THF)$_2$] which can be converted to [Th(OC$_6$H$_3$Me$_2$-2,6)$_4$(py)$_2$]. The effect of the nature of the halide on the reaction of U(IV) halides with KOC$_6$H$_3$Bu$^t_2$-2,6 in THF has been thoroughly investigated.[325] Reaction with [UCl$_4$] progressed via [(THF)$_4$K][U(OAr)$_2$Cl$_3$] and [ClU(OAr)$_3$] to finally yield [U(OAr)$_4$]. With [UI$_4$(NCMe)$_4$] products [I$_2$U(OAr)$_2$(THF)], [IU(OAr)$_3$], and [U(OAr)$_4$] were obtained with two, three, and four equivalents of KOAr. The compounds [Br$_2$U(OAr)$_2$(THF)], obtained from [UBr$_4$(NCMe)$_4$], and [I$_2$U(OAr)$_2$(THF)] could be readily desolvated.[325] Studies failed to detect ligand redistribution reactions between halides [UX$_4$] (X = Cl, Br, I) or [X$_2$U(OC$_6$H$_3$Bu$^t_2$-2,6)$_2$] (X = Br, I) and [U(OC$_6$H$_3$Bu$^t_2$-2,6)$_4$]. The above results clearly show how the coordination number of the adducts of tetra-aryloxides is extremely sensitive to the bulk of the aryloxide ligands.

Treatment of [U{N(SiMe$_3$)$_2$}$_3$] with 2,6-diisopropylphenol has been shown to generate the corresponding U(III) aryloxide [U$_2$(O-η^6-Ar)$_2$(OAr)$_4$] which in the solid state is held together by π-arene interactions but is monomeric in benzene and forms an adduct with THF.[326] The oxidation chemistry of the complex [U(OC$_6$H$_3$Bu$^t_2$-2,6)$_3$] has been extensively investigated.[327] The entire halide series [XU(OAr)$_3$] (X = F, Cl, Br, I) was obtained with reagents such as AgPF$_6$, PCl$_5$, CBr$_4$, and I$_2$. Reaction with CI$_4$ was found to yield [I$_2$U(OAr)$_2$]. Dimeric compounds [(ArO)$_3$U(μ_2-E)U(OAr)$_3$] (E = O, S) were formed from nitrogen oxides and Ph$_3$PS respectively. Interestingly molecular oxygen produced [U(OAr)$_4$] *via* ligand redistribution reactions.[327]

One entry into the organometallic chemistry of thorium aryloxides can be achieved by treatment of the bromides [ThBr$_4$(THF)$_4$] or [Cp*ThBr$_3$(THF)$_3$] with KOC$_6$H$_3$Bu$^t_2$-2,6 to initially produce [Th(OAr)$_2$Br$_2$(THF)$_2$] or [Cp*ThBr$_2$(OAr)] and [Cp*ThBr(OAr)$_2$] respectively.[224,225] Treatment of these compounds with Grignard reagents produces the corresponding alkyls. Besides undergoing cyclometallation chemistry (Section 5.1) the dialkyls [Th(OC$_6$H$_3$Bu$^t_2$-2,6)$_2$(CH$_2$SiMe$_3$)$_2$] and [Cp*Th(OC$_6$H$_3$Bu$^t_2$-2,6)(CH$_2$SiMe$_3$)$_2$] react with H$_2$ to produce the hydride clusters [Th$_3$(μ_3-H)$_2$(μ_2-H)$_4$(OAr)$_6$] and [Cp$^*_3$Th$_3$(μ_3-H)$_2$(μ_2-H)$_4$(OAr)$_3$] respectively.[224,225] Although both compounds demonstrated activity for olefin hydrogenation, only the Cp* compound polymerized ethylene when activated with [PhMe$_2$NH][B(C$_6$F$_5$)$_4$].

The compound [U(O)$_2$Cl$_2$] has been reported to react directly with phenols to produce the corresponding bis-aryloxides.[328] More recently the reaction of amido compounds [{(Me$_3$Si)$_2$N}$_2$U(O)$_2$] and [(thf)$_2$Na]$_2$[{(Me$_3$Si)$_2$N}$_4$U(O)$_2$] towards phenols has been shown to generate new oxo, aryloxides.[329] Isolated species included neutral [U(O)$_2$(OC$_6$H$_3$Pr$^i_2$-2,6)$_2$(py)$_3$] and [(thf)$_3$Na(μ_2-OAr)$_2$(μ_2-O)$_2$U(OAr)$_2$] (OAr = OC$_6$H$_3$Me$_2$-2,6; Table 6.20). The mono-oxo compound [Cp$^*_2$U(O)(OC$_6$H$_3$Pr$^i_2$-2,6)] is produced by treatment of [Cp$^*_2$U(OAr)(THF)] with pyridine *N*-oxide.[330]

6.2.5   Group 4 Metal Aryloxides

Group 4 metal aryloxides have been synthesized by a plethora of methods employing metal halides, alkyls, alkoxides, acetylacetonates,[331] and dialkylamides as substrates. Initially interest in these compounds focussed on their synthesis,[332-336] simple coordination chemistry[337,338] and underlying structure (Tables 6.21–6.23). As with the corresponding alkoxides, the potential use of group 4 metal aryloxides in sol–gel and thermal[339] processes for metal oxide synthesis has been an important recent development. Very recently the hydroquinone ligand has been used to generate covalent three-dimensional titanium(IV)–aryloxide networks.[340] Interestingly the simple phenoxide [Cp$_2$Ti(OPh)$_2$] has been shown to possess anti-tumour activity.[341] The use of aryloxide ligands has been important in allowing the isolation and study of a variety of group 4 metal compounds containing terminal oxo[342] and imido functional groups. The five-coordinate imido derivatives [Ti(=NR)(OAr)$_2$(L)$_2$] have been made via amine elimination from species such as [(ArO)$_2$Ti(NHPh)$_2$],[343] azobenzene cleavage,[344,345] and substitution of preformed imido, chlorides.[346] The nature of the aryloxide ligands can control whether terminal or bridging imido species are produced.[346] Four coordinate titanium imido compounds containing amidinate and aryloxide ligands have also been generated.[347]

The use of aryloxide ligands attached to zirconium and in particular titanium for carrying out stoichiometric and catalytic transformations has resulted in the chemistry of a large number of organometallic derivatives being studied. In many cases there are parallel structural motifs and reactivity patterns between bis-aryloxide derivatives [(ArO)$_2$MX$_2$] (Tables 6.21–6.23) and analogous bis-cyclopentadiene analogues [Cp$_2$MX$_2$].[348] Related chemistry has also been developed for [(RO)$_2$Ti][349] and [(R$_3$SiO)$_2$Ti][350] fragments. This apparent mimicry has been ascribed to underlying bonding similarities (isolobal analogy) between the Cp and OAr ligands when the aryloxide is sufficiently bulky to maintain a terminal bonding situation.[351] "Hybrid" species [Cp(ArO)TiX$_2$] have also been isolated for titanium.[352-355] Despite the isolobal analogy there are definite electronic differences between Cp and ArO ligation. This can be seen in the structural parameters for pseudo-tetrahedral derivatives [(X)(Y)TiZ$_2$] (X, Y = Cp or ArO; Z = Cl or Me) of titanium.[352,356] In both the chloride and alkyl series it can be seen that replacement of Cp by ArO results in a more electron-deficient metal centre and a corresponding shortening of Ti–Cl and Ti–Me bond lengths. The one electron reduction of the dichlorides leads to the corresponding chloro-bridged d^1–d^1 dimers [(X)(Y)Ti(μ-Cl)$_2$Ti(X)(Y)] (Table 6.21). The Ti–Ti distance is exquisitely sensitive to the nature of the terminal ligands. In the paramagnetic bis(cyclopentadiene) compound the long Ti–Ti distance of 3.95 Å (av.) rules out any significant direct bonding between the two metal centres. This distance shortens by a dramatic 1 Å for the bis-aryloxide and results in a diamagnetic material.[357] The mixed Cp–OAr species has a Ti–Ti distance that is almost exactly intermediate. This argues strongly against the concept of an on/off metal–metal single bond. Instead the data suggests that the Ti–Ti distance and interaction is controlled by the structural parameters for the Ti(μ-Cl)$_2$Ti bridge imposed by the terminal ligands; shorter Ti–Cl and larger Cl–Ti–Cl parameters as Cp is replaced by ArO.

Table 6.21 Titanium aryloxides

Compound	Aryloxide	Bond length (Å) M–O	Bond angle (°) M–O–Ar	Ref.
	Homoleptic aryloxides and adducts			
[Ti$_2$(μ_2-OAr)$_2$(OAr)$_6$(HOAr)$_2$]	OC$_6$H$_5$	1.84 (1)	169	i
HOAr hydrogen bonded *trans* to μ_2-OAr,		1.79 (1)	175	
edge shared bis-octahedron.		1.88 (1)	127	
		2.044 (1), 2.026 (1)		
[Ti(OAr)$_4$]	μ_2-OC$_6$H$_5$			
tetrahedral Ti	OC$_6$H$_3$Pr$^i_2$-2,6	1.781 (3)	166	ii
		1.780 (3)	164	
[Ti(OAr)$_4$]	OC$_6$H$_4$Bu^t-2	1.779 (3)	152	iii
tetrahedral Ti				
[Ti(OAr)$_4$]	OC$_6$HMe$_4$-2,3,5,6	1.79 (2)	149	iii
tetrahedral Ti		1.76 (2)	170	
		1.78 (2)	162	
		1.76 (2)	149	
[Ti(OAr)$_4$]	OC$_6$H$_3$Ph$_2$-2,6	1.790 (4)	161	iv
tetrahedral Ti		1.792 (4)	162	
		1.793 (4)	161	
		1.794 (4)	161	
[Ti$_2$(μ_2-OAr)$_2$(OAr)$_4$]	OC$_6$H$_3$Me$_2$-2,6	1.823 (2)	155	v
		1.830 (2)	146	
	μ_2-OC$_6$H$_3$Me$_2$-2,6	2.009 (2), 2.009 (2)		
trans-[Ti(OAr)$_2$(dmpe)$_2$]	OC$_6$H$_5$	1.891 (6)	180	vi
		1.930 (6)	180	

(continued overleaf)

[i] G.W. Svetich and A.A. Voge, *Acta Crystallogr. B*, **28**, 1760 (1972).

[ii] L.D. Durfee, S.L. Latesky, I.P. Rothwell, J.C. Huffman, and K. Folting. *Inorg. Chem.*, **24**, 4569 (1985).

[iii] R.T. Toth and D.W. Stephan, *Can. J. Chem.*, **69**, 172 (1991).

[iv] E.A. Johnson and I.P. Rothwell, unpublished results.

[v] R. Minhas, R. Duchateau, S. Gambarotta, and C. Bensimon, *Inorg. Chem.*, **31**, 4933 (1992).

[vi] R.J. Morris and G.S. Girolami, *Inorg. Chem.*, **29**, 4167 (1990).

539

Table 6.21 (*Continued*)

Compound	Aryloxide	Bond length (Å) M–O	Bond angle (°) M–O–Ar	Ref.
cis-[Ti(OAr)$_2$(bipy)$_2$] two independent molecules, 6-coord. Ti	OC$_6$H$_3$Pr$^i_2$-2,6	1.896 (5) 1.897 (5) 1.880 (5) 1.881 (4)	156 154 155 156	vii
[Ti(OAr)$_2$(py-4Ph)$_3$] tbp, *trans* axial py, equatorial py one electron reduced	OC$_6$H$_3$Pr$^i_2$-2,6	1.865 (4)	166	viii
[Ti$_3$(μ_2-OAr)$_4$(OAr)$_5$(TMEDA)$_2$]	OC$_6$H$_5$	1.883 (9) 1.802 (11) 1.891 (9)	161 180 175	v
	μ_2-OC$_6$H$_5$	2.058 (7), 2.081 (7) 2.022 (7), 2.103 (7)		
[TMEDA-H][Ti(OAr)$_2$Cl$_2$(TMEDA)]	OC$_6$H$_3$Bu$^t_2$-3,5	1.889 (4) 1.900 (4)	158 144	v
[(THF)$_2$Na(μ_2-OAr)$_2$Ti(OAr)$_2$] tetrahedral Ti	OC$_6$H$_3$Pr$^i_2$-2,6	1.87 (1) 1.86 (1)	143 156	ix
	μ_2-OC$_6$H$_3$Pr$^i_2$-2,6	1.95 (1), 2.31 (1) 1.96 (1), 2.30 (1)		
[NaTi(μ_2-OAr)$_2$(OAr)$_2$(py)]	OC$_6$H$_3$Me$_2$-2,6	1.904 (4) 1.871 (4)	149 154	x
	μ_2-OC$_6$H$_3$Me$_2$-2,6	Ti–O–Na 2.081 (3), 2.271 (4) 1.983 (4), 2.225 (4)		
Mixed aryloxy halides				
[Ti(OAr)Cl$_2$(THF)$_2$]	OC$_6$H$_3$Bu$^t_2$-2,6	1.807 (8)	174	xi
[Ti(OAr)Cl$_3$] tetrahedral Ti	OC$_6$H$_3$Bu$^t_2$-2,6	1.748 (4)	167	xii

Compound	OAr	Ti–O (Å)	Ti–O–C (°)	Ref.
[Ti₂(μ₂-OAr)₂(OAr)₂Cl₄]₂ edge-shared bis-octahedron	OC₆H₅, μ₂-OC₆H₅	1.744 (10), 1.911 (9), 2.122 (9)	166	xiii
6-coord. Ti [Ti(OAr)₂Cl₂(THF)₂]	OC₆H₃Me₂-2,6	1.788 (11), 1.789 (10)	170, 170	xiv
[Ti(OAr)Cl₃(THF)]	OC₆H₃Me₂-2,6	1.762 (5)	173	xv
[Ti(μ₂-Cl)(OAr)₂]₂	OC₆H₃Ph₂-2,6	1.820 (2), 1.805 (2), 1.825 (2), 1.819 (2)	154, 152, 147, 155	xvi
[Ti(OAr)₂Cl₂] tetrahedral Ti	OC₆H₃Ph₂-2,6	1.726 (2)	168	xvii
[Ti(OAr)₂Cl₂] tetrahedral Ti	OC₆H₃Me₂-2,6	1.734 (7), 1.736 (8)	167, 169	xviii
[Ti(OAr)₂Cl₂] 6-coord. Ti, trans O, cis Cl	OC₆H₄(c-CCOCH₂CMe₂N)-2	1.857 (2)	140	xix
[Ti(OAr)₂Cl(THF)] 6-coord. Ti, trans O, Cl cis to THF, two molecules	OC₆H₄(c-CCOCH₂CMe₂N)-2	1.860 (4), 1.951 (4), 1.876 (4), 1.951 (5)	140, 143, 138, 142	xix

[vii] L.D. Durfee, P.E. Fanwick, I.P. Rothwell, K. Folting, and J.C. Huffman, *J. Am. Chem. Soc.*, **109**, 4720 (1987).

[viii] L.D. Durfee, J.E. Hill, J.L. Kerschner, P.E. Fanwick, and I.P. Rothwell, *Inorg. Chem.*, **28**, 3095 (1989).

[ix] L.D. Durfee, S.L. Latesky, I.P. Rothwell, J.C. Huffman, and K. Folting, *Inorg. Chem.*, **24**, 4569 (1985).

[x] L.D. Durfee, P.E. Fanwick, and I.P. Rothwell, *Angew. Chem., Int. Ed. Engl.*, **27**, 1181 (1988).

[xi] M. Mazzanti, C. Floriani, A. Chiesi-Villa, and C. Guastini, *J. Chem. Soc., Dalton Trans.*, 1793 (1989).

[xii] L. Matilainen, M. Klinga, and M. Leskela, *Polyhedron*, **15**, 153 (1996).

[xiii] K. Watenpaugh and C.N. Caughlan, *Inorg. Chem.*, **5**, 1782 (1966).

[xiv] N. Kanehisa, Y. Kai, N. Kasai, H. Yasuda, Y. Nakayama, K. Takei, and A. Nakamura, *Chem. Lett.*, 2167 (1990).

[xv] H. Yasuda, Y. Nakayama, K. Takei, A. Nakamura, Y. Kai, and N. Kanehisa, *J. Organomet. Chem.*, **473**, 105 (1994).

[xvi] J.E. Hill, J.M. Nash, P.E. Fanwick, and I.P. Rothwell, *Polyhedron*, **9**, 1617 (1990).

[xvii] J.R. Dilworth, J. Hanich, M. Krestel, J. Beck, and J. Strahle, *J. Organomet. Chem.*, **315**, C9 (1986).

[xviii] S. Waratuke, I.P. Rothwell, and P.E. Fanwick, unpublished results.

[xix] P.G. Cozzi, C. Floriani, A. Chiesi-Villa, and C. Rizzoli, *Inorg. Chem.*, **34**, 2921 (1995).

(continued overleaf)

Table 6.21 (*Continued*)

Compound	Aryloxide	Bond length (Å) M–O	Bond angle (°) M–O–Ar	Ref.
[Ti(OAr)₂Cl₂] octahedral Ti, *cis* Cl and OMe	η²-OC₆H₄OMe-2	1.858 (2)	124	xx
[Ti(μ-Cl)(OAr)Cl₂]₂ edge-shared bis-octahedron, O *trans* to μ-Cl	η²-OC₆H₄OMe-2	1.862 (3), 1.804 (2)	122, 127	xx
[Ti(OAr)₃I] tetrahedral Ti	OC₆H₃Buᵗ₂-2,6	1.810 (9), 1.802 (7), 1.782 (8)	159, 155, 158	xxi
[Ti(OAr)Cl(μ₂-Cl)₂]ₙ 6-coord. Ti, polymeric structure	OC₆H₅	1.746 (4)	165	xxii
Mixed aryloxy alkoxides, amides, etc.				
[(PrⁱO)(PrⁱOH)(ArO)₂Ti(μ₂-OAr)}₂] edge-shared bis-octahedron, one OAr and PrⁱOH *trans* to μ₂-OAr	OC₆F₅ μ₂-OC₆F₅	1.894 (8), 1.833 (7) 2.050 (7), 2.164 (7)	139, 151	xxiii
[{(PrⁱO)₂(ArO)Ti(μ₂-OPrⁱ)}₂] edge-shared bis-octahedron, one OAr and PrⁱOH *trans* to μ₂-OPrⁱ	OC₆H₂Me₃-2,4,6			xxiv
[{(PrⁱO)(ArO)₂Ti(μ₂-OAr)}₂] edge-shared bis-octahedron, one OAr chelated through ortho-F	OC₆H₃F₂-2,6 μ₂-OC₆H₃F₂-2,6	1.802 (5), 1.867 (6) 2.043 (5), 2.054 (5)	169, 126	xxiv
[Ti(OAr)₂(OCH₂SiMe₃)₂]	OC₆H₃Np₂-2,6	1.790 (4), 1.815 (5)	164, 151	xxv
[Ti(OAr)₂(acac)₂] 6-coord. Ti	OC₆H₃Prⁱ₂-2,6	1.834 (5)	162	xxvi
[Ti(OAr)₂(NMe₂)₂] tetrahedral Ti	OC₆H₂Buᵗ₃-2,4,6	1.813 (4), 1.860 (3)	171, 152	xxvii
[Ti(OAr)₂(NMe₂)₂] tetrahedral Ti	OC₆H₃Buᵗ₂-2,6	1.808 (2), 1.828 (2)	177, 175	xxviii

[Ti(OAr)$_2$(NMe$_2$)$_2$]	OC$_6$H$_2$Bu$^t_2$-2,4-Ph-6	1.862 (5)	143	xxix
tetrahedral Ti		1.844 (6)	154	
[Ti(OAr)$_2$(NHPh)$_2$]	OC$_6$H$_3$Ph$_2$-2,6	1.786 (2)	159	xxx,
tetrahedral Ti		1.789 (2)	147	xxxi
[Ti(OAr)$_2$Cl{N(SnMe$_3$)$_2$}]	OC$_6$H$_3$Ph$_2$-2,6	1.805 (1)	161	xxxii
tetrahedral Ti		1.808 (1)	160	
[Ti(OAr)$_2$(py)$_2$]$_2${μ_2-4,4'-bipy}	OC$_6$H$_3$Pr$^i_2$-2,6	1.883 (7)	143	vii
		1.891 (7)	144	
[CpRu(CO)$_2$-Ti(OAr)$_2$(NMe$_2$)]	OC$_6$H$_3$Me$_2$-2,6	1.814 (5)	153	xxxiii
		1.791 (5)	156	
Oxo and imido aryloxides				
[Ti(OAr)$_2$(=NAr')(η^2-ArN=NAr)]	OC$_6$H$_3$Ph$_2$-2,6	1.878 (4)	150	xxxiv
η^2-benzocinnoline		1.852 (4)	150	

[xx]C.-N. Kuo, T.-Y. Huang, M.-Y. Shao, and H.-M. Gau, *Inorg. Chim. Acta.*, **293**, 12 (1999).

[xxi]S.L. Latesky, J. Keddington, A.K. McMullen, I.P. Rothwell, and J.C. Huffman, *Inorg. Chem.*, **24**, 995 (1985).

[xxii]S. Troyanov, A. Pisarevsky, and Yu. T. Struchkov, *J. Organomet. Chem.*, **494**, C4 (1995).

[xxiii]C. Campbell, S.G. Bott, R. Larsen, and W.G. van der Sluys, *Inorg. Chem.*, **33**, 4950 (1994).

[xxiv]S.C. James, N.C. Norman, and A.G. Orpen, *Acta Crystallog. C*, **54**, 1261 (1998).

[xxv]M.G. Thorn and I.P. Rothwell, unpublished results.

[xxvi]P.H. Bird, A.R. Fraser, and C.F. Lau, *Inorg. Chem.*, **12**, 1322 (1973).

[xxvii]R.A. Jones, J.G. Hefner, and T.C. Wright, *Polyhedron*, **3**, 1121 (1984).

[xxviii]V.M. Visciglio, P.E. Fanwick, and I.P. Rothwell, *Acta Crystallog. C*, **50**, 896 (1994).

[xxix]V.M. Visciglio, P.E. Fanwick, and I.P. Rothwell, *Inorg. Chim. Acta*, **211**, 203 (1993).

[xxx]J.E. Hill, R.D. Profilet, P.E. Fanwick, and I.P. Rothwell, *Angew. Chem., Int. Ed. Engl.*, **29**, 664 (1990).

[xxxi]C.H. Zambrano, R.D. Profilet, J.E. Hill, P.E. Fanwick, and I.P. Rothwell, *Polyhedron*, **12**, 689 (1993).

[xxxii]J.R. Dilworth, J. Hanich, M. Krestel, J. Beck, and J. Strahle, *J. Organomet. Chem.*, **315**, C9 (1986).

[xxxiii]W.J. Sartain and J.P. Selegue, *Organometallics*, **8**, 2153 (1989).

[xxxiv]J.E. Hill, P.E. Fanwick, and I.P. Rothwell, *Inorg. Chem.*, **30**, 1143 (1991).

(continued overleaf)

Table 6.21 (*Continued*)

Compound	Aryloxide	Bond length (Å) M–O	Bond angle (°) M–O–Ar	Ref.
[Ti(OAr)$_2$(η^2-PhN=NPh)(py)$_2$] two independent molecules, mono-η^2-azobenzene	OC$_6$H$_3$Pr$^i_2$-2,6	1.836 (5) 1.852 (6) 1.845 (5)	169 169 176	lvi
[Ti$_4$(μ_3-O)$_4$(OAr)$_4${μ-(CO)$_9$Co$_3$(μ_3-CCO$_2$)}$_4$]	OC$_6$H$_5$	1.788 (7) (av.)	145–165	xxxv
[Ti(OAr)$_2$(O)(4-pyrrolidinopyridine)$_2$] tbp Ti, *trans* axial py ligands	OC$_6$H$_3$Pr$^i_2$-2,6	1.863 (6) 1.879 (6)	167 172	xxxvi
[Ti(OAr)$_2$(=NPh)(4-pyrrolidinopyridine)$_2$] tbp Ti, *trans* axial py ligands	OC$_6$H$_3$Pr$^i_2$-2,6	1.867 (3) 1.880 (3)	174 172	xxxi
[Ti(OAr)$_2$(=NC$_6$H$_3$Me$_2$-2,6)(py)$_2$]	OC$_6$H$_3$Pr$^i_2$-2,6			xxxvii
[Ti(OAr)$_2$(μ-NBu^t)]$_2$	OC$_6$H$_3$Pr$^i_2$-2,6			xxxvii
[Ti(μ_2-O)(OAr)$_2$(4-pyrrolidinopyridine)]$_2$	OC$_6$H$_3$Pr$^i_2$-2,6	1.833 (2) 1.817 (2)	176 156	xxxviii
[Ti(OAr)$_2$(=NPh)(3,4,7,8-Me$_4$-1,10-phen)]	OC$_6$H$_3$Ph$_2$-2,6	1.882 (3) 1.931 (3)	150 125	xxxi
Non-Cp organometallic aryloxides				
[Ti(OAr)$_2$(Ph)$_2$] tetrahedral Ti	OC$_6$H$_3$Ph$_2$-2,6	1.794 (3) 1.797 (3)	153 162	xxxix
[Ti(OAr)$_3$(CH$_2$SiMe$_3$)] tetrahedral Ti	OC$_6$H$_3$Ph$_2$-2,6	1.806 (2) 1.802 (2) 1.797 (2)	179 164 170	xxxix

[Ti(OAr)$_3$Bu^t] tetrahedral Ti	OC$_6$H$_3$Pr$^i_2$-2,6	1.786 (2) 1.799 (2) 1.803 (2)	167 153 155	xl
[Ti(OAr)$_3$Me] tetrahedral Ti	OC$_6$H$_3$Ph$_2$-2,6	1.793 (3) 1.798 (2) 1.793 (3)	160 170 163	xli
[Ti(OAr)$_2$Me$_2$] tetrahedral Ti	OC$_6$H$_3$Ph$_2$-2,6	1.795 (1) 1.791 (1)	163 165	xli
[Ti(OAr)$_2$(CH$_2$Ph)$_2$] tetrahedral Ti	OC$_6$H$_3$Ph$_2$-2,6	1.797 (3) 1.784 (3)	161 168	xli
[Ti(OAr)$_3$(CH$_2$CH=CH$_2$)] σ-allyl, tetrahedral Ti	OC$_6$H$_3$Ph$_2$-2,6	1.798 (2) 1.799 (2) 1.792 (2)	161 171 161	xlii
[Ti$^+$(OAr)$_2$(CH$_2$Ph)(η^6-C$_6$H$_5$CH$_2$CH$_2$B$^-$(C$_6$F$_5$)$_3$]	OC$_6$H$_3$Ph$_2$-2,6	1.761 (5) 1.795 (4)	159 159	xli
[Ti(OAr)$_3$(H)(PMe$_3$)] tbp, axial O and P	OC$_6$H$_3$Pr$^i_2$-2,6	1.810 (3) 1.801 (2) 1.850 (2)	177 170 143	xl
[Ti(OAr)(BH$_4$)$_2$(PMe$_3$)$_2$] 7-coord. Ti	OC$_6$H$_3$Pr$^i_2$-2,6	1.804 (3)	178	xl
[(H$_4$B)$_2$Ti(μ_2-OAr)$_2$Ti(BH$_4$)$_2$] 8-coord. Ti	μ_2-OC$_6$H$_3$Pr$^i_2$-2,6	2.005 (2), 2.003 (2)		xl
[Ti(OAr)$_3$(BH$_4$)] 6-coord. Ti	OC$_6$H$_3$Pr$^i_2$-2,6	1.774 (6) 1.781 (6) 1.781 (6)	173 175 172	xl

(continued overleaf)

xxxv Xinjian Lei, Maoyu Shang, and T.P. Fehlner, *Organometallics*, **15**, 3779 (1996).
xxxvi J.E. Hill, P.E. Fanwick, and I.P. Rothwell, *Inorg. Chem.*, **28**, 3602 (1989).
xxxvii P. Collier, A.J. Blake, and P. Mountford, *J. Chem. Soc., Dalton Trans.*, 2911 (1997).
xxxviii J.E. Hill, P.E. Fanwick, and I.P. Rothwell, *Acta Crystallog. C*, **47**, 541 (1991).
xxxix R.W. Chesnut, L.D. Durfee, P.E. Fanwick, I.P. Rothwell, K. Folting, and J.C. Huffman, *Polyhedron*, **6**, 2019 (1987).
xl H. Noth and M. Schmidt, *Organometallics*, **14**, 4601 (1995).
xli M.G. Thorn, Z.C. Etheridge, P.E. Fanwick, and I.P. Rothwell, *J. Organomet. Chem.*, **591**, 148 (1999).
xlii M.G. Thorn, J.S. Vilardo, J. Lee, B. Hanna, P.E. Fanwick, and I.P. Rothwell, *Organometallics*, in press.

Table 6.21 (*Continued*)

Compound	Aryloxide	Bond length (Å) M–O	Bond angle (°) M–O–Ar	Ref.
[Ti(OAr)₂(C₄Et₄)] two independent molecules, titanacyclopentadiene ring	OC₆H₃Ph₂-2,6	1.788 (6) 1.828 (6) 1.804 (6) 1.806 (6)	166 145 156 176	xliii xliv
[Ti(OAr)₂(CH₂)₄] two independent molecules, titanacyclopentane ring	OC₆H₃Ph₂-2,6	1.804 (1) 1.804 (1) 1.817 (1) 1.799 (1)	156 159 151 158	xlv, xlvi
[Ti(OAr)₂(CHPhCH₂CH₂CHPh)] 2,5-diphenyl-titanacyclopentane	OC₆H₃Ph₂-2,6	1.801 (3) 1.804 (3)	149 161	xlvi
[Ti(OAr)₂[CH₂CH(C₄H₈)CHCH₂}]	OC₆HPh₄-2,3,5,6	1.802 (3)	149	xlvi
[Ti(OAr)₂[C(SiMe₂Ph)₂C₆H₈C(PMe₃)}] titanacyclopentene ring	OC₆H₃Pⁱ₂-2,6	1.830 (6) 1.861 (6)	164 159	xlvii
[Ti(OAr)₂(η⁴-CH₂CMeCMeCH₂)] π-bound 2,3-dimethylbutadiene	OC₆H₃Ph₂-2,6	1.804 (2) 1.843 (2) 1.843 (2)	151 144 144	xlviii
[Ti(OAr)₂(η²-C₂H₄)(PMe₃)] pseudo-tetrahedral Ti	OC₆H₃Ph₂-2,6	1.835 (1) 1.838 (1)	161 161	xlix, 1
[Ti(OAr)₂(η²-Ph₂CO)(PMe₃)] pseudo-tetrahedral Ti	OC₆H₃Ph₂-2,6	1.817 (5) 1.839 (5)	159 168	xlix, 1
[(ArO)₂Ti(μ₂-η²-NCMe)₂Ti(OAr)₂]	OC₆H₃Ph₂-2,6	1.819 (4) 1.799 (4)	173 160	xliv
[(ArO)₂Ti(C₄Et₄CPh₂O)] 3,4,5,6-tetraethyl-7,7-diphenyl-1-oxa-2-titanacyclohepta-3,5-diene ring	OC₆H₃Ph₂-2,6	1.804 (1) 1.814 (1)	163 161	xliv

(continued overleaf)

Compound	OAr			Ref.
[(ArO)₂Ti(η²-BuᵗN=CC₄Et₄CPh₂O)] η²-iminoacyl within metallacycle ring	$OC_6H_3Ph_2$-2,6	1.818 (4)	169	xliii
		1.896 (4)	152	
[Ti(OAr)₂(NCPhCHPhNCH₂Ph)] 2,5-diaza-1-titanacyclopent-2-ene	$OC_6H_3Ph_2$-2,6	1.816 (2)	162	xliv
		1.829 (2)	152	
[Ti(OAr)₂{BuᵗNC(CH₂Ph)=C(CH₂Ph)Nxy}] xy = 2,6-xylyl folded 2,5-diazacyclopentene ring	$OC_6H_3Pr^i_2$-2,6	1.845 (4)	154	li
		1.790 (4)	174	
[Ti(OAr)₂{OC(CH₂Ph)=C(CH₂Ph)NBuᵗ}] folded 2-oxa-5-azacyclopentene ring.	$OC_6H_3Pr^i_2$-2,6	1.792 (3)	172	li
		1.814 (3)	171	
[Ti(OAr)₂{PhNC(CH₂SiMe₃)=C(C(CH₂SiMe₃)=NPh)NPh}] folded 2,5-diazacyclopentene ring	$OC_6H_3Ph_2$-2,6	1.856 (7)	146	li
		1.811 (7)	175	
[{Ti(OAr)₂(CH₂CHMeCMe₂)}₂(μ₂-O)] tetrahedral Ti linked by oxo bridge	$OC_6H_3Ph_2$-2,6	1.793 (3)	161	xlvi
		1.795 (2)	168	
[Ti(OC₆H₃Buᵗ-CMe₂CH₂)(OAr)(CH₂SiMe₃)(py)] tbp Ti	$OC_6H_3Bu^t_2$-2,6	1.850 (4)	176	lii
	$OC_6H_3Bu^t$-CMe₂CH₂	1.810 (4)	146	
[Ti(OAr)₂(η²-BuᵗN=CCH₂Ph)(CH₂Ph)] mono-η²-iminoacyl	$OC_6H_3Pr^i_2$-2,6	1.816 (3)	169	liii
		1.811 (3)	163	
[Ti(OAr)₂(η²-PhN=CCH₂SiMe₃)(CH₂SiMe₃)] mono-η²-iminoacyl	$OC_6H_3Pr^i_2$-2,6	1.850 (4)	140	liii
		1.821 (4)	153	

xliii J.E. Hill, P.E. Fanwick, and I.P. Rothwell, Organometallics, 9, 2211 (1990).

xliv J.E. Hill, G. Balaich, P.E. Fanwick, and I.P. Rothwell, Organometallics, 12, 2911 (1993).

xlv J.E. Hill, P.E. Fanwick, and I.P. Rothwell, Organometallics, 10, 15 (1991).

xlvi M.G. Thorn, J.E. Hill, S.A. Waratuke, E.S. Johnson, P.E. Fanwick, and I.P. Rothwell, J. Am. Chem. Soc., 119, 8630 (1997).

xlvii G.J. Balaich, P.E. Fanwick, and I.P. Rothwell, Organometallics, 13, 4117 (1994).

xlviii J.E. Hill, G.J. Balaich, P.E. Fanwick, and I.P. Rothwell, Organometallics, 10, 3428 (1991).

xlix J.E. Hill, P.E. Fanwick, and I.P. Rothwell, Organometallics, 11, 1771 (1992).

l M.G. Thorn, J.E. Hill, S.A. Waratuke, E.S. Johnson, P.E. Fanwick, and I.P. Rothwell, J. Am. Chem. Soc., 119, 8630 (1997).

li L.R. Chamberlain, L.D. Durfee, P.E. Fanwick, L.M. Kobriger, S.L. Latesky, A.K. McMullen, B.D. Steffey, I.P. Rothwell, K. Folting, and J.C. Huffman, J. Am. Chem. Soc., 109, 6068 (1987).

lii S.L. Latesky, A.K. McMullen, I.P. Rothwell, and J.C. Huffman, J. Am. Chem. Soc., 107, 5981 (1985).

liii L.R. Chamberlain, L.D. Durfee, P.E. Fanwick, L. Kobriger, S.L. Latesky, A.K. McMullen, I.P. Rothwell, K. Folting, J.C. Huffman, W.E. Streib, and R. Wang, J. Am. Chem. Soc., 109, 390 (1987).

Table 6.21 (*Continued*)

Compound	Aryloxide	Bond length (Å) M–O	Bond angle (°) M–O–Ar	Ref.
[Ti(OAr)$_2$(η^2-BuN=CC$_4$Et$_4$)(py)] mono-η^2-iminoacyl	OC$_6$H$_3$Ph$_2$-2,6	1.861 (2) 1.841 (2)	157 154	liv
[Ti(OAr)$_2$\{η^2-Bu^{t}N=C(CH$_2$Ph)$_2$\}(py-4Ph)] mono-η^2-imine	OC$_6$H$_3$Pr$^i_2$-2,6	1.829 (4) 1.848 (3)	174 155	lv, lvi
[Ti(OAr)$_2$\{η^2-Bu^{t}N=C(CH$_2$Ph)$_2$\}(Py-4Et)] mono-η^2-imine	OC$_6$H$_3$Pr$^i_2$-2,6	1.840 (2) 1.854 (2)	175 162	lvi
[Ti(OAr)$_2$\{(xy)NCMeCMeN(xy)\}] xy = 2,6-xylyl	OC$_6$H$_3$Ph$_2$-2,6	1.811 (2) 1.864 (2)	175 161	lvii
[Ti(OAr)$_2$\{(xy)NCMe$_2$C(Et)C(Et)\}] xy = 2,6-xylyl	OC$_6$H$_3$Ph$_2$-2,6	1.841 (1) 1.829 (1)	167 151	lvii
[Ti(OAr)(OC$_6$H$_3$Ph-η^1-C$_6$H$_5$)\{N(xy)CHMe$_2$\}] xy = 2,6-xylyl	OC$_6$H$_3$Ph$_2$-2,6 OC$_6$H$_3$Ph-η^1-C$_6$H$_5$	1.804 (1) 1.829 (1)	164 130	lvii
[Ti(OAr)(OAr')\{NBu^t(C$_9$H$_{15}$)\}]	OC$_6$H$_3$Ph-2-C$_6$H$_4$(η^2-CNBu^t)-6	1.902 (2) 1.858 (2)	166 143	lvii
Cyclopentadienyl aryloxides				
[CP$_2$Ti(OAr)$_2$]	OC$_6$H$_2$Cl$_3$-2,4,6	1.946 (?)	141	lviii
[CP$_2$Ti(OAr)$_2$]	OC$_6$H$_4$Me-4	1.907 (3)	139	lix
[CP$_2$Ti(OAr)]	OC$_6$H$_2$Bu$^t_2$-2,6-Me-4	1.892 (2)	142	lx
[CP$_2$Ti(OAr)Cl]	OC$_6$H$_4$Me-2	1.865 (?)	141	lxi
[Cp(C$_5$H$_3$Me-Pr^i-1,3)Ti(OAr)Cl]	OC$_6$H$_4$Cl-2	1.885 (5)	143	lxii
[Cp(C$_5$H$_3$Me-Pr^i-1,3)Ti(OAr)Cl]	OC$_6$H$_3$Me$_2$-2,6	1.885 (5)	145	lxii
[Cp(C$_5$H$_3$Me-Pr^i-1,3)Ti(OAr)(OAr')]	OC$_6$H$_3$Me$_2$-2,6 OC$_6$H$_4$Cl-2	1.946 (6) 1.914 (6)	143 147	lxii
[Cp(C$_5$H$_3$Me-CH$_2$CMe$_2$Ph-3)Ti(OAr)Cl]	OC$_6$H$_3$Me$_2$-2,6	1.862 (5)	150	lxiii

Compound	Ligand		Value	Ref.
[Cp(C₅H₃Me-Pr^i-1,2)Ti(OAr)Cl] two molecules	OC$_6$H$_3$Me$_2$-2,6	147	1.883 (3)	lxiv
		145	1.870 (3)	lxv
[(ansa-Cp)Ti(OAr)$_2$]	2-naphtholato	146	1.903 (6)	lxv
2,3-bis(3-methyl-cyclopentadienyl)butane [(ansa-Cp)Ti(OAr)$_2$].cyclohexane	2-naphtholato	154	1.895 (5)	lxv
2,3-bis(3-methyl-cyclopentadienyl)butane [(ansa-Cp)Ti(OAr)$_2$]	OC$_6$H$_4$Cl-4	138	1.911 (7)	lxvi
		138	1.906 (6)	lxvi
1,1'-(hexamethyltrisilane-1,3-diyl)-bis-Cp	OC$_6$H$_4$Me-4	132	1.941 (6)	lxvii
[Cp$_2$Ti{O-η^6-Ar-Cr(CO)$_3$}$_2$]		144	1.912 (5)	lxvii
[Cp$_2$Ti{O-η^6-Ar-Cr(CO)$_3$}Br]	OC$_6$H$_4$Me-4	143	1.903 (6)	lxvii
[Cp(Cp-ArO)TiCl]	c-C$_5$H$_4$(CEt$_2$C$_6$H$_4$O-2)	136	1.879 (6)	lxviii
[Cp-ArO)Ti(CH$_2$Ph)$_2$]	OC$_6$H$_3$Me-4-(η^5-C$_5$Me$_4$)-2	127	1.851 (7)	lxix
[CpTi(OAr)$_3$]	OC$_6$H$_3$Pr$^i_2$-2,6	152	1.80 (2)	lxx
		145	1.79 (2)	
		163	1.80 (2)	

lxiv J.E. Hill, P.E. Fanwick, and I.P. Rothwell, *Organometallics*, **9**, 2211 (1990).

lxv L.D. Durfee, P.E. Fanwick, I.P. Rothwell, K. Folting, and J.C. Huffman, *J. Am. Chem. Soc.*, **109**, 4720 (1987).

lxvi L.D. Durfee, J.E. Hill, P.E. Fanwick, and I.P. Rothwell, *Organometallics*, **9**, 75 (1990).

lxvii M.G. Thorn, P.E. Fanwick, and I.P. Rothwell, *Organometallics*, **18**, 4442 (1999).

lxviii Yang Qingchuan, Jin Xianglin, Xu Xiaojie, Li Genpei, Tang Youqi, and Chen Shoushan, *Sci. Sin., Ser. B (Engl. Ed.)*, **25**, 356 (1982).

lxix B.S. Kalirai, J.-D. Foulon, T.A. Hamor, C.J. Jones, P.D. Beer, and S.P. Fricker, *Polyhedron*, **10**, 1847 (1991).

lxx B. Cetinkaya, P.B. Hitchcock, M.F. Lappert, S. Torroni, J.L. Atwood, W.E. Hunter, and M.J. Zaworotko, *J. Organomet. Chem.*, **188**, C31 (1980).

lxxi C. Lecomte, Y. Dusausoy, and J. Protas, *C.R. Acad. Sci., Ser. C*, **280**, 813 (1975).

lxxii J. Besancon, S. Top, J. Tirouflet, Y. Dusausoy, C. Lecomte, and J. Protas, *J. Organomet. Chem.*, **127**, 153 (1977).

lxxiii C. Lecomte, Y. Dusausoy, J. Protas, J. Tirouflet, and A. Dormond, *J. Organomet. Chem.*, **73**, 67 (1977).

lxxiv J. Besancon, J. Szymoniak, C. Moise, L. Toupet, and B. Trimaille, *J. Organomet. Chem.*, **491**, 31 (1995).

lxxv S.C. Sutton, M.H. Nantz, and S.R. Parkin, *Organometallics*, **12**, 2248 (1993).

lxxvi Y. Wang, Shan-Sheng Xu, Xiu-Zhong Zhou, Xin-Kai Yao, and Hong-Gen Wang, *Huaxue Xuebao (Acta Chim. Sinica) (Chin.)* **49**, 751 (1991).

lxxvii Ting-Yu Huang, Chi-Tain Chen, and Han-Mou Gau, *J. Organomet. Chem.*, **489**, 63 (1995).

lxxviii Yanlong Qian, Jiling Huang, Xiaoping Chen, Guisheng Li, Weichun Chen, Bihua Li, Xianglin Jin, and Qingchuan Yang, *Polyhedron*, **13**, 1105 (1994).

lxxix You-Xian Chen, Peng-Fei Fu, C.L. Stern, and T.J. Marks, *Organometallics*, **16**, 5958 (1997).

lxxx A.V. Firth and D.W. Stephan, *Inorg. Chem.*, **37**, 4732 (1998).

(continued overleaf)

Table 6.21 (*Continued*)

Compound	Aryloxide	Bond length (Å) M–O	Bond angle (°) M–O–Ar	Ref.
[Cp*Ti(OAr)$_3$]	OC$_6$HF$_4$-2,3,5,6	1.926 (3)	165	lxxi
		1.867 (3)	137	
		1.830 (3)	157	
[Cp*Ti(OAr)Cl$_2$] *ortho*-PPh$_2$ bound	η^2-OC$_6$H$_3$Bu^t-6-PPh$_2$-2	1.860 (5)	135	lxxii
[Cp*Ti(OAr)Cl$_2$]	OC$_6$H$_3$Me$_2$-2,6	1.785 (2)	162	lxxiii
[CpTi(OAr)Cl$_2$]	OC$_6$H$_3$Pr$^i_2$-2,6	1.760 (4)	163	lxxiv
[Cp*Ti(OAr)Cl$_2$]	OC$_6$H$_3$Pr$^i_2$-2,6	1.772 (3)	173	lxxiv
[Cp*Ti(OAr)Cl$_2$]	OC$_6$H$_3$Me$_2$-2,6	1.785 (2)	162	lxxiv
[CpTi(OAr)Cl$_2$]	OC$_6$H$_3$(CH$_2$CH=CH$_2$)-2-Me-6	1.780 (1)	155	xlii
[CpTi(OAr)Cl$_2$]	OC$_6$H$_2$Ph-2-Bu$^t_2$-4,6	1.785 (2)	160	xlii
[CpTi(OAr)Cl$_2$]	OC$_6$H$_2$Np-2-Bu$^t_2$-4,6	1.786 (1)	159	xlii
[CpTi(OAr)Cl$_2$]	OC$_6$H$_1$(indenyl)-2-Bu$^t_2$-4,6	1.785 (2)	159	xlii
[CpTi(OAr)Cl$_2$]	OC$_6$HPh$_4$-2,3,5,6	1.796 (2)	150	xlii
[CpTi(OAr)Cl$_2$]	OC$_6$HNp$_2$-2,6-Me$_2$-3,5	1.780 (4)	153	xlii
[CpTi(OAr)Cl$_2$]	OC$_6$HNp$_2$-2,6-Bu$^t_2$-3,5	1.774 (3)	164	xlii
[Cp*Ti(OAr)Cl$_2$]	OC$_6$HPh$_2$-2,6-Bu$^t_2$-3,5	1.804 (2)	177	xlii
[Cp*Ti(OAr)Cl$_2$]	OC$_6$HPh$_4$-2,3,5,6	1.817 (2)	158	xlii
[Cp(OAr)Ti(μ-Cl)]$_2$	OC$_6$H$_2$Np-2-Bu$^t_2$-4,6	1.817 (2)	167	xlii
[Cp(OAr)Ti(μ-Cl)]$_2$	OC$_6$H$_1$(indenyl)-2-Bu$^t_2$-4,6	1.819 (2)	155	xlii
[CpTi(OAr)Me$_2$]	OC$_6$H$_2$Np-2-Bu$^t_2$-4,6	1.815 (2)	159	lxxv
[CpTi(OAr)Me$_2$]	OC$_6$H$_2$(tetrahydronaphthyl)-2-Bu$^t_2$-4,6	1.811 (2)	160	xlii
[CpTi(OAr)(C$_6$F$_5$){CH$_2$B(C$_6$F$_5$)$_2$}]	OC$_6$HPh$_2$-2,6-Me$_2$-3,5	1.770 (2)	176	lxxv
[(C$_5$H$_3$Bu$^t_2$-1,3)Ti(OAr)Cl$_2$]	OC$_6$H$_3$Pr$^i_2$-2,6	1.773 (2)	163	lxxiv
[CpTi(OAr)$_2$(HNC$_6$H$_3$Pr$^i_2$-2,6)]	OC$_6$H$_3$Pr$^i_2$-2,6	1.844 (6)	147	lxx
		1.850 (6)	147	
[CpTi(OAr)$_2$(SPh)]	OC$_6$H$_3$Pr$^i_2$-2,6	1.80 (2)	150	lxx
		1.81 (1)	141	

[CpTi(OAr)$_2$(HPPh)]	OC$_6$H$_3$Pr$^i_2$-2,6	1.805 (5) 1.819 (5)	156 172	lxx
[Cp*Ti(OAr)$_2$]$_2$(μ-O) almost linear Ti–O–Ti	OC$_6$HF$_4$-2,3,5,6	1.838 (3) 1.888 (3) 1.844 (3) 1.875 (3)	161 144 162 145	lxxi
[CpTi(OAr)(Cl)(SPh)]	OC$_6$H$_3$Pr$^i_2$-2,6	1.80 (2)	151	lxxvi
[CpTi(OAr)(μ-η^2-S=CHMe)]$_2$	OC$_6$H$_3$Pr$^t_2$-2,6	1.814 (7) 1.816 (7)	162 157	lxxvi
[CpTi(OAr)(SCH$_2$Ph)$_2$]	OC$_6$H$_3$Pr$^i_2$-2,6	1.796 (3)	157	lxxvi
[{CpTi(OAr)(μ-S)}$_2$]	OC$_6$H$_3$Pr$^i_2$-2,6	1.802 (8) 1.819 (8)	156 154	lxxvi
[CpTi(OAr)(SCH$_2$C$_6$H$_4$CH$_2$S)]	OC$_6$H$_3$Pr$^i_2$-2,6	1.793 (5)	162	lxxvi
[CpTi(OAr)(SEt)$_2$] two independent molecules	OC$_6$H$_3$Pr$^i_2$-2,6	1.802 (8) 1.809 (8)	155 159	lxxvi
[{CpTi(OAr)}$_3$(μ_3-S)$_3$(TiCp)] three CpTi(OAr) units capped by a CpTi unit	OC$_6$H$_3$Pr$^i_2$-2,6	1.82 (1) 1.81 (1) 1.79 (1)	152 152 157	lxxvi
[{CpTi(OAr)(μ-Cl)}$_2$]	OC$_6$H$_3$Pr$^i_2$-2,6	1.808 (8) 1.808 (8)	158 156	lxxvii
[{Cp*Ti(OAr)(μ-Cl)}$_2$]	OC$_6$H$_3$Pr$^i_2$-2,6	1.825 (6)	176	lxxvii

[lxx] M.J. Sarsfield, S.W. Ewart, T.L. Tremblay, A.W. Roszak, and M.C. Baird, *J. Chem. Soc., Dalton Trans.*, 3097 (1997).

[lxxi] C.A. Willoughby, R.R. Duff Jr, W.M. Davis, and S.L. Buchwald, *Organometallics*, **15**, 472 (1996).

[lxxiii] P. Gomez-Sal, A. Martin, M. Mena, P. Royo, and R. Serrano, *J. Organomet. Chem.*, **419**, 77 (1991).

[lxxiv] K. Nomura, N. Naga, M. Miki, K. Yanagi, and A. Imai, *Organometallics*, **17**, 2152 (1998).

[lxxv] M.G. Thorn, J.S. Vilardo, P.E. Fanwick, and I.P. Rothwell, *J. Chem. Soc., Chem. Commun.*, 2427 (1998).

[lxxvi] A.V. Firth and D.W. Stephan, *Organometallics*, **16**, 2183 (1997).

[lxxvii] A.V. Firth and D.W. Stephan, *Inorg. Chem.*, **37**, 4726 (1998).

(continued overleaf)

Table 6.21 (*Continued*)

Compound	Aryloxide	Bond length (Å) M–O	Bond angle (°) M–O–Ar	Ref.
[CpTi(OAr)(Cl)]$_2$(μ-O)	OC$_6$H$_3$Pr$^i_2$-2,6	1.80 (1)	147	lxxvii
[CpTi(OAr)$_2$]$_2$(μ-S)	OC$_6$H$_3$Pr$^i_2$-2,6	1.84 (1)	151	lxxvii
		1.80 (1)	153	
		1.80 (1)	153	
		1.81 (1)	149	
[CpTi(OAr)]$_2$(μ-S)(μ-S$_2$)	OC$_6$H$_3$Pr$^i_2$-2,6	1.813 (7)	150	lxxvii
		1.816 (7)	159	
[Cp*Ti(OAr)(S$_5$)]	OC$_6$H$_3$Pr$^i_2$-2,6	1.795 (6)	167	lxxvii
[{CpTi(OAr)(μ-SPh)}$_2$]	OC$_6$H$_3$Pr$^i_2$-2,6	1.828 (2)	161	lxxvii
		1.822 (2)	155	
[CpTi(OAr)(μ_2-O)]$_4$	OC$_6$H$_2$Me$_3$-2,4,6	1.841 (3)	145	lxxviii
		1.840 (4)	142	
		1.815 (4)	154	
		1.818 (4)	147	
Di- and poly-aryloxides				
[Ti(di-OAr)Cl$_2$]	CH$_2$(2-OC$_6$H$_3$Bu^t-6-Me-4)$_2$	1.742 (4)	156	lxxix
		1.761 (4)	144	
[Ti(di-OAr)Me$_2$]	CH$_2$(2-OC$_6$H$_3$Bu^t-6-Me-4)$_2$	1.799 (3)	146	lxxix
[Ti(di-OAr)(BH$_4$)$_2$]	CH$_2$(2-OC$_6$H$_3$Bu^t-6-Me-4)$_2$	1.761 (3)	159	lxxx
		1.791 (3)	143	
[Ti(di-OAr)$_2$]	N(2-OC$_6$H$_3$Bu$^t_2$-4,6)$_2$	1.899 (?)	122	lxxxi
6-coord. Ti		1.872 (?)	126	
[(di-OAr)(OPr^i)Ti(μ_2-OPr^i)$_2$Ti(OPr^i)(di-OAr)]	S(OC$_6$H$_2$Me-4-Bu^t-6)$_2$	1.899 (2)	133	lxxxii
6-coord. Ti		1.885 (2)	132	
[Ti(di-OAr)(Cl)(C$_6$H$_4$CH$_2$NMe$_2$)]	S(OC$_6$H$_2$Me-4-Bu^t-6)$_2$	1.831 (3)	136	lxxxiii
6-coord. Ti		1.879 (3)	135	lxxxii

Compound	Ligand			Ref.
[Ti(di-OAr)Cl₂].THF	*N,N′*-ethylene-bis(salicylideneiminato)	1.835 (5)	137	lxxxiv
6-coord. Ti				
[Ti(di-OAr)Cl(THF)]	*N,N′*-ethylene-bis(salicylideneiminato)	1.826 (?)	148	lxxxv
[Ti(di-OAr)]₂	{CH₂N(BH₃)CH₂(2-OC₆H₄)}₂	1.82 (1)	144	lxxxvi
ligand formed by hydroboration of salen		1.99 (2)	110	lxxxvii
[Ti(di-OAr)Cl₂]	*N,N′*-ethylene-bis(3-Buᵗ-5-Me-salicylideneiminato)	1.816 (4)	140	lxxxviii
6-coord. Ti, *trans* Cl		1.820 (5)	141	
[Ti(di-OAr)Cl(py)].THF	*N,N′*-ethylene-bis(salicylideneiminato)	1.899 (8)	140	lxxxix
6-coord. Ti, Cl *trans* to py		1.921 (6)	136	
[Ti(di-OAr)Cl]₂(μ₂-O)	(R,R)-1,2-diaminocyclohexane-*N,N′*-bis(salicylideneaminato)	1.867 (6)	137	xc
linear oxo bridge, 6-coord. Ti		1.859 (6)	137	
		1.868 (4)	137	
		1.871 (4)	131	
trans-[Ti(di-OAr)(Me)₂]	*N,N′*-ethylene-bis(salicylideneiminato)	1.83 (1)	136	xci
6-coord. Ti		1.86 (1)	135	

lxxviii U. Thewalt and K. Doppert, *J. Organomet. Chem.*, **320**, 177 (1987).

lxxix C. Floriani, F. Corazza, W. Lesueur, A. Chiesi-Villa, and C. Guastini, *Angew. Chem. Int. Ed. Engl.*, **28**, 66 (1989).

lxxx F. Corazza, C. Floriani, A. Chiesi-Villa, and C. Guastini, *Inorg. Chem.*, **30**, 145 (1991).

lxxxi A. Caneschi, A. Dei, and D. Gatteschi, *J. Chem. Soc., Chem. Commun.*, 630 (1992).

lxxxii S. Fokken, T.P. Spaniol, Hak-Chul Kang, W. Massa, and Jun Okuda, *Organometallics*, **15**, 5069 (1996).

lxxxiii L. Porri, A. Ripa, P. Colombo, E. Miano, S. Capelli, and S.V. Meille, *J. Organomet. Chem.*, **514**, 213 (1996).

lxxxiv G. Gilli, D.W.J. Cruickshank, R.L. Beddoes, and O.S. Mills, *Acta Crystallogr. B*, **28**, 1889 (1972).

lxxxv Zhang Shiwei, Lai Luhua, Shao Meicheng, and Tang Youqi, *Wuli Huaxue Xuebao (Acta Phys. -Chim. Sin.)*, **1**, 335 (1985).

lxxxvi G. Dell'Amico, F. Marchetti, and C. Floriani, *J. Chem. Soc., Dalton Trans.*, 2197 (1982).

lxxxvii G. Fachinetti, C. Floriani, M. Mellini, and S. Merlino, *J. Chem. Soc., Chem. Commun.*, 300 (1976).

lxxxviii T. Repo, M. Klinga, M. Leskela, P. Pietikainen, and G. Brunow, *Acta Crystallogr. C*, **52**, 2742 (1996).

lxxxix M. Pasquali, F. Marchetti, A. Landi, and C. Floriani, *J. Chem. Soc., Dalton Trans.*, 545 (1978).

xc T. Aoyama, S. Ohba, Y. Saito, C. Sasaki, M. Kojima, J. Fujita, and K. Nakajima, *Acta Crystallogr. C*, **44**, 1309 (1988).

(continued overleaf)

Table 6.21 *(Continued)*

Compound	Aryloxide	Bond length (Å) M–O	Bond angle (°) M–O–Ar	Ref.
[Ti(di-OAr)(C$_6$H$_2$Me$_3$-2,4,6)] ligand formed by addition of mesityl to imine of salen	(2-mesityl)-*N,N'*-ethylenebis(salicylideneiminato)	1.874 (3) 1.822 (3)	141 138	xci
6-coord. Ti	*N,N'*-ethylene-bis(salicylideneiminato)[(R)-O$_2$CCH(CH$_2$Ph)O)]	1.899 (3) 1.859 (4)	139 138	xcii
6-coord. Ti	(R,R)-1,2-Diaminocyclohexane-*N,N'*-bis(salicylideneaminato) [Ti(di-OAr)(oxalate)]	1.860 (1) 1.832 (2)	140 139	xcii
6-coord. Ti, Ph *trans* to THF	*N,N'*-ethylene-bis(salicylideneiminato) [Ti(di-OAr)(Ph)(thf)]	1.85 (1) 1.89 (1)	138 140	xciii
[[Ti(di-OAr)}$_3$ (μ_2-O)$_2$(thf)$_2$][BPh$_4$]$_2$ linear trimeric cation	*N,N'*-ethylene-bis(salicylideneiminato)	1.850 (7)–1.906 (7)		xciv
[[Ti(di-OAr)}$_4$(μ_2-O)$_2$(thf)$_2$][BPH$_4$]$_2$ linear tetrameric cation	*N,N'*-ethylene-bis(salicylideneiminato)	1.860 (7)–1.897 (6)		xciv
[[Ti(di-OAr)}$_4$(μ_2-O)$_3$][BPH$_4$]$_2$ linear polymeric cation	*N,N'*-ethylene-bis(salicylideneiminato)	1.865 (5)–1.967 (5)		xciv
[[Ti(tri-OAr)][BPh$_4$] 1,4,7-tricyclononane with N-pendant aryloxides, *fac*-arrangement of O and N donor atoms	c-(CH$_2$)$_6${NCH$_2$C$_6$H$_3$Bu^t-4-O}$_3$	1.833 (6) 1.824 (7) 1.828 (7)	140 141 139	xcv
[Ti(tetra-OAr)] 6-coord. Ti	(OC$_6$H$_4$CH$_2$)$_2$NCH$_2$CH$_2$(CH$_2$C$_6$H$_4$O)$_2$	1.903 (2) 1.902 (2) 1.848 (2) 1.847 (2)	140 139 134 138	xcvi

[xci] C. Floriani, E. Solari, F. Corazza, A. Chiesi-Villa, and C. Guastini, *Angew. Chem., Int. Ed. Engl.*, **28**, 64 (1989).
[xcii] K.M. Carroll, J. Schwartz and D.M. Ho, *Inorg. Chem.*, **33**, 2707 (1994).
[xciii] E. Solari, C. Floriani, A. Chiesi-Villa, and C. Rizzoli, *J. Chem. Soc., Dalton Trans.*, 367 (1992).
[xciv] F. Franceschi, E. Gallo, E. Solari, C. Floriani, A. Chiesi-Villa, C. Rizzoli, N. Re, and A. Sgamellotti, *Chem. Eur. J.*, **2**, 1466 (1996).
[xcv] U. Auerbach, T. Weyhermuller, K. Weighardt, B. Nuber, E. Bill, C. Butzlaff, and A.X. Trautwein, *Inorg. Chem.*, **32**, 508 (1993).
[xcvi] H. Hefele, E. Ludwig, W. Bansse, E. Uhlemann, Th. Lugger, E. Hahn, and H. Mehner, *Z. Anorg. Allg. Chem.*, **621**, 671 (1995).

Table 6.22 Zirconium aryloxides

Compound	Aryloxide	Bond length (Å) M–O	Bond angle (°) M–O–Ar	Ref.
Aryloxy adducts, halides, amides etc.				
[Zr(OAr)$_2$Cl$_2$(THF)]	OC$_6$H$_3$Me$_2$-2,6	1.906 (9) 1.904 (8)	167 171	i
[Zr(OAr)$_2$Cl$_2$] 6-coord. Zr, *trans* O, *cis* Cl	OC$_6$H$_4$(c-CCOCHPhCHMeN)-2	1.992 (3) 1.986 (4)	143 145	ii
[Zr(OAr)$_2$Cl$_2$] 6-coord. Zr, *cis* arrangement of Cl, *trans* O, two molecules	OC$_6$H$_4$(CH=NMe)-2	1.972 (4) 1.996 (5) 1.993 (5) 1.983 (4)	144 144 144 144	iii
[Zr(di-OAr)Cl$_2$(THF)] 7-coord. Zr, *trans* arrangement of Cl	N,N'-o-phenylene-bis(salicylaldiminato)	2.015 (2) 2.051 (3)	143 142	iii
[Zr(OAr)$_2$(NHPh)$_2$] 4-coord. Zr	OC$_6$H$_3$Bu^t-2,6	1.941 (7) 1.893 (7)	171 175	iv
[Zr(OAr)$_2$(=NPh)(4-pyrrolidinopyridine)$_2$]	OC$_6$H$_3$Bu^t-2,6	2.047 (7) 2.030 (6)	176 174	iv, v
[Zr(OAr)$_3$(pyrazolylborate)] 6-coord. Zr	OC$_6$H$_3$Me$_2$-2,6	1.967 (3) 1.973 (4) 1.978 (4)	175 177 174	vi
[Zr(OAr)$_2$Cl(pyrazolylborate)] 6-coord. Zr	OC$_6$H$_3$Me$_2$-2,6	1.949 (3) 1.948 (3)	172.3 176.8	vii

[i] H. Yasuda, Y. Nakayama, K. Takei, A. Nakamura, Y. Kai, and N. Kanehisa, *J. Organomet. Chem.*, **473**, 105 (1994).
[ii] P.G. Cozzi, C. Floriani, A. Chiesi-Villa, and C. Rizzoli, *Inorg. Chem.*, **34**, 2921 (1995).
[iii] F. Corazza, E. Solari, C. Floriani, A. Chiesi-Villa, and C. Guastini, *J. Chem. Soc., Dalton Trans.*, 1335 (1990).
[iv] C.H. Zambrano, R.D. Profilet, J.E. Hill, P.E. Fanwick, and I.P. Rothwell, *Polyhedron*, **12**, 689 (1993).
[v] R.D. Profilet, C.H. Zambrano, P.E. Fanwick, J.J. Nash, and I.P. Rothwell, *Inorg. Chem.*, **29**, 4362 (1990).
[vi] R.A. Kresinski, L. Isam, T.A. Hamor, C.J. Jones, and J.A. McCleverty, *J. Chem. Soc., Dalton Trans.*, 1835 (1991).
[vii] R.A. Kresinski, T.A. Hamor, C.J. Jones, and J.A. McCleverty, *J. Chem. Soc., Dalton Trans.*, 603 (1991).

(*continued overleaf*)

Table 6.22 (*Continued*)

Compound	Aryloxide	Bond length (Å) M–O	Bond angle (°) M–O–Ar	Ref.
[Zr(OAr)₃(pyrazolylborate)]	OC₆H₄NO₂-2	1.978 (1)	161	vi, viii
[Zr(OAr)₂{OC(CH₂Ph)=C(CH₂Ph)Nxy)] folded 2-oxa-5-azacyclopentene ring, 4-coord. Zr	OC₆H₃Buᵗ₂-2,6	1.929 (5) 1.967 (5)	170 170	ix, x
[Zr(OAr)₂(xyNCMe=CMeNxy)] folded 2,5-diazacyclopentene ring, 4-coord. Zr	OC₆H₃Buᵗ₂-2,6	2.009 (4) 1.951 (4)	157 152	x, xi
[Zr(OAr)₂(OCHMe-py-CHMeO)(py)] 6-coord. Zr	OC₆H₃Buᵗ₂-2,6	2.006 (5) 2.015 (5)	173 179	xii, xiii
[Zr(OAr)₂{threo-OCH(CH₂Ph)-py-CH(CH₂Ph)O}] 5-coord. Zr	OC₆H₃Buᵗ₂-2,6	1.982 (3) 1.989 (3)	148 146	xiii
[Zr(di-OAr)₂(DMF)]	4-amino-N,N'-disalicylidene-1,2-phenylene-diaminato	2.086 (4) 2.088 (4) 2.075 (4) 2.092 (3)	134 135 138 135	xiv

Non-Cp organometallic aryloxides

Compound	Aryloxide	Bond length (Å) M–O	Bond angle (°) M–O–Ar	Ref.
[Zr(OAr)(CH₂Ph)₃] 4-coord. Zr, some η²-benzyl bonding	OC₆H₃Buᵗ₂-2,6	1.942 (9)	165	xv
[Zr(OAr)(CH₂Ph-4F)₃] 4-coord. Zr, some η²-benzyl bonding	OC₆H₃Buᵗ₂-2,6	1.934 (3)	169	xv
[Zr(OAr)(OAr')(CH₂Ph)₂] 4-coord. Zr	OC₆H₃Buᵗ₂-2,6 OC₆H₂Buᵗ₂-2,6-OMe-4	1.903 (4) 1.936 (4) 1.936 (4)	175 175 175	xv
[Zr(OAr)₂(CH₂Ph)₂] 6-coord. Zr, two molecules	OC₆H₄(c-CCOCH₂CMe₂N)-2	2.001 (3) 2.007 (4)	142 145	xvi
[Zr(OAr)₂Me₂(4,4'-Me₂bipy)] distorted octahedral Zr	OC₆H₃Buᵗ₂-2,6	1.983 (3) 1.990 (3)	161 154	xvii

[Zr(Ind-OAr)₂] 1-indenylphenoxide	OC₆H₂(η^5-indenyl)-2-Bu^t-4,6	2.015 (2)	128	xviii
[Zr$^+$(OAr)₂(CH₂Ph)(η^6-C₆H₅CH₂B$^-$(C₆F₅)₃]	OC₆H₃Ph-2,6	1.893 (3) 1.900 (3)	165 171	xix
[Zr$^+$(OAr)(OC₆H₃Bu^t-CMe₂CH₂)(η^6-C₆H₅CH₂B$^-$(C₆F₅)₃]	OC₆H₃Bu$^t_2$-2,6 OC₆H₃Bu^t-CMe₂CH₂	1.920 (2) 1.922 (2)	167 139	xix
[Zr(OAr)₂(η^2-Bu^{t}N=CCH₂Ph)₂] η^2-iminoacyl, 6-coord. Zr	OC₆H₃Bu$^t_2$-2,6	2.027 (2)	173	ix, xx
[Zr(OAr)(η^2-Bu^{t}N=CCH₂Ph)₃] η^2-iminoacyl, 7-coord. Zr	OC₆H₃Bu$^t_2$-2,6	2.058 (2)	153	xx
[Zr(OAr)₂(η^2-Bu^{t}N=CCH₃)(η^2-NPhNHPh)] η^2-iminoacyl, 5-coord. Zr	OC₆H₃Bu$^t_2$-2,6	2.004 (8) 2.015 (9)	161 156	xxi

[viii] R.A. Kresinski, T.A. Hamor, L. Isam, C.J. Jones, and J.A. McCleverty, *Polyhedron*, **8**, 845 (1989).

[ix] A.K. McMullen, I.P. Rothwell, and J.C. Huffman, *J. Am. Chem. Soc.*, **107**, 1072 (1985).

[x] L.R. Chamberlain, L.D. Durfee, P.E. Fanwick, L.M. Kobriger, S.L. Latesky, A.K. McMullen, B.D. Steffey, I.P. Rothwell, K. Folting, and J.C. Huffman *J. Am. Chem. Soc.*, **109**, 6068 (1987).

[xi] S.L. Latesky, A.K. McMullen, G.P. Niccolai, I.P. Rothwell, and J.C. Huffman, *Organometallics*, **4**, 1896 (1985).

[xii] P.E. Fanwick, L.M. Kobriger, A.K. McMullen, and I.P. Rothwell, *J. Am. Chem. Soc.*, **108**, 8095 (1986).

[xiii] C.H. Zambrano, A.K. McMullen, L.M. Kobriger, P.E. Fanwick, and I.P. Rothwell, *J. Am. Chem. Soc.*, **112**, 6565 (1990).

[xiv] M.L. Illingsworth and A.L. Rheingold, *Inorg. Chem.*, **26**, 4312 (1987).

[xv] S.L. Latesky, A.K. McMullen, G.P. Niccolai, I.P. Rothwell, and J.C. Huffman, *Organometallics*, **4**, 902 (1985).

[xvi] P.G. Cozzi, E. Gallo, C. Floriani, A. Chiesi-Villa, and C. Rizzoli, *Organometallics*, **14**, 4994 (1995).

[xvii] L.M. Kobriger, A.K. McMullen, P.E. Fanwick, and I.P. Rothwell, *Polyhedron*, **8**, 77 (1989).

[xviii] M.G. Thorn, P.E. Fanwick, R.W. Chesnut, and I.P. Rothwell, *J. Chem. Soc., Chem. Commun.*, 2543 (1999).

[xix] M.G. Thorn, Z.C. Etheridge, P.E. Fanwick, and I.P. Rothwell, *J. Organomet. Chem.*, **591**, 148 (1999).

[xx] L.R. Chamberlain, L.D. Durfee, P.E. Fanwick, L. Kobriger, S.L. Latesky, A.K. McMullen, I.P. Rothwell, K. Folting, J.C. Huffman, W.E. Streib, and R. Wang, *J. Am. Chem. Soc.*, **109**, 390 (1987).

[xxi] C.H. Zambrano, P.E. Fanwick, and I.P. Rothwell, *Organometallics*, **13**, 1174 (1994).

(continued overleaf)

Table 6.22 *(Continued)*

Compound	Aryloxide	Bond length (Å) M–O	Bond angle (°) M–O–Ar	Ref.
[Zr(di-OAr)(BH$_4$)$_2$(THF)]	CH$_2$(OC$_6$H$_3$Bu^t-2-Me-4)$_2$	1.922 (3) 1.937 (4)	157 144	xxii
[(cot)(OAr)Zr(μ_2-H)$_2$Zr(OAr)(COT)]	OC$_6$H$_3$Bu$^t_2$-2,6	1.983 (2)		xxiii
[(cot)(OAr)Zr(μ_2-H$_2$CO)Zr(OAr)(COT)]	OC$_6$H$_3$Bu$^t_2$-2,6	1.983 (4)		xxiii
Cyclopentadienyl aryloxides				
[Cp$_2$Zr{(μ_2-η^6-OAr)Cr(CO)$_3$}$_2$] Cr π-bound to phenoxide nucleus	μ_2-η^6-OC$_6$H$_5$	1.992 (6)	154	xxiv
[Cp(pyrazolylborate)Zr(OAr)$_2$]	OC$_6$H$_4$Ph-2	1.987 (5) 2.015 (4)	153 140	vii
[Cp$_2$Zr(OAr)OCPh-W(CO)$_5$]	OC$_6$H$_5$	1.971 (6)	155	xxv
[Cp$_2$Zr(OAr)(CH$_2$Ph)]	OC$_6$H$_2$Me$_2$-2,4-(α-Me-c-C$_6$H$_{10}$)-6	1.981 (6)	165	xxvi
[Cp$_2$Zr(OAr)(η^2-Bu^{t}N=CCH$_2$Ph)] η^2-iminoacyl	OC$_6$H$_3$Bu$^t_2$-2-Me-6	2.079 (3)	155	xxvi
[Cp$_2$(ArO)Zr(μ_2-H)(μ_2-NBu^t)IrCp]	OC$_6$H$_4$Me-4	2.083 (4)	144	xxvii xxviii
[Cp$_2$Zr(OAr)(S$_2$CNMe$_2$)]	OC$_6$H$_5$	2.023 (8)	160	xxix
[Cp$_2$Zr(OAr)$_2$]	OC$_6$F$_5$	1.991 (2)	164	xxx
[Cp$_2$Zr(OAr)$_2$]	OC$_6$H$_4$(PPh$_2$)-2	1.979 (7) 2.004 (1)	160 144	xxxi
[Cp$_2$Zr(OAr)Cl]	OC$_6$H$_4$(PPh$_2$)-2	1.993 (6)	144	xxxi
[Cp$_2$Zr(OAr)$_2$]	OC$_6$H$_5$	2.01 (1)	147	xxxii
[Cp*$_2$Zr(OAr)$_2$]	OC$_6$H$_5$	1.989 (3)	173	xxxii
[Cp$_2$Zr(OAr)(μ_2-O)Zr(OAr)Cp$_2$]	OC$_6$H$_4$Cl-4	2.00 (2) 1.99 (2)	153 177	xxxiii
[Cp$_2$Zr(OAr)(μ_2-O)Zr(OAr)Cp$_2$]	OC$_6$H$_5$	2.073 (7)	153	xxxiv
[(C$_5$H$_4$Me)$_2$Zr(OAr)$_2$]	OC$_6$H$_3$Cl$_2$-2,6	2.03 (1)	148	xxxv

[Cp$_2$Zr(OAr)] 18-electron Zr, two 5-membered metallacycles	OC$_6$H$_4$(N=CHCSiMe$_3$=CSiMe$_3$)-2	2.144 (2)	123	xxxvi
[Cp$_2$Zr(OAr)]$_2$ two Zr bridged by N	OC$_6$H$_4$(CH=N)-2	2.101 (4)	140	xxxvi
[Cp*$_2$ZrCl(*p*-quinone)/ZrClCp*$_2$]	*p*-quinone	1.973 (3) 1.974 (3)	158 157	xxxvii
[CpZr(OAr)(CH$_2$Ph)$_2$]	OC$_6$H$_2$Ph-2-Bu$^t_2$-4,6	1.954 (2)	159	xxxviii
[CpZr(OAr)(η^2-Bu^{t}NCCH$_2$Ph)(CH$_2$Ph)] η^2-iminoacyl	OC$_6$H$_2$Ph-2-Bu$^t_2$-4,6	2.009 (5)	157	xxxviii
[CpZr(OAr)(η^2-Bu^{t}NCCH$_2$Ph)$_2$] η^2-iminoacyl	OC$_6$H$_2$Ph-2-Bu$^t_2$-4,6	2.056 (2)	148	xxxviii
[CpZr(OAr)(η^2-Bu^{t}NCCH$_2$Ph)$_2$] η^2-iminoacyl	OC$_6$H$_2$Np-2-Bu$^t_2$-4,6	2.048 (2)	152	xxxviii

xxiiF. Corazza, C. Floriani, A. Chiesi-Villa, and C. Guastini, *Inorg. Chem.*, **30**, 145 (1991).

xxiiiP. Berno, C. Floriani, A. Chiesi-Villa, and C. Guastini, *J. Chem. Soc., Chem. Commun.*, 109 (1991).

xxivJ.A. Heppert, T.J. Boyle, and F. Takusagawa, *Organometallics*, **8**, 461 (1989).

xxvG. Erker, U. Dorf, R. Lecht, M.T. Ashby, M. Aulbach, R. Schlund, C. Kruger, and R. Mynott, *Organometallics*, **8**, 2037 (1989).

xxviB.D. Steffey, Nhan Truong, D.E. Chebi, J.L. Kerschner, P.E. Fanwick, and I.P. Rothwell, *Polyhedron*, **9**, 839 (1990).

xxviiA.M. Baranger, F.J. Hollander, and R.G. Bergman, *J. Am. Chem. Soc.*, **115**, 7890 (1993).

xxviiiA.M. Baranger and R.G. Bergman, *J. Am. Chem. Soc.*, **116**, 3822 (1994).

xxixD.A. Femec, T.L. Groy and R.C. Fay, *Acta Crystallog. C*, **47**, 1811 (1991).

xxxJ.I. Amor, N.C. Burton, T. Cuenca, P. Gomez-Sal, and P. Royo, *J. Organomet. Chem.*, **485**, 153 (1995).

xxxiL. Miquel, M. Basso-Bert, R. Choukroun, R. Madhouni, B. Eichhorn, M. Sanchez, M.-R. Mazieres, and J. Jaud, *J. Organomet. Chem.*, **490**, 21 (1995).

xxxiiW.A. Howard, T.M. Trnka, and G. Parkin, *Inorg. Chem.*, **34**, 5900 (1995).

xxxiiiYang Qingchuan, Jin Xianglin, Xu Xiaojie, Li Genpei, Tang Youqi and Chen Shoushan, *Sci. Sin., Ser. B (Engl. Ed.)*, **25**, 356 (1982).

xxxivChang Wenrui, Dai Jinbi, and Chen Shoushan, *Jiegou Huaxue (J. Struct. Chem.)*, **1**, 73 (1982).

xxxvDai Jinbi, Lou Meizhen, Zhang Jiping, and Chen Shoushan, *Jiegou Huaxue (J. Struct. Chem.)*, **1**, 63 (1982).

xxxviP. Arndt, C. Lefeber, R. Kempe, and U. Rosenthal, *Chem. Ber.*, **129**, 207 (1996).

xxxviiA. Kunzel, M. Sokolow, Feng-Quan Liu, H.W. Roesky, M. Noltemeyer, H.-G. Schmidt, and I. Uson, *J. Chem. Soc., Dalton Trans.*, 913 (1996).

xxxviiiM.G. Thorn, J.T. Lee, P.E. Fanwick, and I.P. Rothwell, unpublished results.

Table 6.23   Hafnium aryloxides

Compound	Aryloxide	Bond length (Å) M–O	Bond angle (°) M–O–Ar	Ref.
[Hf(OAr)₃Cl]$[Hf(OAr)_3Cl]$	$OC_6H_3Bu^t_2$-2,6	1.925 (2)	152	i, ii
tetrahedral Hf		1.917 (3)	156	
		1.938 (3)	160	
$[Cp_2Hf(OAr)_2]$	$OC_6H_3Cl_2$-2,6	2.004 (7)	147	iii
pseudo-tetrahedral Hf		2.022 (8)	149	
$[Hf(OAr)_2(CH_2py\text{-}6\text{-}Me)_2]$	$OC_6H_3Bu^t_2$-2,6	1.996 (3)	173	iv
6-coord. Hf		2.190 (1)	175	
$[Hf(OAr)_2(\eta^2\text{-}PhN{=}CMe)_2]$	$OC_6H_3Bu^t_2$-2,6	1.978 (4)	171	v
6-coord. Hf				
$[Hf(di\text{-}OAr)Cl_2(THF)]$	N,N'-o-phenylene-	2.008 (3)	143	vi
7-coord. Hf, *trans* arrangement of Cl	bis(salicylaldiminato)	2.048 (4)	140	
$[Hf(OAr)_2(CH_2Ph)(THF)][BPh_4]$	$OC_6H_4(c\text{-}$	2.01 (1)	146	vii
6-coord. Hf cation	$CCOCH_2CMe_2N)$-2	1.99 (1)	137	

ⁱL.R. Chamberlain, J.C. Huffman, J. Keddington, and I.P. Rothwell, *J. Chem. Soc., Chem. Commun.*, 805 (1982).
ⁱⁱS.L. Latesky, J. Keddington, A.K. McMullen, I.P. Rothwell, and J.C. Huffman, *Inorg. Chem.*, **24**, 995 (1985).
ⁱⁱⁱDou Shiqi and Chen Shoushan, *Gaodeng Xuexiao Huaxue Xuebao (Chem. J. Chin. Uni.)*, **5**, 812 (1984).
ⁱᵛS.M. Beshouri, P.E. Fanwick, I.P. Rothwell, and J.C. Huffman, *Organometallics*, **6**, 2498 (1987).
ᵛL.R. Chamberlain, L.D. Durfee, P.E. Fanwick, L. Kobriger, S.L. Latesky, A.K. McMullen, I.P. Rothwell, K. Folting, J.C. Huffman, W.E. Streib, and R. Wang, *J. Am. Chem. Soc.*, **109**, 390 (1987).
ᵛⁱF. Corazza, E. Solari, C. Floriani, A. Chiesi-Villa, and C. Guastini, *J. Chem. Soc., Dalton Trans.*, 1335 (1990).
ᵛⁱⁱP.G. Cozzi, E. Gallo, C. Floriani, A. Chiesi-Villa, and C. Rizzoli, *Organometallics*, **14**, 4994 (1995).

The dichlorides [(ArO)₂TiCl₂] are an important class of starting materials for entry into the organometallic chemistry of titanium.[358] For 2,6-disubstituted-phenoxides, structural studies show a monomeric, pseudo-tetrahedral geometry in the solid state (Table 6.21). The dichlorides themselves can be used as Lewis acid catalysts for Diels–Alder reactions.[359] Careful alkylation can lead to the corresponding dialkyls, [(ArO)₂TiR₂] (R = Me, benzyl).[360,361] An alternative approach involves addition of the parent phenol to the alkyl substrate [TiR₄] in hydrocarbon solvent. For R = Me, Ph these alkyls are generated *in situ* whereas for R = CH₂Ph, CH₂SiMe₃ the relatively stable tetra-alkyls can be isolated. The stoichiometry of the product depends strongly on reaction conditions, the bulk of the aryloxide and nature of the alkyl group.

In the case of zirconium and hafnium the lack of accessible dichlorides means that alkyl derivatives are generated *via* the corresponding homoleptic alkyl intermediates. With bulky 2,6-di-*tert*-butylphenoxide the tris-benzyls [(ArO)ZrR₃] can be isolated and studied.[134] The dialkyls [(ArO)₂MR₂] (M = Zr, Hf; R = Me, benzyl) can also be isolated for bulky aryloxides. In the case of 2,6-diphenylphenols the monomethyl species [(ArO)₃ZrMe] are readily formed by simple addition of parent phenol to the dimethyl compound.

The thermal stability of the dialkyls $[(ArO)_2MR_2]$ is a function of all three variables: metal, aryloxide, and alkyl. With 2,6-di-*tert*-butylphenoxide the bis-benzyls for all three metals undergo elimination of toluene and formation of cyclometallated products generated by activation of tert-*butyl* CH bonds (see Section 5.1).

The migratory insertion of organic isocyanides ($RN\equiv C$) into the metal–alkyl bonds of $[(ArO)_xMR_{4-x}]$ species has been extensively investigated.[362] The initial product of these reactions is typically the corresponding η^2-iminoacyl derivatives. The structural and spectroscopic properties of these derivatives has been reviewed and discussed in terms of the bonding of the iminoacyl function.[363] In the case of titanium the mono-η^2-iminoacyls $[(ArO)_2Ti(\eta^2$-$R'NCR)(R)]$ react with pyridine to generate the η^2-imines $[(ArO)_2Ti(\eta^2$-$R'NCR_2)(py)]$.[364,365] The structural parameters for these compounds fit an aza-titanacyclopropane bonding picture. Further reaction with nitrogen heterocycles leads to elimination of imine (eneamine tautomers) and generation of low valent titanium aryloxides.[366] For simple pyridine, radical-induced coupling of py groups leads to a dimeric derivative of Ti(III). In the case of 4-phenylpyridine, a five-coordinate, mononuclear compound $[(ArO)_2Ti(py-4Ph)_3]$ is produced. Structural studies rule out formulation as a d^2–Ti(II)-containing molecule as the equatorial pyridine (tbp) has clearly undergone one electron reduction.[367] The octahedral species $[(ArO)_2Ti(bipy)_2]$ has a *cis* arrangement of aryloxide groups. Here again it is probable that significant amounts of electron density are residing on the bipy ligands. This uncertainty in formal metal oxidation state is absent in the compound *trans*-$[Ti(OAr)_2(dmpe)_2]$ where the supporting donor ligands are redox innocent.[368]

A series of bis-η^2-iminoacyl compounds $[(ArO)_2Zr(\eta^2$-$R'NCR)_2]$ has been isolated by insertion of two equivalents of isocyanide with the corresponding di-alkyls (Table 6.22). Subsequent thermolysis leads to an intramolecular coupling with formation of enediamido derivatives.[369,370] Related titanium compounds are generated directly without isolation of the intermediate bis-η^2-iminoacyl. A non-planar conformation for the di-aza-metallacyclopent-3-ene ring has been shown both in the solid state and solution. Variable-temperature NMR studies have allowed measurement of the barriers to flipping of these folded rings. The formation of the C=C double bond by coupling at zirconium metal centres has been kinetically studied and shown to be a first order, intramolecular event. The use of differing aryl-substituted isocyanides showed the coupling reaction to be accelerated by electron-withdrawing substituents.[371]

The reaction of mixed alkyl, aryloxides of the group 4 metals with CO has not yielded simple insertion products. However, final products are clearly the result of subsequent reactivity of η^2-acyl intermediates. The compound $[(ArO)_2Ti(\eta^2$-$R'NCR)(R)]$ undergoes reaction with CO to produce the non-planar oxa-aza-metallacycle $[(ArO)_2Ti\{OC(R)=C(R)N(R')\}]$.[362] Carbonylation of $[(ArO)_2MR_2]$ in the presence of pyridine remarkably leads to pyridine-di-methoxide complexes. The *threo* and *erythro* isomeric mixture can be rationalized on the basis of nucleophilic attack of acyl groups on coordinated pyridine ligands.[372,373]

The insertion chemistry of the di-alkyls $[Cp(ArO)Zr(CH_2Ph)_2]$ with organic isocyanides also leads to mono- and bis-η^2-iminoacyl derivatives. Their spectroscopic and structural parameters are very similar to the corresponding bis-aryloxides.[374]

The reaction of the bis-benzyls $[(ArO)_2M(CH_2Ph)_2]$ with the Lewis acidic $[B(C_6F_5)_3]$ leads to the zwitterionic species $[(ArO)_2M^+(CH_2Ph)\{(\eta^6$-$C_6H_5CH_2)B^-(C_6F_5)_3\}]$.[375] Structural studies show the M–OAr distances to be among the shortest for Ti and

Zr aryloxides, reflecting the high electrophilicity of the metal centres. The addition of alkenes or alkynes to the titanium species leads to cationic mono-insertion products in which chelation *via* the original benzyl phenyl ring occurs with displacement of the boron anion. The corresponding titanium di-methyl substrates form cationic species with $[B(C_6F_5)_3]$ which decompose to a mixture of $[(ArO)_2Ti(C_6F_5)Me]$ and $[MeB(C_6F_5)_2]$.[361] In the presence of ethylene or propylene, polymerization by the cationic methyl species occurs to yield polymers (low poly-dispersities) whose molecular weights are strongly dependent on the nature of the aryloxide ligand. The polypropylene formed is atactic with predominantly vinylidene end groups (1,2-insertion with β-hydrogen elimination termination step).[361]

Chelating diaryloxo derivatives of titanium have been applied to the polymerization of olefins.[53–57] The nature of the bridging ligand has been shown to strongly affect the activity of catalyst precursors $[(\text{di-ArO})TiX_2]$ activated with MAO. For example the highest activity for the syndiospecific polymerization of styrene was found to be for the sulfur-bridged ligand.[376] Coordination of the sulfur atom is believed to be important and was demonstrated in one case by crystallographic studies.[55]

The derivatives $[Cp'(ArO)TiX_2]$ (ArO = 2,6-dialkylphenoxide; X = Cl, Me, O_3SCF_3)[377] and $[Cp^*Ti(OC_6F_5)X_2]$[378] have been shown to be active catalyst precursors for the polymerization of olefins. Reaction of $[Cp^*TiMe_2(OC_6F_5)]$ with the Lewis acid $[B(C_6F_5)_3]$ leads to $[Cp^*TiMe(OC_6F_5)(\mu\text{-Me})B(C_6F_5)_3]$ which is in equilibrium with its ion pairs. In contrast $[Cp^*Ti(OC_6F_5)_2][MeB(C_6F_5)_3]$ exists in solution as separated ion pairs.[378] In the case of the compounds $[Cp(ArO)TiMe_2]$, (ArO = 2,6-diarylphenoxide) activation with $[B(C_6F_5)_3]$ again leads to thermally unstable cationic species. In this case the use of chiral, *ortho*-(1-naphthyl)phenoxides has allowed the molecular dynamics of these species to be studied in detail by low-temperature NMR.[356] Decomposition in this case leads to elimination of methane and formation of species $[Cp(ArO)Ti\{CH_2B(C_6F_5)_2\}(C_6F_5)]$.

The reduction of the dichlorides $[(ArO)_2TiCl_2]$ in the presence of unsaturated organic substrates can lead to a variety of stable metallacyclic compounds.[351,379–381] The reduction can be achieved using sodium amalgam or by alkylation with two equivalents of *n*-BuLi. These are interesting molecules in their own right as well as being key intermediates in a number of catalytic cycles. The addition of ketones or imines to titanacyclopentadiene or titanacyclopentane species can each lead to simple ring expansion. However, in a number of cases elimination of alkyne or olefin respectively can occur. This indicates that metallacycle formation is a reversible process (see below). The titanacyclopentadiene species generated from alkynes and di-ynes will carry out the cyclotrimerization of alkynes to arenes as well as the selective catalytic $(2+2+2)$ cycloaddition of olefins with alkynes to produce the 1,3-cyclohexadiene nucleus.[382] The observed regio- and stereochemistry of the 1,3-cyclohexadienes is controlled by the structure and isomerization of intermediate titana-norbornene compounds generated by pseudo-Diels–Alder addition of olefin to titanacyclopentadiene rings. The titanacyclopentane rings formed by coupling of olefins exhibit much greater thermal stability than their titanocene counterparts.[351] In solution the titanacyclopentane formed by coupling of ethylene demonstrates fluxionality consistent with facile fragmentation to a bis(ethylene) species on the NMR time scale. The structure of the metallacycle formed from styrene has a *trans*-2,5-regiochemistry in the solid state. However, the slow, catalytic dimerization of styrene by this species in solution produces a dimer that can only arise *via* the 2,4-regioisomer.[351] Hence in solution a rapidly interconverting isomeric mixture of metallacycles are present. The addition of donor ligands such as PMe_3 to these titanacyclopentanes produces titanacyclopropane and one equivalent of olefin. The titana-bicycle produced from 1,7-octadiene initially has a *cis* configuration but in solution will rapidly produce the *trans* species. At 100°C this compound will catalyse the cyclization of 1,7-octadiene to 2-methyl-methylenecyclohexane.[351]

Although β-hydrogen abstraction from five-membered titanacycle rings is slow, expansion to a seven-membered ring containing β-hydrogens can lead to facile abstraction/elimination pathways. This principle underlies a number of highly selective coupling and cross-coupling reactions that can be catalysed by titanium aryloxide compounds. Hence, a mixture of 2,3-dimethylbutadiene and α-olefins can be selectively dimerized by the titanacyclopent-3-ene compound $[(ArO)_2Ti(MCH_2CMe{=}CMeCH_2)]$ to form 1,4-hexadienes.[383] A related coupling of 1,3-*cyclo*-octadiene produces 3-vinyl-*cyclo*-octenes.[358] Both reactions are overall 1,4-hydrovinylations of dienes and in the case of 1,3-*cyclo*-octadiene the reaction has been shown to occur in a *cis* fashion. The key to the selectivity hinges on initially formed 2-vinyl-titanacyclopentanes which undergo facile ring expansion *via* 1,3-allylic shifts to produce titanacyclohept-3-ene rings. Abstraction from the β-position originating from the olefin produces the observed products.

In the case of 1,3-cyclohexadiene, addition to any of the titanacyclopentane compounds or to a mixture of $\{[(ArO)_2TiCl_2]/2n\text{-}BuLi\}$ leads to the rapid formation of a non-Diels–Alder dimer. Following complete dimerization, subsequent isomerization via metal mediated 1,5-hydrogen shifts occurs as well as coupling of dimers to higher oligomers.[358] The regio- and in particular the exclusive *threo* stereochemistry of the initially produced dimer is due to the coupling of 1,3-cyclohexadiene to produce a *cis-anti-cis*-titanacycle. An allylic shift followed by β-hydrogen abstraction/elimination process leads to the *threo* isomer. A logical extension of this work is the cross-coupling of olefins with 1,3-cyclohexadiene. Again the products can be rationalized using the sequence of coupling to a five-membered ring, expansion via 1,3 shifts and β-abstraction/elimination.[358]

6.2.6   Group 5 Metal Aryloxides

For vanadium, the use of available lower valent halides has allowed the isolation of simple aryloxides with the metal in a variety of formal oxidation states (Table 6.24). Important mononuclear examples include $[V(OC_6H_3Me_2\text{-}2,6)_3(py)_2]$[384] and square planar $[V(OC_6H_2Bu^t_2\text{-}2,6\text{-}Me\text{-}4)_2(py)_2]$.[385] A variety of vanadium(v) aryloxides $[(X)V(OAr)_3]$ have also been isolated containing oxo $(X = O)$,[386,387] and imido $(X = NR)$[388–390] ligands.

Many early examples of niobium and tantalum aryloxide compounds were obtained *via* the pentahalides by reaction with either the parent phenol or alkali metal aryloxides.[391–395] The homoleptic aryloxides, monomeric trigonal bipyramidal for bulky ligands, *e.g.* $[Nb(OC_6H_3Me_2\text{-}2,6)_5]$[396] and edge-shared bi-octahedra for small ligands, *e.g.* $[Ta_2(\mu_2\text{-}OC_6H_4Me\text{-}4)_2(OC_6H_4Me\text{-}4)_8]$[397] have been used as precursors for lower valent derivatives. The mixed chloro, aryloxides $[M(OAr)_xCl_{5-x}]$ (Tables 6.24–6.25) are an important group of starting materials. Structural studies typically show a square pyramidal geometry for bulky aryloxides, *e.g.* $[Ta(OC_6H_3Bu^t_2\text{-}2,6)_xCl_{5-x}]$ $(x = 2, 3;$ axial OAr), but the compounds $[M(\mu_2\text{-}Cl)(OC_6H_3Pr^i_2\text{-}2,6)_2Cl_2]_2$ (M = Nb, Ta) are

Table 6.24 Vanadium aryloxides

Compound	Aryloxide	Bond length (Å) M–O	Bond angle (°) M–O–Ar	Ref.
Homoleptic aryloxides and adducts				
[V₂(μ₂-OAr)₂(OAr)₄(THF)] 4- and 5-coord. V	OC₆H₃Me₂-2,6	1.818 (4), 1.836 (4), 1.842 (4), 1.831 (3)	154, 140, 158, 135	i
	μ₂-OC₆H₃Me₂-2,6	1.999 (4), 1.972 (3), 1.926 (3), 2.084 (4)		
[V(OAr)₃(py)₂] tbp V, axial py	OC₆H₃Me₂-2,6	1.870 (3), 1.861 (3), 1.869 (3)	137, 150, 131	i
[V(OAr)₂(py)₂] square planar V	OC₆H₂Bu$^t_2$-2,6-Me-4	1.916 (2)	156	ii
[V(OAr)₂(TMEDA)] 6-coord. V, chelating 2OMe phenoxides	OC₆H₄OMe-2	2.068 (4), 2.105 (3)	121, 123	ii
[V(OAr)₂(TMEDA)] distorted 4-coord. V	OC₆H₃Ph₂-2,6	2.000 (3), 2.038 (3)	142, 138	ii
[V(OAr)₂(py)₃] square pyramidal V, axial py	OC₆H₃Ph₂-2,6	2.025 (4)	146	ii
[(DME)₃Li₃(μ₂-OAr)₆V] octahedral V	μ₂-OC₆H₅	Li–O–V 1.878 (3), 2.007 (1), 1.902 (3), 2.021 (1)		iii
[(THF)Li(μ₂-OAr)₂V(OAr)₂] two independent molecules, 4-coord. V	OC₆H₃Pr$^i_2$-2,6	1.820 (2)	146	iii
	μ₂-OC₆H₃Pr$^i_2$-2,6	1.843 (2)	146	

Compound	Ligand	Distances (Å)		Angle	Ref.
[(thf)Li(μ₂-OAr)₂V(μ₂-(OAr)₂Li(THF)] square planar V	μ₂-OC₆H₃Pr$^i_2$-2,6	1.839 (2), 1.854 (2)		138 140	iv
		Li–O–V			
		1.842 (7), 1.957 (3)			
		1.850 (7), 1.895 (2)			
		1.846 (7), 1.934 (3)			
		1.864 (7), 1.931 (2)			
[(12-crown-4)Li][V(OAr)₄] 12-crown-4 solvate, crystallographic S₄ symmetry	OC₆H₃Pr$^i_2$-2,6	1.865 (3)		142	iii
		Li–O–V			
		1.830 (7), 2.005 (2)			
		1.811 (7), 2.034 (3)			
		1.835 (7), 2.034 (2)			
		1.835 (6), 2.010 (2)			
[PPh₄][V₃(μ₂-OAr)₆] terminally bound S, 6-coord. V, linear confacial tris-octahedron	OC₆H₄S-2	2.009 (3), 1.933 (3)			v
		2.001 (3), 1.983 (3)			
		1.994 (3), 1.985 (3)			
		2.025 (3), 1.972 (3)			
		2.006 (3), 1.984 (3)			
		2.017 (3), 1.963 (3)			

Mixed Aryloxy Halides, Amides etc.

Compound	Ligand	Distances (Å)		Angle	Ref.
[V(OAr)Cl₂(THF)₂]	OC₆H₃Me₂-2,6	1.803 (5)		136	vi

(continued overleaf)

[i] S. Gambarotta, F. van Bolhuis, and M.Y. Chiang, *Inorg. Chem.*, **26**, 4301 (1987).

[ii] R.K. Minhas, J.J.H. Edema, S. Gambarotta, and A. Meetsma, *J. Am. Chem. Soc.*, **115**, 6710 (1993).

[iii] W.C.A. Wilisch, M.J. Scott, and W.H. Armstrong, *Inorg. Chem.*, **27**, 4333 (1988).

[iv] M.J. Scott, W.C.A. Wilisch, and W.H. Armstrong, *J. Am. Chem. Soc.*, **112**, 2429 (1990).

[v] Beisheng Kang, Linghong Weng, Hanqin Liu, Daxu Wu, Liangren Huang, Cenzhong Lu, Jinghua Cai, Xuetai Chen, and Jiazi Lu, *Inorg. Chem.*, **29**, 4873 (1990).

[vi] M. Mazzanti, C. Floriani, A. Chiesi-Villa, and C. Guastini, *J. Chem. Soc., Dalton Trans.*, 1793 (1989).

Table 6.24 (*Continued*)

Compound	Aryloxide	Bond length (Å) M–O	Bond angle (°) M–O–Ar	Ref.
[V(OAr)₂Cl(THF)₂] tbp, axial THF	OC₆H₃Pr$^i_2$-2,6	1.865 (3)		vii
[V(OAr)₂Cl(THF)] 6-coord. V, *trans* O, Cl *trans* to THF	OC₆H₄(*c*-CCOCH₂CMe₂N)-2	1.893 (4), 1.927 (3)	134, 134	viii
[V(L)(OAr)(THF)] L = *N,N'*-ethylenebis(acetylacetoneiminato), 6-coord. V	OC₆H₅	1.881 (3)	133	ix

Oxo and imido aryloxides

Compound	Aryloxide	Bond length (Å) M–O	Bond angle (°) M–O–Ar	Ref.
[V(O)(OAr)₃] pyramidal V	OC₆H₃Pr$^i_2$-2,6	1.761 (4), 1.761 (2)	136, 166	vii
[{V(O)₂(OAr)₂}₂{Li(THF)₂}₂] tetrahedral V	OC₆H₃Pr$^i_2$-2,6	1.810 (9), 1.812 (8), 1.791 (8), 1.798 (7)		xv
[{HB(Me₂pz)₃}V(O)(OAr)₂] 6-coord. V	OC₆H₄Br-4	1.822 (4), 1.857 (4)	151, 129	x
Na[V(O)(OAr)] 5-coord. V, trianionic ligand, both N deprotonated	(OC₆H₄-2)CON(C₆H₄)NCO(py)]	1.887 (4)	120	xi
[V(O)(OAr)(bipy)] 6-coord. V	(2-OC₆H₃Me-4)N=N(2-C₆H₄COO)	1.895 (5)	127	xii
[V₂(μ₂-O)₂(O)₂(OAr)] 6-coord. V, both N of OAr bound but alkoxide not	OC₆H₄{CH₂NH(CH₂)₂NH(CH₂)₂OH}-2	1.905 (1)	132	xiii
[V(=NC₆H₃Pr$^i_2$-2,6)(OAr)(CH₂Ph)₂] pyramidal V	OC₆H₃Pr$^i_2$-2,6	1.746 (4)		xiv
[V(=NNMe)(OAr)₃] pyramidal V	OC₆H₃Pr$^i_2$-2,6	1.783 (3), 1.794 (2), 1.801 (3)		xv

Compound	Ligand	Bond lengths (Å)	Page	Ref.
[V(OAr)₃-N-Li(THF)₃] tetrahedral V	OC₆H₃Pr^{i}₂-2,6	1.828 (4), 1.827 (4); 1.830 (4)		xv
[{V(μ₂-OAr)(OAr)Cl(=NC₆H₄Me-4)}₂] tbp V, imido *trans* to μ₂-OAr	OC₆H₃Me₂-2,6; μ₂-OC₆H₃Me₂-2,6	1.797 (3); 1.888 (3), 2.283 (3)	132	xvi
Di- and poly-aryloxides				
[V(O)(di-OAr)] square planar V, axial oxo	{CH₂N=CMeCH₂C=O(C₆H₄O)}₂	1.946 (6); 1.916 (6)	122; 128	xvii
[NH₄][V(di-OAr)(O)] one phenoxide coordinated, one protonated, 6-coord. V	(CH₂)₂{NHCH(COO)(OC₆H₄-2)₂	1.950 (2)	130	xviii
K[V(O)₂(di-OAr)] 5-coord. V	(OC₆H₄-2)N=N(OC₆H₄-2)	1.953 (2); 1.912 (2)	120; 136	xix
[V₂(μ₂-O)(O)₂(di-OAr)₂] 5-coord. V	(OC₆H₄-2)N=N(OC₆H₄-2)	1.851 (3); 1.832 (3)	126; 127	xix

(continued overleaf)

[vii] R.A. Henderson, D.L. Hughes, Z. Janas, R.L Richards, P. Sobota, and S. Szafert, *J. Organomet. Chem.*, **554**, 195 (1998).

[viii] P.G. Cozzi, C. Floriani, A. Chiesi-Villa, and C. Rizzoli, *Inorg. Chem.*, **34**, 2921 (1995).

[ix] M. Mazzanti, C. Floriani, A. Chiesi-Villa, and C. Guastini, *Inorg. Chem.*, **25**, 4158 (1986).

[x] S. Holmes and C.J. Carrano, *Inorg. Chem.*, **30**, 1231 (1991).

[xi] T.A. Kabanos, A.D. Keramidas, A.B. Papaioannou, and A. Terzis, *J. Chem. Soc., Chem. Commun.*, 643 (1993).

[xii] J. Chakravarty, S. Dutta, S.K. Chandra, P. Basu, and A. Chakravorty, *Inorg. Chem.*, **32**, 4249 (1993).

[xiii] G.J. Colpas, B.J. Hamstra, J.W. Kampf, and V.L. Pecoraro, *Inorg. Chem.*, **33**, 4669 (1994).

[xiv] V.J. Murphy and H. Turner, *Organometallics*, **16**, 2495 (1997).

[xv] R.A. Henderson, Z. Janas, L.B. Jerzykiewicz, R.L. Richards, and P. Sobota, *Inorg. Chim. Acta.*, **285**, 178 (1999).

[xvi] D.D. Devore, J.D. Lichtenhan, F. Takusagawa, and E.A. Maatta, *J. Am. Chem. Soc.*, **109**, 7408 (1987).

[xvii] N.A. Bailey, D.E. Fenton, C.A. Phillips, U. Casellato, S. Tamburini, P.A. Vigato, and R. Graziani, *Inorg. Chim. Acta*, **109**, 91 (1985).

[xviii] P.E. Riley, V.L. Pecoraro, C.J. Carrano, J.A. Bonadies, and K.N. Raymond, *Inorg. Chem.*, **25**, 154 (1986).

[xix] S. Dutta, P. Basu, and A. Chakravorty, *Inorg. Chem.*, **32**, 5343 (1993).

Table 6.24 *(Continued)*

Compound	Aryloxide	Bond length (Å) M–O	Bond angle (°) M–O–Ar	Ref.
[V(O)(di-OAr)] 6-coord. V, one py bound	(OC₆H₄-2)CH₂N(CH₂py)(CH₂)₂N(CH₂py)(OC₆H₄-2)	1.920 (2)	132	xx
		1.926 (3)	131	
Chloro-(bis(2-hydroxy-5-methyl-3-*t*-butylphenyl)methane-*O,O'*)-bis(tetrahydrofuran-*O*)-vanadium(III)		1.832 (4)	132	xxi
		1.847 (4)	129	
[V(di-OAr)(O)Cl] tetrahedral V	(OC₆H₃Buᵗ-2-Me-4)₂CH₂	1.752 (4)	142	xxii
		1.734 (3)	154	
[{V(di-OAr)(O)}₂(μ₂-O)] 5-coord. V	(OC₆H₄-2)N=N(OC₆H₄-2)	1.839 (3)	126	xxiii
		1.831 (4)	127	xxiv
[V(di-OAr)(O)(OMe)(HOMe)] 6-coord. V	(OC₆H₄-2)N=N(OC₆H₄-2)	1.939 (3)	120	xxiv
		1.870 (3)	128	
K₂[V(O)(di-OAr)] Square pyramidal V, two forms, tetra-anionic ligand	Et₂C{C(O)N(2-OC₆H₄)}₂	1.954 (4)	113	xxv
		1.944 (4)	114	
		1.937 (5)	113	
		1.938 (3)	112	
[V(di-OAr)][PF₆] 6-coord. V	{CH₂N(CH₂py)(CH₂C₆H₄O-2)}₂	1.884 (3)	129	xxvi
		1.891 (3)	128	
[V(O)(di-OAr)] 6-coord. V	CH₂{CH₂N(CH₂py)(CH₂C₆H₄O-2)}{CH₂N(Me)(CH₂C₆H₄O-2)}	1.905 (5)	135	xxvii
		1.928 (5)	127	

[V(O)(di-OAr)] 6-coord. V, both OAr, both N and alkoxide bound	$OC_6H_4\{CH_4NH(CH_2)_2N(2\text{-}OC_6H_4)(CH_2)_2O\}\text{-}2$	1.910 (7) 1.840 (3)	136 138	xiii
[V(tri-OAr)][BPh$_4$] octahedral V	$c\text{-}(CH_2)_6\{NCH_2\text{-}2\text{-}OC_6H_3Bu^t\text{-}4\}_3$	1.834 (5) 1.809 (5) 1.831 (4)	136 137 137	xxviii
[V(tetra-OAr)] octahedral V	$(2\text{-}OC_6H_4CH_2)_2N(CH_2)_2N(CH_2C_6H_4O\text{-}2)_2$	1.848 (4) 1.877 (4) 1.872 (4) 1.853 (4)	138 131 129 140	xxix

[xx] A. Neves, A.S. Ceccatto, C. Erasmus-Buhr, S. Gehring, W. Haase, H. Paulus, O.R. Nascimento, and A.A. Batista, *J. Chem. Soc., Chem. Commun.*, 1782 (1993).

[xxi] M. Mazzanti, C. Floriani, A. Chiesi-Villa, and C. Guastini, *J. Chem. Soc., Dalton Trans.*, 1793 (1989).

[xxii] P.J. Toscano, E.J. Schermerhorn, C. Dettelbach, D. Macherone, and J. Zubieta, *J. Chem. Soc., Chem. Commun.*, 933 (1991).

[xxiii] J. Chakravarty, S. Dutta, and A. Chakravorty, *J. Chem. Soc., Dalton Trans.*, 2857 (1993).

[xxiv] E. Ludwig, H. Hefele, U. Schilde, and E. Uhlemann, *Z. Anorg. Allg. Chem.*, **620**, 346 (1994).

[xxv] A.S. Borovik, T.M. Dewey, and K.N. Raymond, *Inorg. Chem.*, **32**, 413 (1993).

[xxvi] A. Neves, A.S. Ceccato, S.M.D. Erthal, I. Vencato, B. Nuber, and J. Weiss, *Inorg. Chim. Acta*, **187**, 119 (1991).

[xxvii] A. Neves, I. Vencato, and Y.P. Mascarenhas, *Acta Crystallog. C*, **50**, 1417 (1994).

[xxviii] U. Auerbach, B.S.P.C. Della Vedova, K. Wieghardt, B. Nuber, and J. Weiss, *J. Chem. Soc., Chem. Commun.*, 1004 (1990).

[xxix] A. Neves, A.S. Ceccato, I. Vencato, Y.P. Mascarenhas, and C. Erasmus-Buhr, *J. Chem. Soc., Chem. Commun.*, 652 (1992).

Table 6.25 Niobium aryloxides

Compound	Aryloxide	Bond length (Å) M–O	Bond angle (°) M–O–Ar	Ref.
	Homoleptic aryloxides and adducts			
[Nb(OAr)₅] tbp Nb	OC₆H₃Me₂-2,6	1.901 (2)	167	i
		1.893 (2)	145	
		1.903 (2)	144	
		1.890 (2)	169	
		1.910 (2)	157	
trans-[Nb(OAr)₂(dmpe)₂] distorted octahedron	OC₆H₄Me-4	2.023 (3)	156	i
		2.022 (3)	171	
[Nb(OAr)₂(dmpe)₂]⁺ [Nb(OAr)₆]⁻ *trans* OAr in cation, 6-coord. Nb	OC₆H₃Me₂-3,5	1.917 (2)	177	i, ii
		1.96 (1)	143	
		1.95 (1)	150	
		1.98 (1)	146	
		1.95 (1)	161	
		1.94 (1)	153	
		1.95 (1)	149	
	Mixed aryloxy halides, amides, etc.			
[Nb(OAr)₂Cl₃] square pyramidal Nb, axial OAr	OC₆HPh₄-2,3,5,6	1.820 (3)	163	iii
		1.870 (3)	151	
[Nb(OAr)₂Cl₃] square pyramidal Nb, axial OAr	OC₆HPh₂-2,6-Buᵗ₂-3,5	1.809 (1)	178	iv
		1.881 (1)	141	
[Nb(OAr)₂I₃] tbp Nb, axial OAr	OC₆HPh₄-2,3,5,6	1.879 (4)	179	v
[Nb(μ₂-Cl)(OAr)₂Cl₂]₂ edge-shared bis-octahedron, OAr *trans* to bridging chloride groups, 6-coord. Nb	OC₆H₃Prⁱ₂-2,6	1.816 (7)	173	vi
		1.816 (7)	169	
[Nb(OAr)Cl₄(THF)] distorted octahedron, THF *trans* to OAr	OC₆H₃Me₂-2,6	1.819 (8)	180	vii

Compound / geometry	OAr			Ref.
cis,mer-[Nb(OAr)₂Cl₃(THF)] distorted octahedron	OC₆H₃Me₂-2,6	1.829 (6)	173	vii,
cis,mer-[Nb(OAr)₂Cl₃(THF)] distorted octahedron	OC₆H₃Ph₂-2,6	1.854 (6)	170	viii
cis,mer-[Nb(OAr)₂Cl₃(THF)] distorted octahedron		1.855 (5)	146	viii
		1.883 (5)	154	
cis,mer-[Nb(OAr)₂Cl₃(py)] distorted octahedron	OC₆H₃Pr$^i_2$-2,6	1.832 (3)	173	vi
		1.852 (3)	175	
cis,mer-[Nb(OAr)₂Cl₃(PMe₂Ph)] distorted octahedron	OC₆H₃Pr$^i_2$-2,6	1.83 (2)	173	vi
		1.82 (2)	172	
trans,mer-[Nb(OAr)₃Cl₂(PMe₂Ph)] distorted octahedron	OC₆H₃Pr$^i_2$-2,6	1.913 (2)	160	vi
		1.868 (2)	176	
		1.894 (2)	165	
all trans-[Nb(OAr)₂Cl₂(py)₂] octahedron	OC₆H₃Pr$^i_2$-2,6	1.903 (3)	164	ix
		1.889 (2)	164	
all trans-[Nb(OAr)₂Cl₂(PMe₂Ph)₂] octahedron	OC₆H₃Pr$^i_2$-2,6	1.901 (2)	177	ix
all trans-[Nb(OAr)₂Cl₂(PPh₃)₂] octahedron	OC₆H₃Pr$^i_2$-2,6	1.892 (7)	174	ix
[Nb(NMe₂)₃(OAr)₂] tbp Nb, equatorial OAr	OC₆H₂Bu$^t_2$-2,4-Ph-6	1.925 (2)	172	x
		1.923 (2)	169	

[i] T.W. Coffindaffer, B.D. Steffy, I.P. Rothwell, K. Folting, J.C. Huffman, and W.E. Streib, J. Am. Chem. Soc., 111, 4742 (1989).

[ii] T.W. Coffindaffer, I.P Rothwell, K. Folting, J.C. Huffman, and W.E. Streib, J. Chem. Soc., Chem. Commun., 1519 (1985).

[iii] M.A. Lockwood, M.C. Potyen, B.D. Steffey, P.E. Fanwick, and I.P. Rothwell, Polyhedron, 14, 3293 (1995).

[iv] J.S. Vilardo, M.M. Salberg, J.R. Parker, P.E. Fanwick, and I.P. Rothwell, Inorg. Chim. Acta, 299, 135 (2000).

[v] S.W. Schweiger, P.E. Fanwick, and I.P. Rothwell, results to be published.

[vi] J.R. Clark, A.L. Pulverenti, P.E. Fanwick, M. Sigalas, O. Eisenstein, and I.P. Rothwell, Inorg. Chem., 36, 3623 (1997).

[vii] H. Yasuda, Y. Nakayama, K. Takei, A. Nakamura, Y. Kai, and N. Kanehisa, J. Organomet. Chem., 473, 105 (1994).

[viii] N. Kanehisa, Y. Kai, N. Kasai, H. Yasuda, Y. Nakayama, K. Takei, and A. Nakamura, Chem. Lett., 2167 (1990).

[ix] S.W. Schweiger, E.S. Freeman, J.R. Clark, M.C. Potyen, P.E. Fanwick, and I.P. Rothwell, Inorg. Chim. Acta, in press.

[x] V.M. Visciglio, P.E. Fanwick, and I.P. Rothwell, Inorg. Chim. Acta, 211, 203 (1993).

(continued overleaf)

Table 6.25 (*Continued*)

Compound	Aryloxide	Bond length (Å) M–O	Bond angle (°) M–O–Ar	Ref.
	Oxo and imido aryloxides			
[Nb(O)(OAr)$_3$] pseudo-tetrahedral, terminal oxo	OC$_6$H$_3$Ph$_2$-2,6	1.890 (9) 1.878 (7) 1.921 (8)	154 155 146	xi
[Nb(μ_2-O)(OAr)$_3$]$_2$ oxo-bridges, 5-coord. Nb	OC$_6$H$_3$Pr$^i_2$-2,6	1.877 (3) 1.880 (3) 1.874 (3)	166 174 165	xii
[Nb(=NMe)(OAr)$_3$(NHMe$_2$)] 5-coord. Nb	OC$_6$H$_3$Ph$_2$-2,6	1.965 (5) 1.958 (4) 1.940 (4)	139 163 156	xiii
	Organometallic aryloxides			
[Nb(OAr)$_2$(CH$_2$C$_6$H$_4$Me-4)$_3$] tbp Nb, axial OAr	OC$_6$H$_3$Ph$_2$-2,6	1.915 (2) 1.917 (2)	178 168	xiv
[Nb(OAr)$_2$Cl$_2$(CH$_2$SiMe$_3$)] tbp Nb, axial OAr	OC$_6$HPh$_4$-2,3,5,6	1.880 (3) 1.892 (3)	172 160	v
[Nb(OAr)$_2$(CH$_2$SiMe$_3$)$_3$] tbp Nb, axial OAr	OC$_6$HPh$_4$-2,3,5,6	1.926 (2) 1.932 (2)	156 170	v
[Nb(OAr)$_2$(CH$_2$SiMe$_3$)$_2$] tetrahedral Nb	OC$_6$HPh$_4$-2,3,5,6	1.902 (2) 1.917 (2)	159 154	v
[Nb(OAr)(CO)$_2$(dmpe)$_2$] 7-coord. Nb	OC$_6$H$_3$Me$_2$-3,5	2.181 (4)	160	i, ii
[(ArO)(Me$_3$SiCH$_2$)Nb(μ-CSiMe$_3$)]$_2$ 1,3-dimetallallabutadiene core, tetrahedral Nb, anti-OAr	OC$_6$H$_3$Ph$_2$-2,6	1.909 (3)	167	xv
[Nb(OC$_6$H$_3$Ph-C$_6$H$_4$)(OAr)$_2$Cl] tbp, axial OAr and O of cyclometallated OAr	OC$_6$H$_3$Ph-η^1-C$_6$H$_4$ OC$_6$H$_2$Ph$_2$-2,6-Bu$^i_2$-3,5	1.872 (3) 1.898 (3)	141 164	xvi
[CpNb(OAr)Cl$_3$] square pyramidal, axial Cp		1.918 (3) 1.872 (4)	131 155	iv

Compound	OAr	d(Nb–O)	angle	Ref.
$[Nb(OC_6H_3Ph-\eta^4-C_6H_7)(OAr)_2]$ three legged piano stool, chelating 2-cyclohexadiene	$OC_6H_3Ph_2$-2,6	1.932 (9)	148	xvii
	$OC_6H_3Ph-\eta^4-C_6H_7$	1.928 (9)	166	
		1.923 (9)	139	
$[Nb(OAr)_2Cl(\eta^4-C_6H_8)]$ three legged piano stool	$OC_6H_3Pr^i_2$-2,6	1.897 (2)	152	xviii
$[Nb(OAr)_3(\eta^4-C_6H_8)]$ three legged piano stool	$OC_6H_3Pr^i_2$-2,6	1.886 (6)	163	xviii
		1.906 (6)	151	
		1.898 (7)	150	
$[Nb(OC_6H_3Pr^i-\eta^2-CMe=CH_2)(OAr_2)(thf)]$	$OC_6H_3Pr^i_2$-2,6	1.875 (2)	158	xix
		1.967 (2)	138	
	$OC_6H_3Pr^i-\eta^2-CMe=CH_2$	1.949 (2)	120	
$[Nb(OC_6H_3Pr^i-\eta^2-CMe=CH_2)(OAr)_2(CNBu^t)_2]$	$OC_6H_3Pr^i_2$-2,6	2.057 (5)	137	xx
		1.900 (4)	171	
	$OC_6H_3Pr^i-\eta^2-CMe=CH_2$	1.949 (2)	123	
$[Nb(OC_6H_3Pr^i-CMe_2)(OAr)_3]$ cyclometallated OAr	$OC_6H_3Pr^i_2$-2,6	1.893 (2)	176	xx
		1.897 (2)	143	
		1.874 (2)	144	
	$OC_6H_3Pr^i-CMe_2$	1.952 (2)	121	
$[Nb(OC_6H_3Pr^i-CMeCH_2CPh=CPh)(OAr)_2]$	$OC_6H_3Pr^i_2$-2,6	1.882 (4)	164	xix
		1.873 (4)	164	
	$OC_6H_3Pr^i-CMeCH_2CPh=CPh$	1.943 (4)	123	

[xi] J.S. Yu, P.E. Fanwick, and I.P. Rothwell, *Acta, Crystallog. C*, **48**, 1759 (1992).

[xii] V.M. Visciglio, P.E. Fanwick, and I.P. Rothwell, *Acta, Crystallog. C*, **50**, 900 (1994).

[xiii] R.W. Chesnut, P.E. Fanwick, and I.P. Rothwell, *Inorg. Chem.*, **27**, 752 (1988).

[xiv] R.W. Chesnut, G.G. Jacob, J.S. Yu, P.E. Fanwick, and I.P. Rothwell, *Organometallics*, **10**, 321 (1991).

[xv] E. Torrez, R.D. Profilet, P.E. Fanwick, and I.P. Rothwell, *Acta, Crystallog. C*, **50**, 902 (1994).

[xvi] R.B. Chesnut, J.S. Yu, P.E. Fanwick, I.P. Rothwell, and J.C. Huffman, *Polyhedron*, **9**, 1051 (1990).

[xvii] B.D. Steffey, R.W. Chesnut, J.L. Kerschner, P.J. Pellechia, P.E. Fanwick, and I.P. Rothwell, *J. Am. Chem. Soc.*, **111**, 378 (1989).

[xviii] V.M. Visciglio, J.R. Chark, M.T. Nguyen, D.R. Mulford, P.E. Fanwick, and I.P. Rothwell, *J. Am. Chem. Soc.*, **119**, 3490 (1997).

[xix] J.S. Yu, P.E. Fanwick, and I.P. Rothwell, *J. Am. Chem. Soc.*, **112**, 8171 (1990).

[xx] J.S. Yu, L. Felter, M.C. Potyen, J.R. Clark, V.M. Visciglio, P.E. Fanwick, and I.P. Rothwell, *Organometallics*, **15**, 4443 (1996).

dimeric with chloro bridges.[398] Although the tribromide [Ta(OC$_6$HPh$_4$-2,3,5,6)$_2$Br$_3$] is also square pyramidal, the iodide [Nb(OC$_6$HPh$_4$-2,3,5,6)$_2$I$_3$] is trigonal bipyramidal with axial OAr. There is an extensive series of six-coordinate adducts with O,[399] N,[400] and P[398] donor ligands. These d^0-adducts are distorted from strict octahedral and this distortion has been analysed.[398] A more pronounced distortion is present in the dihydrides [Ta(OAr)$_2$Cl(H)$_2$(PMePh$_2$)] and [Ta(OAr)$_3$(H)$_2$(PMe$_2$Ph)] (see below).[401]

Treatment of the dimeric [TaCl$_3$(THF)$_2$]$_2$(μ-N$_2$) with six equivalents of [LiOC$_6$H$_3$Pr$^i_2$-2,6] was found to lead to the compound [(THF)(ArO)$_3$Ta(NN)Ta(OAr)$_3$(THF)].[402] The structure contains a hydrazido(4-) ligand with an N–N distance of 1.32 (1) Å.

The alkylation of chloro, aryloxides of niobium and tantalum with Grignard or lithium reagents leads to a wide range of mixed alkyl, aryloxides. Alkylation of the metal–chloride bonds occurs first, but subsequent displacement of aryloxide groups can occur with excess alkylating agent.[403] With very bulky aryloxide ligands, alkylation can lead to alkylidene derivatives via an α-hydrogen abstraction process. It is also possible to photochemically generate corresponding alkylidene species. Mechanistic studies show that irradiation into alkyl-to-metal charge transfer bands leads to transient alkyl radicals that abstract the adjacent α-hydrogen.[203]

A series of tris-aryloxide neopentylidene compounds [Ta(=CHCMe$_3$)(OAr)$_3$(THF)] (OAr = OC$_6$H$_3$Pr$^i_2$-2,6, OC$_6$H$_3$Me$_2$-2,6) have been obtained by treating [Ta(=CHCMe$_3$)Cl$_3$(THF)$_2$] with LiOAr.[404] They react with a variety of olefins to produce isolable tantalacyclobutane derivatives, e.g. styrene produces [Ta(OC$_6$H$_3$Pr$^i_2$-2,6)$_3$\{CH(Ph)CH(Bu^t)CH$_2$\}] whose structure shows the Bu^t substituent at the β-position. An equilibrium between the tantalacyclobutane and alkylidene/THF adduct was observed accounting for both formation of new alkylidenes (e.g. [Ta(=CHSiMe$_3$)(OAr)$_3$(THF)] from CH$_2$=CHSiMe$_3$) and tantalacyclobutane isomerization. Addition of ethylene generates the simplest tantalacyclobutane [Ta(OC$_6$H$_3$Pr$^i_2$-2,6)$_3$(CH$_2$CH$_2$CH$_2$)] which forms a pyridine adduct without fragmenting. The compound [Ta(OC$_6$H$_3$Pr$^i_2$-2,6)$_3$\{CHBu^{t}CH(C$_5$H$_8$CH)\}] (structurally characterized) formed from norbornene acts as a living polymerization catalyst for the formation of polynorbornene.[404] Kinetic studies show that the rate determining step is ring opening of the metallacycle.

Treatment of the compound [(Me$_3$SiCH$_2$)$_2$Ta(μ-CSiMe$_3$)$_2$Ta(CH$_2$SiMe$_3$)$_2$] with bulky phenols leads to a series of substitution products in which the central 1,3-dimetallacyclobutadiene core remains intact.[136] The relative rates of substitution of the first alkyl group by a series of phenols have been obtained by competition reactions. In the case of 2,6-diarylphenols the relative rate of substitution decreases as the bulk of the meta-substituent increases.[405] This is argued to be a consequence of a decrease in the conformational flexibility of the ortho-phenyl rings hence leading to a bulkier ligand.

Mixed hydrido, aryloxides of niobium and tantalum are important reagents. The addition of bulky phenols to [Cp*Ta(PMe$_3$)(H)$_2$(η^2-CH$_2$PMe$_2$)] has been shown to lead to [Cp*Ta(H)$_2$(OAr)$_2$] derivatives.[153] Treatment of [Ta(OAr)(R)(Cl)(η^6-C$_6$Me$_6$)] precursors with [LiBEt$_3$H] leads to [Ta(OAr)(R)(H)(η^6-C$_6$Me$_6$)] species.[406] The hydrogenolysis of mixed alkyl, aryloxides of tantalum in the presence of phosphine donor ligands can produce the corresponding hydrides (Scheme 6.8).[401] It is also possible to generate hydride compounds by addition of [Bu$_3$SnH] to chloro, aryloxides of tantalum (Scheme 6.8). In the case of niobium this reaction does not lead to stable hydrides but to the d^1-species all-trans-[Nb(OAr)$_2$Cl$_2$(PR$_3$)$_2$].[407] The corresponding

Table 6.26 Tantalum aryloxides

Compound	Aryloxide	Bond length (Å) M–O	Bond angle (°) M–O–Ar	Reference
	Homoleptic aryloxides and adducts			
[Ta$_2$(μ_2-OAr)$_2$(OAr)$_8$] 6-coord. Ta	OC$_6$H$_4$Me-4	1.836 (10)	156	i
		1.898 (10)	166	
	μ_2-OC$_6$H$_4$Me-4	1.925 (9)	148	
		1.924 (9)	144	
		2.084 (8)	112	
	Mixed aryloxy halides, amides, hydrides, etc.			
[Ta(OAr)$_2$Cl$_3$] square pyramidal Ta, axial ArO	OC$_6$H$_3$Bu$^t_2$-2,6	1.872 (5)	157	ii
		1.836 (4)	172	
[Ta(OAr)$_3$Cl$_2$] square pyramidal Ta, axial ArO	OC$_6$H$_3$Bu$^t_2$-2,6	1.90 (2)	148	iii
		1.90 (2)	153	
		1.83 (2)	162	
[Ta(OAr)$_3$Cl$_2$] square pyramidal Ta, axial ArO	OC$_6$HPh$_2$-2,6-Me$_2$-3,5	1.803 (3)	168	iv
		1.862 (3)	151	
[Ta(OAr)$_3$Cl$_2$] square pyramidal Ta, axial ArO	OC$_6$HPh$_2$-2,6-Bu$^t_2$-3,5	1.810 (2)	178	iv
		1.883 (2)	141	
[Ta(OAr)$_2$Br$_3$] square pyramidal Ta, axial ArO	OC$_6$HPh$_4$-2,3,5,6	1.865 (3)	153	v
		1.825 (3)	162	
[Ta(OAr)$_3$Br$_2$] square pyramidal Ta, axial ArO	OC$_6$H$_3$Pr$^i_2$-2,6	1.839 (6)	152	v
		1.840 (7)	161	
		1.861 (7)	171	

[i] L.N. Lewis and M.F. Garbauskas, *Inorg. Chem.*, **24**, 363 (1985).
[ii] L.R. Chamberlain, I.P. Rothwell, and J.C. Huffman, *Inorg. Chem.*, **23**, 2575 (1984).
[iii] G.R. Clark, A.J. Nielson, and C.E.F. Rickard, *Polyhedron*, **6**, 1765 (1987).
[iv] J.S. Vilardo, M.A. Lockwood, L.G. Hanson, J.R. Clark, B.C. Parkin, P.E. Fanwick, and I.P. Rothwell, *J. Chem. Soc., Dalton Trans.*, 3353 (1997).
[v] E.S. Freeman, P.E. Fanwick, and I.P. Rothwell, results to be published.

(continued overleaf)

Table 6.26 (Continued)

Compound	Aryloxide	Bond length (Å) M–O	Bond angle (°) M–O–Ar	Reference
[Ta(μ₂-Cl)(OAr)₂Cl₂]₂ edge-shared bis-octahedron, OAr *trans* to bridging chloride groups, 6-coord. Nb	OC₆H₃Pr^i₂-2,6	1.815 (12) 1.791 (14)	173 171	vi
cis,mer-[Ta(OAr)₂Cl₃(py)] distorted octahedron	OC₆H₃Pr^i₂-2,6	1.841 (4) 1.848 (5)	174 174	vi
cis,mer-[Ta(OAr)₂Cl₃(η¹-dcpm)] distorted octahedron	OC₆H₃Pr^i₂-2,6	1.855 (7) 1.889 (6)	171 180	vii
all trans-[Ta(OAr)₂Cl₂(py)₂] octahedron	OC₆H₃Pr^i₂-2,6	1.906 (3) 1.896 (3)	164 164	viii
all trans-[Ta(OAr)₂Br₂(PMe₂Ph)₂] octahedron	OC₆H₃Pr^i₂-2,6	1.900 (4)	175	viii
all trans-[Ta(OAr)₂Cl₂(PMe₂Ph)₂] octahedron	OC₆H₂Pr^i₂-2,6-Br-4	1.909 (3)	175	viii
[Ta₂(μ₂-OMe)₂(OAr)₄(OMe)₄] 6-coord. Ta	OC₆H₃Pr^i₂-2,6	1.888 (4) 1.919 (4)	179 161	ix
[Ta(OAr)(NMe₂)₄]	OC₆H₂(tetrahydronaphthyl)-2-Bu^t₂-4,6	1.954 (3)	162	xxvi
[Ta(OAr)(NMe₂)₄]	OC₆H₂Ph₂-2,6-Bu^t₂-3,5	1.950 (3)	172	x
[Ta(OAr)(OAr')(NMe₂)₃]	OC₆H₂(tetrahydronaphthyl)-2-Bu^t₂-4,6 OC₆HPh₄-2,3,5,6	1.948 (3) 2.012 (3)	155 159	xxvi
[Ta(OAr)Cl₂(=NC₆H₃Pr^i-2,6)(py)₂]	OC₆H₃Me₂-2,6	1.905 (5)	145	xi
[Ta(OAr)₃(xyNCbz=CbzNxy)] tbp Ta, axial OAr and N	OC₆H₃Me₂-2,6	1.904 (6) 1.898 (5) 1.880 (6)	171 163 167	xii, xiii
[Ta(OAr)₂(=NC₆H₃Pr^i₂-2,6)(RNC₆H₃Pr^i₂-2,6)]	OC₆H₃Me₂-2,6	1.916 (5) 1.923 (5)	150 150	xii, xiv
[Ta₂Cl₃(O)(OAr)₅]	OC₆H₃Pr^i₂-2,6	1.81 (2) 1.83 (1) 1.77 (2) 1.99 (2) 2.19 (2)	163 159 161 174 162	iii

Compound	ArO	Ta–O (Å)	Ta–O–C (°)	Ref.
[(THF)(ArO)₃Ta(=NN=)Ta(OAr)₃(THF)]	$OC_6H_3Pr^i_2$-2,6	1.945 (5) / 1.911 (5) / 1.933 (5)	147 / 175 / 160	xv
[Ta(OAr)₂(H)Cl₂(PMe₂Ph)₂] pentagonal bipyramidal Ta, axial O	$OC_6H_3Pr^i_2$-2,6	1.902 (3)	175	xvi
[Ta(OAr)₂(H)₂Cl(PMe₂Ph)₂] pentagonal bipyramidal Ta, axial O	$OC_6H_3Pr^i_2$-2,6	1.899 (5) / 1.901 (5)	174 / 171	xvii
[Ta(OAr)₂(H)₃(PMe₂Ph)₂] pentagonal bipyramidal Ta, axial O	$OC_6H_3Cy_2$-2,6	1.901 (6)	172	xvii, xvi
[Ta(OAr)₂(H)₂Cl(PMePh₂)] highly distorted 6-coord. Ta	$OC_6H_3Bu^t_2$-2,6	1.896 (3) / 1.888 (3)	165 / 168	xvi
[Ta(OAr)₃(H)₂(PMe₂Ph)] highly distorted 6-coord. Ta	$OC_6H_3Pr^i_2$-2,6	1.907 (3) / 1.912 (4) / 1.897 (3)	161 / 154 / 170	xvi xviii

Organometallic aryloxides

Compound	ArO	Ta–O (Å)	Ta–O–C (°)	Ref.
[Ta(OAr)₂Me₃] tbp Ta, axial ArO	$OC_6H_3Bu^t_2$-2,6	1.930 (6) / 1.945 (6)	145 / 155	xix

[vi] J.R. Clark, A.L. Pulverenti, P.E. Fanwick, M. Sigalas, O. Eisenstein, and I.P. Rothwell, *Inorg. Chem.*, **36**, 3623 (1997).

[vii] P.N. Riley, J.C. Clark, and P.E. Fanwick, *Inorg. Chim. Acta*, **288**, 35 (1999).

[viii] S.W. Schweiger, E.S. Freeman, J.R. Clark, M.C. Potyen, P.E. Fanwick, and I.P. Rothwell, *Inorg. Chim. Acta*, in press.

[ix] R. Wang, K. Folting, J.C. Huffman, L.R. Chamberlain, and I.P. Rothwell, *Inorg. Chim. Acta*, **120**, 81 (1986).

[x] J.S. Vilardo, M.M. Salberg, J.R. Parker, P.E. Fanwick, and I.P. Rothwell, *Inorg. Chim. Acta*, **299**, 135 (2000).

[xi] Yuan-Wei Chao, P.A. Wexler, and D.E. Wigley, *Inorg. Chem.*, **28**, 3860 (1989).

[xii] L.R. Chamberlain, I.P. Rothwell, and J.C. Huffman, *J. Chem. Soc., Chem. Commun.*, 1203 (1986).

[xiii] L.R. Chamberlain, L.D. Durfee, P.E. Fanwick, L.M. Kobriger, S.L. Latesky, A.K. McMullen, B.D. Steffey, I.P. Rothwell, K. Folting, and J.C. Huffman, *J. Am. Chem. Soc.*, **109**, 6068 (1987).

[xiv] L.R. Chamberlain, B.D. Steffey, I.P. Rothwell, and J.C. Huffman, *Polyhedron*, **8**, 341 (1989).

[xv] R.R. Schrock, M. Wesolek, A.H. Liu, K.C. Wallace, and J.C. Dewan, *Inorg. Chem.*, **27**, 2050 (1988).

[xvi] B.C. Parkin, J.C. Clark, V.M. Visciglio, P.E. Fanwick, and I.P. Rothwell, *Organometallics.*, **14**, 3002 (1995).

[xvii] B.C. Ankianiec, P.E. Fanwick, and I.P. Rothwell, *J. Am. Chem. Soc.*, **113**, 4710 (1991).

[xviii] V.M. Visciglio, P.E. Fanwick, and I.P. Rowthwell, *J. Chem. Soc., Chem. Commun.*, 1505 (1992).

[xix] L. Chamberlain, J. Keddington, I.P. Rothwell, and J.C. Huffman, *Organometallics*, **1**, 1538 (1982).

(continued overleaf)

Table 6.26 (*Continued*)

Compound	Aryloxide	Bond length (Å) M–O	Bond angle (°) M–O–Ar	Reference
[Ta(OAr)$_2$Me$_3$] tbp Ta, axial ArO	OC$_6$H$_2$Bu$^t_2$-2,6-OMe-4	1.925 (4)	153	xx
[Ta(OAr)$_2$(CH$_2$Ph)$_3$] tbp Ta, axial ArO	OC$_6$H$_3$Me$_2$-2,6	1.919 (4) 1.887 (4)	151 159	xxi
[Ta(OAr)$_2$(CH$_2$C$_6$H$_4$-4Me)$_3$] tbp Ta, axial ArO	OC$_6$H$_3$Ph$_2$-2,6	1.903 (4) 1.906 (4)	178 167	xxii
[Ta(OAr)$_2$(CH$_2$SiMe$_3$)$_3$] tbp Ta, axial ArO, two molecules	OC$_6$HPh$_2$-2,6-Me$_2$-3,5	1.89 (1) 1.90 (1) 1.93 (1) 1.91 (1)	178 165 175 164	iv
[Ta(OAr)$_2$(CH$_2$SiMe$_3$)$_3$] tbp Ta, axial ArO	OC$_6$HPh$_2$-2,6-Pr$^i_2$-3,5	1.922 (3) 1.908 (3)	152 150	iv
[Ta(OAr)$_3$(Ph)$_2$] tbp Ta, axial ArO	OC$_6$H$_3$Me$_2$-2,6	1.848 (5) 1.879 (4)	180 176	xxiii
[Ta(OAr)$_3$(Ph)$_2$] tbp Ta, axial ArO	OC$_6$H$_3$Pr$^i_2$-2,6	1.852 (3) 1.881 (3) 1.880 (3)	153 165 166	xxiii
[Ta(OAr)$_3$(CH$_2$SiMe$_3$)$_2$] tbp Ta, axial ArO	OC$_6$H$_3$Ph$_2$-2,6	1.930 (3) 1.912 (3) 1.918 (3)	153 159 152	iv
[Ta(OAr)$_4$Me] square pyramidal Ta, axial Me	OC$_6$H$_3$Me$_2$-2,6	1.899 (4)	158	xxi
[Ta(OAr)$_2$Cl$_2$(C$_6$H$_{11}$)] Ta-cyclohexyl, square pyramidal Ta, axial C, cis-O	OC$_6$H$_3$Ph$_2$-2,6	1.850 (3) 1.911 (3)	163 145	xxiv
[Ta(OAr)$_2$Cl$_2$(C$_5$H$_9$)] Ta-cyclopentyl, square pyramidal Ta, axial C, cis-O	OC$_6$H$_3$Ph$_2$-2,6	1.907 (4) 1.854 (4)	144 164	xxiv

(continued overleaf)

Compound	OAr	Ta–O (Å)		Ref.
$[Ta(OAr)_2Cl_2(CH_2C_6H_5)]$ tbp Ta, axial ArO	OC_6HPh_4-2,3,5,6	1.900 (3)	156	xxiv
$[Ta(OAr)_2Cl(CH_2C_6H_5)_2]$ tbp Ta, axial ArO	OC_6HPh_4-2,3,5,6	1.885 (3)	167	xxiv
		1.894 (4)	176	xxv
		1.894 (4)	177	
$[Ta(Ind\text{-}OAr)(NMe_2)_3]$ 1-indenylphenoxide	$OC_6H_2(\eta^1\text{-indenyl})$-2-$Bu^t_2$-4,6	2.025 (7)	123	xxv
$[Ta(Ind\text{-}OAr)(4\text{-}Ph\text{-}py)Cl_3]$ 1-indenylphenoxide	$OC_6H_2(\eta^5\text{-indenyl})$-2-$Bu^t_2$-4,6	1.968 (3)	129	xxvi
$[Ta(OAr)_2(=CHSiMe_3)(CH_2SiMe_3)]$ tetrahedral Ta	$OC_6H_3Bu^t_2$-2,6	1.85 (2)	159	xxvii
		1.85 (2)	169	xxi
$[Ta(OAr)_3\{CHBu^tCH(C_5H_8CH)\}]$ tantalacyclobutane formed from norbornene	$OC_6H_3Pr^i_2$-2,6	1.904 (9)	159	xxviii
		1.909 (8)	164	
		1.917 (8)	161	
$[Ta(OAr)_3\{CH(Ph)CH(Bu^t)CH_2\}]$ tantalacyclobutane formed from styrene	$OC_6H_3Pr^i_2$-2,6	1.820 (14)	155	xxix
		1.864 (15)	146	xxviii
		1.893 (13)	157	
$[(ArO)(Me_3SiCH_2)Ta(\mu\text{-}CSiMe_3)Ta(CH_2SiMe_3)_2]$ 1,3-dimetallabutadiene core, tetrahedral Ta	$OC_6H_3Ph_2$-2,6	1.885 (6)	175	xxx
$[(ArO)(Me_3SiCH_2)_2Ta(\mu\text{-}CSiMe_3)_2Ta(CH_2SiMe_3)_2]$ 1,3-dimetallabutadiene core, tetrahedral Ta	$OC_6H_3(1\text{-naphthyl})_2$-2,6	1.897 (4)	178	xxxi

[xx] L. Chamberlain, I.P. Rothwell, and J.C. Huffman, *J. Am. Chem. Soc.*, **108**, 1538 (1986).

[xxi] L.R. Chamberlain, I.P. Rothwell, K. Folting, and J.C. Huffman, *J. Chem. Soc., Dalton Trans.*, 155 (1987).

[xxii] R.W. Chesnut, G.G. Jacob, J.S. Yu, P.E. Fanwick, and I.P. Rothwell, *Organometallics*, **10**, 321 (1991).

[xxiii] B.D. Steffey, P.E. Fanwick, and I.P. Rothwell, *Polyhedron*, **9**, 963 (1990).

[xxiv] S.W. Schweiger, P.E. Fanwick, and I.P. Rothwell, results to be published.

[xxv] M.G. Thorn, P.E. Fanwick, R.W. Chesnut, and I.P. Rothwell, *J. Chem. Soc., Chem. Comm.*, 2543 (1999).

[xxvi] M.G. Thorn, P.E. Fanwick, R.W. Chesnut, and I.P. Rothwell, results to be published.

[xxvii] L. Chamberlain, I.P. Rothwell, and J.C. Huffman, *J. Am. Chem. Soc.*, **104**, 7338 (1982).

[xxviii] K.C. Wallace, A.H. Liu, J.C. Dewan, and R.R. Schrock, *J. Am. Chem. Soc.*, **110**, 4964 (1988).

[xxix] K.C. Wallace, J.C. Dewan, and R.R. Schrock, *Organometallics*, **5**, 2162 (1986).

[xxx] P.E. Fanwick, A.E. Ogilvy, and I.P. Rothwell, *Organometallics*, **6**, 73 (1987).

[xxxi] P.N. Riley, M.G. Thorn, J.S. Vilardo, M.A. Lockwood, P.E. Fanwick, and I.P. Rothwell, *Organometallics*, **18**, 3016 (1999).

Table 6.26 (*Continued*)

Compound	Aryloxide	Bond length (Å) M–O	Bond angle (°) M–O–Ar	Reference
[(ArO)₂Ta(μ-CSiMe₃)₂Ta(CH₂SiMe₃)₂] 1,3-dimetallabutadiene core, tetrahedral Ta	OC₆H₃Bu^t₂-2,6	1.904 (4)	170	xxx
[Ta(OC₆H₃Bu^t-CMe₂CH₂)(OAr)(Ph)₂] tbp Ta, axial ArO	OC₆H₃Bu^t₂-2,6	1.909 (4)	150	xxxii
	OC₆H₃Bu^t-CMe₂CH₂	1.926 (4)	144	xxxiii
[Ta(OAr)(OC₆H₃Ph-C₆H₄)(H)Cl(PMe₃)₂] pentagonal bipyramidal Ta, axial O	OC₆H₃Ph-2,6	1.910 (3)	167	xxxiv
	OC₆H₃Ph-η¹-C₆H₄	1.947 (3)	131	
[Ta(OC₆H₃Ph-C₆H₄)(OAr)₂Me] tbp Ta, axial ArO	OC₆H₃Ph₂-2,6	1.899 (4)	170	xxxv
	OC₆H₃Ph-η¹-C₆H₄	1.883 (4)	147	xxxvi
		1.909 (4)	141	
[Ta(OC₆H₃Ph-C₆H₄)(OAr)₂Bu^i] tbp Ta, axial ArO	OC₆H₃Ph₂-2,6	1.89 (3)	161	xxxvi
	OC₆H₃Ph-η¹-C₆H₅	1.91 (3)	137	
		1.913 (3)	137	
[Ta(OC₆H₃Ph-C₆H₄)₂(OAr)] tbp Ta, axial metallated OAr	OC₆H₃Ph₂-2,6	1.903 (3)	147	xxxvii
	OC₆H₃Ph-η¹-C₆H₅	1.905 (2)	137	
		1.897 (2)	140	
[Ta(OC₆H₃Bu^t-CMe₂CH₂)₂Cl] tbp Ta, axial ArO	OC₆H₃Bu^t-CMe₂CH₂	1.87 (1)	145	xxxvii
		1.87 (1)	145	
[Ta(OAr)₂(η²-xyNCMe)₂(Me)]	OC₆H₃Me₂-2,6	1.913 (4)	169	xii
		1.938 (4)	168	xxxviii
[Ta(OAr)₃(C₄Et₄)]	OC₆H₃Pr^i₂-2,6	1.845 (5)	175	xxxix
		1.921 (5)	171	xl
		1.858 (4)	165	
[Ta(OAr)₃(=CMe-CMe=CHBu^t)(py)]	OC₆H₃Pr^i₂-2,6	1.94 (1)	141	xli
		1.93 (1)	157	
		1.88 (1)	157	
[Ta(OAr)₃(CMe=CMe-Cvinyl-CH₂)]	OC₆H₃Pr^i₂-2,6	1.917 (7)	152	xli
		1.857 (7)	172	
		1.860 (7)	159	
[Ta(OAr)₂(NBu^t)(NBu^t-CPh=CPh₂)]	OC₆H₃Me₂-2,6	1.931 (10)	153	xlii
		1.915 (10)	156	

Compound	OAr	Bond length (Å)	Angle (°)	Ref.
[Ta(OAr)$_2$Cl(η^6-C$_6$Me$_6$)]	OC$_6$H$_3$Pr$^i_2$-2,6	1.887 (5) / 1.935 (5)	163 / 147	xliii
[Ta(OAr)$_2$Cl(η^6-C$_6$H$_3$Bu$^t_3$)]	OC$_6$H$_3$Pr$^i_2$-2,6	1.934 (2) / 1.894 (2)	162 / 175	xliv
[Ta(OAr)$_2$(η^6-C$_6$Et$_6$)]	OC$_6$H$_3$Pr$^i_2$-2,6	1.917 (4) / 1.916 (4)	160 / 158	xlv, xlvi
[Ta(OAr)Cl$_2$(η^6-C$_6$Me$_6$)]	OC$_6$H$_3$Pr$^i_2$-2,6	1.866 (3)	170	xlvii
[Ta(OAr)$_3$(η^4-C$_6$H$_8$)] three legged piano stool	OC$_6$H$_3$Pr$^i_2$-2,6	1.888 (3) / 1.889 (3) / 1.902 (3)	165 / 150 / 151	xlviii
[Ta(OAr)Et$_2$(η^6-C$_6$Me$_6$)]	OC$_6$H$_3$Pr$^i_2$-2,6	1.912 (3)	177	xlix, l

[xxxii] L. Chamberlain, J. Keddington, I.P. Rothwell, and J.C. Huffman, *Organometallics*, **1**, 1538 (1982).

[xxxiii] L.R. Chamberlain, J.L. Kerschner, A.P. Rothwell, I.P. Rothwell, and J.C. Huffman, *J. Am. Chem. Soc.*, **109**, 6471 (1987).

[xxxiv] D.R. Mulford, J.R. Clark, S.W. Schweiger, P.E. Fanwick, and I.P. Rothwell, *Organometallics*, **18**, 4448 (1999).

[xxxv] R.W. Chesnut, B.D. Steffey, I.P. Rothwell, and J.C. Huffman, *Polyhedron*, **7**, 753 (1988).

[xxxvi] R.W. Chesnut, J.S. Yu, P.E. Fanwick, I.P. Rothwell, and J.C. Huffman, *Polyhedron*, **9**, 1051 (1990).

[xxxvii] B.D. Steffey, L.R. Chamberlain, R.W. Chesnut, D.E. Chebi, P.E. Fanwick, and I.P. Rothwell, *Organometallics*, **8**, 1419 (1989).

[xxxviii] L.R. Chamberlain, L.D. Durfee, P.E. Fanwick, L. Kobriger, S.L. Latesky, A.K. McMullen, I.P. Rothwell, K. Folting, J.C. Huffman, W.E. Streib, and R. Wang, *J. Am. Chem. Soc.*, **109**, 390 (1987).

[xxxix] J.R. Strickler, P.A. Wexler, and D.E. Wigley, *Organometallics*, **7**, 2067 (1988).

[xl] J.R. Strickler, P.A. Wexler, and D.E. Wigley, *Organometallics*, **10**, 118 (1991).

[xli] K.C. Wallace, A.H. Liu, W.M. Davis, and R.R. Schrock, *Organometallics*, **8**, 644 (1989).

[xlii] L.R. Chamberlain, B.D. Steffey, I.P. Rothwell, and J.C. Huffman, *Polyhedron*, **8**, 341 (1989).

[xliii] M.A. Bruck, A.S. Copenhaver, and D.E. Wigley, *J. Am. Chem. Soc.*, **109**, 6525 (1987).

[xliv] D.P. Smith, J.R. Stickler, S.D. Gray, M.A. Bruck, R.S. Holmes, and D.E. Wigley, *Organometallics*, **11**, 1275 (1992).

[xlv] P.A. Wexler and D.E. Wigley, *J. Chem. Soc., Chem. Commun.*, 664 (1989).

[xlvi] P.A. Wexler, D.E. Wigley, J.B. Koerner, and T.A. Albright, *Organometallics*, **10**, 2319 (1991).

[xlvii] D.J. Arney, P.A. Wexler, and D.E. Wigley, *Organometallics*, **9**, 1282 (1990).

[xlviii] V.M. Visciglio, J.R. Clark, M.T. Nguyen, D.R. Mulford, P.E. Fanwick, and I.P. Rothwell, *J. Am. Chem. Soc.*, **119**, 3490 (1997).

[xlix] D.J. Arney, P.A. Fox, M.A. Bruck, and D.E. Wigley, *Organometallics*, **16**, 3421 (1997).

[l] D.J. Arney, M.A. Bruck, and D.E. Wigley, *Organometallics*, **10**, 3947 (1991).

(continued overleaf)

Table 6.26 (*Continued*)

Compound	Aryloxide	Bond length (Å) M–O	Bond angle (°) M–O–Ar	Reference
[Ta(OAr)$_3$(OCH$_2$Ph)(CPh=CPhCPh=O)]	OC$_6$H$_3$Pr$^i_2$-2,6	1.915 (2) 1.918 (2)	173 159	li
[TaCl(OAr)$_2$(η^2-2,4,6-Bu$^t_3$-py)]	OC$_6$H$_3$Pr$^i_2$-2,6	1.860 (6) 1.877 (6)	165 162	lii, liii
[Ta(OC$_6$H$_3$MeBu^t-CMe$_2$CH$_2$)Cl$_2$(=CH$_2$Bu^t)]	OC$_6$H$_3$Bu^t-CMe$_2$CH$_2$	1.852 (3)	148	liv
[Ta(OAr)$_3$(η^2-PhC≡CPh)]	OC$_6$H$_3$Pr$^i_2$-2,6	1.889 (2) 1.912 (2) 1.869 (2)	157 148 159	lv
[Ta(OAr)$_3$(η^2-6-Mequin)(PMe$_3$)]	OC$_6$H$_3$Pr$^i_2$-2,6	1.943 (7) 1.904 (7) 1.894 (7)	156 158 174	lvi
[Ta(OAr)$_2$Cl(=CBu^{t}CH=CHCBu^t)]	OC$_6$H$_3$Pr$^i_2$-2,6	1.89 (3) 1.89 (3)	173 157	lvii
[Ta(OAr)$_2$(=NBu^t-CHCHBu^t-CHCHBu^t)]	OC$_6$H$_3$Pr$^i_2$-2,6	1.877 (7) 1.909 (8)	160 155	lvi
[Ta(OAr)$_2$Cl$_2${η^2-ArNC(PMe$_3$)H}]	OC$_6$H$_3$Pr$^i_2$-2,6	1.896 (3) 1.894 (3)	149 150	lviii
[Ta(OAr)$_2$Cl$_2${(Bu^{t}NC)$_3$H}]	OC$_6$H$_3$Pr$^i_2$-2,6	1.919 (2) 1.854 (2)	159 165	lviii
[Cp*Ta(OC$_6$H$_3$Me–CH$_2$)(CH$_2$SiMe$_3$)$_2$] mono-cyclometallated OAr	OC$_6$H$_3$Me–CH$_2$	1.971 (3)	124	lix
[Cp*Ta(OAr)Cl$_3$] square pyramidal Ta, axial Cp*	OC$_6$H$_3$Ph$_2$-2,6-Bu$^t_2$-3,5	1.902 (3)	150	x
[Ta(OAr)$_2$(η^2-PhCH=CH$_2$)(Cl)(PMe$_3$)] Square pyramid, *trans* O and axial η^2-PhCH=CH$_2$	OC$_6$H$_3$Ph$_2$-2,6	1.895 (3) 1.915 (3)	175 166	lx
[Ta(OAr)$_2$(η^2-EtCCEt)Cl(PMe$_3$)]	OC$_6$H$_3$Ph-2,6	1.926 (2) 1.913 (2)	161 171	xxxiv
[Ta(OAr)(OC$_6$H$_3$Bu^{t}CMe$_2$CH$_2$-η^2- C=Nxy)(PhCHCH-η^2-C=Nxy)Cl]	OC$_6$H$_3$Bu$^t_2$-2,6 OC$_6$H$_3$Bu^{t}CMe$_2$CH$_2$-η^2-C=NCH$_2$Ph	1.936 (7) 1.960 (8)	153 149	xxxiv

Compound	OAr			Ref
[Ta(OAr)(OC₆H₃Prⁱ-η²-CMe=CH₂)Cl(PEt₃)₂]	OC₆H₃Prⁱ₂-2,6	1.922 (5)	178	xxxiv
	OC₆H₃Prⁱ-η²-CMe=CH₂	1.981 (5)	123	lx
[Ta(OC₆H₃Buᵗ-CMe₂CH₂)(OAr)(Cl)(CH₂CH₂Ph)] Tbp, axial O (cyclometallated) and Cl	OC₆H₃Buᵗ-2,6	1.863 (4)	178	xxxiv
	OC₆H₃Buᵗ-CMe₂CH₂	1.873 (4)	145	
[Ta(OAr)(OC₆H₃Buᵗ-η¹-CMe₂CH₂)Cl(CH=CHPh)]	OC₆H₃Buᵗ-2,6	1.863 (3)	175	lxi
	OC₆H₃Buᵗ-η¹-CMe₂CH₂	1.871 (3)	147	
[Ta(OAr)₃(η²-C,N,-6-mequin)(PMe₃)] square planar, axial η²-C,N	OC₆H₃Prⁱ₂-2,6	1.904 (6)	158	lxi
		1.943 (6)	156	
		1.894 (5)	175	
[Ta(OAr)₂Cl(η²-C,N-6-mequin)(OEt₂)] tbp, axial η²-C,N and OEt₂	OC₆H₃Prⁱ₂-2,6	1.870 (9)	153	lxi
		1.869 (8)	175	
[Ta(OAr)₂Et(η²-C,N-NC₅H₂Buᵗ₃-2,4,6] 5-coord. Ta	OC₆H₃Prⁱ₂-2,6	1.884 (5)	159	lxii
		1.898 (5)	150	
[Ta(OAr)₂(=NCBuᵗ=CHCBuᵗ=CHCHBuᵗ)] tetrahedral Ta	OC₆H₃Prⁱ₂-2,6	1.877 (6)	160	lxii
		1.909 (6)	155	
[Ta(OAr)₂(μ₂-NCBuᵗ=CHCBuᵗ=CH)]₂ 5-coord. Ta	OC₆H₃Prⁱ₂-2,6	1.905 (3)	169	lxii
		1.907 (3)	161	
[Ta(OAr)₂(=NCBuᵗ=CHCBuᵗ=CHCPhBuᵗ)] tetrahedral Ta	OC₆H₃Prⁱ₂-2,6	1.914 (3)	149	lxiii
		1.875 (3)	175	

[lii] J.R. Strickler, M.A. Bruck, P.A. Wexler, and D.E. Wigley, Organometallics, 9, 266 (1990).

[liii] J.R. Strickler, M.A. Bruck, and D.E. Wigley, J. Am. Chem. Soc., 112, 2814 (1990).

[liv] D.P. Smith, J.R. Strickler, S.D. Gray, M.A. Bruck, R.S. Holmes, and D.E. Wigley, Organometallics, 11, 1275 (1992).

[lv] A.-S. Baley, Y. Chauvin, D. Commereuc, and P.B. Hitchcock, New J. Chem. (Nouv. J. Chim.), 15, 609 (1991).

[lvi] J.R. Strickler, P.A. Wexler, and D.E. Wigley, Organometallics, 10, 118 (1991).

[lvii] S.D. Gray, D.P. Smith, M.A. Bruck, and D.E. Wigley, J. Am. Chem. Soc., 114, 5462 (1992).

[lviii] D.P. Smith, J.R. Strickler, S.D. Gray, M.A. Bruck, R.S. Holmes, and D.E. Wigley, Organometallics, 11, 1275 (1992).

[lix] J.R. Clark, P.E. Fanwick, and I.P. Rothwell, J. Chem. Soc., Chem. Commun., 1233 (1993).

[lx] I. de Castro, M.V. Galakhov, M. Gomez, P. Gomez-Sal, A. Martin, and P. Royo, J. Organomet. Chem., 514, 51 (1996).

[lxi] J.R. Clark, P.E. Fanwick, and I.P. Rothwell, J. Chem. Soc., Chem. Commun., 5531 (1995).

[lxii] K.D. Allen, M.A. Bruck, S.D. Gray, R.P. Kingsborough, D.P. Smith, K.J. Weller, and D.E. Wigley, Polyhedron, 14, 3315 (1995).

[lxii] S.D. Gray, K.J. Weller, M.A. Bruck, P.M. Briggs, and D.E. Wigley, J. Am. Chem. Soc., 117, 10678 (1995).

[lxiii] K.J. Weller, S.D. Gray, P.M. Briggs, and D.E. Wigley, Organometallics, 14, 5588 (1995).

Table 6.27 Chromium aryloxides

Compound	Aryloxide	Bond length (Å) M–O	Bond angle (°) M–O–Ar	Reference
	Homoleptic aryloxides and adducts			
trans-[Cr(OAr)₂(THF)₂] square planar Cr	OC₆H₂Bu^t₂-2,6-Me-4	1.948 (2)	148	i
trans-[Cr(OAr)₂(py)₂] square planar Cr	OC₆H₂Bu^t₃-2,4,6	1.952 (2)	143	ii
[Cr(OAr)₃] 6-coord. Cr, mer-O	OC₆H₄(2-py)-2	1.922 (6), 1.922 (5), 1.957 (6)	122, 121, 123	iii
[Cr(OAr)₂][PF₆] 6-coord. Cr, cis-O	OC₆H₄(2-bipy)-2	1.895 (10), 1.882 (10)	123, 123	iii
[(THF)₂(μ₃-OAr)Li(μ₃-OAr)₃Cr(μ₂-OAr)]₂ two CrLi₃O₄ cubes linked by two (μ₂-OAr) between Cr atoms	μ₂-OC₆H₅ μ₃-OC₆H₅	2.007 (4), 2.006 (4) Li–O–Cr 1.99 (1), 2.050 (4) 1.97 (1), 2.365 (4) 1.97 (1), 2.061 (4)		iv
[(THF)₂Li(μ₂-OAr)₂Cr(μ₂-OAr)₂Cr(μ₂-OAr)₂Li(THF)₂]μ₂-OC₆H₃Me₂-2,6 tetrahedral Cr		2.004 (3) Cr–O–Cr 1.945 (9), 1.987 (3) Li–O–Cr 1.92 (1), 1.988 (3)		iv

Compound	Bridging ligand	Distances		Ref.
[(TMEDA)Na(μ_2-OAr)$_2$]$_2$Cr 4-coord. Cr	μ_2-OC$_6$H$_3$Me$_2$-2,6	Na–O–Cr 2.250 (5), 1.998 (4) 2.247 (5), 2.000 (4) 2.255 (5), 1.988 (4) 2.264 (5), 1.997 (4)		v
[(py)$_4$Na$_4$Cr$_2$(μ_3-OAr)$_8$] octahedron with axial Cr	μ_3-OC$_6$H$_5$	Na–O–Cr 2.360 (4)–2.625 (3) 1.974 (3)–1.987 (3)		v
[(THF)$_4$Na$_4$Cr$_2$(μ_3-OAr)$_8$] octahedral cluster, apical Cr, four equatorial Na	μ_3-OC$_6$H$_5$	Na–O–Cr 2.348 (2)–2.532 (2) 1.985 (2)–2.000 (2)		vi
[(TMEDA)$_4$Na$_4$(μ_3-OAr)$_{10}$Cr$_4$(μ_3-O)$_3$]	μ_3-OC$_6$H$_5$	2.366 (av.)		vii
Mixed ligand aryloxides				
[(acac)$_2$Cr(μ_2-OAr)$_2$Cr(acac)$_2$] 6-coord. Cr	μ_2-OC$_6$H$_4$Me-4	1.974 (2), 1.992 (2)		viii
[(acac)$_2$Cr(μ_2-OAr)$_2$Cr(acac)$_2$] 6-coord. Cr	μ_2-OC$_6$H$_5$	1.977 (5), 1.980 (4)		viii
Li$_6$[BrCr(η^2-O,C-OC$_6$H$_4$)$_4$CrBr] Cr$_2^{4+}$ core	η^2-O,C-OC$_6$H$_4$	2.098 (8) 2.079 (7)	110 110	ix

[i] J.J.H. Edema, S. Gambarotta, F. van Bolhuis, W.J.J. Smeets, and A.L. Spek, *Inorg. Chem.*, **28**, 1407 (1989).

[ii] D. Meyer, J.A. Osborn, and M. Wesolek, *Polyhedron*, **9**, 1311 (1990).

[iii] D.A. Bardwell, D. Black, J.C. Jeffery, E. Schatz, and M.D. Ward, *J. Chem. Soc. Dalton Trans.*, 2321 (1993).

[iv] J.J.H. Edema, A. Meetsma, S. Gambarotta, S.I. Khan, W.J.J. Smeets, and A.L. Spek, *Inorg. Chem.*, **30**, 3639 (1991).

[v] J.J.H. Edema, S. Gambarotta, F. van Bolhuis, and A.L. Spek, *J. Am. Chem. Soc.*, **111**, 2142 (1989).

[vi] J.J.H. Edema, S. Gambarotta, F. van Bolhuis, W.J.J. Smeets, and A.L. Spek, *Inorg. Chem.*, **28**, 1407 (1989).

[vii] J.J.H. Edema, S. Gambarotta, W.J.J. Smeets, and A.L. Spek, *Inorg. Chem.*, **30**, 1380 (1991).

[viii] M. Nakahanada, T. Fujihara, A. Fuyuhiro, and S. Kaizaki, *Inorg. Chem.*, **31**, 1315 (1992).

[ix] F.A. Cotton and S.A. Koch, *Inorg. Chem.*, **17**, 2021 (1978).

(*continued overleaf*)

Table 6.27 (*Continued*)

Compound	Aryloxide	Bond length (Å) M–O	Bond angle (°) M–O–Ar	Reference
[(tpp)Cr(OAr)(THF)] meso-tetraphenylporphyrin, 6-coord. Cr	OC$_6$H$_5$	1.94 (1)		x
	Di- and poly-aryloxides			
[Cr$_4$(μ-F)$_4$(di-OAr)$_4$] tetramer with linear Cr–F–Cr, 6-coord. Cr	{CH$_2$NH(CH$_2$-2-OC$_6$H$_4$)}$_2$	1.919 (3) 1.925 (3) 1.933 (3) 1.886 (3)	126 123 126 129	xi
[{Cr(di-OAr)}$_2$(μ-F)(μ-OEt)]	c-(CH$_2$)$_6${NH-2-OC$_6$H$_3$Bu^t-4}$_2$	1.930 (4) 1.954 (3)	125 126	xi
[Cr(μ-di-OAr)Cl]$_2$ 6-coord. Cr, one OAr bridging	μ-{CH$_2$NH(CH$_2$-2-OC$_6$H$_4$)}$_2$	1.903 (2), 2.024 (2) 2.022 (2)	133	xi
[Cr$_2$(μ-di-OAr)$_2$(py)$_4$] 6-coord. Cr, two Cr centres linked by di-OAr	{CH$_2$NHCO(2-OC$_6$H$_2$Cl$_2$-4,6)}$_2$	1.916 (8) 1.952 (8)	126 127	xii
[Cr(tri-OAr)]$_2$[Mn(OH$_2$)$_6$][BPh$_4$]$_2$	c-(CH$_2$)$_6${NCH$_2$C$_6$H$_3$Bu^t-4-O}$_3$	1.956 (5) 1.945 (5) 1.949 (5)	130 129 128	xiii
[Cr(tri-OAr)] 1,4,7-tricyclononane with N-pendant aryloxides, *fac*-arrangement of O and N donor atoms	c-(CH$_2$)$_6${NCH$_2$-2-OC$_6$H$_3$Bu^t-4}$_3$	1.937 (5)	128	xiv

[x] A.L. Balch, L. Latos-Grazynski, B.C. Noll, M.M. Olmstead, and E.P. Zovinka, *Inorg. Chem.*, **31**, 1148 (1992).

[xi] A. Bottcher, H. Elias, J. Glerup, M. Neuburger, C.E. Olsen, H. Paulus, J. Springborg, and M. Zehnder. *Acta Chem. Scand.*, **48**, 967 (1994).

[xii] T.J. Collins, B.D. Santarsiero, and G.H. Spies, *J. Chem. Soc., Chem. Commun.*, 681 (1983).

[xiii] U. Auerbach, C. Stockheim, T. Weyhermuller, K. Wieghardt, and B. Nuber, *Angew. Chem., Int. Ed. Engl.* **32**, 714 (1993).

[xiv] U. Auerbach, T. Weyhermuller, K. Wieghardt, B. Nuber, E. Bill, C. Butzlaff, and A.X. Trautwein, *Inorg. Chem.*, **32**, 508 (1993).

Table 6.28 Molybdenum aryloxides

Compound	Aryloxide	Bond length (Å) M–O	Bond angle (°) M–O–Ar	Reference
	Homoleptic aryloxides and adducts			
trans-[Mo(OAr)$_4$(HNMe$_2$)$_2$]	OC$_6$H$_4$Ph-2	1.963 (3)	139	i
		1.973 (3)	138	
[Mo$_2$(OAr)$_4$(PMe$_3$)$_4$]	OC$_6$F$_5$	2.136 (5)	138	ii
Mo$_2^{4+}$ core		2.055 (5)	134	
		2.055 (6)	133	
		2.117 (5)	137	
[Mo$_2$(OAr)$_4$(HNMe$_2$)$_4$]	OC$_6$F$_5$	2.06 (1)	140	iii
Mo$_2^{4+}$ core		2.08 (1)	140	
[Mo$_2$(OAr)$_6$(HNMe$_2$)$_2$]	OC$_6$H$_4$Ph-2	1.979 (6)	125	i
Mo$_2^{6+}$ core		1.961 (7)	136	
		1.886 (5)	168	
		1.900 (4)	158	
		1.997 (7)	123	
		1.970 (6)	128	
[Me$_2$NH$_2$][(ArO)$_2$(HNMe$_2$)Mo(μ_2-OAr)$_3$Mo(OAr)$_2$(HNMe$_2$)]	OC$_6$H$_4$Me-4	2.026 (6)	138	iv, v
		2.050 (6)	138	
		2.067 (6)	135	
		2.017 (6)	141	
confacial bi-octahedron, Mo$_2^{6+}$ core	μ_2-OC$_6$H$_4$Me-4	2.118 (6), 2.084 (6)		
		2.089 (6), 2.110 (6)		
		2.112 (6), 2.144 (6)		

[i] M.J. Bartos, C.E. Kriley, J.S. Yu, J.L. Kerschner, P.E. Fanwick, and I.P. Rothwell, *Polyhedron*, **8**, 1971 (1989).

[ii] F.A. Cotton, and K.J. Wiesinger, *Inorg. Chem.*, **30**, 750 (1991).

[iii] R.G. Abbott, F.A. Cotton, and L.R. Falvello, *Inorg. Chem.*, **29**, 514 (1990).

[iv] T.W. Coffindaffer, I.P. Rothwell, and J.C. Huffman, *Inorg. Chem.*, **22**, 3178 (1983).

[v] T.W. Coffindaffer, G.P. Niccolai, D. Powell, I.P. Rothwell, and J.C. Huffman, *J. Am. Chem. Soc.*, **107**, 3572 (1985).

(*continued overleaf*)

Table 6.28 *(Continued)*

Compound	Aryloxide	Bond length (Å) M–O	Bond angle (°) M–O–Ar	Reference
	Mixed aryloxy halides, amides' etc.			
[(ArO)₂Cl₂Mo(μ_2-OAr)₂MoCl₂(OAr)₂] 6-coord. Mo, Cl *trans* to μ-O	OC₆H₅	1.822 (8) 1.806 (7)	166 155	vi
[(2,2,2-Cryptand)Na]₂[Mo₆(OAr)₆Cl₈] hexagonal cluster with terminal OAr	μ_2-OC₆H₅ OC₆H₅	2.059 (8), 2.014 (8) 2.091 (4)	136	vii
1,2-[Mo₂(OAr)₄(OPrⁱ)₂] two independent molecules	OC₆H₃Me₂-2,6	1.902 (4) 1.897 (4) 1.898 (4) 1.903 (4)	157 155 148 145	viii
1,2-[Mo₂(OAr)₄(NMe₂)₂] *anti* rotamer, Mo₂⁶⁺ core	OC₆H₃Me₂-2,6	1.961 (3) 1.926 (3)	115 148	ix
1,2-[Mo₂(OAr)₂(NMe₂)₄] *gauche* rotamer, Mo₂⁶⁺ core	OC₆H₃Buᵗ-2-Me-6	2.003 (7)	126	x
[Mo₂(OAr)₄(O₂CNMe₂)₂] Mo₂⁶⁺ core	OC₆H₃Me₂-2,6	1.971 (7) 1.941 (5)	140 125	ix
[Mo₂(OAr)₄(μ_2-OAr)₂(μ_2-NMe₂)₂(HNMe₂)₂] confacial bi-octahedron, Mo₂⁷⁺ core	OC₆H₃Me₂-3,5 μ_2-OC₆F₅	1.953 (5) 1.988 (4) 1.951 (4) 1.960 (4) 1.971 (4) 2.118 (4), 2.109 (4) 2.140 (4), 2.121 (4)	131 143 140 142 145	v
[Mo₂(OAr)₄(NMe₂)₂(HNMe₂)₂] two forms	OC₆F₅	2.060 (4) 2.073 (4) 2.048 (4) 2.072 (4)	134 133 134 126	iii

Compound	OAr	Mo–O (Å)	Mo–O–C (°)	Ref.
[Mo₂(OAr)₅(NMe₂)(HNMe₂)₂]	OC₆F₅	2.027 (7) 2.003 (7) 2.009 (5) 1.965 (7) 1.995 (6)	142 127 132 145 139	iii
[{HB(mppz)₃}Mo(OAr)(NO)Cl] 6-coord. Mo, OAr *trans* to N(pz)	OC₆H₅	1.903 (4)	137	xi
[NEt₄][Mo(OAr)₂] 6-coord. Mo, *trans* N	OC₆H₄-(2-CH₂NC₆H₄S)	2.030 (7) 2.016 (8)	125 129	xii
Oxo and imido aryloxides				
[Mo(OAr)₂(O)₂] tetrahedral Mo, two molecules	OC₆HPh₂-2,6-Me₂-3,5	1.843 (2) 1.866 (3) 1.844 (2) 1.868 (3)	134 151 133 150	xiii
[Mo(OAr)₂(O)₂] 6-coord. Mo, *cis* di-oxo	OC₆H₄(pyrazolyl)-2	1.957 (3)	128	xiv
[H₂NMe₂]₂[Mo₂(μ₂-O)₂(O)₂(OAr)₄]	OC₆F₅	2.035 (8) 2.012 (8)	124 123	xv
[(pyrazolylborate)Mo(O)(OAr)₂]	OC₆H₄Me-4	1.954 (3) 1.928 (3)	131 146	xvi
[Mo(OAr)₂{NB(Me₂)pz}(O)]	OC₆H₄Cl-4	1.959 (3) 1.934 (3)	132 145	xvii

[vi] B. Kamenar and M. Penavic, *J. Chem. Soc., Dalton Trans.*, 356 (1977).

[vii] N. Perchenek and A. Simon, *Z. Anorg. Allg. Chem.*, **619**, 103 (1993).

[viii] T.W. Coffindaffer, I.P. Rothwell, and J.C. Huffman, *Inorg. Chem.*, **22**, 2906 (1983).

[ix] T.W. Coffindaffer, I.P. Rothwell, and J.C. Huffman, *Inorg. Chem.*, **23**, 1433 (1984).

[x] T.W. Coffindaffer, I.P. Rothwell, and J.C. Huffman, *Inorg. Chem.*, **24**, 1643 (1985).

[xi] M. Cano, J.V. Heras, A. Monge, E. Pinilla, C.J. Jones, and J.A. McCleverty, *J. Chem. Soc., Dalton Trans.*, 1555 (1994).

[xii] O.A. Rajan, J.T. Spence, C. Leman, M. Minelli, M. Sato, J.H. Enemark, P.M.H. Kroneck, and K. Sulger, *Inorg. Chem.*, **22**, 3065 (1983).

[xiii] L.E. Turner, P.E. Fanwick, and I.P. Rothwell, unpublished results.

[xiv] T.N. Sorrell, D.J. Ellis, and E.H. Cheesman, *Inorg. Chim. Acta*, **113**, 1 (1986).

[xv] R.G. Abbott, F.A. Cotton, and L.R. Falvello, *Inorg. Chem.*, **29**, 514 (1990).

[xvi] Chaung-Sheng J. Chang, T.J. Pecci, M.D. Carducci, and J.H. Enemark, *Inorg. Chem.*, **32**, 4106 (1993).

[xvii] Chaung-Sheng J. Chang, T.J. Pecci, M.S. Carducci, and J.H. Enemark, *Acta Crystallog. C*, **48**, 1096 (1992).

(continued overleaf)

Table 6.28 *(Continued)*

Compound	Aryloxide	Bond length (Å) M–O	Bond angle (°) M–O–A	Reference
[{HB(Me$_2$)pz}Mo(O)(OAr)$_2$] 6-coord. Mo, OAr *trans* to N(pz)	OC$_6$H$_5$	1.938 (3) 1.929 (3)	134 136	xviii
[Mo(OAr)(=NBu^t)$_2$Cl] 5-coord. Mo	OC$_6$H$_2$(OMe)-3,5-{1'-Np-2'-C(O)=CHPPh$_3$}-2	2.003 (6)	135	xix
	Organometallic aryloxides			
1,2-[Mo$_2$(OAr)$_2$(CH$_2$SiMe$_3$)$_4$] *anti* rotamer, Mo$_2^{6+}$ core	OC$_6$H$_3$Me$_2$-2,6	1.910 (4)	137	xx
[(ArO)(Me$_3$SiCH$_2$)(py)Mo]$_2$(μ_2-H)(μ_2-CSiMe$_3$)] Mo$_2^{8+}$ core	OC$_6$H$_3$Me$_2$-2,6	1.96 (1) 1.99 (1)	135 136	xx
[Mo(OAr)$_4$(η^2-PhCCPh)] tbp Mo, alkyne lies in equatorial plane	OC$_6$H$_3$Pr$^i_2$-2,6	2.004 (10) 1.876 (10)	144 160	xxi
[Mo(OC$_6$H$_3$Ph-C$_6$H$_4$)(OAr)$_2$(HNMe$_2$)] square pyramidal Mo	OC$_6$H$_3$Ph$_2$-2,6 OC$_6$H$_3$Ph-η^1-C$_6$H$_4$	1.919 (2) 1.976 (2) 1.962 (2)	136 143 131	xxii xxiii
[Mo(OC$_6$H$_3$Ph-C$_6$H$_4$)(OAr)$_2$(py)] square pyramidal Mo	OC$_6$H$_3$Ph$_2$-2,6 OC$_6$H$_3$Ph-η^1-C$_6$H$_4$	1.964 (3) 1.946 (3) 1.939 (3)	139 134 135	xxiii
[Mo(OC$_6$H$_3$Ph-η^6-C$_6$H$_5$)(H)(PMePh$_2$)$_2$] four legged piano stool	OC$_6$H$_3$Ph-η^6-C$_6$H$_5$	2.164 (5)	118	xxi xxiv
	Di-and poly-aryloxides			
[Mo$_2$(di-OAr)$_2$(NMe$_2$)$_2$]	CH$_2$(2-OC$_6$H$_3$Bu^t-6-Me-4)$_2$	1.917 (7) 1.921 (7)	165 139	xxv
[{HB(Me$_2$)pz}Mo(O)$_2$](μ-di-OAr) two 6-coord. Mo each bound to one OAr	(OC$_6$H$_4$-2)SS(2-OC$_6$H$_4$)	1.920 (4)	133	xxvi

[Mo(Hdi-OAr)₂] 6-coord. Mo, only one aryloxide bound, both N deprotonated and bound		$(2\text{-HOC}_6\text{H}_4)\text{NC}_6\text{H}_4\text{N}(2\text{-OC}_6\text{H}_4)$	1.982 (6) 1.967 (5)	132 129	xxvii
[Mo(di-OAr)(O)₂] two molecules, 6-coord. Mo, *trans* OAr		$\{\text{CH}_2\text{NH}(\text{CH}_2\text{-2-OC}_6\text{H}_3\text{Bu}^t\text{-6})\}_2$	1.946 (10) 1.948 (10) 1.909 (10) 1.951 (10)	137 139 134 125	xxviii
[Mo(di-OAr)(O)₂] 6-coord. Mo, *trans* OAr		$\{\text{CH}_2\text{NMe}(\text{CH}_2\text{-2-OC}_6\text{H}_4)\}_2$	1.934 (1)	137	xxix
[Mo(di-OAr)(O)₂] 6-coord. Mo, *trans* OAr		$\{\text{N}(\text{CH}_2\text{-2-OC}_6\text{H}_4)_2(\text{CH}_2\text{CH}_2\text{NMe}_2)\}$	1.953 (3) 1.967 (2)	127 120	xxix
[Mo(di-OAr)(O)₂] 6-coord. Mo, *trans* OAr		$\{\text{N}(\text{CH}_2\text{-2-OC}_6\text{H}_4\text{Bu}^t\text{-4})_2(\text{CH}_2\text{CH}_2\text{NMe}_2)\}$	1.977 (5) 1.954 (5)	130 127	xxix
[Mo(di-OAr)(O)₂] 6-coord. Mo, *trans* OAr		$\{\text{N}(\text{CH}_2\text{-2-OC}_6\text{H}_4)_2(\text{CH}_2\text{py})\}$	1.941 (2) 1.970 (2)	131 131	xxix
[Mo(di-OAr)(O)₂] 6-coord. Mo, *trans* OAr		$\{\text{CH}_2\text{NMe}(2\text{-OC}_6\text{H}_4)\}_2$	1.970 (2) 1.956 (2)	120 122	xxix

[xviii] C.A. Kipke, W.E. Cleland Jr, S.A. Roberts, and J.H. Enemark, *Acta Crystallog. C*, **45**, 870 (1989).

[xix] J. Sundermeyer, K. Weber, H. Werner, N. Mahr, G. Bringmann, and O. Schupp, *J. Organomet. Chem.*, **444**, C37 (1993).

[xx] T.W. Coffindaffer, I.P. Rothwell, and J.C. Huffman, *J. Chem. Soc., Chem. Comm.*, 1249 (1983).

[xxi] E.C. Walborsky, D.E. Wigley, E. Roland, J.C. Dewan, and R.R. Schrock, *Inorg. Chem.*, **26**, 1615 (1987).

[xxii] J.L. Kerschner, P.E. Fanwick, and I.P. Rothwell, *J. Am. Chem. Soc.*, **109**, 5840 (1987).

[xxiii] J.L. Kerschner, J.S. Yu, P.E. Fanwick, I.P. Rothwell, and J.C. Huffman, *Organometallics*, **8**, 1414 (1989).

[xxiv] J.L. Kerschner, E.M. Torres, P.E. Fanwick, I.P. Rothwell, and J.C. Huffman, *Organometallics*, **8**, 1424 (1989).

[xxv] M.H. Chisholm, I.P. Parkin, K. Folting, E.B. Lubkovsky, and W.E. Streib, *J. Chem. Soc., Chem. Commun.*, 1673 (1991).

[xxvi] S.A. Roberts, G.P. Darsey, W.E. Cleland Jr, and J.H. Enemark, *Inorg. Chim. Acta*, **154**, 95 (1988).

[xxvii] S.F. Gheller, T.W. Hambley, M.R. Snow, K.S. Murray, and A.G. Wedd, *Aust. J. Chem.*, **37**, 911 (1984).

[xxviii] P. Subramanian, J.T. Spence, R. Ortega, and J.H. Enemark, *Inorg. Chem.*, **23**, 2564 (1984).

[xxix] C.J. Hinshaw, G. Peng, R. Singh, J.T. Spence, J.H. Enemark, M. Bruck, J. Kristofzski, S.L. Merbs, R.B. Ortega, and P.A. Wexler, *Inorg. Chem.*, **28**, 4483 (1989).

591

Table 6.29 Tungsten aryloxides

Compound	Aryloxide	Bond length (Å) M–O	Bond angle (°) M–O–Ar	Reference
	Homoleptic aryloxides and adducts			
[W(OAr)$_6$] octahedral W	OC$_6$H$_4$Me-4	1.887 (3)	147	i
		1.896 (3)	145	
		1.889 (3)	137	
		1.913 (3)	138	
		1.891 (3)	146	
		1.894 (4)	140	
[NEt$_4$][W(OAr)$_6$] octahedral W anion	OC$_6$H$_5$	1.955 (4)	137	ii
		1.931 (5)	142	
		1.893 (5)	151	
		1.956 (5)	136	
		1.941 (4)	139	
		1.920 (4)	138	
[(THF)$_2$Li(μ_2-OAr)$_2$W(OAr)$_4$] octahedral W	OC$_6$H$_5$	1.915 (6)	139	ii
		1.921 (7)	138	
		1.928 (6)	138	
		1.906 (6)	144	
	μ_2-OC$_6$H$_5$	1.92 (2), 2.011 (7)		
		1.98 (2), 2.012 (7)		
[W(OAr)$_4$] distorted square planar geometry	OC$_6$H$_3$Pr$^i_2$-2,6	1.851 (6)	154	iii, iv
		1.866 (5)	155	
		1.851 (5)	155	
		1.849 (5)	159	
[W(OAr)$_4$] distorted square planar geometry	OC$_6$H$_3$Me$_2$-2,6	1.843 (4)	161	iv
trans-[W(OAr)$_4$(H$_2$NBu^t)$_2$]	OC$_6$H$_4$Ph-2	1.972 (3)	139	v
		1.963 (3)	140	

Compound	OAr			Ref
[W₂(OAr)₆(HNMe₂)₂] nearly eclipsed, *anti*-N atoms, W₂⁶⁺ core	OC₆F₅	1.917(8)	157	vi
		1.95(1)	150	
		1.992(9)	134	

$$\text{Mixed aryloxy halides}$$

Compound	OAr			Ref
[W(OAr)Cl₅]	OC₆H₃Me₂-2,6	1.82(2)	180	vii
trans-[W(OAr)₂Cl₄]	OC₆H₃Me₂-2,6	1.860(5)	180	vii
mer-[W(OAr)₃Cl₃]	OC₆H₃Pr$^i_2$-2,6	1.832(2)	173	viii
		1.848(2)	156	
		1.836(2)	167	
cis-[W(OAr)₂Cl₄]	OC₆H₃Ph₂-2,6	1.824(6)	142	viii
		1.855(7)	149	
cis-[W(OAr)₂Cl₄]	OC₆H₃Ph₂-2,6	1.851(3)	144	ix
		1.840(3)	148	
cis-[W(OAr)₂Cl₄]	OC₆H₃Pr$^i_2$-2,6	1.814(8)	169	x
		1.786(9)	177	
trans-[W(OAr)₄Cl₂]	OC₆H₅	1.83(1)	155	xi
		1.90(1)	166	
		1.82(1)	155	
		1.72(1)	155	

[i] W. Clegg, R.J. Errington, P. Kraxner, and C. Redshaw, *J. Chem. Soc., Dalton Trans.*, 1431 (1992).
[ii] J.I. Davies, J.F. Gibson, A.C. Skapski, G. Wilkinson, and W.-K. Wong, *Polyhedron*, **1**, 641 (1982).
[iii] M.L. Listemann, J.C. Dewan, and R.R. Schrock, *J. Am. Chem. Soc.*, **107**, 7207 (1985).
[iv] M.L. Listemann, R.R. Schrock, J.C. Dewan, and R.M. Kolodziej, *Inorg. Chem.*, **27**, 264 (1988).
[v] J.L. Kerschner, J.S. Yu, P.E. Fanwick, I.P. Rothwell, and J.C. Huffman, *Organometallics*, **8**, 1414 (1989).
[vi] R.G. Abbott, F.A. Cotton, and L.R. Falvello, *Inorg. Chem.*, **29**, 514 (1990).
[vii] N. Kanehisa, Y. Kai, N. Kasai, H. Yasuda, Y. Nakayama, and A. Nakamura, *Bull. Chem. Soc. Jpn.*, **65**, 1197 (1992).
[viii] F. Quignard, M. Leconte, J.-M. Basset, Leh-Yeh Hsu, J.J. Alexander, and S.G. Shore, *Inorg. Chem.*, **26**, 4272 (1987).
[ix] J.L. Kerschner, P.E. Fanwick, I.P. Rothwell, and J.C. Huffman, *Inorg. Chem.*, **28**, 780 (1989).
[x] H. Yasuda, Y. Nakayama, K. Takei, A. Nakamura, Y. Kai, and N. Kanehisa, *J. Organomet. Chem.*, **473**, 105 (1994).
[xi] L.B. Handy and C.K. Fair, *Inorg. Nucl. Chem. Lett.*, **11**, 497 (1975).

(continued overleaf)

Table 6.29 (*Continued*)

Compound	Aryloxide	Bond length (Å) M–O	Bond angle (°) M–O–A	Reference
[W(OAr)₄Cl] square pyramid, axial Cl	OC₆H₃Me₂-2,6	1.855 (7)	176	x
trans-[W(OAr)₂Cl₃(THF)]	OC₆H₃Ph₂-2,6	1.855 (5), 1.857 (5)	154, 152	ix
[W(OAr)₄Cl(OEt₂)] Cl trans to ether	OC₆H₂Cl₃-2,4,6	1.853 (9), 1.964 (10)	165, 140	xii
trans-[W(OAr)₂Cl₃(PMe₂Ph)]	OC₆H₃Ph₂-2,6	1.877 (4), 1.853 (4)	157, 166	xi
all trans-[W(OAr)₂Cl₂(PMe₂Ph)₂]	OC₆H₃Ph₂-2,6	1.966 (4)	140	xi
all trans-[W(OAr)₂Cl₂(PMePh₂)₂]	OC₆H₄Me-4	1.848 (5), 1.840 (5)	166, 172	xiii
all trans-[W(OAr)₂Cl₂(PMePh₂)₂]	OC₆H₃Me₂-2,6	1.854 (5)	180	xiv
	Mixed aryloxy amides, alkoxides, hydrides, etc.			
[W(OAr)₄(NMe₂)₂] octahedral W	OC₆H₂(CF₃)₃-2,4,6	1.949 (5), 1.936 (6), 1.971 (5), 1.945 (6)	151, 147, 136, 133	xv
1,2-[W₂(OAr)₂(NMe₂)₄] gauche rotamer, W₂⁶⁺ core	OC₆H₂Buᵗ₂-2,4-Ph-6	1.90 (2), 1.98 (2)	157, 114	xvi
[W(OAr)₂(OCMe₂CMe₂O)₂] 6-coord. W, cis-OAr	OC₆H₅	1.884 (5), 1.916 (6)	145, 139	xvii
[W(OAr)(H)₃(PMe₃)₄] 8-coord. W	OC₆H₅	2.129 (8)	142	xviii
	Oxo, imido and related aryloxides			
trans-[W(O)(OAr)₂Cl₂] Square pyramid, axial W=O	OC₆H₃Br₂-2,6	1.860 (2), 1.854 (2)	139, 165	xix

Compound	OAr	Bond length		Ref
[W(OAr)₂(=NBuᵗ)₂] tetrahedral W	OC₆H₃Ph₂-2,6	1.920 (2)	126	ix
		1.906 (2)	148	
[W(OAr)₂)(=NPh)₂] tetrahedral W, two molecules	OC₆HPh₄-2,3,5,6	1.897 (6)	141	xx
		1.929 (6)	131	
		1.900 (6)	141	
		1.942 (6)	124	
[W(OAr)₂(=O)₂(PMe₃)] tbp, axial ArO and P	OC₆HPh₄-2,3,5,6	1.943 (3)	137	xxi
		1.902 (3)	177	
[W(OAr)₂(=NPh)(=Ntol)(PMe₃)] tbp, axial ArO and P	OC₆HPh₄-2,3,5,6	1.965 (4)	171	xxi
		2.010 (3)	134	
[W(OAr)₂(=O)(=NPh)(OPMe₃)] square pyramid, axial W=O	OC₆HPh₄-2,3,5,6	1.966 (4)	125	xxi
		1.973 (4)	125	
[W(OAr)₂(=S)₂] tetrahedral W	OC₆HPh₄-2,3,5,6	1.852 (2)	147	xxi
		1.855 (3)	146	
[W(OAr)₂(=Se₂)] tetrahedral W	OC₆HPh₄-2,3,5,6	1.836 (4)	149	xxi
		1.858 (3)	146	
[W(OAr)₂(=Se)(Ph)(SePh)] square pyramidal, axial W=Se	OC₆HPh₄-2,3,5,6	1.929 (6)	142	xxi
		1.875 (6)	152	

xii R.M. Kolodziej, R.R. Schrock, and J.C. Dewan, *Inorg. Chem.*, **28**, 1243 (1989).

xiii L.M. Atagi and J.M. Mayer, *Angew. Chem., Int. Ed. Engl.*, **32**, 439 (1993).

xiv M.R. Lentz, P.E. Fanwick, and I.P. Rothwell, unpublished results.

xv H.W. Roesky, M. Scholz, and M. Noltemeyer, *Chem, Ber.*, **123**, 2303 (1990).

xvi V.M. Visciglio, P.E. Fanwick, and I.P. Rothwell, *Inorg. Chim. Acta*, **211**, 203 (1993).

xvii A. Lehtonen and R. Sillanpaa, *J. Chem. Soc., Dalton Trans.*, 2701 (1995).

xviii K.W. Chiu, R.A. Jones, G. Wilkinson, A.M.R. Galas, M.B. Hursthouse, and K.M.A. Malik, *J. Chem. Soc., Dalton Trans.*, 1204 (1981).

xix W.A. Nugent, J. Feldman, and J.C. Calabrese, *J. Am. Chem. Soc.*, **117**, 8992 (1995).

xx M.A. Lockwood, P.E. Fanwick, O. Eisenstein, and I.P. Rothwell, *J. Am. Chem. Soc.*, **118**, 2762 (1996).

xxi M.A. Lockwood, J.S. Vilardo, P.E. Fanwick, and I.P. Rothwell, unpublished results.

(continued overleaf)

Table 6.29 (*Continued*)

Compound	Aryloxide	Bond length (Å) M–O	Bond angle (°) M–O–A	Reference
[W(OAr)(=NPh)Cl$_3$] 6-coord. W, W≡N *trans* to W-N	OC$_6$H$_2$(CH$_2$NMe$_2$)$_2$-2,6-Me-4	1.880 (3)	136	xxii
[(ArO)$_3$W(μ-N)$_2$W(OAr)$_3$] tbp, axial OAr and N	OC$_6$H$_3$Pr$^i_2$-2,6	1.891 (3) 1.888 (3) 1.886 (3)	155 168 145	xxiii

Organometallic aryloxides

Compound	Aryloxide	Bond length (Å) M–O	Bond angle (°) M–O–A	Reference
[W(OAr)$_3$(C$_3$Et$_3$)] tbp, planar tungstacyclobutadiene	OC$_6$H$_3$Pr$^i_2$-2,6	1.885 (6) 2.008 (6) 1.979 (6)	152 131 135	xxiv
[W(OAr)$_2$(C$_4$Et$_4$)] folded tungstacyclopentatriene	OC$_6$H$_3$Ph$_2$-2,6	1.854 (6) 1.927 (5)	154 136	xxv xxvi
[W(OAr)$_2$(η^2-C$_2$Et$_2$)$_2$] parallel alkyne units	OC$_6$H$_3$Ph$_2$-2,6	1.959 (3) 1.960 (3)	131 136	xxv
[W(OAr)$_2$Cl$_2$(η^2-C$_2$Et$_2$)] square planar, axial η^2-C$_2$Et$_2$	OC$_6$H$_3$Ph$_2$-2,6	1.838 (4) 1.853 (4)	156 151	xxvi
[Cp*W(OAr)(η^2-C$_2$Me$_2$)$_2$]	OC$_6$H$_5$	2.072 (7)	137	xxvii
[Cp*W(OAr)Me$_2$]$_2$(μ_2-N$_2$)	OC$_6$F$_5$	2.079 (8)	142	xxviii
[W(O)(OAr)$_2$(=CHBu^t)(PMePh$_2$)] tbp, axial OAr and P	OC$_6$H$_3$Ph$_2$-2,6	1.993 (5) 1.957 (6)	129 157	xxix
[W(O-η^1-C$_6$H$_4$)(H)$_2$(PMe$_3$)$_4$]	O-η^1-C$_6$H$_4$	2.177 (5)	94	xxx xxxi
[W(OC$_6$H$_2$Me$_2$-2-CH$_2$)(H)$_2$(PMe$_3$)$_4$]	OC$_6$H$_2$Me$_2$-2-CH$_2$	2.145 (8)	118	xxx
[W(OC$_6$H$_3$Ph-η^1-C$_6$H$_4$)$_2$(C$_4$Et$_4$)]	OC$_6$H$_3$Ph-η^1-C$_6$H$_4$	1.861 (9) 1.932 (9)	124 137	xxvi
[W(OC$_6$H$_3$Ph-η^1-C$_6$H$_4$)$_2$(C$_6$Et$_6$)]	OC$_6$H$_3$Ph-η^1-C$_6$H$_4$	1.941 (2) 1.931 (2)	149 125	xxvi

[W(OC$_6$H$_3$Ph-η^1-C$_6$H$_4$)$_2$(PMePh$_2$)$_2$] 6-coord. W, *trans* O, *cis* P	OC$_6$H$_3$Ph-η^1-C$_6$H$_4$	1.88 (1) 1.87 (1)	141 144	xxxii xxiii
[W(OC$_6$H$_3$Ph-η^1-C$_6$H$_4$)$_2$(py)$_2$] 6-coord. W, *trans* O, *cis* py	OC$_6$H$_3$Ph-η^1-C$_6$H$_4$	1.91 (1) 1.93 (1)	142 144	xxxiv xxxv
[W(OC$_6$H$_3$Ph-η^1-C$_6$H$_4$)$_2$(η^2-C$_2$H$_4$)(PMePh$_2$)] octahedral W, *trans* O, *cis* C	OC$_6$H$_3$Ph-η^1-C$_6$H$_4$	1.867 (5) 1.870 (5)	144 142	xxxv
[W(O)(OC$_6$H$_3$Ph-η^1-C$_6$H$_4$)(PMePh$_2$)$_3$]	OC$_6$H$_3$Ph-η^1-C$_6$H$_4$	1.88 (1) 1.87 (1)	141 144	xxxvi
[W(OC$_6$H$_3$Ph-η^6-C$_6$H$_5$)(OAr)(PMePh$_2$)]	OC$_6$H$_3$Ph-2,6 OC$_6$H$_3$Ph-η^6-C$_6$H$_5$	2.001 (6) 2.033 (6)	127 118	xxxii xxxiii
[W(OC$_6$H$_3$Ph-η^6-C$_6$H$_5$)(OAr)(dppm)]	OC$_6$H$_3$Ph-2,6 OC$_6$H$_3$Ph-η^6-C$_6$H$_5$	2.038 (5) 1.986 (5)	119 127	xxxvii xxxiv

(continued overleaf)

[xxii] P.A. van der Schaaf, J. Boersma, W.J.J. Smeets, A.L. Spek, and G. van Koten, *Inorg. Chem.*, **32**, 5108 (1993).

[xxiii] T.P. Pollagi, J. Manna, S.J. Geib, and M.D. Hopkins, *Inorg. Chim. Acta*, **243**, 177 (1996).

[xxiv] M.R. Churchill, J.W. Ziller, J.H. Freudenberger, and R.R. Schrock, *Organometallics*, **3**, 1554 (1984).

[xxv] J.L. Kerschner, P.E. Fanwick, and I.P. Rothwell, *J. Am. Chem. Soc.*, **110**, 8235 (1988).

[xxvi] C.E. Kriley, J.L. Kerschner, P.E. Fanwick, and I.P. Rothwell, *Organometallics*, **12**, 2051 (1993).

[xxvii] M.B. O'Regan, M.G. Vale, J.F. Payack, and R.R. Schrock, *Inorg. Chem.*, **31**, 1112 (1992).

[xxviii] M.B. O'Regan, A.H. Liu, W.C. Finch, R.R. Schrock, and W.M. Davis, *J. Am. Chem. Soc.*, **112**, 4331 (1990).

[xxix] M.B. O'Donoghue, R.R. Schrock, A.M. LaPointe, and W.M. Davis, *Organometallics*, **15**, 1334 (1996).

[xxx] D. Rabinovich, R. Zelman, and G. Parkin, *J. Am. Chem. Soc.*, **112**, 9632 (1990).

[xxxi] D. Rabinovich, R. Zelman, and G. Parkin, *J. Am. Chem. Soc.*, **114**, 4611 (1992).

[xxxii] J.L. Kerschner, I.P. Rothwell, J.C. Huffman, and W.E. Streib, *Organometallics*, **7**, 1871 (1988).

[xxxiii] J.L. Kerschner, P.E. Fanwick, I.P. Rothwell, and J.C. Huffman, *Organometallics*, **8**, 1431 (1989).

[xxxiv] C.E. Kriley, P.E. Fanwick, and I.P. Rothwell, *J. Am. Chem. Soc.*, **116**, 5225 (1994).

[xxxv] P.E. Fanwick, I.P. Rothwell, and C.E. Kriley, *Polyhedron*, **15**, 2403 (1996).

[xxxvi] J.L. Kerschner, C.E. Kriley, P.E. Fanwick, and I.P. Rothwell, *Acta Crystallog. C*, **50**, 1193 (1994).

[xxxvii] J.L. Kerschner, P.E. Fanwick, and I.P. Rothwell, *J. Am. Chem. Soc.*, **110**, 8235 (1988).

Table 6.29 *(Continued)*

Compound	Aryloxide	Bond length (Å) M–O	Bond angle (°) M–O–A	Reference
[W(OC₆HPh₃-η⁶-C₆H₅)(OAr)(PMe₃)]	OC₆HPh₄-2,3,5,6	1.990 (3)	130	xxxviii
	OC₆HPh₃-η⁶-C₆H₅	2.028 (3)	118	
[W(OC₆HPh₃-η⁶-C₆H₅)(OAr))PBu₃)]	OC₆HPh₄-2,3,5,6	1.987 (4)	134	xxxviii
	OC₆HPh₃-η⁶-C₆H₅	2.025 (4)	118	
[W(OC₆HPh₃-η⁶-C₆H₅)(OAr)(PMe₂Ph)]	OC₆HPh₄-2,3,5,6	1.966 (9)	131	xxxix
	OC₆HPh₃-η⁶-C₆H₅	2.022 (9)	120	
[W(OC₆HPh₃-η⁶-C₆H₅)(OC₆HPh₃-η²-C₆H₉)(PMe₂Ph)] chelating arene and cyclohexene rings	OC₆HPh₃-η⁶-C₆H₅	2.130 (4)	125	xxxix
	OC₆HPh₃-η²-C₆H₉	2.066 (4)	118	
[W(OC₆H₃Ph-η⁶-C₆H₅)(H)(PMePh₂)₂]	OC₆H₃Ph-η⁶-C₆H₅	2.163 (9)	119	xl
[W(OC₆H₃Ph-η⁶-C₆H₅)(Cl)(PMe₂Ph)₂]	OC₆H₃Ph-η⁶-C₆H₅	2.128 (4)	122	xl
[W(OC₆H₃Ph-η⁶-C₆H₅)(Cl)(dppm)]	OC₆H₃Ph-η⁶-C₆H₅	2.135 (8)	121	xl
[NEt₄]₃[(CO)₃W(μ₂-OAr)₃W(CO)₃] confacial bis-octahedral anion	μ₂-OC₆H₅	2.210 (6)–2.234 (5)		xli
[NEt₄][W(OAr)(CO)₅]	OC₆H₃Ph₂-2,6	2.165 (9)	128	xlii
		2.175 (9)	124	
[NEt₄][(CO)₅W(μ-OAr)Cr(CO)₃]	μ₂-η⁶-C₆H₅	2.35 (3)	128	xliii
[NEt₄][W(OAr)(CO)₄(PMePh₂)]	OC₆H₅	2.191 (6)	130	xlii
[NEt₄][W(OAr)(CO)₅]	OC₆H₅	2.182 (19)	131	xliv
		2.205 (21)	134	
[NEt₄][W(OAr)(CO)₅]0.5(H₂O)	OC₆H₅	2.224 (15)	132	xliv
[NEt₄][W(OAr)(CO)₅]0.5(H₂O)	OC₆H₅	2.191 (14)	130	xliv
[W(OC₆HPh₃-η¹-C₆H₄)(OAr)(η¹-PhN=CHPh)(PhNCH₃Ph)] tbp, axial O (cyclometallated OAr) and N (amide)	OC₆HPh₄-2,3,5,6	1.922 (4)	166	xlv
	OC₆HPh₃-η¹-C₆H₄	1.961 (4)	131	

Complex	OAr ligand			Ref.
[W(OC$_6$HPh$_3$-η^1-C$_6$H$_4$-c-C$_5$H$_8$C-O)(OAr)(η^2-c-C$_5$H$_8$C=O)]		1.819 (9)	151	xlv
[W(O)(OAr)$_2${OCPh(C$_6$H$_4$)CHPh$_2$}]	OC$_6$HPh$_3$-2,3,5,6	1.942 (9)	141	xlv
	OC$_6$HPh$_3$-η^1-C$_6$H$_4$-c-C$_5$H$_8$C-O	1.911 (5)	149	
	OC$_6$HPh$_4$-2,3,5,6	1.845 (4)	172	xlvi
[W$_3$(di-OAr)$_3$(μ-O)$_3$(O)$_3$]	S(2-OC$_6$H$_4$)$_2$	1.90 (1)	124	
		2.02 (1)	129	
		1.87 (1)	140	
		1.99 (1)	131	
		1.84 (1)	134	
		1.87 (1)	131	
[W$_2$(μ_2-H)(μ_2-NMe$_2$)(di-OAr)(di-OAr-H)(NMe$_2$)(HNMe$_2$)]	CH$_2$(2-OC$_6$H$_3$Bu^t-6-Me-4)$_2$	1.939 (7)	149	xlvii
one cyclometallated di-OAr	CH(2-OC$_6$H$_3$Bu^t-6-Me-4)$_2$	1.998 (7)	148	xlviii
		1.967 (7)	121	
		1.984 (7)	120	

xxxviii M.A. Lockwood, P.E. Fanwick, and I.P. Rothwell, unpublished results.

xxxix M.A. Lockwood, P.E. Fanwick, and I.P. Rothwell, *Polyhedron*, **14**, 3363 (1995).

xl J.L. Kerschner, E.M. Torres, P.E. Fanwick, I.P. Rothwell, and J.C. Huffman, *Organometallics*, **8**, 1424 (1989).

xli D.J. Darensbourg, K.M. Sanchez, and J.H. Reibenspies, *Inorg. Chem.*, **27**, 3269 (1988).

xlii D.J. Darensbourg, B.L. Mueller, C.J. Bischoff, S.S. Chojnacki, and J.H. Reibenspies, *Inorg. Chem.*, **30**, 2418 (1991).

xliii D.J. Darensbourg, B.L. Mueller, and J.H. Reibenspies, *J. Organomet. Chem.*, **451**, 83 (1993).

xliv D.J. Darensbourg, K.M. Sanchez, J.H. Reibenspies, and A.L. Rheingold, *J. Am. Chem. Soc.*, **111**, 7094 (1989).

xlv M.A. Lockwood, P.E. Fanwick, and I.P. Rothwell, *J. Chem. Soc., Chem. Commun.*, 2013 (1996).

xlvi P. Berges, W. Hinrichs, A. Holzmann, J. Wiese and G. Klar, *J. Chem. Res.*, **10**, 201 (1986).

xlvii M.H. Chisholm, I.P. Parkin, K. Folting, E.B. Lubkovsky, and W.E. Streib, *J. Chem. Soc., Chem. Commun.*, 1673 (1991).

xlviii M.H. Chisholm, J-H. Huang, J.C. Huffman, and I.P. Parkin, *Inorg. Chem.*, **36**, 1642 (1997).

Table 6.30 Simple aryloxides of manganese

Compound	Aryloxide	Bond length (Å) M–O	Bond angle (°) M–O–Ar	Reference
[Mn(OAr)₂(NCMe)₂] tetrahedral, 4-coord. Mn	OC₆H₂Bu$^t_3$-2,4,6	1.910 (6)	179	i
[(ArO)Mn(μ-OAr)₂Mn(OAr)] 3-coord. Mn	OC₆H₂Bu$^t_3$-2,4,6 μ₂-OC₆H₂Bu$^t_3$-2,4,6	1.867 (4), 1.879 (4) 2.058 (4), 2.041 (4) 2.054 (4), 2.045 (4)	167 179	ii
[Mn(OAr)₂(THF)₃]	OC₆H₂(CF₃)₃-2,4,6	1.997 (5) 1.993 (5)	162 179	iii
[Mn(OAr)₃].0.5CH₂Cl₂.0.5C₆H₁₄ mer-MnO₃N₃	OC₆H₄(2'-py)-2	1.870 (6) 1.898 (6) 1.906 (6)	119 123 117	iv
[Mn(NBu^t)₃(OAr)] tetrahedral Mn	OC₆F₅	1.896 (2)	120	v
[(bipy)OAr)Mn(μ₂-OAr)₂Mn(OAr)(bipy)]	OC₆H₂Cl₃-2,4,6 μ₂-OC₆H₂Cl₃-2,4,6	2.045 (3), 1.996 (3) 2.123 (3), 2.173 (3) 2.139 (3), 2.125 (3)	132 137	vi
[(bipy)(OAr)Mn(μ₂-OAr)₂Mn(OAr)(bipy)]	OC₆H₂Br₃-2,4,6 μ₂-OC₆H₂Br₃-2,4,6	2.039 (5) 2.024 (5) 2.160 (5), 2.178 (5) 2.198 (5), 2.120 (5)	138 138	vi
[(PhMe₂CCH₂)Mn(μ-OAr)₂Mn(CH₂CMe₂Ph)] 3-coord. Mn	μ₂-OC₆H₂Bu$^t_3$-2,4,6	2.05 (8), 2.073 (4)		vii

600

Compound / description	OAr	M–O distance (Å)		Ref.
[(OC)₃Mn(C₄H₄N)Mn(CO₃)(OAr)] picrate chelated to one Mn; pyrrole π-bound to one Mn and σ-bound to the other	OC$_6$H$_2$(NO$_2$)$_3$-2,4,6	1.989 (8)	127	viii
[Mn$_4$(μ-OAr)$_6$][ClO$_4$]$_2$·3MeCN·2Et$_2$O chain structure with two O between each pair of Mn atoms	μ$_2$-OC$_6$H$_4$(2'-1,10-phen)-2	2.084 (8)–2.176 (8)		ix
[Mn$_4$(μ-OAr)$_6$][BPh$_4$]$_2$·2MeCN·2.5Et$_2$O chain structure with two O between each pair of Mn atoms	μ$_2$-OC$_6$H$_4$(2'-bipy)-2	2.073 (7)–2.162 (7)		x
[Mn$_2$(μ-OAr)$_3$(MeCN)][PF$_6$]$_2$·Et$_2$O edge-shared bis-octahedron, MeCN trans to μ-O	μ$_2$-OC$_6$H$_4$(2'-bipy)-2	2.088 (4), 2143 (4) 2.239 (4), 1.924 (4)		x

[i] D. Meyer, J.A. Osborn, and M. Wesolek, *Polyhedron*, **9**, 1311 (1990).

[ii] R.A. Bartlett, J.J. Ellison, P.P. Power, and S.C. Shoner, *Inorg. Chem.*, **30**, 2888 (1991).

[iii] H.W. Roesky, M. Scholz, and M. Noltemeyer, *Chem. Ber.*, **123**, 2303 (1990).

[iv] D.A. Bardwell, J.C. Jeffery, and M.D. Ward, *Inorg. Chim. Acta*, **236**, 125 (1995).

[v] A.A. Danopoulos, G. Wilkinson, T.K.N. Sweet, and M.B. Hursthouse, *J. Chem. Soc., Dalton Trans.*, 1037 (1994).

[vi] M. Wesolek, D. Meyer, J.A. Osborn, A. De Cian, J. Fischer, A. Derory, P. Legoll, and M. Drillon, *Angew. Chem., Int. Ed. Engl.*, **33**, 1592 (1994).

[vii] R.A. Jones, S.U. Koschmieder, and C.M. Nunn, *Inorg. Chem.*, **27**, 4524 (1988).

[viii] V.G. Andrianov, Yu. T. Struchkov, N.I. Pyshnograeva, V.N. Setkina, and D.N. Kursanov, *J. Organomet. Chem.*, **206**, 177 (1981).

[ix] J.C. Jeffery, P. Thornton, and M.D. Ward, *Inorg. Chem.*, **33**, 3612 (1994).

[x] D.A. Bardwell, J.C. Jeffery, and M.D. Ward, *J. Chem. Soc., Dalton Trans.*, 3071 (1995).

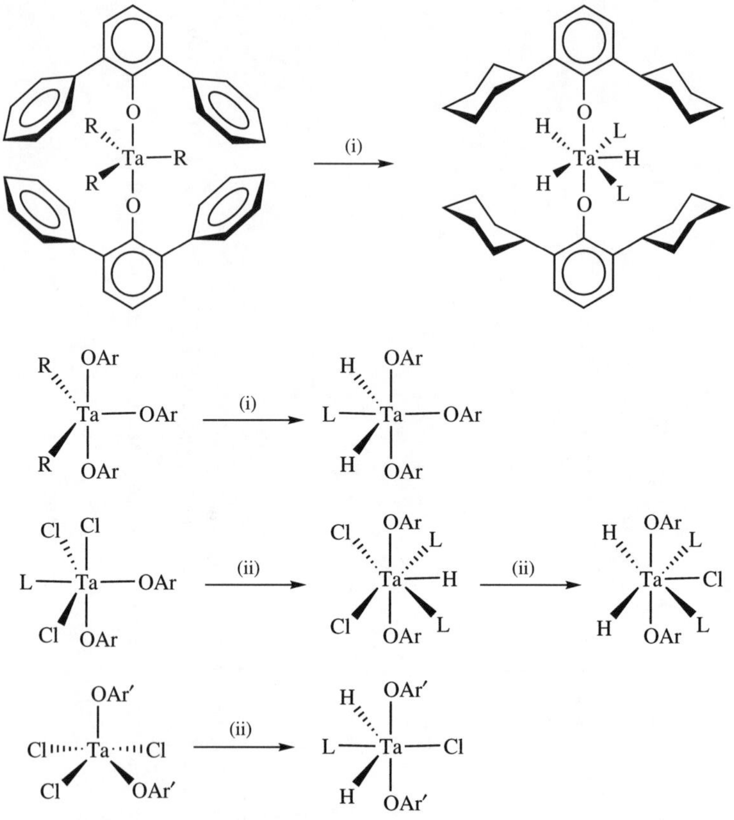

OAr = 2,6-diisopropyl, OAr′ = 2,6-di-tert-butylphenoxide,
L = tertiary phosphine, R = $CH_2C_6H_4Me$-4.

Reagents: (i) L, H_2 (1200 psi), 80° C; (ii) Bu_3^nSnH, L.

Scheme 6.8

Table 6.31   Simple aryloxides of technecium

Compound	Aryloxide	Bond length (Å) M–O	Bond angle (°) M–O–Ar	Reference
[Tc(OAr)$_3$]	OC$_6$H$_4$PPh$_2$-2	2.000 (3)	123	i
mer-TcO$_3$P$_3$ core		1.990 (3)	124	
		2.074 (3)	121	
[Tc(O)Cl(OAr)$_2$]	OC$_6$H$_4$PPh$_2$-2	1.985 (6)	123	ii
cis-O(Ar), trans ArO–Re=O		1.998 (6)	127	

[i] C. Bolzati, F. Refosco, F. Tisato, G. Bandoli, and A. Dolmella, *Inorg. Chim. Acta*, **201**, 7 (1992).
[ii] C. Bolzati, F. Tisato, F. Refosco, and G. Bandoli, *Inorg. Chim. Acta*, **247**, 125 (1996).

Table 6.32 Simple aryloxides of rhenium

Compound	Aryloxide	Bond length (Å) M–O	Bond angle (°) M–O–Ar	Reference
[Re(OAr)$_4$] square planar, 4-coord. Re	OC$_6$H$_3$Pr$_2^i$-2,6	1.864 (4) 1.866 (4) 1.879 (4) 1.867 (4)	138 138 142 144	i, ii
trans-[Re(OAr)$_4$(PMe$_3$)$_2$]	OC$_6$H$_5$	1.992 (9)	133	iii
[Re(OAr)(O)(MeC≡CMe)$_2$] psuedo-tetrahedral	OC$_6$H$_5$	1.938 (9) 1.966 (14)	141 125	iv
fac-[Re(OAr)(CO)$_3$(dppe)]	OC$_6$H$_5$	2.127 (4)	131	v
[Re(OAr)$_3$(=CHBu^t)(NC$_6$H$_3$Pr$_2^i$-2,6)] square pyramidal Re, apical alkylidene	OC$_6$H$_3$Cl$_2$-2,6	1.98 (1) 2.00 (1) 2.00 (1)	129 135 132	vi
[Re(OAr)$_3$(=CHBu^t)(NC$_6$H$_3$Pr$_2^i$-2,6)(THF)]	OC$_6$F$_5$	2.012 (5) 1.991 (5) 1.985 (5)	142 154 128	vi
[(HBpz$_3$)Re(OAr)Cl(py)] 6-coord. Re	OC$_6$H$_5$	2.004 (4)	131	vii
fac,cis-[Re(OAr)(CO)$_3$(PPh$_3$)$_2$]	OC$_6$H$_4$Me-4	2.144 (9) 2.121 (9)	132 132	viii, ix

(continued overleaf)

[i] I.M. Gardiner, M.A. Bruck, and D.E. Wigley, *Inorg. Chem.*, **28**, 1769 (1989).
[ii] I.M. Gardiner, M.A. Bruck, P.A. Wexler, and D.E. Wigley, *Inorg. Chem.*, **28**, 3688 (1989).
[iii] P.G. Edwards, G. Wilkinson, M.B. Hursthouse, and K.M.A. Malik, *J. Chem. Soc., Dalton Trans.*, 2467 (1980).
[iv] T.K.G. Erikson, J.C. Bryan, and J.M. Mayer, *Organometallics*, 7, 1930 (1988).
[v] S.K. Mandal, D.M. Ho, and M. Orchin, *Inorg. Chem.*, **30**, 2244 (1991).
[vi] M.H. Schofield, R.R. Schrock and L.Y. Park, *Organometallics*, **10**, 1844 (1991).
[vii] N. Brown and J.M. Mayer, *Organometallics*, **14**, 2951 (1995).
[viii] R.D. Simpson and R.G. Bergman, *Angew. Chem., Int. Ed. Engl.*, **31**, 220 (1992).
[ix] R.D. Simpson and R.G. Bergman, *Organometallics*, **12**, 781 (1993).

Table 6.32 *(Continued)*

Compound	Aryloxide	Bond length (Å) M–O	Bond angle (°) M–O–Ar	Reference
[NEt$_4$][Re$_3$(μ-H)$_3$(μ-OAr)(CO)$_{10}$] triangular cluster	μ_2-OC$_6$H$_5$	2.176 (5), 2.176 (5)		x
[NEt$_4$][Re$_3$(μ-OAr)$_3$(CO)$_6$] confacial bi-octahedron	μ_2-OC$_6$H$_5$	2.11 (1)–2.15 (1)		x
[NEt$_4$][Re(μ-OH)(μ-OAr)$_2$(CO)$_6$] confacial bi-octahedron	μ_2-OC$_6$H$_5$	2.156 (8) (av.)		xi
[(Bu^{t}N)Br(mes)Re(μ-NBu^t)(μ-O)Re(OC$_6$H$_2$Me$_2$-CH$_2$)(NBu^t)] cyclometallated mesityl oxide	OC$_6$H$_2$Me$_2$-CH$_2$	1.96 (1)	117	xii
[NBu$_4$][Re(O)Cl(OAr)] trans ArO–Re=O	OC$_6$H$_4$PPh$_2$-2	2.016 (6)	127	xiii
[AsPh$_4$][Re(O)$_2$(OAr)$_2$].0.5EtOH.0.5Me$_2$CO Trans O=Re=O, cis-O(Ar)	OC$_6$H$_4$PPh$_2$-2	2.12 (1) 2.13 (1)	122 119	xiii
[Re(O)Cl(OAr)$_2$] cis-O(Ar), trans ArO–Re=O	OC$_6$H$_4$PPh$_2$-2	2.013 (9) 2.032 (9)	124 127	xiii, xiv
[Re(O)Cl(OAr)(HNC$_6$H$_4$PPh$_2$-2)] trans ArO–Re=O	OC$_6$H$_4$PPh$_2$-2	2.04 (1)	129	xiii
fac-cis-[Re(O)(di-OAr)(OAr)] cis-P,P	OC$_6$H$_4$PPh$_2$-2 (OC$_6$H$_4$-2)$_2$PPh	1.994 (3) 2.050 (3) 2.026 (3)	123 119 127	xv
fac-cis-[Re(NPh)(di-OAr)Cl(PPh$_3$)].2CHCl$_3$ cis-P,P	(OC$_6$H$_4$-2)$_2$PPh	2.050 (3) 2.050 (3)	122 126	xv

[x] T. Beringhelli, G. Ciani, G. D'Alfonso, A. Sironi, and M. Freni, *J. Chem. Soc, Dalton Trans.*, 1507 (1985).

[xi] Chenghua Jiang, Yuh-Sheng Wen, Ling-Kang Liu, T.S.A. Hor, and Yaw Kai Yan, *Organometallics*, **17**, 173 (1998).

[xii] A. Gutierrez, G. Wilkinson, B. Hussain-Bates, and M.B. Hursthouse, *Polyhedron*, **9**, 2081 (1990).

[xiii] C. Bolzati, F. Tisato, F. Refosco, G. Bandoli, and A. Dolmella, *Inorg. Chem.*, **35**, 6221 (1996).

[xiv] F. Loiseau, Y. Lucchese, M. Dartiguenave, F. Belanger-Gariepy, and A.L. Beauchamp, *Acta Crystallog. C*, **52**, 1968 (1996).

[xv] Hongyan Luo, I. Setyawati, S.J. Rettig, and C. Orvig, *Inorg. Chem.*, **34**, 2287 (1995).

Table 6.33 Simple aryloxides of iron

Compound	Aryloxide	Bond length (Å) M–O	Bond angle (°) M–O–Ar	Reference
[(ArO)Fe(μ_2-OAr)$_2$Fe(OAr)] 3-coord. Fe	OC$_6$H$_2$Bu$^t_3$-2,4,6 μ_2-OC$_6$H$_2$Bu$^t_3$-2,4,6	1.822 (5) 1.823 (5) 2.022 (6), 2.001 (6) 2.011 (6), 2.023 (6)	162 151	i
[Fe(OAr)$_2$] *cis*-O(Ar), octahedral Fe	OC$_6$H$_4$(2'-1,10-phen-)2	1.893 (3) 1.884 (3)	131 133	ii
[NEt$_4$][Fe(OAr)$_4$] tetrahedral anion	OC$_6$HMe$_4$-2,3,5,6	1.847 (4) 1.852 (4) 1.826 (4) 1.860 (4)	142 137 137 131	iii
[PPh$_4$][Fe(OAr)$_4$] tetrahedral anion	OC$_6$H$_2$Cl$_3$-2,4,6	1.863 (3) 1.867 (3) 1.875 (3) 1.859 (3)	132 134 129 141	iii
[(ArO)Fe(μ-C$_6$H$_2$Me$_3$-2,4,6)$_2$Fe(OAr)] bridging mesityl group	OC$_6$H$_2$Bu$^t_2$-2,6-Me-4	1.946 (3) 1.945 (2)	128 127	iv
[{HB(3,5-Pr$^i_2$pz)$_3$}Fe(OAr)]	OC$_6$F$_5$	1.875 (5)	133	v
[{HB(3,5-Pr$^i_2$pz)$_3$}Fe(OAr)$_2$(MeCN)].MeCN	OC$_6$H$_4$Me-4	1.887 (1) 1.897 (2)	142 142	v
[NEt$_4$]$_2$[(S)$_2$Mo(μ-S)$_2$Fe(OAr)$_2$] 4-coord. Fe	OC$_6$H$_5$	1.93 (2) 1.86 (2)	136 133	vi

[i] R.A. Bartlett, J.J. Ellison, P.P. Power, and S.C. Shoner, *Inorg. Chem.*, **30**, 2888 (1991).

[ii] J.C. Jeffery, C.S.G. Moore, E. Psillakis, and M.D. Ward, *Polyhedron*, **14**, 599 (1995).

[iii] S.A. Koch, and M. Millar, *J. Am. Chem. Soc.*, **104**, 5255 (1982).

[iv] H. Muller, W. Seidel, and H. Gorls, *Z. Anorg. Allg. Chem.*, **622**, 1968 (1996).

[v] M. Ito, H. Amagai, H. Fukui, N. Kitajima, and Y. Moro-oka. *Bull. Chem. Soc. Jpn.*, **69**, 1937 (1996).

[vi] B.-K. Teo, M.R. Antonio, R.H. Tieckelmann, H.C. Silvis, and B.A. Averill, *J. Am. Chem. Soc.*, **104**, 6126 (1982).

(continued overleaf)

Table 6.33 *(Continued)*

Compound	Aryloxide	Bond length (Å) M–O	Bond angle (°) M–O–Ar	Reference
[NEt$_4$]$_2$[Fe$_4$(μ_3-S)$_4$(OAr)$_4$] cubic cluster	OC$_6$H$_5$	1.848 (6) 1.889 (6) 1.865 (6) 1.860 (6)	133 123 129 132	vii
[NEt$_4$]$_3$[(Meida)VFe$_3$(μ_3-S)$_4$(OAr)$_4$] cubic cluster	OC$_6$H$_4$Me-4	1.87 (1) 1.894 (9) 1.89 (1)	135 136 131	viii
[PPh$_4$]$_2$[Fe$_4$(μ_3-S)$_4$(OAr)$_2$Cl$_2$] cubic cluster	OC$_6$H$_5$	2.057 (9)	107	ix
[PPh$_4$]$_2$[Fe$_4$S$_4$(OAr)$_2$(SPh)$_2$] cubic cluster	OC$_6$H$_4$Me-4	1.996 (9)	126	ix
[K(222)][Fe(porph)(OAr)] square pyramidal Fe(II)	OC$_6$H$_5$	1.937 (4)	137	x
[Fe(porph)(OAr)] square pyramidal Fe(III)	OC$_6$H$_3$Cl$_2$-2,6	1.869 (2)	133	xi
[PPh$_4$]$_2$[Fe$_2$(μ_2-S)$_2$(OAr)$_4$] psuedo-tetrahedral Fe	OC$_6$H$_4$Me-4	1.86 (1) 1.87 (1) 1.87 (1) 1.88 (1)	130 143 134 135	xii
[NEt$_4$]$_3$[Fe$_6$(μ_4-S)(OAr)$_4$] rhombic dodecahedral cluster	OC$_6$H$_4$OMe-4	1.88 (2) 1.88 (2) 1.87 (2) 1.88 (2) 1.85 (2) 1.86 (2)	130 129 132 130 143 130	xiii

Compound	OAr			Ref
[NEt$_4$]$_3$[Fe$_6$(μ_4-S)$_4$(OAr)$_4$][Mo(CO$_3$)$_2$] rhombic dodecahedral cluster	OC$_6$H$_4$OMe-4	1.865 (7) 1.831 (7) 1.850 (7)	130 146 133	xiii
[NEt$_4$]$_3$[Fe$_6$(μ_4-S)$_4$(OAr)$_4$][W(CO$_3$)$_2$] rhombic dodecahedral cluster	OC$_6$H$_4$Me-4	1.821 (8) 1.848 (8) 1.859 (8)	154 135 130	xiii
[NEt$_4$]$_3$[Fe$_6$(μ_4-S)$_4$(OAr)$_4$][Mo(CO$_3$)$_2$] rhombic dodecahedral cluster	OC$_6$H$_4$(COMe)-4	1.913 (8) 1.923 (8) 1.909 (8)	146 134 134	xiii
[Fe$_4$(μ_4-S)$_4$(OAr)$_2$(CNBu^t)$_6$] cubic cluster	OC$_6$H$_4$Me-4	1.871 (7) 1.864 (8)	140 133	xiv

[vii] W.E. Cleland, D.A. Holtman, M. Sabat, J.A. Ibers, G.C. DeFotis, and B.A. Averill, *J. Am. Chem. Soc.*, **105**, 6021 (1983).

[viii] Jiesheng Huang, S. Mukerjee, B.M. Segal, H. Akashi, Jian Zhou, and R.H. Holm, *J. Am. Chem. Soc.*, **119**, 8662 (1997).

[ix] M.G. Kanatzidis, N.C. Baenziger, D. Coucouvanis, A. Simopoulos, and A. Kostikas, *J. Am. Chem. Soc.*, **106**, 4500 (1984).

[x] H. Nasri, J. Fischer, R. Weiss, E. Bill, and A. Trautwein, *J. Am. Chem. Soc.*, **109**, 2549 (1987).

[xi] A.M. Helms, W.D. Jones, and G.L. McLendon, *J. Coord. Chem.*, **23**, 351 (1991).

[xii] A. Salifoglou, A. Simopoulos, A. Kostikas, R.W. Dunham, M.G. Kanatzidis, and D. Coucouvanis, *Inorg. Chem.*, **27**, 3394 (1988).

[xiii] S.A. Al-Ahmad, A. Salifoglou, M.G. Kanatzidis, W.R. Dunham, and D. Coucouvanis, *Inorg. Chem.*, **29**, 927 (1990).

[xiv] C. Goh, J.A. Weigel, and R.H. Holm, *Inorg. Chem.*, **33**, 4861 (1994).

Table 6.34 Simple aryloxides of ruthenium

Compound	Aryloxide	Bond length (Å) M–O	Bond angle (°) M–O–Ar	Reference
cis-[Ru(OAr)(H)(PMe₃)₄]	OC₆H₄Me-4	2.152 (3)	133	i
		2.143 (3)	135	
cis-[Ru(OAr)(H)(PMe₃)₄] two independent molecules	OC₆H₄Me-4	2.161 (6)	131	ii
		2.145 (6)	136	
[Ru(OAr)(OAr')(CO)(PPh₃)₂] cis-O(Ar), trans-P	OC₆H₄NO-2	2.080 (2)	111	iii
		2.129 (1)	136	
[Ru(porphyrin)(OAr)₂]₂(μ-O)	OC₆H₄Me-4	1.95 (1)	132	iv
		1.94 (1)	128	
[Ru(OAr)(H)(CO)(PMe₃)₃] CO trans to OAr	OC₆H₄Me-4	2.108 (6)	133	i
[(η³-allyl)Ru(OAr)(PMe₃)₃]	OC₆H₅	2.192 (5)	139	v
[Ru(OC₆H₄-η³-C₃H₄)(PMe₃)₃]	OC₆H₄-η³-C₃H₄	2.124 (5)	120	v
[Ru(OAr)(PMe₃)₃(Me₂P-o-η¹-C₆H₃Me-4)] two molecules, metallated P-OAr	OC₆H₄Me-4	2.161 (4)	141	vi
		2.152 (4)	142	
trans-[Ru(OAr)(H)(dmpe)₂]	OC₆H₄Me-4	2.239 (2)		vii
[Cp*Ru(OAr)(HOAr)] phenol hydrogen bonded to chelated O	OC₆H₄PPh₂-2	2.133 (2)	121	viii
[Cp*Ru(μ₂-OAr)(μ₂-OMe)RuCp*]	μ₂-OC₆H₄Buᵗ-2,4	2.085 (3)		ix
[Ru(OAr)(bipy)₂][PF₆].MeCN octahedral Ru	OC₆H₄(2'py)-2	2.064 (4)	117	x
[Ru(OAr)₂][PF₆].MeCN octahedral Ru, cis-O	OC₆H₄(2'-bipy)-2	1.965 (2)	123	x
		1.947 (3)	124	
[Ru(OAr)₂(CO)₂] all cis octahedral, N-bound	2-NO-1-naphtholato	2.064 (3)	112	xi
		2.086 (3)	112	

[Ru(OAr)₂(CO)(py-CMeO-4)] cis-O, N-bound	2-NO-1-naphtholato	2.06 (1) 2.07 (1)	112 110	xi
[η⁶-C₆H₂Me₄)Ru(di-OAr)] three legged piano stool, two molecules	(MeO)₂C₆H₃-P(2-OC₆H₃OMe-6)₂	2.047 (5)–2.063 (5)		xii
[η⁶-C₆H₄MePrⁱ)Ru(di-OAr)] three legged piano stool	(MeO)₂C₆H₃-P(2-OC₆H₃OMe-6)₂	2.060 (5) 2.065 (5)	121 120	xii
[η⁶-C₆H₄MePrⁱ)RuCl(OAr)] three legged piano stool	(MeO)₂C₆H₃-PPh(2-OC₆H₃OMe-6)	2.069 (6)	121	xii
[η⁶-C₆H₂Me₄)RuCl(OAr)] three legged piano stool	OC₆H₃OMe-6-PPh₂-2	2.059 (3)	120	xii
[Ru(OAr)Br(PPh₃)₂] cyclometallated diazo ligand, trans P	OC₆H₄Me-4-N=NC₆H₄	2.112 (5)	113	xiii
[(H)₂Ru₆(CO)₁₅[P(OMe)₃]{μ₂-O-η²-C₆H₄}] raft cluster, metallated C bridges two metals, phenoxy ring π-binds to another	μ₂-O-η²-C₆H₄	2.182 (3), 2.143 (2)		xiv
[(H)₂Ru₆(CO)₁₆(μ₂-O-η²-c₆H₄)] raft cluster, metallated C bridges two metals, phenoxy ring π-binds to another	μ₂-O-η²-C₆H₄	2.131 (2), 2.173 (2)		xv

[i] J.F. Hartwig, R.A. Andersen, and R.G. Bergman, *Organometallics*, **10**, 1875 (1991).

[ii] K. Osakada, K. Ohshiro, and A. Yamamoto, *Organometallics*, **10**, 404 (1991).

[iii] M. Pizzotti, C. Crotti, and F. Demartin, *J. Chem. Soc., Dalton Trans.*, 735 (1984).

[iv] J.P. Collman, C.E. Barnes, P.J. Brothers, T.J. Collins, T. Ozawa, J.C. Gallucci, and J.A. Ibers, *J. Am. Chem. Soc.*, **106**, 5151 (1984).

[v] M. Hirano, N. Kurata, T. Marumo, and S. Komiya, *Organometallics*, **17**, 501 (1998).

[vi] J.F. Hartwig, R.G. Bergman, and R.A. Andersen, *J. Organomet. Chem.*, **394**, 417 (1995).

[vii] M.J. Burn, M.G. Fickes, F.J. Hollander, and R.G. Bergman, *Organometallics*, **14**, 137 (1995).

[viii] M. Canestrari, B. Chaudret, F. Dahan, Yong-Sheng Huang, R. Poilblanc, Tag-Chong Kim, and M. Sanchez, *J. Chem. Soc., Dalton Trans.*, 1179 (1990).

[ix] K. Bucken, U. Koelle, R. Pasch, and B. Ganter, *Organometallics*, **15**, 3095 (1996).

[x] B.M. Holligan, J.C. Jeffery, M.K. Norgett, E. Schatz, and M.D. Ward, *J. Chem. Soc., Dalton Trans.*, 3345 (1992).

[xi] K. Ka-Hong Lee and Wing-Tak Wong, *J. Chem. Soc., Dalton Trans.*, 2987 (1997).

[xii] Y. Yamamoto, R. Sato, F. Matsuo, C. Sudoh, and T. Igoshi, *Inorg. Chem.*, **35**, 2329 (1996).

[xiii] G.K. Lahiri, S. Bhattacharya, M. Mukherjee, A.K. Mukherjee, and A. Chakravorty, *Inorg. Chem.*, **26**, 3359 (1987).

[xiv] S. Bhaduri, K. Sharma, H. Khwaja, and P.G. Jones, *J. Organomet. Chem.*, **412**, 169 (1991).

[xv] D.S. Bohle and H. Vahrenkamp, *Angew. Chem., Int. Ed. Engl.*, **29**, 198 (1990).

(continued overleaf)

Table 6.34 *(Continued)*

Compound	Aryloxide	Bond length (Å) M–O	Bond angle (°) M–O–Ar	Reference
[(H)$_2$Ru$_6$(CO)$_{16}$(μ_2-O-η^2-C$_6$H$_3$OMe-4)] raft cluster, metallated C bridges two metals, phenoxy ring π-binds to another	μ_2-O-η^2-C$_6$H$_3$OMe-4	2.132 (5), 2.159 (5)		xvi
[(Ph$_3$PAu)(H)Ru$_6$(CO)$_{16}$(μ_2-O-η^2-C$_6$H$_3$OMe-4)] raft cluster, metallated C bridges two metals, phenoxy ring π-binds to another	μ_2-O-η^2-C$_6$H$_3$OMe-4	2.14 (2), 2.15 (2)		xvi
[Ru$_3$(CO)$_8$(μ_2-η^2-OAr)$_2$] aryloxides both chelate through Cl and bridge one edge of triangular cluster	μ_2-OC$_6$H$_4$Cl-2	2.110 (4), 2.163 (4); 2.093 (4), 2.161 (4)		xvii
[Ru$_4$(μ_3-OAr)$_2$(μ_2-Cl)$_2$(CO)$_{10}$]	μ_3-OC$_6$H$_5$	2.144 (1)–2.218 (1)		xviii
[Ru$_4$(μ_3-OAr)$_2$(μ_2-OAr)(μ_2-Cl)(CO)$_{10}$] two molecules	μ_2-OC$_6$H$_4$OMe-2; μ_3-OC$_6$H$_4$OMe-2	2.029 (7)–2.230 (6); 2.121 (7)–2.218 (7)		xix

[xvi] M.P. Cifuentes, T.P. Jeynes, M.G. Humphrey, B.W. Skelton, and A.H. White, *J. Chem. Soc., Dalton Trans.*, 925 (1994).
[xvii] D.J. Darensbourg, B. Fontal, S.S. Chojnacki, K.K. Klausmeyer, and J.H. Reibenspies, *Inorg. Chem.*, **33**, 3526 (1994).
[xviii] S. Bhaduri, N. Sapre, K. Sharma, P.G. Jones, and G. Carpenter, *J. Chem. Soc., Dalton Trans.*, 1305 (1990).
[xix] T.P. Jeynes, M.P. Cifuentes, M.G. Humphrey, G.A. Koutsantonis, and C.L. Raston, *J. Organomet. Chem.*, **476**, 133 (1994).

Table 6.35   Simple aryloxides of osmium

Compound	Aryloxide	Bond length (Å) M–O	Bond angle (°) M–O–Ar	Ref.
[Os(OAr)(H)(CO)(PPr$_3^i$)$_2$] chelated aryloxide, H *trans* to bound Cl, *trans* P	OC$_6$Cl$_5$	2.124 (3)	123	i
[Os(porphyrin)(OAr)$_2$]	OC$_6$H$_5$	1.938 (2)	128	ii

[i]M.A. Tena, O. Nurnberg, and H. Werner, *Chem. Ber.*, **126**, 1597 (1993).
[ii]Ming-Chu Cheng, Chih-Chieh Wang, and Yu Wang, *Inorg. Chem.*, **31**, 5220 (1992).

tantalum species [Ta(OAr)$_2$Cl$_2$(L)$_2$] (L = various P and N donor ligands) are produced by sodium amalgam reduction of the trichlorides.[400] Two general structural motifs are found for the isolated hydrides of tantalum. The seven-coordinate compounds adopt pentagonal bipyramidal geometry with *trans*, axial aryloxide ligands. Spectroscopic studies show that the molecules are stereochemically rigid.[401] There is no evidence that the colourless trihydride compounds exist as η^2-H$_2$ species. With very bulky aryloxides or with three aryloxide ligands, six-coordinate hydrides are produced. There is a dramatic bending of the two hydride ligands towards the donor phosphine ligand. This can be accounted for by an increase in π-bonding between the Cl or OAr ligand *trans* to P upon bending the hydride ligands.[398]

It can be seen that during the hydrogenolysis of alkyls containing 2,6-diphenylphenoxide ligands, the intramolecular hydrogenation of the *ortho*-phenyl rings takes place.[31] These hydride species of tantalum and in particular niobium will carry out the catalytic hydrogenation of simple arenes as well as aryl-phosphines.[408,409] Polynuclear aromatic hydrocarbons are hydrogenated faster than simple benzenes.[410] The catalysts exhibit a high degree of regioselectivity and have been shown to carry out the *all-cis* hydrogenation of many arenes.[411] The most active catalysts for arene hydrogenation are the solutions generated by hydrogenolysis of either the isolated alkyls [Nb(OAr)$_2$R$_3$] or the mixture [Nb(OAr)$_2$Cl$_3$/3n-BuLi]. This reactivity has also been developed into a process for the bulk synthesis of cyclohexylphosphines from their aryl counterparts.[408] Studies have shown the reaction proceeds in a predominantly *all-cis* fashion with no intermediate cyclohexadienyl or cyclohexenylphosphines being detected by ^{31}P NMR spectroscopy.

The two-electron reduction of mixed chloro, aryloxides can lead to cyclometallated derivatives *via* CH bond activation by low valent intermediates (Section 5.1). However, reduction in the presence of unsaturated organic substrates can lead to interesting compounds. The presence of the π-donating aryloxide ligands results in highly reducing d^2-metal fragments that can strongly bind and activate the organic substrate.[412–416] Reduction in the presence of alkynes can lead to mono-alkyne complexes, tantalacyclocyclopentadiene rings or η^6-arene derivatives, hence mapping out the overall cyclotrimerization process.[417] The alkyne complex [Ta(OC$_6$H$_3$Pr$_2^i$-2,6)$_3$(η^2-PhC≡CPh)][418] has tantalacyclopropene character and reacts with ketones to produce 2-oxa-tantalacyclopent-4-ene rings. With aldehydes, coupling is followed by a hydride transfer to a second aldehyde to produce an alkoxide species, *e.g.* [Ta(OC$_6$H$_3$Pr$_2^i$-2,6)$_3$(OCH$_2$Ph)(CPh=CPhCPh=O)] from benzaldehyde.[419] Addition of nitriles to the alkyne complex produces adducts which thermolyse to produce metallacyclic enamines *via* imine intermediates.[420] Reduction of [Ta(OC$_6$H$_3$Pr$^i_2$-2,6)$_2$Cl$_3$(OEt$_2$)] in the presence of Bu^{t}C≡CH produces the α,α'-disubstituted tantalacyclopentadiene.[417] Thermolysis leads to the α,β-metallacycle *via* fragmentation and exchange with free alkyne. Further reaction with alkyne produces the arene complex [Ta(OC$_6$H$_3$Pr$^i_2$-2,6)$_2$Cl(η^6-C$_6$H$_3$Bu$^t_3$-1,3,5)]. Structural studies show the arene to be folded leading to a 7-metallanorbornadiene bonding description. The alkylation chemistry of this and related arene compounds [(η^6-C$_6$Me$_6$)Ta(OAr)$_2$Cl] and [(η^6-C$_6$Me$_6$)Ta(OAr)Cl$_2$][421] has been investigated and the resulting alkyls shown to be thermally stable.[406] The compound [(η^6-C$_6$Me$_6$)Ta(OAr)Et$_2$] shows no evidence for *agostic* interactions in the solid state.

Addition of nitriles to the metallacyclopentadiene complex generates an equivalent of pyridine, which remains strongly η^2-C,N bound to the metal centre, *e.g.* in the complex [TaCl(OC$_6$H$_3$Pr$^i_2$-2,6)$_2$(η^2-2,4,6-Bu$^t_3$-py)].[417] The η^2-C,N bonding activates the pyridine ligand towards ring opening. The phenyl compound [Ta(Ph)(OC$_6$H$_3$Pr$^i_2$-2,6)$_2$(η^2-2,4,6-Bu$^t_3$-py)] thermally converts to the ring-opened complex [Ta(OC$_6$H$_3$Pr$^i_2$-2,6)$_2$(=NCBu^t=CHCBu^t=CHCPhBu^t)].[422] A mechanistic study of the reaction with substituted Ta–C$_6$H$_4$–X groups showed first-order kinetics $\rho = -0.58(10)$ for the aryl migration. Further mechanistic studies of the pyridine cleavage reaction have been carried out in the context of modelling key steps within hydrodenitrogenation catalysis.[415]

A series of 1,3-cyclohexadiene derivatives have also been obtained by reduction of chloro, aryloxides of niobium and tantalum.[416] These compounds are potential intermediates within the catalytic cycle of arene hydrogenation. Structural studies of derivatives such as [M(OAr)$_3$(η^4-C$_6$H$_8$)] (M = Nb, Ta; OAr = OC$_6$H$_3$Pr$^i_2$-2,6) show a metallanorbornene bonding picture. These compounds are not only precursors for arene hydrogenation, but will also catalyse both disproportionation and hydrogenation of 1,3-cyclohexadiene.[416]

6.2.7 Group 6 Metal Aryloxides

There is a diverse chemistry associated with the aryloxide ligand attached to the group 6 metals. Binuclear derivatives of all three metals are known with a range of metal–metal bond orders. In the case of chromium a number of mixed metal clusters containing the group 1 metals have been structurally characterized. In a series of formally Cr(II) compounds the extent of Cr–Cr bonding appears to vary depending on the nature of the alkali metal.[423,424] Compounds containing the unbridged M$_2^{6+}$ and M$_2^{4+}$ cores (M = Mo, W) have been synthesized and shown to contain structural parameters, spectra, and reactivity consistent with the presence of metal–metal triple and quadruple bonds respectively.[425–429] A series of confacial bis(octahedral) anions [Mo$_2$(OAr)$_7$]$^-$ have been isolated for small aryloxides and shown to exhibit contact shifted NMR spectra. Analysis of the temperature dependence of these NMR spectra has allowed the energy gap between the diamagnetic ground state and paramagnetic excited state for the Mo$_2^{6+}$ core to be determined.[425]

The mononuclear aryloxide chemistry of these metals spans a wide range of formal oxidation states. This is particularly true for tungsten, where the aryloxide ligand has been coordinated to this metal centre in its highest oxidation state, *e.g.* [W(OC$_6$H$_4$Me-4)$_6$],[430] low valent carbonyl compounds, *e.g.* [NEt$_4$][W(OPh)(CO)$_5$],[431] and intermediate redox states, *e.g.* $[W(OC_6H_3Me_2-2,6)Cl_5]$,[432] $[W(OC_6H_3Pr^i_2-2,6)_4]$,[433] $[W(OC_6H_3Ph-\eta^6-C_6H_5)(H)(PMePh_2)_2]$,[434] and $[W(OC_6H_3Ph-\eta^6-C_6H_5)(OAr)(dppm)]$.[29,435]

The reduction of the tetrachloride $[W(OC_6H_3Ph_2-2,6)_2Cl_4]$ in the presence of phosphine ligands leads to either bis-cyclometallated compounds, *e.g.* $[W(OC_6H_3Ph-\eta^1-C_6H_4)_2(PMePh_2)_2]$ or else to produce the deep-green W(II) species $[W(OC_6H_3Ph-\eta^6-C_6H_5)(OAr)(PMePh_2)]$.[436,437] The latter compound can be thermally converted to the former with the elimination of H_2. The adducts $[W(OC_6H_3Ph-\eta^1-C_6H_4)_2L_2]$ (L_2 = a variety of pyridines and bipyridines) are paramagnetic in solution. The temperature dependence of the NMR spectra has been used to determine the singlet–triplet energy gap for the d^2-W(IV) metal centre and a pyridine ligand π-acidity scale developed.[438] The use of metallation-resistant 2,3,5,6-tetraphenylphenoxide ligands gives the corresponding, thermally stable η^6-arene derivatives in higher yield and with a variety of different phosphine ligands.[439] Structural studies of these π-arene compounds show the strongly metal-bound arene ring to be reduced, leading to structural parameters consistent with a bonding picture (Scheme 6.9). These compounds can carry out the four-electron reduction of a variety of small molecules including the four-electron cleavage of the N=N double bond in azobenzenes (Scheme 6.9).[440] Mechanistic studies show that the process occurs at a single metal centre, and the pathway has been analysed theoretically.[441] With suitable ketones and aldehydes, products derived by insertion of the carbonyl function into a W–C(metallanorbornadiene) bond are observed.[442]

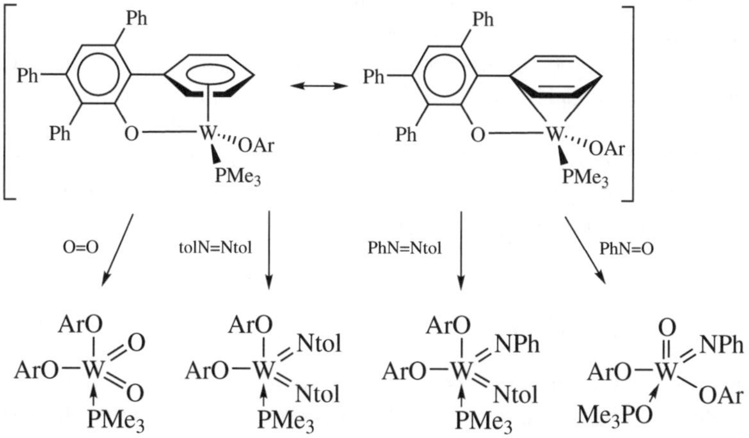

Scheme 6.9

The aryloxide ligand has played an important part in the development of the olefin/alkyne metathesis chemistry associated with molybdenum and tungsten. Many early catalyst systems consisted of Mo/W aryloxide compounds "activated" in the presence of suitable substrates.[443] Examples include the metathesis of olefins by $[W(OAr)_xCl_{6-x}]$ precursors treated with $[EtAlCl_2]$ and related reagents.[444–447] In one study of the metathesis of 2-pentene catalysed by $[W(OAr)_4Cl_2]$ it was shown that electron-withdrawing groups in the 4-position of the phenoxide increase the activity considerably. A linear free-energy relationship was observed over a limited range of substituents (Cl, Br, H, Me, OMe). The presence of methyl substituents in the 2- and 6-positions on the phenoxide ring was also found to lead to considerable rate enhancement.[444] Electro-chemical studies of the precursors have also been carried out.[201] Later studies on the ring-opening polymerization of dicyclopentadiene by precursors $[W(O)(OAr)_x Cl_{4-x}]$ activated with tin hydrides showed similar electronic effects. The activity of the precursors was correlated with the $W(IV)-W(V)$ reduction potentials.[448] The combination of the precursor $trans$-$[W(O)(OC_6H_3Br_2$-$2,6)_2Cl_2]$ activated by $[Et_4Pb]$ has been shown to be a catalyst for the stereoselective ring closing of chiral dienes.[449]

The alkylation of the bis(aryloxide) $[W(OC_6H_3Ph_2$-$2,6)_2Cl_4]$ with $LiCH_2CMe_3$ leads to cyclometallated alkylidene compounds $[W(OC_6H_3Ph$-η^1-$C_6H_4)(OC_6H_3Ph_2$-$2,6)(=CHCMe_3)(OR_2)]$.[227] These compounds will catalyse the metathesis of 2-pentene with high stereoselectivity, polymerize 1-methyl-norbornene to 100% cis, 100% head-to-tail, syndiotactic polymer, and cyclize a variety of functionalized dienes.

The most important development in this area has been the careful mechanistic studies of well-defined, d^0-alkylidene(aryloxide) and alkylidyne(aryloxide) catalysts.[450] The isolated alkylidyne compound $[W(\equiv CCMe_3)(OC_6H_3Pr^i_2$-$2,6)_3]$ reacts with internal alkynes to generate stable tungstabutadiene compounds, $e.g.$ structurally characterized $[W(OC_6H_3Pr^i_2$-$2,6)_3(C_3Et_3)]$.[228] This complex will metathesize alkynes at rates dependent on the rate of fragmentation of the intermediate metallacycles. The corresponding 2,6-dimethylphenoxide species can be generated from dinuclear $[W_2(OC_6H_3Me_2$-$2,6)_6]$ by treatment with alkynes, effectively triple bond metathesis involving a $W\equiv W$ bond.[451] The aryloxide in this case was not bulky enough to generate an active metathesis catalyst, whereas the bulkier 2,6-di-$tert$-butylphenoxide led to a cyclometallated, alkylidene compound. The metallacyclic compound $[W(OC_6H_3Pr^i_2$-$2,6)_3(C_3HBu^t_2)]$ (obtained from $Bu^tC\equiv CH$) undergoes elimination of phenol (deprotonation of the β-CH bond) and formation of $[W(OC_6H_3Pr^i_2$-$2,6)_2(C_3Bu^t_2)]$ which forms a pyridine adduct.[452] An important class of alkene metathesis catalysts are the imido-alkylidene complexes $[M(=NR)(=CHR')(OR)_2]$ (M = Mo, W) typically referred to as "Schrock-type" catalysts. The activity of these catalysts is strongly dependent on the electronic and steric properties of the alkoxide/aryloxide ancillary ligands.[266,453] The discrete catalyst $[W(O)(OC_6H_3Ph_2$-$2,6)_2(=CHBu^t)(PMePh_2)]$ has been structurally characterized.[454]

6.2.8 Group 7 Metal Aryloxides

The vast majority of the studies of manganese aryloxides have involved delineating how ligand architecture affects the aggregation and coordination geometry about the metal. Synthetic strategies for coordination compounds, $e.g.$ $[Mn\{OC_6H_4(2'$-$py)$-$2\}_3]$[455] which contains a mer-MnO_3N_3 core typically utilize starting materials such as manganese acetate in protic solvents. For simple aryloxides, extensive use of the bis(amido) compound $[Mn\{N(SiMe_3)_2\}_2]$ has been made. Early work showed that a variety of extremely oxygen sensitive bis(aryloxides) could be obtained.[456] The complex $[Mn(OC_6H_3Bu^t_2$-$2,6)_2]$ was reported to form the compound (6-I) when exposed to O_2.

Later work showed that the $OC_6H_2Bu^t_3$-$2,4,6$ derivative was dimeric, $[(ArO)Mn(\mu$-$OAr)_2Mn(OAr)]$ with three-coordinate Mn.[457] The tetrahedral adduct $[Mn(OC_6H_2Bu^t_3$-$2,4,6)_2(NCMe)_2]$ was obtained directly from the corresponding chloride (Eq. 6.30).[108]

$$Bu^t \quad\quad\quad Bu^t$$

$$O=\!\!\!\!=\!\!\!\!=\!\!\!\!=O$$

$$Bu^t \quad\quad\quad Bu^t$$

(6-I)

The chemistry of rhenium aryloxides is significant. The homoleptic rhenium aryloxides [Re(OAr)$_4$] (OAr = OC$_6$H$_3$Pr$^i_2$-2,6, OC$_6$H$_3$Me$_2$-2,6) have been obtained from [ReCl$_4$(THF)$_2$] and the corresponding lithium aryloxide.[458,459] Cyclic voltammetric studies show these square planar d^3-species can undergo one-electron reduction. As with other transition metal systems (Section 4.2.3) these aryloxides are more difficult to reduce (more negative potential) than their halide counterparts.[458] The 2,6-dimethylphenoxide compound reacts with alkynes to form adducts [(RCCR)Re(OAr)$_4$].

Treatment of the bis(alkyne) compounds [Re(O)(I)(RC≡CR)$_2$] with TlOPh leads to the corresponding phenoxide.[460] The structure of the 2-butyne, phenoxide showed a pseudo-tetrahedral coordination sphere. The orientation of the phenoxide ligand was interpreted in terms of minimizing π-antibonding interactions. Unlike corresponding alkoxides, the phenoxide complex did not participate in any insertion chemistry. The tris(alkyne) [Re(I)(RC≡CR)$_3$] is converted into the corresponding phenoxide by simply treating with phenol.[461]

The importance of group 5 and 6 metal alkylidene and alkylidyne compounds in metathesis reactions has led to an exploration of related rhenium chemistry.[462] A series of alkylidene compounds [Re(OAr)$_3$(=CHCMe$_3$)(NC$_6$H$_3$Pr^i-2,6)] were obtained by reacting dimeric [Re(CCMe$_3$)(NHAr)Cl$_3$]$_2$ with KOAr.[463] The solid-state structure of the OC$_6$H$_3$Cl$_2$-2,6 derivative showed the alkylidene unit to occupy an apical site within an essentially square pyramidal geometry. The vacant site trans to the alkylidene group is occupied in the adduct [Re(OC$_6$F$_5$)$_3$(=CHCMe$_3$)(NC$_6$H$_3$Pr^i-2,6)(THF)]. The lack of metathesis activity for these compounds was ascribed to the binding of olefins in a similar fashion, precluding $2+2$ cycloaddition. Addition of pyridine to [Re(OAr)$_3$(=CHBu^t)(NC$_6$H$_3$Pr^i-2,6)] yielded the alkylidyne [Re(OAr)$_2$(CCMe$_3$)(NC$_6$H$_3$Pr^i-2,6)(py)] with elimination of phenol.[463] The alkylidene [Re(OAr)Cl$_2$(=CHCMe$_3$)(NC$_6$H$_3$Pr^i-2,6)] undergoes a similar elimination with pyridine, and forms an active olefin metathesis catalyst when activated with GaBr$_3$.

Some important examples of phenyl migration to rhenium-oxo groups have been studied in detail.[464,465] The tris(pyrazolylborate) compound [(HBpz$_3$)Re(O)(Ph)Cl] undergoes photolysis in the presence of donor ligands to yield the d^4-phenoxide [(HBpz$_3$)Re(OPh)Cl(py)]. Mechanistic studies including using the labelled compounds [(HBpz$_3$)Re(^{16}O)(C$_6$H$_5$)Cl] and [(HBpz$_3$)Re(^{18}O)(C$_6$D$_5$)Cl] showed the reaction to be an intramolecular 1,2-migration.[465] In related studies the triflate (OTf) compound [(HBpz$_3$)ReO(Ph)(OSMe$_2$)][OTf] undergoes phenyl to oxo migration to yield [(HBpz$_3$)ReO(OPh)(OSMe$_2$)][OTf] which reversibly loses Me$_2$S. The reactions proceed via the key intermediate [(HBpz$_3$)Re(O)$_2$(Ph)][OTf] which undergoes phenyl to oxo migration at 0°C to yield [(HBpz$_3$)Re(O)(OPh)][OTf].[464]

The structurally characterized compound *fac,cis*-[Re(OC$_6$H$_4$Me-4)(CO)$_3$(PPh$_3$)$_2$] is produced by addition of 4-methylphenol to the corresponding methyl compound.[466] The insertion chemistry of this and related alkoxides has been discussed (Section 5.2).

6.2.9 Group 8 Metal Aryloxides

As with other first row d-block metals, the aryloxide bond is an important component of many coordination compounds of iron formed with polydentate ligands. Simple aryloxide ligands have been applied mainly to the synthesis of binary compounds and adducts, as supporting ligands in (typically sulfur) cluster compounds and as the axial group in iron porphyrin compounds (Table 6.33). Treatment of the diethyl [FeEt$_2$(bipy)$_2$] with phenols HOC$_6$H$_4$X-4 was found to yield the corresponding bis(aryloxides).[467] Kinetic studies supported a pathway involving initial dissociation of bipy, coordination of phenol, and then elimination of ethane. The homoleptic iron(II) aryloxide [(ArO)Fe(μ_2-OAr)$_2$Fe(OAr)] (OAr = OC$_6$H$_2$Bu$_3^t$-2,4,6) was obtained from the corresponding bis(trimethylsilyl)amide.[457] Treatment of FeCl$_3$ with LiOAr in EtOH was found to produce the iron(III) anions [Fe(OAr)$_4$]$^-$ (Eq. 6.31, OAr = OC$_6$HMe$_4$-2,3,5,6; OC$_6$H$_2$Cl$_3$-2,4,6).[109] The bonding in these anions was of interest as a model for tyrosine binding. Interestingly the anions undergo reversible one-electron reductions at -1.32 V and -0.45 V (*vs* SCE) respectively, showing how substituents on the phenoxide ring can influence metal-based redox processes.

A large number of high-spin, five-coordinate Fe(III) complexes [(porphyrin)Fe(OAr)] have been isolated and studied.[468,469] Cyclic voltammetric studies show that the Fe(II)/Fe(III) couple lies at a significantly negative potential.[470] A reasonable correlation between the reduction potential and Hammett σ-values for p-substituents on the aryloxide ligand was found. The high-spin Fe(II) anion [(porphyrin)Fe(OPh)]$^-$ has been isolated and structurally characterized.[471] At low temperatures formation of low-spin adducts occurs with ligands such as pyridine and 1-methylimidazole.[472] The reaction of [(porphyrin)Fe(Ar)] compounds with O$_2$ has been investigated. Initial insertion into the Fe–Ar bond was believed to be followed by rapid homolysis.[142] The aryloxide chemistry of ruthenium[473] and osmium[474] porphyrin compounds has also been investigated.

Aryloxide ligands also play an important role in the area of biologically relevant iron–sulfur cluster chemistry.[475–477] The cubic cluster [Fe$_4$(μ_2-S)$_4$(OAr)$_4$]$^{2-}$ shows significant spin delocalization onto the phenoxide nucleus.[478] The first and second reduction potentials for the [Fe$_4$(μ_2-S)$_4$]$^{2+}$ core were found to undergo significant negative shifts (more difficult to reduce) upon replacing arenethiolates by aryloxide ligands. An important class of rhombic dodecahedral clusters [Fe$_6$(μ_4-S)$_4$(OAr)$_4$]$^{3-}$ has been isolated and extensively studied.[479,480]

Ruthenium chemistry typically finds aryloxide ligands attached to the metal in low oxidation states. The extensive number of carbonyl aryloxide compounds reported for this metal exemplifies this (Table 6.34). The cluster compound [(H)$_2$Ru$_6$(CO)$_{16}$(μ_2-O-η^2-C$_6$H$_4$)] and derivatives contain the metallated phenoxide bridging two metal centres as well as η^6-bound to another.

6.2.10   Group 9 Metal Aryloxides

Aryloxide coordination compounds of Co(II) and Co(III) typically adopt octahedral geometries (Table 6.36). In a number of these derivatives the metal coordination sphere is completed by chelation of *ortho*-substituents on the aryloxide, *e.g.* in [Co(OC$_6$H$_2$Cl$_3$-2,4,6)$_2$L$_2$] (L = imidazole, L$_2$ = TMEDA) compounds where Co–Cl interactions are present.[481,482] The complex [Co(OPh)(PPh$_3$)$_3$][483] is tetrahedral while treatment of

Table 6.36   Simple aryloxides of cobalt

Compound	Aryloxide	Bond length (Å) M–O	Bond angle (°) M–O–Ar	Ref.
[Co(OAr)(PPh$_3$)$_3$] tetrahedral Co	OC$_6$H$_5$	1.900 (9)	138	i
[Co(OAr)(PPh$_3$)$_3$] tbp Co, axial O and P	OC$_6$H$_2$Me-4-Bu^t-6-PPh$_2$-2	2.001 (2)	120	ii
[Co(OAr)$_2$(*N*-methylimidazole)$_2$] 6-coord. Co, Cl bound	OC$_6$H$_2$Cl$_3$-2,4,6	1.898 (8) 1.889 (8)	131 133	iii
[Co(OAr)$_2$(TMEDA)] 6-coord. Co, *trans*-O, Cl bound	OC$_6$H$_2$Cl$_3$-2,4,6	1.919 (3) 1.929 (3)	130 129	iv
[Co(OAr)$_2$(N-methylimidazole)$_2$] 6-coord. Co, *trans* O(Ar), *cis* N	OC$_6$H$_4$(NO$_2$)-2	1.972 (2) 1.978 (2)	132 127	v
[Co(OAr)$_3$] 6-coord. Co, *mer*-N$_3$O$_3$	OC$_6$H$_4$(2′py)-2	1.83 (1) 1.885 (9) 1.83 (1)	121 117 121	vi
[Co(OAr)$_2$][PF$_6$].MeCN *cis*-CoO$_2$N$_4$ core	μ_2-OC$_6$H$_4$(2′-bipy)-2	1.886 (4) 1.869 (3)	120 125	vii
[HCo(OAr)(O$_2$CMe)(PMe$_3$)$_2$] metallated formyl phenoxide	OC$_6$H$_2$Bu^t-6-Me-4-η^1-CO	1.981 (3)	111	viii
[ClCo(OAr)(O$_2$CMe(PMe$_3$)$_2$] metallated formyl phenoxide	OC$_6$H$_2$Bu^t-6-Me-4-η^1-CO	1.977 (2)	113	ix
[BrCo(OAr)(O$_2$CMe)(PMe$_3$)$_2$] metallated formyl phenoxide	OC$_6$H$_2$Bu^t-6-Me-4-η^1-CO	1.978 (4)	113	ix
[ICo(OAr)$_2$(O$_2$CMe)(PMe$_3$)$_2$] metallated formyl phenoxide	OC$_6$H$_2$Bu^t-6-Me-4-η^1-CO	1.965 (2)	110	ix

[i]Y. Hayashi, T. Yamamoto, A. Yamamoto, S. Komiya, and Y. Kushi, *J. Am. Chem. Soc.*, **108**, 385 (1986).

[ii]H.-F. Klein, A. Brand, and G. Cordier, *Z. Naturforsch. B*, **53**, 307 (1998).

[iii]M.B. Cingi, A.M.M. Lanfredi, A. Tiripicchio, J. Reedijk, and R. van Landschoot, *Inorg. Chim. Acta*, **39**, 181 (1980).

[iv]M. Basturkmen, D. Kisakurek, W.L. Driessen, S. Gorter, and J. Reedijk, *Inorg, Chim. Acta*, **271**, 19 (1988).

[v]R.G. Little, *Acta Crystallogr. B*, **35**, 2398 (1979).

[vi]P. Ganis, A. Saporito, A. Vitagliano, and G. Valle, *Inorg. Chim. Acta*, **142**, 75 (1988).

[vii]J.C. Jeffery, E. Schatz, and M.D. Ward, *J. Chem. Soc., Dalton Trans.*, 1921 (1992).

[viii]H.-F. Klein, S. Haller, Hongjian Sun, Xiaoyan Li, T. Jung, C. Rohr, U. Florke, and H.-J. Haupt, *Z. Naturforsch. B*, **53**, 587 (1998).

[ix]H.-F. Klein, S. Haller, Hongjian Sun, Xiaoyan Li, T. Jung, C. Rohr, U. Florke, and H.-J. Haupt, *Z. Naturforsch. B*, **53**, 856 (1998).

Table 6.37   Simple aryloxides of rhodium

Compound	Aryloxide	Bond length (Å) M–O	Bond angle (°) M–O–Ar	Ref.
[Rh(OAr)(PMe$_3$)$_3$](HOAr) square planar Rh, hydrogen bonded HOAr	OC$_6$H$_4$Me-4	2.124 (8)	122	i
trans-[Rh(OAr)(CO)(PPh$_3$)$_2$] square planar Rh	OC$_6$H$_5$	2.044 (2)	126	ii
[Rh(OAr)(PPh$_3$)$_3$] square planar Rh	OC$_6$H$_5$	2.091 (3)	123	iii
trans-[Rh(OAr)(CO)(PPh$_3$)$_2$] square planar Rh	OC$_6$H$_4$NO$_2$-4	2.069 (2)	130	iii
[(dippe)Rh(μ-H)(μ-OAr)Rh(dippe)] diisopropylphosphinoethane	μ_2-OC$_6$H$_5$	2.117 (1), 2.114 (2)		iv
[(ArO)Rh(N$_2$)Rh(OAr)] bridging dinitrogen, square planar Rh	OC$_6$H$_2$(CH$_2$PBu$_2^t$)$_2$-2,6-Me-4	2.041 (2)	99	v
[(CO)$_2$Rh(μ-OAr)$_2$Rh(CO)$_2$]	μ_2-OC$_6$H$_3$Ph$_2$-2,6	2.063 (3), 2.067 (4) 2.083 (3), 2.101 (3)		vi
[Rh(OAr)(CO)(PPh$_3$)] square planar Rh, trans P	OC$_6$H$_4$PPh$_2$-2	2.045 (4)	120	vii

[i] S.E. Kegley, C.J. Schaverien, J.H. Freudenberger, R.G. Bergman, S.P. Nolan, and C.D. Hoff, *J. Am. Chem. Soc.*, **109**, 6563 (1987).
[ii] K.A. Bernard, M.R. Churchill, T.S. Janik, and J.D. Atwood, *Organometallics*, **9**, 12 (1990).
[iii] V.F. Kuznetsov, G.P.A. Yap, C. Bensimon, and H. Alper, *Inorg. Chim, Acta*, **280**, 172 (1998).
[iv] M.D. Fryzuk, May-Ling Jang, T. Jones, and F.W.B. Einstein, *Can. J. Chem.*, **64**, 174 (1986).
[v] M.E. van der Boom, Shyh-Yeon Liou, Y. Ben-David, L.J.W. Shimon, and D. Milstein, *J. Am. Chem. Soc.*, **120**, 6531 (1998).
[vi] D.E. Chebi, P.E. Fanwick, and I.P. Rothwell, *Polyhedron*, **9**, 969 (1990).
[vii] L. Dahlenburg, K. Herbst, and M. Kuhnlein, *Z. Anorg. Allg. Chem.*, **623**, 250 (1997).

Wilkinson's catalyst with NaOAr yields square-planar [Rh(OAr)(PPh$_3$)$_3$].[484,485] The isomorphous pair of carbonyls *trans*-[M(OPh)(CO)(PPh$_3$)$_2$] (M = Rh[486] and Ir[237]) are also obtained from the corresponding chlorides. The treatment of the rhodium compound with MeI yields PhOMe whereas elimination does not take place from the iridium compound.

The hydroxy compound [Cp*Ir(OH)(Ph)(PMe$_3$)] reacts with phenol to produce the corresponding phenoxide.[487] A related compound [Cp*Ir(OC$_6$H$_4$CF$_3$-4)(Me)(PMe$_3$)] can be produced by addition of the parent phenol to the methylene complex [Cp*Ir(CH$_2$)(PMe$_3$)], formally addition of the ArO–H bond across the Ir=CH$_2$ double bond.[488] The oxidative addition of ArOH bonds to [Ir(COD)(PMe$_3$)$_3$]Cl yield the Ir(III) aryloxide *mer*-[Ir(OAr)Cl(H)(PMe$_3$)$_3$]. Attempts to observe interactions between Rh or Ir and the ortho-phenyl rings of 2,6-diphenylphenoxide ligands failed. Only simple

Table 6.38   Simple aryloxides of iridium

Compound	Aryloxide	Bond length (Å) M–O	Bond angle (°) M–O–Ar	Ref.
trans-[Ir(OAr)(CO)(PPh$_3$)$_2$] square planar Ir	OC$_6$H$_5$	2.049 (4)	127	i
trans-[Ir(OAr)(CO)(PPh$_3$)$_2$] square planar Ir	OC$_6$F$_5$	2.058 (3)	135	ii
mer-[Ir(OAr)Cl(H)(PMe$_3$)$_3$].0.5CH$_2$Cl$_2$ octahedral Ir	OC$_6$H$_3$Me$_2$-3,5	2.109 (5)	130	iii
[Cp*Ir(OAr)(Ph)(PMe$_3$)]	OC$_6$H$_5$	2.04 (1)	125	iv
trans-[Ir(OAr)(CO){P(p-tolyl)$_3$](HOAr)} square planar Ir, non hydrogen bonded HOAr	OC$_6$H$_4$Me-4	2.06 (7)	122	v
[(COD)Ir(η^5-HOAr)][(COD)Ir(OAr)$_2$] square planar anion	OC$_6$H$_3$Ph$_2$-2,6	2.08 (1) 2.09 (1)	126 131	vi
[Ir(OAr)(OAr′)(MeCN)] octahedral Ir, trans O	PBu$_2^t$(2-OC$_6$H$_3$OMe-6) CH$_2$CMe$_2$PBu^t(2-OC$_6$H$_3$OMe-6)	2.062 (7) 2.050 (6)	119 119	vii

[i]W.M. Rees, M.R. Churchill, J.C. Fettinger, and J.D. Atwood, *Organometallics*, **4**, 2179 (1985).
[ii]M.R. Churchill, J.C. Fettinger, W.M. Rees, and J.D. Atwood, *J. Organomet. Chem.*, **308**, 361 (1986).
[iii]F.T. Ladipo, M. Kooti, and J.S. Merola, *Inorg. Chem.*, **32**, 1681 (1993).
[iv]K.A. Woerpel and R.G. Bergman, *J. Am. Chem. Soc.*, **115**, 7888 (1993).
[v]C.A. Miller, T.S. Janik, C.H. Lake, L.M. Toomey, M.R. Churchill, and J.D. Atwood, *Organometallics*, **13**, 5080 (1994).
[vi]D.E. Chebi, P.E. Fanwick, and I.P. Rothwell, *Polyhedron*, **9**, 969 (1990).
[vii]H.D. Empsall, P.N. Heys, W.S. McDonald, M.C. Norton, and B.L. Shaw, *J. Chem. Soc., Dalton Trans.*, 1119 (1978).

aryloxide bonding, *e.g.* in [(CO)$_2$Rh(μ-OAr)$_2$Rh(CO)$_2$] or sometimes η^5-bound cyclohexadienonyl groups as in [(COD)Ir(η^5-HOAr)][(COD)Ir(OAr)$_2$] were observed.[102]

6.2.11   Group 10 Metal Aryloxides

The aryloxide chemistry of the group 10 metals is dominated by compounds of general formula [M(OAr)(X)L$_2$] where L can be a variety of N or P donor ligands and X can be groups such as Me or even OAr. By far the most extensive method of introducing the M–OAr group is by treatment of alkyl precursors with phenol. A consistent observation is that generation of the aryloxide [L$_n$M(OAr)] is followed by formation of adducts [L$_n$M(OAr)(HOAr)] where the phenolic proton is hydrogen bonded to the highly nucleophilic aryloxide oxygen atom. The electron-rich nature of the aryloxide group

Table 6.39   Simple aryloxides of nickel

Compound	Aryloxide	Bond length (Å) M–O	Bond angle (°) M–O–Ar	Ref.
trans-[Ni(OAr)Me(PMe$_3$)$_2$](HOAr) square planar Ni, hydrogen bonded HOAr	OC$_6$H$_5$	1.932 (5)	122	i
trans-[Ni(OAr)(H)(PBz$_3$)$_2$](HOAr) square planar Ni, hydrogen bonded HOAr	OC$_6$H$_5$	1.949 (7)	123	ii
[Ni(OAr)$_2$(py)$_3$] square pyramidal Ni, *trans* basal ArO	OC$_6$H$_2$Cl$_3$-2,4,6	2.014 (4)	130	iii
[Ni(OAr)(trithianonane)(OH$_2$)][OAr] 6-cord. Ni, *cis*-O, chelating OAr	OC$_6$H$_2$(NO$_2$)$_3$-2,4,6	2.030 (5)	130	iv
[Ni(OAr)$_2$(OPMe$_3$)$_2$] 6-cord. Ni, *all-trans*, chelating OAr	OC$_6$H$_2$(NO$_2$)$_2$-2,6-Me-4	1.974 (2)	125	v
trans-[Ni(OAr)$_2$(PMe$_3$)$_2$] square planar Ni	OC$_6$H$_3$Bu^t-2-Me-6	1.861 (3)	127	vi
trans-[Ni(OAr)$_2$(PMe$_3$)$_2$] square planar Ni	OC$_6$H$_3$Bu^t-2-Me-4	1.857 (3)	123	vii
[{Ni(OAr)(HOAr)}$_2$][PF$_6$]$_2$ two octahedral Ni units held together by ArOH–OAr hydrogen bonding and π-stacking	OC$_6$H$_4$(2'-bipy)-2	2.06 (2) 2.04 (2)	123 123	viii
[Ni(OAr)(PMe$_3$)$_3$] chelated via acyl group	OC$_6$H$_2$Bu^t-6-Me-4-CO-2	2.042 (3)	111	ix
[{Ni(μ_2-OAr)(PEt$_3$)}$_2$] chelating 2-naphtholato ligand	μ_2-OC$_{10}$H$_6$(CF$_2$CF$_2$)	1.919 (4), 1.940 (4)		x
[Ni(OAr)(dmpe)] square planar Ni	OC$_6$H$_4$-CMe$_2$CH$_2$	1.872 (3)	127	xi
[Ni(OAr)$_2$].2MeCO OAr derived from P{C$_6$H$_2$(OMe)$_3$-2,4,6}$_3$, *trans*, square planar NiO$_2$P$_2$	(MeO)$_3$C$_6$H$_2$-P-2-OC$_6$H$_2$(OMe)$_2$	1.856 (5)	120	xii
[Ni(OAr)$_2$][BF$_4$].2MeCO OAr derived from P{C$_6$H$_2$(OMe)$_3$-2,4,6}$_3$, octahedron with *trans* chelating OMe, *cis*-O(Ar)	(MeO)$_3$C$_6$H$_2$-P-2-OC$_6$H$_2$(OMe)$_2$	1.918 (5) 1.918 (5)	120 120	xii
[Cp*Ni(OAr)(PEt$_3$)(HOAr')] HOC$_6$H$_3$Me$_2$-2,6 hydrogen bonded to OAr	OC$_6$H$_4$Me-4	1.909 (2)	123	xiii
[Cp*Ni(OAr)(PEt$_3$)] two molecules	OC$_6$H$_4$Me-4	1.979 (3) 2.007 (4)	122 123	xiv

Table 6.39   (*Continued*)

Compound	Aryloxide	Bond length (Å) M–O	Bond angle (°) M–O–Ar	Ref.
[Ni$_2$(μ-OAr)$_2$(DMF)$_2$(OH$_2$)$_2$][BPh$_4$]$_2$.4DMF edge-shared bis-octahedron, axial OH$_2$	μ_2-OC$_6$H$_4$(2'-bipy)-2	1.993 (4), 2.121 (4)		xv
[Ni$_2$(μ-OAr)$_2$(DMF)$_6$][BPh$_4$]$_2$.2Et$_2$O edge-shared bis-octahedron, axial py	μ_2-OC$_6$H$_4$(2'-py)-2	2.003 (3), 2.051 (3)		xvi

[i]Yong-Joo Kim, Kohtaro Osakada, A. Takenaka, and A. Yamamoto, *J. Am. Chem. Soc.*, **112**, 1096 (1990).

[ii]A.L. Seligson, R.L. Cowan, and W.C. Trogler, *Inorg. Chem.*, **30**, 3371 (1991).

[iii]M.F. Richardson, G. Wulfsberg, R. Marlow, S. Zaghonni, D. McCorkle, K. Shadid, J. Gagliardi Jr, and B. Farris, *Inorg. Chem.*, **32**, 1913 (1993).

[iv]J.A.R. Hartmann and S.P. Cooper, *Inorg. Chim. Acta*, **111**, L43 (1986).

[v]H.-F. Klein, A. Dal, S. Hartmann, U. Florke, and H.-J. Haupt, *Inorg. Chim. Acta*, **287**, 199 (1999).

[vi]H.-F. Klein, A. Dal, T. Jung, S. Braun, C. Rohr, U. Florke, and H.-J. Haupt, *Eur. J. Inorg. Chem.*, 621 (1998).

[vii]H.-F. Klein, A. Dal, T. Jung, C. Rohr, U. Florke, and H.-J. Haupt, *Eur. J. Inorg. Chem.*, 2027 (1998).

[viii]B.M. Holligan, J.C. Jeffery, and M.D. Ward, *J. Chem. Soc., Dalton Trans.*, 3337 (1992).

[ix]H.-F. Klein, A. Bickelhaupt, B. Hammerschmitt, U. Florke, and H.-J. Haupt, *Organometallics*, **13**, 2944 (1994).

[x]M.A. Bennett M. Glewis, D.C.R. Hockless, and E. Wenger, *J. Chem. Soc., Dalton Trans.*, 3105 (1997).

[xi]K. Koo, G.L. Hillhouse, and A.L. Rheingold, *Organometallics*, **14**, 456 (1995).

[xii]K.R. Dunbar, Jui-Sui Sun, and A. Quillevere, *Inorg. Chem.*, **33**, 3598 (1994).

[xiii]P.L. Holland, R.A. Andersen, R.G. Bergman, J. Huang, and S.P. Nolan, *J. Am. Chem. Soc.*, **119**, 12800 (1997).

[xiv]P.L. Holland, M.E. Smith, R.A. Andersen, and R.G. Bergman, *J. Am. Chem. Soc.*, **119**, 12815 (1997).

[xv]D.A. Bardwell, J.C. Jeffery, and M.D. Ward, *J. Chem. Soc., Dalton Trans.*, 3071 (1995).

[xvi]D.A. Bardwell, J.C. Jeffery, and M.D. Ward, *Inorg, Chim. Acta*, **236**, 125 (1995).

accounts for its reaction with electrophiles such as CO_2 to generate insertion products (Section 5.2). Hence treatment of *trans*-[NiMe$_2$(PMe$_3$)$_2$] or *trans*-[PdR$_2$(PR$_3$)$_2$] (R = Me, Et) with *para*-substituted phenols yields the corresponding compounds *trans*-[MMe(OAr)(HOAr)(PR$_3$)$_2$] (M = Ni, Pd).[489] Both compounds for R = Me, Ar = Ph as well as the non-phenol adduct of Pd were crystallographically characterized. In the case of *cis*-[PtMe$_2$(PMe$_3$)$_2$] initial conversion to *cis*-[PtMe{OCH(CF$_3$)$_2$}(PMe$_3$)$_2$] followed by addition of phenol led to the simple phenoxide *cis*-[Pt(OPh)Me(PMe$_3$)$_2$] prior to formation of the phenol adduct.[240] The one-electron oxidation of [PdMe(OPh)(dmpe)] results in cleavage of the Pd–OPh bond.[490]

Other related alkyl substrates include [PtMe$_2$(bipy)],[491] [PdMe$_2$(NMe$_2$CH$_2$C$_6$H$_4$PPh$_2$)],[492] and [PdMe$_2$(TMEDA)][493] which all form mono-aryloxides. The product [Pt(OPh)Me(bipy)] reacts with MeI to form [Pt(OPh)(I)Me$_2$(bipy)]

Table 6.40 Simple aryloxides of palladium

Compound	Aryloxide	Bond length (Å) M–O	Bond angle (°) M–O–Ar	Ref.
[Pd(OAr)₂] trans-O(Ar)	OC₆H₄(2′py)-2	1.979 (2)	117	i
[(bipy)Pd(OAr)₂] square planar Pd	OC₆H₅	1.996 (7) 1.983 (6)	121 120	ii
[(bipy)Pd(OAr)₂] square planar Pd	OC₆F₅	2.013 (3) 2.013 (4)	123 125	ii
[(TMEDA)Pd(OAr)(Me)] square planar Pd	OC₆H₅	2.024 (3)	123	iii
[(TMEDA)Pd(OAr)(Me)](HOAr) square planar Ph, hydrogen bonded HOAr	OC₆H₅	2.037 (2)	122	iii
[(TMEDA)Pd(OAr)(Me)] square planar Pd	OC₆H₄(NO₂)-4	2.029 (4)	126	iii
[(TMEDA)Pd(OAr)(Me)](HOAr) square planar Ph, hydrogen bonded HOAr	OC₆H₄(NO₂)-4	2.043 (2)	125	iii
trans-[Pd(OAr)(H)(PCy₃)₂](HOAr) square planar Pd, hydrogen bonded HOAr	OC₆H₅	2.135 (2)	121	iv, v
trans-[Pd(OAr)(H)(PCy₃)₂](HOAr) square planar Pd, hydrogen bonded HOAr	OC₆F₅	2.181 (2)	126	iv, v
trans-[Pd(OAr)Me(PMe₃)₂] square planar Pd	OC₆H₅	2.106 (3)	127	vi
trans-[Pd(OAr)Me(PMe₃)₂](HOAr) square planar Pd, hydrogen bonded HOAr	OC₆H₅	2.134 (3)	122	vi
trans-[(Pd(OAr)Me(PMe₃)₂](CF₃CHPhOH) square planar Pd, hydrogen bonded alcohol	OC₆H₅	2.107 (2)	124	vi
[(bipy)Pd(OAr){OCH(CF₃)₂}](HOAr) square planar Pd, hydrogen bonded HOAr	OC₆H₅	1.997 (5)	120	vii
trans-[Pd(OAr)₂(pyrrolidine)₂] square planar Pd, intermolecular hydrogen bonding between ArO and pyrrolidine	OC₆H₅	2.003 (2)	121	viii
trans-[Pd(OAr)₂(pyrrolidine)₂](HOAr)₂ square planar Pd, two hydrogen bonded HOAr	OC₆H₅	2.018 (2)	120	viii

	OAr		Angle	Ref.
[Pd(OAr)(C$_6$H$_3${CH$_2$NMe$_2$}$_2$-2,6)] square planar Pd, ArO *trans* to metallated aryl ring	OC$_6$H$_5$	2.12 (2)	125	viii
[Pd(OAr)(C$_6$H$_3${CH$_2$NMe$_2$}$_2$-2,6)](HOAr) square planar Pd, ArO *trans* to metallated aryl ring, hydrogen bonded HOAr	OC$_6$H$_5$	2.139 (4)	128	viii
[Pd(PPh$_2$C$_6$H$_4$CH$_2$NMe$_2$)(OAr)Me] square planar Pd, ArO *trans* to P	OC$_6$H$_5$	2.088 (5)	123	ix
[Pd(PPh$_2$C$_6$H$_4$CH$_2$NMe$_2$)(OAr)Me](HOAr) square planar Pd, ArO *trans* to P, hydrogen bonded HOAr	OC$_6$H$_5$	2.103 (2)	122	ix
cis-[Pd(OAr)Me(dmpe)] square planar Pd	OC$_6$H$_5$	2.098 (6)	124	x
[(ArO)Pd(O$_2$CCF$_3$)] square planar Pd, two molecules	OC$_6$H$_2$(CH$_2$PBu$^t_2$)$_2$-2,6-Me-4	1.981 (2) 1.987 (2)	103 101	xi
[Pd(OAr)$_2$].3.73CH$_2$Cl$_2$.H$_2$O OAr derived from P[C$_6$H$_2$(OMe)$_3$-2,4,6]$_3$, *trans*, square planar PdO$_2$P$_2$	(MeO)$_3$C$_6$H$_2$-P-2-OC$_6$H$_2$(OMe)$_2$	2.047 (7) 2.047 (8)	121 117	xii

[i] P. Ganis, A. Saporito, A. Vitagliano, and G. Valle, *Inorg. Chim. Acta*, **142**, 75 (1988).

[ii] G.M. Kapteijn, D.M. Grove, H. Kooijman, W.J.J. Smeets, A.L. Spek, and G. van Koten, *Inorg. Chem.*, **35**, 526 (1996).

[iii] G.M. Kapteijn, A. Dervisi, D.M. Grove, H. Kooijman, M.T. Lakin, A.L. Spek, and G. van Koten, *J. Am. Chem. Soc.*, **117**, 10939 (1995).

[iv] D. Braga, P. Sabatino C. Di Bugno, P. Leoni, and M. Pasquali, *J. Organomet. Chem.*, **334**, C46 (1987).

[v] C. Di Bugno, M. Pasquali, P. Leoni, P. Sabatino, and D. Braga, *Inorg. Chem.*, **28**, 1390 (1989).

[vi] Yong-Joo Kim, Kohtaro Osakada, A. Takenaka, and A. Yamamoto, *J. Am. Chem. Soc.*, **112**, 1096 (1990).

[vii] G.M. Kapteijn, D.M. Grove, G. van Koten, W.J.J. Smeets, and A.L. Spek, *Inorg. Chim. Acta*, **207**, 131 (1993).

[viii] P.L. Alsters, P.J. Baesjou, M.D. Janssen, H. Kooijman, A. Sicherer-Roetman, A.L. Spek, and G. van Koten, *Organometallics*, **11**, 4124 (1992).

[ix] M. Kapteijn, M.P.R. Spee, D.M. Grove, H. Kooijman, A.L. Spek, and G. van Koten, *Organometallics*, **15**, 1405 (1996).

[x] A.L. Seligson, R.L. Cowan, and W.C. Trogler, *Inorg. Chem.*, **30**, 3371 (1991).

[xi] M.E. van der Boom, Shyh-Yeon Liou, Y. Ben-David, L.J.W. Shimon, and D. Milstein, *J. Am. Chem. Soc.*, **120**, 6531 (1998).

[xii] Jui-Sui Sun, C.E. Uzelmeier, D.L. Ward, and K.R. Dunbar, *Polyhedron*, **17**, 2049 (1998).

(continued overleaf)

Table 6.40 (Continued)

Compound	Aryloxide	Bond length (Å) M–O	Bond angle (°) M–O–Ar	Ref.
[Pd(OAr)(C$_6$H$_4$CH$_2$NMe$_2$-2)] square planar Pd, *trans* N	OC$_6$H$_4$CH$_2$NMe$_2$-2	2.098 (2)	118	xiii
[Pd(OAr)Cl].CH$_2$Cl$_2$ square planar Pd	OC$_6$H$_4$(2′-bipy)-2	1.939 (6)	125	xiv
[Pd(OAr)(PMe$_3$)$_3$(HOCHCF$_3$Ph)] non-clelating OAr, HOCHCF$_3$Ph hydrogen bonded to O	OC$_6$H$_4$CH$_2$CH=CH$_2$-2	2.129 (3)	122	xv
[Pd(OAr)$_2$].CHCl$_3$ square planar Pd, *cis* O, chelating 2-naphthoxide	OC$_{10}$H$_6$-PPh(Pr^i)	2.046 (1) 2.051 (1)	118 118	xvi
[Pd(OAr)$_2$]	OC$_6$H$_3$(2′-py-Bu^t-4)-2	1.966 (1)	113	xvii
[Pd(OAr)$_2$] square planar PdO$_2$N$_2$	OC$_6$H$_3$(2′-py-Me-2)-2	2.017 (2) 2.007 (2)	110 112	xvii
[Pd(OAr)$_2$] square planar PdO$_2$N$_2$	OC$_6$H$_3$(2′-py-CH$_2$CH$_2$Ph-2)-2	2.000 (3) 2.011 (3)	112 111	xvii

[xiii]P.L. Alsters, H.T. Teunissen, J. Boersma A.L. Spek, and G. van Koten, *Organometallics*, **12**, 4691 (1993).

[xiv]D. Bardwell, D. Black, J.C. Jeffery, and M.D. Ward, *Polyhedron*, **17**, 1577 (1993).

[xv]Youn-Joo Kim, Jae-Young Lee, and Kohtaro Osakada, *J. Organomet. Chem.* **558**, 41 (1998).

[xvi]J. Heinicke, R. Kadyrov, M.K. Kindermann, M. Kloss, A. Fischer, and P.G. Jones, *Chem. Ber.*, **129**, 1061 (1996).

[xvii]C.A. Otter, D.A. Bardwell, S.M. Couchman, J.C. Jeffery, J.P. Maher, and M.D. Ward, *Polyhedron*, **17** 211 (1998)

Table 6.41   Simple aryloxides of platinum

Compound	Aryloxide	Bond length (Å) M–O	Bond angle (°) M–O–Ar	Ref.
trans-[Pt(OAr)(H)(PEt$_3$)$_2$] square planar Pt	OC$_6$H$_5$	2.098 (9)	124	i
[Pt(OAr)Cl(η^2-vinyl-N,N-dimethylaniline)] pseudo-square planar Pd	OC$_6$F$_5$	2.017 (5)	123	ii
cis-[Pt(OAr)Me(PMe$_3$)$_2$]	OC$_6$H$_5$	2.128 (4)	120	iii
[NBu$_4$][Pt(OAr)(CO)(C$_6$F$_5$)] square planar Pt, ArO cis to CO	OC$_6$H$_4$(NO$_2$)-4	2.070 (4)	125	iv
trans-[Pt(OAr)(H)(Pbz$_3$)$_2$] square planar Pt	OC$_6$H$_5$	2.130 (6)	124	v
[Pt(OAr)$_2$].3EtCN OAr derived from P{C$_6$H$_2$(OMe)$_3$-2,4,6}$_3$, trans, square planar PtO$_2$P$_2$	(MeO)$_3$C$_6$H$_2$-P- 2-OC$_6$H$_2$(OMe)$_2$	2.044 (5) 2.047 (8)	119 119	vi
[Pt(OAr)Me(PMe$_3$)] chelating OAr, O trans to P	OC$_6$H$_4$- η^2-CH$_2$CH=CH$_2$-2	2.05 (2)	128	vii
[Pt(OAr)Cl].CH$_2$Cl$_2$ square planar Pt, two molecules	OC$_6$H$_4$(2'bipy)-2	1.95 (1) 1.94 (1)	126 126	viii
[Pt(OAr)Me(bipy)] square planar Pt	OC$_6$H$_5$	2.001 (5)	122	ix
[Pt(OAr)(I)Me$_2$(bipy)] octahedral Pt, I trans to Me	OC$_6$H$_5$	2.014 (5)	127	ix
[Pt(OAr)Me(PPh$_2$C$_6$F$_5$)] square planar Pt, Me trans to O	OC$_6$F$_4$(PPh$_2$)-2	2.12 (1)	117	x

[i]R.L. Cowan and W.C. Trogler, *J. Am. Chem. Soc.*, **111**, 4750 (1989).

[ii]M.K. Cooper, N.J. Hair, and D.W. Yaniuk, *J. Organomet. Chem.*, **150**, 157 (1978).

[iii]K. Osakada, Yong-Joo Kim, and A. Yamamoto, *J. Organomet. Chem.*, **382**, 303 (1990).

[iv]J. Ruiz, V. Rodriguez, G. Lopez, P.A. Chaloner, and P.B. Hitchcock, *Organometallics*, **15**, 1662 (1996).

[v]A.L. Seligson, R.L. Cowan, and W.C. Trogler, *Inorg. Chem.*, **30**, 3371 (1991).

[vi]Jui-Sui Sun, C.E. Uzelmeier, D.L. Ward, and K.R. Dunbar, *Polyhedron*, **17**, 2049 (1998).

[vii]Yong-Joo Kim, Jae-Young Lee, and Kohtaro Osakada, *J. Organomet. Chem.*, **558**, 41 (1998).

[viii]D.A. Bardwell, J.G. Crossley, J.C. Jeffery, A.G. Orpen, E. Psillakis, E.E.M. Tilley, and M.D. Ward, *Polyhedron*, **13**, 2291 (1994).

[ix]G.M. Kapteijn, M.D. Meijer, D.M. Grove, N. Veldman, A.L. Spek, and G. van Koten, *Inorg. Chim. Acta*, **264**, 211 (1997).

[x]S. Park, M. Pontier-Johnson, and D.M. Roundhill, *Inorg. Chem.*, **29**, 2689 (1990).

structurally characterized with I trans to Me but present as two isomers in solution.[491] Treatment of cis-[PdMe$_2$(PMe$_3$)$_2$] with 2-allylphenol yields trans-[Pd(OC$_6$H$_4$CH$_2$CH=CH$_2$-2)Me(PMe$_3$)$_2$] in which the allyl group is unbound.[32] However, treatment of the alkoxide cis-[PtMe{OCH(CF$_3$)$_2$}{HOCH(CF$_3$)$_2$}(PMe$_3$)$_2$] with 2-allylphenol yielded [Pt(OC$_6$H$_4$-η^2-CH$_2$CH=CH$_2$-2)Me(PMe$_3$)] with the chelating olefin trans to the Pt–Me bond.

The use of alkoxide or amide intermediates to form aryloxides is also quite common. Reaction of [NiMe(OMe)(PMe$_3$)]$_2$ with acidic phenols in the presence of PMe$_3$ yields trans-[NiMe(OAr)(PMe$_3$)$_2$].[494] The bulky amido compound [PdCl{N(SiMe$_3$)$_2$}(TMEDA)] reacts with phenols to yield [PdCl(OAr)(TMEDA)].[495]

Hydrido, phenoxides have been isolated by treatment of $trans$-$[Ni(H)Cl(PR_3)_2]$ (R = Pr^i, Cy, CH_2Ph) and $trans$-$[Pt(H)(NO_3)(PEt_3)_2]$ with NaOPh.[243,244]

Bis(aryloxides) of nickel have been obtained by metathetical exchange reactions as well as addition of phenols to $[Ni(PMe_3)_4]$.[496,497] In the case of palladium, treatment of $[Pd(O_2CMe)_2]$ with two equivalents of NaOAr in the presence of chelating diamine ligands yields $[Pd(OAr)_2(L-L)]$ (L–L = bipy, TMEDA, etc.).[498] The platinum compounds $[Pt(OAr)_2(dppe)]$ react with CO to form intermediate bis(aryloxycarbonyl) complexes which finally produce $ArOCO_2Ar$, CO_2, and $[Pt(CO)_2(dppe)]$.[238] Mechanistic studies show that the overall deoxygenation of one of the aryloxide ligands proceeds via a benzyne intermediate.

Treatment of $[Pd_2Cl_2(L-L)_2]$ (L–L = dppm, dmpm) with Na or KOAr leads to the corresponding Pd(I)–Pd(I) bis(aryloxides).[499] Reaction with CO led to insertion into the Pd–Pd bond.

The allyl compounds $[(\eta^3\text{-allyl})M(\mu_2\text{-}OAr)_2M(\eta^3\text{-allyl})]$ (M = Ni,[500] Pd[501]) have been obtained by treating $[(\eta^3\text{-allyl})Ni(\mu_2\text{-}Br)_2Ni(\eta^3\text{-allyl})]$ and $[(\eta^3\text{-allyl})Pd(\mu_2\text{-}Cl)_2Pd(\eta^3\text{-allyl})]$ with Li/NaOAr and $[NBu_4^n][OH]$/phenol respectively. The nickel compounds were found to exist as a mixture of $cis/trans$ isomers. The equilibration of isomers, either by allyl rotation or opening up of an aryloxide bridge, can occur on the NMR time scale depending on the nature of the aryloxide ligand.[500] The OC_6F_5, $OC_6H_3(CF_3)_2$-3,5 and $OC_6H_2(CF_3)_3$-2,4,6 derivatives initiate the rapid polymerization of 1,3-cyclohexadiene and 1,3-butadiene to high-molecular-weight 1,4-linked polymers. The palladium complexes, OAr = OC_6H_4X-4 (X = H, Me, Cl, Br, NO_2) react with PPh_3 to form $[(\eta^3\text{-allyl})Pd(OAr)(PPh_3)]$.[501]

6.2.12   Group 11 Metal Aryloxides

An important aspect of copper aryloxide chemistry is the homogeneous copper/amine catalysed oxidative coupling of phenols. For example the repeated C–O coupling of 2,6-dimethylphenol leads to a poly(phenylene-ether) which is an extensively used plastic owing to its high chemical and thermal stability.[502,503] Typically the reactions are carried out using either Cu(II) salts (e.g. chloride or nitrate) or Cu(I) compounds in the presence of base, amine (e.g. N-methylimidazole) and oxygen as the oxidant. The mechanism of the reaction is complex, but has received both theoretical[504,505] and experimental[506] study. Kinetics show the reaction obeys simple saturation kinetics with respect to phenol, with phenol oxidation possibly being the rate-determining step.[507] The theoretical and experimental data support a mechanism in which dinuclear phenolate-bridged copper(II) species act as intermediates affording phenoxonium cations after a double one-electron transfer. A number of simple coordination compounds of Cu(II) aryloxides have been isolated (Table 6.42) typically with electron-withdrawing substituents, e.g. $[Cu(OC_6H_3Cl_2\text{-}2,6)_2(N\text{-methylimidazole})_2]$,[508] or chelating aryloxides, e.g. $[Cu(OC_6H_3Ph\text{-}4\text{-}CH_2NEt_2\text{-}2)_2]$[509] and $[Cu(\mu_2\text{-}OAr)(OAr)]_2$ (OAr = $OC_6H_3(2'\text{-py-}Bu^t\text{-}4)\text{-}2)$.[510]

Copper(I) aryloxides can be formed by treating CuCl with NaOAr[511] or by adding phenols to organo-copper(I) compounds such as mesityl-copper or $[MeCu(PPh_3)]$.[512] In the absence of donor ligands the $[Cu(OAr)]_n$ derivatives are sparingly soluble and adopted structures were assumed on the basis of known Cu(I) alkyls and alkoxides. However, since 1996 the homoleptic compounds $[Cu(\mu_2\text{-}OC_6H_3Ph_2\text{-}2,6)]_4$ and

Table 6.42 Simple aryloxides of copper

Compound	Aryloxide	Bond length (Å) M–O	Bond angle (°) M–O–Ar	Ref.
[Cu(μ_2-OAr)]$_4$ planar Cu$_4$O$_4$ core, 2-coord. Cu	μ_2-OC$_6$H$_3$Ph$_2$-2,6	1.834 (7)–1.865 (7)		i
[Cu(μ_2-OAr)]$_4$ distorted cubic Cu$_4$O$_4$ core, 4-coord. Cu	μ_3-OC$_6$H$_4$-η^2-CH$_2$CH=CH$_2$-2	1.980 (5), 1.927 (5) 2.747 (5)		ii
[Cu(μ_2-OAr)(OAr)]$_2$ tbp Cu with axial O and μ_2-O	OC$_6$H$_3$(2'-py-Bu^t-4)-2	1.889 (2)	122	iii
[Cu(OAr)$_2$(imidazole)$_2$].H$_2$O all trans, Cl bound	μ_2-OC$_6$H$_3$(2'-py-Bu^t-4)-2 OC$_6$H$_2$Cl$_3$-2,4,6	1.915 (1), 2.193 (2) 1.97 (1)	130	iv
[Cu(OAr)$_2$(N-methylimidazole)$_2$] all trans, Cl bound	OC$_6$H$_3$Cl$_2$-2,6	1.942 (4)	126	v
[(THF)(tfd)Cu(μ_2-OAr)$_2$Cu(tfd)(THF)] tfd = 1,1,1-trifluoro-acac, square pyramidal Cu, axial THF	μ_2-OC$_6$F$_5$	1.952 (3), 1.960 (4)		vi
[Cu(OAr)$_2$] square planar Cu, trans CuO$_2$N$_2$, chelating OAr	OC$_6$H$_3$Ph-4-CH$_2$NMe$_2$-2	1.884 (3) 1.897 (3)	127 118	vii
[Cu(OAr)$_2$] square planar Cu, trans CuO$_2$N$_2$, chelating OAr	OC$_6$H$_3$Ph-4-CH$_2$NEt$_2$-2	1.898 (2)	119	viii
[Cu(μ_2-OAr)(O$_2$CMe)]$_2$ 5-coord. Cu, bidentate acetate, chelating OAr	μ_2-OC$_6$H$_3$Ph-4-CH$_2$NMe$_2$-2	1.926 (2), 1.971 (2)		vii
[Cu(μ_2-OAr)(O$_2$CMe)(HOMe)]$_2$ 5-coord. Cu, monodentate acetate, chelating OAr	μ_2-OC$_6$H$_3$Ph-4-CH$_2$NMe$_2$-2	1.987 (5), 1.985 (5)		vii
[Cu$_2$(μ_2-OAr)$_2$(μ_2-O$_2$CMe)][PF$_6$] square pyramidal Cu, axial μ_2-O	μ_2-OC$_6$H$_4$(2'-1,10-phen)-2	1.942 (2), 1.889 (3)		ix

(continued overleaf)

Table 6.42 (*Continued*)

Compound	Aryloxide	Bond length (Å) M–O	Bond angle (°) M–O–Ar	Ref.
[Cu(OAr)(CNC$_6$H$_4$Me-4)$_2$] trigonal planar 3-coord. Cu	OC$_6$H$_3$Bu$^t_2$-2,6	1.917 (3)	134	x, xi
[Cu(α-diimine)$_2$][Cu(OAr)$_2$] 2-coord. Cu anion	OC$_6$H$_3$Me$_2$-2,6	1.806 (6) 1.798 (8)	137 135	x, xi
[Cu(OAr)(dppe)]$_2$(μ-dppe) 5-coord. Cu	OC$_6$H$_5$	2.023 (5)	128	xi
trans-[Cu(OAr)$_2$(py)$_2$]	OC$_6$H$_2$Cl$_3$-2,4,6	1.909 (2)	126	xii, xiii
[Cu(OAr)(terpy)(ClO$_4$)] square pyramidal Cu with axial ClO$_4$	OC$_6$H$_4$(NO$_2$)-4	1.872 (3)	132	xiv
[(CO)Cu(μ_2-OAr)$_2$Cu(CO)] trigonal planar Cu	μ_2-OC$_6$H$_3$Ph$_2$-2,6	1.953 (7), 1.965 (7) 1.995 (7), 1.966 (7)		i
[(CO)Cu(μ_2-OAr)$_2$Cu(CO)] trigonal planar Cu	μ_2-OC$_6$H$_2$Bu$^t_3$-2,4,6	1.974 (5), 1.984 (5)		xv
[(CO)Cu(μ_2-OAr)$_2$Cu(CO)] trigonal planar Cu	μ_2-OC$_6$H$_2$Me$_3$-2,4,6	1.95 (2), 1.97 (2)		xv
[Cu$_2$(μ_2-OAr)(OAr)$_2$(H$_2$NCH$_2$CH$_2$NH$_2$)$_2$] distorted square pyramidal Cu, hydrogen bonded	OC$_6$H$_5$ μ_2-OC$_6$H$_5$	1.947 (4) 1.926 (3), 2.265 (4)	129	xvi
[Cu$_4$(μ_3-OAr)$_4$(PPh$_3$)$_4$] Cu$_4$O$_4$ cube	μ_3-OC$_6$H$_5$	2.05 (2)–2.26 (21)		xvii
[(Ph$_3$P)$_2$Cu(μ_2-OAr)$_2$Cu(PPh$_3$)$_2$] tetrahedral Cu, two molecules	μ_2-OC$_6$H$_5$	2.13 (2), 2.12 (2) 2.32 (2), 2.03 (2)		xviii
[(Ph$_3$P)Cu(μ_2-OAr)$_2$Cu(PPh$_3$)$_2$] tetrahedral and trigonal planar Cu	μ_2-OC$_6$H$_5$	2.11 (1), 2.00 (1) 2.10 (1), 1.95 (1)		xvii
[(Ph$_3$P)Cu(μ_2-OAr)$_2$Cu(PPh$_3$)$_2$] tetrahedral and trigonal planar Cu	μ_2-OC$_6$H$_4$Me-4	2.071 (6), 1.986 (6) 2.097 (5), 1.952 (5)		xix

| [(RNC)$_2$Cu(μ_2-OAr)$_2$Cu(CNR)$_2$] R = C$_6$H$_4$Me-4, tetrahedral Cu | μ_2-OC$_6$H$_5$ | 2.066 (4), 2.082 (4) | xx |
| [(alkyne)Cu(μ_2-OAr)$_2$Cu(alkyne)] alkyne = 3,3,6-tetramethyl-1-thia-4-c-heptyne | μ_2-OC$_6$H$_5$ | 1.945 (2) | xxi |

[i] C. Lopes, M. Håkansson, and S. Jagner, *Inorg. Chem.*, **36**, 3232 (1997).

[ii] M. Håkansson, C. Lopes, and S. Jagner, *Organometallics*, **17**, 210 (1998).

[iii] C.A. Otter, D.A. Bardwell, S.M. Couchman, J.C. Jeffery, J.P. Maher, and M.D. Ward, *Polyhedron*, **17**, 211 (1998).

[iv] R.Y. Wong, K.J. Palmer, and Y. Tomimatsu, *Acta. Crystallogr. B*, **32**, 567 (1976).

[v] M. Basturkmen, D. Kisakurek, W.L. Driessen, S. Gorter, and J. Reedijk, *Inorg. Chim. Acta*, **271**, 19 (1988).

[vi] W. Bidell, V. Shklover, and H. Berke, *Inorg. Chem.*, **31**, 5561 (1992).

[vii] Y.-L. Lee, W.J. Burke, and R.B. VonDreele, *Acta. Crystallogr. C*, **43**, 209 (1987).

[viii] F. Connac, N. Habaddi, Y. Lucchese, M. Dartiguenave, L. Lamande, M. Sanchez, M. Simard, and A.L. Beauchamp, *Inorg. Chim. Acta*, **256**, 107 (1997).

[ix] B.M. Holligan, J.C. Jeffery, and M.D. Ward, *J. Chem. Soc., Dalton Trans.*, 3337 (1992).

[x] P. Fiaschi, C. Floriani, M. Pasquali, A. Chiesi-Villa, and C. Guastini, *J. Chem. Soc., Chem. Commun.*, 888 (1984).

[xi] P. Fiaschi, C. Floriani, M. Pasquali, A. Chiesi-Villa, and C. Guastini, *Inorg. Chem.*, **25**, 462 (1986).

[xii] M.F.C. Ladd and D.H.G. Perrins, *Acta Crystallogr. B*, **36**, 2260 (1980).

[xiii] J.R. Marengo-Rullan and R.D. Willett, *Acta Crystallog. C*, **42**, 1487 (1986).

[xiv] Whei-Lu Kwik, *J. Coord. Chem.*, **19**, 279 (1988).

[xv] C. Lopes, M. Håkansson, and S. Jagner, *New. J. Chem.*, **21**, 1113 (1997).

[xvi] F. Calderazzo, F. Marchetti, G. Dell'Amico, G. Pelizzi, and A. Colligiani, *J. Chem. Soc., Dalton Trans.*, 1419 (1980).

[xvii] C. Lopes, M. Håkansson, and S. Jagner, *Inorg. Chim. Acta*, **254**, 361 (1997).

[xviii] K. Osakada, T. Takizawa, M. Tanaka, and T. Yamamoto, *J. Organomet. Chem.*, **473**, 359 (1994).

[xix] W.J. Evans, R.E. Golden, and J.W. Ziller, *Acta Crystallog. C*, **50**, 1005 (1994).

[xx] M. Pasquali, P. Fiaschi, C. Floriani, and A.G. Manfredotti, *J. Chem. Soc., Chem. Commun.*, 197 (1983).

[xxi] F. Olbrich, U. Behrens, G. Groger, and E. Weiss, *J. Organomet. Chem.*, **448**, C10 (1993).

Table 6.43 Simple aryloxides of silver

Compound	Aryloxide	Bond length (Å) M–O	Bond angle (°) M–O–Ar	Ref.
[Ag(OAr)] polymeric structure, tbp silver, Ag–phenyl carbon interaction	$OC_6H_2Cl_3$-2,4,6	2.317 (3)–2.543 (3)		i
[(AgOAr)(PPh$_3$)$_2$] trigonal planar Ag with capping Cl	$OC_6H_2Cl_3$-2,4,6	2.235 (4)	125	ii

[i]G. Smith, E.J. O'Reilly, B.J. Reynolds, C.H.L. Kennard, and T.C.W. Mak, *J. Organomet. Chem.*, **331**, 275 (1987).
[ii]G. Wulfsberg, D. Jackson, W. Ilsley, S. Dou, A. Weiss, and J. Gagliardi, *Z. Naturforsch. A*, **47**, 75 (1992).

Table 6.44 Simple aryloxides of gold

Compound	Aryloxide	Bond length (Å) M–O	Bond angle (°) M–O–Ar	Ref.
[Au(OAr)(PPh$_3$)]	OC_6Cl_5	2.047 (5)	131	i
[Au(OAr)(PPh$_3$)]	OC_6H_4Br-2	2.025 (8)	126	ii
		2.019 (8)	129	
[Au(OAr)(PPh$_3$)](HOAr)	OC_6H_4Me-4	2.033 (9)	125	iii
cis-[Au(OAr)Me$_2$(PPh$_3$)] square planar Au	OC_6H_5	2.09 (1)	123	iv

[i]L.G. Kuz'mina and Yu.T. Struchkov, *Koord. Khim.*, **14**, 1262 (1988).
[ii]L.G. Kuz'mina, O.Yu Burtseva, N.V. Dvortsova, M.A. Porai-Koshits, and E.I. Smyslova, *Koord. Khim.*, **15**, 773 (1989).
[iii]L.G. Kuz'mina, Yu.T. Struchkov, and E.I. Smyslova, *Koord. Khim.*, **15**, 368 (1989).
[iv]T. Sone, M. Iwata, N. Kasuga, and S. Komiya, *Chem. Lett.*, 1949 (1991).

[Cu(μ_3-OC_6H_4-η^2-$CH_2CH=CH_2$-2)]$_4$ have been obtained from mesityl-copper and fully characterized.[513,514] The 2,6-diphenylphenoxide adopts a planar cluster with two-coordinate copper. In the 2-allylphenoxide, chelation of the olefinic group leads to four-coordinate copper. The cubic cluster contains μ_3-aryloxide oxygens with one very long Cu–O distance. In the presence of suitable ligands, more manageable, discrete Cu(I) aryloxides are formed. One common structural motif is the dimeric [L$_n$Cu(μ_2-OAr)$_2$CuL$_m$] species. Both three- and four-coordinate copper are observed, depending on the nature of L and the steric requirements of Ar. Examples include the carbonyls [(CO)Cu(μ_2-OAr)$_2$Cu(CO)] (ArO $= OC_6H_3Ph_2$-2,6,[513] $OC_6H_2Bu^t_3$-2,4,6, and $OC_6H_2Me_3$-2,4,6[515]) and isocyanide [(RNC)$_2$Cu(μ_2-OPh)$_2$Cu(CNR)$_2$][516] (R $= C_6H_4Me$-4). The monomeric, three-coordinate complex [Cu($OC_6H_2Bu^t_3$-2,4,6)(CNC$_6H_4$Me-4)$_2$] forms for the bulkier aryloxide.[511,517] The ligand PPh$_3$ can form a variety of structures ranging from cubic [Cu$_4$(μ_3-OPh)$_4$(PPh$_3$)$_4$][518] to dinuclear [(Ph$_3$P)$_2$Cu(μ_2-OPh)$_2$Cu(PPh$_3$)$_2$].[512] Interestingly two examples of mixed three-four-coordinate compounds [(Ph$_3$P)Cu(μ_2-OAr)$_2$Cu(PPh$_3$)$_2$] (OAr $=$ OPh,[518] OC_6H_4Me-4[519]) have been characterized.

The two-coordinate, linear anion $[Cu(OC_6H_3Me_2-2,6)_2]^-$ contains very short Cu–OAr distances of 1.806 (6) Å and 1.798 (8) Å. This terminal aryloxide distance elongates dramatically to 2.023 (5) Å in five-coordinate $[Cu(OAr)(dppe)]_2(\mu\text{-dppe})$.[511]

Besides their insertion chemistry (Section 5.2) Cu(I) aryloxides have been shown to form tetra(aryl) ortho-carbonates with CCl_4,[520] and $(ArO)_2CS$ from CS_2.[521]

The chemistry of silver(I) and gold(I) aryloxides is not as important as that of copper. The reaction of the gold(III) iodide cis-$[AuMe_2(I)(PPh_3)]$ with KOAr produces the corresponding Au(III) aryloxides which form hydrogen-bonded adducts with phenols.[522] The bis(aryloxide) $Na[Au(OPh)_2(C_6H_4NO_2-2)_2]$ is formed from the corresponding chloride and NaOPh.[523]

6.2.13   Group 12 Metal Aryloxides

Interest in the structures of monomeric aryloxides of zinc and cadmium was initially aroused by the observation that $[Zn(OC_6H_2Bu^t_3-2,4,6)_2(THF)_2]$ formed a distorted tetrahedral geometry[524,525] while $[Cd(OC_6H_3Bu^t_2-2,6)_2(THF)_2]$ formed a *trans*, square-planar geometry in the solid state. More recently this interest has been heightened by the observation that discrete zinc aryloxides containing bulky *ortho*-substituents act as homogeneous catalysts for the copolymerization of epoxides with carbon dioxide,[250] a reaction that can be achieved with other zinc catalysts[526–528] and offers potential for the utilization of CO_2. Mechanistic studies of the reactivity indicate that epoxide ring opening requires a vacant site at the metal centre whereas attack at the electrophilic carbon of CO_2 occurs directly from the nucleophilic oxygen bound to zinc. Hence aryloxides containing bulky *ortho*-substituents will initiate epoxide ring opening but not attack CO_2. The initially formed Zn–O–CHR–CH_2–OAr (from an α-olefin epoxide) now has an exposed nucleophilic Zn–O bond which can attack CO_2.[250] The details of the insertion chemistry of zinc and cadmium aryloxides is discussed in Section 5.2.2.

The bis(aryloxides) $[M(OAr)_2(L)_n]$ (M = Zn, Cd) can be readily synthesized by addition of phenols to $[M\{N(SiMe_3)_2\}_2]$ in the presence of ligands L. With the bulky 2,6-di-*tert*-butylphenoxide, phosphine ligands form trigonal planar species such as $[Zn(OC_6H_3Bu^t_2-2,6)_2(PR_3)]$ $(PR_3 = PMe_2Ph, PCy_3)$.[529] A large number of four-coordinate adducts $[M(OAr)_2(L)_2]$ $(L = OEt_2, THF, THT)$ have also been characterized. The zinc compounds are invariably distorted tetrahedral,[250] whereas both tetrahedral and *trans*, square planar geometries are possible for cadmium derivatives (Tables 6.45 and 6.46).[251,530] Although $[Zn(OC_6H_2Me_3-2,4,6)_2(py)_2]$ is four-coordinate, cadmium forms trigonal bipyramidal $[Cd(OC_6H_3Bu^t_2-2,6)_2(py)_3]$ with equatorial aryloxides. The addition of $KOC_6H_3Bu^t_2-2,6$ to THF solutions of $[M(OC_6H_3Bu^t_2-2,6)_2(THF)_2]$ produces the three-coordinate anions $[M(OC_6H_3Bu^t_2-2,6)_3]$ (M = Zn, Cd).[531] More complex anionic zinc species are formed with less bulky sodium aryloxides.[532]

A variety of donor ligand-free, dimeric zinc aryloxides of general formula $[(X)Zn(\mu_2\text{-}OAr)_2Zn(X)]$ (ArO = bulky aryloxide; X = alkyl[533,534] or amide[535]) are known.

6.2.14   Group 13 Metal Aryloxides

Aryloxides are very important ancillary ligands in the chemistry of the group 13 metals. Particular interest in the structure and bonding (see Section 4.2) of aluminium aryloxides

Table 6.45   Simple aryloxides of zinc

Compound	Aryloxide	Bond length (Å) M–O	Bond angle (°) M–O–Ar	Ref.
[Zn(μ_2-OAr)$_3$Zn(μ_2-OAr)$_3$Zn] central ZnO$_6$ and outer ZnO$_3$P$_3$ units	μ_2-OC$_6$H$_4$PPh$_2$-2	2.12 (3)–2.148 (3)		i
[Zn(OAr)$_2$(PMePh$_2$)] trigonal planar Zn	OC$_6$H$_3$Bu$^t_2$-2,6	1.844 (4) 1.864 (4)	136 124	ii
[Zn(OAr)$_2$(PCy$_3$)] trigonal planar Zn	OC$_6$H$_3$Bu$^t_2$-2,6	1.869 (4) 1.875 (4)	138 141	ii
[K(THF)$_6$][Zn(OAr)$_3$] trigonal planar Zn	OC$_6$H$_3$Bu$^t_2$-2,6	1.89 (2) 1.86 (2) 1.85 (2)	126 139 134	iii
[Zn(OAr)$_2$(THF)$_2$] distorted tetrahedral Zn	OC$_6$H$_2$Bu$^t_3$-2,4,6	1.889 (7) 1.885 (6)	126 133	iv, v
[Zn(OAr)$_2$(OEt$_2$)$_2$] distorted tetrahedral Zn	OC$_6$H$_3$Ph$_2$-2,6	1.886 (6)	127	vi
[Zn(OAr)$_2$(thf)$_2$] distorted tetrahedral Zn	OC$_6$H$_3$Ph$_2$-2,6	1.864 (4) 1.864 (4)	122 133	vi
[Zn(OAr)$_2$(thf)$_2$] distorted tetrahedral Zn	OC$_6$H$_3$Pr$^i_2$-2,6	1.85 (1) 1.86 (1)	127 127	vi
[Zn(OAr)$_2$(thf)$_2$] distorted tetrahedral Zn	OC$_6$H$_3$Bu$^t_2$-2,6	1.873 (5) 1.878 (5)	134 131	vi
[Zn(OAr)$_2$(PC)$_2$] propylene carbonate, distorted tetrahedral Zn	OC$_6$H$_3$Bu$^t_2$-2,6	1.85 (1)	155	vi
[Zn(OAr)$_2$(py)$_2$] distorted tetrahedral Zn	OC$_6$H$_2$Me$_3$-2,4,6	1.885 (4)	137	vi
[Na(μ_2-OAr)$_3$Zn(OH$_2$)] tetrahedral Zn, Na π-bound to *ortho*-phenyl rings	μ_2-OC$_6$H$_3$Ph$_2$-2,6	Zn–O–Na 1.954 (5), 2.301 (6)		vii
[(thf)$_2$Na(ArO)ClZn(μ_2-OAr)$_2$Zn(OAr)(thf)] tetrahedral Zn, Na π-bound to ArO nucleus and bound to Cl	OC$_6$H$_3$Pr$^i_2$-2,6 μ_2-OC$_6$H$_3$Pr$^i_2$-2,6	1.84 (1) 1.90 (1) 1.95 (1), 2.03 (1) 1.98 (1), 1.98 (1)	133 130	vii
[Zn(OAr)(OAr′)]	OC$_6$H$_3$Br$_2$-2,4-η^3-O,N,N- OC$_6$H$_2$Bu$^t_2$-py-pic	1.982 (11) 1.914 (9)	139 121	viii
[Et$_3$NH][Zn(OAr)Cl$_2$] tetrahedral Zn, cation hydrogen bonded to chelated O(Ar)	OC$_6$H$_3$Ph-4-CH$_2$NEt$_2$-2	1.947 (2)	119	ix
[PhZn(μ_3-OAr)Zn(pac)$_2$]$_2$ pac = 1,1,1-Me$_3$-acac, 4- and 6-coord. Zn	μ_3-OC$_6$H$_5$	2.04 (1), 2.17 (2), 2.20 (1)		x
[EtZn(μ_2-OAr)$_2$ZnEt] trigonal planar Zn	μ_2-OC$_6$H$_3$Bu$^t_2$-2,6	1.970 (1), 1.990 (1)		xi
[Me$_3$SiCH$_2$Zn(μ_2-OAr)$_2$ZnCH$_2$SiMe$_3$] trigonal planar Zn	μ_2-OC$_6$H$_2$Bu$^t_3$-2,4,6	1.958 (4), 2.021 (4)		xii

Table 6.45   (*Continued*)

Compound	Aryloxide	Bond length (Å) M–O	Bond angle (°) M–O–Ar	Ref.
[Me$_3$SiCH$_2$Zn(μ_2-OAr)$_2$ZnCH$_2$SiMe$_3$] trigonal planar Zn	μ_2-OC$_6$H$_3$Pr$^i_2$-2,6	1.98 (1), 1.93 (1) 1.94 (1), 1.94 (1)		xii
[{Me$_3$SiCH$_2$Zn(μ_2-OAr)$_2$}$_2$Zn] trigonal planar and tetrahedral Zn	μ_2-OC$_6$H$_3$Pr$^i_2$-2,6	1.950 (2)–1.985 (2)		xii
[{(Me$_3$Si)$_2$N}Zn(μ_2-OAr)$_2$Zn{N(SiMe$_3$)$_2$}] trigonal planar Zn	μ_2-OC$_6$H$_3$Pr$^i_2$-2,6	1.951 (5), 1.944 (5) 1.951 (5), 1.962 (4)		xii
[(pz)Zn(OAr)] pyrazolylborate, tetrahedral Zn	OC$_6$H$_4$NO$_2$-4	1.860 (2)	132	xiv
[Zn(OAr)$_2$(BPh$_2$)][BPh$_4$] could be considered an [(ArO)$_2$BPh$_2$] anion	OC$_6$H$_4$(2′-bipy)-2	2.125 (5), 2.136 (5)		xv

[i]D. Weiss, A. Schier, and H. Schmidbaur, *Z. Naturforsch. B*, **53**, 1307 (1998).

[ii]D.J. Darensbourg, M.S. Zimmer, P. Rainey, and D.L. Larkins, *Inorg. Chem.*, **37**, 2852 (1998).

[iii]D.J. Darensbourg, S.A. Niezgoda, J.D. Draper, and J.H. Reibenspies, *Inorg. Chem.*, **38**, 1356 (1999).

[iv]R.L. Geerts, J.C. Huffmann, and K.G. Caulton, *Inorg. Chem.*, **25**, 590 (1986).

[v]R.L. Geerts, J.C. Huffmann, and K.G. Caulton, *Inorg. Chem.*, **25**, 1803 (1986).

[vi]D.J. Darensbourg, M.W. Holtcamp, G.E. Struck, M.S. Zimmer, S.A. Niezgoda, P. Rainey, J.B. Robertson, J.D. Draper, and J.H. Reibenspies, *J. Am. Chem. Soc*, **121**, 107 (1999).

[vii]D.J. Darensbourg, J.C. Yoder, G.E. Struck, M.W. Holtcamp, J.D. Draper, and J H. Reibenspies, *Inorg. Chim. Acta*, **274**, 115 (1998).

[viii]A. Abufarag and H. Vahrenkamp, *Inorg. Chem.*, **34**, 3279 (1995).

[ix]F. Connac, N. Habaddi, Y. Lucchese, M. Dartiguenave, L. Lamande, M. Sanchez, M. Simard, and A.L. Beauchamp, *Inorg. Chim. Acta*, **256**, 107 (1997).

[x]J. Boersma, A.L. Spek, and J.G. Noltes, *J. Organomet. Chem.* **81**, 7 (1974).

[xi]M. Parvez, G.L. BergStresser, and H.G. Richey, *Acta. Crystallog. C.*, **48**, 641 (1992).

[xii]M.M. Olmstead, P.P. Power, and S.C. Shoner, *J. Am. Chem. Soc.*, **113**, 3379 (1991).

[xiii]H. Grutzmacher, M. Steiner, H. Pritzkow, L. Zsolnai, G. Huttner, and A. Sebald, *Chem. Ber.*, **125**, 2199 (1992).

[xiv]R. Walz, K. Weis, M. Ruf, and H. Vahrenkamp, *Chem. Ber.*, **130**, 975 (1997).

[xv]D.A. Bardwell, J.C. Jeffery, and M.D. Ward, *Inorg. Chim. Acta*, **241**, 125 (1996).

has been stimulated by the widespread use of these derivatives to carry out important organic transformations.[13,536] This reactivity is a direct consequence of the Lewis acidity of the aluminium aryloxides. Early work demonstrated that aluminium phenoxide carried out the *ortho*-alkylation of phenols by olefins.[537,538] More recently the use of bulky aryloxide ligands attached to aluminium to generate monomeric, "designer" Lewis acid catalysts has been pioneered by Yamamoto and co-workers.[13] Important examples include the compounds [Al(Me)(OC$_6$H$_2$Bu$^t_2$-2,6-X-4)$_2$] (X = Me, acronym "MAD"[539,540] and X = Br, "MABR"[541]), [Al(Me)(OC$_6$H$_3$Ph$_2$-2,6)$_2$] ("MAPH"[542]) and [Al(OC$_6$H$_3$Ph$_2$-2,6)$_3$] ("ATPH"[543]). These compounds are all easily obtained by adding

Table 6.46 Simple aryloxides of cadmium

Compound	Aryloxide	Bond length (Å) M–O	Bond angle (°) M–O–Ar	Reference
[K(THF)$_6$][Cd(OAr)$_3$] trigonal planar Cd	OC$_6$H$_3$Bu$^t_2$-2,6	2.078 (3), 2.081 (3), 2.057 (4)	154, 123, 159	i
trans-[Cd(OAr)$_2$(THF)$_2$] square planar Cd, room temp and *−80°C data	OC$_6$H$_3$Bu$^t_2$-2,6	2.058 (4), 2.068(3)*	127, 126*	ii, iii
trans-[Cd(OAr)$_2$(THT)$_2$] square planar Cd	OC$_6$H$_3$Bu$^t_2$-2,6	2.102 (6)	125	iv
[Cd(OAr)$_2$(THF)$_2$] distorted tetrahedral Cd	OC$_6$H$_3$Ph$_2$-2,6	2.073 (5), 2.074 (5)	130, 120	iv
[Cd(OAr)$_2$(THT)$_2$] distorted tetrahedral Cd	OC$_6$H$_3$Ph$_2$-2,6	2.12 (2)	133	iii
trans-[Cd(OAr)$_2$(pc)$_2$] pc = propylene carbonate, square planar Cd	OC$_6$H$_3$Bu$^t_2$-2,6	2.038 (8)	128	iii
[Cd(OAr)$_2$(py)$_3$] tbp, axial py	OC$_6$H$_3$Bu$^t_2$-2,6	2.193 (5), 2.190 (5)	165, 165	iii
[Cd(μ-OAr)$_2$Cd(HOAr)(DMF)$_2$][BPh$_4$]$_2$.2DMF edge-shared bis-octahedron, non-equiv. Cd	μ_2-OC$_6$H$_4$(2'-bipy)-2	2.226 (4), 2.310 (4), 2.249 (5), 2.238 (4)		v

[i] D.J. Darensbourg, S.A. Niezgoda, J.D. Draper, and J.H. Reibenspies, *Inorg. Chem.*, **38**, 1356 (1999).
[ii] C. Goel, M.Y. Chiang, and W.E. Buhro, *J. Am. Chem. Soc.*, **112**, 6724 (1990).
[iii] D.J. Darensbourg, S.A. Niezgoda, J.D. Draper, and J.H. Reibenspies, *J. Am. Chem. Soc.*, **120**, 4690 (1998).
[iv] D.J. Darensbourg, S.A. Niezgoda, J.H. Reibenspies, and J.D. Draper, *Inorg. Chem.*, **36**, 5686 (1997).
[v] D.A. Bardwell, J.C. Jeffery, and M.D. Ward, *J. Chem. Soc., Dalton Trans.*, 3071 (1995).

Table 6.47 Simple aryloxides of mercury

Compound	Aryloxide	Bond length (Å) M–O	Bond angle (°) M–O–Ar	Ref.
[Hg(OAr)(Ph)]	OC$_6$H$_3$Cl-2-Br-4	2.006 (8)	129	i
[(1,4,8,11-tetrathiacyclo-tetradecane)Hg(OAr)$_2$]	OC$_6$H$_2$(NO$_2$)$_3$-2,4,6	2.532 (9)	121	ii

[i]L.G. Kuz'mina, N.G. Bokii, Yu.T. Struchkov, D.N. Kravtsov, and L.S. Golovchenko, *Zh. Strukt. Khim.*, **14**, 508 (1973).
[ii]M. Herceg, B. Matkovic, D. Sevdic, D. Matkovic-Cologovic, and A. Nagl, *Croat. Chem. Acta*, **57**, 609 (1984).

the corresponding phenol to AlMe$_3$ and can be generated *in situ* for use. The steric bulk of the aryloxides plays an important role in discriminating between various oxygen donor ligands that can bind (and sometimes be activated) by the aluminium metal centre. As an example MAD demonstrated >99:1 selectivity for the ether MeOCH$_2$CH$_2$CH$_2$Ph over EtOCH$_2$CH$_2$CH$_2$Ph.[13] The reduction of a mixture of the ketones PhCOMe and PhCOBu^t with [Bu$_2^i$AlH] in the presence of MAD led mainly to the more hindered alcohol owing to selective binding (removal) of the smaller ketone.[544] MAD can also be used to discriminate between different esters[545] while ATPH can be used to selectively bind aldehydes.[546] In the last case the less hindered aldehyde is bound to aluminium and can be functionalized by Diels–Alder reactions. The unbound, more bulky aldehyde will react with alkylating agents such as LiBu^n.[547] Hence the smaller, bound substrate is electronically activated but sterically protected.

These bulky aluminium aryloxides will also promote a variety of carbon–carbon bond forming reactions with a high degree of regio and stereoselectivity. Examples include Michael addition to α,β-unsaturated ketones, Diels–Alder additions, and Claisen rearrangements. In the case of Diels–Alder reaction, ATPH promoted the *exo*-selective condensation of α,β-unsaturated ketones with dienophiles.[548] The Claisen rearrangement can be catalysed by ATPH and its more Lewis acidic 4-bromo derivative.[549] A series of chiral aluminium aryloxides were also synthesized and have been applied to asymmetric Claisen rearrangements,[13] aldol reactions,[550] and aldehyde alkylations.[8]

Compounds such as [Al(R)(OC$_6$H$_3$Bu$_2^t$-2,6)$_2$] will carry out the polymerization of methyl and ethyl methacrylate when activated with *t*-BuLi.[551] The nature of the alkyl group was found to affect the tacticity of the polymer. The tetraphenylporphinato aluminium phenoxide has been shown to be active for the ring opening polymerization of epoxides and lactones.[552,553] The reaction is accelerated by the presence of reagents such as MAD.[554] An adduct [Al(OC$_6$H$_2$Bu$_2^t$-2,6-Me-4)Me$_2$(methylmethacrylate)] relevant to this reactivity has been isolated and structurally characterized.[555]

The synthesis of group 13 metal aryloxides can proceed directly from the metal although use of halides and in particular alkyl precursors is much more common. The homoleptic tris(aryloxides) can be monomeric with bulky, *e.g.* trigonal [Al(OC$_6$H$_2$Bu$_2^t$-2,6-Me-4)$_3$],[556,557] or chelating, *e.g.* octahedral [M(OC$_6$H$_4$(oxazole)-2)$_3$] (M = Al, Ga, In).[558] With smaller aryloxides dimeric structures such as [(ArO)$_2$Al(μ_2-ArO)$_2$Al(OAr)$_2$] (OAr = OC$_6$H$_3$Me$_2$-2,6) are obtained.[559] The metal(I) aryloxides [M(μ_2-OC$_6$H$_2$(CF$_3$)$_3$-2,4,6)$_2$M] (M = In,[132] Tl[560]) contain two-coordinate metal centres.

A large number of mixed alkyl, aryloxides of all four metals have been obtained by reacting [MR$_3$]$_n$ with phenols. Monomeric bis(aryloxides) of aluminium, *e.g.*

$[Al(OC_6H_2Bu^t_2-2,6-Me-4)_2Me]^{561}$ and $[CpAl(OC_6H_2Bu^t_2-2,6-Me-4)_2]$,[562] have been characterized. In the case of mono(aryloxides) a ubiquitous stoichiometry is $[(R)_2M(\mu_2-OAr)_2M(R)_2]$ (M = Al, Ga, In, Tl) although the metal coordination number varies from four with simple aryloxides to five for chelating ligands (Tables 6.48–6.51). In the presence of donor ligands, tetrahedral adducts of aluminium aryloxides are very common, *e.g.* $[Al(OC_6H_3Pr^i_2-2,6)_3(py)]$,[563] $[Al(OC_6H_2Bu^t_2-2,6-Me-4)_2(H)(OEt_2)]$,[557] $[Al(OC_6H_2Bu^t_2-2,6-Me-4)Me_2(NH_3)]$,[564] and $[Al(OC_6H_2Bu^t_3-2,4,6)Cl_2(OEt_2)]$.[565] With smaller ligands, five-coordinate species are possible, *e.g.* trigonal bipyramidal $[Al(OC_6H_3Pr^i_2-2,6)_2(H)(THF)_2]$.[566] Particularly important given their intermediacy in a variety of organic reactions (see above) are adducts formed with ketones and related carbonyl compounds.[567]

6.2.15   Group 14 Metal Aryloxides

The "germylene" and "stannylene" aryloxides $[M(OC_6H_2Bu^t_2-2,6-Me-4)_2]$ (M = Ge, Sn) can be obtained by treatment of $[M\{N(SiMe_3)_2\}_2]$ with phenol.[568] An intermediate $[Sn(OC_6H_2Bu^t_2-2,6-Me-4)\{N(SiMe_3)_2\}]$ has been isolated and structurally characterized.[569] All of these molecules are V-shaped with O–M–O angles of less than 100°. They will act as two-electron donors to metal fragments, *e.g.* to $[Fe(CO)_4]$.[570] Addition of $N_3C(O)OAr$ to $[Ge\{N(SiMe_3)_2\}_2]$ was found to lead to phenoxides such as $[Ge(OPh)(CNO)\{N(SiMe_3)(C_6H_2Me_3)\}_2]$.[571] Tin bis(aryloxides) have also been reported to be produced by addition of phenols to $[(C_5H_4Me)_2Sn]$[572] and $[Sn(acac)_2]$.[573]

Aryloxides of Ge(IV) and Sn(IV) can be obtained by reacting the tetrahalides with LiOAr or reacting $[M(NMe_2)_4]$ with phenols.[105] The bonding in these derivatives (Section 4.2) as well as the sometimes facile activation of arene CH bonds at Sn(IV) metal centres (Section 5.1) has been discussed above. "Hypervalent" anions such as $[Me_3Sn(OC_6H_3Me_2-2,6)_2]^-$ have been characterized and their bonding analysed.[574] Although $[SnR_4]$ compounds do not react with phenols under normal conditions, the "hypervalently activated" $[Me_2N(CH_2)_3SnPh_3]$ will undergo stepwise elimination of benzene and formation of corresponding mono and bis(phenoxides) with phenol.[575]

6.2.16   Group 15 Metal Aryloxides

The synthesis of antimony(III)[576,577] and bismuth(III) aryloxides can be achieved by reacting the trichlorides with either phenols or group 1 metal aryloxides or by treating trialkyls with phenolic reagents, typically containing electron-withdrawing substituents. In one case using $[NaOC_6H_2(CF_3)_3-2,4,6]$ the reaction failed owing to C–F bond activation by bismuth.[578] The homoleptic $[Bi(OC_6H_3Me_2-2,6)_3]$[579] is obtained *via* the chloride and is a distorted pyramidal monomer (Table 6.56). Dimeric intermediates such as $[Bi_2(\mu-OC_6H_3Me_2-2,6)_2Cl_4(THF)_2]$ have been isolated.[580] The pentafluorophenoxide (obtained from $[BiPh_3]$) is dimeric, with the electrophilic metal centre coordinating molecules such as toluene and THF.[581,582] Further reaction with $NaOC_6F_5$ leads to polymeric mixed-metal aryloxides.[583] The compound $[BiEt_3]$ reacts slowly with HOPh and HOC_6F_5 to form a mono-aryloxide, which is polymeric in the solid state.[584] Aryloxides of antimony(V) and bismuth(V) can be obtained from $[MPh_5]$ substrates (Tables 6.55 and 6.56). Alternatively the dihalides $[X_2BiPh_3]$ (X = Cl, Br) can be substituted with NaOAr reagents.[585] Isolated species such as $[Bi(OC_6F_5)(Br)Ph_3]$ (which undergoes

Table 6.48 Aluminium aryloxides

Compound	Aryloxide	Bond length (Å) M–O	Bond angle (°) M–O–Ar	Ref.
	Homoleptic aryloxides and adducts			
[Al(OAr)$_3$] trigonal Al	OC$_6$H$_2$Bu$^t_2$-2,6-Me-4	1.657 (6) 1.640 (5) 1.647 (7)	175 178 179	i, ii
[(ArO)$_2$Al(μ_2-ArO)$_2$Al(OAr)$_2$] tetrahedral Al	OC$_6$H$_3$Me$_2$-2,6	1.686 (2) 1.685 (2)	142 150	iii
	μ_2-OC$_6$H$_3$Me$_2$-2,6	1.835 (2), 1.833 (2)		
[Al(OAr)$_3$] 6-coord. Al, mer-O$_3$	OC$_6$H$_4$(oxazole)-2	1.877 (2) 1.844 (2) 1.838 (2)	133 136 133	iv
[Al(OAr)$_3$] 6-coord. Al, mer-O$_3$	OC$_6$H$_3$(oxazole)-2-Me-6	1.849 (2) 1.847 (2) 1.847 (2)	127 134 135	iv
[Al(OAr)$_3$] 6-coord. Al, mer-O$_3$	OC$_6$H$_3$(oxazole)-2-Br-4	1.839 (2) 1.876 (2) 1.839 (2)	136 123 132	v
[Al(OAr)$_3$] tbp with axial N, one O and two O,N-bound ligands	OC$_6$H$_2$(CH$_2$NMe$_2$)$_2$-2,6-Me-4	1.716 (3) 1.754 (3) 1.764 (3)	169 135 133	vi
[Al(OAr)$_3$(py)] tetrahedral Al	OC$_6$H$_3$Pr$^i_2$-2,6	1.702 (3) 1.711 (3) 1.702 (4)	143 141 144	vii

[i] M.D. Healy and A.R. Barron, *Angew. Chem., Int. Ed. Engl.*, **31**, 921 (1992).
[ii] M.D. Healy, M.R. Mason, P.W. Gravelle, S.G. Bott, and A.R. Barron, *J. Chem. Soc., Dalton Trans.*, 441 (1993).
[iii] R. Benn, E. Janssen, H. Lehmkuhl, A. Rufinska, K. Angermund, P. Betz, R. Goddard, and C. Kruger, *J. Organomet. Chem.*, **411**, 37 (1991).
[iv] H.R. Hoveyda, V. Karunaratne, S.J. Rettig, and C. Orvig, *Inorg. Chem.*, **31**, 5408 (1992).
[v] H.R. Hoveyda, C. Orvig, and S.J. Rettig, *Acta Crystallogr. C*, **50**, 1906 (1994).
[vi] M.P. Hogerheide, M. Wesseling, J.T.B.H. Jastrzebski, J. Boersma, H. Kooijman, A.L. Spek, and G. van Koten, *Organometallics*, **14**, 4483 (1995).
[vii] M.D. Healy, J.W. Ziller, and A.R. Barron, *J. Am. Chem. Soc.*, **112**, 2949 (1990).

(continued overleaf)

Table 6.48 *(Continued)*

Compound	Aryloxide	Bond length (Å) M–O	Bond angle (°) M–O–Ar	Ref.
[Al(OAr)$_3$(4-t-butylcyclohexanone)]	OC$_6$H$_2$Bu$^t_2$-2,6-Me-4	1.717 (5) 1.716 (5) 1.691 (5)	147 137 164	ii
[Na(OH$_2$)]$_2$[Al(tri-OAr)]$_4$[ClO$_4$]$_2$ 6-coord. Al, three aryloxides bridging to Na in AlNaAl unit	MeC(CH$_2$NHCH$_2$-2-OC$_6$H$_3$Br-4)$_3$	Al–O–Na 1.821 (5) 1.827 (4) 1.827 (5) 1.836 (5), 2.336 (5) 1.843 (4), 2.585 (5) 1.849 (5), 2.334 (5)	130 125 130	viii
[Al(di-OAr)(OAr)] 5-coord. Al, two molecules	OC$_6$H$_2$Me$_3$-2,4,6 salen	1.737 (3) 1.741 (2) 1.798 (2) 1.787 (3) 1.797 (2) 1.791 (3)	135 140 134 133 133 132	ix
[Al(tri-OAr)(py)] tbp Al	N(C$_6$H$_4$O-2)$_3$	1.762 (5) 1.768 (5) 1.764 (5)	115 117 116	x
[Hquinuclidine][Al(tri-OAr)(OH)] tbp Al	N(C$_6$H$_4$O-2)$_3$	1.778 (2) 1.772 (2) 1.796 (2)	120 120 121	x
[Al$_2$(tri-OAr)$_2$] each ligand has one bridging OAr and two terminal, N-coordinated, 5-coord. Al	N(C$_6$H$_4$O-2)$_3$	1.74 (2) 1.75 (1)	115 115	x
[Al(H-tri-OAr)][ClO$_4$] octahedral Al, two O and N bonded to proton	N[(CH$_2$)$_2$NHCH$_2$-2-OC$_6$H$_3$Cl-4]$_3$	1.852 (4), 1.849 (4) 1.819 (3) 1.838 (3) 1.824 (3)	132 134 132	xi

Compound	OAr	Al–O (Å)	Angle (°)	Ref.
[(ArO)₂Al(μ-OAr)₂Mg(THF)₂(μ-OAr)₂Al(OAr)₂] octahedral Mg attached to two tetrahedral Al centres	OC₆H₅	1.71 (2) 1.68 (2) 1.68 (2) 1.74 (2)	144 143 153 141	xii
	μ₂-OC₆H₅	Mg–O–Al 2.12 (av.), 1.76 (av.)		

Mixed aryloxy halides, alkoxides, amides, etc.

Compound	OAr	Al–O (Å)	Angle (°)	Ref.
[Al(OAr)Cl₂(OEt₂)] tetrahedral Al	OC₆H₂Bu$^t_3$-2,4,6	1.699 (2)	153	xiii
[Al(OAr)Cl₂(OEt₂)] tetrahedral Al	OC₆H₃Bu$^t_2$-2,6	1.700 (2)	155	xiv
[Al(OAr)₂(H)(H₂NBu^t)]	OC₆H₂Bu$^t_2$-2,6-Me-4	1.735 (5) 1.710 (5)	169 164	ii
[Al(OAr)₂(H)(OEt₂)]	OC₆H₂Bu$^t_2$-2,6-Me-4	1.704 (3) 1.711 (3)	165 163	ii
[Al(OAr)₂(H)(THF)₂] tbp with axial THF	OC₆H₃Pr$^i_2$-2,6	1.752 (2)	138	xv
[Al(H)₂(OAr)(NMe₃)] [Al(μ-H)(H)(OAr)(NMe₃)]₂ monomer and dimer co-crystallize	OC₆H₂Bu$^t_2$-2,6-Me-4	1.762 (4) 1.774 (3)	147 134	ii
[Al(OAr)₂(PhNNNPh)] tetrahedral Al	OC₆H₂Bu$^t_2$-2,6-Me-4	1.689 (3)	152	xvi

viii Shuang Liu, E. Wong, V. Karunaratne, S.J. Rettig, and C. Orvig, *Inorg. Chem.*, **32**, 1756 (1993).

ix P.L. Gurian, L.K. Cheatham, J.W. Ziller, and A.R. Barron, *J. Chem. Soc., Dalton Trans.*, 1449 (1991).

x E. Muller and H.-B. Burgi, *Helv. Chim. Acta*, **70**, 520 (1987).

xi Shuang Liu, S.J. Rettig, and C. Orvig, *Inorg. Chem.*, **31**, 5400 (1992).

xii J.A. Meese-Marktscheffel, R.E. Cramer, and J.W. Gilje, *Polyhedron*, **13**, 1045 (1994).

xiii S. Schulz, H.W. Roesky, M. Noltemeyer, H.-G. Schmidt, *J. Chem. Soc., Dalton Trans.*, 177 (1995).

xiv M.D. Healy, J.W. Ziller, and A.R. Barron, *Organometallics*, **11**, 3041 (1992).

xv J.P. Campbell and W.L. Gladfelter, *Inorg. Chem.*, **36**, 4094 (1997).

xvi J.T. Leman, J. Braddock-Wilking, A.J. Coolong, and A.R. Barron, *Inorg. Chem.*, **32**, 4324 (1993).

(continued overleaf)

Table 6.48 (*Continued*)

Compound	Aryloxide	Bond length (Å) M–O	Bond angle (°) M–O–Ar	Ref.
[Al(OAr)$_3${O(CH$_2$)$_4$NMe$_2$Et}] zwitterionic, tetrahedral Al	OC$_6$H$_3$Pr$^i_2$-2,6	1.733 (3) 1.747 (3) 1.748 (4)	149 137 139	xv
[Al(OAr)(NC$_5$H$_6$Me$_4$)$_2$] trigonal Al, tetramethylpiperidino ligands	OC$_6$H$_3$Pr$^i_2$-2,6	1.696 (2)	158	xvii
[{(Me$_3$Si)$_3$Si}Al(OAr)$_3$][H$_2$NC$_5$H$_6$Me$_4$] two phenoxides hydrogen bonded to tetramethylpiperidinyl cation	OC$_6$H$_5$	1.736 (3), 1.784 (3) 1.775 (3),		xvii
[2,6Me$_2$-pyH][Al(OAr)$_2$Cl$_2$] tetrahedral Al	OC$_6$H$_3$Bu$^t_2$-2,6	1.705 (2) 1.703 (2)	170 166	xiv
[(ArO)$_2$Al(μ_2-OH)$_2$Al(OAr)$_2$] 6-coord. Al	OC$_6$H$_4$(benzoxazole)-2	1.836 (4) 1.849 (4) 1.845 (4) 1.840 (4)	129 130 127 129	xviii
[(ArO)$_2$Al(μ_2-OEt)$_2$Al(OAr)$_2$]	OC$_6$H$_2$Bu$^t_3$-2,4,6	1.713 (2) 1.723 (2)	142 143	xiii
Organometallic aryloxides				
[CpAl(OAr)$_2$] η^5-bound Cp	OC$_6$H$_2$Bu$^t_2$-2,6-Me-4	1.736 (2)	135	xix
[Al(OAr)$_2$Me] trigonal planar Al	OC$_6$H$_2$Bu$^t_2$-2,6-Me-4	1.685 (2) 1.687 (2)	147 141	xx

Compound	OAr	Al–O (Å)	Angle	Ref.
[Al(OAr)₂(i-Bu)] trigonal planar Al	OC₆H₃Bu$^t_2$-2,6	1.682 (1)	157	xxi
[Al(OAr)(t-Bu)₂]	OC₆H₂Bu$^t_2$-2,6-Me-4	1.702 (1)	135	xxii
[Al(di-OAr)Me] salen	salen	1.710 (2)	129	xvi
5-coord. Al		1.794 (7)	134	
		1.823 (7)	134	
[Al(OAr)₂Me] tbp with *trans* O(Ar), methylsalicylate	OC₆H₄C(OMe)O-2	1.968 (1)		xxiii
[Al(OAr)(OAr')Me] tetrahedral Al, chelating keto group	OC₆H₂Bu$^t_2$-2,6-Me-4	1.706 (4)	158	xxiv
K[Al(OAr)₂Me₂] tetrahedral anion	OC₆H₂Bu^t-2-Me-4-(CMeO)-6	1.765 (5)	130	xxv
	OC₆H₅	1.800 (3)	125	
[NMe₄][Al(OAr)Cl₂Me] tetrahedral Al	OC₆H₂Bu$^t_2$-2,6-Me-4	1.713 (4)	164	xxvi
[Me₂Al(μ₂-OAr)₂AlMe₂] tetrahedral Al	μ₂-OC₆F₅	1.911 (5), 1.911 (5)		xxvii
		1.881 (5), 1.881 (5)		
[(Bu^t)₂Al(μ₂-OAr)₂Al(Bu^t)₂]	μ₂-OC₆H₅	1.875 (2)		xxviii

[xvii] K. Knabel, I. Krossing, H. Noth, H. Schwenk-Kircher, M. Schmidt-Amelunxen, and T. Seifert, *Eur. J. Inorg. Chem.*, 1095 (1998).

[xviii] H.R. Hoveyda, S.J. Rettig, and C. Orvig, *Inorg. Chem.*, **32**, 4909 (1993).

[xix] J.D. Fisher, P.J. Shapiro, P.M.H. Budzelaar, and R.J. Staples, *Inorg. Chem.*, **37**, 1295 (1998).

[xx] A.P. Shreve, R. Mulhaupt, W. Fultz, J.C. Calabrese, W. Robbins, and S.D. Ittel, *Organometallics*, **7**, 409 (1988).

[xxi] R. Benn, E. Janssen, H. Lehmkuhl, A. Rufinska, K. Angermund, P. Betz, R. Goddard, and C. Kruger, *J. Organomet. Chem.*, **411**, 37 (1991).

[xxii] M.A. Petrie, M.M. Olmstead, and P.P. Power, *J. Am. Chem. Soc.*, **113**, 8704 (1991).

[xxiii] J. Lewinski, J. Zachara, B. Mank, and S. Pasynkiewicz, *J. Organomet. Chem.*, **454**, 5 (1993).

[xxiv] M.B. Power, S.G. Bott, E.J. Bishop, K.D. Tierce, J.L. Atwood, and A.R. Barron, *J. Chem. Soc., Dalton Trans.*, 241 (1991).

[xxv] M.J. Zaworotko, C.R. Kerr, and J.L. Atwood, *Organometallics*, **4**, 238 (1985).

[xxvi] M.D. Healy, D.A. Wierda, and A.R. Barron, *Organometallics*, **7**, 2543 (1988).

[xxvii] D.G. Hendershot, R. Kumar, M. Barber, and J.P. Oliver, *Organometallics*, **10**, 1917 (1991).

[xxviii] C.L. Aitken and A.R. Barron, *J. Chem. Cryst.*, **26**, 293 (1996).

[xxix] R. Kumar, M.L. Sierra, V.S.J. de Mel, and J.P. Oliver, *Organometallics*, **9**, 484 (1990).

[xxx] D.G. Hendershot, M. Barber, R. Kumar, and J.P. Oliver, *Organometallics*, **10**, 3302 (1991).

(continued overleaf)

Table 6.48 (*Continued*)

Compound	Aryloxide	Bond length (Å) M–O	Bond angle (°) M–O–Ar	Ref.
[Me₂Al(μ₂-OAr)₂AlMe₂] tetrahedral Al	μ₂-OC₆H₃(CH₂CH=CH₂)-2-Me-6	1.859 (3), 1.862 (3)		xxix
[Me₂Al(μ₂-OAr)₂AlMe₂] 5-coord. Al, two independent molecules	μ₂-OC₆H₄SMe-2	1.865 (6), 1.946 (6); 1.979 (6), 1.365 (6); 1.856 (6), 1.964 (6)		xxx
[(*i*-Bu)₂Al(μ₂-OAr)₂Al(*i*-Bu)₂] 5-coord. Al	μ₂-OC₆H₄SMe-2	1.870 (2), 1.966 (2)		xxx
[(Et)₂Al(μ₂-OAr)₂Al(Et)₂] 5-coord. Al	μ₂-OC₆H₄OMe-2	1.859 (3), 1.952 (3)		xxx
[(Buⁱ)₂Al(μ₂-OAr)₂Al(Buⁱ)₂] 5-coord. Al	μ₂-OC₆H₄OMe-2	1.861 (1), 1.950 (1)		xxx
[(*i*-Bu)₂Al(μ₂-OAr)₂Al(*i*-Bu)₂] tetrahedral Al	μ₂-OC₆H₃Me₂-2,6	1.871 (1), 1.874 (2)		xxxi
[Me₂Al(μ₂-OAr)₂AlMe₂] two molecules, five-coord. Al due to chelating aryloxide	μ₂-OC₆H₃(CMeO)-2-Cl-4	1.852 (2), 1.964 (2); 1.854 (2), 1.975 (2)		xxxii
[Me₂Al(di-OAr)AlMe₂] two tetrahedral Al linked by keto bridge	O=C(2-OC₆H₂Buᵗ-6-Me-4)₂	1.78 (1)	122	xxxiii
[(Buⁱ)₂Al(μ₂-di-OAr)Al(Buⁱ)₂] 2,2'-(1,2-ethynediyl)bis(phenoxide), both O bridge two Al centres	(OC₆H₂Buᵗ₂-2-Me-4)₂(CC)	1.915 (2), 1.912 (2); 1.917 (2), 1.909 (2)		xxxiv
[K(dibenzo-18-crown-6)][Me₃Al(μ₂-OAr)AlMe₃]	μ₂-OC₆H₅	1.901 (9), 1.88 (1)		xxv
[Me(ArO)Al(μ₂-Cl)₂Al(OAr)(Me)]	OC₆H₃Buᵗ₂-2,6	1.679 (2)	152	xxxv

642

Compound	OAr	Al–O (Å)	Angle	Ref.
[Al(OAr)Cl(Me)(NH$_2$Bu^t)]	OC$_6$H$_3$Bu^t-2,6	1.722 (5)	154	xxxv
[Al(OAr)Me$_2$(NH$_2$Bu^t)]	OC$_6$H$_3$Bu^t-2,6	1.750 (4)	154	xxxv
[Al(OAr)Me$_2$(NH$_3$)] tetrahedral Al	OC$_6$H$_2$Bu^t-2,6-Me-4	1.743 (4)	150	xxxvi
[Al(OAr)Me$_2$(PMe$_3$)] tetrahedral Al	OC$_6$H$_2$Bu$^t_2$-2,6-Me-4	1.736 (5)	165	xxxvii
[Al(OAr)Me$_2$(py)] tetrahedral Al	OC$_6$H$_2$Bu$^t_2$-2,6-Me-4	1.740 (4)	156	vii
[Al(OAr)$_2$Me(pyMe$_2$-3,5)] tetrahedral Al	OC$_6$H$_2$Me$_3$-2,4,6	1.714 (6) / 1.722 (7)	146 / 141	vii
[Al(OAr)Et$_2$(MeO$_2$CC$_6$H$_4$Me-4)] tetrahedral Al	OC$_6$H$_2$Bu$^t_2$-2,6-Me-4	1.749 (5)	146	xx
[Al(OAr)Me$_2$(OCPh$_2$)]	OC$_6$H$_2$Bu$^t_2$-2,6-Me-4	1.731 (8)	158	xxxviii
[Al(OAr)$_2$Me(neopentanal)]	OC$_6$H$_2$Bu$^t_2$-2,6-Me-4	1.729 (3) / 1.726 (3)	141 / 131	xxxviii
[Al(OAr)$_2$Me(methylbenzoate)]	OC$_6$H$_2$Bu$^t_2$-2,6-Me-4	1.721 (8) / 1.714 (9)	156 / 153	xxxviii
[Al(OAr)$_2$Me(OCPh$_2$)] tetrahedral Al	OC$_6$H$_2$Bu$^t_2$-2,6-Me-4	1.733 (5) / 1.721 (6)	143 / 161	xxxix
[Al(OAr)Me$_2$(pyMe$_2$-2,6)]	OC$_6$H$_2$Bu$^t_2$-2,6-Me-4	1.744 (2)	158	xl

[xxxi] R. Benn, E. Janssen, H. Lehmkuhl, A. Rufinska, K. Angermund, P. Betz, R. Goddard, and C. Kruger, *J. Organomet. Chem.*, **411**, 37 (1991).

[xxxii] J. Lewinski, J. Zachara, and I. Justyniak, *Organometallics*, **16**, 4597 (1997).

[xxxiii] V. Sharma, M. Simard, and J.D. Wuest, *J. Am. Chem. Soc.*, **114**, 7931 (1992).

[xxxiv] O. Saied, M. Simard, and J.D. Wuest, *Organometallics*, **15**, 2345 (1996).

[xxxv] J.A. Jegier and D.A. Atwood, *Bull. Soc. Chim. France*, **133**, 965 (1996).

[xxxvi] M.D. Healy, J.T. Leman, and A.R. Barron, *J. Am. Chem. Soc.*, **113**, 2776 (1991).

[xxxvii] M.D. Healy, D.A. Wierda, and A.R. Barron, *Organometallics*, **7**, 2543 (1988).

[xxxviii] M.B. Power, S.G. Bott, D.L. Clark, J.L. Atwood, and A.R. Barron, *Organometallics*, **9**, 3086 (1990).

[xxxix] M.B. Power, S.G. Bott, J.L. Atwood, and A.R. Barron, *J. Am. Chem. Soc.*, **112**, 3446 (1990).

[xl] M.D. Healy, J.W. Ziller, and A.R. Barron, *Organometallics*, **10**, 597 (1991).

[xli] M.B. Power, A.W. Apblett, S.G. Bott, J.L. Atwood, and A.R. Barron, *Organometallics*, **9**, 2529 (1990).

(continued overleaf)

Table 6.48 (*Continued*)

Compound	Aryloxide	Bond length (Å) M–O	Bond angle (°) M–O–Ar	Ref.
[Al(OAr)Et₂(H₂NBuᵗ)]	OC₆H₂Buᵗ₂-2,6-Me-4	1.748 (3)	151	xl
[Al(OAr)Et₂(Opy)]	OC₆H₂Buᵗ₂-2,6-Me-4	1.754 (1)	144	xl
[Al(OAr)₂Me(Opy)]	OC₆H₂Buᵗ₂-2,6-Me-4	1.717 (3), 1.739 (4)	168, 175	xl
[Al(OAr)Me₂{1-azabicyclo{222}octane}]	OC₆F₅	1.787 (1)	123	xl
[(Et₂O)Li(μ₂-OAr)Al(OAr)Me(OCMePh₂)]	OC₆H₂Buᵗ₂-2,6-Me-4	1.736 (3)	166	xxxix
4-coord. Al	μ₂-OC₆H₂Buᵗ₂-2,6-Me-4	Li–O–Al 1.910 (7), 1.805 (2), 1.922 (7), 1.808 (2)		
[Al(OAr)Et{O=C(Buᵗ)CH₂CMe(Buᵗ)O}] tetrahedral Al, one alkoxy and one keto bond	OC₆H₂Buᵗ₂-2,6-Me-4	1.732 (5)	135	xlii
[Al(OAr)Me₂(methylmethacrylate)] tetrahedral Al	OC₆H₂Buᵗ₂-2,6-Me-4	1.738 (8)	151	xlii
[AlMe₂(OAr)-AlMe₃] OAr chelated to one Al, an AlMe₃ group bound to the other N	OC₆H₂(CH₂NMe₂)₂-2,6-Me-4	1.750 (3)	131	vi
[AlMe₂(OAr)-(AlMe₃)₂] OAr chelated to one Al, AlMe₃ groups also bound to O and the other N	μ₂-OC₆H₂(CH₂NMe₂)₂-2,6-Me-4	1.855 (2), 1.989 (2)		vi

xliiiM. Akakura, H. Yamamoto, S.G. Bott, and A.R. Barron, *Polyhedron*, **16**, 4389 (1997).

Table 6.49 Gallium aryloxides

Compound	Aryloxide	Bond length (Å) M–O	Bond angle (°) M–O–Ar	Reference
[Ga(OAr)₃] 6-coord. Ga, *mer*-O₃	OC₆H₄(oxazole)-2	1.934 (2) 1.981 (2) 1.929 (2)	132 130 131	i
[Ga(OAr)₃] 6-coord. Ga, *mer*-O₃	OC₆H₄(oxazole)-2-(CH₂CH=CH₂)-6	1.920 (2) 1.938 (2) 1.923 (2)	130 131 130	i
[Ga(OAr)(NC₅H₆Me₄)₂] trigonal Ga, tetramethylpiperidino ligands	OC₆H₅	1.822 (3)	129	ii
[(Buᵗ)₂Ga(μ₂-OAr)₂Ga(Buᵗ)₂]	μ₂-OC₆H₅	2.035 (1)	129	iii
[Et₂Ga(μ₂-OAr)₂GaEt₂]	μ₂-OC₆H₃Ph₂-2,6	1.998 (2), 1.992 (2) 2.034 (2), 1.998 (2)		iv
[Me₂Ga(μ₂-OAr)₂GaMe₂] 5-coord. Ga	μ₂-OC₆H₄OMe-2	1.957 (4), 2.046 (4)		v
[Ga(tri-OAr)] octahedral Ga	MeC(CH₂NHCH₂-2-OC₆H₄)₃	1.916 (2) 1.922 (2) 1.923 (2)	120 119 120	vi
[Mg(OH₂)₆][Ga(di-OAr)]₂	ethylene-bis(*o*-hydroxyphenyl)glycine)	1.899 (1) 1.886 (2)	126 125	vii
[Ga(OAr)₂(OAc)] 6-coord. Ga	OC₆H₄(benzoxazole)-2	1.873 (2)	127	viii
[Ga(H-tri-OAr)][ClO₄] octahedral Ga, one HOAr bound	*c*-(CH₂CH₂N(-2-OC₆H₂Me₂-4,6)₃	1.937 (3) 1.900 (3)	130 131	ix

(continued overleaf)

645

Table 6.49 *(Continued)*

Compound	Aryloxide	Bond length (Å) M–O	Bond angle (°) M–O–Ar	Reference
[Ga(H-tri-OAr)]Cl octahedral Ga, one OAr protonated, not bound	N[(CH₂)₂NHCH₂-2-OC₆H₃Br-4]₃	1.897 (4) 1.904 (4) 1.829 (4)	131	x
[{(Me₃Si)₃Si}Ga(OAr)₃][H₂NC₅H₆Me₄] two phenoxides hydrogen bonded to tetramethylpiperidinyl cation	OC₆H₅	1.853 (4), 1.892 (4)	124	xi

[i] H.R. Hoveyda, V. Karunaratne, S.J. Rettig, and C. Orvig, *Inorg. Chem.*, **31**, 5408 (1992).
[ii] G. Linti, R. Frey, and K. Polborn, *Chem. Ber.*, **127**, 1387 (1994).
[iii] W.M. Cleaver, A.R. Barron, A.R. McGufey, and S.G. Bott, *Polyhedron*, **13**, 2831 (1994).
[iv] M. Webster, D.J. Browning, and J.M. Corker, *Acta Crystallogr. C*, **52**, 2439 (1996).
[v] D.G. Hendershot, M. Barber, R. Kumar, and J.P. Oliver, *Organometallics*, **10**, 3302 (1991).
[vi] Shuang Liu, E. Wong, V. Karunaratne, S.J. Rettig, and C. Orvig, *Inorg. Chem.*, **32**, 1756 (1993).
[vii] P.E. Riley, V.L. Pecoraro, C.J. Carrano and K.N. Raymond, *Inorg. Chem.*, **22**, 3096 (1983).
[viii] H.R. Hoveyda, S.J. Rettig, and C. Orvig, *Inorg. Chem.*, **32**, 4909 (1993).
[ix] D.A. Moore, P.E. Fanwick, and M.J. Welch, *Inorg. Chem.*, **28**, 1504 (1989).
[x] Shuang Liu, S.J. Rettig, and C. Orvig, *Inorg. Chem.*, **31**, 5400 (1992).
[xi] G. Linti, R. Frey, W. Kostler, and H. Urban, *Chem. Ber.*, **129**, 561 (1996).

Table 6.50 Indium aryloxides

Compound	Aryloxide	Bond length (Å) M–O	Bond angle (°) M–O–Ar	Reference
[In(μ_2-OAr)$_2$In] 2-coord. In	μ_2-OC$_6$H$_2$(CF$_3$)$_3$-2,4,6	2.303 (4), 2.332 (4) 2.332 (4), 2.315 (4)		i
[Me$_2$In(μ_2-OAr)$_2$InMe$_2$] 5-coord. In	μ_2-OC$_6$H$_4$(CHO)-2	2.188 (3), 2.383 (3)		ii
[In(tri-OAr)] octahedral In	MeC(CH$_2$NHCH$_2$-2-OC$_6$H$_4$)$_3$	2.107 (3) 2.096 (3) 2.096 (3)	120 114 121	iii
[(NO$_3$)(EtOH)(H$_2$O)Na][In(tri-OAr)] two OAr bridging Na and In	N{(CH$_2$)$_2$NHCH$_2$-2-OC$_6$H$_4$}$_3$	2.174 (3)	122	iv
		Na–O–In 2.434 (5), 2.177 (3) 2.449 (5), 2.155 (4)		
[In(OAr)$_3$] 6-coord. In, mer-O$_3$	OC$_6$H$_4$(oxazole)-2	2.149 (2) 2.103 (2) 2.104 (2)	131 132 132	v

[i] M. Scholz, M. Noltemeyer, and H.W. Roesky, *Angew. Chem., Int. Ed. Engl.*, **28**, 1383 (1989).
[ii] N.W. Alcock, I.A. Degnan, S.M. Roe, and M.G.H. Wallbridge, *J. Organomet. Chem.*, **414**, 285 (1991).
[iii] Shuang Liu, E. Wong, V. Karunaratne, S.J. Rettig, and C. Orvig, *Inorg. Chem.*, **32**, 1756 (1993).
[iv] Shuang Liu, S.J. Rettig, and C. Orvig, *Inorg. Chem.*, **31**, 5400 (1992).
[v] H.R. Hoveyda, V. Karunaratne, S.J. Rettig, and C. Orvig, *Inorg. Chem.*, **31**, 5408 (1992).

Table 6.51   Thallium aryloxides

Compound	Aryloxide	Bond length (Å) M–O	Bond angle (°) M–O–Ar	Ref.
[Tl(μ_2-OAr)$_2$Tl] 2 coord. Tl	μ_2-OC$_6$H$_2$(CF$_3$)$_3$-2,4,6	2.469 (8), 2.46 (1)		i
[Me$_2$Tl(μ_2-OAr)$_2$TlMe$_2$] almost linear Me–Tl–Me	μ_2-OC$_6$H$_5$	2.37 (9), 2.36 (10)		ii
[Me$_2$Tl(μ_2-OAr)$_2$TlMe$_2$] almost linear Me–Tl–Me	μ_2-OC$_6$H$_4$Cl-2	2.43 (1), 2.40 (1)		ii
[(ArO)TlEt$_2$]	OC$_6$H$_4$CHO-2	2.46 (2)	133	iii

[i]H.W. Roesky, M. Scholz, M. Noltemeyer, and F.T. Edelmann, *Inorg. Chem.*, **28**, 3829 (1989).
[ii]P.J. Burke, L.A. Gray, P.J.C. Hayward, R.W. Matthews, M. McPartlin, and D.G. Gillies, *J. Organomet. Chem.*, **136**, C7 (1977).
[iii]G.H.W. Milburn and M.R. Truter, *J. Chem. Soc. A*, 648 (1967).

Table 6.52   Germanium aryloxides

Compound	Aryloxide	Bond length (Å) M–O	Bond angle (°) M–O–Ar	Ref.
[Ge(OAr)$_2$] vee-shaped molecule	OC$_6$H$_2$Bu$_2^t$-2,6-Me-4	1.807 (4)	123	i
[(ArO)$_2$GeFe(CO)$_4$] trigonal Ge, Ge(OAr)$_2$ in equatorial plane of tbp Fe	OC$_6$H$_2$Bu$_2^t$-2,6-Me-4	1.778 (6) 1.776 (6)	126 121	ii
[Ge(OAr)$_3$Cl] tetrahedral Ge	OC$_6$H$_3$Ph$_2$-2,6	1.739 (5) 1.754 (5) 1.744 (5)	125 123 123	iii
[Ge(OAr)$_2$(NMe$_2$)$_2$] tetrahedral Ge	OC$_6$H$_3$Ph$_2$-2,6	1.795 (3) 1.777 (3)	122 121	iii
[Ge(OAr)(CNO){N(SiMe$_3$)(C$_6$H$_2$Me$_3$)}$_2$] tetrahedral Ge	OC$_6$H$_5$	1.772 (2)	130	iv
[Ge(OAr)(CNO){N(SiMe$_3$)(C$_6$H$_3$Pr$_2^i$)}$_2$] tetrahedral Ge	OC$_6$H$_5$	1.777 (4)	132	iv

[i]B. Cetinkaya, I. Gumrukcu, M.F. Lappert, J.L. Atwood, R.D. Rogers, and M.J. Zaworotko, *J. Am. Chem. Soc.*, **102**, 2088 (1980).
[ii]P.B. Hitchcock, M.F. Lappert, S.A. Thomas, A.J. Thorne, A.J. Carty, and N.J. Taylor, *J. Organomet. Chem.*, **315**, 27 (1986).
[iii]G.D. Smith, P.E. Fanwick, and I.P. Rothwell, *Inorg. Chem.*, **29**, 3221 (1990).
[iv]A. Meller, G. Ossig, W. Maringgele, M. Noltemeyer, D. Stalke, R. Herbst-Irmer, S. Freitag, and G.M. Sheldrick, *Z. Naturforsch. B*, **47**, 162 (1992).

Table 6.53   Tin aryloxides

Compound	Aryloxide	Bond length (Å) M–O	Bond angle (°) M–O–Ar	Ref.
[Sn(OAr)$_2$]	OC$_6$H$_2$Bu$_2^t$-2,6-	2.022 (4)	129	i
vee-shaped molecule	Me-4	1.995 (4)	122	
[Sn(OAr)$_2$]	OC$_6$H$_3$Bu$_2^t$-2,6	2.003 (3)	125	ii
vee-shaped molecule		2.044 (3)	117	
[Sn(OAr){N(SiMe$_3$)$_2$}]	OC$_6$H$_2$Bu$_2^t$-2,6-	2.055 (2)	115	iii
vee-shaped molecule	Me-4			
[(ArO)$_2$SnFe(CO)$_4$]	OC$_6$H$_2$Bu$_2^t$-2,6-	1.983 (5)	118	iv
trigonal Sn, Sn(OAr)$_2$ in equatorial plane of tbp Fe	Me-4	1.963 (5)	125	
[Sn(OAr)$_2$(NMe$_2$)$_2$]	OC$_6$H$_2$Bu$_2^t$-2,6-	1.996 (3)	123	v
tetrahedral Sn	Me-4	1.980 (3)	126	
[Li(μ_2-OAr)$_3$Sn]	μ_2-OC$_6$H$_3$Ph$_2$-2,6	Li–O–Sn		vi
some Li-π-arene interactions		1.988 (4), 2.150 (2)		
[Sn(OC$_6$H$_2$Bu$_2^t$-η^1-C$_6$H$_5$)$_2$(HNMe$_2$)]	OC$_6$H$_2$Bu$_2^t$-η^1- C$_6$H$_5$	2.097 (4)	119	vii
tbp Sn, axial O		2.077 (4)	121	
[Sn(O-η^1-C$_{10}$H$_5$Bu^t-2)$_2$(HNMe$_2$)$_2$]	O-η^1- C$_{10}$H$_5$Bu^t-2	2.126 (3)	112	vii
6-coord. Sn, all-trans SnO$_2$C$_2$N$_2$				
[Sn$_2$(μ_2-OC$_6$H$_3$Ph-C$_6$H$_4$)$_2$Cl$_4$]	μ_2-OC$_6$H$_3$Ph- η^1-C$_6$H$_5$	2.045 (2), 2.259 (2) 2.326 (2), 2.035 (2)		viii
[Sn(OC$_6$H$_3$Ph-C$_6$H$_4$)$_2$(HNMe$_2$)$_2$]	OC$_6$H$_3$Ph-η^1- C$_6$H$_5$	2.091 (8)	123	vii
6-coord. Sn, trans O, cis C				
[S(NMe$_2$)$_3$][Me$_3$Sn(OAr)$_2$]	OC$_6$H$_3$Me$_2$-2,6	2.212 (7)	126	ix
tbp anion, trans OAr		2.225 (7)	122	
[S(NMe$_2$)$_3$][Me$_3$Sn(OAr)Cl]	OC$_6$H$_3$Me$_2$-2,6	2.102 (6)	131	ix
tbp anion, trans Cl and OAr				

[i]B. Cetinkaya, I. Gumrukcu, M.F. Lappert, J.L. Atwood, R.D. Rogers, and M.J. Zaworotko, *J. Am. Chem. Soc.*, **102**, 2088 (1980).

[ii]D.M. Barnhart, D.L. Clark, and J.G. Watkin, *Acta Crystallog. C*, **50**, 702 (1994).

[iii]H. Braunschweig, R.W. Chorley, P.B. Hitchcock, and M.F. Lappert, *J. Chem. Soc., Chem. Commun.*, 1311 (1992).

[iv]P.B. Hitchcock, M.F. Lappert, S.A. Thomas, A.J. Thorne, A.J. Carty, and N.J. Taylor, *J. Organomet. Chem.*, **315**, 27 (1986).

[v]G.D. Smith, P.E. Fanwick, and I.P. Rothwell, *Inorg. Chem.*, **29**, 3221 (1990).

[vi]G.D. Smith, P.E. Fanwick, and I.P. Rothwell, *Inorg. Chem.*, **28**, 618 (1989).

[vii]G.D. Smith, V.M. Visciglio, P.E. Fanwick, and I.P. Rothwell, *Organometallics*, **11**, 1064 (1992).

[viii]G.D. Smith, P.E. Fanwick, and I.P. Rothwell, *J. Am. Chem. Soc.*, **111**, 750 (1989).

[ix]M. Suzuki, I.-H. Son, R. Noyori, and H. Masuda, *Organometallics*, **9**, 3043 (1990).

(*continued overleaf*)

Table 6.53   (*Continued*)

Compound	Aryloxide	Bond length (Å) M–O	Bond angle (°) M–O–Ar	Ref.
$[\{Cl_3Sn(\mu_2\text{-}OAr)_2\}_2SnCl_2]$ 5- and 6-coord. Sn	$\mu_2\text{-}OC_6H_4Me\text{-}4$	2.083 (5), 2.128 (5) 2.111 (5), 2.085 (5)		x
$[Sn(OAr)Me_3]$ tetrahedral Sn	$OC_6H_2Bu^t_2\text{-}$ 2,4-PPh$_2$-2	2.015 (2)	136	xi

[x] H. Jolibois, F. Theobald, R. Mercier, and C. Devin, *Inorg. Chim. Acta*, **97**, 119 (1985).
[xi] J. Heinicke, R. Kadyrov, M.K. Kindermann, M. Koesling, and P.G. Jones, *Chem. Ber.*, **129**, 1547 (1996).

Table 6.54   Lead aryloxides

Compound	Aryloxide	Bond length (Å) M–O	Bond angle (°) M–O–Ar	Ref.
$[Pb(OAr)Ph_3]_n$	$OC_6H_3F\text{-}2\text{-}NO\text{-}4$	2.423 (6)	169	i

[i] N.G. Bokii, A.I. Udel'nov, Yu.T. Struchkov, D.N. Kravtsov, and V.M. Pacherskaya, *Zh. Strukt. Khim.*, **18**, 1025 (1977).

Table 6.55   Antimony aryloxides

Compound	Aryloxide	Bond length (Å) M–O	Bond angle (°) M–O–Ar	Ref.
$[Sb(OAr)Ph_4]$	$OC_6H_3Me_2\text{-}2,6$	2.131 (5)	124	i
$[Sb(OAr)Ph_4]$	$OC_6H_4NO_2\text{-}2$	2.222 (7)	124	ii
$[Sb(OAr)Ph_4]$	$OC_6H_4CHO\text{-}2$	2.202 (5)	130	ii
$[Sb(OAr)Ph_4]$	$OC_6H_3Pr^i\text{-}2\text{-}Me\text{-}5$	2.128 (5)	127	i

[i] V.V. Sharutin, O.K. Sharutina, P.E. Osipov, M.A. Pushilin, D.V. Muslin, N.Sh. Lyapina, V.V. Zhidkov, and V.K. Bel'sky, *Zh. Obshch. Khim.*, **67**, 1528 (1997).
[ii] V.V. Sharutin, V.V. Zhidkov, D.V. Muslin, N.Sh. Lyapina, G.K. Fukin, L.N. Zakharov, A.I. Yanovsky, and Yu.T. Struchkov, *Izv. Akad. Nauk. SSSR, Ser. Khim.*, 958 (1995).

ligand exchange in solution) are interesting given the important arylation of phenols which occurs with reagents such as $[Cl_2BiPh_3]$.[9] A number of interesting cluster aryloxides of bismuth have been characterized, with the aggregates typically held together with oxide ligands.

Table 6.56 Bismuth aryloxides

Compound	Aryloxide	Bond length (Å) M–O	Bond angle (°) M–O–Ar	Ref.
[Bi(OAr)₃] pyramidal Bi	OC₆H₃Me₂-2,6	2.091 (5) (av.)	[124 (4) av.]	i
[Bi₂(μ₂-OAr)₂Cl₄(THF)₂] square pyramid Bi, axial Cl	μ₂-OC₆H₃Me₂-2,6	2.211 (5), 2.463 (5)		ii
[Bi₂(μ₂-OAr)₂Cl₄(THF)₂] square pyramid Bi, axial Cl	μ₂-OC₆H₂Me₃-2,4,6	2.187 (3), 2.497 (3)		ii
[NMe₄]₂ [Bi₂(OAr)₆(μ-Cl)₂] square pyramid Bi, axial OAr	OC₆H₃Me₂-2,6	2.154 (3) 2.152 (3) 2.113 (3)	126 126 125	ii
[PPh₄]₂ [Bi(OAr)₂Br₃] square pyramid Bi, axial OAr	OC₆H₂Me₃-2,4,6	2.193 (7) 2.119 (7)	129 128	ii
[Et₂Bi(μ₂-OAr)]∞ chiral helical chains, tbp with axial OAr, equatorial lone pair	μ₂-OC₆H₅	2.382 (7)		iii
[Et₂Bi(μ₂-OAr)]∞ isomorphous with OPh compound	μ₂-OC₆F₅	2.4105 (7)		iii
[Bi(OAr)₂(μ₂-OAr)(η⁶-toluene)]₂.2toluene dimer with π-bound toluene	OC₆F₅ μ₂-OC₆F₅	2.088 (9), 2.147 (8) 2.210 (8), 2.571 (7)		iv,v
[Bi(OAr)₂(μ₂-OAr)(η⁶-toluene)]₂ dimer with π-bound toluene	OC₆F₅ μ₂-OC₆F₅	2.089 (8), 2.136 (8) 2.168 (7), 2.555 (9)		iv
[Bi(OAr)₂(μ₂-OAr)(THF)₂]₂.C₆H₁₄ dimer	OC₆F₅ μ₂-OC₆F₅	2.21 (1) 2.132 (9) 2.198 (9), 2.75 (1)	126 126	iv
[Bi(OAr)₂(μ₂-OAr)(THF)₂]₂ dimer	OC₆F₅ μ₂-OC₆F₅	2.111 (3) 2.151 (4) 2.245 (4), 2.661 (4)	130 130	iv
		Bi–O–Na		
[NaBi(μ₂-OAr)₄(THF)]∞ polymeric chain, square pyramidal Bi, basal THF	μ₂-OC₆F₅ μ₂-OC₆F₅	2.16 (2), 2.41 (2) 2.18 (2), 2.42 (2) 2.31 (1), 2.24 (2) 2.22 (2), 2.35 (2)		vi

(continued overleaf)

651

Table 6.56 *(Continued)*

Compound	Aryloxide	Bond length (Å) M–O	Bond angle (°) M–O–Ar	Ref.
[Na$_4$Bi$_2$(μ_6-O)(OAr)$_8$(THF)$_4$ octahedral arrangement of metal atoms about O	μ_2-OC$_6$F$_5$	occupancy disorder of Na, Bi		vi
[Bi$_6$(μ_3-O)$_3$(μ_2-OAr)$_7$(OAr)$_5$]	OC$_6$H$_3$Cl$_2$-2,6 μ_2-OC$_6$H$_3$Cl$_2$-2,6	2.116 (8)–2.559 (9)		vii
[Bi$_6$(μ_3-O)$_7$(μ_3-OAr)[Bi(μ_2-OAr)$_4$]$_3$] both THF and toluene solvates. Octahedral core facially capped by seven μ_3-O's and one μ_3-OAr	μ_2-OC$_6$F$_5$ μ_3-OC$_6$F$_5$	2.32 (10)–2.35 (8) 2.52 (8)–2.76 (15)		viii
[Bi(OAr)$_2$Ph$_3$] tbp equatorial Ph	OC$_6$F$_5$	2.228 (5) 2.212 (5)	126 126	ix
[Bi(OAr)(Br)Ph$_3$] tbp equatorial Ph	OC$_6$F$_5$	2.235 (9)	130	ix
[Bi(OAr)Ph$_4$] tbp equatorial Ph	OC$_6$F$_5$	2.544 (7)	126	ix
[Bi(OAr)$_2$Ph$_3$] tbp equatorial Ph	OC$_6$Cl$_5$	2.250 (7) 2.244 (7)	124 125	ix
[Bi(OAr)(Br)Ph$_3$] tbp equatorial Ph	OC$_6$Cl$_5$	2.230 (6)	136	ix
[Bi(OAr)Ph$_4$] tbp equatorial Ph	OC$_6$Cl$_5$	2.543 (9)	123	ix

[i] W.J. Evans, J.H. Hain Jr., and J.W. Ziller, *J. Chem. Soc. Chem. Commun.*, 1628 (1989).

[ii] P. Hodge, S.C. James, N.C. Norman, and A.G. Orpen, *J. Chem. Soc. Dalton Trans.*, 4049 (1998).

[iii] K.H. Whitmire, J.C. Hutchison, A.L. McKnight, and C.M. Jones, *J. Chem. Soc. Chem. Commun.*, 1021 (1992).

[iv] C.M. Jones, M.D. Burkart, R.E. Bachman, D.L. Serra, Shiou-Jyh Hwu, and K.H. Whitmire, *Inorg. Chem.*, **32**, 5136 (1993).

[v] C.M. Jones, M.D. Burkart, and K.H. Whitmire, *Angew. Chem., Int. Ed. Engl.*, **31**, 451 (1992).

[vi] J.L. Jolas, S. Hoppe, and K.H. Whitmire, *Inorg. Chem.*, **36**, 3335 (1997).

[vii] S.C. James, N.C. Norman, A.G. Orpen, M.J. Quayle, and U. Weckenmann, *J. Chem. Soc. Dalton Trans.*, 4159 (1996).

[viii] C.M. Jones, M.D. Burkart, and K.H. Whitmire, *J. Chem. Soc. Chem. Commun.*, 1638 (1992).

[ix] S. Hoppe and K.H. Whitmire, *Organometallics*, **17**, 1347 (1998).

REFERENCES

1. a. D.C. Bradley, R.C. Mehrotra, and D.P. Gaur, *Metal Alkoxides*, Academic Press, New York (1978); b. K.C. Malhotra and R.L. Martin, *J. Organomet. Chem.*, **239**, 159 (1982); c. R.C. Mehrotra, *Adv. Inorg. Chem. Radiochem.*, **26**, 269 (1983); d. M.H. Chisholm and I.P. Rothwell, in *Comprehensive Coordination Chemistry* (G. Wilkinson, R.D. Gillard, J.A. McCleverty eds), Vol. 2, Ch. 15.3, Pergamon Press, Oxford (1987); e. H.E. Bryndza and W. Tam, *Chem. Rev.*, **88**, 1163 (1988); f. W.G. Van der Sluys and A.P. Sattelberger, *Chem. Rev.*, **90**, 1027 (1990); g. K.G. Caulton and L.G. Hubert-Pfalzgraf, *Chem. Rev.*, **90**, 969 (1990); h. R.C. Mehrotra, A. Singh, and U.M. Tripathi, *Chem. Rev.*, **91**, 1287 (1991); i. M.D. Healy, M.B. Power, and A.R. Barron, *Coord. Chem. Rev.* **130**, 63 (1994).
2. L. Que, *Coord. Chem. Rev.*, **50**, 73 (1983).
3. J.P. Phillips, *Chem. Rev.*, **56**, 271 (1956).
4. a. K.B. Mertes and J.-M. Lehn, in *Comprehensive Coordination Chemistry* (G. Wilkinson, R.D. Gillard, and J.A. McCleverty eds), Vol. 2, Ch. 21.3, Pergamon Press, Oxford (1987); b. R.S. Vagg, in *Comprehensive Coordination Chemistry* (G. Wilkinson, R.D. Gillard, and J.A. McCleverty eds), Vol. 2, Ch. 20.4, Pergamon Press, Oxford (1987).
5. a. R. Masrhoff, H. Kohler, H. Bohland, and F. Schmeil, *Z. Chem.*, 122 (1965); b. H. Funk and W. Baumann, *Z. Anorg. Allg. Chem.*, **231**, 264 (1937).
6. a. R.M. Roberts and A.A. Khalaff, *Friedel–Crafts Alkylations, A Century of Discovery*, Marcel Dekker, New York (1984); b. K. Dimroth, *Top. Curr. Chem.*, **129**, 99 (1985).
7. P.N. Riley, M.G. Thorn, J.S. Vilardo, M.A. Lockwood, P.E. Fanwick, and I.P. Rothwell, *Organometallics*, **18**, 3016 (1999).
8. S. Saito, T. Kano, K. Hatanaka, and H. Yamamoto, *J. Org. Chem.*, **62**, 5651 (1997).
9. D.H.R. Barton, N.Y. Bhatnagar, J.-C. Blazejewski, B. Charpiot, J.-P. Finet, D.J. Lester, W.B. Motherwell, M.T.B. Papoula, and S.P. Stanforth, *J. Chem. Soc., Perkin Trans. I*, 2657 (1985).
10. D.H.R. Barton, D.M.X. Donnelly, P.J. Guiry, and J.H. Reibenspies, *J. Chem. Soc., Chem. Commun.*, 1110 (1990).
11. J.S. Vilardo, M.A. Lockwood, L.G. Hanson, J.R. Clark, B.C. Parkin, P.E. Fanwick, and I.P. Rothwell, *J. Chem. Soc., Dalton Trans.*, 3353 (1997).
12. F. Hawthorne and D.J. Cram, *J. Am. Chem. Soc.*, **76**, 5859 (1952).
13. S. Saito and H. Yamamoto, *J. Chem. Soc., Chem. Commun.*, 1585 (1997).
14. G.M. Loudon, *Organic Chemistry*, 2nd edn, 925, Benjamin/Cummings, Menlo Park, CA (1988).
15. a. A.R. Katritzky, M. Karelson, and P.A. Harris, *Heterocycles*, **32**, 329 (1991); b. P. Beak, *Acc. Chem. Res.*, **10**, 186 (1977).
16. A.J. Deeming, I.P. Rothwell, M.B. Hursthouse, and J.D.J. Backer-Dirks, *J. Chem. Soc., Dalton Trans.*, 2039 (1981).
17. M.F. Richardson, G. Wulfsberg, R. Marlow, S. Zaghonni, D. McCorkle, K. Shadid, J. Gagliardi Jr, and B. Farris, *Inorg. Chem.*, **32**, 1913 (1993).
18. Hyungkyu Kang, Shuncheng Liu, S.N. Shaikh, T. Nicholson, and J. Zubieta, *Inorg. Chem.*, **28**, 920 (1989).
19. D.G. Hendershot, M. Barber, R. Kumar, and J.P. Oliver, *Organometallics*, **10**, 3302 (1991).
20. C.-N. Kuo, T.-Y. Huang, M.-Y. Shao, and H.-M. Gau, *Inorg. Chim. Acta.*, **293**, 12 (1999).
21. B.S. Kang, L.H. Weng, D.X. Wu, F. Wang, Z. Guo, L.R. Huang, Z.Y. Huang, and H.Q. Liu, *Inorg. Chem.*, **27**, 1128 (1988).
22. H.-F. Klein, A. Dal, S. Hartmann, U. Florke, and H.-J. Haupt, *Inorg. Chim. Acta*, **287**, 199 (1999).
23. T.C. Higgs and C.J. Carrano, *Inorg. Chem.*, **36**, 291 (1997).
24. T.C. Higgs, N.S. Dean, and C.J. Carrano, *Inorg. Chem.*, **37**, 1473 (1998).
25. B.S. Hames and C.J. Carrano, *Inorg. Chem.*, **38**, 4593 (1999).
26. B.S. Hames and C.J. Carrano, *Inorg. Chem.*, **38**, 3562 (1999).
27. I.P. Rothwell, *Acc. Chem. Res.*, **21**, 153 (1988).
28. D. Rabinovich, R. Zelman, and G. Parkin, *J. Am. Chem. Soc.*, **116**, 4611 (1992).

29. J.L. Kerschner, P.E. Fanwick, I.P. Rothwell, and J.C. Huffman, *Organometallics*, **8**, 1431 (1989).
30. J.S. Yu, P.E. Fanwick, and I.P. Rothwell, *J. Am. Chem. Soc.*, **112**, 8171 (1990).
31. M.A. Lockwood, M.C. Potyen, B.D. Steffey, P.E. Fanwick, and I.P. Rothwell, *Polyhedron*, **16**, 3293 (1995).
32. Yong-Joo Kim, Jae-Young Lee, and Kohtaro Osakada, *J. Organomet. Chem.*, **558**, 41 (1998).
33. M. Hirano, N. Kurata, T. Marumo, and S. Komiya, *Organometallics*, **17**, 501 (1998).
34. M. Calligaris and L. Randaccio, in *Comprehensive Coordination Chemistry* (G. Wilkinson, R.D. Gillard, and J.A. McCleverty eds), Vol. 2, Ch. 20.1, Pergamon Press, Oxford (1987).
35. B. Bosnich, *Inorg. Chem.*, **38**, 2554 (1999).
36. A.S. Borovik and L. Que Jr, *J. Am. Chem. Soc.*, **110**, 2345 (1988).
37. C. Fraser, L. Johnston, A.L. Rheingold, B.S. Haggerty, G.K. Williams, J. Whelan, and B. Bosnich, *Inorg. Chem.*, **31**, 1835 (1992).
38. a. C.G. Pierpont and C.W. Lange, *Prog. Inorg. Chem.*, **41**, 331 (1994); b. C.G. Pierpont and R.M. Buchanan, *Coord. Chem. Rev.*, **38**, 45 (1981).
39. J.R. Telford and K.N. Raymond, in *Comprehensive Supramolecular Chemistry*, Vol. 1, 245, Elsevier Science, Oxford (1996).
40. L. Que and R.Y.N. Ho, *Chem. Rev.*, **96**, 2607 (1996).
41. J.G. Jang, D.D. Cox, and L. Que, *J. Am. Chem. Soc.*, **113**, 9200 (1991).
42. a. R.H. Heistand, A.L. Roe, L. Que Jr, *Inorg. Chem.*, **21**, 676 (1982); b. Van An Ung, S.M. Couchman, J.C. Jeffery, J.A. McCleverty, M.D. Ward, F. Totti, and D. Gatteschi, *Inorg. Chem.*, **38**, 365 (1999); c. J.R. Farrell, C.A. Mirkin, I.A. Guzei, L.M. Liable-Sands, and A.L. Rheingold, *Angew. Chem., Int. Ed. Engl.*, **37**, 465 (1998); d. W.J. Evans, M.A. Ansari, and J.W. Ziller, *Polyhedron*, **17**, 299 (1998); e. F.S. McQuillan, T.E. Berridge, Hongli Chen, T.A. Hamor, and C.J. Jones, *Inorg. Chem.*, **37**, 4959 (1998); f. A. Sen, H.A. Stecher, and A.L. Rheingold, *Inorg. Chem.*, **31**, 473 (1992); g. F.S. McQuillan, Hongli Chen, T.A. Hamor, and C.J. Jones, *Polyhedron*, **15**, 3909 (1996); h. Van An Ung, D.A. Bardwell, J.C. Jeffery, J.P. Maher, J.A. McCleverty, M.D. Ward, and A. Williamson, *Inorg. Chem.*, **35**, 5290 (1996); i. A. Kunzel, M. Sokolow, Feng-Quan Liu, H.W. Roesky, M. Noltemeyer, H.-G. Schmidt, and I. Uson, *J. Chem. Soc., Dalton Trans.*, 913 (1996).
43. T.P. Vais, J.M. Tanski, J.M. Pette, E.B. Lobkovsky, and P.T. Wolczanski, *Inorg. Chem.*, **38**, 3394 (1999).
44. D.R. Mulford, P.E. Fanwick, and I.P. Rothwell, *Polyhedron*, **19**, 35 (1999) and references therein.
45. A. van der Linden, C.J. Schaverien, N. Meijboom, C. Ganter, and A.G. Orpen, *J. Am. Chem. Soc.*, **117**, 3008 (1995) and references therein.
46. a. R. Noyori, *Asymmetric Catalysis in Organic Synthesis*, J. Wiley & Sons, New York (1994); b. R. Zimmer and J. Suhrbier, *J. Prakt. Chem.*, 339 (1997).
47. N.W. Eilerts and J.A. Heppert, *Polyhedron*, **22**, 3255 (1995).
48. K.M. Totland, T.J. Boyd, G.G. Lavoie, W.M. Davis, and R.R. Schrock, *Macromolecules*, **29**, 6114 (1996) and references therein.
49. K. Narasaka, *Synthesis*, **1**, 1 (1991).
50. S.S. Zhu, D.R. Cefalo, D.S. La, J.Y. Jamieson, W.M. Davis, A.H. Hoveyda, and R.R. Schrock, *J. Am. Chem. Soc.*, **121**, 8251 (1999).
51. J.B. Alexander, D.S. La, D.R. Cefalo, A.H. Hoveyda, and R.R. Schrock, *J. Am. Chem. Soc.*, **120**, 4041 (1998).
52. D.S. La, J.B. Alexander, D.R. Cefalo, D.D. Graf, A.H. Hoveyda, and R.R. Schrock, *J. Am. Chem. Soc.*, **120**, 9720 (1998).
53. J. Okuda, S. Fokken, H.-C. Kang, and W. Massa, *Chem. Ber.*, **128**, 221 (1995).
54. S. Foken, T.P. Spaniol, J. Okuda, F.G. Sernetz, and R. Muelhaupt, *Organometallics*, **16**, 4240 (1997).
55. S. Fokken, T.P. Spaniol, Hak-Chul Kang, W. Massa, and Jun Okuda, *Organometallics*, **15**, 5069 (1996).

56. Y. Nakayama, K. Watanabe, N. Ueyama, A. Nakamura, A. Harada, and J. Okuda, *Organometallics*, **19**, 2498 (2000).

57. J. Okuda, S. Fokken, H.-C. Kang, and W. Massa, *Polyhedron*, **17**, 943 (1998).

58. F.C. Anson, J.A. Christie, T.C. Collins, R.J. Coots, T.T. Furutani, S.L. Gipson, J.T. Keech, T.E. Krafft, B.D. Santarsiero, and G.H. Spies, *J. Am. Chem. Soc.*, **106**, 4460 (1984).

59. J. Costamagna, *Coord. Chem. Rev.*, **119**, 67 (1992).

60. Y.N. Ito and T. Katsuki, *Bull. Chem. Soc. Jpn.*, **72**, 603 (1999).

61. L. Canali and D.C. Sherrington, *Chem. Soc. Rev.*, **28**, 85 (1999).

62. a. E.N. Jacobsen, in *Catalytic Asymmetric Synthesis* (I. Ojima, ed.), 159–202, VCH, New York (1993); b. T. Katsuki, *Coord. Chem. Rev.*, **140**, 189 (1995); c. T. Linker, *Angew. Chem., Int. Ed. Engl.*, **36**, 2060 (1997); d. T.J. Katsuki, *Mol. Catal. A: Chem.*, **113**, 87 (1996); e. C.T. Dalton, K.M. Ryan, V.M. Wall, C. Bousquet, and D.G. Gilheany, *Top. Catal.*, **5**, 75 (1998).

63. N.H. Pilkington and R. Robson, *Aust. J. Chem.*, **23**, 2225 (1970).

64. H. Okawa and H. Furutachi, *Coord. Chem. Rev.*, **174**, 51 (1998).

65. a. A.J. Atkins, D. Black, A.J. Blake, A. Marin-Becerra, S. Parsons, L. Ruiz-Ramirez, and M. Schröder, *J. Chem. Soc., Chem. Commun.*, 457 (1990).

66. S.K. Mandal, L.K. Thompson, K. Nag, J.-P. Charland, and E.J. Gabe, *Inorg. Chem.*, **26**, 1391 (1987).

67. H. Wada, T. Aono, K.-I. Motoda, M. Ohba, N. Matsumoto, and H. Okawa, *Inorg. Chim. Acta*, **13**, 246 (1996).

68. C. Fraser, L. Johnston, A.L. Rheingold, B.S. Haggerty, G.K. Williams, J. Whelan, and B. Bosnich, *Inorg. Chem.*, **31**, 1835 (1992).

69. H. Okawa, J. Nishio, M. Ohba, M. Tadokoro, N. Matsumoto, M. Koikawa, S. Kida, and D.E. Fenton, *Inorg. Chem.*, **32**, 2949 (1993).

70. M. Tadokoro, H. Sakiyama, N. Matsumoto, M. Kodera, H. Okawa, and S. Kida, *J. Chem. Soc., Dalton Trans.*, 313 (1992).

71. S.S. Tandon, L.K. Thompson, J.N. Bridson, V. McKee, and A.J. Downard, *Inorg. Chem.*, **31**, 4635 (1992).

72. I.A. Kahwa, S. Folkes, D.J. Williams, S.V. Ley, C.A. O'Mahoney, and G.L. Mcpherson, *J. Chem. Soc., Chem. Commun.*, 1531 (1989).

73. D.E. Fenton, *Adv. Inorg. Bioinorg. Mech.*, **2**, 187 (1983).

74. B.F. Hoskins, R. Robson, and D. Vince, *J. Chem. Soc., Chem. Commun.*, 392 (1973).

75. D.F. Cook, D. Cummins, and E.D. McKenzie, *J. Chem. Soc., Dalton Trans.*, 1369 (1976).

76. A. Neves, A.S. Ceccato, I. Vencato, Y.P. Mascarenhas, and C. Erasmus-Buhr, *J. Chem. Soc., Chem. Commun.*, 652 (1992).

77. a. M. Bell, A.J. Edwards, B.F. Hoskins, E.H. Kachab, and R. Robson, *J. Chem. Soc., Chem. Commun.*, 1852 (1987); b. M. Bell, A.J. Edwards, B.F. Hoskins, E.H. Kachab, and R. Robson, *J. Am. Chem. Soc.*, **111**, 3603 (1989); c. A.J. Edwards, B.F. Hoskins, E.H. Kachab, A. Markiewicz, K.S. Murray, and R. Robson, *Inorg. Chem.*, **31**, 3585 (1992).

78. a. M.J. Grannas, B.F. Hoskins, and R. Robson, *J. Chem. Soc., Chem. Commun.*, 1644 (1990); b. M.J. Grannas, B.F. Hoskins, and R. Robson, *Inorg. Chem.*, **33**, 1071 (1994).

79. J. Vicens and V. Bomer, *Calixarenes: a Versatile Class of Macrocyclic Compounds*, Kluwer Academic, Boston, MA (1991).

80. a. C. Wieser, C.B. Dielman, and D. Matt, *Coord. Chem. Rev.*, **165**, 93 (1997); b. D.M. Roundhill, *Prog. Inorg. Chem.*, **43**, 533 (1995).

81. L. Giannini, A. Caselli, E. Solari, C. Floriani, A. Chiesi-Villa, C. Rizzoli, N. Re, and A. Sgamellotti, *J. Am. Chem. Soc.*, **119**, 9709 (1997) and references therein.

82. O.V. Ozerov, F.T. Ladipo, and B.O. Patrick, *J. Am. Chem. Soc.*, **121**, 7941 (1999).

83. A. Ikeda and S. Shinkai, *Chem. Rev.*, **97**, 1713 (1997).

84. W.J. Evans, R.E. Golden, and J.W. Ziller, *Inorg. Chem.*, **32**, 3041 (1993).

85. K.G. Caulton, M.H. Chisholm, S.R. Drake, K. Folting, J.C. Huffman, and W.E. Streib, *Inorg. Chem.*, **32**, 1970 (1993).

86. S.R. Drake, W.E. Streib, M.H. Chisholm, and K.G. Caulton, *Inorg. Chem.*, **29**, 2707 (1990).

87. P.B. Hitchcock, M.F. Lappert, G.A. Lawless, and B. Royo, *J. Chem. Soc., Chem. Commun.*, 1141 (1990).
88. K.G. Caulton, M.H. Chisholm, S.R. Drake, and K. Folting, *J. Chem. Soc., Chem. Commun.*, 1349 (1990).
89. K.G. Caulton, M.H. Chisholm, S.R. Drake, K. Folting, and J.C. Huffman, *Inorg. Chem.*, **32**, 816 (1993).
90. K.S. Mazdivasni, C.T. Lynch, and J.S. Smith, *Inorg. Chem.*, **5**, 342 (1966).
91. G.B. Deacon, S. Nickel, P. MacKinnon, and E.R.T. Tiekink, *Aust. J. Chem.*, **63**, 1245 (1990).
92. G.B. Deacon, P.B. Hitchcock, S.A. Holmes, M.F. Lappert, P. MacKinnon, and R.H. Newnham, *J. Chem. Soc., Chem. Commun.*, 935 (1989).
93. a. M.-U Haque, W. Horne, and S.J. Lyle, *J. Cryst. Spectrosc.*, **21**, 411 (1991); b. H. Schmidbaur, J. Lettenbauer, D.L. Wilkinson, G. Muller, and O. Kumberger, *Z. Naturforsch. B*, **66**, 901 (1991).
94. H. Schmidbaur, O. Kumberger, and J. Riede, *Inorg. Chem.*, **30**, 3101 (1991).
95. L.B. Cole and E.M. Holt, *J. Chem. Soc., Perkin Trans.*, **2**, 1997 (1986).
96. C.L. Spiro, S.L. Lambert, T.J. Smith, E.N. Duesler, R.R. Gagne, and D.N. Hendrickson, *Inorg. Chem.*, **20**, 1229 (1981).
97. M.W. Glenny, A.J. Nielson, and C.E.F. Rickard, *Polyhedron*, **17**, 851 (1998).
98. R.C. Mehrotra and G. Chander, *J. Indian Chem. Soc.*, **39**, 235 (1962).
99. M. Aresta and R.S. Nyholm, *J. Organomet. Chem.*, **56**, 395 (1973).
100. F. Quignard, M. Leconte, J.-M. Basset, Leh-Yeh Hsu, J.J. Alexander, and S.G. Shore, *Inorg. Chem.*, **26**, 4272 (1987).
101. J.R. Dilworth, J. Hanich, M. Krestel, J. Beck, and J. Strahle, *J. Organomet. Chem.*, **315**, 9 (1986).
102. D.E. Chebi, P.E. Fanwick, and I.P. Rothwell, *Polyhedron*, **9**, 969 (1990).
103. F. Calderazzo, F. Marchetti, G. Dell'Amico, G. Pelizzi, and A. Colligiani, *J. Chem. Soc., Dalton Trans.*, 1419 (1980).
104. W.J. Evans, J.M. Olofson, and J.W. Ziller, *Inorg. Chem.*, **28**, 4308 (1989).
105. G.D. Smith, P.E. Fanwick, and I.P. Rothwell, *Inorg. Chem.*, **29**, 3221 (1990).
106. L.D. Durfee, S.L. Latesky, I.P. Rothwell, J.C. Huffman, and K. Folting, *Inorg. Chem.*, **26**, 4569 (1985).
107. M.L. Listemann, R.R. Schrock, J.C. Dewan, and R.M. Kolodziej, *Inorg. Chem.*, **27**, 264 (1988).
108. D. Meyer, J.A. Osborn, and M. Wesolek, *Polyhedron*, **9**, 1311 (1990).
109. S.A. Koch and M. Millar, *J. Am. Chem. Soc.*, **106**, 5255 (1982).
110. G.D. Smith, P.E. Fanwick, and I.P. Rothwell, *Inorg. Chem.*, **28**, 618 (1989).
111. J.J.H. Edema, A. Meetsma, S. Gambarotta, S.I. Khan, W.J.J. Smeets, and A.L. Spek, *Inorg. Chem.*, **30**, 3639 (1991).
112. D.L. Clark, J.G. Watkin, and J.C. Huffman, *Inorg. Chem.*, **31**, 1554 (1992).
113. H. Yasuda, Y. Nakayama, K. Takei, A. Nakamura, Y. Kai, and N. Kanehisa, *J. Organomet. Chem.*, **673**, 105 (1994).
114. Yanlong Qian, Jiling Huang, Xiaoping Chen, Guisheng Li, Weichun Chen, Bihua Li, Xianglin Jin, and Qingchuan Yang, *Polyhedron*, **13**, 1105 (1994).
115. H.D. Empsall, P.N. Heys, W.S. McDonald, M.C. Norton, and B.L. Shaw, *J. Chem. Soc., Dalton Trans.*, 1119 (1978).
116. Jui-Sui Sun, C.E. Uzelmeier, D.L. Ward, and K.R. Dunbar, *Polyhedron*, **17**, 2049 (1998).
117. Y. Yamamoto, R. Sato, F. Matsuo, C. Sudoh, and T. Igoshi, *Inorg. Chem.*, **35**, 2329 (1996).
118. M.E. van der Boom, Shyh-Yeon Liou, Y. Ben-David, L.J.W. Shimon, and D. Milstein, *J. Am. Chem. Soc.*, **120**, 6531 (1998) and references therein.
119. M.F. Lappert, P.P. Power, A.R. Sanger, and R.C. Srivastava, *Metal and Metalloid Amides*, Ellis Horwood, Chichester (1980).
120. H.A. Stecher, A. Sen, and A.L. Rheingold, *Inorg. Chem.*, **27**, 1130 (1988).
121. T.W. Coffindaffer, I.P. Rothwell, and J.C. Huffman, *Inorg. Chem.*, **23**, 1433 (1984).
122. J.R. van den Hende, P.B. Hitchcock, and M.F. Lappert, *J. Chem. Soc., Chem. Commun.*, 1413 (1994).

123. T.W. Coffindaffer, W.M. Westler, and I.P. Rothwell, *Inorg. Chem.*, **26**, 4565 (1985).
124. M.J. Bartos, C.E. Kriley, J.S. Yu, J.L. Kerschner, P.E. Fanwick, and I.P. Rothwell, *Polyhedron*, **8**, 1971 (1989).
125. R.G. Abbott, F.A. Cotton, and L.R. Falvello, *Inorg. Chem.*, **29**, 514 (1990).
126. R.C. Mehrotra and G. Chander, *J. Indian Chem. Soc.*, **39**, 235 (1962).
127. I.D. Verma and R.C. Mehrotra, *J. Indian Chem. Soc.*, **38**, 147 (1961).
128. T.J. Boyle, D.L. Barnes, J.A. Heppert, L. Morales, F. Takusagawa, and J.W. Connolly, *Organometallics*, **11**, 1112 (1992).
129. R.D. Simpson and R.G. Bergman, *Organometallics*, **12**, 781 (1993).
130. D.H. McConville, J.R. Wolf, and R.R. Schrock, *J. Am. Chem. Soc.*, **115**, 4413 (1993).
131. J.C. Calabrese, M.A. Cushing Jr, and S.D. Ittel, *Inorg. Chem.*, **27**, 867 (1988).
132. M. Scholz, M. Noltemeyer, and H.W. Roesky, *Angew. Chem., Int. Ed. Engl.*, **28**, 1383 (1989).
133. S.L. Latesky, A.K. McMullen, I.P. Rothwell, and J.C. Huffman, *J. Am. Chem. Soc.*, **107**, 5981 (1985).
134. S.L. Latesky, A.K. McMullen, G.P. Niccolai, I.P. Rothwell, and J.C. Huffman, *Organometallics*, **6**, 902 (1985).
135. H. Schmidbaur, J. Lettenbauer, D.L. Wilkinson, G. Muller, and O. Kumberger, *Z. Naturforsch. B*, **66**, 901 (1991).
136. P.E. Fanwick, A.E. Ogilvy, and I.P. Rothwell, *Organometallics*, **6**, 73 (1987).
137. K.B. Starowieyski, S. Pasynkiewicz, and M. Skowronska-Ptasinska, *J. Organomet. Chem.*, **90**, C43 (1975).
138. A.P. Shreve, R. Mulhaupt, W. Fultz, J.C. Calabrese, W. Robbins, and S.D. Ittel, *Organometallics*, **7**, 409 (1988).
139. M. Skowronska-Ptasinska, K.B. Starowieyski, S. Pasynkiewicz, and M. Carewska, *J. Organomet. Chem.*, **160**, 403 (1978).
140. T.V. Lubben and P.T. Wolczanski, *J. Am. Chem. Soc.*, **109**, 424 (1987).
141. A.G. Davies and B.P. Roberts, *Acc. Chem. Res.*, **5**, 387 (1972).
142. R.D. Arasasingham, A.L. Balch, R.L. Hart, and L. Lato-Grazynski, *J. Am. Chem. Soc.*, **112**, 7566 (1990).
143. P.L. Alsters, H.T. Teunissen, J. Boersma, A.L. Spek, and G. van Koten, *Organometallics*, **12**, 4691 (1993).
144. K. Koo, G.L. Hillhouse, and A.L. Rheingold, *Organometallics*, **14**, 456 (1995).
145. M.A. Bennett, M. Glewis, D.C.R. Hockless, and E. Wenger, *J. Chem. Soc., Dalton Trans.*, 3105 (1997).
146. S. Park, M. Pontier-Johnson, and D.M. Roundhill, *Inorg. Chem.*, **29**, 2689 (1990).
147. A. Hess, M.R. Horz, L.M. Liable-Sands, M. Louise, D.C. Lindner, A.L. Rheingold, and K.H. Theopold, *Angew. Chem., Int. Ed. Engl.* **38**, 166 (1999).
148. S.N. Brown and J.M. Mayer, *J. Am. Chem. Soc.*, **118**, 12119 (1996) and references therein.
149. S.N. Brown and J.M. Mayer, *Organometallics*, **14**, 2951 (1995).
150. J.C. Huffman, R.L. Geerts, and K.G. Caulton, *J. Cryst. Spectrosc.*, **16**, 541 (1984).
151. S. Brooker, F.T. Edelmann, T. Kottke, H.W. Roesky, G.M. Sheldrick, D. Stalke, and K.H. Whitmire, *J. Chem. Soc., Chem. Commun.*, 144 (1991).
152. E. Solari, S. DeAngelis, C. Floriani, A. Chiesi-Villa, and C. Rizzoli, *J. Chem. Soc., Dalton Trans.*, 2471 (1991).
153. V.C. Gibson and T.P. Kee, *J. Organomet. Chem.*, **444**, 91 (1993).
154. S.M. Beshouri, T.W. Coffindaffer, S. Pirzad, and I.P. Rothwell, *Inorg. Chim. Acta.*, **103**, 111 (1985).
155. P.J. Fagan, K.G. Moloy, and T.J. Marks, *J. Am. Chem. Soc.*, **103**, 6959 (1981).
156. D.C. Bradley, *Nature*, **182**, 1211 (1958).
157. L.M. Jackman, D. Cizmeciyan, P.G. Williard, and M.A. Nichols, *J. Am. Chem. Soc.*, **115**, 6262 (1993).
158. W.J. Evans, R.E. Golden, and J.W. Ziller, *Inorg. Chem.*, **32**, 3041 (1993).
159. K.G. Caulton, M.H. Chisholm, S.R. Drake, K. Folting, J.C. Huffman, and W.E. Streib, *Inorg. Chem.*, **32**, 1970 (1993).
160. J.J.H. Edema, S. Gambarotta, F. van Bolhuis, and A.L. Spek, *J. Am. Chem. Soc.*, **111**, 2142 (1989).

161. J.J.H. Edema, S. Gambarotta, F. van Bolhuis, W.J.J. Smeets, and A.L. Spek, *Inorg. Chem.*, **28**, 1407 (1989).
162. W.J. Evans, R.E. Golden, and J.W. Ziller, *Inorg. Chem.*, **32**, 3041 (1993).
163. K.G. Caulton, M.H. Chisholm, S.R. Drake, and K. Folting, *J. Chem. Soc., Chem. Commun.*, 1349 (1990).
164. S.J. Vilardo, P.E. Fanwick, and I.P. Rothwell, *Polyhedron*, **17**, 769 (1998).
165. L.D. Durfee, P.E. Fanwick, and I.P. Rothwell, *Angew. Chem., Int. Ed. Engl.* **27**, 1181 (1988).
166. D.M. Barnhart, D.L. Clark, J.C. Gordon, J.C. Huffman, R.L. Vincent, J.G. Watkin, and B.D. Zwick, *Inorg. Chem.*, **33**, 3487 (1994).
167. W.G. Van der Sluys, C.J. Burns, J.C. Huffman, and A.P. Sattelberger, *J. Am. Chem. Soc.*, **110**, 5924 (1988).
168. J.A. Heppert, T.J. Boyle, and F. Takusagawa, *Organometallics*, **8**, 461 (1989).
169. S.D. Loren, B.K. Campion, R.H. Heyn, T.D. Tilley, B.E. Bursten, and K.W. Luth, *J. Am. Chem. Soc.*, **111**, 4712 (1989).
170. R.O. Rosette and D.G. Cole-Hamilton, *J. Chem. Soc., Dalton Trans.*, 1618 (1979).
171. M.A. Bennett and T.W. Matheson, *J. Organomet. Chem.* **175**, 87 (1979).
172. B. Cetinkaya, P.B. Hitchcock, M.F. Lappert, and S. Torioni, *J. Organomet. Chem.*, **88**, C31 (1980).
173. V.F. Kuznetsov, G.P.A. Yap, C. Bensimon, and H. Alper, *Inorg. Chim. Acta.*, **280**, 172 (1998).
174. L. Dahlenburg and N. Hoeck, *J. Organomet. Chem.*, **284**, 129 (1985).
175. Analysis of the Cambridge Crystallographic Database.
176. T.W. Coffindaffer, I.P. Rothwell, and J.C. Huffman, *Inorg. Chem.*, **22**, 2906 (1983).
177. M.H. Chisholm, L.-S. Tan, and J.C. Huffman, *J. Am. Chem. Soc.*, **104**, 4879 (1982).
178. H. Yasuda, Y. Nakayama, K. Takei, A. Nakamura, Y. Kai, and N. Kanehisa, *J. Organomet. Chem.*, **473**, 105 (1994).
179. R.R. Schrock and L.J. Guggenberger, *J. Am. Chem. Soc.*, **97**, 6578 (1975).
180. J.L. Kerschner, P.E. Fanwick, I.P. Rothwell, and J.C. Huffman, *Inorg. Chem.*, **28**, 780 (1989).
181. L.M. Atagi and J.M. Mayer, *Angew. Chem., Int. Ed. Engl.*, **32**, 439 (1993).
182. C.E. Kriley, P.E. Fanwick, and I.P. Rothwell, *J. Am. Chem. Soc.*, **116**, 5225 (1994).
183. R. Anwander, *Top. Curr. Chem.*, **179**, 149 (1996).
184. J.M. Mayer, *Comments Inorg. Chem.* **4**, 125 (1988).
185. R.G. Bergman, *Polyhedron*, **14**, 3227 (1995).
186. H.E. Bryndza, P.J. Domaille, W. Tam, L.K. Fong, R.A. Paciello, and J.E. Bercaw, *Polyhedron*, **7**, 1441 (1988).
187. A.R. Barron, *Polyhedron*, **14**, 3197 (1995).
188. P.J. Brothers and P.P. Power, *Adv. Org.-met. Chem.*, **39**, 1 (1996).
189. M.A. Petrie, M.M. Olmstcad, and P.P. Power, *J. Am. Chem. Soc.*, **113**, 8704 (1991).
190. L. Chamberlain, J.C. Huffman, J. Keddington, and I.P. Rothwell, *J. Chem. Soc., Chem. Commun.*, 805 (1982).
191. K. Watenpaugh and C.N. Caughlan, *Inorg. Chem.*, **5**, 1782 (1966).
192. B.D. Steffey, P.E. Fanwick, and I.P. Rothwell, *Polyhedron*, **9**, 963 (1990).
193. J.L. Kerschner, P.E. Fanwick, I.P. Rothwell, and J.C. Huffman, *Inorg. Chem.*, **28**, 780 (1989).
194. W.A. Howard, T.M. Trnka, and G. Parkin, *Inorg. Chem.*, **34**, 5900 (1995).
195. M.H. Chisholm and D.L. Clark, *Comments Inorg. Chem.*, **6**, 23 (1987).
196. T.W. Coffindaffer, I.P. Rothwell, and J.C. Huffman, *Inorg. Chem.*, **22**, 2906 (1983).
197. L.D. Durfee, S.L. Latesky, I.P. Rothwell, J.C. Huffman, and K. Folting, *Inorg. Chem.*, **24**, 4569 (1985).
198. J.L. Kerschner, P.E. Fanwick, I.P. Rothwell, and J.C. Huffman, *Inorg. Chem.*, **28**, 780 (1989).
199. S.M. Beshouri and I.P. Rothwell, *Inorg. Chem.*, **25**, 1962 (1986).
200. R.M. Kolodziej, R.R. Schrock, and J.C. Dewan, *Inorg. Chem.*, **28**, 1243 (1989).
201. A. Bell and D.W. deWolf, *Cyclic Voltammetric Studies of Tungsten(VI) Phenoxide Complexes*, Abstracts of the 199th ACS Meeting, Boston, MA, ACS, Washington, DC (1990).

202. A. Bell, *J. Mol. Catal.* **76**, 165 (1992).
203. L.R. Chamberlain and I.P. Rothwell, *J. Chem. Soc., Dalton Trans.*, 163 (1987).
204. A.J. Deeming, I.P. Rothwell, M.B. Hursthouse, and J.D.J. Backer-Dirks, *J. Chem. Soc., Dalton Trans.*, 2039 (1981).
205. S. Bhaduri, K. Sharma, H. Khwaja, and P.G. Jones, *J. Organomet. Chem.*, **412**, 169 (1991).
206. D.S. Bohle and H. Vahrenkamp, *Angew. Chem., Int. Ed. Engl.*, **29**, 198 (1990).
207. M.P. Cifuentes, T.P. Jeynes, M.G. Humphrey, B.W. Skelton, and A.H. White, *J. Chem. Soc., Dalton Trans.*, 925 (1994).
208. D. Rabinovich, R. Zelman, and G. Parkin, *J. Am. Chem. Soc.*, **114**, 4611 (1992).
209. J.F. Hartwig, R.G. Bergman, and R.A. Andersen, *J. Organomet. Chem.*, **394**, 6499 (1991).
210. J.F. Hartwig, R.A. Andersen, and R.G. Bergman, *J. Am. Chem. Soc.*, **113**, 8171 (1990).
211. H.-F. Klein, S. Haller, Hongjian Sun, Xiaoyan Li, T. Jung, C. Rohr, U. Florke, and H.-J. Haupt, *Z. Naturforsch. B*, **53**, 587 (1998).
212. H.-F. Klein, A. Bickelhaupt, B. Hammerschmitt, U. Florke, and H.-J. Haupt, *Organometallics*, **13**, 2944 (1994).
213. P. Ghosh, N. Bag, and A. Chakravorty, *Organometallics*, **15**, 3042 (1996) and references therein.
214. H. Aneetha, C.R.K. Rao, K.M. Rao, P.S. Zacharias, Xue Feng, T.C.W. Mak, B. Srinivas, and M.Y. Chiang, *J. Chem. Soc., Dalton Trans.*, 1696 (1997).
215. B.D. Steffey, L.R. Chamberlain, R.W. Chesnut, D.E. Chebi, P.E. Fanwick, and I.P. Rothwell, *Organometallics*, **8**, 1419 (1989).
216. J.L. Kerschner, I.P. Rothwell, J.C. Huffman, and W.E. Streib, *Organometallics*, **7**, 1871 (1988).
217. J.S. Yu, P.E. Fanwick, and I.P. Rothwell, *J. Am. Chem. Soc.*, **112**, 8171 (1990).
218. J.S. Yu, L. Felter, M.C. Potyen, J.R. Clark, V.M. Visciglio, P.E. Fanwick, and I.P. Rothwell, *Organometallics*, **15**, 4443 (1996).
219. J.R. Clark, P.E. Fanwick, and I.P. Rothwell, *J. Chem. Soc., Chem. Commun.*, 1233 (1993).
220. D.R. Mulford, J.R. Clark, S.W. Schweiger, P.E. Fanwick, and I.P. Rothwell, *Organometallics.*, **18**, 4448 (1999).
221. R.W. Chesnut, G.G. Jacob, J.S. Yu, P.E. Fanwick, and I.P. Rothwell, *Organometallics*, **10**, 321 (1991).
222. S.L. Latesky, A.K. McMullen, I.P. Rothwell, and J.C. Huffman, *J. Am. Chem. Soc.*, **107**, 5981 (1985).
223. A.R. Johnson, W.M. Davis, and C.C. Cummins, *Organometallics*, **15**, 3825 (1996).
224. R.J. Butcher, D.L. Clark, S.K. Grumbine, B.L. Scott, and J.G. Watkin, *Organometallics*, **15**, 1488 (1996).
225. D.L. Clark, S.K. Grumbine, B.L. Scott, and J.G. Watkin, *Organometallics*, **15**, 949 (1996).
226. L. Chamberlain, I.P. Rothwell, and J.C. Huffman, *J. Am. Chem. Soc.*, **108**, 1538 (1986).
227. F. Lefebvre, M. Leconte, S. Pagano, A. Mutch, and J.-M. Basset, *Polyhedron*, **14**, 3209 (1995).
228. M.R. Churchill, J.W. Ziller, J.H. Freudenberger, and R.R. Schrock, *Organometallics*, **3**, 1554 (1984).
229. L.R. Chamberlain, J.L. Kerschner, A.P. Rothwell, I.P. Rothwell, and J.C. Huffman, *J. Am. Chem. Soc.*, **109**, 6471 (1987).
230. M.E. Thompson, S.M. Baxter, A.R. Bulls, B.J. Burger, M.C. Nolan, B.D. Santarsiero, W.P. Schaeref, and J.E. Bercaw, *J. Am. Chem. Soc.*, **109**, 203 (1987).
231. I. de Castro, M.V. Galakhov, M. Gomez, P. Gomez-Sal, A. Martin, and P. Royo, *J. Organomet. Chem.*, **514**, 51 (1996).
232. M.H. Chisholm, I.P. Parkin, K. Folting, E.B. Lubkovsky, and W.E. Streib, *J. Chem. Soc., Chem. Commun.*, 1673 (1991).
233. M.H. Chisholm, J-H. Huang, J.C. Huffman, and I.P. Parkin, *Inorg. Chem.*, **36**, 1642 (1997).
234. G.D. Smith, V.M. Visciglio, P.E. Fanwick, and I.P. Rothwell, *Organometallics*, **11**, 1064 (1992).
235. R.W. Chesnut, G.G. Jacob, J.S. Yu, P.E. Fanwick, and I.P. Rothwell, *Organometallics*, **10**, 321 (1991).

236. D.W. Docter, P.E. Fanwick, and C.P. Kubiak, *J. Am. Chem. Soc.*, **118**, 4846 (1996).
237. W.M. Rees, M.R. Churchill, J.C. Fettinger, and J.D. Atwood, *Organometallics*, **4**, 2179 (1985).
238. J. Ni and C.P. Kubiak, *Homogeneous Transition Metal Catalyzed Reactions, Advances in Chemistry Series*, **230**, 515 (1992).
239. A.M. Gull, J.M. Blatnak, and C.P. Kubiak, *J. Organomet. Chem.*, **577**, 31 (1999).
240. K. Osakada, Yong-Joo Kim, and A. Yamamoto, *J. Organomet. Chem.*, **382**, 303 (1990).
241. D.J. Darensbourg, K.M. Sanchez, J.H. Reibenspies, and A.L. Rheingold, *J. Am. Chem. Soc.*, **111**, 7094 (1989).
242. D.J. Darensbourg, B.L. Mueller, C.J. Bischoff, S.S. Chojnacki, and J.H. Reibenspies, *Inorg. Chem.*, **30**, 2418 (1991).
243. R.L. Cowan and W.C. Trogler, *J. Am. Chem. Soc.*, **111**, 4750 (1989).
244. A.L. Seligson, R.L. Cowan, and W.C. Trogler, *Inorg. Chem.*, **30**, 3371 (1991).
245. R.D. Simpson and R.G. Bergman, *Organometallics*, **11**, 4306 (1992).
246. S.P. Abraham, N. Narasimhamurthy, M. Nethaji, and A.G. Samuelson, *Inorg. Chem.*, **32**, 1739 (1993).
247. C. Wycliff, A.G. Samuelson, and M. Nethaji, *Inorg. Chem.*, **35**, 5427 (1996) and references therein.
248. N. Narasimhamurthy, A.G. Samuelson, and H. Manohar, *J. Chem. Soc., Chem. Commun.*, 1803 (1989).
249. C. Wycliff, D.S. Bharathi, A.G. Samuelson, and M. Nethaji, *Polyhedron*, **18**, 949 (1999) and references therein.
250. D.J. Darensbourg, M.W. Holtcamp, G.E. Struck, M.S. Zimmer, S.A. Niezgoda, P. Rainey, J.B. Robertson, J.D. Draper, and J.H. Reibenspies, *J. Am. Chem. Soc.*, **121**, 107 (1999).
251. D.J. Darensbourg, S.A. Niezgoda, J.D. Draper, and J.H. Reibenspies, *J. Am. Chem. Soc.*, **120**, 4690 (1998).
252. J.P. Fackler, D.P. Schussler, and H.W. Chen, *Synth. React. Inorg. Met.-Org. Chem.*, **8**, 27 (1978).
253. L.M. Green and D.W. Meek, *Organometallics*, **8**, 659 (1989).
254. A.G. Lee, *J. Chem. Soc. A*, 467 (1970).
255. M.B. Power, S.G. Bott, E.J. Bishop, K.D. Tierce, J.L. Atwood, and A.R. Barron, *J. Chem. Soc., Dalton Trans.*, 241 (1991).
256. J. Lewinski, J. Zachara, and I. Justyniak, *Organometallics*, **16**, 4597 (1997).
257. C.F. Edwards, W.P. Griffith, A.J.P. White, and D.J. Williams, *Polyhedron*, **11**, 2711 (1992).
258. a. J.K. Whitesell, *Chem. Rev.*, **89**, 1581 (1989); b. C. Rosini, L. Franzini, A. Raffaelli, and P. Salvadori, *Synthesis*, 503 (1992); c. K. Mikami and Y. Motoyama, in *Encyclopedia of Reagents for Organic Synthesis* (L.A. Paquette, ed.), Vol. 1, 397–403, Wiley, New York (1995).
259. For the exact nature of a widely used BINOL-Ti system and its applications see M. Terada, Y. Matsumoto, Y. Nakamura, and K. Mikami, *Inorg. Chim. Acta.*, **296**, 267 (1999) and references therein.
260. a. S. Akutagawa, *Appl. Catal. A*, **128**, 171 (1995); b. P.J. Cox, W. Wang, and V. Snieckus, *Tetrahedron Lett.* **33**, 2253 (1992); c. M.R. Dennis and S. Woodward, *J. Chem. Soc., Perkin Trans.*, 1081 (1998).
261. K.B. Simonsen, K.V. Gothelf, and K.A. Jorgensen, *J. Org. Chem.*, **63**, 7536 (1998) and references therein.
262. M.D. Morton, J.A. Heppert, S.D. Dietz, W.H. Huang, D.A. Ellis, T.A. Grant, N.W. Eilerts, D.L. Barnes, F. Takusagawa, and D. VanderVelde, *J. Am. Chem. Soc.*, **115**, 7916 (1993).
263. H.C. Aspinall, J.L.M. Dwyer, N. Greeves, and A. Steiner, *Organometallics*, **18**, 1366 (1999) and references therein.
264. N. Giuseppone, J. Collin, A. Domingos, and I. Santos, *J. Organomet. Chem.* **590**, 248 (1999) and references therein.
265. M. Shibazaki and H. Groger, *Top. Organomet. Chem.* **2**, 199 (1999).
266. R.R. Schrock, *Tetrahedron*, **55**, 8141 (1999).
267. H. Ishitani, T. Kitazawa, and S. Kobayashi, *Tetrahedron Lett.*, **40**, 2161 (1999).

268. K. Ishihara, H. Kurihara, M. Matsumoto, and H. Yamamoto, *J. Am. Chem. Soc.*, **120**, 6920 (1998).

269. a. A.-M. Sapse and P.v.R. Schleyer, Eds. *Lithium Chemistry*; Wiley, New York (1995); b. R.E. Mulvey, *Chem. Soc. Rev.*, **20**, 167 (1991).

270. a. D.B. Collum, *Acc. Chem. Res.*, **25**, 448 (1992); b. K. Gregory, P.v.R. Schleyer, and R. Snaith, *Adv. Inorg. Chem.*, **37**, 47 (1991); c. B. Goldfuss, P.v.R. Schleyer, and F. Hampel, *J. Am. Chem. Soc.*, **118**, 12183 (1996) and references therein.

271. R.E. Dinnebier, M. Pink, J. Sieler, and P.W. Stephens, *Inorg. Chem.*, **36**, 3398 (1997).

272. A.S. Lindsey and H. Jeskey, *Chem. Rev.*, **57**, 583 (1957).

273. R.E. Dinnebier, M. Pink, J. Sieler, P. Norby, and P.W. Stephens, *Inorg. Chem.*, **37**, 4996 (1998).

274. J.S. Vilardo, P.E. Fanwick, and I.P. Rothwell, *Polyhedron*, **17**, 769 (1998).

275. M. Kunert, E. Dinjus, M. Nauck, and J. Sieler, *Chem. Ber.*, **130**, 1461 (1997).

276. W.J. Evans, R.E. Golden, and J.W. Ziller, *Inorg. Chem.*, **32**, 3041 (1993).

277. D.L. Clark, D.R. Click, R.V. Hollis, B.L. Scott, and J.G. Watkin, *Inorg. Chem.*, **37**, 5700 (1998).

278. L.M. Jackman, D. Cizmeciyan, P.G. Williard, and M.A. Nichols, *J. Am. Chem. Soc.*, **115**, 6262 (1993).

279. P.A. vanderSchaaf, J.T.B.H. Jastrzebski, M.P. Hogerheide, W.J.J. Smeets, A.L. Spek, J. Boersma, and G. van Koten, *Inorg. Chem.*, **32**, 4111 (1993).

280. J.C. Huffman, R.L. Geerts, and K.G. Caulton, *J. Cryst. Spectrosc.*, **14**, 541 (1984).

281. L. Matilainen, M. Leskela, and M. Klinga, *J. Chem. Soc., Chem. Commun.*, 421 (1995).

282. L.M. Jackman and B.D. Smith, *J. Am. Chem. Soc.*, **110**, 3829 (1988).

283. L.M. Jackman and E.F. Rakiewicz, *J. Am. Chem. Soc.*, **113**, 1202 (1991).

284. L.M. Jackman and C.W. DeBrosse, *J. Am. Chem. Soc.*, **105**, 4177 (1983).

285. L.M. Jackman, L.M. Scarmoutzos, and C.W. DeBrosse, *J. Am. Chem. Soc.*, **109**, 5355 (1987).

286. T.P. Hanusa, *Chem. Rev.*, **93**, 1023 (1993).

287. K.G. Caulton, M.H. Chisholm, S.R. Drake, K. Folting, J.C. Huffman, and W.E. Streib, *Inorg. Chem.*, **32**, 1970 (1993).

288. K.F. Tesh, T.P. Hanusa, J.C. Huffman, and C.J. Huffman, *Inorg. Chem.*, **31**, 5572 (1992).

289. P.B. Hitchcock, M.F. Lappert, and A. Singh, *J. Chem. Soc., Chem. Commun.*, 1499 (1983).

290. P.B. Hitchcock, M.F. Lappert, R.G. Smith, R.A. Bartlett, and P.P. Power, *J. Chem. Soc., Chem. Commun.*, 10007 (1988).

291. H.J. Heeres, A. Meetsma, J.H. Teuben, and R.D. Rogers, *Organometallics*, **8**, 2637 (1989).

292. H.A. Stecher, A. Sen, and A.L. Rheingold, *Inorg. Chem.*, **27**, 1130 (1988).

293. R.J. Butcher, D.L. Clark, S.K. Grumbine, R.L. VincentHollis, B.L. Scott, and J.G. Watkin, *Inorg. Chem.*, **34**, 5468 (1995).

294. Z. Hou, Y. Zhang, T. Yoshimura, and Y. Wakatsuki, *Organometallics*, **16**, 2963 (1997).

295. Z. Hou, A. Fujita, T. Yoshimura, A. Jesorka, Y. Zhang, H. Yamazaki, and Y. Wakatsuki, *Inorg. Chem.*, **35**, 7190 (1996).

296. G.B. Deacon, Tiecheng Feng, P. MacKinnon, R.H. Newnham, S. Nickel, B.W. Skelton, and A.H. White, *Aust. J. Chem.*, **46**, 387 (1993).

297. G.B. Deacon, B.M. Gatehouse, Q. Shen, G.N. Ward, and E.R.T. Tiekink, *Polyhedron*, **12**, 1289 (1993).

298. W.J. Evans, W.G. McClelland, M.A. Greci, and J.W. Ziller, *Eur. J. Solid State Inorg. Chem.*, **33**, 145 (1996).

299. B. Cetinkaya, P.B. Hitchcock, M.F. Lappert, and R.G. Smith, *J. Chem. Soc., Chem. Commun.*, 932 (1992).

300. W.J. Evans, M.A. Greci, and J.W. Ziller, *J. Chem. Soc., Chem. Commun.*, 2367 (1998).

301. G.B. Deacon and Q. Shen, *J. Organomet. Chem.*, **511**, 1 (1996).

302. G.B. Deacon, C.M. Forsyth, P.C. Junk, B.S. Skelton, and A.H. White, *Chem. Eur. J*, **5**, 1452 (1999).

303. G.B. Deacon, Tiecheng Feng, S. Nickel, M.I. Ogden, and A.H. White, *Aust. J. Chem.*, **45**, 671 (1992).

304. M.P. Hogerheide, J.T.B.H. Jastrzebski, J. Boersma, W.J.J. Smeets, A.L. Spek, and G. van Koten, *Inorg. Chem.*, **33**, 4431 (1994).

305.   D.M. Barnhart, D.L. Clark, J.C. Gordon, J.C. Huffman, R.L. Vincent, J.G. Watkin, and B.D. Zwick, *Inorg. Chem.*, **33**, 3487 (1994).
306.   D.L. Clark, J.C. Gordon, J.G. Watkin, J.C. Huffman, and B.D. Zwick, *Polyhedron*, **15**, 2279 (1996).
307.   J.R. van den Hende, P.B. Hitchcock, S.A. Holmes, M.F. Lappert, WingPor Leung, T.C.W. Mak, and S. Prashar, *J. Chem. Soc., Dalton Trans.*, 1427 (1995).
308.   C. Barberato, E.E. Castellano, G. Vicentini, P.C. Isolani, and M.I.R. Lima, *Acta Crystallogr. C*, **50**, 351 (1994).
309.   G.B. Deacon, T. Feng, P.C. Junk, B.W. Skelton, and A.H. White, *Chem. Ber.*, **130**, 851 (1997).
310.   W.J. Evans, M.A. Greci, and J.W. Ziller, *J. Chem. Soc., Dalton Trans.*, 3035 (1997).
311.   H.C. Aspinall and M. Williams, *Inorg. Chem.*, **35**, 255 (1996).
312.   W.J. Evans, R.E. Golden, and J.W. Ziller, *Inorg. Chem.*, **32**, 3041 (1993).
313.   D.L. Clark, R.V. Hollis, B.L. Scott, and J.G. Watkin, *Inorg. Chem.*, **35**, 667 (1996).
314.   D.L. Clark, G.B. Deacon, Tiecheng Feng, R.V. Hollis, B.L. Scott, B.W. Skelton, J.G. Watkin, and A.H. White, *J. Chem. Soc., Chem. Commun.*, 1729 (1996).
315.   G.B. Deacon, T. Feng, P.C. Junk, B.W. Skelton, and A.H. White, *J. Chem. Soc., Dalton Trans.*, 1181 (1997).
316.   C.J. Schaverien, J.H.G. Frijns, H.J. Heeres, J.R. van den Hende, J.H. Teuben, and A.L. Spek, *J. Chem. Soc., Chem. Commun.*, 642 (1991).
317.   C.J. Schaverien, *Organometallics*, **13**, 69 (1994).
318.   W.E. Piers, E.E. Bunel, and J.E. Bercaw, *J. Organomet. Chem.*, **407**, 51 (1991).
319.   P. Shao, D.J. Berg, and G.W. Bushnell, *Inorg. Chem.*, **33**, 6334 (1994).
320.   D.L. Clark, J.C. Gordon, J.C. Huffman, J.G. Watkin, and B.D. Zwick, *Organometallics*, **13**, 4266 (1994).
321.   P.B. Hitchcock, M.F. Lappert, A. Singh, R.G. Taylor, and D. Brown, *J. Chem. Soc., Chem. Commun.*, 561 (1983).
322.   J.M. Berg, D.L. Clark, J.C. Huffman, D.E. Morris, A.P. Sattelberger, W.E. Streib, W.G. van der Sluys, and J.G. Watkin, *J. Am. Chem. Soc.*, **114**, 10811 (1992).
323.   P.G. Edwards, R.A. Andersen, and A. Zalkin, *J. Am. Chem. Soc.*, **103**, 7792 (1981).
324.   W.G. van der Sluys, A.P. Sattelberger, W.E. Streib, and J.C. Huffman, *Polyhedron*, **8**, 1247 (1989).
325.   S.D. McKee, C.J. Burns, and L.R. Avens, *Inorg. Chem.*, **37**, 4040 (1998).
326.   W.G. van der Sluys, C.J. Burns, J.C. Huffman, and A.P. Sattelberger, *J. Am. Chem. Soc.*, **110**, 5924 (1988).
327.   L.R. Avens, D.M. Barnhart, C.J. Burns, S.D. McKee, and W.H. Smith, *Inorg. Chem.*, **33**, 4245 (1994).
328.   K.C. Malhotra, M. Sharma, and N. Sharma, *Indian J. Chem. A*, **24**, 790 (1985).
329.   D.M. Barnhart, C.J. Burns, N.N. Sauer, and J.G. Watkin, *Inorg. Chem.*, **34**, 4079 (1995).
330.   D.S.J. Arney and C.J. Burns, *J. Am. Chem. Soc.*, **115**, 9840 (1993).
331.   W.J. Evans, M.A. Ansari, and J.W. Ziller, *Polyhedron*, **17**, 299 (1997).
332.   A.W. Duff, R.A. Kamarudin, M.F. Lappert, R.J. Norton, and J. Reginald, *J. Chem. Soc., Dalton Trans.*, 489 (1986).
333.   K.C. Malhotra, N. Sharma, S.S. Bhatt, and S.C. Chaudhry, *J. Indian Chem. Soc.*, **67**, 830 (1990).
334.   L.I. Vyshinskaya, G.A. Vasil'eva, T.A. Vishnyakova, D.V. Muslin, N.Sh. Lyapina, and A.S. Smirnov, *Metalloorg. Khim.*, **2**, 907 (1989).
335.   R. Gupta, A. Singh, and R.C Mehrotra, *Indian J. Chem. A*, **29A**, 596 (1990).
336.   P. Jutzi, T. Redeker, B. Neumann, and H.-G. Stammler, *Organometallics*, **15**, 4153 (1996).
337.   K.C. Malhotra, N. Sharma, S.S. Bhatt, and S.C. Chaudhry, *J. Indian Chem. Soc.*, **71**, 571 (1994).
338.   K.C. Malhotra, N. Sharma, and S.C. Chaudhry, *Transition Met. Chem.*, **6**, 238 (1981).
339.   K.C. Malhotra, N. Sharma, S.S. Bhatt, and S.C. Chaudhry, *J. Therm. Anal.*, **46**, 327 (1996).
340.   T.P. Vaid, J.M. Tanski, J.M. Pette, E.B. Lobkovsky, and P.T. Wolczanski, *Inorg. Chem.*, **38**, 3394 (1999).

341. B.S. Kalirai, J.-D. Foulon, T.A. Hamor, C.J. Jones, P.D. Beer, and S.P. Fricker, *Polyhedron*, **10**, 1847 (1991).
342. J.E. Hill, P.E. Fanwick, and I.P. Rothwell, *Inorg. Chem.*, **28**, 3602 (1989).
343. C.H. Zambrano, R.D. Profilet, J.E. Hill, P.E. Fanwick, and I.P. Rothwell, *Polyhedron*, **12**, 689 (1993).
344. J.E. Hill, R.D. Profilet, P.E. Fanwick, and I.P. Rothwell, *Angew. Chem., Int. Ed. Engl.*, **29**, 664 (1990).
345. J.E. Hill, P.E. Fanwick, and I.P. Rothwell, *Inorg. Chem.*, **30**, 1143 (1991).
346. P.E. Collier, A.J. Blake, and P. Mountford, *J. Chem. Soc., Dalton Trans.*, 2911 (1997).
347. P.J. Stewart, A.J. Blake, and P. Mountford, *Organometallics*, **17**, 3271 (1998).
348. A. Togni and R.L. Halterman (eds), *Metallocenes*, Wiley-VCH, New York, 1998.
349. a. H. Urabi and F. Sato, *J. Am. Chem. Soc.*, **121**, 1245 (1999); b. T. Nakagawa, A. Kasatkin, and F. Sato, *Tetrahedron Lett.*, **36**, 3207 (1995); c. A. Kasatkin, T.Nakagawa, S. Okamoto, and F. Sato, *J. Am. Chem. Soc.*, **117**, 3881 (1995); d. S. Okamoto, A. Kasatkin, P.K. Zubaidha, and F. Sato, *J. Am. Chem. Soc.*, **118**, 2208 (1996); e. K. Suzuki, H. Urabe, and F. Sato, *J. Am. Chem. Soc.*, **118**, 8729 (1996); f. H. Urabe and F. Sato, *J. Org. Chem.*, **61**, 6756 (1996); g. M. Koiwa, G.P.J. Hareau, D. Morizono, and F. Sato, *Tetrahedron Lett.*, **40**, 4199 (1999); h. Y. Takayama, S. Okamoto, and F. Sato, *J. Am. Chem. Soc.*, **121**, 3559 (1999); i. H. Urabe, T. Hamada, and F. Sato, *J. Am. Chem. Soc.*, **121**, 2931 (1999); j. R. Mizojiri, H. Urabe, and F. Sato, *Tetrahedron Lett.*, **40**, 2557 (1999); k. S. Yamaguchi, R.Z. Jin, K. Tamao, and F. Sato, *J. Org. Chem.*, **63**, 10060 (1998); l. S.Y. Cho, J.H. Lee, R.K. Lammi, and J.K. Cha, *J. Org. Chem.*, **62**, 8235 (1997); m. J.H. Lee, Y.G. Kim, J.G. Bae, and J.K. Cha, *J. Org. Chem.*, **61**, 4878 (1996); n. J. Lee, H. Kim, and J.K. Cha, *J. Am. Chem. Soc.*, **118**, 4198 (1996); o. J.H. Lee, C.H. Kang, H.J. Kim, and J.K. Cha, *J. Am. Chem. Soc.*, **118**, 292 (1996).
350. C.J. Covert, A.R. Mayol, and P.T. Wolczanski, *Inorg. Chim. Acta.*, **263**, 263 (1997).
351. M.G. Thorn, J.E. Hill, S.A. Waratuke, E.S. Johnson, P.E. Fanwick, and I.P. Rothwell, *J. Am. Chem. Soc.*, **119**, 8630 (1997).
352. J.S. Vilardo, M.G. Thorn, P.E. Fanwick, and I.P. Rothwell, *J. Chem. Soc., Chem. Commun.*, 2425 (1998).
353. A.V. Firth and D.W. Stephan, *Organometallics*, **16**, 2183 (1997).
354. K. Nomura, N. Naga, M. Miki, K. Yanagi, and A. Imai, *Organometallics*, **17**, 2152 (1998).
355. M.J. Sarsfield, S.W. Ewart, T.L. Tremblay, A.W. Roszak, and M.C. Baird, *J. Chem. Soc., Dalton Trans.*, 3097 (1997).
356. a) M.G. Thorn, J.S. Vilardo, P.E. Fanwick, and I.P. Rothwell, *J. Chem. Soc., Chem. Commun.*, 2427 (1998). b) M.G. Thorn, J.S. Vilardo, J. Lee, B. Hanna, P.L. Fanwick and I.P. Rothwell, *Organometallics*, in press.
357. J.E. Hill, J.M. Nash, P.E. Fanwick, and I.P. Rothwell, *Polyhedron*, **9**, 1617 (1990).
358. S.R. Waratuke, M.G. Thorn, P.E. Fanwick, A.P. Rothwell, and I.P. Rothwell, *J. Am. Chem. Soc.*, **121**, 9111 (1999).
359. B.P. Santora, A.O. Larsen, and M.R. Gagné, *Organometallics*, **17**, 3138 (1998).
360. R.W. Chesnut, L.D. Durfee, P.E. Fanwick, I.P. Rothwell, K. Folting, and J.C. Huffman, *Polyhedron*, **6**, 2019 (1987).
361. M.G. Thorn, Z.C. Etheridge, P.E. Fanwick, and I.P. Rothwell, *J. Organomet. Chem.*, **591**, 148 (1999).
362. L.R. Chamberlain, L.D. Durfee, P.E. Fanwick, L.M. Kobriger, S.L. Latesky, A.K. McMullen, B.D. Steffey, I.P. Rothwell, K. Folting, and J.C. Huffman, *J. Am. Chem. Soc.*, **109**, 6068 (1987).
363. I.P. Rothwell and L.D. Durfee, *Chem. Rev.*, **88**, 1059 (1988).
364. L.D. Durfee, P.E. Fanwick, I.P. Rothwell, K. Folting, and J.C. Huffman, *J. Am. Chem. Soc.*, **109**, 4720 (1987).
365. L.D. Durfee, J.E. Hill, P.E. Fanwick, and I.P. Rothwell, *Organometallics*, **9**, 75 (1990).
366. L.D. Durfee, P.E. Fanwick, I.P. Rothwell, K. Folting, and J.C. Huffman, *J. Am. Chem. Soc.*, **109**, 4720 (1987).
367. L.D. Durfee, J.E. Hill, J.L. Kerschner, P.E. Fanwick, and I.P. Rothwell, *Inorg. Chem.*, **28**, 3095 (1989).
368. R.J. Morris and G.S. Girolami, *Inorg. Chem.*, **29**, 4167 (1990).

369. A.K. McMullen, I.P. Rothwell, and J.C. Huffman, *J. Am. Chem. Soc.*, **107**, 1072 (1985).
370. L.R. Chamberlain, L.D. Durfee, P.E. Fanwick, L. Kobriger, S.L. Latesky, A.K. McMullen, I.P. Rothwell, K. Folting, J.C. Huffman, W.E. Streib, and R. Wang, *J. Am. Chem. Soc.*, **109**, 390 (1987).
371. L.D. Durfee, L.M. Kobriger, A.K. McMullen, and I.P. Rothwell, *J. Am. Chem. Soc.*, **110**, 1463 (1988).
372. P.E. Fanwick, L.M. Kobriger, A.K. McMullen, and I.P. Rothwell, *J. Am. Chem. Soc.*, **108**, 8095 (1986).
373. C.H. Zambrano, A.K. McMullen, L.M. Kobriger, P.E. Fanwick, and I.P. Rothwell, *J. Am. Chem. Soc.*, **112**, 6565 (1990).
374. M.T. Thorn and I.P. Rothwell, unpublished results.
375. M.G. Thorn, Z.C. Etheridge, P.E. Fanwick, and I.P. Rothwell, *Organometallics*, **17**, 3636 (1998).
376. a. J. Okuda and E. Masoud, *Macromol. Chem. Phys.*, **199**, 543 (1998). b. F.G. Sernetz, R. Muelhaupt, S. Fokken, and J. Okuda, *Macromolecules*, **30**, 1562 (1997).
377. K. Nomura, N. Naga, M. Miki, K. Yanagi, and A. Imai, *Organometallics*, **17**, 2152 (1998).
378. M.J. Sarsfield, S.W. Ewart, T.L. Tremblay, A.W. Roszak, and M.C. Baird, *J. Chem. Soc., Dalton Trans.*, 3097 (1997).
379. J.E. Hill, G. Balaich, P.E. Fanwick, and I.P. Rothwell, *Organometallics*, **12**, 2911 (1993).
380. J.E. Hill, P.E. Fanwick, and I.P. Rothwell, *Organometallics*, **9**, 2211 (1990).
381. J.E. Hill, P.E. Fanwick, and I.P. Rothwell, *Organometallics*, **10**, 15 (1991).
382. E.S. Johnson, G.J. Balaich, and I.P. Rothwell, *J. Am. Chem. Soc.*, **119**, 7685 (1997).
383. J.E. Hill, G.J. Balaich, P.E. Fanwick, and I.P. Rothwell, *Organometallics*, **10**, 3428 (1991).
384. S. Gambarotta, F. van Bolhuis, and M.Y. Chiang, *Inorg. Chem.*, **26**, 4301 (1987).
385. R.K. Minhas, J.J.H. Edema, S. Gambarotta, and A. Meetsma, *J. Am. Chem. Soc.*, **115**, 6710 (1993).
386. R.A. Henderson, D.L. Hughes, Z. Janas, R.L. Richards, P. Sobota, and S. Szafert, *J. Organomet. Chem.*, **554**, 195 (1998).
387. W.A. Herrmann, E. Herdtweck, and G. Weichselbaumer, *J. Organomet. Chem.*, **362**, 321 (1989).
388. D.D. Devore, J.D. Lichtenhan, F. Takusagawa, and E.A. Maatta, *J. Am. Chem. Soc.*, **109**, 7408 (1987).
389. R.A. Henderson, Z. Janas, L.B. Jerzykiewicz, R.L. Richards, and P. Sobota, *Inorg. Chim. Acta*, **285**, 178 (1999).
390. V.J. Murphy and H. Turner, *Organometallics*, **16**, 2495 (1997).
391. K.C. Malhotra, U.K. Banerjee, and S.C. Chaudhry, *J. Indian Chem. Soc.*, **57**, 868 (1980).
392. M. Schoenherr and J. Koellner, *Z. Chem.*, **18**, 36 (1978).
393. M. Schoenherr, *Z. Chem.*, **16**, 374 (1976).
394. K.C. Malhotra, B. Bala, N. Sharma, and S.C. Chaudhry, *J. Indian Chem. Soc.*, **70**, 187 (1993).
395. M. Schoenherr, *Z. Chem.*, **20**, 155 (1980).
396. T.W. Coffindaffer, B.D. Steffey, I.P. Rothwell, K. Folting, J.C. Huffman, and W.E. Streib, *J. Am. Chem. Soc.*, **111**, 4742 (1989).
397. L.N. Lewis and M.F. Garbauskas, *Inorg. Chem.*, **24**, 363 (1985).
398. J.R. Clark, A.L. Pulverenti, P.E. Fanwick, M. Sigalas, O. Eisenstein, and I.P. Rothwell, *Inorg. Chem.*, **36**, 3623 (1997).
399. H. Yasuda, Y. Nakayama, K. Takei, A. Nakamura, Y. Kai, and N. Kanehisa, *J. Organomet. Chem.*, **473**, 105 (1994).
400. K.D. Allen, M.A. Bruck, S.D. Gray, R.P. Kingsborough, D.P. Smith, K.J. Weller, and D.E. Wigley, *Polyhedron*, **14**, 3315 (1995).
401. B.C. Parkin, J.C. Clark, V.M. Visciglio, P.E. Fanwick, and I.P. Rothwell, *Organometallics*, **14**, 3002 (1995).
402. R.R. Schrock, M. Wesolek, A.H. Liu, K.C. Wallace, and J.C. Dewan, *Inorg. Chem.*, **27**, 2050 (1988).
403. L.R. Chamberlain, J. Keddington, and I.P. Rothwell, *Organometallics*, **1**, 1098 (1982).
404. K.C. Wallace, A.H. Liu, J.C. Dewan, and R.R. Schrock, *J. Am. Chem. Soc.*, **110**, 4964 (1988).

405. P.N. Riley, M.G. Thorn, J.S. Vilardo, M.A. Lockwood, P.E. Fanwick, and I.P. Rothwell, *Organometallics*, **18**, 3016 (1999).
406. D.J. Arney, M.A. Bruck, and D.E. Wigley, *Organometallics*, **10**, 3947 (1991).
407. D.E. Wigley and I.P. Rothwell, unpublished results.
408. I.P. Rothwell and J.S. Yu, US Patent 5,530,162 (1996); J.S. Yu, and I.P. Rothwell, *J. Chem. Soc., Chem. Commun.*, 632 (1992).
409. M.C. Potyen and I.P. Rothwell, *J. Chem. Soc., Chem. Commun.*, 849 (1995).
410. I.P. Rothwell, *J. Chem. Soc., Chem. Commun.*, 1331 (1997).
411. J.S. Yu, B.C. Ankianiec, M.T. Nguyen, and I.P. Rothwell, *J. Am. Chem. Soc.*, **114**, 1927 (1992).
412. S.D. Gray, D.P. Smith, M.A. Bruck, and D.E. Wigley, *J. Am. Chem. Soc.*, **114**, 5462 (1992).
413. J.R. Strickler, M.A. Bruck, and D.E. Wigley, *J. Am. Chem. Soc.*, **112**, 2814 (1990).
414. K.J. Weller, S.D. Gray, P.M. Briggs, and D.E. Wigley, *Organometallics*, **14**, 5588 (1995).
415. S.D. Gray, K.J. Weller, M.A. Bruck, P.M. Briggs, and D.E. Wigley, *J. Am. Chem. Soc.*, **117**, 10678 (1995).
416. V.M. Visciglio, J.R. Clark, M.T. Nguyen, D.R. Mulford, P.E. Fanwick, and I.P. Rothwell, *J. Am. Chem. Soc.*, **119**, 3490 (1997).
417. D.P. Smith, J.R. Stickler, S.D. Gray, M.A. Bruck, R.S. Holmes, and D.E. Wigley, *Organometallics*, **11**, 1275 (1992).
418. J.R. Strickler, P.A. Wexler, and D.E. Wigley, *Organometallics*, **10**, 118 (1991).
419. J.R. Strickler, M.A. Bruck, P.A. Wexler, and D.E. Wigley, *Organometallics*, **9**, 266 (1990).
420. J.R. Strickler and D.E. Wigley, *Organometallics*, **9**, 1665 (1990).
421. D.J. Arney, P.A. Wexler, and D.E. Wigley, *Organometallics*, **9**, 1282 (1990).
422. K.J. Weller, S.D. Gray, P.M. Briggs, and D.E. Wigley, *Organometallics*, **14**, 5588 (1995).
423. J.J.H. Edema, S. Gambarotta, F. van Bolhuis, W.J.J. Smeets, and A.L. Spek, *Inorg. Chem.*, **28**, 1407 (1989).
424. J.J.H. Edema, S. Gambarotta, F. van Bolhuis, and A.L. Spek, *J. Am. Chem. Soc.*, **111**, 2142 (1989).
425. T.W. Coffindaffer, G.P. Niccolai, D. Powell, I.P. Rothwell, and J.C. Huffman, *J. Am. Chem. Soc.*, **107**, 3572 (1985).
426. T.W. Coffindaffer, I.P. Rothwell, and J.C. Huffman, *Inorg. Chem.*, **22**, 2906 (1983).
427. R.G. Abbott, F.A. Cotton, and L.R. Falvello, *Inorg. Chem.*, **29**, 514 (1990).
428. F.A. Cotton and K.J. Wiesinger, *Inorg. Chem.*, **30**, 750 (1991).
429. V.M. Visciglio, P.E. Fanwick, and I.P. Rothwell, *Inorg. Chim. Acta*, **211**, 203 (1993).
430. W. Clegg, R.J. Errington, P. Kraxner, and C. Redshaw, *J. Chem. Soc., Dalton Trans.*, 1431 (1992).
431. D.J. Darensbourg, K.M. Sanchez, J.H. Reibenspies, and A.L. Rheingold, *J. Am. Chem. Soc.*, **111**, 7094 (1989).
432. N. Kanehisa, Y. Kai, N. Kasai, H. Yasuda, Y. Nakayama, and A. Nakamura, *Bull. Chem. Soc. Jpn.*, **65**, 1197 (1992).
433. M.L. Listemann, R.R. Schrock, J.C. Dewan, and R.M. Kolodziej, *Inorg. Chem.*, **27**, 264 (1988).
434. J.L. Kerschner, E.M. Torres, P.E. Fanwick, I.P. Rothwell, and J.C. Huffman, *Organometallics*, **8**, 1424 (1989).
435. J.L. Kerschner, I.P. Rothwell, J.C. Huffman, and W.E. Streib, *Organometallics*, **7**, 1871 (1988).
436. J.L. Kerschner, I.P. Rothwell, J.C. Huffman, and W.E. Streib, *Organometallics*, **7**, 1871 (1988).
437. J.L. Kerschner, P.E. Fanwick, I.P. Rothwell, and J.C. Huffman, *Organometallics*, **8**, 1431 (1989).
438. C.E. Kriley, P.E. Fanwick, and I.P. Rothwell, *J. Am. Chem. Soc.*, **116**, 5225 (1994).
439. M.A. Lockwood, P.E. Fanwick, and I.P. Rothwell, *Polyhedron*, **14**, 3363 (1995).
440. M.A. Lockwood, P.E. Fanwick, O. Eisenstein, and I.P. Rothwell, *J. Am. Chem. Soc.*, **118**, 2762 (1996).
441. F. Maseras, M.A. Lockwood, O. Eisenstein, and I.P. Rothwell, *J. Am. Chem. Soc.*, **120**, 6598 (1998).

442.   M.A. Lockwood, P.E. Fanwick, and I.P. Rothwell, *J. Chem. Soc., Chem. Commun.*, 2013 (1996).
443.   V. Dragutan, A.T. Balaban, and M. Dimonie, *Olefin Metathesis and Ring-Opening Polymeriazation of Cyclo-Olefins*, Wiley-Interscience, New York (1985).
444.   H.T. Dodd and K.J. Rutt, *J. Mol. Catal.*, **15**, 103 (1982).
445.   F. Quignard, M. Leconte, and J.M. Basset, *J. Mol. Catal.*, **28**, 27 (1985).
446.   H.T. Dodd and K.J. Rutt, *J. Mol. Catal.*, **28**, 33 (1985).
447.   H.T. Dodd and K.J. Rutt, *J. Mol. Catal.*, **47**, 67 (1988).
448.   A. Bell, *ACS Symp. Ser.*, **496**, 121 (1992).
449.   W.A. Nugent, J. Feldman, and J.C. Calabrese, *J. Am. Chem. Soc.*, **117**, 8992 (1995).
450.   R.R. Schrock, *Polyhedron*, **14**, 3177 (1995).
451.   I.A. Latham, L.R. Sita, and R.R. Schrock, *Organometallics*, **5**, 1508 (1986).
452.   J.H. Freudenberger and R.R. Schrock, *Organometallics*, **5**, 1411 (1986).
453.   R.R. Schrock, *Top. Organomet. Chem.*, **1**, 1 (1998).
454.   M.B. O'Donoghue, R.R. Schrock, A.M. LaPointe, and W.M. Davis, *Organometallics*, **15**, 1334 (1996).
455.   D.A. Bardwell, J.C. Jeffery, and M.D. Ward, *Inorg. Chim. Acta*, **236**, 125 (1995).
456.   B. Horvath, R. Moeseler, and E.G. Horvath, *Z. Anorg. Allg. Chem.*, **449**, 41 (1979).
457.   R.A. Bartlett, J.J. Ellison, P.P. Power, and S.C. Shoner, *Inorg. Chem.*, **30**, 2888 (1991).
458.   I.M. Gardiner, M.A. Bruck, and D.E. Wigley, *Inorg. Chem.*, **28**, 1769 (1989).
459.   I.M. Gardiner, M.A. Bruck, P.A. Wexler, and D.E. Wigley, *Inorg. Chem.*, **28**, 3688 (1989).
460.   T.K.G. Erikson, J.C. Bryan, and J.M. Mayer, *Organometallics*, **7**, 1930 (1988).
461.   R.C. Conry and J.M. Mayer, *Organometallics*, **12**, 3179 (1993).
462.   A.M. Lapointe and R.R. Schrock, *Organometallics*, **14**, 1875 (1995) and references therein.
463.   M.H. Schofield, R.R. Schrock, and L.Y. Park, *Organometallics*, **10**, 1844 (1991).
464.   S.N. Brown and J.M. Mayer, *J. Am. Chem. Soc.*, **118**, 12119 (1996) and references therein.
465.   S.N. Brown and J.M. Mayer, *Organometallics*, **14**, 2951 (1995).
466.   R.D. Simpson and R.G. Bergman, *Angew. Chem., Int. Ed. Engl.*, **31**, 220 (1992).
467.   S. Komiya, S. Taneichi, A. Yamamoto, and T. Yamamoto, *Bull. Chem. Soc. Jpn.*, **53**, 673 (1980).
468.   A.M. Helms, W.D. Jones, and G.L. McLendon, *J. Coord. Chem.*, **23**, 351 (1991).
469.   R.D. Arasasingham, A.L. Balch, C.R. Cornman, J.S. De Ropp, K. Eguchi, and G.N. La Mar, *Inorg. Chem.*, **29**, 1847 (1990).
470.   H. Sugimoto, N. Ueda, and M. Mori, *Bull. Chem. Soc. Jpn.*, **55**, 3468 (1982).
471.   H. Nasri, J. Fischer, R. Weiss, E. Bill, and A. Trautwein, *J. Am. Chem. Soc.*, **109**, 2549 (1987).
472.   E.W. Ainscough, A.W. Addison, D. Dolphin, and B.R. James, *J. Am. Chem. Soc.*, **100**, 7585 (1978).
473.   J.P. Collman, C.E. Barnes, P.J. Brothers, T.J. Collins, T. Ozawa, J.C. Gallucci, and J.A. Ibers, *J. Am. Chem. Soc.*, **106**, 5151 (1984).
474.   Ming-Chu Cheng, Chih-Chieh Wang, and Yu Wang, *Inorg. Chem.*, **31**, 5220 (1992).
475.   M.G. Kanatzidis, N.C. Baenziger, D. Coucouvanis, A. Simopoulos, and A. Kostikas, *J. Am. Chem. Soc.*, **106**, 4500 (1984).
476.   A. Salifoglou, A. Simopoulos, A. Kostikas, R.W. Dunham, M.G. Kanatzidis, and D. Coucouvanis, *Inorg. Chem.*, **27**, 3394 (1988).
477.   C. Goh, J.A. Weigel, and R.H. Holm, *Inorg. Chem.*, **33**, 4861 (1994).
478.   W.E. Cleland, D.A. Holtman, M. Sabat, J.A. Ibers, G.C. DeFotis, and B.A. Averill, *J. Am. Chem. Soc.*, **105**, 6021 (1983).
479.   W.E. Cleland and B.A. Averill, *Inorg. Chim. Acta*, **106**, L17 (1985).
480.   S.A. Al-Ahmad, A. Salifoglou, M.G. Kanatzidis, W.R. Dunham, and D. Coucouvanis, *Inorg. Chem.*, **29**, 927 (1990).
481.   M.B. Cingi, A.M.M. Lanfredi, A. Tiripicchio, J. Reedijk, and R. van Landschoot, *Inorg. Chim. Acta*, **39**, 181 (1980).
482.   M. Basturkmen, D. Kisakurek, W.L. Driessen, S. Gorter, and J. Reedijk, *Inorg. Chim. Acta*, **271**, 19 (1988).

483. Y. Hayashi, T. Yamamoto, A. Yamamoto, S. Komiya, and Y. Kushi, *J. Am. Chem. Soc.*, **108**, 385 (1986).
484. V.F. Kuznetsov, G.P.A. Yap, C. Bensimon, and H. Alper, *Inorg. Chim. Acta*, **280**, 172 (1998).
485. S.E. Kegley, C.J. Schaverien, J.H. Freudenberger, R.G. Bergman, S.P. Nolan, and C.D. Hoff, *J. Am. Chem. Soc.*, **109**, 6563 (1987).
486. K.A. Bernard, M.R. Churchill, T.S. Janik, and J.D. Atwood, *Organometallics*, **9**, 12 (1990).
487. K.A. Woerpel and R.G. Bergman, *J. Am. Chem. Soc.*, **115**, 7888 (1993).
488. D.P. Klein and R.G. Bergman, *J. Am. Chem. Soc.*, **111**, 3079 (1989).
489. Yong-Joo Kim, Kohtaro Osakada, A. Takenaka, and A. Yamamoto, *J. Am. Chem. Soc.*, **112**, 1096 (1990).
490. A.L. Seligson and W.C. Trogler, *J. Am. Chem. Soc.*, **114**, 7085 (1992).
491. G.M. Kapteijn, M.D. Meijer, D.M. Grove, N. Veldman, A.L. Spek, and G. van Koten, *Inorg. Chim. Acta*, **264**, 211 (1997).
492. M. Kapteijn, M.P.R. Spee, D.M. Grove, H. Kooijman, A.L. Spek, and G. van Koten, *Organometallics*, **15**, 1405 (1996).
493. G.M. Kapteijn, A. Dervisi, D.M. Grove, H. Kooijman, M.T. Lakin, A.L. Spek, and G. van Koten, *J. Am. Chem. Soc.*, **117**, 10939 (1995).
494. H.F. Klein and T. Weimer, *Inorg. Chim. Acta*, **189**, 267 (1991).
495. Y-J. Kim, J-C. Choi, and K. Osakada, *J. Organomet. Chem.*, **491**, 97 (1995).
496. H.-F. Klein, A. Dal, T. Jung, C. Rohr, U. Florke, and H.-J. Haupt, *Eur. J. Inorg. Chem.*, 2027 (1998).
497. H.-F. Klein, A. Dal, T. Jung, S. Braun, C. Rohr, U. Florke, and H.-J. Haupt, *Eur. J. Inorg. Chem.*, 621 (1998).
498. G.M. Kapteijn, D.M. Grove, H. Kooijman, W.J.J. Smeets, A.L. Spek, and G. van Koten, *Inorg. Chem.*, **35**, 526 (1996).
499. T.E. Krafft, C.I. Hejna, and J.S. Smith, *Inorg. Chem.*, **29**, 2682 (1990).
500. P.D. Hampton, S. Wu, T.M. Alam, and J.P. Claverie, *Organometallics*, **13**, 2066 (1994).
501. J. Ruiz, J.M. Marti, F. Florenciano, and G. Lopez, *Polyhedron*, **18**, 2281 (1999).
502. D. Aycock, V. Abolins, and D.M. White, *Encycl. Polym. Sci. Eng.*, **13**, 1 (1988).
503. A.S. Hay, H.S. Stafford, G.F. Endres, and J.W. Eustance, *J. Am. Chem. Soc.*, **81**, 6335 (1959).
504. P.J. Baesjou, W.L. Driessen, G. Challa, and J. Reedijk, *J. Am. Chem. Soc.*, **119**, 12590 (1997).
505. K. Tuppurainen and J. Ruuskanen, *Chemosphere*, **38**, 1825 (1999).
506. P.J. Baesjou, W.L. Driessen, G. Challa, and J. Reedijk, *Macromolecules*, **32**, 270 (1999) and references therein.
507. P.J. Baesjou, W.L. Driessen, G. Challa, and J. Reedijk, *J. Mol. Catal. A*, **135**, 273 (1998).
508. M. Basturkmen, D. Kisakurek, W.L. Driessen, S. Gorter, and J. Reedijk, *Inorg. Chim. Acta*, **271**, 19 (1988).
509. F. Connac, N. Habaddi, Y. Lucchese, M. Dartiguenave, L. Lamande, M. Sanchez, M. Simard, and A.L. Beauchamp, *Inorg. Chim. Acta*, **256**, 107 (1997).
510. C.A. Otter, D.A. Bardwell, S.M. Couchman, J.C. Jeffery, J.P. Maher, and M.D. Ward, *Polyhedron*, **17**, 211 (1998).
511. P. Fiaschi, C. Floriani, M. Pasquali, A. Chiesi-Villa, and C. Guastini, *Inorg. Chem.*, **25**, 462 (1986).
512. K. Osakada, T. Takizawa, M. Tanaka, and T. Yamamoto, *J. Organomet. Chem.*, **473**, 359 (1994).
513. C. Lopes, M. Håkansson, and S. Jagner, *Inorg. Chem.*, **36**, 3232 (1997).
514. M. Håkansson, C. Lopes, and S. Jagner, *Organometallics*, **17**, 210 (1998).
515. C. Lopes, M. Håkansson, and S. Jagner, *New. J. Chem.*, **21**, 1113 (1997).
516. M. Pasquali, P. Fiaschi, C. Floriani, and A.G. Manfredotti, *J. Chem. Soc., Chem. Commun.*, 197 (1983).
517. P. Fiaschi, C. Floriani, M. Pasquali, A. Chiesi-Villa, and C. Guastini, *J. Chem. Soc., Chem. Commun.*, 888 (1984).
518. C. Lopes, M. Håkansson, and S. Jagner, *Inorg. Chim. Acta*, **254**, 361 (1997).

519. W.J. Evans, R.E. Golden, and J.W. Ziller, *Acta Cryst. C*, **50**, 1005 (1994).
520. J.F. Harrod and P. Van Gheluwe, *Can. J. Chem.*, **57**, 890 (1979).
521. N. Narasimhamurthy and A.G. Samuelson, *Tetrahedron Lett.*, **29**, 827 (1988).
522. T. Sone, M. Iwata, N. Kasuga, and S. Komiya, *Chem. Lett.*, 1949 (1991).
523. J. Vicente, M.D. Delores, F.J. Carrion, and P.G. Jones, *J. Organomet. Chem.*, **508**, 53 (1996).
524. R.L. Geerts, J.C. Huffman, and K.G. Caulton, *Inorg. Chem.*, **25**, 590 (1986).
525. R.L. Geerts, J.C. Huffman, and K.G. Caulton, *Inorg. Chem.*, **25**, 1803 (1986).
526. M.S. Super and E.J. Beckman, *Trends Polymer Sci.*, **5**, 236 (1997).
527. This reaction was first reported using $ZnEt_2/H_2O$ catalysts, see S. Inoue, *Carbon Dioxide as a Source of Carbon* (M. Aresta and G. Forti eds), 331, Reidel, Dordrecht (1987).
528. M. Cheng, E.B. Lobkovsky, and G.W. Coates, *J. Am. Chem. Soc.*, **120**, 11018 (1998).
529. D.J. Darensbourg, M.S. Zimmer, P. Rainey, and D.L. Larkins, *Inorg. Chem.*, **37**, 2852 (1998).
530. D.J. Darensbourg, S.A. Niezgoda, J.H. Reibenspies, and J.D. Draper, *Inorg. Chem.*, **36**, 5686 (1997).
531. D.J. Darensbourg, S.A. Niezgoda, J.D. Draper, and J.H. Reibenspies, *Inorg. Chem.*, **38**, 1356 (1999).
532. D.J. Darensbourg, J.C. Yoder, G.E. Struck, M.W. Holtcamp, J.D. Draper, and J.H. Reibenspies, *Inorg. Chim. Acta*, **274**, 115 (1998).
533. M.M. Olmstead, P.P. Power, and S.C. Shoner, *J. Am. Chem. Soc.*, **113**, 3379 (1991).
534. M. Parvez, G.L. BergStresser, and H.G. Richey Jr, *Acta. Cryst. C*, **48**, 641 (1992).
535. H. Grutzmacher, M. Steiner, H. Pritzkow, L. Zsolnai, G. Huttner, and A. Sebald, *Chem. Ber.*, **125**, 2199 (1992).
536. H. Yamamoto and S. Saito, *Pure Appl. Chem.*, **71**, 239 (1999).
537. Y.B. Kozlikovskii, V.A. Koshchii, and B.V. Chernyaev, *Zh. Org. Khim.*, **22**, 1014 (1986).
538. F.R.J. Willemse, J. Wolters, and E.C. Kooijman, *Recl. Trav. Chim. Pays-Bas*, **90**, 5 (1971).
539. K. Maruoka, T. Itoh, M. Sakurai, K. Nonoshita, and H. Yamamoto, *J. Am. Chem. Soc.*, **110**, 3588 (1988).
540. M.B. Power, S.G. Bott, J.L. Atwood, and A.R. Barron, *J. Am. Chem. Soc.*, **112**, 3446 (1990).
541. K. Maruoka, T. Ooi, and H. Yamamoto, *J. Am. Chem. Soc.*, **111**, 6431 (1989).
542. K. Maruoka, A.B. Conception, N. Murase, M. Oishi, N. Hirayama, and H. Yamamoto, *J. Am. Chem. Soc.*, **115**, 3943 (1993).
543. K. Maruoka, H. Imoto, S. Saito, and H. Yamamoto, *J. Am. Chem. Soc.*, **116**, 4131 (1994).
544. K. Maruoka, Y. Araki, and H. Yamamoto, *J. Am. Chem. Soc.*, **110**, 2650 (1988).
545. K. Maruoka, S. Saito, and H. Yamamoto, *J. Am. Chem. Soc.*, **114**, 1089 (1992).
546. K. Maruoka, S. Saito, and H. Yamamoto, *Synlett*, 439 (1994).
547. K. Maruoka, S. Saito, A.B. Conception, and H. Yamamoto, *J. Am. Chem. Soc.*, **115**, 1183 (1993).
548. K. Maruoka, H. Imoto, and H. Yamamoto, *J. Am. Chem. Soc.*, **116**, 12115 (1994).
549. K. Shimada and H. Yamamoto, *Synlett*, 720 (1996).
550. S. Saito, K. Hatanaka, T. Kano, and H. Yamamoto, *Angew. Chem., Int. Ed. Engl.*, **37**, 3378 (1998).
551. T. Kitayama, T. Hirano, and K. Hatada, *Tetrahedron*, **53**, 15263 (1997).
552. T. Aida and S. Inoue, *Acc. Chem. Res.*, **29**, 39 (1986).
553. T. Yasuda, T. Aida, and S. Inoue, *Bull. Chem. Soc. Jpn.*, **59**, 3931 (1986).
554. M. Kuroki, T. Watanabe, T. Aida, and S. Inoue, *J. Am. Chem. Soc.*, **113**, 5903 (1991).
555. M. Akakura, H. Yamamoto, S.G. Bott, and A.R. Barron, *Polyhedron*, **16**, 4389 (1997).
556. M.D. Healy and A.R. Barron, *Angew. Chem., Int. Ed. Engl.*, **31**, 921 (1992).
557. M.D. Healy, M.R. Mason, P.W. Gravelle, S.G. Bott, and A.R. Barron, *J. Chem. Soc., Dalton Trans.*, 441 (1993).
558. H.R. Hoveyda, V. Karunaratne, S.J. Rettig, and C. Orvig, *Inorg. Chem.*, **31**, 5408 (1992).
559. R. Benn, E. Janssen, H. Lehmkuhl, A. Rufinska, K. Angermund, P. Betz, R. Goddard, and C. Kruger, *J. Organomet. Chem.*, **411**, 37 (1991).
560. H.W. Roesky, M. Scholz, M. Noltemeyer, and F.T. Edelmann, *Inorg. Chem.*, **28**, 3829 (1989).

561. A.P. Shreve, R. Mulhaupt, W. Fultz, J.C. Calabrese, W. Robbins, and S.D. Ittel, *Organometallics*, **7**, 409 (1988).
562. J.D. Fisher, P.J. Shapiro, P.M.H. Budzelaar, and R.J. Staples, *Inorg. Chem.*, **37**, 1295 (1998).
563. M.D. Healy, J.W. Ziller, and A.R. Barron, *J. Am. Chem. Soc.*, **112**, 2949 (1990).
564. M.D. Healy, J.T. Leman, and A.R. Barron, *J. Am. Chem. Soc.*, **113**, 2776 (1991).
565. S. Schulz, H.W. Roesky, M. Noltemeyer, and H.-G. Schmidt, *J. Chem. Soc., Dalton Trans.*, 177 (1995).
566. J.P. Campbell and W.L. Gladfelter, *Inorg. Chem.*, **36**, 4094 (1997).
567. M.B. Power, S.G. Bott, D.L. Clark, J.L. Atwood, and A.R. Barron, *Organometallics*, **9**, 3086 (1990).
568. B. Cetinkaya, I. Gumrukcu, M.F. Lappert, J.L. Atwood, R.D. Rogers, and M.J. Zaworotko, *J. Am. Chem. Soc.*, **102**, 2088 (1980).
569. H. Braunschweig, R.W. Chorley, P.B. Hitchcock, and M.F. Lappert, *J. Chem. Soc., Chem. Commun.*, 1311 (1992).
570. P.B. Hitchcock, M.F. Lappert, S.A. Thomas, A.J. Thorne, A.J. Carty, and N.J. Taylor, *J. Organomet. Chem.*, **315**, 27 (1986).
571. A. Meller, G. Ossig, W. Maringgele, M. Noltemeyer, D. Stalke, R. Herbst-Irmer, S. Freitag, and G.M. Sheldrick, *Z. Naturforsch. B*, **47**, 162 (1992).
572. P.F.R. Ewings and P.G. Harrison, *J. Chem. Soc., Dalton Trans.*, 2015 (1975).
573. I. Wakeshima, T. Suzuki, A. Takemoto, and I. Kijima, *Synth. React. Inorg. Met.-Org. Chem.*, **27**, 787 (1997).
574. M. Suzuki, I.-H. Son, R. Noyori, and H. Masuda, *Organometallics*, **9**, 3043 (1990).
575. D. Dakternieks, G. Dyson, K. Jurkschat, R. Tozer, and E.R.T. Tiekink, *J. Organomet. Chem.*, **458**, 458 (1993).
576. R.C. Sharma and M.K. Rastogi, *Indian J. Chem., A*, **23**, 431 (1984).
577. T. Athar, R. Bohra, and R.C. Mehrotra, *J. Indian Chem. Soc.*, **66**, 189 (1989).
578. K.H. Whitmire, H.W. Roesky, S. Brooker, and G.M. Sheldrick, *J. Organomet. Chem.*, **402**, C4 (1991).
579. W.J. Evans, J.H. Hain Jr, and J.W. Ziller, *J. Chem. Soc., Chem. Commun.*, 1628 (1989).
580. P. Hodge, S.C. James, N.C. Norman, and A.G. Orpen, *J. Chem. Soc., Dalton Trans.*, 4049 (1998).
581. C.M. Jones, M.D. Burkart, R.E. Bachman, D.L. Serra, Shiou-Jyh Hwu, and K.H. Whitmire, *Inorg. Chem.*, **32**, 5136 (1993).
582. C.M. Jones, M.D. Burkart, and K.H. Whitmire, *Angew. Chem., Int. Ed. Engl.*, **31**, 451 (1992).
583. J.L. Jolas, S. Hoppe, and K.H. Whitmire, *Inorg. Chem.*, **36**, 3335 (1997).
584. K.H. Whitmire, J.C. Hutchison, A.L. McKnight, and C.M. Jones, *J. Chem. Soc., Chem. Commun.*, 1021 (1992).
585. S. Hoppe and K.H. Whitmire, *Organometallics*, **17**, 1347 (1998).

7

Industrial Applications

1 INTRODUCTION

Metal alkoxides have found a variety of uses, either as catalysts for a range of organic reactions or as precursors for forming metal oxide films, ceramic materials, or glasses.

The catalytic activity is a manifestation of the chemical lability of metal alkoxides especially their reactivity with hydroxyl-containing molecules. The volatility and solubility in common organic solvents of certain metal alkoxides has made them attractive precursors for depositing pure metal oxides by chemical vapour deposition (MOCVD) or by the sol–gel process. The requirement for heterometal oxides as useful materials in the electronics industry has stimulated research in this field in recent years and led to renewed interest in the preparation and characterization of alkoxides of some of the p-block elements which had previously been neglected. In particular, the discovery of the high Tc copper oxide-based superconducting heterometal oxides has made a tremendous impact on this field.

In the following sections a few examples of these industrial applications are detailed, with an emphasis on publications in recognized scientific journals rather than on the numerous patent applications.

2 METAL OXIDE FILMS

2.1 Deposition of Metal Oxides by MOCVD

In recent years there has been an upsurge in the use of thin films of metal oxides in the electronics industry. A wide range of applications has been developed, from insulators, high dielectric materials, piezoelectric materials, and nonlinear optical materials to high Tc superconductors and fast ion conductors.[1-3] Volatile metal alkoxides are increasingly being used as precursors for MOCVD because they can be prepared in a high state of purity, are easy to handle, and are readily converted by thermolysis to pure metal oxides. Bradley and Faktor first reported a systematic study of the thermal decomposition of zirconium tetra-alkoxides $[Zr(OR)_4]_n$[4] with a detailed study of the kinetics of decomposition of the volatile monomeric tertiary amyloxide.[5] This revealed that ZrO_2 was deposited on a clean glass surface by a hydrolytic process caused by the surface-catalysed dehydration of the tertiary alcohol. At 250°C the overall stoichiometric reaction involved the formation of alkene and water (Eq. 7.1).

$$Zr(OR)_4 \longrightarrow ZrO_2 + 4\text{alkene} + 2H_2O \qquad (7.1)$$

Several years later Mazdiyasni and Lynch[6] reported the deposition of ZrO_2, HfO_2, and yttrium-stabilized ZrO_2 as a continuous process using chemical vapour deposition of the metal alkoxide vapour in an inert gas atmosphere. The oxides of yttrium, dysprosium, and ytterbium were similarly deposited from the metal isopropoxides at 200–300°C under nitrogen.[7] The mechanism of the decomposition of metal alkoxides to metal oxides has been the subject of several studies. Recent research by Nix et al.[8] has shown that $Ti(OPr^i)_4$ begins depositing TiO_2 (on a TiO_2 surface) at low pressure at ca. 177°C with the formation of a mixture of acetone, isopropanol, and propene. At higher temperatures (>490 K) the organic product is propene. The first-order rates of decomposition of adsorbed intermediate at various temperatures gave an activation energy of ca. 85 kJ mol^{-1}.

It needs to be borne in mind that metal alkoxides do not invariably give rise to pure metal oxides by the MOCVD process. For example it has been shown by Chisholm et al.[9] that whereas $Al_2(OBu^t)_6$, $Mo_2(OBu^t)_6$, and $W_2(OBu^t)_6$ do deposit the oxides γ-Al_2O_3, MoO_2, and WO_2 respectively, the cyclohexyloxides of Mo and W behave differently. At ca. 210°C $Mo_2(OC_6H_{11}\text{-}c)_6$ eliminates a mixture of cyclohexanol, cyclohexanone, cyclohexene, and cyclohexane forming a material of composition "$Mo_2C_4O_4$" which is stable up to 550°C but at higher temperatures (660–706°C) this is converted by loss of CO and CO_2 into molybdenum carbide γ-Mo_2C. The decomposition of $W_2(OC_6H_{11}\text{-}c)_6$ follows a similar course at 200–250°C giving "$W_2C_4O_4$" but at higher temperature (800°C) this loses CO forming W metal.

In order to be useful as a precursor for MOCVD a metal alkoxide needs to have a reasonable vapour pressure within a temperature range of room temperature to ca. 100°C, and this requires mononuclear compounds. The oligomeric nature of the lower alkoxides $[M(OR)_x]_n$ (R = Me, Et, etc.) can be overcome by using bulky alkoxides such as the tert-butoxide when the metals are in a high valency state ($x = 4, 5$ or 6), and monomeric species ($n = 1$) are thus obtained. However, the screening effect of tert-alkoxy groups is not capable of producing monomeric alkoxides of the alkali metals, alkaline earths, or the trivalent metals although some trivalent alkoxides give dimeric species $M_2(OR)_6$ which are sufficiently volatile for low-pressure MOCVD.

Two strategies have been developed for enhancing the volatility of metal alkoxides. One approach uses functionalized alkoxides, e.g. OCH_2CH_2X (X = OMe, OBu, NEt$_2$) which may act as chelating ligands thereby preventing oligomerization.[10–12] For example the copper(II) complex $Cu(OC_2H_4NEt_2)_2$ sublimes at 60°C in vacuo.[12] The other approach is to replace CH_3 groups in the tert-alkoxide by CF_3 groups. The electron-attracting effect of the CF_3 groups weakens the donor ability of the oxygen, thus weakening the alkoxide bridges in the oligomer and raising the volatility. The weaker intermolecular forces between CF_3 groups also enhances volatility. The metal becomes more electrophilic when bonded to fluoro alkoxide groups and has a greater tendency to coordinate with additional neutral ligands.

This is well illustrated by the tri-alkoxides of yttrium and the lanthanides where the tert-butoxides are trinuclear species $[M_3(OBu^t)_9(Bu^tOH)_2]$ having moderate volatility.[13] With the hexafluoro-tert-butoxides mononuclear complexes such as $[M\{OCMe(CF_3)_2\}_3(thf)_3]$ and $[M\{OCMe(CF_3)_2\}_3(\text{diglyme})]$ were obtained.[14] These compounds were considerably more volatile than the trinuclear *tert*-butoxides and the diglyme complexes sublimed at *ca.* 120°C *in vacuo* without loss of diglyme. However, it is noteworthy that thermal decomposition of fluorinated alkoxy compounds may produce metal fluoride or oxyfluoride instead of metal oxide although this may be prevented by incorporating oxygen or water vapour into the system.

Purdy and George[15] have obtained the volatile copper compounds $Cu_4\{OC(CF_3)_3\}_7$ (sublimes at room temperature *in vacuo*) and $[CuOC(CF_3)_3]_n$ (sublimes 40–50°C *in vacuo*) and the remarkably volatile barium copper heterometal alkoxide $Ba[Cu\{OCMe(CF_3)_2\}_3]_2$ (sublimes 70–90°C *in vacuo*). The X-ray crystal structure of the barium copper complex shows the presence of a monomeric molecule containing trigonal planar three-coordinated Cu(II) and with the barium atom closely coordinated to four alkoxide oxygens and more weakly by intramolecular interactions with eight fluorines. Other volatile copper compounds are the Cu(I) mixed ligand species $[Cu(OBu^t)_x(OR)_{1-x}]$ (R = $C(CF_3)_3$, $CMe(CF_3)_2$, $CH(CF_3)_2$; $x \sim 0.5$)[16] and the copper(II) mixed alkoxides $[Cu_4(OBu^t)_6\{OC(CF_3)_3\}_2]$.[17a] The X-ray crystal structure of the tetranuclear copper(II) complex showed a linear *tert*-butoxy-bridged species with the two internal copper atoms in distorted tetrahedral coordination and the two outer copper atoms in trigonal planar coordination involving two bridging *tert*-butoxides and one terminal perfluoro-*tert*-butoxide. The MOCVD process has been used for the deposition of ZrO_2,[17b] $PbTiO_3$[17c] and $SrTiO_3$.[17d]

2.2   Spray Coating and Flash Evaporation of Metal Alkoxide Solutions

As an alternative process to MOCVD, which requires a volatile metal alkoxide, there is the technique of depositing a film of metal alkoxide onto a substrate from solution in a volatile solvent followed by thermolysis. Thus oligomeric metal alkoxides $[M(OR)_x]_n$ that may not be appreciably volatile can by suitable choice of the alkyl group R be made soluble in volatile organic solvents. This method is especially useful for depositing a multicomponent heterometal oxide because the stoichiometry can be determined by the initial concentration of each component metal alkoxide. This technique is similar in principle to the sol–gel technique which is favoured for the formation of bulk materials (Section 3).

A number of applications of this technique have been reported. For example, single layer or multi-layer coatings of SiO_2/metal oxide (metal = aluminium, titanium, or zirconium) have been produced from solutions of hydrolysed alkoxysilane/metal alkoxide mixtures for protection of electronic devices.[18] Alkoxysiloxy transition metal complexes $[M\{OSi(OBu^t)_3\}_4]$ (M = Ti, Zr or Hf) have been shown to be single-source precursors for low-temperature (*ca* 150°C) formation of MSi_4O_{10} oxide materials.[19] Thin porous membranes of mixed TiO_2/SiO_2 oxides have been produced from the alkoxides to form the tubular channels of a ceramic cartridge for tangential microfiltration.[20] Other applications are the manufacture of crystalline fine TiO_2 particles,[21] the deposition of nonreflective or selectively reflective films of TiO_2 on glass,[22] and the deposition of a thin film of metal oxide (Al, Zr, Ti, Si, Y, or Ce) on stainless-steel or nickel supports to improve the bonding to γ-Al_2O_3 catalyst.[23] Although small particles of TiO_2 (<100 nm) have been deposited by MOCVD[24] much smaller particles (nanometer-sized) of anatase have been produced by electrostatic spraying of titanium alkoxide solutions.[25] Coatings of yttrium-doped cubic ZrO_2 have been obtained by hydrolysis of zirconium alkoxides with hydrogen peroxide and nitric acid.[26]

3 CERAMICS AND GLASSES

In recent years there has been considerable interest in using metal alkoxides as molecular precursors for preparing ceramic oxide materials and speciality glasses. In particular the sol–gel process has been applied because it provides a relatively low temperature mechanism for producing solid oxide materials in contrast to the traditional "grind and bake" method.

A comprehensive account of the sol–gel technique has recently been published by Brinker and Scherer[27] who discuss the advantages and disadvantages of the technique together with references to a wide range of applications. Essentially the sol–gel process involves the controlled hydrolysis of the metal alkoxide in a suitable organic solvent (water miscible) to form a gel from which solvent is removed leaving finely divided oxide particles which can be compacted and heated to form a ceramic or glass. Most metal alkoxides undergo extremely rapid hydrolysis and chemical additives are often used to control the rate of hydrolysis. These additives are usually polyols, carboxylic acids or β-diketones. By contrast the hydrolysis of tetra-alkoxysilanes $Si(OR)_4$ is so slow that acid or base catalysts are required to accelerate the process.

The sol–gel process is particularly useful for nonvolatile metal alkoxides which cannot be used in the MOCVD process, and for forming heterometal oxides, where control of stoichiometry is important. Since 1980, numerous excellent reviews have been published on the sol–gel chemistry of metal alkoxides[1–3,28–38] to which readers are referred for detailed bibliography. Numerous publications have appeared detailing the preparation of binary metal oxides and heterometal oxides using the sol–gel technique and a few representative examples only are given here.

The binary metal oxides Al_2O_3,[39] Y_2O_3,[40] TiO_2,[41] TiO_2,[42] V_2O_5,[43] Nb_2O_5[44] and Ta_2O_5[45] have attracted considerable interest covering a range of applications. Although tetra-alkoxysilanes are not metal alkoxides they cannot be ignored because they are precursors for the preparation of various forms of silica and as components for many glasses and ceramics. The acid catalysed hydrolysis of tetra-alkoxysilanes is an essential stage in the sol–gel synthesis of quartz and various glasses, and a host of publications[46] have appeared covering a variety of applications.

The requirement for inorganic electronic materials prepared at low temperatures has stimulated enormous activity in the heterometal oxide field. The sol–gel technique using metal alkoxides has been especially effective in producing ferroelectric, piezo-electric, and pyroelectric materials, e.g. $BaTiO_3$;[47] $LiNbO_3$;[48] $LiTaO_3$;[49] $PbTiO_3$;[50] $Pb(Zr,Ti)O_3(PZT)$;[51] $(Pb,La)(Zr,Ti)O_3(PLZT)$;[52] $Pb(Fe,Nb)O_3$;[53] and $Pb(Mg,Nb)O_3$.[54]

Considerable attention has been devoted to the preparation of high Tc superconductors based on the heterometal copper oxide systems. Most reports have been concerned with the deposition of films of the "1,2,3" superconductors $YBa_2Cu_3O_{7-x}$ on a variety of different substrates.[55]

Other superconducting materials deposited by the sol–gel process using metal alkoxide precursors are the "1,2,4" compound $YBa_2Cu_4O_8$,[56] bismuth strontium calcium copper oxides,[57] bismuth lead strontium calcium copper oxide,[58] and thallium calcium barium copper oxide.[59] Other heterometaloxides of interest are lithium and sodium aluminates (β-aluminia),[60] yttrium aluminium oxides ($YAlO_3$, $Y_4Al_2O_9$ and $Y_3Al_5O_{12}$-YAG),[61] and aluminium zirconium oxide.[62] Other applications involve the preparation of refractory oxides such as aluminium silicates (*e.g.* mullite),[63] and magnesium aluminium silicates (*e.g.* cordierite),[64] and fast ion conductors such as sodium zirconium silicon phosphorus oxide ($Na_2Zr_2Si_2PO_{12}$-NASICON)[65] and lithium silicon vanadium oxide.[66]

Another developing area in which metal alkoxide precursors have been important is the sol–gel processing of hybrid inorganic–organic polymers called ceramers[67] and ormosils (organically modified silicates).[68] These hybrid materials appear to combine the hardness of the metal oxide (or silica) with the characteristic toughness of the organic polymer and are used in thick coatings which are abrasion resistant or corrosion resistant.

Alternatively the organic component can be removed by thermolysis or chemical treatment leaving a microporous metal oxide material.[69]

4   METAL ALKOXIDES AS CATALYSTS

Metal alkoxides are used as catalysts in a wide range of homogenous reactions. In some cases a remarkable degree of steroselectivity has been achieved. The following sections are representative of developments in this field.

4.1   Catalytic Polymerization of Alkenes (Olefins)

The use of metal alkoxides, especially those of Ti and Mg, as active components of multicomponent Ziegler–Natta type catalysts has an established place in industry. Many patent applications have been filed over the years but they read rather like cookery recipes and will not be documented here. Anionic polymerization of styrene, butadiene, and isoprene has also received much investigation. The addition of alkali metal alkoxides MOR to alkali metal amides MNH_2 gives rise to a range of complex base initiators whose properties may be modulated by the nature of the metal, alkoxide group, and solvent. Similarly the effect of lithium alkoxides on the rate of polymerization of styrene initiated by *n*-butyl lithium has also been reported.[70–72]

The effect of metal alkoxides on the properties of initiators for the anionic polymerization of vinyl monomers has also been reported for polymethylmethacrylate,[73] polymethacrylonitrile[74] and poly-2-vinyl pyridine.[75] The polymerization esterification of terephthalic acid with glycols is another important industrial process which is catalysed by metal alkoxides, mainly titanium or alkali metal derivatives. Numerous patent applications have appeared over the years.

4.2   Asymmetric Epoxidation of Allylic Alcohols

One of the most important applications of metal alkoxides post-1978 was reported by Katsuki and Sharpless in 1980.[76] They discovered that a mixture of titanium tetraisopropoxide and (+)-diethyl tartrate was a highly efficient enantioselective reagent for the epoxidation of allylic alcohols using *tert*-butyl hydroperoxide as the oxygenating reagent. A wide range of allylic alcohols was converted to the corresponding epoxy-alcohol with enantiomeric excess values (ee) of 90–95%. Subsequent work by Sharpless and co-workers led to versatile routes to optically pure natural products and drugs.[77] Using racemic secondary allylic alcohols it was shown that one enantiomer reacted much faster (100×) than the other giving rise to kinetic resolution by enantioselective epoxidation.[78] Further research showed that a molar ratio of dialkyltartrate:Ti(OPr^i)$_4$ of 1.2–1.5 gave optimum results and with the addition of molecular sieves the process became catalytic with regard to titanium complex (<100%).[79] Although various alkoxides of Ti, Zr, Hf, Nb, and Ta were tried the best results were obtained with Ti(OPr^i)$_4$.

A detailed account of the earlier work was published in 1985.[80] Subsequently much research was devoted to elucidating the mechanism of this fascinating catalytic reaction. Since titanium alkoxides readily undergo alcohol interchange it was not surprising to find that the dialkyltartrate (a 1,2-diol) reacted with Ti(OPr^i)$_4$ to form a titanium diisopropoxide dialkyltartrate complex which was shown to be dimeric in methylene dichloride or pentane solution (Eq. 7.2):

$$2Ti(OPr^i)_4 + 2RCO_2CH(OH)CH(OH)CO_2R$$

$$\longrightarrow Ti_2(OPr^i)_4[OCH(CO_2R)CHO(CO_2R)]_2 + 4Pr^iOH \qquad (7.2)$$

Although a crystal structure has not yet been obtained for a dialkyltartrate complex (alkyl = Et, Pr^i, etc.), the structure was obtained for the (R,R)-*N,N'*-dibenzyl tartramide complex and also for the related titanium monoethoxide diethyltartrate diphenyl hydroxamate complex.[81] In both of these dimeric species the tartrate moiety acts as a chelating and bridging ligand and the alkoxy ligands are present as terminal groups. Both compounds were catalytically active in asymmetric epoxidation. The coordination of the titanium atoms in each complex is distorted octahedral. In the tartramide complex the tartramide ligand is bonded facially through the two diolato oxygens and one of the carbonyl oxygens which is weakly bound (see Fig. 4.36 and Table 4.9). The two terminal isopropoxide groups are *cis*, one being *trans* to the carbonyl oxygen and the other *trans* to the bridging diolato oxygen. Sharpless *et al*. have reported the results of detailed kinetic investigations[82] and considerations of the catalyst structure and mechanism of the asymmetric epoxidation.[83] By means of IR and ^{1}H, ^{13}C, and ^{17}O NMR spectroscopic studies it was deduced that the catalyst Ti$_2$(OPr^i)$_2$(diisopropyltartrate)$_2$ probably has a similar structure to that of the tartramide. It is believed that the catalytic process involves replacement of the isopropoxide groups by exchange with the allylic alcohol and the *tert*-butylhydroperoxide. The weakly bound carbonyl is dissociated to allow the *tert*-butylperoxide to act as a bidentate ligand which brings the C=C moiety of the allylic alkoxide into proximity with an oxygen of the *tert*-butylperoxide group thus facilitating oxygen transfer and giving rise to *tert*-butoxide and epoxyalkoxide ligands. The products Bu^{t}OH and epoxyalcohol are then released by exchange reactions with the reactants Bu^{t}OOH and allylic alcohol.

In a related reaction using a different Ti(OPr^i)$_4$/DIPT catalyst it was found that kinetic resolution of racemic β-hydroxy amines could be effected by enantioselective *N*-oxide formation.[84] The *N*-oxide could be reduced to the chiral (S) β-hydroxyamine using LiAlH$_4$ in THF or by catalytic hydrogenation. In addition 2,3-epoxyalcohols were converted regio- and stereo-selectively to 2,3-epithioalcohols by thiourea in the presence of Ti(OPr^i)$_4$.[85] Titanium tetraisopropoxide has also been used to facilitate the regioselective opening of 2,3-epoxyalcohols and related compounds by a wide range of nucleophiles.[86] Two interesting new developments have been reported. Thus Canali *et al.*[87] have synthesized polytartrate esters by polycondensation of L-(+)-tartaric acid with the diols (HO(CH$_2$)$_n$OH, n = 2, 6, 8 and 12). These optically active polymers were successfully used in the epoxidation of *trans*-hex-2-en-1-ol using titanium isopropoxide and Bu^{t}O$_2$H at 20°C in CH$_2$Cl$_2$ in the presence of molecular sieve (4A). In a different approach some novel alkoxytitanium silsesquioxane derivatives [R$_7$Si$_7$O$_{12}$Ti(OR′)]$_x$ (R = cyclopentyl or cyclohexyl; R′ = Me, Pr^i, Bu^t; x = 1 or 2) were prepared by reaction of the cubic trisilanols R$_7$Si$_7$O$_9$(OH)$_3$ with the titanium alkoxide. The isopropoxide (R′ = Pr^i) which was shown to exhibit a monomer–dimer equilibrium in CDCl$_3$ by ^{13}C NMR studies also displayed good catalytic activity in the epoxidation of cyclohexene. Interestingly the methoxide gave a crystalline methanolate whose structure was determined by X-ray crystallography (see Table 4.9, p. 278). The dimeric molecule is held together by methoxide bridges, and each titanium is coordinated by a molecule of methanol which is also hydrogen bonded to an Si–O–Ti oxygen atom. This compound is a model for the titano–zeolite heterogeneous catalysts for the selective oxidation of organic molecules.[88]

4.3   Catalytic Sulfoxidation of Organosulfur Compounds

The sulfoxidation of organosulfur compounds by *tert*-butyl hydroperoxide has been effected catalytically using titanium alkoxide-based catalyst.[89] A robust titanium alkoxide catalyst was developed[90] from the reaction of a homochiral trialkanolamine N{CH$_2$CHR(OH)}$_3$ (R = H, MePh) with Ti(OPr^i)$_4$ to form a mononuclear isopropoxy titanatrane (Pr^{i}O)Ti{OCHRCH$_2$}$_3$N containing five-coordinated titanium.

However, addition of excess trialkanolamine led to replacement of the terminal isopropoxy group with formation of polynuclear titaniumtrialkanolaminates, which were characterized by NMR and electrospray ionization mass spectrometry (ESIMS). Addition of Bu^{t}OOH to this system generated the mononuclear *tert*-butyl peroxo titanatrane (Bu^{t}OO)Ti{OCHRCH$_2$}$_3$N which appears to be the active catalytic species in sulfoxidation (see Table 4.9, p. 276).[91]

4.4   Metathesis and Polymerization of Alkenes and Alkynes Using Transition Metal Alkoxy Complexes

The metathesis and polymerization of alkenes (olefins) and alkynes (acetylenes) (Eqs 7.3 and 7.4) is of considerable industrial importance, and much research has been devoted to the synthesis of transition metal catalysts.

$$2R'CH{=}CHR'' \longrightarrow R'CH{=}CHR' + R''CH{=}CHR'' \qquad (7.3)$$

$$2R'C{\equiv}CR'' \longrightarrow R'C{\equiv}CR' + R''C{\equiv}CR'' \qquad (7.4)$$

A major contribution to this field has been made in recent years by Schrock and co-workers[92,93] who have developed a wide range of alkylidene and alkylidyne metal complexes in which alkoxy groups play an important role in optimizing the catalytic performance of the metal complex. Complexes of Mo(VI), W(VI), and Re(VII) with the d^0 electronic configuration have proved to be most effective with OBu^t, $OCMe_2CF_3$, $OCMe(CF_3)_2$, $OCH(CF_3)_2$, and $OC_6H_3R_2$-2,6 as the alkoxy groups.

The pseudo-tetrahedral tungsten alkylidene complexes $W(CHR)(NAr)(OR')_2$ (Ar = 2,3-diPr^{i}phenyl) are catalysts for olefin metathesis.[94] The activity of these molecules is sensitive to electronic and steric factors of all three types of ligand. In the case of the neopentylidene (R = Bu^t) complexes, $W(CHBu^t)(NR)(OBu^t)_2$ does not react with ordinary olefins but does react readily with more reactive substrates such as norbornene. On the other hand $W(NHBu^t)(NR)\{OCMe(CF_3)_2\}_2$, containing the more electronegative tertiary alkoxide, is an active metathesis catalyst for ordinary olefins giving turnover rates of *ca.* 10^3 turnovers min^{-1} in a hydrocarbon solvent at 25°C. X-ray crystal structures of some of these complexes have been obtained. The course of the reaction depends on the nature of the olefin RCH=CHR' and in some cases tungstacyclobutane complexes are formed as pseudo-trigonal bipyramidal complexes containing an axial arylimido ligand and an equatorial bent tungstacyclobutane ring (see Table 4.11, p. 307). The corresponding molybdenum complexes are also active towards olefins but less so than the tungsten complexes except towards the more reactive olefins. Interestingly, although $M(CHBu^t)(NR)(OBu^t)_2$ (M = Mo, W) does not react readily with ordinary internal olefins such compounds do react extremely rapidly with the strained double bonds in norbornenes and norbornadienes.[93] Ring-opening metathesis polymerization (ROMP) of norbornene occurs even at −80°C with $W(CHBu^t)(NR)(OBu^t)_2$ giving a "living polymer" containing up to 500 monomer units. The polymer can be released by addition of benzaldehyde to give the oxo metal complex, or by chain transfer with styrene. The polymers produced are essentially monodisperse. The molybdenum complex $Mo(CHBu^t)(NR)(OBu^t)_2$ was particularly effective as a catalyst for the ROMP of functionalized norbornadienes giving monodisperse living polymers.[94] Alkynes (acetylenes) can be metathesized (Eq. 7.5) using the metal alkylidyne complexes $M(CR)(OR')_3$, where M = Mo or W, R = Bu^t, and R' = 2,6-diisopropyl-C_6H_3.[92]

$$2RC{\equiv}CR' \longrightarrow RC{\equiv}CR + R'C{\equiv}CR' \qquad (7.5)$$

Although $W(CBu^t)(OBu^t)_3$ readily metathesizes internal acetylenes the corresponding molybdenum complex is inactive. However, with a more electronegative alkoxy group such as $OCMe(CF_3)_2$ the Mo-complex will catalyse the metathesis of 3-heptyne. In the case of the more reactive tungsten complexes $W(CBu^t)(OR')_3$, the tungstacyclobutadiene complexes $W(C_3R_3)(OR')_3$ were isolated from the metathesis of RC≡CR. X-ray crystallography confirmed that the tungstacyclobutadiene complexes have a distorted trigonal bipyramidal structure containing a planar WC_3 ring. These complexes are active in the metathesis of acetylenes. Some Re(VII) alkylidyne complexes of the type $Re(CR)(NAr)(OR')_2$ have also been shown to react with acetylenes.[95] The neopentylidyne complex $Re(CBu^t)(NR)\{OCMe(CF_3)_2\}_2$ reacted rapidly with 3-hexyne at 25°C giving $Bu^tC{\equiv}CEt$ and the rhenacyclobutadiene complex $Re(C_3Et_3)(NR)\{OCMe(CF_3)_2\}_2$. Acetylene has been polymerized using the tungsten neopentylidene complex $W(CHBu^t)(NR)(OBu^t)_2$.[96] Di-*tert*butyl-capped oligomers were obtained by carrying out the reaction in the presence of the base quinuclidine and adding pivaldehyde to release the oligomer from the metal.

The molybdenum neopentylidene complex $Mo(CHBu^t)(NR)(OBu^t)_2$ is the active catalyst used in a fascinating development for the synthesis of II–VI semiconductor clusters (ZnS, CdS, PbS) and silver and gold nanoclusters of predictable size within microdomains in films of block copolymers prepared by ROMP.[97] Block copolymers of norbornene and a functionalized norbornene that will complex with a metal-containing compound were prepared and characterized as monodisperse materials. The function-alized component (amine, alkoxide, or thiolate) then sequestered the metal and the metallated block copolymer was cast into a film which was subsequently treated with H_2S to convert the metal into the sulfide. The molybdenum complexes have also featured in the development of the synthesis of side-chain liquid crystal polymers by living ROMP.[98,99]

4.5 Miscellaneous Reactions Catalysed by Metal Alkoxides

Alkali metal alkoxides in association with transition metal carbonyls (e.g. $Ni(CO)_4$) have featured in several patents covering the synthesis of methanol from synthesis gas (Eq. 7.6).[100]

$$2H_2 + CO \longrightarrow CH_3OH \tag{7.6}$$

In the absence of hydrogen, a catalytic reaction of carbon monoxide and an alcohol produces alkyl formate (Eq. 7.7). The catalyst involved is an alkali metal alkoxide alone[101] or together with a transition metal carbonyl cluster, e.g. $Ru_3(CO)_{12}$.[102]

$$ROH + CO \longrightarrow HCO_2R \tag{7.7}$$

Another process featuring alkali metal alkoxides is the conversion of methanol to hydrogen and formaldehyde (Eq. 7.8).[103]

$$CH_3OH \longrightarrow H_2 + H_2CO \tag{7.8}$$

Lithium alkoxides play an important role in the oligomeric cyclization of dinitriles in the synthesis of phthalocyanines.[104] The combination of potassium alkoxide-crown ether complex in a hydrocarbon solvent gives a very powerful catalyst for production of vinyl ethers from alcohols and acetylene (Eq. 7.9).[105]

$$ROH + HC{\equiv}CH \longrightarrow ROCH{=}CH_2 \tag{7.9}$$

Sodium alkoxides have also been used for the isomerization of 1,3-dienes in diene polymers[106] and in the aldol condensation of acetone.[107]

Bimetallic oxoalkoxides are excellent catalysts for the ring-opening-polymerization of lactones, oxiranes, and thioranes.[108]

Aluminium alkoxides have long been used as catalysts for the reduction of aldehydes or ketones by secondary alcohols and recently lanthanide alkoxides were reported to act similarly[109] in addition to catalysing the epoxidation of allylic alcohols with tert-butyl hydroperoxide. Aluminium alkoxides have also been used to catalyse the conversion of aldehydes to alkylesters (Tischtchenko reaction).[110] A review[111] gives a fascinating account of the use of heterometal alkoxides in asymmetric catalysis.

Titanium alkoxides feature as catalysts in the esterification of succinic, adipic, azelaic, and sebacic acids with 2-ethyl hexanol[112] and in the production of acetate esters by transesterification of ethyl acetate with the appropriate alcohol.[113] A titanium isopropoxide chiral Schiff-base complex has been reported to catalyse a novel enantioselective reaction of diketene with aldehydes to give 5-hydroxy-3-oxoesters.[114]

Chiral titanium complexes obtained from the reaction of $TiCl_2(OPr^i)_2$ with various chiral diols have been used in a variety of reactions such as the asymmetric Diels–Alder reaction.[115]

Lanthanum alkoxides have been used in the catalysed transhydrocyanation from acetone cyanohydrin to other aldehydes or ketones and also in the reactions of silyl ketene acetals with aldehydes.[116]

5   MISCELLANEOUS APPLICATIONS

In the book on *Metal alkoxides*[117] some other industrial applications of metal alkoxides were described, such as heat-resisting paints, protective coatings, water-repellent agents, and drying agents for inks and paints. Since 1978 many similar reports have appeared, mainly in the patent literature, and a few examples of more recent references will be given here. A review on corrosion prevention discusses the merits of metal alkoxide formulations versus chromate treatments.[118] Heat-resistant electrical wire insulation,[119] cross-linking of epoxy resins,[120] remedial treatment of concrete,[121] fireproofing of wood,[122] and weather-proofing of decorative panels[123] are also featured in the literature. Other applications involve dental fillings,[124] aluminium soap greases,[125] laser recording media,[126] and a wide range of ceramic formulations based on aluminium nitride,[127] aluminium oxynitride,[128] silicon carbide,[129] silicon nitride,[130] titanium carbide, and titanium nitride.[131]

These examples demonstrate the ubiquitous nature of metal alkoxides.

REFERENCES

1. L.G. Hubert-Pfalzgraf, *New. J. Chem.*, **11**, 663 (1987).
2. D.C. Bradley, *Chem. Rev.*, **89**, 1317 (1989); *Phil. Trans. Roy. Soc. London*, **A330**, 167 (1990); *Polyhedron*, **13**, 1111 (1994); D.C. Bradley, in *Fine Chemicals for the Electronics Industry II*, D.J. Ando and M.G. Pellat (eds), 49–59, Royal Society of Chemistry, Cambridge (1991).
3. P. O'Brien, in *Inorganic Materials*. D.W. Bruce and D. O'Hare (eds), Ch. 9, John Wiley & Sons (1992).
4. D.C. Bradley and M.M. Faktor, *J. Appl. Chem.*, **9**, 5425 (1959).
5. D.C. Bradley and M.M. Faktor, *Trans. Faraday Soc.*, **55**, 2117 (1959).
6. K.S. Mazdiyasni and C.T. Lynch, *USA F Tech. Doc. Rept.*, ASD-TDR-63-322 (May 1963); *USAF Tech. Doc. Rept.*, ML-TDR-64-269 (Sept. 1964).
7. K.S. Mazdiyasni, C.T. Lynch, and J.S. Smith, *Inorg. Chem.*, **5**, 342 (1966).
8. Yuan-Min Wu, D.C. Bradley, and R.M. Nix, *Appl. Surface Sci.*, **64**, 21 (1992), and references therein.
9. D.V. Baxter, M.H. Chisholm, V.F. Distasi, and J.A. Klang, *Chem. Mater.*, **3**, 221 (1991).
10. S.C. Goel, K.S. Kramer, P.C. Gibbons, and W.E. Buhro, *Inorg. Chem.*, **28**, 3620 (1989).
11. H.S. Horowitz, S.J. McLain, A.W. Sleight, J.D. Drulinger, P.L. Gai, M.J. Vanhavelaar, J.L. Wagner, B.D. Biggs, and J.J. Poon, *Science*, **243**, 66 (1989).
12. S.C. Goel, K.S. Kramer, M.Y. Chang, and W.E. Buhro, *Polyhedron*, **9**, 611 (1990).

13. D.C. Bradley, H. Chudzynska, M.B. Hursthouse, and M. Motevalli, *Polyhedron*, **10**, 1049 (1991).
14. D.C. Bradley, H. Chudzynska, M.B. Hursthouse, and M. Motevalli, *Polyhedron*, **12**, 1907 (1993); **13**, 7 (1994).
15. A.P. Purdy and C.F. George, *Inorg. Chem.*, **30**, 1970 (1991); A.P. Purdy, US Patent Application, US 828,634 (15 Aug. 1992).
16. M.E. Gross, *J. Electrochem. Soc.*, **138**, 2422 (1991).
17. a. A.P. Purdy, C.F. George, and G.A. Brewer, *Inorg. Chem.*, **31**, 2633 (1992); b. Y. Takahashi, T. Kawae and M. Nasu, *J. Cryst. Growth*, **74**, 409 (1986); c. B.S. Kwak, E.P. Boyd, and A. Erbil, *Appl. Phys. Lett.*, **53**, 1702 (1988); S.L. Swartz, D.A. Seifert, G.T. Noel, and T.R. Shrout, *Ferroelectrics*, **93**, 37 (1989); d. W.A. Feil, B.W. Wessels, L.M. Tonge and T.J. Marks, *J. Appl. Phys.*, **67**, 3858 (1990).
18. L.A. Haluska, K.W. Michael, and L. Tarhay, *US Patent*, US 4,753,856 (28 June 1988).
19. K.W. Terry and T.D. Tilley, *Chem. Mater.*, **3**, 1001 (1991).
20. B. Castelas, European Patent Application, EP 248,748 (9 Dec. 1987).
21. Y. Nishii and T. Shimizu, Jpn. Kokai Tokkyo Koho JP 62,278,125 (3 Dec. 1987) and 62,226,814 (5 Oct. 1987).
22. T. Sugimoto and K. Kenzi, German Offentlich DE 3,828,137 (2 Mar. 1989).
23. K. Izumi, T. Deguchi, M. Murakami, and H. Tanaka, Jpn. Kokai Tokkyo Koho JP 01 75,040 (20 Mar. 1989).
24. K. Okuyama, J.T. Jeung, Y. Kousaka, H.V. Nguyen, J.J. Wu, and R.C. Flanagan, *Chem. Eng. Sci.*, **44**, 1369 (1989).
25. D.G. Park and J.M. Burlitch, *Chem. Mater.*, **4**, 500 (1992).
26. C. Sakurai, T. Fukui, and M. Okuyama, *J. Am. Ceram. Soc.*, **76**, 1061 (1993).
27. C.J. Brinker and G.W. Scherer, *Sol–Gel Science: The Physics and Chemistry of Sol–Gel Processing*, Academic Press, New York (1990). See also L.C. Klein, *Sol–Gel Technology for Thin Films, Fibers, Preforms, Electronics, and Speciality Shapes*, Noyes, Park Ridge, NJ (1988).
28. S. Sakka and K. Kamiya, *J. Non-Cryst. Solids*, **42**, 403 (1980).
29. E. Matijevic, *Acc. Chem. Res.*, **14**, 22 (1981).
30. N.Ya. Turova and M.I. Yanovskaya, *Izv. Akad. Nauk. SSSR, Neorg. Mater.*, **19**, 693 (1983).
31. K.S. Mazdiyasni, *Better Ceramics Through Chemistry*, in G.J. Brinker, D.E. Clark, and D.R. Ulrich (eds), 175–186, Elsevier, New York (1984).
32. J. Livage, *J. Solid State Chem.*, **64**, 322 (1986).
33. J. Livage, M. Henry, and C. Sanchez, *Prog. Solid State Chem.*, Vol. 18, G.M. Rosenblatt and W.L. Worrell (eds), 259–341, Pergamon Press, Oxford (1988); J.D. Mackenzie, *J. Non-Cryst. Solids*, **100**, 162 (1988).
34. C. Sanchez and J. Livage, *New J. Chem.*, **14**, 513 (1990).
35. M.I. Yanovskaya, E.P. Turevskaya, V.G. Kessler, I.E. Obvintseva, and N.Ya. Turova, *Integr. Ferroelectr.*, **1**, 343 (1992).
36. L.G. Hubert-Pfalzgraf, *Appl. Organomet. Chem.*, **6**, 627 (1992).
37. J. Livage, F. Babonneau, and L. Bonhomme-Coury, *Polymer Preprints*, **94**, 246 (1993).
38. C.D. Chandler, C. Roger, and M.J. Hampden-Smith, *Chem. Rev.*, **93**, 1205 (1993).
39. W.L. Olsen and L.J. Baner, in *Better Ceramics Through Chemistry II*, C.J. Brinker, D.E. Clark, and D.R. Ulrich (eds), 187, Materials Research Society, Pittsburgh, PA (1986); L.F. Nazur and L.C. Klein, *J. Am. Ceram. Soc.*, **71**, C85 (1988); H. Uchihashi, N. Tohge, and T. Minami, *Nipp. Seram. Kyoka Gakujutsu Ronbushi (English)*, **97**, 396 (1989); A Ueno, *Seramikkusu (Japanese)*, **24**, 1042 (1989); K. Maeda, F. Muzukami, M. Watanabe, S. Niwa, M. Toba, and K. Shimizu, *Chem. Ind. (London)*, **23**, 807 (1989); T. Ogihara, H. Nakajima, T. Yanagawa, N. Ogata, K. Yoshida, and N. Matsushita, *J. Am. Ceram. Soc.*, **74**, 2263 (1991); S. Rezhui, B.C. Gates, S.L. Barkett, and M.E. Davis, *Chem. Mater.*, **6**, 2390 (1994).
40. L.G. Hubert-Pfalzgraf, O. Poncelet, and J.-C. Daran, in *Better Ceramics Through Chemistry IV*, C.J. Brinker, D.E. Clark, and D.R. Ulrich (eds), Material Research Society, Pittsburgh, PA (1990).
41. B. Fegley and E.A. Barringer, in *Better Ceramics Through Chemistry*, C.J. Brinker, D.E. Clark, and D.R. Ulrich (eds) 187, Elsevier, New York (1984); A. Leaustic, F. Babonneau,

and J. Livage, *Chem. Mater.*, **1**, 248 (1989); M.J. Munoz-Aguado, M. Gregorkiewitz, and H. Larbot, *Mater. Res. Bull.*, **27**, 87 (1992).

42. K. Yamiya, T. Yoko, K. Tanaka, and H. Ito, *Yogyo Kyokarshi (English)*, **95**, 1157 (1987); N. Toghge, A. Matsuda, and T. Minami, *Chem. Express (English)*, **2**, 141 (1987); H. Saito, H. Suzuki, and H. Hisao, *Nipp. Kagaku Kaishi (Japanese)*, **9**, 1571 (1988); K. Kamiya, K. Takahashi, T. Maeda, H. Nasu, and T. Yoko, *J. Eur. Ceram. Soc.*, **7**, 295 (1991); T.L. Wen, V. Herbert, S. Vilminot, and J.C. Bernier, *J. Mater. Sci.*, **26**, 3787 (1991).

43. C. Sanchez, M. Nabavi, and F. Taulelle, in *Better Ceramics Through Chemistry III*, C.J. Brinker, D.E. Clark and D.R. Ulrich (eds), 93, Materials Research Society, Pittsburgh, PA (1988); S. Hioki, T. Ohishi, K. Takahashi, and T. Nakazawa, *Nipp. Seram. Kyokai Gakujutsu Ronbunshi (Japanese)*, **97**, 628 (1989); M. Nabavi, C. Sanchez, and J. Livage, *Eur. J. Solid State Inorg. Chem.*, **28**, 1173 (1991); J. Livage, *Chem. Mater.*, **3**, 578 (1991).

44. E.I. Ko and S.M. Maurer, *J. Chem. Soc., Chem. Commun.*, 1062 (1990).

45. L.A. Silverman, G. Teowee, and D.R. Uhlmann, in *Better Ceramics Through Chemistry II*, C.J. Brinker, D.E. Clark, and D.R. Ulrich (eds), 725, Materials Research Society, Pittsburgh, PA (1986).

46. H. Dislich, *J. Non-Cryst. Solids*, **73**, 599 (1985); B. Unga, F.G. Wihsmann, K. Forkel, M. Nofz, and U. Bessau, *Wiss. Fortschr.*, **36**, 239 (1986); H. Schmidt, G. Philipp, H. Patzelt, and H. Scholze, *Collected Papers of 14th Intl. Glass Congress*, Vol. 2, 429–436, Indian Ceramic Society, Calcutta (1986); S. Sakka and K. Kamiya, *Nihon Reoroji Gakkaishi (Japanese)*, **14**, 9 (1986); Y. Ozaki, K. Hayashi, and T. Kasai, *Kenkyu Hokoku-Asahi Garasu Kogyo Gijutsu Shoreikai (Japanese)*, **49**, 169 (1986); M. Nogami and K. Nagasaki, *J. Mater. Sci. Lett.*, **6**, 1479 (1987); W. Beier, A.A. Goktas, and G.H. Frischat, *J. Non-Cryst. Solids*, **100**, 531 (1988); M. Nogami, K. Nagasaka, K. Kadono, and T. Kishimoto, *J. Non-Cryst. Solids*, **100**, 298 (1988); W. Beier and G.H. Frischat, in *Better Ceramics Through Chemistry III*, C.J. Brinker, D.E. Clark, and D.R. Ulrich (eds), 817, Material Research Society, Pittsburgh, PA (1988); N. Tohge and T. Minami, *Chem. Express*, **3**, 455 (1988); S. Miyashita, Japanese Patent JP 01160834 (23 June 1989); H. Asano and T. Shimizu, Japanese Patent JP 02,120,247 (8 May 1990); N. Yamamoto, T. Goto, and Y. Horiguchi, European Patent Application EP 261,593 (30 Mar 1988); I.M. Thomas, J.G. Wilder, W.H. Lowdermilk, and M.C. Staggs, *Boulder Damage Symposium*, Boulder, CO, Oct. 15–17 (1984).

47. T. Kasai and Y. Ozaki, *Yogyo Kyokaishi (Japanese)*, **95**, 912 (1987); G. Limmer, H. Buerke, R. Kohl, and G. Tomandl, *Sprechsaal (English)*, **121**, 1099 (1988); G. Tomandl, H. Roesch, and A. Stigelschmitt, in *Better Ceramics Through Chemistry III*, C.J. Brinker, D.E. Clark, and D.R. Ulrich (eds), 281, Materials Research Society, Pittsburgh, PA (1988); R.L. Perrier, A. Safari, and R.E. Riman, *Mater. Res. Bull.*, **26**, 1067 (1991); V.W. Day, T.A. Eberspacher, W.G. Klemperer, and S. Liang, *Chem. Mater.*, **7**, 1607 (1995).

48. N. Puyoo-Castaings, F. Duboudin, and J. Ravez, *J. Mater. Res.*, **46**, 557 (1988); D.J. Eichorst and D.A. Payne, in *Better Ceramics Through Chemistry III*, C.J. Brinker, D.E. Clark, and D.R. Ulrich (eds), 773 *Materials Research Society*, Pittsburgh, PA (1988); S. Hirano, T. Hayashi, K. Nosaki, and K. Kazumi, *J. Am. Ceram. Soc.*, **72**, 707 (1989); K. Nashimoto and M.J. Cima, *Mater. Letts.*, **10**, 348 (1991).

49. J.H. Jean, *J. Mater. Sci.*, **25(2A)**, 859 (1990).

50. K.D. Budd, S.K. Dey, and D.A. Payne, in *Better Ceramics Through Chemistry II*, C.J. Brinker, D.E. Clark, and D.R. Ulrich (eds), 711, *Materials Research Society*, Pittsburgh, PA (1986); Y. Hayashi and J.B. Blum, *J. Mater. Sci.*, **22**, 2655 (1987); C.D. Chandler and M.J. Hampden-Smith, *Chem. Mater.*, **4**, 1137 (1992).

51. R.A. Lipeles, D.J. Coleman, and M.S. Leung, in *Better Ceramics Through Chemistry II*, C.J. Brinker, D.E. Clark, and D.R. Ulrich (eds), 665, Materials Research Society, Pittsburgh, PA (1986); R.A. Lipeles and D.J. Coleman, in *Ultrastructure Processing of Advanced Ceramics*, J.D. Mackenzie and D.R. Ulrich (eds), 919, J. Wiley & Sons, New York (1988); T. Ogihara, H. Kaneko, N. Mizutani, and M. Kato, *J. Mater. Sci. Lett.*, **7**, 867 (1988); F. Imoto, S. Hashimoto, and Y. Ogawa, *Shizuoka Daigaku Kokakubu Kenkyu Hokuku (Japanese)*, **39**, 9 (1988).

52. K.D. Budd, S.K. Dey, and D.A. Payne, *Brit. Ceram. Proc.*, **36**, 107 (1985); M.I. Yanovskaya, E.P. Turevskaya, N.Ya. Turova, M.Ya. Dambekalne, N.V. Kolganova, S.A. Ivanov, A.G. Segalla, V.V. Belov, A.V. Novoselova, and Yu N. Venevtsev, *Izv. Akad. Nauk. SSSR, Neorg. Mater.*, **23**, 658 (1987).

53. P. Griesmar, G. Papin, C. Sanchez, and J. Livage, *J. Mater. Sci. Letts.*, **9**, 1288 (1990).

54. P. Ravindranathan, S. Komarneni, A.S. Bhalla, R. Roy, and L.E. Cross, *Ceram. Trans.*, **1**, 182 (1988); F. Chaput, J.-P. Boilot, M. Lejeune, R. Papiernik, and L.G. Hubert-Pfalzgraf, *J. Am. Ceram. Soc.*, **72**, 1355 (1989); L.F. Francis and D.A. Payne, *Proceedings of the 7th IEEE Intl. Symposium on Applied, Ferroelectronics*, Urbana-Champaign, Illinois, 8 June 1980, IEEE, New York, July (1990); D.A. Payne, D.J. Eichorst, L.F. Francis, and J.F. Campion, *Chem. Process. Advan. Mater.*, L.L. Hench and J.K. West (eds), Wiley, New York (1992).

55. S. Shibata, T. Kitagawa, H. Okazaki, T. Kimura, and T. Murakami, *Jpn. Appl. Phys. Part 2*, **27**, L53 (1988); T. Monde, H. Kozuka, and S. Sakka, *Chem. Lett. (2)*, 287 (1988); S.A. Kramer, G. Kordas, G.C. Hilton, and D.R. Van Harligen, *Appl. Phys. Lett.*, **53**, 156 (1988); S. Kramer, K. Wu, and G. Kordas, *J. Electron. Mater.*, **17**, 135 (1988); S. Shibata, T. Kitagawa, H. Okazaki, and T. Kimura, *Jpn. J. Appl. Phys., Part 2*, **27**, L646 (1988); N. El Khokh, R. Papiernik, L.G. Hubert-Pfalzgraf, F. Chaput, and J.P. Boilot, *J. Mater. Sci. Lett.*, **8**, 762 (1989); G. Moore, S. Kramer, and G. Kordas, *Mater. Lett.*, **7**, 415 (1989); E.P. Turevskaya, N.I. Kozlova, N.Ya. Turova, B.A. Popovkin, M.I. Yanovskaya, O.V. Fedoseeva, and Yu.N. Venertsev, *Sverkhprovodimost: Fiz., Khim., Tekh.*, **2**, 30 (1989); M.W. Rupich, B. Lagos, and J.P. Hachey, *Appl. Phys. Lett.*, **55**, 2447 (1989); E.P. Turevskaya, M.I. Yanovskaya, N.Ya. Turova, A.K. Kochetov, and Yu.N. Venevtsev, *Mater. Sci. Forum*, 62 (1990); G.E. Whitwell, J.H. Wandass, F.M. Cambria, and M.F. Antezzo, in *Better Ceramics Through Chemistry IV*, C.J. Brinker, D.E. Clark, and D.R. Ulrich (eds), 929, Materials Research Society, Pittsburgh, PA (1990); D.M. Millar and D.A. Payne, *NIST Spec. Publ.*, 112 (1991); S. Katayama and M. Sekine, *J. Mater. Chem.*, **1**, 1031 (1991); **6**, 1629 (1991); M.W. Rupich, Y.P. Liu, and J. Ibechem, *Appl. Phys. Lett.*, **60**, 1384 (1992).

56. S. Koriyama, T. Ikemachi, T. Kawano, H. Yamanchi and S. Tanaka, *Physica C (Amsterdam)*, 185 (1991); S. Katayama, M. Sekine, H. Fudouzi, and M. Kuwabara, *J. Appl. Phys.*, **71**, 2745 (1992).

57. S. Katayama and M. Sekine, in *Better Ceramics Through Chemistry IV*, C.J. Brinker, D.E. Clark and D.R. Ulrich (eds), 873, Materials Research Society, Pittsburgh, PA (1990); *J. Mater. Res.*, **6**, 31 (1991).

58. K. Matsumura, H. Nobumasa, K. Shimizu, T. Arima, Y. Kitano, M. Tanaka and K. Sushida, *Jpn. J. Appl. Phys., Part 2*, **28**, L1797 (1989).

59. M.R. Teepe, D.S. Kenzer, G.A. Moore and G. Kordas, in *Better Ceramics Through Chemistry IV*, C.J. Brinker, D.E. Clark and D.R. Ulrich (eds), 901, Materials Research Society, Pittsburgh, PA (1990).

60. S. Yamaguchi, K. Terabe, Y. Iguchi, and A. Imai, *Solid State Ionics*, **25**, 171 (1987); P.E. Morgan, *Solid State Ionics*, **27**, 207 (1988); S. Yamaguchi, *Solid State Ionics*, **27**, 207 (1988); C.W. Turner, B.C. Clatworthy, and A.H.Y. Gin, *Adv. Ceram.*, 141 (1989); O. Renoult, J.-P Boilot, and M. Boncoer, *J. Am. Ceram. Soc.*, **77**, 249 (1994).

61. O. Yamaguchi, K. Takeoka, K. Hirota, H. Takano and A. Hayashida, *J. Mater. Sci.*, **27**, 1261 (1992).

62. S. Dick, C. Suhr, J.L. Rehspringer, and M. Daire, *Mater. Sci. Eng. A.* **109**, 227 (1989).

63. M. Suzuki, S. Hiraishi, M. Yoshimura, and S. Somiya, *Yogyo Kyokaishi (Japanese)*, **92**, 320 (1984); Y. Hirata, K. Sakeda, Y. Matsushita, and K. Shimada, *Yogyo Kyokaishi (Japanese)*, **93**, 577 (1985); L.A. Paulick, Y.F. Yu, and T.I. Mah, *Adv. Ceram.*, 121 (1987); J.C. Pouxviel, J.P. Boilot, O. Poncelet, L.G. Hubert-Pfalzgraf, A. Lecomte, A. Dauger, and J.C. Beloeil, *J. Non-Cryst. Solids*, **93**, 277 (1987); H. Suzuki, H. Nagata, Y. Suyama, and H. Saito, *Funtai Oyobi Funmatsu Yakin (Japanese)*, **35**, 211 (1988); J.S. Sparks and D.S. Tucker, *Adv. Ceram. Mater.*, **3**, 509 (1988); P. Colomban, *J. Mater. Sci.*, **24**, 3011 (1989); F. Chaput, A. Lecomte, A. Dauger, and J.P. Boilot, *Chem. Mater.*, **1**, 199 (1989).

64. H. Suzuki, K. Ota, and H. Saito, *Yogyo Kyokaishi (Japanese)*, **95**, 163 (1987); **95**, 170 (1987); J.C. Broudic, S. Vilminot, and J.C. Bernier, *Mater. Sci. Eng., A*, **109**, 253 (1989); K. Maeda, F. Mizukami, S. Miyashita, Shu-ichi Niwa, and M. Toba, *J. Chem. Soc, Chem. Commun.*, 1268 (1990); T. Fukui, C. Sakurai, and M. Okuyama, *Nippon Kagaku Kaishi (Japanese)*, **4**, 281 (1991).

65. J.P. Boilot, *Ann. Chim. Sci. Mater.*, **10**, 305 (1985).

66. J. Kuwano, Y. Naito, and M. Kato, *Yogyo Kyokaishi (Japanese)*, **95**, 176 (1987).

67. G.L. Wilkes, B. Orler, and H. Huang, *Polymer Preprints (Am. Chem. Soc. Div. Polym. Chem.)*, **26**, 300 (1985); D.E. Rodrigues, A.B. Brennan, C. Betrabet, B. Wang, and G.L. Wilkes, *Chem. Mater.*, **4**, 1437 (1992).

68. H. Schmidt and B. Seiferling, in *Better Ceramics Through Chemistry II*, C.J. Brinker, D.E. Clark, and D.R. Ulrich (eds), 739, Materials Research Society, Pittsburgh, PA (1986); W.F. Doyle, B.D. Fabes, J.C. Root, K.D. Simmons, Y.M. Chiang, and D.R. Uhlmann, in *Ultrastructure Processing of Advanced Ceramics*, J.D. Mackenzie and D.R. Ulrich (eds), 953, J. Wiley & Sons, New York (1988).

69. C. Roger and M.J. Hampden-Smith, *J. Mater. Chem.*, **2**, 1111 (1992).

70. G. Coudert, G. Ndebeka, P. Caubere, S. Raynal, S. Lecolier, and S. Boileau, *J. Polym. Sci., Polym. Letts. Edn.*, **16**, 413 (1978).

71. P. Caubere, S. Raynal, G. Ndebeka, and S. Lecolier, *Polymer Preprints (Am. Chem. Soc. Div. Polym. Chem.)*, **23**, 190 (1982).

72. C.A. Ogle, Xiao Li Wang, F.H. Strickler, and B. Gordon, *Polymer Preprints (Am. Chem. Soc. Div. Polym. Chem.)*, **33**, 190 (1992).

73. J. Trekoval, *Coll. Czech. Chem. Comm.*, **42**, 1529 (1977); P. Bayard, R. Fayt, P. Teyssie, B. Vuillemin, P. Heim, and S.K. Varshney, European Patent Application EP 524,054 (20 Jan. 1993).

74. J. Baca, *Makromol. Chem., Rapid Commun.*, **1**, 609 (1980).

75. Q. Jin, D. Dimov, and T. Hogen-Esch, *Polymer Preprints (Am. Chem. Soc. Div. Polym Chem.)*, **32**, 465 (1991).

76. T. Katsuki and K.B. Sharpless, *J. Am. Chem. Soc.*, **102**, 5974 (1980).

77. K.B. Sharpless, C.H. Behrens, T. Katsuki, A.W.M. Lee, V.S. Martin, M. Takatani, S.M. Viti, F.J. Walker, and S.S. Woodward, *Pure Appl. Chem.*, **55**, 589 (1983).

78. V.S. Martin, S.S. Woodward, T. Katsuki, Y. Yamada, M. Ikeda, and K.B. Sharpless, *J. Am. Chem. Soc.*, **103**, 6237 (1981).

79. R.M. Hanson and K.B. Sharpless, *J. Org. Chem.*, **51**, 1922 (1986).

80. M.G. Finn and K.B. Sharpless, in *Asymmetric Synthesis*, Vol. 5, J.D. Morrison (ed.) Ch. 8, Academic Press, New York (1985); see also K.B. Sharpless, *Chem. Britain*, 38, (Jan. 1986).

81. I.D. Williams, S.F. Pedersen, K.B. Sharpless, and S.J. Lippard, *J. Am. Chem. Soc.*, **106**, 6430 (1984).

82. S.S. Woodward, M.G. Finn, and K.B. Sharpless, *J. Am. Chem. Soc.*, **113**, 106 (1991).

83. M.G. Finn and K.B. Sharpless, *J. Am. Chem. Soc.*, **113**, 113 (1991).

84. S. Miyano, L.D.-L. Lu, S.M. Viti, and K.B. Sharpless, *J. Org. Chem.*, **48**, 3608 (1983); **50**, 4350 (1985); M. Hayashi, F. Okamura, T. Toba, N. Oguni, and K.B. Sharpless, *Chem. Letts. (English)*, 547 (1990).

85. Y. Gao and K.B. Sharpless, *J. Org. Chem.*, **53**, 4114 (1988).

86. M. Caron and K.B. Sharpless, *J. Org. Chem.*, **50**, 1557 (1985); J.M. Chong and K.B. Sharpless, *J. Org. Chem.*, **50**, 1560 (1985).

87. L. Canali, J.K. Karjalainen, D.C. Sherrington, and O. Hormi, *Chem. Commun.*, 123 (1997).

88. T. Maschmeyer, M.C. Klunduk, C.M. Martin, D.S. Shephard, J.M. Thomas, and B.F.G. Johnson, *Chem. Commun.*, 1847 (1997); M. Crocker, R.H.M. Herold, and A.G. Orpen, *Chem. Commun.*, 2411 (1997).

89. F. DiFuria, G. Modena, and R. Seraglia, *Synthesis*, 325 (1984); P. Pitchen, E. Dunach, M.N. Deshmukh, and H.B. Kagan, *J. Am. Chem. Soc.*, **106**, 8188 (1984); N. Komatsu, M. Hashizume, T. Sugita, and S. Vemura, *J. Org. Chem.*, **58**, 4529 (1993).

90. F. Di Furia, G. Licini, G. Modena, R. Motterle, and W.A. Nugent, *J. Org. Chem.*, **61**, 5175 (1996).

91. G. Boche, K. Mobus, K. Harms, and M. Marsch, *J. Am. Chem. Soc.*, **118**, 2770 (1996); M. Bonchio, G. Licini, G. Modena, S. Moro, P. Traldi, and W.A. Nugent, *Chem. Commun.*, 869 (1997).

92. R.R. Schrock, *Acc. Chem. Res.*, **19**, 342 (1986).

93. R.R. Schrock, *Acc. Chem. Res.*, **23**, 158 (1990); H.H. Fox, R.R. Schrock, and R. O'Dell, *Organometallics*, **13**, 635 (1994).

94. R.R. Schrock, R.T. DePue, J. Feldman, C.J. Schaverien, J.C. Dewan, and A.H. Liu, *J. Am. Chem. Soc.*, **110**, 1423 (1988); R.R. Schrock, R.T. DePue, J. Feldman, K.B. Yap, D.C. Yang, W.M. Davis, L. Park, M. di Mare, M. Schofield, J. Anhaus, E. Walborsky, E. Evitt, C. Kruger, and P. Betz, *Organometallics*, **9**, 2262 (1990).

95. G.C. Bazan, E. Khosravi, R.R. Schrock, W.J. Feast, V.C. Gibson, M.B. O'Regan, J.K. Thomas, and W.M. Davis, *J. Am. Chem. Soc.*, **112**, 8378 (1990); G.C. Bazan, R.R. Schrock, N.-N. Cho, and V.C. Gibson, *Macromol*, **24**, 4495 (1991).

96. I.A. Weinstock, R.R. Schrock, and W.M. Davis, *J. Am. Chem. Soc.*, **113**, 135 (1991).

97. R. Schlund, R.R. Schrock, and W.E. Crowe, *J. Am. Chem. Soc.*, **111**, 8004 (1989).

98. V. Sankaran, C.C. Cummins, R.R. Schrock, R.E. Cohen, and R.J. Silbey, *J. Am. Chem. Soc.*, **112**, 6858 (1990); Y. Ng Cheong Chan, R.R. Schrock, and R.E. Cohen, *Chem. Mater.*, **4**, 24 (1992); C.C. Cummins, R.R. Schrock, and R.E. Cohen, *Chem. Mater.*, **4**, 27 (1992); Y. Ng Cheong Chan and R.R. Schrock, **5**, 556 (1993).

99. C. Pugh and R.R. Schrock, *Macromol.*, **25**, 6593 (1992); Z. Komiya and R.R. Schrock, *Macromol.*, **26**, 1393 (1993); C. Pugh and R.R. Schrock, *Polymer Preprints (Am. Chem. Soc. Div. Polym. Chem.)*, **34**, 160 (1993).

100. D. Mahajan, R.S. Sapienza, W.A. Sleigeir, and T.E. O'Hara, US Patent 4,935,395 (June 19, 1990); S.T. Sie, E. Drent, and W.W. Jager, European Patent, 306,114 (Mar. 8, 1989).

101. E. Stroefer, Wolf-Karlo Aders, P. Keller, G.W. Rotermund, F.J. Mueller, and W. Steiner, European Patent. 251,112 (Jan. 7, 1988).

102. D.J. Darensbourg, R.L. Gray, and C. Ovalles, *J. Mol. Catalysis*, **41**, 329 (1987).

103. T. Suzuki and H. Toritame, Japanese Patent, 6,379,850 (April 9, 1988).

104. S.W. Oliver and T.D. Smith, *J. Chem. Soc., Perkin Trans. II.* 1579 (1987).

105. D. Steinborn, H. Mosinski, and T. Rosenstock, *J. Organomet. Chem.*, **414**, C45 (1991).

106. W. Kampf and C. Herrmann, European Patent, 86,894 (Aug. 31, 1983).

107. N. Yahagi, *Sogo Shikensho Nenpo (Tokyo Daigaku Kogakubu)*, **38**, 279 (1978); *Chem. Abs.*, 25478V, **93**, 620 (1980).

108. J. Heuschen, R. Jerome, and Ph. Teyssie, *Macromolecules*, **14**, 242 (1981); A. Hamitou, R. Jerome, and Ph. Teyssie, *J Polym. Sci., Polym. Chem. Ed.*, **15**, 1035 (1977).

109. A. Lebrun, J.-L. Namy, and H.B. Kagan, *Tetrahedron Letts.*, **32**, 2355 (1991).

110. G. Preacher, A. Grund, and H. Petsch, German Patent, 3,514,938 (Oct. 30, 1991).

111. M. Shibasaki, H. Sasai, and T. Aria, *Angew. Chem., Int. Ed. Engl.*, **36**, 1236 (1997).

112. I.A. El-Maghy, E.S. Nasr and M.S. El-Samanoudy, *Proceedings of the Conference on Synthetic Lubricants*, Hung. Hydrocarb. Instit. Szazhalombatta, Hungary, 167 (1989); *Chem. Abs.*, 43,567, **113** (1990).

113. M. Hayashi, I. Inoue, and N. Oguni, *J. Chem. Soc., Chem. Commun.*, 341 (1994).

114. K. Narasaka, *Synthesis*, 1 (1993).

115. H. Ohno, A. Mori, and S. Inoue, *Chem. Letts. (Chemical Soc. of Japan).* 375 (1993).

116. Y. Makioka, I. Nakagawa, Y. Taniguchi, K. Takaki, and Y. Fujiwara, *J. Org. Chem.*, **58**, 4771 (1993).

117. D.C. Bradley, R.C. Mehrotra and D.P. Gaur, *Metal Alkoxides*, Academic Press, London, New York and San Francisco (1978).

118. B.R.W. Hinton, *Met. Finish*, **89**, 15 (1991); see also K. Nagakawa, T. Suzue, H. Furuse, and Y. Miyosawa, Japanese Patent 6136,992 (21 Aug. 1986); T. Nakano, Japanese Patents 02,142,861 (31 May, 1990).

119. T. Suzuki and S. Mazaki, Japanese Patents 02,121,204 (9 May, 1990); S Adachi and A Yamamoto, Japanese Patents 02,220,308 (3 Sept. 1990); S Adachi and I Kamioka, Japanese Patents 02,220,309 (3 Sept. 1990).

120. I. Yamashita, Japanese Patents, 62,250,025 (30 Oct. 1987); 63,295,628 (2 Dec, 1988); 02,167,332 (27 Jun. 1990).

121. S. Ohgishi, S. Itoh, H. Ono, and M. Tsugeno, *Semento Gijutsu Nenpo (Japanese)*, **41**, 241 (1987); K. Kira, T. Ine, T. Kojima, and K. Kuboyama, Japanese Patent 63,319,278 (17 Dec. 1988).
122. S. Hanawa, O. Ishii, M. Tanaka, T. Sakata, and S. Ishihavra, Japanese Patent 05,116,107 (14 May, 1993); S. Hanawa, O. Ishii, M. Tanaka, and S. Ishihara, Japanese Patent 05,278,008 (26 Oct. 1993); S. Hanawa, O. Ishii, M. Tanaka, and S. Ishihara, Japanese Patent, 06,000,802 (11 Jan. 1994).
123. Z. Nozaki and K. Yoshimoto, Japanese Patent 01,164,478 (28 Jun. 1988).
124. Tokuyama Soda Co. Ltd., Japanese Patent 6,011,408 (21 Jan. 1985); B.C.M. Patel, UK Patent Application GB 2,257,438 (13 Jan. 1993); GB 2,257,439 (13 Jan. 1993).
125. C.E. Pratt, US Patent 4,557,842 (10 Dec. 1985).
126. T. Aoi, S. Tezuka, M. Shinkai, N. Nanba, and M. Takayama, Japanese Patent 02,312,020 (27 Dec. 1990).
127. K. Shibata, O. Kamura, and K. Sogabe, Japanese Patent 62,221,764 (30 Jan. 1987); Y. Yoneda, H. Bandai, and M. Murata, Japanese Patent 62,246,866 (28 Oct. 1987); K. Izumi, H. Ishii and T. Hata, Japanese Patent, 01,126,276 (18 May, 1989); Y. Watanabe, Y. Yamate, M. Kumagai, M. Yamamoto, E. Arai, and Y. Fujii, Japanese Patent 01,275,418 (6 Nov. 1989); S. Nishikawa, K. Shibata, K. Takeuchi, T. Tanaka, S. Nakano, and K. Kuroki, Japanese Patent 01,319,682 (25 Dec. 1989); M. Okabe, Y. Sakabe, Y. Yoneda, Y. Takeshima, and H. Tamura, Japanese Patent 2,051,407 (21 Feb. 1990).
128. K. Shimada and S. Watanabe, Japanese Patent 63,206,355 (25 Aug. 1988); Y. Onuma, T. Yonezawa, Y. Itsudo, T. Ando, and H. Inoue, Japanese Patent, 0,176,976 (23 Mar. 1989).
129. H. Aoki, T. Suzuki, T. Katahata, M. Haino, G. Nishimura, H. Kaya, T. Isoda, Y. Tashiro, O. Funayama, and M. Arai, European Patent Application EP 332,357 (13 Sept. 1989); K. Hatsukyo, T. Shidara, H. Isaki, A. Nakanori, and T. Kawakami, Japanese Patent 1,257,111 (13 Oct. 1989).
130. M. Sekine and M. Mitomo, Japanese Patent 3,113,018 (14 May 1991).
131. M.A. Janney, German Offentlich DE 3,608,264 (18 Sept. 1986); A Kawakatsu, Japanese Patent 2,145,772 and 2,145,773 (5 Jun. 1990).

Index

Page numbers in *italics* refers to illustrations and tables.